NITROGEN IN
AGRICULTURAL SOILS

AGRONOMY

A Series of Monographs

The American Society of Agronomy and Academic Press published the first six books in this series. The General Editor of Monographs 1 to 6 was A. G. Norman. They are available through Academic Press, Inc., 111 Fifth Avenue, New York, NY 10003.

1. C. EDMUND MARSHALL: The Colloid Chemical of the Silicate Minerals, 1949
2. BYRON T. SHAW, *Editor*: Soil Physical Conditions and Plant Growth, 1952
3. K. D. JACOB: Fertilizer Technology and Resources in the United States, 1953
4. W. H. PIERRE and A. G. NORMAN, *Editors*: Soil and Fertilizer Phosphate in Crop Nutrition, 1953
5. GEORGE F. SPRAGUE, *Editor*: Corn and Corn Improvement, 1955
6. J. LEVITT: The Hardiness of Plants, 1956

The Monographs published since 1957 are available from the American Society of Agronomy, 677 S. Segoe Road, Madison, WI 53711.

7. JAMES N. LUTHIN, *Editor*: Drainage of Agricultural Lands, 1957   *General Editor*, D. E. Gregg
8. FRANKLIN A. COFFMAN, *Editor*: Oats and Oat Improvement   *Managing Editor*, H. L. Hamilton
9. C. A. BLACK, *Editor-in-Chief* and D. D. EVANS, J. L. WHITE, L. E. ENSMINGER, and F. E. CLARK, *Associate Editors*: Methods of Soil Analysis, 1965.
   Part 1—Physical and Mineralogical Properties, Including Statistics of Measurement and Sampling
   A. L. PAGE, *Editor*: Methods of Soil Analysis, 1982
   Part 2—Chemical and Microbiological Properties, Second Edition   *Managing Editor*, R. C. Dinauer
10. W. V. BARTHOLOMEW and F. E. CLARK, *Editors*: Soil Nitrogen, 1965
    (Out of print; replaced by no. 22)   *Managing Editor,* H. L. Hamilton
11. R. M. HAGAN, H. R. HAISE, and T. W. EDMINSTER, *Editors*: Irrigation of Agricultural Lands, 1967   *Managing Editor*, R. C. Dinauer
12. FRED ADAMS, *Editor*: Soil Acidity and Liming, Second Edition, 1984
         Managing Editor, R. C. Dinauer
13. K. S. QUISENBERRY and L. P. REITZ, *Editors*: Wheat and Wheat Improvement, 1967
         Managing Editor, H. L. Hamiiton
14. A. A. HANSON and F. V. JUSKA, *Editors*: Turfgrass Science, 1969
         Managing Editor, H. L. Hamilton
15. CLARENCE H. HANSON, *Editor*: Alfalfa Science and Technology, 1972
         Managing Editor, H. L. Hamilton
16. B. E. Caldwell, *Editor*: Soybeans: Improvement, Production, and Use, 1973
         Managing Editor, H. L. Hamilton
17. JAN VAN SCHILFGAARDE, *Editor*: Drainage for Agriculture, 1974
         Managing Editor, R. C. Dinauer
18. GEORGE F. SPRAGUE, *Editor*: Corn and Corn Improvement, 1977
         Managing Editor, D. A. Fuccillo
19. JACK F. CARTER, *Editor*: Sunflower Science and Technology, 1978
         Managing Editor, D. A. Fuccillo
20. ROBERT C. BUCKNER and L. P. BUSH, *Editors*: Tall Fescue, 1979
         Managing Editor, D. A. Fuccillo
21. M. T. BEATTY, G. W. PETERSEN, and L. D. SWINDALE, *Editors*: Planning the Uses and Management of Land, 1979   *Managing Editor*, R. C. Dinauer
22. F. J. STEVENSON, *Editor*: Nitrogen in Agricultural Soils, 1982
         Managing Editor, R. C. Dinauer
23. H. E. DREGNE and W. O. WILLIS, *Editors*: Dryland Agriculture, 1983
         Managing Editor, D. A. Fuccillo
24. R. J. KOHEL and C. F. LEWIS, *Editors*: Cotton, 1984
         Managing Editor, D. A. Fuccillo
25. N. L. TAYLOR, *Editor*: Clover Science and Technology, 1985
         Managing Editor, D. A. Fuccillo

NITROGEN IN AGRICULTURAL SOILS

FRANK J. STEVENSON, *editor*

Editorial Committee

F. J. Stevenson

J. M. Bremner

R. D. Hauck

D. R. Keeney

Managing Editor: RICHARD C. DINAUER

Assistant Editor: KRISTINE E. GATES

Editor-in-Chief ASA Publications: MATTHIAS STELLY

Number 22 in the series

AGRONOMY

American Society of Agronomy, Inc.
Crop Science Society of America, Inc.
Soil Science Society of America, Inc.
Publisher
Madison, Wisconsin USA

1982

American Society of Agronomy, Inc.
Crop Science Society of America, Inc.
Soil Science Society of America, Inc.
677 South Segoe Road, Madison, Wisconsin 53711 USA

Second Printing 1985

Library of Congress Cataloging in Publication Data

Nitrogen in agricultural soils.

    (Agronomy; no. 22)
    Includes bibliographies and index.
    1. Soils—Nitrogen content. 2. Plants, Effects of nitrogen
on. 3. Crops and soils. 4. Nitrogen fertilizers.
I. Stevenson, F. J. II. Series.
S592.6.N5N566   631.4′16   82-1704
ISBN 0-89118-070-2   AACR2

Printed in the United States of America

CONTENTS

1 Origin and Distribution of Nitrogen in Soil

F. J. STEVENSON

2 Inorganic Forms of Nitrogen in Soil

J. L. YOUNG AND R. W. ALDAG

3   Organic Forms of Soil Nitrogen

F. J. STEVENSON

4   Retention and Fixation of Ammonium and Ammonia in Soils

HANS NOMMIK AND KAAREL VAHTRAS

5   Biochemistry of Ammonification

J. N. LADD AND R. B. JACKSON

6   Mineralization and Immobilization of Soil Nitrogen

S. L. JANSSON AND J. PERSSON

7 Nitrification in Soil

EDWIN L. SCHMIDT

8 Biological Denitrification

M. K. FIRESTONE

9 Gaseous Losses of Nitrogen Other Than Through Denitrification

DARRELL W. NELSON

10 Biological Nitrogen Fixation

U. D. HAVELKA, M. G. BOYLE, AND R. W. F. HARDY

14 Soil Nitrogen Budgets

J. O. LEGG AND J. J. MEISINGER

15 Crop Nitrogen Requirements, Utilization, and Fertilization

R. A. OLSON AND L. T. KURTZ

DEDICATION

GEORGE STANFORD
1916–1981

It is highly appropriate that *Nitrogen in Agricultural Soils* is dedicated to the memory of Dr. George Stanford, whose research career over the past 4 decades centered chiefly on elucidating nitrogen behavior and transformations in soils as a basis for achieving more effective use of nitrogen fertilizers.

Born 7 March 1916 near Pierre, South Dakota, he received his B.S. degree from South Dakota University in 1938. His research career and studies for advanced degrees began in 1939 at Iowa State University. The M.S. degree was received in 1941 and the Ph.D. in 1947. This period included military service as an infantry officer in World War II.

His professional career was spent at Cornell University (1948–50); Iowa State University (1950–55); TVA (1955–60); Hawaiian Sugar Planters Association and Hawaii Agricultural Experiment Station (1960–65); and USDA, Beltsville, Maryland (1965–80). He retired as a research soil scientist from USDA due to ill health that culminated in his death on 28 January 1981.

Dr. Stanford's research accomplishments were numerous and had considerable influence on soil fertility practices at a time when major expansion in fertilizer use was just beginning. His major research accomplishments dealt with ammonium fixation in soils; a rapid greenhouse-laboratory method for evaluating the relative effectiveness of diverse nitrogen, phosphorus, and potassium sources; optimal use of nitrogen fertilizer on sugarcane; development of a nitrogen availability index (extractable ammonia) for rapidly estimating the nitrogen-supplying capacities of soils; denitrification and nitrogen losses from soils; and long-term field experiments using stable nitrogen isotopes.

Throughout his career, Dr. Stanford's fine qualities as a researcher and team leader were often demonstrated. His careful planning, attention to detail, and good judgment contributed greatly to the success of experiments. He directly involved himself in all phases of the work. His ability to assimilate research data and to reduce it to its most significant terms was a trait admired by all who knew him.

Dr. Stanford's research greatly expanded our knowledge of nitrogen in agronomic systems and has resulted in more than 75 publications. He devoted a large share of his life to nitrogen research, and his findings will continue to influence the trends in research on this important element for years to come. His frank and forthright manner left no doubt as to where he stood on critical or controversial issues. His approach to research was realistic and involved a critical examination of the problems, alternate procedures, possibilities for success, and what contribution the research results made to scientific knowledge and to practical solutions. All those who conduct research in the future may do well to follow his example.

GENERAL FOREWORD

Nitrogen in Agricultural Soils is an update of the 1965 edition of *Soil Nitrogen,* ASA Monograph 10, and replaces the depleted supply of *Soil Nitrogen.* The new book incorporates the significant advances made in this field during the past 17 years and is the 22nd monograph in the series *Agronomy* that was started in 1949. The first six volumes were published by Academic Press, Inc., New York. In 1957, the American Society of Agronomy took over publication of the monographs and continued to be the sole publisher through the 18th monograph published in 1977. The Crop Science Society of America and the Soil Science Society of America were invited to participate in the series and have been copublishers since 1977. The monographs represent an important and continuing effort of the associated societies, their officers, and the 11,700 members located in 100 countries to provide mankind worldwide with the most recent information available.

On behalf of the members of the associated societies and myself, I sincerely thank the Editorial Committee members chaired by Dr. F. J. Stevenson for their diligent work, the many authors for their writings, Managing Editor Richard C. Dinauer for his inexhaustible patience in the compilation of the contents of this book, and all others who have contributed directly or indirectly to the accomplishment of this worthy project.

December 1981

MATTHIAS STELLY
Executive Vice President, ASA-CSSA-SSSA
and Editor-in-Chief, ASA Publications

FOREWORD

The nitrogen reactions in soils and the nitrogen nutrition of crops are insufficiently understood. Soil nitrogen derived from organic matter is essential for plant growth and the formation of proteins required by living matter. The principal source of protein for a major portion of the human population is cereal grain. Humans depend on foraging animals to capture plant proteins and provide meat and milk. These basic characteristics of soil nitrogen challenge our interest in the mechanisms and processes by which people obtain food and fiber products.

An increase in the protein content of cereal grains would enhance the nutritional status of the human population. In many soils and for many crops, farmers increase the nutrient supply of nitrogen by the addition of various forms of nitrogen to the soil or by biological dinitrogen fixation with legumes. Plants consume only about 50% of the added nitrogen forms, and there is a strong need to increase this utilization level. The nitrogen cycle in soils follows a complex series of reactions that require continuing study to utilize this nutrient more effectively and to assure an adequate supply for plants. In the overall process of removing barriers to crop productivity, nitrogen supply to plants plays a major role.

We express appreciation and gratitude to Dr. F. J. Stevenson, editor, and his editorial committee, Drs. J. M. Bremner, R. D. Hauck, and D. R. Keeney for their important functions leading to this publication. We acknowledge and thank the authors for their cooperation and efforts and the help of society members who reviewed manuscripts. We are grateful to the Headquarters staff for editorial and production efforts, which makes it possible to place this fine volume in your hands.

November 1981

STERLING R. OLSEN   KENNETH J. FREY   BOBBY A. STEWART
president ASA   *president CSSA*   *president SSSA*

PREFACE

Nitrogen in Agricultural Soils provides an authoritative review of the principles governing the behavior of nitrogen in the soil-plant system. The volume supersedes ASA Monograph 10 *Soil Nitrogen,* published in 1965. Significant advances on all aspects of the subject have been made since that time, and the need had arisen for a compilation and critical analysis of current knowledge. Material contained in the 1965 monograph has been extensively revised and updated, and new chapters have been introduced in response to increasing concern about energy conservation and preservation of the environment. Authors were allowed considerable latitude in developing their topics, with the result that both panoramic and specific views have been presented for each major component of the soil nitrogen cycle.

The volume covers many facets of soil nitrogen, including forms and distribution, biological and nonbiological transformations, gains, losses, and recycling, plant availability and uptake, modeling and transport, pesticide interactions, experimental approaches, and economic implications of restrictions on fertilizer nitrogen use. The field of study is broad and has involved researchers working in many specialized areas. Because of the voluminous literature that has accumulated over the past two decades, an exhaustive coverage of the literature was not always possible, and selection of references has often been rather arbitrary. The editors and authors apologize for omission of important work.

The editorial committee expresses appreciation to the authors and the organizations they represent for cooperation and support. Acknowledgment is given to Richard C. Dinauer, Matthias Stelly, and other members of the Headquarters staff for advice and assistance in editing and preparing the manuscripts for publication. We pay special tribute to George Stanford, author of Chapter 17, whose untimely death occurred while the monograph was in progress. The assistance of J. J. Meisinger in proofreading and indexing Dr. Stanford's chapter is gratefully acknowledged.

August 1981

The Editorial Committee

F. J. STEVENSON, *editor, University of Illinois, Urbana, Illinois*

J. M. BREMNER, *Iowa State University, Ames, Iowa*

R. D. HAUCK, *Tennessee Valley Authority, Muscle Shoals, Alabama*

D. R. KEENEY, *University of Wisconsin, Madison, Wisconsin*

CONTRIBUTORS

Rudolf W. Aldag Dr. sc. agr., Dr. habil., Institut für Bodenkunde, Universität Göttingen, Göttingen, West Germany

James W. Biggar Professor of Water Science, Department of Land, Air, and Water Resources, University of California, Davis, California

John M. Bremner Professor, Department of Agronomy, Iowa State University, Ames, Iowa

Mark G. Boyle Agronomist, Central Research and Development Department, E.I. DuPont DeNemours Company, Wilmington, Delaware

Mary K. Firestone Assistant Professor of Soil Microbiology, Department of Plant and Soil Biology, University of California, Berkeley, California

Cleve A. I. Goring Technical Director, Agricultural Products Department, The Dow Chemical Company, Midland, Michigan

Ralph W. F. Hardy Director, Life Sciences, Central Research and Development Department, E.I. DuPont DeNemours Company, Wilmington, Delaware

Robin F. Harris Professor of Soil Microbiology, Department of Soil Science and Bacteriology, University of Wisconsin, Madison, Wisconsin

Roland D. Hauck Research Soil Scientist, National Fertilizer Development Center, Tennessee Valley Authority, Muscle Shoals, Alabama

U. D. Havelka Research Staff, Central Research and Development Department, E.I. DuPont DeNemours Company, Wilmington, Delaware

Ron B. Jackson Experimental Officer, Division of Soils, CSIRO, Glen Osmond, South Australia, Australia

Sven L. Jansson Professor, Department of Soil Sciences, Swedish University of Agricultural Sciences, Uppsala, Sweden

Dennis R. Keeney Professor and Chairman, Department of Soil Science, University of Wisconsin, Madison, Wisconsin

Lester T. Kurtz Professor of Soil Fertility, Department of Agronomy, University of Illinois, Urbana, Illinois

Jeffrey N. Ladd Senior Principal Research Scientist, Division of Soils, CSIRO, Glen Osmond, South Australia, Australia

Dennis A. Laskowski Research Leader, Environmental Chemistry, Agricultural Products Department, The Dow Chemical Company, Midland, Michigan

J. O. Legg Soil Scientist (retired), Soil Nitrogen and Environmental Chemistry Laboratory, Agricultural Research Service, U.S. Department of Agriculture, Beltsville, Maryland. Currently Adjunct Professor, Agronomy Department, University of Arkansas, Fayetteville, Arkansas

J. J. Meisinger	Soil Scientist, Soil Nitrogen and Environmental Chemistry Laboratory, Agricultural Research Service, U.S. Department of Agriculture, Beltsville, Maryland
Donald R. Nielsen	Professor of Soil and Water Science, Department of Land, Air, and Water Resources, University of California, Davis, California
Darrell W. Nelson	Professor of Agronomy, Department of Agronomy, Purdue University, West Lafayette, Indiana
Hans Nommik	Professor of Soil Chemistry, Department of Forest Soils, Swedish University of Agricultural Sciences, Uppsala, Sweden
Robert A. Olson	Professor of Agronomy, Department of Agronomy, University of Nebraska, Lincoln, Nebraska
J. Persson	Department of Soil Sciences, Swedish University of Agricultural Sciences, Uppsala, Sweden
William H. Patrick, Jr.	Boyd Professor, Laboratory for Wetland Soils and Sediments, Louisiana State University, Baton Rouge, Louisiana
James R. Peterson	Soil Scientist III, Research and Development Department, The Metropolitan Sanitary District of Greater Chicago, Cicero, Illinois
Edwin L. Schmidt	Professor, Department of Soil Science, University of Minnesota, St. Paul, Minnesota
Jay H. Smith	Soil Scientist, Snake River Conservation Research Center, Agricultural Research Service, U.S. Department of Agriculture, Kimberly, Idaho
George Stanford	Research Soil Scientist, Soil Nitrogen and Environmental Chemistry Laboratory, Agricultural Research Service, U.S. Department of Agriculture, Beltsville, Maryland. Deceased 28 January 1981
Frank J. Stevenson	Professor of Soil Chemistry, Department of Agronomy, University of Illinois, Urbana, Illinois
Earl R. Swanson	Professor of Agricultural Economics, Department of Agricultural Economics, University of Illinois, Urbana, Illinois
Kenneth K. Tanji	Professor of Water Science, Department of Land, Air, and Water Resources, University of California, Davis, California
Kaarel Vahtras	Agronomie Licentiat, Department of Soil Sciences, Swedish University of Agricultural Sciences, Uppsala, Sweden
Peter J. Wierenga	Professor of Soil Science, Department of Agronomy, New Mexico State University, Las Cruces, New Mexico
J. L. Young	Research Chemist and Professor of Soil Science, Agricultural Research Service, U.S. Department of Agriculture, and Soil Science Department, Oregon State University, Corvallis, Oregon

CONVERSION FACTORS FOR U. S. AND METRIC UNITS

To convert column 1 into column 2, multiply by	Column 1	Column 2	To convert column 2 into column 1, multiply by
Length			
0.621	kilometer, km	mile, mi	1.609
1.094	meter, m	yard, yd	0.914
0.394	centimeter, cm	inch, in	2.54
Area			
0.386	kilometer^2, km^2	mile^2, mi^2	2.590
247.1	kilometer^2, km^2	acre, acre	0.00405
2.471	hectare, ha	acre, acre	0.405
Volume			
0.00973	meter^3, m^3	acre-inch	102.8
3.532	hectoliter, hl	cubic foot, ft^3	0.2832
2.838	hectoliter, hl	bushel, bu	0.352
0.0284	liter	bushel, bu	35.24
1.057	liter	quart (liquid), qt	0.946
Mass			
1.102	ton (metric)	ton (U.S.)	0.9072
2.205	quintal, q	hundredweight, cwt (short)	0.454
2.205	kilogram, kg	pound, lb	0.454
0.035	gram, g	ounce (avdp), oz	28.35
Pressure			
14.50	bar	lb/inch^2, psi	0.06895
0.9869	bar	atmosphere, atm	1.013
0.9678	kg(weight)/cm^2	atmosphere, atm	1.033
14.22	kg(weight)/cm^2	lb/inch^2, psi	0.07031
14.70	atmosphere, atm	lb/inch^2, psi	0.06805
Yield or Rate			
0.446	ton (metric)/hectare	ton (U.S.)/acre	2.24
0.892	kg/ha	lb/acre	1.12
0.892	quintal/hectare	hundredweight/acre	1.12
Temperature			
$\left(\dfrac{9}{5}\ °C\right) + 32$	Celsius $-17.8C$ $0C$ $100C$	Fahrenheit $0F$ $32F$ $212F$	$\dfrac{5}{9}\ (°F - 32)$
Water Measurement			
8.108	hectare-meters, ha-m	acre-feet	0.1233
97.29	hectare-meters, ha-m	acre-inches	0.01028
0.08108	hectare-centimeters, ha-cm	acre-feet	12.33
0.973	hectare-centimeters, ha-cm	acre-inches	1.028
0.00973	meters^3, m^3	acre-inches	102.8
0.981	hectare-centimeters/hour, ha-cm/hour	feet^3/sec	1.0194
440.3	hectare-centimeters/hour, ha-cm/hour	U.S. gallons/min	0.00227
0.00981	meters^3/hour, m^3/hour	feet^3/sec	101.94
4.403	meters^3/hour, m^3/hour	U.S. gallons/min	0.227

Plant Nutrition Conversion—P and K

P (phosphorus) $\times$ 2.29 = P_2O_5

K (potassium) $\times$ 1.20 = K_2O

1 Origin and Distribution of Nitrogen in Soil

F. J. STEVENSON

University of Illinois
Urbana, Illinois

I. INTRODUCTION

Nitrogen occupies a unique position among the elements essential for plant growth because of the rather large amount required by most agricultural crops. A deficiency of N is shown by yellowing of the leaves and by slow and stunted growth. Other factors being favorable, an adequate supply of N in the soil promotes rapid plant growth and the development of dark-green color in the leaves.

The combined N in soil is largely bound to organic matter and mineral material; in general, only a few kilograms per hectare exist in available mineral forms (as NO_3^- and exchangeable NH_4^+) at any one time. The total amount of N in many soils is appreciable, often exceeding 4,000 kg/ha to the depth of plowing. When land is first placed under cultivation, the N content of the soil usually declines, and a new equilibrium level that is characteristic of the climate, cultural practice, and soil type is established. At equilibrium, any N removed by harvested crops must be compensated for by incorporation of an equivalent amount of newly fixed N into the organic matter.

Systems of agriculture that rely heavily on soil reserves to meet the N requirements of plants cannot long be effective in producing high yields of crops. In the past, biological N_2 fixation was the chief means of supplying N for cultivated crops. Since World War II, N fertilizers have become increasingly available, which when used to augment the N supplied by natural processes can increase yields and improve the quality of crops. A major concern of present-day farmers is the effective management of N fertilizers for maximum efficiency and minimal pollution of the environment (Chapt. 16, this book).

The subject of soil N is broad and deals not only with the distribution and transformation of organic and inorganic forms within the soil but with interrelationships that exist with the atmosphere and biosphere. Many of these aspects are covered in considerable detail in the chapters that follow. In this chapter, an overview is presented on N cycle processes in soil, with major emphasis on factors influencing soil N levels.

Copyright 1982 © ASA-CSSA-SSSA, 677 South Segoe Road, Madison, WI 53711, USA.
Nitrogen in Agricultural Soils—Agronomy Monograph no. 22.

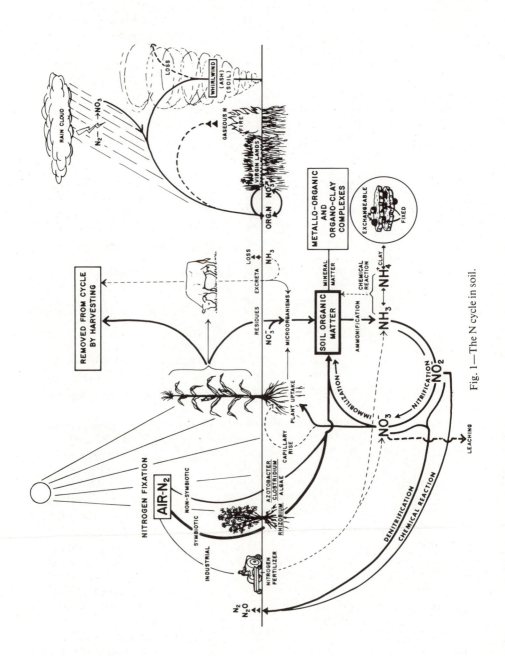

Fig. 1—The N cycle in soil.

II. THE N CYCLE

The N cycle in soil is an integral part of the overall cycle of N in nature, as shown in Fig. 1. In addition to interaction with the total ecosystem, an internal cycle is operative in which N passes from one regulated system to another through a variety of biochemical and chemical processes. It should be noted that a "N cycle" as such does not exist in nature but that any given N atom moves from one form to another in a completely irregular or random fashion.

Gains in soil N occur not only through fixation of molecular N_2 by microorganisms but from the return of NH_4^+ and NO_3^- in rain water; losses occur through crop removal, leaching, and volatilization. The conversion of molecular N_2 to combined forms through *biological N_2 fixation* is of particular interest both theoretically and practically. *Mineralization* is the conversion of organic forms of N to NH_4^+ and NO_3^-. The initial conversion to NH_4^+ is referred to as *ammonification*; the oxidation of this compound to NO_3^- is termed *nitrification.* The utilization of NH_4^+ and NO_3^- by plants and microorganisms constitutes *assimilation* and *immobilization,* respectively. Combined N is ultimately returned to the atmosphere through biological *denitrification,* thereby completing the cycle.

Not all transformations of N in soil are mediated by microorganisms. Ammonia and nitrite (NO_2^-), produced as products of the microbial decomposition of nitrogenous organic materials, are capable of undergoing chemical reactions with organic matter, in some cases leading to the evolution of N gases. Through the physicochemical association of humic materials with mineral matter, metallo-organic and organo-clay complexes are formed, whereby the N compounds are protected against attack by microorganisms. The positively charged NH_4^+ ion undergoes substitution reactions with other cations of the exchange complex and can be fixed by clay minerals.

The basic feature of biological N transformations centers on oxidation and reduction reactions. An entire sequence of events is involved, some of which take place in the cells of microorganisms and some in the tissues of higher plants. Included in the intermediates are gaseous, mineral, and organic forms of N. Oxidation states of N range from $+5$ for NO_3^- to -3 for NH_3 (or NH_4^+). In the oxidized ($+$) state, the outer electrons of N serve to complete the electron shells of other atoms; in the reduced ($-$) state, the electrons required to fill the outer shell of N are supplied by other atoms.

Balance sheets for N are also discussed in Chapt. 16 and 23 (this book). Estimates vary somewhat, but the results indicate that N is being introduced into the biosphere at the rate of about 92 million metric tons/year. The total amount that is denitrified and returned to the atmosphere is about 83 million metric tons per year. The difference (9 metric tons) represents the rate at which fixed N is building up in the biosphere and pedosphere.

As Delwiche (1970) pointed out, the N balance of the earth has been upset through the activities of man. The amount of N fixed industrially has increased steadily since World War II, and some of this N has been carried into lakes and streams in runoff waters from agricultural lands. In some cases, NO_3^--N concentrations in water supplies have approached the level (10 mg/liter) regarded by the U.S. Department of Health as unacceptable for human consumption (see Chapt. 16, this book). The disposal of nitrogenous organic wastes from an expanding human and domestic animal population is also a problem (Chapt. 21, this book). Much of this waste N should be recycled back to the soil, but under present economic conditions it is often cheaper for the farmer to use chemical fertilizers. Extensive research is currently underway in attempts to develop efficient and adequate methods for utilization of organic wastes on soil.

A. Geochemistry of N

Nitrogen is of geological interest because of its occurrence in the four recognized spheres of the earth, namely, lithosphere, atmosphere, hydrosphere, and biosphere. The significance of N arises from the fact that the air over each square foot of the earth's surface contains about 2,700 kg (3 tons) of N and that after C, H, and O, no other element is so intimately associated with the reactions carried out by living organisms.

The approximate inventory of N in the four spheres is given in Table 1. The survey shows that the bulk of the N (about 98%) exists in the lithosphere, and most of the remainder is found in the atmosphere. Molecular N_2 comprises about 78% of the gases in the atmosphere. Other N compounds found in minute amounts include NH_4^+, NO_3^-, nitrous oxide (N_2O), and organically bound N (associated with terrestrial dust).

Many difficulties are encountered in determining the distribution of N in living matter (the biosphere). Unlike other spheres, the biosphere is in a constant state of flux. Also, the matter of the biosphere is not uniformly distributed, and the N contents of different organisms vary widely. The value given (2.8×10^5 Tg) is at best an approximation.

In the ocean (hydrosphere), N occurs as molecular N_2, NH_4^+, NO_2^-, NO_3^- and dissolved and particulate organic matter (Emery et al., 1955). Molecular N_2 is the dominant form, accounting for over 95% of the total.

Table 1—Inventory of N in the four spheres of the earth.

Sphere	Tg of N†
Lithosphere	1.636×10^{11}
Igneous rocks	
a) of the crust	1.0×10^9
b) of the mantle	1.62×10^{11}
Core of the earth	1.3×10^8
Sediments (fossil N)	3.5–5.5×10^8
Coal	1.0×10^5
Sea-bottom organic compounds	5.4×10^5
Terrestrial soils	
a) Organic matter	2.2×10^5
b) Clay-fixed NH_4^+	2.0×10^4
Atmosphere	3.86×10^9
Hydrosphere	2.3×10^7
Biosphere	2.8×10^5

† Tg = terrogram = 10^{12} g or million metric tons. Estimates are from Stevenson (1965), Burns and Hardy (1975), and Söderlund and Svensson (1976).

The N reserve of the ocean can be considered to be in a state of quasiequilibrium. Variations in abundance of the different forms of N occur with depth, season, biological activity, and other factors, but in the long run, the amount of each remains constant. In other words, the total quantity of N in the N reserve at any one time represents a balance between N gains and losses. Gains of N are from the land and atmosphere, from which combined N is carried by rivers and rain. Losses of N occur through deposition of organically bound N in sediments and by bacterial denitrification.

Nitrogen is a common constituent of the lithosphere (soils, sediments, silicate minerals, fossils, and rocks of all types). Many natural products, such as coal, contain N. As can be seen from the inventory given in Table 1, most of the N occurs in association with igneous rocks of the earth's crust and mantle, possibly as NH_4^+ held within the lattice structures of such primary silicate minerals as the micas and feldspars (Stevenson, 1962).

Table 2—Nitrogen in some soil associations of North America, South America, and other areas as tabulated from organic C estimates of Bohn (1976). The values represent amounts to a depth of 1 m.

	Area	Total organic C†	N, $\times 10^3$ Tg		
Association	$\times 10^5$ km²	$\times 10^3$ Tg	Organic	Fixed NH_4^+	Total
North America					
Histosols	13.3	266	8.9	--	8.9
Podzols	31.5	66	5.9	0.7	6.6
Cambisols	9.4	56	5.0	0.6	5.6
Haplic Hastanozems	32.0	51	4.6	0.5	5.1
Eutric Gleysols	10.9	33	3.0	0.3	3.3
Gelic Regosols	15.9	32	2.9	0.3	3.2
Dystric Cambisols	3.6	21	1.9	0.2	2.1
Phaeozems	10.4	21	1.9	0.2	2.1
All others	76.0	119	10.7	1.2	11.9
	203.0	665	44.8	4.0	48.8
South America					
Ferrasols	89.8	108	9.7	1.1	10.8
Dystric Histosols	3.8	76	2.5	--	2.5
Cambisol-Andisols	6.2	35	3.1	0.4	3.5
Cambisols	4.2	26	2.3	0.3	2.6
Acrisol-Xerosol-					
Kastanozems	16.6	13	1.2	0.1	1.3
All others	61.4	43	3.9	0.4	4.3
	182.0	301	22.7	2.3	25.0
Asia, Africa Europe, Oceania					
Histosols	26.0	520	17.3	--	17.3
Cambisols	37.0	280	25.2	2.8	28.0
Podzols	130.0	270	24.3	2.7	27.0
Kastanozems	131.0	210	18.9	2.1	21.0
Chernozems	51.0	200	18.0	2.0	20.0
Ferrasols	141.0	170	15.3	1.7	17.0
Cambisol-Vertisols	28.0	110	9.9	1.1	11.0
All others	282.0	220	19.8	2.2	22.0
	836.0	1,980	148.7	14.6	163.3
Total		2,946	216.2	20.9	237.1

† Tg = terrogram = 10^{12}g or million metric tons.

Table 3—Estimated amounts of N to depths of 15 and 100 cm for the major soil associations of the United States.†

Soil association	Approxi-mate area	Average amount of N per ha		Total N in association	
		To 15 cm	To 1 m	To 15 cm	To 1 m
	10^6 ha	10^3 kg	10^3 kg	10^9 kg	10^9 kg
Brown Forest	72.8	2.8	7.5	203.8	546.0
Red-Yellow	60.7	2.2	4.5	133.5	273.2
Prairie	45.7	3.9	17.9	178.2	818.0
Chermozem and					
Chermozem-like	49.8	5.0	17.9	249.0	891.4
Chestnut	41.3	3.3	12.0	136.3	495.6
Brown	21.0	2.8	9.0	58.8	189.0
			Total	959.6	3213.2

† Adapted from Campbell (1978) as tabulated from data published by Schreiner and Brown (1938).

Estimates for the quantities of N in each pool vary widely, as will be evident from the more extensive coverage of this subject in Chapter 23 (this book). In any event, it can be seen that the amount of N in terrestrial soils is negligible when compared with other pools. The amount held by primary rocks is about 50 times that present in the atmosphere, whereas the amount contained in the latter is nearly 5,000 times that existing in soils.

The values given above for N in terrestrial soils were estimated from Bohn's (1976) tabulation for total organic C in soil associations of the world. The assumption was made that the average C/N ratio was 10 for the mineral soils and 30 for the organic soils (Histosols). An average of 10% of the N in the mineral soils was assumed to occur as clay-fixed NH_4^+. This value may be low, because appreciable amounts of the N in many subsurface soils occur in this form (see Chapt. 2, this book).

A further breakdown of total, organic, and clay-fixed NH_4^+ in soil associations of the world is given in Table 2. Because of their high organic matter and N contents, Histosols are major contributors to the total soil N (28.7×10^3 Tg, or >10% of the total). It should be noted that Histosols are often considerably deeper than the 1-m depth on which the calculations were made. The relatively low amounts of N in the mineral soils of South America are due to the fact that most of them are tropical soils low in organic matter.

Estimated amounts of N to depths of 15 and 100 cm for the major soil associations of the USA are recorded in Table 3. For reasons that will be evident later, most of the N resides in those soils classified as Mollisols (listed as Prairie, Chernozem, and Chestnut soils).

B. Evolutionary Aspects

Events leading to biochemical N transformations and the formation of organically combined N are outlined in Fig. 2. The first event was the formation of an atmosphere enriched with molecular N_2. The second event, which can be considered a preliminary step in the evolution of the biosphere, was associated with the nonbiological formation of organic molecules. The third event was the development of organisms capable of bringing about N transformations.

The various aspects of the origin of N in soil, including a more detailed account of the events illustrated in Fig. 2, are discussed below. Additional information on the geochemistry of N in the terrestrial atmosphere can be obtained from the works of Hutchinson (1944, 1954).

1. GEOLOGICAL HISTORY OF ATMOSPHERIC N

Extensive coverage of this subject is available elsewhere (Hutchinson, 1944, 1954; Fairbridge, 1972), and only the bare essentials are outlined herein.

Our concept of the origin of atmospheric N is closely allied to the mode of origin that we ascribed to the earth, for which two main schools of thought have existed. The first is that the earth (and the solar system as a whole) was formed by the accretion of small solid particles called planetesimals. The second is that the earth was torn from the body of the sun. Astronomical and geochemical evidence is highly in favor of the planetesimal hypothesis. According to this hypothesis, the N in the earth's atmosphere was derived largely from compounds initially occluded or chemically combined with the planetesimals.

It is not known whether all of the N in the present-day atmosphere was formed gradually over geological time or if some of it is a remnant of an original protoatmosphere. According to a widely held theory, the planetesimals were vaporized as they arrived at the surface of the protoplanet; therefore, the temperature was too high to allow for retention of gaseous products. Accordingly, practically all of the N in the atmosphere originated by release of nonvolatile compounds from the interior of the newly formed earth. This could have been brought about by outgassing of igneous rocks as the temperature of the earth increased due to heat generated by compression and decay of radioactive elements (^{234}U, ^{232}U, and ^{40}K). Opponents of this theory maintain that there is too much N in the atmosphere, as well as in sediments and sedimentary rocks, to be accounted for by the simple weathering of igneous materials.

Evidence that the N content of the atmosphere has increased steadily throughout geological time has come from the finding that N is a normal constituent of magmatic gases. Small additions of N have been made to the atmosphere by volatilization of nitrogenous compounds from meteorites during their entry into the earth's atmosphere.

The chemical forms in which N has existed in the atmosphere over geological time is unknown. Many geochemists are of the opinion that the early atmosphere was reducing and that the N was present as NH_3. Others believe that it occurred as diatomic N_2, the form in which it occurs today. According to the NH_3 hypothesis, the early atmosphere contained considerable quantities of hydrogen. Consequently, the atmosphere was reducing, and

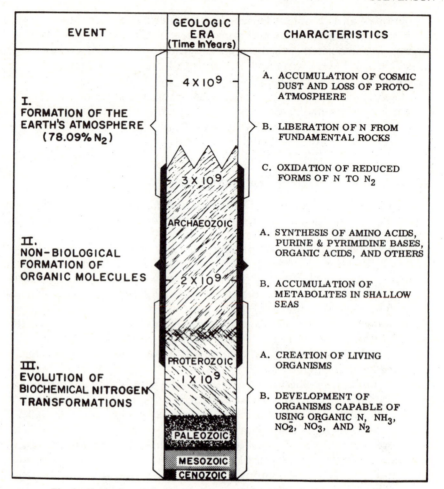

EVENT	GEOLOGIC ERA (Time In Years)	CHARACTERISTICS
I. FORMATION OF THE EARTH'S ATMOSPHERE (78.09% N_2)	4×10^9	A. ACCUMULATION OF COSMIC DUST AND LOSS OF PROTO-ATMOSPHERE
		B. LIBERATION OF N FROM FUNDAMENTAL ROCKS
	3×10^9	C. OXIDATION OF REDUCED FORMS OF N TO N_2
II. NON-BIOLOGICAL FORMATION OF ORGANIC MOLECULES	ARCHAEOZOIC 2×10^9	A. SYNTHESIS OF AMINO ACIDS, PURINE & PYRIMIDINE BASES, ORGANIC ACIDS, AND OTHERS
		B. ACCUMULATION OF METABOLITES IN SHALLOW SEAS
III. EVOLUTION OF BIOCHEMICAL NITROGEN TRANSFORMATIONS	PROTEROZOIC 1×10^9 PALEOZOIC MESOZOIC CENOZOIC	A. CREATION OF LIVING ORGANISMS
		B. DEVELOPMENT OF ORGANISMS CAPABLE OF USING ORGANIC N, NH_3, NO_2^-, NO_3^-, AND N_2

Fig. 2—Evolution of the N cycle (adapted from a drawing by Calvin, 1956).

the N was in a reduced state (as NH_3). With the loss of hydrogen (and other light elements) by diffusion into space and with the appearance of free oxygen in the atmosphere (photochemical dissociation of water vapor; photosynthesis by green plants), the NH_3 was oxidized to elemental N_2.

2. PREBIOLOGICAL FORMATION OF ORGANIC MOLECULES

The prebiological synthesis of organic substances (Fig. 2, Event II) set the stage and established the pattern for the evolution of living organisms, including those that fix dinitrogen gas (N_2).

Organic N compounds (amino acids, peptides, purine and pyrimidine bases, porphyrins) are of paramount importance to the efficient functioning of life processes. From the point of view of science, these compounds, among others, had to be on hand before life arose on this planet.

Reactions leading to the synthesis of biochemically active substances and the development of living organisms are summarized below (Fox, 1963):

CO, CO_2, and/or CH_4 Electricity, heat, solar radiation
NH_3 or N_2 ——————————————————→
H_2O, H_2, H_2S, metals, and H_3PO_4 β rays, γ rays, x rays

Biochemical staples (organic acids,
amino acids, vitamins, carbohydrates,   Heat
pyrimidines, porphyrins, etc.)   ——————→

Macromolecules (proteins,   ————————————→ Organisms
nucleic acids, polysaccharides)

The initial step was the formation of organic molecules from volatile compounds, presumably in an extralithospheric atmosphere. Considerable controversy exists as to whether the initial synthesis took place under reducing or oxidizing conditions (see Kvenvolden, 1974; Dickerson, 1978; Kerr, 1980). Irrespective of the mode of formation, the biochemical staples (amino acids, carbohydrates, porphyrins, pyrimidimes, etc.) are believed to have become concentrated in the shallow seas that formed by condensation of water vapor during cooling of the earth. Following polymerization of active biochemicals into proteins, nucleic acids and polysaccharides, molecular systems (enzymes) developed that had the ability to synthesize protoplasmic constituents from the surrounding medium. The final step was the development of living organisms.

Geochemists have successfully produced biochemical substances in the laboratory under conditions resembling those of the primitive earth (see Kvenvolden, 1974; Dickerson, 1978). For example, amino acids have been synthesized by subjecting such mixtures of gases as CO_2, H_2, H_2O, and NH_3 and CO_2, H_2, H_2O, and N_2 to an electrical discharge.

3. EVOLUTION OF BIOCHEMICAL N TRANSFORMATIONS

The biosphere, that part of the earth that sustains life, developed later than other earth spheres; life was not possible until conditions on the earth's surface were satisfactory. Once living cells became established, the formation of organisms with complex enzyme systems became possible.

Living organisms first emerged during the early history of the earth (Fig. 2). Well-preserved fossils have been identified in limestone formations dating to the Proterozoic era, and structures believed to be those of primitive algae have been observed in Pre-Cambrian limestone. The formation of these organisms was undoubtedly preceded by a long evolutionary cycle. Hutchinson (1954) concluded that life, the oceans, and oxygen appeared concurrently, a hypothesis that implies that the biosphere developed rather early in the earth's history.

Numerous investigators have emphasized that the development of complex living organisms proceeded from heterotrophs to autotrophs and from anaerobes to aerobes (e.g., Oda, 1959). This arrangement suggests that nitrogenous compounds synthesized by nonbiological means (Event II) were the mother substances of primitive N metabolism. With the release of NH_3 from primary organic matter during fermentation by anaerobic organisms, other heterotrophs developed that were capable of utilizing NH_3 (Hutchinson, 1944).

The appearance of oxygen subsequently led to the evolution of organisms capable of oxidizing NH_3 to NO_3^-. As implied above, organisms with different types of metabolisms originated from a common root (a heterotrophic one), and the biochemical functions of autotrophs were derived from systems developed by heterotrophs. On this basis, the oxidizing enzymes of autotrophic nitrifying organisms were inherited from those developed earlier by heterotrophs. It is of interest that many heterotrophic organisms are able to oxidize NH_4^+ to NO_2^- or NO_3^-.

The development of organisms capable of utilizing NO_3^- either by reduction into organic compounds (NO_3^- assimilation) or by the formation of N gases (denitrification) is believed by Hutchinson (1944) to have taken place soon after the appearance of organisms capable of oxidizing NH_3. Wilson (1951) suggested that without these organisms, life would have been threatened by the conversion of NH_3 into highly toxic compounds, whereas the premature development of denitrifying organisms could have led to the depletion of combined N. The enzyme systems of microorganisms that permit the oxygen of NO_3^- to serve as an electron acceptor during respiration were probably adapted from those in which molecular oxygen was the acceptor. Even in modern denitrifying organism, NO_3^- is used as an electron acceptor only in the absence of oxygen.

Concepts regarding the evolution of biological N_2 fixation have been discussed by Imshenetskii (1962), who concluded that the anaerobic *Clostridium pasteurianum* (or spore-forming anaerobes closely related to it) was the most primitive and the earliest of all N_2 fixers. The existence of photosynthetic bacteria capable of fixing N_2 was believed to have developed somewhat later. Only after the appearance of free oxygen did the aerobic N_2-fixing bacteria (e.g., *Azotobacter*) appear. Nitrogen-fixing blue-green algae were believed to have evolved during this same period. According to this sequence, the ability to utilize molecular N_2 originated with heterotrophic organisms. With the later development of photosynthesizing bacteria, other microorganisms developed that were able to fix N_2 without utilizing organic substrates as a source of energy.

The final stage in the evolution of N_2 fixation was the development of a symbiotic relationship between microorganisms and higher plants. Presumably, the organisms involved fixed N_2 while they were living free in the soil, but later, after adaptation with higher plants, they lost this power when grown in culture media.

Symbiotic N_2 fixation is not believed to have developed until the Mesozoic period, presumably in Upper Cretaceous times when the angiosperms dominated the earth's flora. Bond et al. (1956) concluded that the possession of root nodules and the ability to use molecular N_2 were ancient characteristics that dated from a time when only woody flowering plants existed and when conditions were favorable for the initiation of symbiotic associations. Norris (1956) emphasized that the angiosperms originated under conditions resembling the modern wet tropics and that their subsequent history was a progressive adaptation to temperate zones. According to Tutin (1958), the ancestral legume was probably a large tropical tree, and herbaceous annuals were formed according to the sequence; trees → shrubs and woody climbers → perennial herbs → annual herbs.

The development of highly specialized root nodules probably proceeded through several less complex stages before they reached the present degree of organization. The major stages were envisaged by Parker (1957) to be: (i) casual association between free-living N_2-fixing bacteria and the root surfaces of plants; (ii) loose symbiosis on the surface of the root; (iii) entry into, and symbiotic establishment within, the cortical tissues of the root; and (iv) specialization of root tissue, resulting finally in the organized nodule.

III. MECHANISMS BY WHICH N IS ADDED TO SOIL IN NATURE

Several mechanisms have been proposed for the exogenous addition of N to soils under natural conditions, but thus far only fixation of molecular N_2 by microorganisms and accession of NH_4^+ and NO_3^- in atmospheric precipitation are considered important to the soil N cycle. The total amount of atmospheric N returned to land surfaces of the earth has been estimated to be of the order of 194 Tg/year, of which 139 Tg is due to biological N_2 fixation (Burns & Hardy, 1975; Söderlund & Svensson, 1976). This is equivalent to an average annual return of about 12.8 kg of N/ha, or 9.2 kg/ha of biologically fixed N. As one might expect, these amounts are much too low for soils where legumes are grown and too high for most wastelands (e.g., desert regions). An additional 36 Tg of fixed N are added to the oceans each year (Burns & Hardy, 1975).

A. Nitrogen in Atmospheric Precipitation

Combined N, consisting of NH_4^+, NO_2^-, NO_3^-, and organically bound N, is a common constituent of atmospheric precipitation. Nitrite occurs in trace amounts and is usually ignored or included with the NO_3^- determination. The organically bound N is probably associated with terrestrial dust and does not represent a new addition to land masses of the world.

The amount of N added to the soil each year in atmospheric precipitation is normally too small to be of significance in crop production. However, this N may be of considerable importance to the N economy of mature ecosystems, such as undisturbed natural forests and native grasslands. Natural plant communities, unlike domesticated crops, are not subject to continued large losses of N through cropping and grazing, and the N in precipitation serves to restore the small quantities that are lost by leaching and denitrification.

Eriksson (1952) has summarized the measurements for combined N in atmospheric precipitation. For the United States and Europe, the estimates for NH_4^+ plus NO_3^--N range from 0.8 to 22.0 kg of N/ha per year. Many of the higher values may represent analytical or sampling errors, although higher than normal amounts would be expected near highly industrial areas due to burning of fossil fuels.

Tropical air contains 10 to 30% more mineral N than polar air and nearly twice as much as arctic air (Ångstrom & Högberg, 1952). In temperate regions of the earth, the mineral N in precipitation is highest during warm periods of the year, and for any given rainfall, concentrations decrease progressively with duration of precipitation. Rain contains higher quantities of NH_4^+ and NO_3^- than snow, a result that may be due to their greater adsorption in the liquid phase.

Hutchinson (1944) gives the following sources of combined N in atmospheric precipitation:

1) From soil and the ocean
2) From fixation of atmospheric N_2
   a) Electrically
   b) Photochemically
   c) In the trail of meteorites
3) From industrial contamination

Important sources of NH_3 in the atmosphere include volatilization from land surfaces, combustion of fossil fuel, and natural fires. The quantity of N fixed in the trail of meteorites is negligible.

The origin of NO_3^- in atmospheric precipitation is not known with certainty, but formation by electrical discharge during thunderstorm activity has long been a favored theory. However, NO_3^- distribution patterns have seldom correlated with thunderstorm activity. According to Hutchinson (1944), only 10 to 20% of the NO_3^- in precipitation can be accounted for by electrical discharge.

Soils have the ability to absorb atmospheric NH_3, and earlier investigators (notably Liebig) placed considerable emphasis on this process as a means of providing N to plants. The opinion of modern soil scientists is that the process is of little practical significance except under special circumstances. In areas of dense animal populations (e.g., cattle feedlot), redistribution of NH_3 can provide a major input of N to surrounding areas (Hutchinson & Viets, 1969; Elliott et al., 1971). A rapid circulation of NH_3 within a crop canopy has been reported by Denmead et al. (1976), who observed that NH_3 produced near the ground level was almost completely absorbed within the plant cover.

B. Biological N_2 Fixation

Although a vast supply of N occurs in the earth's atmosphere (3.86×10^9 Tg), it is present as an inert gas and cannot be used directly by higher forms of plant and animal life. The covalent triple bond of the N_2 molecule ($N \equiv N$) is highly stable and can be broken chemically only at elevated temperatures and pressures. Nitrogen-fixing microorganisms, on the other hand, perform this seemingly impossible task at ordinary temperatures and pressures.

The total amount of N returned to the earth each year through biological N_2 fixation (land plus ocean) has been estimated at 175 Tg, of which about one half (80 Tg) is contributed by nodulated legumes grown for grain, pasture, hay, and other agricultural purposes (Burns & Hardy, 1975; Hardy & Havelka, 1975; Evans & Barber, 1977). Estimated average rates of biological N_2 fixation for some specific organisms and associations are given in Table 4.

The ability of a few bacteria and blue-green algae to fix molecular N_2 can be regarded as being second in importance only to photosynthesis for the maintenance of life on this earth. The two basic biochemical processes in nature are often considered to be *photosynthesis* and *respiration*; to this list should be added *biological N_2 fixation* and possibly *denitrification*.

The biochemical process of N_2 fixation is responsible for much of the fertility of agricultural soils. Even with the tremendous expansion in facilities for producing fertilizer N since World War II, legumes are still to be regarded as the main source of fixed N for the majority of the world's soils. For many years to come, the N necessary for crop production in underdeveloped countries will have to come from that which is indigenous to the soil or supplied by N_2-fixing microorganisms. Nitrogen fixation in prehistoric times undoubtedly created the combined N that is currently present in many commercially important natural deposits, such as coal, petroleum, and the caliche of the Chilean desert.

The organisms that fix molecular N_2 are conveniently placed into two groups: (i) the nonsymbiotic fixers, or those that fix molecular N_2 apart from a specific host; and (ii) the symbiotic fixers, or those that fix N_2 in association with higher plants.

Table 4—Estimated average rates of biological N_2 fixation for specific organisms and associations (from Evans & Barber, 1977).

Organism or system	N_2 fixed, kg/ha per year
Free-living microorganisms	
Blue-green algae	25
Azotobacter	0.3
Clostridium pasteurianum	0.1–0.5
Plant-algal associations	
Gunnera	12–21
Azollas	313
Lichens	39–84
Legumes	
Soybeans (*Glycine max* L. Merr.)	57–94
Cowpeas (*Vigna, Lespedeza, Phaseolus,* and others)	84
Clover (*Trifolium hybridum* L.)	104–160
Alfalfa (*Medicago sativa* L.)	128–600
Lupines (*Lupinus* sp.)	150–169
Nodulated nonlegumes	
Alnus	40–300
Hippophae	2–179
Ceanothus	60
Coriaria	150

1. NONSYMBIOTIC

Free-living microorganisms capable of utilizing molecular N_2 include a number of blue-green algae of the family Nostocaceae, various photosynthetic bacteria (e.g., *Rhodospirillum*), several aerobic bacteria belonging to the family Azotobacteriaceae (e.g., *Azotobacter, Beijerinckia*), and certain anaerobic bacteria of the Bacillaceae (genus *Clostridium*). A variety of other organisms, including some actinomycetes and fungi, have been reported from time to time to fix molecular N_2, but these claims have yet to be verified, and in any event, the amounts of N fixed would appear to be too small to be of practical significance.

Approximately 40 species of blue-green algae have been reported to fix molecular N_2 in pure culture studies (documented by Stewart, 1970). Those for which fixation has been well established include *Nostoc punctiforme, N. muscorum, N. paludosum, Anabaena variabilis, A. gelatinosa, A. naviculoides, A. humicola, A. cylindrica, A. ambigua, A. fertilissima, Cylindrosphermum licheniforme, C. maius, C. gorakhporense,* and *Aulosira fertilissima* (Fogg, 1947).

The blue-green algae are an archaic group of organisms that have persisted during long epochs of the earth's history. They can be found in practically every environmental situation where sunlight is available for photosynthesis, such as uninhabited wastelands and barren rock surfaces. Their ability to colonize virgin landscapes can be accounted for by the fact that they are completely autotrophic and thereby able to synthesize all of their cellular material from CO_2, molecular N_2, water, and mineral salts. Of additional significance is the fact that they form symbiotic relationships with a variety of other organisms, such as the lichen fungi.

Geographically, lichens are widely distributed over land masses of earth. They make up a considerable portion of the vegetation on the Antarctic continent. Besides being the pioneering plants on virgin landscapes, they bring about the disintegration of rocks to which they are attached, thereby forming soil on which higher plants can get a start. In desert areas of the southwestern United States, they form surface crusts of varying density and cling to surface stones. Crusts of blue-green algae have been found in semiarid soils of eastern Australia and the Great Plains of the United States, where their favorite habitat is the undersurface of translucent pebbles. The initial vegetation on the pumice and ash of Krakatao after the volcanic explosion of 1883, which completely denuded the island of all visible forms of plant life, was a dark-green gelatinous layer containing blue-green algae.

The importance of blue-green algae in supplying combined N to most agricultural soils is probably limited to the initial stages of soil formation. These organisms fix molecular N_2 only in the presence of sunlight. Consequently, their activity is confined almost exclusively to superficial layers of the earth's crust. There is abundant evidence to indicate that blue-green algae are important agents in fixing N_2 in rice (*Oryza sativa* L.) fields.

Classical examples of N_2 fixation by free-living bacteria are by species of the photosynthetic *Rhodospirillum*, the anaerobic heterotroph *Clostridi-*

um, and the aerobic heterotroph *Azotobacter*. To this list can be added *Beijerinckia,* an organism first classified with *Azotobacter* but now relegated to generic rank (Jurgensen & Davey, 1970).

In recent years, the association of N_2-fixing bacteria (*A. paspalum* and *Spirillum lipoferum*) with the roots of corn (*Zea mays* L.) and certain grasses has been established (Hardy & Havelka, 1975).

The requirement of photosynthetic N_2-fixing bacteria for both irradiation and anaerobiosis restricts their activities to shallow, muddy ponds or estuarine muds. They generally are found as a layer that overlies the mud and are covered by a layer of algae. Fixation of N_2 is possible because pigments of the photosynthetic bacteria absorb light rays in the region of the spectra not absorbed by pigments of the overlying algae.

The anaerobic fixer *Clostridium* is universally present in soils, including those too acid for *Azotobacter*. For any given soil, it is more abundant than *Azotobacter,* and it occurs in tropical soils. The normal condition of *Clostridium* is the spore form, vegetative growth occurring only during brief anaerobic periods following rains.

Azotobacter also is widely distributed in soils but is not normally found in acidic soils. The natural habitat of *Beijerinkia* appears to be soils of the tropics and subtropics (Jurgensen & Davey, 1970). This organism is found chiefly in such areas as India, Burma, Java, tropical Africa, northern Australia, and South America.

Under natural soil conditions, the N_2-fixing capabilities of free-living bacteria are greatly restricted. These organisms require a source of available energy, a factor that limits their activities to environments with relatively high organic matter contents. In an early discussion of the subject, Jensen (1950) suggested that many of the estimates for N fixation by nonsymbiotic N_2-fixing bacteria, frequently as high as 20 to 50 kg N/ha per year (18–45 pounds/acre per year), were much too high. He concluded that the level of available organic matter in most soils was too low to support fixation of this magnitude. The consensus of many soil scientists is that no more than about 7 kg of N/ha are added per year to soils of the United States by the combined activities of nonsymbiotic N_2-fixing microorganisms; in semiarid soils, no more than 3 kg of N/ha of N may be fixed per year. On this basis, the amount of N fixed by nonsymbiotic N_2 fixers in soils under intensive cultivation would appear to be too low to have much practical impact. Free-living *Azotobacter* and *Clostridia* may actually fix no more than 0.5 kg of N/ha per year.

Conditions for optimum N_2 fixation by free-living microorganisms include the presence of adequate energy substrates (organic residues, etc.), low levels of available soil N, adequate mineral nutrients, near neutral pH, and suitable moisture (Moore, 1966). In view of the large number of microorgnaisms that have been reported to fix molecular N_2, it would appear that N gains under field conditions result from the cumulative action of numerous organisms fixing rather small amounts of N_2 rather than through fixation by only one or two organisms. The subject of gains in soil N through the activities of free-living N_2-fixing microorganisms has been reviewed by Moore (1966), Jurgensen and Davey (1970), and Burns and Hardy (1975).

Arguments in support of the view that free-living bacteria are not major sources of combined N for plants in most cultivated soils include the following:

1) Free-living N_2 fixers are hetrotrophic and inefficient users of carbohydrates, with only from 1 to 10 mg of N being fixed per gram of carbohydrate used.

2) Heterotrophic N_2 fixers are in severe competition with other bacteria, actinomycetes, and fungi for available organic matter, which is normally in short supply in most soils.

3) Fixation of N_2 is greatly reduced in the presence of readily available combined N. In productive agricultural soils, the level of available N may be sufficiently high to inhibit fixation of N_2.

Substantial gains in N, frequently of the order of 66 to 112 kg/ha per year, have been reported in many soils in the apparent absence of legumes or other plants known to form a symbiotic relationship with N_2-fixing microorganisms (see reviews of Moore, 1966; Jurgensen & Davey, 1970; Burns & Hardy, 1975). Despite the difficulties mentioned below relative to measuring N gains in field soil, it appears rather certain that under special circumstances, significant amounts of combined N can be added through the combined activities of free-living N_2-fixing organisms. Incorporation of organic residues in the soil would be expected to enhance biological N_2 fixation. As noted earlier, a unique opportunity exists in lowland rice for N_2 fixation by blue-green algae.

It is appropriate to mention that considerable difficulty is encountered in evaluating reports for gains in soil N under field conditions. In addition to faulty experimental techniques, lack of statistical control, and errors inherent in measuring small increases in soil N by the Kjeldahl method, the possibility exists that N has accumulated through other means, such as by upward movement of NO_3^- in solution, recycling via plant roots, and accretion from the atmosphere (discussed earlier).

Evidence that nonsymbiotic N_2 fixers often make a significant contribution to the soil N is as follows:

1) More kinds and numbers of N_2-fixing bacteria are present in soil than formally thought. The number of described species of nonsymbiotic fixers has increased greatly in recent years and now number about 100. Although the amount of N_2 fixed by any given species may be small, the combined total for all organisms could be appreciable.

2) The discovery of new N_2 fixers in tropical soils, such as *Beijerinckia*, point to the importance of these organisms in tropical agriculture.

3) The ability to grow crops continuously on the same land for years without N fertilizers (and without growing legumes) is well known and may be due to nonsymbiotic fixation.

4) Appreciable gains in soil N observed for legume-free grass sods suggest extensive fixation in the rhizosphere of crop plants. As noted earlier, recent work has indicated a specialized association of N_2-fixing bacteria with the roots of certain grasses and corn.

The rhizosphere, or that part of the soil near plant roots, would appear to be a particularly favorable site for N_2 fixation because of organic material excreted or sloughed off by roots. In natural plant communities, a relatively low rate of fixation (10 kg of N/ha per year) may be adequate, whereas under intensive farming conditions, fixation of rather large amounts (often >60 kg of N/ha per year) would be required for optimum yields.

2. SYMBIOSIS OF ROOT NODULE BACTERIA WITH LEGUMINOUS PLANTS

The symbiotic partnership between bacteria of the genus *Rhizobium* has had a long and comprehensive development. The importance of this relationship is emphasized by the fact that even with the tremendous expansion in facilities for producing fertilizer N since World War II, legumes are still the main source of fixed N for the majority of the world's soils. On the conservative estimate of an average fixation of 50 kg N/ha for the 30 million ha of legumes planted each year in the United States, a total of 15×10^8 kg of N are fixed annually. For additional information on symbiotic N_2 fixation, see Chapter 10 (this book).

The Leguminosae family contains from 10 to 12 thousand species, most of which are indigenous to the tropics. Thus far, only about 1,200 species have been examined for nodulation, of which about 90% have been found to bear nodules. Approximately 200 species are cultivated by man, and about 50 of these are grown commercially in the United States. Greatest attention has been given to the cultivated legumes, but wild species are of considerable importance for fixation of N_2 in natural ecosystems.

Norris (1956) summarized available information regarding the global distribution of the Leguminosae, from which he compiled the tribal and species distribution. A summary of his tabulation is given in Table 5. The data are undoubtedly obsolete, particularly with respect to the number of species. Nevertheless, the material is adequate for arriving at some generalizations regarding the distribution of the Leguminosae. A smaller number of genera and species of Leguminosae are indigenous to the temperate regions of the earth than to the tropics and subtropics. With the exception of one genus (141 species), the subfamily Mimosoideae is entirely tropical and subtropical, whereas 89 of the 95 genera of plants in the subfamily Caesalpinioideae (>95% of the species) are confined to the tropics and sub-

Table 5—Distribution of the Leguminosae.†

Subfamily	Number of genera and species	Genera and species in tropics and subtropics	Genera and species in temperate regions	Genera occurring in both tropic and temperate zones
Mimosoideae	31–1,341	31–1,200	1–141	1
Caesalpinioideae	95–1,032	89–988	7–44	1
Papilionateae	305–6,514	176–2,430	141–3,084	12
Total	431–8,887	296–4,618	149–3,269	14

† Adapted from Norris (1956). The tabulation is based on early records and does not account for all of the known genera and species of Leguminosae.

tropics. In the subfamily Papilionateae, 141 genera of plants (3,084 species) are located in the temperate regions, and 176 genera (2,430 species) occur in the tropics and subtropics.

The bacterial symbionts, all members of the genus *Rhizobium,* are Gram-negative, nonspore-forming rods that measure 0.5 to 0.9 by 1.2 to 3.0 μm. The following species are generally recognized: *R. meliloti, R. legumi-nosarum, R. phaseoli, R. japonicum, R. lupini,* and *R. trifolii.* A collection of leguminous plants that exhibit specificity for a common *Rhizobium* species is referred to as a *cross-inoculation group.* The validity of these bac-terial-plant associations has been severely challenged because the boundaries between the groups overlap and because some strains of rhizobia form nodules on plants occurring in several different groups. Rates of N_2 fixation by nodulated soybeans, cowpeas, clover, alfalfa, and lupins range from 57 to 600 kg/ha per year, as noted earlier in Table 4.

For the most part, the rhizobia are capable of prolonged independent existence in the soil; however, N_2 fixation takes place only when symbiosis is established with leguminous plants. The maintenance of a satisfactory population of any given *Rhizobium* species in the soil depends largely on the previous occurrence of the appropriate leguminous plant. High acidity, lack of necessary nutrients, poor physical condition of the soil, and attack by bacteriophages contribute to their disappearance. The desirability of legume inoculation to ensure nodulation with host-specific effective rhizobial strains is well known. A discussion of the potential for increasing protein production by legume inoculation has been given by Dawson (1970). The subject of legume seed inoculants has been covered by Roughley (1970).

The presence of sufficient available N in the soil to support the pro-longed growth of plants depresses nodule formation severely, a result that appears to be associated with a low carbohydrate-N ratio in the plant and, consequently, an inadequate supply of carbohydrates to the roots. Numer-ous studies have shown that the quantity of N_2 fixed by rhizobia decreases as the ability of the soil to provide mineral N increases. However, the amount of N released from the soil organic matter is rarely adequate to sup-press N_2 fixation entirely.

3. MICROORGANISMS LIVING IN SYMBIOSIS WITH NONLEGUMINOUS PLANTS

Nitrogen fixation of a nature similar to the symbiotic relationship be-tween the *Rhizobium* and leguminous plants has been demonstrated for many angiosperms, including plants belonging to the families Betulaceae, Casuarinaceae, Coriariaceae, Elaeagnaceae, Myricaceae, Rhamnaceae, and Rosaceae (Becking, 1970). Contrary to popular belief, nodulated non-legumes are not freak plants of limited distribution but are important sources of fixed N for plants in general.

The geographical distribution of the nonleguminous families for which N_2 fixation has been confirmed is outlined in Table 6. Data for the incidence of known species for which N_2 fixation is known to occur were taken from Becking (1970) and Silver (1971) Nodulated nonlegumes occur in 14 genera or 7 families of dicotyledon plants. The 14 genera of nodulated nonlegumes

Table 6—Distribution of nodulated nonlegumes.

Family	Genus	Incidence of nodulating species†	Geographical distribution
Betulaceae	*Alnus*	25/35	Cool regions of the northern hemisphere
Casuarinaceae	*Casuarina*	14/45	Tropics and subtropics, extending from East Africa to the Indian Archipelago, Pacific Islands, and Australia
Coriariaceae	*Coriaria*	12/15	Widely separated regions, chiefly Japan, New Zealand, Central and South America, and the Mediterranean region
Elaeagnaceae	*Elaeagnus*	9/45	Asia, Europe, North America
	Hippophae	1/1	Asia and Europe, from the Himalayas to the Arctic Circle
	Shepherdia	2/3	Confined to North America
Myricaceae	*Myrica*	12/35	Temperate regions of both hemispheres
Rhamnaceae	*Ceanothus*	30/55	Confined to North America
	Discaria	1/10	Temperate, subtropical and tropical regions
Rosaceae	*Cerocarpus*	1/20	Cool regions to temperate zone
	Dryas	3/4	Cool regions of temperate zone
	Purshia	2/2	Cool regions of temperate zone

† Incidence refers to ratio of species bearing nodules to total number of species as reported by Silver (1971). See also Becking (1970).

comprise about 300 plant species, of which about one third have been reported to bear nodules. Very little information is available concerning the organisms responsible for N_2 fixation.

The family Betulaceae, consisting of the birches and alders, is found almost entirely in the cool temperate and arctic zones of the northern hemisphere. Thus far, only the alder (*Alnus*) has been found to bear nodules. Crocker and Major (1955) estimated an annual gain of 50 kg of N/ha by *A. crispa* during colonization of the recessional moraines of Alaskan glaciers. Studies of fossil pollen and tree stumps in peat show that the alder was formerly an abundant tree in the less well-drained parts of England (Bond, 1958). A moderately sized alder tree (*A. glutinosa*) has been found capable of fixing 0.25 to 0.5 kg (0.55–1.10 pounds) of N/ha per year.

Nitrogen fixation has been reported for 14 of the 45 species of Casuarinaceae, the main nonleguminous angiosperm family of nodulating plants occurring in tropical and subtropical areas. Plants of this family are of great ecological significance in the Australian environment.

Twelve of the 15 species of the family Coriariaceae (genus *Coriaria*) have been found to bear nodules. Bond and Montserrat (1958) suggested that the discontinuous distribution of this family indicates that in ancient times, it made a far greater contribution to the supply of fixed N than at present.

The family Elaeagnaceae, consisting of the genera *Elaeagnus, Hippophae,* and *Shepherdia,* is distributed widely in the temperate regions of both hemispheres. The genus *Elaeagnus,* with 45 species (9 of which bear nodules), occurs in Asia, Europe, and North America. *Hippophae,* consisting of a single nodulating species, was a prominent plant in Europe following the Ice Age. *Shepherdia,* a plant confined to North America, consists of three species, two of which nodulate. Crocker and Major (1955) reported that *Shepherdia*, in company with *Alnus,* colonized the moraines of receding glaciers in Alaska.

The family Myricaceae (the galeworts) is distributed widely in the temperate regions of both hemispheres. The family consists of about 35 species of *Myrica,* of which 12 have been found to nodulate. A few species occur in the tropics. Bog myrtle (*M. gale*) may be involved in the fixation of N in acid peats.

The family Rhamnaceae contains 40 genera of trees and shrubs that are spread over most of the globe. However, 30 of the 31 species that have been reported to nodulate occur in *Ceanothus,* a genus of about 55 species confined to North America. More than half of the species are found in the southwestern part of the United States. One species of *Ceanothus* (known locally as deer brush) occurs in practically every plant association in California. In their report on a study of N accumulations during soil development on the Mt. Shasta mudflows in California, Dickson and Crocker (1953) suggested that the build-up may have been due to the symbiosis of N_2-fixing microorganisms with species of *Ceanothus.*

The family Rosaceae occurs typically in cool regions of the temperate zone. Species occurring in three genera have been reported to nodulate.

Scientists have long yearned for the day when N_2-fixing microorganisms could be induced to live in or on the roots of cereal crops, such as corn and wheat. The expectation is that this would permit the farmer to decrease his reliance on N fertilizers without reduction in yields. As noted in the following section, research may ultimately provide a solution to this problem.

4. FUTURE TRENDS AND RESEARCH NEEDS

Considerable attention is currently being given to ways of maximizing biological N_2 fixation as a source of combined N for plants. Interest in this subject has developed from the urgency to solve practical problems related to energy, the environment, and world food requirements. Some goals of research on biological N_2 fixation are as follows (Evans & Barber, 1977):

1) Transfer N_2-fixing genes from bacteria to higher plant cells, thus endowing the plant with the capability for utilizing molecular N_2.
2) Transfer N_2-fixing genes into a beneficial bacterium capable of invading plant cells and establishing an effective N_2-fixing system, such as a nodule.
3) Use protoplast fusion methods to create new symbiotic associations between microorganisms and higher plants.
4) Select or develop by genetic means N_2-fixing bacteria capable of living on the roots of such cereal crops as corn and wheat and providing adequate fixed N for optimum plant growth.
5) Develop by genetic manipulation *Rhizobium* strains that are insensitive to soil NH_4^+ and NO_3^- concentrations that normally inhibit nodulation and N_2 fixation.
6) Develop by use of plant-breeding methods legumes that have in

creased photosynthetic capabilities and, therefore, greater capacities for providing energy to N_2-fixing bacteria in the nodules.

IV. NITROGEN LOSSES FROM SOIL

Quantitative recovery of soil and fertilizer N by growing plants has seldom been achieved under conditions existing in the field. Even under the best circumstances, no more than about two thirds of the N added as fertilizer is accounted for by crop removal during the first growing season. Recoveries considerably less than this are not uncommon (Chapt. 14, this book). Numerous attempts have been made to account for the low recoveries, and it is now known that available mineral forms of N (e.g., NO_3^-), whether added as fertilizer or produced through decay of organic matter, will not remain very long in the soil. Thus, any mineral N not adsorbed by plants or incorporated into the organic matter will be lost by one or more of the processes outlined in other chapters of this book and discussed briefly in the following sections.

A. Volatilization of NH_3

Under suitable conditions, NH_3 can be lost from soils by volatilization. Ammonia entering the atmosphere from world-wide terrestrial sources has been estimated by Söderlund and Svensson (1976) to be as follows: 2 to 6 Tg/year from wild animals, 20 to 35 Tg/year from domestic animals, and 4 to 12 Tg/year from combustion of fossil fuel, giving a total of 26 to 53 Tg N/year. Gaseous loss of NH_3 may account for a large part of the N turnover in a grazing system (Denmead et al., 1974).

Factors affecting NH_3 volatilization from soils have been examined by Denmead et al. (1974, 1976), Mills et al. (1974), Fenn (1975), Ryan and Keeney (1975), and Terman (1979). The facts concerning NH_3 volatilization can be summarized as follows:

1) Losses are of greatest importance on calcareous soils, especially when NH_4^+-containing fertilizers are used. Only slight losses occur in soils of pH 6 to 7, but losses increase markedly as the pH of the soil increases.

2) Losses increase with temperature, and they can be appreciable when neutral or alkaline soils containing NH_4^+ near the surface lose water.

3) Losses are greatest in soils of low CEC, such as sands. Clay and humus adsorb NH_4^+ and prevent its volatilization. In soil with an alkaline reaction, little NH_3 will be lost provided adequate moisture is present.

4) Losses can be high when nitrogenous organic wastes, such as farmyard manures, are permitted to decompose on the soil surface, even a soil that is acid, because of the localized increase in pH resulting from NH_3 formation.

The use of anhydrous NH_3 as fertilizer is widespread. Losses through volatilization are not considered serious as long as the gas is injected well below the surface of the soil and the sorption capacity of the soil is not exceeded.

B. Bacterial Denitrification

In denitrification, NO_3^- and various intermediate reduction products substitute for oxygen as the terminal acceptor of electrons in metabolic reactions. The end products (N_2 and N_2O) are gases, which eventually become part of the atmosphere.

The geochemical significance of denitrification arises from the fact that the process acts as a balance on biochemical N_2 fixation. It is analogous to the relation between photosynthesis and respiration in the C cycle. Just as organically bound C is returned to the atmosphere (as CO_2) through respiration, combined N is returned through denitrification.

Some scientists believe that the reason N_2 is the principal constituent of the earth's atmosphere is because of the continued activity of denitrifying microorganisms throughout geological history. In any event, it is likely that most of the atmospheric N_2 has passed at least once through the denitrification cycle. The annual exchange of N between the atmosphere and the biosphere has been reported to range from 0.017 to 0.034 mg/cm² per year (Hutchinson, 1944). This corresponds to a cycle length of between 44 and 220 million years, or from one tenth to one half of the time span from the Cambrian to the present.

All of the bacteria responsible for denitrification are facultative anaerobes, and they are universally distributed in soils, sediments, and water. Both autotrophic and heterotrophic organisms are involved. The autotrophic denitrifiers include *Micrococcus denitrificans*, a facultative autotroph, and *Thiobacillus denitrificans*, an organism that oxidizes S while reducing NO_3^- under anaerobic conditions. The heterotrophic denitrifers comprise numerous genera. Most of them belong to the genera *Pseudomonas, Micrococcus, Achromobacter,* and *Bacillus.* These organisms are also active in the decay of plant and animal residues in soils and sediments.

Nitrous oxide (N_2O) represents an intermediate in the denitrification process and is normally reduced further to N_2. Consequently, N_2O has only a transitory existence in soil.

Conditions leading to gaseous loss of soil N by bacterial denitrification are as follows:

a. Poor Drainage—Moisture is of importance from the standpoint of its effect on aeration. Denitrification is negligible at moisture levels below about two thirds of the water-holding capacity but is appreciable in flooded soils. The process may occur in anaerobic microenvironments of well-drained soils, such as pores filled with water, the rhizosphere of plant roots, and the immediate vicinity of decomposing plant and animal residues.

b. Temperature at 25°C and Above—Denitrification proceeds at a progressively slower rate at temperatures below 20°C (68°F) and practically ceases at 2°C (36°F).

c. Soil Reaction Near Neutral—Denitrifying bacteria are sensitive to high hydrogen ion concentrations. Their activity in acidic soils ($< pH$ 5) is greatly restricted.

d. Good Supply of Readily Decomposable Organic Matter—The amount of organic matter available to denitrifying microorganisms is generally appreciable in the surface horizon but negligible in the subsoil. Significant amount of soluble organic matter may be found under feedlots as well as in the lower horizons of soils amended with large quantities of organic wastes.

Denitrification can be considered desirable when it occurs below the rooting zone because of reduction in the NO_3^- content of ground water. Denitrifying microorganisms are known to be present at considerable depths in soil, and it is possible that some of the NO_3^- leached into the subsoil may be volatilized before reaching the water table. Meek et al. (1969) concluded that much of the NO_3^- leached into the subsoil in irrigation waters was lost through denitrification. Stewart et al. (1967) found that NO_3^- levels in soil under feedlots decreased sharply with depth and concluded that the decrease was due to denitrification.

C. Leaching

Nitrogen is lost by leaching mainly as NO_3^-, although NH_4^+ may be lost from sandy soils. In intensively cropped soils where fertilizer has not been applied, the loss is greatly reduced, because the NO_3^- content of the soil is lower and less water passes through the soil.

Evidence for N losses through leaching has come from lysimeter experiments. Allison (1955) prepared N balance sheets for a large number of lysimeter experiments conducted in the United States. He reported the following findings:

1) Crops commonly recovered only 50 to 75% of the N that was added or made available from the soil. Low recoveries were usually obtained where large additions of N were made, where the soils were very sandy, and where the crop was not adequate to consume the mineral N.

2) The N content of soils decreased regardless of how much was added as fertilizer unless the soil was kept in uncultivated crops.

3) A large proportion of the N not recovered in the crop was found in the leachate, but substantial unaccounted for losses occurred in most lysimeters. Nitrogen gains were few.

4) The magnitude of the unaccounted for N was largely independent of the form in which the N was supplied, whether as NO_3^-, NH_4^+, or organic N.

5) Unaccounted for N was commonly slightly higher in cropped soils than in fallow soils.

In addition to leaching, considerable N may be lost from the soil as a result of erosion. Sheet erosion is highly selective in that the eroded fraction contains several times more N than the original soil.

D. Chemical Reactions of NO_2^-

Under certain conditions, losses of soil and fertilizer N can occur through chemical reactions involving NO_2^-. Losses by this mechanism have been referred to as being due to "sidetracking of nitrification," as outlined in Chapter 9 (this book).

Nitrite is not usually present in detectable amounts in well-drained neutral or slightly acidic soils. Accumulations occur, however, in calcareous soils, and recent work indicates that this ion often persists, albeit temporarily, when NH_4^+, or NH_4^+-type fertilizers are applied to soil. This NO_2^- accumulation has been attributed to inhibition of nitrification at the NH_4^+ stage. Prsumably, NO_2^--oxidizing organisms (*Nitrobacter*) are more sensitive to NH_3 and an adverse soil reaction than NH_4^+ oxidizers (*Nitrosomonas*).

Four reactions have been suggested for loss of N from soil as a result of the degradation of NO_2^-. These are (i) double decomposition of ammonium nitrite ($NH_4NO_2 \rightarrow 2H_2O + N_2$), (ii) the Van Slyke reaction ($RNH_2 + HNO_2 \rightarrow ROH + H_2O + N_2$), (iii) decomposition of nitrous acid ($3HNO_2 \rightarrow 2NO + HNO_3 + H_2O$), and (iv) chemical reactions with humic and fulvic acids. These reactions are discussed in greater detail in Chapter 9 (this book).

V. FACTORS AFFECTING THE N CONTENT OF SOILS

In undisturbed (uncultivated) soil, the organic matter (and N) content attains a steady-state level (Jenny, 1941) that is governed by the soil-forming factors of climate (cl), topography or relief (r), vegetation and organisms (o), parent material (p), and age or time (t).

$$N = f(cl, r, o, p, t, \ldots) \tag{1}$$

From the above, a set of individual equations can be developed, e.g.,

$$N = f(climate)_{r, o, p, t}, \tag{2}$$

where the subscripts indicate that the remaining factors do not vary.

The numerous combinations under which the soil forming factors operate account for the great variability in the organic matter and N contents of soils even in a very localized area. As suggested by Eq. [2], evaluation of any given factor requires that all other factors remain constant, which seldom, if ever, occurs under natural conditions.

The photosynthetic process is of primary importance in providing the raw material for humus synthesis. By using solar energy plus nutrients derived from the soil, higher plants produce lignin, cellulose, protein, and other organic substances that make up their structures. During decomposition by microorganisms, some of the C is released to the atmosphere as

CO_2, and the remainder becomes part of the soil organic matter. At equilibrium, any loss of C or organic N is balanced by an equivalent gain in newly formed humus through synthesis by microorganisms. Results of studies with ^{14}C-labeled plant residues applied to field soil, summarized by Paul and van Veen (1978), show that for temperate zone soils, approximately one third of the applied C remains behind in the soil after the first year. This residual C has been found to have a mean residence time of about 4 years and to ultimately attain a mean residence time comparable to the native humus.

For all practical purposes, the organic matter content of the soil parallels the N content. It is well known that the C/N ratio of soil organic matter falls within the range of 10 to 12. Thus, factors affecting organic matter content apply equally as well to total N.

Jenny and his co-workers (1930, 1931, 1950, 1960) attempted to evaluate the importance of the soil-forming factors on the N content of the soil by treating each one as an independent variable (Eq. [2]). This approach can be criticized on the grounds that an alteration of any one factor produces changes in one or more of the remaining factors. Notwithstanding, his studies have contributed substantially to our understanding of the factors influencing the N and organic matter contents of soil. According to Jenny (1930), the order of importance of the soil-forming factors in determining the N contents of loamy soils within the United States as a whole is as follows: climate > vegetation > topography = parent material > age.

A. Nitrogen Accumulations during Soil Development (the Time Factor)

Information on the influence of time on soil N levels has come from studies of time sequences (or chronosequences) on mud flows, spoil banks, sand dunes, road cuts, and the moraines of receding glaciers. A detailed account of this work is available elsewhere (Stevenson, 1965).

1. NITROGEN ACCUMULATIONS ON VIRGIN LANDSCAPES

The moraines of Alaskan glaciers offer an excellent opportunity to follow the course of events leading to the establishment of steady-state levels of organic matter in soil using N as an index. The blue-green algae component of lichens, which are usually among the initial colonizers of barren landscapes, are undoubtedly responsible for the early fixation of N. The studies of Crocker and Major (1955) and Crocker and Dickson (1957) show that with the introduction of N into the system, higher forms of plant life become established. A period of plant succession follows, leading eventually to a climax vegetation (spruce forest). The pioneering plants consisted mainly of species of willow and cottonwood, and after about 10 years they were replaced by the alder (*Alnus crispa*) and *Dryas*. These nonlegumes form symbiotic relationships with N_2-fixing microorganisms and dramatic increases in organic matter and N occurred during the 50- to 60-year period they occupied the landscape. In the final spruce stage, no further increases in organic matter or N occurred. Equilibrium levels of N in the soil were attained within about 110 years.

Time functions of N build-up during soil development on the Mt. Shasta mudflows in California (Dickson & Crocker, 1953) bear a striking resemblance to those obtained for the moraines of Alaskan glaciers. However, in the case of the Mt. Shasta mudflows, N accumulated in the mineral soil at a much faster rate (about 3,360 kg/ha in 60 years).

Studies on soil development on other youthful landscapes, such as strip mine spoils (Smith et al., 1971; Caspall, 1975; Hallberg et al., 1978), also show that organic matter (and N) levels increase rapidly during the first few years of soil formation. Equilibrium levels in the surface soil have usually been indicated within a 100-year time period.

Under droughty conditions, rather long periods of time may be required to attain equilibrium levels of N in soil. Thus, Syers et al. (1970) found that organic matter was still increasing after 10,000 years of soil formation on windblown sand in New Zealand. Over extended periods of the order of geological time, the N level may undergo further change due to variations in climate or to alteration in soil composition through pedogenic processes.

An idealized diagram showing the effect of time on the N content of loam soils of the Indo-Gangetic Divide in India, as envisioned by Jenny and Raychaudhuri (1960), is shown in Fig. 3. In this case, steady-state levels of N during soil development on sediments deposited by flood waters during the Asoka period (250 B.C.) were believed to have been attained sometime near the end of the Gupta Dynasty (A.D. 500) and to have remained in this primeval condition for a millennium until the reign of Shah Jahn of Taj Mahal fame (about A.D. 1650), at which time the area was converted to agricultural use. Further changes in soil N levels were subsequently brought about through the activities of man.

Although several reasons have been given for the establishment of equilibrium levels of organic matter in soil, none has provided entirely satisfactory. The following explanations are included:

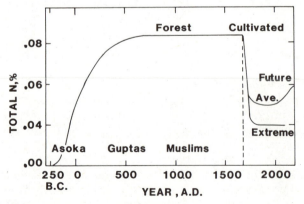

Fig. 3—Idealized N time function for loam soils of the Indo-Gangetic Divide in India as envisioned by Jenny and Raychaudhuri (1960).

1) Organic colloids (e.g., humic acids) that resist attack by microorganisms are produced.
2) Humus is protected from decay through its interaction with mineral matter (e.g., polyvalent cations and clay).
3) A limitation of one or more essential nutrients (N, P, S) places a ceiling on the quantity of stable humus that can be synthesized.

B. Effect of Climate

Climate is the most important single factor that determines the array of plant species at any given location, the quantity of plant material produced, and the intensity of microbial activity in the soil. Consequently, this factor plays a prominent role in determining organic matter levels. If climate is considered in its entirety, a humid climate leads to forest associations and the development of Spodosols (Podzol) and Alfisols (Gray-Brown Podzolic and Red-Yellow Podzolic); a semiarid climate leads to grassland associations and the development of Mollisols (Brunizem, Chernozem, and Chestnut). Grassland soils exceed all other well-aerated soils in organic matter content. Desert, semidesert, and certain tropical soils have the lowest. The profile distribution of N in soils representative of several great soil groups is shown in Fig. 4.

Soils formed under restricted drainage (Histosols and Inceptisols) do not follow a climatic pattern. In these soils, oxygen deficiency prevents complete destruction of organic residues by microorganisms over a wide temperature range.

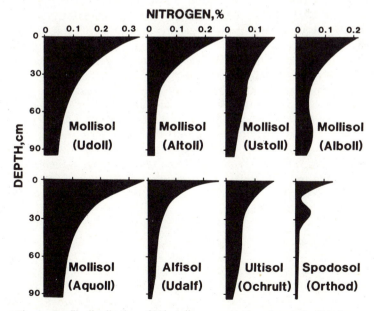

Fig. 4—Profile distribution of N in soils representative of several soil orders.

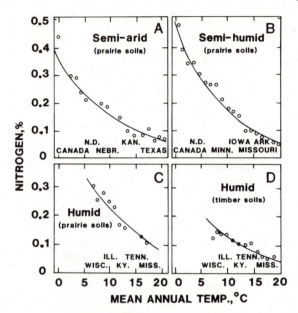

Fig. 5—Average total N content as related to mean annual temperature for soils along north to south transects of the semiarid, semihumid, and humid regions of the central USA (adapted from Jenny, 1930).

Extensive studies were made by Jenny (1930, 1931, 1950, 1960) and Jenny et al. (1948) on the effect of climate on N levels in soil. Jenny's (1930) results for north to south transects of the semiarid, semihumid, and humid regions of the United States are shown in Fig. 5. In each case, the N content of the soil was two to three times lower for each rise of 10°C in mean annual temperature. Whereas relationships derived for U.S. soils cannot be extrapolated directly to other areas (Jenny et al., 1948; Jenny, 1950), it is well known that soils of the warmer climatic zones generally have very low N contents.

Several explanations have been given for the decrease in soil N levels with an increase in mean annual temperature. Jenny (1930, 1931) attempted to relate N levels to the effect of temperature on microbial activity, but this approach can be criticized on the grounds that consideration was not given to temperature effects on photosynthesis (production of raw material for humus synthesis). The differential effect of temperature on the activities of microorganisms and higher plants has been pointed out by Senstius (1958). Specifically, as the temperature is raised, the activities of microorganisms increase more so than does the photsynthetic process of higher plants. Furthermore, the life activities of higher plants start out at lower temperatures than do those of microorganisms (0 vs. 5°C), and the optimum is lower (25 vs. 30°C). On this basis, temperatures below about 25°C should favor the production and preservation of humic substances; higher temperatures should favor destructive processes. Inability to maintain organic N at high levels in tropical soils was attributed by Senstius (1958) to the high activities of microorganisms at the warmer temperatures.

Enders (1943) presented a unique concept concerning the synthesis of genuine stable humus. He concluded that the best soil conditions for the synthesis and preservation of humic substances having high N contents were frequent and abrupt changes in the environment (e.g., humidity and temperature). Consequently, soils formed in harsh continental climates should have higher organic matter and N contents. Harmsen (1951) used this same theory to explain the greater synthesis of humic substances in grassland soils compared with arable land, claiming that in the former the combination of organic substrates in the surface soils and frequent and sharp fluctuations in temperature, moisture, and irradiation leads to greater synthesis of humic substances. According to Harmsen (1951), the extreme surface of the soil (upper few millimeters) is the site of synthesis of humic substances and fixation of N.

The effect of increasing rainfall (moisture component of climate) on soil organic matter content is to promote greater plant growth, and consequently, the production of larger quantities of raw material for humus synthesis. The quantity of vegetable material produced and subsequently returned to the soil can vary from a trace in arid and arctic regions to several metric tons per hectare in warm climates where plant growth occurs throughout the year. Both roots and tops serve as energy sources for humus synthesis.

For grassland soils along a west to east transect of the central United States, a definite correlation exists between the depth of the root system (and thickness of the grass cover) with depth of penetration of organic matter (Fig. 6) and the amount of organic N contained in the surface layer (Fig. 7). Forest soils of the eastern region also show an increase in N with effective rainfall. Total rainfall is not a satisfactory index of soil moisture because of great variations in evaporation. Jenny (see Fig. 7) has used what is known as the NS quotient, which is the ratio of precipitation (in millimeters) to the absolute saturation deficit of the air (in millimeters of Hg).

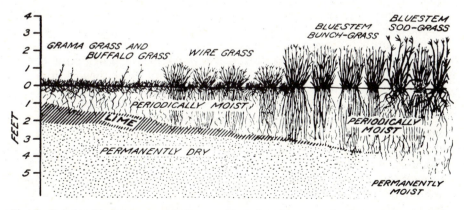

Fig. 6—Relationship between vegetative growth and moisture supply along a west to east transect of the Great Plains (from Shantz, 1923).

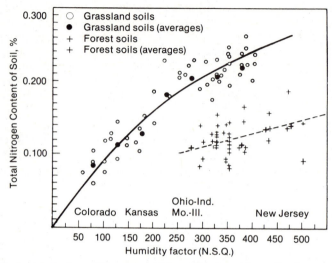

Fig. 7—Relationship between N content and humidity factor for soils along a west to east transect of the central USA (from Jenny, 1930).

C. Vegetation

It is well known that other factors being constant, the organic matter content of grassland soils (e.g., Mollisols) is substantially higher than for forest soils (e.g., Alfisols). Some of the reasons given for this are as follows: (i) larger quantities of raw material for humus synthesis are produced under grass; (ii) nitrification is inhibited in grassland soils, thereby leading to the preservation of N and C; (iii) humus synthesis occurs in the rhizosphere, which is more extensive under grass than under forest vegetation; (iv) inadequate aeration occurs under grass, thereby contributing to organic matter preservation; and (v) the high base status of grassland soils promotes the fixation of NH_3 by lignin (Stevenson, 1965). A combination of several factors is probably involved, with item (iii) being of major importance.

In the case of forest soils, differences in the profile distribution of organic matter and N occur by virtue of the manner in which the leaf litter becomes mixed with mineral matter. In soils formed under deciduous forests on sites that are well drained and well supplied with Ca (Alfisols), the litter becomes well mixed with the mineral layer through the activities of earthworms and faunal organisms. In this case, the top 10 to 15 cm of soil become coated with humus. On the other hand, on sites low in available Ca (Spodosols), the leaf litter does not become mixed with the mineral layer but forms a mat on the soil surface. An organic-rich layer of acid (mor) humus accumulates at the soil surface, and humus accumulates only in the top few centimeters of soil.

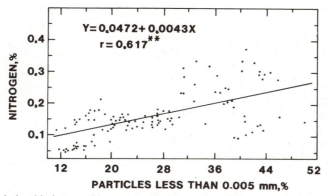

Fig. 8—Relationship between N content and percent of particles <0.005 mm for the surface soils of virgin sod from 13 locations in the Great Plains region of the USA (from Haas et al., 1957).

D. Parent Material

Parent material is effective mainly through its influence on texture. It is a well-known fact that for any given climate zone, provided vegetation and topography are constant, organic matter and N contents depend on textural properties. The fixation of humic substances in the form of organo-mineral complexes serves to preserve organic matter. Thus, heavy-textured soils have higher organic matter contents than loamy soils, which in turn have higher organic matter contents than sandy soils. A typical result showing a positive correlation between organic matter and particles <0.005 mm is given in Fig. 8.

Data reported by Campbell (1978) for the N content of soils from the Canadian prairies show that "Chernozemic" soils developed on glacial till tend to have higher N contents than those developed on lacustrine material, whereas the reverse was found for "Solonetzic" and "Luvisolic" soils of the region.

Organic matter has several characteristics, such as resistance to attack by microorganisms and to removal by chemical extractants, which suggests that it occurs in intimate association with mineral matter. Retention may also be affected by the type of clay mineral present. Montmorillonitic clays, which have high adsorption capacities for organic molecules, are particularly effective in protecting nitrogenous constituents against attack by microorganisms.

E. Topography

Topography, or relief, affects soil organic matter content through its influence on climate, runoff, evaporation, and transpiration. Local variations in topography, such as knolls, slopes, and depressions, modify the plant microclimate, defined by Aandahl (1949) as the climate in the immediate vicinity of the soil profile. Soils occurring in depressions, where the climate is "locally humid," have higher organic matter contents than those occurring on the knolls, where the climate is "locally arid."

Naturally moist and poorly drained soils are usually high in organic matter, because the anaerobic conditions that prevail during wet periods of the year prevent destruction of organic matter. Drainage of organic soils (Histosols) results in subsidence, the lowering of surface elevation. The major cause is enhanced microbial oxidation of the soil organic matter, with an accompanying loss of organic N by conversion to NO_3^- and its loss through leaching and denitrification (Duxbury & Peverly, 1978; Terry, 1980).

F. Effect of Cropping

Marked changes are brought about in the N content of the soil through the activities of man. Usually, but not always, organic N levels decline when soils are first placed under cultivation. For extensive documentation of this work, the reviews of Ensminger and Pearson (1950), Stevenson (1965), and Campbell (1978) are recommended.

Early findings, such as those obtained by Salter and Green (1933) for some rotation experiments at the Ohio Agricultural Experiment Station (Fig. 9) showed that loss of N was least with rotations containing a legume and greatest under continuous row cropping. An item of some significance was that losses were more or less linear over the 31-year period of the experiment. Therefore, unless a change occurred, organic matter content would ultimately reach an absolute minimum, e.g., approach zero. The following equation was derived by Salter and Green (1933) to describe N loss:

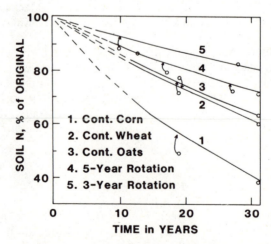

Fig. 9—Changes in the N content of unfertilized soils of the Ohio Agricultural Experimental Station over a 31-year cropping period (from Salter & Green, 1933).

$$N = N_o \exp(-rt), \text{ or } dn/dt = -rN \qquad [3]$$

where N_o is the initial N content and r is the fraction of the N remaining after a single year's cropping.

The conclusion that virtually all of the organic matter (and N) might eventually be lost from soil by cropping caused considerable alarm in agronomic circles. It was feared that unless drastic measures were taken to maintain organic matter reserves, many soils would become unproductive.

Jenny's (1941) early results for losses of soil N over a 60-year period under average farming conditions in the Corn Belt of the United States are in general agreement with the findings of Salter and Green in that cropping caused a decline in organic matter levels. On the other hand, his findings suggested that destruction of organic matter would be far from complete and that a new equilibrium level would ultimately be attained. For soils of the Corn Belt, about 25% of the N was found to be lost the first 20 years, 10% the second 20 years, and 7% the third 20 years.

The validity of Jenny's equilibrium concept has been established in several long-time cropping systems (e.g., see Bartholomew & Kirkham, 1960; Greenland, 1971; Campbell, 1978). Results obtained for the soils at Hays, Kansas, and for the Morrow plots at the University of Illinois are given in Fig. 10 and 11, respectively. In each case, N losses during the early years were much more rapid than later years. In the case of the Morrow plots, losses were greatest with the continuous corn plot and least with the corn-oats-clover rotation plot amended with manure, lime, and phosphate.

Jenny attempted to correct the deficiency in Salter and Green's equation by including a factor for the annual return of N to the soil. Thus, Eq. [3] becomes

$$dN/dt = -rN + A \qquad [4]$$

where A is the annual rate of N addition.

This equation was transformed to the following form by Bartholomew and Kirkham (1960):

$$N = \frac{A}{r} - \left[\frac{A}{r} - N_o\right] \exp(-rt). \qquad [5]$$

A plot of N vs. $\exp(-rt)$ should, therefore, yield a straight line in which the y intercept (A/r) would describe the expected equilibrium value and the term $(N_o - A/r) \exp(-rt)$ the change process. The change in the magnitude of the latter with time provides a measure of the rate of establishment of equilibrium.

Graphical methods were used by Bartholomew and Kirkham (1960) to obtain the constants A and r for the experimental plots of several long-time rotation experiments. In brief, their findings were in agreement with Jenny's observation that equilibrium levels can be expected to be attained within a 50- to 100-year period of cultivation.

Woodruff (1950) envisioned that soil organic matter consisted of com-

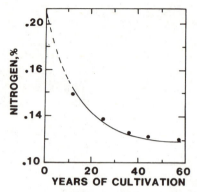

Fig. 10—Decline of soil N content as influenced by years of cropping at Hays, Kansas (from Hobbs & Brown, 1965).

ponents that decomposed at different rates. Thus, Eq. [3] (slightly rearranged) takes the following form:

$$N = N_1 \exp(-r_1 t) + \frac{A_1}{r_1} [1 - \exp(-t_1 t)]$$

$$+ N_2 \exp(-r_2 t) + \frac{A_2}{r_2} [1 + \exp(-r_2 t)] + \text{etc.}$$

[6]

As t approaches infinity, the equation reduces to:

$$N = \frac{A_1}{r} + \frac{A_2}{r} + \ldots \text{etc.}$$

[7]

This expression, as well as Eq. [6], can be simplified to give $N = A/r$, which is the expected equilibrium value.

Crop yields have a potential effect on organic N levels through a feedback effect (greater return of plant residues as yields increase). Russell (1975) used computer-based numerical methods to predict long-term effects of increased yields on N levels for the Morrow plots at the University of Illinois and the Sanborn field at the University of Missouri. The basic equation was:

$$dN/dt = -K_1(t)*N + K_2 + K_3(t)* Y(t)$$

[8]

where $Y(t)$ is plant yield at time t, $K_1(t)$ is the decomposition coefficient, K_2 represents addition to soil organic matter from noncrop sources (including manures), and $K_3(t)$ is a coefficient related to the specific crop at time t. This equation permits estimates to be made of the effect of crop yield within a rotation on soil N levels.

For the Morrow plots, increasing corn yields in a continuous corn system had negligible effects on soil N levels, but strong positive effects were

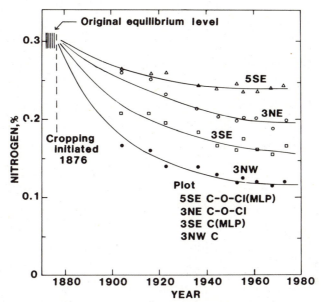

Fig. 11—Effect of long-time rotations on the N content of select soils from the Morrow plots at the University of Illinois. C = corn; O = oats; Cl = clover; MLP = manure-lime-phosphate (from *Nine-Year Report,* Illinois Agric. Exp. Stn., 1938–1947, and unpublished observations; courtesy of S. W. Melsted).

noted for oats and clover. All crops in the Sanborn field had some feedback on soil N levels.

The decline in the N content of the soil when land is cultivated cannot be attributed entirely to a reduction in the quantity of plant residues available for humus synthesis. A temporary increase in respiration rate occurs each time an air-dried soil is wetted (Birch, 1958, 1959), and since considerable amounts of fresh soil are subjected to repeated wetting and drying through cultivation, losses of organic matter by this process would be appreciable. Still another effect of cultivation in stimulating microbial activity may be exposure of organic matter not previously accessible to microorganisms.

For many cultivated soils, particularly the Mollisols, organic matter and N can only be maintained at a level approaching that of the native uncropped soil by inclusion of a sod crop in the rotation or by frequent and heavy applications of manures and crop residues. Increases in soil N levels have been observed by returning previously tilled soil to a grass sod (Jackman, 1964; Clement & Williams, 1967; Giddens et al., 1971; White et al., 1976). A typical result is shown in Fig. 12. Introduction of legumes on soils initially low in N has been reported to lead to increased levels of soil N (see Ensminger & Pearson, 1950). Increases in soil N levels have been observed through zero-tillage also (e.g., Azevedo, 1973; Azevedo & Fernandes, 1973; Fleige & Baeumer, 1974).

In contrast to the work alluded to earlier, indicating the establishment of new equilibrium levels of soil N within a 50- to 100-year period following

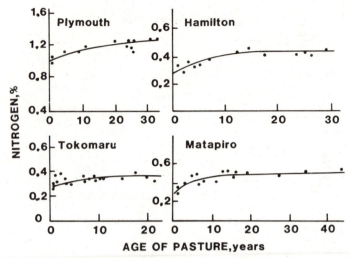

Fig. 12—Changes in the N content of the surface 7.6 cm of soil vs. time after establishment of permanent pasture on four New Zealand soils (from Jackman, 1964).

a change in cultural practice, distinctly linear changes were observed by Rasmussen et al. (1980) for a wheat-fallow cropping sequence on a Pacific Northwest semiarid soil (45-year cropping period). As shown in Fig. 13, linear declines in soil N occurred with or without return of wheat straw residues to the soil; a linear increase occurred when manure was applied at a high rate (22.4 metric tons/ha per year). The greatest loss of N occurred when the wheat straw was burned in the fall. For this soil, well over 100 years will be required for establishment of new equilibrium levels of organic matter and N (Rasmussen et al., 1980).

   Jenkinson and Johnson (1977) demonstrated a pronounced increase in soil N content through long-time applications of manure (35 tons/ha per

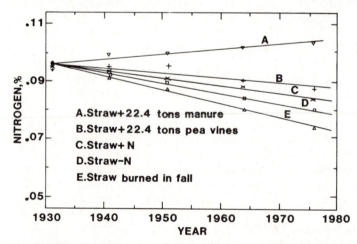

Fig. 13—Soil N levels (30-cm depth) in a silt loam soil near Pendleton, Oregon, as related to crop residue treatment in a wheat-fallow rotation system (from Rasmussen et al., 1980).

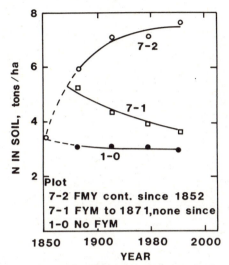

Fig. 14—Nitrogen in the top 23 cm of a soil under continuous barley at the Rothamsted Experimental Station (from Jenkinson & Johnson, 1977).

year) to soil under permanent barley (*Hordeum vulgare* L.) at the Rothamsted Experimental Station in England (Fig. 14). The plots were established in 1852, and equilibrium levels had still not been attained in the manured plot by the time they were sampled 123 years later (in 1975). Only minor changes in soil N levels occurred in the unmanured plot, whereas a slow but continuous decline occurred in a plot receiving farmyard manure annually between 1852 and 1871, but not thereafter. Anderson and Peterson (1973) observed increases in soil N content by long-time applications of manure to a Nebraskan soil (27 metric tons/ha per year).

Results obtained by Larson et al. (1972) indicated that application of plant residues at a rate of 6 tons/ha per year was required for maintenance of organic matter in a Typic Hapludoll soil in Iowa (Marshall silty clay loam); higher application rates led to increases in soil N. Rasmussen et al. (1980) predicted that 5 metric tons of mature crop residue/ha per year was needed to maintain soil organic matter at present levels in a wheat-fallow rotation on Pacific Northwest semiarid soils. For a wheat-fallow system in a cooler climate, a smaller quantity of plant residues may be needed (Black, 1973). Hobbs and Brown (1965) concluded that the application of 56 to 67 metric tons of farmyard manure every 3 years was necessary to prevent further loss of organic matter from some prairie soils of Kansas. In other work, Perepelitsa (1974) observed that "organic fertilizers" applied at rates of 6 metric tons/ha per year or over led to slight increases in the humus content of some Russian soils.

The effectiveness of organic residues in maintaining soil N reserves depends on such factors as rate of application, kind of residue and its N content (C/N ratio), manner of incorporation into the soil, soil characteristics, and seasonal variations in temperature and moisture. Select references on soil management factors affecting organic matter and N levels in the soils of several countries are as follows: Australia (Russell, 1962; Greenland, 1971), Canada (Ridley & Hedlin, 1968, Martel & Paul, 1974; Dubetz et al., 1975, Martel & Deschenes, 1976; Martel & Laverdiere, 1976), France (Morel, 1978), Germany (Fleige & Baeumer, 1974), India (Mathan et al., 1975), Israel (Reinhorn & Avnimelech, 1974), The Netherlands (Kolenbrander, 1974), New Zealand (Jackman, 1964; O'Brien & Stout, 1978), Nigeria (Ayanaba et al., 1976), Portugal (Azevedo, 1973), United Kingdom (Chater & Gasser, 1970; Jenkinson & Johnson, 1977; Jenkinson & Rayner, 1977), United States (Hobbs & Brown, 1965; Oveson, 1966; Larson et al., 1972; Anderson & Peterson, 1973; Black, 1973; Haas et al., 1957; White et al., 1976; Rasmussen et al., 1980), USSR (Perepelitsa, 1974; Koloskova et al., 1979). Factors affecting the fate of organic wastes when they are applied at high disposal rates to soils of the various climatic regions of the United States (arid, cool subhumid and humid, hot humid) are discussed in the volume by Elliott and Stevenson (1977).

VI. SUMMARY

Nitrogen undergoes a wide variety of transformations in soil, most of which involve the organic fraction. Although each process was considered individually, each is affected by others occurring sequentially; in some cases, opposing processes operate simultaneously.

An internal "N cycle" exists in soil apart from the overall cycle of N in nature. Even if N gains and losses are equal, as may occur in a mature ecosystem, the N cycle is not static. Continuous turnover of N occurs through mineralization-immobilization, with transfer of biological decay products into stable humus forms. A complication in interpreting results of ^{15}N measurements is the natural variation in ^{15}N abundance not only between soils but within the various forms of N in any given soil.

Soils vary greatly in their N contents. In undisturbed (uncultivated) soil, the level is governed by the soil-forming factors of climate, topography (or relief), vegetation, parent material, and time. Usually, but not always, the N content of the soil declines and approaches a new equilibrium level during continued cultivation.

Agronomic practices in the future will be geared toward conserving energy and minimizing adverse environmental effect arising from the use of N fertilizers. In the not too distant future, N cycle processes in soil will be continuously monitored by computer analysis of biological data for mineralization, immobilization, nitrification, denitrification, and biological N_2 fixation, thereby providing precise estimates for the quantities of plant available forms of N in the soil at any one time. The internal N cycle will be manipulated so as to protect N from leaching and denitrification while at the same time providing optimum release of organically combined N to the crop. Conventional N fertilizers will continue to be used but in a much more efficient manner than at present. Ultimately, new fertilizers will be developed that will achieve a specific release rate coincident with active uptake by the plant.

LITERATURE CITED

Aandahl, A. R. 1949. The characterization of slope positions and their influence on the total nitrogen content of a few virgin soils of western Iowa. Soil Sci. Soc. Am. Proc. 13:449–454.

Allison, F. E. 1955. The enigma of soil nitrogen balance sheets. Adv. Agron. 7:213–250.

Ångstrom, A., and L. Högberg. 1952. On the content of nitrogen (NH₄-N and NO₃-N) in atmospheric precipitation. Tellus 4:31–42.

Anderson, F. M., and G. A. Peterson. 1973. Effects of continuous corn (*Zea mays* L.), manuring, and nitrogen fertilization on yield and protein content of the grain and on soil nitrogen content. Agron. J. 65:697–700.

Ayanaba, A., S. B. Tuckwell, and D. S. Jenkinson. 1976. The effect of clearing and cropping on the organic reserves and biomass of tropical forest soils. Soil Biol. Biochem. 8:519–525.

Azevedo, A. L. 1973. Evolução do teor em materia orgânica de solos sujeitosa diferentes tratamentos. An. Inst. Super. Agron. Univ. Tec. Lisboa 34:63–114.

Azevedo, A. L., and M. L. V. Fernandes. 1973. Evolucão do teor em materia orgänica de barros castanho-avermelhados sujeitos a um sistema de mobilizacão minima. An. Inst. Super. Agron. Univ. Tec. Lisboa 34:115–137.

Bartholomew, W. V., and D. Kirkham. 1960. Mathematical descriptions and interpretations of culture induced soil nitrogen changes. Int. Congr. Soil Sci. Trans. 7th (Madison, Wis.) II:471–477.

Becking, J. H. 1970. Plant-endophyte symbiosis in non-leguminous plants. Plant Soil 32:611–654.

Birch, H. F. 1958. The effect of soil drying on humus decomposition and nitrogen availability. Plant Soil 10:9–31.

Birch, H. F. 1959. Further observations on humus decomposition and nitrification. Plant Soil 11:262–286.

Black, A. L. 1973. Soil properties associated with crop residue management in a wheat-fallow rotation. Soil Sci. Soc. Am. Proc. 37:943–946.

Bohn, H. L. 1976. Estimate of organic carbon in world soils. Soil Sci. Soc. Am. J. 40:468–470.

Bond, G. 1958. Symbiotic nitrogen fixation by non-legumes. p. 216–231. *In* E. G. Hallsworth (ed.) Nutrition of the legumes. Butterworth, London.

Bond, G., J. T. MacDonnell, and A. H. McCallum. 1956. The nitrogen-nutrition of *Hippophae rhamnoides*. Ann. Bot. (London) 20:501–512.

Bond, G., and P. Montserrat. 1958. Root nodules of *Coriaria*. Nature (London) 182:474–475.

Burns, R. C., and R. W. F. Hardy. 1975. Nitrogen fixation in bacteria and higher plants. Springer-Verlag New York, Inc., New York.

Calvin, M. 1956. Chemical evolution and the origin of life. Am. Sci. 44:248–263.

Campbell, C. A. 1978. Soil organic carbon, nitrogen and fertility. p. 173–271. *In* M. Schnitzer and S. U. Khan (ed.) Soil organic matter. Elsevier North-Holland, Inc., New York.

Caspall, F. C. 1975. Soil development on surface mine spoils in western Illinois. p. 221–228. *In* Proc. Third Symp. on Surface Mining and Reclamation, Vol. II. 21–23 Oct. 1975, National Coal Association and Bituminous Coal Research, Inc. (NCA/BCR), Louisville, Ky.

Chater, M., and J. K. R. Gasser. 1970. Effects of green manuring, farmyard manure, and straw on the organic matter of soil and of green manuring on available nitrogen. J. Soil Sci. 21:127–137.

Clement, C. R., and T. E. Williams. 1967. Leys and soil organic matter. II. The accumulation of nitrogen in soils under different leys. J. Agric. Sci. 69:133–138.

Crocker, R. L., and B. A. Dickson. 1957. Soil development on the recessional moraines of the Herbert and Mendenhall glaciers, southeastern Alaska. J. Ecol. 45:169–185.

Crocker, R. L., and J. Major. 1955. Soil development in relation to vegetation and surface age at Glacier Bay, Alaska. J. Ecol. 43:427–448.

Dawson, R. C. 1970. Potential for increasing protein production by legume inoculation. Plant Soil 32:655–673.

Delwiche, C. C. 1970. The nitrogen cycle. Sci. Am. 223:136–146.

Denmead, O. T., J. R. Freney, and J. R. Simpson. 1976. A closed ammonia cycle within a plant canopy. Soil Biol. Biochem. 8:161–164.

Denmead, O. T., J. R. Simpson, and J. R. Freney. 1974. Ammonia flux into the atmosphere from a grazed pasture. Science 185:609–610.

Dickerson, R. E. 1978. Chemical evolution and the origin of life. Sci. Am. 239:70–86.

Dickson, B. A., and R. L. Crocker. 1953. A chronosequence of soils and vegetation near Mt. Shasta, California. II. The development of the forest floors and the carbon and nitrogen profiles of the soils. J. Soil Sci. 4:142–154.

Dubetz, S., G. C. Kozub, and J. F. Dormaar. 1975. Effects of fertilizer, barnyard manure, and crop residues on irrigated crop yields and soil chemical properties. Can. J. Soil Sci. 55: 481–490.

Duxbury, J. M., and J. H. Peverly. 1978. Nitrogen and phosphorus loss from organic soils. J. Environ. Qual. 7:566–570.

Elliott, L. F., G. E. Schuman, and F. G. Viets, Jr. 1971. Volatilization of nitrogen-containing compounds from beef cattle areas. Soil Sci. Soc. Am. Proc. 35:752–755.

Elliott, L. F., and F. J. Stevenson (ed.). 1977. Soils for management of organic wastes and waste waters. Am. Soc. of Agron., Madison, Wis.

Emery, K. O., W. L. Orr, and S. C. Rittenberg. 1955. Nutrient budgets in the ocean. p. 299–309. *In* Essays of the natural sciences in honor of Captain Allan Hancock. Univ. of Southern California Press, Los Angeles.

Enders, C. 1943. Über den Chemismurs der Huminsäurebilding unter Physiologischen Bedingungen. Biochem. Z. 315:259–292:352–371.

Ensminger, L. E., and R. Pearson. 1950. Soil nitrogen. Adv. Agron. 2:81–111.

Eriksson, E. 1952. Composition of atmospheric precipitation. 1. Nitrogen compounds. Tellus 4:215–232.

Evans, H. J., and L. E. Barber. 1977. Biological nitrogen fixation for food and fiber production. Science 197:332–339.

Fairbridge, R. W. (ed.). 1972. Encyclopedia of geochemistry and environmental sciences: IVA. Van Nostrand Reinhold, New York, p. 795–801, 836, 837, 849.

Fenn, L. B. 1975. Ammonia volatilization from surface applications of ammonium compounds on calcareous soils. III. Effects of mixing low and high loss ammonium compounds. Soil Sci. Soc. Am. Proc. 39:366–368.

Fleige, H., and K. Baeumer. 1974. Effect of zero-tillage on organic carbon and total nitrogen content, and their distribution in different N-fractions in loessial soils. Agro-Ecosystems 1:19–29.

Fogg, G. E. 1947. Nitrogen fixation by blue-green algae. Endeavour 6:172–175.

Fox, S. W. 1963. Prebiological formation of biochemical substances. p. 36–49. *In* I. A. Breger (ed.) Organic geochemistry. Pergamon Press, Inc., Oxford.

Giddens, J., W. E. Adams, and R. W. Dawson. 1971. Nitrogen accumulation in fescuegrass sod. Agron. J. 63:451–454.

Greenland, D. J. 1971. Changes in the nitrogen status of soils under pastures with special reference to the maintenance of the fertility of Australian soils used for growing wheat. Soils Fert. 34:237–251.

Haas, H. J., C. E. Evans, and E. F. Miles. 1957. Nitrogen and carbon changes in Great Plains soils as influenced by cropping and soil treatments. USDA Tech. Bull. 1164:1–111.

Hallberg, G. R., N. C. Wollenhaupt, and G. A. Miller. 1978. A century of soil development in spoil derived from loess in Iowa. Soil Sci. Soc. Am. J. 42:339–343.

Hardy, R. W. F., and V. D. Havelka. 1975. Nitrogen fixation research: A key to world food? Science 188:633–643.

Harmsen, G. W. 1951. Die Bedeutung der Bodenoberfläche für die Humusbildung, Plant Soil 3:110–140.

Hobbs, J. A., and P. L. Brown. 1965. Effects of cropping and management on nitrogen and organic carbon contents of a western Kansas soil. Kansas Agric. Exp. Stn. Tech. Bull. 144:1–37.

Hutchinson, G. E. 1944. Nitrogen in the biogeochemistry of the atmosphere. Am. Sci. 32: 178–195.

Hutchinson, G. E. 1954. The biochemistry of the terestrial atmosphere. p. 371–433. *In* G. P. Kuiper (ed.) The solar system. Vol. II. The earth as a planet. Univ. of Chicago Press, Chicago.

Hutchinson, G. L., and F. G. Viets, Jr. 1969. Nitrogen enrichment of surface water by absorption of ammonia volatilized from cattle feedlots. Science 166:514–515.

Imshenetskii, A. A. 1962. Evolution of biological fixation of nitrogen. Comp. Biochem. Physiol. 4:353–361.

Jackman, R. H. 1964. Accumulation of organic matter in some New Zealand soils under permanent pasture. N. Z. J. Agric. Res. 7:445–479.

Jenkinson, D. S., and A. E. Johnson. 1977. Soil organic matter in the Hoosfield continuous barley experiment. Rothamsted Exp. Stn. report for 1976, Part 2. Harpenden, Herts, England, p. 81–101.

Jenkinson, D. S., and J. H. Rayner. 1977. The turnover of soil organic matter in some of the Rothamsted classical experiments. Soil Sci. 123:298–305.

Jenny, H. 1930. A study on the influence of climate upon the nitrogen and organic matter content of the soil. Missouri Agric. Exp. Stn. Res. Bull. 152:1–66.

Jenny, H. 1931. Soil organic matter-temperature relationships in the eastern United States. Soil Sci. 31:247–252.

Jenny, H. 1941. Factors of soil formation. McGraw-Hill Book Co., New York.

Jenny, H. 1950. Causes of the high nitrogen and organic matter content of certain tropical forest soils. Soil Sci. 69:63–69.

Jenny, H. 1960. Comparison of soil nitrogen and carbon in tropical and temperate regions. Missouri Agric. Exp. Stn. Res. Bull. 765:1–30.

Jenny, H., F. Bingham, and B. Padilla-Saravia. 1948. Nitrogen and organic matter contents of equatorial soils of Colombia, South America. Soil Sci. 66:173–186.

Jenny, H., and S. P. Raychaudhuri. 1960. Effect of climate and cultivation on nitrogen and organic matter reserves in Indian soils. Indian Council Agric. Res., New Delhi, India, p. 1–126.

Jensen, H. L. 1950. A survey of biological nitrogen fixation in relation to the world supply of nitrogen. Int. Congr. Soil Sci., Trans. 4th (Amsterdam) I:165–172.

Jurgensen, M. F., and C. B. Davey. 1970. Nonsymbiotic nitrogen fixing microorganisms in acid soils and the rhizosphere. Soils Fert. 33:435–436.

Kerr, R. A. 1980. Origin of life: New ingredients suggested. Science 210:42–43.

Kolenbrander, G. J. 1974. Efficiency of organic manure in increasing soil organic matter content. Int. Congr. Soil Sci., Trans. 10th (Harne, The Netherlands) II:129–136.

Koloskova, A. V., Ye. S. Mikhaylova, and S. G. Murtazina. 1979. Effect of cultivation on the content and forms of nitrogen in the non-Chernozemic sods of the Volga-Kama forest-steppe. Sov. Soil Sci. 11:142–150.

Kvenvolden, K. 1974. Geochemistry and the origin of life. Halsted Press, New York.

Larson, W. E., C. E. Clapp, W. H. Pierre, and Y. B. Morachan. 1972. Effect of increasing amounts of organic residues on continuous corn. II. Organic carbon, nitrogen, phosphorus, and sulfur. Agron. J. 64:204–208.

Martel, Y. A., and J. M. Deschenes. 1976. Les effects de la mise en culture et de la prairie prolongée sur le carbone, l'azote et la structure de quelques sols du Quebec. Can. J. Soil Sci. 56:373–383.

Martel, Y. A., and M. R. Laverdiere. 1976. Facteurs qui influencent la teneur de la matiere organique et les proprietes d'echange cationique des horizons Ap des sols de grande culture du Quebec. Can. J. Soil Sci. 56:213–221.

Martel, Y. A., and E. A. Paul. 1974. Effects of cultivation on the organic matter of grassland soils as determined by fractionation and radiocarbon dating. Can. J. Soil Sci. 54:419–426.

Mathan, K. K., K. Sankaran, N. Kanakbushani, and K. K. Krishnamoorthy. 1975. Effect of continuous cropping on the organic carbon and total nitrogen content of a black soil. Indian J. Soil Sci. 26:283–285.

Meek, D. B., L. B. Grass, and A. J. MacKenzie. 1969. Applied nitrogen loss in relation to oxygen status of soils. Soil Sci. Soc. Am. Proc. 33:575–578.

Mills, H. A., A. V. Barker, and D. N. Maynard. 1974. Ammonia volatilization from soils. Agron. J. 66:355–358.

Moore, A. W. 1966. Non-symbiotic nitrogen fixation in soil and soil-plant systems. Soils fert. 29:113–128.

Morel, R. 1978. Quelques aspects nouveaux de la dynamique du carbone et de l'azote le sol. Ann. Agron. 29:357–379.

Norris, D. O. 1956. Legumes and the *Rhizobium* symbiosis. Emp. J. Exp. Agric. 24:247–270.

O'Brien, B. J., and J. D. Stout. 1978. Movement and turnover of soil organic matter as indicated by carbon isotope measurements. Soil Biol. Biochem. 10:309–317.

Oda, Y. 1959. Significance of molecular hydrogen metabolism on the transitionary stage from anaerobiosis to aerobiosis. p. 593–605. *In* A. I. Oparin (ed.) The origin of life on the earth. Pergamon Press, London.

Oveson, M. M. 1966. Conservation of soil nitrogen in a wheat-fallow farming practice. Agron. J. 58:444–447.

Parker, C. A. 1957. Evolution of nitrogen-fixing symbiosis in higher plants. Nature (London) 179:593–594.

Paul, E. A., and J. A. van Veen. 1978. The use of tracers to determine the dynamic nature of organic matter. Int. Congr. Soil Sci., Trans. 11th (Edmonton, Canada) III:61–102.

Perepelitsa, V. M. 1974. Role of organic and mineral fertilizers in humus accumulation. Sov. Soil Sci. 6:151–159.

Rasmussen, P. E., R. R. Allmaras, R. C. Rohde, and N. C. Roager, Jr. 1980. Crop residue influences on soil carbon and nitrogen in a wheat-fallow system. Soil Sci. Soc. Am. J. 44: 596–600.

Reinhorn, T., and Y. Avnimelech. 1974. Nitrogen release associated with decrease in soil organic matter in newly cultivated soil. J. Environ. Qual. 3:118–121.

Ridley, A. O., and R. A. Hedlin. 1968. Soil organic matter and crop yields as influenced by frequency of summer fallowing. Can. J. Soil Sci. 48:315–322.

Roughley, R. J. 1970. The preparation and use of legume seed inoculants. Plant Soil 32:675–701.

Ryan, J. A., and D. R. Keeney. 1975. Ammonia volatilization from surface-applied waste water sludge. J. Water Pollut. Control Fed. 47:386–393.

Russell, J. S. 1962. Estimation of the time factor in soil organic matter equilibration under pasture. p. 191–196. *In* Int. Soc. Soil Sci., Trans. Comm. IV, V (Palmerston, N.Z.).

Russell, J. S. 1975. A mathematical treatment of the effect of cropping system on soil organic nitrogen in two long-term sequential experiments. Soil Sci. 120:37–44.

Salter, R. M., and T. C. Green. 1933. Factors affecting the accumulation and loss of nitrogen and organic carbon in cropped soil. J. Am. Soc. Agron. 25:622–630.

Schreiner, O., and B. E. Brown. 1938. Soil nitrogen. p. 361–372. *In* Soils and men: Yearbook of agriculture 1938, U.S. Government Printing Office, Washington, D.C.

Senstius, M. W. 1958. Climax forms of rock-weathering. Am. Sci. 46:355–367.

Shantz, H. L. 1923. The natural vegetation of the Great Plains region. Ann. Assoc. Am. Geog. 13:81–107.

Silver, W. S. 1971. Physiological chemistry of non-leguminous symbiosis. p. 245–281. *In* J. R. Postgate (ed.) The chemistry and biochemistry of nitrogen fixation. Plenum Press, New York.

Smith, R. M., E. H. Tyron, and E. H. Tyner. 1971. Soil development on mine spoil. West Virginia Agric. Exp. Stn. Bull. 604T:1–47.

Söderlund, R., and H. B. Svensson. 1976. The global nitrogen cycle. Ecol. Bull. (Stockholm) 22:23–73.

Stevenson, F. J. 1962. Chemical state of the nitrogen in rocks. Geochim. Cosmochim. Acta 26:797–809.

Stevenson, F. J. 1965. Origin and distribution of nitrogen in soil. *In* W. V. Bartholomew and F. E. Clark (ed.) Soil nitrogen. Agronomy 10:1–42. Am. Soc. of Agron., Madison, Wis.

Stewart, B. A., F. G. Viets, Jr., G. L. Hutchinson, and W. D. Kemper. 1967. Nitrate and other water pollutants under fields and feedlots. Environ. Sci. Technol. 1:736–739.

Stewart, W. D. P. 1970. Algal fixation of atmospheric nitrogen. Plant Soil 32:555–558.

Syers, J. K., J. A. Adams, and T. W. Walker. 1970. Accumulation of organic matter in a chronosequence of soils developed on wind-blown sand in New Zealand. J. Soil Sci. 21: 146–153.

Terman, G. L. 1979. Volatilization losses of nitrogen as ammonia from surface-applied fertilizers, organic amendments, and crop residues. Adv. Agron. 31:189–223.

Terry, R. E. 1980. Nitrogen mineralization in Florida Histosols. Soil Sci. Soc. Am. J. 44: 747–750.

Tutin, T. G. 1958. Classification of the legumes. p. 3–14. *In* E. G. Hallsworth (ed.) Nutrition of the legumes. Butterworth, London.

White, E. M., C. R. Krueger, and R. A. Moore. 1976. Changes in total N, organic matter, available P, and bulk densities of cultivated soils 8 years after tame pastures were established. Agron. J. 68:581–583.

Wilson, P. W. 1951. Biological nitrogen fixation. p. 467–499. *In* C. W. Werkman and P. W. Wilson (ed.) Bacterial physiology. Academic Press, Inc., New York.

Woodruff, C. M. 1949. Estimating the nitrogen delivery from the organic matter determination as reflected by Sanborn Field. Soil Scil Soc. Am. Proc. 14:208–212.

2 Inorganic Forms of Nitrogen in Soil[1]

J. L. YOUNG

Agricultural Research Service, USDA, Oregon State University
Corvallis, Oregon

R. W. ALDAG

Institut für Bodenkunde
Göttingen, West Germany

I. INTRODUCTION

Inorganic N compounds detected in soils number about six, including NO_3^-, NO_2^-, exchangeable NH_4^+, nonexchangeable (mineral-fixed) NH_4^+, dinitrogen gas (N_2), and nitrous oxide (N_2O). Other chemical or biological intermediates, such as nitric oxide (NO), nitrogen dioxide (NO_2), hydroxylamine (NH_2OH), hyponitrous acid (HON=NOH), azide (N_3^-), and even some cyano compounds may also be formed in the soil milieu but have yet to be unequivocally so identified.

Until the late 1950's, the combined total of inorganic N forms was thought to constitute less than one to a few percent at most of the total soil N. Now we know that many soils contain appreciable amounts of N in the form of fixed NH_4^+, particularly in lower horizons. Unfortunately, many textbooks on soils still perpetuate the erroneous concept that practically all the inorganic N in soils occurs as NO_3^- or exchangeable NH_4^+ (e.g., Thompson & Troeh, 1978, p. 235). Although the cyclically available forms (mainly NO_3^- and exchangeable NH_4^+) remain of greatest immediate interest to crop production, other inorganic forms of soil N merit consideration. An expanding litany of environmental issues coupled with world food demands will ensure continuing attention to all forms of soil N, inorganic and organic alike.

This chapter considers mainly post-1965 information on indigenous levels of inorganic N compounds in soil, with emphasis on native mineral-fixed NH_4^+-N, which may constitute over 50% of total N in some subsoils. Reference to all pertinent articles is precluded, but an effort has been made to cite representative papers, including some from less quoted non-English language countries. Scant coverage will be given to transformations, availability, gains, losses, and movement of inorganic forms of soil N since these are subjects of subsequent chapters. Similarly, consideration of methods for water soluble, exchangeable, and especially gaseous forms of inorganic soil N will be minimal; detailed coverage appears in Chapter 13 (this book).

[1] Oregon Agricultural Experiment Station Technical Paper no. 5487, Corvallis, Oregon.

Nitrogen in Agricultural Soils—Agronomy Monograph no. 22.

II. MINERAL-FIXED NH₄⁺-N

Many studies within the past 20 to 25 years have revealed that most mineral soils contain some nonexchangeable NH_4^+ (NH_4^+ held within the lattice structures of silicate minerals).

This entrapped NH_4^+, like K^+, fills interstitial lattice voids formed by hexagonal oxygen rings, and it serves to neutralize negative charges arising from isomorphous substitution (e.g., Mg^{2+} for Al^{3+} in the octahedral and Al^{3+} for Si^{4+} in the tetrahedral lattice layers). This nonexchangeable NH_4^+ resists removal by neutral salt solutions typically used for extracting exchangeable ions; thus, it is considered generally unavailable to plants and microorganisms.

A. General Terminology

Because the terms "fixed NH_4^+" and "fixed N" are used indiscriminantly and sometimes interchangeably, confusion often arises. Biologists and chemists refer to N that is chemically bound to either organic or inorganic materials but releasable by hydrolysis to NH_3 or NH_4^+ as being "fixed." Molecular N combined with H to yield NH_3 or NH_4^+ via autotrophic or symbiotic organisms or by chemical or industrial processes is also called fixed N. Thus, the same unmodified "fixed N" term may evoke quite different images to persons hearing or reading the same paper. Hence there is need to distinguish between "mineral fixed," "organic fixed," "chemically or industrially fixed," and "biologically fixed" NH_3 or NH_4^+ (Bremner, 1965; Young & Cattani, 1962). Because of these terminology problems, Osborne (1976a) recently proposed "intercalated NH_4^+" as a more accurate term for nonexchangeable NH_4^+ held in mineral lattices. Whether entrenched usage of other terms will give way seems doubtful, despite the merits of Osborne's arguments. By editor choice for brevity and uniformity, the short term "fixed NH_4^+" is used hereafter in this chapter in place of both "mineral-fixed NH_4^+-N" and "fixed NH_4^+-N."

B. Origin

The first report of naturally occurring fixed NH_4^+ in soils is usually credited to Rodrigues (1954), who concluded that some tropical soils of the British Caribbean region contained between 282 to 1,920 ppm of fixed NH_4^+. Subsequent studies (Dhariwal & Stevenson, 1958; Bremner, 1959, 1965; Bremner et al., 1967; Mogilevkina, 1969) indicated that these values were erroneously high because of inadequacies in methodology. Nevertheless, Rodrigues' concept that soils contain appreciable amounts of fixed NH_4^+ has gained wide acceptance.

Whether most of this fixed NH_4^+ was incorporated in the parent rock material or whether most of it has been formed and fixed during pedogenetic soil-forming processes remains a debated question. See Stevenson & Dhariwal (1959) and Chapter 1 (this book) for more details about geochemical aspects of soil N. Because of possible interchange involving mineral fixation of NH_4^+ produced during decay of organic matter and concurrent partial utilization of indigenous fixed NH_4^+ by plants and microorganisms, one could expect that part of the fixed NH_4^+ in surface soils is of recent origin. Such an explanation would not account for the relatively large quantities of fixed NH_4^+ in subsurface layers or C horizons largely devoid of plant roots. In nonhumid areas, the NH_4^+ produced through decay of organic matter in surface layers would rarely reach these lower horizons for lack of percolating water even if the NH_4^+ escaped nitrification and were subject to leaching.

Some fixed NH_4^+ found in clay-enriched B horizons of many soils could originate in clay micelles redeposited from the overlying horizons. However, most of the native fixed NH_4^+, particularly in the lower horizons of residual soils, probably has been inherited from in situ parent material. Observations that soil-forming rocks and minerals contain appreciable amounts of native fixed NH_4^+ (Stevenson, 1962; Baur & Wlotzka, 1972; Aldag, 1978; Strathouse et al., 1980) support this probability. Data in Table 1 show that the concentrations of fixed NH_4^+ in several sedimentary rocks (shales) ranged from 210 to 530 ppm; in carbonaceous phyllosilicates from 250 to 1,540 ppm; and in some igneous rocks from 10 to 84 ppm. Values of 8 to 690 ppm for the silicate minerals are of the same order as Schachtschabel (1961) found for illite (294–896 ppm), muscovite (518 ppm), and biotite (84 ppm) and as Young and McNeal (1964) found for 16 specimen

Table 1—Concentration of fixed NH_4^+ in some soil forming rocks and minerals (Stevenson, 1962; Aldag, unpublished data).

Sample[†]	Mineral-fixed NH_4^+-N	
	ppm	as % of N
Igneous rocks		
Granite (6)	11–63	55.9–90.0
Mica peridotite (2)	13–84	61.9–80.8
Others[‡] (5)	10–39	70.0–95.2
Metamorphic rocks		
Slate (1)	378	96.9
Gneiss (2)	17–22	77.3–94.4
Sedimentary rocks[§] (shales)		
Carbonaceous (3)	210–530	12.1–13.2
Noncarbonaceous (3)	330–420	61.7–70.2
Carbonaceous phyllosilicates (9)	250–1,540	30.3–80.6 (Aldag)
Silicate minerals		
Muscovite (3)	10–106	91.0–99.0
Biotite (3)	10–36	88.9–91.0
Illite (1)	690	75.5
Vermiculite (1)	36	90.0
Orthoclase feldspar (4)	8–41	65.9–95.5

† Numeral in parentheses indicates number of samples analyzed.
‡ Includes pegmatite, basalt, porphyry, gabbro, and dunite.
§ See Sullivan et al. (1979) for recent data on several dozen samples.

silicate minerals (2–704 ppm). That some of the fixed NH_4^+ in soils originally existed as NH_3 in the atmosphere and became fixed when washed onto the earth's surface by rain water during early geological times, as proposed by Stevenson and Dhariwal (1959), seems plausible but will be difficult to prove. Baur and Wlotska's (1972) more extensive compilation of natural NH_4^+ abundances in other geological samples includes some discussion of origin and mechanisms of N fixation in rocks, rock-forming minerals, lunar soils, meteorites, and sediments.

C. Methods of Estimation

Many methods have been devised for determining fixed NH_4^+. None has proved entirely satisfactory, mainly because NH_4^+ is produced from labile organic N compounds (amides, amino sugars, etc.) during the recovery of fixed NH_4^+ (Stevenson et al., 1967; Niederbudde & Fischer; 1974). Initially, estimates of fixed NH_4^+ were calculated as the difference between NH_4^+ distilled by NaOH vs. that distilled by KOH (Barshad, 1951; Allison & Roller, 1955). But by 1969, more than 10 methods for determining native fixed NH_4^+ had been published, each of which gave substantially different results (Mogilevkina, 1969). Main features of the more common methods are outlined below.

a. **Method of Rodrigues (1954)**—The soil is treated with an acid mixture of $HF-H_2SO_4$ (4:1, v/v). Considerable heat is generated by the treatment. After removal of HF by further heating with H_2SO_4, the released NH_4^+ is estimated by distillation with alkali. Deamination of organic N compounds gives high results.

b. **Method of Dhariwal and Stevenson (1958)**—A 2-g sample of soil is treated with hot $1N$ KOH to remove exchangeable NH_4^+ and labile organic N compounds. The residue is then washed with $1N$ KCl, and clay minerals are dissolved with 15 ml of a $5N$ $HF-0.75N$ $HCl-0.6N$ H_2SO_4 solution for 12 to 16 hours. Liberated NH_4^+ is recovered by distillation with NaOH.

c. **Method of Bremner (1959)**—A 5-g sample of soil is extracted with 100 ml of a $1N$ $HF-1N$ HCl solution for 24 hours. The mixture is then filtered, and an aliquot of the neutralized filtrate is distilled with pH 8.8 buffer. Exchangeable NH_4^+ is estimated separately and subtracted from the value obtained by HF extraction.

d. **Method of Schachtschabel (1961)**—The soil is first treated with H_2O_2 in the presence of KCl to destroy organic matter. The residue from this treatment is washed with $0.05N$ HCl, water, and ethanol, after which the dried residue is heated with an acid mixture consisting of 50% HF and concentrated H_2SO_4. The liberated NH_4^+ is subsequently distilled with NaOH. Because the H_2O_2 treatment does not completely remove organic matter, a correction is made by the assumption that 3 mg of NH_4^+-N originated from organic matter for each 100 mg C remaining in the H_2O_2-treated residue.

e. Ignition Method of Mogilevkina (1964)—The soil is heated at 400°C for 24 to 72 hours, depending on the organic matter content of the sample. Fixed NH_4^+ is then estimated by Kjeldahl analysis of the residue. The assumptions are that organic N and exchangeable NH_4^+ are quantitatively removed during the heating process, that none of the NH_4^+ released from the organic matter is fixed by clay minerals, and that the heat treatment does not lead to liberation of fixed NH_4^+.

f. Method of Silva and Bremner (1966), Procedure A—Organic matter and exchangeable NH_4^+ are removed by treating the soil (1 g) with alkaline potassium hypobromite (KOBr, 20 ml). After standing for 2 hours, the mixture is boiled for 5 minutes, and the residue is washed with $0.5M$ KCl solution. Mineral residue is dissolved with 20 ml of a $5N$ HF-$1N$ HCl solution, and the liberated NH_4^+ is recovered by distillation with KOH.

g. Method of Silva and Bremner (1966), Procedure B—This method differs from the KOBr-treated residue by heating with the HF–HCl mixture for 30 minutes at 100°C.

Several of these methods for determining fixed NH_4^+ in soil have been compared by Bremner et al. (1967) and Mogilevkina (1969). The results of the two studies, summarized in Table 2, further emphasize the discrepancies between methods for estimating fixed NH_4^+. Although both studies demonstrated that the various methods gave divergent results, agreement was lacking as to which procedure(s) gave the more reliable values. However, method A of Silva and Bremner (1966) seems most widely accepted by investigators so far. The hot KOH pretreatment method (Dhariwal and Stevenson, 1958) is basically the same approach and gives similar results (as concluded by Bremner et al., 1967; Osborne, 1976a; and Young, unpublished data, 1967). More recently, Osborne (1976a) compared four methods and found discrepancies that prompted a proposal to redefine fixed NH_4^+ as intercalated NH_4^+.

Mogilevkina (1969) concluded that a modification of his ignition method gave values comparable with the Silva and Bremner method, but this was not confirmed by Nelson and Bremner (1966). The contention that most fixed NH_4^+ is an artifact (Freney, 1964) also lacks favor or confirmation.

Table 2—Comparison of methods for determining fixed NH_4^+ in soil.

Method	Bremner et al. (1967), avg. for eight U.S. soils	Mogilevkina (1969), avg. for five Russian soils
	ppm	
Rodrigues (1954)	917	137
Dhariwal & Stevenson (1958)	166	133
Bremner (1959)	181	91
Schachtschabel (1961)	316	231
Mogilevkina (1964)	52	69 (72)†
Silva & Bremner Method A (1966)	204	85
Silva & Bremner Method B (1966)	205	91

† Modified value for 1 g of soil rather than the 8 g in the original method.

Aldag (unpublished data, 1976) compared method A of Silva and Bremner (1966) with Bremner's (1959) original method and with that of Schachtschabel (1961). Method A proved most suitable but for safety, economy, and simplified operation was subsequently modified to eliminate mineral dissolution with $5N$ HF-$1N$ HCl. Instead, the mineral residue left after washing with the $0.5N$ KCl solution was digested with concentrated H_2SO_4, as in the usual Kjeldahl procedure. Results were the same.

Ammonium may also occur as an integral part of the structure in such inorganic materials as taranakite. Mineral lattice dissolution techniques as used for intercalated NH_4^+ would liberate but not distinguish such NH_4^+. No differentiation of such NH_4^+, which could also be called "mineral-fixed," has been attempted here.

D. Amounts and Distribution in Soil Profiles

Fixed NH_4^+ at concentrations from a few to several hundred ppm have been reported in soils from countries representing many but not all parts of the world. Range of reported concentrations, together with references and brief notes, are in Table 3, listed generally north to south by continent. Data for several deep sea ocean sediments are included for comparison. Reliability of the data varies depending on the method used, for which readers should check cited papers of interest. Little or no data were found for many large areas, such as Central and South America, Africa, Central and Eastern USSR, Asia, mainland China, Japan, or Southeast Asia.

Reported values for fixed NH_4^+ range from 0 to $>1,000$ ppm, with the lowest values being typical of sandy surface soils and the highest being typical for clayey subsurface horizons. However, the surface layer of some soils will contain as much as 200 ppm N of fixed NH_4^+, equivalent to 750 kg N/ha for a plow layer of 25-cm depth and a soil bulk density of 1.5.

In many soils, the amount of N as fixed NH_4^+ greatly exceeds that present in readily available mineral forms (exchangeable NH_4^+ and NO_3^-). Thus far, highest reported concentrations of fixed NH_4^+ in a *surface* soil layer are for a Maracas soil type developed on micaceous schists and phyllite in Trinidad; the value of 1,162 ppm amounted to 28.5% of the total N (4,648 ppm) as intercalated NH_4^+ (Dalal, 1977). As noted later, these high values require confirmation. High contents of fixed NH_4^+ have also been observed in some soils of Queensland, Australia; the range was from 415 ppm in the surface layer to 1,076 ppm at a depth of 120 cm in one soil (Martin et al., 1970).

For most agricultural soils, from a few to about 10% of the N in the plow layer occurs as fixed NH_4^+. A range of from 3 to 10% has been observed for a wide variety of surface soil types in the North Central and Pacific Northwest regions of the USA (Stevenson, 1959; Walsh & Murdock, 1960; Young, 1962, 1964), in English soils (Bremner, 1959), and in loess-derived soils of Germany (Aldag, 1976; Fleige & Meyer, 1975). Somewhat higher percentages (12–14%) have been reported for several Australian (Martin et al., 1970) and Canadian (Moore, 1965) soils. Still higher percentages are encountered occasionally. For example, 25% of the N in the

Table 3—Amounts of mineral fixed NH_4^+-N reported for various soils and a few marine sediments of the world (north to south by continent).

Location	Conc. range, ppm	Notes	References
North America			
Alberta, Canada	158–330	7 to 14% of N in wide variety of surface soils. Absolute amount *decreased* with depth in 3 profiles but increased as % of N.	Moore (1965)
Saskatchewan, Canada	110–370	7.7 to 13.3% of N in surface 15 cm; up to 58.6% in subsoil, Not affected by cultivation.	Hinman (1964)
Ontario, eastern Canada	12–450	Less fixed NH_4^+ in cropped than in sod soils.	Sowden et al. (1978)
Pacific Northwest, USA	17–138	<1 to 10% of N in broad range of surface soils: 2 to 42% in subsoil horizons. Proportion generally increasing with depth.	Young (1962) Young & McNeal (1964)
Southwest (Nevada) USA	49–79	3.4 to 74.4% of N in low N desert soils of Nevada.	Nishita & Haug (1973)
North Central USA	7–270	4 to 8% in wide range surface soils; lowest in Podzols, highest in silt loams and illite-rich soils. Proportion increased with depth.	Stevenson (1959) Stevenson & Dhariwal (1959) Walsh & Murdock (1960) Keeney & Bremner (1964) Silva & Bremner (1966)
Central America			
Trinidad	378–1,946 (?)	19 to 29% of N in surface soils; 66 to 77% at 1-m depth. High fixed NH_4^+ in micaceous clay, silt, and fine sand of Maracas series formed from micaceous schist and phyllite.	Dalal (1977)
South America			
Peru	140–420	8 to 12% of N in three medium-textured soils (20–40% clay); 420 ppm in 77% clay soil.	Meyer (personal communication, 1979)
Columbia	84–245	3.6 to 17.5% of N in mountain soils on diabase and volcanic ash in humid parts of the Colombian Andes around Cali.	Aldag (personal communication, 1979); for description of soils see Fölster and von Christen (1977)
Western/central Europe			
Sweden	10–17	Podzol profile; low in clay.	Nommik (1967)
Finland	0–623	140 ± 40 ppm for top layers; 400 ± 40 ppm deeper layers. Clay soils (4%), > nonclay soils (9%). Avg. $52 \pm 5\%$ of N in subsoil samples.	Kaila (1966)
Netherlands	161–340	Values decreased after fallowing and cropping	van Schreven (1963)
England	52–252	4 to 8% of N in surface soils; 19 to 45% in subsoils.	Bremner (1959) Bremner & Harada (1959)

(continued on next page)

Table 3—Continued.

Location	Conc. range, ppm	Notes	References
France	25–130	Content positively correlated with amount of clay.	Blanchet et al. (1963) Gouny et al. (1960)
Germany	80–850	80 to 110 ppm in loess soils; 150 to 850 ppm in marsh soils. Content positively correlated with clay.	Schachtschabel (1960, 1961) Mba (1974) Fleige & Meyer (1975) Aldag (1976, 1978)
Bulgaria	36–113	3.3 to 15.6% of N in seven soils.	Ivanov (1962)
Africa			
Tunisia	220–390	20 to 28% of N in surface soils; 30 to 47% in subsurface horizons.	Aldag (1978)
Nigeria	32–220	2 to 6% of N in surface layers; 45 to 63% at depths of 5 to 7 feet.	Moore & Ayeke (1965) Opuwaribo & Odu (1974, 1975)
Libya	150–240	28 to 47% of N in two loess and two sandy desert soils; 92 to 104 ppm in three NE Libyan soils.	Meyer & Aldag (unpublished data, 1979) Mohammed (1979)
Tschad	400–490	14.8 to 17.8% in two Vertisols (30–40% clay), east shore Tschad sea.	Makaini (via Aldag, personal communication, 1979)
Eastern Europe; Mediterranean			
USSR	14–490	2 to 7% of N in surface layers of considerable range of soils representative of the various great soil groups; proportion increased to 10 to 36% at 1-m depth.	Mogilevkina (1964, 1965, 1969) Peterburgskii & Kudeyarov (1966) Smirnov & Fruktova (1963) Koloshova & Shitova (1974) Shkonde et al. (1974)
Lebanon	126–198	Content correlated with amount of clay.	Sayegh & Rehman (1969)
Israel	3–10	About 2 to 25% average of N for 129 samples.	Feigin & Yaalon (1974)
Central Asia; India			
Punjab	10–68	5.5% of N; avg. for 23 surface soils 28 ppm.	Grewal & Kanwar (1967)
Assam	228–465	322 ppm avg. for cultivated soils from different parts of the state.	Tewari et al. (1969)
West Bengal	109–410	5 to 15% of N averages from cultivated surface soils; increased with depth.	Raju & Mukhopadhyay (1973) Singh & Dixit (1972) Prasad et al. (1970)
Utter Pradesh	12–37		
Bihar	90		

(continued on next page)

Table 3—Continued.

Location	Conc. range, ppm	Notes	References
Mainland China;			
Indochina	?	Found no data.	
Pacific rim; Australia; islands			
Taiwan	140–470	10.6 to 32.6% of N in surface layers of nine soils.	Wang et al. (1967)
Philippines	7–428	<1 to 56% of N in 16 soils; slight build-up of fixed NH_4^+ from ammonium sulfate on a clay loam soil after 13 years of long-term fertility trials.	Tilo et al. (1977)
Australia	41–1,076	5 to 90% of N in profiles developed on Permian phyllite (221–1,076 ppm); 5 to 82% in soils from other parent materials.	Martin et al. (1970) Osborne (1976) Black & Waring (1972)
Hawaiian Islands	0–585	6 to 32.9% of N and correlated with total K and mica; 4 to 178 ppm in volcanic ash soils; up to 585 ppm in soils from basalt.	Mikami & Kanehiro (1968)
Marine sediments			
Central Pacific	60–270		Müller (1977)
Atlantic (Northwest Africa)	15–120		
Baltic Sea	300–580	Clays dominated by illites.	

surface layer of an Aridisol (Ustertic Camborthid, Chestnut) soil widely distributed in South Dakota (USA) occurred as fixed NH_4^+; on a profile-depth basis, the fixed NH_4^+-N accounted for nearly 50% of total soil N (Young, 1962). Over 32% of the N in some Taiwan surface soils (Wang et al., 1967) and up to 28% in some Tunisian soils occurred as fixed NH_4^+ (Aldag, 1978).

Figure 1 illustrates a variety of vertical distribution patterns for fixed NH_4^+ in soil profiles representing several great soil groups. Relative and even some absolute increases in fixed NH_4^+ content with depth seem common, and since the organic matter content of most soils decreases sharply with depth, the percentage of total N as fixed NH_4^+ becomes proportionately greater. As much as 90% of the N in some lower soil horizons may occur as fixed NH_4^+ (e.g., in weathered phyllite at >1 m depth, Martin et al., 1970).

The amount of N as fixed NH_4^+ on an area profile-depth basis is often appreciable, as seen in Table 4. Thus, the volume of soil occupied by plant roots may contain 10^2 to $>10^3$ kg fixed NH_4^+-N/ha. Most of the profiles analyzed by Young (1962) contained a total of 470 to 2,387 kg of fixed NH_4^+/ha (depths varied from 0.75 to 1.5 m); those examined by Hinman (1964) ranged from 2,960 to 5,200 kg/ha to a 1.2-m depth; those by Moore and Ayeke (1965) in Nigeria ranged from 1,345 to 2,350 kg/ha to a 1.5-m depth; those by Mogilevkina (1964) ranged from 526 to 1,608 kg/ha (470 to 1,434 pounds/acre) to a 100-cm depth (3.3 feet); and those by Aldag (1978) ranged from 2,250 to 2,600 kg/ha to a 50-cm depth. The exceptional

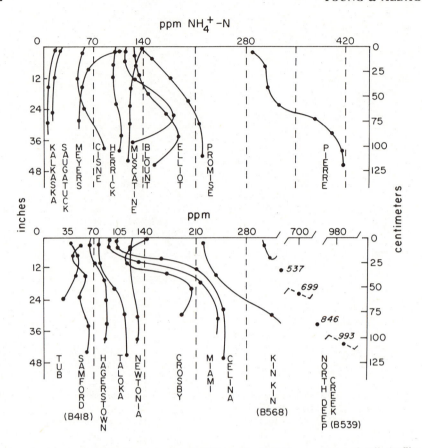

Fig. 1—Distribution of fixed (intercalated) NH$_4$$^+$ with depth in a variety of soil profiles. (Adapted from Stevenson & Dhariwal, 1959; Young, 1962; and Martin et al., 1970.)

Family Level Designations for Soils:

Blount = Fine, illitic, mesic aeric ochra-qualfs

Celina = Fine, mixed, mesic Aeric Hapludolfs

Cisne = Fine, montmorillonitic, mesic Mollic Albaqualfs

Crosby = Fine, mixed, mesic Aeric Ochra-qualfs

Elliott = Fine, mixed, mesic Aquic Argiudolls

Hagerstown = Fine, mixed, mesic Typic Hapludolfs

Herrick = Fine, montmorillonitic, mesic Aquic Argiudolls

Kalkaska = Sandy, mixed, frigid Typic Haplorthods

Meyers (Carney) = Very-fine, montmorillonitic, mesic Typic Chromoxererts

Miami = Fine-loamy, mixed, mesic Typic Hapludolfs

Muscatine = Fine-silty, mixed, mesic Aquic Hapludolls

Newtonia = Fine-silty, mixed, thermic Typic Paleudolls

Pierre = Very-fine, montmorillonitic, mesic, Ustertic Camborthids

Promise = Very-fine, montmorillonitic, mesic Vertic Haplustolls

Saugatuck = Sandy, mixed, mesic, ortstein Aeric Haplaquods

Taloka = Fine, mixed, thermic Mollic Albaqualfs

Tub = Fine, montmorillonitic, mesic calcic Pachic Argixerolls

Undetermined designation for Red Podzolic (Samford B418; North Deep Creek, B539) and Xanthozem (Kin Kin, B568) soils of Australia.

Table 4—Amount of mineral-fixed NH_4^+ and total N for a 15-cm (6 inches) surface and subsurface layer and for top 1 m of several U.S. and Australian soils (adapted from Young, 1962, and Martin et al., 1970).†

Depth interval	Total N	Fixed NH_4^+-N	Total N	Fixed NH_4^+-N
cm		kg/ha		
	Walla Walla		Pierre	
0-15.2	2,150	110 (5)‡	2,590	655(25)‡
76-91.4	605	130(21)	1,215	900(75)
0-100	6,875	810(12)	11,400	5,000(44)
	Nehalem		Chehalis	
0-15.2	8,880	275 (3)	6,500	70 (1)
76-91.4	4,330	270 (6)	1,880	45 (2)
0-100	22,520	1,595 (7)	22,150	335 (1.5)
	Amity			
0-15.2	4,710	220 (5)		
76-91.4	630	140(22)		
0-1C0	13,730	1,000 (7)		
	Jory		Steiwer	
0-15.2	5,740	85(15)	3,205	245 (8)
76-91.4	2,490	70 (3)	785	265(33)
0-100	24,544	465 (2)	10,875	1,670(15)
	Samford (B418)		North Deep Creek	
0-15.2	2,400	120 (5)	7,570	930(12)
76-91.4	315	140(44)	2,460	1,975(80)
0-100	6,310	800(13)	27,035	9,100(34)

† U.S. taxonomy designations: Walla Walla = Typic Haploxeroll; Pierre-Promise = Ustertic Camborthid; Nehalem = Fluventic Haplumbrept; Chehalis = Cumulic Ultic Haploxeroll; Amity = Argiaquic Xeric Argialboll; Jory = Xeric Haplohumult; Steiwer = Ultic Haploxeroll. Samford (B418) and North Deep Creek soils of Australia are Red Podzolics; U.S. designation not determined.
‡ Values in parentheses are fixed NH_4^+-N as % of total N.

amounts of 5,600 to >6,700 kg/ha to a 1-m depth (and >22,400 kg/ha in one N-rich Maracas series, Trinidad), as calculated from Dalal's (1977) data, are so extraordinary as to suggest need for verification.

Generally, large amounts of illite, vermiculites, or micas in parent materials can explain the rather high levels of fixed NH_4^+ in subsoils, e.g., in the Elliott, Promise, and Celina soils, as shown in Fig. 1 (Stevenson, 1960; Stevenson & Dhariwal, 1959); in the Pierre soil developed on shale in South Dakota (Young, 1962); in some soils formed on micaceous-rich Permian phyllite in Australia (Martin et al., 1970) and Trinidad (Dalal, 1977); and in some Hawaiian soils (Mikami & Kanehiro, 1968).

Except as they influence mineralogy, such factors as type of vegetative cover, extent of profile leaching by percolating water, and drainage seem to have little effect on the native fixed NH_4^+ content of the soil (Stevenson & Dhariwal, 1959). A relationship has been noted between K_2O (often used to estimate the percentage of mica) and fixed NH_4^+ in some, though not all, soils (Mogilevkina, 1969; Rich, 1960; Martin et al., 1970). Nevertheless, the amount of fixed NH_4^+ appears to be most closely related to the kinds and amounts of clay minerals present. Micaceous sand and silt fractions from which vermiculitic and illitic clays derive will also contribute, however.

A strong correlation between fixed NH_4^+ and "illites" has been established by Müller (1977) for 121 samples from marine sediments taken out of the Baltic Sea, the Atlantic ocean (off Northwest Africa), and the central Pacific ocean. Clay and clay loam soils generally contain more fixed NH_4^+ than silty loams, which in turn contain more than sandy soils. Relatively low amounts of fixed NH_4^+ in sandy Spodosols (e.g., Kalkaska, Fig. 1) are consistent with their low clay content and highly weathered minerals. However, Spodosols with fixing-type minerals may contain appreciable fixed NH_4^+ (e.g., North Deep Creek, Fig. 1). In the absence of fixing-type clays, organic soils predictably should contain little fixed NH_4^+. From behavior of specimen clay minerals (Young & McNeal, 1964), one would expect the native fixed NH_4^+ content and fixing ability to follow the order: illites > vermiculites $\geq$ smectites > kaolinites. By contrast, vermiculites showed greatest additional fixing capacity.

E. Effect on C/N Ratios

Since Truchot (1875) first expressed the idea of proportionality between C and N, C/N ratios have been considered an important soil characteristic. That soils contain appreciable quantities of fixed NH_4^+ emphasizes that organic C/*total* N ratios can no longer be interpreted as reflecting the character of the soil organic matter, especially in lower soil horizons.

The C/N ratio of the surface soil generally falls within narrow limits, usually about 10 to 12 for cultivated, agriculturally important soils of temperate regions (often a few units higher in forest soils). On the basis of the conventionally used organic C/total N calculation, the ratio often narrows sharply with depth; for subsurface soils, values <5.0 are not uncommon. Since the organic C content of many soils diminishes sharply with depth while the level of fixed NH_4^+ remains fairly constant or increases, the C/N ratio narrows. In upper soil horizons, the C/N ratio is only slightly affected, because there the fixed NH_4^+ represents a relatively small percentage of the total N.

Data for a variety of soils from the North Central region of the USA (Stevenson, 1959) showed that the organic C/total N ratio in eight of nine profiles decreased with depth (from 10.4 to 12.3 in the A1 horizons to 7.0 to 8.0 in most C horizons). Similar changes were observed on Tunisian soils (Aldag, unpublished data), but less consistent changes were found in some U.S. Pacific Northwest soils (Young, 1962). In some tropical soils, Moore and Ayeke (1965) found that the C/total N ratios (at 10 to 18 in the surface layers and 2 to 5 at the lower depths), were shifted slightly to 11 to 19 for the surface but up to 5 to 18 for the subsoil layers by calculating organic C/*organic N* ratios. Divergent results have been reported for some Saskatchewan soils of Canada, where marked widening with depth was noted for the organic C/organic N ratios (range of 9.6 vs. 12.4 in surface layers to 15.5 to 31.5 in the lower layers, Hinman, 1964). Whether inability of the Kjeldahl method to fully measure all total N in the soils (see Stewart & Porter, 1963) could help explain these anomalous results could not be determined from the data given.

Once fixed NH_4^+ data from methods known to yield spurious results have been discounted, it is abundantly clear that use of organic C/total N ratios in relation to the character of soil or sediment organic materials should be discontinued in favor of organic C/organic N ratio.

All narrow soil C/N ratios are not correctable to the 10 to 12 range by accounting for mineral N, however (Young, 1962; Moore & Ayeke, 1965). Extreme examples are Müller's (1977) recent data for several oxic Central Pacific deep sea sediments with organic C/organic N ratios from 5.5 to <2; these show exceptional C depletion vs. relative N conservation. What N-rich organic compounds would survive or be formed to yield these low C/organic N ratios remains uncertain.

F. Stability, Transformation, Fluctuation, and Movement

Only a few examples and limited comment are given on these N topics in anticipation of more extensive coverage in subsequent chapters.

The apparently conflicting results regarding the effect of soil management practices on fixed NH_4^+ content and its availability in soil are less confusing when one views native fixed vs. culturally induced fixed NH_4^+ in light of appropriate influencing factors. Such factors include type of mineral, combined K^+ and NH_4^+ status, degree of lattice weathering, wetting and drying cycles or moisture status, ion position (e.g., whether at feathered edge or blocked interior site), particle size, number and kind of competing ions in fixing sites and in the soil solution, diffusion path length, reaction time allowed for ion equilibrium/exchange, test crop or organism, laboratory (culture plate) or glasshouse (confining pots) vs. field crop evaluation, and chemical methods for analyses.

As often stressed, organic matter is usually lost at a rapid rate when virgin or grassland soils are repeatedly cultivated (Stevenson, 1965). Consequently, the proportion of the total N remaining as fixed NH_4^+ will usually be higher in cropped soils than in virgin soils (Keeney & Bremner, 1964; Stevenson & Dhariwal, 1959). Changes resulting from long-time cropping have been observed, but they have been variable and slight in terms of present-day annual crop needs (Ivanov, 1962; Keeney & Bremner, 1964; Stevenson & Dhariwal, 1959). Similarly, Walsh and Murdock (1963) concluded that even under the most advantageous cropping conditions, very little native fixed NH_4^+ was available to crops, a result attributed to the blocking effect of K^+. On the other hand, van Schreven (1963) concluded that the fixed NH_4^+ content of some recently reclaimed Polder soils was reduced by fallowing and cropping, a result that was attributed to nitrification in the fallow soil and both nitrification and uptake by plants in the cropped soil. Progressive depletion of fixed NH_4^+ in surface compared with subsurface horizons (cf. Miami and Promise soils of Fig. 1) indicates gradual NH_4^+ release and presumably biocycling over geological time.

Because many soils have the ability to fix appreciable amounts of NH_3 or NH_4^+, the continuous application of ammoniacal fertilizers may lead to increases in the fixed NH_4^+ content of the soil. Possible practical consequences are often regarded as negligible at least for any one season. For example, the effect of cultivation (Keeney & Bremner, 1964) or of added mineral N (Black & Waring, 1972; Cheng & Kurtz, 1963; Campbell et al., 1974; Fleige & Capelle, 1975) on the amount of fixed NH_4^+ seemed small. But, sometimes the fixing capacity may be important. Thus, among six soils from Southern New South Wales in Australia, Osborne (1976b) found a heavy-textured gray soil that fixed significant amounts of NH_4^+ after additions of $(NH_4)_2SO_4$. The level of fixed NH_4^+ increased by 55% in the surface and by 100% in the subsoil leading to recommendations against use of NH_4^+ or NH_4^+-forming fertilizers on that high fixing-capacity soil. Smirnov and Furktova (1963) also found higher quantities of fixed NH_4^+ in intensively cultivated and regularly fertilized soils. Likewise, from incubation studies with (Jenkinson, 1965; Raju & Mukhopadhyay, 1976) and without addition of NH_4^+ fertilizer (Aldag, 1976), it is known that appreciable amounts of mineralized or added NH_4^+ can be fixed by the soil minerals. Moreover, under the dessicating effect from volatile anhydrous NH_3 temporarily sweeping sorbed H_2O out of the NH_3 saturation injection zone, even NH_3 can be temporarily trapped or fixed by soil minerals that collapse when dried. Mineral fixation of NH_4^+ or NH_3 by air-dry samples can exceed by several fold the wet fixation from aqua NH_3 (Young & Cattani, 1962).

In brief, availability of native or additionally fixed NH_4^+ varies considerably, as also previously reviewed by Nömmik (1965) and Allison (1973). And although the convenience term "fixed NH_4^+" is often used too broadly and contributes to ambiguities, nature's diversity and range of intergradations make more precise and universally accepted definitions difficult. Hence, it bears reemphasizing that not all NH_4^+ or NH_3 is "fixed" with the same energy, either by mineral or organic components of soils, and that reported results must be interpreted in light of the particular experimental conditions and methods used in each case to evaluate the relative strength of fixation (Young & McNeal, 1964).

III. EXCHANGEABLE AND WATER-SOLUBLE FORMS

Review papers (Harmsen & Kohlenbrander, 1965; Bremner, 1965; Stevenson, 1965) and a continuing flow of primary publications serve to emphasize that only a very small percentage of the total N in soils exists in readily available mineral forms (MacGregor et al., 1974; Robinson, 1975; Smith & Young, 1975; Scharpf & Wehrmann, 1976; Smith et al., 1977; Winner et al., 1976; Osborne, 1977). Agronomically, NO_3^-, NO_2^-, and exchangeable NH_4^+ are considered the most important forms in this category. Other plausible trace or unstable N intermediates generally are neglected as soil constituents.

A. Methods of Determination

Most of the many methods for determination of NO_3^-, NO_2^-, and exchangeable NH_4^+ have been discussed by Bremner (1965) and Bremner and Keeney (1966). More recently, Robinson (1975) described in detail the many parameters of concern in field sampling of soil and subsequent processing, including transport, drying, subsampling, storage, extraction, and determination of soluble soil N. Problems of drying and of obtaining an homogeneous subsample seemed most critical. For best results, immediate extraction of the soil at the field site is recommended together with addition of a preservative to and refrigeration (at $0°C$ or less) of the extract pending further analytical processing (Robinson, 1975; Scharpf & Wehrmann, 1976).

When this chapter was begun, the most recent review pertinent to NO_3^- and NO_2^- determination in soils, solution, and air samples was in a report sponsored by the National Research Council (NRC, 1978). That report also enumerates many hazards and complications of wide ranges in concentrations among various soils (which can change from day to day and also differ for a soil within the same field plot); rapid concentration changes from unchecked microbial action; improper sampling, handling, and storage (considered probable source of most errors in soil NO_3^- analysis); and analytical interferences or other method limitations. Two tables, one for NO_3^- in soils and one for NO_3^- in waters, list 5 and 11 methods, respectively. Each shows applicable range and interferences, along with brief comments and example references. The wet chemical methods are summarized as consisting of two major types: (i) nitration, i.e., the substitution of the $-NO_2$ moiety onto an organic ring compound to form a colored product; and (ii) NO_3^- reduction, usually to NO_2^-, with subsequent colorimetric analysis of the reduced nitrogenous product. Nitration reactions require a strong acid medium (e.g., 50% H_2SO_4), and usually a high temperature. The ion electrode method (e.g., Carlson & Keeney, 1971) requires buffering to a common ionic strength and surmounting interferences from Cl^-, Br^-, I^-, S^{2-}, and NO_2^-.

Although water alone will extract NO_3^- quantitatively from most soils, most methods use a salt solution, such as $CaSO_4$, Na_2SO_4 (Onken & Sundermann, 1977) K_2SO_4, or KCl, often with intent to include NO_2^- and exchangeable NH_4^+, but also to provide clearer extracts. Reduction of NO_3^- to NH_3 with Devarda's alloy and steam distilling in a MgO-buffered medium has been most popular in recent years (consult the 1978 National Research Council report for more narrative comments about the different methods).

A caution regarding ways of expressing NO_3^- concentrations appears in order. Added to the uncertainties about whether data are given as NO_3^- ion or as NO_3^--N is the recent trend of expressing results on the basis of soil solution concentration, that is, assuming all of the nitrate N is dissolved in the soil water (e.g., Broadbent & Carlton, 1978) rather than on the customary basis of soil solids (e.g., mg/kg or ppm of NO_3^--N). "Thus, a soil containing 10 percent water and a nitrate-N concentration of 10 mg/kg would have a soil solution nitrate-N concentration of 100 mg/liter'' (NRC, 1978, p. 136).

For NO_2^-, in contrast to NO_3^-, a simple, highly sensitive, accurate, colorimetric method has long been used. As the anion of weak nitrous acid (HNO_2, pK − 3.36), NO_2^- reacts with sulfanilic acid or sulfanilamide to form a diazonium salt, which in turn couples with substituted naphthyl-amines to form stable red-pink azo dyes having detection limits down to 1 ppm (1 μg N/liter) or less.

Exchangeable NH_4^+ is extracted almost exclusively with KCl solutions (0.5–2N) to ensure displacement of NH_4^+ while maintaining closure of sili-cate clay lattice layers holding fixed NH_4^+. This exchangeable NH_4^+ is most commonly determined by titration after steam distillation with MgO (Bremner, 1965; Bremner & Keeney, 1966). A specific ion electrode for NH_3, claimed as rapid, simple, and precise, could be used according to Ban-wart et al. (1972), although the more widely tested distillation method has held favor. A more rapid, reliable, sensitive probe (to < 30 ppb; HNU Systems Inc., Newton, Mass.) could shift the favor toward more use of the NH_3 electrode (W. Silvester, personal communication, 1980).

For the latest detailed discussion on advances in methods for determin-ing various forms of soil N see Chapter 13 (this book); for laboratory methods of estimating plant available N see Chapter 17 (this book).

B. Amounts and Distribution

Exchangeable and water-soluble mineral N levels at any sampling point or moment are the net product of opposing, yet often concurrent, biological mineralization, assimilation, and immobilization processes, as controlled by the host of almost ceaselessly changing microenvironmental and macro-environmental conditions. Because native mineralized N normally is recycled quickly, these N forms move little from their in situ origin before they are again immobilized (Chapt. 6, 23, this book), assimilated (Chapt. 15, 23, this book), biochemically denitrified (Chapt. 8, 12, this book), or nonbiologically immobilized, fixed, or volatilized (Chapt. 4, 9, 12, this book; Allison, 1973).

Under natural conditions, concentrations generally cycle in the range of a few to a few tens of ppm because of immediate plant and microbial demand, as was thoroughly reviewed by Harmsen and Kolenbrander (1965) and more succinctly covered by Allison (1973). Thus, at times only a few kilograms of soluble N/ha may be available, although during the main vegetative period, levels may rise to 100 or more kg/ha of soluble N (mostly NO_3^-) in fertile systems (Scharpf & Wehrmann, 1976). Typically, in surface layers of soils in temperate climate humid areas, quantities are lowest in winter, increase in spring as mineralization of organic matter commences, decrease in summer due to consumption by plants or lowered mineralization rates as soils dry, and increase once again in fall when plant growth ceases and residues start to decay. Winter levels seldom exceed 10 ppm in unamended soils but may increase fourfold to sixfold during the spring (Harmsen & Kolenbrander, 1965).

In contrast to fixed NH_4^+, the distribution of NO_3^- in soil profiles is not correlated with a specific soil characteristic, such as clay content or particle size, but usually follows the water regime, *except* in soils with significant anion exchange or other sorption sites wherein mobility of NO_3^- could be lessened appreciably (Black & Waring, 1976a, b; Espinoza et al., 1975; Singh & Kanehiro, 1969).

Presumably, with the many possible "sinks" for mineralized N in uncultivated soils, available forms of inorganic N seldom accumulated to more than a few ppm and seldom persisted beyond normal growing seasons. With direct injection of anhydrous- or aqua NH_3, localized concentrations of NH_3 and NH_4^+ reaching several thousand ppm may occur and persist for weeks to months depending on soil temperature and moisture or simultaneous application of effective nitrification inhibitors. Increasingly, NO_3^- accumulations within the rooting zone in amounts ranging from a few to several hundreds ppm are also being found (Linville & Smith, 1971; El-Bassam et al., 1974; MacGregor et al., 1974; Strebel et al., 1975; Scharpf & Wehrmann, 1976; Kick & Lohsse, 1977).

Where soils are not cultivated or not overfertilized, the NO_3^- concentration within the rooting zone normally decreases during the crop growth period (Feigin et al., 1974; Scharpf & Wehrmann, 1976). However, fertilization of continuous corn (*Zea mays* L.) in excess of optimum rates can leave considerable residual NO_3^- in soils after the growing season (Hahne et al., 1977). MacLean (1977) reported similar results with different cropping systems (grasses, corn) but also found that considerable loss of the excess NO_3^--N occurred from early fall to the next spring. Diminished amounts of NO_3^--N in the upper soil layers and increased amounts in the lower layers during the fall-spring period indicated movement of the NO_3^--N. Even when fertilizer N was applied at rates less than required for maximum corn production, Ludwick et al. (1976) observed that NO_3^--N accumulated in a soil profile. In a recent reassessment of long-term (17 and 15 years) N-rate experiments with corn in northwestern Iowa, Jolley and Pierre (1977) found that depth of maximum accumulation of NO_3^- was between 1.2 and 1.5 m; there was no evidence of leaching below 2.4 and 2.7 m in either soil. The accumulations were in addition to appreciable losses (attributed to denitrification) of 37 and 46% at highest N application rates (168 and 134 kg of N/ha).

When the water regime allows (infiltration of precipitation or irrigation exceeding evapotranspiration), large amounts of NO_3^--N have been moved well below 2 m (Pratt et al., 1972; MacGregor, et al., 1974; Schumann et al., 1975). On an untiled Forman clay loam after 15 years of liberally fertilized corn, appreciable amounts of NO_3^- were found to at least the 10-m depth. Movement of the NO_3^- front averaged 1.9 mm/day over the 15-year period (MacGregor et al., 1974).

In fact, sizeable accumulations and transport (Chapt. 11, this book) of leachable N, mainly NO_3^-, are now occurring in soils of developed countries as a result of several management factors. Understandably, growing con-

cerns over existing or potential earth, water, and air contamination from so-
called nonpoint source pollutants, including N compounds, are generating
many nonagricultural (NRC, 1978, 1979) as well as agricultural agency
reports (Foster, 1976). Out of it all should come better inventories of N
forms in soils and recommended management systems for maximum efficiency and minimal pollution (Chapt. 16, this book). (For detailed coverage of N gains, losses, movement, and effects of cultural practices see
Chapt. 11 and 14, this book).

In light of the above reported man-caused perturbations, the recent discoveries of large quantities of so-called natural soil NO_3^- in Texas (Kreitler
& Jones, 1975) and of geological NO_3^- within the deep loess mantle of
southwestern and central Nebraska (Boyce et al., 1976) have added significance. Nitrate N concentrations of 25 to 45 ppm but with values to 85
ppm were found at the 7- to 30-m depth and deeper. The origin of this NO_3^-
and its potential movement into ground water supplies as a result of emerging irrigation and agricultural practices remain unclear.

The origin of several central California soils with very high NO_3^--N
concentrations (several hundred to several thousand ppm and to >15-m
depths) is better established (Sullivan et al., 1979). These westside basin
soils of the San Joaquin Valley apparently derive from mud flow material
carried out of Cretaceous and Tertiary marine sediment formations of the
Diablo Range. Nitrate N concentrations in some mudstone, shale, siltstone,
and even sandstone layers exceeded 1,000 ppm—the highest being >4,750
ppm in a mudstone of the Tulare formation. This geologically stored
soluble N (and Cl^-), along with some fertilizer N, moves readily down the
soil profile with irrigation water, as documented by sizeable quantities
found in the tile-drain effluents (Letey et al., 1977). The potential environmental hazard of this geological N has been further discussed recently by
Strathouse et al. (1980). Other unusual NO_3^- accumulations in soils are
known, e.g., in barren soils of Antarctica (Claridge & Campbell, 1968), and
to amounts of 0.1 to 0.2% in desert soils, as in Northern Chilean areas
(Mueller, 1968). Only future work will show how extensive such similar
natural NO_3^- accumulations may be.

IV. SOIL NITROGEN GASES

As in the aboveground atmosphere, dinitrogen (N_2) is by far the
dominant N gas species within aerated, undisturbed soils. Of the dozen
other possible gas species (Lindsay, 1979, p. 269), only N_2O has been unequivocally detected in soil air under field conditions. Other intermediates
resulting from biologically probable and thermodynamically possible reactions no doubt occur. These transitory intermediates will remain unquanti-
fied until more suitable in situ sampling and analytical techniques are developed and applied. Gaseous NH_3 (Denmead et al., 1974), NO, and NO_2 in
addition to N_2O (Arnold, 1954) have been detected in emissions from field
soils but not in the gases within natural soil profiles. For review of biological or chemical processes where several forms of N gases may be produced

in, interchanged with, or evolved from soils, see Chapters 4, 8, 9, 10, and 12 (this book).

Although it is rarely mentioned in soils literature, elemental or molecular N_2 is also a common inclusion constituent in rocks and rockforming minerals that could be present in soil profiles (Baur & Wlotzka, 1972).

A. Methods of Determination

Various methods for determination of gaseous forms of N up to the year 1962 were exhaustively reviewed by Cheng and Bremner (1965). Notable improvements have since been devised. An example is the rapid, specific, and precise determination of N_2, O_2, Ar, CO_2, CH_4, N_2O, and other gases in soil atmospheres at concentrations ranging from 10 $\mu g/ml$ to 100% (Blackmer & Bremner, 1977).

Many other reports have appeared during the last decade concerning determination of soil gases, primarily by gas chromatography. Chapter 13 (this book) gives a detailed review. For reference to recovery and assay methods (such as grinding under vacuum and volumetric determination) of "inclusion N_2" from rocks or rock-forming minerals, please consult the report of Baur and Wlotzka (1972).

B. Amounts and Distribution

The ambient 78% + N_2 concentration should be appreciably different in air of virgin soils only where physical barriers inhibit ready diffusion or exchange with the aboveground atmosphere. Less O_2 and relatively greater N_2 concentrations may occur in localized microhabitat sites (Greenwood, 1961) or where permanent soil submergence or temporary inundation permit anaerobic conditions to develop (e.g., Yoshida et al., 1975). Amounts of inclusion N_2 in rocks and rock-forming minerals at parts per million by weight (Bauer & Wlotzky, 1972) are somewhat lower but near the same order of magnitude as amounts of fixed NH_4^+ in soils.

Concentrations of N_2O in field soils are orders of magnitude less than N_2, and they fluctuate widely. Mean values reported by Dowdell and Smith (1974) were 20 to 100 ppm (detection limit of method was 20 ppm), but some individual samples contained 1,500 to 6,500 ppm (v/v). Largest values tended to be at deeper depths (30 and 60 cm), during winter, and at lower O_2 concentrations (<8-10%) but not always. Except in a few instances, N_2O was found on any given sampling occasion in only 5 to 15% of the samples; N_2O persisted in the soil for only short periods at a given sampling point and rarely reappeared at that point. Denitrification in anaerobic pockets was believed to be the source since mean N_2O concentration followed the decrease in NO_3^- and NO_2^- and was found where O_2 was present (sometimes at O_2 concentrations near that of the free atmosphere). Burford and Millington (1968) also found high N_2O concentrations in Australian field soils where O_2 existed.

Concentrations of N_2O in some California field soils ranged from 0.36

to 8,000 ppm (v/v), as detected by a gas chromatography technique sensitive to 0.1 ppb (Delwiche & Rolston, 1976). More recently, Focht et al. (1979) reported N_2O levels of < 1 to near 80 μliter/liter in control plot profiles of an irrigated calcareous soil in the Imperial Valley, Calif. Fifty percent of the control plot soils had ≤ 0.8 μliter/liter N_2O; 13% had ≥ 10 μliter/liter, and 4% had > 40 μliter/liter N_2O.

That "NO may be the kinetically feasible intermediate reaction product in the chemical denitrification process" in soils (Lindsay, 1979), though not yet detected in field sites, remains to be evaluated. Similarly, the extent to which N_2O may be generated in virgin field soils during oxidation of NH_4^+ by nitrifying soil organisms as distinct from being a reduction product of NO_3^- via denitrifying microorganisms (Bremner & Blackmer, 1979) also remains to be evaluated.

LITERATURE CITED

Aldag, R. W. 1976. Verfügbarkeit des Stickstoffs in Ackerböden. Bestimmungsprobleme aus der Sicht der Umverteilung der Stickstoff-Bindungsformen durch Bebrütung. Landwirtsch. Forsch., Sonderh. 32/II:91–97.

Aldag, R. W. 1978. Anteile des mineralisch fixierten Ammoniums am Amidstickstoff in Bodenhydrolysaten. Mitt. Dtsch. Bdenkd. Ges. 27:293–302.

Allison, F. E. 1973. Soil organic matter. Elsevier North-Holland, New York. [Especially Chapt. 11, Nonbiological immobilization of nitrogen, p. 206–229.]

Allison, F. E., and E. M. Roller. 1955. A comparison of leaching and distillation procedures for determining fixed ammonium in soils. Soil Sci. 80:349–362.

Arnold, P. W. 1954. Losses of nitrous oxide from soil. J. Soil Sci. 5:116–128.

Banwart, W. L., M. A. Tabatabai, and J. M. Bremner. 1972. Determination of ammonium in soil extracts and water samples by an ammonia electrode. Commun. Soil Sci. Plant Anal. 3:449–458.

Barshad, I. 1951. Cation exchange in soils: I. Ammonium fixation and its relation to potassium fixation and to determination of ammonium exchange capacity. Soil Sci. 72:361–371.

Baur, W. H., and F. Wlotzka. 1972. Nitrogen. p. 7-B-1 through 7-0-3, plus 11 pages of references. In K. H. Wedepohl (ed.) Handbook of geochemistry, Vol. II/1. Springer-Verlag Berlin, Heidelberg.

Black, A. S., and S. A. Waring. 1972. Ammonium fixation and availability in some cereal producing soils in Queensland. Aust. J. Soil Res. 10:197–207.

Black, A. S., and S. A. Waring. 1976a. Nitrate leaching and adsorption in a Krasnozem from Redland Bay, Queensland. I. Leaching of banded ammonium nitrate in a horitcultural rotation. Aust. J. Soil Res. 14:171–180.

Black, A. S., and S. A. Waring. 1976b. Nitrate leaching and adsorption in a Krasnozem from Redland Bay, Queensland. II. Soil factors influencing adsorption. Aust. J. Soil Res. 14:181–188.

Blackmer, A. M., and J. M. Bremner. 1977. Gas chromatographic analysis of soil atmosphere. Soil Sci. Soc. Am. J. 41:908–912.

Blanchet, R., R. Studer, C. Chaumont, and L. LeBlevenec. 1963. Principaux facteurs influencant la retrogradation de l'ammonium dans les conditions naturelles des sols. C. R. Acad. Sci. Paris 256:2223–2225.

Boyce, J. S., J. Muir, A. P. Edwards, E. C. Seim, and R. A. Olson. 1976. Geologic nitrogen in pleistocene loess of Nebraska. J. Environ. Qual. 5:93–96.

Bremner, J. M. 1959. Determination of fixed ammonium in soil. J. Agric. Sci. 52:147–160.

Bremner, J. M. 1965. Inorganic foms of nitrogen. In C. A. Black et al. (ed.) Methods of soil analysis, part 2. Agronomy 9:1179–1237. Am. Soc. of Agron., Madison, Wis.

Bremner, J. M., and A. M. Blackmer. 1979. Effects of acetylene and soil water content on emission of nitrous oxide from soils. Nature (London) 280:381–382.

Bremner, J. M., and T. Harada. 1959. Release of ammonium and organic matter from soil by hydrofluoric acid and effect of hydrofluoric acid treatment on extraction of soil organic matter. J. Agric. Sci. 52:137–160.

Bremner, J. M., and D. R. Keeney. 1966. Determination and isotope-ratio analysis of different forms of nitrogen in soils: 3. Exchangeable ammonium, nitrate, and nitrite by extraction-distillation methods. Soil Sci. Soc. Am. Proc. 30:577–582.

Bremner, J. M., D. W. Nelson, and J. A. Silva. 1967. Comparison and evaluation of methods for determining fixed ammonium in soil. Soil Sci. Soc. Am. Proc. 31:466–472.

Broadbent, F. E., and A. B. Carlton. 1978. Field trials with isotopically labeled nitrogen fertilizer. p. 1–41. *In* D. R. Nielsen and J. G. MacDonald (ed.) Nitrogen in the environment, Vol. 1, Nitrogen behavior in field soils, Academic Press, Inc., New York.

Burford, J. R., and R. J. Millington. 1968. Nitrous oxide in the atmosphere of a red-brown earth. Int. Congr. Soil Sci., Trans. 9th (Adelaide) 2:505–511.

Campbell, C. A., D. W. Stewart, W. Nicholaichuk, and V. O. Biederbeck. 1974. Effects of growing season soil temperature, moisture, and NH_4-N on soil nitrogen. Can. J. Soil Sci. 54:403–412.

Carlson, R. M., and D. R. Keeney. 1971. Specific ion electrodes. Techniques and uses in soil, plant, and water analysis. p. 39–65. *In* L. M. Walsh (ed.) Instrumental methods for analysis of soils and plant tissue. Soil Sci. Soc. of Am., Inc., Madison, Wis.

Cheng, H. H., and J. M. Bremner. 1965. Gaseous forms of nitrogen. *In* C. A. Black et al. (ed.) Methods of soil analysis, part 2. Agronomy 9:1287–1323. Am. Soc. of Agron., Madison, Wis.

Cheng, H. H., and L. T. Kurtz. 1963. Chemical distribution of added nitrogen in soils. Soil Soc. Am. Proc. 27:312–316.

Claridge, G. G. C., and I. B. Campbell. 1968. Origin of nitrate deposit. Nature (London) 217:428.

Dalal, R. C. 1977. Fixed ammonium and carbon-nitrogen ratios of some Trinidad soils. Soil Sci. 124:323–327.

Delwiche, C. C., and D. E. Rolston. 1976. Measurement of small nitrous oxide concentrations by gas chromatography. Soil Sci. Soc. Am. J. 40:324–327.

Denmead, O. T., J. R. Simpson, and J. R. Freney. 1974. Ammonia flux into the atmosphere from a grazed pasture. Science 185:609–610.

Dhariwal, A. P. S., and F. J. Stevenson. 1958. Determination of fixed ammonium of soils. Soil Sci. 86:343–349.

Dowdell, R. J., and K. A. Smith. 1974. Field studies of the soil atmosphere. II. Occurrence of nitrous oxides. J. Soil Sci. 25:231–238.

El-Bassam, N., C. Tietjen, and F. Mertens. 1974. Nitrat-, Nitrit- und Ammoniak-N im Boden und Bodenwasser sowie Aufnahme durch Markstammkohl und Zuckerrüben bei Zufuhr hoher Klärschlammgaben. Landwirtsch. Forsch. Sonderh. 30/II:29–38.

Espinoza, W., R. G. Gast, and R. S. Adams, Jr. 1975. Charge characteristics and nitrate retention by two andepts from south-central Chile. Soil Sci. Soc. Am. Proc. 39:842–846.

Feigin, A., G. Shearer, D. H. Kohl, and B. Commoner. 1974. The amount and nitrogen-15 content of nitrate in soil profiles from two central Illinois fields in a corn-soybean rotation. Soil Sci. Soc. Am. Proc. 38:465–471.

Feigin, A., and D. H. Yaalon. 1974. Non-exchangeable ammonium in soils of Israel and its relation to clay and parent materials. Soil Sci. 25:384–397.

Fleige, H., and A. Capelle. 1975. Bilanz und Umverteilung von markiertem Dünger-N. Mitt. Dtsch. Bodenkd. Ges. 22:375–384.

Fleige, H., and B. Meyer. 1975. Mineralisch fixiertes Ammonium in jungpleistozänen Sedimenten Norddeutschlands und ihren fossilen und holozänen Böden: Ein Indikator für litho-oder pedogenetische Prozesse und Herkünfte? Göttinger Bodenkd. Ber. 34:315–328.

Focht, D. D., L. H. Stolzy, and B. D. Meek. 1979. Sequential reduction of nitrate and nitrous oxide under field conditions as brought about by organic amendments and irrigation management. Soil Biol. Biochem. 11:37–46.

Folster, H., and H. von Christen. 1977. The influence of Quaternary uplift on the altitude zonation of mountain soils on diabase and volcanic ash in humid parts of the Colombian Andes. Catena. 3:233–263.

Foster, S. S. D. 1976. The vulnerability of British groundwater resources to pollution by agricultural leachates. U. K. Tech. Bull. 32:68–91. Ministry of Agric., Fisheries and Food, Inst. of Geological Sciences, Hydrogeological Dep., London.

Freney, J. R. 1964. An evaluation of naturally occurring fixed ammonium in soils. J. Agric. Sci. 63:297–303.

Gouny, P., S. Mériaux, and R. Grosman. 1960. Importance de l'ion ammonium à l'etat non échangeable dans un profil de sol. C. R. Acad. Sci. Paris 251:1418–1420.

Greenwood, D. J. 1961. The effect of oxygen concentration on the decomposition of organic materials in soil. Plant Soil 14:360–376.

Grewal, J. S., and J. S. Kanwar. 1967. Forms of nitrogen in Punjab soils. J. Res. Punjab Agric. Univ. 4:477–480. (From Soils Fert. 31:3855, 1968.)

Hahne, H. C. H., W. Kroontje, and J. A. Lutz, Jr. 1977. Nitrogen fertilization: I. Nitrate accumulation and losses under continuous corn cropping. Soil Sci. Soc. Am. J. 41:562–568.

Harmsen, G. W., and G. J. Kolenbrander. 1965. Soil inorganic nitrogen. In W. V. Bartholomew and F. E. Clark (ed.) Soil nitrogen. Agronomy 10:43–92. Am. Soc. of Agron., Madison, Wis.

Hinman, W. C. 1964. Fixed ammonium in some Saskatchewan soils. Can. J. Soil Sci. 44:151–157.

Ivanov, Ts. 1962. Non-exchangeable absorption of the ammonium cation by some soils. Izv. Dobrudzh. S-stop. Nauch-iszled. Inst. Tolbukhin 3:5–19. (From Soils Fert. 27:30, 1964.)

Jenkinson, D. 1965. Organic matter in soil. New Sci. 27:746–748.

Jolley, V. D., and W. H. Pierre. 1977. Profile accumulation of fertilizer derived nitrate and total nitrogen recovery in two long-term nitrogen-rate experiments with corn. Soil Sci. Soc. Am. J. 41:373–378.

Kaila, A. 1966. Fixed ammonia in some Finnish soils. J. Sci. Agric. Soc. Finl. 38:49–58. (From Soils Fert. 29:2983, 1966.)

Keeney, D. R., and J. M. Bremner. 1964. Effect of cultivation on the nitrogen distribution in soils. Soil Sci. Soc. Am. Proc. 28:653–656.

Kick, H., and H. Lohsse. 1977. Verlagerung von Nährstoffen, insbesondere von Stickstoff, in einer Parabraunerde auf Löess unterlage. Landwirtsch. Forsch. Sonderh. 33/II:147–156.

Koloskova, A. V., and L. I. Shitova. 1974. Nitrogen in tatar soils. Sov. Soil Sci. 6:172–179.

Kreitler, C. W., and D. C. Jones. 1975. Natural soil nitrate: The cause of the nitrate contamination of ground water in Runnels County, Texas. Ground Water 13:53–61.

Letey, J., J. W. Blair, D. Devitt, L. J. Lund, and P. Nash. 1977. Nitrate-nitrogen in effluent from agricultural tile drains in California. Hilgardia 45:239–319.

Lindsay, W. L. 1979. Chemical equilibria in soils. John Wiley & Sons, Inc., New York.

Linville, K. W., and G. E. Smith. 1971. Nitrate content of soil cores from corn plots after repeated nitrogen fertilization. Soil Sci. 112:249–255.

Ludwick, A. E., J. O. Reuss, and E. J. Langin. 1976. Soil nitrates following four years continuous corn and as surveyed in irrigated farm fields of central and eastern Colorado. J. Environ. Qual. 5:82–86.

MacGregor, J. M., G. R. Blake, and S. D. Evans. 1974. Mineral nitrogen movement into subsoils following continued annual fertilization for corn. Soil Sci. Soc. Am. Proc. 38:110–113.

MacLean, A. J. 1977. Movement of nitrate nitrogen with different cropping systems in two soils. Can. J. Soil Sci. 57:27–33.

Martin, A. E., R. J. Gilkes, and J. O. Skjemstad. 1970. Fixed ammonium in soils developed in some Queensland phyllites and its relation to weathering. Aust. J. Soil Res. 8:71–80.

Mba, K. 1974. Boden—und Dunger (N¹⁵)-Stickstoff in Acker-Parabraunerden aus Löess. Diss. Landwirtsch. Fakultät Göttingen, D-16.

Mikami, D. T., and Y. Kanehiro. 1968. Native fixed ammonium in Hawaiian soils. Soil Sci. Soc. Am. Proc. 32:481–485.

Mogilevkina, I. A. 1964. Content of fixed ammonium in some soil types of the USSR and their NH_4^+-fixing capacity. Agrokhimiya 7:26–36. (From Soils Fert. 28:3647, 1965.)

Mogilevkina, I. A. 1969. Comparison of methods of determining fixed ammonium in the soil. Sov. Soil Sci. 16:229–238.

Mohammed, I. H. 1979. Fixed ammonium in Libyan soils and its availability to barley seedlings. Plant Soil 53:1–9.

Moore, A. W. 1965. Fixed ammonium in some Alberta soils. Can. J. Soil Sci. 45:112–115.

Moore, A. W., and C. A. Ayeke. 1965. HF-extractable ammonium nitrogen in four Nigerian soils. Soil Sci. 99:335–338.

Mueller, G. 1968. Genetic histories of nitrate deposits from Antarctica and Chile. Nature (London) 219:1131.

Müller, P. J. 1977. C/N ratios in Pacific deep-sea sediments: Effect of inorganic ammonium and organic nitrogen compounds sorbed by clays. Geochim. Cosmochim. Acta 41:765–776.

National Research Council. 1978. Nitrates: An environmental assessment. National Academy of Sciences, Washington, D.C.

National Research Council. 1979. Ammonia. Subcommittee of Committee on Medical and Biologic Effects of Environmental Pollutants. University Park Press, Baltimore.

Nelson, D. W., and J. M. Bremner. 1966. An evaluation of Mogilevkina's method of determining fixed ammonium in soils. Soil Sci. Soc. Am. Proc. 30:409–411.

Niederbudde, E. A., and W. R. Fischer. 1974. Der Einfluss des Abbaues von orgnischer Substanz auf die NH$_4$-Fixierung und die K-Ca-Austauscheigenschaften smektit- und illithaltiger Böden. Z. Pflanzenernaehr. Bodenkd. 137:6–18.

Nishita, H., and R. M. Haug. 1973. Distribution of different forms of nitrogen in some desert soils. Soil Sci. 116:51–58.

Nommik, H. 1965. Ammonia fixation and other reactions of nonenzymatic immobilization of mineral nitrogen in soil. In W. V. Bartholomew and F. E. Clark (ed.) Soil nitrogen. Agronomy 10:198–258. Am. Soc. of Agron., Madison, Wis.

Nommik, H. 1967. Distribution of forms of nitrogen in a podzolic soil profile from Garpenberg, Central Sweden. J. Soil Sci. 18:301–308.

Onken, A. B., and H. D. Sunderman. 1977. Colorimetric determinations of exchangeable ammonium, urea, nitrate, and nitrite in a single soil extract. Agron. J. 69:49–53.

Opuwaribo, E., and C. T. I. Odu. 1974. Fixed ammonium in Nigerian soils. I. Selection of a method and amounts of native fixed ammonium. J. Soil Sci. 25:256–264.

Opuwaribo, E., and C. T. I. Odu. 1975. Fixed ammonium in Nigerian soils. II. Relationship between native fixed ammonium and some soil characteristics. J. Soil Sci. 26:350–357.

Osborne, G. J. 1976a. The extraction and definition of nonexchangeable or fixed ammonium in some soils from southern New South Wales. Aust. J. Soil Res. 14:373–380.

Osborne, G. J. 1976b. The significance of intercalary ammonium in representative surface and subsoils from southern New South Wales. Aust. J. Soil Res. 14:381–388.

Osborne, G. J. 1977. Chemical fractionation of soil nitrogen in six soils from southern New South Wales. Aust. J. Soil Res. 15:159–165.

Peterburgskii, A. V., and V. N. Kudeyarov. 1966. Fixed ammonium in some soils of the USSR and its availability to plants. Izv. Timiryazevsk. Skh. Akad. 3:72–80. (From Soils Fert. 29:3734, 1966.)

Prasad, B., H. Sinha, and R. N. Prasad. 1970. Forms of ammonium nitrogen in soils of Bihar. J. Indian Soc. Soil Sci. 18:289–295.

Pratt, P. F., W. W. Jones, and V. E. Hunsaker. 1972. Nitrate in deep soil profiles in relation to fertilizer rates and leaching volume. J. Environ. Qual. 1:97–102.

Raju, G. S. N., and A. K. Mukhopadhyay. 1973. Distribution of native fixed ammonium and other forms of nitrogen in different soils of West Bengal. J. Indian Soc. Soil Sci. 21:257–262.

Raju, G. S. N., and A. K. Mukhopadhyay. 1976. Ammonium fixing capacities of West Bengal soils. J. Indian Soc. Soil Sci. 24(3):270–274.

Rich, C. I. 1960. Ammonium fixation by two red-yellow podzolic soils as influenced by interlayer-Al in clay minerals. Int. Congr. Soil Sci., Trans. 7th (Madison, Wis.) 4:468–475.

Robinson, J. B. D. 1975. The soil nitrogen index and its calibration with crop performance to improve fertilizer efficiency on arable soils. Spec. Pub. no. 1. Commonwealth Bureau of Soils, Harpenden, England.

Rodrigues, G. 1954. Fixed ammonia in tropical soils. J. Soil Sci. 5:264–274.

Said, M. B. 1973. Ammonium fixation in the Sudan Gezira soil. Plant Soil 38:9–16.

Sayegh, A. H., and H. Rehman. 1969. Ammonium fixation in alkaline Lebanese soils. Soil Sci. 108:202–208.

Schachtschabel, P. 1960. Fixierter Ammoniumstickstoff in Löess und Marschböden. Int. Congr. Soil Sci., Trans. 7th (Madison, Wis.) 2:22–27.

Schachtschabel, P. 1961. The determination of fixed ammonium in soil. Z. Pflansenernaehr. Dueng. Bodenkd. 93:125–136.

Scharpf, H. C., and J. Wehrmann. 1976. Bedeutung des Mineralstickstoffvorrates des Bodens zu Vegetationsbeginn für die Bemessung der N-Düngung zu Winterweizen. Landwirtsch. Forsch. Sonderh. 32/I:100–114.

Schumann, G. E., T. M. McCalla, K. E. Saxton, and H. T. Knox. 1975. Nitrate movement and its distribution in the soil profile of differentially fertilized watersheds. Soil Sci. Soc. Am. Proc. 39:1192–1197.

Shkonde, E. I., I. Ye Koroleva, and A. P. Shcherbakov. 1974. Reserves and forms of nitrogen in the Chernozems of the east European facies. Sov. Soil Sci. 6:553–558.

Silva, J. A., and J. M. Bremner. 1966. Determination and isotope-ratio analysis of different forms of nitrogen in soils. 5. Fixed ammonium. Soil Sci. Soc. Am. Proc. 30:587–594.

Singh, S. B., and V. K. Dixit. 1972. Native fixed ammonium in a few soils of Uttar Pradesh. J. Indian Soc. Soil Sci. 20:189–191.

Singh, B. R., and Y. Kanehiro. 1969. Adsorption of nitrate in amorphous and kaolinite Hawaiian soil. Soil Sci. Soc. Am. Proc. 33:681–683.

Smirnov, P. M., and N. I. Fruktova. 1963. Non-exchangeable fixation of ammonia by soils. Sov. Soil Sci. 1963:265–272.

Smith, S. J., and L. B. Young. 1975. Distribution of nitrogen forms in virgin and cultivated soils. Soil Sci. 120:354–360.

Smith, S. J., L. B. Young, and G. E. Miller. 1977. Evaluation of soil nitrogen mineralization potentials under modified field conditions. Soil Sci. Soc. Am. J. 41:74–77.

Sowden, F. J., A. A. MacLean, and G. J. Ross. 1978. Native clay-fixed ammonium content, and the fixation of added ammonium of some soils of eastern Canada. Can. J. Soil Sci. 58:27–38.

Stevenson, F. J. 1959. Carbon-nitrogen relationships in soil. Soil Sci. 88:201–208.

Stevenson, F. J. 1960. Some aspects of the distribution of biochemicals in geologic environments. Geochim. Cosmochim. Acta 19:261–271.

Stevenson, F. J. 1962. Chemical state of the nitrogen in rocks. Geochim. Cosmochim. Acta 26:797–809.

Stevenson, F. J. 1965. Origin and distribution of nitrogen in soil. In W. V. Bartholomew and F. E. Clark (ed.) Soil nitrogen. Agronomy 10:1–42. Am. Soc. of Agron., Madison, Wis.

Stevenson, F. J., and A. P. S. Dhariwal. 1959. Distribution of fixed ammonium in soils. Soil Sci. Soc. Am. Proc. 23:121–125.

Stevenson, F. J., G. Kidder, and S. N. Tilo. 1967. Extraction of organic nitrogen and ammonium from soil with hydrofluoric acid. Soil Sci. Soc. Am. Proc. 31:71–76.

Stewart, B. A., and L. K. Porter. 1963. Inability of the Kjeldahl method to fully measure indigenous fixed ammonium in some soils. Soil Sci. Soc. Am. Proc. 27:41–43.

Strathouse, S. M., G. Sposito, P. J. Sullivan, and L. J. Lund. 1980. Geologic nitrogen: A potential geochemical hazard in the San Joaquin Valley, California. J. Environ. Qual. 9:54–60.

Strebel, O., M. Renger, and W. Giesel. 1975. Vertikale Wasserbewegung und Nitratverlagerung unterhalb des Wurzelraums. Mitt. Dtsch. Bodenkd. Ges. 22:277–286.

Sullivan, P. J., G. Sposito, S. M. Strathouse, and C. L. Hansen. 1979. Geologic nitrogen and the occurrence of high nitrate soils in the western San Joaquin Valley, California. Hilgardia 47:15–49.

Tewari, S. N., I. Dutta, and A. K. Deka. 1969. HF-extractable ammonium and ammonium fixing capacity in some soils of Assam. Indian J. Agric. Chem. 2:18–21. (From Soils Fert. 33:32, 1970.)

Thompson, L. M., and F. R. Troeh. 1978. Soils and soil fertility. 4th ed. McGraw-Hill book Co., New York.

Tilo, S. N., N. V. Caramancion, I. J. Manguiat, and E. S. Paterno. 1977. Naturally occurring fixed ammonium in some Philippine soils. Philipp. Agric. 60:413–419.

Truchot, M. 1875. Observations sur la composition les terres arables de l'Auvergne. Ann. Agron. 1:535–551.

van Schreven, D. A. 1963. Nitrogen transformation in the former subaqueous soils of polders recently reclaimed from Lake Ijssel. I. Plant Soil 18:143–162.

Walsh, L. M., and J. T. Murdock. 1960. Native fixed ammonium and fixation of applied ammonium in several Wisconsin soils. Soil Sci. 89:183–193.

Walsh, L. M., and J. T. Murdock. 1963. Recovery of fixed ammonium by corn in greenhouse studies. Soil Sci. Soc. Am. Proc. 27:200–204.

Wang, T. S. C., T.-K. Yang, and S.-Y. Cheng. 1967. Amino acids in subtropical soil hydrolysates. Soil Sci. 103:67–74.

Winner, C., I. Feyerabend, and A. V. Müller. 1976. Untersuchungen über den Gehalt an Nitratstickstoff in einem Bodenprofil und dessen Entzug durch Zuckerrüben. Zucker 29:477–484.

Yoshida, T., Y. Takai, and D. C. Del Rosario. 1975. Molecular nitrogen content in a submerged rice field. Plant Soil 42:653–660.

Young, J. L. 1962. Inorganic soil nitrogen and carbon:nitrogen ratios of some Pacific northwest soils. Soil Sci. 93:397–404.

Young, J. L. 1964. Ammonia and ammonium reactions with some Pacific Northwest soils. Soil Sci. Soc. Am. Proc. 28:339–345.

Young, J. L., and R. A. Cattani. 1962. Mineral fixation of anhydrous NH_3 by air-dry soils. Soil Sci. Soc. Am. Proc. 26:147–152.

Young, J. L., and B. L. McNeal. 1964. Ammonia and ammonium reactions with some layer-silicate minerals. Soil Sci. Soc. Am. Proc. 28:334–339.

3 Organic Forms of Soil Nitrogen

F. J. STEVENSON

University of Illinois
Urbana, Illinois

I. INTRODUCTION

Over 90% of the N in the surface layer of most soils is organically combined. The importance of this organic N from the standpoint of soil fertility has long been recognized, and our knowledge concerning the nature and chemical composition of organic N is extensive. Nevertheless, approximately one half of the soil organic N has not been adequately characterized, and little is known of the chemical bonds linking nitrogenous constituents to other soil components.

The primary objective of this chapter is to present an account of present knowledge of the forms of organic N in soils as a basis for understanding how N behaves in the soil-plant system. An exhaustive review of all aspects of organic N is beyond the scope of the present communication; much additional information can be found in other chapters of this monograph. Previous reviews on the chemistry of soil N include those of Bremner (1965a, 1967), Stevenson and Wagner (1970), Kowalenko (1978), and Parsons and Tinsley (1975).

II. FRACTIONATION OF SOIL N

Most studies on the forms of N in soils have been based on the use of hot mineral acids (or bases) to hydrolyze nitrogenous constituents (e.g., proteins). In a typical procedure, the soil is heated with $3N$ or $6N$ HCl for from 12 to 24 hours, after which the N is separated into the fractions outlined in Table 1. Methods for determining the forms of N in soil hydrolysates have been described in Bremner (1965a), Ferguson and Sowden (1966), Miki et al. (1966a), Fleige et al. (1971), and Shaymukhametov (1977).

The N that is not solubilized by acid hydrolysis is usually referred to as *acid-insoluble N*; that recovered by distillation with MgO is NH_3-*N*. The soluble N not accounted for as NH_3 or in known compounds is the *hydrolyzable unknown (HUN) fraction*. In the older literature, a small amount of N brought down in the MgO precipitate after distillation of NH_3 was called *humin-N*. This fraction is seldom determined and is usually grouped with the HUN fraction.

Nitrogen in Agricultural Soils—Agronomy Monograph no. 22.

Table 1—Fractionation of soil N based on acid hydrolysis.

Form	Definition and method	% of soil N (usual range)
Acid-insoluble N	Nitrogen remaining in soil residue following acid hydrolysis; usually obtained by difference (total soil N-hydrolyzable N).	20–35
NH_3-N	Ammonia recovered from hydrolysate by steam distillation with MgO.	20–35
Amino acid N	Usually determined by the ninhydrin-CO_2 or ninhydrin-NH_3 methods; recent workers have favored the latter.	30–45
Amino sugar N	Steam distillation with phosphate-borate buffer at pH 11.2 and correction for NH_3-N; colorimetric methods are also available (see text); also referred to as hexosamine-N.	5–10
Hydrolyzable unknown N (HUN fraction)	Hydrolyzable N not accounted for as NH_3, amino acids, or amino sugars; part of this N occurs as non-α-amino N in arginine, tryptophan, lysine, and proline.	10–20

The main identifiable organic N compounds in soil hydrolysates are the *amino acids* and *amino sugars.* Soils contain trace quantities of nucleic acids and other nitrogenous biochemicals, but specialized techniques are required for their separation and identification. Only from one third to one half of the organic N in most soils can be accounted for in known compounds.

The procedure for hydrolyzing the soil has not been standardized, and many variations in hydrolytic conditions have been employed. The variables include: (i) type and concentration of acid; (ii) time and temperature of hydrolysis; (iii) ratio of acid to soil; and (iv) pretreatment (Kowalenko, 1978). A two-stage acid hydrolysis procedure has been used by Russian soil scientists, thereby enabling them to subdivide soil N into four fractions: mineral N, easily hydrolyzable N, difficulty hydrolyzable N, and non-hydrolyzable N (Aderikhin & Shcherbakov, 1974; Koloskova & Shitova, 1974; Bunyakina, 1976; references cited by them).

An unusually large amount of the N in soil, usually of the order of 20 to 35%, is recovered as *acid-insoluble N.* At one time it was thought that this fraction was an artifact resulting from the condensation of amino acids with reducing sugars during hydrolysis, but it is now believed that part of this N occurs as a structural component of humic substances. The possibility that artifacts are formed during hydrolysis cannot be dismissed, however, since addition of glucose to soil immediately prior to hydrolysis has been shown to result in increases in the percentage of the soil N as acid-insoluble N and decreases in hydrolyzable NH_3 and amino acid N (Asami & Hara, 1970). Nonenzymatic browning reactions of the type described by Hodge (1953) and Spark (1969) are undoubtedly involved.

Cheng et al. (1975) found that the percentage of the soil N recovered in acid-soluble forms could be increased by pretreating the soil with HF prior to acid hydrolysis. Their results for some surface and subsurface soils are given in Fig. 1. The increased recovery of organic N was due largely to the

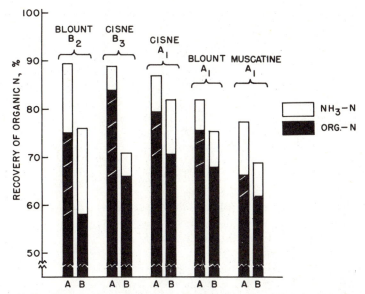

Fig. 1—Recovery of soil organic N as soluble organic compounds and as NH_3 by acid hydrolysis. (*A*) Soil pretreated with HF; (*B*) conventional acid hydrolysis. The broken portion of the bars indicate the increase in recovery of amino acids and other stable organic N compounds (from Cheng et al., 1975).

release of aminio acids and other compounds that were not deaminated by subsequent acid hydrolysis. Other work has shown that the insoluble N can be dissolved with dilute base and partially solubilized by acid hydrolysis, thereby reducing the acid insoluble fraction to about 10% of the total soil N (Freney, 1968; Griffith et al., 1976).

Another unique feature of N fractionation schemes is that a large proportion of the soil N (usually 20 to 30% for surface soils) is recovered as NH_3. Some of the NH_3 is of inorganic origin; part comes from amino sugars and the amino acid amides, asparagine and glutamine. It is also known that NH_3 can arise from the breakdown of certain amino acids during hydrolysis (Bremner, 1965a, 1967). Tryptophan is lost completely; others, such as serine and threonine, are destroyed to a lesser extent. The proportion of the N recovered as NH_3 increases with depth in the soil profile, which can be explained on the basis that the percentage of the N as fixed NH_4^+ also increases with depth (see Chapt. 2, J. L. Young and R. W. Aldag).

An accounting of all potential sources of NH_3 in soil hydrolysates shows that the origin of approximately one half of the NH_3, equivalent to 10 to 12% of the total organic N, is still obscure. Some NH_3 may be derived from complexes formed by fixation reactions of the types described in Chapter 4 (H. Nommik and K. Vahtras). An interesting observation was recently made by Loginow (1967), who reported that humic acids reacted chemically with amino acids in buffered aqueous solutions at 34°C to yield NH_3. Haider et al. (1965) examined the reaction of proteinaceous substances with phenols in the presence of phenoloxidase enzymes. When the products thus formed were subjected to acid hydrolysis, part of the amino acid N was recovered as NH_3.

Aldag (1977) and Aldag and Kickuth (1973) further separated NH_3-N into two fractions: "real amide N" and "pseudo-amide N." The former was obtained by chromatographic assay of soil hydrolysates for aspartic acid and glutamic acid and by assuming that the ratio of NH_3-N to (aspartic acid + glutamic acid)-N was the same for the soil "protein" as for pure proteins (about 0.5). The ratio of pseudo-amide N to real amide N for a variety of soils was found to range from 1.5 to 11.7.

Part of the N that is solubilized by acid hydrolysis occurs in unknown forms (the HUN fraction of Table 1). On the basis of an examination of hydrolysates of 14 proteins, Greenfield (1972) concluded that most of the unidentified N originated from the non-α-amino N present in arginine, tryptophan, lysine, and proline. The non-α-amino N in these amino acids is not included with the amino acid N values as determined by the ninhydrin-NH_3 or ninhydrin-CO_2 methods (discussed later).

The percentage of the soil N accounted for in hydrolyzable unknown compounds (HUN) can be appreciable, often >20% of the total N. The possibility that arginine, tryptophan, lysine, and proline are present in sufficient quantities to account for most of the hydrolyzable unknown N in many soils is remote. Goh and Edmeades (1979) concluded that from one fourth to one half of the hydrolyzable unknown N in the soils they examined occurred as non-α-amino acid N.

A. Distribution of Organic Forms of N in Mineral Soils

Relatively little is known of the factors affecting the distribution of the forms of organic N in soils. Wide variations in N distribution patterns have been observed, but no consistent trend has emerged relating composition to soil properties. As this work is discussed, it should be noted that several problems are encountered in evaluating published data on N distribution patterns in soils. They include inaccuracies in analytical methods, especially for amino acid N. Discrepancies also arise because part of the NH_3 liberated by hydrolysis results from partial destruction of amino sugars, and this has not always been taken into account in estimating NH_3-N and amino sugar N. Still another factor is that some workers have expressed their results in terms of total soil N rather than in terms of an organic N basis. As noted in Chapter 2, significant amounts of the N in many soils occur as fixed NH_4^+, particularly in the subsoil. In some cases, results obtained by different investigators cannot be compared directly due to variations in hydrolysis conditions.

Typical data showing the distribution of the forms of N in soils from different sections of the world are recorded in Table 2. In general, differences in N composition within a similar group of soils are fully as great as for soils having contrasting chemical and physical properties. For the soils shown in Table 2, amino acids accounted for from 13.3 to 52.0% of the N; amino sugars accounted for from 0.8 to 13.8%. The percentage of the N in acid-insoluble forms ranged from 6.9 to 41.3; the range for hydrolyzable unknown forms (HUN) was from 2.4 to 40.2%. Considerable variation also exists in the ratio of amino acid N to amino sugar N (1.9–46.4%). Distribution patterns similar to those recorded in Table 2 have also been observed for soils of Germany (Asmus, 1970) and Japan (Miki et al., 1966b; Iseki, 1968).

Variations in N distribution patterns obtained in individual studies have often been attributed to differences in the physical, chemical, and biological properties of the soils examined. Slight variations in N distribution exist within different size fractions of the soil (Kyuma et al., 1969), and changes are brought about through incubation (Keeney & Bremner, 1966) or when organic matter becomes decomposable through the effect of drying the soil (Hayashi & Harada, 1969; Ahmad et al., 1973).

Sowden's (1977) results for the distribution of N in the surface layers of 40 Canadian soils (see Table 2) show that a higher percentage of the N in the overlying organic layers (LFH and TC horizons) occurred as amino acid N than in the mineral layers (Ah, Ap, and Ae horizons). A plausible explanation for this result is that the organic matter associated with the mineral soil was more highly humified. The percentage of the N in acid-insoluble forms has been found to increase with an increase in "degree of humification" (Yonebayashi et al., 1973), as well as in "degree of decomposition" (Tsukada et al., 1966).

Khan and Sowden (1971) concluded that the main difference between the forms of N in some saline and nonsaline soils of Alberta, Canada, was that a higher proportion of the N in the latter occurred in acid-insoluble forms. Differences have also been observed in N distribution patterns for volcanic vs. nonvolcanic tropical soils, in which case higher percentages of the N in the nonvolcanic soils occurred as acid-insoluble N, NH_3-N, and amino sugar N (Dalal, 1978a, b). In other work, Amara and Stevenson (unpublished observations) found that much lower percentages of the organic N in some Sierra Leone soils occurred as (amino acid + amino sugar)-N than in some typical Illinois soils (averages of 37.0 and 45.9% for the two groups, respectively). One explanation given for these findings is that higher amounts of montmorillonitic-type clays in the Illinois soils resulted in better preservation of protein and carbohydrate constituents. An alternative explanation is that the higher temperatures under Sierra Leone conditions resulted in greater humification of organic matter and more extensive incorporation of amino acid N and amino sugar N into humic compounds. Results obtained by Jorgensen (1967) for three forest soils of North Carolina are unusual in that relatively high percentages of the N in the mineral layers (0–10 cm below the organic horizons) occurred as amino acid N (52.6–55.8%).

Sowden et al. (1977) summarized the N distribution in soils from widely different climatic zones. Included in the comparisons were six soils from arctic and subarctic regions of northern Canada, 62 soils of the cool temperate zones of southern Canada extending from the Atlantic to the Pacific, 6 soils of the dry subtropical Mediterranean region (southern Italy and Israel), and 10 tropical soils from the Caribbean islands of Dominica and St. Vincent. Data presented in Table 3 suggest that higher percentages

Table 2—Distribution of the forms of N in representative surface soils of the world.

Location†	Form of N, % of total soil N					Amino acid N/Amino sugar N ratio	Reference
	Acid insoluble	NH₃	Amino acid	Amino sugar	HUN		
Africa							
Sierra Leone (8)	19.8–24.5	9.3–20.3	22.8–33.3	4.1–13.8	17.4–40.2	2.4– 7.0	Amara & Stevenson‡
Tanzania (4)	14.4–30.7	10.8–21.4	18.6–31.2	5.2–11.5	24.1–36.0	1.9– 4.1	Singh et al. (1978)
Argentina							
Misc. group (4)	22.2–32.7	15.1–21.3	13.3–20.2	2.0–10.9	22.2–32.7	1.2–10.0	Rosell et al. (1978)
Canada							
Representative soils from nine soil orders							
Organic, LFH (6)	15.3 ± 7.3	14.9 ± 2.6	51.0 ± 7.5	7.0 ± 2.7	11.5 avg.	7.3 avg.	Sowden (1977)
Organic, TC (2)	17.3 ± 1.6	14.4 ± 5.7	52.0 ± 2.3	6.4 ± 0.5	9.2 avg.	8.1 avg.	Sowden (1977)
Mineral, Ah (14)	18.3 ± 7.8	18.6 ± 3.4	40.8 ± 8.0	6.7 ± 1.8	15.6 avg.	6.1 avg.	Sowden (1977)
Mineral, Ap (13)	13.4 ± 4.1	21.6 ± 3.2	37.6 ± 4.8	6.7 ± 1.3	20.7 avg.	5.6 avg.	Sowden (1977)
Mineral, Ae (5)	12.5 ± 5.4	23.8 ± 7.6	41.2 ± 9.6	5.9 ± 1.5	16.6 avg.	7.0 avg.	Sowden (1977)
Alberta Gray Wooded (12)	18.4–26.4	13.7–19.0	2.68–33.0	8.8–11.9	17.1–25.0	2.8– 3.3	Khan & Sowden (1971)
Alberta Solonetizic (2)	24.8–25.4	20.7–21.2	34.6–35.8	4.8– 5.2	12.9–13.9	6.7– 7.5	Khan & Sowden (1971)
Quebec cultivated soils (20)	16.2–33.6	15.7–28.2	23.4–37.6	4.8– 9.2	6.9–22.5	2.9– 7.3	Kadirgamathaiyah & MacKenzie (1970)
Germany							
Misc. soils	15.3–18.3	26.4–32.2	22.1–31.2	3.8– 5.0	20.3–29.1	4.9– 8.2	Fleige & Baeumer (1974)
Japan							
Upland soils (3)	9.0–28.0	12.0–24.0	28.0–38.0	5.0–11.0	19.0–29.0	--	Miki et al. (1966b)
Misc. group (3)	21.7–32.6	15.3–24.2	16.9–25.2	1.2– 3.1	19.4–26.6	7.9–14.1	Kyuma et al. (1969)
Paddy soils (3)	11.4–16.8	22.3–22.8	38.8–43.3	7.5– 8.8	20.3–23.4	4.4– 5.7	Hayashi & Harada (1969)

(continued on next page)

Table 2—Continued.

Location[†]	Form of N, % of total soil N					Amino acid N / Amino sugar N ratio	Reference
	Acid insoluble	NH₃	Amino acid	Amino sugar	HUN		
Mediterranean area							
Israel (5)	9.7–24.4	14.4–23.9	30.2–47.9	3.9– 9.3	3.8–27.1	5.2– 7.7	Chen et al. (1977)
Italy (1)	16.5	14.1	46.0	9.2	14.2	5.0	Chen et al. (1977)
Taiwan							
Misc. group (20)	24.4	25.0	37.5	4.8	10.5	7.8	Lin et al. (1973)
United Kingdom							
Arable and pasture (14)	14.0–34.0	16.0–29.0	20.0–39.0	4.0–12.0	13.0–34.0	5.0 avg.	Greenfield (1972)
Forest (8)	14.0–17.0	21.0–37.0	35.0–41.0	4.0– 8.0	15.0–19.0	7.6 avg	Greenfield (1972)
United States							
Illinois (4)	20.6–32.6	15.1–24.4	31.9–40.3	8.7–13.2	2.4–20.0	2.3– 3.8	Amara & Stevenson[‡]
Iowa (20)	18.4–36.7	18.6–29.0	17.8–34.3	3.3– 7.1	17.9–28.9	2.9– 6.3	Keeney & Bremner (1964)
Nebraska (8)	18.0–22.0	18.0–28.0	31.0–46.0	5.0– 9.0	6.0–17.0	4.2– 8.0	Meints & Peterson (1977)
West Indies							
Volcanic (6)	7.6–14.0	19.5–34.3	37.2–49.6	5.6– 7.6	7.6–15.3	5.9– 7.0	Sowden et al. (1976)
Volcanic (4)	11.5–41.3	11.6–17.4	25.4–45.7	0.8– 3.0	4.2–35.0	11.6–46.4	Dalal (1978a)
Nonvolcanic (3)	6.9–22.7	21.5–32.1	20.4–49.8	3.6– 7.9	6.5–20.7	3.9–13.8	Dalal (1978a)

† Numbers in parentheses indicate number of soils analyzed.
‡ Unpublished observations.

Table 3—Nitrogen distribution in soils from widely different climatic zones
(from Sowden et al., 1971).

Climatic zone[†]	Total soil N, %	Form of N, % of total soil N					Amino acid N Amino sugar N ratio
		Acid insoluble	NH₃	Amino acid	Amino sugar	HUN	
Arctic (6)	0.02–0.16	13.9 ± 6.6	32.0 ± 8.0	33.1 ± 9.3	4.5 ± 1.7	16.5	7.4
Cool temperate (82)	0.02–1.06	13.5 ± 6.4	27.5 ± 12.9	35.9 ± 11.5	5.3 ± 2.1	17.8	6.8
Subtropical (6)	0.03–0.30	15.8 ± 4.9	18.0 ± 4.0	41.7 ± 6.8	7.4 ± 2.1	17.1	5.6
Tropical (10)	0.24–1.61	11.1 ± 3.8	24.0 ± 4.5	40.7 ± 8.0	6.7 ± 1.2	17.6	6.1

† Numbers in parentheses indicate number of soils examined.

of the N in soils from the warmer climates occurred as amino acid N and as amino sugar N. Lower percentages occurred as hydrolyzable NH_3. The ratio of amino acid N to amino sugar N was also lower in the tropical and subtropical soils. These findings are in general agreement with results obtained for the Sierra Leone and Illinois soils (discussed above).

1. EFFECT OF CULTIVATION

It is a well-known fact that the N content of most soils declines when land is subjected to intense cultivation (Chapt. 1, F. J. Stevenson). This loss of N is not spread uniformly over all N fractions, although it should be pointed out that neither long-term cropping nor the addition of organic amendments to the soil greatly affects the relative distribution of the forms of N. All forms of soil N, including the acid-insoluble fraction, appear to be biodegradable (Keeney & Bremner, 1964; Ivarson & Schnitzer, 1979).

Data showing the effect of cultivation and cropping systems on the distribution of the forms of N in soil are recorded in Table 4. Cultivation generally leads to a slight increase in the proportion of the N as hydrolyzable NH_3, but this effect is due in part to an increase in the percentage of the soil N as fixed NH_4^+ (Chapt. 2). The proportion of the soil N as amino acid N generally decreases with cultivation, whereas the percentage as amino sugar N changes very little or increases. Khan's (1971) results for two cropping systems (Breton plots of Alberta, Canada) show that the proportion of the soil N as amino acid N and as amino sugar N was significantly greater for a 5-year rotation of grasses and legumes than for a wheat-fallow sequence (Table 4). Khan (1971) found that manure applications resulted in increases in the percentage of the soil N as amino sugars. In contrast, long-term applications of mineral fertilizers had no significant effect on N distribution patterns.

Results obtained for the Morrow plots also indicate that when soils are subject to intensive cultivation, those compounds intimately bound to clay minerals are selectively preserved. These data, summarized in Fig. 2, show that a higher proportion of the organic N in the more heavily cropped soils (those with low N contents) was solubilized when the clay was destroyed with a $5N$ HF–$0.1N$ HCl solution. Thus, it appears that N compounds loosely attached to clay or held between soil aggregates are lost first, followed by those held by strong cohesive forces. Nitrogen compounds held within the lattice structures of clay minerals would be particularly resistant to attack by microorganisms.

Table 4—Effect of cultivation and cropping system on the distribution of the forms of N in soil.

Location[†]	Form of N, % of total soil N					Amino acid N / Amino sugar N ratio	Reference
	Acid insoluble	NH₃	Amino acid	Amino sugar	HUN		
Alberta Canada (Breton plots)[‡]							
Rotation of grains and legumes (6)	21.1	15.1	30.9	10.4	22.6	3.0	Khan (1971)
Wheat-fallow sequence (6)	25.0	17.8	28.8	9.3	19.2	3.1	Khan (1971)
Germany							
Grass sod, cm							
0–5	18.1	26.6	31.2	3.8	20.3	8.2	Fleige & Baeumer (1974)
5–10	15.3	26.6	24.8	4.2	29.1	5.9	Fleige & Baeumer (1974)
10–15	15.9	29.5	26.7	4.6	23.3	5.8	Fleige & Baeumer (1974)
Arable (tilled), cm							
0–5	16.2	32.2	22.1	4.2	25.3	5.3	Fleige & Baeumer (1974)
5–10	16.2	32.2	22.1	4.2	25.3	5.3	Fleige & Baeumer (1974)
10–15	16.2	31.8	22.2	4.2	25.6	5.3	Fleige & Baeumer (1974)
Illinois, USA (Morrow plots)[§]							
Grass border and COCl rotation (2)	20.3	16.6	42.0	10.5	10.7	4.0	Stevenson[¶]
Continuous corn and CO rotation (2)	20.2	16.7	35.0	14.4	13.9	2.4	Stevenson[¶]
Iowa, USA							
Virgin (10)	25.4	22.2	26.5	4.9	21.0	5.4	Keeney & Bremner (1964)
Cultivated (10)	24.0	24.7	23.4	5.4	22.5	4.3	Keeney & Bremner (1964)
Nebraska, USA							
Virgin (4)	20.8	19.8	44.3	7.3	7.8	6.1	Meints & Peterson (1977)
Cultivated (4)	19.3	24.5	35.8	7.0	13.4	5.1	Meints & Peterson (1977)

[†] Numbers in parentheses indicate number of soils analyzed.
[‡] Treatments for each sequence included a control, manure plot, NPKS plot, lime plot, and a P plot.
[§] COCl = corn, oats, clover rotation with lime and P additions; CO = corn-oats rotation.
[¶] Unpublished observations.
[#] Soils of the Ustoll suborder.

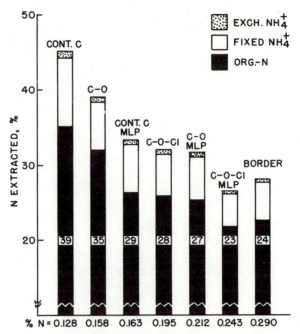

Fig. 2—Organic N and NH_4^+ extracted from the Morrow plot soils with a $2.5N$ HF-$0.1N$ HCl solution. The values in the solid portion of the bars represent the percent recovery of organic N. C = corn, O = oats, Cl = clover, MLP = manure, lime, and phosphate (from Stevenson et al., 1967).

The observation that cultivation has little effect on the percentage distribution of the various forms of N emphasizes that chemical fractionation of soil N following acid hydrolysis is of little practical value as a means of testing soils for available N or for predicting crop yields (Keeney & Bremner, 1964, 1966; Porter et al., 1964; Cornfield, 1957; Cornforth, 1968; Moore & Russell, 1970; Kadirgamathaiyah & MacKenzie, 1970; Giddens et al., 1971; Osborne, 1977).

2. DEPTH DISTRIBUTION PATTERNS

Several studies have dealt with changes in the distribution of the forms of N with depth in the soil profile (e.g., Khan & Sowden, 1971; Meints & Peterson, 1977). Much of this information cannot be interpreted directly in terms of organic N because a significant amount of the subsoil N may occur as fixed NH_4^+. As noted in Chapter 2, the percentage of the soil N as fixed NH_4^+ increases with depth, often $>50\%$ in the B and C horizons. Part of this fixed NH_4^+ will be liberated as NH_3 by acid hydrolysis, and part may remain as a portion of acid-insoluble N. The observation that the proportion of the soil N as hydrolyzable NH_3 increases with depth (e.g., see Meints & Peterson, 1977) can be accounted for by partial release of fixed NH_4^+.

Sowden's (1956) results for amino acid N in the A and B horizons of five Canadian soils, which are recorded in Table 5, suggest that the "pro-

Table 5—Nitrogen accounted for as amino acids in the A and B horizons
of five Canadian soils (from Sowden, 1956).

Soil type	Percentage of total N as amino acids	
	A horizon	B horizon
Black Solonetz (Mollisol)	31.5	19.5
Gray Wooded (Alfisol)	17.1	14.0
Podzol (Spodosol)	22.5	19.9
Podzol (Spodosol)	30.1	22.8
Podzol (Spodosol)	41.8	24.9

tein'' content of the organic matter may be higher in surface soil than in
subsoil. The values shown for the Mollisol and Alfisol are undoubtedly affected by fixed NH_4^+, but Spodosols generally contain very little N in this
form. Recalculation of data published by Nommik (1967) for a Spodosol
profile from Sweden gives the following values for percentage organic N as
amino acids: A0 horizon, 50.8%; A1 horizon, 39.5%; A2 horizon, 30.2%;
B horizon, 27.1%.

An accurate accounting of organic forms of N in the lower soil horizons will require improvements in methods for determining native NH_4^+ as
well as specific organic N forms when they are present in low amounts (e.g.,
amino acid N and amino sugar N). Conventional hydrolysis procedures may
lead to incomplete recoveries of amino acids from clay-rich subsurface soils
(Stevenson & Cheng, 1970).

B. Distribution of the Forms of N in Histosols and Aquatic Sediments

Environmental conditions in aquatic sediments are particularly suitable
for the preservation of microbially produced substances. Thus, the same
forms of N occur in organic soils (Histosols) and lake sediments as in mineral soils, but the N distribution is usually different. In general, higher percentages of the N in Histosols and lake sediments occur in the form of
amino acids, a result that can be explained by less extensive microbial turnover of organic N in aquatic sediments. An exception to this rule occurs for
well-humified peats, where amino acid N values are relatively low, and a
high percentage of the N occurs in the acid-insoluble fraction.

Data presented in Table 6 indicate that the N distribution in Histosols
is similar to that of lake sediments, both of which are formed under waterlogged or partially waterlogged conditions. For the Wisconsin lake sediments, there was a trend toward increasing amino acid N with increasing
level of fertility, which Keeney et al. (1970) attributed to less extensive N
turnover by microorganisms under conditions where O_2 is limited.

As far as Histosols are concerned, the percentage of the total N as
amino acid N is generally higher for mesic peats than for humic peats (see
Table 6). Humic peats are in a more advanced stage of decomposition than
mesic peats; thus, more of the N will have been incorporated into unknown
humus forms. When Histosols are incubated under aerobic or anaerobic

Table 6—Distribution of the forms of N in Histosols and lake sediments.

Distribution†	Acid insoluble	NH₃	Amino acid	Amino sugar	HUN	Reference
Canada						
Humic peat profile	20.3–29.5	11.1–14.7	37.7–43.3	8.7–11.0		6.5–19.4 Sowden et al. (1978)
Mesic peat profile	21.2–24.5	8.6–12.1	46.3–56.3	4.3– 5.8		6.8–17.7 Sowden et al. (1978)
Wisconsin						
Histic materials as collected (9)	4.6–37.0	9.3–23.6	15.9–58.8	3.5–11.5		13.7–30.6 Isirimah & Keeney (1973)
Following incubation (9)	21.6–45.4	13.0–17.1	24.3–39.2	3.9– 4.8		11.6–17.2 Isirimah & Keeney (1973)
Wisconsin						
Oligotrophic (5)	9.4–30.9	13.4–19.1	30.0–38.7	3.5– 6.3		21.4–30.5 Keeney et al. (1970)
Mesotrophic (4)	17.0–30.1	14.4–17.9	33.3–41.9	3.1– 3.6		19.6–24.6 Keeney et al. (1970)
Eutrophic (8)	12.7–21.8	14.3–17.1	34.4–45.1	2.7– 4.4		21.1–26.9 Keeney et al. (1970)

Column header spanning the value columns: **Form of N, % of total N**

Sub-headers within the body: Histosols; Lake sediments

† Numbers in parentheses indicate number of soils analyzed.

conditions in the laboratory, the percentage of the N as amino acid N decreases while the percentage as acid-insoluble N increases (Isirimah & Keeney, 1973).

Results presented in Fig. 3 for the forms of N in the organic ooze of Mud Lake, Florida, show that the percentage of the N as amino acids was substantially higher in the top layer than in the sample at a depth of 1.2 m. In contrast, a higher percentage of the N at the lower depth occurred in acid-insoluble forms. This organic sediment has been formed under warm oxidizing conditions, and the lower percentage of the N in the form of amino acids at 1.2 m can be accounted for by greater turnover of organic matter in the older material, with subsequent incorporation of N into brown nitrogenous polymers (Stevenson, 1974).

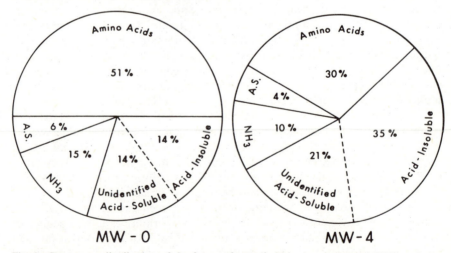

Fig. 3—Percentage distribution of the forms of organic N in the top layer (MW-0) and at a depth of 1.2 m (MW-4) in the organic sediment of Mud Lake, Florida (from Stevenson, 1974).

C. Humic and Fulvic Acids

Humic and fulvic acids contain the same forms of N that are obtained when soils are subjected to acid hydrolysis (Bremner, 1955; Khan & Sowden, 1972; Batsula & Krupskiy, 1974; Aldag, 1977; Rosell et al., 1978; Tsutsuki & Kuwatsuka, 1978). However, N distribution patterns vary somewhat. For example, compared with unfractionated soil, a lower percentage of the N in humic acids occurs as NH_3-N and in hydrolyzable unknown compounds (HUN fraction); a higher percentage occurs as amino acid N and as acid-insoluble N. Results obtained by Rosell et al. (1978) for the distribution of N in some Argentine soils and their humic acids are shown in Fig. 4.

Typical results for the distribution of N in hydrolysates of humic and fulvic acids are given in Table 7. For the preparations analyzed, from 20 to 45% of the N occurred as amino acids, and from 2 to 8% occurred as amino sugars. In comparison with the humic acids, a higher percentage of the N in the fulvic acids occurred as NH_3-N. Also, the fulvic acids contained little, if any, acid-insoluble N, whereas a much higher percentage of the N occurred in hydrolyzable unknown compounds (HUN fraction). The amino acid content of humic acids may decrease with an increase in "degree of humification" (Tsutsuki & Kuwatsuka, 1978).

Data given in Table 7 also show that humic acids extracted from soils with $0.5N$ NaOH have higher N contents than those extracted with $0.1M$ $Na_4P_2O_7$. Furthermore, a higher percentage of the N in alkali-extracted humic acids occurs in the form of amino acids. Ortiz de Serra et al. (1973) found that a higher proportion of the N in fungal humic acids occurred in the form of amino acids compared with soil humic acids.

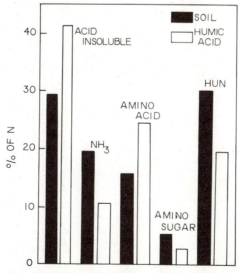

Fig. 4—Comparison of the distribution of the forms of N by acid hydrolysis of soils and their humic acids. *HUN* = hydrolyzable unknown compounds (adapted from data of Rosell et al., 1978).

Table 7—Distribution of the forms of N in humic and fulvic acids.

Extractant†	% N	Acid insoluble	NH₃	Amino acids	Amino sugars	HUN	Amino acid N / Amino sugar N ratio	Reference
				Humic acids				
0.1M Na₄P₂O₇ (6)	1.79–2.63	41.3–59.0	8.8–12.8	19.5–34.5	2.6–5.0	5.2–10.6	3.9–10.1	Bremner (1955)
0.5N NaOH (5)	2.31–3.74	32.6–43.7	8.4–13.7	31.2–44.7	3.4–8.1	4.7– 8.6	3.9–12.5	Bremner (1955)
0.5N NaOH (4)	2.11–2.69	35.9–50.8	8.2–14.0	22.1–26.5	1.8–3.9	16.2–21.8	6.4–14.7	Rosell et al. (1978)
				Fulvic acids				
0.5N NaOH (3)	3.37–3.89	--	15.1–19.3	26.4–34.2	3.6–5.2	41.3–54.9	6.4– 7.3	Khan & Sowden (1972)

† Numbers in parentheses indicate number of soils analyzed.

The fulvic acid fraction of soil organic matter contains a broad spectrum of organic compounds, including amino acids and amino sugars (Forsyth, 1947). Thus far, few attempts have been made to establish the origin of the various organic N compounds in hydrolysates of crude fulvic acid extracts. Differences in N distribution patterns observed for "fulvic acids" from different sources may partially reflect variations in the amounts of "true fulvic acids" that are present relative to nonhumic substances. In some fulvic acid preparations, most of the N appears to be present as amino acids (Otsuka, 1975; Sequi et al., 1975). In others, rather high amounts of NH_3-N have been observed (Stevenson, 1960; Batsula & Krupskiy, 1974).

As was the case for soil hydrolysates, the origin of the NH_3 in acid hydrolysates of humic and fulvic acids is unknown. Kickuth and Scheffer (1976) concluded that the pseudo-amide N (hydrolyzable NH_3 not accounted for as the amino acid amides, asparagine and glutamine) was derived from imino groups ($=NH$) on quinone ring systems. The reaction of amino acids with humic substances (Loginow, 1967) and phenols (Haider et al., 1965; Ladd & Butler, 1966) has been shown to lead to incorporation of N into structures that yield NH_3 when the products are subjected to acid hydrolysis. An alternate source is from condensation products originating from the reaction of amino acids with reducing sugars, as suggested by Stevenson (1960).

As noted previously, as much as one half of the total N in humic substances can be accounted for as amino acid N. Additional quantities of amino acids can be recovered from humic acids by subjecting the acid-hydrolyzed residues to a second hydrolysis with $2.5N$ NaOH (Piper & Posner, 1968, 1972; Griffith et al., 1976). Sodium amalgam reduction of the alkali-treated residue has been shown to lead to a further release of amino acids (Piper & Posner, 1968). Acid hydrolysis would be expected to remove amino acids bound by peptide bonds (I) as well as those linked to quinone rings (II). On the other hand, amino acids bonded directly to phenolic rings (III) would not be released without subsequent alkaline hydrolysis. The effect of alkaline hydrolysis was believed to be due to oxidation of the phenol to the quinone form, with release of the amino acid. A wide variety of amino acids have been observed in alkaline hydrolysates of acid-treated humic acids (Griffith et al., 1976).

It is evident from the above, that part of the acid-insoluble N in humic acids may ocur in the form of N-phenyl amino acids resulting from the bonding between amino groups and aromatic rings (see also Witthauer & Klocking, 1971; references contained therein).

In a study of the incorporation of ^{14}C-labeled glycylglycine into humic acids, Perry and Adams (1972) found that the N-terminal glycine residue

Table 8—Distribution of the forms of N in humic acid as influenced by different
hydrolysis treatment (adapted from Aldag, 1977).

Hydrolysis condition	Form of N, % of total N				
	Acid insoluble	NH₃	Amino acid	Amino sugar	HUN
1) With 6N HCl, 24 hours	38.5	13.8	34.2	2.4	11.0
2) Hydrolysate from (1) further hydrolyzed with 3% H₂O₂-6N HCl	38.5	18.2	35.2	1.2	6.8
3) Residue from one rehydrolyzed with 3% H₂O₂-6N HCl	27.6	21.8	37.8	1.2	11.6

was considerably more resistant to hydrolysis than was the C-terminal residue, indicating that the bond between humic acid and the N-terminal glycine component was more stable to hydrolysis than was the peptide bond. In a subsequent study, Adams and Perry (1973) found that maximum incorporation of amino acids into humic acids occurred at a pH that coincided with the apparent dissociation constants of their α-amino groups. Incorporation took place into forms that were both acid hydrolyzable and nonhydrolyzable, the latter accounting for 10 to 20% of the total. Stepanov (1969) had earlier demonstrated chemical fixation of amino acid N by humic acids.

An interesting result was recently obtained by Aldag (1977), who observed an increase in amino acid N at the expense of the HUN fraction when an acid hydrolysate of humic acid (6N HCl for 24 hours) was subjected to a second hydrolysis with 6N HCl containing 3% H_2O_2. The increased release of amino acids was attributed to the presence of phenolic-amino acid addition products in the hydrolysate (e.g., see structure III). Hydrolysis of the initial acid-insoluble residue with 3% H_2O_2–6N HCl led to a further release of amino acids as well as other N forms.

Results obtained by Aldag (1977) are given in Table 8. The increase in NH_3-N by hydrolysis in the presence of 3% H_2O_2 can be accounted for in part by oxidation of organic N compounds, such as amino sugars.

Part of the N associated with humic acids may exist as peptides or proteins linked to the central core by H bonding (Haworth, 1971). Evidence for N in loosely bound forms has come from the observation that the N content of humic acids can be lowered considerably by passage through a cation exchange resin in the H form (Sowden & Schnitzer, 1967). Protein-rich fractions have been obtained from humic acids by extraction with phenol (Biederbeck & Paul, 1973; McGill & Paul, 1976) and formic acid (Simonart et al., 1967).

Peptide-like substances have been determined in humic acids by chromatographic assay of proteinaceous constituents liberated by cold hydrolysis with concentrated HCl (Bremner, 1955; Cheshire et al., 1967; Piper & Posner, 1968). It has also been demonstrated that amino acids are released through the action of proteolytic enzymes (see Chapt. 5, J. N. Ladd and R. B. Jackson). Infrared spectra of some, but not all, humic acids show absorption bands typical of the peptide linkage (Otsuki & Hanya, 1967; Stevenson & Goh, 1971).

As was the case for soil hydrolysates, some of the N accounted for in hydrolyzable unknown compounds (HUN fraction) may occur as the non-α-amino acid N of such amino acids as arginine, typtophan, lysine, and proline. Kickuth and Scheffer (1976) concluded that hydrolyzable amide groups comprised 50% of the total peripheral amino groups of humic acids and represented the quinoid aspect of humic substances.

In conclusion of this section, it can be said that significant amounts of the N associated with humic and fulvic acids cannot be accounted for in known compounds. This N may occur in the following types of linkages:

1) As a free amino ($-NH_2$) group
2) As an open chain ($-NH-$, $=N-$) group
3) As part of a heterocyclic ring, such as an $-NH-$ of indole and pyrrole or the $-N=$ of pyridine
4) As a bridge constituent linking quinone groups together
5) As an amino acid attached to aromatic rings in such a manner that the intact molecule is not released by acid hydrolysis

D. Distribution and Stabilization of Newly Immobilized N

Soil organic matter is not an inert substance but is constantly changing as a consequence of the activities of microorganisms. Under steady-state conditions, mineralization of native humus is compensated for by synthesis of new humus. During humification, the N of amino acids and other amino compounds is incorporated into the structures of humic and fulvic acids, in which form it is not readily available to plants.

A key to the understanding of the process whereby newly immobilized N is stablized in soils may be provided by consideration of the changes that occur during the humification of plant remains. As noted in the reviews of Flaig et al. (1975) and Haider et al. (1975), the classical lignin-protein theory is now believed to be obsolete. The modern view is that the so-called humic and fulvic acids are formed by a multiple stage process that includes: (i) decomposition of all plant polymers, including lignin, into simple monomers; (ii) metabolism of the monomers by microorganisms with an accompanying increase in the soil biomass; (iii) repeated recycling of the biomass C (and N) with synthesis of new cells; and (iv) concurrent polymerization of reactive monomers into high molecular weight polymers. The consensus is that the polyphenols (quinones) synthesized by microorganisms, together with those liberated from lignin, polymerize alone or in the presence of amino compounds (amino acids, etc.) to form brown-colored polymers. An alternate pathway is by condensation of amino acids and related substances with reducing sugars, according to the Maillard reaction (Hodge, 1953; Spark, 1969). In each case, structural changes take place in the amino acid molecule with incorporation of N into complex polymers and copolymers. Other mechanisms for incorporation of N into humic and fulvic acids include

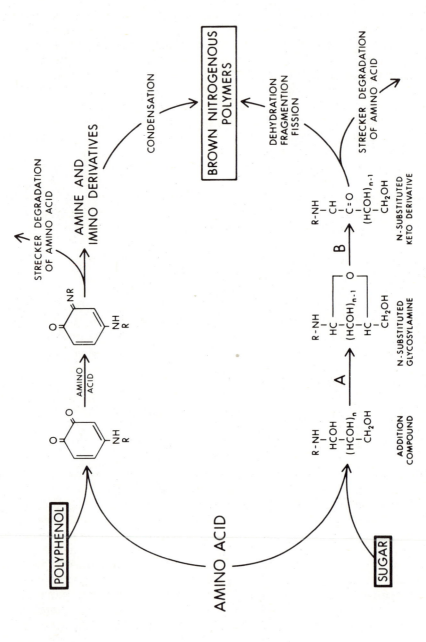

Fig. 5—Suggested scheme for the formation of brown nitrogenous polymers in soils by condensation of amino acids with polyphenols (*upper portion*) and sugars through the Maillard reaction (*lower portion*).

chemical fixation of NH_3 and NO_2^- by lignin and lignin-like substances (see Chapt. 4, H. Nommik and K. Vahtras).

The reaction between amino acids and polyphenols involves simultaneous oxidation of the polyphenol to the quinone form, such as by polyphenol oxidase enzymes. The addition product readily polymerizes to form brown nitrogenous polymers, according to the general sequence shown in Fig. 5 (top part).

In the case of the Maillard reaction, the initial step involves the addition of the amine to the carbonyl group of the sugar, with the formation of an aldosylamine (Fig. 5, bottom). This is followed by the Amadori rearrangement to form the N-substituted keto derivative, which subsequently undergoes dehydration and fragmentation to yield a variety of unsaturated intermediates (Hodge, 1953). In the final stages of browning, the intermediates polymerize into brown polymers and copolymers. The rate of the reaction increases with temperature, pH, and the basicity of the amine. Under laboratory conditions, brown polymers can readily be synthesized from amino acid–sugar mixtures within hours in aqueous solution at 50°C. The significance of nonenzymatic browning during food dehydration has been outlined elsewhere (Hodge, 1953; Spark, 1969).

The net effect of nonbiological reactions involving amino acids and other amino compounds is incorporation of N into the structures of humic and fulvic acids. Thus, whereas amino acids constitute 80% or more of the organic N of microbial tissue (biomass), they account for only about one third of the N of soil humus. As noted earlier, most of the unaccounted for soil N occurs in forms that cannot be accounted for in known compounds by acid hydrolysis. The fraction designated as acid-insoluble N, and part of the N in the NH_3 and hydrolyzable unknown (HUN) fractions, appears to be associated with complex organic pigments.

Information on the composition and fate of newly immobilized N has been provided by studies in which the biomass has been labeled with ^{15}N (Yamashita & Kawada, 1968; Asami, 1971; Ahmad et al., 1973; Kai et al., 1973). The findings of these and other studies can be summarized as follows:

1) Under optimum conditions for microbial activity and in the presence of an available C source, net immobilization of added ^{15}N proceeds rapidly and reaches a maximum at incubation periods as short as 3 days with a simple substrate (e.g., glucose) to as much as 2 months or more for a complex substrate (e.g., mature crop residue) (Kai et al., 1973).

2) At the point of maximum incorporation of N into the biomass, there is a distinct difference in the percentage distribution of the forms of organic N between the newly immobilized N and the native humus N. As noted in Table 9, the immobilized N is higher in amino acid N and in hydrolyzable unknown forms (HUN fraction) but lower in acid-insoluble N; percentages of the N as NH_3-N and as amino sugar N are approximately the same.

3) Following maximum tie-up of N, there is a net release of im-

Table 9—Percentage distribution of native N and applied ^{15}N in various soil organic N
fractions after maximum incorporation of applied N into the biomass
(adapted from Kai et al., 1973).

Source of N and substrate used†	Form of N, % of total N				
	Acid insoluble	NH$_3$	Amino acid	Amino sugar	HUN
Native humus–N	20.8–21.8	11.6–14.3	37.5–40.3	9.8–10.9	12.7–19.3
Applied ^{15}N					
Glucose	4.5	12.6	48.2	9.1	25.6
Cellulose	5.2	13.5	50.2	13.0	18.1
Straw	6.2	13.5	51.7	6.7	21.9

† Incubation times were 3 days, 8 weeks, and 7 days for glucose, cellulose, and straw, respectively.

mobilized ^{15}N to available mineral forms (NH$_4^+$, NO$_3^-$). However, not all N forms are minerlized at the same rate. With time, the percentage of the organic N as amino acids and in hydrolyzable unknown compounds decreases as the percentage as NH$_3$-N and acid insoluble-N increases. These changes are shown in Table 10.

The trends in laboratory incubations (noted above) have also been observed when ^{15}N-labeled fertilizers have been applied to soil in the greenhouse or in the field (Huser, 1971; Legg et al., 1971; Allen et al., 1973; Bunyakina, 1976; Koren'kov et al., 1976; McGill & Paul, 1976). In the case of the field study of Allen et al. (1973), an average of one third of the fertilizer N initially applied was accounted for in the surface soil after the end of the first growing season, the remainder of which had been consumed by plants or lost through leaching and denitrification. Isotope ratio analyses revealed that the residual N had been incorporated into the organic matter. Comparison of the distribution pattern for the fertilizer-derived N with that of the native humus N (Fig. 6) shows that a considerably higher proportion of the fertilizer N occurred in the form of amino acids (59.0 vs. 36.0%) and amino sugars (9.9 vs. 8.0%); lower proportions occurred as hydrolyzable NH$_3$ (10.6 vs. 18.1%), acid-insoluble N (10.3 vs. 21.7%), and hydrolyzable unknown forms (10.2 vs. 16.2%). When the plots were resampled 4 years later, the fertilizer N remaining in the soil, representing one sixth of that

Table 10—Distribution of immobilized ^{15}N after various incubation periods using
glucose as substrate (adapted from Kai et al., 1973).

Source of N and incubation period	Form of N, % of total N				
	Acid insoluble	NH$_3$	Amino acid	Amino sugar	HUN
Native humus–N†	20.8–24.1	12.4–15.5	37.5–38.6	9.1–10.4	12.9–19.3
Applied ^{15}N					
3 days	4.5	12.6	48.2	9.1	25.6
1 week	5.7	12.4	43.5	14.0	24.4
8 weeks	10.7	17.2	41.4	10.7	20.1
20 weeks	8.5	20.0	39.2	9.2	23.1

† Variations reflect changes over a 20-week incubation period.

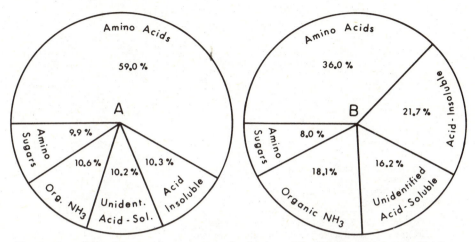

Fig. 6—Comparison between the percentage distribution of fertilizer-derived organic N (*A*) and native humus–N (*B*) (from Allen et al., 1973).

initially applied, had a composition very similar to that of the native humus –N. From these and other studies, it must be concluded that part of the fertilizer added to soils of the Corn Belt section of the United States each year will remain in the soil in organic forms for many decades.

E. Natural Variations in N Isotope Abundance

Slight variations occur in the N isotope composition of natural substances, including soil, peat, coal, petroleum, rocks, minerals, and the proteins of plants and animals (see review of Hauck & Bremner, 1976). These variations result from isotopic effects during biochemical and chemical transformations, such as nitrification (Delwiche & Steyn, 1970) and denitrification (Wellman et al., 1968; Blackmer & Bremner, 1977). The overall effect of these isotope effects in soil is a slight increase in the average ^{15}N content of soil N and its fractions compared with atmospheric N_2.

Natural variations in N isotope abundance are usually expressed in terms of the per mil excess ^{15}N, or delta ^{15}N ($\delta^{15}N$). The equation is:

$$\delta^{15}N = \frac{\text{atom \% } ^{15}N \text{ in sample} - \text{atom \% } ^{15}N \text{ in standard}}{\text{atom \% } ^{15}N \text{ in standard}} \times 1,000.$$

Thus, a $\delta^{15}N$ value of $+10$ indicates that the experimental sample is enriched by 1% compared with the atom % ^{15}N of the standard, usually atmospheric N_2. A negative value indicates that the sample is depleted in ^{15}N relative to the standard.

The work of Meints et al. (1975), Rennie et al. (1976), Shearer et al. (1974), Broadbent et al. (1980), and others has shown that the $\delta^{15}N$ value of the soil generally falls within the range of $+5$ to $+12$, although higher and

Table 11—Variations in the $\delta^{15}N$ content of the various forms of soil nitrogen
(from Cheng et al., 1964).

Form of N	Soil type[†]				
	Grundy sicl	Hayden sil	Austin c	Clarion sil	Glencoe sicl
Total	+16	+7	+5	+3	+2
Hydrolyzable					
Total	+18	+10	+7	+5	+4
Ammonium	+7	+7	+3	+6	+5
Hexosamine	+25	+8	0	+2	−2
Amino acid	+16	+14	+12	+5	+8
Hydroxyamino acid	+19	+11	+8	+7	+3
Nonhydrolyzable	−3	−2	−1	0	−4
N, mineralized[‡]	+6	+2	+1	+1	+1
Fixed ammonium	+6	+6	+4	+2	0

† sicl = silty clay loam, sil = silt loam, c = clay.
‡ Inorganic N produced by incubation of the soil at 30°C and at 50% water-holding
  capacity for 2 weeks.

lower values are by no means rare. For any given soil, variations exist in the
$\delta^{15}N$ value of the various N fractions, as shown in Table 11. Kanazawa and
Yoneyama (1978) found that the ^{15}N abundance of soil amino acids was
higher than for the total soil N.

Natural variations in ^{15}N content may ultimately prove useful in evaluating N cycle processes in soil, such as denitrification. In recent years, the
approach has been used by Kohl et al. (1971) in an attempt to estimate the
relative contribution of soil and fertilizer N to the NO_3^- in surface waters of
an Illinois watershed, the basis of which is that the ^{15}N enrichment of soil-
derived NO_3^- will be higher than that of NO_3^- originating from the fertilizer.
The conclusions of Kohl et al. (1971) generated considerable interest and
were challenged by Hauck et al. (1972) (for a response, see Kohl et al.,
1972). Limitations in using natural variations in ^{15}N content to follow the
fate of fertilizer N in soils have been discussed elsewhere (Focht, 1973;
Hauck, 1973; Blackmer & Bremner, 1977; Broadbent et al., 1980). One
problem in evaluating the contribution from soil is that the ^{15}N content of
mineralized N may not be the same as humus-N. Feigin et al. (1974) found
that the ^{15}N value of the NO_3^- produced on incubation increased with time;
long-term incubations often gave $\delta^{15}N$ values twice as large as those measured after only a short incubation. When the contribution of native humus–
N to the NO_3^- in drainage waters is evaluated, the question naturally arises
as to the proper $\delta^{15}N$ value to select for the soil-derived NO_3^-.

III. AMINO ACIDS

The occurrence of amino acids in soil has been known since the turn of
the century, when Suzuki (1906–1908) reported the presence of aspartic
acid, alanine, aminovaleric acid, and proline in an acid hydrolysate of
humic acid. Bremner's (1967) review shows that by 1917 several other

amino acids had been isolated, including glutamic acid, valine, leucine, iso-leucine, tyrosine, histidine and arginine. As will be shown later, numerous other amino acids have since been identified in soils, many of which are not normal constituents of proteins.

A. Extraction and Quantitative Determination

Amino acids are normally extracted from soil by treatment with hot mineral acids, usually $6N$ HCl under reflux for 16 to 24 hours. Hydrolysis with base is required if tryptophan is to be determined. Conditions for maximum recovery of amino acids have been studied in considerable detail by Bremner (1949), Sowden (1969), Cheng (1975), and Cheng et al. (1975).

The efficiency of conventional acid hydrolysis for recovering amino acids from soils containing fine-grained mineral matter is suspect. It was pointed out earlier that an HF pretreatment may be required for quantitative release of amino acids that are bound to clay minerals and that amino acids chemically attached to humic acids may not be released by acid hydrolysis.

Although several methods have been used for estimating amino acid N in hydrolysates of soil or soil organic matter preparations, most investigators favor methods based on the ninhydrin reaction.

The classical ninhydrin-CO_2 method is highly specific because it requires a free NH_2 group adjacent to a COOH group. However, aspartic acid yields 2 mol of CO_2; proline and hydroxyproline give low values. Bremner (1965b) applied the ninhydrin-NH_3 method to determine amino acid N in soil hydrolysates. Careful control of pH is required, otherwise the NH_3 forms a colored product with ninhydrin as described below.

Methods based on colorimetric analysis of the blue-colored product produced when the ninhydrin reaction is carried out at pH 5.0 have also been used (Stevenson & Cheng, 1970, 1972). This approach is especially useful for subsurface soils and sediments where amino acid levels are low.

A less popular procedure is the nitrous acid method, which is based on the ability of aliphatic amines to react with nitrous acid to form N_2:

$$R-NH_2 + HNO_2 \rightarrow R-OH + H_2O + N_2.$$

The method is subject to interference by many compounds, such as phenols, tannins, lignins, alcohols, and keto acids.

Values recorded for amino acid N based on the analysis of soil hydrolysates must be regarded as minimal. In addition to incomplete extraction and partial destruction of amino acids, part of the N of glutamine and asparagine is liberated as NH_3. Furthermore, some amino acids contain N other than in an NH_2 group adjacent to a COOH group, typical examples of which are arginine, histidine, and lysine.

B. Identification of Amino Acids

Chromatographic techniques have now superseded the more tedious and time-consuming chemical methods for isolating amino acids, and an impressive number of compounds have been detected in soil hydrolysates. The first modern study was conducted by Bremner (1950), who detected the following compounds by paper partition chromatography: glycine, alanine, valine, leucine, isoleucine, serine, threonine, aspartic acid, glutamic acid, phenylalanine, arginine, histidine, lysine, proline, hydroxyproline, α-amino-n-butyric acid, α,ϵ-diaminopimelic acid, β-alanine, and γ-amino butyric acid. A few years later, Stevenson (1954, 1956a) applied the newly developed and more precise technique of ion exchange chromatography. The identifications made by Bremner were confirmed, and in addition, ornithine, cysteine, methionine sulfone, and methionine sulfoxide were identified.

Improvemenets have recently been made in the ultrarapid analysis of amino acids by high-pressure chromatography, and procedures are available for the complete separation of ninhydrin-positive compounds in biological samples containing a multiplicity of constituents (Hamilton, 1968). Separations can be greatly facilitated by use of an automatic amino acid analyzer, several models of which are sold commercially. Gas chromatographic methods are also available and may have advantages for certain applications (Cheng et al., 1975; Pollock et al., 1977).

A rather large number of unidentified ninhydrin-reacting substances have been observed in soil hydrolysates. In the study conducted by Stevenson (1954), 33 amino compounds were isolated, 29 of which were identified. Young and Mortensen (1958) reported 57 ninhydrin-reacting substances in acid hydrolysates of some Ohio soils, only 24 of which were identified. Many of the unidentified compounds occur in minute amounts and can be found only by using special isolation techniques.

It is of interest that many of the amino acids found in soils (specifically, α,ϵ-diaminopimelic acid (IV), ornithine (V), β-alanine (VI), α-amino-n-butyric acid (VII), γ-amino-butyric acid (VIII), 3,4-dihydroxyphenylalanine (IX), and taurine (V)) are not normal constituents of proteins. Several of these amino acids may represent waste products of metabolism; others occur in a variety of natural products synthesized by microorganisms. Ornithine and β-alanine, for example, are constituents of certain antibiotics; the latter is also a component of pantothenic acid, an important vitamin. The occurrence of α,ϵ-diaminopimelic acid is of interest because this amino acid appears to be confined to bacteria, where it occurs as part of the cell wall. It should be noted that thus far, about 140 amino acids and amino acid derivatives have been identified as constituents of living organisms (Uy & Wold, 1977).

$$HOOC-CH-CH_2-CH_2-CH_2-CH-COOH$$
$$\quad\;\; |NH_2 \qquad\qquad\qquad\quad |NH_2$$

$$NH_2-CH_2-CH_2-CH_2-CH-COOH$$
$$\qquad\qquad\qquad\qquad\qquad\quad |NH_2$$

α,ϵ-Diaminopimelic acid   Ornithine

IV   V

$$NH_2-CH_2-CH_2-COOH \qquad CH_3-CH_2-CH-COOH \qquad NH_2-CH_2-CH_2-CH_2-COOH$$
$$\qquad\qquad\qquad\qquad\qquad\qquad\qquad\;\; |NH_2$$

β-Alanine α-Amino-n-butyric acid λ-amino-butyric acid

VI   VII   VIII

$$HO$$
$$HO-\langle\;\rangle-CH_2-CH-COOH \qquad\qquad NH_2-CH_2-CH_2-SO_3H$$
$$\qquad\qquad\qquad\quad |NH_2$$

3,4-dihydroxyphenylanine   Taurine

IX   X

C. Distribution Patterns in Soil

Due to the advanced state of the art of chromatography, no great ingenuity is required to demonstrate the occurrence of amino acids in soil or soil organic matter preparations. Of greater significance than the mere detection of these compounds is a knowledge of precise distribution patterns. Unfortunately, data on this subject are difficult to evaluate because of analytical errors associated with incomplete extraction, losses of specific amino acids during hydrolysis and desalting, possible improper identifications, and others. Sowden (1969) found that most methods used for desalting soil hydrolysates, including treatment on cation exchange resins, led to selective losses of aspartic acid, glutamic acid, and some of the basic amino acids. He found that Fe and Al, which caused the greatest interference during chromatography, could be removed by treatment of the hydrolysate with acetylacetone-chloroform solution with minimal losses of amino acids. Methods for desalting soil hydrolysates have also been evaluated by Pollock and Miyamoto (1971).

A novel desalting procedure has been proposed by Cheng et al. (1975) and is based on the fact that Fe^{3+}, Al^{3+}, and Si^{4+} form polyanion complexes with F^- and are retained as FeF_6^{3-}, AlF_6^{3-}, and SiF_6^{2-} when the hydrolysate is passed through an anion exchange resin in the F^- form.

The extreme variability that has been reported for the amino acid composition of soils is indicated in Table 12, where a tabulation is given of some

Table 12—Percentage of α-amino acid N as acidic, neutral, and basic compounds
for soils of different climatic zones.[†]

Zone[‡]	Distribution of α-amino acid N, %		
	Acid compounds	Neutral compounds	Basic compounds
Temperate (7)	12.6–25.0	66.6–76.2	8.4– 9.8
Semitropical (5)	0.8–10.9	61.2–70.8	10.3–35.5
Tropical (5)	1.2– 6.4	65.0–85.6	8.0–29.1

† Adapted from data published by Almeida et al. (1969), Sowden (1970), Stevenson
  (1956b), Wang et al. (1967), and Young and Mortensen (1958).
‡ Number in brackets indicate number of soils analyzed.

of the early work on the relative distribution of amino acid N as acidic, neutral, and basic compounds in soils of different climatic zones. From 8 to over 35% of the amino acid N in soils has been reported in the form of basic amino acids (lysine, histidine, arginine, and ornithine). Equally divergent results have been recorded for individual amino acids in each group. For example, Stevenson (1956b) and Young and Mortensen (1958) found that lysine was abundant in rather large amounts in soils of the Corn Belt section of the United States, whereas none could be detected by Gupta and Reuszer (1977) for the soils that they examined from this region. Aspartic acid has been reported to constitute practically all of the acidic amino acids in some soils, whereas in others glutamic acid seems to predominate.

In some early studies, Decau (1967, 1969) and Carles and Decau (1960) found that the percentage of the soil amino acids as basic amino acids progressively increased from the colder to the warmer climates. A similar trend can be observed by comparing other data for temperature zone soils (Sowden, 1970; Stevenson, 1956b; Young & Mortensen, 1958) and tropical and subtropical soils (Almeida et al., 1969; Wang et al., 1967). One explanation for this result is that greater microbial activity in the warmer soils results in more extensive turnover of proteinaceous material with selective preservation of basic compounds due to interactions with other soil components.

More recent results by Sowden and his associates (Khan & Sowden, 1971, Chen et al., 1977; Sowden, 1977; Sowden et al., 1976, 1977) have failed to confirm the above mentioned findings. Their data, summarized in Table 13, indicate that it is the acidic amino acids (particularly aspartic) rather than the basic amino acids that predominate in tropical soils. Reasons for the discordant results are unknown but may be due to differences in the nature of the soils examined. Another possibility is that selective losses of amino acids had occurred during extraction and desalting, particularly in some of the earlier investigations.

Studies on the amino acid composition of soils in New Zealand (Goh & Edmeades, 1979) and Japan (Hayashi & Harada, 1969) have given distribution patterns within the ranges shown in Table 13. Lowe (1973) found that amino acid distribution patterns in hydrolysates from the L, F, and H humus horizons of some forest soils were similar to those reported for

Table 13—Distribution of amino acids in soils from several climatic zones
(from the summary of Sowden et al., 1977).†

Amino acids	Climatic zone			
	Arctic	Cool temperate	Sub-tropical	Tropical
	— (N in each amino acid/N in total amino acids) × 100 —			
Acidic				
Aspartic acid	9.1 ± 1.7	12.1 ± 2.3	13.5 ± 4.3	16.3 ± 4.0
Glutamic acid	7.5 ± 1.2	8.7 ± 1.1	8.2 ± 1.0	9.9 ± 1.8
Total acidic	16.6	20.8	21.7	26.2
Basic				
Arginine	11.2 ± 1.2	10.0 ± 1.2	10.0 ± 0.8	8.0 ± 1.6
Histidine	4.5 ± 1.2	4.2 ± 0.9	4.5 ± 0.9	3.9 ± 0.9
Lysine	7.4 ± 0.8	7.6 ± 1.0	8.6 ± 1.0	6.4 ± 0.6
Ornithine	2.6 ± 1.8	2.2 ± 2.2	1.8 ± 2.4	1.6 ± 0.6
Total basic	25.7	24.0	24.9	19.9
Neutral				
Phenylalanine	2.6 ± 0.1	2.3 ± 0.4	2.2 ± 0.4	1.9 ± 0.4
Tyrosine	1.6 ± 0.2	1.1 ± 0.4	1.6 ± 0.5	0.9 ± 0.2
Glycine	10.6 ± 0.5	11.4 ± 0.9	12.2 ± 1.0	12.8 ± 0.4
Alanine	8.0 ± 0.3	8.4 ± 0.8	8.1 ± 0.6	8.5 ± 03
Valine	5.6 ± 0.7	4.8 ± 0.7	4.7 ± 0.4	5.2 ± 0.9
Leucine	5.0 ± 0.5	4.6 ± 0.8	4.6 ± 0.3	4.1 ± 0.5
Isoleucine	3.0 ± 0.1	2.9 ± 0.4	3.1 ± 0.2	2.4 ± 0.5
Serine	5.8 ± 0.9	5.5 ± 0.9	4.9 ± 0.7	5.7 ± 0.5
Threonine	6.2 ± 0.4	5.5 ± 0.5	4.8 ± 0.5	5.1 ± 0.3
Proline	4.4 ± 1.3	3.9 ± 0.7	3.9 ± 1.5	3.7 ± 1.4
Hydroxyproline	1.0 ± 0.6	0.3 ± 0.3	0.2 ± 0.2	0.5 ± 0.3
Total neutral	53.8	50.7	50.3	50.8
Sulfur-containing				
Methionine	0.6 ± 0.1	0.6 ± 0.6	0.5 ± 0.1	0.4 ± 0.2
Cystine	0.8 ± 0.6	0.8 ± 0.6	0.8 ± 1.4	0.4 ± 0.2
Cysteic acid	0.8 ± 0.3	1.1 ± 0.7	0.5 ± 0.4	1.0 ± 0.4
Methionine sulphoxides	0.5 ± 0.3	0.3 ± 0.3	0.2 ± 0.2	0.3 ± 0.1
Total S-containing	2.7	2.8	2.0	2.1
Miscellaneous‡	1.2 ± 0.8	2.1 ± 1.1	0.9 ± 0.7	1.9 ± 0.6

† Data for the arctic and cool temperate zone were from Sowden (1977) and Khan and Sowden (1971), for the subtropical region from Chen et al. (1977), and for the tropical region from Sowden et al. (1976).
‡ Includes allo-isoleucine, γ-NH$_2$-butyric acid, 2-4-diaminobutyric acid, diaminopimelic acid, β-alanine, ethanolamine, and unidentified compounds.

mineral soils. However, with increasing decomposition, the proportion of glutamic acid, proline, and leucine decreased relative to glycine.

Several studies have been conducted to determine the effects of cultivation practices on the amino acid composition of the soil. Stevenson (1956b) examined some soils of the Morrow plots at the University of Illinois and concluded that long-time cultivation (plots established in 1901) without organic matter additions led to an increase in the relative proportion of the amino acid material in the form of basic amino acids (Fig. 7). A similar result was obtained by Yamashita and Akiya (1963) for some Japanese soils.

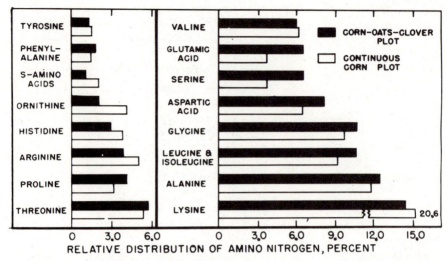

RELATIVE DISTRIBUTION OF AMINO NITROGEN, PERCENT

Fig. 7—Relative distribution of amino acid N in the continuous corn and corn-oats-clover rotation plots of the Morrow plots at the University of Illinois. Relatively higher amounts of the amino acid N in the continuous corn plot occurred as basic amino acids, ornithine, histidine, arginine, and lysine (from Stevenson, 1956b).

On the other hand, Young and Mortensen (1958) failed to detect quantitative differences in amino acid composition of some rotation plots in Ohio. These plots had been in operation for only about 3 decades compared with nearly 7 decades for the Morrow plots; consequently, differences due to cropping would not be as great. Khan (1971) found identical amino acid distribution patterns for the soils of two cropping systems on a "Gray Wooded" soil in Alberta, Canada (plots continuous for 39 years at time of sampling).

Little is known regarding the distribution of individual amino acids in various horizons of the soil profile. Similar amino acid distribution patterns were observed by Sowden (1977) for the A, B, and C horizons of over 30 Canadian soils, and Nommik (1967) noted increases in aspartic acid and decreases in alanine and lysine with depth in a Swedish Spodosol. Decau (1969) observed changes in amino acids for some tropical soil profiles; as depth increased, the molar proportions of β-alanine and γ-aminobutyric acid increased at the expense of leucine and tyrosine. Amino acid levels have served as markers for paleosols in soil profiles (Goh, 1972).

D. Factors Affecting the Distribution of Amino Acids

A variety of complex factors may affect the distribution of amino acids in soils, including synthesis and destruction by the indigenous biota, adsorption by clay minerals, and reactions with quinones and reducing sugars.

The predominant amino acids in soils appear to be those present in the cell walls of microorganisms (Hayashi & Harada, 1969; Kai et al., 1973). Glycine, alanine, aspartic acid, and glutamic acid are often the dominant amino acids in bacterial cells, along with ornithine, lysine, and diamino-pimelic acid. Thus, it appears that much of the amino acid material that accumulates in soil is derived from peptides, mucopeptides, and teichoic acids of microbial cells. The persistence of certain microbially synthesized amino acids has been reported by Wagner and Mutatkar (1968) in a study of the humification of ^{14}C glucose. The highest specific activities were found in those amino compounds known to be constituents of cell walls of microorganisms (alanine, glycine, glutamic acid, and lysine). Glucosamine, an amino sugar found in bacterial and fungal cell walls, also contained large quantities of ^{14}C.

A priori, reactions between amino acids and reducing sugars would be expected to play a significant role in amino acid composition, because it is known that individual amino acids vary greatly in the ease with which they react with polyphenols and sugars. Basic amino acids, for example, undergo browning at considerably higher rates than neutral and acidic amino acids (Hodge, 1953). Thus, if it is assumed that these reactions occur in soils, the more basic compounds, such as lysine, would be affected to a greater extent than other amino acids. Another factor to consider is that the accessory amino group of basic amino acids, when present in peptides, is capable of combining with carbonyl-containing substances, whereas the amino group of neutral and acidic amino acids is inaccessible because of participation in peptide linkages.

Amino acids in soils and sediments may be affected by enzymatic and nonenzymatic browning reactions in several different ways. They could be oxidatively deaminated by Strecker degradation, they could be linked to the addend through a transamination reaction, or they could be degraded in such a manner as to produce a nitrogenous compound whose character is not related to the original amino acid. Since basic amino acids are highly reactive, these compounds could be selectively degraded or preserved, depending on the stability of the amino acid in the addition product. At present, it is not possible to choose between these alternatives. Studies on the browning of amino acids in model systems provide little information on this subject, because these experiments are generally conducted under conditions dissimilar to those found in soils. Nonenzymatic browning reactions, for example, are usually carried out at elevated temperatures to reduce the time required to initiate browning, and there is evidence that these conditions produce results that are qualitatively and quantitatively different from those occurring in natural systems (Spark, 1969). Mention should also be made of the likelihood that in soils, amino acids are linked in some manner to polymeric constituents of humus.

One might expect to find confirmatory evidence for selective preservation (or degradation) of basic amino acids in soils even though the overall effect would be masked by continual resynthesis by microorganisms. Conditions of intense microbial activity, as in tropical soils, should be particularly conducive for mineralization of free or unbound amino acids and peptides and for chemical interactions of the more reactive species. As noted in the previous section, early work suggested greater preservation of basic amino acids in tropical soils, but the effect could not be confirmed in more recent studies.

Adsorption reactions may also be a factor affecting amino acid distribution patterns. Carter and Mitterer (1978) found that organic matter associated with carbonate sediments had a distinctly different amino acid composition than that of noncarbonate sediments. Specifically, the amino acid composition of carbonaceous sediments was characterized by elevated levels of acidic amino acids, aspartic and glutamic acids. According to Carter and Mitterer (1978), carbonate sediments selectively adsorb aspartic acid–enriched organic matter, whereas noncarbonate sediments do not have this property. Greater attention needs to be given to the effect of adsorbing surfaces on the amino acid composition of soils.

E. Amino Acids in Humic and Fulvic Acids

Humic and fulvic acids contain the amino acids found in soil hydrolysates but not necessarily in the same proportions. Differences also exist in the amino acid distribution patterns of humic and fulvic acids. In comparison with fulvic acids, humic acids contain relatively higher amounts of basic amino acids but relatively lower amounts of acidic amino acids (Sowden et al., 1976). This effect can be seen in Table 14, where a compilation of values recorded in the literature for acidic, neutral, and basic amino acids in humic and fulvic acids is given. It is of interest that the values given for the fungal humic acids are within the range shown for the soil humic acids.

The percentage of the amino acid N in humic acids as basic compounds may increase with an increase in "degree of humification," but information on this point is lacking. Changes in amino acid composition may also occur when soils are subjected to long-time cultivation, as noted earlier for amino acids in total soil.

Table 14—Percent molar distribution of amino acids in hydrolysates of the humic acid and fulvic acid fractions of organic matter.

Source or location†	Group			Reference
	Acidic	Neutral	Basic	
Humic acids				
Argentine soils (4)	26.9–27.5	56.6–62.0	11.1–16.4	Rosell et al. (1978)
Canadian soils (1)	18.3–19.8	71.2–74.0	7.0– 9.0	Khan & Sowden (1971)
India (1)	34.4	62.6	3.0	Rudrappa et al. (1972)
Tropical soils (3)	23.6–25.2	65.1–67.8	8.6– 9.7	Sowden et al. (1976)
Carbonate mud (4)	27.5–33.5	59.2–67.3	5.2– 9.2	Carter & Mitterer (1978)
Noncarbonate mud (1)	22.3	68.4	9.3	Carter & Mitterer (1978)
Fungal "humic acids" (4)	22.5–34.0	60.3–73.2	4.3– 7.3	Saiz-Jimenez et al. (1975)
Fulvic acids				
Italian soils (5)	29.6–37.7	56.8–67.7	2.7– 7.5	Guidi et al. (1976)
Tropical soils (2)	43.8–45.6	51.8–51.9	2.7– 4.5	Sowden et al. (1976)
Carbonate mud (4)	34.3–43.5	47.9–58.8	3.5– 8.6	Carter & Mitterer (1978)
Noncarbonate mud (1)	25.4	65.0	9.6	Carter & Mitterer (1978)

† Numbers in brackets refer to number of soils analyzed.

F. Stereochemistry of Amino Acids

The percentage [(D-amino acid $\times$ 100)/(D-amino acid + L-amino acid)] for D-alanine, D-aspartic acid, and D-glutamic acid in soils has been found to be significantly greater than for other protein amino acids (Kawaguchi et al., 1976; Pollock et al., 1977). This is to be expected because these D-amino acids are significant components of bacterial cell walls and metabolites (see Meister, 1965), and it has been demonstrated in ^{14}C studies that cell wall amino acids are selectively preserved in soils (Wagner & Mutatkar, 1968). The significance of the stereochemical nature of amino acids to the humification process has been discussed by Bachelier (1975).

Amino acid racemization-epimerization reactions have been used as a geochronological tool for dating fossil bones, shells, deep sea foraminiferal deposits, and marine sediments. This work has been reviewed by Dungworth (1976) and Schroeder and Bada (1976).

G. Free Amino Acids

Practically all of the proteinaceous material in soil is intimately combined with clay minerals or humus colloids. From the standpoint of N availability to microorganisms and higher plants, free amino acids may be of greater importance. The occurrence of small amounts of amino acids in the soil solution is expected, because these compounds are formed during the conversion of protein N to NH_3 by heterotrophic organisms:

$$\text{Proteins} \rightarrow \text{peptides} \rightarrow \text{amino acids} \rightarrow NH_3.$$

Free amino acids may modify the biological properties of the soil through their effect on specific types or groups of living organisms, as has been observed for lake sediments (Bowen, 1980). Some loss of soil N is possible through leaching of amino acids, but the extent of such losses, if any, is unknown.

The term *free amino acids* has been used to refer to those amino acids in soil that do not exist in peptide linkages. In this sense, the term is a misnomer. Some amino acids may be sorbed by clay in such a manner that they are not removed by the solvents commonly used to extract free amino acids. Sørensen (1972, 1975) found that amino acids metabolites synthesized in soil during periods of intense microbial activity were stabilized by clay.

In addition to serving as a source of N, amino acids may play a role in rock weathering and in pedogenic processes through their ability to form chelate complexes with metal ions.

Solvents used as extractants for free amino acids include distilled water, 80% ethyl alcohol, and dilute aqueous solutions of Ba(OH)$_2$ and ammonium acetate (NH$_4$OAC). Barium hydroxide and NH$_4$OAC appear to be the most efficient, since they remove large quantities of amino acids absorbed on clay particles. Water is rather inefficient in removing amino acids. Paul and Schmidt (1960) found that Ba(OH)$_2$ and NH$_4$OAC extracted from 5 to 25 times more amino acids from soil than 80% ethyl alcohol. A disadvantage of Ba(OH)$_2$ as an extractant is that hydrolysis of organic N compounds may occur during extraction. Accordingly, NH$_4$OAC would appear to be the preferred solvent for recovering "free amino acids" from soil. Sowden and Ivarson (1966) found that extraction of soil with water in the presence of CCl$_4$ increased the quantity of free amino acids by 25 to 100 times, presumably due to the effect of CCl$_4$ on release of amino acids from microbial and other cell material. Use of a complexing agent (e.g., acetylacetone) with CCl$_4$ had little effect on the amounts extracted.

The difficulty of obtaining quantitative release of free amino acids is emphasized by Paul and Schmidt's (1960) finding that only 31 to 83% of the added amino acids could be removed from soil with NH$_4$OAC; the recovery of basic amino acids was somewhat less than that for neutral and acidic compounds. Similar results were obtained by Gilbert and Altman (1966) using 20% ethyl alcohol as extractant. The possibility that amino acids may be held on interlamellar surfaces of expanding lattice clays further complicates the problem of complete extraction (Stevenson & Cheng, 1970; Cheng et al., 1975). Under natural soil conditions, some free amino acids may be held in small voids or micropores and may thereby be inaccessible to microorganisms.

Since amino acids are readily decomposed by microorganisms, they have only an ephemeral existence in soil. Thus, the amounts present in the soil solution at any one time represent a balance between synthesis and destruction by microorganisms. Levels will be highest when microbial activity is intense.

Bremner (1967) has reviewed work done on free amino acids in soil prior to 1965. Studies conducted since that time tend to confirm many of the conclusions of the earlier work, namely that the free amino acid content of the soil is higher in "rhizosphere" soil than in fallow soil and that levels are strongly influenced by weather conditions, moisture status of the soil, type of plant and stage of growth, additions of organic residues, and cultural conditions (Paul & Tu, 1965; Yamashita & Akiya, 1963; Yamashita & Kawada, 1968; Kobo et al., 1970; Ivarson et al., 1970; Sinha, 1972; Kato, 1975; Karamshuk, 1979; Ladd & Paul, 1973; Ktsoyev, 1977).

Levels of free amino acids in soil seldom exceed 2 μg/g, or 4.5 kg/ha plow depth, but they may be sevenfold higher in rhizosphere soil. Ivarson and Sowden (1969, 1970) showed that freezing brought about a 10- to 14-fold increase in free amino acid content. Marked increases have also been observed on prolonged storage of soil at low temperature (Ivarson & Sowden, 1966), as well as by treatment of soil with fungicides (Wainwright & Pugh, 1975; Ladd et al., 1976). On the assumption that NH$_4$OAc-extractable amino acids exist largely in water soluble forms, their concentration in the soil solution would be of the order of 10^{-5} to 10^{-6} M at the 20% moisture level.

In addition to the amino acids mentioned in Bremner's (1967) review, the following free amino acids have now been reported in soil: tryptophan, β-aminoisobutyric acid, and α-aminovaleric acid by Decau (1969) and asparagine, glutamine, and citrulline by Sowden and Ivarson (1968).

There is evidence that healthy plant roots excrete amino acids into the soil. The review of Rovira and McDougall (1967) shows that 21 amino acids have been identified in root exudates, including leucine, isoleucine, valine, γ-aminobutyric acid, glutamine, alanine, β-alanine, asparagine, serine, glutamic acid, aspartic acid, glycine, phenylalanine, threonine, tyrosine, lysine, proline, methionine, and cystathionine. The kinds and amounts of amino acids exuded from roots vary with plant type and maturity and, for any given plant, on environmental conditions affecting growth, such as temperature, light intensity, moisture status of the soil, and availability of nutrient elements.

Most estimates for amino acids exuded by roots have been made with culture solutions under laboratory conditions, and for this reason, the amounts exuded under field conditions cannot be accurately estimated. As indicated above, the free amino acid content of rhizosphere soil is normally many times higher than for nonrhizosphere soil. The difference between the free amino acid content of cropped and uncropped soil cannot be used as an index of the amounts of amino acids decomposed and assimilated by soil microorganisms.

The release of amino acids from roots is an important factor affecting soil microorganisms and, in some cases, growth of the plant. These aspects of the problem, covered in detail by Rovira and McDougall (1967), are beyond the scope of this chapter.

H. State of Amino Acids in Soil

The deliberations of the previous sections indicate that amino acids exist in soils in several different forms, including the following:

1) As free amino acids
    a) In the soil solution
    b) In soil micropores
2) As amino acids, peptides, or proteins bound to clay minerals
    a) On external surfaces
    b) On internal surfaces
3) As amino acids, peptides, or proteins bound to humic colloids
    a) H bonding and van der Waal's forces
    b) In covalent linkage as quinoid-amino acid complexes, as exemplified by the following structures (R = amino acid or peptide).

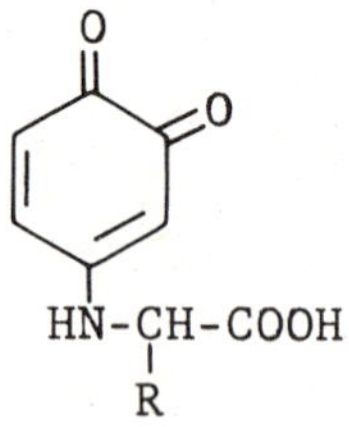

XI 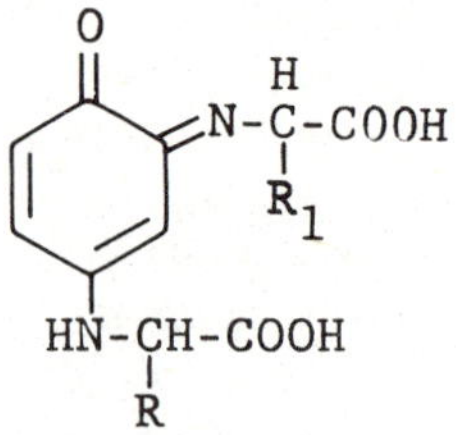   XII 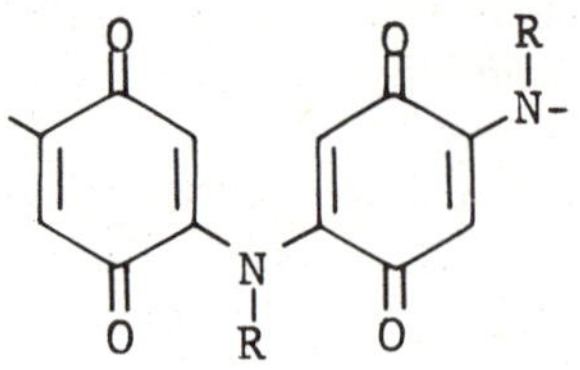   XIII

4) As mucoproteins (amino acids combined with N-acetylhexosamines, uronic acids and other sugars).

5) As a muramic acid–containing mucopeptide derived from bacterial cell walls.

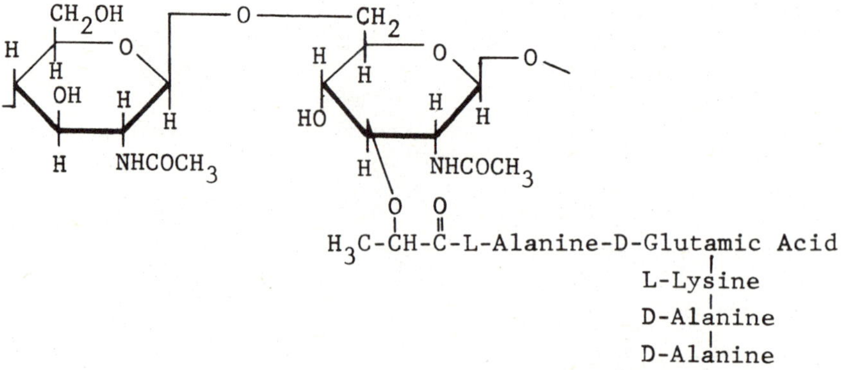

$$H_3C-CH-C-\text{L-Alanine-D-Glutamic Acid}$$
L-Lysine
D-Alanine
D-Alanine

Muramic acid-containing mucopeptide

XIV

6) As teichoic acids (linear polymers of polyol, ribitol or glycerophosphate containing ester-linked alanine).

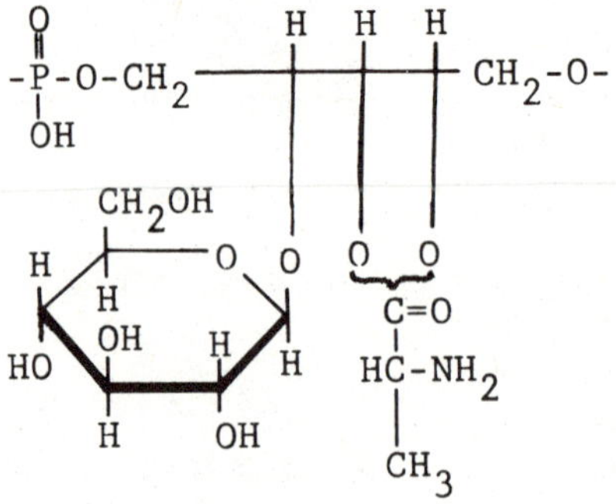

Teichoic acid

XV

Considerable controversy exists as to whether proteins as such occur in significant amounts in soil organic matter. The well-known lignoprotein theory of soil humus has yet to be confirmed, and as mentioned earlier, many investigators believe that the theory in its original form is obsolete.

Failure to account for a significant amount of the soil amino acids as protein or as lignin-protein complexes has led to the conclusion that the amino acids are directly bound to complex polymers formed from phenols or quinones (see no. 3). Bremner (1967) suggested that some of the bound amino acids exists in the form of mucopeptides and teichoic acids (see no. 5 and 6). The latter accounts for as much as 50% of the material in the cell walls of gram-positive bacteria, and as indicated earlier, ^{14}C studies indicate a selective enrichment in cell wall amino acids during the decomposition of organic substrates in soil.

IV. AMINO SUGARS

Amino sugars occur as structural components of a broad group of substances, the mucopolysaccharides, and they have been found in combination with mucopeptides and mucoproteins, as well as with smaller molecules, such as antibiotics. In soil, some of the amino sugar material may exist in the form of an alkali-insoluble polysaccharide referred to as *chitin*. This substance, which is a polymer of N-acetylglucosamine, comprises the cell walls, structural membranes, and skeletal component of fungal mycelia, where it plays a structural role analogous to the cellulose of higher plants. The amino sugars in soil have often been referred to as chitin, but this practice cannot be justified because it is known that soils contain amino sugars other than glucosamine. Furthermore, the capsular material encasing the bodies of many bacteria consists of complex polysaccharides bearing amino sugars, but they cannot be classified as chitin. It is generally assumed that the amino sugars in soil are of microbial origin. Lower members of the animal world, including insects, contain chitin in structural tissue, but it is not known if any of this material persists in soil.

Amino sugars may play the dual role in soil of serving as a source of N for plant growth and promoting good soil structure. The ability of microbial polysaccharides to bind soil particles into aggregates of high stability is well known, and special aggregating properties have sometimes been attributed to nitrogenous polysaccharides (Geoghegan & Brian, 1946; Haworth et al., 1946).

From 5 to 10% of the N in the surface layer of most soils can be accounted for as N-containing carbohydrates or as amino sugars (see II A). Data of Gallali et al. (1975) suggest that a relationship may exist between the percentage of the soil N as amino sugars and Ca^{2+} content, the highest percentages of which are typical of Ca^{2+}-saturated soils. On the other hand, mean values for a broad selection of soils from widely different climatic zones indicate higher percentages for subtropical and tropical soils, which are usually acidic. These relationships are shown in Table 15. The

Table 15—Distribution of amino sugars (glucosamine and galactosamine) in soils from widely different climatic zones (means and standard deviations).†

	Percent of total soil N			Glucosamine/ Galactosamine ratio
Climate zone‡	Amino sugar N	Glucosamine	Galactosamine	
Arctic (6)	4.5 ± 1.7	2.8 ± 0.4	1.7 ± 0.5	1.7
Cool temperate (82)	5.3 ± 2.1	3.4 ± 0.9	1.9 ± 0.7	1.8
Subtropical (6)	7.4 ± 2.1	5.0 ± 1.5	2.4 ± 0.7	2.1
Tropical (10)	6.7 ± 2.1	3.9 ± 0.8	2.8 ± 0.5	1.4

† From Sowden et al. (1977).
‡ Numbers in brackets indicate number of soils analyzed.

discordant results may be due to extensive decomposition and turnover of organic matter in soils of the warmer climates, with selective preservation of amino sugars. The situation is analogous to the observed increase in amino sugar N (relative to total organic N) when soils are subjected to intensive cultivation (see Table 4).

In other work, Namdev and Dube (1973) concluded that applications of urea and herbicides to soil increased amino sugar–N content by stimulating microbial populations that synthesized amino sugars during the decomposition of plant residues.

Early work indicated that the proportion of the soil N as amino sugars may increase with increasing depth in the soil profile, reaching a maximum in the B horizon (Stevenson, 1957; Sowden, 1959). These observations require confirmation, since other studies have failed to show such an effect (Gallali et al., 1975; Sowden, 1977; Singh et al., 1978).

A. Extraction and Quantitative Determination

Amino sugars are normally recovered from soil by hydrolysis with HCl (usually a 3 or 6N solution) for 6 to 9 hours. Some degradation of amino sugars occurs during hydrolysis, and a correction factor must be applied to account for these losses. The liberated amino sugars are then analyzed by the standard colorimetric method of Elson and Morgan or by alkaline distillation. Both methods give similar results (see Stevenson, 1965).

With the Elson-Morgan colorimetric method, the amino sugar is first heated with an alkaline solution of acetylacetone, and then an acid, alcoholic solution of p-dimethylaminobenzaldehyde (Ehrlich's reagent) is added. A chromogen is formed by the first reaction, following which the addition of Ehrlich's reagent produces a red solution.

Many substances, including Fe and mixtures of amino acids and simple sugars, produce colors that interfere with the determination. However, these interfering substances are removed by treatment of the soil hydrolysate on ion exchange resin columns (Stevenson, 1965).

The alkaline distillation method is based on the observation that amino sugars are readily deaminated by heating with alkali. In Bremner's (1965b)

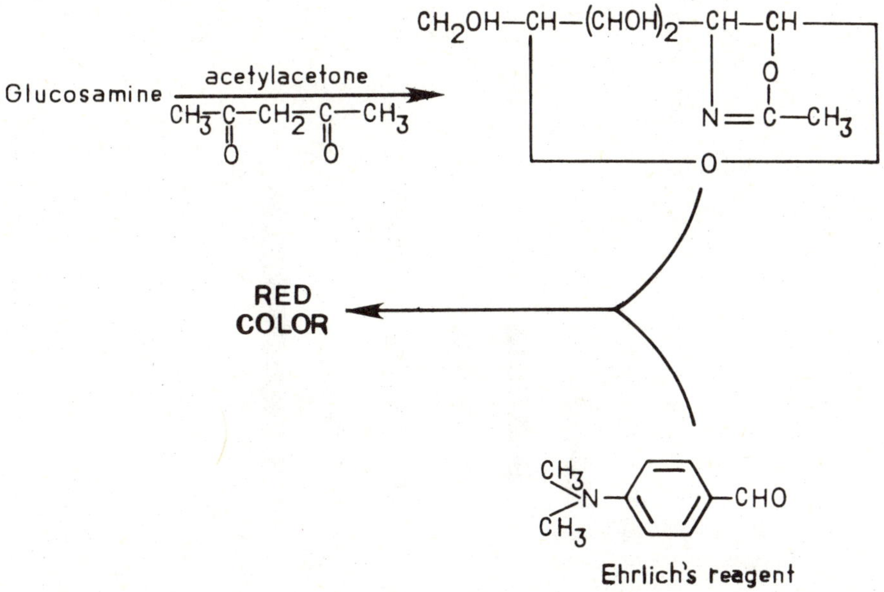

$$CH_2OH-CH-(CHOH)_2-CH-CH-$$

Glucosamine $\xrightarrow[\underset{\underset{O}{\|}\quad\underset{O}{\|}]{CH_3-C-CH_2-C-CH_3}]{acetylacetone}$

$$N=C-CH_3$$

RED COLOR ←

$$\begin{array}{c} CH_3 \\ {>}N{-}\bigcirc{-}CHO \\ CH_3 \end{array}$$

Ehrlich's reagent

distillation procedure, amino sugar N plus NH_3-N is determined by steam distillation with borate buffer of pH 8.7. Preformed NH_3 is then estimated by steam distillation with MgO, a procedure that recovers the NH_3 without deaminating amino sugars. The difference in values obtained by the two distillation procedures is amino sugar N.

B. Isolation of Amino Sugars

Individual amino sugars have been isolated from soil hydrolysates by both paper partition chromatography and ion exchange chromatography. These studies indicate that most of the amino sugar material occurs as D-glucosamine and D-galactosamine, with the former occurring in greatest amounts.

Sowden's (1959) findings indicate that the percentage of the amino sugar N as glucosamine is rather high in acidic soils, an effect that can be explained by enhanced growth of fungi in acidic environments. Unlike bacteria, fungi synthesize glucosamine only.

Variations in the glucosamine/galactosamine ratios for some soils belonging to several suborders are shown in Fig. 8. The highest ratios are found in those soils that are either highly acidic (Spodosol) or have natural pH values in the neutral or slightly alkaline ranges (e.g., Albolls and Ustolls). As suggested above, the high ratio for the highly acidic Spodosol can be accounted for by a high population of fungi. The high ratios for the near neutral or alkaline soils may be due to high populations of actinomycetes. Numbers of these organisms are known to be especially high in soils that are near neutral or slightly alkaline in reaction. Stevenson and Braids

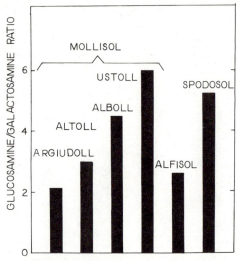

Fig. 8—Variations in the glucosamine/galactosamine ratio in the soils of several soil orders (from Stevenson, unpublished observations).

(1968) observed increases in the glucosamine/galactosamine ratios with increasing depth in some soil profiles.

Muramic acid has been observed in both soil (Miller & Casida, 1970) and bog sediments (Casagrande & Park, 1978). In the case of the study by Miller and Casida (1970), muramic acid levels in soil were 10 to 1,000 times the amounts that could be accounted for by microbial tissue as estimated by the plate-counting technique, but roughly comparable to the amounts expected on the basis of the microscopic counting method.

Skujins and Pukite (1970) reported the presence of N-acetylglucosamine in $2.5M$ acetic acid extracts of soil. Acetylated amino sugars have sometimes been reported in HCl hydrolysates, but this is unlikely, because these compounds are easily broken down by boiling $3N$ or $6N$ HCl, as pointed out by Bremner (1965a, 1967).

By the use of an improved ion exchange chromatographic procedure, a large number of compounds that give color reactions of amino sugars have been isolated from soil (Stevenson, unpublished observations). A typical elution curve showing the separations obtained is given in Fig. 9. In addition to D-glucosamine, and D-galactosamine, D-mannosamine (i) and muramic acid (h) were identified. The compound indicated as no. 6 was tentatively identified as D-fucosamine.

The presence of a wide variety of amino sugars in soil is expected, because over 25 compounds of this class have now been found in products synthesized by microorganisms (Sharon, 1965). In addition to the amino sugars mentioned above, the following 2-amino sugars have been reported: D-gulosamine, D-fucosamine (2-amino-2,6-dideoxy-D-galactose), L-fucosamine (2-amino-2,6-dideoxy-L-galactose), and pneumosamine (2-amino-2,6-dideoxy-L-talose). Other naturally occurring amino sugars include a number of 3-amino-, diamino-, and aminohexuronic sugars.

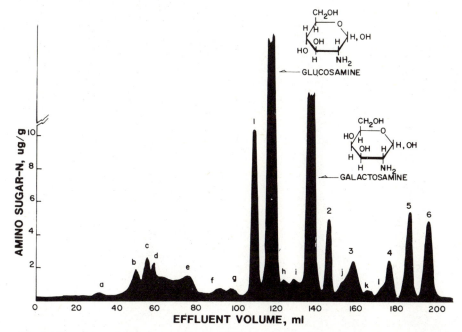

Fig. 9—Separation of amino sugars from a soil hydrolysate by ion exchange chromatography. Note the occurrence of many possible compounds other than glucosamine and galactosamine, six of which occur in rather high amounts (from Stevenson, unpublished observations).

V. OTHER N COMPOUNDS

A wide array of naturally occurring nitrogenous compounds other than amino acids and amino sugars have been found in soil but in very low amounts. They include the nucleic acids and their derivatives, chlorophyll and chlorophyll-degradation products, phospholipids, several amines, and vitamins.

A. Nucleic Acids and Derivatives

Nucleic acids, which occur in the cells of all living organisms, consist of individual mononucleotide units (base-sugar-phosphate) joined by a phosphoric acid ester linkage through the sugar. Two types are known, ribonucleic acid (RNA) and deoxyribonucleic acid (DNA). The two types are identified by the nature of the pentose sugar (ribose or deoxyribose, respectively). Both contain the purine bases adenine and guanine and the pyrimidine base cytosine. In addition, RNA contains the pyrimidine uracil; DNA also contains thymine. Plant DNA contains 5-methyl cytosine.

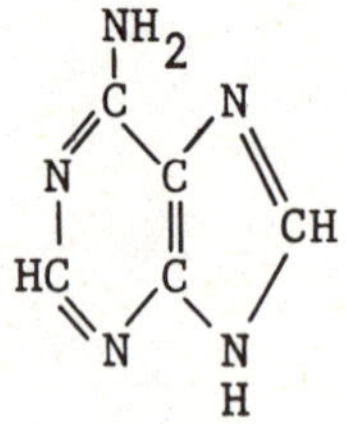

Adenine
XVI

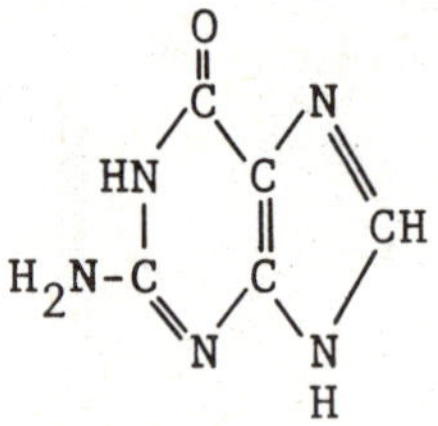

Guanine
XVII

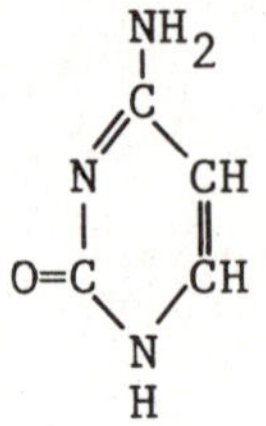

Cytosine
XVIII

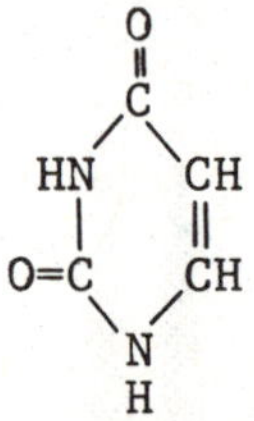

Uracil
XIX

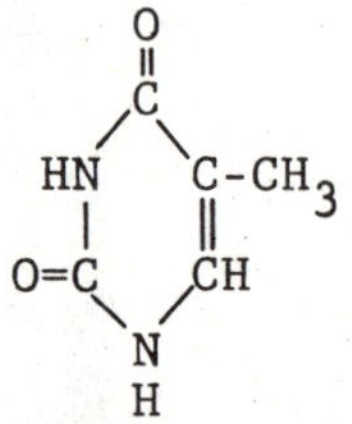

Thymine
XX

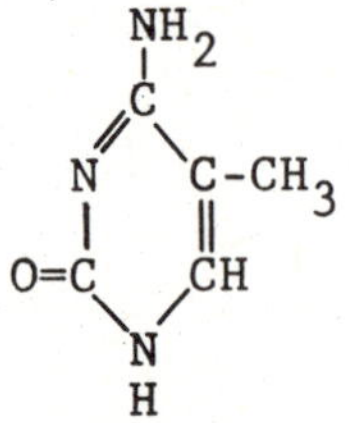

5-Methylcytosine
XXI

The application of chromatographic methods has led to the identification of all of the above mentioned bases in hydrolysates of soil organic matter preparations (Anderson, 1967; Cortez & Schnitzer, 1979). This work indicates that the bases occur as polynucleotides derived primarily from bacterial DNA. The occurrence of 5-methyl cytosine in some, but not all, soils suggests a plant source of some of the DNA (Anderson, 1958; Cortez & Schnitzer, 1979).

In a recent study, Anderson (1970) isolated two pyrimidine nucleotide diphosphates, thymidine-3′:5′ diphosphate and deoxyuridine-3′:5′ diphosphate, from NaOH extracts of soil.

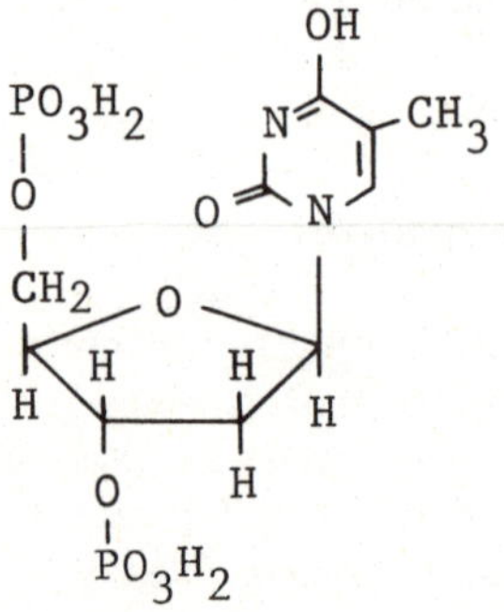

Thymidine-3′:5′ diphosphate
XXII

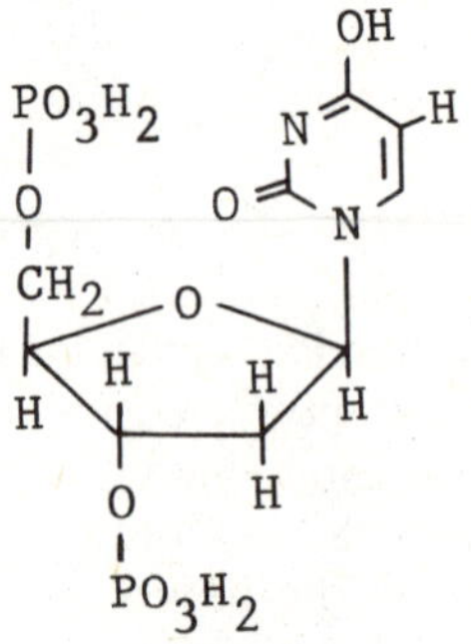

Deoxyuridine-3′:5′ diphosphate
XXIII

Table 16—Percentage of the total soil N, and of the N in the humic acid, fulvic acid, and humin fractions, accounted for as purine and pyrimidine bases.[†]

Soil[‡]	% of N as purine and pyrimidine bases			
	Soil	Humic acid	Fulvic acid	Humin
1 "Chernozemic"	2.1	2.6	3.1	1.7
2 "Chernozemic"	0.9	1.7	2.2	0.4
3 "Chernozemic"	1.3	1.8	2.8	0.8
4 "Chernozemic"	2.3	2.0	6.9	1.3
5 "Chernozemic"	1.4	2.1	1.2	10.4
6 "Regosolic"	4.4	2.1	2.2	7.7
7 "Gleysolic"	0.7	0.4	0.8	0.8
8 "Podzolic"	2.0	1.3	2.9	2.3
9 "Podzolic"	2.8	2.3	6.0	1.6
10 "Podzolic"	3.9	4.9	13.3	2.1
11 "Podzolic"	7.4	10.7	6.9	6.2
12 "Podzolic"	3.1	2.0	3.8	4.1
13 Organic	3.1	2.6	18.6	2.8
14 Organic	1.2	3.7	2.6	1.1
15 Organic	0.2	0.7	0.8	0.2

[†] From Cortez and Schnitzer (1979).
[‡] Equivalents in the Comprehensive Soil Classification system are: "Chernozem," Mollisol; "Regosolic," Entisol; "Podzolic," Spodosol; Organic, Histosol.

The reviews of Anderson (1967), Bremner (1965a, 1967), Kowalenko (1978) and Parsons and Tinsley (1975) show that several nitrogen ring bases other than those shown above have been reported in the literature dating as far back as 1910. These claims require confirmation by modern analytical methods.

The N in purine and pyrimidine bases is usually considered to account for < 1% of the total soil N (see reviews of Bremner, 1965a, 1967; Kowalenko, 1978; Parsons & Tinsley, 1975). Somewhat higher percentages were found by Cortez and Schnitzer (1979). Their date (Table 16) indicated that up to 7.4% of the soil N and up to 18.6% of the N in fulvic acid occurred as purine and pyrimidine bases. On this basis, significant amounts of the unknown N in some soils may occur in nucleic acid bases (see II).

B. Chlorophyll and Chlorophyll Degradation Products

The green color of the landscape largely disappears in the autumn in temperate climate zones, and as Hoyt (1971) pointed out, plants in all climates usually lose their green color before they fall on the soil surface. Accordingly, chlorophyll in its original state is not an obvious residue constituent in most soils. Degradation of chlorophyll is apparently initiated by tissue enzymes and is very rapid (Hoyt, 1971). Nevertheless, significant quantities of chlorophyll or its derivatives are added to the soil each year in plant remains as well as in animal feces (Hoyt, 1966, 1971). The amount of this material that persists depends on a variety of soil conditions, including moisture content and pH. Simonart et al. (1959), using ^{14}C-labeled plant material, found that chlorophyll decomposed more slowly than the cellulose fraction.

Chlorophyll and its derivatives in soil are estimated on the basis of the intensity of absorption peaks near 665 nm in 90% aqueous acetone extracts. A similar procedure has been used for the determination of these compounds in lake and marine sediments (Orr et al., 1958; Hodgson et al., 1968).

A deficiency of O_2 inhibits complete destruction of chlorophyll; thus, poorly drained soils contain larger quantities of chlorophyll-type compounds than do well-drained soils. Cornforth (1969) found that the chlorophyll content of some soils of India was inversely related to acidity, and Gorham's (1959) data showed that the amounts contained in the acidic mor humus layers of forest soils were considerably higher than in the more neutral mull humus layers. The higher chlorophyll content of acidic soils may be due in part to substitution of Fe or Mn for Mg in the molecule (Cornforth, 1969). In other work, Hoyt (1966) found that soil under grass grazed for 100 years containing higher amounts of chlorophyll than soil under grass cut for hay, a result that was attributed to chlorophyll added in the feces.

Hoyt (1967) found that the content of chlorophyll-type compounds in soils was correlated with N uptake by ryegrass (*Lolium multiflorum* Lam.) even though these compounds amounted to <3.4 kg/ha and made an infinitesimal contribution to the total N requirement. Kaszubiak (1976) also found that the potential of N mineralization in soil was closely correlated ($r = 0.96$) with the content of chlorophyll-type compounds.

The geochemical significance of chlorophyll has been discussed by Orr et al. (1958) and Hodgson et al. (1968). Changes produced in chlorophyll A during attack by microorganisms involves loss of Mg to form pheophytin A and then the phytol group to form pheophorbide A (Fig. 10). In sediments, the latter is transformed to a variety of porphyrin-type compounds (Orr et al., 1958). The occurrence of pheophytins and other chlorophyll-type compounds in soil has been reported by Chopra (1976).

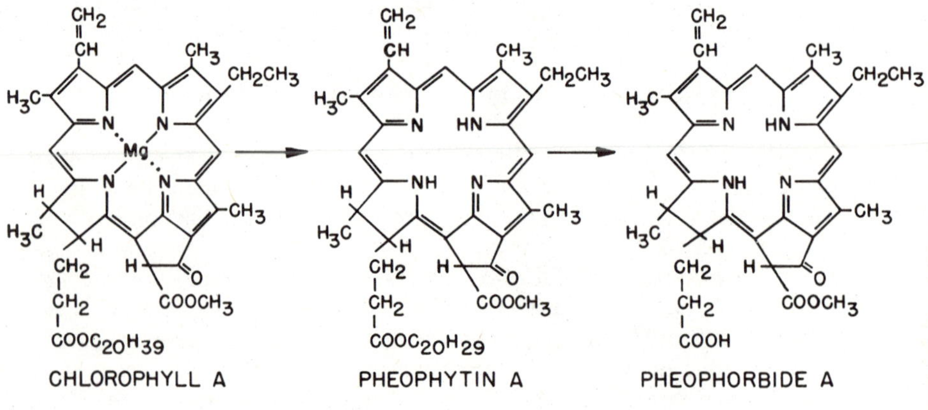

Fig. 10—Initial changes in chlorophyll during decay in soil.

C. Phospholipids

Small amounts of N are extracted from soil with the so-called fat solvents, and it has been established that this N occurs in the form of glycerophosphatides. Hance and Anderson (1963) found glycerophosphate, choline, and ethanolamine in approximate molar ratios of 1:1:0.2, respectively, which indicates that phosphatidyl choline (lecithin) is the most abundant soil phospholipid, followed by phosphatidyl ethanolamine. Microorganisms contain variable amounts of phosphatidyl choline and phosphatidyl ethanolamine, the latter of which comprises over one third of the lipid material in some bacteria.

L-α-Lecithin   Phosphatidyl ethanolamine

XXIV   XXV

D. Amines, Vitamins, and Other Compounds

The review of Bremner (1967) shows that a wide variety of amines and other organic N compounds have been detected in trace amounts in soil or soil extracts, including choline, $CH_2N(CH_3)_3$, ethanolamine, $CH_2OH–CH_2NH_2$, trimethylamine, $(CH_3)N$, urea, $CO(NH_2)_2$, histamine, creatine, allantoin, cyanuric acid, and α-picoline-γ-carboxylic acid.

Histamine   Creatine   Allantoin

XXVI   XXVII   XXVIII

Cyanuric acid   α-Picoline-γ-carboxylic acid

XXIX   XXX

Anaerobic or water-logged conditions are particularly suitable for the formation and preservation of amines in soil (Fujii et al., 1974; references contained therein). The amines identified by Fujii et al. (1974) in plant residues and sand incubated under water-logged conditions were putrescine, cadaverine, methylamine, ethylamine, n-propylamine, and isobutylamine. Temporary accumulations were also noted by aerobic incubations.

$NH_2-(CH_2)_4-NH_2$

Putrecine

XXXI

$NH_2-(CH_2)_5-NH_2$

Cadaverine

XXXII

CH_3-NH_2

Methylamine

XXXIII

$CH_3-CH_2-NH_2$

Ethylamine

XXXIV

$CH_3-CH_2-CH_2-NH_2$

n-Propylamine

XXXV

$\begin{matrix} CH_3 \\ CH_3 \end{matrix}>CH-CH_2-NH_2$

Isobutylamine

XXXVI

Secondary amines, both aliphatic and aromatic, react with nitrous acid to form N-nitrosamines (compounds containing the $>N-N=O$ group), which are carcinogenic and mutagenic at low concentrations. Thus, a potential hazard to the health of man and animals would exist if nitrosamines were formed in soil from pesticide degradation products or from precursors present in manures and sewage sludge. A health hazard would become a reality, however, only if the nitrosamines thus formed were leached into water supplies or taken up by plants used as food by livestock or humans. Trace quantities of nitrosamines have been detected in soils amended with amines (dimethylamine, trimethylamine) and NO_2^- or NO_3^- (Ayanaba et al., 1973; Tate & Alexander, 1974; Mills & Alexander, 1976; Pancholy, 1978), but for the most part, this work has been done under ideal conditions for nitrosamine formation in the laboratory, such as high additions of reactants. Evidence is lacking that the synthesis of nitrosamines in field soil represents a threat to the environment. Mosier and Torbit (1976) were unable to detect N-dimethylnitrosamine and N-diethylnitrosamine in manures even though the necessary precursors were known to be present.

A host of water-soluble, nitrogen-containing B vitamins have also been reported in soils, including biotin, thiamine, nicotinic acid, pantothenic acid, and cobamide coenzyme vitamin B_{12}. These constituents are of special importance, because they may act as growth factors for numerous organisms. Levels of B vitamins in soil are directly related to those factors influencing microbial activity.

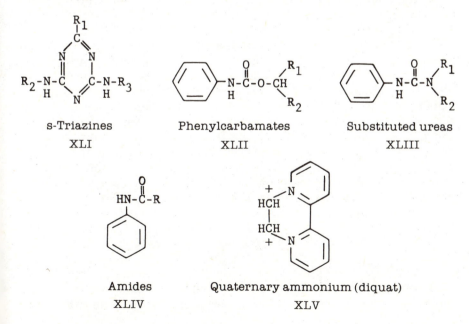

Biotin

XXXVII

Thiamine

XXXVIII

Nicotinic acid

XXXIX

Pantothenic acid

XL

E. Pesticide and Pesticide Degradation Products

Substantial evidence exists to indicate that many pesticides or their partial degradation products can form stable linkages with components of soil organic matter and that such binding increases their persistence in soil (Guenzi, 1974; Kaufman et al., 1976). Many of these pesticides contain N as part of their structures, such as the *s*-triazines, phenylcarbamates, substituted ureas, amides, and quaternary ammonium derivatives. Adsorption by clay minerals also enhances the persistence of many N-containing pesticides in soil.

s-Triazines

XLI

Phenylcarbamates

XLII

Substituted ureas

XLIII

Amides

XLIV

Quaternary ammonium (diquat)

XLV

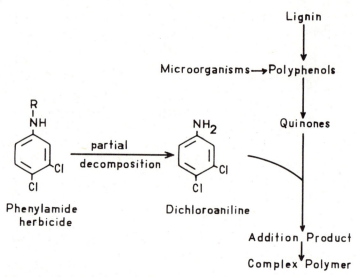

Fig. 11—Postulated chemical reactions leading to stabilization of dichloraniline pesticide residues in soil.

Some pesticides (e.g., *s*-triazines) exhibit a carryover effect in that they are phytotoxic to subsequent crops in a rotation. Paraquat, which has two positive charges and is strongly bound by soil colloids, may persist in soils for years (Kaufman et al., 1976, Chapt. 22).

The microbial metabolism of some phenylamide herbicides results in the formation of reactive chloroaniline residues. These amino compounds can become part of a *pool of precursor molecules* for humus synthesis and, in so doing, may lose their identity (Kaufman et al., 1976, Chapt. 15). The overall reaction is depicted in Fig. 11.

According to Bartha (1971), Bartha and Pramer (1970), and Chiska and Kearney (1970), the bulk of the chloroanilines liberated by partial degradation of phenylamide herbicides (acylanilides, phenylcarbamates, and phenylureas) becomes immobilized in soil by chemical bonding to organic matter. The chemically bound residues could not be recovered by extraction with organic solvents or inorganic salts; partial release was achieved by base and acid hydrolysis (Hsu & Bartha, 1974). The hydrolyzable portion of the humus-bound chloroaniline was believed to be attached as the anil or anilinoquinone and the nonhydrolyzable part as heterocyclic ring structures or in ether linkages. The soil-bound chloroaniline residues are not easily attached by microorganisms (Bartha, 1971; Hsu & Bartha, 1974).

VI. STABILITY OF SOIL ORGANIC N

The high resistance of organic N complexes in soil to microbial attack is of considerable significance to the N balance of the soil, and several theories have been given to account for this phenomenon. A full account of this work is beyond the scope of this chapter (additional information can be obtained from Bremner, 1965a).

Explanations often given to explain the stability of organic N include the following:

1) Proteinaceous constituents (e.g., amino acids, peptides, proteins) are stabilized through their reaction with other organic soil constituents, such as lignins, tannins, quinones, and reducing sugars. Some of the reactions believed to be involved with quinones and reducing sugars have been outlined elsewhere in this chapter (e.g., II D and III F; see also Fig. 5). In the case of lignins, the main reaction is believed to be one involving NH_2 groups of the protein and $C = O$ groups of lignin (see Bremner, 1965a). The high stability of proteins, peptides, and amino acids when linked to aromatic rings has been demonstrated by Verma et al. (1975).

2) Biologically resistant complexes are formed in soil by chemical reactions involving NH_3 or NO_2^- with lignins or humic substances. The complexes thus formed have been shown to be highly resistant to mineralization by soil microorganisms. Reactions involved in the chemical fixation of NH_3 and NO_2^- by soil organic matter are discussed in Chapters 4 (H. Nommik and K. Vahtras) and 9 (D. W. Nelson), respectively.

3) Adsorption of organic N compounds by clay minerals protects the molecule from decomposition. It is well known that the N content of fine-textured soils is higher than coarse-textured soils and that clays, particularly montmorillinitic types, reduce the rate at which proteins and other nitrogenous compounds are decomposed by microorganisms or by proteinase enzymes (see Chapt. 5). Treatment of mineral soils with hydrofluoric acid to decompose clay minerals results in the solubilization of considerable quantities of organic N (see II), which indicates that some of the organic N may be entrapped within the lattice structures of clay minerals.

Recent studies showing the effects of clay minerals on the preservation of nitrogenous metabolites include those of Chichester (1969, 1970), Greaves and Wilson (1970, 1973), and Sørensen (1967, 1972, 1975). For information on the nature of clay-humus complexes, the papers of Emerson (1959), Edwards (1966), Edwards and Bremner (1967), and Greenland (1965, 1971) are recommended.

4) Complexes formed between organic N compounds and polyvalent cations, such as Fe, are biologically stable. One explanation given for the high stability of organic matter in allophanic soils is that reactive groups of humic substances combine with Al in such a way that the surface of the humic molecule no longer provides a suitable fit for enzymes capable of attacking them (Broadbent et al., 1964). Because N is an integral part of the humic molecule, the availability of N in allophanic soils would be restricted by the same mechanism.

5) Some of the organic N occurs in small pores or voids and is physically inaccessible to microorganisms. A typical soil bacterium would be about 0.5 μm in diam and 1 μm or so in length; actinomycetes, fungi; and soil faunal organisms are even larger. Enzymes have the potential for penetrating very small pores, but their movement may be restricted through adsorption.

No single theory can satisfactorily explain the resistance of soil organic N to attack by microorganisms. All of the above mechanisms are undoubtedly important but probably not to the same extent in all soils.

VII. SUMMARY

Despite the excellent progress that has been made since the 1965 ASA Soil Nitrogen Monograph, much more needs to be done before our knowledge of soil N complexes is complete. Also, additional research is required to more fully elucidate the mechanisms whereby nitrogenous substances are bound to organic and inorganic soil colloids. The problems are exceedingly complex, because a variety of chemical and physical processes are involved.

A significant fraction of the organic N in soil can be recovered as simple biochemical compounds, such as amino acids, by hydrolysis procedures, and extensive use has been made of chromatographic techniques in attempts to characterize these compounds. Unfortunately, most of this work cannot be evaluated because of errors associated with incomplete extraction, improper sample preparation, and lack of precision during chromatography. The state of the art has reached the stage where accurate quantitative values are of greater importance than the mere detection of compounds. A wide variety of new separation and analytical techniques are available that should enhance progress in this area.

Nitrogen-15 tracer techniques provide a new dimension to the study of soil organic N and make possible a reassessment of the basic processes involved in synthesis and stabilization of soil N complexes. The research carried out thus far has provided evidence for an internal "N cycle" in soil, a basic feature of which is the transfer of nitrogenous metabolites of microorganisms into stable humus forms.

The nature of as much as one half of the soil organic N will probably remain obscure until more is known about the chemical structures of humic and fulvic acids. Establishment of the composition of these constituents is one of the most challenging problems facing the modern soil scientist. Other major problems remaining to be solved include: (i) nature of clay mineral-organic N complexes, (ii) role of nitrogenous constituents in the binding of metal ions, (iii) resistance of soil N complexes to attack by soil microorganisms, (iv) availability of soil organic N to higher plants, and (v) role of organic N compounds in geochemical and pedogenic processes.

LITERATURE CITED

Adams, W. A., and D. R. Perry. 1973. The effect of pH on the incorporation of amino acids into humic acid extracted from soil. J. Soil Sci. 24:18–25.

Aderikhin, P. G., and A. P. Shcherbakov. 1974. Forms of nitrogen in soils of the Central Black Earth Areas of the USSR. Int. Congr. Soil Sci., Trans. 10th (Aberdeen) III:83–89.

Ahmad, Z., Y. Yahiro, H. Kai, and T. Harada. 1973. Factors affecting immobilization and release of nitrogen in soil and chemical characteristics of the nitrogen newly immobilized. IV. Soil Sci. Plant Nutr. (Tokyo) 19:287–298.

Aldag, R. W. 1977. Relations between pseudo-amide nitrogen and humic acid nitrogen released under different hydrolysis conditions. p. 293–299. *In* Soil organic matter studies. Int. Atomic Energy Agency, Vienna.

Aldag, R. W., and R. Kickuth. 1973. Stickstoff verbindungen im Boden und ihre Beziehung zur Humusdynamik. 2. Z. Pflanzenernaehr. Bodenkd. 136:141–150, 193–202.

Allen, A. L., F. J. Stevenson, and L. T. Kurtz. 1973. Chemical distribution of fertilizer nitrogen in soil as revealed by nitrogen-15 studies. J. Environ. Qual. 2:120–124.

Almeida, L. A. V., R. P. Ricardo, and M. B. M. C. Rouy. 1969. Acides amines de quelques sols typiques des regions tropicales. Agrochimica 13:358–366.

Anderson, G. 1958. Indentification of derivatives of deoxyribonucleic acid in humic acid. Soil Sci. 86:169–174.

Anderson, G. 1967. Nucleic acids, derivatives, and organic phosphates. p. 67–90. *In* A. D. McLaren and G. H. Peterson (ed.) Soil biochemistry. Marcel Dekker, Inc., New York.

Anderson, G. 1970. The isolation of nucleoside diphosphates from alkaline extracts of soil. J. Soil Sci. 21:96–104.

Asami, T. 1971. Immobilization and mineralization of nitrogen compounds in paddy soils, part 5, Distribution of immobilized nitrogen to various organic nitrogen fractions. J. Sci. Soil Manure, Japan 42:103–108. From Soil Sci. Plant Nutr. 18:203. 1972.

Asami, T., and M. Hara. 1970. On the fractionation of soil organic matter after hydrolysis with hydrochloric acid. (In Japanese.) J. Sci. Soil Manure, Japan 41:487–490. From Soil Sci. Plant Nutr. (Tokyo) 17:222. 1971.

Asmus, F. 1970. Effect of different organic manures on the nitrogen content and nitrogen fractions of a sandy brown soil (Rosterde). (In German.) Albrecht-Thaer-Arch. 14:775–782.

Ayanaba, A., W. Verstraete, and M. Alexander. 1973. Formation of dimethylnitrosamine, a carcinogen and mutagen, in soils treated with nitrogen compounds. Soil Sci. Soc. Am. Proc. 37:565–568.

Bachelier, G. 1975. Significance of the stereochemical nature of amino acids in humification processes. (In French.) Rev. Ecol. Biol. Sol 12:383–392.

Bartha, R. 1971. Fate of herbicide-derived chloroanilines in soil. J. Agric. Food Chem. 19:383–387.

Bartha, R., and D. Pramer. 1970. Metabolism of acylanilide herbicides. Adv. Appl. Microbiol. 13:317–341.

Batsula, A. A., and N. K. Krupskiy. 1974. The forms of nitrogen in humus substances of some virgin and developed soils of the left-bank of the Ukraine. Sov. Soil Sci. 6:456–462.

Biederbeck, V. O., and E. A. Paul. 1973. Fractionation of soil humate with phenolic solvents and purification of the nitrogen rich portion with polyvinylpyrrolidone. Soil Sci. 115:357–366.

Blackmer, A. M., and J. M. Bremner. 1977. Nitrogen isotope discrimination in denitrification of nitrate in soils. Soil Biol. Biochem. 9:73–77.

Bowen, S. H. 1980. Detrital nonprotein amino acids are the key to rapid growth of *Tilapia* in Lake Valencia, Venezuela. Science 207:1216–1218.

Bremner, J. M. 1949. Studies on soil organic matter: I. J. Agric. Sci. 39:183–193.

Bremner, J. M. 1950. The amino acid composition of the protein material in soil. Biochem. J. 47:538–542.

Bremner, J. M. 1955. Studies on soil humic acids: I. J. Agric. Sci. 46:247–256.

Bremner, J. M. 1965a. Organic nitrogen in soils. *In* W. V. Bartholomew and F. E. Clark (ed.) Soil nitrogen. Agronomy 10:93–149. Am. Soc. of Agron., Madison, Wis.

Bremner, J. M. 1965b. Organic forms of nitrogen. *In* C. A. Black et al. (ed.) Methods of soil analysis. Agronomy 9:1148–1178. Am. Soc. of Agron., Madison, Wis.

Bremner, J. M. 1967. Nitrogenous compounds. p. 19–66. *In* A. D. McLaren and G. H. Petersen (ed.) Soil biochemistry. Marcel Dekker, Inc., New York.

Broadbent, F. E., R. S. Rauschkolb, K. A. Lewis, and G. Y. Chang. 1980. Spatial variability of nitrogen-15 and total nitrogen in some virgin and cultivated soils. Soil Sci. Soc. Am. J. 44:524–532.

Broadbent, F. E., R. H. Jackman, and J. McNicoll. 1964. Mineralization of carbon and nitrogen in some New Zealand allophanic soils. Soil Sci. 98:118–128.

Bunyakina, R. F. 1976. Fractional composition of nitrogen and transformation of nitrogen fertilizers in the leached Chernozem of the Kuban region (based on pot experiments with N^{15}). Sov. Soil Sci. 8:438–442.

Carles, J., and J. Decau. 1960. Variations in the amino acids of soil hydrolysates. Sci. Proc. Royal Dublin Soc., Ser. 1A:177–182.

Carter, P. W., and R. M. Mitterer. 1978. Amino acid composition of organic matter associated with carbonate and non-carbonate sediments. Geochim. Cosmochim. Acta 42: 1231–1238.

Casagrande, D. J., and K. Park. 1978. Muramic acid levels in bog soils from the Okefenokee swamp. Soil Sci. 125:181–183.

Chen, Y., F. J. Sowden, and M. Schnitzer. 1977. Nitrogen in Mediterranean soils. Agrochimica 21:7–14.

Cheng, C.-N. 1975. Extraction and desalting amino acids from soils and sediments: Evaluation of methods. Soil Biol. Biochem. 7:319–322.

Cheng, C.-N., R. C. Shufeldt, and F. J. Stevenson. 1975. Amino acid analysis of soils and sediments: Extraction and desalting. Soil Biol. Biochem. 7:143–151.

Cheng, H. H., J. M. Bremner, and A. P. Edwards. 1964. Variation of nitrogen-15 abundance in soils. Science 146:1574–1575.

Cheshire, M. V., P. A. Cranwell, C. P. Falshaw, A. J. Floy, and R. D. Haworth. 1967. Structure of humic acids. Tetrahedron 23:1669–1682.

Chichester, F. W. 1969. Nitrogen in organo-mineral sedimentation fractions. Soil Sci. 107: 356–363.

Chichester, F. W. 1970. Transformations of fertilizer nitrogen in soils. II. Total and ^{15}N-labelled nitrogen in soil organic mineral sedimentation and fractions. Plant Soil 33: 437–456.

Chiska, H., and P. C. Kearney. 1970. Metabolism of propanil in soils. J. Agric. Food Chem. 18:854–858.

Chopra, N. M. 1976. Investigations into the fate of plant pigments in some Canadian soils. Soil Sci. 121:103–113.

Cornfield, A. H. 1957. Effect of 8-years fertilizer treatment on the protein-nitrogen content of four cropped soils. J. Sci. Food Agric. 8:509–511.

Cornforth, I. S. 1968. The potential availability of organic nitrogen fractions in some West Indian soils. Exp. Agric. 4:193–201.

Cornforth, I. S. 1969. Chlorophyll compounds and nitrogen availability in West Indian soils. Plant Soil 30:113–116.

Cortez, J., and M. Schnitzer. 1979. Nucleic acid bases in soil and their association with organic and inorganic soil components. Can. J. Soil Sci. 59:277–286.

Dalal, R. C. 1978a. Distribution of organic nitrogen in organic volcanic and nonvolcanic tropical soils. Soil Sci. 125:178–180.

Dalal, R. C. 1978b. The nature and distribution of soil nitrogen in tropical soils. Trop. Agric. 55:369–376.

Decau, J. 1967. Observations sur la repartition comparee des acides amines dans la matiere organique non evoluee et dans l'humus du sol. C. R. Acad. Sci. 264:1836–1839.

Decau, J. 1969. Contribution a l'etude de l'influence des conditions de milieu sur la repartition de l'azote dans le sol. Ann. Agron. 20:35–59, 277–303.

Delwiche, C. C., and P. L. Steyn. 1970. Nitrogen isotope fractionation in soils and microbial reactions. Environ. Sci. Tech. 4:929–935.

Dungworth, G. 1976. Optical configuration and the racemization of amino acids in sediments and in fossils—a review. Chem. Geol. 17:135–153.

Edwards, A. P. 1966. Clay-humus complexes in soil. Int. Congr. Soil Sci., Trans. 10th. (Aberdeen) III:33–39.

Edwards, A. P., and J. M. Bremner. 1967. Microaggregate formation in soils. J. Soil Sci. 18: 64–73.

Emerson, W. W. 1959. The structure of soil crumbs. J. Soil Sci. 10:235–244.

Feigin, A., D. H. Kohl, G. Shearer, and B. Commoner. 1974. Variation in the natural nitrogen-15 abundance in nitrate mineralized during incubation of several Illinois soils. Soil Sci. Soc. Am. Proc. 38:90–95.

Ferguson, W. S., and F. J. Sowden. 1966. A comparison of methods of determining nitrogen fractions in soils. Can. J. Soil Sci. 46:1–6.

Flaig, W., H. Beutelspacher, and E. Rietz. 1975. Chemical composition and physical properties of humic substances. p. 1–211. In J. E. Gieseking (ed.) Soil components: Vol. 1. Springer-Verlag, New York, New York.

Fleige, H., and K. Baeumer. 1974. Effect of zero-tillage on organic carbon and total nitrogen content, and their distribution in different N-fractions in loessial soils. Agro-Ecosystems 1:19–29.

Fleige, H., B. Meyer, and H. Scholz. 1971. Fraktionierung des Boden-Stickstoffs fur N-Haushalts-Bilanzen. Göttinger Bodenkd. Ber. 18:1–37.

Focht, D. D. 1973. Isotope fractionation of ^{15}N and ^{14}N in microbiological nitrogen transformations: A theoretical model. J. Environ. Qual. 2:247–252.

Forsyth, W. G. C. 1947. Studies on the more soluble complexes of soil organic matter: 1. A method of fractionation. Biochem. J. 41:176–181.

Freney, J. R. 1968. The extraction and partial characterization of non-hydrolyzable nitrogen in soil. Int. Congr. Soil Sci., Trans. 9th (Adelaide) III:531–539.

Fujii, K., M. Kobayashi, and E. Takahashi. 1974. Amines in the mixture of plant residues and sand incubated under aerobic and water-logged conditions. Part 5. (In Japanese.) J. Sci. Soil Manure, Japan 43:160–164. From Soil Sci. Plant Nutr. (Tokyo) 20:101–102.

Gallali, T., A. Gluckert, and F. Jacquin. 1975. Etude de la distribution des sucres amines dans la matiere organique des sols. Bull. Ec. Natl. Super. Agron. Ind. Aliment. 17:53–59.

Geoghegan, M. J., and R. C. Brian. 1946. Influence of bacterial polysaccharides on aggregate formation in soils. Nature (London) 158:837.

Giddens, J., R. D. Hauck, W. E. Adams, and R. N. Dawson. 1971. Forms of nitrogen and nitrogen availability in fescuegrass sod. Agron. J. 63:458–460.

Gilbert, R. G., and J. Altman. 1966. Ethanol extraction of free amino acids from soil. Plant Soil 24:229–238.

Goh, K. M. 1972. Amino acid levels as indicators of paleosols in New Zealand soil profiles. Geoderma 7:33–47.

Goh, K. M., and D. C. Edmeades. 1979. Distribution and partial characterisation of acid hydrolysable organic nitrogen in six New Zealand soils. Soil Biol. Biochem. 11:127–132.

Gorham, E. 1959. Chlorophyll derivatives in woodland soils. Soil Sci. 87:258–261.

Greaves, M. P., and M. J. Wilson. 1970. The degradation of nucleic acids and montmorillonite-nucleic-acid complexes by soil microorganisms. Soil Biol. Biochem. 2:257–268.

Greaves, M. P., and M. J. Wilson. 1973. Effects of soil microorganisms on montmorillonite-adenine complexes. Soil Biol. Biochem. 5:275–276.

Greenfield, L. G. 1972. The nature of the organic nitrogen of soils. Plant Soil 36:191–198.

Greenland, D. J. 1965. Interactions between clay and organic compounds in soils: 1 and 2. Soils Fert. 28:412–425; 521–532.

Greenland, D. J. 1971. Interactions between humic and fulvic acids and clays. Soil Sci. 111: 34–41.

Griffith, S. M., F. J. Sowden, and M. Schnitzer. 1976. The alkaline hydrolysis of acid-resistant soil and humic acid residues. Soil Biol. Biochem. 8:529–531.

Guenzi, W. D. (ed.). 1974. Pesticides in soil and water. Soil Sci. Soc. of Am., Madison, Wis.

Guidi, G., G. Petruzzelli, and P. Sequi. 1976. Characterization of amino acid and carbohydrate components in fulvic acid. Can. J. Soil Sci. 56:159–166.

Gupta, U. C., and H. W. Reuszer. 1967. Effect of plant species on the amino acid content and nitrification of soil organic matter. Soil Sci. 104:395–400.

Haider, K., L. R. Frederick, and W. Flaig. 1965. Reaction between amino acid compounds and phenols during oxidation. Plant Soil 22:49–64.

Haider, K., J. P. Martin, and Z. Filip. 1975. Humus biochemistry. p. 195–244. In E. A. Paul and A. D. McLaren (ed.) Soil biochemistry, Vol. 4. Marcel Dekker, Inc., New York.

Hamilton, P. H. 1968. The ion exchange chromatography of urine amino acids. p. B43–B55. In H. A. Sober (ed.) Handbook of biochemistry. The Chemical Rubber Co., Cleveland, Ohio.

Hance, R. J., and G. Anderson. 1963. Identification of hydrolysis products of soil phospholipids. Soil Sci. 96:157–161.

Hauck, R. D. 1973. Nitrogen tracers in nitrogen cycle studies—past use and future needs. J. Environ. Qual. 2:317–327.

Hauck, R. D., W. V. Bartholomew, J. M. Bremner, F. E. Broadbent, H. H. Cheng, A. P. Edwards, D. R. Keeney, J. O. Legg, S. R. Olsen, and L. K. Porter. 1972. Use of variations in natural nitrogen isotope abundance for environmental studies: A questionable approach. Science 177:453–456.

Hauck, R. D., and J. M. Bremner. 1976. Use of tracers for soil and fertilizer nitrogen research. Adv. Agron. 28:219–265.

Haworth, R. D. 1971. The chemical nature of humic acid. Soil Sci. 111:71–79.

Haworth, W. N., F. W. Pinkard, and M. Stacey. 1946. Function of bacterial polysaccharides in soil. Nature (London) 158:836–837.

Hayashi, R., and T. Harada. 1969. Characterization of the organic nitrogen becoming decomposable through the effect of drying of a soil. Soil Sci. Plant Nutr. 15:226–234.

Hodge, J. E. 1953. Chemistry of browning reactions in model systems. J. Agric. Food Chem. 1:928–943.

Hodgson, G. W., B. Hitchen, K. Taguchi, B. L. Baker, and E. Peake. 1968. Geochemistry of porphyrins, chlorins, and polycyclic aromatics in soils, sediments, and sedimentary rocks. Geochim. Cosmochim. Acta 32:737–772.

Hoyt, P. B. 1966. Chlorophyll-type compounds in soil: I and II. Plant Soil 25:167–180; 313–328.

Hoyt, P. B. 1967. Chlorophyll-type compounds in soil: III. Plant Soil 26:5–13.

Hoyt, P. B. 1971. Fate of chlorophyll in soil. Soil Sci. 111:49–53.

Hsu, T.-S., and R. Bartha. 1974. Biodegradation of chloroaniline-humus complexes in soil and in culture solution. Soil Sci. 118:213–220. See also 116:444–452 (1973).

Hüser, R. 1971. Die Umsetzung von N^{15}-Düngern in Problem einer Podzol-Baunerde und ihre Austnutzung durch Einjahrige Koniferen. Plant Soil 35:37–50.

Iseki, A. 1968. The nitrogen components of some typical volcanic ash soils and red-yellow soils in Japan. 1. (In Japanese.) J. Sci. Soil Manure, Japan 39:189–193. From Soil Sci. Plant Nutr. (Tokyo) 15:166.

Isirimah, N. O., and D. R. Keeney. 1973. Nitrogen transformations in aerobic and water-logged Histosols. Soil Sci. 115:123–129.

Ivarson, K. C., and M. Schnitzer. 1979. The biodegradability of the "unknown" soil-nitrogen. Can. J. Soil Sci. 59:59–67.

Ivarson, K. C., and F. J. Sowden. 1966. Effect of freezing on the free amino acids in soil. Can. J. Soil Sci. 46:115–120.

Ivarson, K. C., and F. J. Sowden. 1969. Free amino acid composition of the plant root environment under field conditions. Can. J. Soil Sci. 49:121–127.

Ivarson, K. C., and F. J. Sowden. 1970. Effect of frost action and storage of soil at freezing temperatures on the free amino acids, free sugars, and respiratory activity of soil. Can. J. Soil Sci. 50:191–198.

Ivarson, K. C., F. J. Sowden, and A. R. Mack. 1970. Amino-acid composition of rhizosphere as affected by soil temperature, fertility and growth stage. Can. J. Soil Sci. 50:183–189.

Jorgensen, J. R. 1967. Fractionation of nitrogen in three forest soils. Soil Sci. Am. Proc. 31: 707–708.

Kadirgamathaiyah, S., and A. F. MacKenzie. 1970. A study of soil nitrogen organic fractions and correlation with yield response of Sudan-sorghum hybrid grass in Quebec soils. Plant Soil 33:120–128.

Kai, H., Z. Ahmad, and T. Harada. 1973. Factors affecting immobilization and release of nitrogen in soil and chemical characteristics of the nitrogen newly immobilized. III. Soil Sci. Plant Nutr. (Tokyo) 19:275–286.

Kanazawa, S., and T. Yoneyama. 1978. Determination of ^{15}N abundance of amino acids in soil hydrolysates. Soil Sci. Plant Nutr. 24:153–155.

Karamshuk, Z. P. 1979. Free amino acids in dark Chestnut soil of Northern Kazakhstan. Sov. Soil Sci. 11:745–749.

Kaszubiak, H. 1976. Correlation between determinations of chlorophyll-type compounds, nitrogen availability for plants and number of algae in the soil. Polish J. Soil Sci. 9:47–51. From Soils and Fert. 40(6839):687, 1977.

Kato, T. 1975. On the water-soluble organic matter in paddy soils. 5. Formation of free amino acids in submerged soils. (In Japanese.) J. Sci. Soil Manure, Japan 46:340–342. From Soil Sci. Plant Nutr. (Tokyo) 22:103. 1976.

Kaufman, D. D., G. G. Still, G. D. Paulson, and S. K. Bandal (ed.). 1976. Bound and conjugated pesticide residues. ACS Symp. Ser. 29. Am. Chem. Soc., Washington, D.C.

Kawaguchi, S., H. Kai, and T. Harada. 1976. The occurrence of D-amino acids in soil organic matter and its significance for soil nitrogen metabolism. 1. Methods for the optical resolution of amino acids. (In Japanese.) J. Sci. Soil Manure, Japan 47:243–250. From Soil Sci. Plant Nutr. (Tokyo) 22:358–359. 1976.

Keeney, D. R., and J. M. Bremner. 1964. Effect of cultivation on the nitrogen distribution in soils. Soil Sci. Soc. Am. Proc. 28:653–656.

Keeney, D. R., and J. M. Bremner. 1966. Characterization of mineralizable nitrogen in soils. Soil Sci. Soc. Am. Proc. 30:714–719.

Keeney, D. R., J. G. Konrad, and G. Chesters. 1970. Nitrogen distribution in some Wisconsin lake sediments. J. Water Pollut. Contr. Fed. 42:411–417.

Khan, S. U. 1971. Nitrogen fractions in a gray wooded soil as influenced by long-term cropping systems and fertilizers. Can. J. Soil Sci. 51:431–437.

Khan, S. U., and F. J. Sowden. 1971. Distribution of nitrogen in the black Solonetzic and black Chernozemic soils of Alberta. Can. J. Soil Sci. 51:185–193.

Khan, S. U., and F. J. Sowden. 1972. Distribution of nitrogen in fulvic acid fraction extracted from the black Solonetzic and black Chernozemic soils of Alberta. Can. J. Soil Sci. 52: 116–118.

Kickuth, R., and F. Scheffer. 1976. Constitution and role as plant nutrients of pseudo-amide nitrogen in humic acids. (In German.) Agrochimica 20:373–386.

Kobo, K., H. Wada, and S. Imamura. 1970. Water-soluble organic substances in soils. 2. (In Japanese.) J. Sci. Soil Manure, Japan 41:281–286.

Kohl, D. H., G. B. Shearer, and B. Commoner. 1971. Fertilizer nitrogen: contribution to nitrate in surface water in a corn belt watershed. Science 174:1331–1334.

Kohl, D. H., G. B. Shearer, and B. Commoner. 1972. Response to criticism on use of natural nitrogen isotope abundance for environmental studies. Science 177:454–456.

Koloskova, A. V., and L. I. Shitova. 1974. Nitrogen in Tatar soils. Sov. Soil Sci. 6:172–179.

Koren'kov, D. A., I. A. Lavrova, D. A. Filimonov, and Ye. V. Rudelev. 1976. Transformations of nitrogen fertilizers in soil. 2. Transformations of previously immobilized nitrogen after repeated fertilizer application. Sov. Soil Sci. 8:572–575.

Kowalenko, C. G. 1978. Organic nitrogen, phosphorus, and sulfur in soils. p. 95–136. In M. Schnitzer and S. U. Khan (ed.) Soil organic matter. Elsevier North Holland, Inc., New York.

Ktsoyev, B. K. 1977. Free amino acids in the soils of the northern Caucasus. Sov. Soil Sci. 9: 312–315.

Kyuma, K., A. Hussain, and K. Kawaguchi. 1969. The nature of organic matter in soil organo-mineral complexes. Soil Sci. Plant Nutr. (Tokyo) 15:149–155.

Ladd, J. N., P. G. Brisbane, J. H. A. Butler, and M. Amato. 1976. Studies on soil fumigation. III. Soil Biol. Biochem. 8:225–260.

Ladd, J. N., and J. H. A. Butler. 1966. Comparison of some properties of soil humic acids and synthetic phenolic polymers incorporating amino derivatives. Aust. J. Soil Res. 4:41–54.

Ladd, J. N., and E. A. Paul. 1973. Changes in enzyme activity and distribution of acid-soluble, amino acid-N in soil during nitrogen immobilization and mineralization. Soil Biol. Biochem. 5:825–840.

Legg, J. D., F. W. Chichester, G. Stanford, and W. H. DeMar. 1971. Incorporation of ^{15}N-tagged mineral nitrogen into stable forms of soil organic nitrogen. Soil Sci. Soc. Am. Proc. 35:273–276.

Lin, C. F., A. H. Chang, and C. C. Tseng. 1973. Nitrogen status and nitrogen supplying power of Taiwan soils. (In Chinese.) J. Taiwan Agr. Res. 22:186–203. From Soils Fert. 38(34): 4, 1975.

Loginow, W. 1967. Effect of humic acids on the deamination of amino acids. (In Polish.) Pamiet. Pulawski 29:3–43.

Lowe, L. E. 1973. Amino acid distribution in forest humus layers in British Columbia. Soil Sci. Soc. Am. Proc. 37:569–572.

McGill, W. B., and E. A. Paul. 1976. Fractionation of soil and ^{15}N nitrogen to separate the organic and clay interactions of immobilized N. Can. J. Soil Sci. 56:203–212.

Meints, V. W., L. V. Boone, and L. T. Kurtz. 1975. Natural ^{15}N abundance in soil, leaves, and grain as influenced by long-term additions of fertilizer N at several rates. J. Environ. Qual. 4:486–490.

Meints, V. W., and G. A. Peterson. 1977. The influence of cultivation on the distribution of nitrogen in soils of the Ustoll suborder. Soil Sci. 124:334–342.

Meister, A. 1965. Biochemistry of the amino acids, part 1. Academic Press, Inc., New York, p. 113–118.

Miki, K., Y. Kawato, and T. Mori. 1966a. Fractionation and estimation of organic nitrogen compounds in soil. Tentative method using the Kjeldahl distillation apparatus. (In Japanese.) J. Sci. Soil Manure, Japan 37:542–546. From Soil Sci. Plant Nutr. (Tokyo) 13: 36, 1967.

Miki, K., Y. Kawato, and T. Mori. 1966b. Nitrogen-supplying capacity of upland soils. 3. Distribution of some organic forms of nitrogen in upland soils. Bull. Tokai-Kinki Natn. Agric. Exp. Stn. 15:125–135. From Soils Fert. 30(2411):337, 1967.

Miller, W. N., and L. E. Casida, Jr. 1970. Evidence for muramic acid in soil. Can. J. Microbiol. 16:299–304.

Mills, A. L., and M. Alexander. 1976. Factors affecting dimethylnitrosamine formation in samples of soil and water. J. Environ. Qual. 5:437–440.

Moore, A. W., and J. S. Russell. 1970. Changes in chemical fractions of nitrogen during incubation of soils with histories of large organic matter increase under pasture. Aust. J. Soil Res. 8:21–30.

Mosier, A. R., and S. Torbit. 1976. Synthesis and stability of dimethylnitrosamine in cattle manure. J. Environ. Qual. 5:465–468.

Namdev, K. N., and J. N. Dube. 1973. Residual effect of urea and herbicides on hexosamine content and urease and proteinase activities in a grassland soil. Soil Biol. Biochem. 5:855–859.

Nommik, H. 1967. Distribution of forms of nitrogen in a podzolic soil profile from Garpenberg, central Sweden. J. Soil Sci. 18:301–308.

Orr, W. L., K. O. Emery, and J. R. Grady. 1958. Preservation of chlorophyll derivatives in sediments off Southern California. Bull. Am. Assoc. Petrol. Geol. 42:925–962.

Ortiz de Serra, M. I., F. J. Sowden, and M. Schnitzer. 1973. Distribution of nitrogen in fungal "humic acids". Can. J. Soil Sci. 53:125–127.

Osborne, G. J. 1977. Chemical fractionation of soil nitrogen in six soils from southern New South Wales. Aust. J. Soil Res. 15:159–165.

Otsuka, H. 1975. Accumulated state of humus in the soil profile, and sugar, uronic acid and amino acid contents and amino acid composition in fulvic acids, part 5. (In Japanese.) J. Sci. Soil Manure, Japan 46:138–142. From Soil Sci. Plant Nutr. (Tokyo) 21:420–421.

Otsuki, A., and T. Hanya. 1967. Some precursors of humic acid in recent lake sediments suggested by infrared spectra. Geochim. Cosmochim. Acta 31:1505–1515.

Pancholy, S. K. 1978. Formation of carcinogenic nitrosamines in soils. Soil Biol. Biochem. 10:27–32.

Parsons, J. W., and J. Tinsley. 1975. Nitrogenous substances. p. 263–304. In J. E. Gieseking (ed.) Soil components: Vol. 1. Springer-Verlag, New York, New York.

Paul, E. A., and E. L. Schmidt. 1960. Extraction of free amino acids from soil. Soil Sci. Soc. Am. Proc. 24:195–198.

Paul, E. A., and C. M. Tu. 1965. Alteration of microbial activities, mineral nitrogen, and free amino acid constituents of soils by physical treatment. Plant Soil 22:207–219.

Perry, D. R., and W. A. Adams. 1972. The incorporation of (^{14}C) labelled glycine into extracted humic acid. p. 59–67. In Report Welsh Soils Discussion Group. no. 13. University College, Aberystwyth, U.K.

Piper, T. J., and A. M. Posner. 1968. On the amino acids found in humic acids. Soil Sci. 106:188–192.

Piper, T. J., and A. M. Posner. 1972. Humic acid nitrogen. Plant Soil 36:595–598.

Pollock, G. E., C.-N. Cheng, and S. E. Cronin. 1977. Determination of the D and L isomers of some protein amino acids present in soils. Anal. Chem. 49:2–7.

Pollock, G. E., and A. K. Miyamoto. 1971. A desalting technique for amino acid analysis of use in soil and geochemistry. J. Agric. Food Chem. 19:104–107.

Porter, L. K., B. A. Stewart, and H. J. Haas. 1964. Effects of long-time cropping on hydrolyzable organic nitrogen fractions in some Great Plains soils. Soil Sci. Am. Proc. 28:368–370.

Rennie, D. A., E. A. Paul, and L. E. Johns. 1976. Natural nitrogen-15 abundance of soil and plant samples. Can. J. Soil Sci. 56:43–50.

Rosell, R. A., J. C. Salfeld, and H. Sochtig. 1978. Organic compounds in Argentine soils: 1. Nitrogen distribution in soils and their humic acids. Agrochimica 22:98–105.

Rovira, A. D., and B. M. McDougall. 1967. Microbiological and biochemical aspects of the rhizosphere. p. 417–463. In A. D. McLaren and G. H. Petersen (ed.) Soil biochemistry. Marcel Dekker, Inc., New York.

Rudrappa, T., P. B. Deshpande, and P. V. Monteiro. 1972. Amino acids in humic acid. Mysore J. Agric. Sci. 6:215–218.

Saiz-Jimenez, C., K. Haider, and J. P. Martin. 1975. Anthraquinones and phenols as intermediates in the formation of dark-colored humic acid-like pigments by Eurotium echinulatum. Soil Sci. Soc. Am. Proc. 39:649–653.

Schroeder, R. A., and J. L. Bada. 1976. A review of the geochemical applications of the amino acid racemization reaction. Earth Sci. Rev. 12:347–391.

Sequi, P., G. Guidi, and G. Petruzzelli. 1975. Distribution of amino acid and carbohydrate components in fulvic acid fractionated by polyamide. Can. J. Soil Sci. 55:439–445.

Sharon, N. 1965. Distribution of amino sugars in microorganisms, plants, and invertebrates. p. 1–45. In R. W. Jeanloz and E. A. Balazs (ed.) The amino sugars. Vol. IIA. Academic Press, Inc., New York.

Shaymukhametov, M. Sh. 1977. Improved method for determining amino acid and hexosamine nitrogen in soils. Sov. Soil Sci. 9:503–504.

Shearer, G. B., D. H. Kohl, and B. Commoner. 1974. The precision of determination of the natural abundance of nitrogen-15 in soils, fertilizers, and shelf chemicals. Soil Sci. 118: 308–316.

Simonart, P., L. Batistic, and J. Mayaudon. 1967. Isolation of protein from humic acid extracted from soil. Plant Soil 27:153–161.

Simonart, P., J. Mayaudon, and L. Batistic. 1959. Etude de la decomposition de la matiere organique dans la sol au moyen de carbone radioactif. IV. Plant Soil 11:176–180.

Singh, B. R., A. P. Uriyo, and B. J. Lontu. 1978. Distribution and stability of organic forms of nitrogen in forest soil profiles in Tanzania. Soil Biol. Biochem. 10:105–108.

Sinha, M. K. 1972. Organic matter transformations in soils. III. Nature of amino acids in soils incubated with ^{14}C-tagged oat roots under aerobic and anaerobic conditions. Plant Soil 37:265–271.

Skujins, J., and A. Pukite. 1970. Extraction and determination of N-acetylglucosamine from soil. Soil Biol. Biochem. 2:141–143.

Sørensen, L. H. 1967. Duration of amino acid metabolites in soils during decomposition of carbohydrates. Soil Sci. 104:234–241.

Sørensen, L. H. 1972. Stabilization of newly formed amino acid metabolites in soil by clay minerals. Soil Sci. 114:5–11.

Sørensen, L. H. 1975. The influence of clay on the rate of decay of amino acid metabolites synthesized in soils during decomposition of cellulose. Soil Biol. Biochem. 7:171–177.

Sowden, F. J. 1956. Distribution of amino acids in selected horizons of soil profiles. Soil Sci. 82:491–496.

Sowden, F. J. 1959. Investigations on the amounts of hexosamines found in various soils and methods for their determination. Soil Sci. 88:138–143.

Sowden, F. J. 1969. Effect of hydrolysis time and iron and aluminum removed on the determination of amino compounds in soil. Soil Sci. 107:364–371.

Sowden, F. J. 1970. Extraction of nitrogen-containing organic matter fractions from a brown forest soil. Can. J. Soil Sci. 50:227–232.

Sowden, F. J. 1977. Distribution of nitrogen in representative Canadian soils. Can. J. Soil Sci. 57:445–456.

Sowden, F. J., Y. Chen, and M. Schnitzer. 1977. The nitrogen distribution in soils formed under widely differing climatic conditions. Geochim. Cosmochim. Acta 41:1524–1526.

Sowden, F. J., S. M. Griffith, and M. Schnitzer. 1976. The distribution of nitrogen in some highly organic tropical volcanic soils. Soil Biol. Biochem. 8:55–60.

Sowden, F. J., and K. C. Ivarson. 1966. The "free" amino acids of soil. Can. J. Soil Sci. 46: 109–114.

Sowden, F. J., and K. C. Ivarson. 1968. Determination of asparagine, glutamine and citrulline in aqueous soil extracts with an automatic amino acid analyzer. Can. J. Soil Sci. 48: 349–354.

Sowden, F. J., H. Morita, and M. Levesque. 1978. Organic nitrogen distribution in selected peats and peat fractions. Can. J. Soil Sci. 58:237–249.

Sowden, F. J., and M. Schnitzer. 1967. Nitrogen distribution in illuvial organic matter. Can. J. Soil Sci. 47:111–116.

Spark, A. A. 1969. Role of amino acids in non-enzymatic browning. J. Sci. Food Agric. 20: 308–316.

Stepanov, V. V. 1969. Reaction of humic acids with some nitrogen-containing compounds. Sov. Soil Sci. 2:167–173.

Stevenson, F. J. 1954. Ion exchange chromatography of amino acids in soil hydrolysates. Soil Sci. Soc. Am. Proc. 18:373–377.

Stevenson, F. J. 1956a. Isolation and identification of some amino compounds in soils. Soil Sci. Soc. Am. Proc. 20:201–204.

Stevenson, F. J. 1956b. Effect of some long-time rotations on the amino acid composition of the soil. Soil Sci. Soc. Am. Proc. 20:204–208.

Stevenson, F. J. 1957. Investigations of amino-polysaccharides in soils: I and II. Soil Sci. 83: 113–122; 84:98–106.

Stevenson, F. J. 1960. Chemical nature of the nitrogen in the fulvic acid fraction of soil organic matter. Soil Sci. Soc. Am. Proc. 24:472–477.

Stevenson, F. J. 1965. Amino sugars. In C. A. Black et al. (ed.) Methods of soil analysis. Agronomy 9:1429–1436. Am. Soc. of Agron., Madison, Wis.

Stevenson, F. J. 1974. Nonbiological transformations of amino acids in soils and sediments. p. 701–714. *In* B. Tissot and F. Bienner (ed.) Advances in organic geochemistry. Editions Technips, Paris.

Stevenson, F. J., and O. C. Braids. 1968. Variation in the relative distribution of amino sugars with depth in some soil profiles. Soil Sci. Soc. Am. Proc. 32:598–600.

Stevenson, F. J., and C.-N. Cheng. 1970. Amino acids in sediments: Recovery by acid hydrolysis and quantitative estimation by a colorimetric procedure. Geochim. Cosmochim. Acta 34:77–88.

Stevenson, F. J., and C.-N. Cheng. 1972. Organic geochemistry of the Argentine Basin sediments: carbon-nitrogen relationships and Quaternary correlations. Geochim. Cosmochim. Acta 36:653–671.

Stevenson, F. J., and K. M. Goh. 1971. Infrared spectra of humic acids and related substances. Geochim Cosmochim. Acta 35:471–483.

Stevenson, F. J., G. Kidder, and S. N. Tilo. 1967. Extraction of organic nitrogen and ammonium from soil with hydrofluoric acid. Soil Sci. Soc. Am. Proc. 31:71–76.

Stevenson, F. J., and G. H. Wagner. 1970. Chemistry of nitrogen in soils. p. 125–141. *In* T. L. Willrich and G. E. Smith (ed.) Agricultural practices and water quality. The Iowa State Univ. Press, Ames, Iowa.

Suzuki, S. 1906–1908. Studies on humus formation. Bull. Coll. Agric. (Tokyo) 7:95–101, 419–425, 513–529.

Tate, R. L., and M. Alexander. 1974. Formation of dimethylamine and diethylamine in soil treated with pesticides. Soil Sci. 118:317–321.

Tsukada, T., S. Sugihara, and M. Deguchi. 1966. Nitrogenous compounds and carbohydrates in compost manure. Soil Sci. Plant Nutr. (Tokyo) 12:1–7.

Tsutsuki, K., and S. Kuwatsuka. 1978. Chemical studies of soil humic acids: III and IV. Soil Sci. Plant Nutr. (Tokyo) 24:29–38; 561–570.

Uy, R., and F. Wold. 1977. Posttranslational covalent modification of proteins. Science 198:890–896.

Verma, L., J. P. Martin, and K. Haider. 1975. Decomposition of carbon-14-labeled proteins, peptides and amino acids; free and complexed with humic polymers. Soil Sci. Soc. Am. Proc. 39:279–284.

Wagner, G. H., and V. K. Mutatkar. 1968. Amino components of soil organic matter formed during humification of ^{14}C glucose. Soil Sci. Soc. Am. Proc. 32:683–684.

Wainwright, M., and G. J. F. Pugh. 1975. Changes in the free amino acid content of soil following treatment with fungicides. Soil Biol. Biochem. 7:1–4.

Wang, T. S. C., T.-K. Yang, and S.-Y. Cheng. 1967. Amino acids in subtropical soil hydrolysates. Soil Sci. 103:67–74.

Wellman, B. P., F. O. Cook, and H. R. Krouse. 1968. Nitrogen-15 microbial alteration in abundance. Science 161:269–270.

Witthauer, J., and R. Klocking. 1971. Bundungasrten des Stickstoffs in Huminsauren. Arch. Acker- Pflanzenbau Bodenkd. 15:577–588, 663–670.

Yamashita, T., and T. Akiya. 1963. Amino-acid composition of soil hydrolysates. (In Japanese.) J. Sci. Soil Manure, Japan 34:255–258. From Soil Sci. Plant Nutr. (Tokyo) 10:43, 1964.

Yamashita, T., and C. Kawada. 1968. Studies on the effect of soil sterilization upon the nitrogen composition of tobacco plant cultivated soil. (In Japanese.) J. Sci. Soil Manure, Japan 39:204–209. From Soil Sci. Plant Nutr. (Tokyo) 14:16, 1968.

Yonebayashi, K., K. Kyuma, and K. Kawaguchi. 1973. Readily decomposable organic matter, parts 1–2. (In Japanese.) J. Sci. Soil Manure, Japan 44:327–333, 367–371. From Soil Sci. Plant Nutr. (Tokyo) 20:421–422; 423.

Young, J. L., and J. L. Mortensen. 1958. Soil nitrogen complexes: I. Ohio Agric. Exp. Stn. Res. Circ. 61:1–18.

4   Retention and Fixation of Ammonium and Ammonia in Soils

HANS NOMMIK AND KAAREL VAHTRAS

Swedish University of Agricultural Sciences
Uppsala, Sweden

I. INTRODUCTION

Large quantities of N in the form of NH_4^+ and NH_3 are being used to increase crop production in agriculture. In forestry, too, use of N fertilizers has increased rapidly during the last decade. The reactions involved and the nature of the complexes formed when NH_4^+ and NH_3 are brought into contact with soil materials are, therefore, of interest from both agronomical and theoretical viewpoints.

In the first edition of this monograph, the nature of NH_3 adsorption by clays and organic matter was reviewed by Mortland and Wolcott (1965), and the fixation of NH_4^+ was discussed by Nommik (1965). In the present edition, these two topics are combined into one chapter. Considerations are also given to cation exchange with emphasis on NH_4^+. In reviewing the literature, sources through 1978 have been considered.

II. EXCHANGEABLE BINDING OF NH_4^+ IN SOILS AND CLAY MINERALS

It is of interest that the first quantitative studies on cation adsorption and release in soils were connected with NH_4^+. In 1850 Thompson reported on his discovery that when $(NH_4)_2SO_4$ was added to a soil column and leached with water, much of the added NH_4^+ was retained, while $CaSO_4$ appeared in the soil column. Being interested in these results, the English chemist Way started the first systematic studies in ion exchange, the results of which were published in 1850 and 1852.

After publication of these pioneer works, cation exchange became a subject of intense study. However, it took many decades before the reactions could be satisfactorily described in physical-chemical terms. Theories of cation exchange with special respect to soils have developed contemporaneously with advances in soil chemistry and clay mineralogy. Only a brief account can be given herein. For additional details, the reader is referred to the recent reviews of Bolt (1967), Schuffelen (1972) and Thomas (1977).

A. Cation Adsorption and Exchange

According to the prevailing concepts, soil colloids carry electric charges, both positive and negative, with the negative charges being dominant in most soils. In the lattices of clay minerals the negative charges arise from isomorphous substitution of Si and Al for, usually, Al and Mg, respectively, and from dissociation of H^+ from $-OH$ groups bound to Si and/or Al. In soil organic matter, negative charges originate from the dissociation of COOH and phenolic OH groups. Charges arising from isomorphous substitution represent the so-called permanent charges; those arising from dissociation of functional groups are pH-dependent. The bonding between the charged particle and the counter-ions is largely considered to be of the ion-ion type. According to newer findings, even non-coulombic interactions may exist (see, e.g., Jensen, 1973).

The total of negatively charged sites on a specific soil particle represents its ability to hold positively charged ions (CEC). The binding strength by which cations are held or adsorbed at exchange sites, i.e. their replacing power, is described by a number of factors, such as: the valence and the size of the cation, the charge density and structure of the exchanger material, the nature of the associated anion, the relative concentration of the different cations present (the complementary ion principle), and the water content of the system (the dilution effect).

In general, the binding strength of cations is dependent on valence, effective size, and hydration of the ions. Hydration is inversely proportional to the crystalline radii of the ions (Table 1).

The relative replacing power has often been found to increase in the order corresponding to the lyotropic series. For alkali metal ions, this order is: $Li^+ < Na^+ < K^+ \cong NH_4^+ < Rb^+ < Cs^+$. For alkali earth ions, the order is: $Mg^{2+} < Ca^{2+} < Sr^{2+} < Ba^{2+}$. These series can be considered as "normal orders". The position of the monovalent and divalent ions in a combined series, however, has proved to vary with the nature of the exchange material (composition, structure, and charge distribution) and with the concentration and volume of the surrounding solution. For H^+, the position in relation to the other ions varies even more. In hydrated form, the H_3O^+ ion is of

Table 1—Crystal ion radii, average hydration number, and polarizability of some cations. After Scheffer and Schachtschabel (1966).[†]

	Cation										
	Li^+	Na^+	K^+	NH_4^+	Rb^+	Cs^+	Mg^{2+}	Ca^{2+}	Sr^{2+}	Ba^{2+}	H_3O^+
Radius, Å	0.78	0.98	1.33	1.43	1.49	1.65	0.78	1.06	1.27	1.43	1.45
H_2O/ion	3.3	1.6	1.0	0.7	--	0.4	7.0	5.2	4.7	2.0	--
Polarizability, cm³	0.025	0.170	0.80	--	1.42	2.35	0.10	0.54	0.87	--	--

† Ion radii according to Goldschmidt, the hydration number according to Glueckauf, and polarizability according to Pauling.

the same size as unhydrated K^+ and NH_4^+ and should show somewhat identical exchange properties.

Schachtschabel (1940), studying the release of NH_4^+ from a humic acid and various minerals, observed the following sequence for the relative replacing power of cations:

Kaolinite	$Li^+ < Na^+ < H^+ < K^+ < Mg^{2+} < Ca^{2+}$
Montmorillonite	$Li^+ < Na^+ < K^+ < H^+ < Mg^{2+} < Ca^{2+}$
Muscovite	$Li^+ < Na^+ < Mg^{2+} < Ca^{2+} < K^+ < H^+$
Humic acid	$Li^+ < Na^+ < K^+ < Mg^{2+} < Ca^{2+} < H^+$

The soil colloids of high charge density, i.e. a high charge or CEC per unit of surface area, generally show the greatest preference for highly charged cations. Some results of Schachtschabel (1940) may illustrate this preference. Thus, when Schachtschabel equilibrated samples of humic acid, montmorillonite, kaolinite, and muscovite in an acetate solution of $0.05N$ in both Ca^{2+} and NH_4^+, he found that calcium formed 92, 63, 54, and 6% of the exchangeable cations on these four materials, showing a strong preference for Ca^{2+} by the humic acid, and for NH_4^+ by the muscovite. Ammonium and potassium are preferentially held by micas and vermiculites due to the structure of these minerals (discussed in more detail below).

The anion associated with the added cation can affect the extent of cation exchange. As a rule, cation adsorption increases with increasing strength of adsorption of the salt anion (see Dalal, 1975). Exchanging one cation for another in the presence of a third cation becomes easier as the binding strength of the third cation increases (complementary ion effect). Finally, dilution of a soil:water equilibrium system favors retention of the more highly charged cations, e.g., Ca^{2+} in relation to NH_4^+.

The adsorption affinity of univalent cations to soil colloids usually increases with decreasing size of the hydrated ions. This has been explained qualitatively by assuming that the binding forces are essentially electrostatic (coulombic) and that under ordinary conditions the adsorbed ions are hydrated. The cation with the smallest hydrated radius will, therefore, approach the negative site of attachment more closely and will be held more strongly than a cation with a larger radius. In many cases, however, a selective or preferential adsorption of cations occurs, which cannot be explained entirely by hydration theory. The polarizability and the partial dehydration of cations are considered as the main causes for deviations from the "normal order" of exchange affinity (cf. Shainberg and Kemper, 1967).

When the metal ions are close to the negative charge on the clay surface, a displacement of their positive and negative electrical centers occurs due to deformation of the hydration water shell or to partial dehydration of the cation in question. The ease of polarizability is proportional to the size of the unhydrated ion (see Table 1).

B. Cation Exchange Reactions

Several formulations have been proposed in attempts to describe ion-exchange processes and to predict cation distribution. The formulations used most frequently are those of Vanselow (1932), Gapon (1933), and Donnan (see Wiklander, 1964). They are based on the principle of thermo-dynamic equilibrium between ions in solution and in the adsorbed phase, and obey the mass action law. If monovalent and bivalent ions, like NH_4^+ and Ca^{2+}, are designated as M^+ and M^{2+} and the soil particle as X, the exchange reaction on an equivalent basis is:

$$X_2\text{-}M^{2+} + 2\,M^+ \rightleftharpoons 2\,X\text{-}M^+ + M^{2+}.$$

At equilibrium, the reaction equation above can be written as

$$\frac{(M^+)_i^2 \cdot (M^{2+})_o}{(M^{2+})_i \cdot (M^+)_o^2} = K, \text{ or } \frac{(M^+)_i^2}{(M^{2+})_i} = K\,\frac{(M^+)_o^2}{(M^{2+})_o}$$

where i and o denote the adsorbed and the solution phase, respectively, and K the thermodynamic equilibrium constant. The brackets denote activities of the respective ions. In the above equations, (M^+) and (M^{2+}) refer to the activities of the two ion species. In solution, the activity of an ion is usually defined as $c_M \cdot f_M$, i.e., the product of molar concentration and the activity coefficient of the ion in question. The change of the free energy of the system is related to the thermodynamic constant as $\Delta F^\circ = -RT \ln K$. Cationic exchange reactions are approximately stoichiometric, since the amounts exchanged are chemically equivalent.

The thermodynamic treatment of cation exchange reactions can give valuable information about the reaction mechanisms in ideal systems. In soil systems, however, we deal with exchangers of a complex nature in which exchange is not entirely reversible and hysteresis effects can be caused by a slow approach of equilibrium. The general effect of hysteresis is that the cation introduced first is usually less exchangeable than when added last. In many cases, hysteresis is an indication of the tendency of the ion to be adsorbed in difficulty or nonexchangeable forms (see Marshall, 1964).

Summing up this section, it might be stressed that in most arable soils exchangeable NH_4^+ represents only a fractional part of the total inorganic N pool. The cationic nature of NH_4^+ permits retention of NH_4^+-N by soil colloids in exchangeable form, and in this form the ions are effectively protected against leaching by percolating waters. Losses of NH_4^+-N by leaching will be significant only in soils with extremely low CEC. The agronomic significance of exchangeable NH_4^+ lies in the circumstances that it is available to higher plants and holds a key position in the microbial turnover of N in soil.

III. NONEXCHANGEABLE BINDING (FIXATION) OF NH₄⁺ IN SOILS AND CLAY MINERALS

It is well known that many soils have the ability to bind K^+ and NH_4^+ in such a manner that they cannot readily be replaced by other cations. Much of the basic information on NH_4^+ fixation originates from studies on fixation of K^+. In this review, reference to the literature on K^+ fixation is made when desirable to draw parallels between fixation of these two cations. Earlier work in this field is referred to only to get a desirable background and to have continuity in developments of knowledge with time since the pioneer work by Chaminade and Drouineau (1936).

A. Mechanism of NH₄⁺ Fixation

It is now generally accepted that the soil's capacity to fix K^+ and NH_4^+ is related to the presence of clay minerals of the three-layer or 2:1 type. Each unit cell of these minerals consists of an octahedral Al–O–OH sheet "sandwiched" between two tetrahedral Si–O sheets. As a result of isomorphous substitution of Al^{3+} and Si^{4+} by cations of lower valence during crystallization, the lattice obtains an excess negative charge, which will be balanced by other cations, e.g. Ca^{2+}, Mg^{2+}, K^+, Na^+, H_3O^+, either inside the crystal or outside the structural unit (see section IIA). The magnitude of layer charge plays a dominant role in determining the strength of bonding in the basal plane. The greater the layer charge, the stronger the interlayer bond. In minerals of weak interlayer bonds, such as montmorillonites, cations and water may enter the basal plane, causing the basal spacing to increase, and the mineral to expand or swell. Conversely, with minerals of high layer charge, such as micas, the bond energy is so great that polar molecules cannot enter between the basal planes. The minerals are nonexpanding and the interlayer cations are not exchangeable. For vermiculites, forming intermediates in layer charge, the Ca^{2+}, Mg^{2+} and Na^+ ions in interlayer position are exchangeable, but K^+ and NH_4^+ are not exchangeable by ordinary procedures (Fig. 1).

Early investigations revealed that the saturation of vermiculite and degraded illite (under special conditions also montmorillonite) with NH_4^+ and K^+ leads to collapse of the crystal lattice. The K^+ and NH_4^+ are, thereby, "trapped" between the silica sheets, and are largely withdrawn from exchange reactions (Fig. 2). A number of theories have been proposed to explain the specificity of the different cations to contract or expand the mica lattice, and, thus, their readiness to be fixed between the silica sheets. A discussion of these theories will be limited to a short recapitulation of the previous review by Nommik (1965).

According to the "lattice hole" theory of Page and Baver (1940), cation fixation is related to the size of the cation as well as to the kind of fixing material. The generally accepted idea is that the exposed surface (and

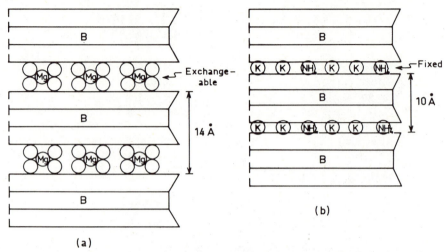

Fig. 1—Schematic picture of the structure of vermiculite (modified after Mitchell, 1977):
(*a*) Expanded vermiculite lattice with Mg^{2+} ions (hydrated) in the interlayer position; and
(*b*) Collapsed lattice of vermiculite after K^+ or NH_4^+ saturation; B = brucite layer.

surfaces between the sheets) of three-layer minerals consists of oxygen ions, arranged hexagonally. The opening within the hexagon is equal to the diameter of an oxygen ion (approx. 2.8 Å). Ions having a diameter of this magnitude (e.g., NH_4^+ and K^+) will fit snugly into the lattice holes and such ions will be held very tightly as they come closer to the negative electrical charges within the crystal. Owing to this, the layers are allowed to approach and be bound together, thus preventing rehydration and re-expansion of the lattice. Cations larger than 2.8 Å cannot enter the cavities and will, therefore, remain more loosely held between the layers and be more accessible for exchange with other cations.

In accordance with the above hypothesis it has been found that the basal spacing of 2:1 lattice clay minerals is controlled by the type of cation occupying the interlayer exchange positions. When Ba^{2+}, Ca^{2+}, Mg^{2+}, Na^+, Li^+, or H^+ is the interlayer cation, the crystal lattice is, at ordinary temperature, in an expanded and hydrated state. With K^+, Rb^+, or Cs^+ in interlayer positions, the crystal lattice becomes contracted and nonexpandable, even when immersed in water. Cations in an expanded crystal lattice are readily replaced by cations which leave the lattice in an expanded state, but only with difficulty by cations which contract the lattice. Likewise, the cations present in a contracted crystal lattice (e.g., NH_4^+, K^+) may be slowly replaced by cations which bring about an expansion of the lattice, but not by cations which contract it. Thus, the interlayer NH_4^+ in a contracted lattice of vermiculite, illite, or montmorillonite can be slowly replaced by such cations as Ca^{2+}, Mg^{2+}, and Na^+, but hardly by K^+. It is evident from this that the concept of cation fixation is somewhat arbitrary, the magnitude of fixation being dependent on the character of the replacing cation. The definition of cation fixation must necessarily include information as to the method used to replace the cation in question.

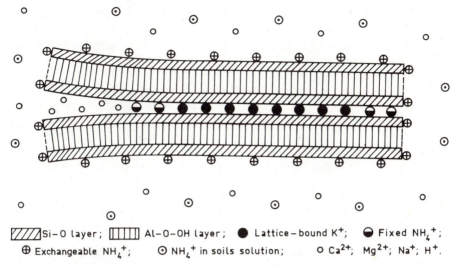

Si–O layer; ▨ Al–O–OH layer; ● Lattice–bound K^+; ◐ Fixed NH_4^+; ⊕ Exchangeable NH_4^+; ⊙ NH_4^+ in soils solution; ○ Ca^{2+}; Mg^{2+}; Na^+; H^+.

Fig. 2—Schematic picture showing the different forms of NH_4^+ on illite (modified after Wiklander, 1958, and Schachtschabel, 1961).

The differential behaviour of 2:1 type clay minerals depends on whether the main part of the negative charge on the lattice originates from the octahedral Al-layer or from the tetrahedral Si-layer. The force of attraction between the positively charged interlayer cations and the negative charges in the lattice will be greatest when the negative charge results from an isomorphous substitution of Al for Si in the tetrahedral layers, giving a shorter distance between the interlayer cations and the negative sites of the lattice. Substitution in tetrahedral Si-sheets accounts for 80 to 90% of the substitution in vermiculites, about 65% or more in illites, and <20% in montmorillonites (see Nommik, 1965).

As regards the different three-layer clay minerals, the vermiculites have the greatest capacity to fix K^+ and NH_4^+. Illite may or may not fix, depending on the degree of weathering and K^+ saturation of the lattice. Montmorillonites do not fix NH_4^+ under moist conditions. Kaolinites are generally considered as nonfixing minerals.

Earlier work in the exchange reactions of K^+ in illites and illitic soils has yielded information indicating that the exchange and displacement behavior of K^+ is due to the existence of several types of exchange positions, each with its own specific exchange constant. On the basis of this information the suggestion has been made that there exists a kind of dynamic equilibrium between K^+ in the soil solution (K_s), exchangeable K^+ (K_e), and fixed K^+ (K_f). A similar equilibrium is likely to occur between similar forms of NH_4^+. In view of the generally overwhelming predominance of K^+ vs. NH_4^+ in soil systems and the similarity of these ions with respect to fixation reactions, Nommik (1957) suggested that the sum of NH_4^+ and K^+ in the respective categories, rather than NH_4^+ alone, should be introduced into the equilibrium equation

$$(NH_4^+ + K^+)_s \rightleftharpoons (NH_4^+ + K^+)_e \rightleftharpoons (NH_4^+ + K^+)_f.$$

With the aid of tracer techniques, Wiklander (1950) and Nommik (1957) showed that only a fractional part of the fixed K^+ or fixed NH_4^+ participates in a real equilibrium with exchangeable ions, indicating that the last stage of the reaction is strongly hysteric. This is in agreement with the finding of Newman and Oliver (1966), who studied the exchange between unlabelled NH_4^+ in hydrobiotite and labelled NH_4^+ in solution.

Studying K^+ in Danish soils, Lamm and Nadafy (1973) presented a somewhat more detailed scheme for the different forms of K^+ in soil in equilibrium with each other. Adapted to NH_4^+, the scheme is as follows:

$$\text{soluble } NH_4^+ \underset{\text{fast}}{\rightleftharpoons} \text{exchangeable } NH_4^+ \underset{\text{slow}}{\rightleftharpoons} \text{intermediate } NH_4^+ \underset{\text{very slow}}{\rightleftharpoons} \text{fixed } NH_4^+.$$

The intermediate K^+ or NH_4^+ ions in the above equilibrium scheme may be considered to occupy interlayer sites of the clay plates which are at a medium stage of collapse, occurring in the transitional zone where the fixed and exchangeable ions meet. The intermediate-NH_4^+ may be considered to be exchangeable by H^+ and K^+. Exchange by other cations (e.g., Ca^{2+} and Na^+) occurs after a gradual opening of the clay lattice, the reaction rate being slow, however.

It may be noted that the presence of interlayer K^+ in clay minerals is always associated with that of fixed NH_4^+ (see Chapter 2). Martin et al. (1970), working with Australian soil profiles, noted that the NH_4^+/K^+ ratio had not changed during the weathering sequence of mica-illite-vermiculite. As concerns the quantitative occurrence of native fixed NH_4^+ in soils and clay minerals, reference should be made to Chapter 2.

B. Methods Used for Studying NH_4^+ Fixation

In studying the NH_4^+-fixing properties of soils and minerals, it is desirable from several points of view to find a procedure which enables a sharp separation of the readily exchangeable NH_4^+ and the NH_4^+ in inter-lattice positions. The cation used to remove the water-soluble and the easily exchangeable NH_4^+ should, if possible, not replace NH_4^+ ions from inter-lattice positions, even on prolonged extraction and leaching. Otherwise, fixation values will be dependent on time and intensity of extraction and will, therefore, not be reproducible. Barshad (1951) proposed that the fixed NH_4^+ should be defined as NH_4^+ which is not replaceable by prolonged extraction and leaching of soil with a K^+-salt solution, and inversely, that fixed K^+ should be defined as the K^+ which is not replaced by NH_4^+. The term "fixed NH_4^+" or "nonexchangeable NH_4^+" will be used this way in the present paper.

In studying the ability of soils to fix NH_4^+, two principally different methods can be used. The first involves treatment of the soil with a known amount of dilute NH_4^+-salt solution and, after prolonged contact, the easily exchangeable portion is removed using a KCl-leaching technique. The difference between the NH_4^+ added and that recovered in the extract represents the NH_4^+ fixed. This method may be used for describing the soils' ability to fix added fertilizer NH_4^+ under cropping conditions.

The second method, which is generally used for estimation of the capacity of the soil to fix NH_4^+, involves a treatment of the soil with excessive amounts of relatively concentrated NH_4^+ solution in order to achieve complete saturation of the fixing minerals with NH_4^+. The fixed NH_4^+ is estimated by one of the following procedures.

1. KJELDAHL-N PROCEDURE

The soil sample, previously saturated with NH_4^+, is leached extensively with a N KCl solution and then analyzed for total N by the Kjeldahl procedure. The increase in N content over the untreated soil (also leached with N KCl), represents the quantity of NH_4^+ fixed (Allison et al., 1951). For soils with a high total N content and/or a low NH_4^+-fixing capacity the method is rather insensitive.

Stewart and Porter (1963), and later Meints and Peterson (1972), demonstrated that the customary Kjeldahl method did not include all of the fixed NH_4^+. Pretreatment of the soil with an HF-acid mixture resulted in increased recovery figures, a finding confirmed by Bremner (1965). The portion of N missed is often negligible compared with the total N content of the soil (Bremner, 1965).

2. ¹⁵N TECHNIQUES

In principle, this is the same procedure as described above, with the difference that the soil is treated with an NH_4^+ salt enriched on N isotope ^{15}N. After KCl leaching, the soil is subjected to determination of both the Kjeldahl N and the isotopic composition of this N fraction. From the data obtained the amount of NH_4^+ fixed is calculated. The method is unquestionably sensitive and gives fully reproducible results (Nommik, 1957).

3. THE ALKALINE DUPLICATE-DISTILLATION PROCEDURE

In this procedure, proposed by Barshad (1951), the fixed NH_4^+ and the NH_4^+-fixing capacity of the soil are estimated from the difference between the NH_4^+ released by distilling the NH_4^+-treated soil (excessive amounts of NH_4^+ salt removed by leaching with alcohol) with NaOH and KOH, respectively. It is assumed that distillation with NaOH removes both exchangeable and interlattice NH_4^+, while distillation with KOH will remove exchangeable NH_4^+ only.

The Barshad procedure has been tested in a number of investigations, showing in some cases an incomplete recovery of fixed NH_4^+. This may be due to: (i) the blocking effect of K^+, either released from the fixing material or present in the added reagents as an impurity; and (ii) the fixation of NH_3 by soil organic matter during alkaline distillation. An additional shortcoming is that the amount of NH_3 released from organic combination is dependent on the rate and duration of distillation and that a part of the native fixed NH_4^+ in the soil may be released on NaOH treatment (see Bremner, 1965; Nommik, 1965).

4. HYDROFLUORIC ACID PROCEDURE

In this method, the NH_4^+-saturated and subsequently KCl-leached soil is treated with an acid mixture containing HF to disintegrate clay minerals and liberate NH_4^+ from interlayer positions. The difference between NH_4^+-N released from the NH_4^+-treated and control samples corresponds to the soils' content of fixed NH_4^+, excluding native fixed NH_4^+.

Bremner (1965) and Silva and Bremner (1966) recommended pretreatment of the soil with alkaline potassium hypobromide (KOBr–KOH) to remove exchangeable NH_4^+ and organic N compounds, which on subsequent treatment with HF may yield NH_4^+. Bremner et al. (1967) considered this method to be especially reliable for the determination of native fixed NH_4^+. Osborne (1976) reported that pretreatment of soil with KOH (Dhariwal & Stevenson, 1958) or KOBr–KOH (Silva & Bremner, 1966) give similar values for native fixed NH_4^+.

Freney (1964) raised the possibility that native fixed NH_4^+ may be a laboratory artifact resulting from the release of labile organic N compounds entrapped in the silicate lattice by the action of HF–HCl. As suggested by Bremner et al. (1967), this is a remote possibility, but it would be difficult to prove otherwise. The high correlation between amounts of native fixed NH_4^+ and total K^+, noted by numerous workers, unquestionably indicates a structural association.

C. Factors Affecting Rate and Magnitude of NH_4^+ Fixation

Several factors other than mineralogical composition are of significance for the capacity of the soils to fix NH_4^+. A brief discussion follows.

1. RATE OF FIXATION

Fixation rate, being controlled mainly by ion diffusion, is highest in periods immediately after NH_4^+ addition and slows down as the equilibrium point is approached (see Nommik, 1965; Dissing-Nielsen, 1971; Sippola et al., 1973). From 60 to 90% of the total fixation may occur within the first few hours (Fig. 3).

2. CONCENTRATION AND VOLUME OF THE NH_4^+ SALT SOLUTION

The amount of NH_4^+ fixed has been generally found to increase with an increase in the amounts of NH_4^+ added (see Nommik, 1965; Black and Waring, 1972; Opuvaribo & Odu, 1974; Sippola et al., 1973; Sowden et al., 1978). Percentage fixation, on the other hand, generally decreases with an increase in the amount of NH_4^+ added (Table 2). Jansson (1958) demonstrated that the amount of NH_4^+ fixed was appreciably decreased by an in-

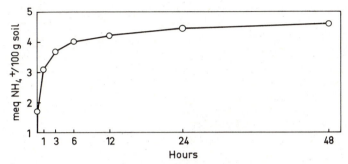

Fig. 3—Influence on fixation of time of reaction between NH_4^+-N and soil; 7.4 meq NH_4^+ added per 100 g soil (Harada and Kutsuna, 1954).

crease in the water/soil ratio used in the fixation tests. This is in variance with the observations by Sowden et al. (1978) showing that water volume had little effect on fixation.

3. TEMPERATURE

In short-term fixation tests, Harada and Kutsuna (1954) and Nommik (1957) demonstrated that within the temperature interval of 0–60°C fixation rate increased with increasing temperature. It has not been shown, however, whether the soils' capacity to fix is affected by temperature.

4. DRYING VS. ALTERNATE WETTING AND DRYING

Drying the soil after the addition of an NH_4^+ salt solution (either in air or at elevated temperatures) has been shown to increase the extent of fixation (Black & Waring, 1972; Blasco & Cornfield, 1966; Opuwaribo & Odu, 1974). Two different reactions may be responsible. First, drying removes water and consequently leads to an increase in the concentration of NH_4^+ in the soil solution. Second, as concerns montmorillonite and montmorillonite-containing soils, drying is a condition necessary for NH_4^+ fixation. In the latter case, drying results in dehydration of interlayer surfaces, thereby making partial contraction of the lattice possible.

Table 2—Relation between the amounts of NH_4^+ added and NH_4^+ fixed in a vermiculite-containing surface soil from central Sweden (Nommik, 1957).†

NH_4^+ added, meq/100 g soil	NH_4^+ fixed, meq/100 g soil	Fixed NH_4^+, % of added NH_4^+
1	0.83	83
2	1.39	70
5	2.17	43
10	3.18	32
20	3.85	19
40	4.48	11

† Soil/water ratio 1:1.

The effect of drying can be illustrated by experimental results reported by Allison et al. (1951), in which a Harpster clay loam (predominating clay mineral being montmorillonite) was able to fix 1.1, 3.5, and 6.2 meq NH_4^+/ 100 g soil under moist, air-dried, and oven-dried (100°C) conditions, respectively. In an investigation of a series of arable montmorillonitic soils from Queensland, Black and Waring (1972) found that fixation of NH_4^+ increased 3- to 10-fold following air-drying.

Alternate drying and wetting may be especially effective in increasing fixation. Thus, Blasco and Cornfield (1966) reported that the average fixation of NH_4^+ by montmorillonite clays was in the ratio 1:12:16 with air-drying, oven-drying (100°C), and five cycles of alternative wetting and oven-drying, respectively. Jansson (1958) showed that in a sandy loam soil from southern Sweden (probably illitic), which under moist conditions fixed approximately 50% of the 100 ppm NH_4^+-N added, fixation was increased to 63% after a single drying of the moistened soil at 30°C, and to 77% after three wettings and dryings. Similar trends in fixation values have been obtained by other workers (see review of Nommik, 1965).

5. FREEZING AND THAWING

Since freezing, like drying, removes water from the system, it may influence the amount of NH_4^+ fixed. Walsh and Murdock (1960, 1963) showed that a treatment for 5 days at −15°C increased fixation in a number of forest soils from 0.08 to 0.14 meq/100 g on the average. According to Nommik (1965), fixation capacity in a vermiculite-containing, NH_4^+-treated clay soil was increased from 3.08 to 3.42 meq/100 g soil by freezing the moist soil for 24 hours at −17°C.

6. PARTICLE SIZE

Barshad (1954a) demonstrated that a decrease in the particle size of vermiculite, either by grinding or exfoliation, led to a decrease in NH_4^+ fixation, as measured by the KCl-leaching procedure. According to McDonnell et al. (1959), nearly all the fixed NH_4^+ in an NH_4^+-saturated vermiculite, amounting to 69 meq/100 g, was made accessible to K^+ replacement by prolonged ball milling. Barshad (1954a) suggested that the increased accessibility accompanying a decrease in particle size is the result of conversion of the internal surfaces of a particle to external ones and of an increase in edge surfaces and fractures through which the replacing and the adsorbed cations can enter or leave. In another work with vermiculite, Bredell and Coleman (1968) reported that silt-size particles had a greater ability to adsorb NH_4^+ and retain it against NaCl displacement than clay-size particles.

In several soil regions, fixation has been shown to be associated with the clay fraction. Nommik (1957) found that the NH_4^+ fixation capacity of a number of Swedish surface soils tended to increase with increasing clay content. The correlation was not close, however, as even clayey silt and fine sand soils often showed considerable fixing capacity. A similar finding was reported by Jansson and Ericsson (1961) for some Scanian and by Kaila (1962, 1966) and Smirnov and Fruktova (1963) for some Finnish and Russian soils, respectively (see also Rich & Lutz, 1965; Schiller & Walicord, 1969; Opuwaribo & Odu, 1974).

7. OTHER CATIONS INCLUDING K⁺

It is evident from the preceding description that besides NH_4^+, even K^+, Rb^+, and Cs^+ may be subject to fixation in 2:1 layer clay minerals. If present, they will compete with NH_4^+ for fixation positions.

The depressive effect of K^+ on fixation of NH_4^+ has been subject to a number of investigations. It has been found that the effect may vary, depending on whether K^+ is added simultaneously, prior to, or after the addition of NH_4^+. In accordance with the discussion above, the addition of K^+ prior to NH_4^+ will depress fixation of the NH_4^+ (see Fig. 4). An addition of K^+ after NH_4^+ will, on the other hand, not appreciably influence the amount of NH_4^+ fixed. Results obtained by Jansson (1958) illustrate this effect (Table 3).

Table 3—Influence of the time of K^+ addition, in relation to the addition of NH_4^+, on the fixation of NH_4^+ (Jansson, 1958).

Addition, meq/100 g soil	% of added NH_4^+ fixed
0.714 meq NH_4^+	42.7
0.714 meq NH_4^+ + 0.714 meq K^+ (simultaneously)	50.0
0.714 meq NH_4^+ + 0.714 meq K^+ (added 2 h after the addition of NH_4^+)	50.4
0.714 meq K^+ + 0.714 meq NH_4^+ (K added 2 h before the addition of NH_4^+)	28.3

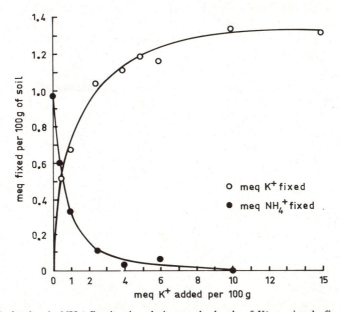

Fig. 4—Reduction in NH_4^+ fixation in relation to the levels of K^+ previously fixed; 10 meq NH_4^+ added per 100 g soil. Fixation period for each cation, 30 min (Stanford and Pierre, 1947).

When NH_4^+ and K^+ are added simultaneously, variable results can be obtained, depending, inter alia, on the type of fixing mineral, the concentration level, and the ratio of NH_4^+ to K^+. Joffe and Levine (1947) found that when equivalent amounts of K^+ and NH_4^+ were added simultaneously to an H-saturated montmorillonite, more K^+ than NH_4^+ was fixed at lower concentrations, the relation being reversed at higher concentrations.

In a study of the effect of increasing amounts of simultaneously added K^+ on fixation of NH_4^+ in a vermiculitic clay soil, Nommik (1957) found that when added in equivalent amounts (1 meq/100 g soil), the soil fixed NH_4^+ and K^+ in the proportion 3.4:1. This is in accordance with the finding of Dissing-Nielsen (1971) and Sippola et al. (1973) for some surface soils of Denmark and Finland, respectively. More recently, Sippola (1976) demonstrated that the ratio NH_4^+/K^+ fixed was lower for surface soils than subsoils. According to Nommik (1965), the preferential fixation of NH_4^+ was characteristic not only of relatively low concentration levels, but also of rather high NH_4^+ and K^+ addition rates (Table 4). One interesting and somewhat unexpected consequence of the preferential fixation of NH_4^+, reported by Nommik (1957) was that within a narrow concentration interval the simultaneous addition of K^+ led to a small but significant increase in the NH_4^+ fixation (Fig. 5).

According to present evidence there generally is a close relationship between the ability of soils to fix NH_4^+ and K^+. Exceptions from this rule have been reported, however. Nommik (1957), thus, found that a vermiculite-containing clay soil fixed 10–15% more NH_4^+ than K^+, and that this was valid for a fairly wide concentration range. According to Wiklander and Andersson (1959), the surface horizon of a soil profile studied fixed about equal amounts of NH_4^+ and K^+; in the subsoil, on the other hand, the NH_4^+ was fixed in considerably greater amounts than K^+. Sippola et al. (1973) found that more K^+ than NH_4^+ was fixed in some Finnish surface soils when the respective ions were added separately.

Numerous investigations have demonstrated that cations which do not contract the crystal lattice may influence the capacity of the minerals to fix NH_4^+. This may be due to competition between NH_4^+ and other cations for exchange positions, resulting in displacement of the equilibrium between exchangeable and fixed NH_4^+ (see III A). Accordingly, addition of a cation of

Table 4—Fixation of NH_4^+ and K^+ in a vermiculite-containing clay, when added singly and when added simultaneously as mixed solutions (Nommik, 1965).

Treatments, per 100 g soil	NH_4^+ fixed, meq	K^+ fixed, meq	$\dfrac{NH_4^+ \text{ fixed}}{K^+ \text{ fixed}}$
1.0 meq NH_4^+	0.80		
1.0 meq K^+		0.69	
1.0 meq NH_4^+ + 1.0 meq K^+	0.89	0.30	3.0
2.5 meq NH_4^+	1.61		
2.5 meq K^+		1.37	
2.5 meq NH_4^+ + 2.5 meq K^+	1.60	0.49	3.3
10.0 meq NH_4^+	3.36		
10.0 meq K^+		3.13	
10.0 meq NH_4^+ + 10.0 meq K^+	2.78	0.98	2.8

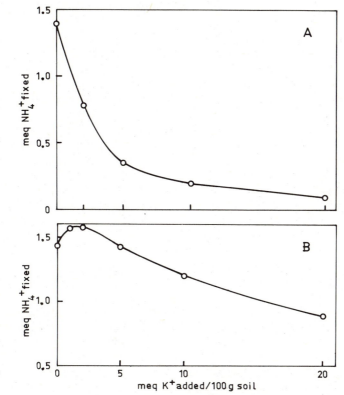

Fig. 5—Effect of K⁺ on fixation of NH₄⁺. 2 meq NH₄⁺ added per 100 g soil: (*A*) K⁺ added 2 days prior to the addition of NH₄⁺; and (*B*) K⁺ and NH₄⁺ added simultaneously in form of mixed solution (Nommik, 1957).

high replacing power will lead to a decrease in exchangeable NH_4^+ and, consequently, to lower fixation of added NH_4^+ or, alternatively, to release of some of the already fixed NH_4^+. The greatest depressive effect on NH_4^+ fixation is exerted by trivalent cations and by H^+ (see Table 5). As concerns Al^{3+}, Rich and Obenshain (1955) demonstrated that the effect was due to the restricted collapse of the mineral lattice on K^+ and NH_4^+ saturation.

Table 5—Fixation of NH₄⁺ by a residual soil from Japan saturated with different cations under moist conditions (Harada & Kutsuna, 1954).

Treatment†	NH₄⁺ fixed	
	meq	% of added NH₄⁺
Untreated	11.9	39.9
Na⁺-saturated	11.2	37.4
K⁺-saturated	0.7	2.3
Ca²⁺-saturated	12.2	40.7
Ba²⁺-saturated	9.7	32.3
Al³⁺-saturated	0.5	1.7
H⁺-saturated	1.7	5.6

† 30 meq NH₄⁺ added per 100 g soil.

8. pH AND DEGREE OF BASE SATURATION

The significance of relative base saturation of soils and clay minerals on their capacity to fix NH_4^+ and K^+ has been the subject of several investigations. Stanford (1948) reported that addition of NaOH or $Ca(OH)_2$, and even of phosphate and fluoride, increased the moist fixation of K^+ by illite, and described this to their ability to remove H^+, Fe^{3+}, and Al^{3+} ions from the interlayer positions. Wiklander and Andersson (1959) demonstrated that the ability of soils to fix NH_4^+ and K^+ was measurably increased by liming (Table 6). Raju and Mukhopadhyay (1975, 1976), working with Indian soils, showed that NH_4^+ fixation increased with increasing pH and base saturation.

Statistical analysis of fixation figures for over 200 surface clay soils from Sweden showed that there was a tendency for NH_4^+ fixation to increase with increasing pH of the soils (Nommik, 1957). Soils with pH values lower than 5.5 generally showed very low fixation. A weak positive correlation between pH and the soil's capacity to fix NH_4^+ was also found by Kaila (1962) for Finnish soils. These observations are in accordance with the findings showing that on H^+ saturation the fixation capacity of the soils is reduced to a minimum (see Nommik, 1965).

The marked effect of H^+ treatment on the fixation of NH_4^+ and K^+ is not fully understood. The high relative replacing power of H^+ is presumably an important factor influencing the equilibrium between $(NH_4^+)_e$ and $(NH_4^+)_f$. Barshad (1954b) considered, however, that the decrease of NH_4^+ fixation in the presence of exchangeable H^+ may be due to the somewhat expanded state in which the adsorbed H^+ leaves the crystal lattice and which, therefore, renders the interlattice NH_4^+ more accessible to a replacing cation.

9. EFFECT OF ASSOCIATED ANIONS

The anion associated with NH_4^+ has been found to influence the capacity for NH_4^+ fixation. This may be ascribed in some cases to the effect of the anion on soil pH. Another possible effect of an associated anion is, according to the Paneth-Fajans-Hahn rule, that an anion which is strongly adsorbed may increase the adsorption of the cation. In this respect the effect of phosphate ions has been given the greatest attention[1] (Leggett, 1958).

---

[1] G. E. Leggett. 1958. Ammonium fixation in soils and minerals. Thesis. Washington State College, Dep. of Agronomy. Pullman, Wash.

Table 6—Effect of liming on the capacity of the surface soil to fix NH_4^+ and K^+. The analyses of soil performed 22 years after the application of lime (Wiklander and Andersson, 1959).†

Lime added, kg CaO/ha	pH_{H_2O}	NH_4^+ fixed, meq	K^+ fixed, meq
0	6.0	1.28	1.31
3,000	6.1	1.24	1.47
6,000	6.4	1.45	1.60
12,000	7.0	1.46	1.69

† 4 meq NH_4^+ and K^+, respectively, added per 100 g soil.

10. EFFECT OF ORGANIC MATTER

In soils the amounts of native fixed NH_4^+ and the NH_4^+-fixing capacity may vary with depth, the fixing capacity usually being higher in subsoil than in surface soil. Reasons for this phenomenon are not known and information on the subject is meager.

Hinman (1966), working with Canadian soils, found that destruction of organic matter with H_2O_2 resulted in an increase in NH_4^+ fixation. The higher the organic matter content the greater was the increase. The author suggested that organic matter either blocked the entry of NH_4^+ to fixing sites or prevented collapse of the basal spacing of the minerals.

According to Porter and Stewart (1970), pretreatment of subsoils or vermiculite with organic compounds containing NH_2 groups caused considerable reduction in the NH_4^+ fixation capacity. Glycerine treatment reduced the amounts of NH_4^+ fixed by as much as 78%. In contrast, organic acids interfered only slightly with NH_4^+ fixation.

11. INFLUENCE OF FIXATION OF NH₄⁺ ON THE CEC OF THE FIXING MATERIAL

It has been observed that, in some soils, the sum of base cations replaceable with NH_4^+ may exceed the CEC as determined by the conventional NH_4^+-saturation procedure (see Nommik, 1965). An acceptable explanation of this discrepancy was given by Bower (1950), who demonstrated that, in some saline and alkaline California soils, the observed excess of exchangeable Na^+ over NH_4^+ maximally sorbed depended on the inability of other cations to replace all of the NH_4^+ in exchange positions. In a subsoil sample Bower found that its content of exchangeable Na^+ was more than 30% greater than its CEC, determined by NH_4^+ saturation.

For Dutch soils, characterized by a high K^+-fixing capacity, Marel (1954) found that the ratio between total exchangeable cations and CEC was, in some cases, as high as 1.30. The same author also noticed that there was a definite correlation between the above ratio and the capacity of the soil to fix K^+. Several workers have demonstrated that the fixation of K^+ by soil is accompanied by a decrease in CEC (see Nommik, 1965, Page et al., 1967).

D. Release of Fixed NH₄⁺ from Soils and Minerals by Different Extraction and Distillation Procedures

As previously described, evidence has accumulated indicating that some form of dynamic equilibrium exists between fixed and exchangeable NH_4^+. Accordingly, NH_4^+ fixation occurs when the concentration of exchangeable NH_4^+ exceeds the equilibrium value for the soil in question and, vice versa, fixed NH_4^+ is released when the exchangeable NH_4^+ drops below this equilibrium value. Consequently, the fixed NH_4^+ may be gradually liberated from soils and clay minerals if the NH_4^+ released is effectively removed from the system, and if no other influencing ions or materials are present.

With regard to the effect of clay mineral type on release of fixed NH_4^+, information is meager and inconsistent. As concerns K^+, it has been reported, inter alia, by Mortland et al. (1957) that fixed K^+ is released comparatively easily from montmorillonites and vermiculites, but not from illites.

The following section deals with the effectiveness of different cations, and different extraction and distillation procedures in releasing fixed NH_4^+. Special attention will be given to the effect of K^+ on replacement.

1. LEACHING WITH DIFFERENT SALT AND ACID SOLUTIONS

In determining NH_4^+ fixation of soils by the Kjeldahl-N procedure Allison et al. (1953b) and Allison and Roller (1955a) found that results can differ greatly, depending on whether exchangeable NH_4^+ is removed by leaching with $1N$ KCl or $CaCl_2$. Leaching with KCl results in consistently higher fixation values than leaching with $CaCl_2$. On considering the effecttiveness of the two cations in expanding the NH_4^+-saturated lattice of the minerals, the authors suggested that $CaCl_2$, unlike KCl, will remove not only the soluble and exchangeable fractions of the soil NH_4^+, but also significant amounts of the fixed NH_4^+.

The release of NH_4^+ from NH_4^+-saturated Montana vermiculite and Wyoming bentonite by treating the samples with NaCl solutions was studied by Hanway et al. (1957). The amount of NH_4^+ extracted from vermiculite increased as the amount of NaCl in the extraction solution was increased (Table 7). At the highest NaCl/mineral ratio as much as 98% of the fixed NH_4^+ was replaced by treatment with the boiling NaCl solution. It was also found that addition of a small amount of K^+ exerted a blocking effect on the release of fixed NH_4^+. At a sufficiently high concentration of K^+ the release of NH_4^+ was entirely blocked. A similar blocking effect has been reported by Leggett and Moodie (1963) and Bredell and Coleman (1968).

According to Scott et al. (1958), more fixed NH_4^+ was released from vermiculite when the NH_4^+ was continuously removed from the extracting solution, either by using alkaline solutions or the successive extraction technique.

The interfering effect of K^+ on the release of fixed NH_4^+ from soils by leaching with different neutral salt solutions has also been studied by Nommik (1957) and Jansson (1958). By using the ^{15}N technique, these authors were able to confirm the results presented above, though it appeared that, in general, successive leaching was less effective in replacing fixed NH_4^+ from soils than from clay minerals.

It is interesting to note that the blocking effect of K^+ on the release of fixed NH_4^+ seems to be of a mutual character. Nommik (1957), thus, demonstrated that the release of fixed and native K^+ from soils by leaching with neutral salt solutions was markedly inhibited by the presence of NH_4^+. Barshad (1954a) and Peech (1948) had earlier shown that NH_4^+ was the least effective cation in replacing native K^+ from soils and minerals.

Table 7—NH₄⁺ extracted from NH₄⁺-saturated Montana vermiculite by boiling for 1 hour in NaCl solutions (Hanway et al., 1957).†

Normality of NaCl	Volume of NaCl solution, ml	NH₄⁺ extracted, meq/100 g
0.1	100	14.4
1	100	49.0
1	200	60.8
2	100	61.1
5	100	73.0
5	250	80.5

† One gram of vermiculite submitted to analyses.

2. ALKALINE DISTILLATION PROCEDURE

Alkaline distillation is a more effective method of releasing fixed NH_4^+ from soils than is leaching. This is due to the fact that by distillation the NH_4^+ released from internal surfaces is continuously removed, leaving the exchangeable NH_4^+ at a low level.

Employing the duplicate distillation procedure of Barshad (1951), Hanway and Scott (1956), and Nommik (1957) demonstrated that the method was as satisfactory as the Kjeldahl procedure. However, the presence of small amounts of K^+ interfered with the replacement of fixed NH_4^+ by NaOH distillation. This question was later studied by Hanway et al. (1957) and Leggett (1958). According to Leggett (1958), approximately 10 times as great a K^+/Na^+ ratio was required to block the release of NH_4^+ from vermiculite as from soils. Nommik (1965) tested the effectiveness of different methods of recovering fixed NH_4^+ from a $^{15}NH_4^+$-saturated vermiculitic clay, and found that the blocking effect of K^+ could be considerably reduced by the addition of large quantities of NaCl to the NaOH distilling solution (Fig. 6).

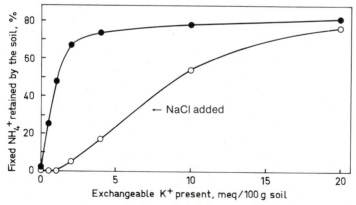

Fig. 6—The percentage of fixed $^{15}NH_4^+$ retained by a vermiculite-containing clay soil after treatment with boiling 0.5N NaOH solution as affected by levels of exchangeable K^+ and by addition of NaCl (10 moles/100 g soil). The soil content of fixed NH_4^+, 4.2 meq/100 g (Nommik, 1965).

Table 8—Amounts of NH_4^+ released from NH_4^+-saturated clay minerals by various treatments, meq per 100 g (Scott et al., 1960a).

Treatment†	NH_4^+ vermiculite	NH_4^+ bentonite	NH_4^+ illite
Kjeldahl digestion	81.4	50.4	25.3
H⁺ resin	80.5	36.7	23.9
Na⁺ resin	80.5	36.3	23.2
NaCl, 1N solution	66.5	33.0	22.4
K⁺ resin	10.7	9.0	22.7
KCl, 1N solution	8.1	5.7	22.1
KOH distillation	2.8	24.6	22.6

† Mineral/resin ratio 1:20. Mineral/salt solution ratio varied between 1:40 and 1:100. The NH_4^+ not liberated by distillation with KOH was considered as fixed.

3. TREATMENT WITH CATION-EXCHANGE RESINS

If a clay mineral or a soil containing fixed NH_4^+ is placed in contact with a cation-exchange resin, this may result in a decrease of the water soluble NH_4^+ as well as NH_4^+ adsorbed on external surfaces of soil particles. This displacement of the equilibrium would reasonably lead to a release of some of the fixed NH_4^+. A procedure based on this principle is frequently used for determining the K⁺-supplying power of soils. The question of the effect of cation-exchange resin on the release of fixed NH_4^+ was studied by Scott et al. (1960a), who found that, with vermiculite, both the H⁺- and Na⁺-saturated resins were effective in this respect (Table 8). As expected, the K⁺ resin was ineffective. The effectiveness of the resin treatment in the release of fixed NH_4^+ was considerably higher for vermiculite and montmorillonite than for illite.

Nommik (1965), working with a heavy clay soil containing 4.91 meq of fixed $^{15}NH_4^+$, demonstrated that the Na⁺ and H⁺ saturated resin, when mixed with the soil in the proportion of 10:1, released 9.2 and 38.1% of the fixed NH_4^+, respectively.

4. TREATMENT WITH SODIUM TETRAPHENYLBORON

Analogous to treatment with cation-exchange resins, addition of a reagent which will reduce the concentration of exchangeable K⁺ and NH_4^+ by precipitation is expected to release some of these cations from fixed positions. Sodium tetraphenylboron (NaBPh₄) has been suggested as a precipitant for K⁺ and NH_4^+ by Gloss (1953). Scott et al. (1960b) used this reagent to release nonexchangeable K⁺ from soils and clay minerals. Essentially all of the K⁺ in vermiculite, 12% of the K⁺ in muscovite, and as much as 47% of the K⁺ in illite was extracted. According to Scott and Reed (1962), the NaBPh₄ treatment of a sample of Grundite-illite resulted in an increase of its capacity to fix NH_4^+ by ca. 35 meq/100 g. Marques and Vidas (1974) reported that NaBPh₄ treatment was effective for removal of fixed NH_4^+ in a number of Mosambique soils.

E. Availability of Fixed NH$_4$$^+$ to Soil Microorganisms

The significance of the fixation reaction as concerns mineral N cycling and biological availability of fertilizer N in arable soils has been subject to a large number of investigations. A brief analysis of the results are given as follows.

In soil, two main microbial processes are responsible for the consumption and removal of the NH$_4$$^+$-N: (i) nitrification, and (ii) assimilation by heterotrophic microflora. As these processes are fundamentally different their relationships to the release of fixed NH$_4$$^+$ will be treated separately.

1. AVAILABILITY OF FIXED NH$_4$$^+$ TO NITRIFIERS

As pointed out previously, fixation and defixation are to be considered as two opposed reactions in a reaction system striving toward equilibrium. Thus, fixed NH$_4$$^+$ will be released from soil when the concentration of water-soluble and exchangeable NH$_4$$^+$ falls below a certain value characteristic of the system. From these considerations it seems likely that nitrification, which involves a consumption of water-soluble and exchangeable NH$_4$$^+$ in the soil, may bring about a release of NH$_4$$^+$ from fixed positions. Several investigators have shown that this is the case.

The first thorough investigations on the availability of fixed NH$_4$$^+$ to nitrifying organisms are those of Bower (1951) and Allison and co-workers (Allison et al., 1951, 1953a, b, c; Allison and Roller, 1955b). The incubations carried out by Bower revealed that 13–28% of the non-exchangeable NH$_4$$^+$ was nitrified over a 4-day period. Prolonged incubation did not result in any further release of fixed NH$_4$$^+$.

Results of other workers are in good agreement with those of Bower. Allison and co-workers found that not more than 15% of the fixed NH$_4$$^+$ in the soils they tested was nitrified by incubation for 4 weeks. Availability of fixed NH$_4$$^+$ was lowest in vermiculite-containing soils and highest in those containing montmorillonite as the fixing mineral (Allison et al., 1953a). Kowalenko and Cameron (1976) and Sowden (1976), working with Eastern Canadian soils, showed that the fixed NH$_4$$^+$ was nitrified only slowly.

Low availability of added NH$_4$$^+$ to nitrifiers in highly fixing Swedish soils has been reported by Nommik (1957) and Jansson (1958), as shown in Fig. 7. According to Nommik, availability of fixed NH$_4$$^+$ was dependent, among other things, on the degree to which the total fixing capacity of the soil was saturated by NH$_4$$^+$. Availability increased with increasing saturation of the soil's NH$_4$$^+$-fixing capacity.

In geological specimens of vermiculite, fixed NH$_4$$^+$ has been shown to be nearly as available to nitrifying organisms as the NH$_4$$^+$ of neutral NH$_4$$^+$-salts (Allison et al., 1953c; Welch and Scott, 1960). Different types of vermiculite appeared, however, to behave rather differently in this respect (Allison et al., 1953c). In a South African vermiculite, availability to nitrifiers was, thus, only 11–16%.

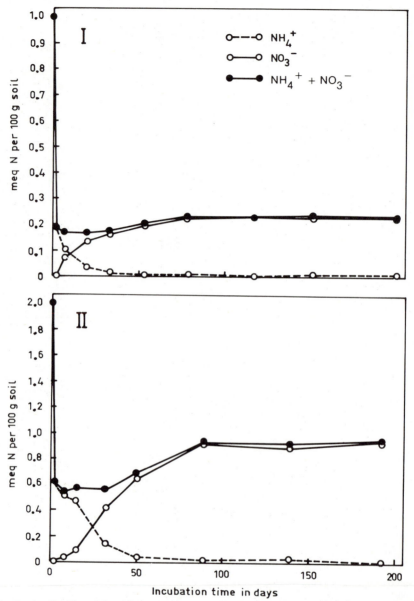

Fig. 7—Availability of fixed NH_4^+ to nitrifying organisms in a highly fixing vermiculitic clay soil. (I) 1 meq NH_4^+ added per 100 g soil. (II) 2 meq NH_4^+ added per 100 g soil (Nommik, 1957).

Axley and Legg (1960) investigated the availability of added NH_4^+ to nitrifiers in soils with different NH_4^+-fixing capacities (both surface soils and subsoils). The percentage of the added NH_4^+ nitrified during incubation showed an unmistakable tendency to decrease with increasing NH_4^+-fixing capacity of the soil. In highly fixing soils, only 10–15% of the added NH_4^+

Table 9—Nitrification of NH_4^+ added to soils with varying fixation capacities
(Axley and Legg, 1960).

Soil	NH_4^+-fixing capacity, meq/100 g	% of added NH_4^+ nitrified[†]
Meyersville A	1.18	46.0
Meyersville B2	7.07	13.5
Meyersville C	8.93	9.6
Chester B2	0.58	57.1
Sassafras B2	0.57	41.8
Hagerstown surface soil	0.70	94.4

† Added at the rate of 100 ppm N.

was recovered as NO_3^- during an incubation period of 50 days (Table 9). Similar results were obtained by Aomine and Higashi (1953b), who studied the nitrification of added NH_4^+ in Japanese soils having various fixing capacities. Investigating an NH_4^+-fixing Belgian soil, Baert-de Bièvre et al. (1961) observed that only NH_4^+-extractable with KCl and NaCl was susceptible to nitrification.

Blasco and Cornfield (1966) reported that nitrification of fixed NH_4^+ during aerobic nitrification for 6 weeks varied from 6 to 88% in different soil clays (Vopěnko & Némec, 1976). Lutz (1966), working with 14 surface soils from the southeastern United States, reported that an average of 20% of the fixed NH_4^+ was nitrified during 4 months of incubation.

Nommik (1966) studied the influence of the pellet size of $(NH_4)_2SO_4$ on the extent of NH_4^+ fixation and the subsequent release of fixed NH_4^+ on nitrification in two highly fixing vermiculitic clays during incubation. Fixation was strongly reduced by increasing the pellet size. Within a 10-week incubation period, up to 37% of the initially fixed NH_4^+ was released on nitrification. For the large pellet size fraction both the nitrification rate and, as a consequence of this, the rate of release of fixed NH_4^+ was low compared with treatments with fine-textural or aqueous sources of $(NH_4)_2SO_4$. No measurable amounts of fixed NH_4^+ were released during incubation as long as more than traces of fertilizer NH_4^+ remained in the soil in exchangeable form. The experimental data indicate, furthermore, that both the extent of fixation and the rate of subsequent release of fixed NH_4^+ were markedly influenced of the soil moisture level.

The low availability of fixed NH_4^+ to nitrifiers, reported by several workers, may be considered inconsistent with the theory of the reversibility of the fixation-defixation reaction. However, it should be borne in mind that in soil systems, K^+ may interfere with defixation of NH_4^+. The interfering effect of K^+ is evidenced by the fact that, contrarily to soils, the availability of fixed NH_4^+ in K^+-free clay minerals, such as vermiculite and montmorillonite, is high and comparable with that of exchangeable NH_4^+. The mechanism of delayed release of fixed NH_4^+ in the nitrification process is at present not fully understood. It might be considered possible that in soils the removal of exchangeable NH_4^+ by nitrification results not only in a release of fixed NH_4^+, but also in a simultaneous liberation of small amounts of the lattice-bound K^+. The latter reaction may be enhanced by H^+ formed as a

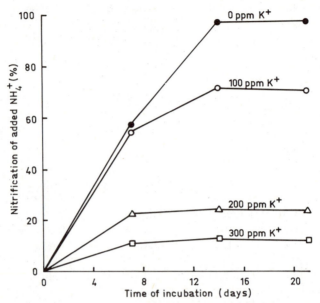

Fig. 8—Nitrification of NH_4^+ in NH_4^+-saturated South Carolina vermiculite as affected by added K^+. The experiment carried out in vermiculite-water suspensions inoculated with nitrifying organisms. Total NH_4^+ content of the vermiculite, 79.5 meq/100 g (Welch and Scott, 1960).

consequence of the transformation of NH_4^+ to NO_3^-. Owing to this, defixation of NH_4^+ may cease at a stage where the concentration of exchangeable K^+ would balance the fixed NH_4^+ at a new equilibrium level.

The interfering effect of K^+ on the effectiveness of nitrifying organisms to release fixed NH_4^+ from vermiculite and montmorillonite has been studied in some detail by Welch and Scott (1960). The results obtained by these workers, as well as by van Schreven (1968), stress the significance of the soils' K^+-status on the availability of fixed NH_4^+ to the nitrification process (Fig. 8).

2. AVAILABILITY OF FIXED NH_4^+ TO HETEROTROPHIC SOIL MICROORGANISMS

In comparison to the nitrifying organisms the heterotrophic microorganisms and higher plants have a high K^+ requirement, which suggests that the two organism groups are different as regards their ability to utilize fixed NH_4^+ in soils.

As concerns the heterotrophic soil microflora, their requirement for assimilable N and K^+ are highly dependent on the supply of available energy, i.e., readily oxidizable organic material. The higher the level of decomposable organic matter, the greater would be the amounts of mineral N and K^+ assimilated by the organisms. In a soil containing excessive amounts of available energy, exchangeable K^+ and NH_4^+ are generally held at low levels, a condition that is favorable for the release of fixed NH_4^+ and K^+ from clay minerals.

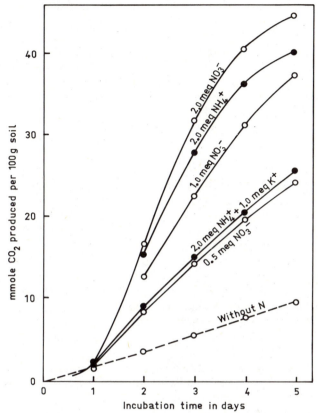

Fig. 9—Availability of NH₄⁺ and NO₃⁻ sources of N to mixed heterotrophic microflora in a highly fixing surface soil in the presence of excess energy (Nommik, 1957).

An incubation technique, implying measurement of CO_2 evolution, has been used by some investigators to study the availability of added NH_4^+ to the heterotrophic microflora in highly fixing soils. Using a vermiculite-containing, heavy clay soil, which under moist conditions fixed about 80–90% of the NH_4^+ added (0.5–2.0 meq NH_4^+/100 g soil), and with glucose as the energy source, Nommik (1957) found that fixed NH_4^+ was approximately two-thirds as effective an N source for microorganisms as NO_3^-. Ability of the microbes to utilize NH_4^+ was markedly reduced by addition of K^+, with only about 10% of the fixed NH_4^+ being available (Fig. 9).

Using the same technique, but different textural fractions of another vermiculite-containing soil, Nommik (1965) reported that in comparison with NO_3^-, the availability of fixed NH_4^+ in the clay, silt, and fine sand fractions was as high as 89, 70, and 50%, respectively.

Similar availability tests, but using ground cereal straw as an energy source, were carried out by Jansson (1958). Fixation markedly reduced the ability of the heterotrophic microflora to utilize the added NH_4^+. According to this investigation, the heterotrophic microflora appeared as a weak and ineffective extractant of fixed NH_4^+, being in some cases equilvalent to and in other cases even weaker than a $1N$ KCl solution.

The apparent disagreement between Nommik's and Jansson's observations was considered to depend on the presence of K^+ in the cereal straw used by Jansson as energy source (see Nommik, 1965). With the knowledge of the blocking effect of K^+ on the release of fixed NH_4^+, the results of Jansson are fully explainable and in accord with previous observations. Control experiments carried out by Jansson with K^+-free straw indicated that the above interpretation probably was correct.

F. Availability of Fixed NH_4^+ to Higher Plants

Although NH_4^+ fixation is a common reaction in many arable soils, our knowledge of its agronomic significance is extremely deficient. Moreover, the information available is rather inconsistent from several points of view. Most studies have been carried out in pot experiments and the results cannot be directly applied to field conditions.

The first attempt to elucidate this problem seems to have been made by Bower (1951), who used a modified Neubauer technique to compare the plant availability of exchangeable and fixed NH_4^+. In two soils containing 3.5 and 4.0 meq of fixed NH_4^+/100 g, respectively, a maximum of only 10% of the fixed NH_4^+ was recovered by barley plants. In a parallel study, from 13 to 28% of the nonexchangeable NH_4^+ was released through nitrification, indicating that plants were unable to utilize fixed NH_4^+ in excess of that made available through nitrification.

Extremely low availability of fixed NH_4^+ has also been reported by Allison et al. (1953a), who demonstrated that millet, grown in the greenhouse, was capable of utilizing only 7% of the NH_4^+ fixed by air-drying and 12% of that fixed by heating. The test soil was pretreated with N KCl to remove exchangeable NH_4^+, followed by a subsequent removal of excess K^+ by leaching with $CaCl_2$ or $MgCl_2$, which may have influenced the accessibility of the fixed NH_4^+. In studies with rice and wheat plants, Aomine and Higashi (1953a) showed that accessibility of fixed NH_4^+ was mainly dependent on the nature of the fixing material.

Considerably higher availability of fixed NH_4^+ to higher plants has been recorded in more recent greenhouse trials. Using a highly fixing, vermiculite-containing, heavy clay soil, Nommik (1957) found that the growth response of oats to NH_4^+-N was only slightly lower ($<10\%$) than that of NO_3^--N. Since analysis of the soil 1 week after the start of the experiment indicated that only 10–20% of the added NH_4^+ could be recovered as exchangeable NH_4^+ or NO_3^-, it was concluded that the oat plants were highly effective in utilizing the fixed NH_4^+ (Fig. 10). Analysis of the crop at an early stage of development indicated, however, that the N added as $Ca(NO_3)_2$ was absorbed considerably more rapidly than N added in the form of $(NH_4)_2SO_4$. In the treatments with $(NH_4)_2SO_4$ the oats absorbed large amounts of N at later stages of development, whereas in the $Ca(NO_3)_2$-treated pots uptake of N was almost completed prior to the heading stage of

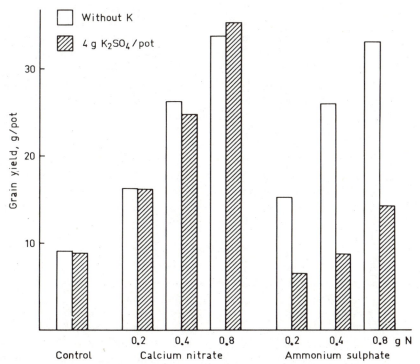

Fig. 10—Yield of oats per pot as influenced by different addition rates of NH_4^+ and NO_3^- in the absence or presence of simultaneously added K^+ fertilizer. Highly fixing clay soil from central Sweden, 6 kg per pot (Nommik, 1957).

the oats. Evidently, fixation resulted in a delayed uptake of added NH_4^+-N (Atanasiu et al., 1968).

In the same pot experiment, fairly large doses of K_2SO_4 did not significantly influence yield and uptake of N by oats when the N was applied in the form of NO_3^-. However, when K^+ was added in combination with $(NH_4)_2SO_4$, uptake of NH_4^+-N by the crop was markedly decreased. It was further shown that the extent of reduction of the N effect was greatly dependent on the K^+ addition rate.

An interesting finding from the above experiment was that the negative effect of K^+ was almost entirely eliminated when K_2SO_4 was added 3 weeks before addition of NH_4^+. This is in accordance with the previous observations on the influence of time of application of K^+ on fixation and release of NH_4^+. Jansson (1960) and Jansson and Ericsson (1961) obtained results which were in good agreement with those described above (see also Shilova, 1969).

Results obtained by Nommik (1957) and Jansson (1958) have been confirmed by investigations carried out in the United States. Thus, Legg and Allison (1959) showed that the uptake of added NH_4^+ by sudangrass from two NH_4^+-fixing subsoils was of the same order as that for a nonfixing surface soil.

Black and Waring (1972) studying the availability of fixed NH_4^+ to wheat crop in pot culture, reported that about 50% of the recently fixed

NH_4^+ was recovered by a single cropping, whereas intense cropping and successive cropping removed almost all of the recently fixed NH_4^+. The availability of added fixed NH_4^+ in montmorillonitic clay was, thus, dependent on the intensity and duration of cropping. Regarding native fixed NH_4^+, insignificant amounts were removed by intensive cropping.

In a study of the availability of NH_4^+ in Maryland soils having different NH_4^+-fixing capacities, Axley and Legg (1960) reported that plant uptake of N from $(NH_4)_2SO_4$ or urea was not greatly affected by the capacity of the soil to fix NH_4^+, unless sufficient K^+ was present to block the release of fixed NH_4^+. Application of K^+ together with NO_3^- had little effect on yield or uptake of N. The authors suggested that NH_4^+ should not be applied with K^+ to soils characteristically having high fixing capacity. The above recommendation is in harmony with the results of greenhouse studies by Walsh and Murdock (1963), who found that N uptake by corn was measurably higher when K^+ and NH_4^+ were applied in separate soil layers than when applied in the same layer.

The high availability of fixed NH_4^+ reported above holds good for cereal crops. The possibility cannot be excluded that different plant species may behave differently. Jansson (1958), thus, found that oats grown after mustard on a fixing soil consumed significant amounts of the tagged fertilizer NH_4^+ not recovered by the mustard. No such residual effect was obtained for oats following another cereal. In a later study, Jansson and Ericsson (1961) demonstrated that oats were considerably more effective in utilizing added NH_4^+ on a fixing soil than mustard; sugarbeets and potatoes took an intermediate position. The authors further reported a marked residual N effect when NH_4^+ was included with a light application of K^+, but not when a large dose of K^+ was given to the first crop. The release of fixed NH_4^+ occurred rather slowly and even five successive crops failed to remove all of the fixed NH_4^+ present.

The readiness with which cereals utilize fixed NH_4^+ in pot experiments need not a priori be valid for crops grown under field conditions. In the field, the density of plant roots per volume of soil is usually much lower than in pots, implying that a considerably smaller part of the exchangeable and even of nonexchangeable K^+ is removed by the activity of the plant roots during the vegetation period. This may certainly influence the rate on which fixed NH_4^+ is released from the clay minerals. Furthermore, the extent of NH_4^+ fixation under field conditions may be considerably different from that induced under laboratory or greenhouse conditions, especially when using large pellet NH_4^+ materials.

Available information on the relative effectiveness of NH_4^+ when applied to fixing soils under field conditions appears extremely meager and inconsistent. In moderately fixing soils, Jansson and Ericsson (1961) recorded small and inconsistent differences in the effect of $(NH_4)_2SO_4$ and $Ca(NO_3)_2$ on spring wheat and sugar beets, the tendency being for NO_3^- to be superior to NH_4^+. According to observations by Kaila and Hänninen (1961), the yield increase obtained by application of NH_4NO_3 limestone to NH_4^+-fixing Finnish soils are not significantly lower than that produced by an equal

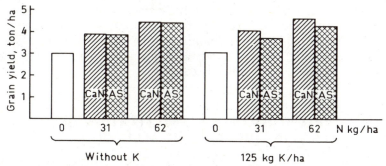

Fig. 11—Effect of $(NH_4)_2SO_4$ (AS) vs. $Ca(NO_3)_2$ (CaN), both in absence of and in combination with K⁺, on the yield of spring cereals. Mean figures of 28 field trials in southern Sweden (J. Ericsson and S. L. Jansson, unpublished data).

amount of N as $Ca(NO_3)_2$. This is supported by the findings of Ericsson and Jansson (unpublished data) from a series of field trials on moderately fixing soils of Southern and Central Sweden (average fixing capacity 40% when 100 ppm of NH_4^+-N added). In combination with K⁺, $Ca(NO_3)_2$ was clearly superior to $(NH_4)_2SO_4$. In treatment with $(NH_4)_2SO_4$ the yield was, in fact, slightly reduced by addition of K⁺ (Fig. 11).

The field data presented by Ericsson and Jansson contained a very large yearly variation. This possibly may be ascribed to differences in the amount and distribution of precipitation during the spring and early summer. Drying of the soil shortly after application of N fertilizers has been found to exert an unfavorable influence on uptake and recovery of NH_4^+-N by the crop. According to Schachtschabel (1961), the general experience in German agricultural practice is that soil N is more available to plants in wet than in dry summers. Schachtschabel suggests that this may be due to easier release of fixed NH_4^+ from the soil during a wet season. According to him, the moisture content influences the degree of expansion of the "degraded" illites and, thus, also the readiness with which the interlattice K⁺ and NH_4^+ can migrate from the interior of the crystal to the peripheral zones of the mineral particle. Kowalenko (1978) suggested that wetting and drying and freezing and thawing cycles may increase the stability of recently fixed NH_4^+.

In summary, it may be concluded that the question of the relative effectiveness of NH_4^+-fertilizers in fixing soils is not satisfactorily elucidated. It is evident, anyhow, that neither the nitrification test nor chemical laboratory tests for determining NH_4^+-fixing capacity of the soil provides a fully reliable measure of the availability and effectiveness of applied NH_4^+-N to field crops. As the great majority of arable surface soils evidently fix much less NH_4^+ than many of the Scandinavian and Japanese soils described in this chapter the agronomic significance of the NH_4^+ fixation should not be over emphasized. It should be taken into account, furthermore, that under cropping conditions the consequences of soils' capacity to fix NH_4^+ are expected to be reduced when using granulated N materials and band application of them.

Finally it should be noted that NH_4^+ fixation should not be considered as an entirely unfavorable reaction from the agronomic point of view. Under certain soil and climatic conditions, the fixation may be a positive factor in preventing losses through leaching and ensuring a more even supply of N throughout the growing season.

IV. RETENTION OF NH_3 AND FIXATION OF NH_3 IN SOIL ORGANIC MATTER

Nitrogen applied to soil as anhydrous NH_3, aqua NH_3, or organic nitrogen materials which decompose to form NH_3 (e.g., urea, calcium cyanamide) is subject to adsorption by organic and mineral fractions of the soil and may be dissolved in soil water. When applied to moist soil, and at proper depth, sorption is nearly complete and rather limited movement occurs from the point of placement.

A number of physical and chemical reactions are responsible for the efficiency of NH_3 retention. They range from weak physical sorption of the type of H-bonding to irreversible incorporation of NH_3 into the soil organic matter. Physically sorbed NH_3 is in a dynamic equilibrium with the gaseous phase and is subject to movement by diffusion. Physical sorption only occurs when there is a positive pressure of NH_3 in the soil atmosphere. As soon as NH_3 pressure decreases, equilibrium is displaced and the physically sorbed NH_3 returns to the gaseous phase. On diffusion through the soil, NH_3 may finally be lost to the atmosphere.

The chemisorbed NH_3 is strongly bound to specific adsorption sites on soil colloids and little migration will occur within the soil. The basic mechanism for chemical binding of NH_3 consists of acquisition of a proton by the NH_3 to form NH_4^+, which is subsequently bound to exchange sites on clay and humus particles. Ammonia may also react irreversibly with organic matter, with formation of various NH_3-organic matter complexes. These complexes are characterized by high chemical stability and high resistance to microbial decomposition.

The aim of the following section is to review current concepts of the reactions responsible for NH_3 retention by the clay and organic fractions of the soil. Various aspects of this topic have previously been summarized by Mortland (1958), Mortland and Wolcott (1965), Broadbent and Stevenson (1966), and Parr and Papendick (1966).

A. General Remarks

Unless the soil is dry or extremely coarse in texture, applied NH_3 is effectively retained by the soil. Hence, high NH_3 concentrations and high pH may be found within the localized retention zone. Immediately after NH_3 application, concentrations ranging from 1,000 to 2,500 ppm N have been reported to occur in a horizontal, cylindrically shaped zone with a diameter of 3 to 6 cm (see Fig. 12 and reviews mentioned above). Concentrations ap-

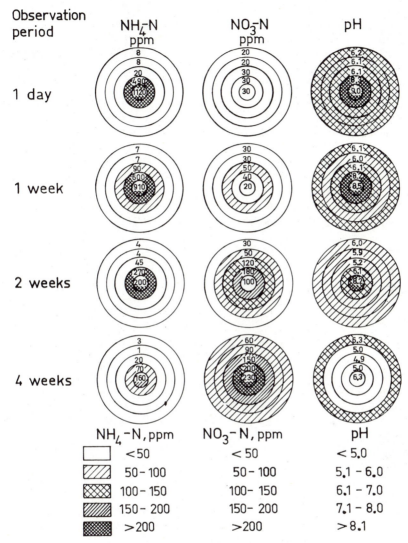

Fig. 12—Distribution of NH_4^+, NO_3^- and pH in soil column after injection of NH_3. Soil characteristics: loamy fine sand; organic C, 3.1%. Ammonia application rate corresponding to 116 kg N/ha at 25-cm spacing. The rings represent distances from the injection line: 0–1.5, 1.5–3.0, 3.0–5.0, 5.0–7.0, and 7.0–9.0 cm, respectively (adapted after Nommik and Nilsson, 1963a).

proaching 1,000 ppm may persist in this zone for a period of several weeks. Through gaseous diffusion and diffusion in solution, the NH_3 distributes radially outwards from the line of injection. According to available data, the ultimate diameter of the localized zone does not exceed, in general, 10–12 cm. This means that at usual field application rates, and with a knife-spacing of 100 cm, a maximum of 20% of the plow layer will be influenced by NH_3 (cf. Stanley & Smith, 1956; Nommik & Nilsson, 1963a; Kurtz & Smith, 1966; Papendick & Parr, 1966).

The volume of the retention zone will vary, depending on the rate of NH_3 application, row space, soil moisture content, buffering capacity, texture, and nitrification capacity etc. The initial distribution pattern is, to a large extent, controlled by the moisture content of the soil, depending both on the high solubility of NH_3 in water and the low diffusion rate of NH_3 in the aqueous phase as compared to the gaseous phase. According to Parr and Papendick (1966), soil moisture functions largely as a temporary repository of NH_3 from which it is subject to slow movement by mass flow with water or by gaseous diffusion away from the zone of high NH_3 concentration.

Under field conditions, moisture content may have an indirect effect on the soils' capacity to retain applied NH_3 by providing physical conditions to ensure rapid and complete sealing of the injection channel (Parr and Papendick, 1966). Hanawalt (1969) reported that a high moisture level was more important in this respect than a high CEC, exchange acidity, and clay and organic matter contents.

In a study using 76 soil samples from 20 different soil profiles of the Pacific Northwest, Young (1964) concluded that total NH_3 retention could be described by the following relationship:

$$\text{ppm } NH_3\text{-N} = 147 + 596\,(\% \text{ organic C}) + 43.4\,(\% \text{ clay}).$$

The equation indicates the importance of the organic fraction on NH_3 retention by soils. The high pH dependent CEC of soil organic matter was largely responsible for this property.

Conditions similar to these arising from the application of NH_3 may also occur in the soil following application of urea (and calcium cyanamide). Common to both types of N materials is the occurrence of local zones of high pH and NH_3 concentration close to the point of placement. However, pH changes with urea are usually not as drastic as with NH_3, and a smaller part of the soil matrix is possibly affected.

A summarized discussion of possible NH_3 retention mechanisms follows. The term retention has been used, thereby, to designate the total of any form of sorption or fixation mechanism in which NH_3 is retained by various mineral and organic soil components, and by soil water.

B. Physical Sorption of NH_3

The NH_3 molecule is very similar to the water molecule in many respects, e.g., both are polar and capable of forming H-bonds. In the case of physical adsorption, the H atoms of NH_3 tend to associate with electronegative atoms of polar molecules, as illustrated by the following equations (see James & Harward, 1964):

$$NH_3 + OX \rightleftharpoons NH_3 \dots OX \qquad [1]$$

$$NH_3 + HOX \rightleftharpoons H_3N \dots HOX \qquad [2]$$

where OX and HOX represents oxide and OH groups on the surfaces of clay minerals or humus colloids.

Ammonia retained by H-bonding is easily replaced by water molecules. The mutual competition between H$_2$O and NH$_3$ molecules for adsorption sites can be described by the law of mass action (James & Harward, 1964). Thus, under soil conditions, NH$_3$ held by this mechanism would be of transitory character, being of importance only in the zones of high NH$_3$ concentration and in the absence of water (Mortland, 1966).

Chao and Kroontje (1960) and Du Plessis and Kroontje (1967) found that at low NH$_3$ concentrations adsorption in soils followed Langmuir's monomolecular adsorption theory. The differential slopes recorded for different sections of the curves were releated to reactions involving proton acception or possibly the formation of multilayers. Sorption isotherms of NH$_3$ on base-saturated clays have been utilized by Mortland (1955) for the determination of specific surface.

A mechanism contributing to adsorption of NH$_3$ on soil colloids, with binding being only slightly more energetic than that of H-bonding, involves coordination between NH$_3$ and exchangeable metal cations. Adsorption of NH$_3$ by cations in the soil system is analogous to hydration of these ions by water. In fact, NH$_3$ and water molecules compete for coordination positions around exchangeable cations and each can displace the other. The following equation describes these interactions (cf. Mortland, 1966):

$$M (H_2O)_n X + n NH_3 \rightleftharpoons M (NH_3)_n X + n H_2O$$

where letter M represents the exchangeable metal ion, X a component of either the organic or inorganic exchange complex, and n the coordination number. The reaction is reversible, implying that the forward reaction occurs on injection of NH$_3$ into soil, and the reverse reaction predominates as NH$_3$ gas diffuses away from the placement zone and water molecules displace NH$_3$ from its coordinating positions. Consequently, under soil conditions, NH$_3$ held by the above mechanism is transitory in nature, involving retention only during the very initial stages of NH$_3$ injection. The stability of complexes of this type is higher for transition metal ions like Cu^{2+} or Fe^{2+} than for alkaline earth and alkali metal ions (Brown & Bartholomew, 1963; Mortland, 1966).

From data on heats of adsorption and adsorption isotherms of NH$_3$ gas on outgased, homoionic soil, Ashworth (1973, 1978) concluded that Ca^{2+}- and Mg^{2+}-saturated soils adsorbed one more NH$_3$ molecule per exchangeable ion than Na$^+$ and K$^+$-saturated soils. The results further suggested that most NH$_3$ is sorbed on these soils through mechanisms not involving coordinate bonding to exchangeable cations. Sorption through mechanisms involving H-bonding was considered negligible.

To the category of physically sorbed NH$_3$ may be included, on various grounds (e.g. reversibility), NH$_3$ dissolved in soil water. The size of this fraction may be considerable and of great significance in controlling NH$_3$ loss during the initial phases of NH$_3$ application in the field.

C. Chemisorption of NH_3

In the determination of chemisorption capacity, the soil is usually subjected to ammoniation with gaseous NH_3 or a water solution of NH_3. Thereafter, the soil is aerated or evacuated to remove physically sorbed NH_3. The amount of NH_3 chemisorbed is obtained as the difference between the total N content of the ammoniated and degased soil and total N content of the corresponding nontreated control.

On extraction of degased soils with dilute acid or salt solutions a further amount of NH_3 is removed. This NH_3 is considered to occupy exchange sites on clay and humus particles, and to occur as NH_4^+. The formation of NH_4^+ involves acquisition of a proton (H^+) by the NH_3 molecule. An acid soil having exchangeable H^+ ions on colloid surfaces may provide the necessary protons:

$$NH_3 + H^+X \rightarrow NH_4^+X,$$

where X is an organic or mineral component of the exchange complex. The NH_4^+, being positively charged, will associate with exchange sites on clay and humus particles. While in this condition, it is protected to a large extent against losses by leaching or by NH_3 volatilization.

According to Mortland (1966), NH_3 may react also with protons provided from OH groups associated with silicon on the edges of clay minerals:

$$\gtreqless Si\text{–}OH + NH_3 \rightleftharpoons \gtreqless Si\text{–}O\text{–}NH_4.$$

The dissociation of these protons is pH-dependent, implying that they become available for NH_4^+ formation only at alkaline pH.

The capacity of the soils to retain NH_3 as exchangeable NH_4^+ is expected to be closely related to pH and buffering capacity. The titratable acidity should, thus, describe satisfactorily the soils' capacity to retain added NH_3. This is valid especially for organic soils showing a high pH-dependent acidity. For clay minerals and mineral soils low in humus, exchange acidity might also be a useful characteristic for this purpose.

The relative importance of clay and organic matter in retaining NH_3 depends on the type of clay mineral as well as the degree of base unsaturation of the colloids. In general, the organic fraction is considerably more effective per unit weight.

D. Fixation of NH_3 in Soil Organic Matter

It is well established that the organic fraction of the soil has the capacity to bind NH_3 in nonexchangeable forms and that the NH_3-organic matter complexes are extremely resistant to microbial decomposition. Neither the mechanism of this reaction, referred to as chemical NH_3 fixation, nor its agronomic consequences are fully understood. Owing to the increasing use of anhydrous NH_3 and NH_3 solutions and of materials which decompose to form NH_3 (e.g., urea), the chemical immobilization of NH_3 by fixation in soil organic matter has become a question of increasing agronomic significance. The literature on this subject up to 1964 has been summarized by Mortland (1958), Mortland and Wolcott (1965), and Nommik (1965).

The term "fixed NH_3", as used in the present paper, refers to NH_3 which is retained by the soil organic matter after intensive extraction and leaching either with diluted mineral acid or neutral salt solutions. Owing to solubilization of soil organic matter on ammoniation by exposure to NH_3, values for the amount of fixation will depend to some extent on the leaching technique used. On mineral acid treatment, some of the solubilized N compounds are precipitated and included in the total N analyses of the extracted residue. On the other hand, with treatments using neutral salts, the solubilized organic N is removed and excluded from the total N figure for the extracted residue. Furthermore, when fixation is calculated by difference between the amounts of NH_3 added and the amounts recovered in the leachate as NH_4^+, soluble forms of organic N will be included with the "fixed NH_3" (Mortland and Wolcott, 1965). These discrepancies should be borne in mind in interpretating the reported data.

1. MECHANISMS OF NH₃ FIXATION

At present, the exact nature of the NH_3-organic matter complexes is not known, although several plausible reaction mechanisms for their formation have been proposed. According to Broadbent and Stevenson (1966), there is no sound basis for assuming a given mechanism as the correct one or that only one mechanism is involved. Reactions of NH_3 with known organic compounds gives us, however, an insight into possible mechanism for the reaction of NH_3 with soil organic matter.

In a comprehensive review, Mortland and Wolcott (1965) have considered the reactions of NH_3 with aromatic and unsaturated alicyclic compounds as well as with carbohydrates. Since then surprisingly little new information has been submitted on this subject.

Groups capable of reacting with NH_3 are primarily present in phenols and quinones, which may arise during the enzymatic break-down of lignin, essential oils, tanning proteins as well as on non-enzymatic alternation of inositol and reducing sugars (see Mortland & Wolcott, 1965). Because lignin is a quantitatively important constituent in all plant materials, it is a widely accepted understanding that lignin derivatives are of considerable importance for fixation of NH_3 in soil organic matter. On the basis of results obtained in investigating humus materials, as well as di- and trihydric phenols and poly-hydric aromatic carboxylic acids, Mattson and Koutler-Andersson (1943) suggested that fixation takes place in the lignin-derived fraction of soil organic matter. The presence of aromatic rings with two or more OH groups was considered essential. They postulated the formation of amino phenols, which on oxidation were transformed to reactive quinone-imino compounds and by subsequent condensation converted to compounds with N incorporated in heterocyclic rings. This theory is supported by findings of

Bennett (1949), who reported that on exposure to oxidative ammoniation the commercial lignin fixed NH_3. Bennett demonstrated, additionally, that ammoniation decreased the number of methoxyl groups. Conversely, methylation reduced the capacity of lignin to fix NH_3 (cf. Jansson, 1960; Valdmaa, 1969).

Flaig (1950) suggested a mechanism for the formation of amorphous substances very similar to soil humic acids through the reaction between NH_3 and quinones in alkaline medium. According to Flaig, fixation of NH_3 by aromatic compounds is dependent on extensive polymerization. On polymerization the N is combined into bridging structures, analogous to those in the phenoxazine dyes.

In an investigation of para-quinone-NH_3 systems, Lindbeck and Young (1965) proposed a fixation mechanism involving a sequential formation of 2.5-diaminohydroquinone and phenazine as transient intermediates (Fig. 13).

Scheffer and Ulrich (1960) reported that aminophenols are readily subject to polymerization with the formation of either phenoazine or phenaziner rings, depending on the orientation of the amino groups. They consider the aminoquinones as probable intermediate oxidation products in humic acid formation.

The reactivity of polyphenols and quinones is due to their tendency to form semiquinone free radicals by univalent oxidation or reduction. The greatly enhanced capacity of soil organic matter to fix NH_3 at alkaline pH thus may be conceived as a consequence of increased formation and polarization of free radicals and accelerated autocatalytic oxidation and polymerization (see Mortland & Wolcott, 1966). The free radicals are considered to be formed also by the action of microbial phenoloxidases and dehydro-

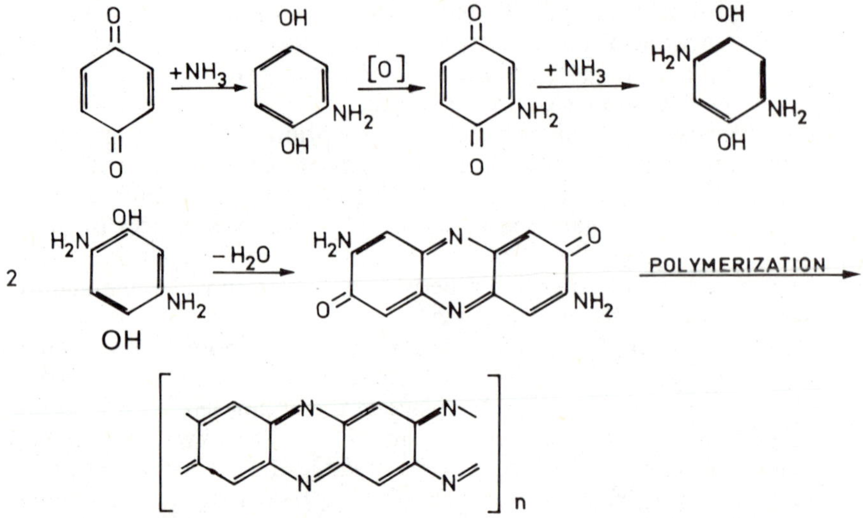

Fig. 13—A mechanism of NH_3 fixation by *p*-quinone proposed by Lindbeck and Young (1965). Cit. Broadbent and Stevenson (1966).

genases (Flaig, 1950). According to Mortland and Wolcott (1965) it is most likely that under field conditions both chemical and enzymatic oxidations are involved.

Even though aromatic compounds and their unsaturated alicyclic counterparts may be primarily responsible for NH_3 fixation in soils, a large number of other biologically active substances may be involved. Mechanisms for their reactions with NH_3 have been studied using simple model compounds. Regarding the reactions of NH_3 with carbohydrates, reducing sugars are primarily involved. The reactions are similar to those described by Hodge (1953) for melanin formation from sugars and amino acids in acid media. Documentation is imperfect as concerns the extent to which the soil carbohydrate contributes to NH_3 fixation under natural soil conditions (Mortland & Wolcott, 1965).

2. CONDITIONS FOR FIXATION

The first information on the ability of soil organic matter to fix NH_3 originated from experimental work carried out with the intention of producing slow-release organic N material appropriate for use as commercial fertilizer. Several German patents for ammoniation of peat were issued around 1930. An extensive study of the fixation of NH_3 in peat under varying conditions of temperature, pressure, length of treatment and moisture content has been conducted by Scholl and Davis (1933). The authors reported that NH_3 treatment of a moss peat at 300°C resulted in a product which showed a N content of 21.7%. Only about one-fifth of this N was in water-soluble form. Economic considerations at the time indicated according to the authors, that commercial production of ammoniated peat was possible.

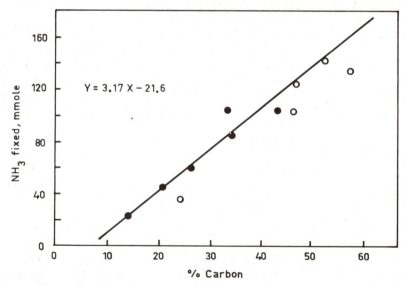

Fig. 14—Relationship between C content and NH_3 fixation in organic soils under aerobic conditions (Burge and Broadbent, 1961).

As concerns soils, their capacity to fix NH_3 in nonexchangeable form may vary considerably. Sohn and Peech (1958) found that among mineral soils, the quantity of NH_3 fixed was highest for acid soils containing large amounts of organic matter. Burge and Broadbent (1961) studied the fixation of N from anhydrous NH_3 by dry organic soils ranging from 14 to 43% C, the quantity of NH_3 fixed was linearly related to C content (Fig. 14). They found that under aerobic conditions one molecule of NH_3 was fixed per 29 C atoms. In treating a series of mineral and organic soils with aqueous NH_3, Nommik and Nilsson (1963b) also noted a close correlationship between NH_3 fixation and organic matter content. They concluded that at NH_3 application rates corresponding to field practice, less than 4% of the NH_3 applied would be fixed by the organic matter in Swedish mineral soils. According to the previously cited report by Young (1964), fixation of NH_3 by the organic fraction of mineral soils amounted to 2 to 28% of the total NH_3 retention. Mattson and Koutler-Andersson (1942, 1943) demonstrated that the NH_3-fixing capacity of the peats was negatively correlated with their degree of humification and thus with their total N content. More recently results of Shoji and Matsumi (1961) and Nyborg (1969) showed no such close relationships.

From the methodological point of view it is important to note that, in mineral soils, the determination of the NH_3 fixation capacity of the organic fraction is generally more complicated than in peat. In mineral soils the increase in the soil's content of Kjeldahl N, following NH_3 treatment, represents the sum of NH_3 fixed by organic matter and NH_4^+ fixed in clay minerals. Several procedures have been proposed to eliminate the interfering effect of the soil mineral fraction. For this purpose, Sohn and Peech (1958) examined the soil as regards its capacity to fix K^+ and suggested that the difference between the amounts of NH_3 and K^+ fixed by the soil would represent a conservative estimate of the NH_3 fixing capacity that can be attributed to the organic fraction. The simplest, even if not fully perfect, way to overcome interference by the mineral fraction is pretreatment of the soil with K^+ prior to exposure to NH_3 (Nommik & Nilsson, 1963b). The imperfectness of the latter technique was demonstrated by data by Nyborg (1969), who showed that fixation in subsoils may be several times greater when soils were treated, air-dry, with gaseous NH_3, rather than in NH_4Cl solution.

In this connection it should be noted that the procedure of estimating the NH_3-fixing capacity of the soil from the increase of soil's content of Kjeldahl N may also for other reasons not give a completely correct answer of the absolute magnitude of the fixation reaction. This is due to the fact that on treatment of the soil with NH_3 there occurs a simultaneous release of NH_3 from organic N combinations, e.g. amides, amino sugars, which are unstable at elevated pH levels. The figures for NH_3-fixation capacity obtained according to the above procedure actually represent net effects between two reactions, one immobilizing and the other releasing NH_3. The above relationship has been demonstrated experimentally by Nommik and Nilsson (1963b) using ^{15}N-labeled NH_3. The tracer technique considerably improves both the detection limit and the accuracy of the NH_3-fixation measurements.

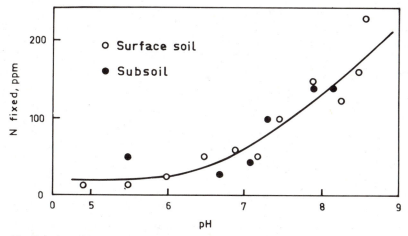

Fig. 15—Relation of NH₃ fixation to pH of a peaty muck. The variation in pH obtained by applying increasing amounts of Ca(OH)₂ (Broadbent et al., 1960).

a. pH Relationships—Chemical reactions to be expected between NH₃ and soil organic matter involve oxygen-containing groups, as COOH, C=O, structures become increasingly polarized and reactive. Even the number of the reactive groups should increase in the alkaline range (see Mortland & Wolcott, 1965). The rate and magnitude of NH₃ fixation is owing to this, strongly dependent upon pH.

In studies on the relationship of NH₃ fixation in organic soils (Shoji & Matsumi, 1961) and forest raw humus (Themlitz, 1958) fixation was found to be insignificant at pH <7 but to increase abruptly above neutratility. Broadbent et al. (1960) and Nommik (1970), by using the ¹⁵N recovery technique, demonstrated measurable fixation of NH₃ even at pH <6 (see Fig. 15, Table 10).

b. Aerobic vs. Anaerobic Fixation—It has been well-established that the fixation of NH₃ in alkaline media involves a simultaneous oxidation reaction, indicated by uptake of O₂. Nommik and Nilsson (1963b) found

Table 10—Fixation of ¹⁵N-labeled NH₃ by Norway spruce raw humus as influenced by liming. Ammonium application rate 100 ppm; reaction period 24 hours; temperature 22°C (Nommik, 1970).

pH	Labeled N recovered		Total organic fixation
	In HCl-leached humus residue	In HCl extract as soluble organic N	
	%		
3.4	1	0	1
4.2	2	1	2
5.5	4	1	5
7.5	10	1	11
10.1	50	3	53
11.3	57	6	63
12.0	54	9	63

a close relationship between NH_3 fixation and O_2 consumption. Investigations by Broadbent et al. (1960) showed, however, that fixation by air-dry peats during a 24-hour exposure was only slightly affected by O_2, the partial pressure of which ranged from 0 to 0.5 atm. In a subsequent paper, Burge and Broadbent (1961) reported that the fixation capacities to NH_3 in a series of peat soils were 106 meq/100 g organic matter in the absence of O_2 and 161 meq when the atmosphere contained 25% O_2 and 75% NH_3. In a forest raw humus at pH 10.3, Nommik (1970) demonstrated that the fixation figures for anaerobic NH_3 treatment were approximately 40–50% lower than those for the corresponding treatments in air. This is in agreement with observations made by Nyborg (1969).

To interpret the role of O_2 in NH_3 fixation, Broadbent et al. (1960) suggested that reactive groupings capable of fixing NH_3, possibly quinones, are already present in soil, explaining fixation on the absence of O_2. When these active groups are consumed, a further fixation of NH_3 will only be possible when new groups, with an appropriate stage of oxidation, are produced during the oxidative treatment (cf. Mortland & Wolcott, 1965).

It is of interest to note, that according to Flaig (1950) the incorporation of NH_3 into synthetic humic acid by alkaline treatment of hydroquinone was closely related to O_2 uptake. The observation was made that the synthetic humic acids formed by autoxidation in NaOH in the absence of O_2, no longer had the capacity to fix NH_3 on subsequent exposure. According to this, the incorporation of NH_3 in humic acid structures could take place only during oxidation and polymerization. This agrees with other work indicating that fixation of NH_3 is increased by simultaneous oxidation, whereas an accomplished oxidation suppresses fixation.

c. Other Factors Affecting Chemical NH_3 Fixation—The extent of chemical fixation is affected by concentration and/or pressure of NH_3, reaction temperature, reaction time, and nature of associated cations and anions.

Fixation generally increases with NH_3 concentration, as illustrated by experimental data of Burge and Broadbent (1961) and Broadbent et al. (1960). Data obtained by Nommik (1970), presented in Table 11, show the effect of increasing levels of $^{15}NH_4Cl$ on fixation in a forest raw humus previously adjusted to pH 11.3 by addition of CaO.

Regarding fixation rate, the reaction at high pH and in the absence of O_2 is essentially complete within 24 hours, although fixation at a greatly reduced rate may continue over considerable periods of time (Mortland & Wolcott, 1965). In an O_2 atmosphere, fixation is more protracted and continues at a decreasing rate for periods of up to several weeks. The period of active fixation is expected to be shortened owing to the pH drop caused by oxidative formation of strongly acidic groups. The relationship between reaction rate and length of reaction period is illustrated by data in Table 12 (Nommik & Nilsson, 1963b). Furthermore, the figures elucidate the discrepancy of fixation figures, obtained either by difference or by ^{15}N recovery. By the difference method fixation attained maximum after 13 days' exposure; with prolonged incubation, fixation showed a definite tendency to

Table 11—Fixation of NH_3 in forest raw humus as influenced by the amounts of ^{15}N-labeled NH_4^+ added. Reaction time 24 hours; pH of the humus adjusted to 11.3 (Nommik, 1970).

NH₄⁺ added	NH₃ fixed	
mmole/g of humus	mmole/g	% of added NH₄⁺
0.1	0.052	52
0.2	0.100	50
0.4	0.136	34
0.8	0.185	23
2.0	0.278	14

Table 12—Aerobic fixation of NH_3 in muck as influenced by length of reaction period. 142 meq ^{15}N-labeled NH_4OH added/100 g soil. Soil pH 9.4. Toluene treated (Nommik and Nilsson, 1963b).

	NH₃ fixed	
Reaction period	Difference method	¹⁵N recovery technique
days	————— meq/100 g soil —————	
1	15.2	17.4
13	30.0	34.6
29	27.7	35.3
51	25.0	34.5

decrease. The tracer data suggested that this was due to the release of nontagged NH_3 by nonenzymatic deamination of the native organic N in the soil.

As pointed out previously an extended aerobic exposure to NH_3 leads to a hydrolytic and oxidative degradation of humic acid polymers and formation of low molecular, water-soluble organic compounds. The persistence in soils of these water-soluble organic compounds, after dissipation of excess NH_3, has not been established. Mangum and Young (1965) concluded that anhydrous NH_3 treatment would seldom cause serious dispersion problems in soils not already predisposed to structural problems. An increased solubility of soil organic matter as a consequence of urea application to forest soil has been reported by Ogner (1972).

Chemical fixation of NH_3 is characteristic not only of the organic matter of soil but of leaf litter, sewage sludge, farmyard manure, cereal straw and sawdust (Mattson & Koutler-Andersson, 1942; Jansson, 1960; Valdmaa, 1969).

3. AMMONIA FIXATION FOLLOWING APPLICATION OF ORGANIC FERTILIZER MATERIALS

As previously suggested, not only anhydrous NH_3 and aqua NH_3, but also urea, N solutions containing urea, and calcium cyanamide may be involved in NH_3 fixation with soil organic matter. For urea, fixation of a magnitude similar to anhydrous NH_3 has been observed (Jung, 1959). However, Nommik and Nilsson (1963b) demonstrated that fixation of comparable magnitude occurred only for low and moderate urea application rates. At high doses, uniformly distributed in the soil, considerably more N was fixed from added aqua NH_3 than from an equivalent amount of urea in aqueous solution. The latter finding was explained by the fact that the activity of the enzyme urease is markedly suppressed when the pH of the medium rises above 8. At pH >8.5, enzymatic breakdown of urea, as a rule, ceases entirely. In urea-treated soils, even at extremely high rates of application, or in close vicinity to the urea pellets, the pH therefore never reaches the level recorded after application of NH_3.

As concerns calcium cyanamide, Themlitz (1958) reported higher N fixation for this compound than for corresponding amount of NH_3. This is in accordance with the findings by Nommik and Nilsson (1963b) and Nommik (1970), who showed that in a cultivated muck and a forest raw humus the extent of N fixation was more than twice as high for calcium cyanamide as for aqua NH_3. The authors suggested that the high fixation observed with cyanamide is probably explained by a retention mechanism, involving the intact cyanamide molecules and not the NH_3 formed through decomposition of cyanamide.

4. CHEMICAL AND BIOLOGICAL RESISTANCE OF NH$_3$-ORGANIC MATTER COMPLEXES

According to the great majority of observations, the NH_3 fixed by soil organic matter is exceedingly resistant to both acid and alkaline treatment. Broadbent et al. (1960) found that some of the NH_3 fixed against extraction with $0.1N$ HCl or acidified $1N$ NaCl was resistant to refluxing for 16 hours with $6N$ HCl. In fractionation of N in two Californian soils treated with ^{15}N-labeled NH_4Cl at pH 9.0, Broadbent and Thenabadu (1967) found that the refluxing with $6N$ HCl for 16 hours released 80 to 90% of both the labeled and native organic N. The NH_4^+-N in the HCl-hydrolyzate comprised, on an average, 30% of total soil N and 77% of the recently fixed N.

In a study of the resistance of native and fixed ^{15}N-labeled nitrogen of a forest raw humus to acid hydrolysis, Nommik (1970) concluded that on refluxing with $6N$ HCl for 8 hours, 54% of the labeled N and 19% of the native organic N resisted the hydrolysis and were recovered in the hydrolysis residue. On the same treatment, 36% of the fixed N and 16% of the native N were released as NH_3.

The varying stability of NH_3-soil organic matter complexes, observed by different workers, may be explained by the heterogenous nature of soil organic matter, and, as a consequence of this, by the existence of varying type of chemical combinations (Broadbent & Stevenson, 1966). The diverging results obtained may depend, additionally, on the conditions under which fixation occurred. Factors such as pH, concentration of NH_3, partial pressure of O_2, and temperature should all be considered in this respect.

Observations of differentiated release of native and recently fixed N from the soil organic matter by acid and/or alkaline digestion may provide a basis for developing methods of detecting the presence of chemically fixed NH_3 in soil organic matter. In combination with tracer techniques, the digestion tests could be used for obtaining evidence whether an accumulation of fertilizer N in the soil organic fraction, following NH_3 application, is a result of chemical or microbial immobilization.

The chemical resistance of NH_3-organic matter complexes indicates a low availability of fixed N to soil heterotrophic microflora and to higher plants. Unfortunately only limited research has been carried out on this subject. Burge and Broadbent (1961), in a pot experiment, using a soil containing ^{15}N-tagged fixed NH_3-N, demonstrated that 4.29% of the tagged and 0.14% of the untagged native nitrogen was removed by a first crop of sudangrass. In a second cutting, only 1.31% of the remaining tagged N was recovered in the plants. The declining release of fixed NH_3 indicates either reversion to increasingly resistant compounds or heterogeneity of the chemical combination of fixed NH_3 (see Mortland & Wolcott, 1965).

Davis et al. (1935) and Nommik (1967) reported that the water-nonextractable fraction of ammoniated peat was exceedingly resistant to biological decomposition, being mineralized and taken up by the crop at about the same rate as the native organic matter of the soil. According to Bremner and Shaw (1954) the acid-insoluble fraction of the ammoniated lignin evidenced a high resistance to mineralization.

5. AGRONOMIC SIGNIFICANCE OF NH₃ FIXATION

As noted above, considerable evidence has accumulated showing the capacity of soil organic matter to react with NH_3 and bind it in a form resistant to chemical hydrolysis and microbial attack. The question as to whether chemical fixation of NH_3 is likely to occur under field conditions, and whether the process alters the effectiveness of NH_3 forms of fertilizer N, is not fully understood, however. The information available is almost exclusively based on data from laboratory investigations. Experimental evidence, showing an accumulation of NH_3-organic matter complexes in field soil, subjected to frequent treatments with NH_3-materials, is still lacking. Theoretically, the reaction is most likely to occur in calcareous, organic soils. The type of organic matter, as well as the moisture content and the nitrification capacity of the soil, are, of course, additional factors influencing the extent of NH_3 fixation.

Owing to the fact that, in row application, the NH_3 is generally distributed only through a fractional part of the top soil (an estimated 1/5–1/20 of the top-soil volume) and that only in the center of the retention zone essentially high pH and NH_3 concentrations will be reached, it seems highly probable that in mineral soils and at moderate rates of application only a small part of the applied NH_3 will be inactivated by reaction with soil organic matter. According to a rough estimation, a mineral soil with a C content of 2% and an NH_3 application rate of 100 kg N per hectare, the extent of immobilization of NH_3 by chemical fixation in soil organic matter will not exceed 5% of the N added. The correctness of this estimation is evidenced by the findings that, in field experiments in general, no significant difference in the effectiveness of N is obtained between solid NH_4^+-salts and anhydrous NH_3.

The possibility cannot be excluded, of course, that NH_3 solubilizes some of the soil organic N and makes it more available, thereby cancelling out the fixation effect.

Regarding urea and calcium cyanamide which decompose in the soil to form NH_3, it is conceivable that under field conditions the pH and the NH_3 concentration in the vicinity of a fertilizer pellet may reach such values that a reaction between NH_3 and organic matter may occur. Use of large-granule materials and localized placement of them may increase the probability of this reaction in originally acid soils.

LITERATURE CITED

Allison, F. E., J. H. Doetsch, and E. M. Roller. 1951. Ammonium fixation and availability in Harpster clay loam. Soil Sci. 72:187–200.

Allison, F. E., J. H. Doetsch, and E. M. Roller. 1953a. Availability of fixed ammonium in soils containing different clay minerals. Soil Sci. 75:373–381.

Allison, F. E., M. Kefauer, and E. M. Roller. 1953b. Ammonium fixation in soils. Soil Sci. Soc. Am. Proc. 17:107–110.

Allison, F. E., E. M. Roller, and J. H. Doetsch. 1953c. Ammonium fixation and availability in vermiculite. Soil Sci. 75:173–180.

Allison, F. E., and E. M. Roller. 1955a. A comparison of leaching and distillation procedures for determining fixed ammonium in soils. Soil Sci. 80:349–362.

Allison, F. E., and E. M. Roller. 1955b. Fixation and release of ammonium ions by clay minerals. Soil Sci. 80:431–441.

Aomine, S., and T. Higashi. 1953a. Studies on the fixation of ammonium in soils. V. Availability of fixed ammonia by crops. J. Sci. Soil Manure, Japan 23:105–108.

Aomine, S., and T. Higashi. 1953b. Studies on the fixation of ammonium in soils. VI. Nitrification of fixed ammonium. J. Sci. Soil Manure, Japan 23:185–188.

Ashworth, J. 1973. Reactions of ammonia with soil. I. Adsorption isotherms and calorimetric heats of adsorption of ammonia gas on homo-ionic soils. J. Soil Sci. 24:104–116.

Ashworth, J. 1978. Reactions of ammonia with soil. II. Sorption of NH_3 on English soils and on Wyoming bentonite. J. Soil Sci. 29:195–205.

Atanasiu, N., A. Westphal, and A. K. Banerjee. 1968. Studien über die Wirkung gedüngten Stickstoffs auf Ertrag und N-Aufnahme der Pflanze bei Böden mit verschiedenem Ammoniumfixierungsvermögen. Agrochimica 12:120–129.

Axley, J. H., and J. O. Legg. 1960. Ammonium fixation in soils and the influence of potassium on nitrogen availability from nitrate and ammonium sources. Soil Sci. 90:151–156.

Baert de Bièvre, M., A. van den Hende, and F. Lox. 1961. Étude de la fixation d'ammonium dans deux sols argileux à l'aide de N^{15}. Pédologie. Symp. int. 2, Appl. sc. nucl. ped., 104–118.

Barshad, I. 1951. Cation exchange in soils. I. Ammonium fixation and its relation to potassium fixation and to determination of ammonium exchange capacity. Soil Sci. 72:361–371.

Barshad, I. 1954a. Cation exchange in micaceous minerals: I. Replaceability of the interlayer cations of vermiculite with ammonium and potassium ions. Soil Sci. 77:463–472.

Barshad, I. 1954b. Cation exchange in micaceous minerals: II. Replaceability of ammonium and potassium from vermiculite, biotite, and montmorillonite. Soil Sci. 78:57–76.

Bennett, E. 1949. Fixation of ammonia by lignin. Soil Sci. 68:399–400.

Black, A. S., and S. A. Waring. 1972. Ammonium fixation and availability in some cereal producing soils in Queensland. Aust. J. Soil Res. 10:197–207.

Blasco, M. L., and A. H. Cornfield. 1966. Fixation of added ammonium and nitrification of fixed ammonium in soil clays. J. Sci. Food Agric. 17:481–484.

Bolt, G. H. 1967. Cation exchange equations used in soil science—A review. Neth. J. Agric. Sci. 15:81–103.

Bower, C. A. 1950. Fixation of ammonium in difficulty exchangeable form under moist conditions by some soils of semiarid region. Soil Sci. 70:375–382.

Bower, C. A. 1951. Availability of ammonium fixed in difficulty exchangeable form by soils of semiarid region. Soil Sci. Soc. Am. Proc. 15:119–122.

Bredell, G. S., and N. T. Coleman. 1968. Factors affecting displacement of fixed ammonium from vermiculite and hydrobiotite. S. Afr. J. Agric. Sci. 11:735–742.

Bremner, J. M. 1965. Inorganic forms of nitrogen. p. 1179–1237. *In* C. A. Black et al. (ed.) Methods of soil analysis, Part 2. Am. Soc. Agron., Inc., Madison, Wis.

Bremner, J. M., W. D. Nelson, and J. A. Silva. 1967. Comparison and evaluation of methods of determining fixed ammonium in soils. Soil Sci. Soc. Am. Proc. 31:466–472.

Bremner, J. M., and K. Shaw. 1954. Studies on the estimation and decomposition of amino sugars in soil. J. Agric. Sci. 44:152–159.

Broadbent, F. E., W. D. Burge, and T. Nakashima. 1960. Factors influencing the reaction between ammonia and soil organic matter. Int. Congr. Soil Sci., Trans. 7th (Madison, Wis.) II:509–516.

Broadbent, F. E., and F. J. Stevenson. 1966. Organic matter interactions. p. 169–187. *In* H. N. McVickar et al. (ed.) Agricultural Anhydrous Ammonia. Technology and Use. Proc. Symp. St. Louis, Mo., 29–30 Sept. 1965. Agric. Ammonia Inst., Memphis, Tenn.; Am. Soc. of Agron., and Soil Sci. Soc. of Am., Madison, Wis.

Broadbent, F. E., and M. W. Thenabadu. 1967. Extraction of ammonia fixed by soil organic matter. Soil Sci. 104:283–288.

Brown, J. M., and W. V. Bartholomew. 1963. Sorption of gaseous ammonia by clay minerals as influenced by sorbed aqueous vapor and exchangeable cations. Soil Sci. Soc. Am. Proc. 27:160–164.

Burge, W. D., and F. E. Broadbent. 1961. Fixation of ammonia by organic soils. Soil Sci. Soc. Am. Proc. 25:199–204.

Chaminade, R., and G. Drouineau. 1936. Recherches sur la mecanique chemique des cations exchangeables. Ann. Agron. 6:677–690.

Chao, T. T., and W. Kroontje. 1960. Ammonia adsorption phenomena in soils. Trans. 7th Int. Congr. Soil Sci. (Madison, Wis.) II:517–522.

Dalal, R. C. 1975. Effect of associated anions on ammonium adsorption by and desorption from soils. J. Agric. Food Chem. 23:684–687.

Davis, R. O. E., R. R. Miller, and W. Scholl. 1935. Nitrification of ammoniated peat and other nitrogen carriers. J. Am. Soc. of Agron. 27:729–737.

Dhariwal, A. P. S., and F. J. Stevenson. 1958. Determination of fixed ammonium in soils. Soil Sci. 86:343–349.

Dissing-Nielsen, J. 1971. Fixation and release of ammonium in Danish soils. (In Danish). Tidsskr. Planteavl. 75:239–255.

Du Plessis, M. C. F., and W. Kroontje. 1967. Characteristics of ammonia adsorption by homo-ionic clays. Soil Sci. Soc. Am. Proc. 31:176–181.

Flaig, W. 1950. Comparative chemical investigations on natural humic compounds and their model substances. Sci. Proc. R. Dublin Soc., Ser. A. 1:149–162.

Freney, J. R. 1964. An evaluation of naturally occurring fixed ammonium in soils. J. Agric. Sci. 63:297–303.

Gapon, E. N. 1933. Theory of exchange adsorption in soils. (In Russian). J. Gen. Chem. USSR. 3:144–152.

Gloss, G. H. 1953. Sodium tetraphenylboron: A new analytical reagent for potassium, ammonium, and some organic nitrogen compounds. Chemist-Analyst 42:50–55.

Hanawalt, R. B. 1969. Soil properties affecting the sorption of atmospheric ammonia. Soil Sci. Soc. Am. Proc. 33:725–729.

Hanway, J. J., and A. D. Scott. 1956. Ammonium fixation and release in certain Iowa soils. Soil Sci. 82:379–386.

Hanway, J. J., A. D. Scott, and G. Stanford. 1957. Replaceability of ammonium fixed in clay minerals as influenced by ammonium or potassium in the extracting solution. Soil Sci. Soc. Am. Proc. 21:29–34.

Harada, T., and K. Kutsuna. 1954. Ammonium fixation by residual soil from crystalline schists at Yahatahama. Bull. Nat. Inst. Agric. Sci. Japan, Ser. B 3:17–41.

Hinman, W. C. 1966. Ammonium fixation in relation to exchangeable K and organic matter content of two Saskatchewan soils. Can. J. Soil Sci. 46:223–225.

Hodge, J. E. 1953. Chemistry of browning reactions in model systems. J. Agric. Food Chem. 1:928–943.

James, D. W., and M. E. Harward. 1964. Competition of NH₃ and H₂O for adsorption sites on clay minerals. Soil Sci. Soc. Am. Proc. 28:636–640.

Jansson, S. L. 1958. Tracer studies on nitrogen transformations in soil with special attention to mineralisation-immobilization relationships. K. Lantbrukshögsk. Ann. 24:101–361.

Jansson, S. L. 1960. On the humus properties of organic manures. II. Potential humus properties. K. Lantbrukshögsk. Ann. 26:135–172.

Jansson, S. L., and J. Ericsson. 1961. Kväve- och kaliumproblem i skånsk växtodling. En preliminär redogörelse. Socker, Handl. 17:9–21.

Jensen, H. E. 1973. Potassium-calcium exchange on a montmorillonite and a kaolinite clay. II. Application of double-layer theory. Agrochimica 17:191–201.

Joffe, J. S., and A. K. Levine. 1947. Fixation of potassium in relation to exchange capacity of soils. II. Associative fixation of other cations, particularly ammonium. Soil Sci. 63:151–158.

Jung, J. 1959. Vergleichende Überprüfung verschiedener Stickstoffverbindungen auf ihre chemische Reaktion mit Rohhumus und die photometrische Erfassung dieses Reaktionseffektes. Z. Pflanzenernähr. Düng. Bodenkd. 85:104–112.

Kaila, A. 1962. Fixation of ammonium in Finnish soils. J. Sci. Agric. Soc. Finl. 34:107–114.

Kaila, A. 1966. Fixed ammonium in Finnish soils. J. Sci. Agric. Soc. Finl. 38:49–58.

Kaila, A., and P. Hänninen. 1961. Fertilizer nitrogen in soil. J. Sci. Agric. Soc. Finl. 33:169–184.

Kowalenko, C. G. 1978. Nitrogen transformations and transport over 17 months in field fallow microplots using ^{15}N. Can. J. Soil Sci. 58:69–76.

Kowalenko, C. G., and D. R. Cameron. 1976. Nitrogen transformation in an incubated soil as affected by combinations of moisture content and temperature and adsorption-fixation of ammonium. Can. J. Soil Sci. 56:63–70.

Kurtz, L. T., and G. E. Smith. 1966. Nitrogen fertility requirements. p. 195–235. In W. H. Pierre et al. (ed.) Advances in corn production. The Iowa State Univ. Press, Ames, Iowa.

Lamm, C. G., and M. H. Nafady. 1973. Plant nutrient availability in soils. III. Studies on potassium in Danish soils. 6. The rate of release and availability of fixed potassium. Agrochimica 17:435–444.

Legg, J. O., and F. E. Allison. 1959. Recovery of N^{15}-tagged nitrogen from ammonium-fixing soils. Soil Sci. Soc. Am. Proc. 23:131–134.

Leggett, G. E., and C. D. Moodie. 1963. The release of fixed ammonium from soils by sodium as affected by small amounts of potassium or ammonium. Soil Sci. Soc. Am. Proc. 27:645–648.

Lindbeck, M. R., and J. L. Young. 1965. Polarography of intermediates in the fixation of nitrogen by p-quinone-aqueous ammonia systems. Anal. Chim. Acta 32:73–80.

Lutz, J. A. 1966. Ammonium and potassium fixation and release in selected soils of southeastern United States. Soil Sci. 102:366–372.

Mangum, D. L., and J. L. Young. 1965. Influence of anhydrous ammonia on solubility of soil organic matter. West Soil Sci. Soc. Abstr. 3–4.

Marques, J. M., and J. A. Vidas. 1974. Ammonium fixation in K depleted soil clays. Agrochimica 18:90–93.

Marshall, C. E. 1964. The physical chemistry and mineralogy of soils. John Wiley & Sons, Inc., New York.

Martin, A. E., R. J. Gilkers, and J. O. Skjemstad. 1970. Fixed ammonium in soils developed on some Queensland phyllites and its relation to weathering. Aust. J. Soil Res. 8:71–80.

Mattson, S., and E. Koutler-Andersson. 1942. The acid-base condition in vegetation, litter and humus: V. Products of partial oxidation and ammonia fixation. Lantbrukshögsk. Ann. 10:284–332.

Mattson, S., and E. Koutler-Andersson. 1943. The acid-base condition in vegetation, litter and humus: VI. Ammonia fixation and humus nitrogen. Lantbrukshögsk. Ann. 11:107–134.

McDonnell, P. M., F. J. Stevenson, and J. M. Bremner. 1959. Release of fixed ammonium from soil by ball milling. Nature (London) 183:1414–1415.

Meints, V. W., and G. A. Peterson. 1972. Further evidence for the inability of the Kjeldahl nitrogen method to fully measure indigenous fixed ammonium nitrogen in subsoils. Soil Sci. Soc. Am. Proc. 36:434–436.

Mitchell, J. K. 1977. Fundamentals of soil behaviours. John Wiley & Sons, Inc., New York.

Mortland, M. M. 1955. Adsorption of ammonia by clays and muck. Soil Sci. 80:11–18.

Mortland, M. M. 1958. Reaction of ammonia in soils. Adv. Agron. 10:325–348.

Mortland, M. M. 1966. Ammonia interactions with soil materials. p. 188–197. *In* H. N. Mc-Vickar et al. (ed.) Agricultural Anhydrous Ammonia. Technology and Use. Proc. Symp. St. Louis, Mo., 29–30 Sept. 1965. Agric. Ammonia Inst., Memphis, Tenn.; Am. Soc. of Agron., and Soil Sci. of Am., Madison, Wis.

Mortland, M. M., K. Lawton, and G. Uehara. 1957. Fixation and release of potassium by some clay minerals. Soil Sci. Soc. Am. Proc. 21:381–384.

Mortland, M. M., and A. R. Wolcott. 1965. Sorption of inorganic nitrogen compounds by soil materials. *In* W. V. Bartholomew and F. E. Clark (ed.) Soil nitrogen. Agronomy 10:150–197. Am. Soc. of Agron., Madison, Wis.

Newman, A. C., and S. Oliver. 1966. Isotopic exchange of fixed ammonium. J. Soil Sci. 17:159–174.

Nommik, H. 1957. Fixation and defixation of ammonium in soils. Acta Agric. Scand. 7:395–436.

Nommik, H. 1965. Ammonium fixation and other reactions involving a nonenzymatic immobilization of mineral nitrogen in soil. *In* W. V. Bartholomew and F. E. Clark (ed.) Soil nitrogen. Agronomy 10:198–258. Am. Soc. of Agron., Madison, Wis.

Nommik, H. 1966. Particle-size effect on the rate of nitrification of nitrogen fertilizer materials with special reference to ammonium-fixing soils. Plant Soil 24:181–200.

Nommik, H. 1967. Ammoniated peat as a nitrogen carrier. Acta Agric. Scand. 17:25–29.

Nommik, H. 1970. Non-exchangeable binding of ammonium and amino nitrogen by Norway spruce raw humus. Plant Soil 33:581–595.

Nommik, H., and K. O. Nilsson. 1963a. Nitrification and movement of anhydrous ammonia in soil. Acta Agric. Scand. 13:205–219.

Nommik, H., and K. O. Nilsson. 1963b. Fixation of ammonia by the organic fraction of the soil. Acta Agric. Scand. 13:371–390.

Nyborg, M. 1969. Fixation of gaseous ammonia by soils. Soil Sci. 107:131–136.

Ogner, G. 1972. Leaching of organic matter from forest soil after fertilization with urea. Medd. Nor. Skogforsöksves. 123:428–440.

Opuwaribo, E., and C. T. I. Odu. 1974. Fixed ammonium in Nigerian soils. I. Selection of a method and amounts of native fixed ammonium. J. Soil Sci. 25:256–264.

Osborne, G. J. 1976. The significance of intercalary ammonium in representative surface and subsoils from southern New South Wales. Aust. J. Soil Res. 14:381–388.

Page, J. B., and L. D. Baver. 1940. Ionic size in relation to fixation of cations by colloidal clay. Soil Sci. Soc. Am. Proc. 4:150–155.

Page, A. L., W. D. Burge, and T. J. Ganje. 1967. Potassium and ammonium fixation by vermiculitic soils. Soil Sci. Soc. Am. Proc. 31:337–341.

Papendick, R. I., and J. F. Parr. 1966. Retention of anhydrous ammonia by soil. III. Dispensing apparatus and resulting ammonia distribution. Soil Sci. 102:193–201.

Parr, J. F., and R. I. Papendick. 1966. Retention of ammonia in soils. p. 213–236. *In* H. N. McVickar et al. (ed.) Agricultural Anhydrous Ammonia. Technology and Use. Proc. Symp. St. Louis, Mo., 29–30 Sept. 1965. Agric. Ammonia Inst. Memphis, Tenn.; Am. Soc. of Agron., and Soil Sci. of Am., Madison, Wis.

Peech, M. 1948. Chemical methods for assessing soil fertility. p. 1–52. *In* H. B. Kitchen (ed.) Diagnostic techniques for soils and crops. Am. Potash Inst., Washington, D.C.

Porter, L. K., and B. A. Stewart. 1970. Organic interferences in the fixation of ammonium by soils and clay minerals. Soil Sci. 100:229–233.

Raju, G. S. N., and A. K. Mukhopadhyay. 1975. Effect of the sequence of addition of potassium and ammonium and preadsorbed cation on fixation of applied ammonium ions in soils. J. Indian Soc. Soil Sci. 23:172–176.

Raju, G. S. N., and A. K. Mukhopadhyay. 1976. Ammonium fixing capacities of West Bengal soils. J. Indian Soc. Soil Sci. 24:270–274.

Rich, C. I., and J. A. Lutz, Jr. 1965. Mineralogical changes associated with ammonium and potassium fixation in soil clays. Soil Sci. Soc. Am. Proc. 29:167–170.

Rich, C. I., and S. S. Obenshain. 1955. Chemical and clay mineral properties of a red-yellow podsolic soil derived from muscovite schist. Soil Sci. Soc. Am. Proc. 19:334–339.

Schachtschabel, P. 1940. Untersuchungen über die Sorption der Tonmineralien und organischen Bodenkolloide. Kolloid-Beih. 51:199–276.

Schachtschabel, P. 1961. Fixierung und Nachlieferung von Kalium- und Ammonium-Ionen. Beurteilung und Bestimmung des Kaliumversorgungsgrades von Böden. Landw. Forschung. Sonderheft 15:29–47.

Scheffer, F., and P. Schachtschabel. 1966. Lehrbuch der Bodenkunde. Ferdinand Enke, Verlag, Stuttgart.

Scheffer, F., and B. Ulrich. 1960. Humus und Humusdüngung. Ferdinand Enke, Verlag, Stuttgart. p. 39–125.

Schiller, H., and A. Walicord. 1969. Das Verhältnis Kaliumfixierung/Ammoniumfixierung in Boden—ein ertragsbestimmender Faktor. Z. Pflanzenernähr. Düng. Bodenkd. 104:119–130.

Scholl, W., and R. O. E. Davis. 1933. Ammoniation of peat for fertilizers. Ind. Eng. Chem. 25:1074–1078.

Schuffelen, A. C. 1972. The cation exchange system of the soil. p. 75–88. In Potassium in soil. 9th Coll. Int. Potash Inst. Proc., Landshut, Germany.

Scott, A. D., J. J. Hanway, and A. P. Edwards. 1958. Replaceability of ammonium in vermiculite with acid solutions. Soil Sci. Soc. Am. Proc. 22:388–392.

Scott, A. D., A. P. Edwards, and J. M. Bremner. 1960a. Removal of fixed ammonium from clay minerals by cation exchange resins. Nature (London) 185:792.

Scott, A. D., R. R. Hunziker, and J. J. Hanway. 1960b. Chemical extraction of potassium from soils and micaceous minerals with solutions containing sodium tetraphenylboron. I. Preliminary experiments. Soil Sci. Soc. Am. Proc. 24:191–194.

Scott, A. D., and M. G. Reed. 1962. Chemical extraction of potassium from soils and micaceous minerals with solutions containing sodium tetraphenylboron. III. Illite. Soil Sci. Soc. Am. Proc. 26:45–48.

Shainberg, I., and W. D. Kemper. 1967. Ion exchange equilibria on montmorillonite. Soil Sci. 103:4–9.

Shilova, E. J. 1969. Fixation of ammonia from various forms of nitrogen fertilizers and utilization of the fertilizer nitrogen by plants. Izv. Timiryazevsk. Skh. Akad. 4:137–143.

Shoji, S., and S. Matsumi. 1961. Chemical characteristics of peat soils. 2. Non-biological fixation of ammonia by peat soils and availability of the fixed ammonia. Res. Bull. Hokkaido Natl. Agric. Exp. Stn. 76:37–41.

Silva, J. A., and J. M. Bremner. 1966. Determination and isotope-ratio analysis of different forms of nitrogen in soils. 5. Fixed ammonium. Soil Sci. Soc. Am. Proc. 30:587–594.

Sippola, J. 1976. Fixation of ammonium and potassium applied simultaneously in Finnish soils. Ann. Agric. Fenn. 15:304–308.

Sippola, J., R. Erviö, and R. Eleveld. 1973. The effects of simultaneous addition of ammonium and potassium on their fixation in some Finnish soils. Ann. Agric. Fenn. 12:185–189.

Smirnov, P. M., and N. I. Fruktova. 1963. Nonexchangeable fixation of ammonia by soils. (In Russian). Pochvovedenie 3:83–93.

Sohn, J. B., and M. Peech. 1958. Retention and fixation of ammonia by soils. Soil Sci. 85:1–9.

Sowden, F. J. 1976. Transformation of nitrogen added as ammonium and manure to soil with a high ammonium-fixing capacity under laboratory conditions. Can. J. Soil Sci. 56:319–331.

Sowden, F. J., A. A. McLean, and J. G. Ross. 1978. Native clay-fixed ammonium content and the fixation of added ammonium of some soils of Eastern Canada. Can. J. Soil Sci. 58:27–38.

Stanford, G. 1948. Fixation of potassium in soils under moist conditions and on drying in relation to type of clay mineral. Soil Sci. Soc. Am. Proc. 12:167–171.

Stanford, G., and W. H. Pierre. 1947. The relation of potassium fixation to ammonium fixation. Soil Sci. Soc. Am. Proc. 11:155–160.

Stanley, F. A., and G. E. Smith. 1956. Effect of soil moisture and depth of application on retention of anhydrous ammonia. Soil Sci. Soc. Am. Proc. 20:557–561.

Stewart, B. A., and L. K. Porter. 1963. Inability of the Kjeldahl method to fully measure indigenous fixed ammonium in some soils. Soil Sci. Soc. Am. Proc. 27:41–43.

Themlitz, R. 1958. Umsetzung verschiedener N-Dünger mit einem durch voraufgegangene Bestandskalkung bzw.—stickstoffdüngung umgewandelten Fichtenrohhumus. Z. Pflanzenernähr. Dung. Bodenkd. 82:165–174.

Thomas, G. W. 1977. Historical developments in soil chemistry: Ion exchange. Soil Sci. Soc. Am. Proc. 41:230–238.

Thompson, H. S. 1850. On the absorbent power of soils. J. R. Agric. Soc. Engl. 11:68–74.

Valdmaa, K. 1969. Ammonia fixation of humus substances in arable soils and organic fertilizers. Lantbrukshögsk. Ann. 35:199–228.

Van Schreven, D. A. 1968. Ammonium fixation and availability of fixed ammonium in some Dutch loam and clay soils. Neth. J. Agric. Sci. 16:91–102.

Van der Marel, H. W. 1954. The amount of exchangeable cations of K-fixing soils. Int. Congr. Soil Sci. Trans., 5th (Leopoldville) 2:300–307.

Vanselow, A. P. 1932. Equilibria of the base exchange reactions of bentonites, permutites, soil colloids and zeolites. Soil Sci. 33:95–113.

Vopènko, L., and A. Nemec. 1976. Fixation and release of fixed ammonium in soil. (In Czech.). Rostl. Výroba 22:1236–1269.

Walsh, L. M., and J. T. Murdock. 1960. Native fixed ammonium and fixation of applied ammonium in several Wisconsin soils. Soil Sci. 89:183–193.

Walsh, L. M., and J. T. Murdock. 1963. Recovery of fixed ammonium by corn in greenhouse studies. Soil Sci. Soc. Am. Proc. 27:200–204.

Way, J. T. 1850. On the power of soils to absorbe manure. J. R. Agric. Soc. Engl. 11:313–379.

Way, J. T. 1852. On the power of soils to absorbe manure. J. R. Agric. Soc. Engl. 13:123–143.

Welch, L. F., and A. D. Scott. 1960. Nitrification of fixed ammonium in clay minerals as affected by added potassium. Soil Sci. 90:79–85.

Wiklander, L. 1950. Fixation of potassium by clays saturated with different cations. Soil Sci. 69:261–268.

Wiklander, L. 1958. The soil. p. 118–169. In W. Ruhland (ed.) Handbuch der Pflanzenphysiologi, Band IV, Springer-Verlag, Berlin-Göttingen-Heidelberg.

Wiklander, L. 1964. Cation and anion exchange phenomena. p. 163–205. In F. E. Bear (ed.) Chemistry of the soil. 2nd ed. Reinhold Publ. Co., New York.

Wiklander, L., and E. Andersson. 1959. Kalkens markeffekt. III. Kemiska undersökningar av ett längvarigt kalkningsförsök på skifte IV vid Lanna. Grundförbättring 12:1–40.

Young, J. L. 1964. Ammonia and ammonium reactions with some Pacific Northwest soils. Soil Sci. Soc. Am. Proc. 28:339–345.

5   Biochemistry of Ammonification

J. N. LADD AND R. B. JACKSON

Division of Soils, CSIRO
Glen Osmond, South Australia, Australia

I. INTRODUCTION

Ammonification denotes the processes by which organic nitrogenous compounds are transformed in enzymically-catalyzed reactions to yield NH_4^+ as a reaction product. Organic N of soils has been partly characterized, usually after chemical hydrolysis, as amino acids, amino sugars, purines, and pyrimidines (Chapt. 3, F. J. Stevenson). Such compounds represent prime organic substrates in deamination reactions. In this chapter, however, we discuss not only the mechanisms and enzymology of deamination reactions, but also the formation of the prime organic substrates by decomposition of materials of high molecular weight. Presumably degradation of the latter involves enzymic hydrolyses of the proteins, aminopolysaccharides, and nucleic acids, which are present in soil in living and dead cells, and as exocellular, stabilized residues.

In addition to organic N of cellular origin, soils may receive organic N as *urea*, a major nitrogenous constituent of the urine of grazing animals, and an applied fertilizer. Decomposition of urea by ureases is an important reaction in the ammonification process and is the subject of an increasing number of studies, to the extent that urease is now the most thoroughly examined of all soil enzymes.

In soils, therefore, substrates which participate directly and indirectly in ammonification processes are of different origins, belong to different chemical classes, and are present in very different microenvironments. In the same way, hydrolases, oxidases, deaminases, and lyases, which act upon these substrates, may originate directly from various plant, animal, and microbial sources. The enzymes may function endocellularly, in dead autolysing cells, free in the soil solution, and when adsorbed to soil colloids. For these reasons soils are unlikely to be rich sources for obtaining pure enzymes in high yields. Also, detailed studies of enzyme properties involving estimates of the kinetic constants of enzymes assayed in whole soils are not especially useful. Nevertheless, limited studies with soils and soil extracts are of importance in determining the states in which enzymes occur in soils, their potential activities towards specific substrates, how they are affected by inhibitors, etc. Such studies, especially of soil proteinases and ureases, are included in this chapter. In situ studies of other enzymes of ammonification, for example, nucleases, and deaminases of amino acids, purines, and pyrimidines are virtually nonexistent.

The biochemical literature reveals that all reactions in the overall process of ammonification have been studied in pure systems in varying detail. The enzymes involved range widely in their action and properties. Many of the enzymes, including some from microbial isolates from soil, have been characterized at the molecular level and reaction mechanisms have often been closely defined. Such enzymes are representatives of those acting in soils and a general review of their properties is presented in this chapter. Included too is some discussion on the behavior of pertinent pure enzymes that have been immobilized on solid support materials. Properties of these insolubilized enzymes may have direct relevance to the properties of enzymes of ammonification in soils, since it is believed that these, in part, may function exocellularly, either adsorbed to, covalently bonded to, or enmeshed within soil colloids. Definition of the microenvironments of exocellular enzymes in soil and determination of the activities of exocellular enzymes under normal conditions of substrate supply are subjects of great interest, and remain as major challenges to soil biochemists.

II. PROTEINS, PEPTIDES, AMIDES, AMIDINES, AND AMINO ACIDS

The formation of NH_4^+ from proteins and peptides requires that initially they are hydrolyzed by a sequence of reactions to form amino acids. Enzymes catalyzing the hydrolysis of peptide bonds are proteinases and peptidases (peptide hydrolases, EC 3.4). Amino acids may also be formed in certain related hydrolytic reactions, involving the cleavage of C–N bonds other than peptide bonds. These reactions are catalyzed by amidohydrolases (EC 3.5.1) and amidinohydrolases (EC 3.5.3). Two enzymes of the amidohydrolase group, asparaginase (EC 3.5.1.1) and glutaminase (EC 3.5.1.2), cleave respectively asparagine and glutamine to release NH_4^+ and the amino acids, aspartate, and glutamate.

Amino acids formed from proteins, peptides, and certain amides are further metabolized, with NH_4^+ release, by reactions catalyzed by amino acid dehydrogenases (EC 1.4.1), amino acid oxidases (EC 1.4.3), or carbon-nitrogen lyases (EC 4.3.1).

The enzymes in each of these classes range widely, both in their origin and in their molecular and catalytic properties. In the following sections these properties will be discussed briefly, with emphasis being given, where appropriate, to those properties which may be especially relevant to the function of the enzymes in the soil environment. More detailed information on the properties of purified enzymes can be found in recent biochemistry texts, such as Metzler (1977) and Wingard et al. (1976).

A. Proteinases and Peptidases

1. CLASSIFICATION, ORIGIN, AND SPECIFICITY

Proteinases are classified according to their mechanisms of catalysis into four major groups, (i) serine proteinases, (ii) SH proteinases, (iii) acid proteinases and (iv) metalloproteinases. Those enzymes which typify their class are usually of animal or plant origin, but several microbial proteinases, mainly from cultures of soil isolates, have also been well-characterized.

Serine proteinases (EC 3.4.21) are inactivated irreversibly by reaction with diisopropylphosphofluoridate; this reagent combines with the OH groups of catalytically-active serine residues, to form O-phosphoserine derivatives. Typical serine proteinases are chymotrypsin, trypsin, subtilisin from *Bacillus subtilis* and other *Bacillus* sp., and proteinases from *Arthrobacter* sp. and *Aspergillus* sp.

Sulfhydryl proteinases (EC 3.4.22) (e.g., papain, ficin) are generally of plant origin, but included in the group are proteinases from *Clostridium histolyticum* (clostripain) and from *Streptococcus* sp. and *Saccharomyces* sp. Enzymes of this group have a cysteine residue at the active site and are typically inactivated by compounds, e.g., *p*-chloromercuribenzoate, which react with SH groups.

Acid proteinases (EC 3.4.23) have pH optima below 5. The group includes pepsin and proteinases from *Saccharomyces* sp., and from various fungi (*Aspergillus, Penicillium, Rhizopus, Endothia*).

Metalloproteinases (EC 3.4.24) require the participation of a metal ion, usually Zn^{2+} in the catalysis. These proteinases, which are true metalloenzymes, have been purified from animal and many microbial sources. The group includes collagenase, thermolysin (from *B. thermoproteolyticus*), and neutral proteinases from *B. subtilis, Proteus aeruginosa, Streptomyces* sp., and *Aspergillus oryzae*.

Some proteinases, e.g., subtilisin, exhibit a broad specificity of action, but most proteinases preferentially hydrolyze specific peptide bonds. Specificity is determined by the nature of the amino acid residues forming the peptide bond to be hydrolyzed. For example, chymotrypsin hydrolyzes peptide bonds that are formed from the COOH groups of hydrophobic amino acids (tyrosine, tryptophane, phenylalanine, or leucine). By contrast, trypsin hydrolyzes bonds formed from the COOH groups of basic amino acids (arginine or lysine). Some proteinases catalyze by analogous reactions the release of NH_4^+ from amides, or the formation of alcohols from esters. Specifically-synthesized amides and esters have proved to be useful substrates in many proteinase assays.

Peptidases (EC 3.4.11-15) include enzymes which specifically hydrolyze dipeptide substrates, and those which split off single amino acids and dipeptide units from the C-terminus and the N-terminus of longer peptide chains. Most of the characterized peptidases are of animal origin.

Within each group, peptidases may be further distinguished by their preferential hydrolyses of peptide bonds formed between specific amino acid residues. Many peptidases are metalloenzymes.

2. REACTION MECHANISMS

Peptide bond hydrolysis is a nucleophilic displacement reaction in which a base, or a basic group at the enzyme's active site, becomes bonded to the electrophilic C atom of the CO group in peptide linkage (Eq. [1]). At the same time, the N atom is displaced, and, simultaneously or subsequently, receives a proton donated by an acid group of the enzyme or by water.

$$\text{enzyme}^- \;+\; R-\overset{\overset{\displaystyle O}{\|}}{\underset{\underset{\displaystyle H}{|}}{C}}-N-R^1 \;+\; H^+ \qquad\qquad\qquad [1]$$

$$\longrightarrow R-\overset{\overset{\displaystyle O}{\|}}{\underset{\underset{\displaystyle \text{enzyme}}{|}}{C}} \quad\overset{OH^-}{\longrightarrow}\quad R-\overset{\overset{\displaystyle O}{\|}}{\underset{\underset{\displaystyle OH}{|}}{C}} \;+\; \text{enzyme}^-$$

$$R^1-NH_2$$

The leaving group, containing the displaced N atom, may be an amino acid or a peptide. The latter, in the presence of the appropriate proteinases and peptidases, will undergo sequential hydrolyses until amino acids only are formed.

Experiments with chymotrypsin, a serine proteinase, suggest that an acyl-enzyme intermediate is formed during hydrolysis. Both serine and histidine residues have been implicated in the reaction mechanism and X-ray diffraction studies have established their presence at the enzyme's active site. It is believed that the $-O^-$ group of the serine residue acts as the base in forming the acyl-enzyme intermediate.

The different specificities of the two typical serine proteinases, chymotrypsin and trypsin, are explained by differences in the nature of other amino acid residues which form the enzymes' substrate-binding sites. In chymotrypsin, the site is composed of amino acids with hydrophobic side chains; whereas in trypsin (which preferentially hydrolyzes peptide bonds between basic amino acid residues) the binding site is negatively charged, due to a COO^- group of an aspartate residue.

The other groups of proteinases lack a serine residue at their active site. The displacing nucleophiles are thought to be a COO^-, in the acid proteinases, and an SH group in the SH proteinases. Of the metal-containing enzymes, the peptidase, carboxypeptidase A has been most intensively studied. Here, a Zn^{2+} is chelated to the enzyme and is so positioned at the enzyme's active site that it greatly augments the electrophilic nature of the C atom in peptide linkage. The displacing nucleophile group is either a COO^- of a glutamate residue, or a water molecule.

3. IMMOBILIZED PROTEINASES

Enzymes may be immobilized by entrapment within, or by bonding to the matrixes of insoluble, porous support materials. Studies of the catalytic behavior of immobilized proteinases were prominent in the early investigations in this field and included the work of McLaren and co-workers (McLaren & Estermann, 1957; Estermann et al., 1959a) on the pH-activity profiles of chymotrypsin, free or adsorbed to kaolinite. Many recent reviews testify to the rapid growth of interest in the properties of immobilized enzymes (for example, Goldstein, 1970; McLaren & Packer, 1970; Goldman et al., 1971; Zaborsky, 1973; Weetall, 1975; Mosbach, 1976; Wingard et al., 1976).

Almost invariably, bonding of proteinases to insoluble support materials decreases their specific activities, especially towards protein substrates. Activities of the enzymes towards low molecular weight substrates are relatively less affected, indicative of less diffusional resistance and steric hindrance. Similarly, steric restrictions may limit the accessibility of the active sites of immobilized proteinases to inhibitors of comparatively low molecular weight. Also, a high molecular weight inhibitor of native trypsin little affected the activities of immobilized trypsin derivatives when assayed against a low molecular weight substrate, but completely inhibited their activities towards protein substrates. Thus, the high molecular weight inhibitor appeared to reach those active sites of the bound enzyme that were also accessible to protein substrates, but did not reach additional sites accessible to low molecular weight substrates only. Again, perhaps in an analogous way, the degradation of immobilized proteinases by exocellular microbial proteinases may be limited to those bound enzyme molecules that are also accessible to protein substrates. Due to steric constraints, immobilized proteinases active against low molecular weight substrates only are expected to be protected from biological attack.

Michaelis constant (K_m) values and pH-activity profiles of immobilized proteinases appear to differ from those of the free enzymes. These effects may result from diffusional resistances, which by comparison with the bulk solution, decrease the concentrations of substrates and increase those of products (including H^+) within the microenvironment of the bound enzymes. Thus, K_m values of bound proteinases are higher than for the free enzymes and the pH optima are displaced towards more alkaline values.

Similarly, ionic interactions between substrates, H^+, and charged enzyme support material will also influence K_m values and the optimal pH for enzyme activity. Thus, pH-activity profiles of proteinases immobilized on polyanionic supports will, when compared with those of free proteinases, be displaced towards more alkaline regions, due to higher H^+ concentrations in the vicinity of the insoluble enzymes (McLaren, 1960; McLaren & Estermann, 1957). Likewise, K_m values appear to increase, since concentrations of the negatively charged protein or peptide substrates within the anionic matrix of the immobilized enzyme are less than those in the bulk solution. By contrast, K_m values decrease and pH-activity profiles are displaced towards more acidic regions when proteinases are bound to polycationic supports. Immobilized proteinases are less vulnerable than free proteinases to biochemical degradation. In addition to protection afforded by steric constraints on the activities of free exocellular proteinases, immobilized proteinases are anchored and are physically separated on and within the support matrix. Thus, they do not autodigest as readily as enzymes free in solution (Rowell et al., 1973).

Also, bonding of proteinases to insoluble support material often enhances their stability against denaturation by heating, alkaline pH, chemical agents such as $8M$ urea, and oxidation. However, there are many instances where the stabilities of proteinases are unaffected or are decreased by immobilization on insoluble supports. The response varies with the enzyme, the nature of the support material, and the type of enzyme-support bonding. The comparative stabilities of the immobilized enzymes are determined primarily by the nature of the amino acid residues bonded to the insoluble carrier and by their role in maintaining the tertiary structure of the enzyme.

B. Proteinases and Peptidases in Soil

1. ASSAY, ACTIVITY, AND SPECIFICITY

Proteins, when added to soils and incubated, are readily decomposed; during incubation, microorganisms proliferate and NH_4^+ is formed. Presumably, the added proteins are initially hydrolyzed by exocellular proteinases. Subsequent metabolism of the amino acid products involves some de novo synthesis and degradation of microbial biomass protein, and the participation of newly-formed microbial exocellular and endocellular proteinases.

Assays of soil proteinase activities are designed to obtain measurements of the rates of hydrolysis of added proteins under conditions where microbial growth and further metabolism of hydrolytic products are eliminated or minimized (Kiss et al., 1975; Skujins, 1967, 1976; Ladd, 1978). Most assays are of several days duration and toluene is usually included as a bacteriostatic agent. Voets and Dedeken (1964) found that toluene-treated soils hydrolyzed gelatin without NH_4^+ release; added amino acids were not deaminated (Voets et al., 1965). Amino acids were also formed and identified in toluene-treated soils incubated without amendment with protein substrates (Dedeken & Voets, 1965). By contrast, Ambroz (1966b) demonstrated NH_4^+ to be present in toluene-treated soils incubated with gelatin for 1 to 2 days. Concentrations of NH_4^+ at the beginning of the incubation, and in control soils without substrate, were not reported so no assessment can be made of the NH_4^+ concentrations, if any, that were due to deamination of amino acids released from the added gelatin. However, provided that toluene effectively inhibits assimilation reactions, some further catabolism (deamination) of amino acid products is unlikely to cause serious errors in those long-term proteinase assays which are based on measurements of the release of amino compounds generally (e.g., compounds reactive with ninhydrin reagents).

Nevertheless, the use of toluene or other bacteriostatic agents may pose other problems in enzyme assays. Not only may the efficiencies of the agents vary with the soil and experimental conditions, but they may exert effects in addition to inhibition of microbial growth (see reviews by Durand, 1965a; Skujins, 1976; Burns, 1978). Short-term assays (< 2 hours) avoid potential problems since assimilation of proteolysis products due to microbial growth is likely to be minimal and bacteriostatic agents can be omitted from the assay reaction mixtures (Ladd & Butler, 1972). Also, with assays of only several hours duration, the reaction conditions necessary for accurate comparisons of potential proteinase activities between soils and soil treatments are more likely to be met (Ladd & Butler, 1972).

Many assays of soil proteinase activities have been described. Substrates have included ovalbumin (Ambroz, 1965), casein (Kuprevich & Shcherbakova, 1961, 1971; Ambroz, 1971; Ladd & Butler, 1972; Mayaudon et al., 1975; Speir et al., 1980), azocasein (Macura & Vagnerova, 1969) haemoglobin (Antoniani et al., 1954; Ladd & Butler, 1972) and gelatin (Hoffmann & Teicher, 1957; Voets & Dedeken, 1964, 1965; Ambroz, 1971; others). Activities have been estimated from the rates of release of amino compounds, as determined with a ninhydrin (Ladd & Butler, 1972) and a copper reagent (Hoffmann & Teicher, 1957), from the formation of colored products (Macura & Vagnerova, 1969), from the formation of TCA—soluble tyrosyl derivatives (Ladd & Butler, 1972; Speir et al., 1980), and from the release of arginine (Voets & Dedeken, 1964).

It might be anticipated that proteinase assays applied to whole soils or crude soil extracts would measure the potential activities of a range of proteinases and that substrate specificities of individual enzymes would be obscured. Certainly, no preferential hydrolysis of specific peptide bonds, based on analyses of protein hydrolysis products, has been reported. Ambroz (1965, 1966a, 1970) however, found that soils differed in their relative rates of hydrolysis of casein, assayed at pH 8.5, and of gelatin, assayed at pH 6.0. Also, casein hydrolysing activities were more sensitive to loss by heating or drying of soils (Ambroz, 1966a, b, 1968). Ladd and Butler (1972) reported that soils differed in their relative activities towards haemoglobin and casein substrates assayed under otherwise identical conditions.

Comparisons of soil proteinase activities reported from different laboratories is a difficult and probably fruitless exercise. Protein hydrolysis is a complex sequence of reactions. Even with a pure enzyme functioning under rigidly controlled assay conditions, a variety of bonds are cleaved at different catalytic rates; and the choice of the initial protein substrate influences both the accessibility of specific peptide bonds and also the nature and concentration of peptide products and substrates in the sequential degradative process. Assays for soil proteinase activities have varied in the choice of substrate, buffer, incubation time, temperature, pH, and methods of product analysis. Under the assay conditions described by Ladd and Butler (1972), the observed proteinase activity was directly related to soil concentration and to assay incubation time. For a limited number of soils, activities towards haemoglobin ranged from 0.032 to 0.083 μmol, and, towards casein, from 0.39 to 1.64 μmol tyrosine equivalents released per gram of soil per hour.

Soils may readily hydrolyze dipeptide derivatives such as benzyloxycarbonyl-phenylalanyl leucine (ZPL), or substituted amides such as N-benzoyl-L-arginine amide (BAA) (Ladd & Butler, 1972). These and similar compounds have proved to be most useful in establishing the substrate specificities of many pure proteinases. Having but one bond to be cleaved by the proteinase, the specificity of attack is more clearly defined, and assay conditions are more readily established under which activity is linearly related to enzyme concentration and assay incubation time. Ladd and Butler (1972) have described such conditions for assaying activities of whole soil suspensions. However, it should be stressed that in the diverse environment of a soil, hydrolysis of these comparatively simple dipeptide substrates of relatively low molecular weight may be by the concerted action of both proteinases and peptidases, and that these enzymes could be both exocellular and endocellular. By contrast, added proteins are presumed to be hydrolyzed by exocellular proteinases only.

N-Benzyloxycarbonyl (Z) derivatives of the dipeptides tested included those containing amino acid residues with either aromatic, acidic, or non-polar aliphatic (long and short) side chains (Ladd & Butler, 1972). Rates of hydrolysis of a given substrate ranged widely with soils of differing pH, clay content, and organic matter content. Nevertheless, soils were consistent in their preferential hydrolysis of those anionic dipeptide derivatives containing amino acid residues with hydrophobic side chains. ZPL was the substrate most readily hydrolyzed by all soils (Table 1). Thus, the use of these anionic substrates demonstrated the presence in soils of enzymes with substrate specificities similar to those of the proteinase, chymotrypsin, or to those of the peptidase, carboxypeptidase A.

By contrast, the rate of hydrolysis of the cationic substrate BAA, (and of casein), relative to that of a Z-dipeptide, varied with each soil (Ladd & Butler, 1972). The ratio of the activities towards BAA and ZPL ranged from 0.2:1 to 0.8:1, respectively. Later studies showed even greater differences, with ratios as high as 15:1 in some topsoils from the New Hebrides (Table 2). The cationic substrate BAA is hydrolyzed by proteinases typified by trypsin and papain.

Table 1—Relative rates of hydrolysis of Z-dipeptides by soils.[†]

Soil[‡]	Z-phenylalanyl leucine hydrolysis, μmol leucine/ g soil per hour	Relative rates of hydrolysis of Z-dipeptides[§]				
		Z-glycyl leucine	Z-glycyl phenyla-lanine	Z-leucyl tyrosine	Z-glutamyl tyrosine	Z-glycyl glycine
Millicent cl	13.9	0.33	0.15	0.13	0.09	0.04
Maitland l	6.9	0.27	0.24	0.16	0.09	0.04
Mt. Crawford sal	2.2	0.11	0.13	0.10	0.02	0.01
Cambrai sa	2.3	0.19	0.18	0.13	0.05	0.01
Cooke Plains sa	2.5	0.16	0.21	0.17	0.08	0.03
Mt. Gambier l	1.8	0.17	0.17	0.16	0.06	0.06

† Adapted from Ladd and Butler (1972).
‡ cl, clay loam; l, loam; sal, sandy loam; sa, sand.
§ Relative to rates of hydrolysis of Z-phenylalanyl leucine (1.00).

Table 2—Hydrolysis of casein, Z-phenylalanyl leucine and benzoyl arginine amide by soils.[†]

Soil	Sample depth, cm	Rates of hydrolysis[‡]		
		Casein	Z-phenylalanyl leucine	N-benzoyl arginine amide
Maitland, S.A., Australia	0–7.5	1.64	6.9	1.3
Cambrai, S.A., Australia	0–7.5	0.24	2.3	0.5
Mt. Gambier, S.A., Australia	0–7.5	0.98	1.8	1.5
Urrbrae, S.A., Australia	0–2.5	0.81	3.8	6.1
Urrbrae, S.A., Australia	2.5–5	0.35	1.6	3.2
Urrbrae, S.A., Australia	5–10	0.07	0.7	2.6
Urrbrae, S.A., Australia	20–25	0.06	0.8	2.3
Bradwell, Sask., Canada	0–25	0.26	4.0	10.3
Efate Island, New Hebrides	0–10	0.25	0.9	14.9
Erromango Island, New Hebrides	0–15	0.6	4.0	18.3
Erromango Island, New Hebrides	15–60	0.0	0.2	6.6
Erromango Island, New Hebrides	60–90	0.0	0.0	1.2
Malekula Island, New Hebrides	0–10	n.d.	7.7	13.9
Malekula Island, New Hebrides	10–50	n.d.	1.0	10.9
Malekula Island, New Hebrides	50–100	n.d.	0.6	10.5

† Adapted from Ladd and Butler (1972), Ladd and Paul (1973), and Butler, Amato and Ladd, unpublished data.

‡ μmol of tyrosine, leucine, and NH_4^+, respectively, released/g soil per hour, from casein, Z-phenylalanyl leucine, and benzoylarginine amide.

It seems likely that soils contain different amounts of proteinases (peptidases) with different substrate preferences. The location of these enzymes within the soil matrix is unknown. Differences in the amounts and proportions of soil proteinases, present either in microbial cells or adsorbed on charged soil colloids, and differences in the interactions of the latter with the cationic and anionic substrates provided, would greatly affect the relative catalytic rates at which substrates are hydrolyzed.

In establishing conditions for assay of ZPL hydrolysis by soils, Ladd and Butler (1972) showed that shaking the soil suspensions increased their activities. Activities of buffered soil suspensions were optimal near pH 8.0 and 60°C. A similar optimal temperature was reported by Hoffmann and Teicher (1957) for gelatin hydrolysis.

Proteinase activities decreased with soil sampling depth (Hoffmann & Teicher, 1957; Amato et al., unpublished data, Table 2), with surface area of rendzina aggregates (Cerna, 1970), and with soil organic matter content (Bei-Bienko, 1970; Franz, 1973). Activities of soils towards casein and ZPL declined more steeply with depth of sampling than did activities towards BAA (Table 2). The rate of hydrolysis of ZPL by topsoils was highly correlated with clay content, CEC, and surface area, but not with organic matter content (Ladd & Butler, 1972). By contrast, activity towards BAA was highly correlated with organic C and N concentrations. Activities towards casein were not related to the capacities of the soils to accumulate inorganic N in incubation tests (Ross & McNeilly, 1975).

2. ORIGIN, STABILITY, AND STATE OF OCCURRENCE

Sources of soil proteinases and peptidases are widespread and variable. A wide range of proteolytic microorganisms can be readily isolated from soils and undoubtedly microorganisms are important contributors to the total activities of these soils when incubated under conditions to promote microbial growth (Ambroz, 1965; Macura & Vagnerova, 1969; Ladd & Paul, 1973; Mayaudon et al., 1975). Nevertheless, seasonal changes in proteinase activities of field soils are not correlated with changes in microbial populations (Franz, 1973; Ladd et al., 1976).

Soil proteinases and peptidases are probably also derived from plant and animal sources, although unequivocal evidence for their participation is lacking. The respective contributions of microbial, plant, and animal proteinases (and peptidases) to the total activities of soils may be expected to be influenced by several factors, including (i) age and type of plant cover, (ii) time of sampling in relation to season and agricultural practice, (iii) pretreatment of the soils before assay, especially the effectiveness with which roots and other plant debris may be removed, (iv) time and conditions of storage, and (v) assay conditions, including the nature of the assay substrate and whether or not bacteriostatic agents are employed.

The acceptance that soil enzymes generally may be of diverse origin is based partly on the belief that in soils some enzyme molecules, irrespective of source, may become stabilized, and hence, may persist for long periods after the original sources have been extensively decomposed. In support of this notion, studies by Simonart and Mayaudon (1961) indicated that more than one-third of a plant protein preparation became directly stabilized when added to soil, and remained protected from degradation by the soil microflora during a 30-day incubation period. It has been hypothesized that in soils protection may be afforded by adsorption of the proteins to clays, or possibly by chemical bonding to resistant organic compounds. Model experiments demonstrate that proteins can be bonded to clays and to complex organic compounds of high molecular weight, and that such reactions generally retard the rates of degradation of the protein by microorganisms, or by specifically added proteinases (Estermann et al., 1959b; Handley, 1961; Bremner, 1965; McLaren and Skujins, 1967; Mayaudon, 1968; Verma et al., 1975; Burns, 1978; others).

Burns (1978) has proposed a model for the location of exoenzymes (and endoenzymes from lysed cells) in the soil microenvironment. Enzymes are considered to be free in solution, and attached to the surfaces of clays and of preformed organic colloids. They may also be entrapped within the interlamellar spaces of clays, and within colloidal humus during formation of the latter. Ladd and Butler (1975) have argued that exocellular enzymes such as proteinases which are active against, and therefore accessible to, added protein substrates of high molecular weight, are likely to have high turnover rates in soils. Proteinases which are free in the soil solution would be vulnerable to degradation, both from autodigestion and from attack by other microbial proteinases newly formed in response to the presence of suitable substrates. Proteinases adsorbed to soil colloids, but still active against protein substrates, would presumably be no longer susceptible to autodigestion. They would, however, remain accessible to attack by other microbial proteinases, possibly at diminished rates, until the latter in turn are adsorbed or degraded. Some proteinase molecules may become entrapped within or adsorbed to soil colloids such that the enzymes are completely stabilized against biological degradation. In such cases the bound enzyme may also have been rendered inaccessible to, and thus inactive towards, substrates of high molecular weight; however, activities towards low molecular weight substrates may be far less affected.

If these hypotheses were correct, rates of hydrolysis of proteins in soils should be markedly influenced by the availability of energy sources for continued de novo synthesis of active proteinases. Activities of exocellular proteinases and peptidases towards appropriate peptides of low molecular weight should be less dependent on enzyme resynthesis, due to an accumulation of protected enzymes. Ladd and Paul (1973) found that the casein-hydrolyzing activity of an incubated soil increased rapidly during a period when numbers of recently formed viable bacteria were decreasing markedly, and when there was relatively intense metabolism of microbial products. Subsequently, as microbial activity decreased due to the depletion of available energy sources, so too did the casein-hydrolyzing activity of the soils. Maximal activity towards casein occurred 4 to 5 days after commencing the incubation and was about 12 times that of the soil initially. After 4 weeks incubation, proteinase activity had declined to <20% of the maximal. Such results are in accord with the above hypothesis.

In the same experiment, activities towards the low molecular weight substrates, ZPL and BAA, were comparatively high initially, possibly due to stabilized enzymes. During incubation, activities increased by about 100% and 50%, respectively, and remained at the new high levels. It is impossible to state whether the increased, comparatively stable, activities towards ZPL and BAA were due to protected, newly formed exocellular proteinases or peptidases. Undoubtedly, the increased activity toward ZPL which occurred soon after commencement of incubation was associated with rapid microbial growth; but ZPL-hydrolyzing activity did not decline, even when cell numbers subsequently decreased. Activity towards BAA increased more slowly throughout the incubation period and was independent of the large rise and decline in the viable bacterial populations. Even so, hydrolyses of these substrates may have been catalyzed, at least partly, by endocellular enzymes; changes in the numbers and activities of those cells causing hydrolysis may not be indicated by changes in the total viable bacterial populations.

Speculations on the location in soil of enzymes hydrolyzing proteins and low molecular weight peptide and amide derivatives are based partly on studies of the properties of proteinases immobilized on solid support materials including clays. Rowell et al. (1973) prepared enzymically active complexes of proteinases by reacting the enzymes with polymerizing *p*-benzoquinone at pH 8.0. These enzyme derivatives were assumed to have properties comparable to those of proteinases which react with, or become entrapped within, humic colloids during the formation of the latter in soils. Ionic bonds, if present in the synthetic complexes, appeared to be of minor importance in linking the proteinases to the aromatic moieties. The model proteinase-humic acid analogues, when compared with the respective free enzymes, were more stable after incubation at 25°C, or at elevated temperatures. Further, the enzyme derivatives exhibited changes in kinetic properties (K_m values, pH optima) consistent with those of enzymes bound to polyanionic supports and assayed with cationic substrates. The specific activities of the complexed enzymes were always less than those of the respective free enzymes, whether proteins or low molecular weight compounds were used as assay substrates. However, there was no consistent indication that the decreases in the specific activities of these enzyme complexes were influenced by the molecular size of the substrates employed.

As yet there has been no direct demonstration that proteinases and peptidases are stabilized in soils by the formation of enzyme-humic matter complexes. Extracts from soils have been shown to be enzymically active towards proteins (Ladd & Paul, 1973; Mayaudon et al., 1975) and peptide derivatives (Ladd, 1972). Humic compounds, also present in the extracts, were largely separated from the proteinases by precipitation of the humic material with salmine (Mayaudon et al., 1975) or with $CaCl_2$ (Ladd, 1972). In the former case, the removal of humic compounds doubled the proteinase specific activity; in the latter case, separation was accomplished without loss of enzymic activity. Ladd (1972) considered that the presence of humic compounds in his active extracts was coincidental.

The extracted soil enzymes active against ZPL appeared to be comparatively stable to autodigestion, to attack by added microbial proteinases, to drying, and to brief heat treatments (Table 3). Thus, the soil extracts, incubated at 1 or 25°C for 10 days, retained all ZPL—hydrolyzing activity, despite the presence of active proteinases (as assayed against protein substrates). When the extracts were incubated for 1 day at 37°C with

Table 3—Stabilities of proteinases in soil extracts.[†]

Treatment of soil extract	Z-phenylalanyl leucine hydrolysis, %[‡]
Freeze-dry	100
Dry at 30°C (90 min)	100
Dry at 50°C (10 min)	91
Incubate solutions at 25°C (10 days)	100
Incubate solutions at 50°C (1 day)	39
Incubate solutions at 50°C (2 days)	20
Incubate solutions at 37°C (1 day)	89
Incubate solutions at 37°C (1 day) + thermolysin	91
Incubate solutions at 37°C (1 day) + pronase	87
Freeze-dry, γ irradiate (10 Mrad)	65

† Adapted from Ladd (1972).
‡ Percentage of activities of untreated extracts.

the added proteinases, thermolysin, or subtilisin, activities of the extracted soil enzymes towards ZPL declined by about 10% only. Activities were unaffected by freeze-drying or by drying at 30°C for 90 min, activities decreased by only 9% when extracts were dried for 10 min at 50°C. Activities towards ZPL decreased by 61% when extracts were incubated at 50°C for 1 day, and by 80%, after 2 days incubation (Ladd, 1972). Whether or not the humic compounds in the extracts afforded some protection to the enzymes from biological attack and from denaturation remains unproven. It is possible that these enzymes may be relatively stable proteins per se, or that they may be protected by other types of compounds in the extracts.

In this latter regard, Mayaudon et al. (1975) demonstrated that the diphenol oxidases present in their soil extracts, were stable to attack by the co-extracted proteinases and by the added microbial proteinase, pronase. Yet diphenol oxidase activity was destroyed after incubation with added lysozyme plus pronase. These workers have suggested that the extracted enzymes were stabilized and protected by the presence of polysaccharides, which also may play a role in stabilizing exocellular enzymes in soils.

It is important to emphasize that the behavior of extracted enzymes may neither accurately reflect their behavior in soils, before extraction; nor the behavior of those enzymes which catalyze similar reactions, but which remain in the extracted soil. The failure to isolate active proteinase-humic matter complexes by extraction does not negate their presence in soils; nor does the characterization of model complexes, analogous to enzyme-humic acid complexes, support the presence of the latter in soils. The locations of soil proteinases and peptidases, and the microenvironments in which they function, have yet to be described.

3. KINETICS

Considering the probable variety of proteinases in soils, and their states of occurrence and reaction conditions, there seems little point in using the activities of whole soils or crude extracts to determine specific values for K_m, V_{max}, etc. Establishment of these kinetic constants (and indeed of the effects of pH, temperature, inhibitors, etc.) are more appropriate when defining the catalytic properties of pure enzymes, which, in the case of proteinases and peptidases, have not as yet been prepared from soils.

Mayaudon et al. (1975) reported that soil extracts, previously treated with salmine and filtered on Sephadex C25 gel to remove humic compounds, optimally hydrolyzed casein at pH 8.5 and at 50°C. The energy of activation (E_a) was 5.8 Kcal mol^{-1} °C^{-1}, calculated in the range 14–37°C; the Michaelis constant (K_m) was 2.2% (wt/vol). Inhibition studies suggested that the casein-hydrolyzing activities of the soil extracts were due to serine proteinases; proteinases of the SH and metalloenzyme classes were not extracted (Table 4). Ladd (1972) showed that the optimal pH and temperature for ZPL hydrolysis by crude soil extracts were about pH 8.0 and 60°C, respectively, each value being slightly higher than the corresponding value (pH 7.0, 50°C) obtained with the extracted soil.

Table 4—Inhibition of proteinases in soil extracts.[†]

Substrate	Inhibitor	Inhibitor concentration	Percentage activity[‡]
Z-Phenylalanyl leucine			
	o-phenanthroline	400 μM	28
	EDTA	20 mM	18
	β-phenylpropionate	4 mM	33
	Mercuric chloride	400 μM	18
	Sodium humate	400 μg/ml	85
	Dimethylsulfoxide	15%	50
Casein	Diisopropylphosphofluoridate	100 μM	61
	Iodoacetamide	10 mM	88
	p-chloromercuribenzoate	10 μM	100
	EDTA	50 mM	100

† Adapted from Ladd and Butler (1972) and Mayaudon et al. (1975).
‡ Percentage of rates observed in absence of inhibitor.

When tested against a range of Z-dipeptide derivatives, enzymes in the soil extract preferentially hydrolyzed those substrates with hydrophobic side chains. The ZPL-hydrolyzing activities of the extracts were inhibited by the metal chelating compounds, EDTA and o-phenanthroline, and by β-phenylpropionate (Table 4). In these respects the enzymes of the soil extracts behaved similarly to the metalloenzyme, carboxypeptidase A (Cunningham, 1965). However, the two enzyme systems differed in their responses to the presence of 15% dimethylsulphoxide (Butler & Ladd, 1969; Ladd, 1972), of NaCl, and of neutralized humic acids (Ladd & Butler; 1969, Ladd, 1972). The soil extract enzymes were far less sensitive to inhibition by humic acids than was carboxypeptidase and, in this respect, they behaved similarly to phaseolain, a plant proteinase of similar substrate specificity (Wells, 1968). The activities of the soil extracts were inhibited 82% by HgCl₂ (400 μM), indicative of the necessity for free SH groups for enzyme catalysis (Ladd, 1972).

C. Amidohydrolases and Amidinohydrolases

Enzymes characterized within this subgroup (EC.3.5) hydrolyze a very wide variety of linear and cyclic amides and amidines, frequently with the release of amino acids, NH_4^+, or urea. Formation of NH_4^+ by deamination of the cyclic amidines, the nucleic acid bases, is discussed elsewhere (section IV), as is the further hydrolysis of urea itself by urease (section V). In this section, discussion is restricted to the properties of two amidohydrolases, asparaginase (EC 3.5.1.1) and glutaminase (EC 3.5.1.2). These, together with urease, are the only enzymes of the entire subgroup whose activities have been assayed in whole soils (Mouraret, 1965; Galstyan, 1973; Galstyan & Saakyan, 1973).

Asparaginase and glutaminase catalyze, respectively, the hydrolyses of the amides, L-asparagine and L-glutamine (Eq. [2] and [3])

$$\text{L-asparagine} + H_2O \rightarrow \text{L-aspartate} + NH_4^+ \qquad [2]$$

$$\text{L-glutamine} + H_2O \rightarrow \text{L-glutamate} + NH_4^+. \qquad [3]$$

Amide hydrolysis is a nucleophilic displacement reaction akin to the hydrolysis of peptide bonds, but differing in that the displaced N atom leaves as an NH_4^+ group.

Asparagine-hydrolyzing activities are widely distributed in animal tissues and in microorganisms, and have been detected in soils. In the latter cases, activities were based on the release of NH_4^+ in toluene-treated soils after incubation with asparagine for 2 to 3 days (Drobnik, 1956; Beck & Poschenrieder, 1963) or for 1 to 5 hours (Mouraret, 1965). Such activities are indicative of catalysis by asparaginase, although its participation remains inconclusive. Nevertheless, asparaginases have been described for a range of microorganisms (Imada et al., 1973), including some commonly found in soil, e.g., *Aspergillus* sp., *Streptomyces griseus, Bacillus* sp., *Pseudomonas fluorescens.* Asparaginase has been highly purified from animal and microbial (*Escherichia coli*) sources (Wriston, 1971). Asparaginases from different sources differ in their molecular and kinetic properties, including substrate range, pH-activity profiles, and response to SH inhibitors.

Glutaminase activity has been demonstrated in toluene-treated soils (Galstyan, 1973; Galstyan & Saakyan, 1973). Both glutamate and NH_4^+ accumulated in soils when incubated with glutamine; assays were based on NH_4^+ formation after incubation for 24 hours. Glutaminase activities were correlated with soil organic C and N contents, and the pHs for optimal activities ranged from 6.8 to 7.2 for different soils.

Glutaminase has been purified from animal, fungal, and bacterial sources, the latter including *Pseudomonas* sp., *Azotobacter agilis,* and especially, *E. coli* (Hartman, 1971). The formation of glutamate from glutamine may be regarded as the specific transfer of the glutamyl group to water, which acts as an acyl acceptor. The enzyme from *E. coli* hydrolyzes a range of glutamyl derivatives, but substrates must have the L-configuration, and have free α-amino and α-carboxyl groups. The carboxyl group at position C-5 is essential.

Glutaminase from *E. coli* is inhibited by heavy metals and by *p*-mercuribenzoate, but is not inhibited by other SH reactive compounds such as iodoacetate or *N*-ethylmaleimide. The enzyme functions optimally at acidic pHs (4–5). The glutaminase purified from the soil bacterium, *A. agilis,* differs from the *E. coli* enzyme in substrate specificity, pH optimum, and sensitivity to inhibition by *p*-mercuribenzoate.

D. Amino Acid Dehydrogenases and Oxidases

Amino acids formed by the hydrolysis of proteins, peptides, and certain amides, may participate further in a variety of reactions, including transamination, decarboxylation (to form amines), racemization, and deamination. Oxidative deamination of amino acids, catalyzed by amino acid dehydrogenases (EC 1.4.1) or amino acid oxidases (EC 1.4.3) to yield α-oxo acids and NH_4^+, is considered principally in this section. The nature of these reactions is indicated by Eq. [4] and [5], respectively.

$$
\underset{\underset{NH_2}{|}}{R-CH-COOH} + NAD^+ \xrightarrow{\overset{NADH\ +H^+}{}} \underset{\underset{NH}{||}}{R-C\ -COOH} \xrightarrow{\overset{H_2O}{}} \underset{\underset{O}{||}}{R-C-COOH} + NH_3 \qquad [4]
$$

$$
\underset{\underset{NH_2}{|}}{R-CH-COOH} + O_2 \xrightarrow{\overset{H_2O_2}{}} \underset{\underset{NH}{||}}{R-C\ -COOH} \xrightarrow{\overset{H_2O}{}} \underset{\underset{O}{||}}{R-C\ -COOH} + NH_3 \qquad [5]
$$

Both reaction mechanisms involve an initial oxidation of the amino acid and the formation of an imino acid as intermediate. Dehydrogenases utilize nicotinamide-adenine dinucleotide (NAD^+) as an H-accepting co-enzyme, whereas the amino acid oxidases are flavoproteins, in which flavin-adenine dinucleotide (FAD) is reduced initially and then reoxidized directly by O_2, with the formation of H_2O_2.

Dehydrogenases, active essentially towards specific amino acids only, have been purified from plant, animal, and microbial sources. In addition, an L-amino acid dehydrogenase of broader specificity and acting on aliphatic amino acids has been characterized. The equilibria of the reactions catalyzed by amino acid dehydrogenases strongly favor production of amino acids from the respective α-oxo acids and reduced NAD.

Most amino acid oxidases exhibit broad specificities of action. The prime constraint imposed by substrate structure on enzyme activity is the configuration about the α-carbon atom. Thus, amino acid oxidases are specific towards either L-amino acids or D-amino acids, but, within each class of substrate, an oxidase may deaminate at varying rates a range of up to 10 amino acids. However, some substrate-specific amino acid oxidases have been described, e.g., D-aspartate or D-glutamate oxidases. The overall oxidative deamination reactions, as catalyzed by the flavin-containing oxidases, are virtually irreversible.

L- and D-amino acid oxidases have been purified from a variety of animal and microbial sources. Kinetic studies, indicative of the mechanism of action of amino acid oxidases, have been reviewed in detail by Bright and Porter (1975). These and other properties of the oxidases have also been discussed by Mahler and Cordes (1971) and by Metzler (1977). Active, immobilized L-amino acid oxidase has been prepared by covalently bonding the enzyme to silica glass particles (Weetall & Baum, 1970).

In addition to the more general, more common, activities of amino acid oxidases, NH_4^+ may be released from specific amino acids by the action of (i) dehydratases acting on L-serine (EC 4.2.1.13), D-serine, or L-threonine (ii) a desulfhydrase (EC 4.4.1.2) acting on homocysteine, and (iii) ammonia lyases acting on aspartate (EC 4.3.1.1) and on other amino acids.

Detailed discussion of these reactions and of the formation of NH_4^+ from specific amines, themselves products of amino acid decarboxylation, is beyond the scope of this chapter.

Ammonium is formed from amino acids when they are incubated with soils under conditions that do not prevent microbial growth. However, toluene-treated soils neither deaminated added amino acids (Voets et al., 1965), nor accumulated NH_4^+ when the soils were incubated with gelatin (Voets & Dedeken, 1964). Amino acids were formed from the added gelatin and from native proteins in the soil (Dedeken & Voets, 1965). Since toluene prevents microbial growth in soils it has been concluded that deamination of amino acids in soils may be due essentially to enzymes associated with live cells (Skujins, 1976). However, the possibility still remains that toluene may directly affect amino acid deaminases, whether they are in live cells, in cell debris, or adsorbed to soil colloids. L-Amino acid oxidase can remain active in an insolubilized state (Weetall & Baum, 1970).

Ladd and Amato (unpublished data) observed that NH_4^+ slowly accumulated in soils when incubated under an atmosphere of chloroform vapor for periods up to 10 days. Under these conditions soil deaminases appeared to remain active for several days at least. However, the locations of the active deaminases, and whether the enzymes belonged to the oxidative deaminase group, and indeed, the nature of the substrates deaminated, remain unknown. Fumigation with chloroform vapor rapidly destroyed soil dehydrogenase and glucose-oxidizing activities, but not soil proteinase activities.

No satisfactory assay has been described for measuring amino acid deaminase activities in soils. Short-term incubation of leucine in soils, untreated with bacteriostatic agents, indicate that in the absence of substantial microbial growth, soil deaminase activities may be low (Ladd & Butler, 1972).

III. AMINOPOLYSACCHARIDES AND AMINO SUGARS

Soil amino sugars are derived mainly from soil microorganisms and the soil fauna. Amino sugars occur in nature largely in the form of polymers, and those providing most of the soil amino sugars are chitin, peptidoglycans, teichoic acids, and other components of bacterial cell walls (reviewed by Parsons, 1981). A variety of amino sugars occur in microbial polysaccharides, but three predominate: glucosamine, galactosamine, and muramic acid; their structures are shown below. These amino sugars commonly occur as their N-acetyl derivatives. The formation of NH_4^+ from amino sugar polymers requires prior hydrolysis to aminomonosaccharides.

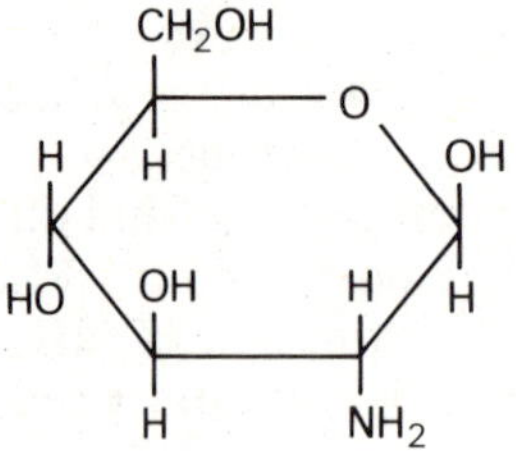

CH₂OH structures — chemical structural diagrams:

$\beta - D - Glucosamine$ $\beta - D - Galactosamine$ $\beta - Muramic\ acid$
$(3 - O - (1 - carboxyethyl) -$
$\beta - D - glucosamine)$

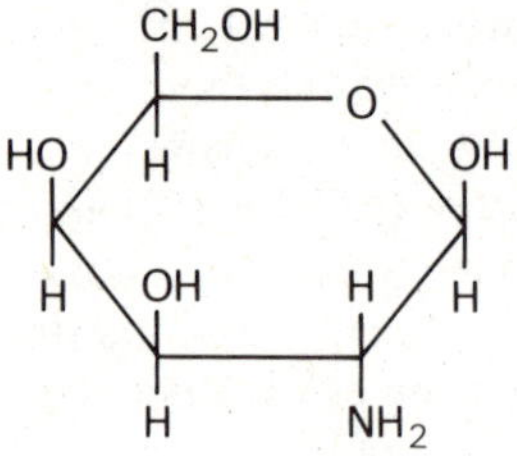

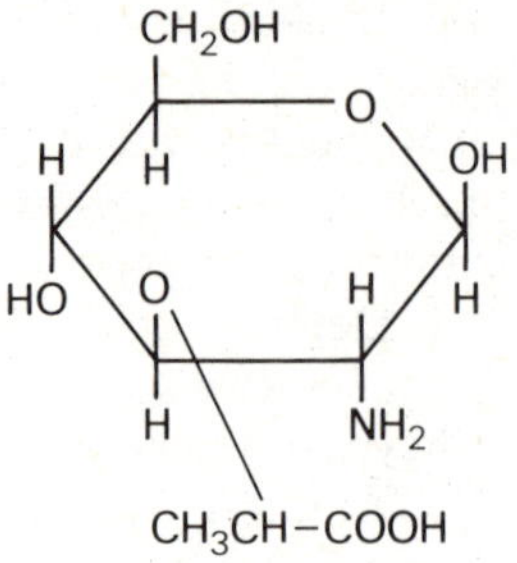

A. Origin and Hydrolysis

1. CHITIN

Complete enzymic hydrolysis of chitin to free N-acetyl-D-glucosamine is effected by a chitinolytic complex consisting of two hydrolases that act consecutively. Chitinase (EC 3.2.1.14) hydrolyzes chitin to N,N'-diacetyl-chitobiose, and this dimer is hydrolyzed to N-acetyl-D-glucosamine by chitobiase (β-N-acetylglucosaminidase, EC 3.2.1.30). The chitinolytic complex is widely distributed among soil bacteria (Clarke & Tracey, 1956), especially the Streptomycetes, and is also produced by many fungi. Chitin is hydrolyzed also by some lysozymes (EC 3.2.1.17). Chitinolytic activity has been found also in protozoans and several soil animals (Tracey & Youatt, 1958). Chitinase activity has been assayed by turbidimetric measurement of chitin disappearance, by following the fall in viscosity due to hydrolysis of soluble substrates such as glycol-chitin and carboxymethylchitin, and by colorimetric measurement of N-acetylglucosamine production; in the latter case chitobiase must also be present since N,N'-diacetyl chitobiose does not react.

No satisfactory method for assaying chitinase activity in soils has been described. The viscometric and turbidimetric methods cannot be applied to soils, and recoveries of N-acetylglucosamine as estimated by the colorimetric method are not always satisfactory (Skujins & Pukite, 1970). Decomposition of chitin added to soils has been monitored by measuring increases in the release of mineral-N (Bremner & Shaw, 1954) or CO_2 (Okafor, 1966a). Such measurements, however, integrate the activities of a large number of enzymes, any one of which may profoundly influence the net result. Recently, ion-exchange and gas-liquid chromatography have been used in estimations of amino sugars, and these techniques may lead to more satisfactory methods for assaying chitinolytic activity in soils.

2. PEPTIDOGLYCANS

Peptidoglycans, which occur in the cell walls of all bacteria except the extreme halophiles, are composed of glycan chains crosslinked by short peptides; the glycan chains consist of alternating β-1, 4 linked units of N-acetylglucosamine and N-acetylmuramic acid. Most of the carboxyl groups of the N-acetylmuramic acid residues are involved in amide bonds with terminal L-alanine residues of the cross-linking peptides. Hydrolysis of either the glycoside or the peptide bonds results in solubilization of the peptidoglycan. Lysozyme hydrolyzes the links between C-1 of N-acetylmuramic acid moieties and C-4 of N-acetylglucosamine moieties, and links between C-1 of N-acetylglucosamine and C-4 of N-acetylmuramic acid moieties are hydrolyzed by *endo*-N-acetylglucosamidinase. The amide bonds between N-acetylmuramoyl residues and the peptide chains are hydrolyzed by N-acetyl-muramoyl-L-alanine amidase (mucopeptide amidohydrolase, EC 3.5.1.28).

3. TEICHOIC ACIDS

Teichoic acids are polymers of either glycerol phosphate or ribitol phosphate to which different sugars, including N-acetylglucosamine and N-acetylgalactosamine, and D-alanine are attached (Archibald, 1974); little is known about the fate of teichoic acids during cell lysis in natural or model systems.

4. LIPOPOLYSACCHARIDES

In Gram-negative bacteria the peptidoglycan layer of the wall is enveloped by a complex of lipopolysaccharide, phospholipid, and protein. These lipopolysaccharides, which comprise 1 to 5% of the dry weight of the cell, have been studied intensively because they determine the antigenic properties of Gram-negative bacteria. Glucosamine and N-acetylgluco-samine occur in all the cell-wall lipopolysaccharides that have been examined; also, at least five other amino sugars have been found, galacto-samine being the most common of these (Luderitz et al., 1968). Little is known about the degradation of these polymers and the release of amino sugars during cell lysis.

B. Stability of Aminopolysaccharides in Soil

Under conditions favouring biomass proliferation, the hexosamine content of soils increases. When incubation is continued without further addition of substrate, the hexosamine content declines slowly (Shields et al., 1973); by contrast, glucosamine and purified chitin added to soils are rapidly decomposed (Bremner & Shaw, 1954; Okafor, 1966a). These observations indicate that structures containing amino sugars are protected to some extent from decomposition. In insect exoskeletons, chitin is closely associated with proteins and lipids that partly protect it from decomposition in soil (Okafor, 1966b). Degradation of chitin in these structures requires concerted action of chitinases and proteolytic enzymes. In the cell walls of fungi, chitin is associated with other polysaccharides (glucans) and hydrolysis by chitinase is limited unless there is concomitant hydrolysis of the glucans by glucanases (Skujins et al., 1965). The cell walls of some fungi contain melanin which inhibits chitinase and glucanase noncompetitively; melanin may also protect chitin by complex formation (Bull, 1970).

Acid hydrolysates of soil humic acids contain amino sugars (see Chapt. 3, F. J. Stevenson). It has been suggested that carbohydrates having free amino groups may become incorporated into humic acids, and this protects them from microbial attack; synthetic polymers produced from polyphenols and aminocarbohydrates decompose in soil much less rapidly than the free aminocarbohydrates (Bondietti et al., 1972).

Most studies on the decomposition of microbial residues in soil have been carried out with fungal material; there is some evidence that cell walls of bacteria, too, may be relatively stable in soils. Hexosamines, and the amino acids associated with peptidoglycans (alanine, glycine, glutamic acid, and lysine) are usually the predominant amino compounds found in acid hydrolysates of soils (Stevenson, 1957); furthermore, after 6 months incubation of ^{14}C-glucose in soil, these particular amino acids had higher specific activities than other amino acids (Wagner & Mutatkar, 1968), and the amount of ^{14}C present in glucosamine was greater than in any other component examined.

Marumoto et al. (1975) incubated isolated cell fractions of *B. subtilis* in sand, and found that the intracellular amino acid fraction was mineralized much faster than the cell wall fraction. When the cell wall fraction was oven-dried, however, mineralization was rapid. It would be interesting to see if drying under field conditions has a similar effect. *N*-acetylmuramic acid must occur in soils, at least in intact bacteria and undegraded cell walls; there are, however, only two reports of its presence in soils. Millar and Casida (1970) found muramic acid in all but one of 33 soils tested; the levels ranged from 19 to 158 μg/g dry soil and were consistent with the total numbers of bacteria found in the soils. In two peat soils muramic acid represented 0.17–0.47% of the total N (Casagrande & Park, 1978).

Various structural features of some peptidoglycans confer resistance to attack by lysozyme (Hayashi et al., 1973) and they could be involved in the persistence of cell wall components in soils. For example: free amino groups in glucosamine residues, the presence of O-acetyl groups, a high degree of peptide crosslinking, and the occurrence of free amino groups in the peptide crosslinks.

C. Hydrolysis of Aminopolysaccharides in Soil

There seem to be no reports about the behavior in soil of the enzymes that catalyze hydrolysis of aminopolysaccharides (aminoglycanhydrolases). Adsorption of aminoglycanhydrolases to soil constituents probably protects them from attack by proteases and, thus, increases their stability. Adsorption, however, may lower catalytic activity by reducing accessibility of substrates, particularly those of high molecular weight and low solubility such as chitin and the peptidoglycans. Adsorption of the chitinolytic complex of a *Streptomyces* sp. by kaolinite resulted in reduction of activity and a shift in optimal pH from 4.7 to 5.7 (Skujins et al., 1974). Chitin rapidly adsorbs chitinase and lysozyme (Skujins et al., 1973), and once this occurs, the insoluble substrate particle may undergo extensive hydrolysis before the enzyme is released into the surroundings; such a mechanism might counteract, to some extent, the limiting effect of the immobility of chitin in soils.

D. Ammonia Production From Amino Sugars

Detailed studies of the formation of NH_4^+ from N-acetylglucosamine in microorganisms have been made only with *B. subtilis* (Bates & Pasternak, 1965), *E. coli* (Rolls & Shuster, 1972; White, 1968), and *Bifidobacterium bifidum* var. *pennsylvanicus* (Veerkamp, 1969). The pathways by which glucosamine and N-acetylglucosamine are degraded by these organisms are shown in Fig. 1.

1. AMINO SUGAR KINASES

The first step leading to NH_4^+ production from glucosamine and N-acetylglucosamine is formation of the respective 6-phosphates by transfer of phosphate groups from ATP, catalyzed by glucosamine kinase (EC 2.7.1.8) and N-acetylglucosamine kinase (EC 2.7.1.59), respectively (Reactions 1 and 2 in Fig. 1). These kinases have been found in *Aerobacter cloacae* (Imanaga, 1957), *B. subtilis* (Bates and Pasternak, 1965), *E. coli* K_{12} (White, 1968), and *Streptococcus pyogenes* (Barkulis, 1966).

2. *N*—ACETYLGLUCOSAMINE 6-PHOSPHATE DEACETYLASE (EC 3.5.1.25)

This enzyme hydrolyzes N-acetylglucosamine 6-phosphate to glucosamine 6-phosphate (Reaction 4 in Fig. 1). Although N-acetylglucosamine deacetylase (EC 3.5.1.33) activity has been demonstrated (Reaction 3 in Fig.

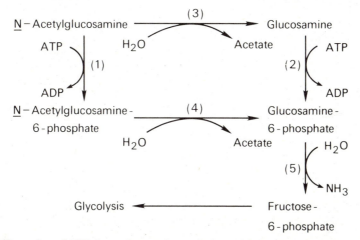

Fig. 1—Formation of NH_3 from glucosamine and N-acetylglucosamine in microorganisms. Enzymes involved: (*1*) N-acetylglucosamine kinase; (*2*) glucosamine kinase; (*3*) N-acetylglucosamine deacetylase; (*4*) N-acetylglucosamine 6-phosphate deacetylase; (*5*) glucosamine 6-phosphate isomerase.

1), deacetylation of *N*-acetylglucosamine 6-phosphate appears to be the physiologically important reaction (White, 1968; Veerkamp, 1969).

3. DEAMINATION OF GLUCOSAMINE 6-PHOSPHATE

The enzyme catalyzing this reaction (Reaction 5 in Fig. 1) has been called glucosamine 6-phosphate deaminase, but because the first step in the reaction is an aldose-ketose isomerization, the Enzyme Commission has named it glucosamine 6-phosphate isomerase (EC 5.3.1.10). A suggested mechanism for the deamination is indicated in Eq. [6] (Noltmann, 1972).

$$
\begin{array}{ccccc}
H-C=O & & H-C-OH & +H_2O & CH_2OH \\
| & \rightleftharpoons & \| & \xrightleftharpoons[-H_2O]{} & | \\
H-C-NH_2 & & C-NH_2 & & HO-C-NH_2 \quad [6] \\
| & & | & & | \\
\text{Glucosamine 6-phosphate} & & & +H_2O & \\
& & & -H_2O & \quad \downarrow\uparrow \\
& & & & CH_2OH \\
& & & & | \quad +NH_3 \\
& & & & C=O \\
& & & & | \\
& & & & \text{Fructose-6-phosphate}
\end{array}
$$

It is not known whether the enzyme also catalyzes the hydrolytic deamination step or whether it occurs spontaneously. The equilibrium of the reaction strongly favours production of NH_4^+ and fructose 6-phosphate, but it is reversible to a slight extent with high concentrations of NH_4^+ and fructose 6-phosphate.

4. DEGRADATION OF *N*-ACETYLGALACTOSAMINE AND *N*-ACETYLMURAMIC ACID

Although *N*-acetylgalactosamine and *N*-acetylmuramic acid are widely distributed in microorganisms, often in amounts comparable with those of *N*-acetylglucosamine, the pathways by which they are degraded to produce NH_4^+ appear to be unexplored.

Hydrolytic removal of the lactyl group from muramic acid would produce glucosamine and, thus, provide a route for NH_4^+ production from muramic acid. Hydrolysis of the ether linkage occurs spontaneously in aqueous solution at pH 12.5, but enzymic removal of the lactyl group has not been reported. The formation of NH_4^+ from galactosamine or *N*-acetyl-galactosamine does not appear to have been reported.

IV. NUCLEIC ACIDS, NUCLEOTIDES, NUCLEOSIDES, PURINES, AND PYRIMIDINES

The conversion of nucleic acid N to NH_4^+ requires the concerted action of a large number of enzymes. Initially nucleic acids must be depolymerized to mononucleotides by nucleases (EC 3.1.4). The nucleotides are dephosphorylated by nucleotidases (EC 3.1.3) yielding nucleosides which are N-glycosides of purines or pyrimidines. Nucleosidases (EC 3.2.2) hydrolyze the nucleosides to purines, pyrimidines, and pentoses.

Purine and pyrimidine N is converted to NH_4^+ by reactions catalyzed mainly by amidohydrolases (EC 3.5.1 and 2) and amidinohydrolases (EC 3.5.3 and 4). Several of these reactions do not release NH_4^+ directly, but split C-N bonds in the heterocyclic rings so ring N atoms become accessible to NH_4^+-producing enzymes. These reactions occur in a wide variety of animals, plants, and microorganisms; in view of the major role of the latter in soils, emphasis will be placed on reactions known in microorganisms.

A. Nucleic Acids

Nucleases catalyze hydrolysis of ester bonds between phosphate groups and pentose units in nucleic acids. Since each phosphate group is linked to two pentose units, the nucleases are classed as phosphodiesterases (EC 3.1.4).

The nucleases are classified into two main groups according to the type of nucleic acid they attack. Ribonucleases (RNases) hydrolyze ribonucleic acids (RNAs), and deoxyribonucleases (DNases) hydrolyze deoxyribonucleic acids (DNAs). Each of these groups may be subdivided on the basis of the site in the nucleic acid chain at which attack occurs. Exonucleases split off mononucleotide units one at a time from the end of the polymer chain, whereas endonucleases attack at many points within the polynucleotide chain.

1. RIBONUCLEASES

Knowledge of these enzymes has expanded greatly in recent years as a result of interest in their enzymology, their use in studies of nucleic acid structure, and attempts to explain the high turnover rate of messenger RNA in bacteria. Bacterial and fungal ribonucleases have been reviewed recently by Datta and Niyogi (1976) and by Uchida and Egami (1971), respectively.

Ribonucleases hydrolyze ribonucleic acids to either 3'- or 5' mononucleotides, depending on whether 5' or 3'ribosylphosphate bonds are split (Fig. 2).

The most systematic studies of ribonucleases have been carried out with *E. coli*. Datta and Niyogi (1976) have described the 12 ribonucleases known in *E. coli*. It is likely that most, if not all of these enzymes are involved in RNA turnover in proliferating cells; their role in autolyzing cells, however, is uncertain. During starvation, four ribonucleases appear to be important in degrading RNA to monocucleotides.

    1) Ribonuclease II (EC 3.1.4.23) is an endonuclease that begins degradation of RNA by splitting bonds between phosphate groups and C-5 of ribose units at various points within the polymer chain. This produces oligonucleotides having terminal 3'-phosphate, and these

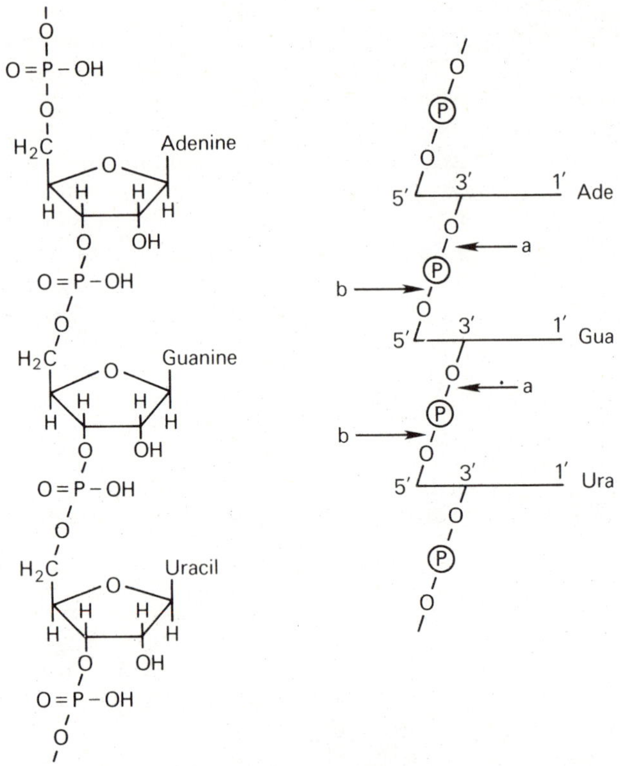

Fig. 2—Portion of a hypothetical RNA molecule and a schematic representation showing points of attack by 3'-ribonucleases (*a*), and 5'-ribonucleases (*b*). P = $-PO_2^--$. Hydrolysis at points *a* yields 5'-nucleotides; hydrolysis at points *b* yields 3'-nucleotides.

are degraded further by RNase II giving 3'-mononucleotides as the final products.

2) Exoribonuclease III (EC 3.1.4.20) hydrolyzes ribosyl 3'-phosphate links so 5'-mononucleotides are produced.

3) Oligonucleotidase (EC 3.1.4.19) is an exonuclease that attacks short oligonucleotides such as those produced by RNase II; the products are 5'-mononucleotides.

4) Polynucleotide phosphorylase (polyribonucleotide nucleotidyltransferase, EC 2.7.7.8) is a nonhydrolytic exonuclease that transfers terminal nucleoside 5'-monophosphate units from polynucleotides to inorganic PO_4^{3-} to produce nucleoside 5'-diphosphates.

The three exonucleases described above act in a processive manner; i.e., a given enzyme molecule hydrolyzes a single polynucleotide chain until a small oligonucleotide remains; only then does the enzyme part from its substrate and begin to attack another polynucleotide. This mechanism of enzyme action differs from the classical Michaelis-Menten concept which requires that the enzyme should separate from the products after each substrate molecule has reacted.

2. DEOXYRIBONUCLEASES

Enzymes that hydrolyze the internucleotide bonds of DNA have been described in many bacteria, but those from *E. coli* have received most attention (reviewed by Lehman, 1971). The *E. coli* exonucleases produce 5′-mononucleotides and small oligonucleotides derived from the ends of polymer chains. *E. coli* endonuclease yields oligonucleotides terminated by 5-phosphoryl groups; after exhaustive digestion of DNA, the average chain length of these oligonucleotides is seven residues.

3. MECHANISMS OF NUCLEASE ACTION

The hydrolysis of bonds between nucleotide groups by phosphoric diesterases is a nucleophilic displacement reaction in which a basic group at the enzyme's active site becomes bonded to the P atom. The O atom of the leaving group (a mono or oligonucleotide) is displaced and receives a proton from an acidic group of the enzyme or from water. In the case of ribonucleases producing 3′-nucleotides, the reaction proceeds in two steps. In the first step the OH group on the 2′ position of the ribose ring is deprotonated by a basic group on the enzyme; the anion, thus generated, attacks the P atom, displaces the O attached to the 5′-C of the next nucleotide unit, and forms a 2′, 3′-cyclic phosphate. In the second step the cyclic phosphate is hydrolyzed to a 3′-phosphate. The hydrolytic reaction is usually slower than the initial displacement reaction so accumulation of the cyclic phosphate is often observed.

With ribonucleases producing 5′-nucleotides, the attacking nucleophile is O^- from water, and the O atom displaced is the one attached to C-3 of a ribose ring. Hydroxyl ion is also the displacing nucleophile in reactions producing 3′- and 5′-nucleotides from DNA; in this case, formation of 2′, 3′-cyclic phosphates is not possible because there is no 2′OH in deoxyribose.

B. Nucleotides and Nucleosides

Mononucleotides produced by the action of nucleases are hydrolyzed by nucleotidases to give nucleosides and inorganic PO_4^{3-} (reviewed by Drummond & Yamamoto, 1971a, b). 5′-Nucleotidase (5′-ribonucleotide phosphohydrolase, EC 3.1.3.5) has been found in many species of microorganisms; it attacks all the common 5′-ribo- and 5′-deoxyribonucleotides.

3′-Ribonucleotides are hydrolyzed by 3′-ribonucleotidases and by ribonucleoside 2′,3′-cyclic phosphate diesterases. The 3′-ribonucleotidases seem to occur mainly in plants, but none has been well characterized. The cyclic phosphate diesterases are widely distributed, and preparations from several microorganisms have been studied in detail. The four 3′-ribonucleotides from RNA and the corresponding 2′,3′-cyclic phosphates are hydrolyzed at similar rates, but the 5′-nucleotides are not attacked.

The *E. coli* and *V. alginolyticus* enzymes are highly specific for ribonucleoside 3′-phosphates and ribonucleoside 2′,3′-cyclic phosphates, but the

B. subtilis enzyme differs in that it also hydrolyzes 3'-deoxyribonucleotides (Shimada & Sugino, 1969). Mononucleotides are also hydrolyzed by some phosphatases. The alkaline phosphatase (EC 3.1.3.1) from *E. coli,* for example, acts on a wide variety of ribo- and deoxyribonucleotides (Reid & Wilson, 1971).

The nucleosides produced by dephosphorylation of mononucleotides are hydrolyzed by nucleoside hydrolases (EC 3.2.2) to the purine or pyrimidine bases, and the pentose components. Nucleosides may also be cleaved by ribo- and deoxyribonucleoside phosphorylases (EC 2.4.2) to give the free bases and ribose 1-phosphate or deoxyribose 1-phosphate; these reactions involve the transfer of a pentosyl group from a nucleoside to inorganic PO_4^{2-}.

C. Nucleases, Nucleotidases, and Nucleosidases in Soil

RNA and DNA added to soils are rapidly and extensively degraded (Greaves & Wilson, 1970); also, pure cultures of many soil microorganisms degrade nucleic acids (Antheunisse, 1972). In most cases, measurement of nuclease activity is based on production of inorganic PO_4^{3-}, so the assay embraces depolymerization of nucleic acids and dephosphorylation of the resulting mononucleotides.

Several authors have suggested that adsorption of nucleic acids by clay minerals in soils may protect them from degradation (Goring & Bartholomew, 1952). There are many reports, however, that nucleic acids are rapidly degraded in soils and inorganic PO_4^{3-} released. Greaves and Wilson (1970) found that complexes of RNA or DNA with montmorillonite were degraded in soil, but X-ray diffraction studies indicated that RNA adsorbed in the central zones of the crystallites was partially protected from enzyme attack. The observed X-ray diffraction patterns, however, could have been caused by adsorption of degradation products such as adenine in the interlayer space of montmorillonite (Greaves & Wilson, 1973).

The only nucleotides that have been found in soil are thymidine 3':5'-diphosphate and deoxyuridine 3':5'-diphosphate, presumably derived from DNA (Anderson, 1970). The lack of other nucleotides indicates that following their formation from nucleic acids, most nucleotides are rapidly degraded or assimilated by microorganisms. Nucleotidases may play only a minor role in the hydrolysis of nucleotides in soil because of the abundance of phosphatases having low specificity. Soil phosphatases have been reviewed in detail by Speir and Ross (1978).

Hydrolysis of nucleosides to the constituent pentoses and purines, or pyrimidines, by nucleosidases has not been studied in soil.

D. Deamination of Nucleotides and Nucleosides

There is no evidence that NH_3 is produced directly from nucleic acids in living organisms, but some nucleotides and nucleosides can be deaminated.

1. NUCLEOTIDES

Adenosine 5'-phosphate (AMP) is hydrolyzed to NH_4^+ and inosine 5'-phosphate (IMP) by AMP deaminase (AMP aminohydrolase, EC 3.5.4.6) which is widely distributed in nature. GMP reductase (NADPH:GMP oxido-reductase, EC 1.6.6.8) catalyzes a reductive deamination of guanosine 5'-phosphate to give NH_4^+ and inosine 5'-phosphate.

2. NUCLEOSIDES

Deaminases that act directly on nucleosides are more widespread than those that act on nucleotides. Adenosine deaminase (adenosine aminohydrolase, EC 3.5.4.4) catalyzes the hydrolysis of adenosine to form NH_4^+ and inosine. The enzyme is widely distributed among animal tissues and microorganisms, and is highly active, the molecular activity being about 10^5 mol adenosine/min. Guanosine deaminase (guanosine aminohydrolase, EC 3.5.4.15) has been prepared from *Pseudomonas convexa,* and cytidine deaminase (cytidine aminohydrolase, EC 3.5.4.5) from *E. coli.*

The significance of deamination of nucleotides and nucleosides in the conversion of nucleic acid N to NH_4^+ is uncertain. The main function of these reactions may be the interconversion of bases in the "salvage" pathways for nucleotide synthesis from preformed bases or nucleosides. Nevertheless, sometimes these reactions participate in bypasses round catabolic reactions for which the usual enzymes are lacking.

E. Catabolism of Purines

1. AEROBIC DEGRADATION OF PURINES

a. Adenine—A wide variety of microorganisms hydrolyze adenine to hypoxanthine and NH_4^+ through the influence of adenine deaminase (EC 3.5.4.2) (Reaction 1, Fig. 3). In some microorganisms lacking adenine deaminase, bypass reactions convert adenine to hypoxanthine or xanthine. One such bypass to hypoxanthine involves conversion of adenine to adenosine, deamination to inosine by adenosine deaminase followed by hydrolysis of inosine to hypoxanthine and ribose. In another bypass, conversion of adenine to adenosine 5'-monophosphate (AMP) occurs prior to deamination; the inosinate produced by deamination of AMP is oxidized to xanthylate which is cleaved to yield xanthine.

b. Guanine—Guanine deaminase (guanine aminohydrolase, EC 3.5.4.3) hydrolyzes guanine to xanthine and NH_3 (Reaction 3, Fig. 3).

c. Hypoxanthine and Xanthine—Hypoxanthine is oxidized to xanthine by xanthine dehydrogenase (xanthine:NAD$^+$ oxidoreductase, EC 1.2.1.37) and by xanthine oxidase (xanthine:oxygen oxidoreductase, EC 1.2.3.2). These enzymes are closely related and all those known contain flavinadenine dinucleotide (FAD), molybdenum, and an iron-sulfur center. The reactions catalyzed may be expressed generally as in Eq. [7],

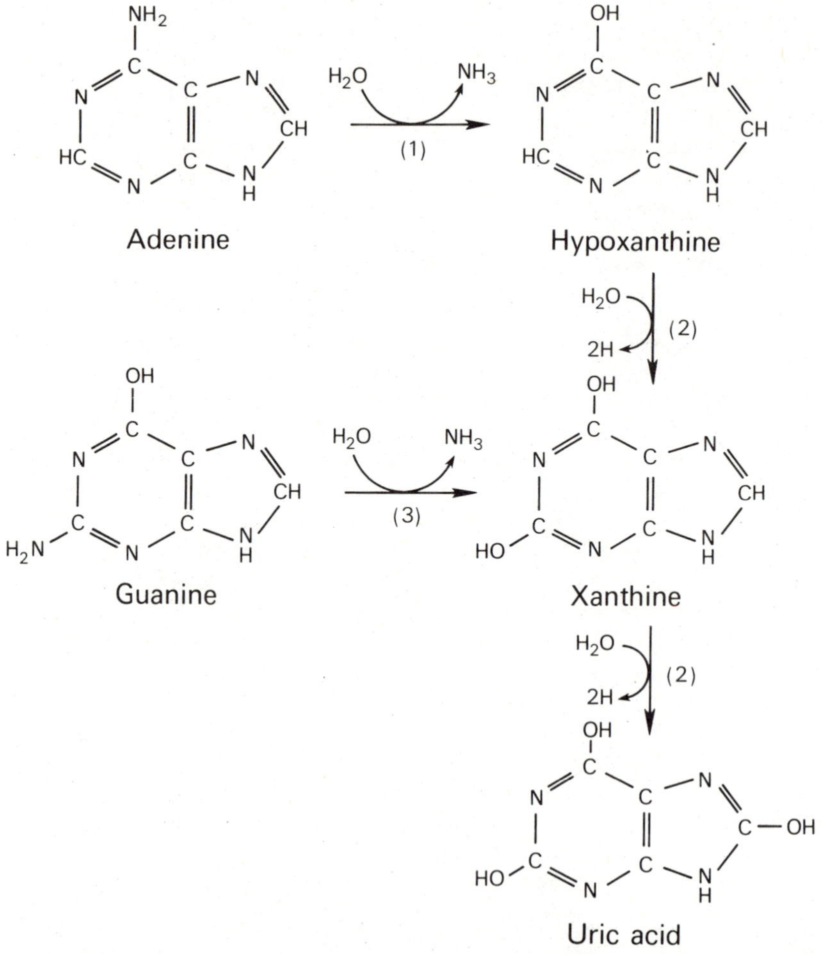

Fig. 3—Conversion of purines to uric acid under aerobic conditions. Enzymes involved: (1) adenine deaminase; (2) xanthine dehydrogenase (3) guanine deaminase.

$$RH + H_2O \xrightarrow[-2H^+]{-2e} ROH \qquad [7]$$

where RH is the substrate. Electrons from the substrate reduce the enzyme and then the reduced enzyme is oxidized by the electron acceptor; the oxygen in the product is derived from water. Enzymes that oxidize xanthine and hypoxanthine are widely distributed among organisms ranging from man to bacteria. Those enzymes purified from microbial species probably use oxidized nicotinamide-adenine dinucleotide (NAD⁺) as the natural electron acceptor under aerobic conditions, and are, therefore, classed as dehydrogenases; in some cases, oxidized nicotinamide-adenine dinucleotide phosphate (NADP⁺), cytochrome c and O_2 can function as electron acceptors, but the reaction rates are usually much slower. Xanthine dehydrogenases

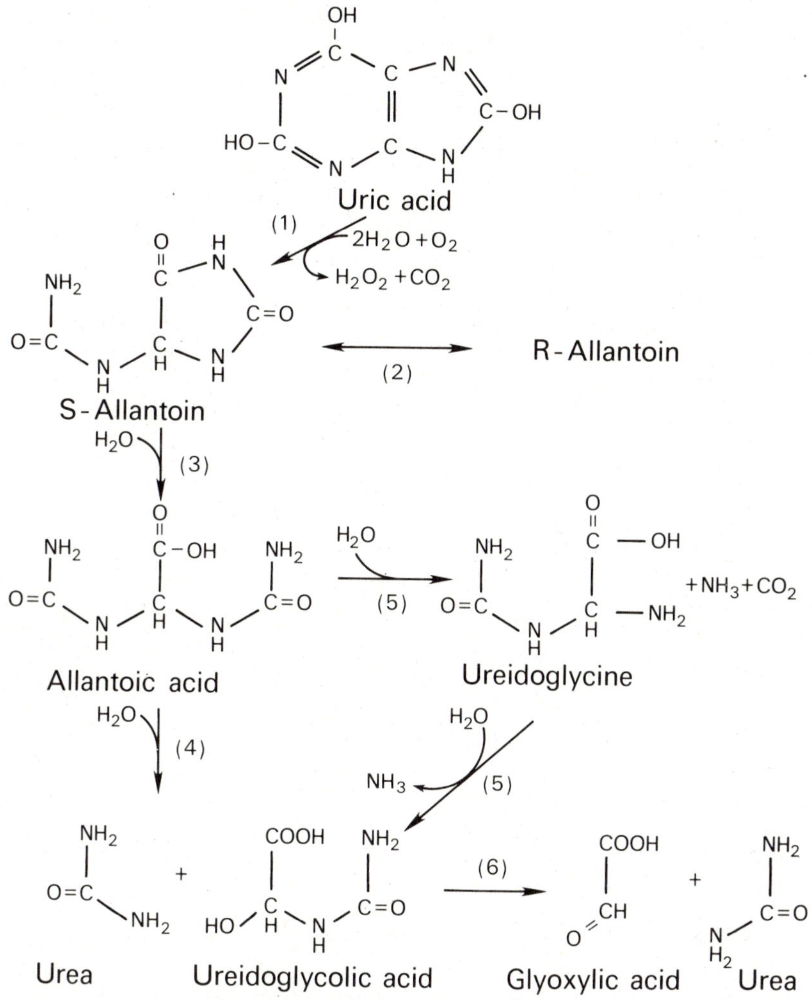

Fig. 4—Pathways of uric acid degradation under aerobic conditions. Enzymes involved: (1) urate oxidase; (2) allantoin racemase; (3) S-allantoinase; (4) allantoicase; (5) allantoate deiminase; (6) ureidoglycolate lyase.

also oxidize xanthine to urate; enzymes from different organisms show variations in specificity towards the two substrates, but in most cases xanthine is oxidized faster than hypoxanthine.

An alternative pathway for the oxidation of hypoxanthine to urate has been demonstrated in some microorganisms; in this pathway, oxidation occurs first at C-8 to give 6, 8-dihydroxypurine, followed by oxidation at C-2 to give urate.

d. Uric Acid—Large quantities of uric acid reach the soil as an end-product of N metabolism in animals and insects, as a constituent of many plants, and as a product of purine degradation by microorganisms. The ability to degrade urate is widespread among soil microorganisms (Antheunisse, 1972); the reactions involved are shown in Fig. 4. Urate is oxidized to allantoin through the action of urate oxidase (urate: oxygen oxidoreductase, EC 1.7.3.3). In contrast to xanthine dehydrogenase, urate oxidase has high specificity both for substrate and electron acceptor; urate is the only known substrate oxidized, and O_2 is the only known electron acceptor. Urate oxidases are widely distributed among animals, microorganisms, and plants. In some species the urate oxidases are firmly bound to structural components, in some they are associated with cell microbodies, while in others the enzymes are soluble.

The reaction catalyzed by urate oxidase includes decarboxylation as well as oxidation, and its mechanism is poorly understood. Experiments with ^{18}O and $H_2{}^{18}O$ indicate that the oxygen in the H_2O_2 produced is derived exclusively from O_2 (Bentley & Neuberger, 1952); therefore, the reaction involves the transfer of two electrons and two protons from each urate ion to O_2. Urate is also oxidized to allantoin by several peroxidase systems, and by the cytochrome oxidase system.

e. Allantoin—Allantoin occurs in many plants possibly as a storage component, in insects, and is the end product of purine catabolism in most mammals other than man and the higher apes. Allantoin occurs in two enantiomorphic forms, the S- and R- forms (Cahn et al., 1956); both forms can be used as a source of C and N by many microbial species.

Allantoinase (allantoin amidohydrolase, EC 3.5.2.5) catalyzes the hydrolysis of an amide bond in allantoin to give allantoate and NH_4^+; this enzyme occurs in many animals, plants, and microorganisms. Some allantoinases act only on S-allantoin, but some microorganisms also possess allantoin racemase (EC 5.1.99.3), so both forms of allantoin are eventually hydrolyzed. On the other hand, some microbial allantoinases hydrolyze both S- and R- allantoin even in the absence of the racemase.

f. Allantoic Acid—Enzymic hydrolysis of allantoate may occur in two ways.

1) Allantoicase (allantoate amidinohydrolase, EC 3.5.3.4) catalyzes the hydrolysis of the bond between the asymmetric C atom and one of the adjoining N atoms to give S-ureidoglycolate and urea (Fig. 4).

2) Allantoate deiminase [allantoate amidinohydrolase (decarboxylating), EC 3.5.3.9] catalyzes hydrolysis of 1 mole of allantoate to 2 mole of NH_4^+, 1 mole of CO_2, and 1 mole of ureidoglycolate (Fig. 4). This reaction takes place in two steps, the first (Eq. [8]) being hydrolysis of one of the peptide bonds giving NH_4^+, CO_2, and ureidoglycine; the second (Eq. [9]) being hydrolysis of ureidoglycine to NH_4^+ and ureidoglycolate:

$$\text{allantoate} + H_2O \rightarrow NH_3 + CO_2 + \text{ureidoglycine} \qquad [8]$$

and

$$\text{ureidoglycine} + H_2O \rightarrow NH_3 + \text{ureidoglycolate.} \qquad [9]$$

g. Ureidoglycolate—The hydrolysis of ureidoglycolate to glyoxylate and urea is catalyzed by ureidoglycolate lyases (EC 4.3.2.3) (Fig. 4).

As a result of the reactions described above, each purine ring and each allantoin molecule gives rise to 2 molecules of urea, or 1 molecule of urea plus 2 molecules of NH_4^+. In the case of adenine and guanine, an additional molecule of NH_4^+ is derived from the amino group attached to the purine ring. The conversion of urea to NH_4^+ and CO_2 is discussed in Section V.

2. METHYLPURINES

The methylxanthines, caffeine (1,3,7-trimethylxanthine), theophylline (1,3-dimethylxanthine), and theobromine (3,7-dimethylxanthine) are widely distributed in nature and occur in high concentrations in the tea, coffee, and cacao plants. These substances can be used as sole sources of C or N by some soil microorganisms including strains of *Pseudomonas putida* and *Ps. fluorescens, Bacillus coagulans,* and *Penicillium roqueforti.*

The degradation of these compounds commences with hydrolytic removal of the methyl groups to give methanol and xanthine (Woolfolk, 1975) (Fig. 5). Further catabolism of xanthine proceeds through the oxidative pathway mediated by xanthine dehydrogenase and urate oxidase.

3. DEGRADATION OF PURINES BY *ENTEROBACTERIACEAE*

Many of the *Enterobacteriaceae*, including several soil isolates, can use a number of purines as sole sources of N, C, and energy. In general, purines are degraded along pathways similar to those found in aerobic microorganisms (Figs. 3 and 4). *Proteus rettgeri* and *Serratia marcescens* oxidize hypoxanthine to urate via 6, 8-dihydroxypurine instead of xanthine. Urate

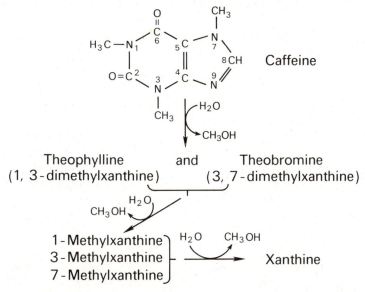

Fig. 5—Degradation of caffeine, theophylline, and theobromine.

oxidase has been demonstrated in several of the *Enterobacteriaceae* (Taupitz-Stahn, 1974). The allantoinases of the *Enterobacteriaceae* differ in some properties from those of other microorganisms and plants; they are activated by Mn^{2+} and by reducing agents, and act on R- and S-allantoin at similar velocities. Allantoicase appears to be absent from the *Enterobacteriaceae*; in these organisms, degradation of allantoate is catalyzed by allantoate deiminase which converts allantoate to ureidoglycolate, NH_4^+, and CO_2. Two pathways for further catabolism of ureidoglycolate have been found. In one, ureidoglycolase catalyzes hydrolysis to give urea and glyoxylate (Fig. 4), whereas in the second, oxalate, NH_4^+, and CO_2 are produced without intermediate formation of urea.

4. ANAEROBIC DEGRADATION OF PURINES

Bacteria capable of decomposing urate under anaerobic conditions are widely distributed in soils. Present knowledge of the anaerobic catabolism of purines is based largely on studies with *Cl. acidiurici* and *Cl. cylindrosporum* which were isolated from soils by enrichment techniques (Barker & Beck, 1942). These clostridia readily decompose guanine, xanthine, urate, and 6, 8-dihydroxypurine; hypoxanthine is attacked more slowly or after a period of adaptation, but adenine is not attacked. Guanine is converted to xanthine by guanine deaminase, and urate is reduced to xanthine by xanthine dehydrogenase.

Degradation of xanthine is brought about by a series of hydrolyses and decarboxylations which produce 2 molecules of NH_4^+ and 2 molecules of CO_2 from each pyrimidine ring, and open the imidazole ring to give formiminoglycine. Further catabolism of formiminoglycine involves intermediate reactions of complex tetrahydrofolate derivatives which are not discussed here. The overall reaction for anaerobic degradation of xanthine is:

$$\text{Xanthine} + 6H_2O \rightarrow 3NH_3 + \text{glycine} + 2CO_2 + \text{formate}.$$

F. Degradation of Pyrimidines

A variety of soil microorganisms can degrade pyrimidines with production of NH_4^+ and other end products such as urea, β-alanine, and β-aminoisobutyrate, from each of which, NH_4^+ can be produced subsequently. Two main routes have been found, one involving oxidation at C-6 of the pyrimidine ring, the other, reduction at this position. Prior to the oxidation or reduction, cytosine is deaminated to give uracil and NH_4^+ by the action of cytosine deaminase (cytosine aminohydrolase, EC 3.5.4.1); this enzyme also deaminates 5-methyl cytosine, producing thymine and NH_4^+ (Fig. 6).

1. OXIDATIVE PATHWAY (FIG. 6)

This pathway was found in bacteria isolated from soil by enrichment culture using various pyrimidines as C and N sources. Organisms that have been isolated in this way include members of the genera *Bacterium, Corynebacterium, Mycobacterium,* and *Nocardia.*

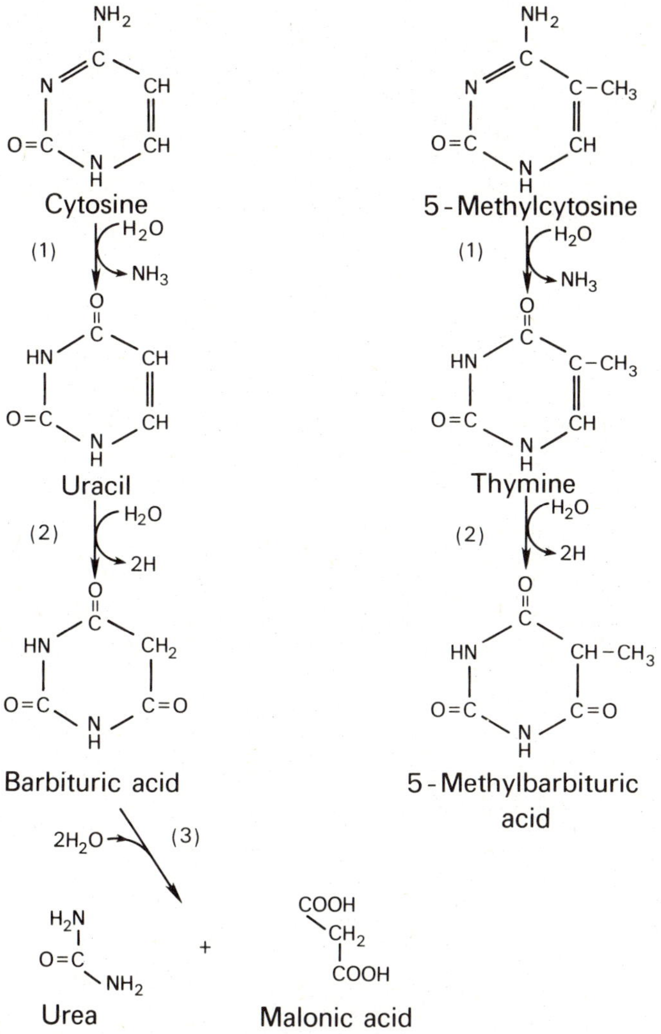

Fig. 6—Oxidative degradation of pyrimidines. Enzymes involved: (1) cytosine deaminase; (2) uracil dehydrogenase; (3) barbiturase.

Uracil dehydrogenase [uracil:(acceptor) oxidoreductase, EC 1.2.99.1] oxidizes uracil and thymine to give barbiturate and 5-methylbarbiturate, respectively; neither NAD^+ nor $NADP^+$ act as electron acceptors in these reactions. Barbiturate is cleaved to urea and malonate by the action of barbiturase (barbiturate amidohydrolase, EC 3.5.2.1); this enzyme does not act on 5-methylbarbiturate, the catabolism of which has not been elucidated.

2. REDUCTIVE PATHWAY (FIG. 7)

The reductive route of pyrimidine degradation was first found in yeasts and has since been demonstrated in a relatively small number of bacteria

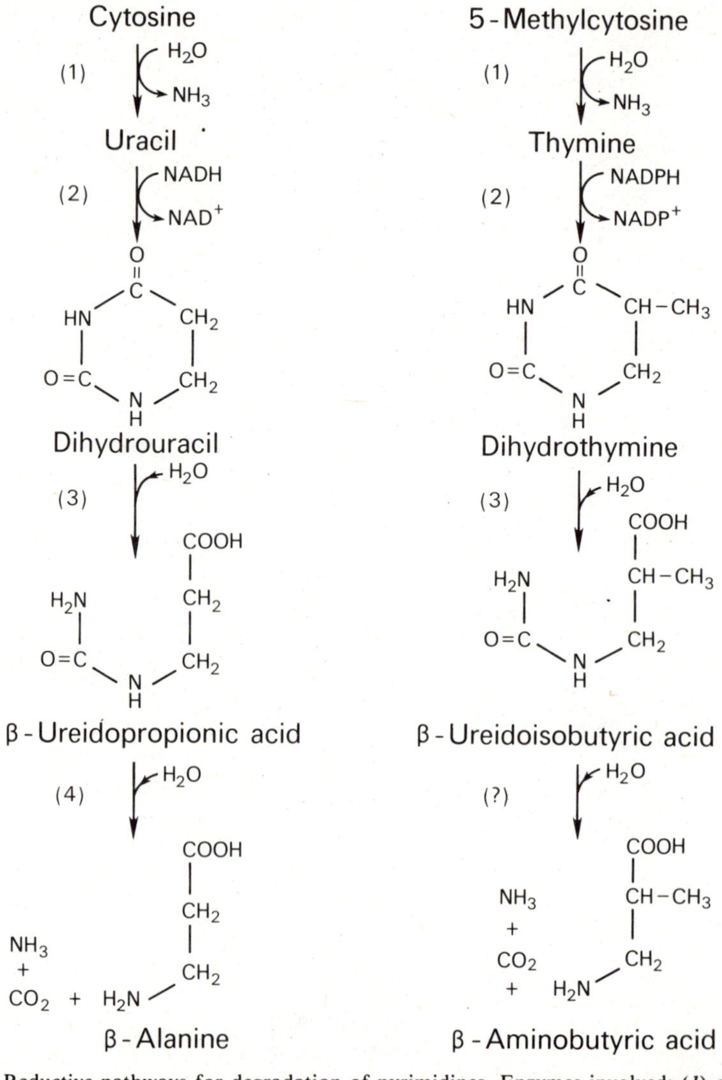

Fig. 7—Reductive pathways for degradation of pyrimidines. Enzymes involved: (*1*) cytosine deaminase; (*2*) dihydrouracil dehydrogenase; (*3*) dihydropyrimidinase; (*4*) β-ureidopropionase.

(listed by Vogels & Van der Drift (1976) and in *Neurospora crassa*. Uracil is hydrogenated by dihydrouracil dehydrogenase (5,6-dihydrouracil:NAD$^+$ oxidoreductase, EC 1.3.1.1), or a similar NADP$^+$ oxidoreductase (EC 1.3.1.2); the latter also catalyzes hydrogenation of thymine.

The reduced pyrimidine ring is opened hydrolytically by dihydropyrimidinase (5,6-dihydropyrimidine amidohydrolase, EC 3.5.2.2) to give *N*-carbamoyl-β-alanine from uracil, and *N*-carbamoyl-β-aminoisobutyrate from thymine. Further hydrolysis of *N*-carbamoyl-β-alanine to NH$_4^+$, CO$_2$, and β-alanine is catalyzed by β-ureidopropionase (*N*-carbamoyl-β-alanine

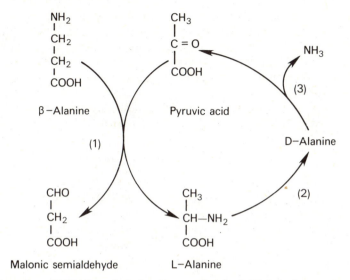

Fig. 8—Formation of NH₄⁺ from β-alanine. Enzymes involved: (*1*) β-alanine: pyruvate amino-transferase; (*2*) alanine racemase; (*3*) D-alanine dehydrogenase.

amidohydrolase, EC 3.5.1.6). Preparations of this enzyme from animal sources also hydrolyze *N*-carbamoyl-β-aminoisobutyrate, but it seems that further catabolism of this compound has not been studied in microorganisms that use the reductive pathway.

Among the microorganisms that degrade pyrimidines by the reductive pathway, only *Candida utilis, Pseudomonas facilis,* and *Ps. aeruginosa* are known to degrade β-alanine and, thus, use the C skeleton of pyrimidines. *Ps. aeruginosa* catabolizes β-alanine through the action of β-alanine:pyruvate aminotransferase (L-alanine:malonate semialdehyde aminotransferase, EC 2.6.1.18) which yields malonic semialdehyde and L-alanine (Robinson and Venables, 1978) (Fig. 8). *Ps. aeruginosa* also catabolizes β-aminoisobutyrate by this route, and with either substrate the catabolism depends upon regeneration of pyruvate from L-alanine by the consecutive action of alanine racemase (EC 5.1.1.1) and D-alanine dehydrogenase.

G. Degradation of Purines and Pyrimidines in Soil

There have been very few studies of the formation of NH₄⁺ from purines and pyrimidines in soils. Adenine complexed with montmorillonite is rapidly degraded in soils (Greaves & Wilson, 1973), but the products have not been identified.

When urate is incubated with soils under conditions that allow microbial growth, it rapidly disappears and NH₄⁺ is formed; in studies using 2-¹⁴C-uric acid, allantoin, allantoic acid, and urea were intermediates in the process (Takahashi et al., 1970). Toluene-treated soils, however, do not produce NH₄⁺ from urate, but allantoin accumulates (Durand, 1961). Similarly, NH_4^+ is formed when allantoin is incubated with soils, but when such soils are subsequently treated with toluene and further incubated with allantoin, allantoate and glyoxylate accumulate. Allantoinase activity is induced in soils by microbial growth at the expense of urate or allantoin, and allantoicase activity is induced when urate, allantoin, or allantoate is the growth substrate.

The degradation of purines and pyrimidines by microorganisms has been reviewed in detail by Vogels and Van der Drift (1976).

V. UREA

In soil, urea is decomposed enzymically to CO_2 and NH_4^+. It has been assumed, probably correctly, that the reaction is catalyzed by soil ureases (Eq.[10]), and the effects of various treatments on the urea-decomposing activities of soils and soil extracts are often related to those obtained with pure ureases from plant and microbial sources:

$$O = C \begin{matrix} NH_2 \\ \\ NH_2 \end{matrix} + 2H_2O = CO_2 + 2NH_3. \qquad [10]$$

However, it is of interest that an alternative pathway of hydrolysis of urea, not involving urease, has been demonstrated in yeast, algae, and in some fungi. In this alternative, two-step reaction sequence, urea is first converted to allophanate by the action of urea carboxylase (urea: CO_2 ligase, EC 6.3.4.6). Allophanate hydrolase (allophanate amidohydrolase, EC 3.5.1.13) then converts allophanate to CO_2 and NH_4^+ (Eq. [11]).

$$O = C \begin{matrix} NH_2 \\ \\ NH_2 \end{matrix} + HCO_3^- + H^+ \xrightarrow[]{ATP \quad ADP + P_{in}} O = C \begin{matrix} NH^{-COOH} \\ \\ NH_2 \end{matrix}$$

$$\xrightarrow{H_2O} 2 CO_2 + 2 NH_3 \qquad [11]$$

The importance in soils of the allophanate pathway for urea degradation has not been demonstrated. Since the initial carboxylation step requires ATP, it is highly probable that the overall reaction sequence would be confined to live cells in soil. The effect of toluene in decreasing rates of urea decomposition in soils may conceivably be due in part to the destruction of ATP in cells and, hence, elimination of this pathway. However, in the absence of any specific information, the tacit assumption is accepted that urea is decomposed in soils by the action of ureases, and all further discussion in this section is restricted to the properties of these enzymes.

A. Ureases

1. REACTION MECHANISM

Ureases (urea amidohydrolases, EC 3.5.1.5) hydrolyze nonpeptide C-N bonds in linear amides. Based on kinetic data, carbamate has been implicated as an obligatory intermediate in a two-step reaction catalyzed by the crystalline enzyme from jack bean (*Canavalia ensiformis*) (Eq. [12]).

$$O = C \begin{smallmatrix} NH_2 \\ \\ NH_2 \end{smallmatrix} + H_2O \longrightarrow \left(O = C \begin{smallmatrix} OH \\ \\ NH_2 \end{smallmatrix} + NH_3 \rightleftharpoons O = C \begin{smallmatrix} ONH_4^+ \\ \\ NH_2 \end{smallmatrix} \right)$$

$$\xrightarrow{\;+\,H_2O\;} H_2CO_3 + 2\,NH_3 \qquad\qquad [12]$$

The reaction is presumed to involve a carbamoyl group transfer from a carbamoyl-enzyme complex to water.

Urease from jack bean contains two essential atoms of bound Ni^{2+} per enzyme molecule (97,000 daltons), but a specific role of the metal ion in the catalysis has yet to be defined. Possibly Ni^{2+} is chelated to amino acid residues and is so positioned at the enzyme's active site, that on formation of the enzyme-substrate complex, the metal enhances the electrophilic nature of the C atom of the $C = O$ group of urea, thus promoting nucleophilic displacement of the N atom. From enzyme ionization constants, SH, NH_4^+, and histidine groups are each thought to be involved at the urease catalytic site.

2. MOLECULAR AND CATALYTIC PROPERTIES

Crystalline jack bean urease has a molecular weight of about 480,000 and consists of polypeptide chain subunits of about 30,000 daltons. The enzyme may dissociate to form active oligomers, or may associate to form either active polymeric isozymes, or active epiisozymes constituted from single active subunits. Urease contains SH groups essential for activity.

Substrate specificity of urease is high. In addition to urea, urease hydrolyzes several urea derivatives, but at far lower rates. Reithel (1971) and Bremner and Mulvaney (1978) have tabulated kinetic data for urea hydrolysis by ureases from jack bean, soy bean, and bacterial sources. Values for K_m, V_{max}, and activation energy vary with enzyme source and preparation, and assay conditions. Maximal catalytic activity of jack bean urease occurs at 65°C and usually in the pH range 6.0 to 7.0. Skujins and McLaren (1967) have measured urease reaction rates at limited water activities. Activity ceased below 60% humidity; at least 1.3 mol of sorbed water per mole of side chain polar groups of urease were required for catalysis.

3. IMMOBILIZED UREASES

Insoluble, but enzymically active, urease derivatives may be formed by adsorbing the enzymes to clays (Durand, 1964) and other inorganic supports, or by binding the enzyme covalently to synthetic organic polymers. Immobilized ureases may exhibit different kinetic properties (V_{max}, K_m) from those of the free enzyme.

The pH-activity profiles of insolubilized and free ureases also may differ. Durand (1964) found that the apparent optimal pH of urease adsorbed to bentonite was greater than that of the free enzyme. This phenomenon was explicable in terms of the increased concentrations of H^+ in the vicinity of the enzyme bound to the polyanionic support, compared with their concentration in the bulk solution. By contrast, urease, bound in an electrically neutral collodion matrix, exhibited an apparently lower pH optimum than did free urease. In this case, the rates of diffusion of NH_4^+ formed in the course of the reaction may influence the steady state pH of the immediate environment of the active site of the immobilized urease, making it more alkaline than that of the external bulk solution.

Enhanced stability to heat, or on storage at room temperatures or lower, is sometimes exhibited by urease derivatives.

4. UREASE INHIBITION

Free ureases are inhibited by heavy metal ions, especially Ag^+ and Hg^{2+}; and by a variety of organic compounds, including some urea derivatives, antibiotics, hydroxylamine, and hydroxamic acids. The effectiveness with which phenylurea derivatives inhibit urease activity is increased by introducing chloro and large hydrophobic substituents into the aromatic ring (Cervelli et al., 1975).

Urease activity is completely lost when either four Ag^+ ions or two hydroxamic acid molecules are bound per urease molecule. Inhibition may be due either to binding at the urease active sites, or to changes induced in the structural configuration of the enzyme. Urease adsorbed to bentonite is less inhibited by Cu^{2+} ions than is free urease. Similarly, urease coupled to acrylic polymers is less sensitive to substrate inhibition.

EDTA inhibits urease by removing essential Ni^{2+} ions.

B. Soil Ureases

1. ASSAY AND ACTIVITY

Methods of assay and factors affecting soil urease activity have been extensively reviewed (Bremner & Mulvaney, 1978; Mulvaney & Bremner, 1981). Most assays are based on the rates of NH_4^+ accumulation in soils when incubated with added urea; others on the rates of CO_2 evolution, or on urea disappearance. Soils are incubated either moist (60% water-holding capacity) or as buffered or unbuffered soil suspensions, often with shaking and usually for several hours. To minimize or eliminate urease activity by proliferating microorganisms, soils may be treated with toluene or sterilized by irradiation. Some recommended assays use untreated soils incubated briefly with urea.

Like other soil enzyme assays, the measured urease activity reflects only the activity of the soil, expressed under the chosen assay conditions; it may not accurately relate to the actual or potential urease activity in the natural soil environment. Nevertheless, some assays are seen to be less suitable than others. For example, measurements of CO_2 evolution from added urea may underestimate urea decomposition in alkaline soils, and measurements of NH_4^+ concentrations may be unreliable indicators of urea decomposition in soils with recent accessions of carbonaceous residues.

The activities of surface soils range widely; comparisons are restricted because of the great variety of assay procedures adopted (Bremner & Mulvaney, 1978). However, significant correlations between soil urease activity and other soil properties, e.g., organic-C content, clay content, pH, and CEC have been reported (McGarity & Myers, 1967; Myers & McGarity, 1968; Silva & Perera, 1971; Dalal, 1975a; Tabatabai, 1977; Zantua et al., 1977; others). The direct relationship between urease activity and organic-C in soils is the most consistently and most commonly found, and is compatible with the view that ureases in soil mainly act exocellularly as stabilized enzymes protected by reaction with soil humic compounds.

2. ORIGIN, STABILITY, AND STATE OF OCCURRENCE

Soil microorganisms which hydrolyze urea include bacteria, actinomycetes, and fungi (Seneca et al., 1962; Roberge & Knowles, 1967; others). Ureolytic bacteria include aerobes, microaerophiles, and anaerobes, and account for a relatively constant proportion of the total bacterial populations in soils, despite large fluctuations in numbers (Roberge & Knowles, 1967; Lloyd & Sheaffe, 1973). Many studies have demonstrated increases in soil urease activity accompanying microbial growth in soils which have been amended with various C substrates (reviewed by Bremner & Mulvaney, 1978). Activity responses to the addition of urea alone to soils has varied from no change in activity (Zantua & Bremner, 1976) to increases (Roberge & Knowles, 1968; Schultz, 1978). Increased urease activities in soils receiving enzyme-saturating concentrations of urea were attributed to induced microbial ureases (Schultz, 1978). Khaziev and Agafarova (1976) concluded that soil ureases were microbial in origin, based on the similarity of Michaelis constants determined for soil and microbial ureases (however, see section V. B. 3).

The contribution of plant ureases to soil urease activity has not been quantified. Their involvement could be inferred from indirect evidence such as the effects of vegetation on soil activities (Pancholy & Rice, 1973; Voets et al., 1974; Cortez et al., 1975; Speir et al., 1980) or the distribution of activities in plant and soil fractions (Speir, 1976, 1977), but the participation of microbial ureases is not excluded. Nevertheless, plants are rich sources of ureases (Reithel, 1971), and exocellular ureases from rice roots have been demonstrated (Mahaptra et al., 1977). In the absence of quantitative evidence, there seems little justification for the widespread, often tacit, assumption that soil ureases are predominantly of microbial origin.

As with other enzymes, sources of urease (including a variety of microbial sources) may vary in importance, not only with the time of sampling and pretreatment of samples, but also with the conditions of assay. For example, the use of toluene-treated or irradiated soils to eliminate activities due to growing microorganisms may either decrease or increase soil urease activity (Thente, 1970; reviews by Ladd, 1978; Bremner & Mulvaney, 1978). It can be surmised that the decreases were due to death of ureolytic microorganisms and/or partial denaturation of soil ureases. Increases in activity may be due to a greater opportunity for enzyme-substrate reaction afforded by rupture of, or permeability changes to, cell membranes. Depending upon irradiation dose, soil type, moisture content, etc., the increases may more than offset decreases due to urease denaturation.

It seems likely that ureases from different sources will have varied opportunities to persist as active enzymes in soils after the death and decomposition of the cells from which they originate (Ladd, 1978). Such differences are indicated from the trends in urease activities with time of incubation of moist soils. Zantua and Bremner (1977) found that urease activities remained constant when moist soils were incubated for 6 months, either aerobically at temperatures ranging from -20 or $40°C$, or anaerobically at 30 or $40°C$ (Table 5). By contrast, other workers have shown urease activities of moist soils to either increase or decrease with incubation. For example, Speir et al. (1980) recorded progressive losses of urease activities of two unplanted New Zealand soils over a period of 5 months. Activity losses in the soils incubated at $25°C$ were greater than those in soils incubated at 10 or $18°C$. In one soil at $25°C$, almost all urease activity had disappeared after 5 months. The same soils planted to rye grass showed less decline, and in some cases a net gain in urease activity, due possibly to direct contributions of enzymes from plant roots and from microorganisms proliferating on plant debris. These results demonstrate that high proportions of the urease activities of the two New Zealand soils, even when unplanted, were attributable to enzymes which were susceptible either to denaturation or to biological degradation.

These results are consistent with those of Voets et al. (1974) who found decreased urease activities with time in soils treated with atrazine to remove vegetation; but they contrast with those of Zantua and Bremner (1975, 1976, 1977) who clearly demonstrated the stability of native ureases in a range of cultivated soils. Not only were the ureases of the latter soils stable to long-term incubation, but also they retained their activity when the soils were either freeze-dried, air-dried, or dried at temperatures up to $60°C$, and when the soils were incubated with the proteolytic enzymes pronase and trypsin (Zantua & Bremner, 1975, 1976, 1977) (Table 5).

Zantua and Bremner (1976, 1977) also showed that jack-bean urease added to these soils, or microbial ureases recently formed in these soils following incubation with added substrates, were not stable (Fig. 9). Of particular interest was that the soils ranged widely in their initial urease ac-

Table 5—Stability of soil ureases.†

Treatment of soil	Effect on urease activity
Dried at 30, 40, 50, or 60°C for 24 hours.	None
Dried at 75°C for 24 hours.	Partial loss of activity
Dried at 105°C for 24 hours.	Complete loss of activity
Autoclaved (120°C) for 2 hours.	Complete loss of activity
Leached with water.	None
Stored at −20, −10, 5, 10, 20, 30, or 40°C for 6 months.	None
Incubated at 30 or 40°C under aerobic or waterlogged conditions for 6 months.	None
Air-dried and stored at 21–23°C for 2 years.	None
Incubated at 30°C after addition of jack bean urease.	Increase followed by decrease to original activity
Incubated at 30°C after treatment with organic materials.	Increase followed by decrease to original activity
Incubated at 30°C after treatment with proteolytic enzymes (pronase or trypsin).	None

† From Bremner and Mulvaney (1978), by courtesy of Academic Press, London.

tivities (i.e., prior to amendment with the plant urease or with the microbial substrates), and that following amendment and incubation for several weeks, urease activities decreased to, and then remained at, their respective initial levels (Zantua & Bremner, 1976, 1977). These results suggest that the ability of soils to retain potentially active ureases varies with the soil; and, of those soils tested, all appeared to have reached their maximal capacity to protect ureases prior to amendment. None (or undetectable amounts) of the

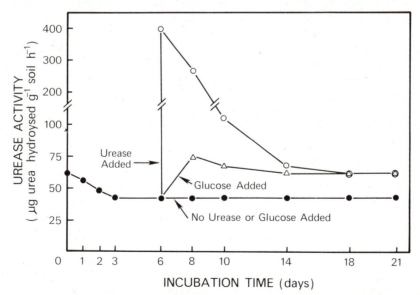

Fig. 9—Urease activity of a soil, incubated at 30°C and 60% WHC. The air dried soil had been remoistened at the commencement of incubation and then, after six days, amended with glucose and urease (from Zantua and Bremner (1977), by courtesy of Pergamon Press, Oxford).

added or newly formed ureases was stabilized within the soils, not even, in the latter case, within residual microbial cells or by association with newly formed, resistant humic compounds. The results with urease, following substrate addition and incubation of soils, are similar to those obtained with protein-hydrolysing enzymes, but they contrast with those obtained when peptides and amides of low molecular weight were used as assay substrates (Ladd & Paul, 1973).

Clearly, the collective experience of a number of workers indicates that soil ureases are, at the time of assay, present in states of varying biological and physico-chemical stability. In some soils the total activities are predominantly due to stabilized ureases, possibly because of a lack of recently added plant residues. Thus, in these soils the proportions of the total activity due to unprotected plant and microbial ureases would be low.

Some workers have found that air-drying of soils may either increase or decrease urease activity (for example, McGarity & Myers, 1967; Cerna, 1968; Thente, 1970; Gould et al., 1973; Speir & Ross, 1975). Zantua and Bremner (1977) reported that air-drying of their soils did not affect urease activities on immediate assay, but after remoistening and incubation of the soils, urease activities decreased to new stable levels within three days (Fig. 10). Air drying appeared to remove the mechanism by which some of the soil ureases were protected from biological decomposition in the field-moist soils. Repeated drying and incubation of the remoistened soils had no further effects on their urease activities. The anticipated increases in the availability of organic substrates, due to drying of the soils, and the growth of microorganisms, during incubation, did not apparently lead to sufficient de novo synthesis of urease to offset destruction of some previously stabilized enzymes.

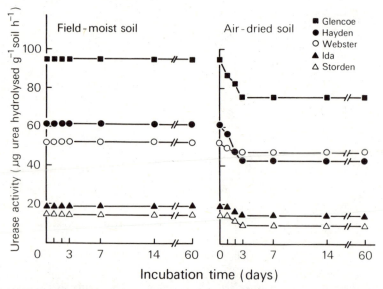

Fig. 10—Effect on urease activity of aerobic incubation (30°C, 60% WHC) of field moist and air-dried soils (from Zantua and Bremner (1977), by courtesy of Pergamon Press, Oxford).

The contribution of viable ureolytic microflora to the total urease activity of soils may be small (Ananthanarayana & Mithyantha, 1970), but the percentage will vary according to the conditions prevailing prior to, and during, the assay (Paulson & Kurtz, 1969; Thente, 1970; Lloyd & Sheaffe, 1973; Zantua & Bremner, 1976). Soil ureases are thought to function mainly exocellularly, however. The generally demonstrated stability of ureases in soils suggests that the enzymes are protected against both biological degradation and, to some extent, physical and chemical denaturation. Bonding of the enzyme to inorganic or organic soil constituents may confer stability (Burns et al., 1972a, 1972b; McLaren, 1975; McLaren et al., 1975; Nannipieri et al., 1974, 1975; Burns, 1977, 1978; Ceccanti et al., 1978). Bonding or entrapment of ureases in a resistant, organic colloidal matrix may, like analogous enzyme-organic derivatives, (i) enhance the rigidity of the enzyme structures and, thus, help stabilize them against denaturation by heating or drying, and (ii) render the ureases inaccessible to attack by soil proteinases, yet may not exclude small molecular weight substrates and products. Such mechanisms are not entirely speculative. Urease-active preparations of soil organic matter, free of clay, are resistant to proteolysis (Burns et al., 1972b) and to heat inactivation (Pettit et al., 1976). Humus-urease complexes, extracted from soil and fractionated according to molecular size, resist proteolysis and thermal denaturation to greater extents than free urease. Complexes of molecular weights $> 10^5$ were the most stable (Ceccanti et al., 1978; Nannipieri et al., 1978) (Table 6).

However, the manner in which urease is bound and the nature of the

Table 6—Stabilities of ureases from jack bean and in soil extracts.

Urease preparation	Treatment	Enzyme activity %[†]	Reference
Jack bean	Incubate at 25°C (7 days)	25	Pettit et al.
Soil extract		72	(1976)
Jack bean	Incubate at 45°C (7 days)	0	
Soil extract		55	
Jack bean	Incubate at 25°C (1 day) + pronase	48	
Soil extract		75	
Jack bean	Incubate at 25°C (5 days) + pronase	12	
Soil extract		75	
Jack bean	Lyophilize	6	
Soil extract		61	
Jack bean	Incubate at 70°C (30 min)	40	Nannipieri
Soil extract		70	et al. (1978)
Fraction AI (mol. wt. $>10^5$)		80	
Fraction AII (mol. wt. $<10^5$)		67	
Soil extract	Incubate at 25°C (4 days) + pronase	90	
Fraction AI		75	
Fraction AII		36	

† Percentages of initial activities or of untreated urease preparations.

organic ligands in humus-urease complexes are unknown. Urease is a Ni metalloenzyme, and both Ni^{2+} and an SH group are essential for urease activity. It seems likely that covalent reactions of the enzyme's SH groups with quinones during humic acid formation, and chelation of enzyme Ni^{2+} by preformed humic polymers, could lead to the formation of some enzymically-inactive complexes, but not necessarily to the exclusion of some active derivatives. Stabilization of extracellular enzymes may also involve enzyme bonding to polysaccharides (Martin & Haider, 1971; Mayaudon et al., 1975) as well as linkage to aromatic polymers.

3. KINETICS

Briggs and Segal (1963) obtained from soil, a crystalline urease-active preparation by acetone precipitation from a filtered, centrifuged, phosphate buffer extract. The crystalline material contained 8.78% N, and could be separated by ultracentrifugation into three components. Urease activity, attributed to a mixture of proteins in the preparation, was maximal at pH 7.1, and was inhibited by heavy metal ions such as Ag^+, Hg^{2+}, and Cu^{2+}. No other reports of crystalline or of highly purified soil ureases have been made, although urease activity has been demonstrated in crude soil extracts (Lloyd, 1975; Pettit et al., 1976) and in partially-purified extracts (Burns et al., 1972a; Ceccanti et al., 1978). Considering the variety of potential sources of soil ureases, and the likelihood of a heterogeneous distribution of the enzymes in soils, it is not surprising that values for a catalytic property range widely for different soils, and that they differ significantly from those of crystalline jack-bean urease with which they are frequently compared.

Soil urease activities increase then remain constant with increasing substrate concentrations (Paulson & Kurtz, 1970; Douglas & Bremner, 1971; Tabatabai & Bremner, 1972; Gould et al., 1973; Tabatabai, 1973; Ardakani et al., 1975; Dalal, 1975a; Pettit et al., 1976; Bremner & Mulvaney, 1978; Beri & Brar, 1978; Beri et al., 1978). Zero order reaction rates for three soils were attained with minimal urea concentrations ranging from 250 to 1,000 μg urea N/g soil (from Bremner & Mulvaney, 1978). An assay recommended by Roberge (1978) employs 3,500 μg urea N/g moist soil. Calculated Michaelis constants (K_m) and maximum reaction velocities (V_{max}) vary with the soil and conditions of assay (Tabatabai, 1973). Values for K_m for ureases of unfractionated soils have ranged from $1.3 \times 10^{-3}M$ (Tabatabai, 1973) to $213 \times 10^{-3}M$ (Paulson & Kurtz, 1970). Values for soil ureases in particle-size fractions also differed, ranging from $1.1 \times 10^{-3}M$ to $4.9 \times 10^{-3}M$; no trend was evident (Tabatabai, 1973).

The K_m value for ureases in a soil extract ($145 \times 10^{-3}M$) exceeded that of the unextracted soil ($52.3 \times 10^{-3}M$) (Pettit et al., 1976). Ureases from soil extracts, fractionated according to molecular size, ranged in K_m values from 8×10^{-3} to $40 \times 10^{-3}M$; again no consistent trend was obtained (Nannipieri et al., 1978) (Table 7). Differences in Michaelis constants may reflect varying rates of diffusion of urea substrate to the active sites of the bound ureases, in which case properties such as the nature and molecular size of the organic ligands may be important. However, in the absence of

Table 7—Kinetic properties of ureases from jack bean and in soil extracts.

Urease preparation	Property		Reference
	K_m value, mM	Optimal pH	
Jack bean	5.5[a]†	7.5[a]	Pettit et al.
	19[b]	6.5[b]‡	(1976)
Soil extract (crude)	145[a]	7.0[a]	
		6.8[b]	
Soil extract (crude)	100[a]	8.4[c]§	Nannipieri et al.
Fraction AI (mol wt >10⁵)	28[a]	7.6[b,c], 8.8[c]	(1978)
Fraction AII (mol wt <10⁵)	35[a]	7.0[b], 7.8[bc]	
Subfraction SI (mol wt >5 × 10⁵)	11[a]	8.2[c]	
Subfraction SII (mol wt 1.3–1.5 × 10⁵)	25[a]	7.5[b,c], 8.8[c]	
Subfraction SIII (mol wt 6–8 × 10⁴)	8[a]	8.7[c]	
Subfraction SIV (mol wt 7–9 × 10⁴)	40[a]	7.0[b], 7.8[b,c]	
Subfraction SV (mol wt 4–5 × 10⁴)	11[a]	7.8[b,c], 8.8[c]	

† a = Assayed with tris buffer.
‡ b = Assayed with phosphate buffer.
§ c = Assayed with borate buffer.

firm evidence, interpretations must be speculative. Differences in K_m values could also be attributed to differences between ureases of different origin (K_m range, 20 to 40 × $10^{-3}M$) or to differences in the structure and charge distribution in the enzyme's active site induced by complex formation.

Paulson and Kurtz (1969, 1970) proposed that in soil, ureases are either in microbial cells or are adsorbed on soil colloids. K_m values for soil ureases decreased as the proportions of soil activity attributable to living microorganisms increased; values increased as the contributions from intact cells decreased. Values of K_m for cellular and adsorbed ureases were calculated to be $0.057M$ and $0.252M$, respectively. The higher K_m values determined for adsorbed ureases are consistent with earlier results (Durand, 1964, 1965b), which indicated that the K_m of urease adsorbed to bentonite was greater than that of free urease.

Soil urease activity has been detected at $-20°$ (Bremner & Zantua, 1975) and at $-33°C$ (Tagliabue, 1958). Activities increase with temperature to a maximum near 70°C, and decline rapidly at 80°C (reviewed by Bremner & Mulvaney, 1978). The activation energies of soil ureases range from 3.90 to 24.5 Kcal/mol (Rachinskii & Pelttser, 1967; Gould et al., 1973; Dalal, 1975a, b). The mean activation energy (21.9 Kcal/mol) of ureases of 15 soils, assayed in the presence of toluene, was significantly higher than that (5.2 Kcal/mol) of ureases from the same soils without toluene (Dalal, 1975b). Free energies of activation of soil ureases, with or without toluene, were similar however. It was concluded that the greater activation energies of soil ureases in the presence of toluene are consistent with the greater energy requirements for the formation of enzyme-substrate complexes in soils, where the free intracellular ureases are eliminated and the active enzymes are those adsorbed or complexed with soil colloids.

The optimum pH for soil urease activity has been reported to be 6.5 to 7.0 (Hofmann & Schmidt, 1953; Pettit et al., 1976), similar to that of jack bean urease (Reithel, 1971; Bremner & Mulvaney, 1978). However, Tabatabai and Bremner (1972) and May and Douglas (1976) found that the pH optimum of soil ureases was 8.8 to 9.0. The pH-activity profile of jack bean urease is influenced by the nature of the buffer and the substrate concentration used in the assay (Bremner & Mulvaney, 1978). The latter may be of greater importance in establishing the optimum pH for soil ureases since the higher pH optima for soil ureases were attained in assays using relatively low concentrations of urea substrate, whereas both phosphate and Tris buffers were used in assays in which high and low pH optima were observed.

Pettit et al. (1976), using both phosphate and Tris buffers, found that the pH optimum for a urease-active soil extract was 6.5 to 7.0, similar to that (6.5) of the unextracted soil (Table 7). By contrast, Nannipieri et al. (1978), using phosphate and borate buffers, showed that the pH optimum of a soil urease extract was 8.3. Further fractionation of the extract yielded active urease-humus complexes of different molecular size, and having different pH-activity profiles. All showed pH optima in the range 8.0 to 8.8, but some fractions exhibited a second optimum at pH 7.0 to 7.5, closer to that (6.5-7.6) of jack bean urease in phosphate buffer (Pettit et al., 1976) (Table 7). The results of Nannipieri et al. (1978) are consistent with the view that the extracted ureases are bonded in polyanionic humic matrixes, which to varying extents, increase the H^+ concentrations within the microenvironment of the enzyme relative to that of the bulk solution. It is of interest that of the subfractions SI to SV, those (SII and SIV) exhibiting the highest K_m values also exhibit the double pH optimum (Table 7). If the higher K_m values were due to a generally more restricted diffusion of substrate to the bound enzyme, and if the pH-activity profiles were influenced by polyanionic ligands, it seems that the structural complexities of the ligands and their charge distributions must vary between the different fractions. However, it is not yet clear how the ureases are complexed in these fractions or the extent to which the molecular sizes of the different components of the fractions are due to the ligands themselves. Since the molecular weight and other properties (including K_m and pH optimum) of crystalline jack bean urease itself can vary according to its preparation and assay conditions, inferences drawn from comparisons of the kinetic properties of soils and soil fractions (containing ureases of diverse origins) and those of the crystalline plant enzyme, must be treated with caution.

4. INHIBITION

Bremner and Mulvaney (1978) and Mulvaney and Bremner (1981) have recently reviewed studies of the inhibition of soil ureases. Unextracted soil ureases are inhibited by the same compounds which inhibit purified plant ureases, but in soil inhibition is less marked. For example, the heavy metal cations, Ag^+, Hg^{2+}, Au^{2+} or Cu^{2+}, in concentrations of < 10 $\mu g/ml$, completely inhibited jack bean urease, but at fivefold greater concentrations they inhibited soil urease on average by only 48, 37, 18, and 14%, respectively (Bremner and Douglas, 1971) (Table 8). The effective concentrations

Table 8—Inhibition of soil urease activity by inorganic and miscellaneous
organic compounds.

Inhibitor	Inhibitor concentration	Inhibition of urease activity, %†	Reference
Silver sulphate	50 µg/g soil	48	Bremner &
Mercuric chloride		37	Douglas (1971)
Gold chloride		18	
Cupric sulphate		14	
Phenylmercuric acetate	50 µg/g soil	67	
p-chloromercuribenzoate		35	
N-Ethylmaleimide		25	
Acetohydroxamic acid		14	
3-(Phenyl)-1,1-dimethylurea	20 µg/g soil	28	Cervelli et al.
3-(p-Chlorophenyl)-1,1-dimethyl urea		32	(1976)
3-(3,4-Dichlorophenyl)-1,1-dimethyl urea		24	
1-(2-Methylcyclohexyl)-3-phenyl urea		29	
Acetohydroxamic acid	100 µg/g soil	17	Gould et al.
Hydroxyurea		16	(1978)
Thiourea		9	
1,3,4-thiadiazole-2,5-dithiol		46	
5-amino-1,3,4-thiadiazole-2-thiol		27	

† Decreased activity, as percentage of urease activity in absence of inhibitor.

of heavy metal ions in soils would be less than that of the applied solutions due to metal ion adsorption reactions. Results of Tabatabai (1977) show that both the order of effectiveness of the metal ions and the extent of inhibition vary with the soil. However, Ag^+ is consistently the most effective (Bremner & Douglas, 1971; Tabatabai, 1977). Its mode of action is believed to be a reaction with enzyme SH groups, but the loci of reaction, even in pure ureases, are unknown. In soils free or unprotected ureases might be anticipated to react with metal ions in the same manner as pure ureases from plant sources. However, stabilized, exocellular soil ureases probably would not be as susceptible as the pure, free ureases to any configurational changes which might result from reactions of heavy metals with SH groups away from the enzymes' active sites. Indeed, if the hypotheses concerning the mechanism of soil urease stabilization are correct (Burns, 1978), such SH groups may have already chemically reacted with soil humic compounds. In such cases metal ions are more likely to inhibit soil ureases by reaction with SH groups at the enzymes' active sites.

Substituted urea derivatives inhibit jack bean urease (Cervelli et al., 1975) and soil ureases (Cervelli et al., 1976, 1977) (Table 8). In all systems, inhibition was of the mixed type. The urea derivatives exhibited different inhibition constants, when derived from assays with the pure plant urease. However, differential adsorption of these inhibitors by soil reduced and obscured their relative effects on soil urease activity. By taking into account the effects of soil adsorption, Cervelli et al. (1977) were able to calculate inhibition constants for each compound for the soil urease systems. Corrected values were greater than those obtained with jack bean urease. Differences in inhibition constants obtained from data with jack bean urease and soil urease assays may be due to differences in the properties of the enzymes themselves, to the effects of enzyme immobilization in soil, or to the different rates of diffusion of substrate, inhibitor, and product to and from the enzymes' active sites.

Other organic inhibitors of plant and soil ureases include hydroxamates (Pugh & Waid, 1969a, b; others) and heterocyclic mercaptans and disulfides (Gould et al., 1978). The latter inhibit by the formation of disulfide bonds between urease and inhibitor, by a thioldisulfide exchange reaction. Some heterocyclic sulfur compounds were powerful inhibitors of jack bean urease but were almost ineffective against soil urease (Gould et al., 1978).

The most effective organic inhibitors of soil ureases are 1:4 benzoquinones and related phenols (Quastel, 1933; Bremner & Douglas, 1971, 1973; Bundy & Bremner, 1973; Gould et al., 1978) (Table 9). Methyl- and halogen-substituted p-benzoquinones were the most inhibitory (Bundy & Bremner, 1973). Quinones probably inhibit urease through reaction with enzyme SH groups (Cecil & McPhee, 1959; Webb, 1966). Dihydric phenols inhibit only after their oxidation to the respective quinones (Quastel, 1933). Inhibition of soil urease activities by p-benzoquinone and hydroquinone was inversely related to the organic C, organic N, clay, and silt contents of 25 surface soils tested and to their CEC (Mulvaney & Bremner, 1978).

Urease activities in soils may be markedly decreased by treatment with methyl bromide (Bhavanandan & Fernando, 1970) or acetone (Lethbridge et al., 1976).

Table 9—Inhibition of soil urease activity by quinones and related compounds.

Inhibitor	Inhibitor concentration	Inhibition of urease activity, %‡	Reference
p-Benzoquinone	50 µg/g soil	70	Bundy &
Methyl-p-benzoquinone		54	Bremner (1973)
2,3-Dimethyl-p-benzoquinone		54	
2,5-Dimethyl-p-benzoquinone		28	
2,6-Dimethyl-p-benzoquinone		30	
Trimethyl-p-benzoquinone		0	
Tetramethyl-p-benzoquinone		0	
2,5-Dichloro-p-benzoquinone		69	
2,6-Dichloro-p-benzoquinone		61	
Trichloro-p-benzoquinone		42	
Tetrachloro-p-benzoquinone		48	
p-Benzoquinone	50 µg/g soil	62	Bremner &
Hydroquinone		64	Douglas (1971)
Catechol		74	
Phenol		42	
4-Chlorophenol		35	
p-Benzoquinone	100 µg/g soil	88	Gould et al.
Hydroquinone		88	(1978)
2,5-Dimethyl-p-benzoquinone		50	
2,6-Dimethyl-p-benzoquinone		45	
Catechol		47	
Phenol		10	

† Decreased activity, as percentage of urease activity in absence of inhibitor.

VI. OTHER COMPOUNDS

Small amounts of simple organic N compounds, such as choline, ethanolamine, and trimethylamine have been found in soils (reviewed by Bremner, 1967).

$(CH_3)_3N-CH_2-CH_2OH$ $H_2N-CH_2-CH_2OH$ $(CH_3)_3N$

Choline   Ethanolamine   Trimethylamine

Plants, microorganisms, and soil animals contain variable amounts of phosphatidyl choline and phosphatidyl ethanolamine, the latter comprising as much as 40% of the total lipid in some bacteria. Trimethylamine occurs in some fungi and higher plants. While it is evident that choline, ethanolamine, and trimethylamine from these and other sources will eventually be degraded by the soil microflora, their contribution to NH_4^+ formation is unknown.

Choline is degraded to glycine in a reaction sequence involving two successive oxidation steps followed by three successive demethylation reactions. Formation of NH_4^+ from glycine is catalyzed by amino acid oxidases and dehydrogenases or by the specific enzyme, glycine dehydrogenase [glycine:NAD$^+$ oxidoreductase (deaminating), EC 1.4.1.10].

Three pathways for NH_4^+ formation from ethanolamine have been found in microorganisms: (i) ethanolamine is converted to NH_4^+ and acetaldehyde by ethanolamine ammonia-lyase (EC 4.3.1.7), (ii) oxidation of ethanolamine to NH_4^+ and glycolaldehyde is catalyzed by ethanolamine oxidase [ethanolamine: oxygen oxidoreductase (deaminating), EC 1.4.3.8], and (iii) ethanolamine is phosphorylated by transfer of PO_4^{3-} from ATP, catalyzed by an amino alcohol-ATP phosphotransferase, and the ethanolamine O-phosphate formed is hydrolyzed by a phospholyase (EC 4.2.99.7) to give NH_4^+, acetaldehyde, and PO_4^{3-}.

Trimethylamine is oxidized to methylamine and formaldehyde, and primary amine dehydrogenase converts methylamine to NH_4^+ and formaldehyde.

Complex nitrogenous compounds such as antibiotics and vitamins are produced by soil microorganisms; also, vitamins and other compounds such as alkaloids and chlorophyll reach the soil as constituents of plant litter. Since the concentrations of these complex substrates in soil are probably small, and their rates of degradation are unknown, the multi-step metabolic pathways leading to NH_4^+ formation are not discussed here.

For the same reasons, the degradation of nitrogenous pesticides are not discussed. Excellent comprehensive reviews of the degradation of these compounds in soil are available (Goring & Hamaker, 1972; Kearney & Kaufman, 1975, 1976).

LITERATURE CITED

Ambroz, Z. 1965. The proteolytic complex decomposing proteins in soil. Rostl. Vyroba 11: 161–170.

Ambroz, Z. 1966a. Some reasons for differences in the occurrences of proteases in soils. Rostl. Vyroba 12:1203–1210.

Ambroz, Z. 1966b. Some notes on the determination of activities of certain proteases in the soil. Sb. Vys. Sk. Zemed Brne Rada A 1:57–62.

Ambroz, Z. 1968. Investigation of the effects of soil structure and associated factors on the activity of proteases. Rostl. Vyroba 14:201–208.

Ambroz, Z. 1970. Factors influencing the distribution of some proteolytic enzymes in soil. Zentralbl. Bakteriol., Parasitenkd., Infektionskr. Hyg., Abt. 2: 125:433–437.

Ambroz, Z. 1971. The determination of the activities of gelatinase and caseinase in the soil. Biol. Sol. 13:28–29.

Ananthanarayana, R., and M. S. Mithyantha. 1970. Urease activity of the soils of M.R.S. (Main Research Station), Hebbal. Mysore J. Agric. Sci. 4:109–111.

Anderson, G. 1970. The isolation of nucleoside diphosphates from alkaline extracts of soil. J. Soil Sci. 21:96–104.

Antheunisse, J. 1972. Decomposition of nucleic acids and some of their degradation products by microorganisms. Antonie van Leeuwenhoek J. Microbiol. Serol. 38:311–327.

Antoniani, C., T. Montanari, and A. Camoriano. 1954. Soil enzymology. I. Cathepsin-like activity. A preliminary note. Ann. Fac. Agric. Univ. Milano 3:99–101.

Archibald, A. R. 1974. The structure, biosynthesis and function of teichoic acid. Adv. Microb. Physiol. 11:53–95.

Ardakani, M. S., M. G. Volz, and A. D. McLaren. 1975. Consecutive steady-state reactions of urea, ammonium and nitrite nitrogen in soil. Can. J. Soil Sci. 55:83–91.

Barker, H. A., and J. V. Beck. 1942. *Clostridium acidi-urici* and *Clostridium cylindrosporum,* organisms fermenting uric acid and some other purines. J. Bacteriol. 43:291–304.

Barkulis, S. S. 1966. *N*-Acetyl-D-glucosamine kinase. I. *Streptococcus pyogenes.* p. 415–420. *In* W. A. Wood (ed.) Methods in enzymology. Vol. IX. Carbohydrate metabolism. Academic Press, New York.

Bates, C. J., and C. A. Pasternak. 1965. Further studies on the regulation of amino sugar metabolism in *Bacillus subtilis.* Biochem. J. 96:147–154.

Beck, T., and H. Poschenrieder. 1963. Experiments on the effect of toluene on the soil microflora. Plant Soil 18:346–357.

Bei-Bienko, N. V. 1970. Effect of mineral nitrogen fertilizers on enzyme activity in soil. Pochvovedenie 2:87–93.

Bentley, R., and A. Neuberger. 1952. The mechanism of the action of uricase. Biochem. J. 52: 694–699.

Beri, V., and S. S. Brar. 1978. Urease activity in subtropical alkaline soils of India. Soil Sci. 126:330–335.

Beri, V., K. P. Goswami, and S. S. Brar. 1978. Urease activity and its Michaelis constant for soil systems. Plant Soil 49:105–115.

Bhavanandan, V. P., and V. Fernando. 1970. Studies on the use of urea as a fertilizer for tea in Ceylon. 2. Urease activity in tea soils. Tea Q. 41:94–106.

Bondietti, E., J. P. Martin, and K. Haider. 1972. Stabilization of amino sugar units in humic-type polymers. Soil Sci. Soc. Am. Proc. 36:597–602.

Bremner, J. M. 1965. Organic nitrogen in soils. *In* W. V. Bartholomew and F. E. Clark (ed.) Soil nitrogen. Agronomy 10:93–149. Am. Soc. Agron., Madison, Wis.

Bremner, J. M. 1967. Nitrogenous compounds. p. 19–66. *In* A. D. McLaren and G. H. Peterson (ed.) Soil biochemistry. Vol. 1. Marcel Dekker Inc., New York.

Bremner, J. M., and L. A. Douglas. 1971. Inhibition of urease activity in soils. Soil Biol. Biochem. 3:297–307.

Bremner, J. M., and L. A. Douglas. 1973. Effects of some urease inhibitors on urea hydrolysis in soils. Soil Sci. Soc. Am. Proc. 37:225–226.

Bremner, J. M., and R. L. Mulvaney. 1978. Urease activity in soils. p. 149–196. *In* R. G. Burns (ed.) Soil enzymes. Academic Press, London.

Bremner, J. M., and K. Shaw. 1954. Studies on the estimation and decomposition of amino-sugars in soil. J. Agr. Sci. 44:152–159.

Bremner, J. M., and M. I. Zantua. 1975. Enzyme activity at subzero temperatures. Soil Biol. Biochem. 7:383–387.

Briggs, M. H., and L. Segal. 1963. Preparation and properties of a free soil enzyme. Life Sci. 1:69–72.

Bright, H. J., and D. J. T. Porter. 1975. Flavoprotein oxidases. p. 421–505. *In* P. D. Boyer (ed.) The enzymes. Vol. XII. Oxidation-Reduction, 3rd ed. Academic Press, New York.

Bull, A. T. 1970. Inhibition of polysaccharases by melanin: enzyme inhibition in relation to mycolysis. Arch. Biochem. Biophys. 137:345–356.

Bundy, L. G., and J. M. Bremner. 1973. Effects of substituted *p*-benzoquinones on urease activity in soils. Soil Biol. Biochem. 5:847–853.

Burns, R. G. 1977. Soil enzymology. Sci. Proc., (Oxford) 64:275–285.

Burns, R. G. 1978. Enzyme activity in soil: some theoretical and practical considerations. p. 295–340. *In* R. G. Burns (ed.) Soil enzymes. Academic Press, London.

Burns, R. G., M. H. El-Sayed, and A. D. McLaren. 1972a. Extraction of a urease-active organo-complex from soil. Soil Biol. Biochem. 4:107–108.

Burns, R. G., A. H. Pukite, and A. D. McLaren. 1972b. Concerning the location and persistence of soil urease. Soil Sci. Soc. Am. Proc. 36:308–311.

Butler, J. H. A., and J. N. Ladd. 1969. The effect of methylation of humic acids on their influence on proteolytic enzyme activity. Aust. J. Soil Res. 7:263–268.

Cahn, R. S., C. K. Ingold, and V. Prelog. 1956. The specification of asymmetric configuration in organic chemistry. Experientia 12:81–94.

Casagrande, D. J., and K. Park. 1978. Muramic acid levels in bog soils from the Okefenokee Swamp. Soil Sci. 125:181–183.

Ceccanti, B., P. Nannipieri, S. Cervelli, and P. Sequi. 1978. Fractionation of humus-urease complexes. Soil Biol. Biochem. 10:39–45.

Cecil, R., and J. R. McPhee. 1959. The sulfur chemistry of proteins. Adv. Protein Chem. 14: 155–389.

Cerna, S. 1968. The influence of desiccation of the structural soil upon the intensity of biochemical reactions. Acta Univ. Carol. Biol. 4:285–288.

Cerna, S. 1970. The influence of crushing of structural aggregates on the activity of hydrolytic enzymes in the soil. Acta Univ. Carol. Biol. 6:461–466.

Cervelli, S., P. Nannipieri, G. Giovannini, and A. Perna. 1975. Jack bean urease inhibition by substituted ureas. Pest Biochem. Physiol. 5:221–225.

Cervelli, S., P. Nannipieri, G. Giovannini, and A. Perna. 1976. Relationships between substituted urea herbicides and soil urease activity. Weed Res. 16:365–368.

Cervelli, S., P. Nannipieri, G. Giovannini, and A. Perna. 1977. Effect of soil on urease inhibition by substituted urea herbicides. Soil Biol. Biochem. 9:393–396.

Clarke, P. H., and M. V. Tracey. 1956. The occurrence of chitinase in some bacteria. J. Gen. Microbiol. 14:188–196.

Cortez, J., G. Billes, and P. Lossaint. 1975. Étude comparative de l'activité biologique des sols sous peuplement arbustifs et herbacés de la garrigue méditerraneéne. II. Activités enzymatiques. Rev. Écol. Biol. Sol. 12:141–156.

Cunningham, L. 1965. The structure and mechanism of proteolytic enzymes. p. 85–188. *In* M. Florkin and E. H. Stotz (ed.) Comprehensive biochemistry. Vol. 16. Elsevier, Amsterdam.

Dalal, R. C. 1975a. Urease activity in some Trinidad soils. Soil Biol. Biochem. 7:5–8.

Dalal, R. C. 1975b. Effect of toluene on the energy barriers in urease activity of soils. Soil Sci. 120:256–260.

Datta, A. K., and S. K. Niyogi. 1976. Biochemistry and physiology of bacterial ribonucleases. Prog. Nucleic Acid Res. Mol. Biol. 17:271–308.

Dedeken, M., and J. P. Voets. 1965. Studies on the metabolism of amino acids in soil. I. Metabolism of glycine, alanine, aspartic acid and glutamic acid. Ann. Inst. Pasteur, Paris Suppl. 109:103–111.

Douglas, L. A., and J. M. Bremner. 1971. A rapid method of evaluating different compounds as inhibitors of urease activity in soils. Soil Biol. Biochem. 3:309–315.

Drobnik, J. 1956. Degradation of asparagine by the enzyme complex of soils. Cesk. Mikrobiol. 1:47.

Drummond, G. I., and M. Yamamoto. 1971a. Nucleotide phosphomonoesterases. p. 337–354. *In* P. D. Boyer (ed.) The enzymes. Vol. IV. Hydrolysis: other C-N bonds, phosphate esters. 3rd ed. Academic Press, New York.

Drummond, G. I., and M. Yamamoto. 1971b. Nucleoside cyclic phosphate diesterases. p. 355–371. *In* P. D. Boyer (ed.) The enzymes, Vol. IV. Hydrolysis: other C-N bonds, phosphate esters. 3rd ed. Academic Press, New York.

Durand, G. 1961. Sur la dégradation des bases puriques et pyrimidiques dans le sol:dégradation aérobie de l'acide urique. C. R. Acad. Sci., Paris 252:1687-1689.

Durand, G. 1964. Changes in urease activity in the presence of bentonite. C. R. Acad. Sci., Paris 259:3397-3400.

Durand, G. 1965a. Enzymes in soil. Rev. Ecol. Biol. Sol. 2:141-205.

Durand, G. 1965b. Enzymatic splitting of urea in the presence of bentonite. Ann. Inst. Pasteur, Paris Suppl. 109:121-132.

Estermann, E. F., E. E. Conn, and A. D. McLaren. 1959a. Influence of pH on particulate and soluble aconitase and glutamic dehydrogenase of *Lupinus albus*. Arch. Biochem. Biophys. 85:103-108.

Estermann, E. F., G. H. Peterson, and A. D. McLaren. 1959b. Digestion of clay-protein, lignin-protein and silica-protein by enzymes and bacteria. Soil Sci. Soc. Am. Proc. 23:31-36.

Franz, G. 1973. Comparative investigations on the enzyme activity of some soils in Nordrhein-Westfalen and Rheinland-Pfalz. Pedobiologia 13:423-436.

Galstyan, A. S. 1973. Formation of easily hydrolysable nitrogen in the soil. Biol. Zh. Arm. 26:15-19.

Galstyan, A. S., and E. G. Saakyan. 1973. Determination of soil glutaminase activity. Dokl. Akad. Nauk SSSR 209:1201-1202.

Goldman, R., L. Goldstein, and E. Katchalski. 1971. Water-insoluble enzyme derivatives and artificial enzyme membranes. p. 1-78. *In* G. R. Stark (ed.) Biochemical aspects of reactions on solid supports. Academic Press, New York.

Goldstein, L. 1970. Water-insoluble derivatives of proteolytic enzymes. p. 935-962. *In* G. Perlmann and L. Lorand (ed.) Methods in enzymology. Vol. XIX. Proteolytic enzymes. Academic Press, New York.

Goring, C. A. I., and W. V. Bartholomew. 1952. Adsorption of mononucleotides, nucleic acids and nucleoproteins by clays. Soil Sci. 74:149-164.

Goring, C. A. I., and J. W. Hamaker (ed.). 1972. Organic chemicals in the soil environment. Vol. 1 and 2. Marcel Dekker, Inc., New York.

Gould, W. D., F. D. Cook, and J. A. Bulat. 1978. Inhibition of urease activity by heterocyclic sulfur compounds. Soil Sci. Soc. Am. J. 42:66-72.

Gould, W. D., F. D. Cook, and G. R. Webster. 1973. Factors affecting urea hydrolysis in several Alberta soils. Plant Soil 38:393-401.

Greaves, M. P., and M. J. Wilson. 1970. The degradation of nucleic acids and montmorillonite-nucleic acid complexes by soil microorganisms. Soil Biol. Biochem. 2:257-268.

Greaves, M. P., and M. J. Wilson. 1973. Effects of soil microorganisms on montmorillonite-adenine complexes. Soil Biol. Biochem. 5:275-276.

Handley, W. R. C. 1961. Further evidence for the importance of residual leaf protein complexes in litter decomposition and the supply of nitrogen for plant growth. Plant Soil 15: 37-73.

Hartman, S. C. 1971. Glutaminase and γ-glutamyltransferases. p. 79-100. *In* P. D. Boyer (ed.) The enzymes. Vol. IV. Hydrolysis: other C–N bonds, phosphate esters. 3rd ed. Academic Press, New York.

Hayashi, H., Y. Araki, and E. Ito. 1973. Occurrence of glucosamine residues with free amino groups in cell wall peptidoglycan from Bacilli as a factor responsible for resistance to lysozyme. J. Bacteriol. 113:592-598.

Hoffmann, G., and K. Teicher. 1957. The enzyme system of our arable soils. VII. Proteases. Z. Pflanzenernaehr. Dueng. Bodenkd. 95:55-63.

Hofmann, E., and W. Schmidt. 1953. Uber das Enzym-system unserer Kulturboden. II. Urease. Biochem. Z. 324:125-127.

Imada, A., S. Igarasi, K. Nakahama, and M. Isono. 1973. Asparaginase and glutaminase activities of microorganisms. J. Gen. Microbiol. 76:85-99.

Imanaga, Y. 1957. Metabolism of D-glucosamine. The formation of D-fructose 6-phosphate. J. Biochem. (Tokyo). 44:69-79.

Kearney, P. C., and D. D. Kaufman (ed.). 1975. Herbicides: chemistry, degradation and mode of action. 2nd ed. Vol. 1. Marcel Dekker, Inc., New York.

Kearney, P. C., and D. D. Kaufman (ed.). 1976. Herbicides: chemistry, degradation and mode of action. 2nd ed. Vol. 2. Marcel Dekker, Inc., New York.

Khaziev, F. Kh., and Y. M. Agafarova. 1976. Michaelis constants of soil enzymes. Pochvovedenie 8:150-157.

Kiss, S., M. Dragan-Bularda, and D. Radulescu. 1975. Biological significance of enzymes accumulated in soil. Adv. Agron. 27:25-87.

Kuprevich, V. F., and T. A. Shcherbakova. 1961. Method for determining proteolytic activity of the soil. Dokl. Akad. Nauk B. SSR 3:133–136.

Kuprevich, V. F., and T. A. Shcherbakova. 1971. Comparative enzymatic activity in diverse types of soil. p. 167–201. *In* A. D. McLaren and J. J. Skujins (ed.) Soil biochemistry. Vol. 2. Marcel Dekker, Inc., New York.

Ladd, J. N. 1972. Properties of proteolytic enzymes extracted from soil. Soil Biol. Biochem. 4:227–237.

Ladd, J. N. 1978. Origin and range of enzymes in soil. p. 51–96. *In* R. G. Burns (ed.) Soil enzymes. Academic Press, London.

Ladd, J. N., P. G. Brisbane, J. H. A. Butler, and M. Amato. 1976. Studies on soil fumigation. III. Effects on enzyme activities, bacterial numbers and extractable ninhydrin reactive compounds. Soil Biol. Biochem. 8:255–260.

Ladd, J. N., and J. H . A. Butler. 1969. Inhibition and stimulation of proteolytic enzyme activities by soil humic acids. Aust. J. Soil Res. 7:253–261.

Ladd, J. N., and J. H. A. Butler. 1972. Short-term assays of soil proteolytic enzyme activities using proteins and dipeptide derivatives as substrates. Soil Biol. Biochem. 4:19–30.

Ladd, J. N., and J. H. A. Butler. 1975. Humus-enzyme systems and synthetic organic polymer-enzyme analogs. p. 143–194. *In* E. A. Paul and A. D. McLaren (ed.) Soil biochemistry. Vol. 4. Marcel Dekker, New York.

Ladd, J. N., and E. A. Paul. 1973. Changes in enzymic activity and distribution of acid-soluble, amino acid-nitrogen in soil during nitrogen immobilization and mineralization. Soil Biol. Biochem. 5:825–840.

Lehman, I. R. 1971. Bacterial deoxyribonucleases. p. 251–270. *In* P. D. Boyer (ed.) The enzymes. Vol. IV. Hydrolysis: other C–N bonds, phosphate esters. 3rd ed. Academic Press, New York.

Lethbridge, G., N. M. Pettit, A. R. J. Smith, and R. G. Burns. 1976. The effect of organic solvents on soil urease activity. Soil Biol. Biochem. 8:449–450.

Lloyd, A. B. 1975. Extraction of urease from soil. Soil Biol. Biochem. 7:357–358.

Lloyd, A. B., and M. J. Sheaffe. 1973. Urease activity in soils. Plant Soil 39:71–80.

Luderitz, O., K. Jann, and R. Wheat. 1968. Somatic and capsular antigens of Gram-negative bacteria. p. 105–228. *In* M. Florkin and E. H. Stotz (ed.) Comprehensive biochemistry. Vol. 26, part A. Elsevier, Amsterdam.

Macura, J., and K. Vagnerova. 1969. Colorimetric determination of the activity of proteolytic enzymes in soil. Rostl. Vyroba 15:173–180.

Mahaptra, B., B. Patnaik, and D. Mishra. 1977. The exocellular urease in rice roots. Curr. Sci. 46:680–681.

Mahler, H. R., and E. H. Cordes. 1971. Biological chemistry. 2nd ed. Harper and Row, New York.

Martin, J. P., and K. Haider. 1971. Microbial activity in relation to soil humus formation. Soil Sci. 111:54–63.

Marumoto, T., H. Kai, T. Yoshida, and T. Harada. 1975. Contribution of microbial cells and their cell walls to an accumulation of soil organic matter becoming decomposable due to drying a soil. Part 4. Soil Sci. Plant Nutr. (Tokyo) 21:193–194.

May, P. B., and L. A. Douglas. 1976. Assay for soil urease activity. Plant Soil 45:301–305.

Mayaudon, J. 1968. Stabilization biologique des proteines ^{14}C dans le sol. p. 177–188. *In* Isotopes and radiation in soil organic matter studies. Int. Atomic Energy Agency, Vienna.

Mayaudon, J., L. Batistic, and J. M. Sarkar. 1975. Properties of proteolytically active extracts from fresh soils. Soil Biol. Biochem. 7:281–286.

McGarity, J. W., and M. G. Myers. 1967. A survey of urease activity in soils of northern New South Wales. Plant Soil 27:217–238.

McLaren, A. D. 1960. Enzyme action in structurally restricted systems. Enzymologia 21:356–364.

McLaren, A. D. 1975. Soil as a system of humus and clay immobilized enzymes. Chem. Scr. 8:97–99.

McLaren, A. D., and E. F. Estermann. 1957. Influence of pH on the activity of chymotrypsin at a solid-liquid interface. Arch. Biochem. Biophys. 68:157–160.

McLaren, A. D., and L. Packer. 1970. Some aspects of enzyme reactions in heterogeneous systems. Adv. Enzymol. Relat. Areas Mol. Biol. 33:245–308.

McLaren, A. D., A. H. Pukite, and I. Barshad. 1975. Isolation of humus with enzymatic activity from soil. Soil Sci. 119:178–180.

McLaren, A. D., and J. J. Skujins. 1967. The physical environment of microorganisms in soils. p. 3-24. *In* T. R. G. Gray and D. Parkinson (ed.) The ecology of soil bacteria. Liverpool Univ. Press.

Metzler, D. E. 1977. Biochemistry (The chemical reactions of living cells). Academic Press, New York.

Millar, W. N., and L. E. Casida, Jr. 1970. Evidence for muramic acid in soil. Can. J. Microbiol. 16:299-304.

Mosbach, K. (ed.). 1976. Methods in enzymology. Vol. XLIV. Immobilized enzymes. Academic Press, New York.

Mouraret, M. 1965. Contribution a l'etude de l'activite des enzymes due sol: L'asparaginase. ORSTOM, Paris.

Mulvaney, R. L., and J. M. Bremner. 1978. Use of *p*-benzoquinone and hydroquinone for retardation of urea hydrolysis in soils. Soil Biol. Biochem. 10:297-302.

Mulvaney, R. L., and J. M. Bremner. 1981. Use of urease and nitrification inhibitors for control of urea transformations in soils. p. 153-196. *In* E. A. Paul and J. N. Ladd (ed.) Soil biochemistry. Vol. 5. Marcel Dekker, Inc., New York.

Myers, M. G., and J. W. McGarity. 1968. The urease activity in profiles of five great soil groups from northern New South Wales. Plant Soil 28:25-37.

Nannipieri, P., B. Ceccanti, S. Cervelli, and P. Sequi. 1974. Use of 0.1M pyrophosphate to extract urease from a Podzol. Soil Biol. Biochem. 6:359-362.

Nannipieri, P., B. Ceccanti, S. Cervelli, and P. Sequi. 1978. Stability and kinetic properties of humus-urease complexes. Soil Biol. Biochem. 10:143-147.

Nannipieri, P., S. Cervelli, and F. Pedrazzini. 1975. Concerning the extraction of enzymically active organic matter from soil. Experientia 31:513-515.

Noltmann, E. A. 1972. Aldose-ketose isomerases. p. 271-354. *In* P. D. Boyer (ed.) The enzymes. Vol. VI. Carboxylation and decarboxylation (nonoxidative), isomerization. 3rd ed. Academic Press, New York.

Okafor, N. 1966a. Estimation of the decomposition of chitin in soil by the method of carbon dioxide release. Soil Sci. 102:140-142.

Okafor, N. 1966b. The ecology of microorganisms on, and the decomposition of, insect wings in the soil. Plant Soil 25:211-237.

Pancholy, S. K., and E. L. Rice. 1973. Soil enzymes in relation to old field succession: amylase, cellulase, invertase, dehydrogenase and urease. Soil Sci. Soc. Am. Proc. 37:47-50.

Parsons, J. W. 1981. Chemistry and distribution of amino sugars in soils and soil organisms. p. 197-227. *In* E. A. Paul and J. N. Ladd (ed.) Soil biochemistry. Vol. 5. Marcel Dekker, Inc., New York.

Paulson, K. N., and L. T. Kurtz. 1969. Locus of urease activity in soil. Soil Sci. Soc. Am. Proc. 33:897-901.

Paulson, K. N., and L. T. Kurtz. 1970. Michaelis constant of soil urease. Soil Sci. Soc. Am. Proc. 34:70-72.

Pettit, N. M., A. R. J. Smith, R. B. Freedman, and R. G. Burns. 1976. Soil urease: activity, stability and kinetic properties. Soil Biol. Biochem. 8:479-484.

Pugh, K. B., and J. S. Waid. 1969a. The influence of hydroxamates on ammonia loss from an acid loamy sand treated with urea. Soil Biol. Biochem. 1:195-206.

Pugh, K. B., and J. S. Waid. 1969b. The influence of hydroxamates on ammonia loss from various soils treated with urea. Soil Biol. Biochem. 1:207-217.

Quastel, J. H. 1933. The action of polyhydric phenols on urease; the influence of thiol compounds. Biochem. J. 27:1116-1122.

Rachinskii, V. V., and A. S. Pelttser. 1967. Effect of temperature on rate of decomposition of urea in soil. Agrokhimiya 10:75-77.

Reid, J. W., and I. B. Wilson. 1971. *E. coli* alkaline phosphatase. p. 373-415. *In* P. D. Boyer (ed.) The enzymes. Vol. IV. Hydrolysis: other C-N bonds, phosphate esters. 3rd ed. Academic Press, New York.

Reithel, F. J. 1971. Ureases. p. 1-21. *In* P. D. Boyer (ed.) The enzymes. Vol. IV. Hydrolysis: other C-N bonds, phosphate esters. 3rd ed. Academic Press, New York.

Roberge, M. R. 1978. Methodology of soil enzyme measurement and extraction. p. 341-370. *In* R. G. Burns (ed.) Soil enzymes. Academic Press, London.

Roberge, M. R., and R. Knowles. 1967. The ureolytic microflora in a black spruce (*Picea mariana* Mill.) humus. Soil Sci. Soc. Am. Proc. 31:76-79.

Roberge, M. R., and R. Knowles. 1968. Urease activity in black spruce humus sterilized by gamma radiation. Soil Sci. Soc. Am. Proc. 32:518-521.

Robinson, E. I., and W. A. Venables. 1978. β-Alanine and β-aminoisobutyrate utilisation by *Pseudomonas aeruginosa*. Soc. Gen. Microbiol: Quarterly 6:25.

Rolls, J. P., and C. W. Shuster. 1972. Amino sugar assimilation by *Escherichia coli*. J. Bacteriol. 112:894–902.

Ross, D. J., and B. A. McNeilly. 1975. Studies of a climosequence of soils in tussock grasslands. 3. Nitrogen mineralization and protease activity. N.Z. J. Sci. 18:361–375.

Rowell, M. J., J. N. Ladd, and E. A. Paul. 1973. Enzymically active complexes of proteases and humic acid analogues. Soil Biol. Biochem. 5:699–703.

Schultz, H. 1978. Ureolytic activity in samples of various arable soil types. Pedobiologia 18: 57–63.

Seneca, H., P. Peer, and R. Nally. 1962. Microbial urease. Nature (London) 193:1106–1107.

Shields, J. A., E. A. Paul, W. E. Lowe, and D. Parkinson. 1973. Turnover of microbial tissue in soil under field conditions. Soil Biol. Biochem. 5:753–764.

Shimada, K., and Y. Sugino. 1969. Cyclic phosphodiesterase having 3′-nucleotidase activity from *Bacillus subtilis*. Biochem. Biophys. Acta 185:367–380.

Silva, C. G., and A. M. A. Perera. 1971. A study of the urease activity in the rubber soils of Ceylon. Quart. J. Rubber Res. Inst. Ceylon 47:30–36.

Simonart, P., and J. Mayaudon. 1961. Humification des proteines ¹⁴C dans le sol. p. 91–103. *In* 2nd Int. Pedologie Symp. Applications des Sciences Nucleaires en Pedologie. Bull. de la Societe Belge de Pedologie.

Skujins, J. J. 1967. Enzymes in soil. p. 371–414. *In* A. D. McLaren and G. H. Peterson (ed.) Soil biochemistry. Vol. 1. Marcel Dekker Inc., New York.

Skujins, J. 1976. Extracellular enzymes in soil. CRC Crit. Rev. Microbiol. 4:383–421.

Skujins, J., and A. D. McLaren. 1967. Enzyme reaction rates at limited water activities. Science 158:1569–1570.

Skujins, J., H. J. Potgieter, and M. Alexander. 1965. Dissolution of fungal cell walls by a Streptomycete chitinase and β-(1→3) glucanase. Arch. Biochem. Biophys. 111:358–364.

Skujins, J., and A. Pukite. 1970. Extraction and determination of *N*-acetylglucosamine from soil. Soil Biol. Biochem. 2:141–143.

Skujins, J., A. Pukite, and A. D. McLaren. 1973. Adsorption and reactions of chitinase and lysozyme on chitin. Mol. Cell. Biochem. 2:221–228.

Skujins, J., A. Pukite, and A. D. McLaren. 1974. Adsorption and activity of chitinase on kaolinite. Soil Biol. Biochem. 6:179–182.

Speir, T. W. 1976. Studies on a climosequence of soils in tussock grasslands. 8. Urease, phosphatase, and sulphatase activities of tussock plant materials and of soil. N.Z. J. Sci. 19: 383–387.

Speir, T. W. 1977. Studies on a climosequence of soils in tussock grasslands. 10. Distribution of urease, phosphatase, and sulphatase activities in soil fractions. N.Z. J. Sci. 20:151–157.

Speir, T. W., R. Lee, E. A. Pansier, and A. Cairns. 1980. A comparison of sulphatase, urease and protease activities in planted and fallow soils in a pot trial. Soil Biol. Biochem. 12: 281–291.

Speir, T. W., and D. J. Ross. 1975. Effects of storage on the activities of protease, urease, phosphatase, and sulphatase in three soils under pasture. N.Z. J. Sci. 18:231–237.

Speir, T. W., and D. J. Ross. 1978. Soil phosphatase and sulphatase. p. 197–250. *In* R. G. Burns (ed.) Soil enzymes. Academic Press, London.

Stevenson, F. J. 1957. Investigations of aminopolysaccharides in soils: 2. Distribution of hexosamines in some soil profiles. Soil Sci. 84:99–106.

Tabatabai, M. A. 1973. Michaelis constants of urease in soils and soil fractions. Soil Sci. Soc. Am. Proc. 37:707–710.

Tabatabai, M. A. 1977. Effects of trace elements on urease activity in soils. Soil Biol. Biochem. 9:9–13.   9:9–13.

Tabatabai, M. A., and J. M. Bremner. 1972. Assay of urease activity in soils. Soil Biol. Biochem. 4:479–487.

Tagliabue, L. 1958. Cryoenzymological research on urease in soils. Chimica Milano 34:448–491.

Takahashi, E., S. Konishi, and K. Hazama. 1970. Studies on the uptake and utilization of uric acid by crop plants. Soil Sci. Plant Nutr. (Tokyo) 16:257.

Taupitz-Stahn, E. L. 1974. Bacterial uricase effects in vitro. Aerztl. Lab. 20:8–12.

Thente, B. 1970. Effects of toluene and high energy radiation on urease activity in soil. Lantbruks hoegsk. Ann. 36:401–418.

Tracey, M. V., and G. Youatt. 1958. Cellulase and chitinase in two species of Australian termites. Enzymologia 19:70–72.

Uchida, T., and F. Egami. 1971. Microbial ribonucleases with special reference to RNases T_1, T_2, N_1, and U_2. p. 205–250. *In* P. D. Boyer (ed.) The enzymes. Vol. IV. Hydrolysis: other C–N bonds, phosphate esters. 3rd ed. Academic Press, New York.

Veerkamp, J. H. 1969. Uptake and metabolism of derivatives of 2-deoxy-2-amino-D-glucose in *Bifidobacterium bifidum* var. *pennsylvanicus.* Arch. Biochem. Biophys. 129:248–256.

Verma, L., J. P. Martin, and K. Haider. 1975. Decomposition of carbon-14-labeled proteins, peptides and amino acids; free and complexed with humic polymers. Soil Sci. Soc. Am. Proc. 39:279–284.

Voets, J. P., and M. Dedeken. 1964. Studies on biological phenomena of proteolysis in soil. Ann. Inst. Pasteur, Paris Suppl. 107:320–329.

Voets, J. P., and M. Dedeken. 1965. Influence of high-frequency and gamma irradiation on the soil microflora and the soil enzymes. Meded. Landbouwhogesch. Opzoekingstn. Staat Gent 30:2037–2049.

Voets, J. P., M. Dedeken, and E. Bessems. 1965. Behavior of some amino acids in gamma irradiated soils. Naturwissenshaften 52:476.

Voets, J. P., P. Meerschman, and W. Verstraete. 1974. Soil microbiological and biochemical effects of long-term atrazine applications. Soil Biol. Biochem. 6:149–152.

Vogels, G. D., and C. Van der Drift. 1976. Degradation of purines and pyrimidines by microorganisms. Bacteriol. Rev. 40:403–468.

Wagner, G., and V. K. Mutatkar. 1968. Amino components of soil organic matter formed during humification of ^{14}C glucose. Soil Sci. Soc. Am. Proc. 32:683–686.

Webb, J. L. 1966. Quinones. p. 421–594. *In* J. L. Webb (ed.) Enzyme and metabolic inhibitors. Vol. 3. Academic Press, New York.

Weetall, H. H. 1975. Immobilized enzymes, antigens, antibodies and peptides, preparation and characterization. Marcel Dekker, Inc., New York.

Weetall, H. H. and G. Baum. 1970. Preparation and characterization of insolubilized L-amino acid oxidase. Biotechnol. Bioeng. 12:399–407.

Wells, J. R. E. 1968. Characterization of three proteolytic enzymes from French beans. Biochim. Biophys. Acta 167:388–398.

White, R. J. 1968. Control of amino sugar metabolism in *Escherichia coli* and isolation of mutants unable to degrade amino sugars. Biochem. J. 106:847–858.

Wingard, L. B., E. Katchalski-Katzir, and L. Goldstein (ed.). 1976. Applied biochemistry and bioengineering. Vol. 1. Immobilized enzyme principles. Academic Press, New York.

Woolfolk, C. A. 1975. Metabolism of *N*-methylpurines by a *Pseudomonas putida* strain isolated by enrichment on caffeine as the sole source of carbon and nitrogen. J. Bacteriol. 123:1088–1106.

Wriston, J. C. 1971. L-Asparaginase. p. 101–121. *In* P. D. Boyer (ed.) The enzymes. Vol. IV. Hydrolysis. 3rd ed. Academic Press, New York.

Zaborsky, O. R. 1973. Immobilized enzymes. CRC Press, Cleveland, Ohio.

Zantua, M. I., and J. M. Bremner. 1975. Preservation of soil samples for assay of urease activity. Soil Biol. Biochem. 7:297–299.

Zantua, M. I., and J. M. Bremner. 1976. Production and persistence of urease activity in soils. Soil Biol. Biochem. 8:369–374.

Zantua, M. I., and J. M. Bremner. 1977. Stability of urease in soils. Soil Biol. Biochem. 9: 135–140.

Zantua, M. I., L. C. Dumenil, and J. M. Bremner. 1977. Relationships between soil urease activity and other soil properties. Soil Sci. Soc. Am. J. 41:350–352.

6 Mineralization and Immobilization of Soil Nitrogen

S. L. JANSSON AND J. PERSSON

Swedish University of Agricultural Sciences
Uppsala, Sweden

I. BACKGROUND

The N supply is one of the general decisive factors in crop production. In many agricultural systems, it is customary to increase the input of N through fertilization. There is an urgent need to improve the N factor of crop production, both by increasing the efficiency with which N is utilized by crops and by limiting N losses to the external environment.

Underlying the various possibilities for improving the N economy in crop production is a complete understanding of N transformation processes, most of which are performed by microorganisms. Any given transformation can affect the final result, the output of useful plant products.

Individual N transformations have been studied extensively within such basic sciences as biochemistry, microbiology, and plant physiology. The result has been the accumulation of considerable information concerning the various transformations, their environmental demands, their mechanisms, the intermediates, and the end products. In contrast, little is known regarding the ecological unity of the transformations—the complete process of plant production on soils.

Studies of single transformations often provide information that is too isolated and specialized to be informative from an ecological point of view. Specialized knowledge on any given N transformation must be supplemented with integrated research taking into account the entire ecosystem with all of its individual transformations, pathways, pools, and interactions.

A. The Processes of Mineralization and Immobilization

In this chapter the ecological function of two individual N transformation processes in soil will be treated, namely, mineralization and immobilization. Both are biochemical in nature, and both are bound to the activities of the organisms making up the heterotrophic biomass (Bartholomew, 1965; Jansson, 1971).

N mineralization is defined as the transformation of N from the organic state into the inorganic forms of NH_4^+ or NH_3. The process is performed by heterotrophic soil organisms that utilize nitrogenous organic substances as an energy source.

N immobilization is defined as the transformation of inorganic N compounds (NH_4^+, NH_3, NO_3^-, NO_2^-) into the organic state. Soil organisms assimilate inorganic N compounds and transform them into organic N constituents of their cells and tissues, the soil biomass.

Plant uptake and assimilation of inorganic N compounds is a variant of immobilization, as is N_2 fixation by autotrophic and heterotrophic soil organisms. In spite of this, N assimilation by plants and N_2 fixation are usually excluded from the definition of immobilization.

B. Relations to the Universal N Cycle

The interaction of individual N-transforming processes in the soil ecosystem leads to a pattern of N pools connected by biochemical pathways along which N is translocated. This functional pattern is commonly known as the N cycle (Campbell, 1978; Jansson, 1971). Both mineralization and immobilization have fundamental functions in the universal N cycle (Bartholomew, 1965; Jansson, 1958).

A major feature of the universal N cycle operating in all ecosystems is the interaction between autotrophic and heterotrophic biological activity. Through photosynthesis, green plants trap and store solar energy in the form of plant tissue. On return to the soil, the plant material is used as a source of energy by heterotrophic microorganisms, and the organic N is transformed back into the simple inorganic compounds originally taken up by the plants. Thereby the cycle becomes closed. Part of the inorganic N is utilized by a new generation of plants.

In this way, a stream of the original solar energy passes through the various ecosystems without any possibility of reuse. In contrast, N and other nutrient elements are repeatedly used and reused by continuous circulation between the autotrophic and heterotrophic phases of the ecosystem.

The process of mineralization plays a key role in this universal N cycle, being responsible for the fundamental transformation of organic N in plant remains back into the simple inorganic forms originally used by the plants in their photosynthetic activities.

Immobilization is an indispensable and integrated prerequisite for mineralization. The heterotrophic phase of the ecosystems, carrying out mineralization, is a living, biological phenomenon. As such, it not only is respiring and mineralizing but is also developing by processes of multiplication, growth, change, and renewal. These processes result in the formation of organic matter, microbial cells, and tissues.

Thus, in all mineralization activities there is a component of immobilization, a renewal of organic matter, and an assimilation of mineral nutrients providing the multiplication, growth, and maintenance of the living and active microbial flora or biomass. This renewal will normally be restricted to a minor part of the organic matter (organic C) utilized by the microbes; the major part will be mineralized into simple inorganic compounds. With regard to the nutrient elements—among them N—the situation is often more complicated.

Soil organisms, chiefly microbes, constitute the living agency carrying out the heterotrophic phase of the N cycle. Mineralization is one of its basic activities. In addition, mineralization is important for functioning of the cycle, for the autotrophic phase proliferating on its mineral end products (NH_4^+ and NH_3).

For the heterotrophic phase, immobilization is as important and determining as mineralization. In dealing with mineralization and its role in the autotrophic phase of the biological cycle, i.e., plant production, one cannot neglect immobilization. This simple fact justifies special consideration of mineralization-immobilization relationships.

C. Partition of the Universal N Cycle Into Three Subcycles

In reality the simple and schematic universal N cycle (Hauck, 1971; Stevenson, 1965) is built up of three interdependent partial cycles having one or more common pathways (Campbell, 1978; Jansson, 1971). These three subcycles may be called the elemental cycle (E), the autotrophic cycle (A), and the heterotrophic cycle (H). They are schematically illustrated in Fig. 1.

The elemental subcycle (E) includes the connection of biological life to the dominating N pool of the earth, the atmosphere. Its specific N pathways are biological N_2 fixation and denitrification under certain environmental conditions (restricted O_2 supply).

The autotrophic subcycle (A) includes the activities of green plants, their photosynthetic binding of solar energy, and the build-up of primary organic N substances.

The heterotrophic subcycle (H) is determined by the activities of heterotrophic microorganisms. The specific ecological characteristic of this cycle is mineralization, energy dissipation from organic matter, whereby the nitrogenous organic substances are converted to NH_3 or NH_4^+. The functioning of all three subcycles is dependent on this mineralized N. Partly, but invariably, this mineralized N will be immobilized in the heterotrophic subcycle, partly taken up by plants of the autotrophic subcycle, and partly nitrified and denitrified in the elemental subcycle.

Thus, mineralization and immobilization are by definition bound to the heterotrophic subcycle. They are the basic functions of the heterotrophic biomass. The two processes work in opposite directions, building up and breaking down organic matter, respectively. The difference between the two processes will be a net effect, net mineralization, or net immobilization. Because the basic activity of heterotrophic organisms is dissipation of organically bound energy, net mineralization will be the normal and dominating reaction. Net mineralization of organic matter (organic C compounds) will always be less than gross mineralization.

The resulting effect of the two opposing processes—suitably expressed as net mineralization or net immobilization—will determine the N supply to

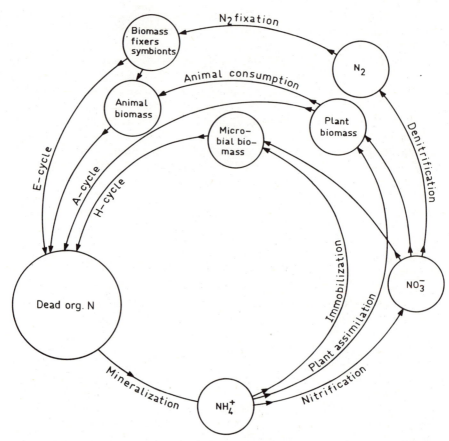

Fig. 1—The universal N cycle divided into its three subcycles: the elemental (*E*), the autotrophic (*A*), and the heterotrophic (*H*).

the other two subcycles and, thus, the production of nonleguminous plants
under nonfertilizer conditions.

In all three subcycles organic N is built up in the biomass, from N_2 in
the elemental cycle and from NH_4^+-N or NO_3^--N in the other two cycles.
Parts of the biomasses are continuously transformed into the dead organic
matter pool, whereby the N becomes mineralized to NH_4^+-N or NH_3-N.

D. Competition Among the N Subcycles

Beginning with mineralization, the three subcycles start a new circulation of N. A situation of competition will be established. Which cycle will
prevail? Which process will prevail—heterotrophic immobilization, plant
uptake, or nitrification?

The NO_3^- pool, formed by nitrification, constitutes a second point of
competition between the three N subcycles (immobilization, plant uptake,
and denitrification).

E. Mineralization-Immobilization Turnover (MIT)

Here our purpose is to follow and discuss on the basis of existing experimental evidence the functioning of the heterotrophic N subcycle, the mineralization-immobilization turnover (MIT), and its interactions with the other two subcycles. The main objective is to provide a better understanding and improved basis of managing soils, crops, and fertilizers for optimum N efficiency and economy. Prevention of N losses to the external environment of the ecosystem concerned is an important aspect of this aim.

It must be stressed that the heterotrophic biomass is in a dynamic, regenerating state. The continuous transfer of mineralized N into organic products of synthesis and of immobilized N back into inorganic decay products—underlying the building up and dying away of the heterotrophic biomass—can be defined as MIT (mineralization-immobilization turnover) (Campbell, 1978; Jansson, 1958; Knowles & Chu, 1969; Ladd & Paul, 1973; Shields et al., 1973; Wallace & Smith, 1954; Westerman & Tucker, 1974).

II. FEATURES AND FUNCTIONS OF MIT

A. Inadequacies of Net Effect Determinations

The net effects of heterotrophic activity—net mineralization or net immobilization of N—are easy and convenient to measure and, in addition, are of immediate and practical importance from the standpoint of plant production. However, in terms of soil biology, biochemistry, and ecology they are not too informative. Specifically, they fail to reflect what is actually going on in the soil (Bartholomew & Hiltbold, 1952; Hiltbold et al., 1951; Jansson, 1958). A small net effect may be the result of low overall biological activity in the ecosystem, or it may be the result of high activity in which the processes work in opposite directions. Differentiation between the alternatives, including estimation of the real scope of the opposing processes, cannot be performed on the basis of net effect determinations. Some idea of the real scope of the basic processes can be obtained by estimating energy dissipation from the system, as revealed, for example, by CO_2 production, O_2 consumption, or enzymatic activities. Regardless of net changes in inorganic N content, large energy dissipation indicates vigorous biological activity, including MIT.

B. Possibilities of Measuring Gross Effects: Usefulness of Tracer Techniques

Demonstration of the magnitude of the two opposing gross processes by analyses for inorganic N is not feasible by ordinary, conventional means. This can be done only by use of tagged N (^{15}N). If, for example, the inorganic N pool of an ecosystem is tagged, immobilization can be demonstrated and determined by the appearance of tagged N in the organic pool regardless of whether net immobilization or net mineralization has occurred. On the other hand, mineralization can be demonstrated by dilution of the tagged inorganic pool with untagged N originating from the organic pool. If two crosswise tagged but otherwise identical decomposition systems are arranged, gross transformations of N can be followed and determined with improved accuracy.

In the long run, the continuous opposing processes constituting MIT will be expected to lead to equilibration between tagged and nontagged N in the inorganic and organic pools. An almost complete equilibration of this kind was demonstrated in early tracer work with a simplified system involving straw decomposition in the absence of soil (Jansson, 1958; Jansson et al., 1955).

In more complicated ecosystems, equilibration will normally not be attained because of the existence of processes and disturbances within the systems that make the pools or phases heterogeneous. Within the inorganic phase, nitrification is a complicating factor causing heterogeneity (presence of both NH_4^+- and NO_3^--N); within the organic phase, humus formation (i.e., accumulation of residual organic substances and aging phenomena) causes heterogeneities that hamper or delay equilibrium. The consequences of nitrification and humus formation on MIT are discussed in sections II G and II M, N, and O, respectively.

Theoretically, experiments with tagged decomposition systems and, especially, systems with their inorganic and organic phases crosswise tagged, may be used for quantification of the opposing gross processes of mineralization and immobilization. Some attempts to calculate the scope of the processes on this basis under different environmental conditions have been made (Hiltbold et al., 1951; Jansson, 1958; Kirkham, 1957; Kirkham & Bartholomew, 1954, 1955). However, the results have not been too encouraging, due undoubtedly to complications indicated above. It is possible, however, that difficulties associated with determination of the scope of the gross processes under different environmental conditions can be overcome using tracer techniques. The task requires more basic knowledge and refined research methods, especially with regard to tracer techniques (Hauck, 1973; Hauck & Bremner, 1976; Hauck & Bystrom, 1970; Jansson, 1966; Paul & van Veen, 1979; Shields et al., 1973).

C. Confusion Caused by MIT

The action of the two opposing processes in soil and in other decomposition media results in the continuous turnover of inorganic N into the organic state and vice versa. If, for example, an experiment is started in which the inorganic phase is tagged (particularly the NH_4^+ pool), biological activity will result in an exchange of tagged N for nontagged N in the inorganic pool and of a corresponding exchange of nontagged N for tagged N in the organic pool.

The existence and manifestations of MIT in tracer experiments have not been fully understood and have caused considerable confusion with regard to the interpretation of tracer data on fertilizer and soil N transformations.

In studies of fertilizer N uptake by crops, the conventional indirect method of determining the difference in crop uptake between the fertilized treatment and an unfertilized control and the direct method of determining the amount of tagged fertilizer N taken up by the crop seldom give identical results. Normally, the indirect method gives higher recoveries than does the direct tracer method (Bartholomew & Hiltbold, 1952; Bartholomew et al., 1950; Dowdell & Webster, 1980; Harmsen & Kolenbrander, 1965; Jansson, 1958, 1971; Kaindl & Haunold, 1965; Terman & Brown, 1968a; Tyler & Broadbent, 1958; Westerman & Kurtz, 1974). The difference has appeared intriguing and has initiated several explanations, of which two are dominating.

A simple and straightforward explanation is that the fertilizer N leads to an increase in net mineralization of soil N, with subsequent consumption of the mineralized N by the crop (Andreeva & Scheglova, 1968; Chichester & Smith, 1978; Filimonov & Rudelev, 1977; Harmsen & Kolenbrander, 1965; Hauck & Bremner, 1976; Kissel & Smith, 1978; Sapozhnikov et al., 1968; Stojanovic & Broadbent, 1956; Tyler & Broadbent, 1958; Turtschin et al., 1960; Westerman & Kurtz, 1974; Zamyatina et al., 1968).

There is reason to believe that N fertilizers would have side effects in soil leading to net mineralization. For example, many N fertilizers can modify the soil pH. This change may affect the composition of the microflora and microbial activities, including net mineralization (Jansson, 1971).

The other main explanation is that the phenomenon is a reflection of MIT (Bartholomew & Hiltbold, 1952; Hauck & Bremner, 1976; Jansson, 1958, 1971; Nommik, 1968; Stewart et al., 1963a; Walker et al., 1956). Tagged fertilizer N will be immobilized at the same time that nontagged soil N is mineralized. These processes will change the isotopic composition of the inorganic N pool from which the crop draws its N but may or may not change the size of the pool. By taking up isotopically diluted pool N, the crop will contain less tagged and more nontagged N than what is indicated by the difference method.

The discrepancy between the two methods of fertilizer N evaluation has also been demonstrated in laboratory incubation experiments, where the size and composition of the inorganic N pool have been followed by chemical analyses for NH_4^+ and NO_3^-. Both interpretations of the difference have been applied to these findings.

D. Priming Effect and Related Phenomena

More than 50 years ago, Löhnis (1926) concluded from experiments with green manure that additions of fresh plant material, rich in readily available energy, stimulated mineralization of indigenous soil organic N (and C). Until the introduction of tracer work on MIT about 20 years later, this observation attracted little attention. Around 1950, however, tracer work with N and C isotopes (Broadbent, 1948; Broadbent & Norman, 1947) seemed to corroborate the results of Löhnis and the new findings attracted considerable interest. The stimulating effect of added energy material on soil organic matter transformations was called "priming action," or "priming effect" (Bingeman et al., 1953; Hallam, 1953; Jenkinson, 1966a; Mortensen, 1963).

The great attention paid to priming is somewhat surprising. It is true that the initial findings of Broadbent and Norman (1947) were quite astonishing, but their findings on the extent of priming were not corroborated in later work. Both positive and negative priming effects have been reported (Broadbent, 1966; Hauck & Bremner, 1976; Jansson, 1971; Jenkinson, 1966; Olson, 1980; Olson et al., 1979), but the majority of the changes have been minor in relation to total biological activity, including MIT.

In principle, small priming effects as defined above are a phenomenon that may often be expected. It is natural that soil microorganisms will react to additions of energetic material (Jansson, 1971). This reaction, stimulating MIT, may include some of the indigenous soil organic matter, that is, give rise to a priming effect. For example, the added material may cause a change in the pattern of microbial species making up the biomass, and this may result in increased or decreased attacks on indigenous soil organic matter.

It should be noted that MIT as such is normally manifested in a way that can erroneously be interpreted as priming. Undoubtedly, misinterpretations of turnover manifestations within the heterotrophic N cycle in soil have occurred. This has led to an exaggeration of the importance of the priming effect in soil biochemistry and has been responsible for much of the interest in priming effect investigations.

E. Fertilizer N and MIT

Another observation in tracer research closely related to priming, as defined above, is that increasing additions of tagged inorganic N (in the form of fertilizers) have led to increases in nontagged inorganic N in the soil and, consequently, in plant uptake of such N (Andreeva & Scheglova, 1968; Broadbent, 1965; Hills et al., 1978; Kissel et al., 1977; Stojanovic & Broadbent; 1956; Turtschin et al., 1960; Westerman & Kurtz, 1973, 1974). It has often been claimed that N fertilizer additions stimulate net mineralization of soil N (Filimonov & Rudelev, 1977; Sapozhnikov et al., 1968; Zamyatina et al., 1968). Physical and chemical causes have been held responsible for the phenomenon (Broadbent & Nakashima, 1971; Heilman, 1975; Laura, 1974, 1975a, b; Westerman & Tucker, 1974); for example, salt effects, including osmosis, pH changes, protolytic action on NH_4^+ formation and of NH_4^+ on nitrogenous bases in soil organic matter, and other side effects produced by N fertilizers. In cropped soils, the phenomenon has been attributed to stimulated root development and enhanced N uptake due to rhizosphere effects (Aleksic et al., 1968; Fried & Broeshart, 1974; Kissel & Smith, 1978; Sapozhnikov et al., 1968).

A different explanation of the phenomenon, however, is that it is a regular feature of MIT (Hauck & Bremner, 1976; Jansson, 1958, 1971; Stewart et al., 1963a; Walker et al., 1956). Provided the soil contains a newly established inorganic pool of tagged N, turnover will cause this pool to lose tagged N by immobilization and to gain nontagged soil N by mineralization. The larger the pool, the more substantial will be the turnover effect. This means that more nontagged soil N will accumulate in the large tagged pool of the fertilized soil than in the small pool of a nonfertilized control. It does not necessarily mean that a real stimulation of mineralization by the tagged N addition will have occurred. Instead, turnover merely means that tagged N is immobilized at the same time as the nontagged N is mineralized.

It is inadequate, however, to assume that this turnover should be stoichiometric, that is, that the amount of tagged N immobilized will be equivalent to the amount of nontagged N mineralized (cf. Nommik, 1968; Riga et al., 1980). There are many reasons against stoichiometric turnover. The organic pool (the biomass) may be in an expanding or decreasing stage, and its pattern of microbial species may be changing. Furthermore, the inorganic pool may be heterogeneous in containing both NH_4^+ and NO_3^-, it may change continuously through nitrification, and it may be diluted with nontagged N as turnover proceeds.

In many investigations where a priming effect has been recorded, immobilization of tagged N has been ignored. In some cases this can be revealed by the data given; in others, the necessary data are missing in the original papers. Scrutinizing the results in the former case has often revealed that the priming effect was associated with increased immobilization of tagged N, a feature of MIT. As already stressed, MIT is stimulated by additions of energetic material along with the (tagged) inorganic N. This means that such additions often increase an apparent priming effect of N fertilizers (Hauck & Bremner, 1976; Jansson, 1971; Taki et al., 1967).

Ever since the stimulating effects of decomposing organic matter and N fertilizers on soil N and C were found in tracer work more than 30 years ago, considerable research has been devoted to these effects. None of the explanations suggested for the phenomena fully explains the observed net effects (Hauck & Bremner, 1976). It is more reasonable to conclude that several mechanisms contribute to a real or apparent priming effect. One must be aware of the fact that MIT is an indispensable feature of all ecosystems where priming effects occur and that the occurrence of MIT will often explain these observations.

It is somewhat disappointing that so much work has been devoted in attempts to register a priming effect. Future tracer research on N transformations should concentrate on extending our knowledge of MIT, its mechanisms, and its quantification under different environmental conditions. When such knowledge becomes available, the puzzling enigmas regarding apparent or real priming effects will be solved.

F. Evaluation of N Fertilizers

MIT may influence the evaluation of fertilizer N where tagged N is used. In most but not all tracer experiments on plant uptake of fertilizer N, a nonfertilized control has been included in the experimental setup, making possible a comparison between the conventional difference method of determining fertilizer N uptake and the direct tracer method. When determined by the tracer method, crop uptake of fertilizer N has normally been lower than when determined by the difference method (Bartholomew et al., 1950; Jansson, 1958; Terman & Brown, 1968a; Tyler & Broadbent, 1958; Westerman & Kurtz, 1973, 1974) more so for NH_4^+ (Broadbent & Tyler, 1962, 1965; Frederick & Broadbent, 1966; Jansson, 1971; Jansson et al., 1955; Kaindl & Haunold, 1965; Ketcheson & Jakovljevic, 1968) than for NO_3^- (when these two sources of plant N can be separated). Also, the difference has been accentuated by the presence of energetic organic material stimulating MIT (Hallam, 1953; Jansson, 1958).

To return to the problems discussed previously, is the difference caused by a real stimulation of soil N mineralization, or is it a feature of MIT? To obtain information on this subject, the experiments must encompass both methods of evaluation and, in addition, include a determination of immobilized tagged N in the organic phase of the soil.

From the viewpoint of the practical fertilizer user (the farmer), who is interested in the simple overall effect of the fertilizer in improving yields, the conventional difference method is still recommended (Jansson, 1966); simple tracer experiments in which MIT is ignored can give misleading results. On the other hand, when a detailed interpretation of the results is desired, the use of tagged fertilizer N is required, and the experiment must fulfill the above requirements. That is, it must include a nonfertilized control and a ^{15}N analysis of the organic phase of soil N.

As an illustration of the necessity of these requirements, it may be added that Sapozhnikov et al. (1968), in a series of pot experiments, applied an interesting split-root technique to find the causes of the increased net mineralization of soil N following the addition of inorganic fertilizer N. Half of the root system of the experimental crop was allowed to develop in a pot compartment containing ^{15}N-fertilized inert sand, half in another compartment filled with unfertilized normal soil. It was found that the fertilizer addition resulted in increased uptake of soil N by the crop and more so for NH_4^+ than for NO_3^- fertilization. Such results seem to exclude MIT as the cause of the phenomenon, and the authors claimed that it was due to fertilizer-stimulated development of the root system. However, this explanation is not fully convincing. Thus, the NH_4^+ treatment gave more soil N in the crop than did the NO_3^- treatment, but this difference corresponded to less uptake of tagged fertilizer N in the former treatment. This result indicates that an exchange had taken place in the substrates and that something unexpected had happened regarding the split-root technique. Had the requirement on the experiment (discussed above) of determining immobilized tagged N in the substrates been fulfilled, the investigation would have gained in clarity and value.

G. MIT and Nitrification

As indicated in Fig. 1, the NH_4^+-N produced in the mineralization process or otherwise accumulated in an NH_4^+ pool in the soil is markedly preferred over NO_3^--N by heterotrophic microorganisms during immobilization. When NO_3^- is available but NH_4^+ missing, the former is utilized (Jansson, 1958).

The preference of heterotrophic microorganisms for NH_4^+ further complicates the problem of evaluating MIT after tagged N has been added. Since NH_4^+ makes up the main and normal inorganic phase of MIT, nitrification will normally be associated with conditions of net mineralization. Further, nitrification will result in withdrawal of inorganic N from MIT; the NO_3^- will stay off the turnover pathway as long as net mineralization conditions persist, though it may be used for plant uptake (Jansson, 1958; Ketcheson & Jakovljevic, 1968).

If under conditions of net mineralization newly added tagged fertilizer N is nitrified or given as NO_3^-, this N may escape MIT. In such cases, plant recovery of tagged N will be high; the difference between the direct and indirect method of determining fertilizer N recovery will be small and less than in nonnitrifying soils.

Such situations have been indicated in many investigations, though they have largely been ignored or overlooked. It must also be emphasized that the preference of heterotrophic microorganisms for NH_4^+ has turned out to be substantial, although not absolute (Jansson, 1971). Simple and fully clear-cut results can seldom be expected in dealing with a highly mixed microflora under the complicated environmental conditions in soil. The ecological importance of NH_4^+ preference has been stressed in recent reviews (Campbell, 1978; Paul & Juma, 1981; Paul & van Veen, 1979).

H. Consequences of N_2 Fixation and Denitrification on MIT

It must be added that N transformations in soil other than those treated in the foregoing sections also affect MIT, its scope, progress, and consequences. Among such processes are N_2 fixation and denitrification, both belonging to the elemental N subcycle (Fig. 1).

Biological N_2 fixation, either by free-living autotrophic and heterotrophic organisms or in symbiosis with higher plants, provides an alternative to N immobilization and decreases the dependence of the ecosystem on N mineralization. At the same time, N mineralization may be enhanced by the ample N supply to the biomass often associated with N_2 fixation.

Denitrification leads to gaseous loss of NO_3^--N that may interfere with immobilization. However, O_2 restrictions leading to denitrification will also restrict microbial activity, including immobilization; under anaerobic conditions, microbial activity is lower than under fully aerobic conditions (Campbell, 1978).

I. MIT Interactions with Plants

The interactions between MIT and the activities of green plants have already been touched on; plants serve as energy sources for heterotrophic microorganisms, and they compete for the mineralization outflow.

A special energy source is provided in the close vicinity of plant roots and root hairs—the rhizosphere. Root exudates contain readily decomposable organic substances, and small parts of the root system are continuously sloughed off and attacked by heterotrophic rhizosphere organisms. Recent ^{14}C investigations (Sauerbeck & Johnen, 1976; Sauerbeck et al., 1976) indicate that this energy supply is large and that MIT is stimulated in the rhizosphere by the ample energy supply (Bartholomew & Clark, 1950; Campbell, 1978; Campbell & Paul, 1978; Legg & Allison, 1960; Legg & Stanford, 1967). The primary reaction is immobilization and, in somewhat later stages, mineralization as well. The increased MIT affects plant recovery of tagged fertilizer N. Compared with conditions in nonplanted soil, the direct method of determining N recovery can be expected to give lower values. At the same time, plant uptake of nontagged soil N will be increased.

J. Effects of Physical and Chemical Soil Factors

Changes in chemical and physical soil properties often affect microbial activity and thereby MIT and its manifestations. Pertinent chemical changes include variations in pH and conductivity, changes in plant nutrients, and changes in the contents of toxic chemicals and heavy metals. Secondary effects of fertilizers, herbicides, and pesticides may cause chemical changes affecting MIT. Chemical NH_4^+ fixation to inorganic soil colloids or NH_3 fixation to organic substances may affect the scope and competitive conditions of the immobilization process.

Physical conditions such as water logging and compaction of the soil may affect aeration and establish more or less anaerobic conditions, generally hampering microbial activity, stimulating NH_4^+ or NH_3 formation and accumulation, and causing NO_3^- losses by denitrification, all of which may have a determining influence on MIT.

Sudden changes in the chemical and physical conditions of the soil caused, for example, by wetting and drying or freezing and thawing may kill microorganisms and may be followed by sudden and temporary flushes of energy stimulating MIT (Campbell, 1978; Ladd et al., 1977b).

K. Energy-Nutrient Relationships

The net effect of MIT—the difference between the two opposing processes of mineralization and immobilization—will be dependent on the energy supply to soil microorganisms and so will the already discussed difference between the indirect and direct method of determining plant uptake

of N that is fully in line with the MIT explanation of this difference (Broadbent & Stevenson, 1966; Hallam, 1953; Jansson, 1958; Jones & Richards,
1977; Kanamori & Yasuda, 1979; McGill et al., 1975; Nommik, 1961). An
ample supply of energetic material to the soil (for example, by glucose and
straw additions) has increased the difference by lowering plant uptake of
tagged N. The findings indicate that the energy supply has stimulated MIT.

This leads to a discussion of the energy-nutrient (N) relationships of
MIT. The supply of decomposable organic matter is an indispensable prerequisite for this process. In functioning as an energy source, decomposable
N-free organic compounds (carbohydrates and lipids, for example, in crop
residues added to the soil) enhance microbial growth. Energy supply is,
however, not the only prerequisite. Since the N-free energetic materials do
not provide N to the growing microbes, the microbes draw on the inorganic
N pool; if this pool is depleted, they will suffer from N deficiency, and their
activities will be limited.

On the other hand, nitrogenous organic matter—proteins—when used
as the source of energy, will lead to net mineralization of N and its accumulation in the inorganic pool of the soil. For optimum conditions, the
energetic material should have a certain E (energy)/N ratio; N mineralization should just meet the N requirements of microorganisms responding to
the energy supply. The proper "diet" of the soil microflora will be a mixture of decomposable nitrogenous and nonnitrogenous substances, giving
no net change in the inorganic N content of the system. Such a well-
balanced condition seldom occurs; different types of organisms have different demands, and the demand on the E/N ratio may change with the ecological development of the biomass (succession of species). Biomass dominated by fungi—in acid soils—are regularly satisfied by a higher E/N ratio
than are biomasses dominated by bacteria.

L. The C/N Ratio: C and N Interdependence

In soil biology, the E/N ratio is not normally used to characterize soil
organic matter or organic soil amendments. Instead, the C/N ratio is used
for this purpose. Well-balanced nutritional conditions of the soil biomass
are in normal arable soils represented by a C/N ratio of about 25.

It should be kept in mind, however, that the C/N ratio is an approximation of the really important parameter—the E/N ratio—and that this approximation may sometimes be misleading. A main cause of this is that
some of the C and N constituents of the organic matter undergoing decomposition are not readily available to microorganisms. They are not easily
mineralized. For example, the N-free lignins of many plant materials and
the low N residual substances of many peat soils are poor energy sources for
most microorganisms. Though they have high C/N ratios, such materials
will not cause any substantial net immobilization of N.

In spite of the above limitation, the C/N ratio has found widespread
practical use in characterizing soil organic matter and organic materials
added to soils. This also applies to their effect on MIT.

In addition, the use of the C/N ratio implies a close interdependence of C and N transformations in biological decomposition.

M. The Phase Concept of Soil Organic Matter

So far we have discussed the two components, or phases, of soil organic matter directly involved in MIT, i.e., the biomass and the readily available energetic material in the form of fresh organic debris of plant, microbial, and animal origin. Normally this debris is not completely decomposed or instantly attacked by soil organisms. Some constituents are more or less resistant to microbial attack, and they accumulate in the soil, often changed or modified by biological or chemical means but far from fully mineralized or transformed into biomass. Such accumulated organic residues make up a considerable part of soil organic matter, thereby constituting a relatively passive phase (Allen et al., 1973; Broadbent & Nakashima, 1965, 1967; Chu & Knowles, 1966; Freney & Simpson, 1969; Jenkinson & Rayner, 1977; Kuo & Bartholomew, 1966; Ladd & Paul, 1973; Nommik, 1961, 1968; Paul, 1976; Stanford et al., 1970; Stewart et al., 1963b). From an ecological point of view, they are the humus substances of the soil (Jansson, 1966, 1971).

With regard to MIT, these organic residues may be excluded from the process. In contrast to fresh biomass debris, they are only slowly and ineffectively attacked by living organisms for energy or nutrient supply. Notwithstanding, this passive phase constitutes an important basic energy and nutrient source for soil biomass due to the considerable and dominating amounts present.

N. Humus Formation and Decay: A Dynamic Phenomenon

The humus content of the soil, the passive organic phase, will be determined by an equilibrium between accumulation of organic residues on the one hand and slow decomposition on the other. This equilibrium will be determined by many factors—among others, soil texture, structure, and chemical composition, climate, cropping system, and crop residue management—of which some are controllable by man (the farmer) and others are not.

In addition, the passive phase of soil organic matter is heterogeneous with regard to origin, composition, and age (Chichester et al., 1975; Chichester & Smith, 1970; Jenkinson & Rayner, 1977; Ladd et al., 1977a, b; Legg et al., 1971; Nommik, 1961; Persson, 1968; Smith et al., 1978; Stanford et al., 1970). Its wide age span has been revealed by radiocarbon dating (Campbell, 1978).

The study of the chemical properties of the passive phase has developed into a special branch of soil chemistry, the study of the humus substances, their extraction, composition, and properties. In their efforts to achieve progress in this area, humus chemists may have overlooked the fact that processes occurring within soil organic matter should primarily be regarded as biological and biochemical phenomena (Jansson, 1967).

There is an urgent need to better characterize and define the different fractions or development stages within the passive phase of soil organic matter and their relations to the biomass and to MIT. It can be inferred that the different fractions are not sharply separated but are successively passing from more active stages into less active ones. They are continuously being attacked by the heterotrophic microflora, with each subsequent fraction being attacked at a steadily diminishing rate. Within the total soil organic matter, the fractions interchange and attain some kind of a dynamic equilibrium or steady state. This will be valid for soil N as well as for soil C. A somewhat specific dynamic equilibrium can be expected to develop in every individual soil.

O. Use of Equilibrium Concepts in Phase Determinations

Postulated interchange and equilibria between different fractions of soil organic matter have been used in tracer work to calculate the size of an active phase of soil organic matter, which is made up of the biomass and the fresh debris of plant, microbial, and animal origin (Jansson, 1958; Nommik, 1968).

The residue formation of soil organic N is a process of stabilization and aging. It can be looked on as a passive alternative to MIT. A primary and central task in the study of soil organic N dynamics will be to find reliable methods for determination of the microbial biomass.

The traditional way of determining the biomass has been to count and estimate the size or weight of individual organisms making up the biomass. Despite several difficulties and disadvantages, it is still used as some kind of standard.

In early remineralization studies of newly immobilized tagged N, Jansson (1958) concluded that the active phase of soil organic N, including the biomass, accounted for up to 10 to 15% of total organic N in arable mineral soils. The remainder presumably made up a biologically passive pool undergoing slow interchange with the active phase.

The general formula for calculating quantities involved in isotopic equilibria was applied to the method for estimating this active phase by using tracer data:

$$A = B(1 - y)/y$$

where y is the experimentally determined net mineralization outflow of the newly immobilized tagged N. If the amount of (newly immobilized) tagged N in the organic state (B) is known, the amount of soil N (A) equally active as the tagged organic N can be calculated.

In long-term experiments, such calculations founded on remineralization data give increasing A values (amounts of soil N in the active phase). In reality, this increase is an expression of stabilization and aging of the tagged N and its equilibration with increasing amounts of passive soil N (Jansson, 1958; Nommik, 1968).

P. "A" and Related Values

The above equation is formally identical with the expression used to calculate the special A value of Fried and Dean (1952), which is a measure of soil nutrient expressed in terms of a standard substance (fertilizer) added to the soil regardless of the transformations this standard may undergo in the soil. Several investigators have calculated A values according to Fried and Dean for soil N on the basis of inorganic ^{15}N fertilizer additions to the soil (Aleksic et al., 1968; Broadbent, 1970; Campbell & Paul, 1978; Dev & Rennie, 1979; van den Hende, 1968; Hills et al., 1978; Kaindl & Haunold, 1965; Legg & Allison, 1959; Legg & Stanford, 1967; Pomares-Garcia & Pratt, 1978; Rennie & Rennie, 1973; Smith & Legg, 1971; Taki & Yamashita, 1967; Terman & Brown, 1968b). The added ^{15}N in these investigations may have been distributed on several of the pathways and pools of soil N in an uncontrolled way. Therefore these calculations would be expected to give undefined results and may even be misleading in the study of transformations involving active soil N.

In contrast to this, an estimation of the active N phase in the soil, including the biomass, must be founded on a standard obtained under strictly defined conditions; tagged N incorporated into active heterotrophic biomass and fresh debris originating from this biomass (Jansson, 1958).

Q. Attempts at Extended Phase Separations

Another new principle in determining the soil biomass is by partial sterilization (e.g., chloroform fumigation), followed by incubation, to determine net mineralization caused by the killed biomass. The size of the original biomass can be calculated on the basis of the mineralization data. Results obtained by this method look interesting and promising (Anderson & Domsch, 1978a, b; Jenkinson, 1966b; Jenkinson & Powlson, 1976a, b).

On the other hand, soil organic matter includes strongly stabilized and very old substances. It has been possible to determine this carbonaceous material by ^{14}C dating (Campbell, 1978; Jenkinson & Rayner, 1977; Paul, 1970; Paul & Juma, 1981) and to calculate the amount of N associated with this old C.

In this way, three biologically interesting phases of soil organic matter can be estimated, namely, the living biomass (by the fumigation method), the fresh debris making up the active phase (by isotope dilution calculations in short-term net mineralization experiments with correction for the separate biomass determination), and the very old and passive material (by ^{14}C dating). The soil N not accounted for by these three estimations makes up a fourth fraction. Paul and Juma (1981) defined this residual fraction as stabilized N with an estimated half-life of about 30 years (cf. Fig. 2). Continued research along these lines would appear fruitful.

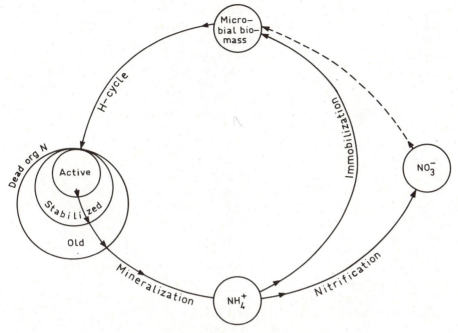

Fig. 2—Tentative fractionation of the dead soil organic N.

III. PROBLEMS AND PROSPECTS

In summary, MIT constitutes an important and characterizing component of the heterotrophic subcycle of N and other nutrients. It regulates in many respects the net outflow of mineralized N and influences the functioning of the other two N subcycles, the autotrophic and the elemental. It is therefore an urgent task to understand and, consequently, to investigate and clarify the dynamics of MIT under different environmental conditions. This is a subject that may be called functional microbiology, in which the soil microflora or biomass is treated as a working unity responsible for important processes, pools, and pathways of the nutrient cycles (the N cycles). In the applications of soil microbiology, for example, in soil fertility and plant production, this functional approach has proved most fruitful.

A general impression of the research and scientific discussion on MIT during the last decades is that the necessary knowledge of the process has not been easily achieved. Progress has been slow and misinterpretations have been frequent. New research openings have not been effectively utilized (Hauck & Bremner, 1976; Jansson, 1971).

A. A New Line in Soil Organic Matter Research

Although some indications on MIT were given before the advent of tracer techniques (Jansson, 1971; Paul & Juma, 1981), understanding of the two opposite but obligately complementary microbial processes of mineralization and immobilization is primarily a result of tracer investigations. Tracer research has added a new dimension to soil microbiology and especially to the role played by the heterotrophic biomass in soil organic matter dynamics. In this connection, the relationship between the total scope of mineralization and immobilization, often with rather limited net results, is of special interest. Although the two processes have been indicated and defined, a solid basis of extensive experimental knowledge is still missing.

Findings obtained thus far suggest that the biomass plays a central role in soil organic matter dynamics, and they stress that there are major difficulties in clarifying these dynamics in an unequivocal way. According to Paul and van Veen (1979), the biomass holds a unique position among the soil organic matter fractions because it functions as both a sink and a transformation station for nutrients (N) and energy in the soil. In other words, the system is complicated and the transformation possibilities manifold.

In spite of the difficulties and confusion, a new and improved theory of soil organic matter dynamics has emerged that is closely related to the functions of the biomass and that stresses the role of MIT. This theory requires a new line of research, starting with the biomass, its size, energy and N requirements, functions, metabolic processes, and left-overs. The traditional soil organic (humus) chemistry approach, which deals with the extraction and purification of fulvic and humic acids and their characterization by purely chemical means, would appear to be inadequate from the standpoint of the new theory. An important task included in the emerging line will be to develop a new, biologically founded analytical technique with regard to soil organic matter.

The first step in such a new analytical approach will be to find a biochemical method for measuring the soil biomass. In this respect, the fumigation method (Jenkinson, 1966b; Jenkinson & Powlson, 1976a, b) or some modification of it (Anderson & Domsch, 1978a, b) appears promising.

Next there is a need to characterize and fractionate dead soil organic matter with regard to its relationships to the biomass. Characterization of an active phase of dead soil organic matter should be given high priority. Such a phase must be closely related to the biomass and intimately interacting with it, serving as its main energy source and receiving its organic outputs, dying organisms, and leftovers. Thus, such an active phase must be strongly associated with MIT. The best way of determining this phase may be to use isotope dilution principles in short-term net mineralization experiments, preferentially including stepwise and crosswise tagging, or "labeled single-treatment experiments" (Fried et al., 1975).

The parts of the active phase that are rejected from MIT will under chemical and biological modifications pass into increasingly recalcitrant stages—for example, by reactions with mineral constituents of the soil, primarily clay colloids. From the ecological point of view, this accumulation of organic residues, increasingly resistant to or physically protected against microbial attack, means humus formation. How properly to characterize this humus formation in unequivocal chemical and ecological terms is still an unsolved problem of great importance. Much of the recalcitrant material is of considerable age, as revealed by radiocarbon dating; however, in a dynamic model, all of it must have a turnover time (Jenkinson & Rayner, 1977). How many fractions of dead soil organic matter should be considered in relation to MIT and how they should be characterized and experimentally determined is still questionable.

B. Inorganic Phase Problems

The inorganic phase of MIT also requires further study. First, one must remember that growth of the biomass may start with assimilation of organic decay products or metabolites, for example, amino acids. If so, the decay and renewal cycle of the biomass will only partly constitute MIT and will not be fully registered in MIT experiments. Such experiments may underestimate the true rate of biomass decay and renewal. This possibility was discussed in early tracer investigations (Jansson, 1958), but so far it has not been further considered.

The inorganic phase of MIT is not homogenous and simple. In the relatively simple case of nonnitrifying and fallow soil, the net mineralization outflow and other NH_4^+ sources will establish and maintain a pool included in MIT. Complications arise because NH_4^+ is immobile and is adsorbed by soil constituents at varying strengths—from free ions in the soil solution to exchangeable ions and ions strongly and permanently fixed by clay colloids. Also, the NH_3 molecule can be captured by chemical reactions to dead organic matter. This means that immobilization does not draw from a homogenous, equilibrated NH_4^+ pool; instead, newly mineralized NH_4^+ in the close vicinity of the active biomass may be preferred. This will make determinations and calculations approximate and uncertain.

Where nitrification occurs, the situation becomes even more complicated. Competition for the mineralization outflow becomes established between the heterotrophic biomass and the nitrifiers. This situation was explored to some extent by Jansson (1958), but since then it has not been given much consideration (cf. Jones & Richards, 1977).

Even more complicated will be the situation where a nitrifying soil carries a growing crop. Plants enter into competition with microorganisms for inorganic N, and in addition, they increase the energy supply and immobilization demand and stimulate MIT by their root exsudates and sloughing off parts of their root system.

Although indicated in early tracer work (Jansson, 1958), the many complications on the inorganic side of MIT have largely been ignored. More experimentation and elucidation is required in the future.

C. Simulation Modeling

The fashion of today in biological and ecological research is to build simulation models of processes, reaction chains, and nutrient cycles. These models are founded on mathematical functions, illustrating the reaction stages, and computer techniques. Mineralization-immobilization turnover has also been subjected to simulation modeling of this kind (Campbell, 1978; McGill et al., 1973; Mehran & Tanji, 1974; Paul, 1976; Paul & Juma, 1981). In these models, problems regarding fractionation of the organic phase discussed above are treated, and the transformation stages are assumed to follow first-order kinetics. The models have yielded outputs compatible with experimental data insofar as such data are available. When adequately described and discussed, simulation modeling is interesting and informative and may be fruitful in stimulating further research.

With veneration for simulation modeling, however, extended experimental evidence is still the most urgent task in soil biology research. With regard to research on MIT, a primary and fundamental need is the building up of a solid basis of experimental evidence comprising all situations provided by the rich variation and interaction of the environmental conditions with regard to soil, climate, energy, nutrient and oxygen supply, etc.

D. Future Role of Tracer Techniques

In future research on MIT, tracer techniques should be emphasized. Their greatest value is that they provide for an extended study of the soil as a reactive unity centered around the active biomass. By application of step-by-step tracer investigations—last but not least ^{15}N investigations—including crosswise tagging and biochemical fractionation, the many complicated and confusing features of MIT as well as organic matter dynamics can be clarified. Tracer techniques would appear to be particularly fruitful in studies of the opposing processes in soil N dynamics and transformations occurring within the organic phase of soil N (for example, humus formation and decay). It is in this area that many of our unsolved biological soil problems are to be found. Hereby an extension from laboratory and pot experiments to field conditions is imperative.

LITERATURE CITED

Aleksic, Z., H. Broeshart, and V. Middelboe. 1968. The effect of nitrogen fertilization on release of soil nitrogen. Plant Soil 29:474–478.

Allen, A. L., F. J. Stevenson, and L. T. Kurtz. 1973. Chemical distribution of residual fertilizer nitrogen in soil as revealed by nitrogen-15 studies. J. Environ. Qual. 2:120–124.

Anderson, J. P. E., and K. H. Domsch. 1978a. Mineralization of bacteria and fungi in chloroform-fumigated soils. Soil Biol. Biochem. 10:207–213.

Anderson, J. P. E., and K. H. Domsch. 1978b. A physiological method for the quantitative measurement of microbial biomass in soils. Soil Biol. Biochem. 10:215–221.

Andreeva, E. A., and G. M. Scheglova. 1968. Uptake of soil nitrogen on application of nitrogen fertilizers and nitrification inhibitors as revealed by greenhouse pot experiments using ^{15}N. Int. Congr. Soil Sci., Trans. 7th (Madison, Wis.) II:523–532.

Bartholomew, W. V. 1965. Mineralization and immobilization of nitrogen in the decomposition of plant and animal residues. In W. V. Bartholomew and F. E. Clark (ed.) Soil nitrogen. Agronomy 10:285–306. Am. Soc. of Agron., Madison, Wis.

Bartholomew, W. V., and F. E. Clark. 1950. Nitrogen transformations in soil in relation to the rhizosphere microflora. Int. Congr. Soil Sci., Trans. 4th (Amsterdam) II:112–113.

Bartholomew, W. V., and A. E. Hiltbold. 1952. Recovery of fertilizer nitrogen by oats in the greenhouse. Soil Sci. 73:193–202.

Bartholomew, W. V., L. B. Nelson, and C. H. Werkman. 1950. The use of the nitrogen isotope N^{15} in field studies with oats. Agron. J. 42:100–103.

Bingeman, C. W., J. E. Varner, and W. P. Martin. 1953. The effect of the addition of organic materials on the decomposition of an organic soil. Soil Sci. Soc. Am. Proc. 17:34–38.

Broadbent, F. E. 1948. Nitrogen release and carbon loss from soil organic matter during decomposition of added plant residues. Soil Sci. Soc. Am. Proc. 12:246–249.

Broadbent, F. E. 1965. Effect of fertilizer nitrogen on the release of soil nitrogen. Soil Sci. Soc. Am. Proc. 29:692–696.

Broadbent, F. E. 1966. Interchange between inorganic and organic nitrogen in soils. Hilgardia 37:6.

Broadbent, F. E. 1970. Variables affecting A values as a measure of soil nitrogen availability. Soil Sci. 110:19–23.

Broadbent, F. E., and T. Nakashima. 1965. Plant recovery of immobilized nitrogen in greenhouse experiments. Soil Sci. Soc. Am. Proc. 29:55–60.

Broadbent, F. E., and T. Nakashima. 1967. Reversion of fertilizer nitrogen in soils. Soil Sci. Soc. Am. Proc. 31:648–652.

Broadbent, F. E., and T. Nakashima. 1971. Effect of added salts on nitrogen mineralization in three California soils. Soil Sci. Soc. Am. Proc. 35:457–460.

Broadbent, F. E., and A. G. Norman. 1947. Some factors affecting the availability of organic nitrogen in soil—a preliminary report. Soil Sci. Soc. Am. Proc. 11:264–267.

Broadbent, F. E., and F. J. Stevenson. 1966. Organic matter interactions. p. 169–187. In M. H. McVickar et al. (ed.) Agricultural anhydrous ammonia technology and use. AAI, ASA, and SSSA, Memphis, Tenn., and Madison, Wis.

Broadbent, F. E., and K. B. Tyler. 1962. Laboratory and greenhouse investigations of nitrogen immobilization. Soil Sci. Soc. Am. Proc. 26:459–462.

Broadbent, F. E., and K. B. Tyler. 1965. Effects of pH on nitrogen immobilization in two California soils. Plant Soil 23:314–322.

Campbell, C. A. 1978. Soil organic carbon, nitrogen and fertility. p. 173–271. In M. Schnitzer and S. U. Kahn (ed.) Soil organic matter. Developments in soil science 8. Elsevier Scientific Publishing Co., New York.

Campbell, C. A., and E. A. Paul. 1978. Effect of fertilizer N and soil moisture on mineralization, N recovery and A values under spring wheat grown in small lysimeters. Can. J. Soil Sci. 58:39–51.

Chichester, F. W., J. O. Legg, and G. Stanford. 1975. Relative mineralization rates of indigenous and recently incorporated ^{15}N-labelled nitrogen. Soil Sci. 120:455–460.

Chichester, F. W., and S. J. Smith. 1970. Transformations of fertilizer nitrogen in soil: 2. Total and N^{15}-labelled nitrogen of soil organo-mineral sedimentation fractions. Plant Soil 33:437–456.

Chichester, F. W., and S. J. Smith. 1978. Disposition of ^{15}N-labeled fertilizer nitrate applied during corn culture in field lysimeters. J. Environ. Qual. 7:227–233.

Chu, J. P.-H., and R. Knowles. 1966. Mineralization and immobilization of nitrogen in bacterial cells and in certain soil organic fractions. Soil Sci. Soc. Am. Proc. 30:210–213.

Dev, G., and D. A. Rennie. 1979. Isotope studies on the comparative efficiency of nitrogenous sources. Aust. J. Soil Res. 17:155–162.

Dowdell, R. J., and C. P. Webster. 1980. A lysimeter study using nitrogen-15 on the uptake of fertilizer nitrogen by perennial ryegrass swards and losses by leaching. J. Soil Sci. 31:65–75.

Filimonov, D. A., and Ye. V. Rudelev. 1977. Transformation of immobilized nitrogen in soil upon application of nitrogen fertilizers. Sov. Soil Sci. 9:588–592.

Frederick, L. R., and F. E. Broadbent. 1966. Biological interactions. p. 198–212. In M. H. McVickar et al. (ed.) Agricultural anhydrous ammonia technology and use. AAI, ASA, and SSSA, Memphis, Tenn., and Madison, Wis.

Freney, J. R., and J. R. Simpson. 1969. The mineralization of nitrogen from some organic fractions in soil. Soil Biol. Biochem. 1:241–251.

Fried, M., and H. Broeshart. 1974. Priming effect of nitrogen fertilizers on soil nitrogen. Soil Sci. Soc. Am. Proc. 38:858.

Fried, M., and L. A. Dean. 1952. A concept concerning the measurement of available soil nutrients. Soil Sci. 73:263–271.

Fried, M., R. J. Soper, and H. Broeshart. 1975. ^{15}N-Labelled single treatment fertility experiments. Agron. J. 67:393–396.

Hallam, M. J. 1953. Influence of moisture, aeration, and substrate on rates of biological transformations of carbon and nitrogen in soil. Iowa State Coll. J. Sci. 27:185–186.

Harmsen, G. W., and G. J. Kolenbrander. 1965. Soil inorganic nitrogen. In W. V. Bartholomew and F. E. Clark (ed.) Soil nitrogen. Agronomy 10:43–92. Am. Soc. of Agron., Madison, Wis.

Hauck, R. D. 1971. Quantative estimates of nitrogen-cycle processes—concepts and review. p. 65–80. In Nitrogen-15 in soil-plant studies. Int. Atomic Energy Agency, Vienna.

Hauck, R. D. 1973. Nitrogen tracers in nitrogen cycle studies—past use and future needs. J. Environ. Qual. 2:317–327.

Hauck, R. D., and J. M. Bremner. 1976. Use of tracers for soil and fertilizer nitrogen research. Adv. Agron. 26:219–266.

Hauck, R. D., and M. Bystrom. 1970. ¹⁵N. A selected bibliography for agricultural scientists. The Iowa State University Press, Ames, Iowa.

Heilman, P. 1975. Effect of added salts on nitrogen release and nitrate levels in forest soils of the Washington Coastal Area. Soil Sci. Soc. Am. Proc. 39:778–782.

van den Hende, A. 1968. L'incorporation d'azote dans la matière organique du sol; possibilités de l'utilisation d'azote-15 lourd. p. 319–326. In The use of isotopes in soil organic matter studies. Int. Atomic Energy Agency, Vienna.

Hills, F. J., F. E. Broadbent, and M. Fried. 1978. Timing and rate of fertilizer nitrogen for sugarbeet related to nitrogen uptake and pollution potential. J. Environ. Qual. 7:368–372.

Hiltbold, A. E., W. V. Bartholomew, and C. H. Werkman. 1951. The use of tracer techniques in the simultaneous measurement of mineralization and immobilization of nitrogen in soil. Soil Sci. Soc. Am. Proc. 15:166–172.

Jansson, S. L. 1958. Tracer studies on nitrogen transformations in soil. Ann. Roy. Agric. Coll. Sweden 24:101–361.

Jansson, S. L. 1966. Nitrogen transformation in soil organic matter. p. 283–296. In The use of isotopes in soil organic matter studies. FAO/IAEA, Pergamon Press, Inc., Oxford.

Jansson, S. L. 1967. Soil organic matter and fertility. p. 1–10. In G. V. Jacks (ed.) Soil chemistry and fertility. Int. Soc. Soil Sci., Trans. Comm. II, IV (Aberdeen, Sept. 1966). The University Press, Aberdeen.

Jansson, S. L. 1971. Use of ¹⁵N in studies of soil nitrogen. p. 129–166. In A. D. McLaren and J. Skujins (ed.) Soil biochemistry II. Marcel Dekker Inc., New York.

Jansson, S. L., M. J. Hallam, and W. V. Bartholomew. 1955. Preferential utilization of ammonium over nitrate by microorganisms in the decomposition of oat straw. Plant Soil 6: 382–390.

Jenkinson, D. S. 1966a. The priming action. p. 199–208. In The use of isotopes in soil organic matter studies. FAO/IAEA, Pergamon Press, Inc., New York.

Jenkinson, D. S. 1966b. Studies on the decomposition of plant material in soil. II. Partial sterilization of soil and the soil biomass. J. Soil Sci. 17:280–302.

Jenkinson, D. S., and D. S. Powlson. 1976a. The effects of biocidal treatments on metabolism in soil. I. Fumigation with chloroform. Soil biol. Biochem. 8:167–177.

Jenkinson, D. S., and D. S. Powlson. 1976b. The effects of biocidal treatments on metabolism in soil. V. A method for measuring soil biomass. Soil Biol. Biochem. 8:209–213.

Jenkinson, D. S., and J. H. Rayner. 1977. The turnover of soil organic matter in some of the Rothamsted classical experiments. Soil Sci. 123:298–305.

Jones, J. M., and B. N. Richards. 1977. Effect of reforestation on turnover of ¹⁵N-labelled nitrate and ammonium in relation to changes in soil microflora. Soil Biol. Biochem. 9: 383–392.

Kaindl, K., and E. Haunold. 1965. Die Aufnahme von N¹⁵-markiertem Stickstoffdünger durch Mais im Gefäss- und Feldversuch. Landwirtsch. Forsch. Sonderh. 19:140–144.

Kanamori, T., and T. Yasuda. 1979. Immobilization, mineralization and the availability of the fertilizer nitrogen during the decomposition of the organic matter applied to the soil. Plant Soil 52:219–227.

Ketcheson, J. W., and M. Jakovljevic. 1968. Transformation of NO_3^- and NH_4^+ in soils. p. 125–130. In Isotopes and radiation in soil organic-matter studies. Int. Atomic Energy Agency, Vienna.

Kirkham, D. 1957. Mathematical aspects of soil nitrogen studies. U.S. Atomic Energy Commission Rep. TID 7512:349–359.

Kirkham, D., and W. V. Bartholomew. 1954. Equations for following nutrient transformations in soil, utilizing tracer data. Soil Sci. Soc. Am. Proc. 18:33–34.

Kirkham, D., and W. V. Bartholomew. 1955. Equations for following nutrient transformations in soil, utilizing tracer data: II. Soil Sci. Soc. Am. Proc. 19:189–192.

Kissel, D. E., and S. J. Smith. 1978. Fate of fertilizer nitrate applied to Coastal bermudagrass on a swelling clay soil. Soil Sci. Soc. Am. J. 42:77–80.

Kissel, D. E., S. J. Smith, W. L. Hargrove, and D. W. Dillow. 1977. Immobilization of ferti-lizer nitrate applied to a swelling clay soil in the field. Soil Sci. Soc. Am. J. 41:346–349.

Knowles, R., and D. T.-H. Chu. 1969. Survival and mineralization and immobilization of ^{15}N-labelled *Serratia* cells in a boreal forest raw humus. Can. J. Microbiol. 15:223–228.

Kuo, M. H., and W. V. Bartholomew. 1966. On the genesis of organic nitrogen in decom-posed plant residue. p. 329–335. *In* The use of isotopes in soil organic matter studies. FAO/IAEA, Pergamon Press, Inc., Oxford.

Ladd, J. N., J. W. Parsons, and M. Amato. 1977a. Studies of nitrogen immobilization and mineralization in calcareous soils. I. Distribution of immobilized nitrogen amongst soil fractions of different particle size and density. Soil Biol. Biochem. 9:309–318.

Ladd, J. N., J. W. Parsons, and M. Amato. 1977b. Studies of nitrogen immobilization and mineralization in calcareous soils. II. Mineralization of immobilized nitrogen from soil fractions of different particle size and density. Soil Biol. Biochem. 9:319–325.

Ladd, J. N., and E. A. Paul. 1973. Changes in enzymatic activity and distribution of acid-soluble amino acid-nitrogen in soil during nitrogen immobilization and mineralization. Soil Biol. Biochem. 5:825–840.

Laura, R. D. 1974. Effect of neutral salts on carbon and nitrogen mineralization of organic matter in soil. Plant Soil 41:113–127.

Laura, R. D. 1975a. On the "priming effect" of ammonium fertilizers. Soil Sci. Soc. Am. Proc. 39:385.

Laura, R. D. 1975b. The "A-value" and the priming effect of nitrogen fertilizers on soil nitro-gen. Soil Sci. Soc. Am. Proc. 39:596.

Legg, J. O., and F. E. Allison. 1959. Recovery of N^{15}-tagged nitrogen from ammonium-fixing soils. Soil Sci. Soc. Am. Proc. 23:131–134.

Legg, J. O., and F. E. Allison. 1960. Role of rhizosphere microorganisms in the uptake of nitrogen by plants. Int. Congr. Soil Sci., Trans. 7th (Madison, Wis.) II:545–550.

Legg, J. O., F. W. Chichester, G. Stanford, and W. H. DeMar. 1971. Incorporation of ^{15}N-tagged mineral nitrogen into stable forms of soil organic nitrogen. Soil Sci. Soc. Am. Proc. 35:273–276.

Legg, J. O., and G. Stanford. 1967. Utilization of soil and fertilizer N by oats in relation to the available N status of soil. Soil Sci. Soc. Am. Proc. 31:215–219.

Löhnis, F. 1926. Nitrogen availability of green manure. Soil Sci. 22:253–290.

McGill, W. B., E. A. Paul, J. A. Shields, and W. E. Lowe. 1973. Turnover of microbial popu-lations and their metabolites in soil. Ecol. Res. Commun. (Stockholm) Bull. 17:293–301.

McGill, W. B., J. A. Shields, and E. A. Paul. 1975. Relation between carbon and nitrogen turnover in soil organic fractions of microbial origin. Soil Biol. Biochem. 7:57–63.

Mehran, M., and K. K. Tanji. 1974. Computer modeling of nitrogen transformations in soil. J. Environ. Qual. 3:391–396.

Mortensen, J. L. 1963. Decomposition of organic matter and mineralization of nitrogen in Brookston silt loam and alfalfa green manure. Plant Soil 19:374–384.

Nommik, H. 1961. Effect of the addition of organic materials and lime on the yield and nitro-gen nutrition of oats. Acta Agric. Scand. 11:211–226.

Nommik, H. 1968. Nitrogen mineralization and turnover in Norway spruce (*Picea Abies* (L.) Karst.) raw humus as influenced by liming. Int. Congr. Soil Sci., Trans. 9th (Adelaide) 11: 533–545.

Olson, R. V. 1980. Fate of tagged nitrogen fertilizer applied to irrigated corn. Soil Sci. Soc. Am. J. 44:514–517.

Olson, R. V., L. S. Murphy, H. C. Moser, and C. W. Swallow. 1979. Fate of tagged fertilizer nitrogen applied to winter wheat. Soil Sci. Soc. Am. J. 43:973–975.

Paul, E. A. 1970. Plant components and soil organic matter. *In* C. Steelink and V. C. Runeckles (ed.) Recent Adv. Phytochem. 3:59–104.

Paul, E. A. 1976. Nitrogen cycling in terrestrial ecosystems. p. 225–243. *In* J. O. Nriagu (ed.) Environmental biochemistry, Vol. I. Ann Arbor Science Publishers, Inc., Ann Arbor, Mich.

Paul, E. A., and N. G. Juma. 1981. Mineralization and immobilization of soil nitrogen by microorganisms. p. 179–199. *In* F. E. Clark and T. Rosswall (ed.) Terrestrial nitrogen cycles. Ecol. Bull. 33 (Stockholm).

Paul, E. A., and J. A. van Veen. 1979. The use of tracers to determine the dynamic nature of organic matter. p. 75–132. *In* J. K. R. Gasser (ed.) Modelling nitrogen from farm wastes. Applied Science Publishers, Ltd, London.

Persson, J. 1968. Biological testing of chemical humus analysis. Ann. Agric. Coll. Sweden 34: 81–217.

Pomares-Garcia, F., and P. F. Pratt. 1978. Recovery of ^{15}N-labeled fertilizer from manured and sludge-amended soil. Soil Sci. Soc. Am. J. 42:717–720.

Rennie, R. J., and D. A. Rennie. 1973. Standard nitrogen versus nitrogen balance criteria for assessing the efficiency of nitrogen sources for barley. Can. J. Soil Sci. 53:73–77.

Riga, A., V. Fischer, and H. J. van Praag. 1980. Fate of fertilizer nitrogen applied to winter wheat as Na $^{15}NO_3$ and $(^{15}NH_4)_2SO_4$ studied in microplots through a four-course rotation: 1. Influence of fertilizer splitting on soil and fertilizer nitrogen. Soil Sci. 130:88–99.

Sapozhnikov, N. A., E. J. Nerterova, I. P. Rusinova, L. B. Sirota, and T. K. Livanova. 1968. The effect of fertilizer nitrogen on plant uptake of nitrogen from different podzolic soils. Int. Congr. Soil Sci., Trans. 9th (Adelaide) II:467–474.

Sauerbeck, D., and B. Johnen. 1976. Der Umsatz von Pflanzenwurzeln im Laufe der Vegetationsperiode und dessen Beitrag zur "Bodenatmung". Z. Pflanzenernaehr. Bodenkd. 139: 315–328.

Sauerbeck, D., B. Johnen, and R. Six. 1976. Atmung, Abbau und Ausscheidungen von Weizenwurzeln im Laufe ihrer Entwicklung. Landwirtsch. Forsch. Sonderh. 32:49–58.

Shields, J. A., E. A. Paul, W. E. Lowe, and D. Parkinson. 1973. Turnover of microbial tissue in soil under field conditions. Soil Biol. Biochem. 5:753–764.

Smith, S. J., F. W. Chichester, and D. E. Kissel. 1978. Residual forms of fertilizer nitrogen in field soils. Soil Sci. 125:165–169.

Smith, S. J., and J. O. Legg. 1971. Reflections on the A value concept of soil nutrient availability. Soil Sci. 112:373–375.

Stanford, G., J. O. Legg, and F. W. Chichester. 1970. Transformations of fertilizer nitrogen in soil. 1. Interpretations based on chemical extractions of labelled and unlabelled nitrogen. Plant Soil 33:425–435.

Stevenson, F. J. 1965. Origin and distribution of nitrogen in soil. In W. V. Bartholomew and F. E. Clark (ed.) Soil nitrogen. Agronomy 10:1–42. Am. Soc. of Agron., Madison, Wis.

Stewart, B. A., D. D. Johnson, and L. K. Porter. 1963a. The availability of fertilizer nitrogen immobilized during decomposition of straw. Soil Sci. Soc. Am. Proc. 27:656–659.

Stewart, B. A., L. K. Porter, and D. D. Johnson. 1963b. Immobilization and mineralization of nitrogen in several organic fractions of soil. Soil Sci. Soc. Am. Proc. 27:302–304.

Stojanovic, B. J., and F. E. Broadbent. 1956. Immobilization and mineralization of nitrogen during decomposition of plant residues in soil. Soil Sci. Soc. Am. Proc. 20:213–218.

Taki, M., and T. Yamashita. 1967. Methods of measuring forms of available soil nitrogen. Hatano Tobacco Exp. Stn. Bull. 60:71–98.

Taki, M., M. Yamashita, and T. Yamashita. 1967. Effect of nitrogen in green manure on tobacco cultivation. 2. Utilization by tobacco plants of the nutrients from applied green manure. Hatano Tobacco Exp. Stn. Bull. 60:61–69.

Terman, G. L., and M. A. Brown. 1968a. Crop recovery of applied fertilizer nitrogen. Plant Soil 29:48–65.

Terman, G. L., and M. A. Brown. 1968b. Uptake of fertilizer and soil nitrogen by ryegrass, as affected by carbonaceous residue. Soil Sci. Soc. Am. Proc. 32:86–90.

Turtschin, F. B., S. N. Bersanjewa, I. A. Koritzkaja, G. G. Shidkick, and G. A. Lobowikowa. 1960. Die Stickstoffumwandlung im Boden nach der Angaben der Untersuchungen des Isotops N^{15}. Int. Congr. Soil Sci., Trans. 7th (Madison, Wis.) II:236–245.

Tyler, K. B., and F. E. Broadbent. 1958. Nitrogen uptake by ryegrass from three tagged ammonium fertilizers. Soil Sci. Soc. Am. Proc. 22:231–234.

Walker, T. W., A. F. R. Adams, and H. D. Orchiston. 1956. Fate of labelled nitrate and ammonium nitrogen when applied to grass and clover grown separately and together. Soil Sci. 81:339–351.

Wallace, A., and R. L. Smith. 1954. Nitrogen interchange during decomposition of orange and avocado tree residues in soil. Soil Sci. 78:231–242.

Westerman, R. L., and L. T. Kurtz. 1973. Priming effect of ^{15}N-labeled fertilizer on soil nitrogen in field experiments. Soil Sci. Soc. Am. Proc. 37:725–727.

Westerman, R. L., and L. T. Kurtz. 1974. Isotopic and nonisotopic estimations of fertilizer nitrogen uptake by sudangrass in field experiments. Soil Sci. Soc. Am. Proc. 38:107–109.

Westerman, R. L., and T. C. Tucker. 1974. Effects of salts plus nitrogen-15-labeled ammonium chloride on mineralization of soil nitrogen, nitrification, and immobilization. Soil Sci. Soc. Am. Proc. 38:602–605.

Zamyatina, V. B., N. I. Borisova, N. M. Varyushkina, S. V. Burtzeva, and L. I. Kirpaneva. 1968. Investigations on the balance and use of ^{15}N-tagged fertilizer nitrogen by plants in soils. Int. Congr. Soil Sci., Trans. 9th (Adelaide) II:513–521.

7 Nitrification in Soil

EDWIN L. SCHMIDT

University of Minnesota
St. Paul, Minnesota

I. INTRODUCTION

The decomposition of proteins, nucleic acids, and other nitrogenous organic substances in soils leads to the release of NH_3 which equilibrates to the ionic species, NH_4^+, in all but highly alkaline soils. In the presence of readily available carbonaceous materials, the NH_4^+ is assimilated rapidly into newly forming microbial biomass. Under the more usual soil circumstances in which microbial development is limited by available C and energy, most of the NH_4^+ is oxidized to NO_3^- as rapidly as it is formed. The process whereby NH_4^+ is oxidized to NO_3^- is referred to as nitrification.

Nitrification occurs worldwide in terrestrial, aquatic, and sedimentary ecosystems. It is of paramount importance to primary productivity, to nutrient cycling, to waste disposal, and to the quality of water. The nitrification process interfaces with other component processes of the N cycle—a feature that contributes to the difficulty of its study, especially in soil.

The strictly biological nature of nitrification was firmly established with the isolation of "nitrifying bacteria" by Winogradsky in the period 1889–1890. He described representatives of two small groups of specialized chemoautotrophic (chemolithotrophic) bacteria and clearly related the metabolism of each to the two corresponding stages of nitrification: the oxidation of NH_4^+ to NO_2^-, and the subsequent oxidation of NO_2^- to NO_3^-. So far as is known, these autotrophic nitrifying bacteria are the principal agents of nitrification throughout nature. A compilation of Winogradsky's works has been published in French (1949). Waksman (1946) commemorated the occasion of Winogradsky's 90th birthday with a highly readable account summarizing his career.

Nitrification has been the subject of recent review from a variety of viewpoints by Painter (1977), Focht and Verstraete (1977) and Belser (1979). There is a wealth of older literature dealing with the chemistry of soil nitrification that should not be ignored. Adequate techniques and procedures to follow the chemistry of the nitrification process were available very early and have been little changed or improved to date. Consequently, a great deal of basic information on soil factors as they influence nitrification appeared decades ago, but is rarely retrieved and apparently destined for periodic repetition. Particularly good summaries and interpretations of

Copyright 1982 © ASA-CSSA-SSSA, 677 South Segoe Road, Madison, WI 53711, USA.
Nitrogen in Agricultural Soils—Agronomy Monograph no. 22.

the older literature are to be found in the texts of Waksman (1932) and Russell (1973) and in earlier editions of those texts.

II. THE PROCESS OF NITRIFICATION IN SOILS

Nitrification is not the only process contributing NO_3^- to the soil, but it is by far the most important. Nitric oxide produced in atmosphere surrounding lightning discharges is converted to NO_3^- by photochemical oxidations, and reaches the earth with rainfall. Annual increments to the entire biosphere by this means are estimated to be about 10^7 tons NO_3^--N (Yung & McElroy, 1979). Nitrification in the 2.3×10^7 acres of agricultural soils of Minnesota alone, for example, probably amounts to about 10^6 tons NO_3^--N annually.

Nitrification takes place in virtually all soils where NH_4^+ is present and conditions are favorable with respect to the major factors of temperature, moisture, pH, and aeration. Oxidation of NO_2^- is more rapid than that of NH_4^+, so that only rarely in natural soils is there more than trace amounts of NO_2^- present. Arable soils in the temperate regions have a fairly constant but low content of NH_4^+, but a variable and often high NO_3^- content (Russell, 1973). The total amount of mineral N (NH_4^+ plus NO_3^-) usually reflects the difference between the rates of NH_4^+ production from ammonification and of NO_3^- removal by plant assimilation, microbial reductions, and leaching. The NO_3^--N contents of soils at a single farm site in Minnesota were found to vary from 8 to 247 $\mu g/g$ on the same date of sampling, and to vary substantially in the same field according to sampling site and sampling time (Schmidt, 1956). The relatively stable background level of NH_4^+ and the great variability of NO_3^- that are consistent features of most soils is illustrated in Fig. 1 for a semiarid coastal soil of Morocco (Chiang et al., 1972). Although the dynamic nature of soil NO_3^- is well known, conclusions based on NO_3^- concentrations in soils at a single sampling time still enter the literature (Lodhi, 1978, for example).

A. Factors Regulating Nitrification in Soils

The main factors which limit nitrification in soil are substrate NH_4^+, O_2, CO_2, pH, and temperature. These factors were discussed extensively by Focht and Verstraete (1977) for nitrifying microorganisms in culture and for nitrification in natural and N-enriched soils. In general, nitrification appears to proceed in soils under a much broader range of conditions than is predictable on the basis of the biochemistry of the process and the physiology of the autotrophic nitrifiers. Probably this merely reflects present lack of knowledge of the nitrifying population involved in any particular nitrification situation, and of the way the main environmental determinants are integrated at the nitrification microsite in the soil.

Cold and wet soils are essentially inactive with respect to nitrification.

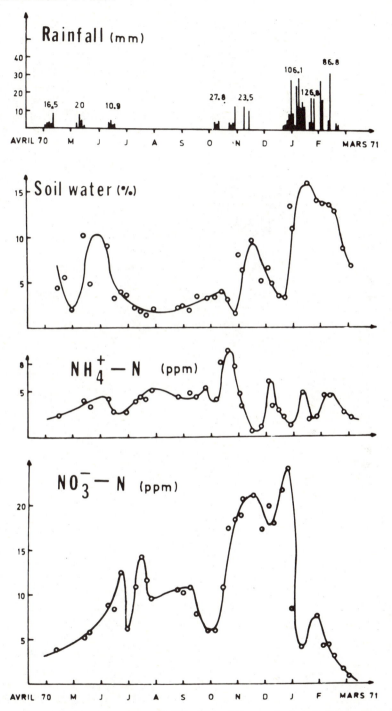

Fig. 1—Seasonal measurements on rainfall, soil moisture, and mineral nitrogen in the cultivated layer of coastal soil of Morocco. Data of Chiang et al. (1972).

Such limitations prevail until the soil warms to approximately 4 or 5°C (Anderson & Boswell, 1964). The optimum temperature for nitrification appears to vary widely among soils. Mahendrappa et al. (1966) reported maximum nitrification rates at 20–25°C for a group of soils from the northwestern United States and at 30–40°C for soils of the southwestern United States. The value of 40°C was found to be the maximum temperature for nitrification in the midwestern United States soils examined by Keeney and Bremner (1967), whereas a tropical Australian soil was found by Myers (1975) to nitrify at temperatures up to 60°C. Nothing is known of possible successional changes within a nitrifying population in response to seasonal increases in temperature—or to fluxes in any of the major parameters—but the data clearly suggest adaptations on the part of some nitrifiers to the temperature regimes of their soil habitats.

The O_2 and HCO_3^- needed for nitrification are provided by the liquid phase of the soil system. Therefore, the amount and composition of the soil water influence nitrification. There is no evidence that assimilable carbon is ever limiting to soil nitrifiers, but O_2 clearly can be. Depletion of O_2 in the soil water is favored by (i) high soil moisture content, which fills soil pores and restricts recharge of O_2 from the gaseous phase; (ii) high soil temperatures, which reduce the solubility of O_2 and increase O_2 demand by heterotrophic microorganisms; and (iii) oxidizable organic matter, which also increases heterotrophic O_2 demand. In organic soils with high total microbial activity, it may well be that nitrification is favored by low temperatures as a result of decreased demand for NH_4^+ by heterotrophs, and increased solubility of O_2. The effect of moderately high moisture levels (pF 1.0–2.0) is to enhance nitrification in most soils so long as aeration is adequate (Fig. 2), whereas in dried soils nitrification varies with soil texture and those properties affecting osmotic pressure (Russell, 1973). Desiccation probably reduces the nitrifying population, but little information is available on this subject. The tenacity of nitrifiers under conditions of high soil temperatures, high solute concentration, and desiccation over a 4 month dry period,

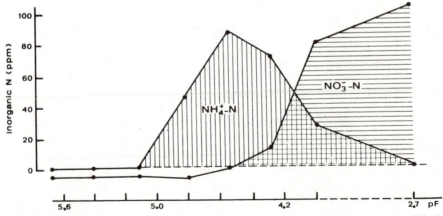

Fig. 2—Influence of soil pF on the accumulation of NH_4^+-N and NO_3^--N in a tropical black clay soil of Senegal (after Dommergues, 1977).

is reflected in the speed with which nitrification is resumed in a Moroccan coastal soil following the first October rainfalls (Fig. 1).

Nitrification proceeds at soil reactions far below the pH limits observed for the nitrifying bacteria in pure culture. This is a long standing anomaly that is no closer to resolution than it was in 1927 when Waksman reviewed the literature to note that nitrification ceased at pH 3.9–4.5 and that nitrifying bacteria isolated from acid soils may be somewhat adapted to such soils. Most observations indicate an arbitrary lower limit for nitrification of pH 4.0, obvious nitrification in the pH 4 to 6 range, and pH independent nitrification in the range 6 to 8 (Weber & Gainey, 1962; Morrill & Dawson, 1967; Chase et al., 1968; Sarathchandra, 1978b). The effect of acidity may be an expression of Al toxicity (Brar & Giddens, 1968). Moreover, a soil sample having a pH of 5.5 as based on the overall measurement may contain microhabitats of pH 7.0 or higher and microhabitats of pH 5.5 or lower. The nature of nitrification at low soil pH and its limitation at extreme acidities should be approached by characterization of the nitrifying population and their microsites.

B. Substrates and Products

Ammonification provides the starting substrate for nitrification when NH_3 equilibrates in the soil as partially represented in the following equations (Smith, 1964):

$$NH_3 + HOH \rightleftharpoons NH_4OH$$

$$NH_4OH \rightleftharpoons NH_4^+ + OH^-$$

$$NH_4OH + \text{H-Soil} \rightleftharpoons NH_4\text{-Soil} + HOH.$$

Ionization of NH_3 depends on pH (pk_a 9.0) so that the percentages of nonionized NH_3 are 0.1, 1.0, 10.0, and 50 at pH 6.0, 7.0, 8.0, and 9.0, respectively. Soil conditions that favor the occurrence of NH_3 (high pH and low CEC) restrict nitrification because of the toxicity of free NH_3. The NO_2^- oxidation stage is most susceptible to inhibition by molecular NH_3 so that applications of urea or anhydrous NH_3 to calcareous soils have resulted in NO_2^- accumulation (Chapman & Liebig, 1952). At normal levels of NH_3 release, where nitrification is linked to ammonification, substrate concentrations are low and rate limiting. Virtually all NH_3 occurs as NH_4^+ bound to the soil exchange complex. Some NH_4^+ is tightly bound at soil exchange sites within the lattices of certain clays, such as illite and vermiculite and generally biologically unavailable, but much of this fixed NH_4^+ is nevertheless nitrified (Jansson, 1958).

Nitrite usually occurs only in trace amounts except under unusual circumstances of high applications of nitrogenous fertilizers combined with high soil pH. Such conditions lead to NO_2^- accumulations in some soils but not in others (Russell, 1973). Nitrite accumulation is of concern since this anion is mobile and highly toxic to many microorganisms. NO_2^- can react with secondary amines in various environments to form the potent nitrosamine carcinogens (Cassens et al., 1978), but formation in soils is highly unlikely in view of the very low concentrations of both reactants. Pancholy (1978) found no evidence of nitrosamines in soils unless both NO_2^- and amines were added.

Nitrate is mobile also, but is less reactive than NO_2^- and commonly accumulates in soils in the range of a few to several hundred $\mu g/g$. Identification of N_2O as a product of NH_4^+ oxidation stemmed from enzymatic studies of *Nitrosomonas europaea* (Falcone et al., 1963; Hooper, 1968). Bremner and Blackmer (1978) showed that N_2O was emitted during nitrification of fertilizer N in several soils under strictly aerobic conditions. Evolution of N_2O was not enhanced by NO_3^- addition, was inhibited by nitrapyrin [2-chloro-6-(trichloromethyl) pyridine], and was absent in sterilized, N-treated control soils. Elkins et al. (1978) subsequently reported work suggestive of nitrification-derived N_2O emission in marine environments.

C. Interactions with Other N Cycle Events

Satisfactory methods for the determination of NO_3^- in soils were available before the nature of nitrification was known, but their use in the field has contributed little to an understanding of the dynamics of the process. The chemical description of nitrification in soil, especially at low NH_4^+ levels are limited by a variety of interactions which may lead to the virtual absence of NO_3^- despite active nitrification. For example, NO_3^- may be nearly absent in actively nitrifying soil if that soil is sampled after a heavy rain. Similarly, roots and associated microorganisms in an aerobic rhizosphere may assimilate NO_3^- as rapidly as formed. When NO_3^- is formed adjacent to anaerobic soil sites, dissimilatory reductions of NO_3^- in those sites present still another type of interaction that may mask nitrification. Possi-

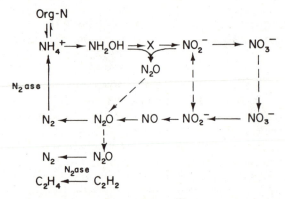

Fig. 3—Possible interrelationships, as shown by dotted lines, among the processes of nitrification (*upper*) denitrification (*middle*) and nitrogen fixation (*bottom*), presented schematically by Knowles (1978).

ble couplings between nitrification, denitrification, and N fixation by way of NO_3^-, NO_2^-, or N_2O are presented in Fig. 3. Nitrate is a N sink in the ocean (Yung & McElroy, 1979), but clearly it is not such in the soil. The many interactions of nitrification with other N transformations in soil pose serious complications, not just to the study of nitrification, but also to the construction of models to predict the rates at which N transformations occur under various environmental conditions.

Interactions between NO_3^- formation and dissimilatory reductions of NO_3^- as both processes go on concurrently have received attention only recently. These have been pictured by Knowles (1978) as a coupling in which nitrification and denitrification occur simultaneously, but in separate microsites on opposite sides of an aerobic-anaerobic interface. The situation may be visualized easily as occurring at soil crumbs where NO_3^- formed at the aggregate surface can diffuse to anaerobic sites within to be denitrified. Direct evidence for concurrent nitrification-nitrate reduction has been provided by Starr et al. (1974) in an NH_4^+ enrichment soil column and by Koike and Hattori (1978) for 0–2 cm coastal sediments. By using a ^{15}N dilution technique to accomplish simultaneous determinations, the latter investigators found that NO_3^- reduction proceeded at a rate of 10^{-2} to 10×10^{-2} μg-atoms of N/g per hour, and that 30% of the NO_3^- reduced was accounted for as N_2 from denitrification reactions in muddy top sediments. A field study by Volz et al. (1975) suggested that NO_2^- produced by reduction of added NO_3^- may have become available for simultaneous reoxidation by NO_2^- oxidizers. Belser (1977) subsequently examined the data of Volz et al. in the light of additional studies with pure cultures and laboratory soil perfusion columns; his findings reinforced the view that the NO_2^- oxidizing populations observed could not have developed from NH_4^+ oxidation alone. The possibility of at least some N cycling between the NO_2^- and NO_3^- states should be considered if NO_2^- oxidizers appear unusually abundant in some soils.

D. Approaches to Soil Nitrification

As noted above, the numerous and often subtle interconversions that occur among the forms of mineral N in soil limit the extent to which NO_3^- determinations alone can be useful as a measure of nitrification. Limitations are particularly severe when the approach is merely an analysis of field samples for NO_3^- alone or NO_3^- together with NH_4^+. Such determinations sometimes are applied to agricultural soils in the spring to obtain a "residual NO_3^-" value as a rough guide to fertilizer application. However, such measurements can only be of value if the soils are not subject to extensive leaching or waterlogging, and their moisture regimes are relatively predictable. Nitrate is often determined on soil samples brought into the laboratory to better isolate the process, and to derive nitrification rates. The NO_3^- formed under incubation conditions standardized with respect to time, temperature, and moisture, may provide useful comparative information on the "nitrifying potential," or "N mineralization potential" of soils

(Stanford & Smith, 1972). Modifications of this approach were in use for routine soil testing, but have been largely discontinued because of the time required for assay and intrinsic difficulties in relating the data to soil management practice.

Nitrification rates are needed not only for the prediction of fertilizer needs in the interest of more efficient use of fertilizer N, but increasingly, also in the development of models designed to simulate the behavior of N in soils (Beek & Frissel, 1973; Reuss & Innis, 1977). The derivation of rates has been largely from NO_3^- measurements only as typified by the approaches of Sabey et al. (1969) and Stanford and Smith (1972), with little direct attention to the microbiology of the nitrifying populations.

Microbiological data on the nitrifiers in relation to the process as carried out in nature would contribute considerably to an understanding of the basis for a particular nitrification rate, and to the reliability of models to predict the rates at which nitrification takes place under various environmental conditions. Thus, it would be desirable to know for a given circumstance what nitrifying microorganisms are represented, what population responses occur, and what growth rates, growth yields, and substrate conversions occur in the soil. As will be noted later, the common simplification of equating *Nitrosomonas* with NH_4^+ oxidation for all circumstances is not justifiable. Other genera and strains of autotrophic nitrifiers may participate and may predominate over *Nitrosomonas* (Belser & Schmidt, 1978a). However, as desirable as information on the microbiological base for nitrification in natural environments may be, such data have not been available because of the complexity of the ecology of the nitrifiers and the inadequacy of available methodology. New approaches are urgently needed to better relate the process of nitrification to the microbiology in soil of the agents responsible for it.

III. MICROBIOLOGICAL BASIS OF NITRIFICATION

Reviews of the early literature (Waksman, 1932; Barritt, 1933; Stephenson, 1939) make it clear that many isolates were claimed as putative agents of nitrification and that a considerable amount of controversy was engendered in relation to those claims. To a limited degree some measure of controversy still exists. This is based on inability to explain the occurrence of nitrification in certain habitats where the presence of the classical autotrophs would seem unlikely—based on their known physiology—or where autotrophic nitrifiers are either not detectable or detectable only at very low numbers.

A. Nitrification by Heterotrophs

Interest in the possibility that heterotrophic microorganisms may contribute to the nitrification process was renewed by the observation of Quastel and Scholefield (1948) that pyruvic oxime was converted to NO_2^- in soil percolation columns. Several bacteria, identified as *Achromobacter* and *Corynebacterium* genera, were reported to be responsible (Quastel et al., 1950, 1952). Jensen (1951) isolated three more groups of hetrotrophic bacteria which formed NO_2^- from the oxime of pyruvic acid, including 24 strains of *Nocardia corallina*, 1 strain of *Agrobacterium* sp., and 3 strains of *Alcaligenes* sp. One *N. corallina* pure culture produced 40 μg NO_2^--N/ml in 8 days with pyruvic oxime as the only N source, and 70 μg NO_2^--N/ml in oxime + peptone medium. All of these bacteria formed only NO_2^- as a nitrification product, but were of interest because of their relatively high activity in this regard.

The first instance of NO_3^- formation by a heterotroph was reported by Schmidt (1954) with the isolation of a soil fungus, *Aspergillus flavus*. When grown in pure culture in a yeast extract-peptone-glucose medium, *A. flavus* formed 25–30 μg NO_3^--N/ml in 7–14 days; the major product was NH_4^+, with NO_2^--N rarely exceeding 1.5 μg/ml. Eylar and Schmidt (1959) carried out a survey of microorganisms isolated from soil to determine the incidence of heterotrophic forms capable for producing NO_2^- or NO_3^-. The ability to form NO_3^- was found to reside almost exclusively in the *A. flavus* isolates. One or more isolates of *A. flavus* were obtained from 8 to 22 soils studied, and isolation was made more commonly from 12 soils of high nitrifying capacity. Various bacteria and fungi, comprising about 7% of all the isolates, were able to form very low (0.2–1.0 μg/ml) concentrations of NO_2^-.

Clearly there are heterotrophic microorganisms, bacteria and fungi, that can be isolated from soil and cultured under conditions that lead to the accumulation of NO_2^- or NO_3^-. There is no evidence from the relatively few studies attempted that any of the heterotrophs that nitrify in culture ever do so in nature. *A. flavus* was shown to produce NO_3^- in pure culture when grown on alfalfa or barnyard manure; the same residues incorporated into sterile soil yielded no NO_3^- after inoculation with *A. flavus* (Schmidt, 1960b). In a nonsterile soil of favorable pH (7.3) amended with 4% alfalfa, inoculation with *A. flavus* had no affect on NO_3^- formation; the fungus grew extensively and NO_3^- was formed, but the NO_3^- was produced by a streptomycin-sensitive component of the soil population and not by *A. flavus* (Schmidt, 1960b). Features of nitrification by heterotrophs as demonstrated in pure culture are not readily extended to the highly competitive, substrate limited circumstances of soil. Pure culture nitrification by *A. flavus* was characterized by a high and narrow pH range, high substrate concentrations poised at C/N ratios <5, extremely high ratios of biomass to nitrification product, and formation of nitrification product after cell synthesis has been completed (Schmidt, 1960a). The same features pertain to other heterotrophic isolates shown to form nitrification products in pure culture. On the other hand, nitrification sometimes occurs in nature under soil circumstances where autotrophic nitrifiers should not grow or cannot be detected (Focht & Verstraete, 1977). Heterotrophic nitrification is commonly invoked to account for nitrification in such instances, but nothing short of isolating a particular heterotroph, demonstrating its potential in pure culture, and then relating its occurrence in the natural environment to the progression of the process in that natural environment, can provide un-equivocal evidence of heterotrophic nitrification.

B. Methane Oxidizing Bacteria

Several CH_4 oxidizing bacteria were shown by Hutton and Zobell (1953) to oxidize NH_4^+ to NO_2^- in a defined medium. Their data indicate that about 20 μg NO_2^--N/ml were produced during 19 days of incubation. The CH_4 oxidizing bacteria, or methylotrophs as frequently designated, oc-cur in soils and waters in aerobic sites in contact with anaerobic, CH_4 generating sites. They are a morphologically diverse group of bacteria, grouped together by a common ability to oxidize CH_4 as their prime C and energy source. The methylotrophic bacteria are aerobic as they incorporate one oxygen from dioxygen into methane by means of a mixed function oxidase (Whittenbury et al., 1970). The recent indication of uncharacterized anaerobic CH_4 oxidizing bacteria in enrichment culture (Panganiban et al., 1979) presents the possibility of a new type of methylotroph whose bio-chemistry and ecology are as yet unassessed. Excellent discussions of the CH_4 oxidizing bacteria are available (Smith & Hoare, 1977; Brock, 1979).

The morphological and biochemical similarities between the obligate CH_4 oxidizers and the autotrophic nitrifying bacteria are noteworthy. Morphologically, both have extensive proliferations of internal membrane assemblies which appear to be invaginations of the cytoplasmic membrane. The membraneous systems are unproven as to function, but are generally considered to bind the enzymes responsible for oxidation of their respective substrates (Smith & Hoare, 1977). Biochemically, both groups have an in-complete tricarboxylic acid cycle and each oxidizes its energy source, CH_4 or NH_4^+, by mixed function oxidases. Methane oxidizers are able to oxidize NH_4^+ to NO_2^- and when they do so the CH_4 mono-oxygenase enzyme com-plex appears to be the same one which oxidizes NH_4^+. Oxidation of NH_4^+ by CH_4 oxidizers is considered by Whittenbury and Kelly (1977) to be an in-stance of "co-oxidation" whereby an organism is able to oxidize a com-pound but is not able to grow on it as a single source of energy. It is not yet clear how useful NH_4^+ oxidation is to a co-oxidizing methylotroph.

Dalton (1977) examined the behavior of soluble extracts of *Methylo-coccus capsulatus* that oxidize CH_4 to methanol and NH_4^+ to NO_2^-. He found that the apparent Km value for NH_4^+ in cells and extracts was very high, which suggests that nitrification would be significant only when either the NH_4^+ concentration was high or the CH_4 concentration was low. Dalton thought it likely that such a situation could prevail in aquatic environments wherein high populations of CH_4 oxidizers may have built up previously; they would then be the major nitrifying organisms. In reality, nothing is known as to what, if any, contribution the CH_4 oxidizers may make to nitrification, but in view of the features shared with the autotrophic nitri-fiers, the possibility is intriguing and bears examination. The possibility of co-oxidation of methane by NH_4^+ oxidizers is also intriguing as an un-explored factor in the survival of the autotrophic nitrifiers in nature.

C. Autotrophic Nitrifiers

The only microorganisms linked directly to nitrification in natural environments at the present time are the Gram negative chemosynthetic autotrophic nitrifying bacteria comprising the family *Nitrobacteriaceae.* A listing is presented in Table 1. The arrangement is that of Watson (1974) as compiled for the 8th edition of *Bergey's Manual,* with the addition of two recently described species of NH_4^+ oxidizers *Nitrosovibrio tenuis* (Harms et al., 1976) and *Nitrosococcus mobilis* (Koops et al., 1976). *Nitrosovibrio* constitutes a new genus.

The brevity of Table 1 is surprising in light of the diversity of environments in which nitrification takes place, but not in the light of the limited research effort focused on the microbiology of nitrification during the past four decades, and the difficulties associated with such study.

Difficulties associated with the study of nitrifiers commence with their isolation in pure culture. Plating, even on strictly inorganic media, is not suitable for their direct isolation from nature because the organic materials introduced with the inoculum permit growth of heterotrophic contaminants. Most successful isolations of autotrophic nitrifiers have been preceded by extensive and careful serial enrichment procedures (Schmidt, 1978). Once isolated, nitrifiers are slow growing and yields are low. Biochemical studies have been hampered by these features and by the susceptibility of nitrifier cultures to contamination. This is especially true for studies with NH_4^+ oxidizers. With few pure culture isolates available, biochemical studies have been limited to *Nitrosomonas* and *Nitrobacter,* and usually to the same few strains of these genera.

Table 1—Listing of chemoautotrophic nitrifiers expanded from Bergey's Manual of Determinative Bacteriology, 8th ed. (Watson, 1974).

Genus	Species	Habitat
	(Oxidize NH_3 to NO_2^-)	
Nitrosomonas	europaea	Soil, water, sewage
Nitrosospira	briensis	Soil
Nitrosococcus	nitrosus	Marine, soil
	oceanus	Marine
	mobilis†	Marine
Nitrosolobus	multiformis	Soil
Nitrosovibrio‡	tenuis‡	Soil
	(Oxidize NO_2^- to NO_3^-)	
Nitrobacter	winogradskyi§	Soil
	(agilis)§	Soil, water
Nitrospina	gracilis	Marine
Nitrococcus	mobilis	Marine

† New species (Koops et al., 1976).
‡ New genus, new species (Harms et al., 1976).
§ *N. winogradskyi* is comprised of at least two serotypes, one of which has been referred to traditionally (prior to the 8th ed. of *Bergey's Manual*) as *N. agilis.* The two serotypes differ somewhat in growth response as well as in antigenicity (Fliermans et al., 1974).

Ecological studies dealing with nitrifiers in soil and other habitats have been restricted by the complexities of the habitat and the specialized biology of the nitrifiers. As a result, essentially nothing is known about two basic aspects of nitrifying populations: the diversity of nitrifiers within a given population, and the in situ nitrifying activities of the individual components of a population. It has been possible only to arrive at rough, statistical estimates of the overall nitrifying population by means of most probable number (MPN) techniques (section VI). Questions with respect to the densities of nitrifying populations in given situations, species and strain composition, successional changes, response to major environmental factors, and rates of cell synthesis and substrate conversions have been approached largely by extrapolation from pure culture studies.

IV. AMMONIUM OXIDIZING BACTERIA IN SOIL

All genera of NH_4^+ oxidizers listed in Table 1 include species isolated from soil. Despite this evidence that various genera could be responsible for NH_4^+ oxidation in a particular soil, the textbook convention of attributing all such oxidations to *Nitrosomonas*, has been widely accepted. The prevailing view that *Nitrosomonas* is the most common NH_4^+ oxidizer in soil, has led to the designation of others as "secondary genera" (Focht & Verstraete, 1977). Recent reports have shown that assumptions relative to the prime role of the genus *Nitrosomonas* do not necessarily hold (Walker, 1978; Belser & Schmidt, 1978a). Relatively little is known of the biological features of genera other than *Nitrosomonas* since nearly all physiological and biochemical studies have focused on *N. europaea*. All genera of NH_4^+ oxidizers, however, appear to be basically similar with respect to energetics, C fixation, and independence of pre-fixed organic compounds.

A. Biochemistry of NH_4^+ Oxidation

Nitrosomonas oxidizes NH_4^+ to NO_2^- according to the equation:

$$NH_4^+ + 1\tfrac{1}{2}\,O_2 \overset{6e}{\rightarrow} NO_2^- + H_2O + 2H^+; \text{F} -65 \text{ kcal.}$$

Energy yielded in this reaction, although small, provides all of the energy for biosynthesis and maintenance. As shown by Hofman and Lees (1953) hydroxylamine (NH_2OH) is a likely intermediate. There is a valency change from N from -3 (NH_4^+) to $+3$ (NO_2^-) which would suggest at least one other intermediate. Nitroxyl (NOH) is usually considered the most probable second intermediate (Nicholas, 1978), although it is unstable and there is no direct evidence in favor of it. The pathway of NH_4^+ oxidation is given in Fig. 4 after Nicholas (1978).

Cell-free extracts of *N. europaea* have low activity so that little is known of the mechanism of the first step of oxidation, NH_4^+ to NH_2OH

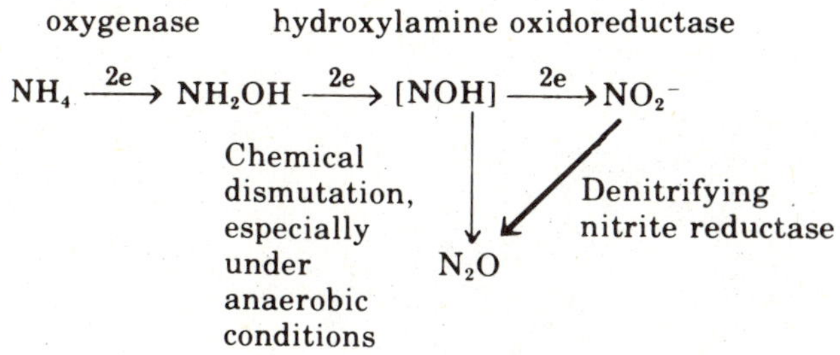

oxygenase   hydroxylamine oxidoreductase

$$NH_4 \xrightarrow{2e} NH_2OH \xrightarrow{2e} [NOH] \xrightarrow{2e} NO_2^-$$

Chemical
dismutation,   Denitrifying
especially   nitrite reductase
under N_2O
anaerobic
conditions

Fig. 4—Overall reaction scheme of the oxidation of NH_4^+ to NO_2^- as depicted by Nicholas (1978).

(Suzuki, 1974). This hydroxylation is an energy requiring step, and if carried out by a conventional mono-oxygenase type of reaction, NADH would be expended without any return in the form of ATP synthesis. Hydroxylamine is readily oxidized by whole cells or by cell extracts. In extracts, the oxidation of NH_2OH to NO_2^- is carried out by the enzyme hydroxylamine oxidoreductase only in the presence of a suitable electron acceptor, indicating that the enzyme had been separated from the terminal oxidase. Two electrons from the dehydrogenation of NH_2OH, forming [NOH], are thought to pass to a membrane-associated terminal electron transport chain involving cytochrome a_2 (Erickson et al., 1972). Synthesis of ATP is coupled to that reaction and reduction of pyridine nucleotide is coupled to ATP hydrolysis (Aleem, 1965). Oxidation of [NOH] to NO_2^- takes place with the net addition of an atom of oxygen derived from O_2. Hooper (1978) has proposed that this step occurs by an internal mixed function oxygenase mechanism. Features of the electron transport components were reviewed by Suzuki (1974), Hooper (1978), and Nicholas (1978).

All enzymes involved in electron transport are thought to remain membrane-bound, even though hydroxylamine oxidoreductase is easily solubilized. Most of the nitrifiers are characterized by an elaborate membrane system which provides extensive sites for enzyme activity and is thought to compensate for the inherently low energy yield of nitrifiable substrates. The membrane system is comprised of folds and convolutions of the cytoplasmic membrane; similar cytomembranes occur in the photosynthetic bacteria and in the CH_4 oxidizing bacteria. Membranes are generally more extensive in the NO_2^- oxidizers of the genus *Nitrobacter,* than in the NH_4^+ oxidizers. Figure 5 is a thin section electron micrograph of an NH_4^+ oxidizing soil isolate showing peripheral membranes occurring as flattened lamellae.

N_2O evolved during NH_4^+ oxidation as depicted in Fig. 4 could arise from interactions of hydroxylamine oxidoreductase and nitrite reductase, influenced by conditions of aerobiosis (Nicholas, 1978). The former enzyme forms the presumed intermediate [NOH] or its dimer hyponitrite which may dismutate chemically under reduced O_2 tensions to N_2O. Nitrite reductase apparently is a dissimilatory enzyme which yields N_2O when O_2 becomes limiting and NO_2^- replaces O_2 as the electron acceptor. Data obtained by Ritchie and Nicholas (1972) with ^{15}N-labeled and unlabeled NH_2OH, and

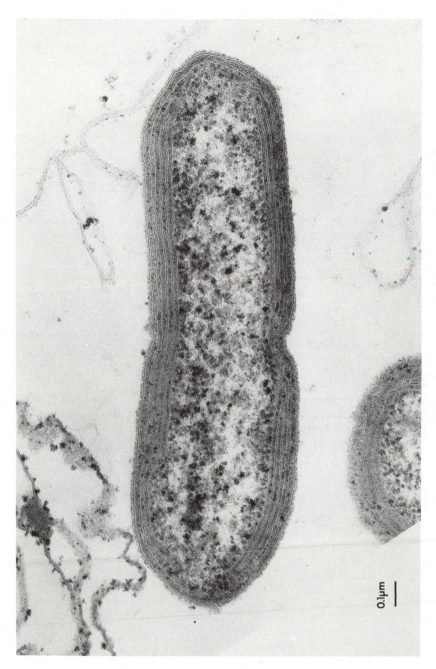

Fig. 5—Thin section electron micrograph of an NH_4^+ oxidizing soil isolate showing extensive cytomembrane system packed against cell wall. (Photograph provided by H. C. Tsien, University of Minnesota).

0.1μm

with ^{15}N-labeled and unlabeled NO_2^- further suggested that some N_2O may form from reduction of NO_2^- via [NOH]. In the light of recent interest in mechanisms leading to the generation of N_2O, more attention should be given in both biochemical and ecological studies to the generation of N_2O by NH_4^+ oxidizers other than *N. europaea,* and to the regulatory role of O_2 tensions.

B. Carbon Metabolism

Nitrifiers can synthesize all of their cell constituents from CO_2. Fixation of C is by way of the Calvin reductive pentose phosphate cycle (ribulose diphosphate cycle). The pathway requires ATP and reducing equivalents. The key reaction is the condensation of CO_2 with ribulose diphosphate (RUDP) to give two molecules of phosphoglyceric acid (C_3). These molecules undergo cyclic transformations to regenerate RUDP, with a gain of one C_3 molecule for every three molecules CO_2 fixed. The three CO_2 molecules which in effect have been converted into one phosphoglyceric acid are subsequently converted in that form to hexoses. Some phosphoglyceric acid also serves for synthesis of alanine and serine (Nicholas, 1978).

Nitrifying bacteria have an incomplete tricarboxylic cycle which further decreases their already limited metabolic versatility (Quayle & Ferenci, 1978). The absent enzyme is α-ketoglutarate dehydrogenase. A consequence of this lesion is that the tricarboxylic acid cycle cannot function as an energy generating mechanism, but operates instead in a purely biosynthetic role.

The sensitivity of NH_4^+ oxidizers to inhibition by trace amounts of numerous organic compounds is well known and was emphasized by Winogradsky (1949, p. 212–238). Growth of these autotrophs in nature in the presence of organic matter has generally been considered to reflect indifference to or tolerance of such organic matter except as it may provide substrate NH_4^+. Studies using labeled compounds have shown, however, that *Nitrosomonas,* at least, will incorporate certain organic compounds readily. In a study of growth stimulation of *N. europaea* by a heterotrophic contaminant, Clark and Schmidt (1966) were able to mimic the stimulation effect by addition of pyruvate to a pure culture of the nitrifier. Pyruvate-2-^{14}C was incorporated readily and became widely distributed in cell fractions. In a later study (Clark & Schmidt, 1967), each of 14 labeled amino acids presented in trace amounts was taken up by growing cells and metabolized. This was true even for those amino acids which were inhibitory at higher concentrations. It seems most probable that nitrifiers can and do utilize many organic metabolites when growing in natural habitats, and respond favorably or adversely depending on the kind and concentrations of the metabolites. In soil, such organic metabolites are likely to be present; B vitamins were found by Schmidt and Starkey (1951), and a wide range of free amino acids were reported by Paul and Schmidt (1961). Assimilation of inhibitory organic substances by NH_4^+ oxidizers may limit nitrification in some circumstances, but this has not been demonstrated. Nor has it been demonstrated that organic molecules may serve as even a partial energy supply for NH_4 oxidizers, since their assimilation is observed only when the organism is utilizing its specific energy source, NH_4^+.

C. Soil Genera

The points have been made that NH_4^+ oxidizers are difficult to isolate from soil, that isolations are attempted infrequently, and that new isolates are usually designated as *Nitrosomonas*. However, there is recent evidence that *Nitrosomonas* is not always the most abundant NH_4^+ oxidizer in soil. Walker (1978) reported the results of isolations that he made since 1960 from a world-wide collection of soils. Of 40 isolates, 17 were *Nitrosolobus,* 11 were *Nitrosospira,* 7 were *Nitrosomonas,* 1 was *Nitrosovibrio*, and 4 were unidentified. This author noted "growing evidence that *Nitrosolobus* and not *Nitrosomonas* are the dominant nitrifiers of agricultural soils." The first examination of diversity within the NH_4^+ oxidizing population of a single soil was carried out by Belser and Schmidt (1978a). They found three genera coexisting in the soil. *Nitrosomonas* and *Nitrosospira* occurred much more commonly than *Nitrosolobus*. This evidence does not support the view of McLaren and Ardakani (1972) that a single species excludes others from a generalized NH_4^+ oxidizing niche (principle of competitive exclusion). Instead, it appears that a rather diverse population of NH_4^+ oxidizers coexist in multiple niches which may vary individually with respect to numerous parameters.

Nitrosomonas isolates are identifiable in pure culture as short rods which exhibit straight-line motility and nitrify in strictly inorganic media. Their morphology is not sufficiently distinctive to permit recognition in mixed culture unless an appropriate fluorescent antibody (FA) is available (Belser & Schmidt, 1978a, b). Not all isolates of *Nitrosomonas* conform to the description of the one recognized species *N. europaea* (Belser & Schmidt, 1978b; Walker, 1978). *Nitrosospira* isolates do have a distinctive morphology due to their tightly would spiral nature (Watson, 1974), but the cells are small and thin ($0.6-0.8 \times 1.5-2.5$ μm), so that their recognition by phase microscopy demands considerable experience. In liquid culture *Nitrosospira* tends to spin and tumble in a circular path, and to grow more slowly than *Nitrosomonas* or *Nitrosolobus* (Belser & Schmidt, 1978a). *Nitrosospira* was found to be the dominant NH_4^+ oxidizer in a Minnesota soil, and to exhibit considerable serotypic diversity (Belser & Schmidt, 1978a, b). The genus *Nitrosolobus* is perhaps the only NH_4^+ oxidizer so distinctive in gross morphology as to be recognizable in enrichment culture by light microscopy. Cells are large ($1.0-1.5$ by $1.0-3.0$ μm) with an irregular lobular appearance and a tumbling motility. The slender nitrifying vibrio *Nitrosovibrio* has thus far been isolated from only two soils (Harms et al., 1976; Walker, 1978).

V. NITRITE OXIDIZING BACTERIA OF SOIL

The generic diversity of NO_2^- oxidizers in soil is more restricted than that of the NH_4^+ oxidizers, since only *Nitrobacter* is associated with soil habitats (Table 1). Two species of *Nitrobacter, N. winogradskyi* and *N. agilis,* were generally accepted as comprising the genus prior to 1974. In revising the *Nitrobacteraceae* for the 8th edition of *Bergey's Manual,* Watson (1974) consolidated these two morphologically similar forms into one species, *N. winogradkyi.*

A. Biochemistry of NO_2^- Oxidation

The oxidation of NO_2^- to NO_3^- by *Nitrobacter* is as follows (Nicholas, 1978):

$$NO_2^- + \tfrac{1}{2} O_2 \xrightarrow{2e} NO_3^-; \quad F, \ -17.8 \ \text{kcal.}$$

With only a 2e shift in oxidation state from $+3$ to $+5$, the reaction poses no obvious problem of intermediates. However, NO_2^- oxidizers are probably among the least favored of bacteria since their high energy requirement for biosynthesis is linked to oxidation of a low energy-yielding substrate. The reaction is carried out by a NO_2^- oxidase system with electrons carried to O_2 via cytochromes leading to the generation of ATP. Part of the ATP is consumed in a reversal of the $NADH_2$-cytochrome c reductase reaction that is conventional for heterotrophs, to account for the generation of reducing power (Aleem et al., 1963). Thus, the synthesis of NADH at the expense of ATP and reduced cytochrome further limits the energy efficiency of NO_2^- oxidizers, but it does solve the problem of providing a reductant with an E_0 value low enough to participate in the reduction of CO_2. All NO_2^- oxidizing and associated ATP synthesizing systems are located in the membrane, and a particularly extensive cytomembrane system is a feature of *Nitrobacter*; the cytomembrane arrangement probably provides the large number of NO_2^- oxidizing sites seemingly needed in order to compensate for the low yielding oxidations, by making possible high turnover rates.

Nitrobacter has an interesting array of assimilatory reductase enzymes. Nicholas (1978) reviewed earlier work in his laboratory which showed that *Nitrobacter* cells incorporated ^{15}N from NH_2OH, NO_2^-, NH_4^+, and NO_3^- into cellular N. Presumably the assimilatory NO_3^- reductase system is the main reductase system operative during normal growth, and accounts also for the inhibitory action of the chlorate ion; NO_3^- reductase reduces chlorate to the toxic chlorite ion.

B. Carbon Metabolism

Carbon dioxide is fixed in NO_2^- oxidizers by the Calvin reductive pentose cycle as in the NH_4^+ oxidizers. Shively et al. (1977) reported that the phage-like polyhedral bodies sometimes observed in electron micrographs of *Nitrobacter* are carboxysomes, which enclose particles of the key CO_2-fixing enzyme ribulose diphosphate carboxylase. The carboxysomes may aid in the survival of *Nitrobacter*, since particles isolated from cells starved of NO_2^- retain a high specific activity for the enzyme (Shively et al., 1977).

Nitrobacter differs substantially from the NH_4^+ oxidizers in that it may be a facultative rather than obligate autotroph. The distinction is based on demonstrations that certain strains can be induced to grow heterotrophically, using organic C and organic energy sources in the absence of NO_2^- (Van Gool & Laudelout, 1966; Smith & Hoare, 1968; Bock, 1976). Acetate, formate, pyruvate, yeast extract-peptone, and casein hydrolyzate, all in low concentrations, are assimilated in the absence of NO_2^-. The heterotrophic mode is clearly not preferred. A *Nitrobacter* strain examined by Bock (1976) required 6 months adaptation before heterotrophic growth commenced, and subsequent growth was extremely slow (generation time 70 hours or more as compared to about 14 hours for autotrophic growth). The long hetrotroph lag, slow heterotrophic growth, and the ability to interchange from autotrophic to heterotrophic growth and vice versa has been confirmed for Bock's strain, and the consistency of its serological properties has been established (P. M. Stanley & E. L. Schmidt, unpublished data). It seems unlikely that *Nitrobacter* may use the heterotrophic mode in soil for other than mere survival. The use of organic substances as auxiliary C and energy sources in the presence of NO_2^- (mixotrophic growth) sometimes stimulates *Nitrobacter* (Bock, 1978). This has interesting implications that should be examined for the natural environment.

The spectrum of NO_2^- oxidizers in soil appears to be limited. Fliermans et al. (1974) prepared fluorescent antibodies specific for the *N.* "agilis" and *N. winodgradskyi* strains and used them to examine 15 pure culture autotrophic NO_2^- oxidizing isolates obtained from soils. All isolates were clearly identifiable by one or the other of the FA preparations. However, the conclusion that the NO_2^- oxidizing population of soils is restricted to these two serotypes is not justified since subsequent isolations made by a technique that avoids enrichment yielded *Nitrobacter* strains belonging to at least five additional serotypes (P. M. Stanley & E. L. Schmidt, unpublished data).

VI. STUDY OF NITRIFYING POPULATIONS OF SOILS

Approaches that are available for the study of nitrifiers are limited by the complexities of the soil microhabitat and the physiology of the nitrifiers. Certainly, the inability to plate out the nitrifiers readily on selective media has repressed progress and interest in the microbiology of nitrification. The plating methods cited by Walker (1975) have not come into widespread usage, probably because of problems associated with slow growth, small colony size, and overgrowth by heterotrophs.

A. Isolation

Isolations are made more easily for the NO_2^- oxidizers than for the NH_4^- oxidizers but are difficult and tedious in any case. Relatively few research groups have invested the time and effort needed to obtain and maintain culture collections of the nitrifiers.

Most isolation attempts are preceded by enrichment, but this approach has distinct disadvantages in diversity studies since a single isolate is likely to achieve dominance during enrichment. This selectivity may or may not favor a significant member of the soil population, and the same procedure may result in the isolation of very similar strains from different soils as reflected in the data of Fliermans et al. (1974). Enrichment as a prelude to isolation is avoidable in the procedures used by Soriano and Walker (1968) in which soil dilutions are plated and individual colonies are subsequently picked for enrichment in liquid culture in attempts to overgrow contaminants. Belser and Schmidt (1978a) also avoided initial enrichment by transferring directly from positive MPN tubes into a dilution isolation series in liquid media. This approach provides the possibility of obtaining isolates from different numerically abundant segments of a given nitrifying population.

B. Most Probable Number (MPN) Enumeration

The usual approach to the enumeration of nitrifiers is indirect, involving some modification of the most probable number (MPN) technique for statistical estimation of nitrifiers present in an inoculum. MPN procedures are time consuming (Matulewich et al., 1975), inherently lacking in precision (Cochran, 1950; Taylor, 1962), and may be selective as a function of medium composition (Belser & Schmidt, 1978b).

The MPN protocol most commonly used for soil is that of Alexander and Clark (1965). An incubation time of 3 weeks at 28°C is required, but a longer incubation apparently is advisable. Matulewich et al. (1975) observed that maximum counts for their MPN medium were not obtained until 20–55 days for NH_4^+ oxidizers, and required more than 100 days for NO_2^- oxidizers. Belser and Schmidt (1978a) compared MPN enumeration with FA enumeration using pure, exponentially growing cultures of three genera of NH_4^+ oxidizers and three MPN media: those of Alexander and Clark (1965) (AC); Soriano and Walker (1968) (SW); and Matulewich et al. (1975) (MSF). The SW medium required the shortest incubation period with an average time of 3.1 weeks, and an average maximum count that was 94% of the FA count. Medium MSF required a slightly longer period (avg 4.2 weeks) and was about as efficient as SW. Medium AC required an average time of 6.1 weeks to obtain a maximum count that was only 60% of the FA count. At the recommended 3 week incubation time, the AC medium gave an average count of only 3.7% of the FA enumeration. The data also showed high variability between duplicate samples on some occasions for all

MPN media, and that a shorter incubation period was required for the *Nitrosomonas* and *Nitrosolobus* strains used, than for the *Nitrosospira*.

The composition of the MPN medium affects not only the counting efficiency and incubation time, but has other effects as well. In the experiments just cited for three genera of NH_4^+ oxidizers and three MPN media, Belser and Schmidt (1978a) found selectivity with respect to all strains evident in each medium.

C. Fluorescent Antibody (FA) Techniques

Fluorescent antibody (FA) or immunofluorescence technique provides a potential capability that is inherent in no other approach: that of studying the population dynamics of a specific microorganism directly in a natural environment in relation to its biochemical activity. The potentials, limitations, and precautions associated with the method were described by Schmidt (1973). The development of FA procedures and their application to the detection of strains of nitrifiers in soil are summarized by Fliermans et al. (1974), Schmidt (1974, 1978), Fliermans and Schmidt (1975), Rennie and Schmidt (1977a, b), and Belser and Schmidt (1978b, c).

Examination of a nitrifying population by FA has given some indication of diversity within that population. Little data are available thus far on NO_2^- oxidizer diversity at the serotype level, but it is clear that multiple serotypes exist among the NH_4^+ oxidizers. Belser and Schmidt (1978b) focused on the nitrifying population of a single soil to examine genus and serotype diversity within the population. The pure culture isolates obtained from this soil included the genera *Nitrosomonas, Nitrosospira,* and *Nitrosolobus.* A suite of 15 FAs was available for serotyping these isolates. Of the 7 *Nitrosomonas* isolates, only 4 could be stained with the 7 *Nitrosomonas* strain-specific FAs available; 6 or 12 *Nitrosospira* strains were reactive to 4 of the 5 *Nitrosospira* FAs available; and the 1 *Nitrosolobus* isolate cross-reacted with 1 of the 3 *Nitrosolobus* FAs on hand. These data illustrate a major problem with respect to the practical application of FA: the antigenic diversity among NH_4^+ oxidizers appears to be so extensive as to require a substantial number of FAs for a given population. The isolation of pure cultures needed for FA preparation is time consuming, and so is the FA preparation, therefore a great deal of effort may be necessary to establish an inventory of all nitrifiers involved in a given environment. A reasonable objective may be to determine the genera and strains that predominate in numbers and activity and to develop the FA reagents capable of their detection and enumeration.

D. Short Term Nitrification Activity

Nitrification rates provide an indirect chemical approach to the activity of the total nitrifying population of a soil. Such information is readily obtained, and may be useful for characterizing soils with respect to N transformations and fertilizer response. Nitrification rates obviously depend on the nitrifying population, but the relationships between rates and nitrifier densities remain to be established. The types of measurements referred to in Section II D to assess a "nitrifying potential" or "N mineralization potential" are examples of long term activity measurements involving incubation of the soil for several weeks to allow equilibration of N transformations in the soil or to note response to an added nitrifiable substrate.

Short term nitrification activity measurements are made over a period of a few hours to about 1 day, and are meant to estimate the nitrifying potential of the indigenous population of a soil at the time of sampling. A known amount of soil is introduced into a solution of buffer and NH_4^+ substrate and the slurry is shaken throughout the assays. Periodic analyses of $NO_2^- + NO_3^-$ yield data for rate determinations under conditions where substrate and oxygen are not limiting. The rate determined is not necessarily that of the soil where O_2, for example, may be limiting. Since the assay is short, population increase is not a complication, and the rate should be a reflection of the enzymic potentials of the standing crop of nitrifiers in the soil at the start of the assay. Substrate conversion rates can be calculated on a per cell basis from pure culture data (Table 3, for example). Hence, if it is assumed that the same per cell activity takes place in the soil, the size of the nitrifying population can be calculated from the short term activity rate measurements.

Knowles et al. (1965) used the activity approach to study nitrification in an aquatic system. They determined oxidation rates for the nitrifying population of water samples. By relating these data to activity measurements made in pure culture where biomass could be determined, they were able to estimate the nitrifier biomass of the water samples. Little use has been made of short term activity measurements for soils, but Sarathchandra (1978a) recently studied nitrification in New Zealand soils in this manner. He also estimated populations by MPN, but observed no relationship between the activity measurements and the MPN counts. This may be another reflection of the statistical uncertainty of MPN enumerations.

VII. REGULATION OF NITRIFYING POPULATIONS IN SOILS

Various factors have an obvious impact on the nitrifying population as a consequence of the nature of the process and the biochemistry of the nitrifiers. Some factors such as NH_4^+, NO_2^-, and O_2 directly involved as reactants or products, or pH and temperature as they affect concentrations and reaction rates have been mentioned here briefly, and discussed in other reviews (Focht & Verstraete, 1977; Belser, 1979). The way in which these and other factors operate to regulate nitrifying populations in soils is not yet known sufficiently well to allow directed manipulation of the nitrification process, but this is a worthwhile objective.

A. Naturally Occurring Inhibitors

Inhibition of growth by organic compounds of various kinds is well documented for nitrifiers in pure or enrichment culture (Gundersen, 1955; Delwiche & Finstein, 1965; Clark & Schmidt, 1967; Hooper & Terry, 1973). Many toxic compounds such as certain amino acids and N bases are released during decomposition of nitrogenous organic matter, but nitrification proceeds vigorously even in sewage and manure wastes so long as major environmental factors are favorable. Evidence that nitrification may be affected by certain decomposition products of organic residues or by metabolites excreted by plants or microorganisms in soil is still only suggestive. Most of the research on the matter has focused on grassland soils, and the reports are highly contradictory.

Grassland soils often appear to have consistently low to normal levels of NH_4^+, but very little NO_3^-. Theron (1951) obtained evidence for the hypothesis that limited occurrence of NO_3^- was due not to immobilization, but to inhibition of the nitrifiers due to substances excreted by grass roots. Clark and Paul (1970) reviewed the various reports both for and against this hypothesis. Much of the evidence for inhibition has derived from the addition of root extracts to nitrifying cultures (Rice, 1964; Munro, 1966; Neal, 1969), but extrapolation of in vitro results to rhizosphere conditions can be misleading. Rice (1964), for example, placed discs saturated with root extract in contact with nitrifiers, or added the extract to liquid medium at the rate of 5%. The presence of inhibitory substances under such conditions is probable, but it is unlikely that similar concentrations of inhibitory substances occur in the rhizosphere. A direct illustration is found in plant exudates prepared by Molina and Rovira (1964). The exudates inhibited *Nitrosomonas* in liquid culture, but MPN numbers of both NH_4^+ oxidizers and NO_2^- oxidizers were increased in the rhizosphere soil of the same plants. Exudates at the root surface in soil are dilute and subject to degradation by rhizosphere bacteria and to adsorption by soil. The exudate effect following concentration and in vitro assay is certainly attributable to a plant effect since not all extracts are inhibitory (Neal, 1969). This is still another of the numerous "potential" effects that so abound in nitrification research.

Effects other than root inhibition may account for an observed absence of NO_3^-. Brar and Giddens (1968) were unable to demonstrate that nitrification was affected by leaching (to remove inhibitors). Increasing soil pH by liming, addition of nitrifiable substrate, or relieving moisture stress has enhanced nitrification where inhibition had been observed (Robinson, 1963; Chase et al., 1968; Neal, 1969). Some of the strongest evidence for specific inhibition by plant exudates was provided by the somewhat different approach of Moore and Waid (1971). They added fresh root washings to a soil that was adjusted to a constant nitrification rate. The most pronounced depression of nitrification was obtained with ryegrass root extracts, an effect that could be repeated by addition of a second increment of extract to the same soil after recovery from the first addition. The authors were conservative in extrapolating their data back to the rhizosphere, but the attractiveness of their approach suggests that further work is desirable.

Rice and Pancholy (1972, 1973) gave evidence in support of the hypothesis that many soils under climax vegetation are low in NO_3^- due to an inhibition of nitrification by climax plant species. They measured NO_3^-, NH_4^+, and nitrifier numbers (MPN) every 2 months on soils taken from three different vegetation types. In each type, soils were taken from two old-field stages of succession and the climax. Nitrate was low in all cases and generally lower under climax than under earlier successional stages. Counts of nitrifiers were commonly at 10^2–10^3/g of old field and 10^1–10^2/g of climax soil.

The data of Rice and Pancholy (1972, 1973) are consistent with the climax inhibition hypothesis, but are by no means conclusive because:

1) Estimates of nitrifier populations by the MPN method suffer from inherently high statistical uncertainty (Section VB).
2) Nitrifier populations of old field soils may differ qualitatively from those of climax soils, with the consequence, due to the selectivity of a single MPN medium (Belser & Schmidt, 1978a), that a predominant nitrifier of the climax soil might not be enumerated.
3) The proposed mechanism of inhibition as involving tannins was not substantiated in the work of Bohlool et al. (1977) who found that pure cultures of nitrifiers were unaffected by concentrated pine bark tannins even at 5 mg/ml.
4) The lower NO_3^- levels in soils under climax as compared to old field vegetation may merely reflect more efficient NO_3^- assimilation by climax plants, or greater vulnerability to losses of NO_3^- through leaching of soils under climax vegetation. Vitousek and Reiners (1975) presented data in support of the latter possibility.

B. Inhibition by Pesticides Added to Soil

Nitrifiers are considered to be among the most sensitive of the physiological groups of bacteria to exotic chemicals that reach the soil. Insecticides, fungicides, fumigants, and herbicides added to soil as part of modern agricultural practice may have undesirable effects on nontarget organisms. The nitrification process is frequently used to assess such effects. Measurements on the effects of chemicals on nitrification have been made in many ways, but those that measure pesticide effects in the soil, as opposed to other media, appear to be most relevant (Atlas et al., 1978).

Effects of pesticides on the transformations of N in soil are considered in Chapter 8 (M. K. Firestone), and only some of the references dealing with nitrifiers and the nitrification process will be cited here. Powlson and Jenkinson (1971), Bremner and Bundy (1974), Ridge (1976), and Rovira (1976) reported data which identified the fumigants carbon disulfide, methyl bromide, and chloropicrin as potent inhibitors of nitrification. Fungicides may also inhibit nitrification, but not at the usual rates of soil application (Wainwright & Pugh, 1973; Ross, 1974; Wainwright, 1977; Ramakrishna et al., 1978).

By far the greatest attention to pesticide-nitrification interactions has been with respect to herbicides. Many such compounds are in widespread use, and monitoring for soil biological effects is an important and continuing task. A recent study by Atlas et al. (1978) discussed the general problems and approaches to monitoring the soil impacts of herbicides, while that of Ratnayake and Audus (1978) illustrated in considerable detail the approaches for evaluating effects specifically on nitrification and nitrifying bacteria. Domsch and Paul (1974) presented their own data on herbicide effects on nitrification, and an excellent analysis of other pertinent results for the period 1960–1973. Virtually all studies concur in finding that applications at field rates of the large majority of herbicides are unlikely to affect nitrification. The urea, carbamate, thiolcarbamate, and aminotriazole herbicides appear to be the groups most likely to influence nitrifier activity in soil.

C. Specific Inhibitors of Nitrification

Direct regulation of nitrification in soil with the nitrifiers as the target organisms became feasible with the introduction of nitrapyrin patented as a specific inhibitor of the NH_4^+ oxidation stage of nitrification (Goring, 1962a, b). Nitrapyrin is distributed by the Dow Chemical Co., Midland, Mich., USA, under the trade name N-Serve. Campbell and Aleem (1965a, b) confirmed the sensitivity of *N. europaea* to inhibition by nitrapyrin at concentrations of 0.2 $\mu g/ml$, and the lack of sensitivty of *Nitrobacter* at concentrations as high as 50 $\mu g/ml$. Relatively little is known about the response of other NH_4^+ oxidizing genera and strains to nitrapyrin or other specific inhibitors.

Inhibition of nitrification as an agricultural practice has as its main objectives the conservation of fertilizer N and the more efficient use of N by the crop. Ideally, with the conversion of NH_3, NH_4^+, and urea fertilizer N to NO_3^- inhibited, losses associated with denitrification and leaching can be mitigated, and economic and environmental benefits will accrue. Numerous chemicals have been used in attempts to achieve these objectives and progress has been summarized in several reviews (Gasser, 1970; Prasad et al., 1971; Hauck, 1972; Huber et al., 1977). The most comprehensive comparative testing of nitrification inhibitors was reported by Bundy and Bremner (1973), who used 24 compounds applied to three soils at the rate of 10 $\mu g/g$ to observe effects on the nitrification of 200 $\mu g/g$ NH_4^+-N. Ten of the compounds had essentially no effect on NO_3^- generation; the others were ranked as follows on the basis of decreasing average effectiveness: 2-chloro-6-(trichloromethyl) pyridine (nitrapyrin), 4-amino-1,2,4-triazole (ATC), sodium or potassium azide, 2,4-diamino-6-trichloromethyl-*s*-triazine (CL-1580), dicyandiamide, 3-chloroacetanilide, 1-amidino-2-thiourea, 2,5-dichloroaniline, phylmercuric acetate, 3-mercapto-1,2,4-triazole or 2-amino-4-chloro-6-methylpyridine (AM), sulfathiazole (ST), sodium diethyldithiocarbamate. The first six gave more than 50% effectiveness as averaged for the three soils.

The effectiveness of nitrification inhibitors is influenced by a number of soil factors. Bundy and Bremner (1973) noted that soil texture was an important factor with effectiveness being greater on light-textured soils. Increasing pH and organic matter content of soils required higher concentrations of nitrapyrin (Goring, 1962a). All compounds studied by Bundy and Bremner (1973) were more effective at 15°C than at 30°C. Nitrapyrin was less effective at 15°C than ATC, whereas the reverse was true at 30°C. Other significant factors are those that affect the degradation of the inhibitor. Nitrapyrin has a short half-life in soil of about 4–22 days so that the duration of nitrification inhibition is a function of application rate, although the delay rate is not linear with concentration (Hughes & Welch, 1970). Degradation probably is both biological and chemical for most inhibitors and both would be temperature dependent. Bremner et al. (1978) demonstrated that aqueous solutions of nitrapyrin undergo rather rapid chemical hydrolysis as a function of temperature. Storage at 30°C resulted in 40% and 90% hydrolysis within 3 and 14 days, respectively. The hydrolytic product of nitrapyrin, 6-chlorodipicolinic acid, was ineffective as a nitrification inhibitor at the rate of 20 ppm.

Some special problems associated with the application of urea as a nitrogen fertilizer may be alleviated by use of nitrification inhibitors. Urea is hydrolyzed rapidly to ammonium carbonate in soil with resultant accumulation of NH_4^+ and an increase in pH; consequences in the form of possible NO_2^- build-up and toxicity, NH_3 loss, and inhibition of NO_2^- oxidation may follow. Bundy and Bremner (1974) found that nitrapyrin, ATC, and CL-1580 were effective in retarding oxidation of NH_4^+ derived from urea. Two soils that accumulated more than 160 ppm NO_2^--N after urea application with no inhibitors yielded little or no NO_2^- when inhibitors were added with urea. The results could be significant in view of the increasing importance of urea as a fertilizer in world agriculture.

Specific inhibitors of nitrification have been useful and are of considerable potential for experimental study of the nitrification process and the ecology and biochemistry of the nitrifying bacteria. Of the many inhibitors that specifically affect NH_4^+ oxidizers (Hooper & Terry, 1973) nitrapyrin has been used most frequently for soil-related research. Bremner et al. (1978) drew attention to certain precautions in the use of aqueous nitrapyrin solutions: storage at low temperature to minimize abiological hydrolysis, and the use of glass bottles with glass rather than rubber (which adsorbs and later releases nitrapyrin), silicone, or polyethylene stoppers. Experimenters concerned with the selective inhibition of the NO_2^- oxidizers are limited to relatively few reagents. The Na^+ or K^+ salt of chlorate was shown by Lees and Simpson (1957) to be a sensitive and specific inhibitor of *Nitrobacter*. This inhibitor appears to have considerable potential for use in nitrification activity determinations (Section VD) so that NO_2^- accumulation can serve as a more sensitive and convenient measure of nitrification rate than NO_3^- accumulation.

VIII. GROWTH OF NITRIFYING BACTERIA IN SOIL

Generalizations as to the densities of nitrifiers in soils are of little value since limiting factors vary strikingly and distinctively with different soils, and since enumeration of nitrifiers is beset with so many difficulties. Nitrifier numbers may be expected to vary from a few hundred to 10^5 per gram in natural (NH_4^+ limited) soils, and from a few thousand to 10^6 or 10^7 per gram in N-amended soils. The number of NO_2^- oxidizers reported for a given soil is commonly lower than for NH_4^+ oxidizers, but there are many exceptions to this as noted by Belser (1977). Difficulties associated with the reliability of population estimates by MPN methods suggest that the most informative of such data are likely to derive from studies comparing the effects of various factors on the nitrifying population of a single soil. In addition, the time-consuming and cumbersome nature of MPN determinations discourage frequent samplings, so that knowledge of the dynamics of nitrifying populations is fragmentary.

A. Growth Rates

Growth of nitrifiers is slow even in pure cultures where nonlimiting and balanced growth conditions should pertain. Generation times of NH_4^+ oxidizers are known mostly in terms of *Nitrosomonas europaea* and range from about 7 to 14 hours (Bock, 1978, Table 2). The genus appears to include the fastest growing forms of NH_4^+ oxidizers. Among other genera generation times of 24 hours for *Nitrosospira briensis* (Watson, 1971), 18 hours for *Nitrosolobus multiformis* (Watson et al., 1971), and 24 hours for *Nitrosovibrio tenuis* (Harms et al., 1976) have been reported. The slow growth of *Nitrosospira* is evident also in Table 2, and in the observation that this soil isolate and another soil strain of *Nitrosospira* took significantly longer to develop maximum MPN counts than strains of other genera (Belser & Schmidt, 1978a). Pure cultures of *Nitrobacter* have generation times of about 8–16 hours (Belser, 1977; Rennie & Schmidt, 1977b; Bock, 1978).

Data on growth rates of nitrifiers in soils are extremely limited. Most attempts that have been made to measure nitrifier numbers as a function of substrate oxidation have been restricted to NO_2^- oxidizers; these are summarized in Table 3. The study of Morrill and Dawson (1962) included MPN counts of NH_4^+ oxidizers in 6 soils during nitrification in soil percolators. Generation times for these populations ranged from 34 hours in a pH 7.6 soil to 104 hours in a pH 6.2 soil. Growth rates were also reported for NO_2^- oxidizing populations during perfusion of the same 6 soils (Table 3). Growth in the pH 6.2 soil was with a generation time of about 55 hours; in the other 5 soils generation times of the NO_2^- oxidizers, unlike those of the NH_4^+ oxidizers, were close to pure culture values (19–24 hours) and independent of pH. A similar value of 26.7 hours (Table 3) was obtained by

Table 2—Growth responses of nitrifying bacteria in pure culture.

	Generation time	Yield	Activity	Counting method[†]	Reference
	hours	cells/μmol $\times 10^6$	pmol/cell per hour		
NH$_4^+$ oxidizers					
Nitrosomonas europaea	12.7[‡]	2.7[‡]	0.020	MPN	Engel & Alexander (1958)
N. europaea	12.0 ± 1.7	5.0 ± 0.3	0.012	FA	Belser & Schmidt (1980)
Nitrosomonas "FH1"	12.8 ± 3.0	2.4 ± 0.2	0.023	FA	Belser & Schmidt (1980)
Nitrosospira "AV2"	20.9 ± 1.5	8.0 ± 0.5	0.004	FA	Belser & Schmidt (1980)
Nitrosolobus "AV3"	16.0 ± 1.0	1.9 ± 0.3	0.023	FA	Belser & Schmidt (1980)
NO$_2^-$ oxidizers					
Nitrobacter sp.	14.4	4.1[‡]	0.011[†]	P-H	Chiang (1969)[¶]
Nitrobacter "Engel"	8.3	4.6	0.018	MPN	Belser (1977)
Nitrobacter "agilis"	16.9	0.96	0.042	MPN	Belser (1977)
N. winogradskyi	15.1[§]	5.1[§]	0.012[§]	FA	Rennie & Schmidt (1977b)
Nitrobacter "agilis"	13.8	5.3[§]	0.009[§]	FA	Rennie & Schmidt (1977b)

† MPN—most probably number, FA—fluorescent antibody; P-H—Petroff-Hausser.
‡ Calculated by Belser (unpublished PhD thesis, University of California Berkeley, 1974) from reference cited.
§ Calculated by Belser and Schmidt (1978c) from reference cited.
¶ Chiang, C. N., unpublished D.Sc. thesis, Universite Catholique de Louvain, Louvain, Belgium, 1969.

Belser (1977) in another study combining MPN counts and soil percolator conditions.

Microhabitat circumstances in the soil percolator are likely to be more artificial than those of the statically incubated soil sample with respect to water relationships and elution of nitrification products. The only growth studies carried out in statically incubated soils were those of Schmidt (1974) and Rennie and Schmidt (1977b). In both studies, enumeration was accomplished by FA techniques. The 1974 study involved a partially sterilized soil reinoculated with a N. winogradskyi strain originally isolated from the same soil; the growth response of the strain was followed by quantitative FA. A generation time of 32 hours was observed for the NO$_2^-$ oxidizer. The 1977 study was with unsterilized soils and gave very long generation times (Table 3). Growth in soil would appear from these data to be surprisingly slow as compared to pure culture conditions, but if such is the case, the reasons for

Table 3—Growth responses of NO_2^- oxidizing populations as calculated during log phase growth in soil in various experimental systems.

Soil	Experimental approach[†]	Generation time	Yield	Activity	Counting method[†]	Reference
		hours	cells/μmol $\times 10^6$	pmol/cell per hour		
Carrot River	a	129[‡]	0.051[§]	0.16[‡]	FA	Rennie & Schmidt (1977b)
Hubbard	a	74[‡]	0.056[‡]	0.18[‡]	FA	Rennie & Schmidt (1977b)
Hanford	b	27	0.137	0.24	MPN	Belser (1977)
Hanford	c	22	0.53	0.07	MPN	Belser (1977)
Bearden/Nb.	d	32	4.0[§]	0.005[§]	FA	Schmidt (1974)
6 Soils	b	20–55	0.031[§]		MPN	Morrill & Dawson (1962, 1967)

† (a) field soil, static incubation; (b) soil percolator; (c) shaken soil suspension; (d) partially sterilized soil reinoculated with *N. winogradskyi*, static incubation.
‡ Calculated by Belser & Schmidt (1978c) from literatured cited.
§ Calculated by Belser (unpublished Ph.D. thesis, University of California, Berkeley, 1974) from literatured cited.

it are unclear. Growth rates of specific nonnitrifying microorganisms in soil are no better known, so that comparisons of NO_2^- oxidizers to other bacteria as to relationships between growth in nature and growth in pure culture, cannot be made. The data obtained thus far by FA techniques have been sufficient only to demonstrate that the method has considerable potential for the study of nitrifying populations directly in soil (Rennie & Schmidt, 1977b; Belser & Schmidt, 1978c). Improvements in both the FA and MPN techniques and more intensive use of both in combination should provide badly needed information on the population ecology of nitrifiers in soil.

B. Yields

Cellular growth yield is easily calculated for pure culture conditions on the basis of the number of cells produced per μmole of NH_4^+ or NO_2^- oxidized, but the enumeration method is important. The growth constants reported in Table 3 for NH_4^+ oxidizers were taken from a study (Belser and Schmidt, 1980) in which counts were obtained by three methods: FA, Petroff-Hausser (P-H) and MPN. Both the P-H counts, which can only be useful in pure culture, and the FA counts were more precise by about a factor of 5 than the MPN counts. For the FA technique the average standard deviation was 8.6% of the count, while for the MPN enumerations it was 42.2% of the count.

The yield data of Table 2 reflect substantial differences among the NH_4^+ oxidizers and emphasize the importance of taking species diversity into account in investigations of nitrifying populations and nitrifying activity. *Nitrosospira,* despite its slow growth, produced the most cells per μmol of NH_4^+ oxidized, but had the lowest activity. The large number of cells formed is compensated by the small size of *Nitrosospira* cells, hence it is possible that yield might be measured better by biomass than by cell number. The largest of the genera, *Nitrosolobus,* formed the least number of cells per unit substrate. These relationships are consistent since *Nitrosomonas* isolate FH 1 was distinctly larger and had a lower cellular growth yield than *N. europaea.*

Yield values reported in Table 2 are fairly consistent for the NO_2^- oxidizers in pure culture. The very low efficiency reported by Belser (1977) for a *Nitrobacter* "agilis" soil isolate was discussed by that author in some detail. Underestimation of cell numbers by MPN, larger unit size of the strain, and slow growth with consequently greater maintenance energy expenditures may all have contributed to the low yield.

The striking feature of the growth yields as compiled for populations of NO_2^- oxidizers in soil (Table 3) is their very low values as compared to pure culture yields, being about 100-fold less under static incubation or soil perfusion conditions, as enumerated by either MPN or FA. It appears that the production of biomass is much less efficient in soil than in pure culture even when nutrients are apparently nonlimiting. A number of factors may have a bearing on the poor yields observed in soil. Enumeration methods are significant as they tend to underestimate the soil population. The statistical uncertainty and selectivity of the MPN approach has been mentioned; to this must be added the likelihood of underestimation due to clumping of cells in association with soil particles during the dilution steps. The FA procedures also tend to underestimate by virtue of their serologic specificity, and as a consequence of inefficient release of cells from soil. Increases in cell size in soil as compared to pure culture also could tend to give lower yield values for the soil situation; little is known of this since so little direct microscopy has been done, but the possibility merits examination. Finally the slow growth rate observed for NO_2^- oxidizers in soil suggests that yields would also be decreased as a result of increased maintenance energy per generation.

The various factors considered above make it appear likely that the growth efficiency in soil may not be quite as low as the few published data indicate. Nevertheless it appears likely that cell yields in soil are significantly less than in pure culture even when O_2, NO_2^-, and pH seem unlimiting. If so, identification of the limiting factors responsible for slow growth and poor yield in soil represents one of the most important problems to be resolved in the ecology of the nitrifiers.

C. Activity

Theoretical considerations underlying the use of k, the oxidation rate per cell, as a growth parameter have been outlined by Belser (1977, 1979). This oxidation rate is not a constant, and has been observed to obey Michaelis-Menten kinetics for *N. europaea* in pure culture (Suzuki et al., 1974; Laudelout et al., 1976). These investigators measured the Michaelis-Menten constant, K_m, and found it to be pH dependent and to range between 1.0 and 2.2 mM in the pH range of 7.5 to 8.0. The pure culture growth studies for the NH_4^+ oxidizers listed in Table 2 were carried out at concentrations well above the K_m for *N. europaea,* and if the same K_m holds roughly for the other genera, the values for k should approximate the maximum oxidation rate per cell.

Nitrosospira, the smallest and slowest growing of the NH_4^+ oxidizers listed in Table 2, also had the lowest activity ($k = 0.004$ pmol/hour per cell) a value which is only 18–45% of the values calculated for the others. If *Nitrosospira* strains were to contribute significantly to nitrification in soil, and approximately the same k were to apply, their population would have to be high. Belser and Schmidt (1978a) found *Nitrosospira* to be the predominant genus in field soil fertilized with NH_4NO_3.

Most of the oxidation rates per cell for pure cultures of the *Nitrobacter* strains of Table 2 are in general agreement. The Hanford soil isolate (*N.* "agilis") of Belser (1977) had an activity more than twice that of the "Engel" strain. While this was attributed to the greater biomass of the Hanford isolate it is also possible that cell numbers were underestimated by the MPN counting procedure used.

The interesting feature of the k values estimated for NO_2^- oxidizing populations in soil (Table 3) is that they are higher by a factor of about 10 than pure culture activities. This is somewhat surprising because the one parameter that would seem most likely to be about the same in soil and pure culture is activity per cell. If k depends on purely enzymatic mechanisms, and if neither O_2 or NO_2^- are limiting, k should be at a maximum in either pure culture or soil. The discrepancy observed in the form of much higher rates of oxidation per cell in the soil has been attributed to underestimation of the NO_2^- oxidizing population (Belser, 1977; Belser & Schmidt, 1978c). While this is a most likely explanation, other factors may pertain. A similar effect would result if cells in soil were larger than in pure culture, or if natural inhibitors present in the soil acted to uncouple substrate oxidation from growth, or if the population were systematically cropped by predators. The data on which k in soil and k in culture are based are still very limited, but thus far at least they pose an intriguing question that could be central to the relationships between the nitrification process in soil and the population dynamics of the microorganisms responsible for the process. If the activity of nitrifiers in soil is maximal, then why do they grow so slowly and produce cells so inefficiently?

IX. CONCLUDING COMMENTS

Agronomic concerns have prompted most of the work on nitrification in soils. The critical role attributed to nitrification in soil fertility centers around the availability of N to the plant. Assimilation of N by plants in unfertilized soil is largely in the NO_3^- form, but soil and fertilizer N converted to this form by nitrification are subject to denitrification and leaching losses. With the agronomic concern to deliver available N to the plant now heightened by demands for increased world food production and decreased energy expenditures, the need for much more detailed knowledge of nitrification is readily apparent.

Knowledge of nitrification is inadequate also to deal with concerns raised by impacts, or possible impacts, of the soil process on the environment. Excess NO_3^- leached out of surface soil into streams and ground water constitutes a threat to water quality (Keeney, 1971) and occasionally to public health (Walton, 1951). Still another area of concern directly linked to nitrification has emerged with the clear demonstration that N_2O can be evolved in the course of nitrification of fertilizer NH_4^+ in well-aerated soil (Bremner & Blackmer, 1978). This points to yet another source of a gas that is considered a threat to the ozone layer of the stratosphere (Crutzen, 1970). There is an obvious need to assess the contribution of nitrification to N_2O evolution from soils and to examine the nature and significance of this unexplored aspect of soil nitrification.

The microbiology of the nitrification process has lagged far behind its chemistry. The reasons for this are based in the difficulties associated with the peculiar physiological features of nitrifying microorganisms, and with the complexities of the natural environments in which they grow. If agronomic and environmental concerns are to be addressed effectively, nitrification research must place much greater emphasis on the microorganisms responsible for the nitrification process. Such research should aim to identify and characterize the major component microorganisms of a given nitrifying population, to estimate the abundance and activity of each major component, and to assess the impact of soil factors on the dynamics of the nitrifying population.

ACKNOWLEDGMENT

Contribution from the Department of Soil Science and the Agricultural Experiment Station, University of Minnesota, St. Paul, MN 55108. The unpublished studies cited here were supported by National Science Foundation grant DEB76-19518.

LITERATURE CITED

Aleem, M. I. H. 1965. Generation of reducing power in chemosynthesis. II. Energy-linked reduction of pyridine nucleotides in the chemoautotroph, *Nitrosomonas europaea,* Biochim. Biophys. Acta 113:216–224.

Aleem, M. I. H., H. Lees, and D. J. D. Nicholas. 1963. Adenosine triphosphate-dependent reduction of nicotinamide adenine dinucleotide by ferro-cytochrome C in chemoautotrophic bacteria. Nature (London) 200:759–761.

Alexander, M., and F. E. Clark. 1965. Nitrifying bacteria. p. 1477–1486. *In* C. A. Black et al. (ed.) Methods of soil analysis. Am. Soc. Agron., Madison, Wis.

Anderson, O. E., and F. C. Boswell. 1964. The influence of low temperature and various concentrations of ammonium nitrate on nitrification in acid soils. Soil Sci. Soc. Am. Proc. 28:525–532.

Atlas, R. M., D. Pramer, and R. Bartha. 1978. Assessment of pesticide effects on non-target soil microorganisms. Soil Biol. Biochem. 10:231–239.

Barritt, N. W. 1933. The nitrification process in soils and biological filters. Ann. Appl. Biol. 20:165-184.

Beek, J., and M. J. Frissel. 1973. Simulation of nitrogen behaviour in soils. Centre for Agricultural Publication and Documentation, Wageningen, The Netherlands.

Belser, L. W. 1977. Nitrate reduction to nitrite, a possible source of nitrite for growth of nitrite-oxidizing bacteria. Appl. Environ. Microbiol. 34:403-410.

Belser, L. W. 1979. Population ecology of nitrifying bacteria. Ann. Res. Microbiol. 33:309-333.

Belser, L. W., and E. L. Schmidt. 1978a. Diversity in the ammonia oxidizing nitrifier population of a soil. Appl. Environ. Microbiol. 36:584-588.

Belser, L. W., and E. L. Schmidt. 1978b. Serological diversity within a terrestrial ammonia-oxidizing population. Appl. Environ. Microbiol. 36:589-593.

Belser, L. W., and E. L. Schmidt. 1978c. Nitrification in soils. p. 348-351. In D. Schlessinger (ed.) Microbiology 1978. Am. Soc. Microbiol. Washington, D.C.

Belser, L. W., and E. L. Schmidt. 1980. Growth and oxidation kinetics of three genera of ammonia oxidizers. FEMS Microbiol. Lett. 7:213-216.

Bock, E. 1976. Growth of Nitrobacter in the presence of organic matter. II. Chemoorganotrophic growth of Nitrobacter agilis. Arch. Microbiol. 108:305-312.

Bock, E. 1978. Lithoautotrophic and chemoautotrophic growth of nitrifying bacteria. p. 310-314. In D. Schlessinger (ed.) Microbiology 1978. Am. Soc. Microbiol. Washington, D.C.

Bohlool, B. B., E. L. Schmidt, and C. Beasley. 1977. Nitrification in the intertidal zone: influence of effluent type and effect of tannin on nitrifiers. Appl. Environ. Microbiol. 34:523-528.

Brar, S. S., and J. Giddens. 1968. Inhibition of nitrification in Bladen grassland soil. Soil Sci. Soc. Am. Proc. 32:821-823.

Bremner, J. M., and L. G. Bundy. 1974. Inhibition of nitrification in soils by volatile sulfur compounds. Soil Biol. Biochem. 6:161-165.

Bremner, J. M., and A. M. Blackmer. 1978. Nitrous oxide: emission from soils during nitrification of fertilizer nitrogen. Science 199:295-296.

Bremner, J. M., A. M. Blackmer, and L. G. Bundy. 1978. Problems in use of nitrapyrin to inhibit nitrification in soils. Soil Biol. Biochem. 10:441-442.

Brock, T. D. 1979. Biology of microorganisms. 3rd ed. Prentice-Hall, Englewood Cliffs, N. J.

Bundy, L. G., and J. M. Bremner. 1973. Inhibition of nitrification in soils. Soil Sci. Soc. Am. Proc. 37:396-398.

Bundy, L. G., and J. M. Bremner. 1974. Effects of nitrification inhibitors on transformations of urea nitrogen in soils. Soil Biol. Biochem. 6:369-376.

Campbell, N. E. R., and M. I. H. Aleem. 1965a. The effect of 2-chloro, 6-(trichloromethyl) pyridine on the chemoautotrophic metabolism of nitrifying bacteria. I. Ammonia and hydroxylamine oxidation by Nitrosomonas. Antonie van Leeuwenhoek; J. Microbiol. Serol. 31:124-136.

Campbell, N. E. R., and M. I. H. Aleem. 1965b. The effect of 2-chloro, 6-(trichloromethyl) pyridine on the chemoautotrophic metabolism of nitrifying bacteria. II. Nitrite oxidation by Nitrobacter. Antonie van Leeuwenhoek; J. Microbiol. Serol. 31:137-144.

Cassens, R. G., T. Ito, M. Lee, and D. Buege. 1978. The use of nitrite in meat. BioScience 28:633-637.

Chapman, H. D., and G. F. Liebig. 1952. Field and laboratory studies of nitrite accumulation in soils. Soil Sci. Soc. Am. Proc. 16:276-282.

Chase, F. E., C. T. Corke, and J. B. Robinson. 1968. Nitrifying bacteria in soils. p. 593-611. In T. R. G. Gray and D. Parkinson (ed.) The ecology of soil bacteria. Liverpool University Press, Liverpool.

Chiang, C., J. Sinnaeve, and G. Dubuisson. 1972. Ecologie microbienne des sols du Maroc 1. Fluctuations saisonnieres. (In French). Ann. Inst. Pasteur, Paris 122:1171-1182.

Clark, C., and E. L. Schmidt. 1966. Effect of mixed culture on Nitrosomonas europaea simulated by uptake and utilization of pyruvate. J. Bacteriol. 91:367-373.

Clark, C., and E. L. Schmidt. 1967. Growth response of Nitrosomonas europaea to amino acids. J. Bacteriol. 93:1302-1308.

Clark, F. E., and E. A. Paul. 1970. The microflora of grassland. Adv. Agron. 22:375-435.

Cochran, W. G. 1950. Estimation of bacterial densities by means of the most probable number. Biometrics 6:105-116.

Crutzen, P. J. 1970. The influence of nitrogen oxides on the atmospheric ozone content. Q. J. Roy. Meterol. Soc. 96:320-325.

Dalton, H. 1977. Ammonia oxidation by the methane oxidising bacterium *Methylococcus capsulatus* strain Bath. Arch. Microbiol. 114:273–279.

Delwiche, C. C., and M. S. Finstein. 1965. Carbon and energy sources for the nitrifying autotroph. Nitrobacter. J. Bacteriol. 90:102–107.

Dommergues, Y. 1977. Biologie du sol. Presses Universitaire de France, Paris.

Domsch, K. H., and W. Paul. 1974. Simulation and experimental analysis of the influence of herbicides on soil nitrification. Arch. Microbiol. 97:283–301.

Elkins, J. W., S. C. Wofsy, M. B. McElroy, C. E. Kolb, and W. A. Kaplan. 1978. Aquatic sources and sinks for nitrous oxide. Nature (London) 275:602–606.

Engel, M. S., and M. Alexander. 1958. Growth and autotrophic metabolism of *Nitrosomonas europaea*. J. Bacteriol. 76:217–222.

Erickson, R. H., A. B. Hooper, and K. R. Terry. 1972. Solubilization and purification of cytochrome a_1 from *Nitrosomonas*. Biochim. Biophys. Acta 283:155–166.

Eylar, O. R., Jr., and E. L. Schmidt. 1959. A survey of heterotrophic microorganisms from soil for ability to form nitrite and nitrate. J. Gen. Microbiol. 20:473–481.

Falcone, A. B., A. L. Shug, and D. J. D. Nicholas. 1963. Some properties of a hydroxylamine oxidase from *Nitrosomonas europaea*. Biochim. Biophys. Acta 77:199–208.

Fliermans, C. B., B. B. Bohlool, and E. L. Schmidt. 1974. Autecological study of the chemoautotroph *Nitrobacter* by immunofluorescence. Appl. Microbiol. 27:124–129.

Fliermans, C. B., and E. L. Schmidt. 1975. Autoradiography and immunofluorescence combined for autecological study of single cell activity with *Nitrobacter* as a model system Appl. Microbiol. 30:676–684.

Focht, D. D., and W. Verstraete. 1977. Biochemical ecology of nitrification and denitrification. Adv. Microbiol. Ecol. 1:135–214.

Gasser, J. K. R. 1970. Nitrification inhibitors—their occurrence, production and effects of their use on crop yields and composition. Soils Fert. 33:547–554.

Goring, C. A. I. 1962a. Control of nitrification by 2-chloro-6-(trichloromethyl) pyridine. Soil Sci. 93:211–218.

Goring, C. A. I. 1962b. Control of nitrification of ammonium fertilizers and urea by 2-chloro-6-(trichloromethyl) pyridine. Soil Sci. 93:431–439.

Gundersen, K. 1955. Effects of B-vitamins and amino acids on nitrification. Physiol. Plant. 8:136–141.

Harms, H., H. P. Koops, and H. Wehrmann. 1976. An ammonia-oxidizing bacterium, *Nitrosovibrio tenuis* nov. gen. nov. sp. Arch. Microbiol. 108:105–111.

Hauck, R. D. 1972. Synthetic slow-release fertilizers and fertilizer amendments. p. 633–690. *In* C. A. I. Goring and J. W. Hamker (ed.) Organic chemicals in the soil environment, Part B. Marcel Dekker, New York.

Hofman, T., and H. Lees. 1953. The biochemistry of the nitrifying organisms. 4. The respiration and intermediary metabolism of *Nitrosomonas*. Biochem. J. 54:579–583.

Hooper, A. B. 1968. A nitrite-reducing enzyme from *Nitrosomonas europaea*. Preliminary characterization with hydroxylamine as electron donor. Biochim. Biophys. Acta 162:49–65.

Hooper, A. B. 1978. Nitrogen oxidation and electron transport in ammonia-oxidizing bacteria. p. 299–304. *In* D. Schlessinger (ed.) Microbiology 1978. Am. Soc. Microbiol. Washington, D.C.

Hooper, A. B., and K. R. Terry. 1973. Specific inhibitors of ammonia oxidation in *Nitrosomonas*. J. Bacteriol. 115:480–485.

Huber, D. M., H. L. Warren, D. W. Nelson, and C. Y. Tsai. 1977. Nitrification inhibitors—new tools for food production. BioScience 27:523–529.

Hughes, T. D., and L. F. Welch. 1970. 2-chloro-6-(trichloromethyl) pyridine as a nitrification inhibitor for anhydrous ammonia applied in different seasons. Agron. J. 62:821–824.

Hutton, W. E., and C. E. Zobell. 1953. Production of nitrite from ammonia by methane oxidizing bacteria. J. Bact. 65:216–219.

Jansson, S. L. 1958. Tracer studies on nitrogen transformation in soil with special attention to mineralization-immobilization relationships. Kungl. Lantbruk. Annaler 24:101–361.

Jensen, H. L. 1951. Nitrification of oxime compounds by h eterotrophic bacteria. J. Gen. Microbiol. 5:360–368.

Keeney, D. R. 1971. Microbiological aspects of the pollution of fresh water with inorganic nutrients. p. 181–200. *In* G. Sykes and F. A. Skinner (ed.) Microbiol aspects of pollution. Soc. Appl. Bacteriol. Academic Press, London.

Keeney, D. R., and J. M. Bremner. 1967. Determination and isotope-ratio analysis of different forms of nitrogen in soils: 6. Mineralizable nitrogen. Soil Sci. Am. Proc. 31:34–39.

Knowles, G., A. L. Downing, and M. J. Barrett. 1965. Determination of kinetic constants for nitrifying bacteria in mixed culture, with the aid of an electronic computer. J. Gen. Microbiol. 38:263–273.

Knowles, R. 1978. Common intermediates of nitrification and denitrification, and the metabolism of nitrous oxide. p. 367–371. *In* D. Schlessinger (ed.) Microbiology 1978. Am. Soc. Microbiol., Washington, D.C.

Koike, I., and A. Hattori. 1978. Simultaneous determinations of nitrification and nitrate reduction in coastal sediments by a ^{15}N dilution technique. Appl. Environ. Microbiol. 35:853–857.

Koops, H. P., H. Harms, and H. Wehrmann. 1976. Isolation of a moderate halophilic ammonia-oxidizing bacterium. *Nitrosococcus mobilis* nov. sp. Arch. Microbiol. 108:277–282.

Laudelout, H., R. Lambert, and M. L. Pham. 1976. Influence du pH et de la pression partielle d'oxygene sur la nitrification. Ann. Microbiol. (Paris). 127A:367–382.

Lees, H., and J. R. Simpson. 1957. The biochemistry of the nitrifying organisms. 5. Nitrite oxidation by *Nitrobacter*. Biochem. J. 65:297–305.

Lodhi, M. A. K. 1978. Comparative inhibition of nitrifiers and nitrification in a forest community as a result of the allelopathic nature of various tree species. Am. J. Bot. 65:1135–1137.

Mahendrappa, J. K., R. L. Smith, and A. T. Christianson. 1966. Nitrifying organisms affected by climatic region in western United States. Soil Sci. Soc. Am. Proc. 30:60–62.

Matulewich, V. A., P. F. Strom, and M. S. Finstein. 1975. Length of incubation for enumerating nitrifying bacteria present in various environments. Appl. Microbiol. 29:265–268.

McLaren, A. D., and M. S. Ardakani. 1972. Competition between species during nitrification in soil. Soil Sci. Soc. Am. Proc. 36:602–605.

Molina, J. A. E., and A. D. Rovira. 1964. The influence of plant roots on autotrophic nitrifying bacteria. Can. J. Microbiol. 10:249–256.

Moore, D. R. E., and J. S. Waid. 1971. The influence of washings of living roots on nitrification. Soil biol. Biochem. 3:69–83.

Morrill, L. G., and J. E. Dawson. 1962. Growth rates of nitrifying chemoautotrophs in soil. J. Bacteriol. 83:205–206.

Morrill, L. G., and J. E. Dawson. 1967. Patterns observed for the oxidation of ammonium to nitrate by soil organisms. Soil Sci. Soc. Am. Proc. 31:757–760.

Munro, P. E. 1966. Inhibition of nitrifiers by grass root extracts. J. Appl. Ecol. 3:231–239.

Myers, R. J. K. 1975. Temperature effects on ammonification and nitrification in a tropical soil. Soil Biol. Biochem. 7:83–86.

Neal, J. L. 1969. Inhibition of nitrifying bacteria by grass and forb root extracts. Can. J. Microbiol. 15:23–28.

Nicholas, D. J. D. 1978. Intermediary metabolism of nitrifying bacteria, with particular reference to nitrogen, carbon, and sulfur compounds. p. 305–309. *In* D. Schlessing (ed.) Microbiology, 1978. Am. Soc. Microbiol. Washington, D.C.

Painter, H. A. 1977. Microbial transformations of inorganic nitrogen. Prog. Water Technol. 8(4/5):3–29.

Pancholy, S. K. 1978. Formation of carcinogenic nitrosamines in soils. Soil Biol. Biochem. 10:27–32.

Panganiban, A. T., Jr., T. E. Patt, W. Hart, and R. S. Hanson. 1979. Oxidation of methane in the absence of oxygen in lake water samples. Appl. Environ. Microbiol. 37:303–309.

Paul, E. A., and E. L. Schmidt. 1961. Formation of free amino acids in rhizosphere and non-rhizosphere soil. Soil Sci. Soc. Am. Proc. 25:359–362.

Powlson, D. S., and D. S. Jenkinson. 1971. Inhibition of nitrification in soil by carbon disulphide from rubber bunds. Soil Biol. Biochem. 3:267–269.

Prasad, R., G. B. Rajale, and B. A. Lakhdive. 1971. Nitrification retarders and slow-release nitrogen fertilizers. Adv. Agron. 23:337–383.

Quastel, J. H., and P. G. Scholefield. 1949. Influence of organic nitrogen compounds on nitrification in soil. Nature (London) 164:1068–1072.

Quastel, J. H., P. G. Scholefield, and J. W. Stevenson. 1950. Oxidation of pyruvic oxime by soil organisms. Nature (London) 166:940–942.

Quastel, J. H., P. G. Scholefield, and J. W. Stevenson. 1952. Oxidation of pyruvic oxime by soil organisms. Biochem. J. 51:278–284.

Quayle, J. R., and T. Ferenci. 1978. Evolutionary aspects of autotrophy. Microbiol. Rev. 42: 251–273.

Ramakrishna, C., V. R. Rao, and N. Sethunathan. 1978. Nitrification in simulated oxidized surface of a flooded soil amended with carbofuran. Soil biol. Biochem. 10:555–556.

Ratnayake, M., and J. L. Audus. 1978. Studies on the effects of herbicides on soil nitrification. II. Pest. Biochem. Physiol. 8:170–185.

Rennie, R. J., and E. L. Schmidt. 1977a. Immunofluorescence studies of Nitrobacter populations in soils. Can. J. Microbiol. 23:1011–1017.

Rennie, R. J., and E. L. Schmidt. 1977b. Autecological and kinetic analysis of competition between strains of Nitrobacter in soils. Ecol. Bull. (Stockholm) 25:431–441.

Reuss, J. O., and G. S. Innis. 1977. A grassland nitrogen flow simulation model. Ecology 58: 379–388.

Rice, E. L. 1964. Inhibition of nitrogen-fixing and nitrifying bacteria by seed plants. Ecology 45:824–837.

Rice, E. L., and S. K. Pancholy. 1972. Inhibition of nitrification by climax vegetation. Am. J. Bot. 59:1033–1040.

Rice, E. L., and S. K. Pancholy. 1973. Inhibition of nitrification by climax ecosystems. II. Additional evidence and possible role of tannins. Am. J. Bot. 60:691–702.

Ridge, E. H. 1976. Studies on soil fumigation-II. Soil Biol. Biochem. 8:249–253.

Ritchie, G. A. F., and D. J. D. Nicholas. 1972. Identification of the sources of nitrous oxide produced by oxidation and reductive processes in Nitrosomonas europaea. Biochem. J 126:1181–1191.

Robinson, J. B. 1963. Nitrification in a New Zealand grassland soil. Plant Soil 19:173–183.

Ross, D. J. 1974. Influence of four pesticides formulations on microbial processes in a New Zealand pasture soil. II. Nitrogen mineralization. N.Z. j. Agric. Res. 17:9–17.

Rovira, A. D. 1976. Studies on soil fumigation—I. Soil Biol. Biochem. 8:241–247.

Russell, E. W. 1973. Soil conditions and plant growth. 10th ed. Longman, London.

Sabey, B. R., L. R. Frederick, and W. V. Bartholomew. 1969. The formation of nitrate from ammonium nitrogen in soils: IV. Use of the delay and maximum rate phases for making quantitative predictions. Soil Sci. Soc. Am. Proc. 33:276–278.

Sarathchandra, S. V. 1978a. Nitrification activities of some New Zealand soils and the effect of some clay types on nitrification. N.Z. J. Agric. Res. 21:615–621.

Sarathchandra, S. V. 1978b. Nitrification activities and the changes in the populations of nitrifying bacteria in soil perfused at two different H-ion concentrations. Plant Soil 50:99–111.

Schmidt, E. L. 1954. Nitrate formation by a soil fungus. Science 119:187–189.

Schmidt, E. L. 1956. Soil nitrification and nitrates in waters. Pub. Health Reports 71:497–503.

Schmidt, E. L. 1960a. Cultural conditions influencing nitrate formation by Aspergillus flavus. J. Bacteriol. 79:553–557.

Schmidt, E. L. 1960b. Nitrate formation by Aspergillus flavus in pure and mixed culture natural environments. Int. Congr. Soil Sci. Trans. 7th (Madison, Wis.) 2:600–607.

Schmidt, E. L. 1973. Fluorescent antibody techniques for the study of microbial ecology. Bull. Ecol. Res. Commun. (Stockholm) 17:67–76.

Schmidt, E. L. 1974. Quantitative autecological study of microorganisms in soil by immunofluorescence. Soil Sci. 118:141–149.

Schmidt, E. L. 1978. Nitrifying microorganisms and their methodology. p. 288–291. In D. Schlessinger (ed.) Microbiology 1978. Am. Soc. Microbiol. Washington, D.C.

Schmidt, E. L., and R. L. Starkey. 1951. Soil microorganisms and plant growth substances. II. Transformations of certain B-vitamins in soil. Soil Sci. 71:221–231.

Shively, J. M., E. Bock, K. Westphal, and G. C. Cannon. 1977. Icosahedral inclusions (carboxysoms) of Nitrobacter agilis. J. Bacteriol. 132:673–675.

Smith, A. J., and D. S. Hoare. 1968. Acetate assimilation by Nitrobacter agilis in relation to its "obligate autotrophy". J. Bacteriol. 95:844–855.

Smith, A. J., and D. S. Hoare. 1977. Specialist phototrophs, lithotrophs, and methylotrophs: a unity among a diversity of procaryotes? Bact. Rev. 41:419–448.

Smith, J. H. 1964. Relationships between soil cation-exchange capacity and the toxicity of ammonia to the nitrification process. Soil Sci. Soc. Am. Proc. 28:640–644.

Soriano, S., and N. Walker. 1968. Isolation of ammonia-oxidizing autotrophic bacteria. J. Appl. Bacteriol. 31:493–497.

Stanford, G., and S. J. Smith. 1972. Nitrogen mineralization potential in soils. Soil Sci. Soc. Am. Proc. 36:465–472.

Starr, J. L., F. E. Broadbent, and D. R. Nielsen. 1974. Nitrogen transformations during continuous leaching. Soil Sci. Soc. Am. Proc. 38:283–289.

Stephenson, M. 1939. Bacterial metabolism. 2nd ed. Longmans, Green, and Co., London.

Suzuki, I. 1974. Mechanisms of inorganic oxidation and energy coupling. Ann. Rev. Microbiol. 28:85–101.

Suzuki, I., U. Dular, and S. C. Kwok. 1974. Ammonia or ammonium ion as substrate for oxidation by *Nitrosomonas europaea* cells and extracts. J. Bacteriol. 120:556–558.

Taylor, J. 1962. The estimation of numbers of bacteria by tenfold dilution series. J. Appl. Bacteriol. 25:54–61.

Theron, J. J. 1951. The influence of plants on the mineralization of nitrogen and the maintenance of organic matter in soil. J. Agric. Sci. 41:289–296.

Van Gool, A. P., and H. Laudelout. 1966. Formate utilization by *Nitrobacter winogradskyi*. Biochim. Biophys. Acta 127:295–301.

Vitousek, P. M., and W. A. Reiners. 1975. Ecosystem succession and nutrient retention: a hypothesis. BioScience 25:376–381.

Volz, M. G., L. W. Belser, M. S. Ardakani, and A. D. McLaren. 1975. Nitrate reduction and nitrite utilization by nitrifiers in an unsaturated Hanford sandy loam. J. Environ. Qual. 4: 179–182.

Wainwright, M. 1977. Effects of fungicides on the microbiology and biochemistry of soils—A review. Z. Planzenernaehr. Bodenkd. 140:587–602.

Wainwright, J., and G. J. F. Pugh. 1973. The effect of three fungicides on nitrification and ammonification. Soil biol. Biochem. 5:577–584.

Waksman, S. A. 1932. Principles of soil microbiology. 2nd ed. Williams & Wilkins, Baltimore.

Waksman, S. A. 1946. Sergei Nikolaevitch Winogradsky. The story of a great bacteriologist. Soil Sci. 62:197–226.

Walker, N. 1975. Nitrification and nitrifying bacteria. p. 133–146. *In* N. Walker (ed.) Soil microbiology. Butterworths, London.

Walker, N. 1978. On the diversity of nitrifiers in nature. p. 346–347. *In* D. Schlessinger (ed.) Microbiology 1978. Am. Soc. Microbiol. Washington, D.C.

Walton, G. 1951. Survey of literature related to infant methemoglobinemia due to nitrate contaminated water. Am. J. Pub. Health 41:986–996.

Watson, S. W. 1971. Reisolation of *Nitrosospira briensis* S. Winogradsky and H. Winogradsky 1933. Arch. Mikrobiol. 75:179–188.

Watson, S. W. 1974. Gram-negative chemolithotrophic bacteria. Family I. p. 450–456. *In* R. E. Buchanan and N. E. Gibbons (ed.) Bergey's manual of determinative bacteriology, 8th ed. The Williams & Wilkins Co., Baltimore.

Watson, S. W., L. B. Graham, C. C. Remsen, and F. W. Valois. 1971. A lobular, ammonia-oxidizing bacterium, *Nitrosolobus multiformis* nov. gen. nov. sp. Arch. Mikrobiol. 76: 183–203.

Weber, D. F., and P. L. Gainey. 1962. Relative sensitivity of nitrifying organisms to hydrogen ions in soils and in solutions. Soil Sci. 94:138–145.

Whittenbury, R., and D. P. Kelly. 1977. Autotrophy: a conceptual phoenix. p. 121–149. *In* B. A. Haddock and W. A. Hamilton (ed.) Microbial energetics. 17th Symp. of the Soc. for Gen. Microbiol. Cambridge University Press. Cambridge.

Whittenbury, R., K. C. Phillips, and J. F. Wilkinson. 1970. Enrichment, isolation and some properties of methane-utilising bacteria. J. Gen. Microbiol. 61:205–218.

Winogradsky, S. N. 1949. Microbiologie du sol: problemes et methodes. Masson et Cie., Paris.

Yung, Y. L., and M. B. McElroy. 1979. Fixation of nitrogen in the prebiotic atmosphere. Science 203:1002–1004.

8 Biological Denitrification

M. K. FIRESTONE

Department of Plant and Soil Biology
University of California, Berkeley

I. INTRODUCTION

Denitrification is the major biological process through which fixed N is returned from the soil to the atmosphere. Even though the central role of denitrification in the N cycle has made it the subject of considerable research, this process remains one of the more poorly understood facets of soil N transformation, from the level of field quantification through fundamental microbial biochemistry. Quantitative estimates of N loss from agricultural soils through denitrification vary tremendously, ranging from 0 to 70% of applied fertilizer N (Rolston et al., 1976, 1979; Craswell, 1978; Kissel & Smith, 1978; Kowalenko, 1978). In a review of field N studies that employed ^{15}N, Hauck (1981) estimated an average fertilizer N deficit of between 25 to 30%. This estimated fertilizer N loss may be due largely to denitrification, but other mechanisms may also be involved (Chapt. 23, Hauck and Tanji). Total N losses due to denitrification in some irrigated California soils studied by Ryden and Lund (1980) were found to range from 95 to 233 kg of N $ha^{-1} \cdot year^{-1}$.

The potential for groundwater or water supply pollution by NO_3^- derived from fertilizer or soil organic matter has stimulated research on denitrification over the last decade. The role of nitrous oxide (N_2O) in stratospheric chemical reactions (Johnston, 1971; Crutzen, 1970) has also generated interest in the process of denitrification. It has been hypothesized that as the amounts of industrially or biologically fixed N_2 used for crop production increase, the production of N_2O due to denitrification could cause significant depletion of the earth's ozone shield (Crutzen & Ehhalt, 1977; McElroy et al., 1977; Pratt et al., 1977; Sze & Rice, 1976), and/or could contribute to a warming of the earth's surface by influencing the radiative budget of the troposphere (Wang et al., 1976). Denitrification in soil produces both N_2O and N_2; hence the process may serve either as a source of N_2O or as a sink for N_2O through reduction of N_2O to N_2. The importance of denitrification in the terrestrial N_2O budget is not yet totally clear, but this question has led to significant research on and increased understanding of not only denitrification but also microbial N metabolism in soil generally.

II. BIOCHEMICAL AND MICROBIOLOGICAL BASIS

A. Definition and Pathway

In the SSSA *Glossary of Soil Science Terms* (1979), denitrification is defined as "the microbial reduction of nitrate or nitrite to gaseous nitrogen either as molecular nitrogen or as an oxide of nitrogen." However, several types of microbial N metabolism including nitrification and NO_3^- reduction to NH_4^+ result in gaseous N oxide production (N_2O, NO, or both) possibly through NO_2^- reduction (Ritchie & Nicholas, 1972; 1974; Bollag & Tung, 1972; Yoshida & Alexander, 1970). To avoid confusion, a more explicit definition seems advisable. Most microbiologists identify denitrification as a respiratory process present in a limited number of bacterial genera. In this process, N oxides serve as terminal electron acceptors for respiratory electron transport leading from a "reduced" electron donating substrate through numerous electron carriers to a more oxidized N oxide. During electron transport to at least several of the N oxides, energy is conserved by electron transport phosphorylation. Denitrifying bacteria can grow in the absence of molecular oxygen (O_2) while reducing NO_3^- or NO_2^- to N_2 and/or N_2O. The reduction of the anion species to only gaseous products (dominantly N_2 and N_2O) and the quantity of gas produced are characteristics that distinguish denitrification from other types of microbial N metabolism.

The exact mechanism of N_2O production during nitrification is not clear; the gas may result from oxidation of hydroxylamine (Hooper & Terry, 1979) or from reduction of NO_2^- via a NO_2^- reductase (Ritchie & Nicholas, 1972, 1974). Blackmer et al. (1980) have suggested that N_2O production during nitrification in soils results from NO_2^- reduction, and it has been shown that lowered O_2 availability enhances N_2O production by NH_4^+ oxidizers in cultures and in soil (Blackmer et al., 1980; Goreau et al., 1980). It has been suggested that nitrifying organisms can shift to a form of anaerobic respiration (or denitrification) when O_2 is scarce (Payne, 1973); however, it has yet to be shown that N oxides can serve as sole terminal electron acceptors for growth of chemoautotrophic NH_4^+ oxidizers. The mechanism by which N_2O is produced during NO_2^- reduction to NH_4^+ (Bollag & Tung, 1972; Yoshida & Alexander, 1970) is even less clear than in nitrification. There are several reports of N_2O production during NO_2^- reduction to NH_4^+ in enteric bacteria (Tiedje, 1981). In at least one of these organisms, *Klebsiella pneumoniae,* N_2O evolution has been reported to be associated with the use of NO_2^- as sole terminal electron acceptor during anaerobic respiration (Hom et al., 1980). Yoshinari (1980) recently reported reduction of N_2O to N_2 by a fermentative organism that reduces NO_3^- to NH_4^+.

The difficulty in distinguishing denitrification from other metabolic processes that in the past have been believed to be quite different, results from expanding knowledge of N metabolism in general. Future results may more clearly delineate differences among these metabolic processes or require expansion of the concept of denitrification. This review will be concerned with denitrification as defined here. In the past decade, other reviewers have covered this topic or aspects of it in greater detail (Focht & Verstraete, 1977; Delwiche & Bryan, 1976; Stouthamer, 1976; Garcia, 1975; Payne, 1973).

The general requirements for denitrification are (i) the presence of bacteria possessing the metabolic capacity; (ii) suitable electron donors such as organic C compounds, reduced S compounds, or molecular hydrogen, H_2; (iii) anaerobic conditions or restricted O_2 availability; and (iv) N oxides, NO_3^-, NO_2^-, NO, or N_2O as terminal electron acceptors.

The pathway of N oxide reduction during denitrification is generally thought to be:

$$(+5) \quad (+3) \quad (+2) \quad (+1) \quad (0)$$
$$NO_3^- \rightarrow NO_2^- \rightarrow NO \rightarrow N_2O \rightarrow N_2.$$

This reductive sequence has been hypothesized for some time, and a significant amount of evidence has accumulated supporting the role of each N oxide in this pathway. Isotope mixing and exchange studies have been used to investigate the role of two of the more "controversial" intermediates, NO and N_2O. Using ^{15}N labeled NO_2^- in studies with *Pseudomonas aeruginosa,* St. John and Hollocher (1977) found no ^{15}N label exchange with nonlabeled pools of NO, and the N_2 product showed no isotopic mixing. These workers concluded that NO was not a free intermediate in denitrification. Firestone et al. (1979a) reported that the ^{13}N label from NO_2^- did show substantial rapid mixing with nonlabeled NO added to cultures of two *Pseudomonas* species. This work indicated that NO was either a central intermediate or was in rapid equilibrium with such an intermediate. There is still considerable disagreement as to the role of NO as an obligatory intermediate (Zumft & Cardenas, 1979; Zumft & Vega, 1979; St. John & Hollocher, 1977; Delwiche & Bryan, 1976). The basis for this uncertainty is discussed further in the sections on NO_2^- and NO reductases. Recent isotope exchange and kinetic evidence argues against involvement of free trans-hyponitrite $(O-N=N-O)^{2-}$ in the reductive pathway (Hollocher et al., 1980).

Exchange studies with ^{15}N in *P. aeruginosa* (St. John & Hollocher, 1977) and with ^{13}N in several cultures and soils (Firestone et al., 1980) indicated that N_2O was a free, obligate intermediate during NO_2^- reduction to N_2. This work, as well as the commonly observed stoichiometric accumulation of N_2O in the presence of acetylene, an N_2O reductase inhibitor (Balderston et al., 1976; Yoshinari & Knowles, 1976), would seemingly establish N_2O as a central intermediate in denitrification. However, recent work by Bryan (1980) casts some doubt on this conclusion. After adding ^{15}N labeled NO_2^- and ^{14}N-N_2O to stationary phase cultures of *Pseudomonas stutzeri,* Bryan determined the pattern of $^{15}N^{15}N$, $^{15}N^{14}N$, and $^{14}N^{14}N$ in the resulting N_2. If the simple reductive sequence $NO_3^- \rightarrow NO_2^- \rightarrow NO \rightarrow N_2O \rightarrow N_2$ was in effect, only $^{15}N^{15}N$ and $^{14}N^{14}N$-N_2 should have resulted. However, a quadratic type of distribution was found, with the dominant form of N_2 being $^{15}N^{14}N$. Bryan (1980) also reported that although *P. aeruginosa* obtains energy during the final reductive step to N_2, it is unable to grow on added N_2O. The implications of these results are not totally clear, but the question of N_2O as a simple intermediate during denitrification may not yet be fully answered.

B. Organisms Involved

The capacity to denitrify has been reported to be present in about 23 genera of bacteria. The listing of denitrifying genera in Table 1 includes 13 genera for which there is confirmed or multiple documentation. There are reports of denitrifying strains of *Corynebacterium, Xanthomonas, Acineobacter* and *Gluconobacter,* most of which are of uncertain taxonomic affiliation. Jeter and Ingraham (1981) have compiled a more complete list and include discussion of denitrifying species in the genera *Chromobacterium, Cytophaga, Neisseria, Simonsiella, Thiomicrospira* and *Thermothrix.* Denitrification appears to be a relatively limited metabolic capability when compared to Hall's list (1978) of 73 bacterial genera capable of dissimilatory reduction of NO_3^- to NO_2^-. The most notable additions to previous compilations by Payne (1976) and Focht and Verstraete (1977) are genera of *Rhizobium, Flavobacterium,* and *Agrobacterium* (Zablotowicz et al., 1978; Pichinoty et al., 1976, 1977). It is also interesting to note the recent preliminary report of a magnetic spirillum apparently capable of microaerophilic denitrification (Escalante-Semerena, 1980). Of the organisms studied, most are capable of reducing NO_3^-, NO_2^- or N_2O as sole terminal electron acceptor. A few strains, such as *Alcaligenes odorans,* cannot utilize NO_3^- (Pichinoty et al., 1976, 1978b; Vangai & Klein, 1974). A few others (generally fluorescent pseudomonads) produce N_2O as the terminal product (Payne & Balderston, 1978; Greenberg & Becker, 1977; Renner & Becker, 1970) or grow very poorly on N_2O (St. John & Hollocher, 1977; Bryan, 1980). There are very few reports of organisms capable of growing on NO

Table 1—Genera of bacteria capable of denitrification.

Genus†	Interesting characteristics of some species
Alcaligenes	Commonly isolated from soils.
Agrobacterium	Some species are plant pathogens.
Azospirillum	Capable of N_2 fixation, commonly associated with grasses.
Bacillus	Thermophilic denitrifiers reported.
Flavobacterium	Denitrifying species recently isolated.
Halobacterium	Requires high salt concentrations for growth.
Hyphomicrobium	Grows on one-carbon substrates.
Paracoccus	Capable of both lithotrophic and heterotrophic growth.
Propionibacterium	Fermentors capable of denitrification.
Pseudomonas	Commonly isolated from soils.
Rhizobium	Capable of N_2 fixation in symbiosis with legumes.
Rhodopseudomonas	Photosynthetic bacteria.
Thiobacillus	Generally grow as chemoautotrophs.

† Reference to many of the genera can be found in text. For further documentation see Jeter and Ingraham (1981), Focht and Verstraete (1977), Garcia (1975), Buchanan and Gibbons (1974).

(Pichinoty et al., 1978a; Ishaque & Aleem, 1973). The capacity to grow on NO may either be uncommon or merely difficult to document because of the inherent toxicity of the compound.

Most denitrifying bacteria are chemoheterotrophs. That is, they use chemical energy sources (not light), and they use organic C compounds as electron donors (reductants) and as sources of cellular C. One photosynthetic bacterium, *Rhodopseudomonas sphaeroides,* has been documented to be a denitrifier, but it is capable of growth as a chemoheterotroph while denitrifying (Satoh, 1977). Several denitrifying bacteria can grow as lithotrophs; *Thiobacillus denitrificans* uses reduced S compounds, and *Paracoccus denitrificans* and several *Alcaligenes* species can use H_2 as an electron donor (Thauer et al., 1977; Buchanan & Gibbons, 1974). *Thiobacillus denitrificans* can also grow as an autotroph (using CO_2 as C source) while denitrifying.

Almost all denitrifiers are aerobic organisms capable of anaerobic growth only in the presence of the N oxides. Respiration using O_2 as an electron acceptor and respiration using oxides of N as acceptors are similar processes in that many of the electron-carrying species used during denitrification are also used during O_2 respiration (Fig. 1). Bacteria of the genus *Propionibacterium* seem to be the only obligately anaerobic fermenters currently known to be capable of denitrification (Thauer et al., 1977; Payne, 1976). Species of this genus have been demonstrated to possess at least some of the cytochromes necessary for electron transport coupled NO_x reduction (Stouthamer, 1976), but the evidence concerning the nature of N oxide reduction by these organisms is not yet complete. Some denitrifying species of the genus *Bacillus* are facultative organisms capable of both fermentation and respiration.

A few N_2-fixing organisms, including *Azospirillum brasilense* (formerly designated *Spirillum lipoferum*) and *Rhizobium japonicum,* also denitrify (Zablotowicz et al., 1978; Eskew et al., 1977; Neyra et al., 1977). Neyra and van Berkum (1977) and Scott and Scott (1978) have reported what appears to be N_2 fixation coupled to denitrification in *Azospirillum* species. However, work by Nelson and Knowles (1978) on O_2 and NO_3^- control of the two processes in *A. brasilense* indicated that significant amounts of N_2 fixation and denitrification do not occur simultaneously. Active reduction of NO_3^- to NO_2^- has been shown to inhibit nitrogenase activity in this organism (Magalhaes et al., 1978). In soils, N_2 fixation activity was found to be detectable only after added NO_3^- was depleted by denitrification (Yoshinari et al., 1977).

It can be very difficult to ascertain which denitrifying organisms are functionally important in soils. In a study of 19 soils, Gamble et al. (1977) found that the bacteria most commonly isolated were of the genera *Pseudomonas* and *Alcaligenes.* Many species of these genera would be expected to grow well on the relatively rich laboratory media most commonly used for isolation from soil. There are at least two major underlying difficulties in determining which bacteria are important as soil denitrifiers: the inherent selectivity of culture media or the almost impossible task of simulating the complex soil environment in culture media and the dual metabolic capability of all denitrifiers. That is, the isolation of a bacterial strain from soil capable of denitrification does not necessarily mean that this organism commonly grows as a denitrifier in the natural environment. The importance of both of these problems is clearly demonstrated in work by Smith and Tiedje (1980). Soil isolates that grew well as denitrifiers on laboratory media did not compete well in soil to which no C was added, whereas the strain that grew most slowly on NO_3^- broth grew most rapidly in saturated soil. These workers also found that the growth or survival capabilities of isolates differed significantly depending on whether the soil was anaerobic (water saturated) or aerobic.

In their study of soils, Gamble et al. (1977) found *Pseudomonas fluorescens* to comprise the major single species isolated. Yet this organism has received very little in-depth study. The denitrifying bacteria most commonly employed for physiological studies are *Pa. denitrificans, Pseudomonas perfectormarinus,* and *P. aeruginosa.* Of these species, only *P. aeruginosa* was found in the soils studies by Gamble et al. (1977). Much of the information concerning the biochemistry of denitrification discussed in this review was obtained using organisms that may not be important in the soil environment. Similarities between species in enzyme characteristics and control most likely exist, but the extrapolation from these studies to field soils must be made with reservation.

C. Cellular Control

In any metabolic process, control can be exerted at either or both of two basic levels: enzyme synthesis or activity of an existing enzyme. In denitrification, several parameters (e.g., O_2 and NO_3^-) control the process at both levels. This section discusses the genetics of denitrification, the conditions controlling gene expression or subsequent enzyme synthesis, and finally, the parameters that affect enzyme activity. It will be apparent that the reduction of NO_3^- to NO_2^- has been much more extensively studied than have subsequent reductive steps. This results in part from the fact that respiratory NO_3^- reduction to NO_2^- is common to many bacteria other than just denitrifiers (Hall, 1978). The characteristics of the dissimilatory nitrate reductases (DNR's) present in denitrifiers are quite similar to those found in NO_3^- respirers. Hence, information gained from studies on NO_3^- respirers is commonly extrapolated to denitrification.

1. GENETICS

Little is known about the genes that control synthesis of N oxide reductases. Mutants in DNR from *P. aeruginosa* have been mapped and partially characterized by van Hartengsveldt et al. (1971; van Hartengsveldt & Stouthamer, 1973). Of the five different types of mutants isolated, four were also found to have lost the ability to reduce NO_3^- to NO_2^- for assimilatory purposes. This implied that *P. aeruginosa* had only one type of NO_3^- reductase that was used for both assimilation and denitrification. It has, however, been more recently determined that of the four mutants unable to assimilate or dissimilate NO_3^-, three contain multiple mutations. The fourth strain was incapable of incorporating Mo into either the assimilatory or dissimilatory NO_3^- reductase (Sias et al., 1980). Isolation of DNR mutants from *Pa. denitrificans* and the characterization of one of these mutants as also being incapable of Mo incorporation have been reported by Burke et al. (1980) and Calder et al. (1980).

Van Hartingsveldt et al. (1971) reported the isolation of five *P. aeruginosa* mutants that were affected in NO_2^- reductase activity. There is preliminary information as to the location of several of these genes on the bacterial genome. Cytochrome spectra indicated that these mutants may have been unable to synthesize the *d*-type cytochrome of NO_2^- reductase.

There are as yet no reports concerning the genetic control of NO or N_2O reducing activity.

2. ENZYME SYNTHESIS

Synthesis of dissimilatory NO_3^- reductase is generally repressed in the presence of O_2 and derepressed in the absence of O_2. In many organisms studied, NO_3^- reductase is readily synthesized when the supply of O_2 is limited (Calder et al., 1980; Zumft & Vega, 1979; Swain et al., 1978), presumably when O_2 demand exceeds rate of supply. Sias and Ingraham (1979) reported that synthesis of DNR by *P. aeruginosa* is fully derepressed when O_2 supply is ≤ 0.02 mM of $O_2 \cdot$liter$^{-1} \cdot$min^{-1} and NO_3^- is present. In fact these "semiaerobic" conditions have been found to be ideal for production of DNR by *Pa. denitrificans* (Calder et al., 1980) and *P. aeruginosa* (Carlson, 1981). Under conditions of limited O_2 availability, aerobic respiration can apparently provide the energy needed for synthesis of new proteins for NO_3^- reduction. In fact, it has been suggested that some denitrifiers cannot grow after an abrupt shift from highly aerobic to strictly anaerobic conditions, presumably due to the absence of any energy generating mechanism for de novo enzyme synthesis (Payne et al., 1971). Conversely, some organisms, such as *T. denitrificans,* seem to require strict anaerobiosis for significant synthesis of DNR to occur (Justin & Kelley, 1978).

Anaerobiosis alone, even in the absence of NO_3^-, is sufficient to cause synthesis of DNR in some organisms, such as *P. perfectomarinus* (Payne et al., 1971; Stouthamer, 1976). However, in many denitrifiers, the presence of NO_3^- is either required for or greatly enhances synthesis of DNR (Calder et al., 1980; Sias & Ingraham, 1979; Stouthamer, 1976). Several other compounds, including NO_2^-, azide, and chlorate, can also serve as inducers or stimulators of DNR synthesis (Carlson, 1981; Calder et al., 1980; Stouthamer, 1976).

The observation that DNR is not normally synthesized in the presence of sufficient O_2 is consistent with a generally observed metabolic pattern in which formation of a reductase for a given electron acceptor is prevented when an alternate terminal acceptor that would give a higher energy yield is present (Stouthamer, 1976). This provides a rationale for O_2 repression of NO_3^- reductase synthesis, but it indicates little about the mechanism of such repression. In his 1976 review, Stouthamer discussed a redox control model in which the oxidation-reduction status of the electron transport chain components, or the terminal NO_3^- reductase, serves as the factor regulating synthesis of the reductases. This model seems to be consistent with recent observations in studies with *Pa. denitrificans* by Burke et al. (1980) and Calder et al. (1980) concerning the role of azide as an inducer of DNR and the effects of mutations on regulation of synthesis.

The regulation of synthesis of the other reductases involved in denitrification is not as well studied as is NO_3^- reductase. Work in the early 1970's with *P. perfectomarinus* indicated that all of the denitrifying enzyme activities were induced simultaneously by the absence of O_2 alone (Payne, 1973; Payne et al., 1971; Payne & Riley, 1969). Since this time, however, much evidence has accumulated from studies with other organisms that suggests that NO_2^- reductase synthesis is not coordinately regulated with DNR synthesis; synthesis of significant quantities of NO_2^- reductase is commonly observed to lag behind that of NO_3^- reductase by several hours (Calder et al., 1980; Zumft & Vega, 1979; Nelson & Knowles, 1978; Swain et al., 1978; Williams et al., 1978). Most of these studies were done by shifting cultures to anaerobic conditions in the presence of NO_3^-. The lag in synthesis of NO_2^- reductase may represent the time required for sufficient accumulation of NO_2^- to cause induction or amplification of synthesis. However, several workers have reported that NO_2^- reductase is not as readily synthesized as is NO_3^- reductase under "semiaerobic" conditions (Calder et al., 1980; Zumft & Vega, 1979; Swain et al., 1978), suggesting that the lag in synthesis of NO_2^- reductase may be due to differential O_2 control. Most of these studies were done using NO_3^- as an inducer, hence it is difficult to delineate the effects of NO_2^- production (or NO_3^- depletion) from those of O_2. In *T. denitrificans,* Justin and Kelley (1978) reported that NO_2^- reductase was synthesized at higher O_2 concentrations in the presence of NO_2^- than was NO_3^- reductase.

The observation that in some organisms the synthesis of NO_2^- reductase lags behind synthesis of NO_3^- reductase may explain the common observation of NO_2^- accumulation in cultures and soils that have been shifted to denitrifying conditions. However, there are other explanations for NO_2^- accumulation, as is discussed later.

Like DNR, N_2O reductase can apparently be synthesized under semiaerobic conditions in some organisms, such as *P. perfectomarinus* and *Alcaligenes* sp. or *Achromobacter* sp. (Payne et al., 1971; Matsubara, 1971). Induction of N_2O reducing activity as a result of growth on NO_3^- is commonly observed (Payne, 1973). Yet, it is not known whether NO_3^- and NO_2^- serve as effective inducers for synthesis of N_2O reductase since the reduction product, N_2O, may, in fact, be the primary inducer. It has been reported that cells grown on N_2O have higher rates of N_2O reduction than cells grown on NO_3^- or NO_2^- (Matsubara, 1971; Payne et al., 1971; Delwiche, 1959). Firestone and Tiedje (1979) reported that synthesis of N_2O reducing activity occurred several hours after synthesis of the other denitrifying enzymes in soils and culture. This pattern may have resulted from accumulation of N_2O and its subsequent induction or amplification of N_2O reductase synthesis. The lag in synthesis of N_2O reducing activity may, in part, explain the transient accumulation of N_2O commonly observed in cultures and soils after a shift to anaerobic conditions.

3. ENZYME ACTIVITY

When actively denitrifying bacteria are shifted to fully aerobic conditions, NO_3^- reduction ceases (John, 1977; Stouthamer, 1976). Inhibition of NO_3^- reduction by O_2 does not seem to be due to a direct effect of O_2 on the enzyme itself; O_2 must function as a terminal electron acceptor for it to affect NO_3^- reduction.

van Hartingsveldt and Stouthamer (1974) reported that when a mutant of *P. aeruginosa*, which was unable to synthesize certain electron transport components aerobically, was shifted from anaerobic to aerobic conditions, the reduction of NO_3^- was largely unaffected. Similarly, it has been suggested that in some cases NO_3^- respiration can continue for a short period after a shift to aerobic conditions until the electron transport chain to O_2 becomes functional (Stouthamer, 1976). There are reports that preformed NO_3^- reductase in cells may be slowly inactivated when cells are incubated aerobically; however, purified NO_3^- reductase from *Klebsiella aerogenes* was not inactivated by exposure to O_2 (Stouthamer, 1976). In studies with *Pa.·denitrificans,* John (1977) found that while O_2 immediately shut down NO_3^- reduction in whole cells, both O_2 and NO_3^- reduction occurred simultaneously in membrane vesicles. The reason for this difference between intact cells and membrane vesicles of *Pa. denitrificans* is unknown.

The presence of the other N oxide intermediates of denitrification does not seem to affect the rate of NO_3^- reduction. The presence of NO_2^- does not affect the rate of NO_3^- reduction in *Pa. denitrificans* (John, 1977). Nitrite, however, may partially uncouple energy generation from NO_3^- reduction (Meijer et al., 1979b).

The effect of O_2 on NO_2^- reducing activity during denitrification appears to be very similar to its effect on NO_3^- reduction. That is, NO_2^- reduction occurs only in the absence of O_2 (John, 1977). As previously mentioned, NO_2^- is commonly observed to accumulate when denitrifying bacteria are grown on NO_3^- (Payne, 1973). Working with *P. perfectomarinus,* Payne and Riley (1969) demonstrated that NO_3^- directly inhibited NO reduction in a partially purified cell fraction. Payne (1973) thus suggests that this inhibitory effect of NO_3^- causes accumulation of NO_2^-. More recently, Betlach (1979), working with a *P. fluorescens* and a *Flavobacterium* sp. grown in the presence of tungstate to eliminate NO_3^- reductase activity, demonstrated that the rate of NO_2^- reduction was not influenced by the presence of NO_3^-. In cells capable of reducing both NO_3^- and NO_2^-, the rate of NO_2^- accumulation seemed to simply reflect the difference in the rates of NO_3^- and NO_2^- reduction.

The presence of O_2 or its use as an electron acceptor also affects N_2O reduction. Work by Betlach (1979) with several cultures indicated that the amount of N_2O produced relative to N_2 increased with increasing O_2 availability, whereas the overall rate of gas production declined. Similar observations as to the effect of O_2 on the gaseous products of denitrification have been made in soils (Firestone et al., 1979b; Cady & Bartholomew, 1961; Nommik, 1956). These results can be interpreted as indicating that the inhibitory effect of O_2 on N_2O reduction is greater than its effect on preceding reductions. Betlach (1979), however, using a kinetic model of the reductive steps in denitrification, suggested that increased N_2O production relative to N_2 would result with increasing O_2 concentration if O_2 inhibited all steps in denitrification equally.

Under anaerobic conditions, N_2O commonly accumulates in cultures (Delwiche & Bryan, 1976; Payne, 1973) as well as in soils (Garcia, 1975). One explanation commonly given for this observation is that NO_3^- (or NO_2^-) is used preferentially over N_2O as an electron acceptor (Payne, 1973; Delwiche, 1959). This explanation is consistent with observations in soils in which the rate of N_2O reduction was found to decline with increasing NO_3^- or NO_2^- (Firestone et al., 1979b; Blackmer & Bremner, 1978). However, Betlach (1979), working with several different denitrifying cultures, found that the presence of NO_3^- or NO_2^- (and its reduction) had no influence on the rate of N_2O reduction. The differences in the effect of NO_3^- or NO_2^- on N_2O reduction in soils vs. cultures may result from differences in the availability of electron donors (C compounds). If the rate of denitrification is being limited by the availability of reducing equivalents (as may be true in some soils), preferential use of NO_3^- or NO_2^- should be apparent. If, however, denitrification is acceptor limited, then both NO_3^- and N_2O might be reduced at maximum rates.

Sørensen et al. (1980) reported that sulfide inhibited the rate of N_2O reduction (as well as NO reduction) in *P. fluorescens*. A number of other environmental parameters have also been shown to cause increased N_2O production relative to N_2 (Table 2); several of these factors (pH, C, temperature) are discussed in later sections on these topics. In some cases, such as the effects of NO_3^-/NO_2^- and sulfide, the accumulation of N_2O seems to result from inhibition of N_2O reduction. For other factors, the mechanism by which N_2O production is increased relative to N_2 is not clear. Betlach (1979), using a kinetic model of denitrification, predicted that any factor that would cause a decline in the overall rate of denitrification would result in increased N_2O production.

Table 2—Summary of factors that affect the proportion of N_2O and N_2 produced during denitrification

Factor†	Effect on N_2O/N_2
NO_3^- concentration	Increasing NO_3^- increases ratio.
NO_2^- concentration	Increasing NO_2^- increases ratio.
O_2 concentration	Increasing O_2 increases ratio.
pH	Decreasing pH increases ratio and enhances effect of NO_3^-.
Sulfide	Increasing sulfide increases ratio.
Carbon	Increasing C availability has been reported to decrease ratio.
Eh	Redox potential changes below 0 mV do not affect ratio (Sørensen et al., 1980).
Enzyme status	Synthesis of (or absence of) N_2O-reducing activity relative to preceding reductases can increase or decrease ratio.

† See text for references.

D. Characteristics of Specific Reductases

1. NITRATE REDUCTASE

In bacteria, NO_3^- is reduced to NO_2^- for two distinct physiological purposes. In one case, NO_3^- is ultimately reduced to NH_4^+ and can serve as the sole source of N for cellular constituents. In the second case, NO_3^- is used as a terminal electron acceptor during respiration to provide energy for growth under anaerobic conditions. During denitrification, the reduction of NO_3^- to NO_2^- serves the second purpose, and the enzyme catalyzing this reductive step is called *dissimilatory nitrate reductase* (DNR), or *nitrate reductase A,* according to Pichinoty's (1973) proposed distinction. The anaerobically produced NO_3^- reductase A was primarily identified by its ability to reduce chlorate to chlorite. For a discussion of NO_3^- reductase A and B, see Stouthamer (1976). The reduction of NO_3^- to NO_2^- is thought to be similar among all organisms capable of NO_3^- respiration whether the ultimate product of the reduction is NO_2^-, N_2O, N_2, or NH_4^+.

The characteristics of purified DNR vary depending on the bacterial species being studied and the method of purification employed. The DNR enzymes isolated from *Escherichia coli, K. aerogenes,* and *Pa. denitrificans* have been relatively thoroughly studied. The enzymes generally consist of multiple subunits, and in some cases a cytochrome *b* copurifies with the DNR (Stouthamer, 1976). The enzymes studied have several characteristics in common; each contains Mo, Fe (both as heme and nonheme Fe), and labile sulfide groups. As previously discussed, mutants incapable of incorporating Mo produce inactive enzymes (see previous section). When tungstate or vanadate are supplied to microbes in place of molybdate, these metals are incorporated into NO_3^- reductase, producing a nonfunctional enzyme (Betlach, 1979; Scott et al., 1979; Stouthamer, 1976). The presence of Fe or Mo chelating or binding agents, such as thiocyanate (chelates Mo) or bathophenthroline (binds Fe), strongly inhibits DNR activity (Stouthamer, 1976). These studies indicate that Mo and Fe are integral components of the enzyme essential for activity, but they give no indication as to whether the metals are directly involved in the electron transfer to NO_3^- by DNR. Electron paramagnetic resonance (EPR) studies have been used to investigate the oxidation state of Mo in oxidized and reduced NO_3^- reductase from several bacteria, including *Pa. denitrificans, E. coli,* and *K. aerogenes.* The results of these EPR studies are summarized by Stouthamer (1976). Some of the EPR data has been interpreted as indicating that the oxidized NO_3^- reductase contains Mo(V) and that Mo(III) is present in the reduced enzyme (Stouthamer, 1976). However, other interpretations of EPR data have been proposed (Bray et al., 1976). Electron paramagnetic resonance studies of the Fe-S centers indicate that the oxidized NO_3^- reductase contains nonheme Fe(III) and that these Fe-S centers are directly involved in electron transfer to NO_3^- (Stouthamer, 1976).

In almost all denitrifying organisms, at least one type of cytochrome *b* has been found to be involved in electron transport to NO_3^- (Stouthamer,

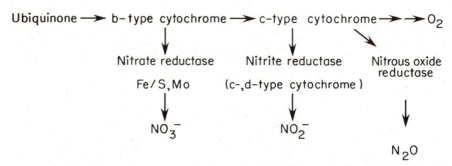

Fig. 1—Hypothetical electron transport scheme for denitrification similar to that proposed for *Pa. denitrificans* by Boogerd et al. (1980), Haddock and Jones (1977), and John and Whatley (1975).

1976). As indicated in Fig. 1, it has been proposed that electron transport to NO_3^- branches off the aerobic electron transport chain at a cytochrome *b* level (Stouthamer, 1980; Haddock & Jones, 1977; Payne, 1976). There is indication that in some organisms, such as *T. denitrificans* (Sawhney & Nicholas, 1978), a cytochrome *c* is involved in electron transport to DNR (Thauer et al., 1977). It should be expected that the specific electron carriers involved in NO_3^- reduction (and denitrification) will vary with different bacteria.

The dissimilatory NO_3^- reductase is membrane bound in most denitrifiers (Haddock & Jones, 1977). There is evidence that in *E. coli* (a NO_3^- respirer), the NO_3^- reductase complex spans the cytoplasmic membrane, and the binding and reduction of NO_3^- occur on the outer side of the membrane (Jones & Garland, 1977; Boxer & Clegg, 1975; Garland et al., 1975). Kristjansson and Hollocher (1979), however, have reported that the NO_3^- binding site in *E. coli* is on the inner aspect of the membrane. Sawada and Satoh (1980) found that NO_3^- reductase is located in the outer region of the membrane in the phototrophic denitrifier *R. sphaeroides*. In *Pa. denitrificans,* the evidence indicates that NO_3^- approaches DNR from the cell cytoplasm and that a NO_3^- carrier transports NO_3^- into the cell (Kristjansson et al., 1978; John, 1977). Kristjansson et al. (1978) propose that NO_3^- (and NO_2^-) enter the cell through facilitated diffusion down a concentration gradient created by internal reduction of the ions.

There are relatively few estimates of the kinetic parameters of purified NO_3^- reductases from denitrifying bacteria or from NO_3^- reducing activities of whole cells. Estimates of K_m values from purified DNR enzymes range from 16 μM for *P. aeruginosa* (Fewson & Nicholas, 1961b) to 1,300 μM for *Paracoccus halodenitrificans* (Rosso et al., 1973). The Michaelis constants for NO_3^- reducing activities in several denitrifying bacteria were reported by Betlach (1979) to be lower than those previously reported for purified reductases (< 15 μM). Estimated K_m values for whole cells may reflect either the kinetics of NO_3^- reduction or NO_3^- transport into the cell.

2. NITRITE REDUCTASE

There appear to be two general types of NO_2^- reductases involved in denitrification. One type, a Cu-containing protein, has been found in *Achromobacter cycloclastes, R. sphaeroides* f. sp. *denitrificans,* and an *Alcaligenes* sp. formerly described as *Pseudomonas denitrificans* (Sawada et al., 1978; Iwasaki & Matsubara, 1972; Iwasaki et al., 1963). The second type of dissimilatory NO_2^- reductase, a cytochrome *cd*, seems to be more common and has been investigated in greater detail. This enzyme consists of two identical subunits, each containing a *c*-type and a *d*-type heme, and has a molecular weight of about 120,000. The characteristics of cytochrome *cd* have been most extensively studied in *Pa. denitrificans* and *P. aeruginosa*; however, the NO_2^- reductase has also been well characterized in *P. perfectomarinus, T. denitrificans* and *Alcaligenes faecalis* (Zumft et al., 1979; Sawhney & Nicholas, 1978; Gudat et al., 1973; Payne, 1973). In at least several of these bacteria, the NO_2^- reductase accepts electrons from the respiratory chain at a level of cytochrome *c* as indicated in Fig. 1 (Bamforth & Quayle, 1978; Payne, 1976; John & Whatley, 1975).

There is considerable disagreement as to the exact location of NO_2^- reductase in the cell. In early work by Cox and Payne (1973) and Matsubara and Iwasaki (1971), NO_2^- reducing activity was found in the soluble fraction of the cell. This was an important point, because the absence of apparent association of the reductase with the cell membrane led a number of workers to conclude that the reduction of NO_2^- was not coupled to ATP formation. More recently, Saraste and Kuronen (1978) found that cytochrome *cd* was associated with the cell membrane of *P. aeruginosa* but that the reductase was easily solubilized by manipulation. In *P. perfectomarinus,* Zumft and Vega (1979) found NO_2^- reductase to occur both in membrane association and in the soluble fraction of the cell. Saraste and Kuronen (1978) reported that cytochrome *cd* was associated with the inner surface of the membrane of *P. aeruginosa*; however, work with the same organism by Wood (1978) indicated a periplasmic location (the outer membrane surface). In two other organisms, a periplasmic location for NO_2^- reductase has also been reported (Sawada & Satoh, 1980; Meijer et al., 1979a). There is general agreement that the enzyme does not occupy a transmembrane location. The cellular location of NO_2^- reductase is of interest for determining whether NO_2^- must be transported into the cell before reduction and for understanding the access of possible competitors or inhibitors to NO_2^- reduction.

In general, the kinetics of NO_2^- reduction in denitrifiers has been found to be relatively complex. Sawada et al. (1978) reported a K_m value of 51 μM NO_2^- for a NO_2^- reductase of the Cu-containing type isolated from a photosynthetic denitrifier. Robinson et al. (1979), in a relatively detailed investigation of the kinetics of cytochrome *cd* from *Pa. denitrificans,* were unable to totally resolve the complexities of the NO_2^- reduction sequence. Among other possible explanations, they suggested that the enzyme may be subject to inhibition by a product of the reaction. The K_m values for NO_2^- in *P.*

aeruginosa were determined by Saraste and Kuronen (1978) for intact cells, cellular supernatant, and the purified enzyme (of the cytochrome *cd* type). In each case, Lineweaver-Burk plots revealed a distinct biphasic character of NO_2^- reduction kinetics, with two K_m value ranges resulting (6 μM and 50 to 85 μM NO_2^-). Saraste and Kuronen postulated several explanations for the biphasic kinetics, including the possibility of interaction between the two NO_2^- binding sites on the dimeric enzyme. Betlach (1979) apparently found simple NO_2^- reduction kinetics in the denitrifying cultures that he studied, reporting K_m values of 12, 6, and 5 μM NO_2^- for an *Alcaligenes* sp., a *Flavobacterium* sp. and *P. fluorescens,* respectively.

The product of NO_2^- reduction by purified NO_2^- reductases has most commonly been found to be NO (LeGall et al., 1979; Zumft et al., 1979; Payne, 1973). However, in several organisms both NO and N_2O are produced by NO_2^- reductase (Zumft & Vega, 1979; Sawhney & Nicholas, 1978; Matsubara & Iwasaki, 1972). Although much evidence does support the single-step, single-electron reduction of NO_2^- to NO by NO_2^- reductase, extrapolations from in vitro, purified reductase studies to the in vivo cell must be made with caution. Considerable controversy still remains as to the in vivo product of NO_2^- reductase and the role of NO as a denitrification intermediate (Zumft & Vega, 1979; St. John & Hollocher, 1977; Delwiche & Bryan, 1976).

The pattern of NO production and reduction in several denitrifying cultures has been found to be consistent with its hypothesized role as an intermediate in denitrification (Betlach, 1979). In NO flux measurements from soils, exhalation of NO has been estimated at 1 kg of $N \cdot ha^{-1} \cdot year^{-1}$ (Gabally & Roy, 1978). However, it is not clear what proportion of the NO commonly encountered in soils results from biological NO_2^- reduction during denitrification. Nitrite is a relatively reactive species in soil, possessing a number of possible chemical fates. The literature on "chemodenitrification" of NO_2^- is extensive and is not reviewed here. Several recent papers have included excellent discussions of NO_2^- reactions and NO formation (Smith & Chalk, 1980a, b; Keeney et al., 1979).

3. NITRIC OXIDE REDUCTASE

Unlike NO_3^- reductase and NO_2^- reductase, the enzyme responsible for NO reduction has not been highly purified or extensively studied. Cell-free fractions from a number of organisms have been found to possess NO reducing activity (Cox & Payne, 1973; Matsubara & Iwasaki, 1971; Miyata, 1971; Payne et al., 1971; Miyata et al., 1969; Fewson & Nicholas, 1961a). In *P. perfectomarinus* and *P. aeruginosa,* the NO reducing activity was found in the soluble fraction (Cox & Payne, 1973; Payne et al., 1971; Fewson & Nicholas, 1961b), whereas in *Alcaligenes* sp., the activity was found associated with the particulate membrane containing fraction (Matsubara & Iwasaki, 1971; Miyata, 1971). As with NO_2^- reductase, the possible soluble nature of NO reducing activity has been interpreted as indicating that NO reduction was not coupled to ATP formation. Work with *P. perfectomarinus* indicated that NO reducing activity was distinct from NO_2^- reductase (Cox & Payne, 1973; Payne et al., 1971); however, in one of the few characterized NO reductases, the activity resulted from cytochrome cd, NO_2^- reductase (Masubara & Iwasaki, 1972). The cytochromes involved in electron transport to NO have not been clearly identified; preliminary evidence indicates the involvement of b-type and c-type cytochromes (Cox & Payne, 1973; Miyata, 1971).

The highly reactive character of the NO molecule has made the isolation and characterization of NO reductases difficult. Nitric oxide exhibits nonspecific binding (and/or reaction) with proteins not associated with in vivo NO reduction (Rowe et al., 1977; Zumft et al., 1979; Matsubara & Iwasaki, 1972). The nature of and, in fact, the unequivocal existence of specific NO reductases remain a major point of controversy in the biochemistry of denitrification (Zumft & Cardenas, 1979).

4. NITROUS OXIDE REDUCTASE

Little is known about N_2O reductase. Significant purification of the enzyme has not yet been accomplished due largely to its labile nature and loss of electron-transport links upon cell breakage (Kristjansson & Hollocher, 1980; Payne et al., 1971; Matsubara & Mori, 1968). The recent description of a cell-free assay system in *Pa. denitrificans* by Kristjansson and Hollocher (1980) may indicate that significant advances in our understanding of this enzyme may be soon forthcoming. In 1973, Cox and Payne reported that N_2O reductase was particle bound in cell-free preparations of *P. perfectomarinus,* whereas Kristjansson and Hollocher (1980) characterize N_2O reductase from *Pa. denitrifcans* as soluble. In *Pa. denitrificans,* N_2O reductase appears to be linked to electron transport at the level of cytochrome c, as indicated in Fig. 1 (Boogerd et al., 1980).

Kristjansson and Hollocher (1980) reported a K_m value for N_2O reduction of 5 μM in cell-free preparations of *Pa. denitrificans*. In intact cells, K_m values of 0.5 μM for *Flavobacterium* sp., 30 to 60 μM for *Alcaligenes* sp., and a value of <100 μM for *P. aeruginosa* have been reported (Betlach, 1980; St. John & Hollocher, 1977; Matsubara & Mori, 1968). Rate measurements of N_2O utilization for kinetic determinations can easily be confounded by gas-liquid phase transfer limitations. This difficulty may explain the range of values reported and would suggest that the lowest estimates may be the most valid. In soil, Yoshinari et al. (1977) reported a N_2O K_m value of 0.7 to 1.0 mM. It is not clear, however, whether these workers attempted to maximize the rate of gas diffusion into the soil solution.

Not all denitrifying bacteria are capable of reducing N_2O to N_2. Several strains, including *Corynebacterium nephridii, Psuedomonas aurefaciens* and *Psuedomonas chlororaphis,* produce N_2O as the terminal product of NO_3^- reduction (Firestone et al., 1979; Greenberg & Becker, 1977; Renner & Becker, 1970). Such organisms, however, appear to be relatively uncommon.

An interesting and technically important characteristic of N_2O reductase is its inhibition by acetylene. Work with cell-free preparations from *Pa. denitrificans* indicates that the inhibitory effect of acetylene on N_2O reduction is on the reductase enzyme itself, not the electron transport system to the reductase, and that acetylene does not compete with the electron donor for a common site on the enzyme (Kristjansson & Hollocher, 1980). In the same investigation, acetylene was shown to be a noncompetative inhibitor with a K_i of 28 μM. The noncompetitive nature of the inhibition (Kristjansson & Hollocher, 1980) should mean that acetylene does not interfere with N_2O binding by the enzyme, that the degree of inhibition is independent of N_2O concentration, and that the inhibition is reversible. It is somewhat surprising that acetylene acts as a noncompetitive inhibitor of N_2O reduction since the structural similarities of the two gases would suggest a dual competition at the active site of the enzyme. The characteristics of acetylene inhibition of N_2O reductase from cell-free studies are consistent with those reported by Yoshinari and Knowles (1976) and Balderston et al. (1976) who originally reported the effectiveness of the inhibition in bacterial cultures of denitrifiers and suggested the application to assays of denitrification in natural systems.

If in the presence of acetylene, N_2O is the sole product of denitrification, then analysis is greatly simplified, because N_2O, unlike N_2, is a minor atmospheric constituent and can be assayed by sensitive gas chromatographic detectors. The basic assumption that the quantity of N_2O produced in the presence of acetylene is a valid measure of the total gaseous N that would be produced in the absence of the inhibitor has been confirmed by ^{13}N and ^{15}N techniques in soils (Ryden et al., 1979a; Smith et al., 1978). Many investigators (including Ryden et al., 1979b; Klemedtsson et al., 1978; Smith et al., 1978; Yoshinari et al., 1977) have now successfully employed this technique to measure denitrification rates in soil. However, significant problems have been encountered in applying the acetylene inhibition technique to soils. Smith et al. (1978) found that relatively high concentrations of acetylene (0.1 atm) were required for effective inhibition at low NO_3^- concentrations. Yeomans and Beauchamp (1978) reported that N_2O reducing activity in soil was not totally inhibited by 1 kPa of acetylene (0.01 atm) over a relatively long incubation (168 hours), thus suggesting the presence of denitrifiers in soil that are not readily inhibited by acetylene. Work by Tam and Knowles (1979) indicated that in the presence of sulfide, acetylene did not effectively block N_2O reduction in soil. However, sulfide did not relieve the acetylene inhibition of N_2O reduction in *P. aeruginosa* or *T. denitrificans*. These authors suggest that the presence of sulfide allows the development of anaerobic microorganisms that require a low redox potential and are not readily inhibited by acetylene (Tam & Knowles, 1979). Possibly the most significant problem in the application of this technique arises from the fact that acetylene is a biologically active compound. Nitrification is one of a number of metabolic processes that is inhibited by acetylene (Walter et al., 1979; Hynes & Knowles, 1978). The application of the acetylene inhibition technique for assaying denitrification in soils should still prove useful despite these problems if incubation is limited to short periods of time and if diffusion of acetylene to active sites in soil is optimized.

E. Energy Conservation During Denitrification

Denitrification supports bacterial life through respiration. When a reduced substrate, such as an organic C compound, donates electrons via carriers to an oxidized N oxide, the resulting decline in free energy is harnessed to the production of ATP, thus conserving a portion of the energy. It is well established that such electron transport phosporylation (ADP → ATP) occurs during denitrification (Payne, 1973). However, the sites of ATP generation during electron transport and the number of ATP's produced during reduction of individual N oxide species are not as well established and are the topic of current research. This discussion of energy production or ATP generation during denitrification focuses on evidence from growth yield studies with information from proton translocation studies mentioned as confirming or contradicting evidence.

The underlying assumption in growth yield studies is that cell mass yield is directly related to ATP production. When the availability of electron acceptors (O_2 or N oxides) is the factor limiting growth, yield studies should give information concerning the relative efficiencies of ATP production during O_2 respiration vs. NO_3^- respiration.

It is generally agreed that on the basis of yield per number of electrons transported, growth on NO_x is less efficient than is growth on O_2; exactly how much less efficient varies depending on the organisms studied, the methods employed, and the electron donor employed. Koike and Hattori (1975a) found that O_2 respiration was about 1.7 times more efficient than NO_3^- reduction to N_2 in supporting growth of an organism described as *P. denitrificans*. Studies with *Pseudomonas stutzeri* by Elliot and Gilmour (1971) also showed a relative efficiency of 1.7 for growth on O_2 vs. NO_3^-. Similar relative efficiencies were reported for proton translocation studies (1.9) by both Kristjansson et al. (1978) and van Verseveld et al. (1977) with *Pa. denitrificans*. Recent growth studies by Bryan (1980), however, indicate slightly less energy yield from denitrification relative to O_2 respiration with *P. aeruginosa* (2.6) and *P. stutzeri* (2.9, relative efficiency for growth on O_2 vs. NO_3^-).

How one interprets the relative growth efficiencies on O_2 and NO_3^- depends on the electron transport scheme hypothesized. For example, if in the scheme proposed for *Pa. denitrificans* shown in Fig. 1, one assumes one site of phosphorylation preceding cytochrome *b*, a second site between cytochrome *b* and *c*, and a third site after cytochrome *c*, available only during O_2 respiration, then the following interpretation could be made. During NO_3^- reduction to NO_2^-, only the first site of ATP generation would be used, whereas during reduction of the other N oxides, the first two sites would be available; this was the basic assumption used by van Verseveld et al. (1977). If this is the case, then 1.6 ATP would be generated for every two electrons used during NO_3^- reduction to N_2, whereas three ATP's per two electrons would result from O_2 reduction. This would make O_2 reduction about 1.9 times more efficient than NO_3^- reduction to N_2 and would be consistent with several of the reports mentioned earlier. However, both Koike and Hattori (1975b) and Bryan (1980) found that the efficiency of growth on NO_3^-, NO_2^-, and N_2O as terminal acceptors was proportional to the oxidation state of the N, indicating that the same number of ATP's were being generated for each two electron reductive step. This is also roughly consistent with proton translocation studies by Kristjansson et al. (1978). The hypothesis that there is a single site of phosphorylation during electron transport to each N oxide and that the location of this site is the same for each N oxide (as proposed by Bryan, 1980) would be consistent with the electron transport scheme proposed by Payne (1976).

III. DENITRIFICATION IN SOIL

A. Carbon Supply

It is well established that denitrification in soil is strongly dependent on the availability of organic compounds as electron donors and as sources of cellular material. The presence of ample C substrate can also cause rapid O_2 consumption and possible depletion in soil microenvironments, thus also indirectly enhancing the potential for denitrification. One might be tempted to conclude that any factor that increased C availability in soil, such as high native organic matter content, organic C amendments, or the presence of plant roots, would increase denitrification. This relationship between C and denitrification is frequently found to be true, but the net effect of C amendments and plant roots on denitrification are not easily predicted.

1. SOIL ORGANIC MATTER

In 1958, Bremner and Shaw (1958a, b) observed the general relationship between soil organic matter content and denitrification. Denitrification could be detected by total N analysis in soils of higher organic matter content and was not measurable in soils with < 1% native soil C. McGarity (1961) also found a close relationship between denitrification and the quantity of native soil organic C. However, it is the availability of this C that is important, not simply the total organic C present in soil. In 1975, Craswell and Martin (1975b) concluded that "if denitrifying capacity is to be accurately predicted in the field, . . . some account of the content of readily decomposable organic matter must be made." That same year, two groups reported procedures for determining available or readily decomposable C. Burford and Bremner (1975) reported that water-soluble C or mineralizable C (determined by CO_2 evolution during a 7-day incubation) were highly correlated with denitrification activity. Working on the assumption that 1 μg of available C was required for production of 1.17 μg of N as N_2O or of 0.99 μg of N as N_2, Burford and Bremner calculated that the amount of water-soluble C found could account for about 71% of the N_2O and N_2 evolved and that the mineralizable C determined was about 1.36 times that required. They concluded that quantification of either mineralizable C or water-soluble organic C was a good index for the capacity of a soil to denitrify. Stanford et al. (1975) found that extractable "glucose-equivalent," determined by boiling soils for 1 hour in $0.01M$ $CaCl_2$, was highly correlated with denitrification potential in soil and provided a reliable basis for predicting a denitrification rate constant based on C availability. It should be noted that both groups worked with air-dried soils and that recent work by Patten et al. (1980) indicated that the routine procedure of drying the soil before use increases the capacity of the soil to denitrify by increasing the availability of C. The higher the temperature of drying and the longer the storage period, the greater was the effect on C availability and denitrification. This indicates that determination of available C would be more useful when applied to field moist soils.

Since denitrification is strongly dependent on organic C availability, denitrification activity might be expected to be high in soils containing a high percentage of organic C. Studies on organic soils and high C marsh soils using laboratory incubation techniques indicate that this may be the case. The organic soils studied by Reddy et al. (1980) and Bartlett et al. (1979) had high potentials for denitrification of added NO_3^-. Terry and Tate (1980) reported an average denitrification rate of 0.7 kg of $N \cdot ha^{-1} \cdot day^{-1}$ in laboratory incubations of a cropped Histosol. Engler and Patrick (1974) also reported high rates of NO_3^- removal (7.6 $ppm \cdot day^{-1}$) by a saltwater marsh soil containing 12% organic C. Although laboratory incubation studies would indicate high denitrification potentials in soils containing large amounts of organic material, there are few field estimates available to confirm these studies. Particularly in flooded organic soils, the availability of NO_3^- may limit denitrification activity. However, the observation that denitrification potential in an Everglade Histosol was enhanced by the presence of plant roots may indicate that even in this organic soil, denitrification was C limited (Terry & Tate, 1980).

2. ORGANIC SOIL AMENDMENTS

There are numerous reports that the addition of organic materials, such as plant residues or manure, greatly increases denitrification activity. The effects of these additions, however, are dependent on the quality of the organic material added. The addition of manure, a readily decomposable C source that is relatively high in N, greatly enhances denitrification both in laboratory and field studies (Rolston et al., 1979; Guenzi et al., 1978). The addition of plant materials to soil has also been found to enhance denitrification potential in laboratory studies (Brar et al., 1978), with cellulose supporting greater rates of denitrification than straw or grass, which was greater than sawdust or lignin (Bremner & Shaw, 1958b). However, Craswell (1978) reported that the addition of wheat straw to cores of a clay soil reduced N loss to denitrification due to increased immobilization of N.

3. RHIZOSPHERE

The effects of plants on denitrification appear to be relatively complex. Work by Woldendorp (1962, 1963) indicated that the presence of plant roots stimulated denitrification by supplying C and causing a localized zone of lowered O_2 concentration. Woldendorp reasoned that root respiration and microbial respiration, enhanced by root excretions, reduced the O_2 tension in the soil near roots; he found that about 67% of the O_2 consumption in his system resulted from roots and about 33% resulted from the root-associated microbes (Woldendorp, 1962, 1963).

Smith and Tiedje (1979a) found that the potential for denitrification was much higher in soil near corn roots (*Zea mays* L.) and decreased rapidly in the first few millimeters of soil away from roots. Bailey (1976) also reported a stimulation of denitrification in the presence of corn roots. In this work, the root containing soil cores were incubated anaerobically with NO_3^- added, so that the observed stimulation was most likely due to increased C availability.

Volz et al. (1976), comparing fallow plots and plots planted to barley (*Hordeum vulgare* L.), found NO_3^- losses in excess of that taken up by the roots and attributed this to NO_3^- reduction and denitrification. During the early stages of an irrigation, barley absorbed over half of the NO_3^- removed from solution, whereas after about 24 hours of ponding, microbes were responsible for about 90% of the NO_3^- disappearance. Carbon availability (by "glucose-equivalent" test) was greater and redox potential was lower in the irrigated (ponded) barley plot. They concluded that in the early stages of an irrigation event, root uptake of NO_3^- was favored by the "aerobic" conditions, whereas after several hours of ponding, the decreased O_2 availability (by ODR probes) in the root zone inhibited root uptake and enhanced microbial NO_3^- utilization.

Stefanson (1972a) reported that the presence of wheat roots (*Triticum aestivum* L.) enhanced denitrification at all water contents below field capacity. Above field capacity, soil denitrification was limited by NO_3^- and C availability. Stefanson recognized the confounding interaction of pore space relationships, native soil organic C availability, and the form of N added with rhizosphere denitrification (Stefanson, 1972a-c, 1976). In soils of lower organic matter content, wheat roots stimulated denitrification of added NO_3^-, whereas no effect was found in soils with higher native C (Stefanson, 1976). However, when N was added as NH_4^+, the presence of roots increased denitrification in higher organic matter soil but decreased denitrification in soil with lower C content (Stefanson, 1972b).

Several groups have now reported that under certain conditions, denitrification is not enhanced by the presence of plant roots. In soil columns to which wastewater was applied, Lance et al. (1978) reported that the presence of barley or bermudagrass (*Cynodon dactylon* L. Pers.) did not increase N loss due to denitrification. In this case, the presence of vegetation did not affect the redox status of the soil in the root zone since it was already reduced by flooding. Guenzi et al. (1978) found that in soils receiving manure additions, the presence of plants decreased the total denitrification, and they attributed this effect to root competition for added NO_3^-.

In 3 years of field studies with barley, Kowalenko and Cameron (1978) found that plants did not influence the amount of N lost (presumably) to denitrification. These workers noted that maximum rates of denitrification roughly coincided with periods of maximum uptake of N by the barley and that the soil profiles in cropped fields were often drier than those in fallow fields.

Smith and Tiedje (1979a) reported that denitrification rates were higher in planted than unplanted soils in the presence of sufficient NO_3^-, but when NO_3^- availability was low, the presence of plants decreased denitrifying activity.

Under some soil conditions, O_2 availability may be decreased in the microenvironment of the rhizosphere. However, in flooded soils the opposite situation can occur. In flooded rice paddies (*Oryza sativa* L.), O_2 availability near rice roots may be higher than in the surrounding soil due to transport of O_2 to the root zone by the plant. The influence of these oxidized zones on N transformations in flooded soils is discussed in Chapter 12 (Patrick).

Thus, plant roots have the potential to influence soil denitrification in a number of ways. Roots can:

1) Provide C, which serves to support a population capable of denitrification and acts as electron donor for NO_3^- reduction when O_2 availability is low.
2) Create anaerobic zones through respiratory O_2 consumption when O_2 supply is sufficiently limited by pore space/water conditions.
3) Create a drier soil near the roots through evapotranspiration, thus increasing the rate of O_2 diffusion in the root zone.
4) Increase O_2 availability near roots of aquatic plants in flooded soils or sediments.
5) Remove NO_3^- through uptake from the potentially active rhizosphere soil.

The net effect of these influences may be positive or negative, or the summation may yield no net effect. Whether the presence of roots increases or decreases denitrification in field soils must depend on the status of a number of other soil parameters, including availability of native soil C, O_2 diffusion controlers, such as pore size and water content, and the rate of NO_3^- supply to the root zone.

4. DEPTH

To some degree the changes in denitrifying activity that occur with increasing depth in soil reflect the distribution of organic matter in the profile. However, in addition to C availability, other soil characteristics that change with depth, such as porosity/permeability, pH, temperature, and water table, may also be important controllers of denitrification.

Kahn and Moore (1968) found that the potential for denitrification was the greatest in the surface samples from six profiles of mineral soils. In a peat soil, high potentials were found to a depth of 100 cm. Kahn and Moore, as well as Bailey and Beauchamp (1973), reported decreasing denitrification activity with increasing depth of sampling when C and NO_3^- were added, indicating that the depth function reflected the presence of populations capable of denitrification. That the populations of denitrifiers commonly decrease with depth was confirmed by most probable number counts (Brar et al., 1978).

By sampling gases at different depths in a Yolo loam field plot planted in ryegrass (*Lolium perenne* L.), Rolston et al. (1976) found patterns of O_2, CO_2, and N_2 that indicated that the zone of denitrification was limited to the surface 60 cm. This pattern was attributed to the presence of C sources and anoxic microsites in the upper horizons.

Cho et al. (1979) determined denitrification rates in profile samples incubated under flooded conditions, finding an exponential decrease in rates with depths down to 150 cm. This decrease directly paralleled microbial activity (indicated by CO_2 production) but not total organic matter content.

Organic matter content of soil usually decreases with depth, but available C may occur in the deeper horizons. Myers and McGarity (1971, 1972) suggested that organic C leaching from the surface layers to a solodized solonetz B horizon supported relatively high denitrification activity. Similarly, Volz et al. (1976) hypothesized that leaching of C from root sources facilitated NO_3^- loss from below the root zone.

Gilliam et al. (1978) found that significant amounts of denitrification could occur at depths of 30 to 75 cm and that it was supported by C indigenous to those depths or by C input from percolating water. These investigators pointed out that C availability in subsoils may be quite variable in soils from different regions and that organic content of surface horizons was not necessarily indicative of subsurface C status. This work also clearly indicated that any profile characteristic that caused impedance of water movement should be expected to increase the potential for denitrification.

5. INFLUENCE ON NITROUS OXIDE PRODUCTION

The availability of C or electron donor seems to influence the proportion of N_2O and N_2 produced. It is generally concluded that increasing C availability decreases the ratio of N_2O/N_2 (Smith & Tiedje, 1979a; Focht & Verstraete, 1977; Nommik, 1956; Delwiche, 1959). However in most of the work on this subject, changes in C availability were accompanied by changes in other factors that may have affected the product composition. The information available concerning C influence on N_2O production, reduction, or both is yet much too fragmentary for meaningful generalization.

The importance of C availability in controlling denitrification is apparent in the relatively extensive literature on native organic matter, C additions to soil, plant roots, and depth in profile. But the regulatory role of C in soil denitrification is tightly intertwined with O_2 regulation of the process. Considering either O_2 or C control of denitrification alone cannot realistically describe the complex process.

B. Oxygen Control

It is relatively well understood that the absence of O_2 or reduced O_2 availability is required for both synthesis and activity of the enzymes involved in denitrification. The expression of this critical controller is relatively easily observed in both artificial and natural soil systems. Yet the control of denitrification by O_2 in natural soil systems is exceedingly difficult to realistically quantify and properly predict or encompass by modeling efforts.

1. BULK SOIL PARAMETERS THAT INFLUENCE OXYGEN CONTROL

That denitrification activity is inversely related to the water content soils is almost universally observed. This simply results from the fact that the supply of O_2 to metabolically active microsites is strongly affected by the amount of water through which the O_2 must diffuse. The rate of diffusion of O_2 through water is about 10^4 times less than through air. Soil parameters that have been used in attempts to describe O_2 control of this process include soil moisture content, soil water potentials, partial pressures of O_2, O_2 concentrations in solution, percentage air-filled porosities, and redox potentials. The variety of techniques used to describe the O_2 status in soil is in itself an impediment to a comparative summary. Oxygen concentrations in soil atmosphere at which denitrification has been observed range from 4 to 17% O_2 (Gilliam et al., 1978; Volz et al., 1976; Stefanson, 1973; Wijler & Delwiche, 1954). Pilot and Patrick (1972) found that the critical air-filled porosities below which denitrification became significant ranged from 11 to 14% in several soils. The finer the soil texture, the higher the critical air-filled porosity. Brandt et al. (1964), using Pt electrodes to amperimetrically measure rates of O_2 diffusion, found that significant denitrification occurred at O_2 diffusion rates <0.2 μg of $O_2 \cdot cm^{-2} \cdot min^{-1}$. Others, such as Burford and Stefanson (1973), have found that soil water contents were useful in predicting the potential for denitrification. However, Craswell and Martin (1974) found no significant denitrification in a well-structured clay soil even with moisture contents up to 90% (by weight); only at moisture contents $>100\%$ did they record significant denitrification. In terms of soil water potentials, Pilot and Patrick (1972) reported that critical moisture tensions at which denitrification occurred in several soils ranged between 20 and 40 cm (corresponding to about 2 and 4 kPa); the finer the soil texture, the higher was the critical moisture tension. However, Craswell (1978) reported a critical moisture pressure of about -10 kPa. Ryden and Lund (1980) reported that peak denitrification rates occurred between suctions of 50 to 100 mbars (5–10 kPa) and that above about 250 mbars (25 kPa) the rates were quite low. They also recorded very low rates of denitrification occurring at 50 mbars (5 kPa), but this was immediately following an irrigation event, and the redox potential of the soil solution had not yet dropped. When the soil was maintained for prolonged periods at around 50 mbars, high rates of denitrification resulted. The various redox potentials at which denitrification has been reported to be significant range between 300 and 650 mV (Ryden & Lund, 1980; Bailey & Beauchamp, 1973).

It is inherent to all of the parameters discussed above that they measure a gross or average characteristic of the soil. They give little direct information as to the O_2 status at microsites in soil. Gilliam et al. (1978) determined moisture content, tension, Eh, O_2 content of atmosphere, water conductivity, and denitrification in columns of several soils of distinct textural character. They found that to some degree, the differences in denitrification rates were not reflected by differences in any of the measured parameters. They concluded that it was the microstructure of soil that was critical.

2. OXYGEN AVAILABILITY AT SOIL MICROSITES

The occurrence of anaerobic microsites has long been used to explain how an O_2 sensitive process can occur in a well-structured aerobic soil. This microsite concept, advanced by Greenwood (1961) and Currie (1961) proposed that anaerobic sites occur within centers of saturated soil aggregates due to impeded diffusion of O_2 and consumption of O_2 by microbes throughout the aggregate. In 1961, Greenwood concluded that water-saturated aggregates with diameters >3 mm could have anaerobic centers. In this work, however, O_2 consumption was stimulated by the addition of glucose. In more recent calculations, Greenwood (1975) concludes that O_2-free zones will not occur in soils having gas-filled pore space >10% and no water-saturated aggregates >9 mm in radius. Greenwood worked from the equation:

$$R = (6DSP_R/M_t)^{1/2}.$$

This defines the maximum spherical radius (R) that an aggregate can possess without having an O_2-free center as a function of D, the diffusion coefficient of O_2 through water-saturated soil, taken to be 1×10^{-5} $cm^2 \cdot sec^{-1}$; S, the solubility of O_2 in the soil water, taken to be 0.028 $ml \cdot ml^{-1} \cdot atm^{-1}$; P_R, the partial pressure of O_2 taken to be 0.21 atm; and M_t, the rate of O_2 uptake. Greenwood considered that M_t, the rate of O_2 consumption, was dependent on the C content of the soil and was enhanced by the presence of roots in soil. He based this value on a literature survey, assuming a cropped soil containing 2% organic C, and determined that the overall rate of O_2 uptake in such a soil would be 4.31×10^{-7} $ml \cdot ml^{-1} \cdot sec^{-1}$. Using this value in the equation given, he determined that only aggregates with the radius >9 mm were likely to contain O_2-free regions. It does, however, seem quite likely that localized regions with O_2 uptake rates significantly greater than this average value would occur in soil. Of the parameters on which this calculation is based, the rate of O_2 consumption must be by far the most variable. Craswell and Martin (1975a) concluded that "the anaerobic pockets of soil postulated by many workers are pockets of intense respiratory activity rather than of passive anaerobiosis." The occurrence of such microsites of intense O_2 consumption is thus largely dependent on localized C availability. Such microsites may occur in rhizosphere soil, in pockets of decomposing plant material, or in small zones of water-saturated soil. Again, the interaction of O_2 and C in controlling denitrification is critical to understanding the process in soil.

3. EXTRAPOLATION OF CELLULAR OXYGEN CONTROL TO SOILS

Extrapolations from pure culture studies to natural soils can be rather equivocal. Certainly the environment experienced by the microbe and its physiological state must be quite different in the two situations. With this important limitation in mind, some of the physiological control characteristics discussed in previous sections are related to control of denitrification in soil.

One commonly considers that the environment experienced by denitrifiers in soil is one of "aerobic" conditions or "anaerobic" conditions. Yet the microenvironment most commonly inhabited by these organisms may be best characterized as one of reduced O_2 availability, somewhere between fully aerobic and fully anaerobic. The observation that NO_3^- reductase is synthesized under "semiaerobic" conditions in several denitrifiers has interesting implications for soils. This enzyme may be present in environments in which O_2 availability is high enough to repress its activity. As a microenvironment becomes more anaerobic, the existing NO_3^- reducing enzymes would be expected to become active. If NO_3^- reductase is synthesized before NO_2^- reductase (due either to a temporal lag or differential O_2 control), then one would expect to find transient NO_2^- accumulations in soils becoming more anaerobic. Accumulations of NO_2^- are not commonly observed in "aerobic" field soils, but such NO_2^- accumulations are found in newly flooded soils. Transient accumulations of NO_2^- have been reported by a number of investigators in newly flooded soil columns and field plots (Volz & Starr, 1977; Volz et al., 1975, 1976; Bailey, 1976; Doner et al., 1974, 1975). The common explanation given for this observation is that NO_3^- respirers outnumber denitrifiers in soil (Volz & Heichel, 1979; Volz & Starr, 1977; Volz et al., 1975, 1976). This may be the case, or it may be that those organisms considered to be capable of only NO_3^- reduction to NO_2^- (in MPN determinations) may in fact be denitrifiers (Gamble et al., 1977; Volz, 1977).

The full suite of denitrification enzymes may be synthesized in aerobic soils under conditions in which they are not fully active. Smith and Tiedje (1979b) found that when O_2 was removed from aerobic soils, active denitrification commenced almost immediately. The rate of denitrification determined in the first hour of anaerobic incubation was not altered by the presence of chloramphenicol, an inhibitor of protein synthesis. This early phase of denitrification represented the maximum potential activity of denitrification enzymes existing in an aerobic soil. The actual denitrification activity in the aerobic soil appeared to be controlled primarily by O_2 availability since NO_3^- and glucose additions did not alter the rate in the first hour of anaerobic incubation. Hence synthesis of denitrifying enzymes seems to occur in soil at levels of O_2 that are inhibitory to enzyme activity.

The observation that some denitrifiers cannot grow after an abrupt shift from highly aerobic to anaerobic conditions may have little meaning for most soils. The removal of O_2 from soil solution would be due to respiratory consumption and should provide conditions under which derepression and synthesis of denitrification enzymes could readily occur.

C. Nitrate Supply

For many years it was commonly concluded that the rate of denitrification in soil was independent of NO_3^- concentration (Focht & Verstraete, 1977). The zero-order kinetics reported may have resulted from the high concentration of NO_3^- employed, commonly 100 mg of NO_3^--N•liter^{-1} or greater (Focht & Chang, 1975). Starr and Parlange (1975) and Stanford et al. (1975) found that denitrification kinetics in soils appeared to be first-order when concentrations of < 40 mg N/liter were employed.

The K_m values estimated for NO_3^- reduction in soils are much higher than those obtained with cultures or purified enzymes, ranging from 0.13 mM NO_3^- for a St. Bernard sandy loam without C amendment (Yoshinari et al., 1977) to 12 mM NO_3^- for a fine sand with glucose added (Bowman & Focht, 1974). However, the kinetic parameters for NO_3^- reduction (or denitrification) in soils are not simple to interpret. Bowman and Focht (1974) recognized that the kinetics of denitrification must reflect both C availability and NO_3^- availability and hence attempts to fit their data to an equation incorporating K_m values for both C and NO_3^-. However, Michaelis-Menton kinetics did not accurately describe rate dependence on C and NO_3^- in an unmixed Hanford soil. Similarly, Doner et al. (1974) proposed a multistep sequence in which the enzyme first reacted with NO_3^- and then with reductant, but the determination of kinetic parameters in this column study was thwarted by the difficulty of measuring the C available from soil organic matter. Kohl et al. (1976) also recognized that the kinetic parameters that they obtained for denitrification most likely reflected the kinetics of C limitation more than those of NO_3^-. These workers concluded that in a system as complex as soil, the kinetic mechanism of a reaction cannot necessarily be determined by examining the goodness of fit of the data to a proposed model.

More recently, Philips et al. (1978) and Reddy et al. (1978) questioned whether such kinetic analyses in soil actually reflect denitrification at all. They concluded that measurements of NO_3^- reduction in flooded soil frequently measured the rate of NO_3^- diffusion from solution into the soil matrix rather than biological denitrification. They suggested that the first-order kinetics for NO_3^- concentration reported by several groups resulted from the concentration dependence of NO_3^- diffusion rather than NO_3^- reduction.

Nitrate reduction during denitrification in soil is an enzyme catalyzed reaction. It is inherent to enzyme catalysis that the finite number of catalytic sites must exhibit concentration dependent response as well as saturation kinetics. It is probable that no one has yet examined the NO_3^- concentration ranges in soils in which NO_3^- reduction would be expected to be first order. As previously mentioned, Betlach (1979) found K_m values to be < 15 μM

NO_3^- for several denitrifying soil isolates. He was not able to determine specific K_m values, however, because the colorimetric assays commonly used to quantify NO_3^- were not sufficiently sensitive to quantify NO_3^- in concentration ranges this low. A NO_3^- concentration of 15 μM would translate to a soil weight concentration of 0.04 ppm in a soil with a water content of 20 g of water/100 g of dry weight soil. If K_m values for NO_3^- reduction in soil reflected those of the biological catalyst, then the first-order concentration range would be below that measurable by traditional colorimetric means and, in fact, below bulk NO_3^- concentrations commonly encountered in fertilized soils. The advent of easily applicable, more sensitive techniques for NO_3^- quantification should yield useful data concerning the kinetics of NO_3^- reduction in soil. High-pressure liquid chromatography with ultraviolet detectors holds promise to be such a technique (Thayer & Huffaker, 1980). Sensitive techniques, coupled with recognition of NO_3^- diffusion phenomenon and C influence, should lead to better understanding of denitrification control in soils. However, the inherent heterogeneity of natural soils in terms of NO_3^- concentration, C availability, aeration, and microbial distribution should make the elucidation of denitrification kinetics in natural soils a challenging topic for some time to come.

D. Effect of Temperature

Ideally, the dependence of denitrification rate on temperature should be described by the Arrhenius equation:

$$\ln v = (-\Delta H^*/RT) + C$$

where v is the velocity, ΔH^* is the activation energy, R is the gas constant, T is temperature in °K, and C is a constant. Thus, denitrification activity would be expected to increase exponentially with increasing temperature temperature within the range of enzyme activity. Usually the minimum temperature for biological activity is several degrees above the freezing point of water, and the maximum temperature is established by thermal denaturation of proteins (Stanier et al., 1976).

Focht (1974) found that Nommik's data (1956) were consistent with the exponential relationship predicted. Others have also reported that the Arrhenius equation is predictive of the relationship between denitrification and temperature (Dawson & Murphy, 1972). But biochemical systems are not well characterized by this exponential function below about 15°C (Ingraham, 1962), and temperatures in this range must be considered to describe denitrification in soil. Craswell (1978) suggests that the interaction of temperature with O_2 control may be quite complex due to the effect of temperature on O_2 solubility and O_2 consumption as well as O_2 diffusion. Decreasing temperature would increase O_2 solubility and decrease O_2 consumption. Misra et al. (1974) found that increasing O_2 concentration had a greater inhibitory effect on denitrification at lower temperature. These problems may explain why data from other investigators does not seem to be consistent with such an exponential relationship (Cho et al., 1979; Keeney et al., 1979). Cho et al. (1979) reported a linear relationship between denitrification intensity and temperature between 2.7 and 20°C.

Bollag et al. (1970) found that four denitrifiers isolated from soil grew best at 30°C (comparing 20, 22, 30, and 37°C), and all grew poorly at 10°C. Characterizing 95 denitrifying isolates from various temperate soils (mean annual temperature, $\leq 20°C$), Gamble et al. (1977) reported that 68% were capable of growth at 4°C, 10% were capable of growth at 41°C, and 22% could grow only at room temperature. From tropical soils (temperature $> 20°C$), none of the 33 isolates could grow at 4°C, 67% grew at 41°C, and 33% grew only at room temperature. These data suggest that denitrifiers are adapted to and capable of growth over a relatively wide soil temperature range.

Reports of minimum temperatures for the occurrence of denitrification in soil range from 2.7 to 10°C (Cho et al., 1979; Craswell, 1978; Bailey, 1976). Craswell (1978) reported that as temperature decreased, the minimum soil water content for denitrification to occur increased. At 10°C, he could not detect denitrification even in effectively flooded soil cores.

Maximum temperatures for denitrification in soil seem to be about 75°C (Keeney et al., 1979; Bremner & Shaw, 1958b). The optimum temperature has been observed to be around 65°C (Bremner & Shaw, 1958b; Nommik, 1956). Focht and Verstraete (1977) suggest that this high optimum results from the presence of thermophilic *Bacillus* species. More recently, however, Keeney et al. (1979) have shown that chemical decomposition reactions of NO_2^- become quite important above 50°C. They hypothesize that in the 50 to 67°C range, thermophilic organisms rapidly reduce NO_3^- to NO_2^-, which then reacts with oxidized-N functional groups on organic compounds to form nitrogenous gases. Thus, the commonly reported optimum temperature for denitrification may well result from stimulation of chemical decomposition of NO_2^- in the range of 50 to 70°C.

In some cases, decreasing soil temperature seems to cause an increasing proportion of N_2O as the product of biological denitrification (Keeney et al., 1979; Bailey, 1976; Nommik, 1956), whereas in other instances, temperature seems to have little effect (Bailey & Beauchamp, 1973).

E. Effect of pH

The pH range and optima for denitrifying organisms appears to be similar to that for heterotrophic organisms generally. In the relatively neutral pH range of soils (pH 6–8), there is little effect of pH on denitrification rates (Burford & Bremner, 1975; Stanford et al., 1975; Khan & Moore, 1968; Wijler & Delwiche, 1954). However, in naturally acid soils, denitrification activity is inhibited. Klemedtsson et al. (1978) found low rates of denitrification in a peat soil of pH 3.5. Since addition of carbon and nutrients did not increase the rate and since increasing the pH to 6.5 strongly stimulated denitrification, they concluded that pH was the limiting factor. Müeller et al. (1980) determined denitrification rates in several horizons from 22 Spodosols and peat soils with pH values ranging from 3.6 to 7.3. Denitrifying activity was highly correlated with soil pH but not with organic C content. Denitrification potential was comparable for soils of similar pH values despite wide differences in soil types and vegetation cover. Rates were higher in B horizons of the Spodosols, presumably reflecting the more neutral pH's of the lower horizons. The average rate of denitrification was much lower in the acid soils than in the soils of pH >5, but denitrification activity was detectable and could account for significant N loss. Several other groups have also reported that significant amounts of denitrification can occur in soils of pH <5 (Gilliam & Gambrell, 1978; Van Cleemput & Patrick, 1974; Ekpete & Cornfield, 1965).

It is interesting to speculate whether the low rates of denitrification that have been found in extremely acid soils (pH 3.5–4.0) result from a few species with low pH tolerances/optima, from a small population in microsites of more hospitable pH, or from a general population with neutral growth optima that function poorly at low pH's. The tremendous increase in denitrification activity associated with neutralization of an acid peat soil (Klemedtsson et al., 1978) would suggest that the latter possibilities may be the case. Most denitrifying isolates studied have neutral growth optima (Bollag et al., 1970; Valera & Alexander, 1961), but then most of these strains were isolated (and hence selected for) on neutral laboratory media.

It is also interesting to speculate whether the limited ability of denitrifiers to function in acid soils results from a direct effect of soil solution pH or from pH-induced deficiencies or toxicities. Since Mo availability decreases with declining pH and since NO_3^- reductase is a molybdo-protein, a Mo deficiency would seem to be a possibility. However, Bremner and Shaw (1958b) found that the addition of neither Mo nor Ca increased denitrification in a soil of pH 3.6. With decreasing pH, NO_2^- formed by NO_3^- reduction would become more toxic, and solubilization of Al or Mn might cause toxicity effects. Preliminary evidence from Bremner and Shaw (1958b) suggested that NO_3^- reduction to NH_4^+ was favored in moderately acid soils compared with slightly alkaline soils. It is possible that increasing acidity may favor NO_3^- reduction to NH_4^+ rather than denitrification.

It has commonly been observed that the proportion of N_2O occurring as product of denitrification increases as pH decreases, with N_2O frequently appearing as the dominant product in acid soils (Blackmer & Bremner, 1978; Nommik, 1956; Wijler & Delwiche, 1954). Blackmer and Bremner (1978) suggested that soil acidity enhances the influence of NO_3^- on the composition of the gaseous products, and work by Firestone et al. (1980) supported this hypothesis. The effect of acidity on N_2O production (or reduction to N_2) appears to be an immediate effect, as opposed to causing a population shift. Tiedje et al. (1981) used ^{13}N to continuously monitor the gaseous products of denitrification from a Brookston soil during acidification from pH 6.7 to 5.2. At pH 6.7, N_2O was a minor product component, but when the pH was lowered to 5.2, N_2O immediately (within 3 min) became the dominant product. The pH shift had little effect on the total rate of gas production (N_2 + N_2O). It may be that N_2O reductase is quite sensitive to pH or that decomposition to N_2O of an NO_x intermediate may be promoted by acidity.

IV. CONCLUDING COMMENTS

In the past 6 years there has been a substantial increase in research on and interest in denitrification. To some degree, our understanding of the process has expanded correspondingly. Increased access to and technology for the stable isotcpe ^{15}N and advent of the acetylene inhibition technique have resulted in relatively reliable estimates of field denitrification. Improved sensitivity through the use of ^{63}Ni electron capture detectors in gas chromatography, the radioactive isotope ^{13}N, and high-pressure liquid chromatography coupled with ultraviolet detection of NO_3^- and NO_2^- has and will yield new information from laboratory and field experiments. At the cellular and subcellular level, new genetic techniques, proton translocation studies, and chemostat techniques should continue to increase understanding of the microbiology and biochemistry of denitrification. The ecology of denitrifiers in soil has been a difficult area to investigate; traditional counting techniques have produced little meaningful information. The use of antibiotic resistant isolates and fluorescent antibody techniques may hold some promise for a better understanding of the relationship of denitrifiers to their soil environment.

There are many aspects of denitrification in which research is needed. Included among these areas are carbon metabolism by denitrifying organisms, the effects of carbon quality on denitrification in soils, the importance of rhizosphere denitrification in field N losses, O_2 control at the cellular and molecular level, and a means of meaningfully quantifying O_2 control in soil microsites.

Increased understanding of the process may never yield useful specific inhibitors, but expanding knowledge should allow improved field management practices involving timing of fertilizer application, form of N fertilizer, timing of irrigation, and tillage and mulching techniques.

To be most useful, the biochemical and microbiological information gained must continuously be related to denitrification in soil, and the reductive process occurring in soil must be integrated with the other segments of the N cycle.

LITERATURE CITED

Bailey, L. D. 1976. Effects of temperature and root on denitrification in a soil. Can. J. Soil Sci. 56:79–87.

Bailey, L. D., and E. G. Beauchamp. 1973. Effects of moisture, added NO_3^-, and macerated roots on NO_3^- transformation and redox potential in surface and subsurface soils. Can. J. Soil Sci. 53:219–230.

Balderston, W. L., B. Sherr, and W. J. Payne. 1976. Blockage by acetylene of nitrous oxide reduction in *Pseudomonas perfectomarinus*. Appl. Environ. Microbiol. 31:504–508.

Bamforth, C. W., and J. R. Quayle. 1978. Aerobic and anaerobic growth of *Paracoccus denitrificans* on methanol. Arch. Microbiol. 119:91–97.

Bartlett, M. S., L. C. Brown, N. B. Hanes, and N. H. Nickerson. 1979. Denitrification in freshwater wetland soil. J. Environ. Qual. 8:460–464.

Betlach, M. R. 1979. Accumulation of intermediates during denitrification—kinetic mechanisms and regulation of assimilatory nitrate uptake. Ph.D. Thesis. Michigan State Univ. Univ. Microfilms. Ann Arbor, Mich. (Diss. Abstr. 80:06083).

Blackmer, A. M., and J. M. Bremner. 1978. Inhibitory effect of nitrate on reduction of N_2O to N_2 by soil microorganisms. Soil Biol. Biochem. 10:187–191.

Blackmer, A. M., J. M. Bremner, and E. L. Schmidt. 1980. Production of nitrous oxide by ammonia-oxidizing chemoautotrophic microorganisms in soil. Appl. Environ. Microbiol. 40:1060–1066.

Bollag, J. M., M. L. Orcutt, and B. Bollag. 1970. Denitrification by isolated soil bacteria under various environmental conditions. Soil Sci. Soc. Am. Proc. 34:875–879.

Bollag, J. M., and G. Tung. 1972. Nitrous oxide release by soil fungi. Soil Biol. Biochem. 4: 271–276.

Boogerd, F. C., H. W. van Versveld, and H. H. Stouthamer. 1980. Electron transport to nitrous oxide in *Paracoccus denitrificans*. FEBS Lett. 113:279–284.

Bowman, R. A., and D. D. Focht. 1974. The influence of glucose and nitrate concentrations upon denitrification rates in sandy soil. Soil Biol. Biochem. 6:297–301.

Boxer, D. H., and R. A. Clegg. 1975. A transmembrane location for the proton translocating reduced ubiquinone—nitrate reductase segment of the respiratory chain of *Escherichia coli*. FEBS Lett. 60:54–57.

Brandt, G. H., A. R. Wolcott, and A. E. Erickson. 1964. Nitrogen transformations in soil as related to structure, moisture, and oxygen diffusion rate. Soil Sci. Soc. Am. Proc. 28:71–75.

Brar, S. S., R. H. Miller, and T. J. Logan. 1978. Some factors affecting denitrification in soils irrigated with wastewater. J. Water Pollut. Control Fed. 50:709–717.

Bray, R. C., S. P. Vincent, D. J. Lowe, R. A. Clegg, and P. B. Garland. 1976. Electron-paramagnetic-resonance studies on the molybdenum of nitrate reductase from *Escherichia coli* K12. Biochem. J. 155:201–203.

Bryan, B. A. 1980. Cell yield and energy characteristics of denitrification with *Pseudomonas stutzeri* and *Pseudomonas aeruginosa*. Ph.D. Thesis. Univ. California, Davis. Univ. Microfilms. Ann Arbor, Mich. (Diss. Abstr. 80:27039).

Bremner, J. M., and K. Shaw. 1958a. Denitrification in soil. I. Methods of investigation. J. Agric. Sci. 51:22–39.

Bremner, J. M., and K. Shaw. 1958b. Denitrification in soil. II. Factor affecting denitrification. J. Agric. Sci. 51:39–52.

Broadbent, F. E., and F. Clark. 1965. Denitrification. *In* W. V. Bartholomew and F. E. Clark (ed.) Soil nitrogen. Agronomy 9:344–359. Am. Soc. of Agron., Madison, Wis.

Buchanan, R. E., and N. E. Gibbons. 1974. Bergey's manual of determinative bacteriology. 8th ed. The Williams & Wilkins Co., Baltimore.

Burford, J. R., and J. M. Bremner. 1975. Relationships between the denitrification capacities of soils and total, water soluble and readily decomposable soil organic matter. Soil Biol. Biochem. 7:389–394.

Burford, J. R., and R. C. Stefanson. 1973. Measurement of gaseous losses of nitrogen from soils. Soil Biol. Biochem. 5:133–141.

Burke, K. A., K. Calder, and J. Lascelles. 1980. Effects of molybdenum and tungsten on induction of nitrate reductase and formate dehydrogenase in wild type and mutant *Paracoccus denitrificans*. Arch. Microbiol. 126:155–159.

Cady, F. B., and W. V. Bartholomew. 1961. Influence of low PO_2 on denitrification processes and products. Soil Sci. Soc. Am. Proc. 25:362–365.

Calder, K., K. A. Burke, and J. Lascelles. 1980. Induction of nitrate reductase and membrane cytochromes in wild type and chlorate-resistant *Paracoccus denitrificans*. Arch. Microbiol. 126:149–153.

Carlson, C. A. 1981. On regulating the synthesis of nitrate reductase. p. 445–450. *In* J. M. Lyons, R. C. Valentine, D. A. Phillips, D. W. Rains, and R. C. Huffaker (ed.) Genetic engineering of symbiotic nitrogen fixation and conservation of fixed nitrogen. Plenum Press, New York.

Cho, C. M., L. Sakdinan, and C. Chang. 1979. Denitrification intensity and capacity of three irrigated Alberta soils. Soil Sci. Soc. Am. J. 43:949–950.

Cox, C. D., and W. J. Payne. 1973. Separation of soluble denitrifying enzymes and cytochromes from *Pseudomonas perfectomarinus*. Can. J. Microbiol. 19:861–872.

Craswell, E. T. 1978. Some factors influencing denitrification and nitrogen immobilization in a clay soil. Soil Biol. Biochem. 10:241–245.

Craswell, E. T., and A. E. Martin. 1974. Effect of moisture content on denitrification in a clay soil. Soil Biol. Biochem. 6:127–129.

Craswell, E. T., and A. E. Martin. 1975a. Isotopic studies of the nitrogen balance in a cracking clay. I. Recovery of added nitrogen from soil and wheat in the glasshouse and gas lysimeter. Aust. J. Soil Res. 13:43–52.

Craswell, E. T., and A. E. Martin. 1975b. Isotope studies of the nitrogen balance in a cracking clay. II. Recovery of Nitrate ^{15}N added to columns of packed soil and microplots growing wheat in the field. Aust. J. Soil Res. 13:53–61.

Crutzen, P. J. 1970. The influence of nitrogen oxides on the atmospheric ozone content. Q. J. R. Meteorol. Soc. 96:320–325.

Crutzen, P. J., and D. H. Ehhalt. 1977. Effects of nitrogen fertilizers and combustion on the stratospheric ozone layer. Ambio 6:112–117.

Currie, J. A. 1961. Gaseous diffusion in porous media. III. Wet granular materials. Br. J. Appl. Phys. 12:275–281.

Dawson, R. N., and K. L. Murphy. 1972. The temperature dependency of biological denitrification. Water Res. 6:71–83.

Delwiche, C. C. 1959. Production and utilization of nitrous oxide by *Pseudomonas denitrificans*. J. Bacteriol. 77:55–59.

Delwiche, C. D., and B. A. Bryan. 1976. Denitrification. Annu. Rev. Microbiol. 30:241–262.

Doner, H. E., M. G. Volz, L. W. Belser, and J.-P. Loken. 1975. Short term nitrate losses and associated microbial populations in soil columns. Soil Biol. Biochem. 7:261–263.

Doner, N. E., M. G. Volz, and A. D. McLaren. 1974. Column studies of denitrification in soil. Soil Biol. Biochem. 6:341–346.

Ekpete, D. M., and A. H. Cornfield. 1965. Effect of pH and addition of organic materials on denitrification losses from soil. Nature (London) 208:1200.

Elliott, R. G., and C. M. Gilmour. 1971. Growth of *Pseudomonas stutzeri* with nitrate and oxygen as terminal electron acceptors. Soil Biol. Biochem. 3:331–335.

Engler, R. M., and W. H. Patrick, Jr. 1974. Nitrate removal from flood water overlying flooded soils and sediments. J. Environ. Qual. 3:409–413.

Escalante-Semerena, J. C., R. P. Blakemore, and R. S. Wolfe. 1980. Nitrate dissimilation under microaerophilic conditions by a magnetic spirillum. Appl. Environ. Microbiol. 40: 429–430.

Eskew, D. L., D. D. Focht, and I. P. Ting. 1977. Nitrogen fixation denitrification, and pleomorphic growth in a highly pigmented *Spirillum lipoferum*. Appl. Environ. Microbiol. 34:582–585.

Fewson, C. A., and D. J. D. Nicholas. 1961a. Nitric oxide reductase from *Pseudomonas aeruginosa*. Biochem. J. 78:9P.

Fewson, C. A., and D. J. D. Nicholas. 1961b. Nitrate reductase from *Pseudomonas aeruginosa*. Biochem. Biophys. Acta. 49:355–349.

Firestone, M. K., R. B. Firestone, and J. M. Tiedje. 1979a. Nitric oxide as an intermediate in denitrification: Evidence from nitrogen-13 isotope exchange. Biochem. Biophys. Res. Commun. 91:10–16.

Firestone, M. K., R. B. Firestone, and J. M. Tiedje. 1980. Nitrous oxide from soil denitrification: Factors controlling its biological production. Science 208:749–751.

Firestone, M. K., M. S. Smith, R. B. Firestone, and J. M. Tiedje. 1979b. The influence of nitrate, nitrite and oxygen on the composition of the gaseous products of denitrification in soil. Soil Sci. Soc. Am. J. 43:1140–1144.

Firestone, M. K., and J. M. Tiedje. 1979. Temporal change in nitrous oxide and dinitrogen from denitrification following onset of anaerobiosis. Appl. Environ. Microbiol. 38:673–679.

Focht, D. D. 1974. The effect of temperature, pH and aeration on the production of nitrous oxide and gaseous nitrogen—a zero-order kinetic model. Soil Sci. 118:173–179.

Focht, D. D., and A. C. Chang. 1975. Nitrification and denitrification processes related to waste water treatment. p. 153–186. *In* D. Perlman (ed.) Advances in applied microbiology, vol. 19. Academic Press, Inc., New York.

Focht, D. D., and W. Verstraete. 1977. Biochemical ecology of nitrification and denitrification. p. 135–214. *In* M. Alexander (ed.) Advances in microbial ecology, Vol. 1. Plenum Press, New York.

Galbally, I. E., and C. R. Roy. 1978. Loss of fixed nitrogen from soils by nitric oxide exhalation. Nature (London) 275:734–735.

Gamble, T. N., M. R. Betlach, and J. M. Tiedje. 1977. Numerically dominant denitrifying bacteria from world soils. Appl. Environ. Microbiol. 33:926–939.

Garcia, J. L. 1975. La denitrification dans les sols. Bull. Inst. Pasteur Paris. 73:167–193.

Garland, P. B., J. A. Downie, and B. A. Haddock. 1975. Proton translocation and the respiratory nitrate reductase of *Escherichia coli*. Biochem. J. 152:547–559.

Gilliam, J. W., S. Dasberg, L. J. Lund, and D. D. Focht. 1978. Denitrification in four California soils: Effect of soil profile characteristics. Soil Sci. Soc. Am. J. 42:61–66.

Gilliam, J. W., and R. P. Gambrell. 1978. Temperature and pH as limiting factors in loss of nitrate from saturated Atlantic Coastal Plain soils. J. Environ. Qual. 7:526–532.

Goreau, T. J., W. A. Kaplan, S. C. Wofsy, M. B. McElroy, F. W. Valois, and S. W. Watson. 1980. Production of NO_2^- and N_2O by nitrifying bacteria at reduced concentrations of oxygen. Appl. Environ. Microbiol. 40:526–532.

Greenberg, E. P., and G. E. Becker. 1977. Nitrous oxide as end product of denitrification by strains of fluorescent pseudomonads. Can. J. Microbiol. 23:903–907.

Greenwood, D. J. 1961. The effect of oxygen concentration on the decomposition of organic materials in soil. Plant Soil 14:360–376.

Greenwood, D. J. 1975. Measurement of soil aeration in soil physical conditions and crop production. Min. Agric. Fish Food, Tech. Bull. no. 29, H.M.S.O., London, p. 261–272.

Gudat, J. C., J. Singh, and D. C. Wharton. 1973. Cytochrome oxidase from *Pseudomonas aeruginosa*. I. Purification and some properties. Biochem. Biophys. Acta 292:376–390.

Guenzi, W. D., W. E. Beard, F. S. Watanabe, S. R. Olsen, and L. K. Porter. 1978. Nitrification and denitrification in cattle manure-amended soil. J. Environ. Qual. 7:196–202.

Haddock, B. A., and C. W. Jones. 1977. Bacterial respiration. Bacteriol. Rev. 41:47–99.

Hall, J. B. 1978. Nitrate-reducing bacteria. p. 296–298. *In* D. Schlessinger (ed.) Microbiology, 1978. Am. Soc. Microbiol., Washington, D.C.

Hauck, R. D. 1981. Nitrogen fertilizer effects in nitrogen cycle processes. p. 551–562. *In* F. E. Clark and T. Rosswall (ed.) Terrestrial nitrogen cycles. Ecol. Bull. 33. Swedish Natural Science Research Council, Stockholm.

Hollocher, T. C., E. Garber, A. J. L. Cooper, and R. E. Reiman. 1980. ^{13}N, ^{15}N isotope and kinetic evidence against hyponitrite as an intermediate in denitrification. J. Biol. Chem. 255:5027–5030.

Hom, S. S. M., H. Hennecke, and K. T. Shanmugm. 1980. Regulation of nitrogenase biosynthesis in *Klebsiella pneumoniae*: Effect of nitrate. J. Gen. Microbiol. 117:169–179.

Hooper, A. B., and K. R. Terry. 1979. Hydroxylamine oxidoreductase of *Nitrosomonas* production of nitric oxide from hydroxylamine. Biochim. Biophys. Acta 571:12–20.

Hynes, R. K., and R. Knowles. 1978. Inhibition by acetylene of ammonia oxidation in *Nitrosomonas europaea*. FEMS Microbiol. Lett. 4:319–321.

Ingraham, J. L. 1962. Temperature relationships. p. 265–296. *In* I. C. Gunsalus and R. Y. Stanier (ed.) The bacteria, Vol. 4. Academic Press, Inc., New York.

Ishaque, M., and M. I. H. Aleem. 1973. Intermediates of denitrification in the chemoautotroph *Thiobacillus denitrificans*. Arch. Microbiol. 94:269–282.

Iwasaki, H., and T. Matsubara. 1972. A nitrite reductase from *Achromobacter cycloclastes*. J. Biochem. (Tokyo) 71:645–652.

Iwasaki, H., S. Shidara, H. Suzuki, and T. Mori. 1963. Studies on denitrification. VII. Further purification and properties of denitrifying enzyme. J. Biochem. (Tokyo) 53:299–303.

Jeter, R. M., and J. L. Ingraham. 1981. The denitrifying procaryotes. p. 000–000. *In* M. P. Starr (ed.) The procaryotes. Springer-Verlag, New York, New York.

John, P. 1977. Aerobic and anaerobic bacterial respiration monitored by electrodes. J. Gen. Microbiol. 98:231–238.

John, P., and F. R. Whatley. 1975. *Paracoccus denitrificans* and the evolutionary origin of the mitochondrion. Nature (London) 254:495–498.

Johnston, H. S. 1971. Reduction of stratospheric ozone by nitrogen oxide catalysts from SST exhaust. Science 173:517–522.

Jones, R. W., and P. B. Garland. 1977. Sites and specificity of the reaction of bipyridyllium compounds with anaerobic respiratory enzymes of *Escherichia coli*: Effects of permeability barriers imposed by the cytoplasmic membrane. Biochem. J. 164:199–211.

Justin, P., and D. P. Kelley. 1978. Metabolic changes in *Thiobacillus denitrificans* accompanying transition from aerobic to anaerobic growth in continuous chemostat culture. J. Gen. Microbiol. 107:131–137.

Keeney, D. R., I. R. Fillery, and G. P. Marx. 1979. Effect of temperature on the gaseous nitrogen products of denitrification in a silt loam soil. Soil Sci. Soc. Am. J. 43:1124–1128.

Khan, M. F. A., and A. W. Moore. 1968. Denitrification capacity of some Alberta soils. Can. J. Soil Sci. 48:89–91.

Kissel, D. E., and S. J. Smith. 1978. Fate of fertilizer nitrate applied to Coastal bermudagrass on a swelling clay soil. Soil Sci. Soc. Am. J. 42:77–80.

Klemedtsson, L., B. H. Svensson, T. Lindberg, and T. Rosswall. 1978. The use of acetylene inhibition of nitrous oxide reductase in quantifying denitrification in soils. Swed. J. Agric. Res. 7:179–185.

Kohl, D. H., F. Vithayathil, P. Whitlow, G. Shearer, and S. H. Chien. 1976. Denitrification kinetics in soil systems: The significance of good fits of data to mathematical forms. Soil Sci. Soc. Am. J. 40:249–253.

Koike, I., and A. Hattori. 1975a. Growth yield of a denitrifying bacterium. *Pseudomonas denitrificans*, under aerobic and denitrifying conditions. J. Gen. Microbiol. 88:1–10.

Koike, I., and A. Hattori. 1975b. Energy yield of denitrification: An estimate from growth yield in continuous cultures of *Pseudomonas denitrificans* under nitrate-, nitrite-, and nitrous oxide-limited conditions. J. Gen. Microbiol. 88:11–19.

Kowalenko, C. G. 1978. Nitrogen transformations and transport over 17 months in field fallow microplots using ^{15}N. Can. J. Soil Sci. 58:69–76.

Kowalenko, C. G., and D. R. Cameron. 1978. Nitrogen transformations in soil-plant systems in three years of field experiments using tracer and non-tracer methods on an ammonium-fixing soil. Can. J. Soil Sci. 58:195–208.

Kristjansson, J. K., and T. C. Hollocher. 1979. Substrate binding site for nitrate reductase of *Escherichia coli* is on the inner aspect of the membrane. J. Bacteriol. 137:1227–1233.

Kristjansson, J. K., and T. C. Hollocher. 1980. First practical assay for soluble nitrous oxide reductase of denitrifying bacteria and a partial kinetic characterization. J. Biol. Chem. 255:704–707.

Kristjansson, J. K., B. Walter, and T. C. Hollocher. 1978. Respiration-dependent proton translocation and transport of nitrate and nitrite in *Paracoccus denitrificans* and other denitrifying bacteria. Biochemistry 23:5014–5019.

Lance, J. C., R. C. Rice, F. D. Whisler. 1978. Effects of vegetation on denitrification and phosphate movement during rapid infiltration of soil columns. J. Water Pollut. Control Fed. 50:2183–2188.

LeGall, J., W. J. Payne, T. V. Morgan, and D. DerVartanian. 1979. On the purification of nitrite reductase from *Thiobacillus denitrificans* and its reaction with nitrite under reducing conditions. Biochem. Biophys. Res. Commun. 87:355–362.

Magalhaes, L. M. S., C. A. Neyra, and J. Döbereiner. 1978. Nitrate and nitrate reductase negative mutants of N_2-fixing *Azospirillum* spp. Arch. Microbiol. 117:247–252.

Matsubara, T. 1971. Studies on denitrification. XIII. Some properties of the N_2O-anaerobically grown cell. J. Biochem. (Tokyo) 69:991–1001.

Matsubara, T., and H. Iwasaki. 1971. Enzymatic steps of dissimilatory nitrite reduction in *Alcaligenes faecalis*. J. Biochem. (Tokyo) 69:859–868.

Matsubara, T., and H. Iwasaki. 1972. Nitric oxide-reducing activity of *Alcaligenes faecalis* cytochrome cd. J. Biochem. (Tokyo) 72:57–64.

Matsubara, T., and T. Mori. 1968. Studies on denitrification. IX. Nitrous oxide, its production and reduction to nitrogen. J. Biochem. (Tokyo) 64:863–871.

McElroy, M. B., S. C. Wofsy, and Y. L. Yung. 1977. The nitrogen cycle: perturbations due to man and their impact on atmospheric N_2O and O_3. Philos. Trans. R. Soc. London Ser. B. 277:159–181.

McGarity, J. W. 1961. Denitrification studies on some South Australian soil. Plant Soil 14:1–21.

Meijer, E. M., J. W. Van Der Zwaan, and A. H. Stouthamer. 1979a. Location of the proton-consuming site in nitrite reduction and stoichiometrics for proton pumping in anaerobically grown *Paracoccus denitrificans*. FEMS Microbiol. Lett. 5:369–372.

Meijer, E. M., J. W. Van Der Zwaan, R. Wever, and A. H. Stouthamer. 1979b. Anaerobic respiration and energy conservation in *Paracoccus denitrificans*. Functioning of iron-sulfur centers and the uncoupling effect of nitrite. Eur. J. Biochem. 96:69–76.

Misra, C., D. R. Nielsen, and J. W. Biggar. 1974. Nitrogen transformations in soil during leaching: II. Steady state nitrification and nitrate reduction. Soil Sci. Soc. Am. Proc. 38:294–299.

Miyata, M. 1971. Studies on denitrification. XIV. The electron donating system in the reduction of nitric oxide and nitrate. J. Biochem. (Tokyo) 70:205–213.

Miyata, M., T. Matsubara, and T. Mori. 1969. Studies on denitrification. XI. Some properties of nitric oxide reductase. J. Biochem. (Tokyo) 66:759–765.

Müller, M. M., V. Sundman, and J. Skujins. 1980. Denitrification in low pH Spodosols and peats determined with the acetylene inhibition method. Appl. Environ. Microbiol. 40: 235–239.

Myers, R. J. K., and J. W. McGarity. 1971. Factors influencing high denitrifying activity in the subsoil of solodized solonetz. Plant Soil 35:145–160.

Myers, R. J. K., and J. W. McGarity. 1972. Denitrification in undisturbed cores from a solodized solonetz B horizon. Plant Soil 37:81–89.

Nelson, L. M., and R. Knowles. 1978. Effect of oxygen and nitrate on nitrogen fixation and denitrification by *Azospirillum brasilense* grown in continuous culture. Can. J. Microbiol. 24:1395–1403.

Neyra, C. A., J. Döbereiner, R. Lalande, and R. Knowles. 1977. Denitrification by N_2-fixing *Spirillum lipoferum*. Can. J. Microbiol. 23:300–305.

Neyra, C. H., and P. van Berkum. 1977. Nitrate reduction and nitrogenase activity in *Spirillum lipoferum*. Can. J. Microbiol. 23:306–310.

Nommik, H. 1956. Investigations on denitrification in soil. Acta Agric. Scand. 6:195–228.

Patten, D. K., J. M. Bremner, and A. M. Blackmer. 1980. Effects of drying and air-dry storage of soils on their capacity for denitrification of nitrate. Soil Sci. Soc. Am. J. 44: 67–70.

Payne, W. J. 1973. Reduction of nitrogenous oxides by microorganisms. Bacteriol. Rev. 37: 409–452.

Payne, W. J. 1976. Denitrification. Trends Biochem. Sci. 1:220–222.

Payne, W. J., and W. L. Balderston. 1978. Denitrification. p. 339–342. *In* D. Schlessinger (ed.) Microbiology, 1978. Am. Soc. Microbiol., Washington, D.C.

Payne, W. J., and P. S. Riley. 1969. Suppression by nitrate of enzymatic reduction of nitric oxide. Proc. Soc. Exp. Biol. Med. 132:258–260.

Payne, W. J., P. S. Riley, and C. D. Cox. 1971. Separate nitrite, nitric oxide, and nitrous oxide reducing fractions from *Pseudomonas perfectomarinus*. J. Bacteriol. 106:356–361.

Phillips, R. E., K. R. Reddy, and W. H. Patrick. 1978. The role of nitrate diffusion in determining the order and rate of denitrification in flooded soil: II. Theoretical analysis and interpretation. Soil Sci. Soc. Am. J. 42:272–278.

Pichinoty, F. 1973. La reduction bacterienne des composes oxygenes mineraux de l'azote. Bull. Inst. Pasteur Paris 71:317–395.

Pichinoty, F., J. Bigliardi-Rouvier, M. Mandel, B. Greenway, G. Metenier, and J.-L. Garcia. 1976. The isolation and properties of a denitrifying bacterium of the genus *Flavobacterium*. Antonie von Leeuwenhoek. J. Microbiol. Serol. 42:349–354.

Pichinoty, F., J.-L. Garcia, M. Mandel, C. Job, and M. Durand. 1978a. Isolation of bacteria that use nitric oxide as a respiratory electron acceptor in anaerobiosis. C. R. Acad. Sci. Ser. D. 286:1403–1405.

Pichinoty, F., M. Mandel, and J.-L. Garcia. 1977. Etude de six souches de *Agrobacterium tumefaciens* et *A. radiobacter*. Ann. Microbiol. (Paris) 128A:303–310.

Pichinoty, F., M. Vernon, M. Mandel, M. Durand, C. Job, and J.-L. Garcia. 1978b. Etude physiologique et taxonomique du genre *Alcaligenes: A. denitrificans, A. odorans* et *A. faecalis*. Can. J. Microbiol. 24:743–753.

Pilot, L., and W. H. Patrick. 1972. Nitrate reduction in soils: effect of soil moisture tension. Soil Sci. 114:312–316.

Pratt, P. F., J. C. Barber, M. L. Corrin, J. Goering, R. D. Hauck, H. S. Johnston, A. Klute, R. Knowles, D. W. Nelson, R. C. Pickett, and E. R. Stephens. 1977. Effect of increased nitrogen fixation on stratospheric ozone. Climatic Change 1:109–135.

Reddy, K. R., W. H. Patrick, and R. E. Phillips. 1978. The role of nitrate diffusion in determining the order of denitrification in flooded soil: I. Experimental results. Soil Sci. Soc. Am. J. 42:268–272.

Reddy, K. R., P. D. Sacco, and D. A. Graets. 1980. Nitrate reduction in an organic soil-water system. J. Environ. Qual. 9:283–288.

Renner, E. D., and G. E. Becker. 1970. Production of nitric oxide and nitrous oxide during denitrification by *Corynebacterium nephridii*. J. Bacteriol. 101:821–826.

Ritchie, G. A. F., and D. J. D. Nicholas. 1972. Identification of the sources of nitrous oxide produced by oxidative and reductive processes in *Nitrosomonas europea*. Biochem. J. 126:1181–1191.

Ritchie, G. A. F., and D. J. D. Nicholas. 1974. The partial characterization of purified nitrite reductase and hydroxylamine oxidase from *Nitrosomonas europaea*. Biochem. J. 138: 471–480.

Robinson, M. K., K. Martinkus, P. J. Kennelly, and R. Timkovich. 1979. Implications of the integrated law for the reactions of *Pseuomonas denitrificans* nitrite reductase. Biochemistry 18:3921–3926.

Rolston, D. E., F. E. Broadbent, and D. A. Goldhamer. 1979. Field measurement of denitrification: II. Mass balance and sampling uncertainty. Soil Sci. Soc. Am. J. 43:703–708.

Rolston, D. E., M. Fried, and D. A. Goldhamer. 1976. Denitrification measured directly from nitrogen and nitrous oxide gas fluxes. Soil Sci. Soc. Am. J. 40:259–266.

Rosso, J. P., P. Forget, and F. Pichinoty. 1973. Les nitrate reductases bacteriennes. Solubilization, purification et proprieties de l'enzyme a de *Micrococcus halodenitrificans*. Biochim. Biophys. Acta 321:443–445.

Rowe, J. J., B. F. Sherr, W. J. Payne, and R. G. Eagon. 1977. A unique nitric oxide-binding complex formed by denitrifying *Pseudomonas aeruginosa*. Biochem. Biophys. Res. Commun. 77:253–258.

Ryden, J. C., and L. J. Lund. 1980. Nature and extent of directly measured denitrification losses from some irrigated vegetable crop production units. Soil sci. Soc. Am. J. 44:505–511.

Ryden, J. C., L. J. Lund, and D. D. Focht. 1979a. Direct measurement of denitrification loss from soils: I. Laboratory evaluation of acetylene inhibition of nitrous oxide reduction. Soil Sci. Soc. Am. J. 43:104–110.

Ryden, J. C., L. J. Lund, J. Letey, and D. D. Focht. 1979b. Direct measurement of denitrification loss from soils: II. Development and application of field methods. Soil Sci. Soc. Am. J. 43:110–118.

St. John, R. T., and T. C. Hollocher. 1977. Nitrogen 15 tracer studies on the pathway of denitrification in *Pseudomonas aeruginosa*. J. Biol. Chem. 252:212–217.

Saraste, M., and T. Kuronen. 1978. Interaction of Pseudomonas cytochrome cd, with the cytoplasmic membrane. Biochim. Biophys. Acta. 513:117–131.

Satoh, T. 1977. Light-activated, inhibited and independent denitrification by a denitrifying phototrophic bacterium. Arch. Microbiol. 115:293–298.

Sawada, E., and T. Satoh. 1980. Periplasmic location of dissimilatory nitrate and nitrite reductases in a denitrifying phototrophic bacterium, *Rhodopseudomonas sphaeroides* forma sp. *denitrificans*. Plant Cell Physiol. 21:205–210.

Sawada, E., T. Satoh, and H. Kitamura. 1978. Purification and properties of a dissimilatory nitrite reductase of a denitrifying phototrophic bacterium. Plant Cell Physiol. 19:1339–1531.

Sawhney, V., and D. J. D. Nicholas. 1978. Sulphide-linked nitrite reductase from *Thiobacillus denitrificans* with cytochrome oxidase activity: purification and properties. J. Gen. Microbiol. 106:119–128.

Scott, D. B., and C. A. Scott. 1978. Nitrate-dependent nitrogenase activity in *Azospirillum* spp. under low oxygen tensions. p. 350–351. *In* J. Dobereiner et al. (ed.) International symposium on the limitations and potentials for biological nitrogen fixation in the tropics. Plenum Press, New York.

Scott, R. H., G. T. Sperl, and J. A. DeMoss. 1979. *In vitro* incorporation of molybdate into demolybdoproteins in *Escherichia coli*. J. Bacteriol. 137:719–726.

Sias, S. R., and J. L. Ingraham. 1979. Isolation and analysis of mutants of *Pseudomonas aeruginosa* unable to assimilate nitrate. Arch. Microbiol. 122:263–270.

Sias, S. R., A. H. Stouthamer, and J. L. Ingraham. 1980. The assimilatory and dissimulatory nitrate reductases of *Pseudomonas aeruginosa* are encoded by different genes. J. Gen. Microbiol. 118:229–234.

Smith, C. J., and P. M. Chalk. 1980a. Gaseous nitrogen evolution during nitrification of ammonia fertilizer and nitrite transformations in soils. Soil Sci. Soc. Am. J. 44:277–282.

Smith, C. J., and P. M. Chalk. 1980b. Fixation and loss of nitrogen in transformations of nitrite in soils. Soil Sci. Soc. Am. J. 44:288–291.

Smith, M. S., M. K. Firestone, and J. M. Tiedje. 1978. The acetylene inhibition method for short-term measurement of soil denitrification and its evlauation using nitrogen-13. Soil Sci. Soc. Am. J. 42:611–615.

Smith, M. S., and J. M. Tiedje. 1979a. The effect of roots on soil denitrification. Soil Sci. Soc. Am. J. 43:951–955.

Smith, M. S., and J. M. Tiedje. 1979b. Phases of denitrification following oxygen depletion in soil. Soil Biol. Biochem. 11:261–267.

Smith, M. S., and J. M. Tiedje. 1980. Growth and survival of antibiotic-resistant denitrifier strains in soil. Can. J. Microbiol. 26:854–856.

Soil Science Society of America, Terminology Committees. 1979. Glossary of soil science terms. Rev. ed. Soil Sci. Soc. of Am., Madison, Wis.

Sørensen, J., J. M. Tiedje, and R. B. Firestone. 1980. Inhibition by sulfide of nitric and nitrous oxide reduction by denitrifying *Pseudomonas fluorescens*. Appl. Environ. Microbiol. 39:105-108.

Stanford, G., R. A. Vander Pol, and S. Dzienia. 1975. Denitrification rates in relation to total and extractable soil carbon. Soil Sci. Soc. Am. Proc. 39:284-289.

Stanier, R. Y., E. A. Adelberg, and J. Ingraham. 1976. The microbial world. Prentice-Hall, Inc., Englewood Cliffs, N.J.

Starr, J. L., and J.-V. Parlange. 1975. Non-linear denitrification kinetics with continuous flow in soil columns. Soil Sci. Soc. Am. Proc. 39:875-880.

Stefanson, R. C. 1972a. Soil denitrification in sealed soil-plant systems. I. Effect of plants, soil water content and soil organic matter content. Plant Soil 33:113-127.

Stefanson, R. C. 1972b. Soil denitrification in sealed soil-plant systems. II. Effect of soil water content and form of applied nitrogen. Plant Soil 37:129-140.

Stefanson, R. C. 1972c. Effect of plant growth and form of nitrogen fertilizer on denitrification from four south Australian soils. Aust. J. Soil Res. 10:183-195.

Stefanson, R. C. 1973. Evolution patterns of nitrous oxide and nitrogen in sealed soil plant systems. Soil Biol. Biochem. 5:167-169.

Stefanson, R. C. 1976. Denitrification from nitrogen fertilizer placed at various depths in the soil-plant system. Soil Sci. 121:353-363.

Stouthamer, A. H. 1976. Biochemistry and genetics of nitrate reductase in bacteria. p. 315-375. *In* A. H. Rose and D. W. Tempest (ed.) Advances in microbial physiology. Academic Press, Inc., New York.

Stouthamer, A. H. 1980. Bioenergetic studies on *Paracoccus denitrificans*. Trends Biochem. Sci. 5:164-166.

Swain, H. M., H. J. Somerville, and J. A. Cole. 1978. Denitrification during growth of *Pseudomonas aeruginosa* on octane. J. Gen. Microbiol. 107:103-112.

Sze, N. D., and H. Rice. 1976. Nitrogen cycle factors contributing to N_2O production from fertilizers. Geophys. Res. Lett. 3:343-346.

Tam, T.-Y., and R. Knowles. 1979. Effect of sulfide and acetylene on nitrous oxide reduction by soil and by *Pseudomonas aeruginosa*. Can. J. Microbiol. 25:1133-1138.

Terry, R. E., and R. L. Tate III. 1980. Denitrification as a pathway for nitrate removal from organic soils. Soil Sci. 129:162-166.

Thauer, R. K., K. Jungermann, and K. Decker. 1977. Energy conservation in chemotrophic anaerobic bacteria. Bacteriol. Rev. 41:100-180.

Thayer, J. R., and R. C. Huffaker. 1980. Determination of nitrate and nitrite by high-pressure liquid chromatography: Comparison with other methods for nitrate determination. Anal. Biochem. 102:110-119.

Tiedje, J. M. 1981. Use of nitrogen-13 and nitrogen-15 in studies on the dissimilatory fate of nitrate. p. 481-497. *In* J. M. Lyons et al. (ed.) Genetic engineering of symbiotic nitrogen fixation and conservation of fixed nitrogen. Plenum Press, New York.

Tiedje, J. M., R. B. Firestone, M. K. Firestone, M. R. Betlach, H. F. Kaspar, and J. Sørensen. 1981. Use of nitrogen-13 in studies of denitrification. p. 295-317. *In* R. A. Krohn and J. W. Root (ed.) Recent developments in biological and chemical research with short-lived isotopes. Advances in chemistry. Am. Chem. Soc., Washington, D.C.

Valera, C. L., and M. Alexander. 1961. Nutrition and physiology of denitrifying bacteria. Plant Soil 15:268-280.

Van Cleemput, O., and W. H. Patrick, Jr. 1974. Nitrate and nitrite reduction in flooded gamma-irradiated soil under controlled pH and redox potential conditions. Soil Biol. Biochem. 6:85-88.

Vangai, S., and D. A. Klein. 1974. Nitrite-dependent dissimilatory microorganisms isolated from Oregon soils. Soil Biol. Biochem. 6:335-339.

van Hartingsveldt, J., M. G. Marinus, and A. H. Stouthamer. 1971. Mutants of *Pseudomonas aeruginosa* blocked in nitrate or nitrite dissimilation. Genetics 67:469-482.

van Hartingsveldt, J., and A. H. Stouthamer. 1973. Mapping and characterization of mutants of *Pseudomonas aeruginosa* affected in nitrate respiration in aerobic or anaerobic growth. J. Gen. Microbiol. 74:97-106.

van Hartingsveldt, J., and A. H. Stouthamer. 1974. Properties of a mutant of *Pseudomonas aeruginosa* affected in aerobic growth. J. Gen. Microbiol. 83:303-310.

van Verseveld, H. W., E. M. Meyer, and A. H. Stouthamer. 1977. Energy conservation during nitrate respiration in *Paracoccus denitrificans.* Arch. Microbiol. 112:17–23.

Volz, M. G. 1977. Denitrifying bacteria can be enumerated in nitrite broth. Soil Sci. Soc. Am. J. 41:549–551.

Volz, M. G., M. S. Ardakani, R. K. Schulz, L. H. Stolzy, and A. D. McLaren. 1976. Soil nitrate loss during irrigation: Enhancement by plant roots. Agron. J. 68:621–627.

Volz, M. G., L. W. Belser, M. S. Ardakani, and A. D. McLaren. 1975. Nitrate reduction and associated microbial populations in a ponded Hanford sandy loam. J. Environ. Qual. 4: 99–102.

Volz, M. B., and G. H. Heichel. 1979. Nitrogen transformations and microbial population dynamics in soil amended with fermentation residue. J. Environ. Qual. 8:434–439.

Volz, M. G., and J. L. Starr. 1977. Nitrate dissimilation and population dynamics of denitrifying bacteria during short term continuous flow. Soil Sci. Soc. Am. J. 41:891–896.

Walter, H. M., D. R. Keeney, and I. R. Fillery. 1979. Inhibition of nitrification by acetylene. Soil Sci. Soc. Am. J. 43:195–196.

Wang, W. C., Y. L. Yung, A. L. Lacis, T. Mo, and J. E. Hanson. 1976. Greenhouse effects due to man-made perturbations of trace gases. Science 194:685–689.

Wijler, J., and C. C. Delwiche. 1954. Investigations on the denitrifying process in soil. Plant Soil 5:155–169.

Williams, D. R., J. J. Rowe, P. Romero, and R. G. Eagon. 1978. Denitrifying *Pseudomonas aeruginosa*: Some parameters of growth and active transport. Appl. Environ. Microbiol. 36:257–263.

Woldendorp, J. W. 1962. The quantitative influence of the rhizosphere on denitrification. Plant Soil 17:267–270.

Woldendorp, J. W. 1963. The influence of living plants on denitrification. Meded. Landbouwhogesch. Wageningen 63:1–100.

Wood, P. D. 1978. Periplasmic location of the terminal reductase in nitrite respiration. FEBS Lett. 92:214–218.

Yeomans, J. C., and E. G. Beauchamp. 1978. Limited inhibition of nitrous oxide reduction in soil in the presence of acetylene. Soil Biol. Biochem. 10:517–519.

Yoshida, T., and M. Alexander. 1970. Nitrous oxide formation by *Nitrosomonas europaea* and heterotrophic microorganisms. Soil Sci. Soc. Am. Proc. 34:880–882.

Yoshinari, T. 1980. N_2O reduction by *Vibrio succinogenes.* Appl. Environ. Microbiol. 39:81–84.

Yoshinari, T., R. Hynes, and R. Knowles. 1977. Acetylene inhibition of nitrous oxide reduction and measurement of denitrification and nitrogen fixation in soil. Soil Biol. Biochem. 9:177–183.

Yoshinari, T., and R. Knowles. 1976. Acetylene inhibition of nitrous oxide reduction by denitrifying bacteria. Biochem. Biophys. Res. Commun. 69:705–710.

Zablotowicz, R. M., D. L. Eskew, and D. D. Focht. 1978. Denitrification in *Rhizobium.* Can. J. Microbiol. 24:757–760.

Zumft, W. G., and J. Cardenas. 1979. The inorganic biochemistry of nitrogen bioenergetic processes. Naturwissenschaften 66:81–88.

Zumft, W. G., B. F. Sherr, and W. J. Payne. 1979. A reappraisal of the nitric oxide-binding protein of denitrifying *Pseudomonas.* Biochem. Biophys. Res. Commun. 88:1230–1236.

Zumft, W. G., and J. M. Vega. 1979. Reduction of nitrate to nitrous oxide by a cytoplasmic membrane fraction from the marine denitrifier *Pseudomonas perfectomarinus.* Biochim. Biophys. Acta 548:484–499.

9 Gaseous Losses of Nitrogen Other Than Through Denitrification[1]

DARRELL W. NELSON

Purdue University
West Lafayette, Indiana

I. INTRODUCTION

Research on processes leading to gaseous loss of N from soils has been stimulated by the accumulation of evidence indicating that crop recovery of fertilizer N is often $< 50\%$ of that applied and that a substantial amount of N added to soils is apparently volatilized during the growing season. Nitrogen balance experiments conducted with lysimeters have suggested that N deficits (presumed to result from volatile N losses) are of the order of 15 to 20% (Allison, 1955, 1966, 1973). Furthermore, experiments involving use of ^{15}N-labeled fertilizers have confirmed that a significant proportion of applied N is lost from soils during cropping (for reviews, see Broadbent & Clark, 1965; Allison, 1973; Makarov & Makarov, 1976). Bacterial denitrification may account for some of the N deficits. However, other mechanisms of gaseous N loss, such as NH_3 volatilization and NO_2^- reactions in soil, probably account for a substantial proportion of the N deficits observed in many experiments (Allison, 1965, 1966, 1973; Broadbent & Clark, 1965; Woldendorp, 1968).

II. AMMONIA LOSS FROM SOILS

During the past 50 years, a large number of studies have been conducted to determine the magnitude of NH_3 losses following applications of N fertilizers to soil and to define the soil and environmental factors affecting NH_3 volatilization from soils. Volatilization can occur whenever free NH_3 is present near the soil surface. Ammonia concentrations in the soil solution or soil atmosphere are markedly increased by applying ammoniacal fertilizers or decomposable organic materials to neutral or alkaline soils or by concentrating an alkaline ammoniacal fertilizer in a limited volume of soil. It has been recognized for more than a century that NH_3 volatilization is a pathway for N loss from soils, but only in recent years

[1] A contribution of the Indiana Agricultural Experiment Station, Purdue University, West Lafayette, IN 47907. Journal Paper no. 7822.

Nitrogen in Agricultural Soils—Agronomy Monograph no. 22.

have careful estimates of NH_3 loss rates from soils been made. Table 1 gives a summary of results from field studies of NH_3 losses following surface applications of fertilizers. The amounts of NH_3 volatilized are small when N fertilizers are incorporated into the soil, and NH_3 losses are normally low ($\leq 15\%$ of applied N) when ammoniacal fertilizers are surface applied to acidic or neutral soils. However, large amounts of NH_3 may be evolved on addition of N fertilizers or decomposable waste materials (animal manure, sewage sludge) to the surface of alkaline soils. Ammonia volatilization not only represents a mechanism for gaseous N loss from soils but also is a means for N enrichment of surface waters. Hutchinson and Viets (1969) have shown that NH_3 volatilized from a feedlot was rapidly removed from the atmosphere by water in a nearby lake, thereby serving as a source of N for algal growth.

Terman (1979) has provided a comprehensive review of NH_3 volatilization studies. For convenience, discussion of NH_3 losses from soil are subdivided into those associated with anhydrous NH_3 application and those following addition of NH_4^+ and NH_4^+-forming solid or liquid fertilizers.

A. Ammonia Volatilization Following Ammonium Fertilization of Soils

A wide variety of NH_4^+ and NH_4^+-forming compounds are used as N fertilizers in crop production [e.g., $(NH_4)_2SO_4$, $(NH_4)_2HPO_4$, NH_4NO_3, NH_4Cl, urea, NH_4OH]. Except for NH_4OH, all these materials may be applied as dry salts or liquid fertilizers formed by dissolving or suspending the salts in water. Aqua ammonia (NH_4OH) is produced by dissolving NH_3 in water. However, the behavior of the fertilizer after incorporation into moist soil is similar regardless of whether it is applied as a dry salt or as a liquid. Materials such as $(NH_4)_2SO_4$ or NH_4NO_3 dissolve in the soil solution, and the NH_4^+ ions produced interact with the cation exchange complex, resulting in electrostatic binding of the ions to clay and organic matter. Urea dissolves in the soil solution and undergoes enzymatic hydrolysis to $(NH_4)_2CO_3$, which immediately dissociates to NH_4^+ and CO_2. Ammonium hydroxide behaves similarly to NH_4^+ salts except that some added N may be chemisorbed by soil organic matter because of the high pH of the fertilizer. Ammonium ions in the soil solution also enter into equilibrium reactions with NH_3 as discussed below.

1. AMMONIA-AMMONIUM EQUILIBRIUM IN SOILS

Reactions described in Eq. [1] and [2] occur when NH_4^+ salts are added to or formed in noncalcareous soils:

$$NH_4^+ \rightleftharpoons NH_{3(aq)} + H^+ \qquad [1]$$

$$\frac{[NH_{3(aq)}][H^+]}{[NH_4^+]} = K = 10^{-9.5} \qquad [2]$$

Table 1—A summary of ammonia losses as measured in the field.

Fertilizer added		Soil		% of added N evolved as NH_3	Reference
Type	Amount, kg N/ha	Texture	pH		
Urea	50	Loamy sand	7.7	22	Nommik (1966)
NH_4NO_3	50			17	
Urea	200	Forest litter	4.3	25	Nommik (1973)
$(NH_4)_2SO_4$	280	Clay†	7.6	50	Hargrove et al. (1977)
$(NH_4)_2SO_4$	280	Clay	7.6	35	Kissel et al. (1977)
Urea	500	Forest litter	4.1	3.5	Overrein (1968)
Urea	100	Forest litter	--	7	Volk (1970)
Urea	100	Grass sod	--	20	
		Fine sandy loam	5.6‡	40	Volk (1959)
		Fine sandy loam	5.8§	9	
$(NH_4)_2SO_4$	150	Silt loam	6.3	4	Kresge & Satchel (1960)
Urea	150	Silt loam	6.3	19	
$(NH_4)_2SO_4$	100	Clay loam¶	7.1	3	Ventura & Yoshida (1977)
Urea	100			8	
$(NH_4)_2SO_4$	90	Clay¶	7.0	7	Mikkelsen et al. (1978)
Urea	90			6	

† Ammonium sulfate applied to surface of a grass sod.
‡ Soil CEC was 1.5 meq/100 g.
§ Soil CEC was 11.5 meq/100 g.
¶ Fertilizer broadcast on surface of flood water.

$$\log \frac{[NH_{3(aq)}]}{[NH_4^+]} = -9.5 + pH \qquad [3]$$

where $NH_{3(aq)}$ is the concentration of NH_3 in solution. At pH values of 5, 7, and 9, approximately 0.0036, 0.36, and 36%, respectively, of the total ammoniacal N in the soil solution is present as $NH_{3(aq)}$. The rate of NH_3 loss from a solution is a function of the partial pressure difference between $NH_{3(aq)}$ and NH_3 in the atmosphere (P_{NH_3}) above the solution. At equilibrium, the amount of $NH_{3(aq)}$ is related to the P_{NH_3} in the atmosphere by the Henry Constant (K_H) according to Eq. [4].

$$[NH_{3(aq)}] = K_H P_{NH_3} \qquad [4]$$

Increase in the $NH_{3(aq)}$ concentration by addition of NH_4^+ or increase in pH will result in a change in the equilibrium between $NH_{3(aq)}$ and P_{NH_3} with resultant loss of NH_3 to the atmosphere as indicated in Eq. [5]:

$$NH_{3(aq)} \rightarrow NH_{3(air)}. \qquad [5]$$

The rate of NH_3 loss from an aqueous solution is directly related to the partial pressure difference between NH_3 in the air and NH_3 dissolved in water. Because the NH_3 concentration in the air is low and relatively constant, the rate of NH_3 volatilization from solutions is directly related to the $NH_{3(aq)}$ concentration that is determined by solution pH and NH_4^+ concentration. During NH_3 volatilization, the pH of unbuffered solutions decrease because of the H^+ formed as NH_4^+ is converted to $NH_{3(aq)}$ (Eq. [1]). As a result, Vlek and Stumpe (1978) found that only about 38% of the N in an $(NH_4)_2SO_4$ solution adjusted to pH 9 was lost before NH_3 volatilization ceased because of low pH. Vlek and Stumpe emphasized that loss of NH_3 from a solution is accompanied by an equivalent loss of titratable alkalinity (HCO_3^-, CO_3^{2-}) and that depletion of alkalinity terminates NH_3 volatilization. This results from the interrelationship between water pH and the $CO_2-H_2CO_3$ system in surface waters (Eq. [6]–[8]):

$$CO_2 + H_2O \rightleftharpoons H^+ + HCO_3^- \qquad\qquad [6]$$

$$\frac{[H^+][HCO_3^-]}{[CO_2]} = K = 10^{-7.81} \qquad\qquad [7]$$

$$pH = 7.81 + \log \frac{[CO_2]}{[HCO_3^-]}. \qquad\qquad [8]$$

Similar conclusions were reached by Mikkelsen et al. (1978), who found that the pH and NH_3 volatilization rate of rice flood water were determined by the dissolved $CO_2-H_2CO_3-HCO_3^--CO_3^{2-}$ equilibria. Water pH values changed diurnally with high values (9.5–10) at midday resulting from algal assimilation of carbonates and low values (6.5–8.0) at night resulting from CO_2 released by algal respiration. Avnimelech and Laher (1977) established that the buffering capacity of a solution is an important variable in determining NH_3 volatilization, because highly buffered systems will continue NH_3 evolution even though large amounts of H^+ are released during the reaction. Other factors of importance influencing NH_3 loss from solutions are the partial pressure of NH_3 in the atmosphere, the pH of the solution, and the NH_4^+ concentration in the solution (Avnimelech & Laher, 1977).

Fenn and Kissel (1973) postulated that application of NH_4^+ compounds to the surface of a calcareous soil results in the formation of $(NH_4)_2CO_3$ through the reaction:

$$X(NH_4)_zY + NCaCO_3(s) = N(NH_4)_2CO_3 + Ca_nY_x \qquad\qquad [9]$$

where Y refers to the anion associated with the NH_4^+ cation, N and X refer to the equation stoichiometry, and n, x, and z are dependent on the valences of the anions and cations. Ammonium carbonate then decomposes according to Eq. [10]:

$$(NH_4)_2CO_3 + H_2O = 2NH_3\uparrow + H_2O + CO_2\uparrow. \qquad\qquad [10]$$
$$\uparrow \downarrow$$
$$2NH_4OH$$

If Ca_nY_x is insoluble, the reaction favors formation of more $(NH_4)_2CO_3$ and NH_4OH. If no insoluble precipitate is formed, no appreciable amount of $(NH_4)_2CO_3$ will be formed. Fenn and Kissel (1973) were able to show that the NH_4^+ salts that produced the highest loss of NH_3 were those that formed insoluble precipitates with $Ca(F^-, SO_4^{2-}, HPO_4^{2-})$, whereas other salts having soluble reaction products with Ca (NO_3^-, Cl^-, I^-) gave low NH_3 losses. Ammonium carbonate formed from hydrolysis of urea in calcareous soil reacts as described in Eq. [10].

Feagley and Hossner (1978) established that because of the high pH values required to form $(NH_4)_2CO_3$ in soil, it is unlikely that NH_3 evolution from a calcareous soil proceeds through an $(NH_4)_2CO_3$ intermediate. Instead, Feagley and Hossner postulated that an NH_4HCO_3 intermediate is involved. Equations [11], [12], and [13] indicate the sequence of reactions to be expected when $(NH_4)_2SO_4$ is added to the surface of a calcareous soil:

$$CaCO_3 + (NH_4)_2SO_4 = Ca^{2+} + 2OH^- + CaSO_4 + 2NH_4HCO_3 \quad [11]$$

$$2NH_4HCO_3 = 2NH_3\uparrow + 2CO_2\uparrow + 2H_2O. \quad [12]$$

The Ca^{2+} and OH^- produced during hydrolysis of $CaCO_3$ may react with $(NH_4)_2SO_4$ according to:

$$Ca^{2+} + 2OH^- + (NH_4)_2SO_4 = CaSO_4 + 2NH_3\uparrow + 2H_2O. \quad [13]$$

Combining Eq. [11], [12], and [13] gives the overall reaction of $(NH_4)_2SO_4$ with $CaCO_3$:

$$CaCO_3 + (NH_4)_2SO_4 = 2NH_3\uparrow + CO_2\uparrow + H_2O + CaSO_4. \quad [14]$$

Equation [14] is similar to that given by Fenn and Kissel (1973).

2. EFFECT OF pH ON AMMONIA LOSS

Numerous studies have demonstrated that the loss of NH_3 following the addition of urea or NH_4^+ salts to soil increases with the increase in soil pH (Martin & Chapman, 1951; Wahhab et al., 1957; Volk, 1959; Ernst & Massey, 1960; Kresge & Satchell, 1960; Chao & Kroontje, 1964; du Plessis & Kroontje, 1964; Blasco & Cornfield, 1966; Loftis & Scarsbrook, 1969; Watkins et al., 1972). However, significant amounts of NH_3 may be lost at soil pH values as low as 5.5 if large amounts of urea or NH_4^+ salts are surface applied or if high incubation temperatures are used (Blasco & Cornfield, 1966; de Plessis & Kroontje, 1964; Ernst & Massey, 1960). de Plessis and Kroontje (1964) attempted to predict the extent of NH_3 volatilization based on knowledge of soil pH and the calculated proportion of ammoniacal N present as $NH_{3(aq)}$ (see Eq. [1] and [2]). However, their predicted NH_3 losses were about six times lower than those actually found although a definite relationship existed between predicted and measured NH_3 losses.

It has generally been observed that higher NH_3 losses occur following addition of NH_4^+ salts to calcareous soils compared with noncalcareous soils (Kahn & Haque, 1965; Fenn & Kissel, 1974; Terry et al., 1978). Furthermore, the addition of $CaCO_3$ or other alkaline materials to soil increases the proportion of added NH_4^+ evolved as NH_3 (Ernst & Massey, 1960; Matocha, 1976; Raison & McGarity, 1978).

Soil pH is a significant factor in influencing the proportion of added urea or NH_4^+-N volatilized as NH_3 because of the effect of H^+ concentration on the NH_4^+–NH_3 equilibrium in soils. However, under certain conditions, other factors such as soil texture or cation exchange capacity (CEC) may be primary factors in NH_4^+ loss, as suggested by substantial NH_3 evolution from acid soils (Ernst & Massey, 1960; Blasco & Cornfield, 1966). The pH in the solution immediately surrounding urea or NH_4^+ salt granules may be much more important in determining NH_3 losses than the soil pH.

3. EFFECT OF TEMPERATURE ON AMMONIA LOSS

A number of investigators have reported that the loss of NH_3 following the application of urea or NH_4^+ salts to soil increases with increasing temperature up to about 45°C (Martin & Chapman, 1951; Wahhab et al., 1957; Volk, 1959; Ernst & Massey, 1960; Gasser, 1964; Baligar & Patil, 1968a; Watkins et al., 1972; Prasad, 1976). Vlek and Stumpe (1978) showed that NH_3 volatilization from solutions increased with increase in temperature up to 46°C. However, Fenn and Kissel (1974) established that NH_3 volatilization resulting from application to calcareous soils of NH_4^+ salts that form insoluble precipitates with Ca^{2+} was only slightly increased by temperature. Increasing temperature from 12 to 32°C markedly increased the amount of NH_3 volatilized following treatment of calcareous soils with NH_4^+ salts that do not form insoluble Ca precipitates (e.g., NH_4NO_3).

The effect of temperature on the rate of NH_3 volatilization from soil or solutions can be explained at least in part by an increase in the equilibrium constant (K in Eq. [2]) with increasing temperature (Vlek & Stumpe, 1978), which results in a higher proportion of ammoniacal N being present as $NH_{3(aq)}$. Higher temperatures may also increase the speed of NH_3 diffusion from the soil and allow a more rapid conversion from $NH_{3(aq)}$ to $NH_{3(air)}$.

4. EFFECT OF AMMONIUM CONCENTRATION ON AMMONIA LOSS

The relationship between NH_3 volatilization from soil and the amount of urea or NH_4^+ salt added has been studied a number of times. It has normally been found that the amount of NH_3 evolved increases with increasing rate of NH_4^+-N added but that the proportion of added N volatilized as NH_3 is constant (Martin & Chapman, 1951; Chao & Kroontje, 1964; Hargrove et al., 1977; Vlek & Stumpe, 1978). Other investigators have reported that the proportion of added NH_4^+ or urea N evolved as NH_3 increases with the fertilizer addition rate (Wahhab et al., 1957; Overrein & Moe, 1967; Baligar & Patil, 1968b). Fenn and Kissel (1974) observed that the percentage of added $(NH_4)_2SO_4$-N evolved as NH_3-N from calcareous soils increased with increasing addition rate, whereas the proportion of NH_4NO_3-N volatilized as NH_3 was not affected by addition rate. They concluded that the addition rate of NH_4^+ salts that form soluble precipitates by reaction with $CaCO_3$ is of critical importance in determining both the total NH_3 losses and the proportion of added NH_4^+-N evolved as NH_3.

The amount of urea or NH_4^+ added to soil must have a direct effect on

NH_3 evolved as predicted in Eq. [1] if all other factors are held constant. The increasing proportion of added N volatilized as NH_3 with increasing NH_4^+ or urea addition rate may result from elevated soil pH values obtained from application of high amounts of urea, NH_4OH, or $(NH_4)_2HPO_4$ or may result from interactions of $(NH_4)_2SO_4$ and NH_4HPO_4 with $CaCO_3$ as described by Fenn and Kissel (1974).

5. EFFECT OF SOIL CHARACTERISTICS ON AMMONIA LOSS

A number of investigators have observed that if all other conditions are similar, more of the NH_4^+-N added to coarse-textured (sandy) soils is volatilized as NH_3 compared with that from fine-textured soils (Wahhab et al., 1957; Gasser, 1964; Verma & Sarkar, 1974; Fenn & Kissel, 1976; Gandhi & Paliwal, 1976; Terry et al., 1978). Most reports suggest that the effect of soil texture on NH_3 volatilization is due to the higher CEC of fine-textured soils. With a high CEC, a greater proportion of added NH_4^+ would be present on the exchange complex, and less $NH_{3(aq)}$ would be present in the soil solution:

$$NH_{4(exc)}^+ \rightleftharpoons NH_{4(aq)}^+ \rightleftharpoons NH_{3(aq)}.$$ [15]

Gasser (1964) concluded that CEC was the most important factor governing the loss of NH_3 following application of urea and $(NH_4)_2SO_4$ to soils.

There have been reports that incorporation of organic residues with urea decreases gaseous loss of urea-N as NH_3 (Tripathi & Benarjee, 1955; Tripathi, 1958). However, other workers have observed that the addition of decomposable organic materials in the zone of urea incorporation accelerates volatilization of urea (Moe, 1967; Khan & Rashid, 1971; Rashid, 1977). On the other hand, Verma and Sarkar (1974) found that the addition of 5% by weight of farmyard manure did not affect the loss of NH_3 following urea application to a calcareous loam soil. The inconsistent effects of organic residue addition on NH_3 volatilization may result from two factors associated with waste application. First, organic N in waste may be mineralized with liberation of NH_4^+, which will increase the $NH_{3(aq)}$ concentration in the soil solution. Second, the waste materials may have some CEC, may have a high C/N ratio (which promotes immobilization of NH_4^+-N), or may be acidic. All of these waste characteristics will lower the $NH_{3(aq)}$ concentration. The net effect on NH_3 volatilization may depend on the nature and rate of residue added.

A number of investigators agree that the addition of urea to soil covered with a grass sod, plant residues, or an organic mulch will result in greater NH_3 losses than application to bare soil (Volk, 1959, 1970; Meyer et al., 1961; Watson et al., 1962; Watkins et al., 1972). These findings likely result from hydrolysis of urea on the surfaces of leaves or residue particles and subsequent volatilization of NH_3. Due to the limited CEC of residues, much less NH_4^+ would be retained on exchange sites than would be the case for hydrolysis on the soil surface. Consequently, NH_3 volatilization would be high because of the high $NH_{4(aq)}^+$ concentration. Kresge and Satchell

(1960) and Volk (1959), however, reported that NH_3 losses were lower when urea was top-dressed on grass sod than when it was applied to bare soil if the soil had a low CEC.

Moe (1967) established that the NH_3 volatilization rate following urea application to moderately acid soils was stimulated by the addition of urease to increase the rate of urea hydrolysis and was reduced when a urease inhibitor, p-chloromercuribenzoate, was added. He speculated that the use of a specific urease inhibitor in conjunction with urea applications may significantly reduce NH_3 losses in the field.

Gadhi and Paliwal (1976) found that the loss of NH_3 following the addition of urea and $(NH_4)_2SO_4$ to NaCl-amended soils increased with increasing salinity over the electrical conductivity (EC) range of 1.1 to 50 mmho/cm. At the highest salinity level, NH_3 volatilization was three times higher than that in soils not treated with NaCl. About 35% of added N was volatilized at EC values of 50 mmho/cm irrespective of soil type and type of fertilized added.

The nature of the cations on the exchange complex affect NH_3 volatilization from soil. Martin and Chapman (1951) observed higher NH_3 losses from K^+- and Na^+-saturated soils compared with those saturated with Ca^{2+} and Mg^{2+}, presumably because of the higher pH values in K^+- and Na^+-saturated soils. In addition, they reported that a neutral Mg^{2+}-saturated soil lost considerably more NH_3 than a Ca^{2+}-saturated soil when $(NH_4)_2SO_4$ was added but not when NH_4OH was applied. On the other hand, du Plessis and Kroontje (1964) reported that more NH_3 was volatilized on the addition of NH_4^+ to Ca^{2+}-saturated clay compared with a Mg^{2+}-saturated clay. However, at a high partial pressure of CO_2 in the system, the reverse effect was observed. They concluded that at high CO_2 levels, a more complete precipitation of $CaCO_3$ occurs, and the Mg^{2+} saturated soil would therefore have a higher pH and greater NH_3 loss.

6. EFFECT OF SOIL MOISTURE CONTENT AND MOISTURE LOSS ON AMMONIA VOLATILIZATION

In studies on the effect of initial soil moisture content on NH_3 volatilization where drying of the soil occurred during the incubation, a wide variety of results have been obtained. Some workers have reported that NH_3 losses increase with increasing soil moisture contents up to field capacity (Volk, 1959, 1966; Ernst & Massey, 1960; Kresge & Satchell, 1960; Baligar & Patil, 1968a). Similar effects were noted by Martin and Chapman (1951) but only at high temperatures. These results have been attributed to (i) small NH_3 losses at low moisture contents due to a lack of water necessary to dissolve added urea and (ii) large NH_3 losses at high moisture contents because of the long period of time required to bring soils to an air-dry state where NH_3 losses are minimal. Other workers have found that NH_3 losses decrease with increasing initial soil moisture contents when sufficient water is present to dissolve added N fertilizers (Martin & Chapman, 1951; Wahhab et al., 1957; Fenn & Escarzaga, 1976; Prasad, 1976). However, drying of soil was minimal in several of these studies. Without drying, the $NH_{3(aq)}$ concentration is in part a function of initial soil moisture, with low moisture content soils having higher $NH_{4(aq)}^+$ and $NH_{3(aq)}$ levels.

Jewitt (1942) reported that the initial soil moisture content did not greatly affect NH_3 volatilization from $(NH_4)_2SO_4$-treated alkaline soil but that the loss of NH_3 was related to soil moisture loss. Similar results were reported by Fenn and Escarzaga (1977) in studies of NH_3 loss from sand. Martin and Chapman (1951) found that NH_3 loss was promoted when dry air was circulated over NH_4^+-treated calcareous soils but that no NH_3 was volatilized when moist air was circulated over the samples. They concluded that moisture loss was a requirement for NH_3 volatilization. Several investigators have reported that the loss of NH_3 and water from soils are directly related (Wahhab et al., 1957; Chao & Kroontje, 1964). In attempting to quantitate the relationship between water loss and NH_3 volatilization, Chao and Kroontje (1964) observed that loss of water and NH_3 volatilization from NH_4OH-treated soils followed different functions. Water loss rate was constant with time, whereas NH_3 volatilization rate decreased with time. However, Ernst and Massey (1960) established that NH_3 volatilization and water loss were not related following addition of urea to an acid soil. They observed that substantial NH_3 loss occurred without drying but that greater NH_3 volatilization took place when slow drying of soil occurred. Very rapid drying of soil resulted in low losses of applied urea N. Terry et al. (1978) established that NH_3 loss following surface application of sewage sludge to a neutral soil was similar in samples losing markedly different amounts of water.

The relative humidity (RH) of air flowing over the surface has a marked effect on the water loss rate from soils. In laboratory experiments, the rate of water loss from soils has been shown to be inversely related to the RH of the air used to purge samples but that NH_3 loss rate was largely independent of RH (Ernst & Massey, 1960; Chao & Kroontje, 1964; Terman et al., 1968; Terry et al., 1978). These authors have shown that significant NH_3 volatilization from soils occurs when 100% RH air is flowing over samples and water loss is insignificant. Ernst and Massey (1960) emphasized that NH_3 volatilization following urea application is depressed by rapid soil drying induced by purging with low RH air, whereas Terry et al., (1978) suggested that rapid drying of soil increases the effective $NH_{4(aq)}^+$ concentration in the soil solution and prevents nitrification of NH_4^+-N, thereby increasing NH_3 losses. Furthermore, Terry et al. (1978) showed that slow drying of soils decreases NH_3 volatilization, because nitrification resulted in decreased NH_4^+ concentration and increased H^+ in the soil solution. Hargrove et al. (1977) observed that the rate of NH_3 volatilization following application of $(NH_4)_2SO_4$ to the surface of a relatively dry calcareous soil peaked each day at 0600 to 0800 hours and decreased to a minimum at 1800 to 2000 hours. They demonstrated that high NH_3 loss rates were related to high atmospheric RH values and speculated that moist air is required to dissolve $(NH_4)_2SO_4$, to allow reaction to NH_4^+ with $CaCO_3$ with subsequent decomposition of the reaction product, or to displace NH_3 that may be sorbed on soil surfaces.

Losses of NH_3 following irrigation are greater when $(NH_4)_2SO_4$, urea, and NH_4NO_3 are applied to the surface of initially wet calcareous soils compared with their addition to initially dry soils (Nommik, 1966; Fenn & Escarzaga, 1977). Fenn and Escarzaga (1976) observed that losses of NH_3 are greater when $(NH_4)_2SO_4$ is added as a solution to air-dry soil compared with its application as a solution to wet soil or as a dry salt to air-dry soil. Losses of NH_3 were less when urea or NH_4NO_3 was applied to the surface of wet soil compared with applications to air-dry soil. Terman et al. (1968) emphasized that drying mildly acidic soils to which urea solutions had been surface applied markedly increased NH_3 volatilization. Volk (1966) found that no NH_3 volatilization occurred when urea was added to dry soil but that application of urea to soil maintained in a wet condition resulted in large NH_3 losses.

It is clear that NH_3 volatilization from soil can occur without concurrent loss of water. However, in most cases loss of water from soil promotes NH_3 evolution by increasing the $NH_{3(aq)}$ concentration in the soil solution and preventing nitrification of NH_4^+. Rapid drying of soil may lead to reduced losses of NH_3 from top-dressed urea prills because of the moisture requirements for dissolution of the fertilizer and hydrolysis of urea. However, rapid drying of soils treated with $(NH_4)_2SO_4$ solutions normally leads to greater NH_3 losses than are obtained by slow drying or maintenance of the initial soil moisture level. High initial soil moisture contents tend to reduce NH_3 losses when soils are not dried during the study. However, the highest amounts of volatilized NH_3 are normally obtained with high moisture contents (below saturation) when soils are allowed to dry. Essentially no NH_3 losses are observed when dry fertilizer is added to soils of low moisture content or when fertilizer solutions are added to very dry soils.

7. EFFECT OF AERATION RATE ON AMMONIA LOSS

A number of workers have reported that the rate of NH_3 volatilization from urea- and $(NH_4)_2SO_4$-treated soils increases with an increase in the rate of air flow over the samples (Chao & Kroontje, 1964; Overrein & Moe, 1967; Watkins et al., 1972; Kissel et al., 1977; Terry et al., 1978; Vlek & Stumpe, 1978). Recent work has established that the maximum rates of NH_3 volatilization and the largest total amounts of NH_3 evolved are attained in volatilization chambers having 15 to 20 air exchanges per minute (Kissel et al., 1977; Vlek & Stumpe, 1978), equivalent to wind speeds of 0.1 to 0.3 km/hour. Ammonia volatilization rates appear to be directly proportional to air flow at exchange rates below 10 volumes per minute (Overrein & Moe, 1967; Watkins et al., 1972; Kissel et al., 1977; Terry et al., 1978; Vlek & Stumpe, 1978). Unfortunately, many of the laboratory studies evaluating NH_3 losses resulting from fertilizer applications to soils have used unrealistically low air exchange rates that in no way simulate the rapid air exchange occurring in the field. For example, Kissel et al., (1977) observed that wind flow over soil surfaces in Texas exceeded 0.26 km/hour (the wind speed giving maximum NH_3 volatilization rates) >98% of the time. It is doubtful that a lack of air movement limits NH_3 volatilization in the field, but NH_3

loss in many laboratory studies may have been underestimated because the air exchange used was too low.

Rapid movement of air across the soil surface promotes NH_3 volatilization by maintaining a low partial pressure of NH_3 in the atmosphere adjacent to the soil, thus, permitting rapid diffusion of NH_3 from the soil in response to a large partial pressure gradient. Furthermore, rapid movement of low RH air across the soil surface can promote rapid loss of soil moisture, thereby stimulating NH_3 evolution (Terry et al., 1978).

8. EFFECT OF FERTILIZER INCORPORATION ON AMMONIA LOSS FROM SOILS

A number of workers have demonstrated that much more NH_3 is volatilized from surface applications of urea or NH_4^+ salts than from incorporation of these fertilizers into soil (Ernst & Massey, 1960; Khan & Haque, 1965; Overrein & Moe, 1967; Baligar & Patil, 1968b; Terman et al., 1968; Fenn & Kissel, 1976; Terry et al., 1978). Furthermore, Macrae & Ancajas (1970) and Mikkelsen et al. (1978) established that placement of N fertilizers into rice soils before flooding markedly reduced NH_3 volatilization compared with broadcasting the fertilizer in the flood water. Ernst and Massey (1960) demonstrated that NH_3 losses from a near neutral soil were reduced by 75% when urea was mixed with the top 3.8 cm instead of applied to the surface. Similar results were reported by Khan and Haque (1965). However, Terman et al. (1968) were able to achieve only a 25 to 50% reduction in NH_3 loss from urea and NH_4^+ fertilizers when they were thoroughly mixed with an acidic silt loam soil in greenhouse pots compared with surface application. Overrein and Moe (1967) demonstrated that increasing the depth of soil over a urea band decreased NH_3 volatilization but observed substantial losses even when a 2.5-cm layer of a sandy soil (pH 5.7) was placed over the fertilizer. Lower NH_3 losses were obtained if the overlying soil was moist rather than nearly air-dry. Fenn and Kissel (1976) found that elimination of NH_3 volatilization from fertilizer bands required a much deeper layer of sandy soil (pH > 7) than clay soil (pH > 7). A cover of dry soil (pH > 7) over banded $(NH_4)_2SO_4$ or NH_4NO_3 was more effective in minimizing NH_3 volatilization than was wet soil (pH > 7), because dry soil more efficiently sorbed NH_3 and water vapor diffusing from the underlying fertilizer-treated moist soil.

9. COMPARISON OF AMMONIA VOLATILIZATION FROM DIFFERENT NITROGEN FERTILIZERS

With neutral and acidic soils, surface applications of such alkaline fertilizers as urea or NH_4OH lead to larger NH_3 losses than application of neutral or acidic fertilizers such as $(NH_4)_2SO_4$ or $NH_4H_2PO_4$ (Terman et al., 1968; Watkins et al., 1972; Matocha, 1976). Mixing neutral NH_4^+ salts or acidifying reagents such as H_3PO_4 or $NH_4H_2PO_4$ with urea prior to their addition to the soil surface markedly reduces NH_3 volatilization (Terman et al., 1968; Bremner & Douglas, 1971; Watkins et al., 1972). Application of amide compounds such as formamide or the addition of formamide-urea mixtures to the surface of acidic soils has also been shown to result in substantial NH_3 losses (Terman et al., 1968; Loftis & Scarsbrook, 1969). Coating urea prills with S markedly reduces NH_3 volatilization from soils following surface application because of the slow release of urea from coated prills (Matocha, 1976; Prasad, 1976). More NH_3 is volatilized when S-coated urea is incorporated into an acidic soil than when it is applied to the surface due to more rapid decomposition of the coating when it is mixed with soil (Matocha, 1976).

Losses of NH_3 are normally higher from the addition of urea to calcareous soils compared with the application of NH_4^+ salts (Nommik, 1966; Gandhi & Paliwal, 1976; Prasad, 1976). Fenn and Kissel (1973, 1974) have shown that volatilization of NH_3 is much more pronounced when NH_4^+ compounds that react with $CaCO_3$ to form insoluble precipitates $[(NH_4)_2SO_4, NH_4F, (NH_4)_2HPO_4]$ are added to calcareous soil compared with the addition of NH_4^+ compounds that do not form insoluble precipitates with Ca^{2+} (NH_4NO_3, NH_4Cl, NH_4I). Apparently, formation of insoluble Ca precipitates in soils leads to concurrent formation and subsequent decomposition of $(NH_4)_2CO_3$). Inclusion of $NH_4H_2PO_4$ at a rate of 30% of the N is a fertilizer mixture markedly reduced NH_3 losses when $(NH_4)_2SO_4$ or NH_4F mixtures were added to calcareous soil (Fenn, 1975). Mixing NH_4NO_3 with urea reduced NH_3 volatilization relative to application of urea alone (Kresge & Satchell, 1960).

Nommik (1973) observed that NH_3 volatilization was initially much lower when large urea pellets rather than prills were applied to a forest soil. However, after 30 days, cumulative NH_3 losses from prills or from varying sizes of urea pellets were about the same. Watkins et al. (1972) also showed that pellet size had little influence on NH_3 volatilization following urea application to forest soils. However, Volk (1961) found that NH_3 losses from large pellets may exceed that from small pellets when urea is applied to bare soil or turf.

10. MODERN METHODOLOGY TO STUDY AMMONIA LOSSES FROM SOILS

Most studies of NH_3 volatilization from soils have employed an air stream drawn over the surface of soil contained in a vessel. The vessel may range from an Erlenmeyer flask in the laboratory to a box placed over a field site. Ambient NH_3 is normally removed from inflowing air, and NH_3 volatilized from soil present in the container is estimated by trapping it in an acid solution and back titrating with acid or determining NH_3 by steam distillation or colorimetric methods. A number of authors have pointed out that these methods do not give a true estimate of NH_3 losses occurring under field conditions. Watkins et al. (1972) found that air movement in "cuvette" methods is much lower than that under natural conditions, thereby minimizing NH_3 losses. Kissel et al. (1977) emphasized that field studies using vessels placed over the soil surface create artificial conditions over and around the area being sampled. Containment of soil in vessels also makes study of factors affecting NH_3 volatilization difficult and almost ensures that the absolute loss of NH_3 will be different from that under natural conditions. Nommik (1973) suggested that air flow across the surface of field soils contained in vessels is normally much lower than that experienced under natural conditions, which results in higher soil moisture levels and considerably lower NH_3 losses than those attained with unconfined soils.

Nommik (1973) proposed that NH_3 volatilization could be accurately assessed under field conditions through the addition of ^{15}N-labeled urea to vertically isolated microplots and subsequent recovery of ^{15}N from the soil. This method requires that no leaching or denitrification occur during the experimental period. The technique is costly and requires careful field sampling and accurate quantitation of ^{15}N in the soil. Nommik suggested that the ^{15}N recovery method is the most accurate for assessing NH_3 volatilization in the field, because the losses are assessed under natural conditions with unchanged aeration, temperature, and soil moisture levels.

A micrometerological technique for estimating NH_3 volatilization from fertilizer-treated or grazed fields has been described (Denmead et al., 1974, 1977). The method involves calculation of the aerial transport of NH_3 across the downwind edge of a treated field from measurements of wind speed, wind direction, and NH_3 concentrations in the atmosphere at various heights above the soil surface. Although the method requires a relatively large experimental area and accurate determination of several micrometerological variables, NH_3 losses as small as 1 kg of N/ha can be detected with a maximum error of 20% (Denmead et al., 1977).

Kissel et al. (1977) described a device for field measurement of NH_3 loss that involved passage of air over a soil microplot contained in a volatilization chamber. The system proposed is different from the containment methods previously used in that the lid of the chamber is closed only for short intervals during the NH_3 loss measurement period (10 minutes every 3 hours), and air flows over the soil only when the chamber lid is closed. Total NH_3 volatilized is calculated by integrating the rate of NH_3 losses over the period of measurement. During the period between individual measurements the chamber lid is open, allowing the soil to be exposed to natural environmental conditions. Kissel et al. (1977) suggested that NH_3 losses measured with their system are similar to those occurring under natural conditions.

B. Ammonia Losses from Soils Following Anhydrous Ammonia Application

Anhydrous NH_3 differs chemically and physically from other N fertilizers during and after application, because it changes from a liquid to a gaseous state during injection into soil. As a gas, NH_3 has the tendency to escape to the atmosphere; however, it is normally retained in soil due to its reactions with soil components. A detailed discussion of mechanisms for NH_3 retention in soil is given in Chapter 4 (Nommik and Vahtras).

A number of factors affect the proportion of anhydrous NH_3 volatilized after addition to soils. Soil moisture content has been shown to have a variety of effects on N loss following NH_3 application to soil. Stanley and

Smith (1956) reported that NH_3 losses were highest at low moisture contents, decreased at intermediate moisture levels, and then increased as the soil reached field capacity. Greater NH_3 losses were observed from some soils at field capacity when they were treated with NH_3 relative to similar soils that were air-dry when treated. However, McDowell and Smith (1958) reported that the addition of NH_3 to air-dry soils led to larger N losses than the addition of NH_3 to moist soils, and Jackson and Chang (1947) reported that soil moisture content had no effect on NH_3 losses except in a sandy soil where losses were higher from dry soil. The size of the initial NH_3 retention zone decreases as soil moisture content increases (Blue & Eno, 1954; Nommik & Nilsson, 1963) presumably because the rate of NH_3 diffusion through water is much less than through air. Parr and Papendick (1966b) found that the initial capacity of a soil to retain NH_3 increased with increasing moisture content but that after aeration of soils, more NH_3 was retained by those that were initially air-dry than those that were initially moist. They also demonstrated that some of the NH_3 retained by a dry soil could be displaced by water vapor during aeration with moist air. This finding suggests that NH_3 and water molecules compete for sorption sites on soil components. Parr and Papendick (1966a) concluded that soil moisture functions as a temporary repository for NH_3 that would otherwise be retained in soil and that soil moisture content at the time of NH_3 application is not a critical factor in NH_3 retention in soil. However, moisture is an important component influencing the soil physical conditions necessary for rapid sealing of the injection channel. Soils having intermediate moisture contents usually have the least loss of NH_3 from injection channels, because soils that are excessively wet or dry normally do not seal well.

Several investigators have shown that substantially higher losses of NH_3 occur following application of anhydrous NH_3 to sandy soils relative to fine-textured soils (Blue & Eno, 1954; McDowell & Smith, 1958) probably because sands have fewer sorption and reaction sites than medium-textured soils, and sands allow rapid diffusion of NH_3 immediately after fertilizer injection (Parr & Papendick, 1966a). However, under certain conditions sandy soils have the capacity to retain as much NH_3 as silt loam soils; NH_3 retention depends on factors such as type of clay minerals, organic matter content, and soil moisture content in addition to texture (Jackson & Chang, 1947; Parr & Papendick, 1966a). It has been shown that regardless of soil type, placement of anhydrous NH_3 at a depth of 5 to 13 cm results in minimal NH_3 losses to the atmosphere provided that the injection channel is sealed (Parr & Papendick, 1966a). This finding suggests that most soils are efficient sorbers for NH_3.

Retention of NH_3 by soils occurs over a wide range in pH values up to 10 (Jackson & Chang, 1947; Mortland, 1955). Ammonia is more efficiently retained by clay minerals at acid pH values, but organic matter chemisorbs NH_3 to the greatest extent under alkaline conditions (see Chapt. 4). Losses of NH_3 from anhydrous NH_3-treated soils is not related to pH if the fertilizer is applied properly.

The capacity of soils to retain NH_3 tends to increase with increasing CEC (Chao & Kroontje, 1960; Parr & Papendick, 1966a). The capacity of a pure clay system to sorb NH_3 is directly related to CEC (James & Harward, 1964). However, some soils have the capacity to retain much higher amounts of NH_3 than predicted by CEC (Brown & Bartholomew, 1962; Parr & Papendick, 1966a). This suggests that although CEC is important in determining the NH_3 retention capacity of soil, other factors (e.g., organic matter content) greatly influence the degree to which NH_3 is held by a soil. Therefore, the amount of NH_3 retained by a series of soils may not be highly correlated with CEC.

Research conducted to date indicates that losses of N during and immediately following application of anhydrous NH_3 are minimal if the fertilizer is applied at an appropriate depth and if soil moisture conditions are such that rapid closure of the injection channel occurs.

III. GASEOUS NITROGEN LOSS FROM SOILS THROUGH NITRITE REACTIONS

Unexplained N losses have been observed by a number of investigators during studies of nitrification in soils. These losses of N have often been associated with accumulation of NO_2^-, and several studies have provided strong presumptive evidence that significant gaseous loss of fertilizer N can occur through chemical reactions of NO_2^- formed by nitrification of NH_4^+ and NH_4^+-forming fertilizers in acidic or mildly acidic soils. Vine (1962) has used the phrase "side-tracking of nitrification," and Clark (1962) has used the term "chemodenitrification" to designate the processes responsible for gaseous loss of N from soils through chemical reactions of NO_2^-. The term *chemodenitrification* has gained considerable acceptance, and for lack of a better term, it will be adopted here.

Nitrite accumulation in soil has been reported during nitrification of urea (Chapman & Liebig, 1952; Court et al., 1974; Hauck & Stephenson, 1965; Wetselaar et al., 1972), NH_4^+ salts (Chapman & Liebig, 1952; Justice & Smith, 1962; Hauck & Stephenson, 1965; Wetselaar et al., 1972) and anhydrous or aqua NH_3 (Wetselaar et al., 1972; Chalk et al., 1975). Furthermore, high concentrations of NO_2^- have been observed in NO_3^--treated soils during biological denitrification (Wijler & Delwiche, 1954; Cady & Bartholomew, 1960; Cooper & Smith, 1963). The presence of NO_2^- provides a mechanism for gaseous N losses from soil because of the tendency of NO_2^- to react with soil components to form N gases (e.g., N_2, N_2O, NO, NO_2). A number of investigators have observed large N deficits during nitrification of urea or NH_4^+ fertilizers in soil and have attributed these N losses to chemical reactions of NO_2^- (Gerretsen & de Hoop, 1957; Soulides & Clark, 1958; Clark et al., 1960; Clark, 1952; Hauck & Stephenson, 1965; Steen & Stojanovic, 1971; Smith & Chalk, 1978). Vine (1962) also observed that a portion of added NH_4^+-N was evolved as N_2O and N_2 during nitrification, but he suggested that N gases were produced as a result of formation and decomposition of NH_2OH. Further evidence for involvement of NO_2^- in gaseous N losses from soils is provided by studies of biological denitrification. A number of investigators have observed formation of NO in anaerobic soils that accumulate NO_2^- (Wijler & Delwiche, 1954; Cady & Bartholomew, 1960; Cooper & Smith, 1963). The formation of NO has been attributed to self-decomposition of HNO_2 formed in acidic soils during denitrification.

Allison (1965, 1966, 1973), Broadbent and Clark (1965), Harmsen and Kolenbrander (1965), and Woldendorp (1968) have reviewed the evidence concerning chemical reactions of NO_2^- as pathways for N loss from soils. The evidence currently available can be summarized as follows:

1) Substantial gaseous loss of N from soils has been observed under conditions that are not conducive to microbial denitrification or volatilization of NH_3.
2) Experiments involving incubation of soil under aerobic conditions after addition of NO_3^- or NH_4^+ have shown that recovery of NO_3^- after incubation is often much higher than the recovery of NH_4^+.
3) Significant gaseous loss of fertilizer N has been observed under conditions that lead to accumulation of NO_2^- (e.g., after application of high rates of NH_4^+ or NH_4^+-forming fertilizers).
4) Nitrite added to sterilized or unsterilized acidic soils is rapidly decomposed with formation of gaseous forms of N.

The deduction from these observations that substantial gaseous loss of N from soils may occur through chemical decomposition of NO_2^- is based on the fact that NO_2^- is the only intermediate that has been detected in studies of nitrification in soils. Although a number of studies of NO_2^- reactions in soils have been conducted, the overall role of these reactions in the N budget of field soils is unclear.

Bremner and Blackmer (1978) have shown that small amounts of N_2O are emitted during nitrification of urea and NH_4^+ fertilizers in aerobic soils. The emission of N_2O is not the result of biological denitrification or NO_2^- reactions but appears to be a "side-tracking" reaction during the oxidation of NH_4^+ to NO_2^-. No N_2O is emitted when urea is added to sterile soils or to soils treated with a nitrification inhibitor, and the amount of N_2O emitted is related to the NH_4^+-N nitrified.

Other investigators have proposed that chemical reactions of NH_2OH and NO_3^- result in gaseous N losses from soil (Wetselaar, 1961; Vine, 1962; Thomas & Kissel, 1970). However, there is little current evidence to suggest that these reactions are important mechanisms for N loss from soils.

Loss of N from soils as N_2O, NO, and NO_2 resulting from NO_2^- decomposition or other reactions involving inorganic N not only decreases the efficiency of applied fertilizer but also promotes environmental contamination. Nitric oxide and NO_2 are recognized air pollutants. Although NO and NO_2 emissions from soil will not in themselves create air quality problems, they may contribute to the overall nitrogen oxide burden in the atmosphere. In fact, Robinson and Robbins (1970) estimated that the amounts of NO and NO_2 produced by natural processes far exceed those emitted from man's activities. However, it is important to recognize that soils also readily sorb NO and NO_2 from atmospheres enriched with these gases (Prather et al., 1973a, b; Nelson & Bremner, 1970a; Miyamota et al., 1974; Smith & Mayfield, 1978).

Nitrous oxide produced during reactions of inorganic forms of N in soils may also pose a threat to the environment through involvement in the catalytic destruction of stratospheric ozone (CAST, 1976; Crutzen, 1976). However, the amounts of N_2O produced in the soil by mechanisms other than nitrification and biological denitrification are believed to be small. Therefore, little significance has been given to chemical decomposition of NO_2^- or NH_2OH in soils as mechanisms for formation of N_2O. However, Bremner and Blackmer (1978) have pointed out that with high NH_4^+ applications, the amounts of N_2O produced during nitrification could represent an important component of total N_2O emissions from the earth's surface.

A. Nitrite-Nitrous Acid Equilibria in Soils

Nitrous acid rather than the NO_2^- ion is likely involved in most chemo-denitrification reactions leading to gaseous loss of N from soils. The NO_2^--HNO_2 equilibrium in aqueous systems is given in Eq. [16] and [17]:

$$HNO_2 \rightleftharpoons H^+ + NO_2^- \qquad [16]$$

$$\frac{[H^+][NO_2^-]}{[HNO_2]} = 6.0 \times 10^{-4}. \qquad [17]$$

In an aqueous system at pH 5, only about 1.6% of the NO_2^--N is present as HNO_2, whereas at pH vlaues of 4 and 3, HNO_2 comprises 14 and 63%, respectively, of the NO_2^--N present. This suggests that NO_2^- reactions involving HNO_2 occur only under conditions of low pH. However, the water films surrounding clay particles are often 100 times more acidic than the bulk soil solution (McLaren & Estermann, 1957; Harter & Ahlrichs, 1967), and the proportion of NO_2^--N present in a soil as HNO_2 may be markedly higher than that predicted by pH measurements in soil-water mixtures.

B. Factors Affecting Nitrite Instability in Soils

Numerous investigators have demonstrated that NO_2^- is rapidly decomposed when it is added to acidic or mildly acidic soils (Clark et al., 1960; Reuss & Smith, 1965; Bremner & Nelson, 1968; Nelson & Bremner, 1969; Jones & Hedlin, 1970; Bollag et al., 1973). Most workers have found that the rate and extent of NO_2^- decomposition in soils are inversely related to soil pH (Reuss & Smith, 1965; Bremner & Nelson, 1968; Nelson & Bremner, 1969; Wullstein, 1969; Bollag et al., 1973). Furthermore, several investigations have shown that NO_2^- is not decomposed or is decomposed very slowly when it is added to alkaline soils (Tyler & Broadbent, 1960; Meek & MacKenzie, 1965; Reuss & Smith, 1965; Bremner & Nelson, 1968; Nelson & Bremner, 1969; Bollag et al., 1973). The rate of decomposition of NO_2^- in soils is greatly reduced by the addition of $CaCO_3$ (Tyler & Broadbent, 1960; Nelson & Bremner, 1969; Wullstein, 1969).

Air-drying acidic and neutral soils markedly promotes NO_2^- decomposition (Clark et al., 1960; Nelson & Bremner, 1969). Nelson and Bremner (1969) found that air-drying NO_2^--treated calcareous soils resulted in significant loss of N, whereas Jones and Hedlin (1970) observed little loss of NO_2^--N when a soil (pH 8) was air-dried.

Several investigators have observed that the rate of NO_2^- decomposition in soil increases with the increase in soil organic matter content (Smith & Clark, 1960; Nelson & Bremner, 1969). Clark & Beard (1960) found that a pretreatment with H_2O_2 to remove organic matter reduced the ability of a soil to decompose NO_2^-, and a similar effect has been obtained by ignition (Nelson & Bremner, 1969).

Decomposition of NO_2^- in acidic soils (pH < 6.4) is rapid (approximately one half of the added N can be lost in 2 hours) (Nelson & Bremner, 1969) and follows first-order rate kinetics for a given NO_2^--N addition rate (Nelson & Bremner, 1969; Laudelout et al., 1977). However, Nelson and Bremner (1969) observed that the proportion of added N lost decreased with increasing NO_2^--N addition over the range from 10 to 1,000 $\mu g/g$ of soil.

The rate of NO_2^- decomposition in soil decreases with increasing moisture content (Nelson & Bremner, 1969; Laudelout et al., 1977) and increases with increasing temperature (Nelson & Bremner, 1969; Laudelout et al., 1977). Sterilization of soils has little effect on NO_2^- decomposition (Tyler & Broadbent, 1960; Reuss & Smith, 1965; Nelson & Bremner, 1969; Bollag et al., 1973), but it prevents biological reduction of NO_2^- when neutral and alkaline soils are incubated under anaerobic conditions (Bollag et al., 1973).

C. Gaseous Products of Nitrite Reactions in Soils

A variety of N gases are evolved from soils treated with NO_2^-. Nitric oxide (NO) has not been identified as a product of NO_2^- decomposition in soils under aerobic conditions, but it has been detected when the atmospheres above NO_2^--treated soils contained little O_2 (Smith & Clark, 1960; Reuss & Smith, 1965; Nelson & Bremner, 1970a; Bulla et al., 1970; Bollag et al., 1973; Van Cleemput et al., 1976; Chalamet & Bardin, 1977). This likely results from the fact that NO is readily oxidized to nitrogen dioxide (NO_2) in the presence of O_2 (Eq. [18]) and thus should not be present under aerobic conditions:

$$2\,NO + O_2 \rightleftharpoons 2\,NO_2. \qquad [18]$$

Nitric oxide has also been detected following treatment of nonsterile, acidic soils with NO_3^- under anaerobic conditions (Wijler & Delwiche, 1954; Cady & Bartholomew, 1960; Cooper & Smith, 1963). The detection of NO under these conditions has been attributed to chemical decomposition of HNO_2 produced by microbial reduction of NO_3^-.

Several investigators (Gerretson & de Hoop, 1957; Reuss & Smith, 1965; Bremner & Nelson, 1968; Nelson & Bremner, 1970a; Bollag et al., 1973; Smith & Chalk, 1980a, b) have reported detection of NO_2 as a product of NO_2^- decomposition in acidic soils. However, Smith and Clark (1960) and Tyler and Broadbent (1960) were unable to detect NO_2 by gas chromatographic or mass spectrometric analysis of aerobic atmospheres above acidic soils treated with NO_2^-. Several investigators (Reuss & Smith, 1965; Bremner & Nelson, 1968; Nelson & Bremner, 1970a) have demonstrated that the failure of Smith and Clark (1960) and Tyler and Broadbent (1960) to detect NO_2 was due to their use of closed, aerobic systems, which encouraged sorption of evolved NO_2 by moist soil through the following reactions:

$$2 NO_2 \rightleftharpoons N_2O_4 \qquad\qquad [19]$$

$$N_2O_4 + H_2O \rightleftharpoons HNO_2 + HNO_3. \qquad\qquad [20]$$

Several investigators have observed that NO_2 is evolved on treatment of neutral and alkaline soils with NO_2^- (Reuss & Smith, 1965; Bremner & Nelson, 1968; Bulla et al., 1970; Nelson & Bremner, 1970a; Smith & Chalk, 1980a, b). Steen and Stojanovic (1971) reported that significant amounts of N oxides (probably NO_2) were evolved during nitrification of urea and $(NH_4)_2SO_4$ or following the addition of NO_2^- to a calcareous loam soil.

Dinitrogen (N_2) is a major product of NO_2^- decomposition in acidic soils under anaerobic conditions (Smith & Clark, 1960; Tyler & Broadbent, 1960; Bulla et al., 1970; Nelson & Bremner, 1970a; Bollag et al., 1973; Van Cleemput et al., 1976; Smith & Chalk, 1980a, b) or aerobic conditions (Reuss & Smith, 1965; Nelson & Bremner, 1970a). In addition, N_2 is evolved following the addition of NO_2^- to sterile alkaline soils (Bulla et al., 1970; Nelson & Bremner, 1970a; Van Cleemput et al., 1976; Smith & Chalk, 1980a, b).

Small amounts of N_2O are evolved on treatment of sterile acidic and alkaline soils with NO_2^- (Smith & Clark, 1960; Meek & MacKenzie, 1965; Reuss & Smith, 1965; Bulla et al., 1970; Nelson & Bremner, 1970a; Bollag et al., 1973; Smith & Chalk, 1980a, b). However, Van Cleemput et al. (1976) failed to detect formation of N_2O on the addition of NO_2^- to sterile soil slurries maintained under anaerobic conditions.

For most soils, a higher proportion of added NO_2^--N is converted to $(NO + NO_2)$-N than to N_2 (Reuss & Smith, 1965; Bulla et al., 1970; Nelson & Bremner, 1970a; Bollag et al., 1973). The $(NO + NO_2)$-N/N_2 ratio normally varies from about 2:1 at soil pH values of 5.0 to 5.8 to about 1:1 at pH values above 5.8. However, the addition of NO_2^- to some soils with high organic matter contents leads to the formation of greater amounts of N_2 than of $(NO + NO_2)$-N (Reuss & Smith, 1965; Bulla et al., 1970; Nelson & Bremner, 1970a). Van Cleemput et al. (1976) observed significant amounts of N_2 produced, but no NO evolved, when sterile, anaerobic soil slurries maintained above pH 6 were treated with NO_2^-.

During studies of reaction products of HNO_2 with lignin, Stevenson and Swaby (1963) observed formation of a previously unobserved N gas. This gas was later found to be methyl nitrite (Edwards & Bremner, 1966; Stevenson & Kirkman, 1964; Stevenson & Swaby, 1964), presumably formed through nitrosation of phenolic ethers (methoxyl groups) in lignin, demethoxylation with formation of methyl alcohol, and esterification of methyl alcohol with HNO_2 to yield the gas (Stevenson & Swaby, 1964). Methyl nitrite is formed under strongly acidic conditions when lignin is treated with HNO_2 and has not been observed when soils are treated with NO_2^-; however, Smith and Chalk (1980b) suggested that some of the $^{15}NO_2^-$-N not recovered after addition to soils may have been lost as methyl nitrite.

D. Mechanisms for Gaseous Loss of Nitrite N from Soils

Investigators have proposed a variety of NO_2^- reactions leading to gaseous loss of N from soils. Described below are major mechanisms that may play a role in chemodenitrification.

1. SELF-DECOMPOSITION OF NITROUS ACID

Nitrous acid is produced when NO_2^- is added to or formed in acidic soils and undergoes spontaneous decomposition as indicated in Eq. [21] (Suzawa et al., 1955; Nelson & Bremner, 1970a). The rate of decomposition increases with decrease in soil pH because of the higher proportion of HNO_2 present at low pH:

$$2 HNO_2 \rightleftharpoons NO + NO_2 + H_2O. \tag{21}$$

In closed, anaerobic systems, the NO_2 produced is sorbed by moist soil with the formation of HNO_2 and HNO_3 as indicated in Eq. [19] and [20]. The overall result of HNO_2 self-decomposition in a closed, anaerobic system is as follows (Nelson & Bremner, 1970a):

$$3 HNO_2 \rightleftharpoons 2 NO + HNO_3 + H_2O. \tag{22}$$

This equation suggests that two thirds of the HNO_2-N decomposed should be present as NO-N in the head space of the incubation vessel and one-third as NO_3^--N in the soil.

In closed, aerobic systems, both the NO and NO_2 produced are sorbed by moist soil. Nitric oxide is first oxidized to NO_2 by O_2, and the NO_2 present in the system is rapidly sorbed by reactions indicated in Eq. [19] and [20]. The overall result of HNO_2 decomposition in a closed, aerobic system is (Nelson & Bremner, 1970a):

$$2 HNO_2 + O_2 \rightleftharpoons 2HNO_3. \tag{23}$$

In open (aerobic) systems, the NO and NO_2 formed by self-decomposition of HNO_2 can escape to the atmosphere, but some NO_3^- is formed by oxidation of NO to NO_2 and hydration of NO_2 during diffusion of these gases from the reaction medium. The amount of NO_3^- formed will depend on the rate of diffusion of these gases through the reaction medium and will

Table 2—Nitrogenous gases evolved on treatment of soils with $NaNO_2$ solution (25°C) (Nelson & Bremner, 1970a).†

Soil				Recovery of NO_2^--N (%)					
Type‡	pH	% of C	Treatment time, days	As NO_2^--N	As NO_3^--N	As NO_2-N	As N_2	As NO_2-N§	Total¶
Pershing sil	5.1	1.79	1	16	18	39	24	1	97
			2	11	18	44	24	1	97
			3	5	20	47	26	1	98
Clyde sil	5.5	4.30	1	11	11	36	38	1	96
			2	3	11	40	42	1	96
			3	2	11	40	42	1	96
Muscatine sicl	6.0	2.25	1	40	7	28	23	1	98
			2	29	8	35	26	1	98
Nicollet l	7.0	2.27	1	59	1	20	18	1	98
			2	49	1	26	22	1	98
			3	41	1	29	27	1	98

† Soil samples (20 g) were treated with 6 ml of $NaNO_2$ solution containing 8 mg of NO_2^--N. Treatments were performed in sealed gas analysis units (helium-oxygen atmosphere) with $KMnO_4$ solution in center chamber.
‡ Sil = silty loam; sicl = silty clay loam; l = loam.
§ Determined by analysis of $KMnO_4$ solution.
¶ Recovery as $(NO_2^- + NO_3^- + NO_2 + N_2 + N_2O)$-N.

therefore depend on the nature of the reaction medium and the geometry of the reaction vessel.

The proportion of added NO_2^- evolved as $(NO + NO_2)$-N increases with decrease in soil pH (Reuss & Smith, 1965; Bremner & Nelson, 1968; Nelson & Bremner, 1970a; Bollag et al., 1973), but it is not significantly related to soil organic matter content (Bremner & Nelson, 1968; Nelson & Bremner, 1970a) (see Table 2). Significant amounts of $(NO + NO_2)$-N are evolved on treatment of neutral and alkaline soils (pH < 7.5) with NO_2^- (Nelson & Bremner, 1970a) (see Table 2), whereas the addition of NO_2^- to buffer solutions at pH values of 6 or greater does not lead to $(NO + NO_2)$-N production (Bremner & Nelson, 1968). The amounts of $(NO + NO_2)$-N formed by treatment of soils with NO_2^- in pH 5 or 6 buffer are practically identical to the amounts formed by self-decomposition of HNO_2 in the pure buffers (Bremner & Nelson, 1968). These findings suggest that $(NO + NO_2)$-N evolution following NO_2^- addition to soil results from self-decomposition of HNO_2 and is not promoted by inorganic or organic soil constituents. Furthermore, to account for $(NO + NO_2)$-N evolution from soils with pH values > 6.0, soils must contain regions having much lower pH values than that measured in soil-water slurries. Support for this explanation is provided by investigations showing that the pH at the surfaces of soil clay minerals is approximately 2 pH units lower than the pH of the bulk soil solution (McLaren & Estermann, 1957; Harter & Ahlrichs, 1967).

Although HNO_2 decomposition occurs in acidic and neutral soils, several investigators have questioned the importance of this reaction in regard to gaseous loss of N from soil (Broadbent & Clark, 1965; Allison, 1966; Broadbent & Stevenson, 1966). Broadbent and Clark (1965) concluded that little NO would form in neutral soils and that NO produced in aerated, acidic soils would be oxidized and hydrated to HNO_3 before it could escape to the atmosphere. It is possible that spuriously high amounts of $(NO + NO_2)$-N are formed following NO_2^- addition to soils in a closed system containing an alkaline permanganate solution (to absorb NO and NO_2), because the trap is a highly efficient sink for evolved N oxides and may stimulate $(NO + NO_2)$-N production. In undisturbed soil, it is possible that N oxides may be sorbed by soil before they escape into the atmosphere. However, Nelson and Bremner (1970a) collected large amounts of $(NO + NO_2)$-N from an air stream passing over the surface of soils treated with NO_2^-. Furthermore, a variety of investigators have reported $(NO + NO_2)$-N emissions from untreated soils and from soils during nitrification of urea and NH_4^+ fertilizers (Steen & Stojanovic, 1971; Kim, 1973).

Nitrous acid decomposition is much slower in anaerobic systems with no provision for trapping evolved N oxides compared with aerobic systems or anaerobic systems with alkaline permanganate traps for sorbing N oxides (Reuss & Smith, 1965; Nelson & Bremner, 1970a). This probably results from an accumulation of NO in the atmosphere above the soil and subsequent repression of HNO_2 decomposition as would be suggested by Eq. [22]. Decomposition of HNO_2 appears to be an important mechanism for loss of NO_2^- from acidic and neutral soils under aerobic and anaerobic conditions.

2. REACTION OF NITROUS ACID WITH TRANSITION METAL CATIONS

Wullstein and Gilmour (1964) postulated that metallic cations play an important role in HNO_2 decomposition in soils, and this theory has received considerable attention in reviews of N transformations in soil (Broadbent & Clark, 1965; Allison, 1966; Wullstein, 1967; Hauck, 1958; Woldendorp, 1968). The experimental basis for their theory was work indicating that extraction of kaolinite and an Oregon soil with $1M$ NaCl reduced the amount of NO formed when these materials were treated with NO_2^- under mildly acidic conditions and that $1M$ NaCl extracts of these materials contained metal ions that promoted NO_2^- decomposition. Wullstein and Gilmour concluded that Cu ions in the kaolinite extract and Mn ions in the soil extract were responsible for NO_2^- destruction and that reduced states of the transition metals Cu, Fe, and Mn and certain Al salts promote NO_2^- decomposition as given below:

$$Mn^{2+} + HNO_2 + H^+ = Mn^{3+} + NO + H_2O \qquad [24]$$

$$Fe^{2+} + HNO_2 + H^+ = Fe^{3+} + NO + H_2O. \qquad [25]$$

However, Nelson and Bremner (1970b) found that extraction of Iowa soils with $1M$ NaCl did not reduce NO_2^- decomposition and that $1M$ NaCl extracts of these soils did not promote NO_2^- destruction. Nelson and Bremner (1970b) attributed this divergence as a lack of adequate pH control in the experiments conducted by Wullstein and Gilmour, because the latter did not use buffered systems.

Studies of HNO_2 decomposition by transition metal cations in buffer solutions (pH 5) indicated that only Fe^{2+}, Sn^{2+}, and Cu^+ promote formation of NO (Nelson & Bremner, 1970b). Theoretical calculations and solution studies established that Mn^{2+} and Al^{3+} cannot reduce HNO_2 to NO. Detailed studies in buffered solutions indicated that the Fe^{2+}, Cu^+, and Sn^{2+} concentrations must be at least 0.001, 0.005, and $0.001M$, respectively, before significant decomposition of HNO_2 occurs (Nelson & Bremner, 1970b).

Nelson and Bremner (1970b) reported that $(NO + NO_2)$-N accounted for $< 50\%$ of the N volatilized on treatment of HNO_2 with Fe^{2+}, Sn^{2+} or Cu^+ at pH 6. Wullstein and Gilmour (1966) had previously demonstrated that as much as 15% of added NO_2^--N was evolved as N_2 at pH values of 3.0 and 6.4 in the presence of 5,000 mg/liter of Fe^{2+} but that no N_2O was produced under these conditions. Chao and Kroontje (1966) found that, under aerobic conditions, NO_2^--N reduction to NO by Fe^{2+} was decreased as pH was increased from 5 to 6. In a vacuum, NO and trace amounts of N_2O were produced during reaction of NO_2^--N with Fe^{2+}, but N_2 was not formed. In anaerobic studies of NO_2^- decomposition by Fe^{2+} in the pH range from 6 to 8, Moraghan and Buresh (1977) found that large amounts of N_2O and smaller amounts of N_2 were produced. The addition of Cu^{2+} catalyzed the decomposition of NO_2^- by Fe^{2+} at pH 6 but had no effect at pH 8. Quantitative reduction of NO_2^- to N_2O and N_2 by Fe^{2+} was obtained at pH 8. Nitrous oxide was reduced to N_2 by Fe^{2+} at pH 9 in the presence of Cu^{2+}. Chalamet and Bardin (1977) found that N_2O and N_2 are produced on the addition of NO_2^- to anaerobic soils containing large amounts of Fe^{2+}.

The possibility that metallic cations play a significant role in chemodenitrification seems unlikely, because soils should not contain significant amounts of Fe^{2+}, Sn^{2+}, or Cu^+ under conditions required for microbial oxidation of NH_4^+ to NO_2^-. The possibility that Fe^{2+} may promote decomposition of NO_2^- formed by microbial reduction of NO_3^- in waterlogged soils deserves serous consideration, because significant amounts of Fe^{2+} are often present in anaerobic soils (Nhung & Ponnamperuma, 1966). However, there does not appear to be any report in the literature of simultaneous occurrence of significant amounts of Fe^{2+} and NO_2^- in waterlogged soils. In fact, several investigations have shown that NO_3^- has a sparing effect on the reduction of Fe^{3+} to Fe^{2+} (Alexander, 1964; Turner & Patrick, 1968).

3. REACTION OF NITROUS ACID WITH CLAY MINERALS

Few investigators have studied the reaction of HNO_2 with clay minerals. Wullstein and Gilmour (1964) observed that kaolinite decomposed NO_2^- to NO even after exchangeable transition metals were removed by NaCl extraction. Mortland (1965) found that NO was chemically sorbed on montmorillonite and nontronite in the absence of air when the exchange complex was saturated with certain transition metal ions and that N_2O was formed when NO was sorbed by Fe^{3+}-saturated nontronite. Studies conducted in buffer solution at pH 5 have shown that a wide variety of pure primary, secondary, and clay minerals commonly found in soils do not promote decomposition of NO_2^- (Nelson & Bremner, 1970b). Based on current knowledge, it seems unlikely that soil minerals play a role in chemodenitrification reactions.

4. REACTIONS OF NITRITE WITH AMMONIUM AND HYDROXYLAMINE

Allison (1963) pointed out that both NH_4^+ and NO_2^- are often present in soils treated with NH_4^+ or NH_4^+-forming fertilizers and postulated that significant gaseous loss of N from soils can occur by chemical decomposition of NH_4NO_2 (Eq. [26]):

$$NO_2^- + NH_4^+ \rightleftharpoons NH_4NO_2 \rightarrow N_2 + H_2O. \qquad [26]$$

Nitrite is known to react slowly with NH_4^+ under acidic conditions, but there is no evidence to suggest that NH_4^+ and NO_2^- react under neutral or alkaline conditions.

Gerretsen and de Hoop (1957) claimed that at pH values between 4.0 and 5.5, large amounts of N_2 and smaller amounts of NO were evolved from solutions containing high concentrations of NO_2^- and NH_4^+. Smith and Clark (1960) found that NH_4^+ and NO_2^- reacted to liberate N_2 from solutions if the atmospheres above the reaction solution contained no O_2, reactant concentrations were high, solution pH values were < 5.2. However, Nelson (D. W. Nelson, unpublished observations) observed no effect of NH_4^+ addition on decomposition of NO_2^- in buffer solutions having pH values varying from 4 to 8 irrespective of reactant concentration.

Allison (1963) and Ewing and Bauer (1966) have discussed the conditions under which NH_4NO_2 decomposition is likely to occur in soils and concluded that this reaction will take place when NO_2^- and NH_4^+ are present simultaneously in acidic soils. Steen and Stojanovic (1971) reported that simultaneous addition of NH_4^+ and NO_2^- to a soil at pH 7.5 markedly increased the amount of NO produced relative to the addition of only NO_2^-. However, numerous other workers (Nommik, 1956; Smith & Clark, 1960; Bremner & Nelson, 1968) have failed to detect NH_4NO_2 decomposition during incubation of soils treated with NH_4^+ and NO_2^-. For example, ^{15}N-tracer studies by Nommik (1956) showed that none of the N_2 evolved from soils treated with NH_4^+ and NO_2^- was formed from NH_4NO_2 decomposition.

Wahhab and Uddin (1954) obtained evidence that NH_4NO_2 decomposition occurred when neutral and alkaline soils containing NH_4^+ and NO_2^- were air-dried but concluded that N loss through volatilization of NH_3 and self-decomposition of HNO_2 on air-drying was much larger than N loss through NH_4NO_2 decomposition. Bremner and Nelson (1968) conducted extensive studies that confirmed Wahhab and Uddin's earlier work. They concluded that NH_4NO_2 decomposition does not occur during incubation or air-drying of acidic soils containing NH_4^+ and NO_2^- but that some NH_4NO_2 decomposition occurs when light-textured, neutral, and alkaline soils treated with NH_4^+ and NO_2^- are air-dried. Jones and Hedlin (1970) have reported similar results.

Arnold (1954) found that large amounts of N_2O were evolved when NH_2OH was added to soils and suggested that gaseous loss of soil N may occur through reaction of HNO_2 and NH_2OH, as indicated in Eq. [27]:

$$NH_2OH + HNO_2 \rightleftharpoons N_2O + 2H_2O. \qquad [27]$$

Wijler and Delwiche (1954) and Vine (1962) also suggested that decomposition of NH_2OH by HNO_2 may account for gaseous N loss from soil. Work by Nelson (1978) has established that NH_2OH is quantitatively decomposed by HNO_2 at pH 5 with production of a large amount of N_2O and a small amount of N_2. Bremner et al. (1980) found that the addition of NO_2^- to NH_2OH-treated soils had a small effect on the amounts of N_2O evolved, because large volumes of N_2O were produced in the absence of NO_2^-.

It appears unlikely that the reaction of NO_2^- with NH_4^+ or NH_2OH is of significance in regard to gaseous loss of N from soils. The only situation in which NH_4NO_2 decomposition may be important is during air-drying of neutral or alkaline soils containing high concentrations of NH_4^+ and NO_2^-. There is no evidence that chemodenitrification occurs to any extent under such conditions. The importance of the reaction between HNO_2 and NH_2OH in N loss from soil is even more questionable, because NH_2OH has not been detected in soils.

5. REACTION OF NITROUS ACID WITH COMPOUNDS CONTAINING FREE AMINO GROUPS

It has been known for many years that compounds containing free amino groups (amino acids, urea, amines, etc.) will react with HNO_2 under acidic conditions to liberate N_2 according to Eq. [28]:

$$R \cdot NH_2 + HNO_2 \rightleftharpoons R \cdot OH + N_2 + H_2O. \qquad [28]$$

This reaction is the basis for the HNO_2 method for estimation of free amino groups in proteins and is often referred to as the Van Slyke reaction. Early workers postulated that the Van Slyke reaction was an important mechanism for N loss from soils and other natural systems. In recent years, Reuss & Smith (1965) and Smith & Chalk (1980b) suggested that at least part of the N_2 produced on treatment of acidic soils with NO_2^- is formed by a "Van Slyke type reaction involving labile NH_2 groups in soil organic matter." Smith and chalk (1980b) found that the atom % ^{15}N of N_2 evolved from some soils is less than the percentage of ^{15}N of added NO_2^- indicating that a portion of the N_2 originated from indigenous soil N. However, studies of the reactions of NO_2^- with amino acids and urea in buffer solutions at various pH values have indicated that these reactions are not likely to occur to any significant extent in soils (Allison & Doetsch, 1951; Allison et al., 1952; Sabbe & Reed, 1964). Interestingly, Nelson (D. W. Nelson, unpublished observations) found that the addition of large amounts of glucosamine promoted NO_2^- decomposition in buffer solutions and soil if the pH of the medium was 5 or lower but that the addition of high amounts of alanine or urea did not promote NO_2^- decomposition during incubation or air-drying of soils having a range in pH values.

It would appear that the Van Slyke reaction is of limited importance as a mechanism for gaseous N loss from soils. At pH values where the Van Slyke reaction occurs (pH < 5), reactions leading to NO_2^- formation are slow, and any NO_2^- produced is likely to be lost by self-decomposition of HNO_2. Furthermore, even if HNO_2 is formed, its rate of reaction with amino groups is slow in the presence of air (Broadbent & Clark, 1965).

6. REACTION OF NITROUS ACID WITH SOIL ORGANIC MATTER

A number of investigators have noted a relationship between the rate of NO_2^- decomposition and the organic matter level in soil. Smith and Clark (1960) and Reuss and Smith (1965) noted that soils high in organic matter decomposed NO_2^- to N_2 more extensively than did a soil low in organic matter. Clark and Beard (1960) found that pretreatment of a soil with H_2O_2 to remove organic matter reduced its ability to decompose NO_2^-. Similarly, Nelson and Bremner (1969) observed that removal of organic matter by ignition markedly reduced the ability of several soils to destroy NO_2^-. Clark and Beard (1960) noted that decomposition of NO_2^- in quartz sand was promoted by the addition of peptone or alfalfa meal. It has also been observed that lignin and humic acid preparations reduce HNO_2 to N_2 and N_2O under the conditions of the Van Slyke method of estimating free amino groups (Bremner, 1957; Stevenson & Swaby, 1964).

Table 3—Effects of various organic materials on nitrite decomposition at pH 5 (Bremner & Nelson, 1968).†

Materials	Effect‡
High molecular weight	
Polysaccharides	None or slight
Proteins	None
Nucleic acids	None or slight
Lipids	None or slight
Polyphenols (lignins, tannins)	Marked or very marked
Humic acids	Marked
Low molecular weight	
Amino acids	None
Amino sugars	Slight
Amides	None
Purines and pyrimidines	None or slight
Carboxylic acids	None
Sugars and alcohols	None
Aromatic compounds (nonphenolic)	None
Phenols, monohydric	Marked or very marked
Phenols, dihydric	Very marked
Phenols, trihydric	Very marked

† One milligram of NO_2^--N (as $NaNO_2$) was added to 20 ml of pH 5 Na acetate buffer (0.5M), which contained 0.1 g of high molecular weight material or was 0.025M with respect to low molecular weight material, and the mixture was shaken slowly for 24 hours at 25°C.

‡ Effect on NO_2^- decomposition as determined by analysis for NO_2^--N after 24 hours. Where effect is desribed as slight, the recovery of NO_2^--N was not much lower than the recovery obtained when no organic material was added (88–90%). Where effect is described as marked, the recovery of NO_2^- did not exceed 45%; where effect is described as very marked, the recovery did not exceed 5% and was usually zero.

These findings led Bremner and Nelson (1968) to undertake extensive studies to determine the organic substances that promote NO_2^- decomposition and N_2 production in soils treated with NO_2^-. The ability of a wide variety of organic materials, which have been isolated from soil or are believed to contribute substantially to soil organic matter, to decompose NO_2^- in buffer solutions at pH 5 was evaluated (Table 3). Bremner and Nelson (1968) concluded that phenolic constituents in soil organic matter are largely, if not entirely, responsible for the formation of N_2 and N_2O observed on treatment of soils with NO_2^-. Furthermore, the reactivity of substituted and unsubstituted phenols toward NO_2^- at pH 5 was closely related to their ease of nitrosation, and only phenols tested that did not react with NO_2^- at pH 5 were those that cannot form nitroso derivatives (e.g., 2,4,6-trinitrophenol).

Figure 1 presents likely mechanisms for the reactions of phenols with NO_2^- under acidic conditions (Kainz & Huber, 1959). Mechanism I involves formation of a p-nitrosophenol, tautomerization of this product to a quinone monoxime, and formation of N_2 and N_2O by reaction of the oxime with HNO_2. A similar mechanism was proposed by Bremner (1957) for the formation of N_2 and N_2O during treatment of lignin and humic acid preparations with NO_2^- under strongly acidic conditions. Nelson (D. W. Nelson, unpublished observations) obtained evidence that mechanism I may occur under mildly acidic conditions by establishing that p-nitrosophenol and oximes promoted NO_2^- decomposition at pH 5. Further, evidence was obtained from experiments with $Na^{15}NO_2$ that showed that reaction of p-nitrosophenol with NO_2^- at pH 5 led to formation of N_2 and N_2O and loss of nitroso-N. Mechanism II involves formation of an o-nitrosophenol and production of N_2 through decomposition of the diazo group in the diazonium compound formed by reaction of this o-nitrosophenol with HNO_2. Evidence that mechanisms II may occur under mildly acidic conditions was indicated by the finding that p-substituted phenols and nitrosobenzene promoted NO_2^- decomposition at pH 5. A number of studies have shown that diazo compounds are formed by treatment of certain phenols and nitrosobenzene with NO_2^- under acidic conditions (Morel & Sisley, 1927; Philpot & Small, 1938) and that the diazo compounds decompose at room temperature in acidic, aqueous media with formation of N_2 (DeTar & Ballentine, 1956).

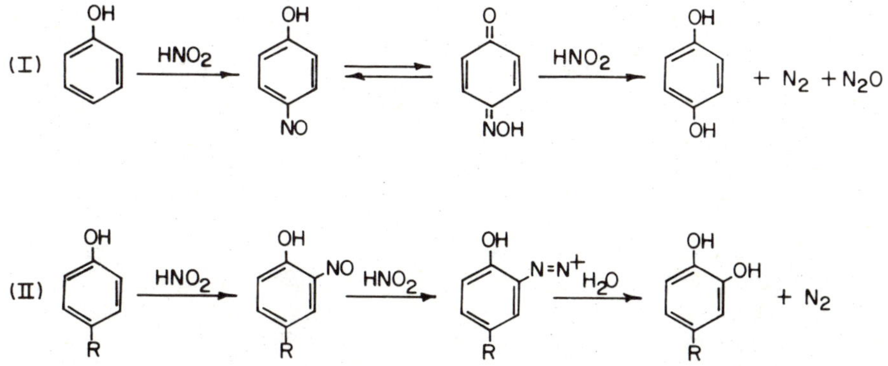

Fig. 1—Reactions of phenols with nitrous acid.

Bremner (1957) found that substantial amounts of N were fixed when humic acid or lignin was treated with NO_2^- under strongly acidic conditions and that only about one third of the fixed N could be recovered by hydrolysis with $6N$ HCl. Führ and Bremner (1964a) investigated the possibility that soils have the ability to fix N on treatment with NO_2^-. They found that the addition of NO_2^- to soils with pH values ranging from 3 to 7 led to fixation of 10 to 28% of added N and conversion of 33 to 79% of NO_2^--N to gaseous forms. Later work established that the amount of added NO_2^--N that was fixed increased with a decrease in soil pH, increased with an increase in soil organic matter, and increased with an increase in NO_2^- concentration (Bremner & Führ, 1966; Führ & Bremner, 1964b; Nelson & Bremner, 1969; Smith & Chalk, 1980b). The lignin-derived fraction of soil organic matter is responsible for NO_2^- fixation, and the mechanism for fixation involves formation of nitroso groups on phenolic rings (Bremner & Führ, 1966). Nelson and Bremner (1969) demonstrated that a wide variety of soils having pH values as high as 7.8 are capable of fixing NO_2^-, that airdrying promotes NO_2^- fixation, that addition of $CaCO_3$ markedly reduces NO_2^- fixation, that NO_2^- fixation is a relatively slow process requiring 96 hours of incubation to reach a maximum value, and that NO_2^- fixation is unaffected by soil moisture content and sterilization. Smith and Chalk (1979) found that fixed NO_2^--N was resistant to mineralization but was more available than indigeous organic N.

E. Importance of Nitrite Reactions in Nitrogen Losses from Soils

Although it has been amply demonstrated that the addition of NO_2^- to soils results in gaseous loss of N, the importance of NO_2^- reactions in N loss during nitrification and denitrification is much in doubt. It seems unlikely that NO_2^- reactions account for significant gaseous N loss from neutral and alkaline soils that accumulate NO_2^- during nitrification, because NO_2^- is relatively unreactive at high pH and little N_2 or $(NO + NO_2)$-N is evolved on treating soils having pH values >7.0 with NO_2^-. Furthermore, Bundy and Bremner (1974) have shown that the N deficits observed during nitrification of urea in alkaline soils that accumulate large amounts of NO_2^- are not reduced if a nitrification inhibitor is used to eliminate NO_2^- accumulation. However, Smith and Chalk (1980a) found significant amounts of N_2, $NO + NO_2$, and N_2O evolved during nitrification of NH_4OH in alkaline soils. Air-drying of neutral alkaline soils containing appreciable amounts of NO_2^- will enhance NO_2^- fixation, $(NO + NO_2)$-N evolution through self-decomposition of HNO_2, and N_2 evolution resulting from reaction of HNO_2 with phenolic components in soil organic matter or decomposition of NH_4NO_2.

Loss of N through NO_2^- reactions is much more likely during nitrification in acidic soils. A number of investigators have shown that NO_2^- accumulates in the periphery of NH_3, NH_4OH, and urea bands in acid soils and in the microenvironment surrounding the prills after urea addition to soil (Smith & Chalk, 1980a). Nitrite ions accumulate because of a combina-

Table 4—Recovery of nitrite N added to pH 5 buffer containing various materials
(25°C) (Nelson & Bremner, 1970a).†

Material‡	Amount, g	Recovery of NO_2^--N after 24 hours (%)						
		As NO_2^--N	As NO_3^--N	As NO_2-N	As N_2	As N_2O-N	As fixed N	Total§
None (control)		82	3	15	0	0	0	100
Glencoe sic, 8.9% C	20	44	3	15	21	1	15	99
Clyde sic, 4.3% C	20	59	4	14	12	1	9	99
Nicollet l, 2.3% C	20	68	3	15	7	<1	6	99
Buckner sa, 0.3% C	20	78	3	15	1	<1	2	99
Ignited Glencoe	20	83	3	14	0	0	0	100
Ignited Clyde	20	82	3	15	0	0	0	100
Ignited Nicollet	20	82	3	14	0	0	0	99
Oxidized Clyde	20	82	3	15	0	0	0	100
Oxidized Nicollet	20	82	3	14	0	0	0	99
Quartz sand	20	83	3	14	0	0	0	100
Clay	20	82	3	15	0	0	0	100
Humic acid	1	49	6	16	19	1	9	100
Lignin	1	23	8	13	27	1	27	99

† Fourteen milliliters of pH 5 Na acetate buffer (4M) containing material specified was treated with 6 ml of $NaNO_2$ solution containing 8 mg of ^{15}N-enriched NO_2^--N. Treatments were performed in sealed gas analysis units (helium-oxygen atmosphere) containing $KMnO_4$ solution in center chamber.

‡ Ignited soils were soils heated at 700°C for 4 hours. Oxidized soils were soils pretreated with KOBr-KOH solution. Clay was mixture of equal parts (by weight) of kaolinite, illite, montmorillonite, and vermiculite, Sic = silty clay; sil = silt loam; l = loam; sa = sand.

§ Recovery as $(NO_2^- + NO_3^- + NO_2 + N_2 + N_2O + fixed)$-N.

tion of high pH and high NH_4^+ concentrations in the fertilizer band or adjacent to the urea prill, which inhibit the activity of *Nitrobacter* sp. responsible for oxidation of NO_2^- to NO_3^- in soils. It appears plausible that NO_2^- diffuses from the alkaline zone surrounding the band or prill into the surrounding acid soil. Once NO_2^- is present in acid soil, HNO_2 is formed immediately and reacts with soil organic matter to liberate N_2 and to be fixed or to undergo self-decomposition with evolution of $(NO + NO_2)$-N. Evidence for these conclusions is given in Table 4, which shows that the amount of added NO_2^--N decomposed in soil and the amount fixed or converted to N_2 at pH 5 increased with increase in soil organic matter content. Like soils, the organic materials tested in the work reported in Table 4 (humic acid and lignin) promoted NO_2^- decomposition, fixed NO_2-N, and converted NO_2^--N to N_2 and N_2O, whereas the soils pretreated to remove organic matter and the inorganic materials tested (quartz sand and clay minerals) did not interact with added NO_2^-. Furthermore, none of the materials tested promoted NO_2 formation at pH 5, which suggests that pH is the soil factor determining NO_2 production through self-decomposition of HNO_2. Much additional research is needed to discover if NO_2^- accumulation and reactions in soil surrounding fertilizer bands and urea prills lead to significant gaseous loss of N from soils. Nitrite accumulations or reactions are not likely when acidic fertilizers such as $(NH_4)_2SO_4$ or NH_4NO_3 are applied to acid soils.

IV. GASEOUS NITROGEN LOSS FROM SOILS THROUGH REACTIONS OF NITRATE AND HYDROXYLAMINE

Wetselaar (1961) observed that an Australian soil stored in a dry state outdoors lost a substantial proportion of its NO_3^- during hot, dry weather. Subsequently, Thomas and Kissel (1970) found that considerable amounts of NO_3^- were lost when soils containing moderate levels of exchangeable Al^{3+} or H^+ were heated at temperatures $>55°C$. They attributed the NO_3^- loss to volatilization of HNO_3 as indicated in Eq. [29]:

$$Ca(NO_3)_2 + Al(H_2O)_6\text{-soil} \rightarrow Ca\text{-soil-}Al(OH)_2(H_2O)_4 + 2\,HNO_3\uparrow. \qquad [29]$$

Thomas and Kissel (1970) concluded that volatilization of HNO_3 from soils is only possible with high levels of soil acidity. The temperatures required for NO_3^- losses are higher than those normally encountered in the field, and large losses of HNO_3 are to be expected only during oven-drying of soil samples.

The chemical reduction of NO_3^- to N_2O and N_2 by Fe^{2+} in the presence of small amounts of Cu^{2+} have been reported at pH values of 7 to 9 (Gunderloy et al., 1970; Buresh & Moraghan, 1976). Buresh and Moraghan found that conditions favoring NO_3^- reduction to N_2O and N_2 were an atmosphere free of O_2, a Fe^{2+}–NO_3^- mole ratio of 12, a pH of 8 to 8.5, and the presence of 10 mg/liter Cu^{2+}. At pH values of 9 and 10, NO_3^- was reduced by Fe^{2+} to NH_4^+ and NO_2^-. Buresh and Moraghan (1976) suggested that reduction of $NO3^-$ to N gases may occur on the addition of NO_3^- to waterlogged soils because of the presence of high Fe^{2+} concentrations in such soils.

Several investigators have reported that NH_2OH cannot be recovered soon after its addition to soils and that added NH_2OH is not converted to NO_2^- or NO_3^- (Duesberg & Buehrer, 1954; Bremner & Shaw, 1958). Bremner and Shaw (1958) speculated that NH_2OH reacts with higher oxides of Mn and Fe to produce gaseous N compounds. Arnold (1954) and Nommik (1956) found that N_2O and N_2, respectively, were evolved when NH_2OH was added to moist soils. In studies with $^{15}NH_2OH$, Nelson (1978) demonstrated that from 10 to 56% of NH_2OH-N was fixed by organic matter when it was added to a variety of soils. Furthermore, substantial amounts of N_2 and N_2O, but no NO, were evolved from soils treated with NH_2OH. Additional studies conducted in pH 5 buffer demonstrated that MnO_2 and NO_2^- quantitatively decompose NH_2OH with N_2O as the predominant product, whereas Fe^{2+} reacted with NH_2OH to produce N_2O and N_2 in a 2.5:1 ratio. Bremner et al. (1980) found that evolution of N_2O by chemical decomposition of NH_2OH in soil is more rapid than production of N_2 and that N_2O formation is highly correlated with exchangeable and oxidized Mn in soils.

There is little likelihood that chemical reactions of NO_3^- or NH_2OH lead to significant loss of gaseous N from soils. Conditions favoring NO_3^- loss from soil by chemical reactions are rarely encountered in the field.

Hydroxylamine has not been detected in soils, and there is no evidence that NH_2OH reactions are important in chemodenitrification. However, such evidence is difficult to obtain if NH_2OH is decomposed rapidly in soils.

V. MANAGEMENT TECHNIQUES TO MINIMIZE GASEOUS NITROGEN LOSSES

Direct loss of NH_3 during and immediately following anhydrous NH_3 application to soils may be essentially eliminated by proper management. Anhydrous NH_3 should be injected at an appropriate depth into the soil as determined by soil type. Coarse-textured soils require deeper placement to avoid NH_3 losses than do fine-textured soils. Application should be timed to coincide with proper soil moisture content and soil physical conditions that give rapid and complete closure of the injection channel. The combination of anhydrous NH_3 application rate, spacing of the injector knives, and applicator speed should be selected to give maximum efficiency of NH_3 retention by the soil.

Losses of NH_3 following applications of dry or liquid ammoniacal fertilizers to soils may be minimized by a number of techniques. Surface applications of fertilizer solutions containing free NH_3 (e.g., NH_4OH) should not be used on any soils. Surface applications of alkaline ammoniacal fertilizers [e.g., urea, $(NH_4)_2HPO_4$] without incorporation are also discouraged. However, if surface applications are required, use of $(NH_4)_2SO_4$ or NH_4NO_3 on neutral and acidic soils is usually recommended. Ammonia losses resulting from surface applications can be minimized if the fertilizer is applied to moist soils (not saturated) immediately prior to rainfall during periods of relatively low temperature. If N fertilizers are surface applied to calcareous soils, use of NH_4NO_3 is recommended, because high NH_3 losses occur with $(NH_4)_2SO_4$ and urea. Proper incorporation immediately following broadcast application of dry or liquid N-containing fertilizers or injection of liquid fertilizers will minimize NH_3 losses from soil. Placement of NH_4^+ fertilizers in the soil immediately prior to flooding will give much lower NH_3 losses from rice fields than broadcasting fertilizer on the water surface after flooding.

Control of gaseous N losses resulting from chemodenitrification reactions in soil is difficult. One approach to minimize chemodenitrification is to manage soils such that rapid nitrification occurs and situations where NO_2^- may temporarily accumulate (e.g., high applications of ammoniacal fertilizers on alkaline soils, banding alkaline fertilizers in neutral or acid soils) are avoided. Apparently there is little loss of N if N fertilizers are broadcast at moderate rates and mixed throughout the upper few cm of soil Alternatively, nitrification inhibitors may be useful in completely stopping nitrification or markedly slowing down the rate of NH_4^+ oxidation. Interestingly, several workers have demonstrated substantial yield benefits resulting from reduced N losses obtained from use of nitrification inhibitors (Nelson & Huber, 1979; Warren et al., 1975). The reduced N losses obtained from application of nitrification inhibitors with ammoniacal fertilizers may, in part, result from elimination of chemodenitrification. Gaseous N losses from NO_2^- reactions may also be reduced through use of controlled-release fertilizers, such as sulfur-coated urea, urea-formaldehyde, or isobutylenediurea. However, the increased cost of these materials may exceed the value of any reduction in N loss expected from their use.

LITERATURE CITED

Alexander, M. 1964. Biochemical ecology of soil microorganisms. A. Rev. Microbiol. 18:217–252.

Allison, F. E. 1955. The enigma of soil nitrogen balance sheets. Adv. Agron. 7:213–250.

Allison, F. E. 1963. Losses of gaseous nitrogen from soils by chemical mechanisms involving nitrous acid and nitrites. Soil Sci. 96:404–409.

Allison, F. E. 1965. Evaluation of incoming and outgoing processes that affect soil nitrogen. In W. V. Bartholomew and F. E. Clark (ed.) Soil nitrogen. Agronomy 10:573–606. Am. Soc. of Agron., Madison, Wis.

Allison, F. E. 1966. The fate of nitrogen applied to soils. Adv. Agron. 18:219–258.

Allison, F. E. 1973. Soil organic matter and its role in crop production. Elsevier Scientific Publishing Co., New York.

Allison, F. E., and J. H. Doetsch. 1951. Nitrogen gas production by the reaction of nitrites with amino acids in slightly acidic media. Soil Sci. Soc. Am. Proc. 15:163–166.

Allison, F. E., J. H. Doetsch, and L. D. Sterling. 1952. Nitrogen gas formation by interaction of nitrites and amino acids. Soil Sci. 74:311–314.

Arnold, P. W. 1954. Losses of nitrous oxide from soils. J. Soil Sci. 5:116–126.

Avnimelech, Y., and M. Laher. 1977. Ammonia volatilization from soils: Equilibrium considerations. Soil Sci. Soc. Am. J. 41:1080–1084.

Baligar, V. C., and S. V. Patil. 1968a. Effect of initial soil moisture and temperature on volatile losses of ammonia with the application of urea to soils. Mysore J. Agric. Sci. II:85–93.

Baligar, V. C., and S. V. Patil. 1968b. Volatile losses of ammonia as influenced by rate and methods of urea application to soils. Indian J. Agron. 13:230–233.

Blasco, M. L., and A. H. Cornfield. 1966. Volatilization of nitrogen as ammonia from acid soils. Nature (London) 212:1279–1280.

Blue, W. G., and C. F. Eno. 1954. Distribution and retention of anhydrous ammonia in sandy soils. Soil Sci. Soc. Am. Proc. 12:157–164.

Bollag, J. M., S. Drzymala, and L. T. Kardos. 1973. Biological versus chemical nitrite decomposition in soil. Soil Sci. 116:44–50.

Bremner, J. M. 1957. Studies on soil humic acids. II. Observations on the estimations of free amino groups. Reactions of humic acid and lignin preparations with nitrous acid. J. Agric. Sci. 48:352–353.

Bremner, J. M., and A. M. Blackmer. 1978. Nitrous oxide: Emission from soils during nitrification of fertilizer nitrogen. Science 199:295–296.

Bremner, J. M., A. M. Blackmer, and S. A. Waring. 1980. Formation of nitrous oxide and dinitrogen by chemical decomposition of hydroxylamine in soils. Soil Biol. Biochem. 12:263–269.

Bremner, J. M., and L. A. Douglas. 1971. Decomposition of urea phosphate in soils. Soil Sci. Soc. Am. Proc. 35:575–578.

Bremner, J. M., and F. Führ. 1966. Tracer studies of the reaction of soil organic matter with nitrite. p. 337–346. In The use of isotopes in soil organic matter studies. Pergamon Press, Inc., Elmsford, N.Y.

Bremner, J. M., and D. W. Nelson. 1968. Chemical decomposition of nitrite in soils. Int. Congr. Soil Sci. Trans., 9th (Adelaide) II:495–503.

Bremner, J. M., and K. Shaw. 1958. Denitrification in soil. J. Agric. Sci. 51:22–39.

Broadbent, F. E., and F. E. Clark. 1965. Denitrification. In W. V. Bartholomew and F. E. Clark (ed.) Soil nitrogen. Agronomy 10:344–359. Am. Soc. of Agron., Madison, Wis.

Broadbent, F. E., and F. J. Stevenson. 1966. Organic matter interactions. p. 169–187. In M. H. McVicker et al. (ed.) Agricultural anhydrous ammonia technology and use. Agric. Ammonia Inst., Memphis, Tenn.

Brown, J. M., and W. V. Bartholomew. 1962. Sorption of anhydrous ammonia by dry clay systems. Soil Sci. Soc. Am. Proc. 26:258–262.

Bulla, L. A., C. M. Gilmour, and W. B. Bollen. 1970. Non-biological reduction of nitrite in soil. Nature (London) 225:664.

Bundy, L. G., and J. M. Bremner. 1974. Effects of nitrification inhibitors on transformaitons of urea nitrogen in soils. Soil Biol. Biochem. 6:369–376.

Buresh, R. J., and J. T. Moraghan. 1976. Chemical reduction of nitrate by ferrous iron. J. Environ. Qual. 5:320–325.

Cady, F. B., and W. V. Bartholomew. 1960. Sequential products of anaerobic denitrification in Norfolk soil material. Soil Sci. Soc. Am. Proc. 24:477–482.

Chalamet, A., and R. Bardin. 1977. Action des ions ferreux sur la reduction de l'acide nitreux dans les sols hydromorphes. Soil Biol. Biochem. 9:281–285.

Chalk, P. M., D. R. Keeney, and L. M. Walsh. 1975. Crop recovery and nitrification of fall and spring applied anhydrous ammonia. Agron. J. 67:33–37.

Chao, T. T., and W. Kroontje. 1960. Ammonia adsorption phenomena in soils. Int. Cong. Soil Sci. Trans., 7th (Madison, Wis.) III:517–522.

Chao, T. T., and W. Kroontje. 1964. Relationship between ammonia volatilization, ammonia concentration, and water evaporation. Soil Sci. Soc. Am. Proc. 28:393–395.

Chao, T. T., and W. Kroontje. 1966. Inorganic nitrogen transformations through the oxidation and reduction of iron. Soil Sci. Soc. Am. Proc. 30:193–196.

Chapman, H. D., and G. F. Liebig. 1952. Field and laboratory studies of nitrite accumulation in soils. Soil Sci. Soc. Am. Proc. 16:276–282.

Clark, F. E. 1962. Losses of nitrogen accompanying nitrification. p. 173–176. In Int. Soc. Soil Sci., Trans. Comm. IV, V (Palmerston, N.Z.).

Clark, F. E., and W. E. Beard. 1960. Influence of organic matter on volatile losses of nitrogen from soil. Int. Congr. Soil Sci. Trans. 7th (Madison, Wis.) III:501–503.

Clark, F. E., W. E. Beard, and D. H. Smith. 1960. Dissimilar nitrifying capacities of soils in relation to losses of applied nitrogen. Soil Sci. Soc. Am. Proc. 24:50–54.

Cooper, G. S., and R. L. Smith. 1963. Sequence of products formed during denitrification in some diverse western soils. Soil Sci. Soc. Am. Proc. 27:659–662.

Council for Agricultural Science and Technology. 1976. Effect of increased nitrogen fixation on stratospheric ozone. Report 53, Iowa State Univ., Ames.

Court, M. M., R. C. Stephen, and J. S. Waid. 1974. Nitrite toxicity arising from the use of urea as a fertilizer. Nature (London) 194:1263–1265.

Crutzen, P. M. 1976. Upper limits on atmospheric ozone reduction following increased application of fixed nitrogen to soil. Geophys. Res. Lett. 3:169.

Denmead, O. T., J. R. Simpson, and J. R. Freney. 1974. Ammonia flux into the atmosphere from a grazed pasture. Science 185:609–610.

Denmead, O. T., J. R. Simpson, and J. R. Freney. 1977. A direct field measurement of ammonia emission after injection of anhydrous ammonia. Soil Sci. Soc. Am. J. 41:1001–1004.

DeTar, D. F., and A. R. Ballentine. 1956. The mechanism of diazonium salt reactions. II. A redetermination of the rates of thermal decomposition of six diazonium salts in aqueous solution. Am. Chem. Soc. J. 78:3916–3920.

Duesberg, P. C., and T. F. Buehrer. 1954. Effect of ammonia and its oxidation products on rate of nitrification and plant growth. Soil Sci. 78:37–49.

du Plessis, M. C. F., and W. Kroontje. 1964. The relationship between pH and ammonia equilibria in soil. Soil Sci. Soc. Am. Proc. 28:751–754.

Edwards, A. P., and J. M. Bremner. 1966. Formation of methyl nitrite in reaction of lignin with nitrous acid. p. 347–348. In The use of isotopes in soil organic matter studies. Pergamon Press, Inc., Elmsford, N.Y.

Ernst, J. W., and H. F. Massey. 1960. The effects of several factors on volatilization of ammonia formed from urea in soil. Soil Sci. Soc. Am. Proc. 24:87–90.

Ewing, G. J., and N. Bauer. 1966. An evaluation of nitrogen losses from the soil due to the reaction of ammonium ions with nitrous acid. Soil Sci. 102:64–69.

Feagley, S. E., and L. R. Hossner. 1978. Ammonia volatilization reactions mechanisms between ammonium sulfate and carbonate systems. Soil Sci. Soc. Am. J. 42:364–367.

Fenn, L. B. 1975. Ammonia volatilization from surface applications of ammonium compounds on calcareous soils: III. Effects of mixing low and high loss ammonium compounds. Soil Sci. Soc. Am. Proc. 39:366–368.

Fenn, L. B., and R. Escarzaga. 1976. Ammonia volatilization from surface applications of ammonium compounds on calcareous soils: V. Soil water content and method of nitrogen application. Soil Sci. Soc. Am. J. 40:537–541.

Fenn, L. B., and R. Escarzaga. 1977. Ammonia volatilization from surface applications of ammonium compounds to calcareous soils: VI. Effects of initial soil water content and quantity of applied water. Soil Sci. Soc. Am. J. 41:358–362.

Fenn, L. B., and D. E. Kissel. 1973. Ammonia volatilization from surface applications of ammonium compounds on calcareous soils: I. General theory. Soil Sci. Soc. Am. Proc. 37: 855–859.

Fenn, L. B., and D. E. Kissel. 1974. Ammonia volatilization from surface applications of ammonium compounds on calcareous soils: II. Effects of temperature and rate of ammonium nitrogen application. Soil Sci. Soc. Am. Proc. 38:606–610.

Fenn, L. B., and D. E. Kissel. 1976. The influence of cation exchange capacity and depth of incorporation on ammonia volatilization from ammonium compounds applied to calcareous soils. Soil Sci. Soc. Am. J. 40:394–398.

Führ, F., and J. M. Bremner. 1964a. Untersuchungen zur Fixierung des Nitrit-Stickstoffs durch die organische Masse des Bodens. Landwirtsch. Forsch. 11:43–51.

Führ, F., and J. M. Bremner. 1964b. Beeinflussende Faktoren in der Fixierung des Nitrit-Stickstoffs durch die organische Masse des Bodens. Atompraxis 10:109–113.

Gandhi, A. P., and K. V. Paliwal. 1976. Mineralization and gaseous loss of nitrogen from urea and ammonium sulfate in salt-affected soil. Plant Soil 45:247–255.

Gasser, J. K. R. 1964. Some factors affecting the losses of ammonia from urea and ammonium sulfate applied to soils. J. Soil Sci. 15:258–272.

Gerretsen, F. C., and H. de Hoop. 1957. Nitrogen losses during nitrification in solutions and acid sandy soils. Can. J. Microbiol. 3:359–380.

Gunderloy, F. C., R. I. Wagner, and V. H. Dayan. 1970. Development of a chemical denitrification process. EPA Water Pollution Cont. Res. Ser. Rep. WPCR 17010 EEZ 10/70. Environmental Protection Agency, Cincinnati, Ohio.

Hargrove, W. L., D. E. Kissel, and L. B. Fenn. 1977. Field measurements of ammonia volatilization from surface applications of ammonium salts to a calcareous soil. Agron. J. 69:473–476.

Harmsen, G. W., and G. J. Kolenbrander. 1965. Soil inorganic nitrogen. In W. V. Bartholomew and F. E. Clark (ed.) Soil nitrogen. Agronomy 10:43–92. Am. Soc. of Agron., Madison, Wis.

Harter, R. D., and J. L. Ahlrichs. 1967. Determination of clay surface acidity by infrared spectroscopy. Soil Sci. Soc. Am. Proc. 31:30–33.

Hauck, R. D. 1968. Soil and fertilizer nitrogen—A review of recent work and commentary. Int. Congr. Soil Sci., Trans. 9th (Adelaide) II:475–486.

Hauck, R. D., and J. M. Stephenson. 1965. Nitrification of nitrogen fertilizers. Effect of nitrogen source, size and pH of the granule, and concentration. J. Agric. Food Chem. 13: 486–492.

Hutchinson, G. W., and F. G. Viets, Jr. 1969. Nitrogen enrichment of surface water by adsorption of ammonia volatilized from cattle feed lots. Science 166:514–515.

Jackson, J. L., and S. C. Chang. 1947. Anhydrous ammonia retention by soils as influence by depth of application, soil texture, moisture content, pH value, and tilth. J. Am. Soc. Agron. 39:623–633.

James, D. W., and M. E. Harward. 1964. Competition of ammonia and water for adsorption on clay minerals. Soil Sci. Soc. Am. Proc. 28:636–640.

Jewitt, T. N. 1942. Loss of ammonia from ammonium sulfate applied to alkaline soils. Soil Sci. 54:401–409.

Jones, R. W., and R. A. Hedlin. 1970. Nitrite instability in three Manitoba soils. Can. J. Soil Sci. 50:339–345.

Justice, J. K., and R. L. Smith. 1962. Nitrification of ammonium sulfate in a calcareous soil as influenced by combinations of moisture, temperature, and levels of added nitrogen. Soil Sci. Soc. Am. Proc. 26:246–250.

Kainz, G., and H. Huber. 1959. Anomalous reactions in amino nitrogen determination. V. The anomaly of phenols. Mikrochim. Acta 1959:891–902.

Khan, D. H., and M. Z. Haque. 1965. Volatilization loss of nitrogen from urea added to some soils of East Pakistan. J. Sci. Food Agric. 16:725–729.

Khan, D. H., and G. H. Rashid. 1971. Losses of nitrogen from urea in some soils of East Pakistan. Exp. Agric. 7:107–112.

Kim, C. M. 1973. Influence of vegetation types on the intensity of ammonia and nitrogen dioxide liberation from soil. Soil Biol. Biochem. 5:163–166.

Kissel, D. E., H. L. Brewer, and G. F. Arkin. 1977. Design and test of a field sampler for ammonia volatilization. Soil Sci. Soc. Am. J. 41:1133–1138.

Kresge, C. B., and D. P. Satchell. 1960. Gaseous loss of ammonia from nitrogen fertilizers applied to soil. Agron. J. 52:104–107.

Laudelout, H., L. Germain, P. F. Chabalier, and C. H. Chiang. 1977. Computer simulation of loss of fertilizer nitrogen through chemical decomposition of nitrite. J. Soil Sci. 28: 329–339.

Loftis, J. R., and C. E. Scarsbrook. 1969. Ammonia volatilization from ratios of formamide and urea solutions in soils. Agron. J. 61:725–727.

Macrae, I. C., and R. Ancajas. 1970. Volatilization of ammonia from submerged tropical soils. Plant Soil 33:97–103.

Makarov, B. N., and N. B. Makarov. 1976. Gaseous nitrogen losses from soil and fertilizer. Sov. Soil Sci. 8:692–704.

Martin, J. P., and H. D. Chapman. 1951. Volatilizations of ammonia from surface fertilized soils. Soil Sci. 71:25–34.

Matocha, J. E. 1976. Ammonia volatilization and nitrogen utilization from sulfur-coated ureas and conventional nitrogen fertilizers. Soil Sci. Soc. Am. J. 40:597–601.

McDowell, L. L., and G. E. Smith. 1958. The retention and reactions of anhydrous ammonia on different soil types. Soil Sci. Soc. Am. Proc. 22:38–42.

McLaren, A. D., and E. Estermann. 1957. Influence of pH on the activity of chymotrypsin at solid-liquid interface. Arch. Biochem. Biophys. 68:157–160.

Meek, B. D., and A. J. MacKenzie. 1965. The effect of nitrite and organic matter on aerobic gaseous losses of nitrogen from a calcareous soil. Soil Sci. Soc. Am. Proc. 29:176–178.

Meyer, R. D., R. A. Olson, and H. F. Rhoades. 1961. Ammonia losses from fertilized Nebraska soils. Agron. J. 53:241–244.

Mikkelsen, D. S., S. K. De Datta, and W. N. Obcemea. 1978. Ammonia volatilization losses from flooded rice soils. Soil Sci. Soc. Am. J. 42:725–730.

Miyamoto, S., R. J. Prather, and H. L. Bohn. 1974. Nitric oxide sorption by calcareous soils: II. Effect of moisture on capacity, rate, and sorption products. Soil Sci. Soc. Am. Proc. 38:71–74.

Moe, P. G. 1967. Nitrogen losses from urea as affected by altering soil urease activity. Soil Sci. Soc. Am. Proc. 31:380–382.

Moraghan, J. T., and R. J. Buresh. 1977. Chemical reduction of nitrite and nitrous acid by ferrous iron. Soil Sci. Soc. Am. J. 41:47–50.

Morel, A., and P. Sisley. 1927. Constitution of the azo compounds of the silk fibers. Soc. Chim. Fr. Bull. 41:1217–1224.

Mortland, M. M. 1955. Absorption of ammonia by clays and muck. Soil Sci. 80:11–18.

Mortland, M. M. 1965. Nitric oxide adsorption by clay minerals. Soil Sci. Soc. Am. Proc. 29: 514–519.

Nelson, D. W. 1978. Transformations of hydroxylamine in soils. Proc. Indiana Acad. Sci. 87: 409–413.

Nelson, D. W., and J. M. Bremner. 1969. Factors affecting chemical transformations of nitrite in soils. Soil Biol. Biochem. 1:229–239.

Nelson, D. W., and J. M. Bremner. 1970a. Gaseous products of nitrite decomposition in soil. Soil Biol. Biochem. 2:203–215.

Nelson, D. W., and J. M. Bremner. 1970b. Role of soil minerals and metallic cations in nitrite decomposition and chemodenitrification in soil. Soil Biol. Biochem. 2:1–8.

Nelson, D. W., and D. M. Huber. 1980. Performance of nitrification inhibitors in the Midwest(east). p. 75–88. In J. J. Meisinger (ed.) Nitrification inhibitors—potentials and limitations. ASA Spec. Pub. no. 38, Madison, Wis.

Nhung, M. M., and F. N. Ponnamperuma. 1966. Effects of calcium carbonate, manganese dioxide, ferric hydroxide, and prolonged flooding on chemical and electrochemical changes and growth of rice in a flooded acid sulfate soil. Soil Sci. 103:29–41.

Nommik, H. 1956. Investigations on denitrification in soil. Acta Agric. Scand. 6:195–226.

Nommik, H. 1966. Use of micro-plot technique for studying gaseous loss of ammonia from added nitrogen materials under field conditions. Acta Agric. Scand. 16:147–154.

Nommik, H. 1973. Assessment of volatilization loss of ammonia from surface-applied urea on forest soil by N^{15} recovery. Plant Soil 38:589–603.

Nommik, H., and K. O. Nilsson. 1963. Nitrification and movement of anhydrous ammonia in soil. Acta Agric. Scand. 13:205–219.

Overrein, L. M. 1968. Lysimeter studies of tracer nitrogen in forest soil: I. Nitrogen losses by leaching and volatilization after addition of urea-N^{15}. Soil Sci. 106:280–290.

Overrein, L. M., and P. G. Moe. 1967. Factors affecting urea hydrolysis and ammonia volatilization in soil. Soil Sci. Soc. Am. Proc. 31:57–61.

Parr, J. F., Jr., and R. I. Papendick. 1966a. Retention of ammonia in soils. p. 213–236. In M. C. McVickar et al. (ed.) Agricultural anhydrous ammonia technology and use. Agric. Ammonia Inst., Memphis, Tenn.

Parr, J. F., and R. I. Papendick. 1966b. Retention of anhydrous ammonia by soil: II. Effect of ammonia concentration and soil moisture. Soil Sci. 10:109–119.

Philpot, J. S., and P. A. Small. 1938. The action of nitrous acid on phenols. Biochem. J. 32: 534–541.

Prasad, M. 1976. Gaseous loss of ammonia from sulfur-coated urea, ammonium sulfate, and urea applied to calcareous soil (pH 7.3). Soil Sci. Soc. Am. J. 40:130–134.

Prather, R. J., S. Miyamoto, and H. L. Bohn. 1973a. Sorption of nitrogen dioxide by calcareous soils. Soil Sci. Soc. Am. Proc. 37:860–863.

Prather, R. J., S. Miyamoto, and H. L. Bohn. 1973b. Nitric oxide sorption by calcareous soils: I. Capacity, rate, and sorption products in air dry soils. Soil Sci. Soc. Am. Proc. 37:877–879.

Raison, R. J., and J. W. McGarity. 1978. Effect of plant ash on nitrogen fertilizer transformations and ammonia volatilization. Soil Sci. Soc. Am. J. 42:140–143.

Rashid, G. H. 1977. The volatilization losses of nitrogen from added urea in some soils of Bangladesh. Plant Soil 48:549–556.

Reuss, J. O., and R. L. Smith. 1965. Chemical reactions of nitrites in acid soils. Soil Sci. Soc. Am. Proc. 29:267–270.

Robinson, E., and R. C. Robbins. 1970. Gaseous nitrogen compound pollutants from urban and natural sources. J. Air Pollut. Control Assoc. 20:303–306.

Sabbe, W. E., and L. W. Reed. 1964. Investigations concerning nitrogen loss through chemical reactions involving urea and nitrite. Soil Sci. Soc. Am. Proc. 28:478–481.

Smith, C. J., and P. M. Chalk. 1978. Losses of nitrogenous gases from soils during nitrification of ammonia fertilizers. p. 483–490. In A. R. Ferguson et al. (ed.) Plant nutrition 1978. New Zealand D.S.I.R. Information Series no. 134, The Government Printer, Wellington, New Zealand.

Smith, C. J., and P. M. Chalk. 1979. Mineralization of nitrite fixed by soil organic matter. Soil Biol. Biochem. 11:515–519.

Smith, C. J., and P. M. Chalk. 1980a. Gaseous nitrogen evolution during nitrification of ammonia fertilizer and nitrite transformations in soils. Soil Sci. Soc. Am. J. 44:277–282.

Smith, C. J., and P. M. Chalk. 1980b. Fixation and loss of nitrogen during transformations of nitrite in soil. Soil Sci. Soc. Am. J. 44:288–291.

Smith, D. H., and F. E. Clark. 1960. Volatile losses of nitrogen from acid or neutral soils or solution containing nitrite and ammonium ions. Soil Sci. 90:86–92.

Smith, E. A., and C. I. Mayfield. 1978. Effects of nitrogen dioxide on selected soil processes. Water Air Soil Pollut. 9:33–43.

Soulides, D. A., and F. E. Clark. 1958. Nitrification in grassland soils. Soil Sci. Soc. Am. Proc. 22:308–311.

Stanley, F. A., and G. E. Smith. 1956. Effect of soil moisture and depth of application on retention of anhydrous ammonia. Soil Sci. Soc. Am. Proc. 20:557–561.

Steen, W. C., and B. J. Stojanovic. 1971. Nitric oxide volatilization from a calcareous soil and model aqueous systems. Soil Sci. Soc. Am. Proc. 35:277–282.

Stevenson, F. J., and M. A. Kirkman. 1964. Identification of methyl nitrite in the reaction product of nitrous acid with lignin. Nature (London) 201:107.

Stevenson, F. J., and R. J. Swaby. 1963. Occurrence of a previously unobserved nitrogen gas in the reaction product of nitrous acid and lignin. Nature (London) 199:97–98.

Stevenson, F. J., and R. J. Swaby. 1964. Nitrosation of soil organic matter: I. Nature of gases evolved during nitrous acid treatment of lignin and humic substances. Soil Sci. Soc. Am. Proc. 23:773–777.

Suzawa, T., M. Honda, O. Manabe, and H. Hijama. 1955. Decomposition of nitrous acid in aqueous solution. Chemical Soc. of Japan. Ind. Chem. Sect. J. 58:744–746.

Terman, G. L. 1979. Volatilization losses of nitrogen as ammonia from surface-applied fertilizers, organic amendments, and crop residues. Adv. Agron. 31:189–223.

Terman, G. L., J. F. Parr, and S. E. Allen. 1968. Recovery of nitrogen by corn from solid fertilizers and solutions. J. Agric. Food Chem. 16:685–690.

Terry, R. E., D. W. Nelson, L. E. Sommers, and G. J. Meyer. 1978. Ammonia volatilization from wastewater sludge applied to soil. J. Water Pollut. Control Fed. 50:2657–2665.

Thomas, G. W., and D. E. Kissel. 1970. Nitrate volatilization from soils. Soil Sci. Soc. Am. Proc. 34:828–830.

Tripathi, O. N. 1958. Role of energy-rich materials as conservators in presence of nitrogenous fertilizers in Sagar soil. Indian Acad. Sci. Proc. 27A:215–220.

Tripathi, O. N., and S. N. Benarjee. 1955. Decomposition of urea as such and in the presence of glucose in Sagar soil. Indian Acad. Sci. Proc. 24A:257–262.

Turner, F. T., and W. H. Patrick. 1968. Chemical changes in waterlogged soils as result of oxygen depletion. Int. Congr. Soil Sci., Trans. 9th (Adelaide) IV:53–65.

Tyler, K. B., and F. E. Broadbent. 1960. Nitrite transformations in California soils. Soil Sci. Soc. Am. Proc. 24:279–282.

Van Cleemput, O., W. H. Patrick, Jr., and R. C. McIlhenny. 1976. Nitrite decomposition in flooded soil under different pH and redox potential conditions. Soil Sci. Soc. Am. J. 40: 55–60.

Ventura, W. B., and T. Yoshida. 1977. Ammonia volatilization from a flooded tropical soil. Plant Soil 46:521–531.

Verma, R. N. Singh, and M. C. Sarkar. 1974. Some soil properties affecting loss of nitrogen from urea due to ammonia volatilization. J. Indian Soc. Soil Sci. 22:80–83.

Vine, H. 1962. Some measurements of release and fixation of nitrogen in soil of natural structure. Plant Soil 17:109–130.

Vlek, P. L. G., and J. M. Stumpe. 1978. Effect of solution chemistry and environmental conditions on ammonia volatilization losses from aqueous systems. Soil Sci. Soc. Am. J. 42: 416–421.

Volk, G. M. 1959. Volatile loss of ammonia following surface application of urea to turf or bare soils. Agron. J. 51:746–749.

Volk, G. M. 1961. Gaseous loss of ammonia from surface-applied nitrogenous fertilizers. J. Agric. Food Chem. 9:280–283.

Volk, G. M. 1966. Efficiency of urea as affected by method of application, soil moisture, and lime. Agron. J. 58:249–252.

Volk, G. M. 1970. Gaseous loss of ammonia from prilled urea applied to slash pine. Soil Sci. Soc. Am. Proc. 34:513–516.

Wahhab, A., M. S. Randhawo, and S. Q. Alam. 1957. Loss of ammonium sulfate under different conditions when applied to soils. Soil Sci. 84:249–255.

Wahhab, A., and F. Uddin. 1954. Loss of nitrogen through reaction of ammonium and nitrite ions. Soil Sci. 78:119–126.

Warren, H. L., D. M. Huber, and D. W. Nelson, and O. W. Mann. 1975. Stalk rot incidence and yield of corn as affected by inhibiting nitrification of fall-applied ammonium. Agron. J. 67:655–660.

Watkins, S. H., R. F. Strand, D. S. De Bell, and J. Esch, Jr. 1972. Factors influencing ammonia losses from urea applied to northwestern forest soils. Soil Sci. Soc. Am. Proc. 36: 354–357.

Watson, G. A., T. T. Chin, and P. W. Wong. 1962. Loss of ammonia by volatilization from surface dressings of urea in Hevea cultivations. J. Rubber Res. Inst. Malays. 17:77–90.

Wetselaar, R. 1961. Nitrate distribution in tropical soils. I. Possible causes of nitrate accumulation near the surface after a long dry period. Plant Soil 15:110–120.

Wetselaar, R., J. B. Passioura, and B. R. Singh. 1972. Consequences of banding nitrogen fertilizers in soil. I. Effects on nitrification. Plant Soil 36:159–175.

Wijler, J., and C. C. Delwiche. 1954. Investigations on the denitrifying process in soils. Plant Soil 5:155–169.

Woldendorp, J. W. 1968. Losses of soil nitrogen. Stikstuf 12:32–38.

Wullstein, L. H. 1967. Soil nitrogen volatilization. Agric. Sci. Rev. Coop. State Res. Ser. U.S. Dep. Agric. 5:8–13.

Wullstein, L. H. 1969. Reduction of nitrite deficits by alkaline earth carbonates. Soil Sci. 108: 222–226.

Wullstein, L. H., and C. M. Gilmour. 1964. Non-enzymatic gaseous loss of nitrite from clay and soil systems. Soil Sci. 97:426–430.

Wullstein, L. H., and C. M. Gilmour. 1966. Non-enzymatic formation of nitrogen gas. Nature (London) 210:1150–1151.

10   Biological Nitrogen Fixation

U. D. HAVELKA, M. G. BOYLE, AND R. W. F. HARDY

Central Research and Development Department
E. I. du Pont de Nemours and Company
Wilmington, Delaware

I. GENERAL INTRODUCTION

Although molecular nitrogen (N_2) composes almost 80% of the earth's atmosphere, N is increasingly scarce in forms that living organisms can assimilate. The reason for this scarcity is the large amount of energy required to cleave the two triple-bond atoms of the N_2 molecule. Only after these triple bonds are broken can the single atoms combine with hydrogen or oxygen to form ammonia (NH_3) or nitrate (NO_3^-) compounds that plants can use.

Biological N_2 fixation in nature is a unique property possessed by only a few genera of prokaryotic organisms that contain the genetic information to synthesize the enzyme nitrogenase. Nitrogenase catalyzes the conversion of N_2 to NH_3 under mild temperature and normal atmospheric pressures. This is contrasted to the commercial Haber-Bosch process that requires high temperatures and pressures.

The microorganisms that possess the ability to fix N_2 include bacteria and blue-green algae (cyanobacteria) that may be free living or symbiotic or that may form associative relationships with higher plants. Some may function in either mode. They range from strict anaerobes to microaerophiles to obligate aerobes; a few are photosynthetic. The most well-known and most agriculturally important relationship is the root nodule–forming symbiosis between *Rhizobium* species and some legumes. Another important root nodule symbiosis involves the actinomycete *Frankia* and forest species such as the alder (*Alnus* sp.). The association of the photosynthetic blue-green alga *Anabaena* within leaf cavities of waterferns of the genus *Azolla* is important in Asian rice (*Oryza sativa* L.) paddy culture. There are great expectations for future potential from associations such as *Azospirillum lipoferum,* which grows on the surface of roots of a tropical grass, *Digitaria,* corn (*Zea mays* L.), and wheat (*Triticum aestivum* L.).

Nitrogenases have been extracted from many diverse diazotrophs, representing the various physiological types (aerobes, anaerobes, microaerophilic bacteria, photosynthetic bacteria, blue-green algae, and nodule bacteria), and all are similar (Chen et al., 1973; Eady & Postgate, 1974).

A. Extent of N_2 Fixation

The legume/*Rhizobium* symbioses were discovered about 100 years ago (Hellriegel, 1886). Further investigation revealed that these *Rhizobium* spp. had the ability to infect roots of legumes, form nodules, and conduct a symbiotic relationship with their host plant. These intimate legume/*Rhizobium* relationships had the capacity to synthesize all or part of their N requirements from N_2 captured from the soil atmosphere. Estimates have been made that these symbioses fix 90×10^6 metric tons of N annually (Hardy & Holsten, 1972). This is about twice the N in chemical fertilizers and more than half of all biological fixation per year. Total annual global N_2 fixation was estimated by Burns and Hardy (1975) at about 175×10^6 metric tons. Lower estimates were achieved by a 1976 conference at Uppsala—122×10^6 metric tons (Burris, 1980) and 100×10^6 metric tons (Delwich, 1970).

B. Nitrogenase Enzyme

Nitrogenase is composed of two soluble proteins. One contains molybdenum (Mo) and nonheme iron (Fe) and is called the *Mo-Fe protein.* The other protein contains nonheme iron and is designated the *Fe protein.*

The Mo-Fe protein has a molecular weight of 200 to 250,000 and contains two Mo, 28 to 34 nonheme Fe, and 26 to 28 acid-labile sulfide atoms. This protein is a tetramer composed of duplicate copies of two different but similar subunits. The Fe protein is a dimer composed of two identical subunits and has a molecular weight of 55 to 65,000. It contains four nonheme Fe atoms and four acid-labile sulfides (Eady & Postgate, 1974).

Both proteins are required for nitrogenase activity, with a ratio of one or two Fe proteins for each Mo-Fe protein. The commonality of nitrogenase is illustrated by the fact that in many instances an Fe protein isolated from one organism can be recombined with a Mo-Fe protein isolated from a different organism to produce a functional nitrogenase.

Both proteins are extremely O_2 labile (Bulen & LeComte, 1966). Catalytic activity is lost after only a few seconds of exposure to air. The Mo-Fe and Fe proteins are classified as Fe-S proteins. Their Fe is contained in $Fe_4S^*_4$ clusters, although the Mo-Fe protein may contain $Fe_2S^*_2$ clusters as well (Mortenson & Thorneley, 1979).

C. Nitrogenase Reaction

Catalytic reduction of one molecule of N_2 to two molecules of NH_3 requires the addition of six protons (H +) and six electrons. The source of the electrons in nature is either ferredoxin, an Fe-S protein, or flavodoxin. Electrons are transferred from ferredoxin to the Fe protein and then to the Mo-Fe protein. However, electrons alone are insufficient for nitrogenase turnover. The system requires energy in the form of adenosine-5'-triphosphate

(ATP) to facilitate the transfer of electrons between the two component proteins. The Fe protein hydrolyzes four molecules of Mg·ATP per two electrons transferred.

Nitrogenase is a versatile enzyme and can serve as a reductase for a variety of substrates. In the absence of N_2, nitrogenase reduces protons (H+) to molecular hydrogen (H_2), but even in the presence of atmospheric concentrations of N_2, H+ is reduced and H_2 evolved (Bulen et al., 1965). This reaction competes for electrons with N_2 fixation and is discussed further in section III C3.

Nitrogenase can also reduce a number of triple bonds containing substrates other than N_2. For example, N_3^- to NH_3 + N_2, CN^- to NH_3 + CH_4, and C_2H_2 to C_2H_4 are all catalyzed by nitrogenase (Hardy et al., 1971). These reactions require both proteins and are coupled to Mg·ATP hydrolysis (Mortenson, 1966).

The acetylene reduction reaction, although quite unphysiological, has proved to be an extremely useful technique. The acetylene-ethylene assay developed from this reaction is widely used in measurement of N_2 fixation. This assay is discussed in section III D4.

D. Nitrogenase Regulation

The ability of a diazotroph to produce nitrogenase depends on the presence and expression of a linear array of about 17 structural nitrogen fixation (nif) genes. This genetic information is required not only for the synthesis of nitrogenase proteins but also for the proteins required for their assembly; for incorporation of Mo, Fe, and S; for electron transfer; for nitrogenase protection from O_2; and for nitrogenase regulation that is dependent on fixed N_2. The last regulatory system prevents the expression of the N_2-fixing system when the organism has an adequate supply of fixed N.

Rhizobia do not incorporate most of the NH_3 from N_2 fixation but excrete it to the legume cells where it is transformed into molecules for uptake by the legume transport system.

II. MICROBIOLOGY OF N_2 FIXATION

A. Introduction

The biological fixation of N_2 is restricted to a relatively small but diverse group of prokaryotic organisms referred to collectively as the diazotrophs. The diazotrophs include a wide variety of bacteria and some of the blue-green algae. These N_2-fixing prokaryotes may be broadly divided into autotrophs and heterotrophs (Fig. 1) according to their usual C source. In addition, they may be subdivided and designated as free living or symbiotic according to the growth habitats in which they may be found when fixing N_2 (Fig. 1). Examples of the bacterial diazotrophs include the predominantly symbiotic heterotrophs of the genera *Rhizobium* and *Frankia* (actinomy-

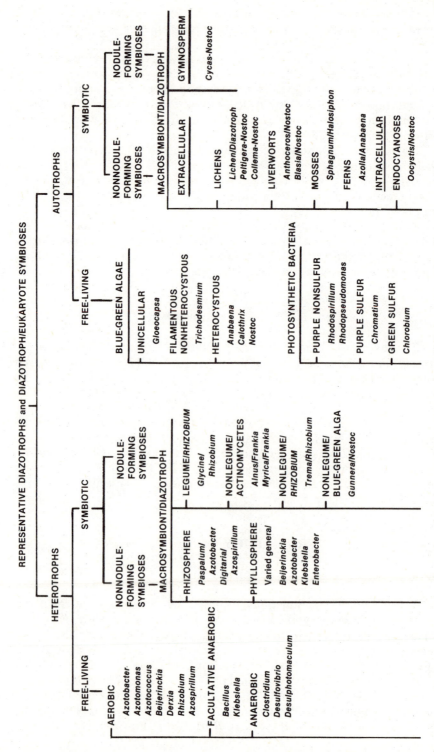

Fig. 1—Classification of representative genera of diazotrophs and associated eukaryotes.

cetes), various free-living heterotrophs such as the azotobacters, and the free-living autotrophs such as the photosynthetic bacterium *Rhodospirillum rubrum*. The blue-green algae, which are primarily autotrophic, have diazotrophic representatives in both the free-living and symbiotic forms. The more common diazotrophs of the blue-green are heterocystous algae, such as *Nostoc*, although some nonheterocystous types and one unicellular type, *Gloeocapsa*, have been demonstrated to fix N_2. The plant forms found in symbioses with blue-green algae include fungi, liverworts, ferns, gymnosperms, and angiosperms.

In general, diazotrophs in the free-living state are in their original habitat, whereas symbiotic interactions represent the colonization by selected diazotrophs of favorable alternative microenvironments within and around plants. Since biological N_2 fixation is energy demanding and sensitive to O_2 and combined forms of N, it is favored by conditions of high available C, low partial pressure of O_2 (pO_2), and low available combined N. Such conditions are met to varying degrees in the plant microenvironments of the root nodule, the rhizosphere, and the phyllosphere. However, plant/diazotroph interactions differ markedly in the degree of interdependence between microorganisms and the host plant.

In root nodule–forming systems, such as the legume/*Rhizobium* system, the plant/diazotroph interaction is highly evolved, and adaptations to the host plant and the bacterium have developed to accommodate the diazotroph within the root. The host plant and bacteria exchange C and N via vascular connections, and the relationship constitutes a clear symbiosis. For nodule-inducing diazotrophs, such as rhizobia, the symbiotic mode of N_2 fixation is the usual one. In nonnodule-forming associations, favorable plant microenvironments serve as alternative niches for otherwise free-living diazotrophs. Such associations, which have been referred to as associative symbioses (Stewart et al., 1979), involve selected rhizosphere and phyllosphere bacteria as well as certain blue-green algae. In general, they do not display the vascular continuity of nodule-forming symbioses. The magnitude of the contributions of rhizosphere and phyllosphere diazotrophs to the host plant is uncertain, and in this review such systems are referred to simply as *associations*. In the blue-green alga/plant system, the diazotroph association with the plant host is more intimate than a rhizosphere or phyllosphere association but apparently less elaborate than a legume/*Rhizobium* symbiosis. The nonnodule-forming alga/plant interactions will be referred to as *associative symbioses*.

With the demonstration that certain rhizobia can have an active nitrogenase when grown free from the host (Keister, 1975; Kurz & LaRue, 1975; McComb et al., 1975; Pagan et al., 1975; Tjepkema & Evans, 1975), it became clear that although the *Rhizobium* symbiosis is commonly symbiotic, it is not obligately symbiotic. Further, with the recent demonstration that actinomycetes of nonlegume root nodules could fix N_2 in the free-living state (Tjepkema et al., 1980), representatives of all major groups of diazotrophs have been shown to be able to fix N_2 when in the free-living state. Nonetheless, the functional divisions of diazotrophs into symbiotic (both nodule forming and nonnodule forming) and free living, based on the usual in situ relationships, are useful in describing the diazotrophs and assessing the relative contribution of the different types of biological N_2 fixation. In this section, a brief general survey of the microbiology of N_2 fixation is presented, including the usual habitat, prominent metabolic characteristics, and relative importance of the various diazotrophs. For extensive reviews of the diazotrophs, see Quispel (1974), Stewart (1975), Nutman (1976), Hardy and Silver (1977), and Hardy and Gibson (1977).

B. Free-Living Diazotrophs

1. FREE-LIVING AUTOTROPHS

The autotrophic asymbiotic diazotrophs that have been studied are primarily photoautotrophs, although some diazotroph species of the genera *Desulfovibrio* and *Methanobacterium* can exist in a chemoautotrophic mode. The phototrophs utilize light as their energy source and include the photosynthetic bacteria and the blue-green algae.

a. Photosynthetic Bacteria—The photosynthetic bacteria are aquatic gram-negative organisms found in both marine and fresh-water environments. The three main groups are the green sulfur bacteria (Chlorobiaceae), the purple sulfur bacteria (Chromatiaceae, formerly Thiorhodaceae), and the purple nonsulfur bacteria (Rhodospirillaceae, formerly Athiorhodaceae). In bacterial photosynthesis, water does not serve as the electron donor. In the green and purple sulfur bacteria, which are photolithotrophs, hydrogen sulfide (H_2S) is commonly utilized as the electron donor with CO_2 as their normal C source. In contrast, the purple nonsulfur bacteria are photoorganotrophic and cannot use H_2S as the electron donor. They require organic substrates as the electron donor and primarily use organic substrates as their C source. All bacterial photosynthesis operates under anaerobic conditions, although the purple nonsulfur bacteria are facultative anaerobes and can grow under microaerobic to aerobic conditions as chemoorganotrophs.

The O_2 requirements and metabolic patterns demonstrated by the photosynthetic bacteria are a reflection of the ecological niche that they normally occupy (Pfennig, 1978). The characteristics required are the presence of light, limited O_2 to anaerobiosis, and organic matter or sulfides. Such conditions are found in lower layers of bodies of water that normally remain stratified into aerobic and anaerobic layers. The green sulfur bacteria are the lowest layer of phototrophs and are found in the low-light, sulfide-rich environment of the sediments. The purple sulfur bacteria are found in the lower layers of bodies of water slightly above the green sulfur bacteria. The environment is relatively sulfide rich but variable. At the border of aerobic and anaerobic conditions, the purple nonsulfur bacteria thrive, since they possess the ability to be anaerobic phototrophs in the light and anaerobic chemotrophs in the dark.

The ability of a photosynthetic bacterium to fix molecular N_2 was first demonstrated by Kamen and Gest (1949) in *Rs. rubrum*. This discovery followed the observation that both N_2 and NH_4^+ could inhibit the photoevolution of H_2 in this species. Initially, it was thought that the photoevolution of H_2 resulted strictly from the activity of a hydrogenase. It was later realized that photoevolution of H_2 was just another manifestation of nitrogenase activity. The nature of these relationships is discussed more fully in Yoch (1978).

Following the discovery of N_2 fixation in *Rs. rubrum,* other nonsulfur purple bacteria were identified as N_2 fixers, including *Rhodopseudomonas sphaeroides, Rhodopseudomonas capsulata, Rhodopseudomonas gelatinosa,* and *Rhodopseudomonas palustris.*

Among the other photosynthetic bacteria, representatives of the purple sulfur bacteria and the green sulfur bacteria have been reported to fix nitrogen (Lindstrom et al., 1950). Although N fixation by the photosynthetic bacteria proceeds optimally only in the light, a low rate of N_2 fixation in the dark can occur, presumably utilizing photosynthate accumulated in the light period as an energy source.

Those photosynthetic bacteria that can function as diazotrophs are highly adapted in view of their ability to fix both C and N_2. However, despite these adaptive assets, the ecological niche they occupy is restricted, and, therefore, the contribution to N_2 fixation on a global scale is uncertain. Little quantitative data are available, but it appears that even though photosynthetic bacterial diazotrophs are important in the long-term N status of aquatic systems, they are not currently of major agricultural significance.

b. Blue-Green Algae—The other major group of autotrophic diazotrophs that can function asymbiotically are the blue-green algae (also referred to as the Cyanophyceae or Myxophyceae). The habitats in which they are found are extremely diverse, and they have been reported in a wide variety of microaerobic to aerobic environments, including aquatic and terrestrial habitats. However, Brock (1973) notes that the habitats in which they prevail or are very abundant are more restricted due to competition with eukaryotes. Terrestrially, they are found at or just below the soil surface (Jurgensen & Davey, 1968) and are common in moderate reducing environments, such as rice paddies. Of the aquatic environments, fresh-water habitats are more common than marine. In fresh-water habitats, mesotrophic waters (intermediate in nutrient availability) and eutrophic waters (high in nutrients but variable in O_2 tension) appear to be preferred over oligotrophic habitats (well-mixed waters with high O_2 levels but low in dissolved nutrients). Stewart (1977) has cited several factors governing the distribution and activities of the blue-green algae in terrestrial systems. From these studies, it was concluded that desiccation is the most limiting factor and that the pH level (optimum is neutral to slightly alkaline) is an important influence in both aquatic and terrestrial systems.

The blue-green algae are unique among prokaryotes in that they exhibit oxygenic photosynthesis, which is characteristic of higher plants. In oxygenic photosynthesis, water serves as the electron donor, and the ultimate electron acceptor, O_2, is produced. The blue-green algae contain chlorophyll a and the efficient accessory pigment phycobilin and thus can survive in low light environments. They employ oxidative phosphorylation and cyclic photophosphorylation to generate ATP; however, they are able to grow under conditions of low O_2 concentration. The ability to survive in low light and low pO_2 environments also contributes to the wide distribution of the blue-green algae.

The three main groups of the blue-green algae are (i) the unicellular, (ii) the filamentous nonheterocystous, and (iii) the filamentous heterocystous. Diazotrophic representatives have been found in all groups, although the ability to fix N_2 is more typical of the heterocystous forms. Among the non-heterocystous blue-green algae, the marine organism *Trichodesmium* has been frequently cited as a diazotroph. Other genera include *Plectonema* and *Oscillatoria*. The only known unicellular diazotroph, *Gloeocapsa,* was first demonstrated to fix N_2 by Wyatt and Silvey (1969). Diazotrophic blue-green algae are found both free living and in a variety of symbiotic associations with other organisms. Asymbiotic examples of heterocystous and unicellular blue-green algae are discussed here, and blue-green algae associative symbioses are discussed in section II C4a.

i. Heterocystous Blue-Green Algae

The morphology and physiology of the N_2-fixing blue-green algae differ among the various algal types. Under aerobic conditions, the heterocystous types conduct some or all of their N_2 fixation in the heterocysts, which are specialized vegetative cells (Stewart, 1977). These specialized cells protect against O_2 damage to nitrogenase through morphological and physiological adaptations.

Morphologically, nitrogenase in heterocysts is protected from O_2 damage by a multilayer envelope outside the cell wall that surrounds the cell and reduces gas diffusion. Physiologically, the heterocyst site provides nitrogenase protection from O_2 damage by its lack of the water-splitting, O_2-evolving pigment system referred to as *photosystem II* (Donze et al., 1972). In contrast, photosystem II is characteristically found in the vegetative cells of blue-green algae. Thus, heterocystous algae protect the nitrogenase sites by compartmentalization away from the sites of O_2 evolution.

ii. Unicellular Blue-Green Algae

Gloeocapsa is unique among the blue-green algae in that it is the only unicellular blue-green alga known to be able to fix N_2. It is an aerobic soil organism, but N_2-fixing activity is usually higher at lower O_2 levels, such as 0.1 atm (Gallon et al., 1975). In culture, it has been shown to be very light sensitive and bleaches at light intensities > 500 lm/m^2 (Rippka et al., 1971). As opposed to the cell-free extracts of other algal nitrogenases, crude viscous extracts of *Gloeocapsa* are stable in air as long as the flasks containing the cultures are not disturbed by shaking (Gallon et al., 1972).

The status of *Gloeocapsa* metabolically as a photosynthetic aerobic diazotroph and morphologically as a unicellular alga that does not differentiate into specialized N_2-fixing heterocysts poses particular problems for protection of its nitrogenase system from O_2 damage. Several adaptions appear to be employed by *Gloeocapsa* to avoid O_2 damage. The nitrogenase of *Gloeocapsa* has been reported by Gallon et al. (1972) as particulate and as such may be protected from O_2 damage in a manner similar to that suggested for *Azotobacter*. Gallon et al. (1975) de-emphasized the physical separation mechanism and suggested that *Gloeocapsa* protects its nitrogenase by temporal separation of maximum O_2-producing and N_2-fixing activities. Support for this was obtained in batch cultures that demonstrated highest nitrogenase activity during the period of maximal cell division and relatively low photosynthetic O_2 evolution.

iii. Contribution of Free-Living Blue-Green Algae

The free-living blue-green algae of direct agricultural significance are found in terrestrial habitats, including the specialized soil environment of the paddy field as well as in temperate and tropical soils. Though widely distributed in all soils, they are more abundant in tropical soils. The genera of blue-green algae found in paddy soils varies with location, but representative genera include *Anabaena, Nostoc, Calothrix,* and *Gloeotrichia.* In Indian rice fields, the genera *Aulosira, Anabaena, Cylindrospermum,* and *Scytonema* are commonly found (Singh, 1961). In Japan and Southeast Asia, Kobayashi et al. (1967) have reported that *Tolypothrix, Cylindrospermum, Calothrix Anabaena, Nostoc,* and *Anabaenopsis* are common.

The contribution of blue-green algae to N_2 fixation in the paddy field environment has been reported to be equivalent to or exceeding the requirement for rice production. However, such reports are based largely on greenhouse rather than field studies. Field studies based on N balance and acetylene reduction have indicated a potential fixation of 70 kg of $N_2 \cdot ha^{-1} \cdot year^{-1}$. However, this amount is based on the entire diazotrophic population, which includes the heterotrophic rhizosphere bacteria as well as the free-living and symbiotic blue-green algae (Balandreau et al., 1976). In addition to the contribution of fixed N_2, the blue-green algae may aid rice production by reducing sulfide injury.

2. FREE-LIVING HETEROTROPHS

Dinitrogen-fixing free-living heterotrophic bacteria may be divided on the basis of their O_2 requirement into aerobes, facultative anaerobes, and strict anaerobes. Mulder (1975) cited representatives of each group, including (i) the aerobic azotobacters; (ii) the facultatively anaerobic klebsiella (formerly referred to as the aerobacters), some facultative bacilli; (iii) most of the anaerobic saccharolytic clostridia; and (iv) anaerobic sulfate reducers, such as *Desulphovibrio* and *Desulphotomaculum*. The ability to fix N_2 is the unifying characteristic of the group. As in all N_2 fixation, a source of energy and reductant is required. These heterotrophic bacteria use organic compounds as their electron donor and oxidation reduction reactions as their energy source. The aerobic members of the heterotrophic bacterial diazotrophs obtain their energy from partial degradation by a fermentative pathway followed by respiration, that is, the oxidative degradation of organic compounds ultimately to CO_2 and water. The anaerobes commonly utilize carbohydrates as fuel molecules and degrade them by solely fermentative pathways, which leads to less complete breakdown. The facultative anaerobes are capable of growth under both anaerobic and aerobic conditions but function anaerobically as diazotrophs. For tabulations of the free-living bacterial diazotrophs, see Knowles (1978) and La Rue (1977).

a. Aerobic N_2 Fixation—Most of the aerobic heterotrophic diazotrophs are members of the family Azotobacteraceae (Mulder & Brotonegro, 1974) although bacteria of other families, such as the Corynebacteraceae, Mycobacteraceae, Achromobacteraceae, and Pseudomonadaceae, have been reported to fix N_2. The Azotobacters are typically found in neutral to alkaline soils, although there are exceptions. *Azotobacter beijerinckii* and *Derxia* are found over a wide range of pH and often in lower pH environments. In soils, the most abundant *Azotobacter* is *Azotobacter chroococcum,* and it has worldwide distribution. The azotobacters are mesophilic in their temperature requirement, with optimum growth at approximately 30°C and a pH minimum of 6.0 for N_2 fixation (LaRue, 1977). *Beijerinckia* is commonly found in the tropical and subtropical regions but are essentially absent from temperate areas (Mulder & Brotonegro, 1974). They have been reported to fix N_2 over a pH range of 3 to 10 (LaRue, 1977).

The anaerobic diazotrophs avoid the negative effects of O_2 on N_2 fixation simply by the absence of O_2 in their natural environment. Similarly facultative organisms that can fix N_2 do so in general only under anaerobic conditions (Grau & Wilson, 1962). However, since the aerobic bacterial diazotrophs require O_2 for their respiratory-type metabolism, special mechanisms for the protection of nitrogenase are required. Several mechanisms have been suggested (Mulder & Brotonegro, 1974). First, in *Azotobacter,* the nitrogenase (as isolated in cell-free crude extracts) is associated with membranes and as such may be protected from ready diffusion of gases. Similarly, slime production and bacterial clump formation may reduce O_2 diffusion. Postgate (1974) has suggested that conformational and respiratory protection mechanisms may moderate negative O_2 effects. Conformational protection refers to changes in the conformation of the nitrogenase enzyme, which leads to a protected but nonfunctioning nitrogenase. This mechanism is viewed as a backup to respiratory protection, which is the maintenance of high levels of respiration to utilize potentially damaging O_2. Such a mechanism leads to a reduced efficiency of N_2 fixation because increasing amounts of C are consumed to produce a given amount of fixed N_2. In addition to the above methods, aerobic diazotrophs commonly live in environments that contain sufficient pO_2 to support aerobic metabolism but less than the atmospheric pO_2 that can hinder N_2 fixation. It has been reported that N_2 fixation is maximal under relatively low pO_2 (Mulder & Brotonegro, 1974).

b. Limitations to N_2 Fixation of Free-Living Heterotrophs—The different catabolic pathways employed by the various bacteria for the production of energy and reductant place different limitations on N_2 fixation by these organisms. The aerobic bacterial diazotrophs employ oxidative (respiratory chain) phosphorylation to more completely capture the energy of available oxidizable materials. As such, their energy limitation would be the availability of oxidizable fuel molecules. However, the aerobic diazotrophs must expend energy on protection of nitrogenase against O_2 damage. The anaerobic diazotrophs utilize the fermentative pathway to provide ATP and reductant through substrate level phosphorylations, i.e., the oxidation of a substrate with the concommittant capture of the liberated energy into phosphorylated compounds. This method is less efficient than oxidative phosphorylation, and thus diazotrophs employing anaerobic catabolism must process large quantities of reduced C to ensure that ATP is not limiting N_2 fixation. Any excess electrons may be consumed by the reduction of pyridine nucleotides or by the evolution of H_2 gas. In the case of the photosynthetic sulfur bacteria, which commonly utilize H_2S as the electron donor, there is cyclic photophosphorylation with a cytochrome-type electron transport system for ATP production. Here the supply of reductant rather than ATP could ultimately limit N_2 fixation.

Microaerophilic growth may be the most ideal in that ATP and reductant are produced by oxidative pathways, and cellular expenditures for O_2 protection are not maximal. Under such conditions, N_2 fixation is optimal, and the need for protection of nitrogenase is relatively low. Such conditions may lead to a build-up of reduced pyridine nucleotides and acetyl–CoA. These compounds may be stored by the cell as poly-β-hydroxybutyrate, which may later be used as an energy source.

The agricultural significance of asymbiotic N_2 fixation by free-living heterotrophic bacteria is variable. When conditions such as high C/N ratio in the soil favor population increases of the heterotrophic diazotrophs, their contribution may become significant. Such conditions may be artificially induced by the addition of plant residues such as wheat straw or naturally occur in the microenvironments of the rhizosphere and phyllosphere (sections II C4b and II C4c). In any case, the contribution of fixed N_2 by these free-living heterotrophs is indirect since they do not excrete fixed N_2 and must be decomposed and mineralized before the nitrogenous compounds are released.

C. Symbiotic N_2 Fixation

1. TYPES OF NODULE-FORMING SYMBIOSES

Symbiotic N_2 fixation involving nodule formation has been identified in both leguminous and nonleguminous plants. In the legume system, the microsymbionts are of the genus *Rhizobium,* whereas in most nonlegume N_2 fixation systems, the microsymbionts are actinomycetes (thread bacteria), usually of the genus *Frankia*. One nonlegume in which a *Rhizobium* species has been confirmed as the microsymbiont is the woody dicot *Trema cannabina* var. *scabra* (Trinick, 1973). Additional nonlegume, nodule-forming symbioses involve blue-green algae as the endophyte. In the case of the herbaceous dicot *Gunnera,* a stem symbiosis is formed with species of the genus *Nostoc*, whereas in the gymnosperm *Cycas,* root nodules containing *Nostoc* or *Anabaena* are formed. Although actinomycetes remain the predominant symbiotic diazotroph in nonleguminous higher plants, the discovery of the *Trema* symbiosis suggests that *Rhizobium* and other diazotrophs associated with plants may be more widely distributed than previously realized.

2. NONLEGUME SYMBIOSES

The major biological source of reduced N_2 in agricultural soils is broadly recognized as the legume/*Rhizobium* symbiosis. However, in nonagricultural lands, such as wooded areas, uncultivated fields, wetlands, sandy soils, and disturbed soils, certain woody dicots nodulated by species of the actinomycete *Frankia* represent the major biological source of reduced N_2 (Torrey, 1978). As more systematic investigations are conducted, the frequency or occurrence of actinomycete nodulation has become better defined (Bond, 1976). Tabulations and discussions of these actinomycete nodulated or actinorhizal dicots are presented in Torrey (1978), Becking (1975, 1977), Bond (1976), and Silvester (1977). It has been noted (Bond, 1976; Becking, 1975) that nodulation is more prevalent in certain species than others. For example, *Alnus* is almost universally nodulated, whereas *Rubus* has only one observed nodulated species. Representative examples include tree species within the genera *Alnus* (alders) and *Casuarina* as well as smaller species such as sweetgale (*Myrica gale*), sweetfern (*Comptonia peregrina* L. J. M. Coult.), bayberry (*Myrica pennsylvanicum*), and various shrubs of the genus *Ceanothus*. In addition to the actinorhizal dicot/*Frankia* symbiosis, other confirmed examples of nonlegume, nodule–forming symbioses are the *Trema/Rhizobium* symbiosis (section II C2b), and the stem symbiosis of *Gunnera/Nostoc* (section II C2c), which are discussed as noted.

a. Nonlegume/Actinomycete Symbioses—Two types of actinomycete-induced nodules are recognized: the *Alnus* type, in which coralloid root structures develop, and the *Myrica* type, in which nodule terminal lobes give rise to a normal but negatively geotropic root that grows almost vertically upward.

The infection process of actinomycetal symbioses, though not as thoroughly investigated as the legume/*Rhizobium* symbiosis, displays some similarities. Torrey (1978) reported on several structural studies of actinomycete nodulation and noted that the mode of infection was consistently root hair invasion. Also, although many root hairs are deformed at a nodule site, only one is invaded. Following infection, the endophyte can be observed in the root enclosed in a host-produced polysaccharide sheath. Ultimately, the endophyte invades cortical cells and establishes a nodule.

Some success in the isolation of the actinomycete microsymbiont has been achieved. Callaham et al. (1978), using sweetfern root nodules, reported the isolation and culture of an actinomycete endophyte. This endophyte successfully reinfected sweetfern seedling roots and produced effective (acetylene-reducing) nodules, thereby satisfying Koch's postulates. More recently, Tjepkema et al. (1980) reported the successful culture of the actinomycete *Frankia* CPI1 and demonstrated in vitro $N_2[C_2H_2]$ fixation. These new developments will aid in the elucidation of the nature of the relationship.

b. *Trema cannabina/Rhizobium* **Symbiosis**—*Trema cannabina* is a woody nonlegume of the family Ulmaceae and the order Urticales. It has been reported as the only nonleguminous angiosperm to be nodulated by *Rhizobium* (Trinick, 1976). However, earlier less substantiated descriptions of *Rhizobium* symbioses with nonlegumes were reported by Sabet (1946), Mostafa and Mahmoud (1951), and Clawson (1972).[1]

Trinick (1975) first reported collecting *Trema* as a volunteer species in between rows of tea (*Camellia sinensis*) in a New Guinea plantation, where it grew to a height of 0.5 to 1.0 m. When mature, however, *Trema* may reach a height of approximately 15 m. The root nodules are found on the tap root and lateral root systems. The isolated endophyte demonstrated the bacteriological features of a slow-growing *Rhizobium*. In addition, the host range of the *Trema* endophyte appears to be restricted to legumes that normally would be nodulated by a cowpea-type *Rhizobium*.

Structurally, the nodule-forming symbiosis of *Trema cannabina* display similarities to both legume and nonlegume nodules. Within the nodule, the rod shape of the bacterium in pure culture assumes a bacteroid form as in legume nodules. However, the central vascular bundle of the nodule of the *Trema cannabina* root nodule exemplifies the type of vascular organization found in actinomycetal root nodules of nonlegumes. Becking (1977) interprets nodules developed in this manner to be simply modified lateral roots. In summary, the *Trema cannabina/Rhizobium* symbiosis is currently an uncommon association but suggests the possibility that angiosperm/diazotroph symbioses are probably more common than formerly realized.

c. Gunnera/Nostoc Stem Symbiosis—Species of the genus *Gunnera* are herbaceous dicots of the family Haloragaceae, which live symbiotically with an intracellular N_2-fixing endophyte. In this symbiosis, morphological modification of the invaded stem tissue is substantial, and the resultant N_2-fixing structures are referred to as *glands,* or *domatia,* by some authors (Millbank, 1974; Becking, 1976), whereas another describes the structures as *nodules* (Silvester, 1976). Reviews and descriptions of the *Gunnera/ Nostoc* symbiosis are found in Silvester (1976) and Becking (1976, 1977), and an overview of these works is presented here.

Gunnera species range in size from a small herb up to a stout plant approximately 1.5 m tall. They are found in wet environments and in their natural habitat invariably contain a blue-green algal endophyte of the family Nostocaceae (Becking, 1977). *Gunnera* spp. have been found in South America, New Zealand, Southeast Asia, and South Africa (Becking, 1977). The *Nostoc* endophyte is found within special stem glands near the base of the leaves. Algal growth begins intercellularly in the mucous glands, but it is ultimately intracellular. In the *Gunnera* symbiosis, the *Nostoc* endophyte

---

[1] M. A. Clawson, R. B. Farnsworth, and M. Hammond. 1972. New species of nodulated nonlegumes on range and forest soils. Agron. Abstr., p. 138.

displays several adaptations (Silvester, 1976). Morphologically, the endophyte has a much higher heterocyst frequency relative to free-living algae. Physiologically, N_2 fixation is enhanced in the light but proceeds also in the dark; the pigment phycocyanin is not found in the endophytic *Nostoc*; ^{14}C is not significantly incorporated into *Nostoc* glands; and the photosystem II inhibitor DCMU does not inhibit the stimulatory effect of light on endophytic *Nostoc* nitrogenase activity. The interpretation of these results is that in the symbiotic association with *Gunnera,* the normally autotrophic *Nostoc* may function heterotrophically depending on the host for carbohydrates.

Estimates of the potential contribution of the *Gunnera/Nostoc* symbiosis vary among investigators. Dinitrogen fixation activity of the *Nostoc* endophyte of *Gunnera arenaria* was first clearly established by Silvester and Smith (1969) using $^{15}N_2$ and acetylene reduction techniques. These authors estimated a maximum N_2 fixation rate for *Gunnera dentata* of 72 kg of N_2 fixed•ha^{-1}•year^{-1}. Becking (1976, 1977) using *Gunnera macrophylla* under Indonesian conditions estimated much lower fixation rates of approximately 10 to 20 kg of N_2 fixed•ha^{-1}•year^{-1}. Regardless of differences in estimates of N_2 fixed, it is clear that the symbiosis can add a substantial supply of reduced N_2 to the ecosystem in which it is found.

d. N_2 Fixation in Coralloid Roots of Cycads—Various genera of the gymnosperm family Cycadaceae have been observed to possess dichotomously branching or coralloid roots that may become infected with N_2-fixing blue-green algae such as *Nostoc* or *Anabaena* (Millbank, 1974). These gymnosperms are primarily found in tropical and subtropical environments (Silvester, 1977). Although the mechanism of infection is not well defined, it has been shown to be an effective N_2-fixing system. Using $^{15}N_2$ in a greenhouse study of *Macrozamia* nodules, Bergerson et al. (1965) reported fixation of 5.2 μg of N•g^{-1} of fresh nodules•hour^{-1}. On an area basis, Halliday and Pate (1976) calculated 19 kg of N•ha^{-1}•year^{-1} by *Macrozamia riedlei*. Physiologically, the endophyte of *Cycas* displays some similarities to the blue-green algal endophytes of other associative symbioses. As in the *Azolla/Anabaena* (Silvester, 1976) and *Gunnera/Nostoc* (Silvester, 1977) symbiosis, photosystem II is not active.

3. LEGUME/*RHIZOBIUM* SYMBIOSIS

The legume/*Rhizobium* symbiosis is an intensively studied system with investigations including a range of laboratory and field studies. This symbiosis is briefly mentioned here with a detailed discussion of the legume/*Rhizobium* symbiosis in section III.

Rhizobia are gram-negative rods, metabolically chemoorganotrophs that grow best at 25 to 30°C on complex media. They often contain poly-β-hydroxybutrate storage granules. A general division of *Rhizobium* based on growth patterns into slow growers and fast growers has been recognized by *Bergey's Manual of Determinative Bacteriology* (Buchanan & Gibbons, 1974). Fast growers lower the pH of a yeast mannitol agar and have a mean generation time of 2 to 4 hours, shorter than slow growers (6 to 8 hours), which produce an alkaline endpoint on the same medium (Vincent, 1977). Fast growers include *Rhizobium leguminosarum, Rhizobium trifolii, Rhizobium phaseoli,* and *Rhizobium meliloti.* The slow growers include *Rhizobium lupini* and *Rhizobium japonicum.* Another important classification of symbiotic diazotrophs is based on the host range and specificity. Thus, on the basis of normal host/diazotroph relationships, legumes have been commonly divided into seven infective or cross-inoculation groups. These groups are discussed in section III E.

4. NONNODULE-FORMING SYMBIOSES

Nonnodule-forming symbioses include associations of the rhizosphere and phyllosphere bacteria with various host plants and the associative symbioses of blue-green algae with various higher plants. The associative blue-green algal diazotrophs are found in both intracellular and extracellular associations and, like their free-living counterparts, may release organic N, thus serving as a source of fixed N_2 to the macrosymbiont. The bacterial diazotroph associations are primarily extracellular organisms found in the surface environments of plant roots and leaves. Their fixed N_2 contribution is indirect in that the organism must die, and following degradation, fixed N_2 is released. In spite of the less direct linkage between the macrosymbiont and the fixed N_2 source, bacterial associations may prove to be of agricultural significance to grass crops since nodule-forming symbioses in angiosperms are restricted to the dicots. The blue-green algal association with the waterfern *Azolla* is also agriculturally significant. A discussion of these associations, along with brief descriptions of other associative symbioses, follow.

a. Associative Symbioses of Blue-Green Algae—Intracellularly, blue-green algae may form endocyanoses in which one blue-green alga lives within another. Extracellularly, blue-green algal associations also have been observed with a range of lower plants. These include associative symbioses of blue-green algae with achlorophyllous fungi to form lichens as well as with lower plants such as the Bryophytes (mosses and liverworts) and the Pteridophytes (ferns). Alga/Bryophyte symbioses and endocyanoses are not discussed here.

i. Blue-Green Alga/Fungus Symbioses—Lichens

In the symbiosis of algae with fungi to form lichens, both partners are microorganisms. The blue-green alga is surrounded by the hyphae of the fungi and could be referred to as the *endophyte,* or *microsymbiont.* However, by convention, since they are symbionts of approximately the same size, the term *mycobiont* is used to refer to the fungal partner, and *phycobiont* refers to the algal partner. Of the approximately 17,000 lichen species identified, only 8% have a blue-green alga as the phycobiont (Stewart et al., 1979). Nonetheless, their N contribution may be significant in a range of environments from artic tundra (Alexander, 1975) to temperate soils (Hitch & Stewart, 1973). Common genera of the lichens that have been widely investigated are *Peltigera* and *Collema,* both of which are found with a phycobiont of the *Nostoc* genus. The lichens are found in a variety of harsh surface habitats. Thus, they are found on rock surfaces, soil, tree bark, arctic tundra, and so on.

In lichen symbioses, as in the free-living state, blue-green algae release substantial amounts of fixed N_2. In addition to this fixed N_2, the mycobiont may receive biotin and thiamine from the phycobiont. In the symbiosis, the phycobiont benefits by the surrounding presence of the mycobiont, which reduces dessication and may provide a more favorable low pO_2 environment for the alga.

ii. Azolla/Anabaena Associative Symbiosis

The *Azolla/Anabaena* associative symbiosis was reviewed by Peters et al. (1980). Morphologically, the *Azolla* sporophyte stem is a branched floating structure that bears bilobed leaves and true roots. The dorsal leaf lobe is aerial and contains a cavity on the ventral surface in which the blue-green alga *Anabaena* is found. There are no vascular connections to the leaf; however, multicellular hairs with transfer cell–type ultrastructure have been observed (Peters et al., 1978). In association with the waterfern, the frequency of heterocysts increases. Biological N_2 fixation demonstrates a light dependence under microaerophilic conditions; however, it will occur at increasingly reduced rates under aerobic light and aerobic dark conditions, respectively. With $^{15}N_2$, it was demonstrated that isolated *Anabaena* releases up to 60% of N_2 fixed mostly as NH_3. The *Azolla/Anabaena* associative symbiosis is agriculturally significant and has been used successfully as a green manure crop.

b. Rhizosphere Habitat—The rhizosphere habitat is recognized as a microenvironment substantially different from the surrounding soil. The biology of the rhizosphere has been reviewed by Rovira and Davey (1971), and some salient points from this discussion are reviewed here. The term *rhizosphere* refers to a zone of soil proximal to and under the influence of the root in which the microorganisms are more abundant than the surrounding soil. The boundaries of this zone are not clearly defined, although this zone is characterized by a gradient of microbiological activity that decreases with distance from the root. Distinctive terms such as outer rhizosphere, inner rhizosphere, and rhizoplane have been proposed for specific applications. However, the more general term rhizosphere is more common and is used here.

The microbial population of the rhizosphere includes bacteria, fungi, actinomycetes, nematodes, and to some extent, the algae and protozoa. The location of microorganisms on or near the root varies. Also, the extent of the rhizosphere effect, that is, the extent of increase in microbial activity due to association with plant roots, may be associated with species (Rouatt & Katznelson, 1961) or, as in wheat, with the stage of plant growth (Riviere, 1960). Increases in microbial activity are both quantitative and qualitative. Quantitative increases are readily detected and are assayed by microorganism counts following appropriate washing procedures. In addition to increased numbers of microorganisms, the rhizosphere is also characterized by changes in the dominant types of microorganisms. In particular, with regard to bacteria, roots selectively enhance the Gram-negative species (Rovira & Davey, 1971), which demonstrate high growth rates and strong growth response to supplied amino acids and glucose. In general, a nutritionally more fastidious bacterium would be able to thrive more readily in the nutritionally enriched environment of the rhizosphere. This enrichment is due to increased organic matter from decaying roots as well as root exudates from active roots. The root exudates may include sugars, amino acids, organic acids, and, to some extent, vitamins and enzymes.

i. Rhizosphere N_2 Fixation

Among the other microbial inhabitants, diazotrophs have been found in the rhizosphere region of certain plants. Such an association with crop plants, particularly the grasses, would be of great promise if the N_2 fixed were substantial and available. The early work on plant-bacterial associations by Döbereiner et al. emphasized the *Paspalum notatum/ Azotobacter paspali* associative symbiosis and was summarized by Döbereiner and Day (1976). The association required weeks or months to become established and had a maximal $N_2[C_2H_2]$ reduction activity of approximately 0.25 kg of N/day. This activity appeared to be linked to photosynthate availability and closely root associated as indicated by minimal activity reduction following vigorous root washing. A subsequent report of the association of rhizosphere bacteria with roots of several tropical grasses (Döbereiner & Day, 1975) suggested maximal N_2 fixation rates of 1 kg of $N \cdot ha^{-1} \cdot day^{-1}$ and perhaps higher. Such high potential fixation rates focused attention on these rhizosphere bacteria, and more characterization studies followed. Döbereiner and Day (1976) reported the responsible diazotroph in their system to be a member of the genus *Spirillum*. Further taxonomic studies of the N_2-fixing *Spirillum* were reported by Krieg and Tarrand (1978). They concluded that *Spirillum lipoferum* should be put into a new genus *Azospirillum* and divided into two groups. Group I includes strains of the species *Azospirillum brasilense,* and Group II includes strains of the species *Azospirillum lipoferum*. Both types displayed a primarily respiratory type of metabolism, although Group II strains were able to grow anaerobically to some extent.

Research on plant root associations with *Azospirillum* spp. continues to be active and promising. However, investigators differ in their assessments of the agronomic potential of such associative N_2 fixation as an alternative to fertilizer N or simply a supplementary N source. Significant contributions to the plant growth of foxtail millet (*Setaria italica* L. Beauv.) and corn in terms of total leaf N and plant dry matter have been demonstrated in greenhouse studies (Cohen, et al., 1980). In a field study, Smith et al. (1976, 1977) reported significant increases in the dry matter production of pearl millet (*Pennisetum americanum*) and guineagrass (*Panicum maximum* Jacq.), which had been inoculated with *A. lipoferum* and fertilized at low levels. However, duplication of results in some cases has been difficult (van Berkum & Bohlool, 1980).

In addition to the potential N contribution, *A. brasilense* has been reported to produce growth hormones that increased the lateral root development of pearl millet grown in inoculated solution culture (Tien et al., 1979). Currently, it appears that selected rhizosphere bacteria may improve plant productivity by serving as a supplementary N source and possibly a stimulatory hormone source. More research on such factors as host range and host-plant specificity may help define the potential role of *Azospirillum* spp. in specific crop plant/bacterial associations.

c. Phyllosphere Habitat—The leaf surface microenvironment, or phyllosphere, is an additional specialized plant habitat in which microorganisms may thrive. A broad general review of the characteristics of leaf surfaces and leaf surface microorganisms is contained in Preece and Dickinson (1971). Some of the general points reviewed are noted here.

The *phyllosphere,* which is also referred to as the *phylloplane,* refers to the outer skin, or cuticle, of the leaf. The distinction between the terms phylloplane and phyllosphere is not as great as that between the analogous terms rhizoplane and rhizosphere. The physical extent of the leaf microhabitat is generally not as substantial as the root microhabitat and much more variable in terms of temperature and moisture. Here the term *phyllosphere* is used to emphasize the concept of a microhabitat as opposed to a physical location.

In the strict sense, the description phyllosphere microorganism refers to noninvasive organisms that live on but not within the leaf (Menna, 1971). Their food supply consists of leaf exudates, dead cell materials, and detritus that may collect on the leaf. Structurally, the microhabitat of the cuticular membrane consists of any epicuticular waxes and the cuticle below. It is nonliving and noncellular. The cuticle itself is a matrix of the biopolymer cutin with embedded cuticular waxes as well as cellulose and pectins at the junction of the cuticular membrane and epidermal wall. The physical topography of the leaf surface varies widely with type of venation, the presence or absence of trichomes, and the presence and type of epicuticular waxes. This intricate but variable microenvironment serves as a potential niche that may be colonized by a variety of microorganisms.

i. Phyllosphere N_2 Fixation

Among the other microflora of the phyllosphere, certain diazotrophic organisms have been identified. They include both the free-living and symbiotic blue-green algal diazotrophs, as well as the heterotrophic bacterial diazotrophs (Ruinen, 1975). The bacterial diazotrophs observed in the phyllosphere include representatives of the aerobic, anaerobic, and facultative heterotrophs, although Silver (1977) has noted that the most commonly isolated types are facultative enterics, such as the *Klebsiella* species. The associated host plants are from various genera and geographical locations, and phyllosphere N_2 fixation has been reported from temperate to tropical locations. However, the more consistent high-moisture conditions in the tropics make it a more favorable environment.

The actual sites within the phyllosphere in which the diazotrophs may develop include the leaf surface itself and leaf enclosures such as the leaf sheath of grasses (Ruinen, 1971). The morphology of the grass leaf sheath provides a natural catchment for water collection and subsequent microbial growth. The sources of moisture on leaf surfaces are both internal and external to the plant. The internal plant sources are the exudation by guttation and transpiration, whereas externally, dew and rain provide moisture. These exudates and leachates are high in carbohydrates and low in N. Such a high C/N ratio environment favors the development of nutritionally less exacting bacteria in general and diazotrophs in particular.

Experimentally, N_2 fixation has been demonstrated in the phyllosphere by Kjeldahl, $^{15}N_2$, and acetylene reduction techniques. However, the contribution of N_2 fixation in the phyllosphere to the combined N of local ecosystems is uncertain. Partially, this is attributable to the difficulty of establishing quantitatively the population density of the diazotrophic organisms in question. The level of the phyllosphere microbial population varies with changes in external influences such as light, moisture, and temperature conditions as well as the plant metabolic status, age, and production of exudates. Nonetheless, there are reports of significant contributions of combined N to plant ecosystems by phyllosphere diazotrophs. These include contributions to the aerial plant parts as well as the suggestion of Ruinen (1971) that rhizosphere fixation may be enhanced.

III. PHYSIOLOGY AND AGRONOMY OF N_2 FIXATION— LEGUME/*RHIZOBIUM* SYMBIOSIS

A. Introduction

The fixation of atmospheric N_2 by the legume/*Rhizobium* symbiosis is an integrated process in which the host plant (macrosymbiont) supplies the bacterium (microsymbiont) with an energy supply, and the bacterium supplies the plant with reduced N_2. There are apparently genetic and/or chemical signals in the cell walls or other sites on the bacteria. The coupling of these systems is believed to account for the specificity. The plant unites with the bacterium in providing a suitable milieu (the nodule) for these mutually beneficial transformations to take place. After the initial invasion of the root hair of the host plant, the rhizobia multiply in the corticular cells where the nodule is developed. The nodule contains the compartmentalized colonies of rhizobia, called *bacteroids,* which in turn contain the enzyme nitrogenase, responsible for the catalysis of N_2 to NH_3. However, though nitrogenase is extremely O_2 labile, the reaction requires O_2 for the production of ATP. Nitrogenase is protected from destruction by O_2 and by an ingenious protein, leghemoglobin, which regulates the flux of O_2 to the bacteroid for its required respiratory processes. The plant (nodule tissue) conducts amination and amino transfer reactions (supplementing those of the rhizobia) to rapidly remove the NH_3 via the xylem stream.

Most of the discussions in this section pertain to the symbiosis of the soybean (*Glycine max* L. Merr.) and *R. japonicum,* although the information in many cases can apply to the other *Rhizobium* species.

B. Nodulation Process

The observation that certain species of *Rhizobium* were able to infect and induce nodulation of only specific species of legumes gave rise to the classification system of "cross-inoculation groups" (Fred et al., 1932). However, a certain degree of promiscuity exists within several *Rhizobium* species. See the discussion in section III E.

An increasing body of evidence supports the theory that certain plant proteins, called *lectins,* bind with carbohydrates found in the cell walls of the bacteria (Bohlool & Schmidt, 1974). Recently reported data support the idea that lectins are involved in the infection process. Lectins have been isolated from soybean seed that bind preferentially to *Rhizobium* strains that normally infect those plants.

Early stages of the development of the symbiosis are described by Vincent (1980). Prior to infection, a colonization of rhizobia occurs at the root surface. The bacteria approach possible infection sites in a polar fashion at right angles to the plane of attachment. It is at this point that the protein/carbohydrate interaction is believed to occur, which results in a specific recognition event that results in legume/*Rhizobium* specificity (Dazzo, 1980). The mechanism of this binding event is not clearly defined. Rhizobial presence near the root hairs causes root hair branching and marked curling, which signals that invasion is possible. The bacteria penetrate the primary cell wall, frequently at the point of curling, then cause an invagination (Nutman, 1956) of the inner cell wall and plasmalemma, leading to the inward growing infection thread. The bacteria lie embedded end to end in a polysaccharide matrix (Goodchild & Bergersen, 1966). The rhizobia divide continually as the infection thread continues through several layers of the host cortex.

The development of *Rhizobium*-directed polyploid meristem centers within the cortex is the first step in the development of the nodule meristem. Further development involves penetration of new tissue by the infection thread, with subsequent release of bacteria through a thin-unwalled section of the infection thread into the host cells. A period of rapid multiplication of the bacteria follows, and the young disomatic cells become tightly packed with bacteria, which then undergo morphological and metabolic changes to become "bacteroids" (Fig. 2). Biochemical changes in the bacteroids produce the enzyme systems necessary for the symbiotic N_2-fixing relationship. Also, at this time, the symbionts cooperate in producing the leghemoglobin that protects the O_2-labile nitrogenase and, at the same time, provides just enough O_2 for respiratory activity in the bacteroids.

Generally, only one strain of *Rhizobium* will be found per nodule, although a given root can form nodules of more than one *Rhizobium* strain.

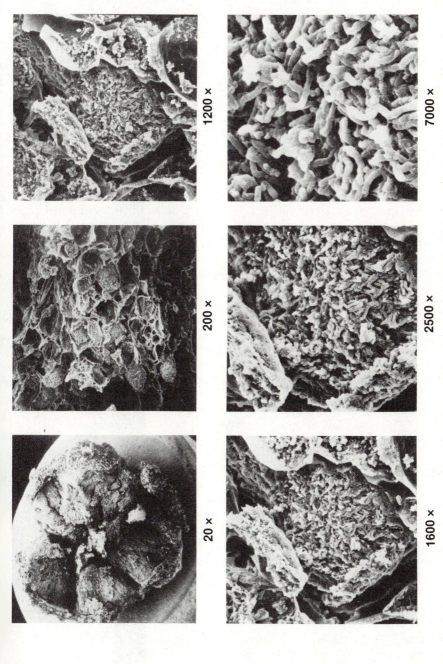

Fig. 2—Scanning electron micrograph of a freeze-cleaved soybean nodule at a low magnification showing one half of the nodule to high magnifications, which show up to 10,000 *Rhizobium* bacteroids within each host cell. Large amounts of energy are consumed within these bacteroids, and N_2 is reduced to NH_3, as well as protons to H_2. (Hardy et al , 1980).

Several native *Rhizobium* strains may exist in the field. These native *Rhizobium* strains compete with introduced strains. In a mixed inoculum, the infection frequency is not necessarily related to its proportion in the initial inoculum. Infection frequency of different strains is affected by temperature in various ways. If planting dates are adjusted, it is possible to favor inoculation by superior strains.

There are lines of leguminous species that differ genetically in their susceptibility to inoculation by various rhizobial strains. A mutant homozygous single recessive factor, *no*, has been bred into several legume species, yielding isogenic lines differing only in the nodulating characteristic. These lines are useful for studying nodulation and N_2 fixation in field situations.

Legumes growing in soils with high amounts of fixed N_2 nodulate poorly and fix little N_2. Fixation of N_2 in high N soils does not appear to be regulated at the nitrogenase synthesis level but at the level of nodulation (Brill, 1980). Nitrate was thought by Tanner and Anderson (1964) to inhibit nodule initiation by catalytic destruction of IAA by NO_2^-, both of which, they said, were produced by the bacteria. However, studies with NO_3^- reductase deficient mutants (Gibson et al., 1977) showed delayed nodule initiation by NO_3^-. If NO_2^- is responsible for IAA destruction, then the NO_2^- is in the plant. Nodulation is apparently inhibited by either NO_3^- itself or a plant metabolite of NO_3^-.

A NO_3^--induced carbohydrate deficiency in the roots is suggested as the cause of delayed nodulation (Gibson et al., 1977). Certain levels of fixed N_2 in root tissue may inactivate *Rhizobium*-binding lectins (Brill, 1980; Dazzo, 1980).

The myoglobin-like leghemoglobin is the O_2 buffering protein that delivers O_2 to the bacteroids for respiration, but because of its high affinity for O_2, it prevents high concentrations of O_2 from reaching and inactivating the nitrogenase. Since leghemoglobin is found only in the legume root nodule, the question has been asked, "What protects nitrogenase in nonlegume nodules?"

It has been suggested that polyphenol oxidases protect nitrogenase in nonlegume nodules, such as *Trema cannabina,* infected by cowpea rhizobia that do not contain leghemoglobin (Coventry et al., 1970). Leghemoglobin maintains the flow of O_2 into the system of air passages found in the central part of the nodule where uniform O_2 pressure is maintained at the surface of the cells (Wittenburg, 1980). Fixation of N_2 in the intact nodule appears to be O_2 limited and increases rapidly in response to increased O_2 pressure. However, if exposed for several hours to low rhizosphere O_2 concentration, the system can adapt and recovers to its initial rate of N_2 fixation (Criswell et al., 1976b).

Globin is induced by the plant, but the possibility that the bacterium contributes to the synthesis of the heme component has also been examined. There is more than one species of leghemoglobin in nodule plant cells, and their ratios change during developmental stages (Brill, 1980).

For further reading on leghemoglobin, see Dilworth and Appleby (1977), Dilworth (1980), and Wittenburg (1980).

C. Energy Relationships in N_2 Fixation

Large amounts of energy are consumed by N_2-fixing reactions whether the process is synthetic or natural. The legume/*Rhizobium* symbiosis requires about 10 kg of carbohydrates·kg^{-1} of N_2 fixed, documenting the high-energy requirement of the process. Biological N_2 fixation consumes the equivalent of 25 to 28 molecules of ATP for each molecule of N_2 fixed. This high energy requirement for N_2 fixation in legumes such as soybeans is coupled, unfortunately, with a C_3 or photosynthetically inefficient system.

1. FACTORS LIMITING N_2 FIXATION

Since N_2 fixation is an energy-intensive process, it is apparent that factors controlling rates of photosynthesis or the distribution of photosynthate in the plant would significantly affect N_2 fixation. These factors may be biological, environmental, or nutritional. Biological factors may be genetic or source-sink relationships. Environmental and nutritional factors normally affect N_2 fixation through photosynthate production, but some environmental factors may exert a direct influence on N_2 fixation itself. Environmental factors would include light, temperature, water relations, mineral nutrition, canopy pCO_2 and pO_2, soil pH, rhizosphere pO_2, and rhizosphere temperatures. A summary of factors that affect N_2 fixation is shown in Table 1.

a. Evidence That Photosynthesis Limits N_2 Fixation—There is a body of evidence that documents the high rate of flow of photosynthate to the root systems and actively functioning legume nodules. Several studies document the high sink strength of legume nodules actively fixing N_2.

Small and Leonard (1969) demonstrated that functional nodules are active sinks for newly formed photosynthate from the tops of pea (*Pisum sativum* L.) and subterranean clover (*Trifolium subterraneum* L.). Four hours after a 15-min exposure of the tops to $^{14}CO_2$, 5 and 9% of the fixed ^{14}C was found in the nodules of the two plants, respectively. Addition of combined N in a concentration inhibiting to N_2 fixation caused a decrease in nodule sink strength.

The partitioning of photosynthate within the plant influences nodule development and subsequent N_2 fixation. Minchin and Pate (1973) found that in vegetative pea and lupin (*Lupinus*), 74 and 71% of the fixed C is translocated to the roots. Of this, 47 and 40% is respired to the soil, and 16 and 9% is recycled to the stem as amino compounds produced by the nodules, respectively. Brun (1976) reported that 20 hours after pulse feeding nodulated, 5-week-old soybean plants with $^{14}CO_2$, the label was distributed between shoots, roots, and nodules in a ratio of 76, 17, and 7%, respectively.

Bach et al. (1958) studied the distribution of ^{14}C-labeled photosynthate in nodulated soybeans and found a high proportion distributed to nodules, where it was divided among organic acids, amino acids, and carbohydrates. However, during the rapid seed development stage of soybeans, roots and nodules accumulate very little photosynthate, and N_2 fixation declines rapidly (Hume & Criswell, 1973).

The genetics of the legume in question could have an influence on the

Table 1—Summary of factors that affect symbiotic N_2 fixation (Silver & Hardy, 1976).

	Effect on N_2 fixation		
Factors	Increase	None	Decrease
Photosynthate available to nodule			
Light quantity	Day Long days Supplemental light		Night Short days Shading Lodging
Source size	Additional foliage Low planting density		Defoliation High planting density
CO_2/O_2 ratio around canopy	Increased pCO_2 Decreased pO_2		Decreased pCO_2 (?) Increased pO_2
Photosynthetic type	C_4 (?)		C_3
Competitive sinks	Removal of reproductive structures		Development of reproductive structures
Translocation			Girdling
Water			Low water potential
Fixed N_2			Nitrogen fertilization
Fixed N_2	Starter N	Some forms of organic N Rhizobia constitutive for nitrogenase (?)	High levels of fixed N_2 as NO_3, NH_4, or urea applied to soil, or foliage at early or mid development
Stress			
Water	Irrigation		Drought
Temperature		> 20C < 30 C	> 30°C and < 20°C
Soil pH			< pH 5.0
Rhizosphere pO_2		Adaptation from 5 to 30%	< 5%, > 50%
Nodule sink activity			
Shoot (reproductive) Sink activity for fixed N_2			Arrested reproductive growth by 5% O_2
Bacterial strain	†	†	†
Plant variety	†	†	†
Management practices Inoculant and inoculation techniques Fixed N_2 Organic matter Plant variety–bacterial strain interaction	†	†	†

† Any of these effects may be obtained depending on the particular situation.

photosynthesis rate and subsequent N_2 fixation. Dornhoff and Shibles (1970) found significant differences in photosynthetic rates among 20 genotypes tested in the field at a range of CO_2 concentrations.

These studies document the high photosynthetic flux to the nodules of plants that are actively fixing N_2. There is further indirect evidence that factors that affect the rate of photosynthesis also affect rates of N_2 fixation.

i. Light Intensity

Diurnal effects on specific N_2-fixing activity of soybean and alder nodules have been observed, with maxima occurring near the period of maximum light intensity (Hardy et al., 1968; Wheeler, 1969; Sloger et al., 1975). Fifty percent shading imposed at the end of flowering in soybeans decreased N_2 fixation from 125 to 91 kg of $N \cdot ha^{-1} \cdot season^{-1}$, whereas supplemental light increased N_2 fixation to 165 kg of $N \cdot ha^{-1} \cdot season^{-1}$. Nodule specific N_2-fixing activity was similarly affected (Ham et al., 1976). Mague and Burris (1972) reported a diurnal cycle of acetylene reduction by intact soybean plants dependent on light intensity and temperature. In the soybean subjected to variable light treatments, the ATP/ADP ratio in the nodule is directly related to N_2 fixation (Ching et al., 1975), as is the ATP/ADP ratio and bacteroid N_2-fixing activity in the presence of increasing amounts of the inhibitor N-phenylimidazole (Appleby et al., 1975).

One-day exposure to darkness of Chippewa soybeans caused a 50% decrease in acetylene reduction (Ching et al., 1975), a loss of activity attributed to a decrease in photosynthate availability. They found the dark treatment decreased nodule sucrose, ATP, and ATP/ADP ratio by 60, 70, and 44%, respectively.

Subterranean clover plants grown under low and high light intensities for 20 days following inoculation were transferred to alternate conditions. Nitrogenase activity of plants placed into low light decreased by 40%, whereas a 50% increase in activity was recorded for plants placed into high light (Gibson, 1976a). Furthermore, nodule weights of plants placed into low light remained static, whereas those placed into high light increased significantly within 3 days (Gibson, 1976a).

ii. Source Size

The effect of four factors—defoliation, additional foliage, plant density, and lodging—related to source size have been measured.

Fixation of N_2 by defoliated plants was reduced (Hardy et al., 1968; Moustafa, 1969) by partial defoliation (removal of two leaflets from each soybean leaf after flowering), causing a fixation reduced from 125 to 100 kg of N/ha (Ham et al., 1976).

When source size was increased by grafting a second top to a single root system (Streeter, 1973), specific N_2-fixing activity of nodules was increased by up to 100% for a short time period.

Fixation of N_2 per plant increased in inverse proportion to plant density, but N_2 fixed $\cdot ha^{-1} \cdot season^{-1}$ is almost independent of planting density (Hardy & Havelka, 1976).

Biological $N_2[C_2H_2]$ fixation by lodged soybeans was less than one-third that of unlodged plants, 82 vs. 275 mg/plant for lodged and unlodged plants, respectively. This major decrease was produced by the loss of the exponential phase within 3 or 4 days following lodging (Hardy & Havelka, 1976).

iii. Competitive Sinks

Pod removal from every other node of soybeans at the end of flowering increased N_2 fixation by presumably making more photosynthate available to the nodules (Ham et al., 1976).

The metabolic demand during rapid pod fill becomes a strong competitor of nodules for photosynthate. Fixation of N_2 virtually terminates at mid pod fill.

Lawrie and Wheeler (1974) observed a 60% decline in ^{14}C photosynthate accumulation and nitrogenase activity in pea nodules between flowering and fruiting, even though total plant photosynthesis doubled during that period.

Time courses of N_2 fixation by soybeans, peanuts (*Arachis hypogaea* L.), and peas have been determined for several locations and cultivars (Criswell et al., 1976; Hardy et al., 1968, 1971, 1973; Hardy & Havelka, 1975; Harper, 1974, 1976; Lawn & Brun, 1974; Mague & Burris, 1972; Shivashankar et al., 1976; Sloger et al., 1975; Thibodeau & Jaworski, 1975; Weber et al., 1971). The N_2 fixation profiles are all similar, with maximum $N_2[C_2H_2]$ fixation occurring in the postanthesis period and the exponential phase terminating during the mid bean-filling stage when the sink demand of the reproductive phase increases greatly.

iv. Photosynthate Translocation—Girdling

Stem-girdled nonlegumes show a rapid decline in N_2-fixing activity (Wheeler, 1971), with >50% decline within 2 hours of girdling. These results suggest the close coupling of photosynthate movement to the nodule and N_2-fixing activity and are consistent with the rapid response in specific N_2-fixing activity to CO_2 enrichment. Exposure of the soybean foliage canopy to 1,500 ppm of CO_2 for only 6 hours increased specific activity by >70% (Havelka & Hardy, 1976).

v. Moisture Stress

Decreased photosynthesis resulting from low water potential resulted in decreased $N_2[C_2H_2]$ fixation activity (Huang et al., 1975a, b). Carbon dioxide enrichment of the foliage canopy overcame some of this decrease of N_2 fixation activity. Water stress as well as waterlogging can seriously reduce N_2 fixation by depressing the activity of existing nodules and reducing nodulation (Sprent, 1976).

b. Effects of Canopy pCO_2 and pO_2 on N_2 Fixation—Fixation of N_2 by laboratory-grown alfalfa (*Medicago sativa* L.) and red clover (*Trifolium pratense* L.) was substantially increased when grown under high canopy pCO_2 (Wilson et al., 1933). Greenhouse experiments on peas (Phillips et al.,

1976) and soybeans (Shivashankar et al., 1976) involving elevated canopy pCO_2 resulted in large increases in N_2 fixation.

The effect of elevated pCO_2 on N_2 fixation is attributed to an increased energy level in the nodules (Hardy & Havelka, 1975; Quebedeaux et al., 1975). The increased energy level of the nodules results from an increased rate of photosynthesis presumably because of decreased photorespiration.

Increased $N_2[C_2H_2]$ fixation of field-grown soybeans was first reported by Hardy and Havelka (1973).[2] Soybeans (cv. Kent) grown in open-top chambers in a field situation were exposed to a canopy pCO_2 concentration of 800 to 1,200 ppm during the daylight hours. Carbon dioxide enrichment began at 38 days of age (5 days before flowering) and continued until senescence. Fixation of $N_2[C_2H_2]$ on a per plant basis was increased fivefold (842 vs. 167 mg) by canopy CO_2 enrichment. Hardy and Havelka (1976) attributed this large increase in $N_2[C_2H_2]$ fixation to (i) a doubling of the nodule mass, (ii) a doubling of the average nodule specific N_2-fixing activity (unit of $N_2[C_2H_2]$ fixed per unit of nodule mass), and (iii) an 8-day extension of the critical exponential phase (section III D 4b). More $N_2[C_2H_2]$ fixation occurred during the 13th week of age by CO_2-enriched plants than by the control plants during their entire life cycle. Reproductive yield was almost doubled by CO_2 enrichment (Fig. 3).

Other field CO_2-enrichment experiments conducted since 1973 on several varieties of soybeans, corroborate and complement the above results (U.D. Havelka & R. W. F. Hardy, unpublished results).

Carbon dioxide enrichment studies of other field-grown legumes have been conducted on peas and peanuts (Hardy et al., 1977). Fixation of $N_2[C_2H_2]$ was substantially increased by CO_2 enrichment (1,500 ppm of the foliage canopy) 8 hours daily from early vegetative period to maturity.

In the case of peas, $N_2[C_2H_2]$ fixed was increased from 87.5 to 128 mg/plant. As in the case of soybeans, this $N_2[C_2H_2]$ fixation increase was the result of increased nodule specific activity, nodule mass, and delayed senescence (increased period of N_2 fixation activity). The $N_2[C_2H_2]$ fixation increase was accompanied by a 53% dry seed yield increase from 3,446 to 5,262 kg/ha.

With peanuts grown under CO_2 enrichment, $N_2[C_2H_2]$ fixation activity increased by 59%, from 157 to 248 mg/plant. This increase was due to increased nodule mass and delayed senescence of the nodules. Specific nodule activity was not increased. However, normal nodule specific activity of peanuts is two to three times that of soybeans. Dry yield of nuts was increased by only 30%. The nut yield in the CO_2-enriched chambers did not reach full potential because pegging was restricted by the increased (44%) vegetative mass.

Fixation of N_2 by pinto beans (*Phaseolus vulgaris*) is normally poor (Burton, 1976). Carbon dioxide enrichment of beans (U. D. Havelka & R. W. F. Hardy, unpublished data) resulted in a $N_2[C_2H_2]$ fixation increase of

---

[2] R. W. F. Hardy and U. D. Havelka. 1973. Symbiotic N_2 fixation: Multifold enhancement by CO_2-enrichment of field-grown soybeans. Plant Physiol. Abstr. 51(Suppl.):35.

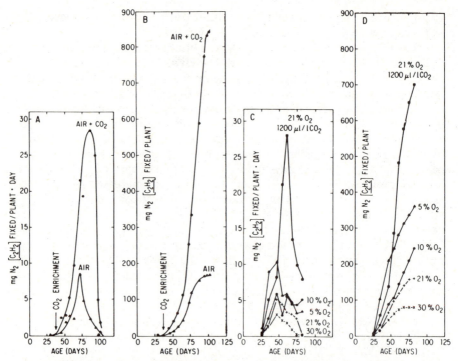

Fig. 3—Photosynthate as a major limiting factor for N_2 fixation based on effect of altered canopy CO_2/O_2 ratio on N_2 fixation. In A and B, CO_2 enrichment to 800 to 1,200 μliters/liter of field-grown soybeans occurred in open-top enclosures from 40 days of age to senescence. For C and D, CO_2 enrichment to 1,200 μliters/liter and O_2 alteration to 5, 10, and 30% at 300 μliters/liter of CO_2 are compared with air for laboratory-grown soybeans.

46%. Dry bean yield increase was substantial, 4,660 vs. 2,930 kg/ha for ambient controls.

It has been suggested that the major effect of increased CO_2 was to decrease photorespiration. Two other experiments lend support to this hypothesis.

Hardy and Havelka (1977) examined the effect of various concentrations of canopy pCO_2, from 300 to 1,500 ppm, in a field situation on soybean $N_2[C_2H_2]$ fixation throughout the entire growing season. Fixation of $N_2[C_2H_2]$ per plant per season increased linearly from 300 to 1,000 ppm of CO_2 and curvilinearly from 1,000 to 1,500 ppm of CO_2.

This type of relationship would be expected for the CO_2/O_2 competition characteristic of photorespiration. In the second experiment, the CO_2/O_2 ratio around the canopy was altered by either CO_2 enrichment at a constant pO_2 of 0.21 atm (as in the previous experiments), by O_2 depletion, or by O_2 enrichment at a constant CO_2 concentration of 300 ppm (Quebedeaux et al., 1975; Hardy et al., 1977). During vegetative growth, including the early flowering stage, there was a close correlation between CO_2/O_2 ratio and $N_2[C_2H_2]$ fixation (Fig. 3). This further supported the hypothesis that the effect of CO_2 enrichment was due to decreased photorespiration and increased photosynthate available to the nodule.

When previously untreated, flowering field-grown soybean plants were exposed to elevated pCO_2 concentration, specific $N_2[C_2H_2]$-fixing activity (unit of $N_2[C_2H_2]$ fixed per unit of nodule mass) increased in a few hours (Hardy & Havelka, 1976).

These experiments can lead to the conclusion only that photosynthate is the most limiting factor in N_2 fixation. Carbon dioxide enrichment is obviously not a practical method of increasing N_2 fixation or field crop yield, but the use of CO_2 enrichment has proved to be a useful tool in assessing the significance of increased photosynthesis. The practical approaches to the use of this knowledge are increased photosynthesis or restricted photorespiration via genetic or chemical manipulation (Hardy et al., 1976, 1977, 1978, 1980).

c. Factors Other Than Photosynthesis That Limit N_2 Fixation—The limiting factors discussed up to this time have been mainly concerned with photosynthesis and the plant component of the symbiosis.

Strains of the rhizobial component of the symbiosis differ in their efficiency of N_2 fixation. For example, three strains of exceptionally efficient soybean rhizobia produce nodules that fix N_2 at about twice the rate of normal strains (Gibson, 1977). The inefficient strains form as much nodule tissue as the efficient ones, thus expending energy needlessly. In fact, inefficient strains form large nodules but fix little or no N_2. Efficient strains of *Rhizobium* also vary in their compatibility with different host cultivars (Burton, 1976; Hardy & Havelka, 1976).

Temperature, soil pH and rhizosphere pO_2, nutritional factors, and soil NO_3^- also affect N_2 fixation and are discussed in the following sections.

i. Temperature Effects

Temperature effects on N_2 fixation may be manifested through either root or shoot effects.

Low root temperatures of both temperate and tropical legumes have been found to cause a greater degree of nodular development, which could compensate for the lower specific N_2-fixing activity at low temperatures (Gibson, 1976a, b). Fixation of $N_2[C_2H_2]$ of growthroom-grown soybean plants exposed to varying rhizosphere temperatures showed a rapid decline and recovery after cooling (24–14°C) and rewarming (14–24°C) but a slow decline and very slow recovery following heating (24–34°C) and cooling (34–24°C) (Hardy et al., 1977).

Low soil temperatures retard root hair infection as well as nodule development and N_2 fixation. Different *Rhizobium* strains and different legume species interact differently. Temperatures >28°C can affect the symbiosis, but here again, the inhibition varies with the host plant and *Rhizobium* strain (Gibson, 1976a, 1977).

ii. Soil pH and Rhizosphere pO_2

Soil pH is probably one of the most important external factors affecting the symbiosis. Some legumes, such as lucerne (*Medicago sativa* L. ssp. *sativa*) and black medic (*Medicago lupulina* L.), grow best at neutral pH or slightly alkaline, whereas serradella (*Ornithopus sativus* Brot.) nodulates at pH 5 or even lower (Gibson, 1977). Part of the adverse effect of low soil pH can be attributed to the poor survival of rhizobia under acid conditions. Large differences in soil pH tolerance between nodulated legumes is due, in part, to the ability of the rhizobia to grow and survive in acid media (Gibson, 1977).

It has been shown that low soil pH is inhibitory to root hair infection and early nodulation in peas and lucerne. Low soil pH may also reduce the supply of nutrients required for nodulation. Liming acid soils that presented problems to the symbiosis greatly improved N_2 fixation.

Previous reports and reviews suggest that low rhizosphere pO_2's are extremely detrimental to N_2 fixation (Bergersen, 1962; Gibson, 1976a, 1977); Hardy et al., 1977). Bergersen (1962) reported substantial increases in N_2 fixation of excised nodulated roots of soybeans subjected to supra-atmospheric pO_2 and large decreases when subjected to subatmospheric pO_2.

Hardy et al. (1977) reported on the effect of altered pO_2 on $N_2[C_2H_2]$ fixation of excised nodulated roots of field-grown soybeans assayed six times during the complete growth cycle. Inhibitions of up to 87% occurred from ≥ 0.06 to ≤ 0.21 atm pO_2, and stimulations of up to 91% occurred from ≥ 0.26 to ≤ 0.41 atm pO_2 (pO_2 for maximum stimulations varied with the age of the plant), whereas inhibitions occurred from ≥ 0.65 to ≤ 0.90 atm pO_2.

However, in the intact soybean system at least, there appears to be an O_2 control system that adapts to an altered pO_2. Long-term studies with intact growthroom-grown soybean plants (Hardy et al., 1977) revealed the intact systems ability to recover $N_2[C_2H_2]$–fixing ability after a 4- to 24-hour exposure to pO_2's of 0.06 atm. Returning the rhizosphere to ambient pO_2 caused a 58% decrease in $N_2[C_2H_2]$ fixation, but activity returned to normal after 4 to 24 hours. Lower pO_2's (0.02–0.03 atm) resulted in an incomplete adaptation. Adaptation was also less complete and slower at cool temperatures (14–19°C) than at ambient (24°C) or hot (29–34°C) temperatures.

If the intact plant was substrate depleted, the initial inhibition response to lowered rhizosphere pO_2 was much greater, whereas an elevated rhizosphere pO_2 substantially increased N_2 fixation. Hardy et al. (1977) suggest that the soybean nodule contains a sophisticated O_2 control mechanism that enables the legume/*Rhizobium* system to adapt to varying pO_2's not only to maintain an orderly rate of N_2 fixation but also to conserve carbohydrates when appropriate.

iii. Nutritional Factors and Soil Nitrate Effects

Nutritional factors also affect N_2 fixation through their effect on either the host or the rhizobia. Molybdenum deficiency is commonly the most important trace element deficiency. In general, the effects of mineral factors on N_2 fixation would be pronounced only under poor cultural management.

That fixed N_2 inhibits N_2 fixation is well documented (Fred et al., 1932; Hardy & Havelka, 1976; Gibson, 1976a).

Translocation of ^{14}C-labeled photosynthate to pea, subterranean clover, and soybean nodules declined after plants growing in N-free media were fed NO_3^- (Gibson, 1976a). Nodulated soybean plants growing in N-free media exhibited a decline in nitrogenase activity within 48 hours of receiving NO_3^-. After 10 days treatment and following elution of the NO_3^-, nitrogenase specific activity and nodule development resumed. No breakdown of nodule integrity was observed by electron microscopy 4 days after treatment.

Decrease of N_2 fixation activity in pea nodules has been correlated with decreases in nodule leghemoglobin content following treatment with NH_4NO_3 (Bisseling et al., 1978). Also in peas, darkening can lead to an apparently energy-related leghemoglobin degradation (Roponen, 1970). The possible role of the nodule NO_2^- concentration in mediating these effects is uncertain. Although NO_3^- clearly affects nitrogenase activity, and NO_2^- can convert leghemoglobin to a form unable to transport O_2, NO_2^- has not been isolated from nodules in sufficient quantity to implicate it as an intermediate in the NO_3^- effect (Dilworth, 1980). Whether NO_3^- effects on N_2 fixation are the result of competition for photosynthate or a more direct NO_3^- metabolite effect on module metabolism, or some combination thereof, is yet to be resolved.

2. AMMONIA ASSIMILATION AND NITROGEN TRANSPORT IN NODULATED LEGUME/*RHIZOBIUM* SYSTEMS

Several reports of analyses of bleeding sap of nodulated legumes have indicated that asparagine is the primary transport form of N from the nodulated root to the aerial portion of the plant (Clauss et al., 1964; Pate & Wallace, 1964; Pate et al., 1965; Streeter, 1972) beginning with the period of high N_2-fixing activity.

However, Matsumoto et al. (1976), Ishisuka (1977), McClure and Israel (1979), and Streeter (1979) have reported that the ureides, allantoin, and allantoic acid—not asparagine—are the major constituents of stem exudate during periods of high N_2-fixing activity. The correlation of N_2 fixation and ureide transport has led to the proposal that ureide content of xylem sap may be used as an indicator of N_2-fixing activity (McClure et al., 1980).

Ammonia assimilation in legume nodules is believed to be centered in the plant cytosol and not in the bacteroid. Enzyme activities in the bacteroid are too low to account for removal of all NH_3 produced by N_2 fixation. Specific activities of enzymes in the assimilation chain rise dramatically when nitrogenase activity rises. No concurrent rise is seen in the specific activity of bacteroid enzymes (Boland et al., 1980).

Boland et al. (1980) proposed a pathway for NH_3 assimilation, metabolism, and transport of N compounds in several nodulated, N_2-fixing legumes. Ammonia reaches the plant cytosol and is converted to glutamine via glutamine synthetase, with subsequent production of glutamate by glutamate synthase. Further transformation is the synthesis sequence aspartate to asparagine from oxaloacetate, with glutamate and glutamine as N donors. The enzymes aspartate aminotransferase and asparagine synthetase are involved, and their specific activities rise in a manner consistent with increased N_2 fixation. Other transformations include synthesis of various other amino acids and plant proteins. In the light of new information, another pathway in some legumes, including soybeans, may involve de novo purine synthesis. Subsequent oxidative decomposition yields allantoin and allantoic acid as the translocated forms. Their exact metabolism is still uncertain.

3. ENERGY COSTS AND H_2 EVOLUTION

The energy cost of N_2 fixation to the plant component is both large and difficult to assess. Energy, in the form of photosynthate, is required for the root hair infection, nodule development, and bacteroid and nodule maintenance.

Fixation of N_2 requires ATP and reductant for nitrogenase reactions and organic acids for use of the nodule cells to synthesize glutamate, glutamine, aspartate, aspragine, ureides, and other amino acids from NH_3. This energy, in the form of photosynthetic C, is the contribution of the host, whereas NH_3 is furnished to the host by the bacteria.

Bacteroids are known to metabolize various C compounds supplied by the plant cell (Bergersen, 1974). Little knowledge is available as to the particular form of C compound furnished by the plant cell (Evans & Barber, 1977).

The cost of N_2 fixation would be expected to be relatively expensive in the early nodulation stage and decline as maximum nodule mass is attained. On a theoretical basis, 4 kg of carbohydrates are required to fix 1 kg of N_2 (Hardy & Havelka, 1975). This includes only ATP and reductant requirements for nitrogenase based on in vitro measurements. When nodule production and maintenance are considered, the cost goes up to 8 to 17 kg of carbohydrates/kg of N_2 fixed (Minchin & Pate, 1973; Mahon, 1977; Ryle et al., 1977; Hardy et al., 1977; Heytler & Hardy, 1979, personal communication).

Briefly, the nitrogenase reaction involves its two subunits, the Mo-Fe protein and the Fe protein. Either of two electron-transferring proteins, ferredoxin or flavodoxin, is used to transfer six electrons per molecule of N_2 reduced to the Fe protein. Unfortunately, these electrons alone are not sufficient for nitrogenase turnover. Two Mg•ATP's per electron transferred are bound to the reduced Fe protein, making the electrons more strongly reducing and enabling the reduction of the Mo-Fe protein with the energy of ATP released by its hydrolysis to ADP and inorganic phosphate. The six electrons transferred have an equivalent cost of nine ATP's, making this an expensive reaction. The total energy cost based on the above stoichiometry is 21 molecules of ATP for each molecule of N_2 fixed.

Unfortunately, the situation becomes even more distressing. It has been known for some time that soybean nodules evolve hydrogen (H_2) (Hoch et al., 1960; Bergerson, 1963). Later it was discovered (Bulen et al., 1965; Hardy et al., 1965) that nitrogenase, in the absence of N_2, irreversibly reduces protons (H⁺) to molecular H₂. However, even at saturating concentrations of N_2 (0.8 atm), nitrogenase continues to reduce protons (Bulen et al., 1965). In vitro, the extent of H_2 evolution is dependent on the component ratio, electron supply, and ATP concentration (Hageman & Burris, 1978; Ljones & Burris, 1972; Silverstein & Bulen, 1970). High ATP concentration favors N_2 reduction (Silverstein & Bulen, 1970). At saturating concentrations of N_2, about 25% of the electrons are wasted in the H_2 evolution reaction (Bulen & Le Comte, 1966; Burris, 1977; Burns & Hardy, 1972; Mortenson, 1966). When this 25% loss is taken into consideration, the cost of fixing a molecule of N_2 goes up to 28 ATP's.

The efficiency of the system could be improved if the energy in the H_2 evolved could be recaptured. Phelps and Wilson (1941) described a hydrogenase isolated from pea nodules. Dixon (1968, 1972), on investigating this system, found that oxidation of H_2 could produce 2 ATP's per two electrons transported, or the electrons can reduce nitrogenase (Dixon, 1972; Emerich et al., 1979). Since the original wasteful H_2 evolution system cost the system six ATP's, the hydrogenase allows the recouping of only 33% of the original lost to H_2 evolution.

Measurements by Schubert and Evans (1976) of H_2 evolution determined the nodular efficiency of H_2 metabolism of most legumes to range from 0.40 to 0.61, with cowpeas, at 0.99, being an exception. Nodular efficiency is defined as 1 minus the ratio of the electrons evolved as H_2 to the total electrons transferred to N_2 and H_2. Theoretically, the nodular efficiency would be 100% if all the H_2 electrons were recaptured.

Research with various *Rhizobium* strains possessing Hup⁺ (those responsible for uptake hydrogenase) and Hup⁻ genes indicates increased N_2 fixation is possible with symbioses involving bacterial strains with positive H_2 uptake genes in cowpeas (*Vigna sinensis*) and soybeans (Schubert et al., 1977; Evans et al., 1980) and soybeans (Albrect et al., 1979). The value of these systems remains to be further elucidated. For example, for relevant comparisons, the efficient N_2-fixing isogenic strains of rhizobia, differing only in H_2 uptake genes, will need to be field tested. If, indeed, an efficient H_2 conserving strain of *Rhizobium* is found or developed, the problem is only addressed. The introduced *Rhizobium* must be able to heavily nodulate legumes in soils containing populations of highly competitive less efficient rhizobia.

4. NITRATE EFFECTS ON N₂ FIXATION

The inhibitory effects of NO_3^- on nodulation and N_2 fixation are well documented (Fred et al., 1932; Ham et al., 1976; Hardy & Havelka, 1975; Harper, 1976; Gibson et al., 1977). "Photosynthate deprivation" in the nodules is said to occur because C is being used in the NO_3^- assimilatory pathway (Oghoghorie & Pate, 1971), and a C source for N_2 fixation is limited. Other suggestions as to the cases of the NO_3^- effect were proposed. Nitrite could complex with leghemoglobin, thus destroying its O_2-regulating activity, and NO_2^- could inhibit nitrogenase activity (Rigaud et al., 1973; Kennedy et al., 1975).

Gibson et al. (1977) used strain 32H1 and four NR (nitrate reductase) deficient mutants to test the photosynthate deprivation of nodules by NO_2^-. Siratro plants (*Macroptilium atropurpureum* DC. Urb.) nodulated by the five strains were fed NO_3^-, and $N_2[C_2H_2]$-fixing activity was measured 24 and 48 hours later. The $N_2[C_2H_2]$-fixing activity and nodule growth were decreased in all treatments. Nitrogenase specific activity declined in all cases. Nitrate reductase activity was high in strain 32H1 nodules but was undetectable in nodules of the four NR deficient mutant strains. These results suggested to them that NO_3^- effects on N_2 fixation are due to effects other than those of NO_2^- and could support the photosynthate deprivation theory.

Harper's (1978) hydroponic experiments, with tungstate as an inhibitor of NO_3^- reductase activity, confirm an apparent indirect effect of NO_3^- metabolism on nodulation and N_2 fixation. The mobilization of carbohydrates for NO_3^- metabolism appeared to deprive nodules of carbohydrates for growth and N_2 fixation. The claim that well-nodulated soybean plants receiving high amounts of photosynthate prefer to fix their own N (Hardy & Havelka, 1975) may corroborate the photosynthate deprivation theory (Gibson et al., 1977). Regardless of the reasons for the turning off of biological N_2 fixation in soils containing high amounts of fixed N_2, this loss of N_2-fixing power results in the loss of a potential N-conserving opportunity.

D. Methodology of N_2 Fixation Measurement

1. INTRODUCTION

The investigation of physiological-biochemical processes requires suitable methods of measurement. Several methods of measuring rates of N_2 fixation have been employed, and excellent techniques are currently available (Burns & Hardy, 1975; Burris, 1974; Criswell et al., 1976a; Hardy et al., 1973; Hardy & Holsten, 1977).

Only methods best suited for agronomic research are discussed. These methods involve (i) nodule number and mass and leghemoglobin concentration, (ii) total N analysis and ^{15}N tracer methods, and (iii) reduction of alternative substrates (commonly called the acetylene reduction method).

The acetylene reduction assay has become, however, the most popular method of measuring rates of N_2 fixation both in the laboratory and the field. Since its introduction, this technique has enabled the unleashing of a deluge of N_2 fixation knowledge at all organizational levels.

2. NODULE NUMBER AND MASS AND LEGHEMOGLOBIN CONTENT

Measurements of nodule number and mass and leghemoglobin concentration are relatively simple to do but, at the same time, the least quantitative of available methods (Criswell et al., 1976a).

Hematin content, determined as pyridine hemochromogen (Virtanen et al., 1947b), which consists of largely leghemoglobin and legmethemoglobin

(Virtanen et al., 1947a), parallels the amount of N_2 fixed in peas, horse beans (*Vicia faba* L.), and soybeans inoculated with different bacterial strains. Methods and modifications of extracting hematin from legume nodules have been described and used (Harper & Cooper, 1971; Jordan & Gerrard, 1951; Keilin & Wang, 1945; Virtanen et al., 1947a). Leghemoglobin content only approximates quantitative measurement of N_2 fixation. Leghemoglobin presence in the nodule does not always reflect N_2-fixing activity. Green nodules from field-grown soybeans fix $N_2[C_2H_2]$ at reduced rates compared with red nodules). Nodules infected with various ineffective rhizobial strains lacked N_2 fixation, although leghemoglobin was present (Schwinghamer et al., 1970). The relationship between leghemoglobin concentration and nitrogenase $N_2[C_2H_2]$ activity changes with soybean plant age (Hardy et al., 1971; Johnson & Hume, 1973).

3. NITROGEN ANALYSIS

A. Total N Analysis by Kjeldahl—Kjeldahl analysis may be used to assess the total N in plant samples. This method does not distinguish between N from N_2 fixed or from combined N, thus appropriate controls must be used to determine the proportion of N derived from N_2 fixed.

One control method utilizes either uninoculated-unnodulated plants of the specific legume or a nonnodulating isoline of the specific legume.

In field experiments, controls employed low N fertility, low native effective *Rhizobium* strain populations, uninoculated seed (Bell & Nutman, 1971; Vest, 1971), or inoculation with ineffective *Rhizobium* strains, or massive soil applications of ineffective rhizobial strains to suppress indigenous *Rhizobium* strains (Bell & Nutman, 1971). In such comparisons, it was assumed that the controls and active N_2-fixing plants were using soil N at similar rates, an assumption that would likely underestimate N_2 fixation (Criswell et al., 1976a). Similar underestimates may be made in the case of using nonnodulating isolines as controls, which, with their greater root proliferation (Weber, 1966), take up more soil N than their N_2-fixing isolines.

One promising approach using total N determination would involve a screening method in a breeding program. Total N production of progeny effectively nodulated by an efficient *Rhizobium* strain would be a selection criterion for high N_2 fixation. The same approach could be used to select for efficient rhizobia.

b. Nitrogen-15 Analysis—The ^{15}N technique, first practiced in the early 1940's (Burris & Miller, 1941), eliminates the need to correct for N other than N_2. Nitrogenase neither differentiates between $^{15}N_2$ and $^{14}N_2$ nor catalyzes an exchange between $^{15}N_2$ and $^{14}N_2$. The primary detection system for $^{15}N_2$ is mass spectrometry (Burris, 1972; Burris & Wilson, 1957), but optical emission spectrometry has also been used.

In ^{15}N-fertility studies in soybeans, ^{15}N-enriched fertilizer is used. The basis of this method, termed the ^{15}N A-value method (Fried & Broeshart, 1975; Fried & Middleboe, 1977), is the fact that when ^{15}N is supplied to the soil at a low level, the legume/*Rhizobium* symbiosis will dilute the ^{15}N content in contrast to a reference crop that cannot fix N_2. The ideal reference crop is a nonnodulating isoline of the legume under study. In the absence of the "ideal" reference crop, a nonlegume may be used if appropriate coefficients are determined to account for morphological and biochemical differences between species (Phillips, 1980).

4. ALTERNATIVE SUBSTRATES: ACETYLENE–ETHYLENE REDUCTION ASSAY

The discoveries that nitrogenase has the capability of reducing substrates other than N_2 provided an excellent measurement method. The demonstration of nitrogenase's reduction of acetylene to ethylene by Dilworth (1966) in Australia and Schöllhorn and Burris (1966) at the University of Wisconsin led to the development of the acetylene-ethylene assay technique (Koch & Evans, 1966; Hardy et al., 1968; Hardy & Knight, 1967). This method of N_2 fixation measurement is based on reduction of acetylene to ethylene, followed by gas chromatographic separation of acetylene and ethylene coupled with their assay by the flame ionization technique. The acetylene serves as an internal standard. The equipment used in this assay is relatively inexpensive and readily attainable. The only objection is the precaution necessary in handling the explosive acetylene. The method is nondestructive, and the sample can be retrieved for other tests or allowed to continue to grow after the assay, since both the substrate, acetylene, and the product (ethylene) are relatively inocuous gases.

Data obtained with the results are identified with the bracketed formula "μg $N_2[C_2H_2]$ fixed." The two parameters used to quantitate the results of the acetylene–ethylene assay are the ratio of C_2H_2/N_2 reduced and the K_m value for acetylene. A saturating concentration of acetylene consumes almost all of the electrons transferred by nitrogenase in in vitro studies, and ATP-dependent H_2 evolution is almost eliminated, but a saturating concentration of N_2 consumes only about 75% of the electrons transferred, and the rest are evolved as H_2 (Burns & Hardy, 1975). Hydrogenases interfere with conducting such experiments in vivo; therefore the theoretical factor of 3:1 is commonly used to quantitate the conversion of acetylene reduced to N_2 fixed (Criswell et al., 1976a), although the ratio of 4:1 has been proposed as nearer the proper value (Burris, 1980). Detailed examination of stoichiometry in the presence of an uptake hydrogenase gives a value of 3.6:1 (P. G. Heytler, personal communication).

Although a saturating concentration of acetylene is 0.05 to 0.10 atm and equivalent to 0.80 atm of N_2, there are cases (in situ continual or repeating assay) where a lower concentration of acetylene is desired to avoid potential toxic effects. In these cases, the K_m value for acetylene is used to adjust the reaction velocity of V_{max}. Although literature values for this K_m are of the order of 0.005 atm, these were measured under Ar/O_2 atmospheres. The apparent K_m value determined in the presence of air must be used; this has been measured as 0.29 atm (P. G. Heytler & R. W. F. Hardy, personal communication).

The analysis of ethylene and the internal standard acetylene are done on a flame ionization gas chromatograph, for which a variety of columns are suitable (Hardy et al., 1973). Injected sample size can be 50 to 200 μliters. Standard samples of 1, 10, 100, 1,000, and 10,000 ppm of ethylene are used for standardization. Experience has shown that peak height is as satisfactory a measurement as area.

The formula for converting moles of acetylene reduced to N_2 fixed is given by Hardy and Holsten (1977):

grams N_2 [C_2H_2] fixed per hour•sample

$$= 28 \left[\left(\frac{e - b - i}{s}\right) \cdot c \cdot r \cdot v \cdot \frac{1}{t} \cdot \frac{1}{f}\right]$$

where e, b, i, and s, are peak height, or area, for acetylene in an analyzed sample of 50 μliters from, respectively, (i) experimental sample incubated with acetylene, (ii) experimental sample preincubated in the absence of acetylene (for ethylene background), (iii) incubation chamber with acetylene but without the sample (for ethylene impurity), and (iv) ethylene standard; c is concentration of ethylene in standard expressed as moles per liter at STP; r is ratio of peak height of internal standard in incubation chamber without sample to peak height in experimental incubation chamber with sample; v is volume of incubation chamber in liters at STP; t is time of incubation in hours; f is conversion factor for moles of acetylene reduced to moles of N_2 fixed; and 28 is the molecular weight of N_2.

The assay requires a gas-tight incubation chamber to contain the sample during the incubation stage. Serum bottles, vials, tubes, Mason jars, disposable syringes, plastic bags, large containers for intact potted plants, and plastic pots and bottles in which plants are growing have been used. Also, metal cylinders may be driven into the soil around nodulated plants in the field in situ (Criswell et al., 1976a), or plants may be grown in open-bottom chambers in the field (Sinclair, 1978).

a. Excised Nodulated Root N_2[C_2H_2] Fixation Assay—Plants being grown for N_2 fixation assays should be spaced as uniformly as possible so as to minimize variability among individual plants. Environmental factors known to limit N_2 fixation should be carefully controlled to the fullest extent possible. Samples should consist of three to six plants with four replications per individual treatment. For a seasonal profile, assays should begin at 10 to 20 days of age and continue weekly or every 10 days or so until the terminal stage is approached. The last two or three assays should be done every 5 days to avoid missing the termination phase.

Continuing assays should be done under similar weather conditions and at the same time period each day to minimize abnormal environmental effects. Unless the field area where the plants are growing is adjacent to the laboratory, the entire operation (exclusive of gas chromatographic analysis) should be done in the field (Havelka & Hardy, 1976). Polyethylene bags (4 ml, 150 by 225 mm) make suitable incubation chambers. Each bag should be marked with a sealing site that will result in a predetermined known volume suitable for the size of the nodulated root sample when the bag is filled with the incubation gas mixture.

Roots should not be dug from excessively wet soils or washed, because excess moisture on the nodule surface results in depressed rates of acetylene reduction (Hardy et al., 1968; LaRue & Kurz, 1973; Minchin & Pate, 1975; Schwinghamer et al., 1970; Van Straten & Schmidt, 1975).

Plants should be carefully dug to minimize nodule loss, excess soil carefully removed, the roots cut off from the tops, and the excised roots carefully packed into the bag. The bag should be manually compressed to eliminate most of the air and heat sealed (Model 210, Quick Seal Impulse Sealer, National Instrument Co., Baltimore, Md.).

The bag is then inflated to a constant pressure with a 1:9 mixture of acetylene (the acetylene used should be the highest grade purity available) and Ar/O_2 (80:20). Purification of acetylene uses traps of soda-asbestos to remove CO_2, water to remove acetone, concentrated H_2SO_4 to remove PH_3, and H_2S and water to remove SO_2. The CO_2 trap may be omitted in nodule assays. The 1:9 mixture of acetylene and Ar/O_2 may be prepared with a gas proportionator (Gas Proportionator 665 with 602 and 603 tubes, Matheson Gas Products, E. Rutherford, N.J.). The addition of 0.1 atm of acetylene to the incubation chamber produces the equivalent saturation of nitrogenase as 0.8 atm of N_2. However, because the acetylene concentration used greatly exceeds its K_m value and because acetylene is a noncompetitive inhibitor of N_2 reduction (Hwang et al., 1973), air may be used with little error. The bag may be filled with air and 10%, by volume, of acetylene added.

After a short, 30-min incubation period to prevent any possibility of derepression of nitrogenase under acetylene (Hardy & Holsten, 1977), an aliquot of gas is transferred into an evacuated blood serum sample tube (3200 TGS Vacutainer tube, Becton Dickinson, Rutherford, N.J.) for storage until analysis for acetylene. Needle holders and receivers facilitate gas sampling (Vacutainer Needle Holder Combination 3200 HN, Beckton Dickinson, Rutherford, N.J.). The entire operation should be conducted at the temperature etc. appropriate for the sample.

The analysis for ethylene and acetylene is done on a flame ionization gas chromatograph using a variety of column packings (Hardy et al., 1973). Additional details of analysis have been discussed in the preceding section.

b. $N_2(C_2H_2]$ Fixation Profile—Profiles of N_2 fixation activity in field-grown soybeans, peas, peanuts, and other legumes have been measured and have been found to be similar (Hardy & Havelka, 1977). Plants were harvested throughout the complete growth cycle, and N_2 $[C_2H_2]$-fixing activity was measured. Activity was expressed as $N_2[C_2H_2]$ fixed per plant per day, as well as $N_2[C_2H_2]$ fixed per plant. Another useful measurement is nodule species $N_2[N_2H_2]$ fixation activity, which relates units of $N_2[C_2H_2]$ fixed per unit nodule weight, expressed either on a fresh weight or dry weight basis.

More than 90% of the N_2 is fixed during the last half of the growth cycle, the reproductive growth stage (Hardy et al., 1971; Weber et al.,

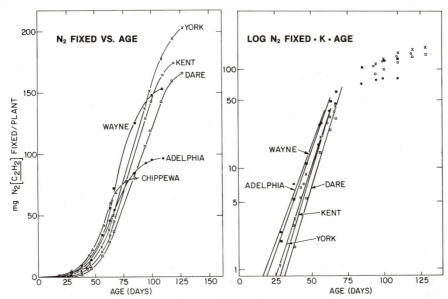

Fig. 4—Age: The $N_2[C_2H_2]$-fixing activity profiles for several varieties of *Glycine max* of various maturity dates and flowering characteristics (Hardy et al., 1973).

1971). Semilogarithmic plots show an exponential phase in the development of N_2-fixing activity and the amount of N_2 fixed. Three time frames are used to describe the N_2 fixation profile: (i) the initiation stage, or period when N_2-fixing activity begins; (ii) the exponential phase, the period when the amount of total N fixed per day increases by 7 to 10%, so that total N_2 fixed doubles every 7 to 10 days; and (iii) the age at termination of exponential phase when N_2 fixation slows. The total amount of N_2 fixed per season is an integrated value obtained from the seasonal profile. See Fig. 4 for N_2 fixation profiles of five soybean varieties.

There is a case of host/*Rhizobium* interaction where Clark soybeans inoculated with *R. japonicum* strain 35 form large normal-appearing nodules but fail to fix N_2 for about 6 weeks. Then, after this lag period, the plants begin to fix N_2 efficiently and grow normally (Vest et al., 1973).

Indeterminate varieties of soybeans can secure 25 to 50% of their N requirement from N_2 fixation, with most being fixed during the reproductive phase. Fixation of N_2 increases rapidly after flowering and ceases rapidly during seed filling (Weber et al., 1971). The shape of N_2 fixation profiles is determined by such factors as growth habit, i.e., determinate or indeterminate, maturity class, and environmental factors. Examination of N_2 fixation profiles reveals that early maturing varieties initiate N_2 fixation sooner (15–20 days) and terminate sooner (55–65 days) than later maturing varieties (Hardy et al., 1973). The delayed exponential phase of later maturing varieties allows them to fix up to twice as much N_2 as the early varieties. Determinate soybeans are reported to fix more N during the vegetative-flowering growth phase (up to 20%) and can fix as much as 75 to 90% of their total N requirements from N_2 (Israel, 1978).

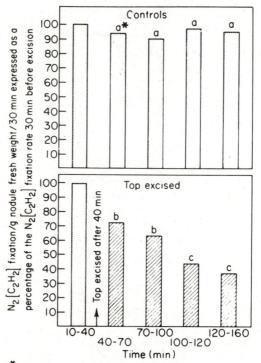

Fig. 5—The $N_2[C_2H_2]$ fixation measured by the intact plant-incubation pot assay over 3 hours and comparison with plants decapitated at 40 min (Hardy et al., 1977).

c. Intact Plant and in situ Field Assays—A formidable objection to the excised root $N_2[C_2H_2]$ fixation assay just described is the extremely large number of plants that must be evaluated over a time period in order to define the $N_2[C_2H_2]$ fixation profile. Another objection to the excised root technique is the significant reduction of $N_2[C_2H_2]$ fixation activity 30 min following decapitation, which may result in a 20 to 30% underestimation of $N_2[C_2H_2]$ fixation activity (Fig. 5), (Hardy et al., 1977).

Intact plant assays eliminate the large requirement for plants, since repeated assays may be done over the growth cycle on the same plant. Successive $N_2[C_2H_2]$ fixation assays of intact plants may more accurately reflect the N_2 fixation status of the plants (Criswell et al., 1976a).

The entire plant system may be enclosed in the incubation chamber (Fishbeck et al., 1973; LaRue & Kurz, 1973; Lie, 1971; Sinclair, 1973). More generally, however, in plant growthroom and greenhouse N_2 fixation studies, the pot in which the plant is growing is sealed, and the rhizosphere is subjected to incubation gases. Meanwhile, the aerial portion of the plant may be subjected to a modified environment and different measurements, such as photosynthesis and transpiration, made (Fig. 6) (Quebedeaux et al., 1975).

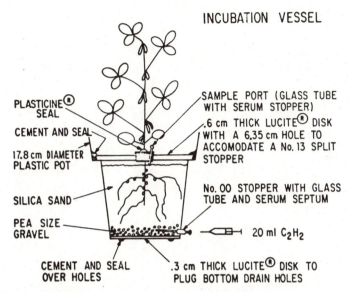

INCUBATION VESSEL

PLASTICINE® SEAL

CEMENT AND SEAL

17.8 cm DIAMETER PLASTIC POT

SILICA SAND

PEA SIZE GRAVEL

CEMENT AND SEAL OVER HOLES

SAMPLE PORT (GLASS TUBE WITH SERUM STOPPER)

.6 cm THICK LUCITE® DISK WITH A 6.35 cm HOLE TO ACCOMODATE A No. 13 SPLIT STOPPER

No. 00 STOPPER WITH GLASS TUBE AND SERUM SEPTUM

20 ml C_2H_2

.3 cm THICK LUCITE® DISK TO PLUG BOTTOM DRAIN HOLES

Fig. 6—The pot technique for measurement of N_2 fixation by intact soybean plants. This technique allows separate treatments of the aerial and root parts of the plant (Quebedeaux et al., 1975) and has been used in our laboratory to assess effects of an altered CO_2/O_2 ratio around canopy and an altered pO_2 around rhizosphere (Criswell et al., 1976a).

A continuous flow field technique, in situ, has been described in which metal cylinders are driven into the soil to enclose the nodulated root zone (Balandreau & Dommergues, 1973; Hardy et al., 1977).

Ideally, determination of N_2-fixing activity would involve accurate monitoring of plants growing in the natural field situation over the entire growing season with minimum perturbation of the environment and metabolism of the plant.

An open cylinder, whose diameter and length are dependent on the width and depth of the prospective nodulated root zone, is buried in the soil (Fig. 7). Plants are seeded into this cylinder. The top of the metal or plastic cylinder is fitted with a method for sealing, with openings for gas inlet and exit ports. A mixture of subsaturating (0.2%) acetylene and air is continually forced through metal probes whose exit ports are located well below the nodulated root zone. A proportion of the introduced gas flows upward through the nodulated root zone. An amount of acetylene proportional to the N_2-fixing activity of the nodules is reduced to ethylene. The flow out of the exit port is measured. A Michaelis-Menten relationship is used to determine total $N_2[C_2H_2]$ fixation rate. Relationships for calculation of $N_2[C_2H_2]$ fixation rates are described for this similar methodology (Hardy et al., 1977).

Sinclair (1978) reported on a system designed for in situ $N_2[C_2H_2]$ fixation measurements. Soybean plants grown in the chambers were not visually distinguishable from other field-grown plants. Exposures to low acetylene

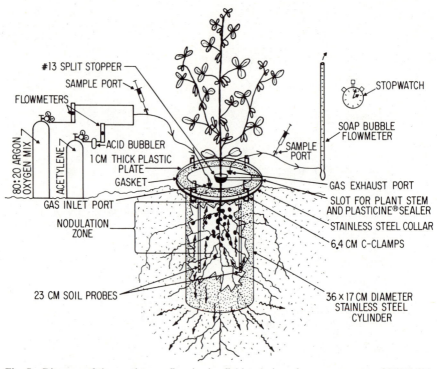

Fig. 7—Diagram of the continuous flow in situ field technique for measurement of $N_2[C_2H_2]$ fixation (Hardy et al., 1977).

concentrations for up to 2 weeks did not result in detectable visual or physiological changes. A computer controlled gas flow and recorded gas chromatograph analyses. Data from a single plant were consistent, but a large degree of plant-to-plant variability was found within one variety.

E. *Rhizobium* Classification

The family Leguminosae is composed of > 12,000 species of plants, of which < 100 species are used in food production. These belong chiefly to the subfamily Papilionatae, most of which have the ability to form N_2-fixing root nodule symbioses with bacteria of the genus *Rhizobium*. The ability to form nodules is a relatively specific interaction between a specific bacteria and groups of related host legumes. This specificity of interaction is the basis of a classification system of *Rhizobium* species and legume groups called cross-inoculation groups (Table 2). Groups of legumes that can be infected by a single *Rhizobium* species are classified in the same inoculation group. There are six cross-inoculation groups of single species and an additionl seventh group, the cowpea miscellany. The *Rhizobium* spp. associated with the cowpea miscellany cross-inoculate a wide variety of legumes from many different genera.

Table 2—Cross-inoculation groups of legumes and associated *Rhizobium* spp.

Group common name	*Rhizobium* species	*Rhizobium* growth rate	Host(s) nodulated
Soybean	*R. japonicum*	Slow	*Glycine* max.
Lupini	*R. lupini*	Slow	*Lupinus, Ornithopus* spp.
Clover	*R. trifolii*	Fast	*Trifolium* spp.
Peas and vetch	*R. leguminosarum*	Fast	*Pisum, Vicia, Lathurus, Lens* spp.
Bean	*R. phaseoli*	Fast	*Phaseolus vulgaris Phaseolus coccineus*
Alfalfa	*R. meliloti*	Fast	*Medicago, Melilotus, Trigonella*
Cowpea miscellany	*Rhizobium* spp.	Variable	*Vigna, Lespedeza, Archis, Stylosanthes, Desmodium, Cajanus, Crotalaria Pueraria*, etc.

Rhizobium species are further divided into classifications called strains and are numbered according to where they were isolated (Burton, 1979). Rhizobia are classified into cross-inoculation groups on the basis of the nodulation assay. The rhizobial isolate is tested on several legumes to determine into which grouping the isolate belongs.

Cross-inoculation grouping on the basis of susceptibility to nodulation by a specific species of *Rhizobium* has been an important tool of the inoculant manufacturer. However, within several of the cross-inoculation groups, particularly in lupin, cowpea, and soybean, and also in the clover and pea groups, promiscuity occurs. A further complication is illustrated by the alfalfa grouping (Vincent, 1974). One genus of this group is nodulated by virtually all of the strains in this group. The other two are nodulated by different strains, neither of which will inoculate the other.

Napoli et al. (1980) described four methods of identifying *Rhizobium* strains based on their surface properties: (i) fingerprint analysis of the sugar composition of *Rhizobium* lipopolysaccharides, (ii) reaction of *Rhizobium* with lipopolysaccharide-specific antibodies, (iii) reaction of *Rhizobium* with flagella-specific antibodies, and (iv) reaction with *Rhizobium* bacteriophage.

Serology, a method of identifying and characterizing the various *Rhizobium* strains, is used to study competition among *Rhizobium* strains. Isolation of *Rhizobium* is not required, because antisera are added to crushed nodule suspensions. This technique has been used successfully to identify *Rhizobium* strains in large-scale field experiments.

Highly efficient strains of *Rhizobium* have been isolated and cultured. Unfortunately, these efficient strains cannot compete with native populations in the soil. It has been reported that an inoculum rate of 1,000 times the native population would be needed to infect 50% of the nodules (Weaver & Frederick, 1974). Clearly, the apparent competitive advantage of native rhizobia must be understood and overcome before the development of super-efficient N_2-fixing strains can become a useful tool in agriculture.

F. Legume/*Rhizobium* Interactions

The family Leguminosae comprises over 12,000 species, but < 100 are used for food production. Of these, grain legumes (soybeans, peas, dry beans [*Phaseolus* spp.], and peanuts) are the most economically important. However, the forages also form an important economic group.

Legumes are large N users. The production of a bushel of soybeans requires approximately 1.82 kg (4 pounds) of N in the seed and another 0.91 kg (2 pounds) in the vegetative portion of the plant (Havelka & Hardy, 1976). Biological N_2 fixation can account for 25 to 80% of this required N. On a world-wide basis, this presents a great potential for conservation of the fossil fuel energy used to manufacture N fertilizers.

Burton (1976) reported that in the United States, soybeans accounted for 80% of the inoculant sold. Forage legumes accounted for 14% of the bacteria sold, and the common grain legumes, field beans (*Phaseolus vulgaris*), mung beans (*Vigna radiata* L. Wilczek var. radiata), peanuts, peas, and broadbeans (*Vicia faba* L.) accounted for < 1% of the inoculant sold.

Legume plant breeders are interested in producing new high-yielding varieties through the incorporation of desirable traits into new genotypes. Unfortunately, the breeding processes are not usually coupled with measurements of N_2-fixing ability or compatibility with highly efficient and ubiquitous N_2-fixing strains. Thus, when new varieties of legume species are developed, efficient compatible species of *Rhizobium* must be found. Plants vary genetically in their affinity for *Rhizobium* species, with some plant varieties being totally resistant toward infection and nodulation.

There are three main possibilities in the host plant/*Rhizobium* interaction: (i) nonnodulation vs. nodulation, (ii) ineffective vs. effective nodulation, and (iii) inefficient N_2-fixing activity vs. efficient N_2-fixing activity.

In plants that are ineffectively nodulated, the completion of the nodule-forming process has been interrupted. The nodules are small with white, green, or pink interiors. The host plant shows typical N deficiency symptoms.

Certain plants may be effectively nodulated, but the N_2-fixing system does not efficiently fix N_2. The roots are well infected with large normal-appearing nodules. The insides of the nodules are white, green, or slightly pink. Only nodules with slightly pink interiors are able to fix small amounts of N_2. The host plants appear chlorotic and exhibit typical N deficiency symptoms.

Plants that are effectively nodulated with efficient N_2-fixing bacteria show good growth characteristics with a dark green foliage color. The tap roots are heavily clustered with large nodules that, upon being split open, are a pink to beefsteak color.

In soybeans, the nodulating patterns of *Rhizobium* strains have been classified into two categories (Wright, 1925): Type A, those strains that cause nodulation clustered around the tap root; and Type B, those strains that induce nodules on lateral roots. The Type A strains were found to be associated with effective host/strain combinations, whereas Type B nodules were usually associated with ineffective host/strain combinations.

Nodule mass (Nutman, 1959) is a more important criterium than nodule number and is related to the host and not the bacterium. However, plants inoculated with effective strains showed variations in nodule number and mass that were not correlated with yield.

In the case of legume species, genetic differences exist among varieties in their ability to form efficient symbiotic relationships with various strains of *Rhizobium*. *Rhizobium* strains vary in their ability to infect the host. Several strains may be able to infect the host, but the most competitive is not necessarily and, unfortunately, not usually the most effective.

The ability to infect a host in the presence of other strains is termed *competitiveness*. Competitiveness is the ability of a *Rhizobium* strain to almost exclusively infect and nodulate roots of host plants in the presence of other highly infective *Rhizobium* strains.

G. Seed Inoculant Technology

Inoculation of legume seeds with rhizobia began soon after Hellriegel's (1886) discovery that nodulated legumes fix N_2, and Beijerinck's (1888) isolation of rhizobia as the causal agent in the root nodules. Use of laboratory produced cultures for legume seed inoculation began soon after this, but these procedures were not immediately successful. With time, different carriers for the rhizobia were developed, with pulverized peat coming into use in the 1920's. "The selection of effective strains of rhizobia for specific leguminous species was one of the key developments and continues to be one of the major responsibilities of the inoculant manufacturer today" (Burton, 1979). For further information on the history and development of the inoculant industry, see Fred et al. (1932), Burton (1967), Date and Roughley (1977), and Burton (1979).

1. STRAIN SELECTION

The responsibility of the inoculant manufacturer is to supply the farmer with high quality inoculum with several required characteristics. The primary requirement is selection of the appropriate strains.

The strains selected should form efficient N_2-fixing nodules on the roots of a range of species and varieties of host plants and supply a large proportion of their N requirements. They should be able to form a large proportion of efficient N_2-fixing nodules on roots of their host plants in the presence of highly competitive, but inefficient bacteria. They should be able to nodulate host plants over a large range of soil temperatures, pH, withstand antagonistic soil microorganisms, and persist in the soil after initial inoculation. A further benefit would be the ability to fix large quantities of N_2 in the presence of large amounts of soil N. Strains of *Rhizobium* that have the ability to infect, nodulate, and effectively fix N_2 in soils containing high residual N are desirable from an energy conservation standpoint. Some such strains have been identified (Burton, 1976), however their performance is not consistent from year to year. The reasons for this inconsistency are not clear. The interaction between N_2 fixation and soil N may be a host problem (Gibson et al., 1977).

2. CULTURE OF RHIZOBIA

The purpose of culturing rhizobia is to provide a high-count broth of effective rhizobia suitable for preparing a strong inoculant capable of quickly affecting vigorous nodulation.

Rhizobia are divided into two groups based on rapidity of growth. Slow growers require about twice the time to attain comparable growth as fast growers, i.e., 8 to 10 days vs. 3 to 5 days. Fast growers include cultures of *R. meliloti, R. trifolii, R. leguminosarum,* and *R. phaseoli.* Slow growers include *R. japonicum, R. lupini,* and cowpea miscellany. Growth rates may be narrowed somewhat by judicious manipulation of the required nutrients in the growth media.

A basic yeast extract is used as the culture media, and sucrose is the C source. More rapid growth of rhizobia is obtained with vigorous aeration. Since rhizobia are slow growing, high inoculum levels of the culture media are desirable.

Equipment used in growing rhizobia varies from small bottles to large automated fermenters equipped with mechanical agitators and aerators (Date & Roughley, 1977). Aeration should provide 1 liter of air/min for each 20 liters of medium. Most rhizobia grow best at temperatures of 28 to 30°C. A cell population of 4 or 5×10^9 ml can be achieved in 96 hours with a 1.0% inoculum with a good medium and proper aeration. Increasing the inoculum can substantially decrease the incubation time. Broth culture for preparing inoculum should contain a minimum of 1×10^9 viable rhizobia/ml.

3. INOCULANT MANUFACTURE

Rhizobia should increase in number 10-fold in 3 to 5 weeks in a good carrier medium. A carrier medium should be nontoxic, absorbant, and easily milled, and should be sterilized, should adhere to seed, and should be available in good supply at moderate cost.

Peat is the most widely used carrier medium. Peat characteristics vary widely according to origin, thus tests of its suitability must be made on each different source used. Demilco peat is commonly used in the United States. The wet peat is screened to remove foreign material, shredded, and flash-dried by hot air from 60% moisture to 8 or 9% in a few seconds. The hot peat is run through a hammer mill to the desired particle size and then bagged until use. The peat must be rapidly heated, dried, and cooled so that degradation does not occur. The normally acid peat must be neutralized to a pH of 6.5 to 7.0 with $CaCO_3$.

The peat carrier is sterilized to assure good growth and survival of bacteria in the carrier medium. The death rate of rhizobia is high in unsterilized media. Methods of sterilization vary with the facilities available, type of packaging, and number of cultures being prepared.

Sterilization methods include flash-drying, autoclaving, gamma radiation, and chemical. Flash-drying is used in the United States, and results are as satisfactory as autoclaving peat for 4 hours at 121°C. Gamma radiation is the most convenient. If radiation facilities are not available, autoclaving is the only recommended alternative.

The inoculum is prepared by spraying a 1×10^9 rhizobia/ml broth–$CaCO_3$ suspension over pulverized, heat-treated peat as it is being mixed. The addition of finely ground $CaCO_3$ neutralizes the acid in the broth and assures a final pH of 6.8 in the peat-base inoculum.

The inoculant is spread in thin layers and allowed to incubate for 48 to 72 hours to allow for uniform moisture dispersal and the heat of wetting to disperse.

In addition to maintenance of optimal moisture levels, proper attention must be paid to aeration and storage temperatures. The mixture is then fine milled and packaged in polyethylene bags. When the packaged inoculum is kept at 25°C, the rhizobia reach their peak numbers in 2 to 3 or 4 to 5 weeks depending on their growth group. After this period, storage at 4°C or lower will maintain a high count of viable rhizobia for 12 months or longer. Without refrigeration, the rhizobia do not maintain their viability for long periods of time.

There should be certain standards for an inoculum relating to the number of viable rhizobia per seed for effective inoculation. These numbers would be low under controlled conditions in the greenhouse or laboratory but much higher in the field. Obviously, the standards must be related to the rhizobial count that is reasonable to obtain under present commercial techniques.

H. Inoculant Application

Two methods of applying inoculum are employed. The direct method involves coating the seed with the rhizobia; the indirect method involves sowing inoculated seed and concomitantly applying the rhizobia to the soil by some other means.

Direct inoculation consists of slurry inoculation and pelleting. Slurry inoculation consists of applying a water suspension of the inoculum or applying the inoculum into moistened seed and mixing thoroughly. However, as the seed dries, the inoculum tends to fall off. This can be avoided to some extent by adding a sticking agent, such as gum arabic or cellulose compounds. Care must be exercised to avoid use of agents toxic to rhizobia.

With the pelleting method, the legume seeds are coated with a mixture of inoculum and various forms of calcium carbonate, dolomite, clay minerals, phosphates, titanium dioxide, soil, humus, talc, and activated charcoal. Various glues are used as adhesives to coat the seed with the pelleting mixture.

In some cases, direct application of inoculum to the seed is not the best means of inoculation. Some species of legumes lift the seed coat out of the ground, thus separating rhizobia from legume roots. Certain soil conditions may dictate the use of strong chemical treatments applied to the seed that would kill the rhizobia, or it may be necessary to apply the bacteria in amounts greater than can be applied on the surface of the seed alone. This may involve very small seed or cases where a high population of undesirable rhizobia present severe competitive problems. Slurry inoculation of fragile seed such as peanuts may decrease germination. Also, the new air planters depend on smooth untreated seed for successful operation.

In general, soil inoculants allow a much greater quantity of bacteria to be introduced into the soil than could be implanted on the seed. This can be important in promoting prompt, effective nodulation. Furthermore, soil-applied inoculants are not affected by toxic fungicides applied with the seed. Soil inoculants are easily applied with delivery systems readily obtainable for most makes of planting equipment. Finally, soil inoculants may be applied to growing legumes that have not nodulated.

There are three types of soil-applied inoculants: the granular peat-base material, the inert cone coated with peat-base inoculants and clays, and the frozen *Rhizobium* culture.

The granular base material is designed for metering through equipment used for application of granular fungicides or herbicides at 4 to 5 kg/ha. When this method is used, large quantities of rhizobia in peat granules are metered into the seed furrow into close proximity to the seed by means of an applicator box and delivery system other than on the seed.

The inert-core, peat-base inoculants are mixed with the planting seed. The frozen liquid concentrate is thawed, diluted with water, and metered into the seed furrow at 15 to 30 liters/ha using a liquid fertilizer attachment.

Under most conditions, the use of an inoculant slurry to adequately coat seed with viable rhizobia before planting is sufficient for good nodulation. Combination seed treatments are sometimes employed when fungicide and inoculation (and perhaps Mo) are necessary.

To plant legumes in acidic soils, the seed is coated with an adhesive mixture of cellulose gums or gum arabic and peat inoculant. This mixture is then rolled in finely pulverized limestone until all the seed are coated. This is called a *lime pellet*.

Seed to be sown in neutral or slightly acidic soils that harbor high populations of native, ineffective rhizobia may receive multiple coatings of a gum arabic–inoculant slurry. This high inoculant rate per seed enables the introduced rhizobia to compete successfully with native rhizobia and to survive better in dry, hot soils following planting. Various experiments have been conducted to test the feasibility of adding nutrients to pelleted seed to stimulate either plant root growth or rhizobial growth and subsequent nodulation. Results of the experiments are not conclusive.

There is an increasing demand for the availability of preinoculated seed to save time at planting. This means that the seed needs to be inoculated with large counts of rhizobia, equal to the rhizobial count obtained with the "at planting inoculation" procedure, at the time the seed are planted. Successful methodology remains to be devised. Some method of seed pelleting probably offers the best avenue for seed preinoculation.

Several grain legumes, such as dry beans, chickpeas (*Cicer arietinum* L.), peanuts, and dry peas, occupy substantial acreages, but only a small portion of the acreage involved receives inoculation. The low benefits received do not encourage increased inoculation efforts. Research remains to be done to determine the causes of the extremely poor N_2 fixation by these crops.

IV. FUTURE APPLICATIONS

Knowledge of the symbiotic N_2-fixing system is increasing at a rapid pace. The prospects of extending N_2 fixation to cereal crops, such as wheat and corn, and other grasses is exciting. However, the main thrust of N_2 fixation should continue to unravel the intricacies of the legume-*Rhizobium* system. Much progress remains to be made in improving this system.

The problem of control of nitrogenase activity in soils possessing high amounts of fixed N_2 must be addressed. Native soil N should be conserved for non-N_2-fixing crops. The relative efficiencies of NO_3^- reduction and symbiotic N_2 fixation must be evaluated and compared. Because of fossil fuel considerations, N input from N_2 fixation may be the desired N source.

Energy in the form of photosynthate has been shown to limit N_2 fixation in the field. Further research is necessary to define the significance of the induction and control of enzymes of the NH_3 transport system in the rate limiting process. The ureide transport systems have only recently been proposed.

Some of the energy lost due to nitrogenase catalyzed proton reduction during N_2 fixation may be recaptured by a bacterial hydrogenase. Field experiments with an isogenic line of *Rhizobium,* with and without the H_2 uptake genes, should be evaluated to assess the significance of this system.

The development of superior legume/*Rhizobium* strain combinations that function in the laboratory and greenhouse will be a futile exercise unless they work in the field. The basis of specificity and the infection process is just beginning to be elucidated. As this knowledge increases, hopefully methods of ensuring the success of the superior strains in the field will be devised.

LITERATURE CITED

Albrecht, S. L., R. J. Maier, F. J. Hanus, A. Russell, W. Emerich, and H. J. Evans. 1979. Hydrogenase in *Rhizobium japonicum* increases nitrogen fixation by nodulated soybeans. Science 203:1255–1257.

Alexander, V. 1975. Nitrogen fixation by blue-green algae in polar and sub-polar regions. p. 1975–188. *In* W. D. P. Stewart (ed.) Nitrogen fixation by free-living micro-organisms. Cambridge University Press, New York.

Appleby, C. A., G. L. Turner, and P. K. Manicol. 1975. Involvement of oxyleghaemoglobin and cytochrome P-450 in an efficient oxidative phosphorylation pathway which supports nitrogen fixation in *Rhizobium.* Biochim. Biophys. Acta 387:461–474.

Bach, M. K., W. K. Magee, and R. H. Burris. 1958. Translocation of photosynthetic products to soybean nodules and their role in nitrogen fixation. Plant Physiol. 33:118–124.

Balandreau, J., and Y. Dommergues. 1973. Assaying nitrogenase (C_2H_2) activity in the field. Bull. Ecol. Res. Commun. (Stockholm) 17:247.

Balandreau, J. P., G. Rinaudo, M. M. Oumarov, and Y. R. Dommergues. 1976. Asymbiotic N_2 fixation in paddy soils. p. 611–628. *In* W. E. Newton and C. J. Nyman (ed.) Proceedings of the First International Symposium on Nitrogen Fixation, Vol. II, Pullman, Wash. 3–7 June 1974. Washington State University Press, Pullman.

Becking, J. H. 1975. Root nodules in non-legumes. p. 507–566. *In* J. G. Torrey and D. T. Clarkson (ed.) The development and function of roots. Academic Press, Inc., New York.

Becking, J. H. 1976. Nitrogen fixation in some natural ecosystems in Indonesia. p. 539–560. *In* P. S. Nutman (ed.) Symbiotic nitrogen fixation in plants. Cambridge University Press, New York.

Becking, J. H. 1977. Dinitrogen-fixing associations in higher plant other than legumes. p. 185–275. *In* R. W. F. Hardy and W. S. Silver (ed.) A treatise on dinitrogen fixation. Section III. Biology. John Wiley & Sons, Inc., New York.

Beijerinck, M. W. 1888. Die bakterien der papilionaceen knöllchen. Bot. Ztg. 46:726–804.

Bell, F., and P. S. Nutman. 1971. Experiments on nitrogen fixation by nodulated legumes. Plant Soil (Special Vol.), p. 231–264.

Bergersen, F. J. 1962. The effects of partial pressure of oxygen upon restoration and nitrogen fixation by soybean root nodule. J. Gen. Microbiol. 29:113–125.

Begersen, F. J. 1963. The relationship between hydrogen evolution, hydrogen exchange, nitrogen fixation, and applied oxygen tensions in soybean root nodules. Aust. J. Biol. Sci. 16: 669–680.

Bergersen, F. J. 1974. Formation and function of bacteriods. p. 473–498. *In* A. Quispel (ed.) The biology of nitrogen fixation. North-Holland Publishing Co., Amsterdam.

Bergersen, F. J., G. S. Kennedy, and W. Whitmann. 1965. Nitrogen fixation in the coralloid roots of *Macrozamia comminis* L. Johnson. Aust. J. Biol. Sci. 18:1135–1142.

Bisseling, R., R. C. VandenBos, and A. VanKammen. 1978. The effect of ammonium nitrate on the synthesis of nitrogenase and the concentration of leghemoglobin in pea root nodules induced by *Rhizobium leguminosarum*. Biochim. Biophys. Acta 539:1–11.

Bohlool, B. B., and E. L. Schmidt. 1974. Lectins: A possible basis for specificity in the *Rhizobium* legume root nodule symbiosis. Science 185:269–271.

Boland, M. J., K. J. F. Farnden, and J. G. Robertson. 1980. Ammonia assimilation in nitrogen-fixing legume nodules. p. 33–52. *In* W. E. Newton and W. H. Orme-Johnson (ed.) Nitrogen fixation, Vol. II. University Park Press, Baltimore.

Bond, G. 1976. The results of the IBP survey of root-nodule formation in non-leguminous angiosperms. p. 443–474. *In* P. S. Nutman (ed.) Symbiotic nitrogen fixation in plants. Cambridge University Press, New York.

Brill, W. J. 1980. Biochemical aspects of nitrogen fixation. Microbiol. Rev. 44:449–467.

Brock, T. D. 1973. Evolutionary and ecological aspects of the Cyanophytes. p. 487–500. *In* N. G. Carr and B. A. Whitton (ed.) The biology of the blue-green algae. University of California Press, Berkeley.

Brun, W. A. 1976. The relation of N_2 fixation to photosynthesis. p. 135–150. *In* L. D. Hill (ed.) World soybean research conference. The Interstate Printers and Publishers, Inc., Danville, Ill.

Buchanan, R. E., and N. E. Gibbons (ed.) 1974. Bergey's manual of determinative bacteriology. Williams & Wilkins Co., Baltimore. p. 52.

Bulen, W. A., R. C. Burns, and J. R. LeComte. 1965. Nitrogen fixation: Hydrosulfite as electron donor with cell-free preparations of *Azotobacter vinelandii* and *Rhodospirillum rubrum*. Proc. Natl. Acad. Sci. U.S.A. 53:532–538.

Bulen, W. A., and J. R. LeComte. 1966. The nitrogenase system from *Azotobacter*: Two enzyme requirements for N_2 reduction, ATP-dependent H_2 evolution and ATP hydrolysis. Proc. Natl. Acad. Sci. U.S.A. 56:979–986.

Burns, R. C., and R. W. F. Hardy. 1972. Purification of nitrogenase and crystallization of its Mo-Fe protein. Methods Enzymol. 24B:480–496.

Burns, R. C., and R. W. F. Hardy. 1975. Nitrogen fixation in bacteria and higher plants. Springer-Verlag New York, New York.

Burris, R. H. 1972. Nitrogen fixation—assay methods and techniques. Methods Enzymol. 24: 415–431.

Burris, R. H. 1974. Methodology. p. 9–31. *In* A. Quispel (ed.) The biology of nitrogen fixation. North-Holland Publishing Co., Amsterdam.

Burris, R. H. 1977. The energetics of N_2 fixation. p. 273–289. *In* A Mitsui et al. (ed.) Biological solar energy conversion. Academic Press, Inc., New York.

Burris, R. H. 1980. The global nitrogen budget-science or seance? p. 7–16. *In* W. E. Newton and W. H. Orme-Johnson (ed.) Nitrogen fixation, Vol. I. University Park Press, Baltimore.

Burris, R. H., and C. E. Miller. 1941. Application of N^{15} to the study of biological nitrogen fixation. Science 93:114–115.

Burris, R. H., and P. W. Wilson. 1957. Methods for measurement of nitrogen fixation. Methods Enzymol. 4:355–356.

Burton, J. C. 1967. *Rhizobium* culture and use. p. 1–33. *In* H. J. Peppler (ed.) Microbial technology. Van Nostrand Reinhold Co., New York.

Burton, J. C. 1976. Pragmatic aspects of the *Rhizobium*: leguminous plant associations. p. 429–446. *In* W. E. Newton and C. J. Nyman (ed.) Symposium on dinitrogen fixation. Washington State University Press, Pullman.

Burton, J. C. 1979. Rhizobium species. p. 29–54. *In* Microbial technology, Vol. I. 2nd ed. Academic Press, Inc., New York.

Callaham, D., P. DelTredici, and J. G. Torrey. 1978. Isolation and cultivation in vitro of the actinomycete causing root nodulation in *Comptonia*. Science 199:899–902.

Chen, J. S., J. S. Multani, and L. E. Mortenson. 1973. Structural investigation of nitrogenase components from *Clostridium pasteurianum* and comparison with similar components of other organisms. Biochim. Biophys. Acta 310:54–59.

Ching, T. M., S. Hedtke, S. A. Russell, and H. J. Evans. 1975. Energy state and dinitrogen fixation in soybean nodules of dark grown plants. Plant Physiol. 55:796–798.

Clauss, H., D. C. Mortimer, and P. R. Gorham. 1964. Timecourse study of translocation of the products of photosynthesis in soybean leaves. Plant Physiol. 34:269–273.

Cohen, E., Y. Okon, J. Kigel, I. Nur, and Y. Henis. 1980. Increase in dry weight and total nitrogen content in *Zea mays* and *Setaria italica* associated with nitrogen-fixing *Azospirillum* spp. Plant Physiol. 66:746–749.

Coventry, D. R., M. J. Trinick, and C. A. Appleby. 1976. Search for a leghaemoglobin-like compound in root nodules of *Trema cannabina Lour.* Biochim. Biophys. Acta 420:105–111.

Criswell, J. G., R. W. F. Hardy, and U. D. Havelka. 1976a. Nitrogen fixation in soybeans: Measurement techniques and examples of applications. p. 108–134. *In* L. D. Hill (ed.) World soybean research conference. The Interstate Printers and Publishers, Inc., Danville, Ill.

Criswell, J. G., U. D. Havelka, B. Quebedeaux, and R. W. F. Hardy. 1976b. Adaptation of nitrogen fixation by intact soybean nodules to altered rhizosphere pO_2. Plant Physiol. 58:622–625.

Date, R. A., and R. J. Roughley. 1977. Preparation of legume seed inoculants. p. 243–275. *In* R. W. F. Hardy and A. H. Gibson (ed.) A treatise on dinitrogen fixation. Section IV. agronomy and ecology. John Wiley & Sons, Inc., New York.

Dazzo, F. B. 1980. Determinants of host specificity in the *Rhizobium*-clover symbiosis. *In* W. E. Newton and W. H. Orme-Johnson (ed.) Nitrogen fixation, Vol. II. University Park Press, Baltimore.

Delwich, C. C. 1970. The nitrogen cycle. Sci. Am. 223:136–146.

Dilworth, M. J. 1966. Acetylene reduction by nitrogen fixing preparations from *Clostridium pasteurianum*. Biochim. Biophys. Acta 127:285.

Dilworth, M. J. 1980. Host and rhizobium contributions to the physiology of legume nodules. p. 3–31. *In* W. E. Newton, and W. H. Orme-Johnson (ed.) Nitrogen fixation, Vol. II. University Park Press, Baltimore.

Dilworth, M. J., and C. A. Appleby. 1977. Leghemoglobin and *Rhizobium* hemoproteins. p. 691–764. *In* R. W. F. Hardy et al. (ed.) A treatise on dinitrogen fixation. John Wiley & sons, Inc., New York.

Dixon, R. O. D. 1968. Hydrogenase in pea root nodule bacteroids. Arch. Mikrobiol. 62:272–283.

Dixon, R. O. D. 1972. Hydrogenase in legume root nodule bacteroids: Occurrence and properties. Arch. Mikrobiol. 85:193–201.

Döbereiner, J., and J. M. Day. 1975. Nitrogen fixation in the rhizosphere of tropical grasses. p. 39–56. *In* W. D. P. Stewart (ed.) Nitrogen fixation by free-living micro-organisms. Cambridge University Press, New York.

Döbereiner, J., and J. M. Day. 1976. Associative symbioses in tropical grasses: characterization of microorganisms and dinitrogen-fixing sites. p. 518–538. *In* W. E. Newton and C. J. Nyman (ed.) Proceedings of the First International Symposium on Nitrogen Fixation, Vol. II, Pullman, Wash. 3–7 June 1974. Washington State University Press, Pullman.

Donze, M., J. Haveman, and P. Schierick. 1972. Absence of photosystem 2 in heterocysts of the blue-green alga, *Anabaena*. Biochim. Biophys. Acta 256:157-161.

Dornhoff, G. M., and R. M. Shibles. 1970. Varietal differences in net photosynthesis of soybean leaves. Crop Sci. 10:42-45.

Eady, R. R., and J. R. Postgate. 1974. Nitrogenase. Nature (London) 249:805-810.

Emerich, D. W., T. Ruiz-Argueso, T. M. Ching, and H. J. Evans. 1979. Hydrogen-dependent nitrogenase activity and ATP formation in *Rhizobium japonicum* bacteroids. J. Bacteriol. 137:153-160.

Evans, H. J., and L. Barber. 1977. Biological nitrogen fixation for food and fiber production. Science 197:332-339.

Evans, H. J., D. W. Emerich, T. Ruiz-Argueso, R. J. Maier, and S. L. Albrect. 1980. Hydrogen metabolism in the legume-*Rhizobium* symbiosis. p. 69-86. *In* W. E. Newton and W. H. Orme-Johnson (ed.) Nitrogen fixation, Vol. II. University Park Press, Baltimore.

Fishbeck, K., H. J. Evans, and L. L. Boersma. 1973. Measurement of nitrogenase activity of intact legume symbionts *in situ* using the acetylene reduction assay. Agron. J. 65:429-433.

Fred, E. B., I. L. Baldwin, and E. McCoy. 1932. Root nodule bacteria and leguminous plants. University of Wisconsin Press, Madison.

Fried, M., and H. Broeshart. 1975. An independent measurement of the amount of nitrogen fixed by a legume crop. Plant Soil 43:707-711.

Fried, M., and V. Middlelboe. 1977. Measurement of amount of nitrogen fixed by a legume crop. Plant Soil 47:713-715.

Gallon, J. R., W. G. W. Kurz, and T. A. LaRue. 1975. The physiology of nitrogen fixation by a *Gloeocapsa* sp. p. 159-173. *In* W. D. P. Stewart (ed.) Nitrogen fixation by free living micro-organisms. Cambridge University Press, New York.

Gallon, J. R., T. A. LaRue, and W. G. W. Kurz. 1972. Characteristics of nitrogenase activity in broken cell preparations of the blue-green alga *Gloeocapsa* sp. LB795. Can. J. Microbiol. 18:329-332.

Gibson, A. H. 1976a. Recovery and compensation by nodulated legumes to environmental stress. p. 385-403. *In* P. S. Nutman (ed.) Symbiotic nitrogen fixation in plants. Cambridge University Press, New York.

Gibson, A. H. 1976b. Limitation to dinitrogen fixation by legumes. p. 400-428. *In* W. E. Newton and C. J. Nutman (ed.) Proceedings of the First International Symposium on Nitrogen Fixation, Vol. II, Pullman, Wash. 3-7 June 1974. Washington State University Press, Pullman.

Gibson, A. H. 1977. The influence of the environment and managerial practices on the legume-*Rhizobium* symbiosis. p. 393-450. *In* R. W. F. Hardy and A. H. Gibson (ed.) A treatise on dinitrogen fixation. Section IV. Agronomy and ecology. John Wiley & Sons, Inc., New York.

Gibson, A. H., W. R. Scowcroft, and J. R. Pagan. 1977. Nitrogen fixation in plants: An expanding horizon? p. 387-417. *In* W. H. Newton et al. (ed.) Recent developments in nitrogen fixation. Academic Press, Inc., New York.

Goodchild, D. J., and F. J. Bergersen. 1966. Electron microscopy of the infection and subsequent development of soybean nodule cells. J. Bacteriol. 92:204.

Grau, F. H., and P. W. Wilson. 1962. Physiology of nitrogen fixation by *Bacillus polymyxa*. J. Bacteriol. 83:490-496.

Hageman, R. V., and R. H. Burris. 1978. Nitrogenase and nitrogenase reductase associate and dissociate with each catalytic cycle. Proc. Natl. Acad. Sci. U.S.A. 75:2699-2702.

Halliday, J., and J. S. Pate. 1976. Symbiotic nitrogen fixation by coralloid roots of the Cyad *Macrozamia riedlei*. Physiological characteristics and ecological significance. Aust. J. Plant Physiol. 3:349-358.

Ham, G. E., R. J. Lawn, and W. A. Brun. 1976. Influence of inoculation, nitrogen fertilizers and photosynthetic source-sink manipulations on field-grown soybeans. p. 239-253. *In* P. S. Nutman (ed.) Symbiotic nitrogen fixation in plants. Cambridge University Press, New York.

Hardy, R. W. F., R. C. Burns, R. R. Hebert, R. D. Holsten, and E. K. Jackson. 1971. Biological nitrogen fixation: A key to world protein. Plant Soil (Special Vol.), p. 561-590.

Hardy, R. W. F., R. C. Burns, and R. D. Holsten. 1973. Applications of the acetylene-ethylene assay for measurement of nitrogen fixation. Soil Biol. Biochem. 5:47-81.

Hardy, R. W. F., J. G. Criswell, and U. D. Havelka. 1977. Investigations of possible limitations of nitrogen fixation by legumes: (1) methodology, (2) identification, and (3) assessment of significance. p. 451-467. *In* W. E. Newton et al. (ed) Nitrogen fixation. Academic Press, Inc., New York.

Hardy, R. W. F., and A. H. Gibson (ed.). 1977. A treatise on dinitrogen fixation. Section IV. Agronomy and ecology. John Wiley & Sons, Inc., New York.

Hardy, R. W. F., and U. D. Havelka. 1975. Nitrogen fixation research: A key to world food. Science 188:633–643.

Hardy, R. W. F., and U. D. Havelka. 1976. Photosynthate as a major factor in limiting nitrogen fixation by field-grown legumes with emphasis on soybeans. p. 421–439. In P. S. Nutman (ed.) Symbiotic nitrogen fixation in plants. Cambridge University Press, New York.

Hardy, R. W. F., and U. D. Havelka. 1977. Possible routes to increase the conversion of solar energy to food and feed by grain legumes and cereal grains (crop production): CO_2 and N_2 fixation, foliar fertilization, and assimilate partitioning. p. 299–322. In A. Mitsui et al. (ed.) Biological solar energy conversion, Academic Press, Inc., New York.

Hardy, R. W. F., U. D. Havelka, and P. G. Heytler. 1980. Nitrogen input with emphasis on N_2 fixation in soybeans. p. 57–71. In Proceedings of the World Soybean Research Conference II, Raleigh, N.C. 24–29 March 1979. Westview Press, Boulder, Co.

Hardy, R. W. F., U. D. Havelka, and B. Quebedeaux. 1976. Opportunities for improved seed yield and protein production: N_2 fixation, CO_2 fixation, and oxygen control of reproductive growth. p. 196–228. In Genetic improvement of seed protein. National Academy of Sciences, Washington, D.C.

Hardy, R. W. F., U. D. Havelka, and B. Quebedeaux. 1978. The opportunity and significance of alteration of ribulose 1,5-bisphosphate carboxylase activity in crop production. p. 165–178. In H. W. Siegelman and G. Hind (ed.) Photosynthetic carbon assimilation. Plenum Press, New York.

Hardy, R. W. F., and R. D. Holsten. 1972. Global nitrogen cycling: Pools, evolution, transformations, transfers, quantitation, and research needs. p. 87–132. In L. J. Guarraia and R. K. Ballentine (ed.) The aquatic environment: Microbial transformations and water management implications. Environmental Protection Agency, EPA 430/G-73-008, U.S. Government Printing Office, Washington, D.C.

Hardy, R. W. F., and R. D. Holsten. 1977. Methods for measurement of N_2 fixation, p. 451–486. In R. W. F. Hardy et al. (ed.) A treatise on dinitrogen fixation, John Wiley & Sons, Inc., New York.

Hardy, R. W. F., R. D. Holsten, E. K. Jackson, R. C. Burns. 1968. The acetylene-ethylene assay for N_2 fixation: Laboratory and field evaluation. Plant Physiol. 43:1185–1207.

Hardy, R. W. F., E. Knight, Jr., and A. J. D'Eustachio. 1965. An energy-dependent hydrogen evolution from dithionite in nitrogen-fixing extracts of Clostridium pasteurianum. Biochem. Biophys. Res. Commun. 20:539–544.

Hardy, R. W. F., and E. Knight, Jr. 1967. ATP-dependent reduction of azide and HCN by H_2-fixing enzymes of Azotobacter vinelandii and Clostridium pasteuranium. Biochim. Biophys. Acta 139:69–90.

Hardy, R. W. F., and W. S. Silver (ed.). 1977. A treatise on dinitrogen fixation. Section III. Biology, John Wiley & Sons, Inc., New York.

Harper, J. E. 1974. Soil and symbiotic nitrogen requirements for optimum soybean production. Crop Sci. 14:255–260.

Harper, J. E. 1976. Contribution of dinitrogen and soil or fertilizer nitrogen to soybean (Glycine max L. Merr.) production. p. 101–107. In L. D. Hill (ed.) World soybean research conference. The Interstate Printers and Publishers, Inc., Danville, Ill.

Harper, J. E. 1978. Nitrogen inputs into soybean and directions for future research. p. 11–12. In T. A. LaRue (ed.) Selecting and breeding legumes for enhanced nitrogen fixation. Boyce Thompson Institute at Cornell, Ithaca, N.Y.

Harper, J. E., and R. L. Cooper. 1971. Nodulation response of soybeans (Glycine max L. Merr.) to application rate and placement of combined nitrogen. Crop Sci. 11:438–440.

Havelka, U. D., and R. W. F. Hardy. 1976. Legume N fixation as a problem in carbon nutrition. p. 456–475. In W. E. Newton and C. J. Nyman (ed.) Proceedings of the First International Symposium on Nitrogen Fixation, Vol. II, Pullman, Wash. 3–7 June 1974. Washington State University Press, Pullman.

Hitch, C. J. B., and W. D. P. Stewart. 1973. Nitrogen fixation by lichens in Scotland. New Phytol. 72:509–524.

Hellriegel, H. 1886. Welche Stickstoffquellen stehen der Pflanze zu Gebote? Z. Ver. Ruebenzucker Ind. Dtsch. Reichs 38:863–877.

Hoch, G. E., K. C. Schneider, and R. H. Burris. 1960. Hydrogen evolution and exchange and conversion of N_2O to N_2 by soybean root nodules. Biochim. Biophys. Acta 37:273–279.

Huang, C. Y., J. S. Boyer, and L. N. Vanderhoef. 1975a. Acetylene reduction (nitrogen fixation) and metabolic activities of soybean having various leaf and nodule water potentials. Plant Physiol. 56:228–232.

Huang, C. Y., J. S. Boyer, and L. N. Vanderhoef. 1975b. Limitation of acetylene reduction (nitrogen fixation) by photosynthesis in soybean having various leaf and water potentials. Plant Physiol. 56:228–232.

Hume, D. J., and J. G. Criswell. 1973. Varietal differences in net photosynthesis of soybean leaves. Crop Sci. 13:519–524.

Hwang, J. C., C. H. Chen, and R. H. Burris. 1973. Inhibition of nitrogenase-catalyzed reduction. Biochim. Biophys. Acta 292:256–270.

Ishisuka, J. 1977. Function of symbiotically fixed nitrogen for grain production in soybean. p. 618–624. *In* Proceedings of the International Sem. Soil Environment Fertility Management in Intense Agriculture. Society of the Science of Soil and Manure, Japan, Tokyo.

Israel, D. 1978. Nitrogen fixation during soybean development and strategies for genetic selection. p. 12. *In* T. A. LaRue (ed.) Selecting and breeding legumes for enhanced nitrogen fixation. Boyce Thompson Institute at Cornell, Ithaca, N.Y.

Johnson, H. S., and D. J. Hume. 1973. Comparison of nitrogen fixation estimates in soybeans by nodule weight, leghemoglobin content, and acetylene reduction. Can. J. Microbiol. 19: 1165–1168.

Jordan, D. C., and E. H. Garrard. 1951. Studies on the legume root nodule bacteria. Can. J. Bot. 29:360–372.

Jurgenson, M. F., and C. B. Davey. 1968. Nitrogen fixing blue-green algae in acid forest and nursery soils. Can. J. Microbiol. 14:1179–1183.

Kamen, M. D., and H. Gest. 1949. Evidence for a nitrogenase system in the photosynthetic bacterium *Rhodospirillum rubrum*. Science 109:560.

Keilin, D., and Y. L. Wang. 1945. Haemoglobin in the root nodules of leghuminous plants. Nature (London) 155:227–229.

Kennedy, I. R., J. Riguaud, and J. C. Trinchaut. 1975. Nitrate reductase from bacteroids of *Rhizobium japonicum*: Enzyme characteristics and possible interaction with nitrogen fixation. Biochim. Biophys. Acta 397:24–35.

Keister, D. L. 1975. Acetylene reduction by pure cultures of *Rhizobia*. J. Bacteriol. 123:1265–1268.

Knowles, R. 1978. Free living bacteria. p. 25–40. *In* J. Döbereiner et al. (ed.) Limitations and potentials for biological nitrogen fixation in the tropics. Plenum Press, New York.

Kobayashi, M., E. Takahashi, and K. Kawaguchi. 1967. Distribution of nitrogen fixing microorganisms in paddy soils of south east Asia. Soil Sci. 104:113–118.

Koch, B., and H. J. Evans. 1966. Reduction of acetylene to ethylene by soybean nodules. Plant Physiol. 41:1748–1750.

Krieg, W. R., and J. J. Tarrand. 1978. Taxonomy of the root-associated nitrogen fixing bacterium *Spirillum lipoferum*. p. 317–333. *In* J. Döbereiner et al. (ed.) Limitations and potentials for biological nitrogen fixation in the tropics. Plenum Press, New York.

Kurz, W. G. W., and T. A. LaRue. 1975. Nitrogenase activity in rhizobia in absence of plant host. Nature (London) 256:407–408.

LaRue, T. A., and W. G. W. Kurz. 1973. Estimation of nitrogenase in intact legumes. Can. J. Microbiol. 19:304–305.

LaRue, T. A. 1977. The bacteria. p. 19–62. *In* R. W. F. Hardy, and W. S. Silver (ed.) A treatise on dinitrogen fixation. Section III. Biology. John Wiley & Sons, Inc., New York.

Lawn, R. J., and W. A. Brun. 1974. Symbiotic nitrogen fixation in soybeans. I. Effect of photosynthetic source-sink manipulation. Crop Sci. 14:22–25.

Lawrie, A. C., and C. T. Wheeler. 1974. The effects of flowering and fruit formation on the supply of photosynthetic assimilates to the nodules of *Pisum sativum* L. in relation to the fixation of nitrogen. New Phytol. 73:1119–1127.

Lie, T. A. 1971. Symbiotic nitrogen fixation under stress conditions. Plant Soil (Special Vol.), p. 118–127.

Lindstrom, E. S., S. R. Trove, and P. W. Wilson. 1950. Nitrogen fixation by the green and purple bacteria. Science 112:197–198.

Ljones, T., and R. H. Burris. 1972. ATP hydrolysis and electron transfer in the nitrogenase reaction with different combinations of the iron protein and the molybdenum-iron protein. Biochim. Biophys. Acta 275:93–101.

Mague, T. H., and R. H. Burris. 1972. Reduction of acetylene and nitrogen by field-grown soybeans. New Phytol. 71:275–286.

Mahon, J. D. 1977. Respiration and the energy requirement for nitrogen fixation in nodulated pea roots. Plant Physiol. 60:817–821.

Matsumoto, T., Y. Yamamoto, and M. Yatazawa. 1976. Role of root nodules in the nitrogen nutrition of soybeans. II. Fluctuation in allantoin concentration of the bleeding sap. J. Sci. Soil Manure Japan 47:463–469.

McClure, P. R., and D. W. Israel. 1979. Transport of nitrogen in the xylem of soybean plants. Plant Physiol. 64:411–416.

mcClure, R. R., D. W. Israel, and R. J. Volk. 1980. Evaluation of the relative ureide content of xylem sap as an indicator of N_2 fixation in soybeans. Plant Physiol. 66:720–725.

McComb, J. A., J. Elliot, and M. J. Dilworth. 1975. Acetylene reduction by *Rhizobium* in pure culture. Nature (London) 256:409–410.

Menna, M. E. di. 1971. The mycroflora of leaves of pasture plants in New Zealand. p. 159–174. *In* T. F. Preece and C. H. Dickinson (ed.) Ecology of leaf surface micro-organisms. Academic Press, Inc., New York.

Millbank, J. W. 1974. Associations with blue-green algae. p. 238–264. *In* A. Quispel (ed.) The biology of nitrogen fixation. North-Holland Publishing Co., Amsterdam.

Minchin, F. R., and J. S. Pate. 1975. Effects of water, aeration, and salt regime on nitrogen fixation in a nodulated legume-definition of an optimum root environment. J. Exp. Bot. 26:60–69.

Minchin, F. R., and J. S. Pate. 1973. The carbon balance of a legume and the functional economy of its root nodules. J. Exp. Bot. 24:259–271.

Mortensen, L. E. 1966. Components of cell-free extract of *Clostridium pasteurianum* required for ATP-dependent H_2 evolution from dithionite and for N_2 fixation. Biochim. Biophys. Acta 127:18–25.

Mortenson, L. E., and R. N. F. Throneley. 1979. Structure and function of nitrogenase. Annu. Rev. Biochem. 48:387–418.

Mostafa, M. A., and M. Z. Mahmond. 1951. Bacterial isolates from root nodules of Zygophyllaceae. Nature (London) 167:446–447.

Moustafa, E. 1969. Use of acetylene reduction to study the effect of nitrogen fertilizer and defoliation on nitrogen fixation by field-grown white clover. N.Z. J. Agric. Res. 12:691–696.

Mulder, E. G. 1975. Physiology and ecology of free living, nitrogen fixing bacteria. p. 3–28. *In* W. D. P. Stewart (ed.) Nitrogen fixation by free living micro-organisms. Cambridge University Press, New York.

Mulder, E. G., and S. Brotonegro. 1974. Free-living heterotrophic nitrogen fixing bacteria. p. 37–85. *In* A. Quispel (ed.) The biology of nitrogen fixation. North Holland Publishing Co., Amsterdam.

Napoli, C., R. Sanders, R. Carlson, and P. Albersheim. 1980. Host-symbiont interactions: Recognizing *Rhizobium*. p. 189–203. *In* W. E. Newton and W. H. Orme-Johnson (ed.) Nitrogen fixation, Vol. II. University Park Press, Baltimore.

Nutman, P. S. 1956. The influence of the legume on root-nodule symbiosis. Biol. Rev. Cambridge Philos. Soc. 31:109–151.

Nutman, P. S. 1959. Sources of incompatability affecting nitrogen fixation in legume symbiosis. Symp. Soc. Exp. Biol. 13:42–58.

Nutman, P. S. (ed.). 1976. Symbiotic nitrogen fixation in plants. Cambridge University Press, New York.

Oghoghorie, C. G. O., and J. S. Pate. 1971. The nitrate stress syndrome of the nodulated field pea (*Pisum arvense* L.). Plant Soil (Special Vol., p. 185–202.

Pagan, J. D., J. J. Child, W. R. Scowcroft, and A. H. Gibson. 1975. Nitrogen fixation by *Rhizobium* cultured on a defined medium. Nature (London) 256:406–407.

Pate, J. S., J. Walder, and W. Wallace. 1965. Nitrogen-containing compounds of the shoot system of *Pisum arvense* L. Ann. Bot. New Ser. 29:475–493.

Pate, J. S., and W. Wallace. 1964. Movement of assimilated nitrogen from the root system of the field pea (*Pisum arvense* L.). Ann. Bot. New Ser. 28:83–99.

Peters, G. A., R. E. Toia, Jr., N. J. Levine, and D. Raweed. 1978. The *Azolla–Anabaena Azollae* relationship. VI. Morphological aspects of the association. New Phytol. 80:583–593.

Peters, G. A., T. B. Ray, B. C. Mayne, and R. E. Toia, Jr. 1980. *Azolla-Anabaena* association: Morphological and physiological studies. p. 293–309. *In* W. E. Newton and W. H. Orme-Johnson (ed.) Nitrogen fixation, Vol. II. University Park Press, Baltimore.

Pfennig, N. 1978. General physiology and ecology of the photosynthetic bacteria. p. 3–18. *In* R. K. Clayton and W. R. Sistrom (ed.) The photosynthetic bacteria. Plenum Press, New York.

Phelps, A. S., and P. W. Wilson. 1941. Occurrence of hydrogenase in nitrogen fixing organisms. Proc. Soc. Exp. Biol. Med. 47:473–476.

Phillips, D. A. 1980. Efficiency of symbiotic nitrogen fixation in legumes. Annu. Rev. Plant Physiol. 31:29–49.

Phillips, D. A., K. D. Newell, S. A. Hassell, and C. S. Felling. 1976. The effect of CO_2 enrichment on root nodule development and symbiotic N_2 reduction in *Pisum sativum* L. Am. J. Bot. 63:356–362.

Postgate, J. 1974. Prerequisites for biological nitrogen fixation in free-living heterotrophic bacteria. p. 663–686. *In* A. Quispel (ed.) The biology of nitrogen fixation. North-Holland Publishing Co., Amsterdam.

Preece, T. F., and C. H. Dickinson. 1971. Ecology of leaf surface microorganisms. Academic Press, Inc., New York.

Quebedeaux, B., U. D. Havelka, K. Livak, and R. W. F. Hardy. 1975. Effect of altered pO_2 in the aerial part of soybean on symbiotic N_2 fixation. Plant Physiol. 56:761–764.

Quispel, A. (ed.). 1974. The biology of nitrogen fixation. North-Holland Publishing Co., Amsterdam.

Rigaud, J., R. J. Bergersen, G. L. Turner, and R. M. Daniel. 1973. Nitrate-dependent anaerobic acetylene reduction and nitrogen fixation by soybean bacteroids. J. Gen. Microbiol. 77:137–144.

Rippka, R. A., A. Nielson, R. Kienesaura, and G. Cohen-Bazire. 1971. Nitrogen fixation by unicellular blue-green algae. Arch. Mikrobiol. 76:341–348.

Riviere, J. 1960. Etude de la rhizosphere du ble. Ann. Agron. 11:397–440.

Roponen, I. 1970. The effect of darkness on the leghemoglobin content and amino acid levels of root nodules of pea plants. Physiol. Plant 23:452–460.

Rouatt, J. W., and H. Katznelson. 1961. A study of the bacteria on the root surface and in the rhizosphere soil of crop plants. J. Appl. Bacteriol. 24:164–171.

Rovira, A. D., and C. B. Davey. 1971. Biology of the rhizosphere. p. 153–204. *In* E. W. Carson (ed.) The plant root and its environment. The University Press of Virginia, Charlottesville.

Ruinen, J. 1971. The grass sheath as a site for nitrogen fixation. p. 567–579. *In* T. F. Preece and C. H. Dickinson (ed.) Ecology of leaf surface micro-organisms. Academic Press, Inc., New York.

Ruinen, J. 1975. Nitrogen fixation in the phyllosphere. p. 85–100. W. D. P. Stewart (ed.) Nitrogen fixation by free-living micro-organisms. Cambridge University Press, New York.

Ryle, G. J. A., C. E. Powell, and A. J. Gordon. 1977. p. 321. *In* D. O. Hall et al. (ed.) Fourth International Congress of Photosynthesis, Reading. 4–9 Sept. 1977. Biochemical Society, London.

Sabet, Y. S. 1946. Bacterial root nodules in the Zygophyclaceae. Nature (London) 157:656–657.

Schöllhorn, R., and R. H. Burris. 1966. Study of intermediates in nitrogen fixation. 50th Annual Meeting, Atlantic City, N.J. Fed. Proc. 25:710.

Schubert, K. R., J. A. Engelke, S. A. Russell, and H. J. Evans. 1977. Hydrogen reactions of nodulated leguminous plants. I. Effect of rhizobial strain and plant age. Plant Physiol. 60:651–654.

Schubert, K. R., and H. J. Evans. 1976. Hydrogen evolution: A major factor affecting the efficiency of nitrogen fixation in nodulated symbionts. Proc. Natl. Acad. Sci. U.S.A. 73:1207–1211.

Schwinghamer, E. A., H. J. Evans, and M. D. Dawson. 1970. Evaluation of effectiveness of mutant strains of rhizobium by acetylene reduction relative to other criteria of N_2 fixation. Plant Soil 33:192–212.

Shivashankar, K., K. Vlassak, and J. Livens. 1976. A comparison of the effect of straw incorporation and CO_2 enrichment on the growth, nitrogen fixation, and yield of soybeans. J. Agric. Sci. 87:181–185.

Silver, W. S. 1977. Foliar associations in higher plants. p. 153–184. *In* R. W. F. Hardy and W. S. Silver (ed.) A treatise on dinitrogen fixation. Section III. Biology. John Wiley & Sons, Inc., New York.

Silver, W. S., and R. W. F. Hardy. 1976. Newer developments in biological dinitrogen fixation of possible relevance to forage production. p. 1–36. *In* C. S. Hoveland (ed.) Biological N fixation in forage-livestock systems. ASA Spec. Pub. No. 28, Madison, Wis.

Silverstein, R., and W. A. Bulen. 1970. Kinetic studies of the nitrogenase catalyzed hydrogen evolution and nitrogen reduction reactions. Biochemistry 9:3809–3815.

Silvester, W. B. 1976. Ecological and economic significance of the nonlegume symbiosis. p. 489-506. *In* W. E. Newton and C. J. Nyman (ed.) Proceedings of the First International Symposium on Nitrogen Fixation, 3-7 June 1974. Washington State University Press, Pullman.

Silvester, W. B. 1977. Dinitrogen fixation by plant associations excluding legumes. *In* R. W. F. Hardy and A. H. Gibson (ed.) A treatise on dinitrogen fixation. Section IV. Agronomy and ecology. John Wiley & Sons, Inc., New York.

Silvester, W. B., and D. R. Smith. 1969. Nitrogen fixation by *Gunnera-Nostoc* symbiosis. Nature (London) 224:1231.

Sinclair, A. G. 1973. Non-destructive acetylene reduction assay of nitrogen applied to white clover plants growing in soil. N.Z. J. Agric. Res. 16:263-270.

Sinclair, T. R. 1978. Measurements of nitrogen fixation rates in soybean by *in situ* acetylene reduction. p. 7-8. *In* T. A. LaRue (ed.) Selection and breeding legumes for enhanced nitrogen fixation. Boyce Thompson Institute at Cornell, Ithaca, N.Y.

Singh, R. N. 1961. Role of blue-green algae in the nitrogen economy of Indian agriculture. Indian Council of Agricultural Research, New Delhi.

Sloger, C., D. Bezdicek, R. Milberg, and N. Boonkerd. 1975. Seasonal and diurnal variations in $N_2[C_2H_2]$-fixing activity in field soybeans. p. 271-284. *In* W. D. P. Stewart (ed.) Nitrogen fixation in free-living micro-organisms. Cambridge University Press, New York.

Small, J. G., and O. A. Leonard. 1969. Translocation of C^{14}-labelled photosynthate in nodulated legumes as influenced by nitrate-nitrogen. Am. J. Bot. 56:187-194.

Smith, R. L., J. H. Bouton, S. C. Shank, K. H. Quesenberry, M. E. Tyler, M. H. Gaskens, and R. C. Littell. 1976. Nitrogen fixation in grasses inoculated with *Spirillum lipoferum*. Science 193:1003-1005.

Smith, R. L., J. H. Bouton, S. C. Schank, and K. H. Quesenberry. 1977. Yield increases of tropical grain and forage grasses after inoculation with *Spirillum lipoferum* in Florida. p. 307-311. *In* A. Ayanaba and P. J. Dart (ed.) Biological nitrogen fixation in farming systems of the tropics. John Wiley & Sons, Inc., New York.

Sprent, J. I. 1976. Nitrogen fixation by legumes subjected to water and light stresses. p. 405-420. *In* P. S. Nutman (ed.) Symbiotic nitrogen fixation in plants. Cambridge University Press, New York.

Stewart, W. D. P. (ed.). 1975. Nitrogen fixation by free-living microorganisms. Cambridge University Press, New York.

Stewart, W. D. P. 1977. Blue-green algae. p. 63-123. *In* R. W. F. Hardy and W. S. Silver (ed.) A treatise on dinitrogen fixation. Section III. Biology. John Wiley & Sons, Inc., New York.

Stewart, W. D. P., P. Rowell, and C. M. Lockhart. 1979. Associations of nitrogen-fixing prokaryotes with higher and lower plants. p. 45-66. *In* E. J. Hewitt and C. V. Cutting (ed.) Nitrogen assimilation of plants. Academic Press, Inc., New York.

Streeter, J. G. 1972. Nitrogen nutrition of field-grown soybean plants. I. Seasonal variations in soil nitrogen and nitrogen composition of stem exudate. Agron. J. 64:311-314.

Streeter, J. G. 1973. Growth of two shoots on a single root as a technique for studying physiological factors limiting the rate of nitrogen fixation by nodulated legumes. Plant Physiol. 48(Suppl.):34.

Streeter, J. G. 1979. Allantoin and allantoic acid in tissues and stem exudate from field-grown soybean plants. Plant Physiol. 63:478-480.

Tanner, J. W., and I. C. Anderson. 1964. External effect of combined nitrogen on nodulation. Plant Physiol. 39:1039-1043.

Thibodeau, P. S., and E. G. Jaworski. 1975. Patterns of nitrogen utilization in the soybean. Planta 127:133-147.

Tien, T. M., M. H. Gaskins, D. H. Hubbell. 1979. Plant growth substances produced by *Azospirillum brasilense* and their effect of the growth of pearl millet (*Pennisetum americanum* L.). Appl. Environ. Microbiol. 37:1016-1024.

Tjepkema, J. D., and H. J. Evans. 1975. Nitrogen fixation by free-living *Rhizobium* in a defined liquid medium. Biochem. Biophys. Rev. Commun. 65:625-628.

Tjepkema, J. D., W. Ormerod, and J. G. Torrey. 1980. Vesicle formation and acetylene reduction activity in *Frankia* sp. CPI1 cultured in a defined media. Nature (London) 287:633-635.

Torrey, J. G. 1978. Nitrogen fixation by acetinomycete-nodulated angiosperms. BioScience 28:586-592.

Trinick, M. J. 1973. Symbiosis between *Rhizobium* and the non-legume *Trema aspera*. Nature (London) 244:459-460.

Trinick, M. J. 1976. *Rhizobium* symbiosis with a non-legume. p. 507–517. *In* W. E. Newton and C. J. Nyman (ed.) Proceeding of the First International Symposium of Nitrogen Fixation, Vol. II, Pullman, Wash. 3–7 June 1974. Washington State University Press, Pullman.

van Berkum, P., and B. B. Bohlool. 1980. Evaluation of nitrogen fixation by bacteria in association with roots of tropical grasses. Microbiol. Rev. 44:419–517.

Van Straten, J., and E. L. Schmidt. 1975. Action of water in depressing acetylene reduction by detached nodules. Appl. Microbiol. 29:432–434.

Vest, G. 1971. Nitrogen increases in a non-nodulating soybean genotype with nodulating genotypes. Agron. J. 63:356–359.

Vest, G., D. F. Weber, and C. Sloger. 1973. Nodulation and nitrogen fixation. p. 353–390. *In* B. E. Caldwell (ed.) Soybeans: Improvement, production, and uses. Agronomy 16: 353–390. Am. Soc. of Agron., Madison, Wis.

Vincent, J. M. 1974. Root-nodule symbiosis with *Rhizobium*. p. 265–341. *In* A. Quispel (ed.) Biology of nitrogen fixation. North-Holland Publishing Co., Amsterdam.

Vincent, J. M. 1977. Rhizobium: General microbiology. p. 277–366. *In* R. W. F. Hardy and W. S. Silver (ed.) A treatise on dinitrogen fixation. Section III. Biology. John Wiley & Sons, Inc., New York.

Vincent, J. M. 1980. Factors controlling the legume-*Rhizobium* symbiosis. p. 103–129. *In* W. E. Newton and W. H. Orme-Johnson (ed.) Nitrogen fixation, Vol. II. University Park Press, Baltimore.

Virtanen, A. I., J. Erkama, and H. Linkola. 1947a. On the relation between nitrogen fixation and leghaemoglobin content of leguminous root nodules. Acta Chem. Scand. 1:861–870.

Virtanen, A. I., J. Jorma, H. Linkola, and A. Linnasalmi. 1947b. On the relation between nitrogen fixation and leghaemoglobin content of leguminous root nodules. II. Acta Chem. Scand. 1:90–111.

Weaver, R. W., and L. R. Fredrick. 1974. Effect of inoculum rate on competitive nodulation of *Glycine max* L. Merrill. II. Field studies. Agric. J. 66:233–236.

Weber, C. R. 1966. Nodulating and nonnodulating soybean isolines. Agron. J. 58:43–46.

Weber, D. F., B. E. Caldwell, C. Sloger, and G. H. Vest. 1971. Some USDA Studies on the soybean-rhizobium symbiosis. Plant Soil (Special Vol.), p. 293–304.

Wheeler, C. T. 1969. The diurnal fluctuation in nitrogen fixation in nodules of *Alnus glutinosa* and *Myrica gale*. New Phytol. 68:675–682.

Wheeler, C. T. 1971. The causation of the diurnal changes in nitrogen fixation in the nodules of *Alnus glutinosa*. New Phytol. 70:487–495.

Wilson, P. W., E. B. Fred, and M. R. Salmon. 1933. Relation between carbon dioxide and elemental nitrogen assimilation in leguminous plants. Soil Sci. 35:145–165.

Wittenburg, J. B. 1980. Utilization of leghemoglobin-bound oxygen by *Rhizobium* bacteroids. p. 53–67. *In* W. E. Newton, and W. H. Orme-Johnson (ed.) Nitrogen fixation, Vol. II. University Park Press, Baltimore.

Wright, W. H. 1925. The nodule bacteria of soybeans. I. Bacteriology of the strains. Soil Sci. 20:95–129.

Wyatt, J. T., and J. K. G. Silvey. 1969. Nitrogen fixation by *Gloeocapsa*. Science 165:908–909.

Yoch, D. C. 1978. Nitrogen fixation and hydrogen metabolism by the photosynthetic bacteria. p. 657–676. *In* R. K. Clayton and W. R. Sistrom (ed.) The photosynthetic bacteria. Plenum Press, New York.

11 Nitrogen Transport Processes in Soil

D. R. NIELSEN AND J. W. BIGGAR

University of California
Davis, California

P. J. WIERENGA

New Mexico State University
Las Cruces, Nex Mexico

I. INTRODUCTION

Our understanding of NO_3^- behavior in agricultural soils has improved markedly within the last decade due to a combination of several issues and events. First, attempts to increase food and fiber production by the application of fertilizers and the inefficiencies in N use were recognized as having caused excessive amounts of N to leave agricultural fields. Losses through leaching of NO_3^- into surface and ground waters (e.g., Strebel et al., 1975) and losses through denitrification and volatilization of N_2, N_2O, and NH_3 (e.g., Ryden, 1981) have been reported. Second, it was realized that the quality of the environment was at stake. In several areas (e.g., California and Israel) underground waters for municipal use were impaired by high NO_3^- levels. More information was demanded by society about levels of NO_3^- in ground waters and their relationship to the management of agricultural fields. The management and disposal of nitrogenous wastes were also being questioned. On a global basis, because of the fact that N_2O reacts and destroys ozone in the stratosphere, concern arose as to the possibility that levels of ozone would be decreased sufficiently to cause marked changes in ultraviolet light reaching the earth's surface. Third, the trade-offs between energy, water, and fertilizer were made more acute with the realization that fossil fuel energy sources are finite. Fourth, the development of computer science and its technology provided a quantum jump in our ability to begin to model the linkages between chemical, physical, and biological processes that ultimately control the behavior of NO_3^- in environment (e.g., Frissel & van Veen, 1981). With these and other issues, the impetus was there to learn more about soil N. Through financial support of national and international agencies responsible for food production and management of the environment, research on soil NO_3^- was accelerated in the 1970's.

At the beginning of the last decade, crop and soil scientists depending primarily on observations taken within the top soil and above the soil surface were not able to account for the fate of all the fertilizer applied to soil. Each research discipline began to look more closely at the kinetics of the processes they deemed important to explain soil and crop N behavior. Soil microbiologists examined Michaelis-Menten type of reactions in open soil systems through which water leached (McLaren, 1969) instead of in the closed systems traditionally conducted in laboratory incubation studies. Soil chemists departed from equilibrium exchange isotherms in favor of nonequilibrium kinetic reactions for ascertaining the reactions of solutes with soil particle surfaces. Soil physicists explained solute behavior in soils on the basis of pore water velocity distributions occurring at different soil water contents and soil water fluxes. Soil fertility and plant nutrition experts paid attention to long-range transport of solutes within plants together with their attendant biochemical processes. All of these, including numerous other related efforts not identified above, were largely deterministic or parametric in nature. That is, models for various nitrogeneous constituents in the soil or crop were arrived at through consideration of underlying principles (e.g., conservation of mass, momentum) and related through parameters obtained by experimentation. Once model parameters (e.g., chemical or biological reaction coefficients, soil hydraulic conductivity) are determined, deterministic models always yield the same output for a given input. The fundamental assumption with deterministic equations is that it is possible to ascertain values of the parameters that are indeed applicable over domains of space and time in which the processes are occurring. At the end of the last decade when such deterministic efforts were made by agricultural scientists to explain, monitor, or predict NO_3^- retention and movement in field soils (not laboratory column or small field plots), they were, in general, dismayed and disappointed in the divergence and uncertainty of their results, which they attributed to the spatial and temporal variability of field soils. It is now being recognized that deterministic models must give way to stochastic models or at least to stochastic parameters included in the deterministic models. A stochastic model is one whose solutions are predictable in a statistical sense. Agricultural scientists, by necessity, are becoming more aware of the advances made using time and space series analysis, Monte Carlo simulations, regionalized variable analysis, pertubation theory, and other techniques not normally included in the principles advanced by Fisher (1956). The objective of this chapter is to provide illustrative examples of the deterministic models presently being used to describe N behavior in soil profiles and to suggest stochastic methods already available in other scientific disciplines that should, in our opinion, be further developed and used routinely in the agricultural sciences if we expect to fully understand and manage effectively N in our land and water resources.

II. DETERMINISTIC ANALYSES

In this section we summarize the deterministic technology developed to date on water and solute transport in soils. Examples of the underlying equations will be presented together with experimental measurements taken in the laboratory or in the field. In each case, it was tacitly assumed that each set of experimental observations was adequate to unambiguously ascertain the values of the soil parameters.

A. Soil Water Movement

Darcy's equation written for the steady-state flux of soil water J_w (cm^3 cm^{-2} day^{-1}) by Richards (1931) is

$$J_w = -K(\theta) \frac{dH}{dz} \tag{1}$$

where dH/dz is the hydraulic gradient and K the hydraulic conductivity, which is strongly dependent on the soil water content θ (cm^3/cm^3) (see Fig. 1). Because the hydraulic head H (cm) is considered to be merely the sum of the gravitational potential head and the matric potential head h, the value of h may depend on the techniques by which it is measured. $K(\theta)$ is dependent on the quality of soil water and the particular distribution of solutes associated with the soil matrix. Transient soil water behavior under isothermal conditions, neglecting transport in the gaseous phase, is generally described by

$$\frac{\partial \theta}{\partial t} = \frac{\partial}{\partial z} \left[K(\theta) \frac{\partial H}{\partial z} \right] + S \tag{2}$$

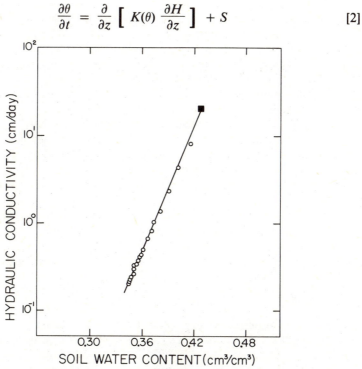

Fig. 1—Soil hydraulic conductivity K as a function of soil water content θ (Nielsen et al., 1973).

where t is time and S is a source or sink. Philip (1957) greatly advanced the understanding of infiltration by solving Eq. [2] for a specific set of conditions. His work, coupled with the later work of Parlange (1972), Kutilek (1980), and many others who provided numerical solutions for a broader set of conditions (e.g., Hanks and Bowers, 1962; de Wit & van Keulen, 1972), furnishes an adequate foundation on which to explain most soil water problems. The utility of this foundation for soil-plant-water investigations hinges on the evaluation of the parameter K and may be complicated by hysteresis (Vachaud & Thony, 1971), the fact that the soil may swell and shrink as θ changes (Bouma et al., 1977), and the fact that the kinetics of root water uptake remain largely empirical (Feddes et al., 1976). Without such complications, there has been reasonable agreement between theory and experiment, even for two- or three-dimensional water flow in the vicinity of water emitters for trickle irrigation (e.g., Bresler et al., 1971).

B. Soil Solute Movement

Similar to Eq. [1], the isothermal, steady state flux of a soil solute J_s (g cm^{-2} day^{-1}) is described by

$$J_s = -\theta D \frac{dc}{dz} + J_w c \qquad [3]$$

where c is the concentration in the soil water (g/cm^3) and D the apparent diffusion coefficient. Equation [3] is fraught with difficulties inasmuch as inside a soil, both c and J_w are spatial averages over an ensemble of soil pores. Experimentally, each remains somewhat ambiguous, with their values depending on the method of measurement and the size of the ensemble that is sampled. Hence, neither their product nor the gradient dc/dz in Eq. [3] is unique. Further difficulties stem from the fact that soil particle surfaces have a net charge density that influences the distribution of solutes within each soil pore, which in turn is modified further by the soil water content and the soil water flux. Because of these and related difficulties, the parameter D in Eq. [3] has been the subject of numerous investigations and, in general, has a unique value only in relation to a particular set of conditions.

Because soil solute movement rarely exists in the steady-state condition, many descriptions of solute behavior are obtained from solutions of the following equation:

$$\frac{\partial \varrho s}{\partial t} + \frac{\partial \theta c}{\partial t} = \frac{\partial}{\partial z}\left(\theta D \frac{\partial c}{\partial z}\right) - \frac{\partial J_w c}{\partial z} + \phi \qquad [4]$$

where ϱ is the soil bulk density (g/cm^3), s the solute associated with the soil particles [g/(g soil)], and ϕ is an irreversible source or sink of solute (g cm^{-3} day^{-1}). Day (1956) used Eq. [4] without its first and last terms to describe the leaching of a simple inorganic salt through sand under steady-state

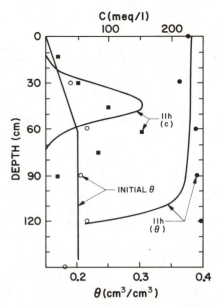

Fig. 2—Measured and theoretical distributions of solutes c and soil water content θ for a leaching time of 11 hours (Warrick et al., 1971).

water flux conditions. Figure 2 shows how well the movement of a solute being leached through a field soil is described by an equation similar to that used by Day for infiltration when J_w is a function of time. The value of D was chosen such that the observations matched the solution of the equation.

The value of D has been observed in laboratory experiments to vary with both θ (Smiles et al., 1978) and average pore water velocity $v (v = J_w/\theta)$ (Fried & Combarnous, 1971). An extensive set of field observations (Biggar & Nielsen, 1976) shows that

$$D = 0.6 + 2.93 \, v^{1.11} \tag{5}$$

for a recent alluvial soil being leached at near saturation. The first term on the right-hand side of Eq. [5] is attributed to molecular diffusion, whereas the last term accounts for the mixing of the solutes not accounted for by using values of J_w and c averaged over unknown ensembles of soil pores in Eq. [3] and [4]. The last term is often simplified to βv in numerical simulations of leaching, with β called the dispersion length (Frissel & Reiniger, 1974) or dispersivity (Freeze & Cherry, 1979).

The reversible process of sorption and exchange of solutes on the solid soil phase as well as chemical precipitation and dissolution described by the first term of Eq. [4] takes the form

$$\partial \varrho s/\partial t = f(s, c, t) \tag{6}$$

in the case of kinetic processes (Lapidus & Amundson, 1952). For instantaneous equilibrium processes we have

$$s = g(c). \tag{7}$$

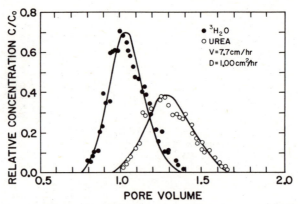

Fig. 3—Experimentally measured and theoretical breakthrough curves of ^3H_2O and urea to ascertain the value of the distribution coefficient R (Wagenet et al., 1977).

When the kinetic form is used, its parameters will implicitly and necessarily include any diffusion-limited transport of the solute to the reactive surfaces inasmuch as no other term appears in Eq. [4]. On the other hand, if Eq. [7] is used, the transport of solute to reactive surfaces associated with those pores in which the soil water is relatively stagnant is usually described by an auxiliary equation (Coats & Smith, 1964) and has been described as the mobile-immobile miscible displacement model by van Genuchten and Wierenga (1976).

An example of the simplest case is given in Fig. 3 for the leaching of ^{15}N-enriched urea through a laboratory column of silty clay loam. The sorption-desorption for urea followed Eq. [7] with $g(c) = Rc$ (where $R = R'\varrho/\theta$), assuming there was no stagnant water within the column, and the value of D was identical for both urea and tritiated water (^3H_2O). Another

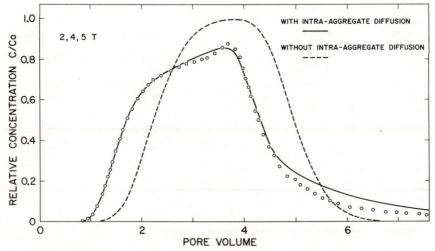

Fig. 4—Measured 2,4,5-T effluent curves calculated with and without immobile water (intra-aggregate diffusion) (van Genuchten et al., 1977).

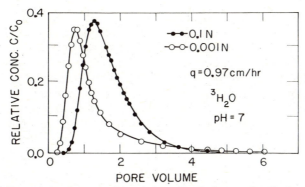

Fig. 5—The effect of the concentration of the soil solution on the displacement of ^3H_2O through an oxisol soil (Nkedi-Kizza, 1979).

example for the leaching of a pulse of ^{14}C-enriched 2,4,5-T (2,4,5-trichloro-phenoxyacetic acid) through a laboratory column of clay loam is given in Fig. 4. In this case, the experimental results could be described by the equilibrium isotherm $g(c) = Rc$ only with the aid of an auxiliary equation that allowed 2,4,5-T to diffuse into the stagnant or immobile water that was estimated to be equal to 6% of the total water in the soil.

In a recent effort (Nkedi-Kizza, 1979), the displacements of ^{45}Ca, ^{36}Cl, and ^3H_2O through laboratory columns of an oxisol at different chemical concentrations, flow velocities, pH values, and soil particle aggregates were described using Eq. [6], which accounted for both kinetic and instantaneous exchange sites (Selim et al., 1976; Cameron et al., 1977), and using Eq. [7], which accounted for mobile-immobile water (van Genuchten & Wierenga,

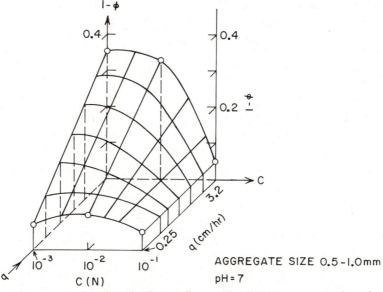

Fig. 6—Immobile water as a function of soil water flux q and soil solution concentration c for a soil composed of 0.5- to 1.0-mm aggregates (Nkedi-Kizza, 1979).

1976). An example of the influence of concentration on ^3H_2O movement and retention is given in Fig. 5. At the more dilute concentration, the volume of immobile water is greater. Figure 6 shows that the fraction of immobile water is directly proportional to the soil water flux and inversely proportional to the chemical concentration. Nkedi-Kizza (1979) concluded that the concepts of instantaneous exchange in the presence of immobile-mobile water and two-site kinetic-equilibrium exchange were essentially equivalent, with neither providing an advantage over the other. Both require obtaining the values of several parameters through the matching of experimental observations, and both yield a match between measurements and theory of equal quality.

For anion exclusion, the concept of immobile water was found attractive to quantitatively identify the average distance that an anion is excluded from negatively charged soil particles during the leaching process (Krupp et al., 1972; Bresler, 1973). If ^3H_2O is neither attracted nor repelled from the soil particle surfaces, the separation between tritiated water and an anion as shown in Fig. 7 is a direct measure of the exclusion distance, which is a function of soil water content and leaching rate.

The last term of Eq. [4], descriptive of irreversible sources or sinks, includes (i) the decay or generation of radioactive elements, (ii) microbial transformations, (iii) absorption or exudation by plant roots, and (iv) chemical precipitation or dissolution. We shall discuss only the first three that are more important to NO_3^- movement. For reaction (i), ϕ is simply $\pm \lambda \theta c$ where λ is the decay or generation constant. For reaction (ii), the complex relationships between microbial growth kinetics and substrate environments can only be approximated (Bazin et al., 1976). Probably the most useful conceptual basis for ϕ stems from the Monod (1942) function that relates the specific growth rate of a microbial population to the concentration of a limiting nutrient in the substrate. For soils being leached, ϕ may be considered (McLaren, 1970) as

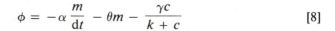

$$\phi = -\alpha \frac{m}{dt} - \theta m - \frac{\gamma c}{k + c} \qquad [8]$$

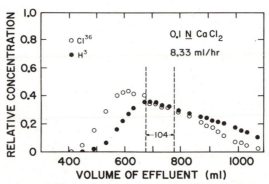

Fig. 7—Breakthrough curves for ^{36}Cl and ^3H_2O whose separation distance indicates anion exclusion (Krupp et al., 1972).

where m is the biomass, and α, β, γ, and k are constants associated with microbial growth, maintenance without growth, and transformation of a solute independent of growth and maintenance. Because of the lack of information on the microbiological population and the difficulty of solving Eq. [4] with Eq. [8] substituted for ϕ, data from most leaching experiments have been compared with solutions of Eq. [4], assuming that ϕ is either a zero or first-order reaction. Under certain conditions, Eq. [8] may be approximated by a zero or first-order reaction. Macura and Kunc (1965), Cho (1971), and McLaren (1969) provided theoretical developments that accelerated us to our present-day technology for laboratory soil columns. As an example, Fig. 8 shows the effluent concentrations of ^{15}N-urea, ^{15}NH$_4^+$, and ^{15}NO$_3^-$ for a column of silty clay loam leached with a pulse of ^{15}N-enriched urea under steady-state, unsaturated leaching conditions where the soil atmosphere was controlled at 20% O$_2$. Three consecutive equations [4] were written for urea, NH$_4^+$, and NO$_3^-$, i.e.,

$$\phi = -k_1 c_1,$$

$$\phi = k_1 c_1 - k_2 c_2, \qquad\qquad [9]$$

$$\phi = k_2 c_2 - k_3 c_3$$

where c_1, c_2 and c_3 were the concentrations of urea, NH$_4^+$ and NO$_3^-$, respectively, and k_i is the assumed first-order coefficients for the hydrolysis of urea, the oxidation of NH$_4^+$, and the reduction of NO$_3^-$. The solutions of

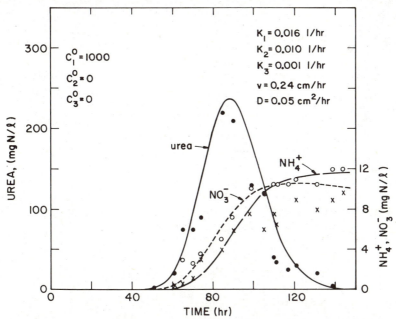

Fig. 8—Experimental and theoretical effluent curves of urea, NH$_4^+$, and NO$_3^-$ stemming from a pulse of ^{15}N-urea leached through an unsaturated soil with its gaseous phase controlled at 20% O$_2$ (Wagenet et al., 1977).

the three equations, which also considered the sorption of urea and the exchange of NH_4^+, were found to be useful tools for studying the microbial transformations in the presence of different levels of aeration.

Sink or source terms for solutes moving into or out of plant roots are somewhat more complex than those above because of the distribution, growth, and activity of the root system, which are all dependent on microenvironmental conditions not explicitly accounted for in Eq. [4]. Equations of the form of Eq. [8] have been used successfully for specific experimental conditions (Nye & Tinker, 1977). Reviews of NO_3^- absorption by plants using equations similar to Eq. [8] (Tanji & Mehran, 1979) and the attendant water absorption by plants (Belmans et al., 1979) indicate that all efforts remain somewhat empirical because of our inability to model root behavior that depends on countless rate-limiting processes that control the growth and development of plants.

The time-dependent nature of θ and J_w in Eq. [4] can be somewhat relaxed by assuming steady-state water flow conditions. By disregarding pertubations in θ and J_w caused by intermittent applications of water to the soil surface (analogous to irrigation and rainfall events). Wierenga (1977) and Beese and Wierenga (1980) showed that temporal and spatial averages of θ and J_w simplified the formulation and solution of Eq. [4] without impairing the prediction of solutes leaching from soil profiles. Further simplifications are possible and are being advocated for field use where irregular infiltration and evaporation events occur and where observations of the plant-soil-water environment are sufficiently limited to preclude analyses stemming from Eq. [4]. The most common lumped parameter equation appropriate for steady-state leaching is

$$L = Ic_w c_s^{-1} \qquad [10]$$

where L is the rate at which water is leaching from the soil profile (cm^3 cm^{-2} $year^{-1}$), I the rate of infiltration into the profile (cm^3 cm^{-2} $year^{-1}$), c_w the concentration of the infiltrating water [$g/(cm^3$ water)], and c_s the concentration of the water leaching from the profile (U.S. Salinity Laboratory Staff, 1954). Rose et al. (1979, 1981) and Dayananda et al. (1980) have recently provided similar lumped parameter models for interpreting the dynamics of leaching in field soils for solutes that are absorbed by plant roots and for solutes that exchange or react with the soil particle surfaces. For example, for a solute that does not react with soil particles, and if one assumes that the solute is not absorbed, the equation that describes the rate at which the average solute content $\bar{c}$ within a profile to depth z changes is

$$\frac{d\bar{c}}{dt} = (Ic_w - Lc_s)(z\bar{\theta})^{-1} \qquad [11]$$

where $\bar{\theta}$ is the average soil water content.

Solutions of Eq. [11] or the more complex Eq. [4] have been difficult to evaluate with some level of confidence because of spatial and temporal fluctuations in the values of the terms in these equations. Such fluctuations have generally been treated statistically, if one assumes that observations were spatially or temporally independent and stemmed from a known frequency distribution. However, their impact on the solution of such equations as Eq. [11] have not been treated on the basis of observations that are spatially or temporally dependent.

III. STOCHASTIC ANALYSES

Although the deterministic equations in the previous section provide a conceptual framework for understanding the movement and retention of nitrogenous constituents in the soil profile, their utility to monitor or manage soil N hinges on how well their parameters represent an entire field. Allison (1955, 1966) spoke to the enigma of soil N balance sheets, the dearth of accurate field data, and the impossibility of accounting for all soil gains and losses of N in a single experiment. Kundler (1970), reviewing ^{15}N-labeled fertilizer research, reported that from 10 to 30% of the fertilizer N was lost or unaccounted for in balance sheet studies. Today our technology continues to improve, but we still face large uncertainties in balance sheet studies even for small field plots. For entire fields under a given cultural practice, the technology remains largely undeveloped primarily because of our inability to properly sample and analyze the results by taking into account the spatial and temporal variability of the observations. The following examples illustrate the uncertainty of observations using standard techniques.

In California (Pratt, 1979) an experiment was designed to examine the behavior of N fertilizer applied to a maize (*Zea mays* L.) crop under three different irrigation regimes (Pratt, 1979). Briefly, the experiment consisted of 12 plots that were individually irrigated units. Within each plot, N fertilizer was applied to four subplots as ^{15}N-depleted ammonium sulphate at 0, 90, 180, and 360 kg/ha. Fertilizer was applied in a single application each year. The irrigation treatments corresponded approximately to 1/3, 3/3, and 5/3 of the normal evapotranspiration (ET) requirements of the maize crop. Water was applied by sprinkler irrigation at 2-week intervals during the growing season. The irrigation treatments corresponded to cumulative totals of 20, 60, and 100 cm of water applied in 7 irrigations each year. In addition, before the maize was planted, each plot received an application of irrigation water that contributed to the water flux below the root zone during the growing season.

Porous ceramic suction probes for extracting soil solution samples and tensiometers for measuring soil water pressure were placed at the 30-, 60-, 120-, 180-, 240-, and 300-cm depths. Duplicate probes were located at each depth in 24 of the 48 subplots, and all 48 subplots had duplicate probes at 240- and 300-cm depths. Soil solution samples were collected every 2 weeks during the growing season and less frequently during intervening months.

Figure 9 depicts soil NO_3^--N in the 360 kg of N fertilizer subplots for the 1/3 and 5/3 ET irrigation treatments. Soil solution samples obtained from the suction probes and values given in the figure are means of two ob-

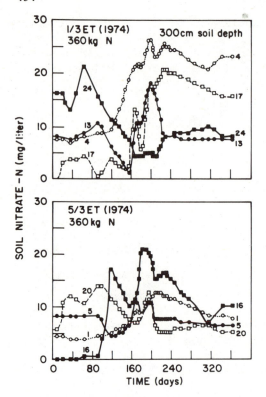

Fig. 9—Soil solution NO₃⁻-N obtained from probes at the 300-cm depth for the 1/3 and 5/3 ET treatments receiving 360 kg of N in 1974. Numbers indicate subplots (Nielsen et al., 1979).

servations. The curves manifest considerable variability between identically treated subplots. Inspection of soil NO_3^--N curves for all the treatment subplots showed that it is impossible to distinguish these curves on the basis of fertilizer application, even within plots given the same irrigation treatment. The erratic behavior of soil NO_3^--N is a consequence of both measurement error and the spatial variability of the soil.

Table 1 shows the amount of inorganic N in the 3-m soil profile after maize harvest for the years 1974 through 1976 (Broadbent & Carlton, 1979). These values are the mean of eight soil cores. Although the mean values are more or less consistent with levels of fertilizer application, the coefficients

Table 1—Inorganic N in soil profile (0–3 m) after maize harvest in three successive years.†

	Nitrogen fertilizer application (kg/ha)							
	0		90		180		360	
Year	Mean	C.V.	Mean	C.V.	Mean	C.V.	Mean	C.V.
	kg N/ha	%	kg N/ha	%	kg N/ha	%	kg N/ha	%
1974	123	43.8	104	41.0	153	11.7	331	22.4
1975	120	11.8	138	4.8	139	11.2	409	19.5
1976	109	4.2	116	11.6	107	7.5	394	48.9

† Plots were irrigated with water equivalent to that evaporated and transpired during the growing season.

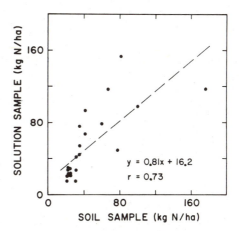

Fig. 10—Correlation between solution and soil methods for determining soluble N in a field soil (Broadbent & Carlton, 1979).

$$y = 0.81x + 16.2$$
$$r = 0.73$$

of variation ranging from about 4 to 50% manifest the spatial variability of the soil and the uncertainty of the observations.

The question often asked is "Which is more reliable, a soil solution sample extracted in situ, or an extract of a soil sample taken from the field?" Figure 10 shows a comparison of the N values obtained in October 1975. Soil solution values converted to kilograms of N/per hectare necessitated their multiplication by the average soil water content at the time the sample was extracted. On the other hand, to estimate the concentration of the soil solution from the soil sample analysis, the uncertainty of the soil water content enters the calculation. Although the data in Fig. 10 shows that the results of the two methods are related, the reliability of both are equally suspect.

A second example of the variability of field soils is shown in Fig. 11, where the leaching of a pulse of ^3H_2O and chloride applied on the soil surface is monitored by samples of the soil solution extracted through suction probes located at the depths of 33.5, 46.0, and 63.5 cm. The curves at the 33.5- and 63.5-cm (a & c) depths appear at approximately the same leaching times, but those at the 46.0-cm depth (b) appear much later. Had the leaching taken place uniformly, the appearance of the curves would have become progressively later with soil depth. Similar field observations that are now frequently reported in soils research are attributed to spatial variability or to the difficulty of obtaining representative samples due to the concept that water moves preferentially through a limited area or limited set of soil pores.

It has been known for at least a century that water and solutes are not displaced uniformly through field soils (Lawes et al., 1881). The literature is replete with such evidence (e.g., Means & Holmes, 1900; Schlichter, 1905; Cunningham & Cooke, 1958; Miller et al., 1965; Wild, 1972; Kissel et al., 1974; Addiscott & Cox, 1976; Wild & Babiker, 1976; Misra & Mishra, 1977; Jones, 1976; Quissenberry & Phillips, 1976; Bouma et al., 1977; Bouma & Wosten, 1979; Scotter, 1978; Cameron et al., 1979; Rose et al., 1981). Reasons for the relatively incomplete displacement are based on the presence of cracks and fissures, clay lenses, worm holes, root channels, textural varia-

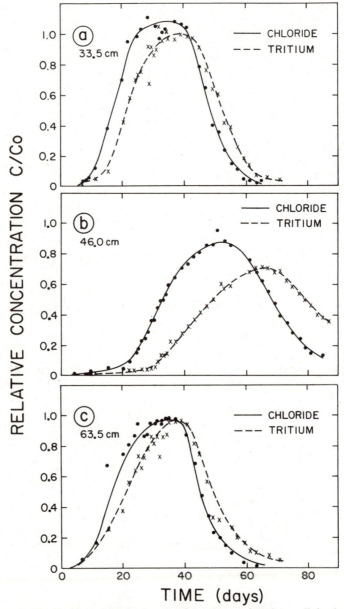

Fig. 11—Relative chloride and $^{3}H_{2}O$ concentrations measured at three soil depths during the leaching of a field soil (Van De Pol et al., 1977).

tions, density and viscosity variations of the soil solution, and others. Soil solution suction probes when placed in the vicinity of a clay lense compared with a root channel, will yield different quality samples depending on factors such as the soil water content and extraction rate. Similarly, soil samples removed from the field and then analyzed for NO_3^- will yield different solute concentrations depending on the position and size of the sample relative to the clay lense, root channel, or other microenvironmental condition.

How many samples are needed to characterize an area, or how large of an area is characterized by a single sample? The latter can be answered by using the autocorrelogram that discloses intervals of distance at which observations are repetitive and also gives the magnitude of the intervals at which observations are no longer correlated. Whittle (1954) and Petrova (1978) used autocorrelograms to express changes in field measured soil properties.

The autocorrelogram of a spatial series is calculated from

$$r(h) = \frac{\text{autocov}(x, x+h)}{\sqrt{\text{var}(x)} \sqrt{\text{var}(x+h)}} \qquad [12]$$

where r is the autocorrelation function and h the lag or distance between observations in the series. If, for the minimum sampling interval as well as all other distances (h), the value of r does not significantly differ from zero within a given field, one may conclude that the observations are spatially independent. Under such circumstances, the locations of the sampling sites within the field should be selected randomly, with the number of samples chosen on the basis of their frequency distribution and the level of desired probability for a specified fiducial limit. On the other hand, if r is not zero but gradually decreases from one (at $h = 0$) as h increases, the observations are spatially dependent, with the scale of observation λ being defined by

$$r = \exp(-h/\lambda). \qquad [13]$$

Figure 12 contains an autocorrelogram calculated for solute concentration measured at 7.6-m intervals along a 750-m transect in a water table (Gelhar et al., 1980) underneath a field planted to maize (Fig. 13). The data exhibit a scale of observation equal to 30.2 m. Sampling at intervals much <30 m is costly and does not add appreciably to our knowledge of the solute in the

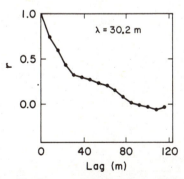

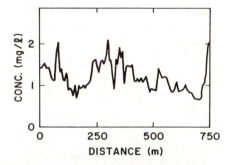

Fig. 12—An autocorrelogram calculated from the solute concentration distribution shown in Fig. 13 (Gelhar et al., 1980).

Fig. 13—Solute concentration observations in a water table along a 750-m transect (Gelhar et al., 1980).

water table. Sampling at distances somewhat > 30 m allows poorer interpolations between measured locations.

Sampling the soil or soil solution with the intent of better understanding or managing the crop and the soil environment should be done on the basis of a knowledge of the scale of observation for each parameter being assessed. Scales of observation for each parameter should be compatible. For example, if the flux J_w in Eq. [1] does not have the same scale of observation as the hydraulic gradient dH/dz, the value of the hydraulic conductivity K derived from their ratio will be physically as well as spatially ambiguous. Similarly, the generally poor agreement between solution vs. soil sample data given in Fig. 10 may be attributed to the comparison of two parameters having two different scales of observations. Such consideration of scales of observations is at its infancy in the agricultural science and yet must be pursued if we wish to solve the enigma of balance sheets in soil N.

A second method of ascertaining the number of samples required for a given precision and accuracy utilizing their spatial dependence is based on the concept of a regionalized variable (Matheron, 1971). The kriging technique, which is derived from this concept, is an optimum interpolator that uses the variogram stemming from a set of observations on a particular field to estimate additional values in the neighborhood of measured values as well as to assess the probable error associated with estimated variances. In order to krige, a variogram is constructed by plotting the semivariance γ against the lag h. The semivariance of a set of observations Y is defined as one-half the variance of the difference $(y_{x+h} - y_x)$:

$$\gamma (h) = \tfrac{1}{2}\mathrm{var}(y_{x+h} - y_x) \tag{14}$$

where y_x and y_{x+h} are deviations around a mean of zero. The y_x and y_{x+h} are actually obtained by subtracting the mean of observations $\overline{Y}$ from each observation Y_x and Y_{x+h} at locations x and $(x+h)$, which are separated by the distance h. A plot of an idealized semivariogram is shown in Fig. 14. It is clear from the figure that as the distance h becomes infinitesimally small, the value of semivariance $\gamma(h)$ becomes zero. On the other hand, as h becomes larger, the value of semivariance approaches the value of variance.

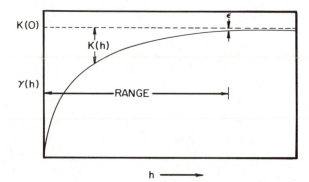

Fig. 14—An idealized variogram.

The value of semivariance may not exactly become equal to variance unless the distance h becomes sufficiently large. The value of h at which the semi-variance approaches to within an arbitrarily selected small value ϵ is called the range. Beyond this range, the observed values are considered to be independent of each other and therefore not capable of being kriged. Moreover, beyond the above range, normal statistical methods used for values independent of sampling position apply.

A kriged, or estimated, value is found by attributing weights i to its neighbors' values that are the known measured values. For kriging to be the best linear unbiased estimator, two constraints apply. First the estimation must be unbiased:

$$E[Y_o^* - Y_o] = 0, \qquad [15]$$

and the square average error should be a minimum:

$$\text{Var}[Y_o^* - Y_o] = \min, \qquad [16]$$

where Y_o^* in this study is the estimation of the value at one point (point to be kriged), and Y_o is the true value of the observation point x_o. The kriged value is computed from

$$Y_o^* = \sum_{i=1}^{n} \lambda_i Y_i \qquad [17]$$

where Y_i are the measured values at locations x_i.

The associated estimation of kriging variance σ^2 is computed from

$$\sigma_K^2 = \sum_{i=1}^{n} \lambda_i \gamma(x_i - x_o) + \mu \qquad [18]$$

where μ is the Lagrange multiplier. Many variants of this so-called simple kriging process exist according to the number of neighbors considered, the presence of a trend, etc. For additional details see Journel and Huijbregts (1978) and Volpi and Gambolati (1978). We shall present two illustrative examples.

Figure 15 is a variogram calculated from 78 observations of soil salinity stemming from a 30-cm deep by 100-cm wide soil profile under a bed of pepper plants (*Capsicum annuum* L.) sampled on a grid. The peppers grown under drip irrigation in a desert environment are easily stressed as a result of lack of water or a build-up of soil salinity. Because of soil spatial variability, observations of salinity within a given soil profile as well as differences between irrigation treatments are often erratic and difficult to interpret. The same kinds of results are typically manifested by NO_3^- concentrations within soil profiles. With the aid of the above variogram, 356 additional values of the solute concentration were kriged within the soil pro-

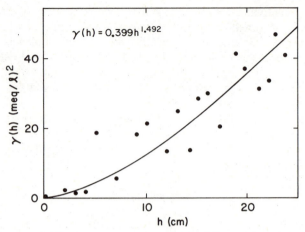

Fig. 15—Variogram of soil solute concentration calculated from observations taken in a grid within a soil profile (P. J. Wierenga, unpublished data).

file and contours of isosalinity drawn as shown in Fig. 16. From the contours, the centroid of solute mass is easily computed, and hence all the observations of salinity are coalesced into a single value. The centroid can be observed as a function of time for a given treatment or compared with those of other irrigation treatments. The technique uses all observations, avoids the dilemma of sampling in or between plant rows, and can be used to judge whether more or fewer observations are required relative for a desired precision.

A second example of kriging is that of predicting the infiltration rate at various locations within an agricultural field. Leaching losses, which are directly related to infiltration and the hydraulic conductivity of the soil, are known to be nonuniform across a field and, hence, are difficult to ascertain. Within a 1-ha experimental site, Vieira et al., (1981) made 1,280 observations of steady-state infiltration V_o in an irregular grid pattern. With

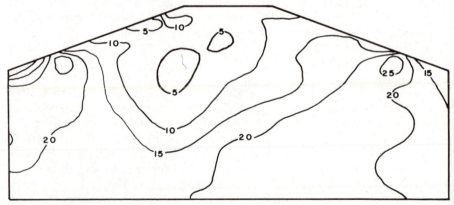

Fig. 16—Contours of isosalinity drawn through 78 measured and 367 kriged observations (P. J. Wierenga, unpublished data).

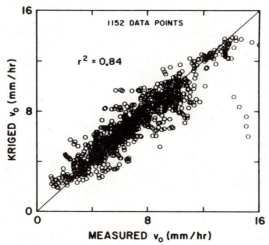

Fig. 17—Kriged values of the steady-state infiltration rate vs. measured values. Kriged values were based on 128 observations of V_o (Vieira et al., 1981).

the use of autocorrelograms and variograms, they showed that only 128 observations of V_O could serve as an adequate basis to predict the infiltration rate at all locations within the experimental site. Figure 17 is a plot of the kriged values of V_o vs. the measured values of V_o using only 128 observations to predict the values at 1,150 other locations within the field. The uncertainty of the predictions was of the same order as the uncertainty of each of the original measurements.

A third method to more thoroughly analyze a set of field observations is to examine the spectral density of the autocorrelation function (Eq. [12]). The spectral density function is useful in identifying periodicities in data associated with cyclic events such as harvesting, cultivating, irrigation or rainfall, daily or seasonal temperatures, etc., or cyclic variations of soil properties over horizontal or vertical distances. The spectral density function $P(f)$ is calculated from $r(h)$

$$P(f) = 2 \int_0^\infty r(h) \cos(2\pi f h) \, dh. \qquad [19]$$

The peaks in this function indicate which frequencies in the observations have the most variance. Gelhar et al. analyzed the solute distribution within a field by sampling the soil at 30-cm intervals to a depth of 22 m (Fig. 18).[2] The spectrum $P(f)$ shown in Fig. 19 has a peak at 0.37 m^{-1}, which indicates a period of 2.7 m. Hence, the fluctuations of solute concentration in Fig. 18 could possibly be indicative of annual cycles of leaching associated with irrigation in a desert environment.

Spectral analyses can be made for various forms of soil N sampled at a given location as a function of time or at a given time as a function of location. For example, the potential denitrification occurring in the lower and frequently wetter locations within the bottom of furrows could be examined. Similarly, the behavior of soil N within and between furrows

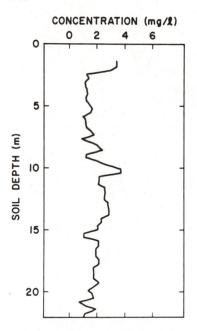

Fig. 18—Concentration distribution of solutes measured at 30-cm intervals (Gelhar et al., 1980).

could be monitored more accurately using spectral analysis. Or the fluctuations of NO_3^- in drainage water discharging into surface waters could be more thoroughly analyzed with spectral analysis.

Most field research on crop response to fertilizer N and the attendant potential leaching of NO_3^- has been conducted on small field plots and in field-located lysimeters. By extending the above three methods we anticipate future experiments that will treat an entire field managed agriculturally as a single unit. Instead of dividing the field into plots to receive different fertilizer treatments, we would treat the entire field (or large sections of it) at the same level of fertilization. And, instead of partitioning the variance into those components associated with each of the treatments, we would analyze the spatial variances and covariances from samples taken from

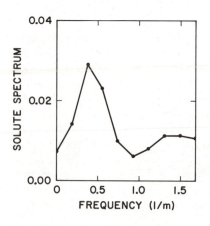

Fig. 19—Spectral density function of the solute distribution of Fig. 18 (Gelhar et al., 1980).

across the entire field. Spatial cross correlations would be made between, for example, the yield of crop vs. soil NO_3^- or ^{15}N in the crop vs. that in the soil (Davis, 1973). In addition, we anticipate observations of the soil (or the crop) that are difficult or expensive to make to be related to other more easily obtained observations through co-kriging (Journel & Huijbregts, 1978).

The above methods illustrated how a given set of field observations could be analyzed more effectively than by merely calculating mean values and standard deviations. Similarly, equations such as Eq. [2], [4], and [11] can be written as mixed deterministic-stochastic models with their parameters, and/or their dependent variables written as the sum of their expected mean and a random function associated with the uncertainties of a spatially and temporally variable field (e.g., Kado, 1971; Gelhar, 1974; Tang & Pinder, 1979). For example, Eq. [4] written without its first and last terms:

$$\frac{\partial c}{\partial t} = D \frac{\partial^2 c}{\partial x^2} - \frac{\partial vc}{\partial z}, \qquad [20]$$

would be solved assuming that

$$c = \bar{c} + c',$$
$$v = \bar{v} + v', \text{ and} \qquad [21]$$
$$D = \bar{D} + D'$$

where the bar and prime refer to mean and pertubated values, respectively. If Eq. [21] is substituted into [20] and the products of pertubation terms are neglected, two equations can be solved—one for the expected means and one for the pertubations. Solutions of such equations will allow us to assess the confidence we have in their predictions as well as the suitability of the technique used to match experimental values with those of theory. Moreover, the results will aid in the design of future experiments to ascertain the soil properties or the behavior of soil N. Such analyses for leaching field soils are beginning to appear in the literature (Bresler & Dagan, 1979).

The description of statistical variations of soil water properties encountered in the field may be simplified by introducing the concept of similar media (Miller & Miller, 1956). The consequences of the concept are that at given soil water contents, the soil water pressure head h is related to an average value h_m by

$$h_m = \alpha_r h_r \qquad [22]$$

where α_r is a scaling factor appropriate for location r. For the hydraulic conductivity K, the corresponding relation is

$$K_m = K_r \alpha_r^{-2}. \qquad [23]$$

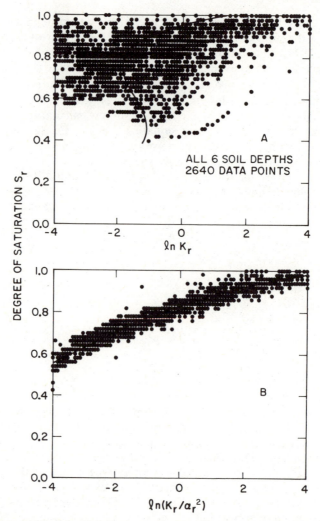

Fig. 20—Measured (A) and scaled (B) observations of hydraulic conductivity K vs. the degree of water saturation measured in a 150-ha field (Warrick et al., 1977).

Figure 20A shows 2,640 observations of K vs. relative soil water content for 6 soil depths within a 150-ha field. Figure 20B shows the same observations scaled using Eq. [23]. The advantage of scaling is that it allows an average curve to characterize the entire field, with the soil heterogeneity being described by a single stochastic parameter α_r (Peck et al., 1977; Warrick et al., 1977; Simmons et al., 1979). Warrick and Amoozegar-Fard (1979) have shown how infiltration and drainage can be described using scaling factors that were assumed to be spatially independent. Examination of the spatial dependence of the scaling factors and their usage in solutions of Eq. [2] remain for future research. Moreover, concepts of similar media for other soil properties remain untested.

In conclusion, it should be pointed out that classical Fisher statistics is relatively inexpensive for field investigations because of generally fewer number of observations required compared with those analyzed geostatistically. Geostatistical analyses provide a great deal more information albeit more expensive. The cost, however, will be tempered if we do not disregard the advantage of combining the theory of regionalized variables with our present concepts of soil classification. Variograms or other modes of describing the spatial variance structure of parameters descriptive of physical, chemical, and biological processes that contribute to the behavior of soil N should become an integral part of the data base for each soil mapping unit.

LITERATURE CITED

Addiscott, T. M., and D. Cox. 1976. Winter leaching of nitrate from autumn-applied calcium nitrate, ammonium sulphate, urea, and sulphur-coated urea in bare soil. J. Agric. Sci. 87: 381–389.

Allison, F. E. 1955. The enigma of soil nitrogen balance sheets. Adv. Agron. 7:213–250.

Allison, F. E. 1966. The fate of nitrogen applied to soils. Adv. Agron. 18:219–258.

Bazin, M. J., P. T. Saunders, and J. I. Prosser. 1976. Models of microbial interactions in the soil. CRC Critical Rev. Microbiol. p. 463–498.

Beese, F., and P. J. Wierenga. 1980. Solute transport through soil with absorption and root water uptake computed with a transient and a constant flux model. Soil Sci. 129:245–252.

Belmans, C., J. Feyen, and D. Hillel. 1979. An attempt at experimental validation of macroscopic scale models of soil moisture extraction by roots. Soil Sci. 127:174–186.

Biggar, J. W., and D. R. Nielsen. 1976. Spatial variability of the leaching characteristics of a field soil. Water Resour. Res. 12:78–84.

Bouma, J., A. Jongerius, O. Boersma, A. Jager, and D. Schoonderbeek. 1977. The function of different types of macropores during statistical flow through four swelling soil horizons. Soil Sci. Soc. Am. J. 41:945–950.

Bouma, J., and J. H. M. Wosten. 1979. How patterns during extended saturated flow in two undisturbed swelling clay soils with different macrostructures. Soil Sci. Soc. Am. J. 43: 16–21.

Bresler, E. 1973. Anion exclusion and coupling effects in nonsteady transport through unsaturated soils: I. Theory. Soil Sci. Soc. Am. Proc. 37:663–669.

Bresler, E., and G. Dagan. 1979. Solute dispersion in unsaturated heterogeneous soil at field scale: II. Applications. Soil Sci. Soc. Am. J. 43:467–472.

Bresler, E., J. Heller, N. Diner, I. Ben-Asher, A. Brandt, and D. Goldberg. 1971. Infiltration from a trickle source: II. Experimental data and theoretical predictions. Soil Sci. Soc. Am. Proc. 35:683–689.

Broadbent, F. E., and C. A. B. Carlton. 1979. Methodology for field trials with ^{15}N-depleted fertilizer. p. 33–56. In P. F. Pratt (principal investigator) Nitrate in effluents from irrigated lands. Final Report to the National Science Foundation, Univ. of California, Riverside.

Cameron, D. R., and A. Klute. 1977. Convective dispersive solute transport with a combined equilibrium and kinetic adsorption model. Water Resour. Res. 13:183–188.

Cameron, D. R., C. G. Kowalenko, and C. A. Campbell. 1979. Factors affecting nitrate nitrogen and chloride leaching variability in a field plot. Soil Sci. Soc. Am. J. 43:455–457.

Cho, C. M. 1971. Convective transport of ammonium with nitrification in soil. Can. J. Soil Sci. 51:339–350.

Coats, K.H., and B. D. Smith. 1964. Dead-end pore volume and dispersion in porous media. Soc. Petrol. Eng. J. 4:73–84.

Cunningham, R. K., and G. W. Cooke. 1958. Soil nitrogen. II. Changes in levels of inorganic nitrogen in a clay loam soil caused by fertilizer application by leaching and uptake by grass. J. Sci. Food Agric. 9:317–324.

Davis, J. C. 1973. Statistics and data analysis in geology. John Wiley & Sons, Inc., New York.

Day, P. R. 1956. Dispersion of a moving salt water boundary advancing through a saturated sand. Trans. Am. Geophys. Union 37:595–601.

Dayananda, P. W. A., F. P. W. Winteringham, C. W. Rose, and J. Y. Parlange. 1980. Leaching of a sorbed solute: a model for peak concentration displacement. Irrigation Sci. 1: 169–175.

de Wit, C. T., and H. van Keulen. 1972. Simulation of transport processes in soils. Centre for Agric. Publ. and Docum., Wageningen.

Feddes, R. A., P. Kowalik, S. P. Neuman, and E. Bresler. 1976. Finite difference and element simulation of field water uptake by plants. Hydrol. Sci. Bull. XXI 13:81–98.

Fisher, R. A. 1956. Statistical methods and scientific inference. Oliver and Boyd, Edinburgh.

Freeze, R. A., and J. A. Cherry. 1979. groundwater. Prentice-Hall, Inc., Englewood, N.J.

Fried, J. J., and M. A. Combarnous. 1971. Dispersion in porous media. Adv. Hydrosci. 7: 170–282.

Frissel, M. J., and P. Reiniger. 1974. Simulation of accumulation and leaching in soils. Centre for Agric. Publ. and Docum., Wageningen.

Frissel, M. J., and J. A. van Veen. 1981. Simulation model for nitrogen immobilization and mineralization. p. 359–381. In I. K. Iskander (ed.) Modeling wastewater renovation: land treatment. John Wiley & Sons, Inc., New York.

Gelhar, L. W. 1974. Stochastic analysis of phreatic aquifers. Water Res. 10:539–545.

Gelhar, L. W., P. J. Wierenga, C. J. Duffy, K. R. Rehfeldt, R. B. Senn, M. Simonett, T.-C. Yeh, A. L. Gutjahr, W. R. Strong, and A. Bustamante. 1980. Irrigation return flow studies at San Acacia, New Mexico: Monitoring modeling and variability. Technical Progress Report no. H-3, New Mexico Institute of Mining and Technology, Socorro.

Hanks, R., and S. A. Bowers. 1962. Numerical solution of the moisture flow equation into layered soils. Soil Sci. Soc. Am. Proc. 26:530–534.

Jones, M. J. 1976. Water movement and nitrate leaching in a Nigerian savannah soil. Exp. Agric. 12:69–79.

Journel, A. G., and Ch. J. Huijbregts. 1978. Mining geostatistics. Academic Press, Inc., New York.

Kado, T. 1971. Stochastic analysis of dispersion models. Ph.D. Thesis. Univ. of Texas, Austin (Diss. Abstr. 72-19617).

Kissel, D. E., J. T. Ritchie, and E. Burnett. 1974. Nitrate and chloride leaching in a swelling clay soil. J. Environ. Qual. 3:401–404.

Krupp, H. K., J. W. Biggar, and D. R. Nielsen. 1972. Relative flow rates of salt and water in soils. Soil Sci. Soc. Am. Proc. 36:412–417.

Kundler, P. 1970. Utilization, fixation and loss of fertilizer nitrogen (review of international results of the last 10 years basic research). Albrecht-Thaer-Arch. 14:191–210.

Kutilek, M. 1980. Constant rainfall infiltration. J. Hydrol. 45:289–303.

Lapidus, L., and N. R. Amundson. 1952. Mathematics of adsorption in beds. 6. The effect of longitudinal diffusion in ion exchange and chromatographic columns. J. Phys. Chem. 56:984–988.

Lawes, J. B., J. H. Gilbert, and R. Warington. 1881. On the amount and composition of the rain and drainage waters collected at Rothamsted. J. R. Agric. Soc. Engl. Ser. 2, 17:249–179, 311–350.

Macura, J., and F. Kunc. 1965. Continuous flow method in microbiology. V. Nitrification. Folia Microbiol. (Prague) 10:125–134.

Matheron, G. 1971. The theory of regionalized variables and its applications. Les Cashiers due centre de Morphologie Mathematique, Fas. 5., C. G. Fontainbleau.

McLaren, A. D. 1969. Steady state studies in nitrification in soil: Theoretical considerations. Soil Sci. Soc. Am. Proc. 33:273–276.

McLaren, A. D. 1970. Temporal and vectorial reactions of nitrogen in soil: A review. Can. J. Soil Sci. 50:97–109.

Means, T. H., and J. G. Holmes. 1900. Soil survey around Fresno, California. In U.S. House Document 107, USDA, Field Operations Div. of Soils, no. 526, p. 333–384.

Miller, E. E., and R. D. Miller. 1956. Physical theory for capillary flow phenomena. J. Appl. Phys. 27:324–332.

Miller, R. J., J. W. Biggar, and D. R. Nielsen. 1965. Chloride displacement in Panoche clay loam in relation to water movement and distribution. Water Resour. Res. 1:63–72.

Misra, C., and B. K. Mishra. 1977. Miscible displacement of nitrate and chloride under field conditions. Soil Sci. Soc. Am. J. 41:496–499.

Monod, J. 1942. Recherches sur la croissance des cultures bacteriennes, Hermann et Cie, Paris.

Nielsen, D. R., J. W. Biggar, and Y. Barrada. 1979. Water and solute movement in field soils. p. 165–183. *In* Isotopes and radiation techniques in research on soil-plant relationships. Int. Atomic Energy Agency, Vienna.

Nielsen, D. R., J. W. Biggar, and K. T. Ehr. 1973. Spatial variability of field-measured soil-water properties. Hilgardia 42:215–259.

Nkedi-Kizza, P. 1979. Ion exchange in aggregated porous media during miscible displacement. Ph.D. Thesis. Univ. of California, Davis (Diss. Abstr. 80-16777).

Nye, P. H., and P. B. Tinker. 1977. Solute movement in the soil-root system. Studies in ecology 4. Univ. of California Press, Berkeley.

Parlange, J. Y. 1972. Theory of water movement in soils, 8. One-dimensional infiltration with constant flux at the surface. Soil Sci. 114:1–4.

Peck, A. J., R. J. Luxmoore, and J. L. Stolzy. 1977. Effects of spacial variability of soil hydraulic properties in water budget modeling. Water Resour. Res. 13:348–354.

Petrova, M. V. 1978. Some questions of measurement of accuracy and placement of soil moisture sensors. (In Russian.) Pochvovedeniye 7:144–150.

Philip, J. R. 1957. Numerical solution of equations of the diffusion type with diffusivity concentration—dependent. II. Aust. J. Phys. 10:29–42.

Pratt, P. F. (principal investigator). 1979. Nitrate in effluents from irrigated lands. Final Report to the National Science Foundation, Univ. of California, Riverside.

Quissenberry, V. L., and R. E. Phillips. 1976. Percolation of surface-applied water in the field. Soil Sci. Soc. Am. J. 40:484–489.

Richards, L. A. 1931. Capillary conduction of liquid through porous materials. Physic. 1:318–333.

Rose, C. W., F. W. Chichester, J. R. Williams, and J. T. Ritchie. 1982. A contribution to simplified models of field solute transport. J. Environ. Qual. 11:146–150.

Rose, C. W., P. W. A. Dayananda, D. R. Nielsen, and J. W. Biggar. 1979. Long-term solute dynamics and hydrology in irrigated slowly permeable soils. Irrig. Sci. 1:77–87.

Ryden, J. C. 1981. Gaseous nitrogen losses. p. 277–312. *In* I. K. Iskander (ed.) Modeling wastewater renovation: land treatment. John Wiley & Sons, Inc., New York.

Schlichter, C. S. 1905. Field measurements of the rate of movement of underground waters. U.S. Geological Survey, Water Supply and Irrigation Paper no. 140, Washington, D.C.

Scotter, D. R. 1978. Preferential solute movement through larger soil voids. I. Some computations using simple theory. Aust. J. Soil Res. 16:257–267.

Selim, H. M., J. M. Davidson, and R. S. Mansell. 1976. Evaluation of a two-site adsorption-desorption model for describing solute transport in soil. p. 444–448. *In* Proc. Summer Computer Simulation Conf., Washington, D.C. Simulation Councils, Inc., La Jolla, Calif.

Simmons, C. S., D. R. Nielsen, and J. W. Biggar. 1979. Scaling of field-measured soil water properties. Hilgardia 47:44–174.

Smiles, D. E., J. R. Philip, J. H. Knight, and D. E. Elrick. 1978. Hydrodynamic dispersion during absorption of water by soil. Soil Sci. Soc. Am. J. 42:229–234.

Strebel, O., M. Renger, and W. Giesel. 1975. Vertical water movement and nitrate displacement below the root zone. Mitt. Dtsch. Bodenkd. Ges. 22:277–286.

Tang, D. H., and G. F. Pinder. 1979. Analysis of mass transport with uncertain physical parameters. Water Resour. Res. 15:1147–1155.

Tanji, K. K., and M. Mehran. 1979. Conceptual and dynamic models for nitrogen in irrigated croplands. p. 555–646. *In* P. F. Pratt (principal investigator) Nitrate in effluents from irrigated lands. Final Report to the National Science Foundation, Univ. of California, Riverside.

U.S. Salinity Laboratory Staff. 1954. Diagnosis and improvement of saline and alkali soils. USDA Agric. Handbook 60, U.S. Government Printing Office, Washington, D.C.

Vachaud, G., and J. L. Thony. 1971. Hysteresis effects during infiltration and redistribution in a soil column at different water contents. Water Resour. Res. 7:111–127.

Van De Pol, R. M., P. J. Wierenga, and D. R. Nielsen. 1977. Solute movement in a field soil. Soil Sci. Soc. Am. J. 41:10–13.

van Genuchten, M. T., and P. J. Wierenga. 1976. Mass transfer studies in sorbing media: I. Analytical solutions. Soil Sci. Soc. Am. J. 40:473–480.

van Genuchten, M. T., P. J. Wierenga, and G. A. O'Connor. 1977. Mass transfer studies in sorbing porous media: III. Experimental evaluation with 2,4,5-T. Soil Sci. Soc. Am. J. 41:278–285.

Vieira, S., D. R. Nielsen, and J. W. Biggar. 1981. Spatial variability of field measured infiltration rate. Soil Sci. Soc. Am. J. 45:1040–1048.

Volpi, G., and G. Gambolati. 1978. On the use of main trend for the Kriging technique in hydrology. Adv. Water Res. 1:345–349.

Wagenet, R. T., S. W. Biggar, and D. R. Nielsen. 1977. Tracing the transformation of urea fertilizer during leaching. Soil Sci. Soc. Am. J. 41:896–902.

Warrick, A. W., and A. Amoozegar-Fard. 1979. Infiltration and drainage calculations using spatially scaled hydraulic properties. Water Resour. Res. 15:1116–1120.

Warrick, A. W., J. W. Biggar, and D. R. Nielsen. 1971. Simultaneous solute and water transfer for an unsaturated soil. Water Resour. Res. 7:1216–1225.

Warrick, A. W., G. J. Mullen, and D. R. Nielsen. 1977. Scaling field-measured soil hydraulic properties using a similar media concept. Water Resour. Res. 13:355–362.

Whittle, P. 1954. On stationary processes in the plane. Biometrika 41:434–449.

Wierenga, P. J. 1977. Solute distribution profiles computed with steady-state and transient water movement models. Soil Sci. Soc. Am. J. 41:1050–1055.

Wild, A. 1972. Nitrate leaching under bare fallow at a site in northern Nigeria. J. Soil Sci. 23:315–324.

Wild, A., and I. A. Babiker. 1976. The asymetric leaching pattern of nitrate and chloride in a loamy sand under field conditions.

12   Nitrogen Transformations in Submerged Soils

W. H. PATRICK, JR.

Center for Wetland Resources
Louisiana State University
Baton Rouge, Louisiana

I. INTRODUCTION

The amount of the earth's surface that is submerged exceeds by far the area of drained land. Rice fields, lake and stream bottoms, estuaries, and ocean bottoms are all characterized by having a layer of soil or sediment covered with water. In addition, some upland areas undergo periodic saturation or submergence. Nitrogen transformations in submerged (waterlogged, flooded) soils and sediments are markedly different from those taking place in drained soils and affect not only the amount of N available for biological activity, but also influence global N budgets. In order to properly deal with these N transformations it is necessary to understand the biological and chemical conditions that exist in submerged soils and sediments.

II. PROPERTIES OF SUBMERGED SOILS THAT AFFECT NITROGEN BEHAVIOR

The major characteristic of a submerged soil that results in it being different insofar as N transformations are concerned is the depletion of O_2 throughout most of the root zone. The overlying water column usually contains dissolved O_2 which usually moves a short distance into the soil before it is depleted. The thickness of the oxidized surface layer is determined by the net effect of O_2 consumption rate in the soil and O_2 supply from the flood water; a high consumption rate results in a thin oxidized surface layer, 1 mm or so in thickness, while a low consumption rate results in a thicker oxidized layer, which may be up to several centimeters. Because of the higher O_2 demand, a soil with an appreciable supply of readily decomposable organic matter usually has a thin oxidized layer. If the amount of energy source in the soil and in the water column is sufficient, O_2 may not reach the soil surface and there will not be an aerobic surface layer. This condition is usually temporary, however, since the continuous supply of O_2 from the atmosphere will in time satisfy the O_2 demand of the water column and O_2 will diffuse into the soil surface for a short distance. The reduced soil has been shown to consume O_2 at a rate several times that of an oxidized soil (Patrick & Sturgis, 1955; Patrick, 1960).

The differentiation of a submerged soil into two distinct zones based on O_2 penetration was first described by Pearsall and Mortimer (1939), although Shioiri and Mitsui (1935) had earlier noted a thin yellowish layer due to oxidized iron when a highly reduced soil was exposed to the atmosphere (Mitsui, 1954). Pearsall and Mortimer noted that the surface few centimeters of a submerged soil could be differentiated from the underlying layer by the presence of NO_3^- instead of NH_4^+, ferric Fe instead of ferrous Fe, and a high or positive redox potential instead of a low or more negative redox potential. Later, Alberda (1953) demonstrated the existence of these two layers with oxidation-reduction potential measurements made at 2-mm intervals from the aerated flood water down into the reduced soil layer.

The presence of a thin aerobic soil layer in close proximity to an anaerobic soil layer has important effects on N transformations as well as other important soil reactions. Nitrate is not readily denitrified in the aerobic layer but is denitrified in the anaerobic layer, while NH_4^+-N will be converted to NO_3^--N in the aerobic layer but not in the anaerobic layer. Several of the N processes taking place in submerged soils can be explained on the basis of movement of these two forms of N from one zone to the other and the biological reactions they subsequently undergo.

The oxidizing rhizosphere of aquatic plants also provides an aerobic-anaerobic interface (Sturgis, 1936; van Raalte, 1941; Armstrong, 1964) that affects N transformation and other processes. Plants that have evolved to grow in an aquatic environment transport O_2 from the atmosphere to the roots through an internal porous structure (aerenchyma). This alternate O_2 supply is necessary for root respiration since the root zone is usually devoid of O_2. Some of the O_2 reaching the root diffuses through the root cells into the surrounding soil, creating a thin aerobic zone surrounded by a much more extensive anaerobic zone. This interface is apparently important for both nitrification-denitrification reactions and N_2 fixation.

As long as O_2 is present in the soil, NO_3^--N and other oxidized components of the soil do not undergo extensive biological and chemical reduction. After O_2 has disappeared from a submerged soil the need for electron acceptors by facultative anaerobic and true anaerobic organisms results in the reduction of NO_3^--N, higher oxides of Mn, hydrated ferric oxide, SO_4^{2-}, and even CO_2. Nitrate and MnO_2 are reduced at fairly high redox potentials, whereas SO_4^{2-} and CO_2 are reduced only under the strictly anaerobic conditions associated with extremely low potentials. Ferric oxide reduction is intermediate. Thus, when O_2 is depleted following submergence, the reduction of the oxidized inorganic soil components is at least somewhat sequential; NO_3^--N and manganic Mn, followed in order by reduction of Fe^{3+} to Fe^{2+}, SO_4^{2-} to S^{2-}, and lastly, CO_2 to CH_4. Takai and Kamura (1966), Turner and Patrick (1968), Connell and Patrick (1968), Martens and Berner (1974), and others have presented evidence for the sequential reduction of O_2, NO_3^--N, manganic Mn, Fe^{3+}, SO_4^{2-}, and CO_2 in submerged soils. Oftentimes one component is not completely reduced before reduction of

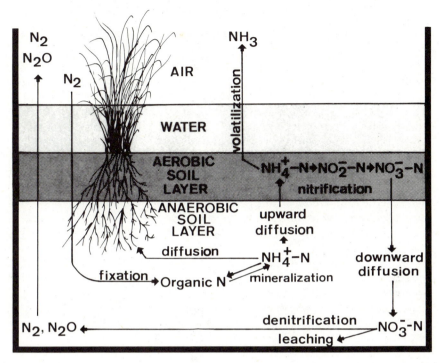

Fig. 1—A schematic outline of the redox zones in a submerged soil showing some of the N transformations.

the next component commences. Nitrate and Mn reduction usually begins before O_2 has completely disappeared. Reduction of the other oxidized components, however, is inhibited by either O_2 or NO_3^--N and consequently soils that contain these oxidants will not produce Fe^{2+}, S^{2-}, or CH_4. Also, when O_2 or NO_3^--N moves into a reduced soil zone containing Fe^{2+} or S^{2-} rapid chemical reduction of the O_2 and biological reduction of NO_3^--N occurs.

Although both O_2 and NO_3^--N are important in retarding reduction, they are usually present in such low amounts that they are rapidly reduced and consequently disappear soon after waterlogging. Hydrated ferric oxide and MnO_2, on the other hand, are usually present in much higher quantities in soils. The Fe system, in particular, generally has a large reserve of solid phase reducible matter. The oxidation and reduction of these compounds are cyclic—that is, they are reduced when the soil is waterlogged and reoxidized when the soil is subsequently drained and exposed to O_2. Consequently, active Fe and Mn compounds serve as buffers against the development of reducing conditions in the soil (Takahashi, 1960). Nitrogen reactions are also cyclic, in a sense, since under aerobic conditions all of the N reactions are proceeding toward the NO_3^--N form while under anaerobic conditions NO_3^--N is converted to more reduced forms, usually N_2 and N_2O, and sometimes NH_4^+. Unlike Mn and Fe, however, some of the N is lost from the soil as it undergoes cyclic oxidation-reduction reactions.

III. NITROGEN TRANSFORMATIONS IN SUBMERGED SOILS

The forms of N present in submerged soils are generally similar to those of upland soils with some well-recognized differences such as in the relative contents of NO_3^--N and NH_4^+-N. Nitrogen transformations, however, differ markedly under the different conditions. This difference in behavior of N in upland and submerged soils is due largely to the difference in activity of the microorganisms functioning under aerobic and anaerobic conditions. In drained soil, most N transformations are carried out by aerobic or sometimes facultative anaerobic bacteria while in submerged soils facultative anaerobic and true anaerobic bacteria predominate. As in aerobic soils the major microbial processes involving N in submerged soil are mineralization, immobilization, nitrification, denitrification, and N fixation. Submergence of the soil modifies these processes considerably, however.

A unique characteristic of flooded soils is that all of these processes can be occurring at the same time because of the presence of both aerobic and anaerobic zones. Thus, a flooded soil is a very complex system insofar as N transformations are concerned and it is usually very difficult to evaluate the significance of any single reaction.

A. Mineralization and Immobilization

Anaerobic decomposition of organic N compounds to inorganic forms in submerged soils can be differentiated from aerobic decomposition in two important aspects: first, by its much slower rate, and second, by the nature of the end products. Microbial decomposition of organic matter in a well-drained soil is accomplished by a wide group of microorganisms in which fungi play a prominent role. Respiration by these organisms is associated with high energy release, and decomposition of substrate and synthesis of cell substances proceed at a rapid rate. As a result, most components forming the bulk of the added organic matter are rapidly converted to CO_2, while those less susceptible to microbial attack persist. As cell synthesis proceeds, a heavy demand is placed upon mineral nutrients, particularly N, except when highly nitrogenous materials are added.

Anaerobic decomposition, on the other hand, is largely dependent on the activities of anaerobic bacteria. Facultative anaerobes will also assist in decomposition as long as alternative electron acceptors are present. The anaerobic SO_4^{2-} reducers are important in coastal wetlands that receive SO_4^{2-} in sea water. Since anaerobic bacteria operate at a much lower energy level, they are much less efficient than aerobic microflora. Consequently, the processes of both decomposition and resynthesis are much slower in a submerged soil than in a well-drained soil. This is in agreement with the general observation that greater accumulation of plant residues occurs in bog and marsh soils.

Tenney and Waksman (1930) conducted a comprehensive study of the decomposition rate of various chemical constituents of a variety of plant material under both aerobic and anaerobic conditions and found that anaerobic decomposition was slower than aerobic decomposition regardless of the type of plant material used. The difference in the rates of aerobic and anaerobic decomposition was even more pronounced for organic nitrogenous complexes. For corn stalks high in water-soluble substances, including reducing sugars and N compounds, 20% of the total material was lost under anaerobic conditions after 27 days, whereas 37% disappeared during the same period of time under aerobic conditions. Relatively slower rates of decomposition were observed with rye straw, which has a lower content of soluble carbohydrates and a higher lignin content. About 17% of the rye straw decomposed in 66 days and 29% in 143 days in the aerobic system, whereas under anaerobic conditions 7 and 22% were decomposed in 84 and 163 days, respectively. The three principal chemical constituents of plants, cellulose, hemicellulose, and lignin, also decomposed much more slowly under anaerobic than under aerobic conditions. A similar decomposition pattern was noted for alfalfa in spite of its relatively high N content. This indicates that even a high N content did not cause rapid decomposition under anaerobic conditions. This was true not only for the total plant material, but also for the individual plant constituents. It is noteworthy that accumulation of proteins was greater under anaerobic conditions than under aerobic conditions in all cases. This may have been due to the more economic use of N, to the small loss of NH_3, or to the lesser decomposition of synthesized protein in the anaerobic system.

The characteristic features of anaerobic bacterial degradation of organic matter in submerged soils are: (i) incomplete decomposition of carbohydrate into CH_4, organic acids, H_2, and CO_2 with consequent low energy yield (Alexander, 1961); (ii) low energy of fermentation, resulting in the synthesis of fewer microbial cells per unit of organic C degraded; for example, only 2 to 5% of the substrate C is assimilated by anaerobic bacteria compared with about 30 to 40% assimilation by fungi in the aerobic system (Alexander, 1961); and (iii) low N requirement of the anaerobic metabolism leading to a more rapid release of NH_4^+ ions than would ordinarily be expected on the basis of high C/N ratios of decomposing plant material and the slow rate of anaerobic decomposition of plant residues.

The greater release of NH_4^+ during anaerobic decomposition is extremely important for lowland rice culture. Acharya (1935) showed that the amount of N required for decomposition of rice straw was higher under aerobic conditions than under anaerobic conditions. Conversely, the net release of inorganic N to the solution phase was higher in the anaerobic than in the aerobic system. Quantitatively, the N released by rice straw decomposing under anaerobic conditions was about 5 to 6 times higher than that released under aerobic conditions. Sircar et al. (1940) presented similar evidence and reported that inorganic N release from decomposing rice straw occurred at a higher C/N ratio under anaerobic conditions. Further, they noted that a minimum N concentration in rice straw of 1.7 to 1.9% was necessary for net accumulation of mineral N, whereas only 0.45 to 0.50% was sufficient under anaerobic conditions. In a field study in California,

Williams et al. (1968) concluded that the N requirement for the decomposition of rice straw in submerged soils was one-third (0.54 vs. 1.5%) the average concentration of N required for aerobic decomposition of plant residues. de Geus (1954) attributed the response of rice to added organic matter to the ready release of NH_4^+ during anaerobic decomposition. Waring and Bremner (1964) observed a more rapid release of inorganic N under waterlogged than under aerobic conditions in a number of soils. Quantitatively, they showed that in a 2-week period, for every 1 part of inorganic N produced aerobically 1.23 parts were produced under waterlogged conditions.

Another feature of anaerobic mineralization of N is its sensitivity to pH. The anaerobic bacterial flora operate most efficiently around neutral pH. This may explain why the liming of paddy soils in Japan has sometimes been claimed to cause an appreciable increase in ammonification (Mitsui, 1954). Nitrogen mineralization under submerged conditions is usually not limited as much as would be expected by adverse pH conditions, however, since submergence tends to drive the pH of both acid and alkaline soils toward the neutral point (Ponnamperuma, 1972), and submerged soils have a narrower range of pH values than do drained soils.

Studies with ^{15}N show that under almost all soil conditions mineralization and immobilization are proceeding at the same time. Insofar as availability of soil N is concerned the imbalance between these two processes or the net mineralization or net immobilization is the significant factor. It is possible to have little or no net effect even though the simultaneous processes of mineralization and immobilization are going on vigorously. The resulting net effect usually depends on the ratio between the carbonaceous material (energy source) and the N in substances undergoing decomposition (Jansson, 1966). The aeration status of the soil also plays a role in determining if net mineralization or net immobilization will occur.

B. Nitrification-Denitrification

Nitrification is completely inhibited in a submerged soil if no free O_2 exists. However, some nitrification can take place in certain parts of a submerged soil, since O_2 is usually present in the water column and in the surface layer of soil as well as in the aerobic root rhizosphere of aquatic plants. Submerged soils and other submerged sediment systems are characterized by having aerobic zones where nitrification can occur in close proximity to anaerobic zones where denitrification can occur. This condition has important implications for fertilizer N utilization by crops and for the overall biogeochemical N cycle.

Denitrification in O_2-deficient soils is covered in detail in Chapter 8 (M. K. Firestone) and the discussion of denitrification in this chapter will be as a part of the overall nitrification-denitrification process that resulted in the loss of N from submerged soil. The two major sites for these reactions are the surface layer of soil and the root rhizosphere, where aerobic and anaerobic conditions occur in close proximity.

When a soil containing an energy source such as plant residue is submerged, NO_3^--N present in the soil appears to be largely lost in gaseous form. Many earlier workers assumed that this NO_3^--N was converted to NH_4^+-N, since an NH_4^+-N buildup generally accompanied the disappearance of NO_3^--N after flooding. Experiments conducted in recent years utilizing $^{15}NO_3^--N$, however, have shown that most of the NO_3^--N added to a submerged soil is converted to N_2 or N_2O with little assimilatory NO_3^--N reduction occurring (Broadbent & Stojanovic, 1952; Wijler & Delwiche, 1954; Cady & Bartholomew, 1960). It is obvious from these studies that restriction of the O_2 supply by waterlogging results in a much higher demand for NO_3^--N as an electron acceptor than as a nutrient source of N. The addition of an energy source to the soil has been found by most workers to increase NO_3^- reduction and denitrification.

Although conversion to N_2 and N_2O appears to be the likely fate of NO_3^--N present in a soil at the time of submergence, in some cases reduction all the way to NH_4^+-N occurs. Stanford et al. (1975a) found substantial reduction of $^{15}NO_3^--N$ to NH_4^+-N and incorporation of $^{15}NO_3^--N$ into soil organic matter in glucose-amended, anaerobic soils. In a subsequent study, Stanford et al. (1975b) reported that the extent of NO_3^--N reduction to NH_4^+-N relative to denitrification increased with an increase in the C:N ratio. Reduction of $^{15}NO_3^--N$ to NH_4^+-N preceded microbial assimilation of labeled N. The results suggested that NH_4^+-N rather than NO_3^--N was the primary form of N assimilated. Buresh and Patrick (1978) confirmed the results of Stanford and found that glucose was not necessary if enough soil organic matter was present to create intense reducing conditions. A 20-day preincubation under Ar before the NO_3^--N was added caused about one fifth of the NO_3^--N to be reduced to NH_4^+-N.

Conversion of NO_3^--N to organic N and NH_4^+-N and denitrification can occur simultaneously in lake sediments (Chen et al., 1972; Keeney et al., 1971). Chen et al. (1972) observed 7% and 3% conversion of added $^{15}NO_3^--N$ to ^{15}N-labeled organic N and NH_4^+-N, respectively, after a 2-day anaerobic incubation of calcareous lake sediments. They attributed the presence of ^{15}N-labeled organic N to (i) assimilatory NO_3^- immobilization with subsequent rapid mineralization, or (ii) fixation of $^{15}N_2$ formed via denitrification. Under aerobic conditions, 10.5% of the applied $^{15}NO_3^--N$ was recovered in the organic N fraction but negligible ^{15}N appeared in the NH_4^+-N form. Mineralization of ^{15}N-labeled organic N to NH_4^+-N was assumed to be absent.

The importance of aerobic-anaerobic profile differentiation in the studies of N transformations in submerged soils was perhaps first realized in Japan. In laboratory experiments, Shiori and Mitsui (1935) observed considerable losses of N from applied $(NH_4)_2SO_4$ as incubation proceeded. The losses were apparently too great to be accounted for by direct volatilization or by other possible ways of N loss known at that time. Further experimentation to determine the effect of stirring on NH_4^+ accumulation under waterlogged conditions led them to believe that even in a submerged soil NH_4^+-N could be nitrified to NO_3^--N if applied to the soil surface. They hypothesized that NO_3^--N thus formed moves down into the anaerobic zone by percolation or by diffusion, and is subsequently denitrified biologically and possibly chemically to gaseous N. This process was subsequently confirmed by Pearsall (1950), Shioiri and Tanada (1954), and Mitsui (1954).

Numerous studies indicate that these losses of N through nitrification and subsequent denitrification will largely account for the low recovery of ammoniacal fertilizers by rice in submerged soils (IAEA, 1966). Abichandani and Patnaik (1955) estimated these losses to be 20–40% of the applied N in India, while losses of 30–50% were reported in Japan (Mitsui, 1954). Other reports indicate crop recovery of $(NH_4)_2SO_4$ will be as low as 40% (Basak et al., 1957; De & Digar, 1955; Yamane, 1957). According to Amer (1960), the recovery of applied $(NH_4)_2SO_4$ decreased with increased fertilizer levels. Nelson and Hauck (1965) indicated that even under the best conditions losses of about 10 to 20% are unavoidable.

Application of fertilizer NH_4^+-N to flooded rice field by deep placement has been shown to be superior to surface application in a number of reports (Mitsui, 1954; Mikkelsen & Finfrock, 1957; Patrick et al., 1967; Yanagisawa & Takahashi, 1964; DeDatta & Venkateswarlu, 1968). Quantitative recovery of applied fertilizer N in the soil-plant system under field conditions was determined by Patrick and Reddy (1976b) and Reddy and Patrick (1976). Their studies indicated that when labeled $(NH_4)_2SO_4$ was applied by deep placement, 49% of the applied N was recovered in the plant, 26% of the applied N remained in the soil either in the roots or incorporated in soil organic matter, and 25% was lost from the system. It was suggested that at least part of the N losses were due to NH_4^+-N diffusing from the anaerobic layer to the aerobic layer where it underwent nitrification and subsequent denitrification.

The importance of the nitrification-denitrification sequence has also been recognized by researchers working with nonagricultural systems. Chen et al. (1972) showed that nitrification and subsequent denitrification functioned simultaneously in lake sediments incubated under air in the laboratory. Graetz et al. (1973) observed that during anoxic conditions NH_4^+-N was released to the water at a relatively constant rate. Aeration caused rapid nitrification and the NO_3^- thus formed in the overlying water decreased with time, possibly due to diffusion into the highly reduced sediment and denitrification. Kemp and Mudrochova (1972) working with Lake Ontario sediments, indicated that N was returned to the atmosphere through nitrification-denitrification processes. Billen (1975) showed that nitrification of NH_4^+-N accounts for the increase in NO_3^- levels in the upper layers of the sediment. When this NO_3^- moved below the 3-cm layer it was lost by denitrification. Brujewicz and Zaitseva (1972), working with deep ocean sediments, reported the distribution of NH_4^+-N was low in the overlying water and the surface oxidized sediment layer, while high NH_4^+ levels were found in the lower sediment layers. These results suggest that the same nitrification-denitrification process occurs in deep ocean sediments as in submerged soils and unstratified lakes.

Tracer studies showing the pathway of the nitrification-denitrification reaction have been carried out by several investigators. Tusneem and Patrick (1971), Broadbent and Tusneem (1971), Patrick and Tusneem

(1972), Manguiat and Yoshida (1973), and Yoshida and Padre (1974) reported that substantial N losses occurred in soils receiving N as $(NH_4)_2SO_4$ and maintained in a flooded condition throughout the experimental period. These studies demonstrated that nitrification and subsequent denitrification reactions were functioning simultaneously in a flooded system and were controlling N loss. In other recent studies, Patrick and DeLaune (1972), Patrick and Gotoh (1974), Takai and Uehara (1973), and Patrick and Reddy (1976a) showed that more N can be lost from a submerged soil than is actually present as NH_4^+ in the aerobic soil layer at any one time.

The sequential processes functioning in a natural submerged soil or a water-sediment system are NH_4^+ diffusion from the anaerobic layer to the surface aerobic layer → nitrification → NO_3^- diffusion from the aerobic layer to the anaerobic layer → denitrification. These four processes will occur continuously if O_2 is present at the sediment surface and if there is enough organic matter to create reducing conditions in the sediment and to provide NH_4^+. It is important to know at what rates these processes are functioning and which of these processes is limiting the overall rate. The kinetics of each process were investigated by Reddy et al. (1980). Their data showed NH_4^+ diffusion rate (0.216 cm^2/day) and nitrification rate (2.07 μg/g per day) functioning slower than NO_3^- diffusion (1.33 cm^2/day) and denitrification (0.315/day) in a submerged Crowley silt loam soil. Nitrate diffusion and denitrification processes apparently did not slow down the nitrification-denitrification sequence in the soil studied. Ammonium diffusion is dependent on nitrification to remove NH_4^+ and to establish a concentration gradient of NH_4^+. If nitrification does not occur, there will be little NH_4^+-N movement. No NH_4^+-N loss was observed in soils incubated under an O_2 free atmosphere (Patrick & Reddy, 1976a).

Nitrogen losses in submerged soils and sediments can be greatly influenced by the presence of plants. The magnitude of N loss will be reduced by plant uptake of inorganic N. Conversely, N losses will be enhanced by the presence of the aerobic layer around the root zone. Woldendorp (1968) suggested that plant roots can accelerate denitrification in the rhizosphere of aerobic soils by taking up O_2 and by secreting organic substances which can serve as hydrogen donors in the denitrification process. Data obtained by workers at IRRI (1964) have indicated the presence of nitrifying bacteria around the root zone. At present, no information is available on the activity of nitrifying organisms in the oxidized rhizosphere of predominately reduced soils. It is speculated, however, that the high root density of rice plants in submerged soil can increase the total area of the aerobic zone and create a favorable condition for nitrification to occur. These same reactions can occur in the aerobic root rhizosphere of swamp and marsh plants.

C. Nitrogen Fixation

Nitrogen fixation has been shown to be a significant process in submerged soils. Organic N tends to accumulate in submerged soils and wetlands in higher concentration than in drained soil. Conversion of drained land areas to flooded conditions invariably results in higher N content, while the reverse process results in loss of N. In subsidence agricultural systems, where grain is grown every year with no added fertilizer, yields of lowland rice are usually higher than yields of upland grain crops. Part of this yield difference can be attributed to a better water supply in submerged soils, but part of the difference is likely due to greater N fixation in the submerged soil. This section deals with the significance of fixation in the various parts of a submerged soil and the factors affecting fixation. For a more comprehensive treatment see Buresh et al. (1980).

Nitrogen fixation by blue-green algae in the water column and on the surface of submerged soils is known to be an important source of N in flooded systems (De, 1939; Singh, 1961; Watanabe et al., 1978). Nitrogen fixation can occur also in other parts of the plant-soil-water system, although relative contributions of these other types of fixation are poorly understood. Watanabe et al. (1978) found that the N-fixing activity in a rice field was related to the activity and biomass of blue-green algae. Yoshida and Ancajas (1973) speculated that shading from the rice plants decreased algal growth and, thus, N-fixing activity. Nitrogen fixation in the overlying water of a rice field was greater in the dry season than in the wet season (Yoshida & Ancajas, 1973) and in unfertilized plots than in those fertilized with NPK (Watanabe et al., 1978).

In addition to free-living blue-green algae, *Anabaena azollae* in symbiotic association with the freshwater fern *Azolla* fixes N. *Azolla* is widely distributed in paddy fields and water bodies within tropical and temperate regions (Moore, 1969). The algae inhabits the leaf cavity of *Azolla,* and it obtains a degree of physical protection and probably metabolites from the fern (Moore, 1969; Peters & Mayne, 1974). *Azolla* containing the algae is able to assimilate atmospheric N. Peters and Mayne (1974) confirmed by the acetylene reduction technique that *Anabaena* was the agent of nitrogenase activity. The N-fixing activity of *Anabaena* associated with *Azolla* nearly covering the surface of a plant-free flooded field reached a maximum of about 1 kg N/ha per day 20 days after inoculation (IRRI, 1975). The *Azolla-Anabaena* association is of agronomic interest as a green manure and possible partial substitute for N fertilizer in rice fields (Moore, 1969; Watanabe, 1978).

The aerobic surface layer of soil is also a site of active N fixation. Magdoff and Bouldin (1970) reported substantially greater N fixation in the surface layer compared to lower layers in a flooded, cellulose enriched soil-sand media. These authors hypothesized that the products of anaerobic cellulose decomposition move from the anaerobic to the aerobic zone where they serve as substrates for aerobic heterotrophic N fixers such as *Azotobacter.* Rice and Paul (1972), on the other hand, concluded that products of plant residue decomposition in the aerobic soil layer were utilized as energy sources by anaerobic N fixers in the lower layer. Rice et al. (1967) found high rates of fixation in a thin layer of saturated rice straw-amended soil with a large aerobic-anaerobic interfacial area. Waughman (1976) suggested that the presence of both an aerobic and anaerobic soil zone can stimulate nitrogenase activity.

Nitrogen-fixing activity has been reported in the anaerobic zone of aquatic (Howard et al., 1970; Keirn & Brezonik, 1971; Macgregor et al., 1973), estuarine (Brooks et al., 1971; Herbert, 1975), and salt marsh sediments (Jones, 1974; Marsho et al., 1975), and rice soil (MacRae & Castro, 1967). The organisms responsible for N fixation in the anaerobic layer appear to be predominantly *Clostridium* (Rice et al., 1967; Brooks et al., 1971; Rice & Paul, 1972), but *Desulfovibrio* (Herbert, 1975), *Klebsiella,* and *Enterobacter* (Werner et al., 1973) may also be important.

Nitrogen fixation decreases with depth down the soil profile in estuarine (Brooks et al., 1971) and aquatic sediments (Keirn & Brezonik, 1971) and salt marsh soils (Marsho et al., 1975; Whitney et al., 1975). Brooks et al. (1971) found that the largest N-fixing activity in an estuarine sediment occurred in the top 2- to 5-cm section. Teal et al. (1979) found that the rates of N fixation by rhizosphere and free-living bacteria were highest near the surface of salt marshes. Bacterial N_2 fixation was more than 10 times larger than fixation by algae. Casselman et al. (1981) have shown that highest N fixation as measured by nitrogenase activity occurs in the surface layer of vegetated salt marsh soils. Nitrogen fixation profiles followed the root distribution of the marsh plants.

Although N fixation by free-living bacteria in unamended soil is low compared to fixation by blue-green algae, bacteria occupy a larger fraction of the flooded habitat than algae and over a period the amount fixed by bacteria may be significant (Jones, 1974; Whitney et al., 1975). Availability of an energy source appears to be the primary factor limiting nonphotosynthetic N-fixing bacteria (Stewart, 1969; Hanson, 1977a).

Associations between N-fixing bacteria and the roots of various non-nodulated plants have been verified with the acetylene reduction method. Döbereiner et al. (1972) reported that the association of *Azotobacter paspali* with the roots of a tropical grass, *Paspalum notatum,* fixed an estimated 90 kg N/ha per year. De-Polli et al. (1977) confirmed both N fixation by bacteria associated with the roots of tropical grasses and plant assimilation of the fixed N by the $^{15}N_2$ reduction method.

Nonsymbiotic N fixation by bacteria has also been found in the rhizosphere of rice (Rinaudo et al., 1971; Yoshida & Ancajas, 1971; Yoshida & Ancajas, 1973; Dommergues et al., 1973) and fresh water (Bristow, 1974; Silver & Jump, 1975), marine (Patriquin & Knowles, 1972), and salt marsh angiosperms (Jones, 1974). Yoshida and Ancajas (1973) found much higher N fixation in soil planted with rice than in unplanted soil, and Lee et al. (1977a) observed higher N fixation in the soil of planted areas of a rice field than within the soil of unplanted areas between plant rows.

Larger populations of N-fixing bacteria are observed in rhizosphere than nonrhizosphere soil in rice (Balandreau et al., 1975) and *Thalassia* (Patriquin & Knowles, 1972), and nitrogenase activity in a salt marsh has been shown to approximate the root distribution of *Spartina alterniflora* (Hanson, 1977b). Researchers have suggested that root exudates of rice (Yoshida & Ancajas, 1973), mangroves (Zuberer & Silver, 1978), and *Spartina alterniflora* (Hanson, 1977b) supply N-fixing rhizosphere bacteria with an energy source. Balandreau et al. (1975) speculated that the response of N fixation in the rice rhizosphere to illumination results from root exudation promoted by increased plant photosynthesis. Root exudates are known to stimulate microbial growth and activity in the rhizosphere of plants (Rovira, 1965).

Atmospheric N is transported by the rice plant to its roots where it diffuses into the rhizosphere for fixation by bacteria (Yoshida & Broadbent, 1975). The rate of N gas transport by the rice plant is greatest at the heading and flowering stage (Yoshida & Broadbent, 1975), the growth stage which corresponds to greatest N-fixing activity (Yoshida & Ancajas, 1973). *Spartina alterniflora* is also capable of transporting air from the atmosphere to its roots (Teal & Kanwisher, 1966).

Submerged soils with plants producing large amounts of root exudates are located in regions with a sunny climate where the ecosystems favor maximum N fixation (Dommergues et al., 1973). Rice fields and coastal wetlands are such ecosystems. Although N fixation by bacteria in the rhizosphere is possibly less than that of blue-green algae in a rice field, it is nonetheless quantitatively significant (Knowles, 1977).

The evidence is strong that N accumulates under excess water, low O_2 conditions. However, the greater N accumulation in wetlands compared to drained soils is not altogether the result of increased N fixation, but to slower mineralization and subsequent lower losses of N through leaching and/or denitrification. The slow mineralization rate in wetlands was discussed earlier in this chapter.

D. Ammonium Volatilization

Nitrogen can also be lost from submerged soils through volatilization of NH_3. Major factors that favor NH_3 volatilization are high NH_3 concentration (especially in the overlying floodwater), high pH, and high temperature. The CEC of the soil is also a factor, since it affects the distribution of NH_4^+ between the solution and solid phases. It has generally been considered that under most submerged soil conditions NH_3 volatilization does not account for large amounts of gaseous N going into the atmosphere. However, recent research has indicated that a reassessment of the significance of NH_3 volatilization in flooded rice fields may be in order (Bouldin & Alimagmo, 1976; Mikkelsen et al., 1978; Vlek & Craswell, 1979). These researchers found that significant amounts of N can be lost by this mechanism.

Urea is now the main form of fertilizer N for rice and NH_3 volatilization following hydrolysis of urea may be a greater problem than loss from $(NH_4)_2SO_4$. This loss would be enhanced by a pH increase in the water column, which usually occurs in natural waters in the middle of the day when CO_2 in the water column is largely depleted as a result of photosynthesis (Park et al., 1958). In a greenhouse experiment Vlek and Craswell (1979) found as much as one half of the NH_3 produced from urea was lost from the submerged soil in a 3-week incubation. Nitrogen loss from $(NH_4)_2SO_4$ was only about one-fifth the loss from urea. In a field experiment in the Philippines, Mikkelsen et al. (1978) measured a loss of one-fifth of the surface applied $(NH_4)_2SO_4$ from a near-neutral pH soil. Carbon dioxide depletion due to algae growth in the floodwater was sufficient to raise the water pH to above 9; on an acid soil where algae did not grow well, the water pH did not rise past the neutral point and little topdressed N was lost. MacRae and Ancajas (1970) and Ventura and Yoshida (1977) reported similar results with losses from $(NH_4)_2SO_4$ being low and losses from urea as high as 19% in the extreme case. Most studies also show that the loss is greatest immediately after application.

It is apparent from the above studies that NH_3 volatilization from submerged soils is quite variable but that it can be significant under certain conditions. The amount and form of N added, the pH of the soil and water, the temperature of the water, and the CEC of the soil are all important factors in determining the amount of N lost.

IV. MANAGEMENT OF SUBMERGED SOIL TO MINIMIZE NITROGEN LOSS

Management practices can be used that will minimize losses of both native soil N and applied fertilizer N in submerged soils planted to rice. These practices take advantage of anaerobic soil condition to prevent oxidation of NH_4^+ to NO_3^-. The greatest efficiency of N fertilizer use is usually obtained when NH_4^+ fertilizers or NH_4^+-forming materials such as urea are placed several centimeters deep in the soil and the soil flooded immediately and maintained in a reduced condition for most of the growing season. On the other hand, NH_4^+ fertilizer applied to the soil surface or in the flood water is subject to oxidation to NO_3^- which in turn can be denitrified and lost. Where NH_3 volatilization is likely to be a problem, NH_3 loss will be lower if the fertilizer is placed beneath the soil surface. Deep placement of NH_4^+ fertilizer to minimize N loss was first utilized in Japan as a result of the work of Shiori, Mitsui and others (Mitsui, 1954). Mikkelson and Finfrock (1957) in California showed that this technique was adapted to mechanized rice farming using anhydrous NH_3 or aqua NH_3 as the source of N. Patrick et al. (1967) also found better utilization of N with deep-placed fertilizer as compared to top dressed material. The key factor in maintaining soil conditions that will minimize N loss from denitrification is to keep the soil in an anaerobic condition so that nitrification cannot take place. This is done by keeping the soil flooded throughout the growing season so that O_2 will not enter the soil.

Just as N loss can be minimized by maintaining the soil in a flooded condition, N loss can be accelerated by alternating periods of flooding and draining. In the rare cases where more available N is present in the soil than can be utilized by the rice crop, a worthwhile management practice is to flood and drain the soil several times during the growing season to stimulate nitrification and subsequent denitrification. Such a condition is likely to exist only where new land is cleared for rice cultivation or where legume-grass pastures are converted to rice.

LITERATURE CITED

Abichandani, C. T., and S. T. Patnaik. 1955. Mineralizing action of lime on soil nitrogen in waterlogged rice soils. Int. Rice Comm. News Lett. 13:11–13.

Acharya, C. N. 1935. Studies on the anaerobic decomposition of plant materials. I. Anaerobic decomposition of rice straw. Biochem. J. 29:528–541.

Alberda, T. 1953. Growth and root development of lowland rice and its relation to oxygen supply. Plant Soil 5:1–28.

Alexander, M. 1961. Introduction to soil microbiology. John Wiley & Sons, Inc., New York.

Amer, F. 1960. Evaluation of dry sub-surface and wet surface ammonium sulfate application for rice. Plant Soil 13:47–54.

Armstrong, W. 1964. Oxygen diffusion from the roots of some British bog plants. Nature (London) 204:801.

Balandreau, J. P., G. Rinaudo, I. Fares-Hamad, and Y. R. Dommergues. 1975. Nitrogen fixation in the rhizosphere of rice plants. p. 57–70. In W. D. P. Stewart (ed.) Nitrogen fixation by free-living microorganisms. Cambridge Univ. Press, Cambridge.

Basak, M. N., T. Dutt, and D. K. Nag. 1957. Effect of mixed forms of nitrogen on yield. J. Indian Soc. Soil Sci. 5:55–63.

Billen, G. 1975. Nitrification in the Scheldt Estuary (Belgium and the Netherlands). Estuarine Coastal Mar. Sci. 3:79–89.

Bouldin, D. R., and B. V. Alimagmo. 1976. NH₃ volatilization losses from IRRI paddies following broadcast application of fertilizer nitrogen. Internal report. Int. Rice Res. Inst., Philippines.

Bristow, J. M. 1974. Nitrogen fixation in the rhizosphere of freshwater angiosperms. Can. J. Bot. 52:217–221.

Broadbent, F. E., and M. E. Tusneem. 1971. Losses of nitrogen from some flooded soils in tracer experiments. Soil Sci. Soc. Am. Proc. 35:922–926.

Broadbent, F. E., and B. F. Stojanovic. 1952. The effect of partial pressure of oxygen on some soil nitrogen transformations. Soil Sci. Soc. Am. Proc. 16:359–363.

Brooks, R. H., P. L. Brezonik, H. D. Putnam, and M. A. Keirn. 1971. Nitrogen fixation in an estuarine environment: The Waccasassa on the Florida Gulf Coast. Limnol. Oceanogr. 16:701–710.

Brujewicz, S. V., and E. D. Zaitseva. 1972. Chemistry of sediments of the northwestern Pacific Ocean. p. 1–121. In Problems of sea chemistry. Translated from Russian. Akademiya Nauk SSSR. Trudy Inst. Okeanol. Vol. XLII. 1960. U.S. Dep. Commerce. Ind. Natl. Sci. Doc. Cent., New Delhi.

Buresh, R. J., and W. H. Patrick, Jr. 1978. Nitrate reduction to ammonium in anaerobic soil. Soil Sci. Soc. Am. J. 42:913–918.

Buresh, R. R., M. E. Casselman, and W. H. Patrick, Jr. 1980. Nitrogen fixation in flooded soil systems, A review. Adv. Agron. 33:149–192.

Cady, F. B., and W. V. Bartholomew. 1960. Sequential products of anaerobic denitrification in Norfolk soil material. Soil Sci. Soc. Am. Proc. 24:477–482.

Casselman, M. E., W. H. Patrick, Jr., and R. D. Delaune. 1981. Nitrogen fixation in Gulf Coast salt marsh. Soil Sci. Soc. Am. J. 45:51–56.

Chen, R. L., D. R. Keeney, D. A. Graetz, and A. J. Holding. 1972. Denitrification and nitrate reduction in Wisconsin lake sediments. J. Environ. Qual. 1:158–162.

Connell, W. E., and W. H. Patrick, Jr. 1968. Sulfate reduction in soil: Effect of redox potential and pH. Science 159:86–87.

De, P. K. 1939. The role of blue-green algae in nitrogen fixation in rice fields. Proc. Roy. Soc. Lond. Ser. B. 127:121–139.

De, P. K., and S. Digar. 1955. Influence of the rice crop on the loss of nitrogen gas from waterlogged soils. J. Agric. Sci. 45:280–282.

DeDatta, S. K., and J. Venkateswarlu. 1968. Uptake of fertilizer phosphorus and nitrogen from different methods of application by lowland rice growing on major Indian soils. 9th Int. Congr. Soil Sci. Trans. (Adelaide, Australia) 4:9–18.

de Geus, J. G. 1954. Means of increasing rice production. Centre d'Etude de l'Azote, Geneva, Switzerland.

De-Polli, H., E. Matsui, J. Döbereiner, and E. Salati. 1977. Confirmation of nitrogen fixation in two tropical grasses by ¹⁵N₂ incorporation. Soil Biol. Biochem. 9:119–123.

Döbereiner, J., J. M. Day, and P. J. Dart. 1972. Nitrogenase activity and oxygen sensitivity of the *Paspalum notatum—Azotobacter paspali* association. J. Gen Microbiol. 71:103–116.

Dommergues, Y., J. Balandreau, G. Rinaudo, and P. Weinhard. 1973. Non-symbiotic nitrogen fixation in the rhizospheres of rice, maize, and different tropical grasses. Soil Biol. Biochem. 5:83–89.

Graetz, D. A., D. R. Keeney, and R. B. Aspiras. 1973. Eh status of lake sediment-water systems in relation to nitrogen transformations. Limnol. Oceanogr. 18:908–917.

Hanson, R. B. 1977a. Comparison of nitrogen fixation activity in tall and short *Spartina alterniflora* salt marsh soils. Appl. Environ. Microbiol. 33:596–602.

Hanson, R. B. 1977b. Nitrogen fixation (acetylene reduction) in a salt marsh amended with sewage sludge and organic carbon and nitrogen compounds. Appl. Environ. Microbiol. 33:846–852.

Herbert, R. A. 1975. Heterotrophic nitrogen fixation in shallow estuarine sediments. J. Exp. Mar. Biol. Ecol. 18:215–225.

Howard, D. L., J. I. Frea, R. M. Pfister, and P. R. Dugan. 1970. Biological nitrogen fixation in Lake Erie. Science 169:61–62.

International Atomic Energy Agency. 1966. IAEA. Tech. Rep. Ser. 55.

International Rice Research Institute. 1964. IRRI. Annual Report. p. 241–256.

International Rice Research Institute. 1975. IRRI. Annual Report. Los Baños, Philippines.

Jansson, S. L. 1966. Use of isotopes in soil organic matter studies. p. 415–422. *In* Report of the FAO/IAEA Technical Meeting, Brunswick-Völkenrode, 1963. Pergamon Press, New York.

Jones, K. 1974. Nitrogen fixation in a salt marsh. J. Ecol. 62:553–565.

Keeney, D. R., R. L. Chen, and D. A. Graetz. 1971. Importance of denitrification and nitrate reduction in sediments to the nitrogen budgets of lakes. Nature (London) 233:66–67.

Keirn, M. A., and P. L. Brezonik. 1971. Nitrogen fixation by bacteria in Lake Mize, Florida, and in some lacustrine sediments. Limnol. Oceanogr. 16:720–731.

Kemp, A. L. W., and A. Mudrochova. 1972. Distribution and forms of nitrogen in a Lake Ontario sediment core. Limnol. Oceanogr. 17:855–867.

Knowles, R. 1977. The significance of asymbiotic dinitrogen fixation by bacteria. p. 33–83. *In* R. W. F. Hardy and A. H. Gibson (ed.) A treatise on dinitrogen fixation Section IV: Agronomy and ecology. John Wiley & Sons, Inc., New York.

Lee, K. K., B. Alimagno, and T. Yoshida. 1977a. Field technique using the acetylene reduction method to assay nitrogenase activity and its association with the rice rhizosphere. Plant Soil 47:519–526.

Macgregor, A. N., D. R. Keeney, and K. L. Chen. 1973. Nitrogen fixation in lake sediments: Contribution to nitrogen budget of Lake Mendota. Environ. Lett. 4:21–26.

MacRae, I. C., and R. Ancajas. 1970. Volatilization of ammonia from submerged tropical soils. Plant Soil 33:97–103.

MacRae, I. C., and T. F. Castro. 1967. Nitrogen fixation in some tropical rice soils. Soil Sci. 103:277–280.

Magdoff, F. R., and D. R. Bouldin. 1970. Nitrogen fixation in submerged soil-sand-energy material media and the aerobic-anaerobic interface. Plant Soil 33:49–61.

Manguiat, I. J., and T. Yoshida. 1973. Nitrogen transformation of ammonium sulfate and alanine in submerged Maahas Clay. Soil Sci. Plant Nutr. (Tokyo) 19:95–102.

Marsho, T. V., R. P. Burchard, and R. Fleming. 1975. Nitrogen fixation in the Rhode River estuary of Chesapeake Bay. Can. J. Microbiol. 21:1348–1356.

Martens, C. S., and R. A. Berner. 1974. Methane production in the interstitial water of sulfate depleted marine sediments. Science 185:1167–1169.

Mikkelsen, D. S., S. K. De Datta, and W. N. Obcemea. 1978. Ammonia volatilization losses from flooded rice soils. Soil Sci. Soc. Am. J. 42:725–730.

Mikkelsen, D. S., and D. C. Finfrock. 1957. Availability of ammonical nitrogen to lowland rice as influenced by fertilizer placement. Agron. J. 49:296–300.

Mitsui, S. 1954. Inorganic nutrition, fertilization, and amelioration for lowland rice. Yokendo Ltd., Tokyo.

Moore, A. W. 1969. *Azolla*: Biology and agronomic significance. Bot. Rev. 35:17–33.

Nelson, L. B., and R. D. Hauck. 1965. Nitrogen fertilizers: Progress and problems. Agric. Sci. Rev., USDA 3:38–47.

Park, K., D. W. Hood, and H. T. Odum. 1958. Diurnal pH variations in Texas bays and its application of primary production estimates. Pub. Inst. Mar. Sci. Univ. of Texas. 5:47–64.

Patrick, W. H., Jr. 1960. Nitrate reduction rates in a submerged soil as affected by redox potential. 7th Int. Congr. Soil Sci. Trans. (Madison, Wisconsin). II:494–500.

Patrick, W. H., Jr., and R. D. DeLaune. 1972. Characterization of the oxidized and reduced zones in flooded soil. Soil Sci. Soc. Am. Proc. 36:573–576.

Patrick, W. H., Jr., and S. Gotoh. 1974. The role of oxygen in nitrogen loss from flooded soils. Soil Sci. 118:78–81.

Patrick, W. H., Jr., F. J. Peterson, J. E. Seaholm, M. D. Faulkner, and R. J. Miears. 1967. Placement of nitrogen fertilizers for rice. Louisiana State Univ., Agric. Exp. Stn. Bull. 619:3–19.

Patrick, W. H., Jr., and K. R. Reddy. 1976a. Nitrification-denitrification reaction in flooded soils and sediments: Dependence on oxygen supply and ammonium diffusion. J. Environ. Qual. 5:469–472.

Patrick, W. H., Jr., and K. R. Reddy. 1976b. Fate of fertilizer nitrogen loss in a flooded soil. Soil Sci. Soc. Am. J. 40:678–681.

Patrick, W. H., Jr., and M. B. Sturgis. 1955. Concentration and movement of oxygen as related to absorption of ammonium and nitrate nitrogen by rice. Soil Sci. Soc. Am. Proc. 19:59–62.

Patrick, W. H., Jr., and M. E. Tusneem. 1972. Nitrogen loss from flooded soil. Ecology 53: 735–737.

Patriquin, D., and R. Knowles. 1972. Nitrogen fixation in the rhizosphere of marine angiosperms. Mar. Biol. 16:49–58.

Pearsall, W. H. 1950. The investigation of wet soils and its agricultural implications. Emp. J. Exp. Agric. 18:289–298.

Pearsall, W. H., and C. H. Mortimer. 1939. Oxidation-reduction potentials in waterlogged soils, natural waters and muds. J. Ecol. 27:483–501.

Peters, G. A., and B. C. Mayne. 1974. The *Azolle: Anabaena azollae* relationships: II. Localization of nitrogenase activity as assayed by acetylene reduction. Plant Physiol. 53:820–824.

Ponnamperuma, F. N. 1972. The chemistry of submerged soils. Adv. Agron. 24:29–96.

Reddy, K. R., and W. H. Patrick, Jr. 1976. Yield and nitrogen utilization by rice as affected by method and time of application of labelled nitrogen. Agron. J. 68:905–969.

Reddy, K. R., W. H. Patrick, and R. E. Phillips. 1980. Evaluation of selected processes controlling nitrogen loss in flooded soil. Soil Sci. Soc. Am. J. 44:1241–1246.

Rice, W. A., and E. A. Paul. 1972. The organisms and biological processes involved in asymbiotic nitrogen fixation in waterlogged soil amended with straw. Can. J. Microbiol. 18: 715–723.

Rice, W. A., E. A. Paul, and L. R. Wetter. 1967. The role of anaerobiosis in asymbiotic nitrogen fixation. Can. J. Microbiol. 13:829–836.

Rinaudo, G., J. Balandreau, and Y. Dommergues. 1971. Algal and bacterial non-symbiotic nitrogen fixation in paddy soils. Plant Soil. Spec. Vol.: 471–479.

Rovira, A. D. 1965. Interactions between plant roots and soil microorganisms. Annu. Rev. Microbiol. 19:241–266.

Shioiri, M., and S. Mitsui. 1935. J. Sci. Soil Manure, Japan 9:261–268 (quoted by Mitsui, 1954).

Shioiri, M., and T. Tanada. 1954. The chemistry of paddy soils in Japan. Minist. Agric. For., Tokyo.

Silver, W. S., and A. Jump. 1975. Nitrogen fixation associated with vascular aquatic macrophytes. p. 121–125. *In* W. D. P. Stewart (ed.) Nitrogen fixation by free-living microorganisms. Cambridge Univ. Press, Cambridge.

Singh, R. N. 1961. Role of blue-green algae in nitrogen economy in Indian agriculture. Ind. Council Agric. Res., New Delhi.

Sircar, S. S. G., S. C. De, and H. D. Bhowmick. 1940. Microbiological decomposition of plant materials. 1. Ind. J. Agric. Sci. 10:119–157.

Stanford, G., J. O. Legg, S. Dzienia, and E. C. Simpson, Jr. 1975a. Denitrification and associated nitrogen transformations in soils. Soil Sci. 120:147–152.

Stanford, G., J. O. Legg, and T. E. Staley. 1975b. Fate of ^{15}N-labelled nitrate in soils under anaerobic conditions. p.677–673. *In* E. R. Klein and P. D. Klein (ed.) Proc. 2nd Int. Conf. on Stable Isotopes. Argonne Natl. Lab., Argonne, Ill.

Stewart, W. D. P. 1969. Biological and ecological aspects of nitrogen fixation by free-living microorganisms. Proc. Roy. Soc. Lond. Ser. B. 172:367–388.

Sturgis, M. B. 1936. Changes in the oxidation-reduction in soils as related to the physical properties of the soil and the growth of rice. Louisiana State Univ. Agric. Exp. Stn. Bull. 271.

Takahashi, J. 1960. Review of investigations on physiological diseases of rice. Part I. Int. Rice comm. News Lett. 9:1–6.

Takai, Y., and T. Kamura. 1966. The mechanism of reduction in waterlogged paddy soils. Folia Microbiol. (Prague) 11:304–313.

Takai, Y., and Y. Uehara. 1973. Nitrification and denitrification in the surface layer of submerged soil. Part I. Oxidation-reduction condition, nitrogen transformation and bacterial flora in the surface and deeper layers of submerged soils. J. Sci. Soil Manure, Japan 44: 463–502.

Teal, J. M., and J. W. Kanwisher. 1966. Gas transport in the marsh grass, *Spartina alterniflora*. J. Exp. Bot. 17:355–361.

Teal, J. M., I. Valiela, and D. Berlo. 1979. Nitrogen fixation by rhizosphere and free-living bacteria in salt marsh sediments. Limnol. Oceanogr. 24:126–132.

Tenney, F. G., and S. A. Waksman. 1930. Composition of natural organic materials and their decomposition in the soil. V. Decomposition of various chemical constituents in plant materials under anaerobic conditions. Soil Sci. 30:143–160.

Turner, F. T., and W. H. Patrick, Jr. 1968. Chemical changes in waterlogged soils as a result of oxygen depletion. 9th Int. Cong. Soil Sci. Trans. (Adelaide, Australia) 4:53–63.

Tusneem, M. E., and W. H. Patrick, Jr. 1971. Nitrogen transformations in waterlogged soil. Louisiana State Univ. Agric. Exp. Stn. Bull. 657:1–75.

van Raalte, M. H. 1941. On the oxygen supply of rice roots. Ann. Bot. Gard. Buitenzorg. 54:13–34.

Ventura, W. B., and T. Yoshida. 1977. Ammonia volatilization from a flooded tropical soil. Plant Soil 46:521–531.

Vlek, P. L. G., and E. T. Craswell. 1979. Effect of nitrogen source and management on ammonia volatilization losses from flooded rice-soil systems. Soil Sci. Soc. Am. J. 43:352–358.

Waring, S. A., and J. M. Bremner. 1964. Ammonium production in soil under waterlogged conditions as an index of nitrogen availability. Nature (London) 201:951–952.

Watanabe, I. 1978. Biological nitrogen fixation in rice soils. p. 465–478. *In* Soils and rice. Int. Rice Res. Inst. Los Banos, Philippines.

Watanabe, I., K. K. Lee, and B. V. Alimagno. 1978. Seasonal changes of N_2-fixing rate in rice field assayed by in situ acetylene reduction technique. I. Experiments in long-term fertility plots. Soil Sci. Plant Nutr. (Tokyo) 24:1–13.

Waughman, G. J. 1976. Investigations of nitrogenase activity in rheotrophic peat. Can. J. Microbiol. 22:1561–1566.

Werner, D., H. J. Evans, R. J. Seidler. 1973. Facultatively anaerobic nitrogen-fixing bacteria from the marine environment. Can. J. Microbiol. 20:59–64.

Whitney, D. E., G. M. Woodwell, and R. W. Howarth. 1975. Nitrogen fixation in Flax Pond: A Long Island salt marsh. Limnol. Oceanogr. 20:640–643.

Wijler, J., and C. C. Delwiche. 1954. Investigations on the denitrifying process in soil. Plant Soil 5:155–169.

Williams, W. A., D. S. Mikkelsen, K. E. Mueller, and J. E. Ruckman. 1968. Nitrogen immobilization by rice straw incorporated in lowland rice production. Plant Soil 28:49–60.

Woldendorp, J. W. 1968. Losses of soil nitrogen. Stikstof 12:32–46.

Yamane, I. 1957. Nitrate reduction and denitrification in flooded soils. Soil Plant Food (Tokyo) 3:100–103.

Yanagisawa, M., and J. Takahashi. 1964. Studies on the factors related to the productivity of paddy soil in Japan with special reference to the nutrition of rice plant. (In Japanese). Bull. Nat. Inst. Agric. Sci. Tokyo. Ser. B. 14:41–171.

Yoshida, Y., and R. R. Ancajas. 1971. Nitrogen fixation by bacteria in the root zone of rice. Soil Sci. Soc. Am. Proc. 35:156–158.

Yoshida, T., and R. R. Ancajas. 1973. Nitrogen-fixing activity in upland and flooded rice fields. Soil Sci. Soc. Am. Proc. 37:42–46.

Yoshida, Y., and F. E. Broadbent. 1975. Movement of atmospheric nitrogen in rice plants. Soil Sci. 120:288–291.

Yoshida, T., and B. C. Padre. 1974. Nitrification and denitrification in submerged Maahas clay soil. Soil Sci. Plant Nutr. (Tokyo) 20:241–247.

Zuberer, D. A., and W. S. Silver. 1978. Biological dinitrogen fixation (acetylene reduction) associated with Florida mangroves. Appl. Environ. Microbiol. 35:567–575.

13 Advances in Methodology for Research on Nitrogen Transformations in Soils

J. M. BREMNER

Iowa State University
Ames, Iowa

R. D. HAUCK

Tennessee Valley Authority
Muscle Shoals, Alabama

I. INTRODUCTION

For meaningful research in any field, it is essential to have effective research techniques, and most of the progress made during the past 30 years in research on the nature and transformations of N in soils has resulted from advances in methodology for such research. For example, most of our knowledge concerning the organic N in soils resulted from development of paper and ion exchange chromatographic methods of detecting and determining amino acids, amino sugars, and purine and pyrimidine bases, and much of the information currently available concerning production of gaseous forms of N in soils resulted from the introduction of gas chromatographic methods for determining nitrogenous gases.

Comprehensive reviews of the methodology available for research on N transformations in soils were published in 1965 (Bremner, 1965a–e; Cheng & Bremner, 1965; Stevenson, 1965a, b). The purpose of this chapter is to review the major advances in methodology for soil N research since that time. The methods currently available for assessment of soil N availability will not be reviewed because these methods are discussed in Chapter 17 (G. Stanford) and in a recent article by Keeney (1981). Space limitations have made it impossible to cite many papers relating to methodology published since 1965, but references to most of these papers can be found in reviews cited. The extent of the literature relating to methodology is indicated by the recent publication of a 702-page monograph dealing solely with the methods now available for studying biological N_2 fixation (Bergersen, 1980).

Nitrogen in Agricultural Soils—Agronomy Monograph no. 22.

II. DETERMINATION OF DIFFERENT FORMS OF NITROGEN

A. Total Nitrogen

The methods currently available for determination of total N in soils have been discussed in recent reviews by Nelson and Sommers (1980) and Bremner and Mulvaney (1981). The methods recommended in these reviews are semimicro modifications of macro-Kjeldahl procedures described by Bremner (1965a) and include procedures that permit quantitative recovery of NO_3^--N and NO_2^--N. The fact that the Kjeldahl procedures commonly used for total N analysis of soils do not effect quantitative recovery of NO_3^-- or NO_2^--N deserves emphasis here, because many studies of N transformations in soils (particularly ^{15}N-tracer studies of denitrification) have been vitiated by failure to recognize this limitation of customary Kjeldahl procedures.

The most noteworthy developments in Kjeldahl analysis since 1965 have been the introduction of heated Al blocks for Kjeldahl digestion and the use of NH_3 electrodes for determination of NH_4^+ in Kjeldahl digests.

Nelson and Sommers (1972), Schuman et al. (1973), Gallaher et al. (1976) and Douglas et al. (1980) have described the use of block digester techniques for Kjeldahl digestion of soils. In these techniques, digestion is performed in Pyrex tubes placed in holes drilled in an Al block that is heated by an electric hot plate or by internal electric heaters connected to a temperature control. Aluminum block digesters with internal heaters and temperature controls are supplied by Tecator Inc. (Boulder, Colo.), Technicon Industrial Systems (Tarrytown, N.Y.), Labconco Corporation (Kansas City, Mo.), and Brinkmann Instruments, Inc. (Westbury, N.Y.). Evidence that the Tecator and Technicon digesters are satisfactory for soil analysis has been reported by Schuman et al. (1973) and Nelson and Sommers (1980), respectively. These block digesters have significant advantages over the digestion stands normally used for Kjeldahl analysis because they provide better temperature control during digestion, require less attention and fume hood space, and allow simultaneous digestion of 40 to 100 samples.

The use of NH_3 electrodes for determination of NH_4^+ in Kjeldahl analysis resulted from research by Orion Research Inc. (Cambridge, Mass.) leading to development of a highly specific electrode for determination of NH_3 (Orion Model 95-10 ammonia electrode). Bremner and Tabatabai (1972) demonstrated that this electrode was satisfactory for determination of NH_4^+ in Kjeldahl digests of soils (see also Gallaher et al., 1976), and other workers have shown that it can be used for determination of NH_4^+ in Kjeldahl digests of plant materials (Eastin, 1976; Gallaher et al., 1976).

Ammonia electrodes with an internal reference electrode are now available from Orion Research Inc., Lazar Research Laboratories, Inc. (Los Angeles), and Electronic Instruments Ltd. (Chertsey, Surrey, England). They respond to the activity of NH_3 in solutions made alkaline (pH > 11) to convert NH_4^+ to NH_3 and can be used in conjunction with any sensitive pH-mV meter. Their attraction is that they allow direct analysis of Kjeldahl digests for NH_4^+ and thereby eliminate the need for the customary distillation step before NH_4^+ analysis.

Gallaher et al. (1976) have described a semiautomated Kjeldahl procedure for determination of total N in soils and plant materials in which an NH_3 electrode is used for direct determination of NH_4^+ in digests obtained using an Al block digester. This procedure permits analysis of more than 100 samples in a normal working day and seems well suited for routine Kjeldahl analysis of soils.

Several workers have described the use of autoanalyzer systems for colorimetric determination of NH_4^+ in Kjeldahl digests of soils (e.g., Henzell et al., 1968; Holz & Kremers, 1971; Schuman et al., 1973; Skjemstad & Reeve, 1976). In these methods, NH_4^+ in the digest is separated from other digest constituents by distillation or dialysis and determined by a procedure involving measurement of the color of the indophenol complex formed by the Berthelot reaction.

A considerable proportion of the N in some soils (particularly subsoils) is in the form of NH_4^+ trapped in the lattices of clay minerals. Several studies have indicated that this fixed NH_4^+-N is usually recovered satisfactorily by Kjeldahl procedures commonly used for total N analysis of soils (Bremner, 1959, 1960; Bremner & Harada, 1959; Ahlrichs, 1965; Nelson & Bremner, 1966; Keeney & Bremner, 1967a; Mogilevkina, 1970), but investigations reported by Stewart and Porter (1963), Keeney and Bremner (1967a), and Meints and Peterson (1972) have shown that certain subsoils contain fixed NH_4^+-N that is not recovered quantitatively by these procedures. Total N in such subsoils can be determined by use of a Kjeldahl procedure involving pretreatment of the soil sample with HF to destroy clay minerals before Kjeldahl digestion (Keeney & Bremner, 1967a).

One of the most interesting advances in methodology for total N analysis during the past decade has been the introduction of the Kjel-Foss Automatic Analyzer, an automated instrument for rapid Kjeldahl analysis of agricultural and food products. This instrument was developed by N. Foss Electric of Hillerod, Denmark, and it is marketed in the United States by Foss America, Inc., Fishkill, N.Y. It involves the use of H_2O_2 to reduce the time required for conversion of total N to NH_4^+ by the Kjeldahl digestion technique, and it permits 120 to 180 analyses per day. This instrument has proved very satisfactory for determination of total N in a wide variety of materials, including plants, animal feeds, fertilizers, and NBS standards (Oberrieth & Mermelstein, 1974; Wall et al., 1975; Noel, 1976; Larson & Peterson, 1979; Bjarno, 1980), and it clearly merits evaluation for routine determination of total N in soils.

Attempts to determine total N in soils by the Dumas combustion technique have shown that, although customary Dumas and Kjeldahl methods give similar total N values with mineral soils, Dumas methods give considerably higher values with soils containing substantial amounts of organic matter (Dyck & McKibbin, 1935; Bremner & Shaw, 1958; Stewart et al., 1963; Bremner, 1965a; Morris et al., 1968). Stewart et al. (1963, 1964) demonstrated that the higher values obtained by the Dumas method with such soils were due to incomplete combustion, which resulted in methane (CH_4) being formed and measured with the N_2 produced by combustion. They also showed that, when the Coleman Model 29 Nitrogen Analyzer (an automated Dumas combustion instrument) was modified to achieve complete combustion and eliminate interference by CH_4, the results obtained when it was used for total N analysis of organic and mineral soils agreed closely with those obtained by Kjeldahl analysis (Stewart et al., 1964).

The Coleman Model 29 Analyzer was not designed for analysis of soils or other heterogeneous materials having low N contents, and its use for total N analysis of such materials is complicated by sampling problems imposed by sample-size limitations. To overcome this difficulty and eliminate the need for fine grinding of samples and use of a microbalance in analysis of heterogeneous materials having low N contents, the Coleman Company (now Coleman Instruments Division, Perkin-Elmer Corporation, 2000 York Road, Oak Brook, Ill.) developed the Model 29A Nitrogen Analyzer, which permits analysis of relatively large samples of heterogeneous materials. Keeney and Bremner (1967a) found that this instrument could be adapted satisfactorily for total N analysis of both mineral and organic soils. They also found that when modified to improve combustion, the Model 29A Analyzer gave nearly quantitative (96–98%) recovery of NO_3^-- or NO_2^--N added to soils but did not give complete recovery of total N in two subsoils containing substantial amounts of fixed NH_4^+-N.

The Coleman automated Dumas combustion instruments evaluated by Stewart et al. (1964) and Keeney and Bremner (1967a) have the attraction that they require very little bench space compared with the equipment needed for total N analysis by Kjeldahl methods, but they have the disadvantage that they require skilled personnel for satisfactory use and permit only 30 to 40 analyses per day. Other automated Dumas combustion instruments have been developed by several companies in the United States, Europe, and Japan, but most of these instruments have been designed for analysis of pure compounds having high N contents, and there is no published evidence that any of them can be adapted successfully for total N analysis of soils. It seems likely, however, that the Leco UO-14SP Nitrogen Determinator (an automated Dumas combustion instrument developed by the Laboratory Equipment Corporation, 3000 Lakeside Ave., St. Joseph, Mo.) could be adapted satisfactorily for soil analysis because Wong and Kemp (1977) found that this instrument was satisfactory for total N analysis of sediments, and the problems in total N analysis of sediments are similar to those encountered in soil analysis. There is a clear need for an evaluation of this instrument for total N analysis of soils because it is simple and rapid, and Wong and Kemp (1977) found that it permitted more than 60 analyses of sediments in a normal working day.

Other automated instruments that merit consideration for total N analysis of soils are the Leco TN-15 Nitrogen Determinator and the Leco TC-36 Nitrogen/Oxygen Analyzer recently developed by the Laboratory Equipment Corporation and the automated N analyzers recently introduced by Antek Instruments, Inc. Houston. The Leco TN-15 and Leco TC-36 instruments resemble the Leco UO-14SP instrument in that they involve conversion of total N to N_2 by fusing the sample in a graphite crucible at high temperature in an inert (He) atmosphere and determination of N_2 by a gas chromatographic procedure involving use of a thermal conductivity detector. The Antek instruments involve pyrolysis of the sample in an O_2-Ar atmosphere to oxidize N to NO, which is determined by using a chemiluminescent detector to measure the light emitted by metastable NO_2 produced by mixing NO with O_3. These instruments have been used to determine N in petroleum fractions (Drushel, 1977) and to measure gaseous loss of N from plants during transpiration (Stutte & Weiland, 1978), but they have not been evaluated for total N analysis of soils.

B. Inorganic Forms of Nitrogen

Keeney and Nelson (1981) have recently updated Bremner's (1965b) review of methods for determination of inorganic forms of N in soils. Comparison of these reviews shows that there have been no major advances in methodology for inorganic N analysis during the past 15 years and that the following methods recommended by Bremner (1965b) remain the methods of choice for most research on N transformations in soils: the steam distillation methods of determining exchangeable NH_4^+, NO_2^-, and NO_3^- described by Bremner and Keeney (1965, 1966); the KOBr-HF method of determining nonexchangeable (fixed) NH_4^+ proposed by Silva and Bremner (1966); and the colorimetric method of determining NO_2^- described by Bremner (1965b), which involves use of the Griess-Ilosvay procedure as modified by Barnes and Folkard (1951).

The methods of determining exchangeable NH_4^+, NO_2^-, and NO_3^- proposed by Bremner and Keeney (1966) involve extraction of the soil sample with $2M$ KCl and analysis of the extract by steam distillation techniques involving the use of MgO and Devarda alloy for reduction of NO_2^- and NO_3^- to NH_3. They have been used very extensively during the past 15 years, because besides being simple, rapid, and precise, they are applicable to colored extracts, require only one extraction of the soil sample, and permit isotope ratio analysis of inorganic forms of N. Substantial amounts of phosphate or silicate can interfere with reduction of NO_3^- to NH_3 by MgO and Devarda alloy (Bremner & Keeney, 1965; Freney, 1971), but this problem can be overcome by use of $FeSO_4$ instead of Devarda alloy (Bremner & Bundy, 1973). There is evidence that modifications of these methods in which the extraction step is omitted (Keeney & Bremner, 1966) are not satisfactory for soils containing high levels of alkali-labile organic N compounds (Robinson, 1973; Sahrawat & Ponnamperuma, 1978) or exchangeable Mg^{2+} (Clausen et al., 1980).

Studies to evaluate various methods proposed for determination of nonexchangeable (fixed) NH_4^+ in soils have shown that the methods proposed by Rodrigues (1954), Schachtschabel (1960, 1961), and Mogilevkina (1964) have serious defects (Bremner, 1959; Nelson & Bremner, 1966; Bremner et al., 1967) and that the values obtained by these methods are greatly different from those obtained by the methods proposed by Dhariwal and Stevenson (1958), Bremner (1959) and Silva and Bremner (1966). Bremner et al. (1967) found that the method of Silva and Bremner (1966) gave slightly higher values than the methods proposed by Dhariwal and Stevenson (1958) and Bremner (1959) and reported evidence that it was the most reliable of nine methods compared in their work. They pointed out, however, that there is no way of establishing that this, or any other, method of determining nonexchangeable NH_4^+ in soils gives accurate results and drew attention to possible defects of the Silva-Bremner method.

Keeney and Nelson (1981) have recently discussed the problems associated with the use of colorimetric methods for determination of NH_4^+ and NO_3^- in soils and have recommended methods designed to minimize these problems. In these methods, NH_4^+, NO_2^-, and NO_3^- are extracted by $2M$ KCl and determined by direct analysis of the extract by indophenol blue, Griess-Ilosvay, and Cd reduction methods, respectively. These methods are similar to procedures now commonly used in automated analysis schemes for determination of inorganic forms of N in natural waters. They are sensitive and precise, and they are not subject to significant interference by common soil constituents. The method for determination of NH_4^+ by the indophenol blue reaction is a modification of a method proposed by Kempers (1974). The method for determination of NO_3^- involves use of a copperized Cd column for reduction of NO_3^- to NO_2^- and is similar to methods described by Henriksen and Selmer-Olsen (1970) and Jackson et al. (1975).

One of the most interesting developments in methodology for determination of inorganic forms of N during the past 15 years has been the introduction of ion electrodes for determination of NH_4^+, NO_3^-, and NO_2^-. These electrodes have received considerable attention, but they have not gained significant acceptance for soil analysis. Keeney and Nelson (1981) have discussed their merits and defects in a recent review of literature relating to their use for determination of inorganic forms of N in soils and other natural materials and have cautioned against their adoption without careful evaluation of their suitability for a particular application.

The most specific of the electrodes currently available for determination of inorganic forms of N are the NH_3 electrodes discussed in section IIA, which respond to the activity of NH_3 in solutions made alkaline (pH > 11) to convert NH_4^+ to NH_3. Banwart et al. (1972) demonstrated that the Orion Model 95-100 NH_3 electrode was satisfactory for determination of NH_4^+ in soil extracts and water samples, and other workers have shown that NH_3 electrodes supplied by Orion Research Inc. or by Electronic Instruments Ltd. can be used for determination of NH_4^+ in fresh and saline waters, sewage effluents, and acid extracts of fresh silage samples (Thomas & Booth, 1973; Beckett & Wilson, 1974; Byrne & McCormack, 1978). Besides being highly specific (only volatile amines interfere), these NH_3 electrodes have the advantage that they are not subject to significant interference by substances known to be present in soils. In contrast, the NO_3^- electrodes currently available are not highly specific, and they are subject to interference by anions commonly found in soil extracts. Numerous studies to evaluate NO_3^- electrodes and reduce interference by these anions have been reported (Keeney & Nelson, 1981), but many of these studies have given conflicting results, and none of the NO_3^- electrode methods thus far described seems likely to be satisfactory for all soils. As noted by Keeney and Nelson (1981), the NO_3^- electrodes currently available require continual restandardization and lack the sensitivity needed for satisfactory analysis of many soil extracts. These criticisms apply to the NO_2^- electrode developed by Orion Research, Inc., which responds to NO_x (NO, NO_2, N_2O_3, N_2O_4) produced by acidification of solutions containing NO_2^-. This electrode gives satisfactory recovery of NO_2^- added to soil extracts and water samples (Tabatabai, 1974), but it is much less sensitive than colorimetric methods available for determination of NO_2^-, and it would seem to have no potential value for determination of the small amounts of NO_2^- normally present in soils, plants, or natural waters.

Siegel (1980) has recently described a method of determining NO_3^- in soil extracts that involves the use of an NH_3 electrode to determine the NH_3 produced by treatment of an aliquot of extract with Devarda alloy and NaOH to reduce NO_3^-. This method is a modification of a method developed by Mertens et al. (1975) for determination of NO_3^- in natural waters. It is subject to interference by NO_2^-.

C. Organic Forms of Nitrogen

Although the chemical nature of most of the organic N in soils is still obscure, there is good evidence that from 20 to 40% of the total N in most surface soils is in the form of combined amino acids and that from 5 to 10% is in the form of combined hexosamines (2-amino sugars) (for references, see Chapt. 3, F. J. Stevenson; Bremner, 1967; Parsons & Tinsley, 1975; Kowalenko, 1978; Stevenson, 1981). Also, recent work by Cortez and Schnitzer (1979) has indicated that from 1 to 7% of the total N in most mineral soils is in the form of nucleic acid bases (purines and pyrimidines).

Methods proposed for determination of organic forms of N in soils have been discussed in articles by Bremner (1965c) and Stevenson (1981). They involve the use of hot mineral acid (usually 6N HCl) to hydrolyze the nitrogenous organic materials in soil and subsequent analysis of the hydrolysate by chromatographic, steam distillation, or colorimetric techniques. Ion exchange chromatography has been used to identify and estimate amino acids and hexosamines in soil hydrolysates (for references, see Sowden et al., 1977; Stevenson, 1981; Chapter 3), and both ion exchange chromatography and ion exclusion chromatography have been used to identify and estimate purines and pyrimidines in hydrolysates of soils or soil fractions (Cortez & Schnitzer, 1979). However, applications of these techniques has been very limited, and most research since 1965 to characterize organic N in soils and obtain estimates of amino acid N and hexosamine N has been performed by N fractionation techniques similar to that described by Bremner (1965c), which involves the use of rapid steam distillation techniques to estimate amino acid N, hexosamine N, and ammonium N released by acid hydrolysis of soil (see Table 1, Chapt. 3).

Although only trace amounts of the organic N in most soils is in the form of urea, research to develop methods for determination of urea in soils has been stimulated by the rapidly increasing importance of urea as a N fertilizer. The methods now available for this analysis have been discussed in a recent article by Bremner (1981). The most sensitive and precise of these methods are the colorimetric procedure described by Douglas and Bremner (1970) and the two enzymatic procedures described by Keeney and Bremner (1967b). The procedure of Douglas and Bremner (1970) involves extraction of the soil sample with $2M$ KCl containing a urease inhibitor (phenylmercuric acetate) and analysis of the extract by a colorimetric method based on the reaction of urea with diacetyl monoxime and thiosemicarbazide in the presence of H_3PO_4 and H_2SO_4. Douglas et al. (1978) have described an automated modification of this method, and Mulvaney and Bremner (1979) have described a modification to eliminate problems caused by impurities in some batches of H_3PO_4. The two enzymatic procedures proposed by Keeney and Bremner (1967b) involve use of urease to convert urea-N to NH_4^+-N. In one, the soil sample is extracted with $2M$ KCl, and an aliquot of the extract is treated with K phosphate buffer (pH 8) and urease. In the other, the soil sample is treated directly with pH 8 K phosphate buffer and urease. In both, the NH_4^+ produced through enzymatic hydrolysis of urea by urease is separated by a rapid steam distillation technique and determined by titration of the distillate with standard acid. Besides being specific and precise, these enzymatic procedures have the advantage that they permit isotope ratio analysis of urea-N in ^{15}N-tracer studies of urea transformations in soils.

D. Gaseous Forms of Nitrogen

Bremner and Blackmer (1981) have recently reviewed the methods available for determination of dinitrogen (N_2), nitrous oxide (N_2O), nitric oxide (NO), nitrogen dioxide (NO_2), ammonia (NH_3), and volatile amines in research on N transformations in soils. Comparison of their review with a 1965 review of the same topic by Cheng and Bremner (1965) shows that there have been striking advances in the methodology for determination of gaseous forms of N in soil research during the past 15 years and that these advances have resulted largely from the introduction of gas chromatographic (GC) methods of analysis.

Bremner and Blackmer (1981) have discussed GC methods currently available for determination of N_2, N_2O, and NO in laboratory research on denitrification and other processes leading to formation of these gases in soils and have drawn attention to the important advantages of the method proposed by Blackmer and Bremner (1977), which involves use of an ultrasonic detector and two columns of Porapak Q at different temperatures. This method permits determination of O_2, Ar, and CO_2 as well as N_2, N_2O, and NO, and it has a combination of features (sensitivity, specificity, durability, versatility, etc.) that make it well suited for research on denitrification in soils and for routine determination of the major constituents of soil atmospheres.

Numerous problems have been encountered in attempts to determine NO and NO_2 in gaseous mixtures containing these gases (Cheng & Bremner, 1965; Bethea & Meador, 1969; Allen, 1973). One major problem is that NO is oxidized by O_2 to NO_2. This reaction is very rapid at high NO concentrations, but its rate decreases rapidly as the concentration of NO decreases. Another serious problem is that NO_2 is sorbed by many materials used to collect, store, and transfer gas samples and reacts with many materials used as column packings for GC analysis. Nitric oxide also reacts with several materials used to pack GC columns, but it is not as reactive as NO_2.

Numerous GC methods have been proposed for detection or estimation of NO_2, but most have been designed for analysis of simple gaseous mixtures containing high concentrations of NO_2, and none permit determination of small amounts of NO_2 in gaseous mixtures as complex as soil atmospheres. Moreover, it has been found that NO_2 can vitiate GC determination of NO by reacting with materials commonly used to pack GC columns for N gas analysis (Bremner & Blackmer, 1981). It is noteworthy that none of the GC methods thus far proposed for determination of NO in soil atmospheres allows determination of NO in the presence of large amounts of N_2 or appreciable amounts of O_2 or Ar.

Several workers have used alkaline or acidic $KMnO_4$ solutions to trap NO and NO_2 evolved from soils incubated in closed systems and have determined $(NO + NO_2)$-N by analysis of these solutions for $(NO_2^- + NO_3^-)$-N or NO_3^--N (for references, see Bremner & Blackmer, 1981).

Galbally and Roy (1978) have described a chamber method for field measurement of emissions of NO and NO_2 from soils that involves use of a Mylar-lined chamber and a highly sensitive chemiluminescent detector developed by Galbally (1977) for estimation of NO and $NO_x(NO + NO_2)$ in surface air. Similar detectors have been used extensively for NO and NO_x analysis in air pollution research during the past 5 years. They originated from research by Fontijn et al. (1970) leading to development of a sensitive detector permitting use of the chemiluminescent reaction of NO with ozone (O_3) for determination of small amounts of NO in air. Subsequent research (Stevens & Hodgeson, 1973; Black & Sigsby, 1974) increased the sensitivity of this method of determining NO and led to development of methods of converting NO_2 to NO for determination of $NO + NO_2$ in air by chemiluminescent detectors. These chemiluminescent detector techniques are subject to several interferences (Winer et al., 1974; Matthews et al., 1977; Folsom & Courtney, 1979), but they are much more sensitive than other methods proposed for determination of NO or NO_2 (they can detect < 1 ppb of NO in air), and there seems little doubt that they will prove valuable for research to assess emissions of NO and NO_2 from soils. Oxidation of NO to NO_2 by atmospheric O_2 is not a problem in determination of NO in air, because the concentration of NO in air is so low that oxidation of NO to NO_2 is very slow (Galbally & Roy, 1978).

Kim (1973) measured emissions of NO_2 and NH_3 from Korean forest soils by a chamber technique in which Petri dishes containing $0.01N$ NaOH and $0.05N$ H_2SO_4 were used to trap NO_2 and NH_3 evolved from soil covered by a plastic hood.

Research by atmospheric scientists during the early 1970s created international concern that N fertilization of soils may lead to a significant increase in the atmospheric concentration of N_2O and thereby cause partial destruction of the stratospheric O_3 layer (Crutzen & Ehhalt, 1977) and add to the greenhouse effect expected to result from the continuing increase in the atmospheric concentration of CO_2 (Yung et al., 1976). This concern has stimulated extensive research to assess N_2O emissions from soils and the effects of N fertilizers on these emissions. For such research, it is essential to have a highly sensitive method of determining N_2O because air normally contains only about 300 ppb (vol/vol) of this gas. None of the methods commonly used for determination of N_2, N_2O, and NO in research on denitrification in soils has the sensitivity needed for direct N_2O analysis of air.

Although several methods, including mass spectrometry and infrared spectroscopy, have been proposed for determination of N_2O in air and other gaseous mixtures containing small amounts of N_2O, most N_2O analyses of air during the past 5 years have been performed by GC methods similar to that described by Rasmussen et al. (1976a), which involves use of a heated (350°C) ^{63}Ni electron capture detector. These methods originated from work showing that when operated in the pulse sampling mode and at a high temperature (ca. 350°C), a ^{63}Ni electron capture detector is highly sensitive to N_2O (Wentworth et al., 1971; Wentworth & Freeman, 1973). They are much simpler than previous GC methods for N_2O analysis of air because they do not require steps to concentrate N_2O in the sample under analysis, and they are so sensitive that they permit direct analysis of as little as 1 ml of air. The column packings used for N_2O analysis of air by GC-electron capture detector methods have included Porapak Q, Porapak R, Porapak Q in series with Porapak R, Porapak QS, Porasil B, Porasil C, and Carbosieve B (for references, see Bremner & Blackmer, 1981). Most workers currently favor Porapak Q and use Ar containing 5% methane as carrier gas.

A collaborative study organized and reported by Hughes et al. (1978) showed that widely divergent results were obtained when two gas samples containing ca. 260 and 300 ppb (vol/vol) of N_2O in N_2 were analyzed by 15 laboratories using GC-electron capture detector methods of determining N_2O. This divergence in results has not been explained, but there seems little doubt that it resulted at least partly from the unavailability of reliable gas standards for calibration of GC methods for N_2O analysis of air. This problem may be alleviated by research to prepare such standards now being undertaken by the National Bureau of Standards.

The major disadvantages of GC-electron capture detector methods of determining N_2O have been that they require frequent calibration and are subject to interference by water vapor and fluorocarbons. Recent work by Mosier and Mack (1980) indicates that these interferences can be eliminated by use of a GC procedure that permits a column backflush to remove water vapor and fluorocarbons.

Blackmer and Bremner[1] have described a GC-ultrasonic detector method of determining N_2O in air and soil atmospheres that does not require frequent recalibration, is not subject to interferences observed using

GC–electron capture detector methods, and has the important advantage that it permits use of the Xe in air as an internal standard. This method involves quantitative removal of N_2O and Xe from the air sample by Porapak Q cooled to $-135°C$ and subsequent determination of these gases by a GC technique in which N_2O and Xe are separated by a short column of Porapak Q at $65°C$ and determined by an ultrasonic detector. It requires a larger volume of air than do GC–electron capture detector methods because the ultrasonic detector is not as sensitive as the electron capture detector toward N_2O, but it does not require a temperature-controlled laboratory. (GC–electron capture detector methods have this requirement because the sensitivity of the electron capture detector is markedly affected by the laboratory temperature.)

Although several techniques have been proposed for field assessment of N_2O emissions from soils, field studies of these emissions have been performed almost exclusively by chamber methods. Two types of chamber methods have been used. In one, air is drawn through a chamber inserted into the soil surface, and N_2O in the air leaving the chamber is trapped by molecular sieve and subsequently recovered and determined by a GC–thermal conductivity detector technique (Dowdell & Crees, 1974; Guthrie & Duxbury, 1978; Ryden et al., 1978). In the other, a chamber is inserted into (or placed over) the soil surface, and the air within the chamber is sampled at intervals for N_2O analysis by a GC technique involving use of an electron capture or ultrasonic detector (McKenney et al., 1978; Roy, 1979; Matthias et al., 1980). Denmead (1979) has described chamber methods of measuring N_2O emissions from soils that involve use of an infrared technique for N_2O analysis of air passing through a cylindrical chamber driven into soil. This technique is not as sensitive as GC methods currently available for determination of N_2O in air, and it requires use of a series of traps to remove interfering gases (water vapor, CO, and CO_2).

It is appropriate to refer here to methods currently available for N_2O analysis of water samples because recent work (Guthrie & Duxbury, 1978; Dowdell et al., 1979) indicates that significant amounts of N_2O are removed from soils in drainage water. Two types of methods have been proposed for N_2O analysis of water samples. In one, the sample is equilibrated with air or N_2 in a sealed container, and its N_2O content is calculated from N_2O analysis of the gas phase by a GC–electron capture detector technique (Cicerone et al., 1978; Dowdell et al., 1979). In the other, N_2O is stripped from the sample, adsorbed on molecular sieve, and subsequently desorbed and determined by gas chromatography using a thermal conductivity or electron capture detector (Cohen, 1977; Guthrie & Duxbury, 1978). Rasmussen et al. (1976b), Kaplan et al. (1978), and Elkins (1980) have described methods for N_2O analysis of water samples based on the static gas-partitioning technique developed by McAuliffe (1969, 1971). These methods involve equilibration of the water sample with He, and N_2O analysis of the gas phase by a GC–electron capture detector method.

---

[1] A. M. Blackmer and J. M. Bremner. 1978. Determination of nitrous oxide in air. Agron. Abst. 1978, p. 137.

No GC techniques permitting direct analysis of soil atmospheres for NH_3 and volatile amines have been reported, but GC methods have been developed for NH_4^+ and amine analysis of solutions, and these methods may prove useful for analysis of acidic solutions used to sorb and estimate NH_3 and volatile amines evolved from soils in closed systems.

Two types of methods have been proposed for GC analysis of amine solutions. In one, the solution is analyzed by GC techniques in which an alkaline precolumn or column packing is used to volatilize the amines under analysis (Andre & Mosier, 1973; Mosier et al., 1973). In the other, the amines under analysis are converted to halogenated derivatives that are separated and detected by GC techniques involving use of an electron capture detector, which is highly sensitive toward halogenated compounds (Mosier et al., 1973; Hoshika, 1977). The compounds used to convert amines to halogenated derivatives for GC analysis have included pentafluorobenzaldehyde, pentafluorobenzoylchloride, and 1-fluoro-2,4-dinitrobenzene.

Mosier et al. (1973) used GC methods of the two types described to analyze acidic solutions used to trap NH_3 and volatile amines evolved from cattle feedlots and found that they permitted detection and identification of nine basic compounds in these traps. The compounds identified included NH_3, methylamine, dimethylamine, and trimethylamine.

Hoshika (1977) has described a very sensitive and selective GC method of detecting and identifying lower aliphatic amines. It involves conversion of these amines to F-containing Schiff bases by reaction with pentafluorobenzaldehyde and separation and detection of these bases by a GC technique involving use of a heated (250°C) electron capture (^{63}Ni) detector. This method permits detection of very small (picogram) amounts of methylamine, ethylamine, and other primary amines in the presence of substantial amounts of NH_3, dimethylamine, and diethylamine.

Many workers have used acid solutions to trap NH_3 evolved from soils incubated in closed systems and have determined NH_3 by analyzing these solutions for NH_4^+ by distillation-titration techniques. Such methods of determining NH_3 are subject to interference by volatile amines (Elliott et al., 1971).

Methods for field assessment of emissions of NH_3 from soils have been described or discussed in articles by Denmead et al. (1974, 1976, 1977, 1978), Kissel et al. (1977), Mikkelsen and DeDatta (1978), Mikkelsen et al. (1978), Vlek and Stumpe (1978), Hargrove and Kissel (1979), Hauck (1979), Vlek (1979), Vlek and Craswell (1979), and Bouwmeester and Vlek (1981). Most of these methods involve use of acid to trap NH_3 evolved from soil covered by a metal or plastic chamber. Two types of chamber techniques have been used. In one, the NH_3 evolved is collected in an acid trap placed under the chamber (static technique). In the other, NH_3-free air is passed through the chamber, and the NH_3 evolved from the soil under the chamber is collected by passing the air emerging from the chamber through an acid trap. The major criticism of these techniques (particularly the static techniques) is that they do not allow normal air movement across the soil surface or normal diurnal fluctuations in soil temperature and moisture content. Denmead et al. (1974, 1976, 1977, 1978) have described a micrometeorological method of studying NH_3 emissions from soils that is not open to this criticism. This method is similar to one that has been used extensively in micrometeorology to measure rates of gas and vapor exchange above natural surfaces, and it does not impose unnatural conditions on the area under study. However, it is based on assumptions that are difficult to validate and has the disadvantages that it requires expensive equipment for accurate determination of wind speed and other micrometeorological variables and is applicable only to rather large experimental areas of good fetch.

Determination of NH_3 in air is complicated by interference by particleborne NH_4^+ ions. Ferm (1979) has suggested that this problem can be solved by use of a method of determining NH_3 based on the fact that, when ambient air passes through a tube, gas molecules diffuse much more rapidly than particles to the tube wall. In this method, air is sucked for 24 hours through a glass tube coated with oxalic acid (to sorb NH_3), and this coating is subsequently dissolved in water for NH_4^+ analysis by a NH_3 electrode technique. This method may prove useful for NH_3 analysis in assessment of NH_3 emissions from soils by micrometeorological techniques.

Several workers (e.g., Sigsby et al., 1973; Aneja et al., 1978; Baumgartner et al., 1979) have described sensitive and rapid methods for NH_3 analysis of air involving use of a chemiluminescent detector for determination of the NO produced through oxidation of NH_3 by a thermal catalytic converter. These methods have not been evaluated for determination of NH_3 in research to assess emissions of NH_3 from soils, but there seems little doubt that they will be adopted for such research.

Hoell et al. (1980) have recently reported use of the Infrared Heterodyne Radiometer (a passive remote sensing instrument that uses the sun as a source) to study the vertical distribution of atmospheric ammonia, and Levine et al. (1980) have described use of this instrument to study changes in atmospheric NH_3 profiles after application of NH_4NO_3 fertilizer.

III. TRACER TECHNIQUES

A. Stable N Techniques

Use of the stable N isotopes ^{14}N and ^{15}N in investigations of N transformations in soils has increased dramatically during the past decade. About 1,000 articles and monographs that reported the use of ^{15}N in agricultural and agriculturally related research were published before 1970 (Hauck & Bystrom, 1970); about 37% of these were published between 1965 and 1969. More than 1,000 additional studies in the agricultural and environmental sciences using N tracers have been published since 1970. This rapid recent growth in N tracer use is the result of (i) the availability of relatively large amounts of N tracer materials at costs severalfold less than they were 10 to 15 years ago; (ii) improved instrumentation for N isotope-ratio analysis; (iii) development of rapid, precise, and accurate steam distillation methods for determining different N forms and for preparing samples for N isotope-ratio analysis; (iv) the need for more accurate quantitative measurements of N transformations in the biosphere that led to the adoption of tracer techniques; and (v) increased funding for purchase of the materials and facilities needed to use N tracer techniques. Public concern for preserving the quality of the environment and for energy conservation stimulated administrative support for research directed toward maximizing the efficiency of N use in agriculture and elucidating the effect of increased levels of fixed N in the biosphere on the environment.

Included among recent reviews of the scope and nature of ^{15}N use in studies of N cycle processes are those of Kundler (1970), Hauck (1971, 1973), Koren'kov et al. (1973), Hauck and Bremner (1976), and Bremner (1977). We focus here on important recent advances relevant to N tracer use in studies of N in the soil environment. The most important of these advances are the large-scale production of ^{15}N-depleted and ^{15}N-enriched materials, the development of optical emission spectroscopy for routine N isotope-ratio analysis, and the development of automatic mass spectrometers.

The availability of kilogram amounts of N tracer materials at relatively low cost ($47 to $99/g of ^{15}N during 1979) makes practical large-scale field experiments in which plot size does not dictate the kind of crop to be grown nor the management practices to be used. Most of the work involving ^{15}N use in soil-plant studies formerly was conducted in greenhouse pots or small field plots (1 to a few square meters in size) cropped to grasses or small grains. Recently, field plots as large as 75 m^2 each that were cropped to corn (*Zea mays* L.) have been used. In addition, the availability of large amounts of labeled N fertilizers has increased the scope of coordinated field trials (Joint FAO/IAEA, 1974, 1978) and N balance studies in a variety of crop management systems (Hauck & Kilmer, 1975). It had been evident for about 3 decades that the price of ^{15}N was high (usually $> $300/g of ^{15}N) because production was low, and that demand was low because the price was high. Recognizing this paradox, the (then) U.S. Atomic Energy Agency (now U.S. Department of Energy) initiated the ICONS (Isotopes of Carbon, Oxygen, Nitrogen, and Sulfur) program at its Los Alamos Scientific Laboratory in New Mexico. A program objective was to stimulate stable isotope use by making available lower cost materials. Apparently in response to the ICONS program, other primary producers of ^{15}N lowered their prices of N tracer materials. At Los Alamos, ^{14}N and ^{15}N are concentrated by means of fractional distillation of NO at liquid N$_2$ temperatures. This cryogenic process simultaneously separates the O$_2$ isotopes. Of the several isotopic species of NO molecules separated, $^{14}N^{16}O$ is produced in largest amounts; this is reacted with H$_2$ to form ^{15}N-depleted NH$_3$ (containing about 30 to 100 ppm ^{15}N compared with about 3,660 ppm ^{15}N in NH$_3$ of natural isotopic composition). Hundreds of kilograms of $(^{14}NH_4)_2SO_4$ and $^{14}NH_4^{14}NO_3$ have been produced and are available at prices about 30% lower than those of the respective ^{15}N-enriched compounds of comparable tracer value. To date, the main users of ^{15}N-depleted materials have been the U.S. Department of Agriculture (SEA-AR), the Tennessee Valley Authority in cooperative research with land-grant universities, and the University of California. These organizations have used about 2,000 kg of materials since 1972.

Mass spectroscopy has been the method chosen by most investigators for determining the isotopic composition of N tracer–labeled samples. However, in addition to the high cost of mass spectrometers and maintenance problems associated with their use, the need for about 1 mg of N for routine N isotope-ratio analysis can be a major limitation of their use for some studies. Broida and Chapman (1958) developed a relatively low-cost microprocedure using optical emission spectroscopy for determining the N isotopic composition of N_2 and NO. The commercial development of the optical emission spectrometer followed in 1968 (Meier, 1976), and since that time several models have been developed by manufacturers in the German Democratic Republic (GDR) and Japan. The analytical method is based on the fact that the wavelength of light emitted from N_2 molecules is affected by the nuclear mass of their constituent atoms. In practice, the gas is enclosed in an electrodeless discharge chamber and excited by microwave energy. Light in the ultraviolet range is emitted and resolved in a monochromator, after which the light intensities at each wavelength corresponding to the three molecular species of N_2 are measured. The main advantages of this method of determining N isotope ratios are the lower instrumentation costs (about one third of the cost for lower price mass spectrometers), lower operating and maintenance costs, and low sample requirement (< 10 μg of N). Acceptable ^{15}N determinations (0.370 ± 0.005 atom % ^{15}N for natural abundance standards) have been obtained from as little as 0.2 μg of N through use of a He-Xe mixture to sustain emission in the discharge tube (Goleb & Middelboe, 1968). The main disadvantage of optical emission spectroscopy is its lack of sensitivity and precision. Optical emission spectrometers are insensitive to ^{15}N concentrations < 0.36 atom % and usually are 10 to 100 times less precise than mass spectrometers, characteristics that either preclude their use for measuring N isotope ratios of ^{15}N-depleted samples or for studying variations in natural N isotope abundance.

Several approaches to automating either type of instrument have been taken, ranging from automatic sequential introduction of prepared N_2 samples into the analyzer to an automatic procedure that converts total sample N to N_2 before measuring its N isotope ratio. Of these developments, two are of particular significance to agricultural and environmental research because of their capacity for rapid and routine analysis of large numbers of samples.

The Isonitromat 5200[2] is an optical emission spectrometer developed by VEB Statron (GDR). The sample usually presented to the instrument is in the form of a solution of NH_4Cl (at least 0.6 ml containing preferably 400 μg of N/ml). As many as 380 samples contained in glass vials of 1-ml capacity can be stored in the instrument awaiting analysis. Each vial is moved in sequence to a sampler, and by means of a microperistaltic pump and appropriate gas and liquid flow regulators, the sample is sucked into

---

[2] Mention of a trade name does not constitute endorsement of the product by TVA nor exclusion of a similar product that also may be suitable.

the sampler, purged with He to expel air, and reacted with alkaline hypobromite reagent to form N_2. The sample preparation system can be modified for the direct conversion of NH_4^+-N in Kjeldahl digests or of amino acid–N to N_2. After conversion of sample N to N_2, the He-N_2 mixture is separated from the liquid phase, dried with silica gel, and introduced at a pressure of 15 torr into the analyzer, where the isotope ratio of the N_2 is determined.

The automated mass spectrometer developed at the Los Alamos Scientific Laboratory (McInteer & Montoya, 1980) also has a sample preparation system that converts NH_4^+-N to N_2 through alkaline hypobromite oxidation. The NH_4^+ sample (either solid or in solution) is placed into a tiny polyethylene vial (total capacity, about 150 μliters), 260 vials per sample tray. By means of an x-y plotter, the sample tray is positioned by computer command in such a way that each vial is connected in its programmed sequence to the sample preparation system. The system is then purged free of air with a condensable gas (Freon-12 or CO_2 have been used), hypobromite reagent is added, and the resultant N_2 is allowed to flow under reduced pressure through a liquid N_2 trap (eliminating the purging gas and gaseous impurities formed during the reaction). After purification, the N_2 pressure is reduced to about 30 torr, and the N_2 is allowed to enter the analyzer, where its isotope ratio is automatically determined.

Both automated systems described use flushing gases free of N_2 rather than vacuum to eliminate contaminant air. Because of its greater sensitivity to impurities and greater precision (an order of magnitude greater than for the emission spectrometer), the mass spectrometer requires sample N_2 virtually free of contaminants. This is better achieved by using a condensable purge gas that can be removed, along with impurities in it, by a liquid N_2 trap rather than an inert gas (He) that is admitted into the analyzer with sample N_2, along with impurities that may not be removed in the sample preparation train. Sample size for the emission spectrometer must be > 120 μg of N (200 μg/ml) and < 600 μg of N (1,000 μg/ml) with best results obtained using a concentration of 400 μg of N/ml, and with a ^{15}N concentration of at least 0.36 but no more than 30 atom %. For the automated mass spectrometer, sample concentration is adjusted to about 1 μg of N/μliter for convenience; isotope ratio determinations usually are made on 25 to 50 μg of N with a precision of at least $\pm$ 0.002 atom % ^{15}N in the full range of N isotopic composition. Analyses can be performed on samples containing < 10 μg of N but with a threefold to fivefold loss in precision. The automated emission spectrometer and mass spectrometer are capable of analyzing 18.5 and 30 samples/hour, respectively. Currently, a few laboratories that do not have automated instruments offer analytical services on a limited basis to investigators who do not have facilities for making N isotope-ratio determinations. Development of automatic instruments for N isotope-ratio analysis will foster the establishment of additional service laboratories.

As can be seen from the sample requirements discussed, both types of automated instruments, but especially the mass spectrometer, can be of great value to investigators who need N isotope-ratio analyses of N in dilute aqueous solutions, such as those obtained from surface and groundwaters, tile drain effluents, and soil extracts. Facilities for the rapid, routine N isotope-ratio analysis of small amounts of N will undoubtedly lead to increased use of N tracer techniques by limnologists, oceanographers, and other environmental scientists, by plant physiologists, and by soil scientists studying the distribution of tracer N among soil organic fractions and N transformations in soil microsites.

The capability for making N isotope-ratio determinations on microgram quantities of N now permits investigators to use convenient microprocedures for sample preparation that were not practical to use when preparing samples for routine analysis by conventional mass spectrometers that required samples containing 0.5 to 4 mg of N. Included among these techniques are the micro-Dumas procedures for preparing N_2 for optical emission spectroscopy (Fiedler & Proksch, 1975), modified Conway microdiffusion methods for determining various forms of inorganic N as NH_3 liberated from solution at ambient temperatures (Bremner, 1965b; Hauck, 1981), and use of paper chromatography (Kato, 1975) or thin-layer gel chromatography (Muhammad & Kumazawa, 1975, 1976) for the separation of amino acids.

Many techniques have been developed to meet the special requirements of a particular study. These include the coupling of a dry combustion or a pyrolysis process to a mass spectrometer or emission spectrometer for determining total N and N isotope ratio on the same sample, the coupling of a gas chromatograph to a mass spectrometer for separating the nitrogenous components of air before determining their N isotope ratios, the use of nuclear magnetic resonance spectroscopy for measuring the N content of and ^{15}N distribution within single kernels of seed or for following the change in fungal cell N components over time, and the use of a high-intensity laser beam for N isotope-ratio analysis by Raman scattering. Mueller (1965) and Jones (1969) suggested several new ways in which ^{14}N and ^{15}N could be used. Techniques currently being developed in several laboratories should considerably extend the scope of stable N tracer use in agricultural research.

B. Nitrogen-13 Techniques

Of the four known radionuclides of N, only ^{13}N has a half-life sufficiently long (10.05 min) for practical use in studies of biological systems. Although there are severe limitations to its use, ^{13}N nevertheless provides an extremely sensitive means of measuring the pathways of N transformations in biological systems within very short time intervals. The radionuclide has been produced as $^{13}N_2$ by the $^{12}C(d,n)^{13}N$ reaction using thin graphite platelets, granular charcoal, or flowing CO_2 as the target for deuteron bombardment, by the $^{13}C(p,n)^{13}N$ reaction using proton bombardment of powdered charcoal, or by the $^{14}N(p,d)^{13}N$ reaction using proton bombardment of melamine powder. It has been produced as $^{13}NO_2$ by the $^{16}O(p,a)^{13}N$ reaction using O_2 at high pressure or water as targets. The $^{13}NO_2$ thus produced is readily converted to $^{13}NO_2^-$, $^{13}NO_3^-$, or $^{13}NH_4^+$, as required.

The first use of ^{13}N (as $^{13}N_2$) as a biological tracer was by Ruben et al. (1940) to study the potential for N_2 fixation by barley plants (*Hordeum vulgare* L.), this study being published about 1 year before Burris and Miller (1941) reported the first use of $^{15}N_2$ to study N_2 fixation. Over 2 decades passed before ^{13}N was again used in research on biological systems (by Nicholas et al., 1961). To date, we are aware of only 10 published studies in which ^{13}N was used to study N transformations in soils—five concerned with evaluating the potential for N_2 fixation in nonleguminous plants, bacteria, or bacterial extracts and for determining the first reaction products of N_2 fixation in blue-green algae and five concerned with denitrification transformation rates and products (e.g., Firestone et al., 1979). Such work is necessarily restricted to those investigators with access to a cyclotron and associated facilities. Because of its sensitivity, the ^{13}N technique can be used in reaction rate studies within short time intervals, studies that cannot satisfactorily be made using stable N techniques. On the other hand, because N isotope-ratio determinations can now be made conveniently on microgram quantities of N, stable N techniques will be used in increasingly more sophisticated applications. Both stable N isotope and N radioisotope techniques give promise of new opportunities for refining our approaches to investigating N in plant-soil systems.

C. Use of Variations in Natural Nitrogen-15 Abundance

The concentration of ^{15}N (i.e., the ratio of ^{14}N to ^{15}N) in natural and synthetic substances is not constant. Slight variations in ^{15}N abundance occur because of isotope effects during the biological or chemical transformation of one N form into another. An isotope effect is defined here as the effect of nuclear characteristics other than atomic number on the nonnuclear chemical and physical properties of isotopes that lead to variations in the expression of these properties. Such expression may be observed as a shift in the wavelength of absorbed or emitted light, as a change in the rate of diffusion or reaction or as a change in the distribution of isotopes among different chemical forms during reaction. Because isotope effects result in only slight changes in the distribution of isotopes during N transformations, investigators using N tracers in studies of N transformations in natural systems need rarely be concerned about significant error from these effects. However, measuring the result of slight differences in reaction rates of ^{14}N and ^{15}N is the basis for using variations in natural ^{15}N abundance to study N transformations in biological systems (where N tracer has not been added).

Over 75 papers have been published on the occurrence of slight variations in the N isotope compositions of animal products, atmosphere, crude oils, coal, plant tissues, rain, rocks, minerals, soils, soil organic fractions, various organic compounds and substances, and other materials (for selected references, see Hauck & Bystrom, 1970; Hauck, 1973). Fractionation of the N isotopes (isotope effects leading to changes in isotopic composition) during biological processes has been observed in (i) nitrification of NH_4^+ resulting in NO_2^- and NO_3^- that are depleted in ^{15}N and substrate NH_4^+ enriched in ^{15}N, and (ii) denitrification of NO_3^- resulting in gaseous N slightly depleted in ^{15}N and substrate NO_3^- slightly enriched in ^{15}N (for references, see Hauck & Bremner, 1976). The extent of ^{15}N enrichment or depletion is determined in part by the degree of change in the substrate/product ratio in the N pool undergoing transformation. Thus, during reaction, an infinite number of N isotopic compositions of substrate and product are possible. Moreover, where nitrification and denitrification occur together or sequentially in the same system, NO_3^- as a product of nitrification is depleted in ^{15}N, whereas as a substrate for denitrification, it becomes enriched in ^{15}N. The N isotopic compositions of substrate and product in the total soil inorganic pool at any moment also are affected by the rates of their removal and replenishment through mineralization-immobilization reactions, external additions, and removals through plant uptake, leaching, and other means. In addition, the isotopic composition of N in the inorganic and organic pools can change significantly during NH_4^+ assimilation by heterotrophic microorganisms or through differential sorption of ^{14}N and ^{15}N on soil exchange complexes.

For most soil-plant studies, it is impossible in practice to continually monitor the changes in N isotopic composition of the different N forms of interest. Nevertheless, several investigators have attempted to make quantitative estimates of N transformation activities in soils using as a basis the changes in N isotopic composition of soil N constituents. Attempts have been made to measure the fractional contribution of soil N and fertilizer N to NO_3^- concentrations in waters, movement of fertilizer N within soil, and uptake of fertilizer N by corn, cereals, and legumes and to obtain evidence of N_2 fixation by legumes and transfer of legume-fixed N_2 to nonlegumes.

The information obtained from these studies is largely qualitative, often of a type that could more readily have been obtained by other means. For example, attempts have been made to estimate the amount of fertilizer N that is taken up by plants by comparing the N isotopic compositions of fertilized and unfertilized plants. The rationale is that because nitrification of NH_4^+ or NH_4^+-forming fertilizers results in NO_3^--N that is slightly depleted in ^{15}N, plants that take up this NO_3^--N will form tissue that is depleted in ^{15}N. It is assumed that the higher the fertilizer N application rate, the greater the degree of ^{15}N depletion in the plant tissue. Published data indicate that the amount of fertilizer N applied can be correlated with a decrease in ^{15}N concentration of plant tissue (within the range of natural ^{15}N abundance) but that the amount of N taken up cannot be accurately estimated. On the other hand, measuring the increase in dry matter production or total N of treated plants over controls also gives presumptive evidence of fertilizer N uptake and, in addition, gives a quantitative measure of uptake.

Naturally occurring or culturally added nitrogenous substances with slightly higher or lower ^{15}N concentrations than that of a standard such as atmospheric N_2 can be considered as tracer materials with low levels of ^{15}N enrichment or depletion. Whether they can be used to produce data that can be interpreted accurately depends on whether the substance selected as the tracer has an N isotopic composition that is sufficiently different from that of other N in the system under study so that it can be recognized as the tracer and accurately measured. Measurement of the tracer usually is complicated by inherent variation or change in its N isotopic composition and in that of other N components that interact with the tracer material (for an analysis of some of the problems inherent in the use of substances of natural N isotopic composition as tracers, see Hauck, 1971;[3] Hauck, 1973). This problem is obviated when using N tracers with an enrichment or depletion in ^{15}N markedly outside the range of natural ^{15}N abundance (i.e., the ^{15}N concentration of the tracer is sufficiently high, or low, that after dilution it remains measurably higher, or lower, than nontracer N in the system under study).

Proponents of using measurements of natural ^{15}N concentrations find this approach attractive because it theoretically permits study of N transformations and movements in ecosystems without perturbation from addition of fertilizer N and associated management practices. In our view, using natural ^{15}N abundance measurements for this purpose may provide useful information in simple systems containing only two major sources of N with comparatively wide differences in N isotopic composition (within the range of natural ^{15}N abundance) that do not change appreciably during the period of study. However, the many sources of error in sample collection and processing and in data interpretation militate against use of measuring variations in natural ^{15}N abundance as a means of obtaining reliable information about N cycle processes in complex biological systems.

IV. METHODS FOR ASSAY OF THE ACTIVITY OF ENZYMES CAUSING NITROGEN TRANSFORMATIONS IN SOILS

Only two enzymes, namely nitrogenase and urease, have received substantial attention in recent research to develop methods of assaying the activity of enzymes involved in N transformations in soils. Nitrogenase reduces N_2 to NH_3 ($N_2 + 6H \rightarrow 2NH_3$), and urease hydrolyses urea to NH_3 and CO_2 ($NH_2CONH_2 + H_2O \rightarrow 2NH_3 + CO_2$). The interest in assay of nitrogenase activity has been stimulated by the key role of this enzyme in biological N_2 fixation. The interest in assay of urease activity has been stimulated by the rapidly increasing importance of urea as a fertilizer in world agriculture and by the accumulation of evidence that the fate and performance of this fertilizer is greatly influenced by soil urease.

---

[3] R. D. Hauck. 1971. Evaluation of a proposed method of determining the extent to which fertilizer nitrogen may contribute to nitrate contents in waters. 10th Plant Nutrient Hearings, Illinois Pollution Control Board, Edwardsville, Ill.

A. Nitrogenase Activity

One of the most important advances in methodology for soil N research since 1965 has been the introduction of the acetylene reduction method of assaying nitrogenase activity. This method originated from the finding that the nitrogenase enzyme responsible for reduction of atmospheric N_2 to NH_3 also reduces acetylene to ethylene ($C_2H_2 + 2H \rightarrow C_2H_4$) (Dilworth, 1966; Schollhorn & Burris, 1966, 1967). Hardy et al. (1968) developed a sensitive gas chromatographic procedure involving use of a flame ionization detector for measurement of reduction of acetylene to ethylene and described use of this procedure for assay of nitrogenase activity in soil and soil-plant studies (see also Hardy et al., 1973). Many variations of the acetylene reduction method of assaying nitrogenase activity have been described (Ham, 1977; Smith, 1977; Bergersen, 1980), and several hundred papers reporting use of this method for research on biological N_2 fixation have been published during the past decade. Problems in its use are discussed in section V.

B. Urease Activity

Numerous methods have been proposed for assay of urease activity in soils (Bremner & Mulvaney, 1978; Ladd, 1978). Most involve determination of the NH_4^+ released on incubation of toluene-treated soil with buffered urea solution. Others involve estimation of the urea decomposed or the CO_2 released on incubation of soil with urea. Most of the methods proposed must be considered empirical because no studies to evaluate them have been reported. The methods that have been most thoroughly evaluated are the buffer method proposed by Tabatabai and Bremner (1972) and the nonbuffer method proposed by Zantua and Bremner (1975). The latter method is essentially a scaled-down version of the method proposed by Douglas and Bremner (1971; the only significant difference is that toluene is omitted. It involves determination of the amount of urea hydrolyzed on incubation of the soil sample with urea at 37°C for 5 hours, urea hydrolysis being estimated by colorimetric determination of urea in the extract obtained by shaking the incubated soil sample with $2M$ KCl containing a urease inhibitor (phenylmercuric acetate) and filtering the resulting suspension. The buffer method of Tabatabai and Bremner (1972) involves determination of the NH_4^+ released on incubation of the soil sample with THAM (tris–H_2SO_4) buffer (pH 9.0), urea, and toluene at 37°C for 2 hours, NH_4^+ release being determined by shaking the incubated soil sample with $2.5M$ KCl containing a urease inhibitor (Ag_2SO_4) and by steam-distilling an aliquot of the resulting soil suspension with MgO. Both methods give precise results, but the buffer method gives markedly higher values than the nonbuffer method and detects urease activity that does not occur when soils are treated with urea in the absence of buffer. The nonbuffer method provides a good index of the ability of soils to hydrolyze urea under natural conditions (Zantua & Bremner, 1975), and its results are not affected by inclusion of toluene.

It should be emphasized that the choice of method for assay of urease activity in soils should depend on the purpose of the assay because this has frequently been overlooked in research on soil urease. If the purpose of the assay is to obtain an index of the ability of the urease in the soil sample under study to hydrolyze urea under natural conditions, the nonbuffer method described is obviously superior to the buffer method. If the purpose is to detect urease in soils or soil fractions, the buffer method should be preferred because it detects urease activity not detected by the nonbuffer method.

C. Other Enzymes

Methods proposed for assay of proteinase, peptidase, nuclease, and amidase activity in soils are discussed in Chapter 5 (J. N. Ladd and R. B. Jackson) and in reviews by Ladd (1978) and Roberge (1978). Most of these methods have not been evaluated and are open to a variety of criticisms. This does not hold for the methods of assaying soil proteolytic activity proposed by Ladd and Butler (1972) or the procedure for assay of soil amidase activity proposed by Frankenberger and Tabatabai (1980).

V. METHODS FOR RESEARCH ON BIOLOGICAL NITROGEN FIXATION

The very substantial increase in support for research on biological N_2 fixation during the past 15 years has resulted in important advances in the methodology for such research. These advances have been discussed at length in a recent monograph devoted entirely to methods of studying biological N_2 fixation (Bergersen, 1980), and no attempt is made to discuss them in any detail here.

The method that has been used most extensively in N_2 fixation research since 1965 is the acetylene reduction method of assaying nitrogenase activity described in section IV A. This method gained rapid acceptance because it is simple, sensitive, and inexpensive compared with the $^{15}N_2$ method of studying N_2 fixation and can be used to study N_2 fixation under field conditions. It is, nevertheless, an indirect method of measuring N_2 fixation, and its results must be compared with those obtained by a direct method of measuring N_2 fixation before they can safely be interpreted in terms of N_2 fixed.

Many users of the acetylene reduction method have calculated N_2 fixed from acetylene reduced by using the theoretical conversion factor (acetylene reduced/N_2 fixed) of 3.0. Burris (1972, 1974) strongly criticized this practice and emphasized the importance of establishing valid conversion factors for N_2-fixing systems studied by the acetylene reduction technique. He also stressed that reduction of $^{15}N_2$ is the primary method of choice for validation of data obtained by the acetylene reduction technique and by other indirect methods of studying N_2 fixation and that for valid interpretation of acetylene reduction data in terms of N_2 fixation, it is essential to measure acetylene and $^{15}N_2$ reduction under identical conditions. Despite Burris's admonitions, few comparisons of the acetylene and $^{15}N_2$ reduction methods of studying N_2 fixation have been reported (for references, see Bremner, 1977; Hudd et al., 1980; Roskoski, 1981), and most conversion factors adopted have not been established by comparing these methods under identical conditions. Most experimentally determined conversion factors have ranged from 3.0 to 6.3, but values as high as 25 have been reported for soils under waterlogged or anaerobic conditions (for references, see Bremner, 1977).

Hudd et al. (1980) compared acetylene and $^{15}N_2$ reduction by nodules of field beans (*Vicia faba*) and found that the factor for calculation of N_2 fixed from acetylene reduced (5.75) was about twice the theoretical factor (3.0). They concluded that this discrepancy arose because with acetylene, all the electrons available to nitrogenase were used to form ethylene, whereas only about half the electron supply was used to fix N_2, the remainder being consumed in the production of H_2. This conclusion was based on work by Schubert and Evans (1976) indicating that for a wide range of legumes, only 40 to 60% of the electron flow to nitrogenase was transferred to N_2, the remainder being lost through H_2 evolution.

Bremner (1977) has emphasized the need for controls in use of the acetylene reduction technique to check that the N_2-fixing system under study does not produce ethylene in the absence of acetylene and does not take up ethylene produced by reduction of acetylene. There is a clear need for such controls because it is now well established that soils and other natural systems can take up ethylene and can produce ethylene in the absence of acetylene and that materials used to fabricate sealed systems for studies of N_2 fixation by the acetylene reduction technique can both sorb and evolve ethylene (Kavanagh & Postgate, 1970; Flett et al., 1975; DeBont & Mulder, 1976; Bremner, 1977; Watanabe & de Guzman, 1980; Culbertson et al., 1981).

Problems have been encountered in application of the acetylene reduction technique to flooded soils and aquatic systems (Flett et al., 1975, 1976; Lee & Watanabe, 1977) and through use of long periods of incubation with acetylene in application of this technique (David & Fay, 1977). Also, it has been demonstrated that acetylene inhibits both nitrification in soils and reduction of N_2O to N_2 by soil microorganisms (see section VI).

Although ^{15}N-enriched NH_4^+ and NO_3^- compounds have been used in tracer studies of N_2 fixation (for examples, see Fried & Broeshart, 1975; Johnson et al., 1975; Legg & Sloger, 1975; Ham, 1978; Rennie et al., 1978), most ^{15}N-tracer studies of N_2 fixation have involved exposure of the N_2-fixing system to $^{15}N_2$ in a gas-tight container. Burris (1972, 1974) has provided detailed descriptions of techniques for exposing N_2-fixing systems to $^{15}N_2$ and for generating and purifying $^{15}N_2$ for use in these techniques. The importance of using $^{15}N_2$ purified to remove NH_3 and N oxides deserves emphasis because failure to use purified $^{15}N_2$ has been a common source of error in ^{15}N-tracer studies of biological N_2 fixation.

Although the first use of $^{13}N_2$ to study N_2 fixation (Ruben et al., 1940) preceded the first use of $^{15}N_2$ to study this process (Burris & Miller, 1941), very few studies of N_2 fixation involving the use of $^{13}N_2$ have been reported during the past 40 years (Bremner, 1977; Hauck, 1979). This is not difficult to explain, because although ^{13}N-tracer techniques are considerably more sensitive than ^{15}N-tracer techniques, $^{13}N_2$ has a half-life of only 10.05 min, and very expensive and sophisticated equipment is required for preparation and purification of this gas for N_2 fixation research.

Three methods using N tracer techniques have been proposed for field measurement of biological N_2 fixation. All three are based on assumptions that require validation. One involves labeling of soil organic matter with ^{15}N by use of glucose to stimulate immobilization of ^{15}N-enriched inorganic N added to soil. Legg and Sloger (1975) used this approach and estimated the amount of N_2 fixed by soybeans (*Glycine max* L. Merr.) from isotope dilution calculations in which it was assumed that the immobilized ^{15}N mineralized during plant growth was a measure of mineralizable soil N and that the fractions of soil organic matter that released plant-available N during plant growth were uniformly labeled.

Another N-tracer technique proposed for field measurement of N_2 fixation involves application of low rates of ^{15}N-labeled fertilizer to leguminous plants and high rates to nonleguminous plants and calculation of A values (defined as the amounts of available soil N) for each plant species (Fried & Broeshart, 1975). The amount of N_2 fixed is calculated as [(A value of N_2-fixing system $-$ A value of nonfixing system) $\times$ Percentage of fertilizer N taken up by the legume]. As noted by Hauck (1979), this approach is based on the following assumptions:

1) The legume and nonlegume have essentially identical rooting patterns (i.e., they extract the same amount of N at the same time from each portion of their expanding root systems).
2) The amount of soil N taken up by either plant species is independent of the amount of fertilizer N applied.
3) The A value is a constant, unaffected by application of fertilizer N.

The validity of these assumptions is questionable (Hauck & Bremner, 1976; Hauck, 1979).

Rennie et al. (1976) proposed a tracer method of estimating N_2 fixation under field conditions without use of ^{15}N-enriched fertilizer. This method involves measurement of natural variations in ^{15}N abundance in legumes and nonlegumes and is similar to the method proposed by Legg and Sloger (1975) in that soil can be regarded as a tracer material having an extremely low level of ^{15}N enrichment. Amarger et al. (1977) modified the calculations in this method to take account of N isotope fractionation during plant uptake of N from the atmosphere and other N sources. Problems in interpretation of data obtained by such methods have been discussed in articles by Hauck (1973), Hauck and Bremner (1976), and Bremner (1977).

Weber (1966) estimated symbiotic fixation of N_2 by soybeans from the difference in total plant N between nodulating and nonnodulating soybean isolines, and Ham et al. (1975) used this approach to measure the depressant effect of fertilizer N on N_2 fixation. Use of this method involves the assumption that both isolines have identical soil N uptake patterns.

VI. METHODS FOR RESEARCH ON DENITRIFICATION

Methods for research on denitrification in soils have been discussed in recent articles by Rolston et al. (1976), Focht (1978), Hauck (1979), Bremner and Blackmer (1981), Knowles (1981), and Tiedje et al. (1981). They include methods involving the use of [15]N- and ^{13}N-tracer techniques discussed in section III and/or GC techniques for determination of N_2, N_2O, and NO (gaseous products of denitrification) discussed in section II D. Several methods proposed have rather obvious limitations and are not discussed here. These include measurement of disappearance of NO_3^- or N_2O, measurement of N_2/Ar ratios, and measurement of N gas production by manometric techniques (for discussions of these methods, see Knowles, 1981).

Research on denitrification is greatly complicated by the fact that the major product of this process under most conditions (N_2) is the major constituent of air. For this reason, most research on denitrification in soils during the past decade has been conducted by incubating NO_3^--treated soils in closed systems under N_2-free atmospheres that permit GC determination of N_2, N_2O, and NO produced by denitrification. Most workers have used He or Ar to establish N_2-free atmospheres, but $He-O_2$ and $Ar-O_2$ mixtures have been employed in several investigations. This method of studying denitrification has proved very valuable for laboratory research on denitrification in soils, but because it requires establishment of a N_2-free atmosphere, it cannot be used to study denitrification in soils under natural conditions. However, chamber techniques have been developed that permit detection of ^{15}N-labeled N_2 evolved from soils into a natural atmosphere through denitrification of ^{15}N-labeled NO_3^- (McGarity & Hauck, 1969; Rolston et al., 1978, 1979). These techniques require use of NO_3^- highly enriched with ^{15}N and are complicated and expensive, but they permit direct measurement of denitrification under field conditions. Requirements and problems in use of such techniques have been discussed by Hauck (1979).

Although both [15]N- and ^{13}N-tracer techniques have been used in recent research on denitrification in soils, studies using ^{13}N have received considerably more attention than those using [15]N. Tiedje et al. (1979, 1981) have described methods for production, purification, and detection of ^{13}N in such studies and have reviewed the results of recent research on denitrification in soils involving use of [13]N. The major advantages of ^{13}N for research on denitrification are that it allows direct measurement of N_2 in any atmosphere and provides such a sensitive measure of denitrification activities that it permits research that is not possible using ^{15}N, including isotope exchange experiments at low substrate concentrations and very short-term measurements of rates of denitrification. However, the short half-life of ^{13}N severely limits the type of research that can be performed with this isotope, because all data collection must be completed within 2 to 3 hours from the time of production of the ^{13}N used. This means that research with this isotope must be performed close to the accelerator used to produce it, that the ^{13}N substrate must be mixed rapidly with the sample, that the analytical techniques adopted must be rapid, and that it is difficult to analyze many replicates or treatments in an experimental period. Moreover, it has not thus far been found possible to use ^{13}N for measurement of the rates of denitrification in soils under natural conditions. As noted by Tiedje et al. (1981), ^{13}N- and ^{15}N-tracer methods of studying denitrification should be considered complementary because each method is uniquely suited for studies of certain aspects of this process.

The technique that has received the most attention by researchers on denitrification during the past 5 years is the so-called acetylene inhibition method, which originated from a report by Federova et al. (1973) that acetylene inhibited reduction of N_2O to N_2 by denitrifying microorganisms. This report has been confirmed by studies of the effects of acetylene on denitrification by pure cultures of denitrifying microorganisms (Balderston et al., 1976; Yoshinari & Knowles, 1976) and on denitrification in soils and sediments (Klemedtsson et al., 1977; Yoshinari et al., 1977; Smith et al., 1978; Tiedje et al., 1978; Chan & Knowles, 1979; Knowles, 1979; Ryden et al., 1979a, b; Van Raalte & Patriquin, 1979), and several of these studies have indicated that denitrification in soils under laboratory or field conditions can be measured by relatively simple methods involving the use of gas chromatography to determine the amount of N_2O evolved when acetylene is used to inhibit microbial reduction of N_2O produced by denitrification of NO_3^- in soils incubated in closed systems. However, other studies have shown that the C_2H_2 inhibition method of studying denitrification has several limitations, and that data obtained by this method must be interpreted with considerable caution. For example, recent work has shown that acetylene inhibits oxidation of NH_4^+ by nitrifying microorganisms (Hynes & Knowles, 1978) and is a potent inhibitor of nitrification in soils (Bremner & Blackmer, 1979; Walter et al., 1979). Moreover, there is evidence that acetylene promotes denitrification of NO_3^- by soil microorganisms (Germon, 1980a, b) and that the ability of acetylene to inhibit reduction of N_2O by denitrifying microorganisms in soils depends on the time of exposure of these microorganisms to acetylene (Yeomans & Beauchamp, 1978) and the NO_3^- (Smith et al., 1978; Tiedje et al., 1978) and sulfide (Tam & Knowles, 1979; Sörensen et al., 1980) content of the soil under study. Also, there seems no doubt that application of the acetylene inhibition method of studying denitrification to flooded systems will be complicated by the problems encountered in application of the acetylene reduction method of studying biological N_2 fixation to such systems (see section V). Other problems that must be considered in application of the acetylene inhibition method of studying denitrification are the susceptibility of acetylene to oxidation by both aerobic and anaerobic microorganisms (Watanabe & de Guzman, 1980; Culbertson et al., 1981) and the solubility of N_2O in water.

As noted in section II D, concern about the potential adverse effects of an increase in the atmospheric concentration of N_2O has stimulated research to assess the contributions of soils and N fertilizers to atmospheric N_2O and has resulted in the development of chamber techniques for field measurement of N_2O emissions from soils. These techniques initially seemed very promising for assessment of gaseous losses of soil and fertilizer N as N_2O produced via denitrification of NO_3^- because it was generally assumed that the N_2O evolved from soils is produced exclusively through reduction of NO_3^- by denitrifying microorganisms. However, recent work has indicated that this assumption is not valid and that much of the N_2O evolved from soils treated with NH_4^+ or NH_4^+-producing compounds commonly used as N fertilizers is produced by nitrifying microorganisms during oxidation of NH_4^+ to NO_3^- (Bremner & Blackmer, 1979).

LITERATURE CITED

Ahlrichs, L. E. 1965. Ammonium fixation and release on some Minnesota soils. Ph.D. Thesis. Univ. of Minnesota (Libr. Congr. Card no. Mic. 66-8853). Univ. Microfilms. Ann Arbor, Mich. (Diss. Abstr. 27B:661).

Allen, J. D. 1973. A review of methods of analysis for oxides of nitrogen. J. Inst. Fuel 46: 123–133.

Amarger, N., A. Mariotti, and F. Mariotti. 1977. An attempt at estimating the rate of symbiotic fixation of nitrogen in lupine by natural isotopic tracing (^{15}N). C. R. Acad. Sci., Ser. D. 284:2179–2182.

Andre, C. E., and A. R. Mosier. 1973. Precolumn inlet system for the gas chromatographic analysis of trace quantities of short-chain aliphatic amines. Anal. Chem. 45:1971–1973.

Aneja, V. P., E. P. Stahel, H. H. Rogers, A. M. Witherspoon, and W. W. Heck. 1978. Calibration and performance of a thermal converter in continuous atmospheric monitoring of ammonia. Anal. Chem. 50:1705–1708.

Balderston, W. L., B. Sherr, and W. J. Payne. 1976. Blockage by acetylene of nitrous oxide reduction in *Pseudomonas perfectomarinus*. Appl. Environ. Microbiol. 31:504–508.

Banwart, W. L., M. A. Tabatabai, and J. M. Bremner. 1972. Determination of ammonium in soil extracts and water samples by an ammonia electrode. Commun. Soil Sci. Plant Anal. 3:449–458.

Barnes, H., and A. R. Folkard. 1951. The determination of nitrites. Analyst (London) 76: 599–603.

Baumgartner, R., W. A. McClenny, and R. K. Stevens. 1979. Optimized chemiluminescence system for measuring atmospheric ammonia. EPA-600/2-79-028, U.S. Environmental Protection Agency, Research Triangle, N.C.

Beckett, M. J., and A. L. Wilson. 1974. The manual determination of ammonia in fresh waters using an ammonia-sensitive membrane electrode. Water Resour. Res. 8:333–340.

Bergersen, F. J. (ed.). 1980. Methods for evaluating biological nitrogen fixation. John Wiley & Sons, Inc., New York.

Bethea, R. M., and M. C. Meador. 1969. Gas chromatographic analysis of reactive gases in air. J. Chromatogr. Sci. 7:655–664.

Bjarno, O.-C. 1980. Kjel-Foss automatic analysis using an antimony-based catalyst: Collaborative study. J. Assoc. Off. Anal. Chem. 63:657–663.

Black, F. M., and J. E. Sigsby. 1974. Chemiluminescent method for NO and NO_x (NO + NO_2) analysis. Environ. Sci. Technol. 8:149–152.

Blackmer, A. M., and J. M. Bremner. 1977. Gas chromatographic analysis of soil atmospheres. Soil Sci. Soc. Am. J. 41:908–912.

Bouwmeester, R. J. B., and P. L. G. Vlek. 1981. Rate control of ammonia volatilization from rice paddies. Atmos. Environ. 15:131–140.

Bremner, J. M. 1959. Determination of fixed ammonium in soil. J. Agric. Sci. 52:147–160.

Bremner, J. M. 1960. Determination of nitrogen in soils by the Kjeldahl method. J. Agric. Sci. 55:11–33.

Bremner, J. M. 1965a. Total nitrogen. In C. A. Black et al. (ed.) Methods of soil analysis, Part 2. Agronomy 9:1149–1178. Am. Soc. of Agron., Madison, Wis.

Bremner, J. M. 1965b. Inorganic forms of nitrogen. In C. A. Black et al. (ed.) Methods of soil analysis. Part 2. Agronomy 9:1179–1237. Am. Soc. of Agron., Madison, Wis.

Bremner, J. M. 1965c. Organic forms of nitrogen. In C. A. Black et al. (ed.) Methods of soil analysis, Part 2. Agronomy 9:1238–1255. Am. Soc. of Agron., Madison, Wis.

Bremner, J. M. 1965d. Isotope-ratio analysis of nitrogen in nitrogen-15 tracer investigations. In C. A. Black et al. (ed.) Methods of soil analysis, Part 2. Agronomy 9:1256–1286. Am. Soc. of Agron., Madison, Wis.

Bremner, J. M. 1965e. Nitrogen availability indexes. In C. A. Black et al. (ed.) Methods of soil analysis, Part 2. Agronomy 9:1324–1345. Am. Soc. of Agron., Madison, Wis.

Bremner, J. M. 1967. Nitrogenous constituents. p. 19–66. In A. D. McLaren and G. H. Peterson (ed.) Soil biochemistry. Marcel Dekker, Inc., New York.

Bremner, J. M. 1977. Use of nitrogen-tracer techniques for research on nitrogen fixation. p. 335–352. In A. Ayanaba and P. J. Dart (ed.) Biological nitrogen fixation in farming systems of the tropics. John Wiley & Sons, Inc., New York.

Bremner, J. M. 1981. Urea. In A. L. Page et al. (ed.) Methods of soil analysis, Part 2. Rev. ed. Am. Soc. of Agron., Madison, Wis. (In press.)

Bremner, J. M., and A. M. Blackmer. 1979. Effects of acetylene and soil water content on emission of nitrous oxide from soils. Nature (London) 280:380–381.

Bremner, J. M., and A. M. Blackmer. 1981. Composition of soil atmospheres. In A. L. Page et al. (ed.) Methods of soil analysis, Part 2. Rev. ed. Am. Soc. of Agron., Madison, Wis. (In press.)

Bremner, J. M., and L. G. Bundy. 1973. Use of ferrous hydroxide for determination of nitrate in soil extracts. Commun. Soil Sci. Plant Anal. 4:285–291.

Bremner, J. M., and T. Harada. 1959. Release of ammonium and organic matter from soil by hydrofluoric acid and effect of hydrofluoric acid treatment on extraction of soil organic matter by neutral and alkaline reagents. J. Agric. Sci. 52:137–146.

Bremner, J. M., and D. R. Keeney. 1965. Steam distillation methods for determination of ammonium, nitrate and nitrite. Anal. Chim. Acta 32:485–495.

Bremner, J. M., and D. R. Keeney. 1966. Determination and isotope-ratio analysis of different forms of nitrogen in soils: 3. Exchangeable ammonium, nitrate, and nitrite by extraction-distillation methods. Soil Sci. Soc. Am. Proc. 30:577–582.

Bremner, J. M., and R. L. Mulvaney. 1978. Urease activity in soils. p. 149–196. In R. G. Burns (ed.) Soil enzymes. Academic Press, Inc., New York.

Bremner, J. M., and C. S. Mulvaney. 1981. Total nitrogen. In A. L. Page et al. (ed.) Methods of soil analysis, Part 2. Rev. ed. Am. Soc. of Agron., Madison, Wis. (In press.)

Bremner, J. M., D. W. Nelson, and J. A. Silva. 1967. Comparison and evaluation of methods of determining fixed ammonium in soils. Soil Sci. Soc. Am. Proc. 31:466–472.

Bremner, J. M., and K. Shaw. 1958. Denitrification in soil. I. Methods of investigation. J. Agric. Sci. 51:22–39.

Bremner, J. M., and M. A. Tabatabai. 1972. Use of an ammonia electrode for determination of ammonium in Kjeldahl analysis of soils. Commun. Soil Sci. Plant Anal. 3:159–165.

Broida, H. P., and M. W. Chapman. 1958. Stable nitrogen isotope analysis by optical spectroscopy. Anal. Chem. 30:2049–2055.

Burris, R. H. 1972. Nitrogen fixation—assay methods and techniques. p. 415–431. In S. P. Colowick and N. O. Kaplan (ed.) Methods in enzymology, Vol. 24. Academic Press, Inc., New York.

Burris, R. H. 1974. Methodology. p. 9–33. In A. Quispel (ed.) The biology of nitrogen fixation. Elsevier North-Holland Inc., New York.

Burris, R. H., and C. E. Miller. 1941. Application of N^{15} to the study of biological nitrogen fixation. Science 93:114–115.

Byrne, E., and S. McCormack. 1978. Determination of ammonium nitrogen in silage samples by an ammonia electrode. Commun. Soil Sci. Plant Anal. 9:667–684.

Chan, Y. K., and R. Knowles. 1979. Measurement of denitrification in two freshwater sediments by an in situ acetylene inhibition method. Appl. Environ. Microbiol. 37: 1067–1072.

Cheng, H. H., and J. M. Bremner. 1965. Gaseous forms of nitrogen. In C. A. Black et al. (ed.) Methods of soil analysis, Part 2. Agronomy 9:1287–1323. Am. Soc. of Agron., Madison, Wis.

Cicerone, R. J., J. D. Shetter, D. H. Stedman, T. J. Kelly, and S. C. Liu. 1978. Atmospheric N_2O: Measurements to determine its sources, sinks and variations. J. Geophys. Res. 83: 3042–3050.

Clausen, C., B. R. Bock, G. A. Peterson, and R. A. Olson. 1980. The magnesium problem in nitrate determination by steam distillation. Soil Sci. Soc. Am. J. 44:1326–1327.

Cohen, Y. 1977. Shipboard measurement of dissolved nitrous oxide in seawater by electron capture gas chromatography. Anal. Chem. 49:1238–1240.

Cortez, J., and M. Schnitzer. 1979. Purines and pyrimidines in soils and humic substances. Soil Sci. Soc. Am. J. 43:958–961.

Crutzen, P. J., and D. H. Ehhalt. 1977. Effects of nitrogen fertilizers and combustion on the stratospheric ozone layer. Ambio 6:112–117.

Culbertson, C. W., A. J. B. Zehnder, and R. S. Oremland. 1981. Anaerobic oxidation of acetylene by estuarine sediments and enrichment cultures. Appl. Environ. Microbiol. 41: 396–403.

David, K. A. V., and P. Fay. 1977. Effects of long-term treatment with acetylene on nitrogen-fixing microorganisms. Appl. Environ. Microbiol. 34:640–646.

DeBont, J. A. M., and E. G. Mulder. 1976. Invalidity of the acetylene reduction assay in alkane-utilizing, nitrogen-fixing bacteria. Appl. Environ. Microbiol. 31:640–647.

Denmead, O. T. 1979. Chamber systems for measuring nitrous oxide emission from soils in the field. Soil Sci. Soc. Am. J. 43:89–95.

Denmead, O. T., J. R. Freney, and J. R. Simpson. 1976. A closed ammonia cycle within a plant canopy. Soil Biol. Biochem. 8:161–164.

Denmead, O. T., R. Nulsen, and G. W. Thurtell. 1978. Ammonia exchange over a corn crop. Soil Sci. Soc. Am. J. 42:840–842.

Denmead, O. T., J. R. Simpson, and J. R. Freney. 1974. Ammonia flux into the atmosphere from a grazed pasture. Science 185:609–610.

Denmead, O. T., J. R. Simpson, and J. R. Freney. 1977. A direct field measurement of ammonia emission after injection of anhydrous ammonia. Soil Sci. Soc. Am. J. 41:1001–1004.

Dhariwal, A. P. S., and F. J. Stevenson. 1958. Determination of fixed ammonium in soils. Soil Sci. 86:343–349.

Dilworth, M. J. 1966. Acetylene reduction by nitrogen-fixing preparations from *Clostridium pasteurianum*. Biochim. Biophys. Acta 127:285–294.

Douglas, L. A., and J. M. Bremner. 1970. Extraction and colorimetric determination of urea in soils. Soil Sci. Soc. Am. Proc. 34:859–862.

Douglas, L. A., and J. M. Bremner. 1971. A rapid method of evaluating different compounds as inhibitors of urease activity in soils. Soil Biol. Biochem. 3:309–315.

Douglas, L. A., A. Riazi, and C. J. Smith. 1980. A semi-micro method for determining total nitrogen in soils and plant material containing nitrite and nitrate. Soil Sci. Soc. Am. J. 44:431–433.

Douglas, L. A., H. Sochtig, and W. Flaig. 1978. Colorimetric determination of urea in soil extracts using an automated system. Soil Sci. Soc. Am. J. 42:291–292.

Dowdell, R. J., J. R. Burford, and R. Crees. 1979. Losses of nitrous oxide dissolved in drainage water from agricultural land. Nature (London) 278:342–343.

Dowdell, R. J., and R. Crees. 1974. Measurement of the nitrous oxide content of the atmosphere. Lab. Pract. 23:488–489.

Drushel, H. V. 1977. Determination of nitrogen in petroleum fractions by combustion with chemiluminescent detection of nitric oxide. Anal. Chem. 49:932–939.

Dyck, A. W. J., and R. R. McKibbin. 1935. The non-protein nature of a fraction of soil organic nitrogen. Can. J. Res. 13:264–268.

Eastin, E. F. 1976. Use of an ammonia electrode for total nitrogen determination in plants. Commun. Soil Sci. Plant Anal. 7:477–481.

Elkins, J. W. 1980. Determination of dissolved nitrous oxide in aquatic systems by gas chromatography using electron-capture detection and multiple phase equilibration. Anal. Chem. 52:263–267.

Elliott, L. F., G. E. Schuman, and F. G. Viets, Jr. 1971. Volatilization of nitrogen-containing compounds from beef cattle areas. Soil Sci. Soc. Am. Proc. 35:752–755.

Federova, R. I., E. I. Milekhina, and N. I. Il'yukhina. 1973. Evaluation of the method of "gas metabolism" for detecting extraterrestrial life. Identification of nitrogen-fixing microorganisms. (In Russian.) Izv. Akad. Nauk SSSR Ser. Biol. 6:797–806.

Ferm, M. 1979. Method for determination of atmospheric ammonia. Atmos. Environ. 13: 1385–1393.

Fiedler, R., and G. Proksch. 1975. The determination of nitrogen-15 by emission and mass spectrometry in biochemical analysis: A review. Anal. Chim. Acta 78:1–62.

Firestone, M. K., R. B. Firestone, and J. M. Tiedje. 1979. Nitric oxide as an intermediate in denitrification: Evidence from nitrogen-13 isotope exchange. Biochem. Biophys. Res. Commun. 91:10–16.

Flett, R. J., R. D. Hamilton, and N. E. R. Campbell. 1976. Aquatic acetylene-reduction techniques: solution to several problems. Can. J. Microbiol. 22:43–51.

Flett, R. J., J. W. M. Rudd, and R. D. Hamilton. 1975. Acetylene reduction assays for nitrogen fixation in fresh waters: A note of caution. Appl. Microbiol. 29:580–583.

Focht, D. D. 1978. Methods for analysis of denitrification in soils. p. 433–490. In D. R. Nielsen and J. G. MacDonald (ed.) Nitrogen in the environment, Vol. 2. Academic Press, Inc., New York.

Folsom, B. A., and C. W. Courtney. 1979. Accuracy of chemiluminescent analyzers measuring nitric oxide in stack gases. J. Air. Pollut. Control Assoc. 29:1166–1169.

Fontijn, A., A. J. Sabadell, and R. J. Ronco. 1970. Homogeneous chemiluminescent measurement of nitric oxide with ozone. Implications for continuous selective monitoring of gaseous air pollutants. Anal. Chem. 42:575–579.

Frankenberger, W. T., Jr., and M. A. Tabatabai. 1980. Amidase activity in soils: I. Method of assay. Soil Sci. Soc. Am. J. 44:282–287.

Freney, J. R. 1971. Phosphate interference in the Devarda's alloy reduction method for nitrate. Commun. Soil Sci. Plant Anal. 2:479–484.

Fried, M., and H. Broeshart. 1975. An independent measure of the amount of nitrogen fixed by a legume crop. Plant Soil 43:707–711.

Galbally, I. E. 1977. Measurement of nitrogen oxides in the background atmosphere. p. 179–185. In Air pollution measurement techniques. Spec. Environmental Report no. 10, World Meteorological Organization, Geneva.

Galbally, I. E., and C. R. Roy. 1978. Loss of fixed nitrogen from soils by nitric oxide exhalation. Nature (London) 275:734–735.

Gallaher, R. N., C. O. Weldon, and F. C. Boswell. 1976. A semiautomated procedure for total nitrogen in plant and soil samples. Soil Sci. Soc. Am. J. 40:887–889.

Germon, J. C. 1980a. Étude quantitative de la dénitrification biologique dans le sol à l'aide de l'acetylene. I. Application à différents sols. Ann. Microbiol. (Inst. Pasteur) 131B:69–80.

Germon, J. C. 1980b. Étude quantitative de la dénitrification biologique dans le sol à l'aide de l'acétylène. II. Évolution de l'effet inhibiteur de l'acétylène sur la N2O-réductase; incidence de l'acétylène sur la vitesse de dénitrification et sur la reórganisation de l'azote nitrique. Ann. Microbiol. (Inst. Pasteur) 131B:81–90.

Goleb, J. A., and V. Middelboe. 1968. Optical nitrogen-15 analysis of small nitrogen samples with a mixture of helium and xenon to sustain the discharge in an electrodeless tube. Anal. Chim. Acta 43:229–234.

Guthrie, T. F., and J. M. Duxbury. 1978. Nitrogen mineralization and denitrification in organic soils. Soil Sci. Soc. Am. J. 42:908–912.

Ham, G. E. 1977. The acetylene-ethylene assay and other measures of nitrogen fixation in field experiments. p. 325–334. In A. Ayanaba and P. J. Dart (ed.) Biological nitrogen fixation in farming systems of the tropics. John Wiley & Sons, Inc., New York.

Ham, G. E. 1978. Use of 15-N in evaluating symbiotic nitrogen fixation of field grown soybeans. p. 151–162. In Isotopes in biological nitrogen fixation. International Atomic Energy Agency, Vienna.

Ham, G. E., I. E. Liener, S. D. Evans, R. D. Frazier, and W. W. Nelson. 1975. Yield and composition of soybean seed as affected by N and S fertilizer. Agron. J. 67:293–297.

Hardy, R. W. F., R. C. Burns, and R. D. Holsten. 1973. Applications of the acetylene-ethylene assay for measurement of nitrogen fixation. Soil Biol. Biochem. 5:47–81.

Hardy, R. W. F., R. D. Holsten, E. K. Jackson, and R. C. Burns. 1968. The acetylene-ethylene assay for N2 fixation: Laboratory and field evaluation. Plant Physiol. 43:1185–1207.

Hargrove, W. L., and D. E. Kissel. 1979. Ammonia volatilization from surface applications of urea in the field and laboratory. Soil Sci. Soc. Am. J. 43:359–363.

Hauck, R. D. 1971. Quantitative estimates of nitrogen-cycle processes—concepts and review. p. 65–80. In Nitrogen-15 in soil-plant studies. International Atomic Energy Agency, Vienna.

Hauck, R. D. 1973. Nitrogen tracers in nitrogen cycle studies—past use and future needs. J. Environ. Qual. 2:317–327.

Hauck, R. D. 1979. Methods for studying N transformations in paddy soils: Review and comments. p. 73-93. *In* Proc. Nitrogen and Rice Symposium, Los Banos. 18-21 Sept. 1978. The International Rice Research Institute, Los Banos, Philippines.

Hauck, R. D. 1981. Isotope-ratio analysis in investigations using stable nitrogen tracers. *In* A. L. Page et al. (ed.) Methods of soil analysis, Part 2. Rev. ed. Am. Soc. of Agron., Madison, Wis. (In press.)

Hauck, R. D., and J. M. Bremner. 1976. Use of tracers for soil and fertilizer nitrogen research. Adv. Agron. 28:219-266.

Hauck, R. D., and M. Bystrom. 1970. ^{15}N—a selected bibliography for agricultural scientists. Iowa State University Press, Ames.

Hauck, R. D., and V. J. Kilmer. 1975. Cooperative research between the Tennessee Valley Authority and land-grant universities on nitrogen fertilizer and water quality. p. 655-660. *In* E. R. Klein and P. D. Klein (ed.) Proc. Second Int. Conf. on Stable Isotopes, Oak Brook, Ill. 20-23 Oct. 1975. Argonne Natl. Laboratory, Argonne, Ill.

Henriksen, H., and A. A. Selmer-Olsen. 1970. Automatic methods for determining nitrate and nitrite in water and soil extracts. Analyst (London) 95:514-518.

Henzell, E. F., I. Vallis, and J. E. Lindquist. 1968. Automatic colorimetric methods for the determination of nitrogen in digests and extracts of soils. Int. Congr. Soil Sci. Trans. 9th (Adelaide, Australia) III:513-520.

Hoell, J. M., C. N. Harward, and B. S. Williams. 1980. Remote infrared heterodyne radiometer measurements of atmospheric ammonia profiles. Geophys. Res. Lett. 7:313-316.

Holz, F., and H. Kremers. 1971. Die automatische Bestimmung des Stickstoffs als Indophenolgrün in Böden und Pflanzen. Landwirtsch. Forsch. Sonderh. 26:177-191 (Chem. Abstr. 76:58119f).

Hoshika, Y. 1977. Gas chromatographic determination of lower aliphatic primary amines as their fluorine-containing Schiff bases. Anal. Chem. 49:541-543.

Hudd, G. H., C. P. Lloyd-Jones, and D. G. Hill-Cottingham. 1980. Comparison of acetylene-reduction and nitrogen-15 techniques for the determination of nitrogen fixation by field bean (*Vicia faba*) nodules. Physiol. Plant 48:111-115.

Hughes, E. E., W. D. Dorko, and J. K. Taylor. 1978. Evaluation of methodology for analysis of halocarbons in the upper atmosphere: Phase 1. NBSIR 78-1480. National Bureau of Standards, U.S. Government Printing Office, Washington, D.C.

Hynes, R. K., and R. Knowles. 1978. Inhibition by acetylene of ammonia oxidation in *Nitrosomonas europaea*. FEMS Microbiol. Lett. 4:319-321.

Jackson, W. A., C. E. Frost, and D. M. Hildreth. 1975. Versatile multi-range analytical manifold for automated analysis of nitrate-nitrogen. Soil Sci. Soc. Am. Proc. 39:592-593.

Johnson, J. W., L. F. Welch, and L. T. Kurtz. 1975. Environmental implications of N fixation by soybeans. J. Environ. Qual. 4:303-306.

Joint FAO/IAEA Division of Atomic Energy in Food and Agriculture. 1974. Isotope studies on wheat fertilization: Results of a four-year coordinated research programme of the Joint FAO/IAEA Division of Atomic Energy in Food and Agriculture. Vienna, International Atomic Energy Agency. Technical Reports Series no. 157.

Joint FAO/IAEA Division of Atomic Energy in Food and Agriculture. 1978. Isotope studies on rice fertilization: Results of a five-year coordinated research programme of the Joint FAO/IAEA Division of Atomic Energy in Food and Agriculture using nitrogen-15-labelled fertilizers. Vienna, International Atomic Energy Agency. Technical Reports Series no. 181.

Jones, J. R. 1969. Prospects for stable isotopes. Applications involving nuclear magnetic resonance spectrometry. J. Labelled Compd. 5:305-311.

Kaplan, W. A., J. W. Elkins, C. E. Kolb, M. B. McElroy, S. C. Wofsy, and A. P. Duran. 1978. Nitrous oxide in fresh water systems: An estimate for the yield of atmospheric N_2O associated with disposal of human waste. Pure Appl. Geophys. 116:423-437.

Kato, T. 1975. Determination of ^{15}N abundance in amino acids separated by paper chromatography. (In Japanese.) J. Sci. Soil Manure, Japan 46:66-67.

Kavanagh, E. P., and J. R. Postgate. 1970. Absorption and release of hydrocarbons by rubber closures: A source of error in some biological assays. Lab. Pract. 19:159-160.

Keeney, D. R. 1981. Nitrogen availability indexes. *In* A. L. Page et al. (ed.) Methods of soil analysis, Part 2. Rev. ed. Am. Soc. of Agron., Madison, Wis. (In press.)

Keeney, D. R., and J. M. Bremner. 1966. Determination and isotope-ratio analysis of different forms of nitrogen in soils. 4. Exchangeable ammonium, nitrate and nitrite by direct-distillation methods. Soil Sci. Soc. Am. Proc. 30:583-587.

Keeney, D. R., and J. M. Bremner. 1967a. Use of the Coleman Model 29A Analyzer for total nitrogen analysis of soils. Soil Sci. 104:358–363.

Keeney, D. R., and J. M. Bremner. 1967b. Determination and isotope-ratio analysis of different forms of nitrogen in soils: 7. Urea. Soil Sci. Soc. Am. Proc. 31:317–321.

Keeney, D. R., and D. W. Nelson. 1981. Inorganic forms of nitrogen. In A. L. Page et al. (ed.) Methods of soil analysis, Part 2. Rev. ed. Am. Soc. of Agron., Madison, Wis. (In press.)

Kempers, A. J. 1974. Determination of sub-microquantities of ammonium and nitrates in soils with phenol, sodium nitroprusside and hypochlorite. Geoderma 12:201–206.

Kim, C. M. 1973. Influence of vegetation types on the intensity of ammonia and nitrogen dioxide liberation from soil. Soil Biol. Biochem. 5:163–166.

Kissel, D. E., H. L. Brewer, and G. F. Arkin. 1977. Design and test of a field sampler for ammonia volatilization. Soil Sci. Soc. Am. J. 41:1133–1137.

Klemedtsson, L., B. H. Svensson, T. Lindberg, and T. Rosswall. 1977. The use of acetylene inhibition of nitrous oxide reductase in quantifying denitrification in soils. Swed. J. Agric. Res. 7:179–185.

Knowles, R. 1979. Denitrification, acetylene reduction and methane metabolism in lake sediment exposed to acetylene. Appl. Environ. Microbiol. 38:486–493.

Knowles, R. 1981. Denitrification. p. 323–369. In E. A. Paul and J. N. Ladd (ed.) Soil biochemistry, Vol. 5. Marcel Dekker, Inc., New York.

Koren'kov, D. A., I. A. Lavrova, and L. I. Romanyuk. 1973. State and prospects for studies using the nitrogen-15 stable isotope in agronomic chemistry. p. 5–45. In A. Koren'kov et al. (ed.). Use of the stable isotope ^{15}N in agricultural research. Kolos, Moscow.

Kowalenko, C. G. 1978. Organic nitrogen, phosphorus and sulfur in soils. p. 15–136. In M. Schnitzer and S. U. Khan (ed.) Soil organic matter. Elsevier Scientific Publishing Co., New York.

Kundler, P. 1970. Utilization, fixation, and loss of fertilizer nitrogen (review of international results of the last 10 years basic research). Albrecht-Thaer-Arch. 14:191–210.

Ladd, J. N. 1978. Origin and range of enzymes in soil. p. 51–96. In R. G. Burns (ed.) Soil enzymes. Academic Press, Inc., New York.

Ladd, J. N., and J. H. A. Butler. 1972. Short-term assays of soil proteolytic enzyme activities using proteins and dipeptide derivatives as substrates. Soil Biol. Biochem. 4:19–30.

Larson, S. L., and H. P. Peterson. 1979. Semiautomated determination of nitrogen in nitrate-containing fertilizers. Anal. Chem. 51:2414–2415.

Lee, K. K., and I. Watanabe. 1977. Problems of the acetylene reduction technique applied to water-saturated paddy soils. Appl. Environ. Microbiol. 34:654–660.

Legg, J. O., and C. Sloger. 1975. A tracer method for determining symbiotic nitrogen fixation in field studies. p. 661–667. In Proc. of the Second Int. Conference on Stable Isotopes, Oakbrook, Ill. Argonne National Laboratory, U.S. Department of Energy.

Levine, J. S., T. R. Augustsson, and J. M. Hoell. 1980. The vertical distribution of tropospheric ammonia. Geophys. Res. Lett. 7:317–320.

Matthews, R. D., R. F. Sawyer, and R. W. Schefer. 1977. Interferences in chemiluminescent measurement of NO and NO₂ emissions from combustion systems. Environ. Sci. Technol. 11:1092–1096.

Matthias, A. D., A. M. Blackmer, and J. M. Bremner. 1980. A simple chamber technique for field measurement of emissions of nitrous oxide from soils. J. Environ. Qual. 9:251–256.

McAuliffe, C. 1969. Determination of dissolved hydrocarbons in subsurface brines. Chem. Geol. 4:225–233.

McAuliffe, C. 1971. GC determination of solutes by multiple phase equilibration. Chem. Tech. 1:46–51.

McGarity, J. W., and R. D. Hauck. 1969. An aerometric apparatus for the evaluation of gaseous nitrogen transformations in field soils. Soil Sci. 108:335–344.

McInteer, B. B., and J. G. Montoya. 1980. Automation of a mass spectrometer for nitrogen isotope analysis. U.S. Dep. of Energy Rep. no. LA-UR-80-245, Los Alamos, N.M.

McKenney, D. J., D. L. Wade, and W. I. Findlay. 1978. Rates of N_2O evolution from N-fertilized soil. Geophys. Res. Lett. 5:777–780.

Meier, G. 1976. ^{15}N analysis with NOI-5 analyzer. p. 10.1–10.13. In Interregional training course on the use of ^{15}N in soils research, Leipzig. 21 Sept. to 22 Oct. 1976. Academy of Sciences of GDR, Leipzig.

Meints, V. W., and G. A. Peterson. 1972. Further evidence for the inability of the Kjeldahl total nitrogen method to fully measure indigenous fixed ammonium nitrogen in subsoils. Soil Sci. Soc. Am. Proc. 36:434–436.

Mertens, J., P. Van der Winkel, and D. L. Mässart. 1975. Determination of nitrate in water with an ammonia probe. Anal. Chem. 47:522–526.

Mikkelsen, D. S., and S. K. DeDatta. 1978. Ammonia volatilization from flooded rice soils. Soil Sci. Soc. Am. J. 43:630.

Mikkelsen, D. S., S. K. DeDatta, and W. N. Obcemea. 1978. Ammonia volatilization losses from flooded rice soils. Soil Sci. Soc. Am. J. 42:725–730.

Mogilevkina, I. A. 1964. Fixation of ammonium in the soil and method of determining it. Sov. Soil Sci. 2:185–196.

Mogilevkina, I. A. 1970. Testing the applicability of the Kjeldahl method for the determination of total nitrogen in soils with a high content of fixed ammonium. (In Russian). Pochvovedeniye 1970:139–143 (Chem. Abstr. 74:12147u).

Morris, G. F., R. B. Carson, D. A. Shearer, and W. T. Jopkiewicz. 1968. Comparison of the automatic Dumas (Coleman Model 29A Nitrogen Analyzer II) and Kjeldahl methods for the determination of total nitrogen in agricultural materials. J. Assoc. Off. Anal. Chem. 51:216–219.

Mosier, A. R., C. E. Andre, and F. G. Viets, Jr. 1973. Identification of aliphatic amines volatilized from cattle feedyard. Environ. Sci. Technol. 7:642–644.

Mosier, A. R., and L. Mack. 1980. Gas chromatographic system for precise, rapid analysis of N_2O. Soil Sci. Soc. Am. J. 44:1121–1123.

Mueller, G. 1965. Special methods of stable isotope analysis. Kernenergie 8:265–283.

Muhammad, S., and K. Kumazawa. 1975. Use of emission spectrometry to trace ^{15}N-labeled ammonium and nitrate nitrogen in amino acids of rice panicle. p. 674–682. In Proc. of the Second International Conference on Stable Isotopes, Oakbrook, Ill. Argonne National Laboratory, U.S. Dep. of Energy.

Muhammad, S., and K. Kumazawa. 1976. Nitrogen distribution studies by emission spectrometry in free amino acids and amides of developing panicle of rice. Z. Pflanzenernaehr. Bodenkd. 5:529–536.

Mulvaney, R. L., and J. M. Bremner. 1979. A modified diacetyl monoxime method for colorimetric determination of urea in soil extracts. Commun. Soil Sci. Plant Anal. 10: 1163–1170.

Nelson, D. W., and J. M. Bremner. 1966. An evaluation of Mogilevkina's method of determining fixed ammonium in soils. Soil Sci. Soc. Am. Proc. 30:409–410.

Nelson, D. W., and L. E. Sommers. 1972. A simple digestion procedure for estimation of total nitrogen in soils and sediments. J. Environ. Qual. 1:423–425.

Nelson, D. W., and L. E. Sommers. 1980. Total nitrogen analysis for soil and plant tissues. J. Assoc. Off. Anal. Chem. 63:770–778.

Nicholas, D. J. D., D. J. Silvester, and J. F. Fowler. 1961. Use of radioactive nitrogen in studying nitrogen fixation in bacterial cells and their extracts. Nature (London) 189:634–636.

Noel, R. J. 1976. Collaborative study of an automated method for the determination of crude protein in animal feeds. J. Assoc. Off. Anal. Chem. 59:141–147.

Oberrieth, R., and N. H. Mermelstein. 1974. Instrument automates, accelerates nitrogen determinations. Food Technol. (Chicago) 28(6):40–43.

Parsons, J. W., and J. Tinsley. 1975. Nitrogenous substances. p. 263–304. In J. E. Gieseking (ed.) Soil components, Vol. 1. Organic components. Springer-Verlag New York, New York.

Rasmussen, R. A., J. Krasnec, and D. Pierotti. 1976a. N_2O analysis in the atmosphere via electron capture-gas chromatography. Geophys. Res. Lett. 3:615–618.

Rasmussen, R. A., D. Pierotti, J. Krasnec, and B. Halter. 1976b. Trip report on the cruise of the Alpha Helix research vessel from San Diego, California to San Martin, Peru. Washington State Univ., Pullman.

Rennie, D. A., E. A. Paul, and L. E. Johns. 1976. Natural nitrogen-15 abundance of soil and plant samples. Can. J. Soil Sci. 56:43–50.

Rennie, J. R., D. A. Rennie, and M. Fried. 1978. Concepts of 15-N usage in dinitrogen fixation studies. p. 107–133. In Isotopes in biological dinitrogen fixation. International Atomic Energy Agency, Vienna.

Roberge, M. R. 1978. Methodology of soil enzyme measurement and extraction. p. 341–370. In R. G. Burns (ed.) Soil Enzymes. Academic Press, Inc., New York.

Robinson, J. B. D. 1973. The preservation, unaltered, of mineral nitrogen in tropical soils and soil extracts. Plant Soil 27:32–80.

Rodrigues, G. 1954. Fixed ammonia in tropical soils. J. Soil Sci. 5:264–274.

Rolston, D. E., F. E. Broadbent, and D. A. Goldhamer. 1979. Field measurement of denitrification: II. Mass balance and sampling uncertainty. Soil Sci. Soc. Am. J. 43:703–708.

Rolston, D. E., M. Fried, and D. A. Goldhamer. 1976. Denitrification measured directly from nitrogen and nitrous oxide gas fluxes. Soil Sci. Soc. Am. J. 40:259–266.

Rolston, D. E., D. L. Hoffman, and D. W. Toy. 1978. Field measurement of denitrification: I. Flux of N_2 and N_2O. Soil Sci. Soc. Am. J. 42:863–869.

Roskoski, J. P. 1981. Comparative C_2H_2 reduction and $^{15}N_2$ fixation in deciduous wood litter. Soil Biol. Biochem. 13:83–85.

Roy, C. R. 1979. Atmospheric nitrous oxide in the mid-latitudes of the southern hemisphere. J. Geophys. Res. 84:3711–3718.

Ruben, S., W. Z. Hassid, and M. D. Kamen. 1940. Radioactive nitrogen in the study of N_2 fixation by non-leguminous plants. Science 91:578–579.

Ryden, J. C., L. C. Lund, and D. D. Focht. 1978. Direct in-field measurement of nitrous oxide flux from soils. Soil Sci. Soc. Am. J. 42:731–737.

Ryden, J. C., L. J. Lund, and D. D. Focht. 1979a. Direct measurement of denitrification loss from soils: I. Laboratory evaluation of acetylene inhibition of nitrous oxide reduction. Soil Sci. Soc. Am. J. 43:104–110.

Ryden, J. C., L. J. Lund, J. Letey, and D. D. Focht. 1979b. Direct measurement of denitrification loss from soils: II. Development and application of field methods. Soil Sci. Soc. Am. J. 43:110–118.

Sahrawat, K. L., and F. N. Ponnamperuma. 1978. Measurement of exchangeable NH_4^+ in tropical rice soils. Soil Sci. Soc. Am. J. 42:282–283.

Schachtschabel, P. 1960. Fixierter Ammoniumstickstoff in Löss- und Marschböden. Int. Congr. Soil Sci., Trans. 7th (Madison) 2:22–27.

Schachtschabel, P. 1961. Bestimmung des fixierten Ammoniums im Boden. Z. Pflanzenernaehr. Dueng. Bodenkd. 93:125–136.

Schollhorn, R., and R. H. Burris. 1966. Studies of intermediates in nitrogen fixation. Fed. Proc. Fed. Am. Soc. Exp. Biol. 25:710.

Schollhorn, R., and R. H. Burris. 1967. Acetylene as a competitive inhibitor of N_2 fixation. Proc. Natl. Acad. Sci. U.S.A. 58:213–216.

Schubert, K. R., and H. J. Evans. 1976. Hydrogen evolution: A major factor affecting the efficiency of nitrogen fixation in nodulated symbionts. Proc. Natl. Acad. Sci. U.S.A. 73: 1207–1211.

Schuman, T. E., M. A. Stanley, and D. Knudsen. 1973. Automated total nitrogen analysis of soil and plant samples. Soil Sci. Soc. Am. Proc. 37:480–481.

Siegel, R. S. 1980. Determination of nitrate and exchangeable ammonium in soil extracts by an ammonia electrode. Soil Sci. Soc. Am. J. 44:943–947.

Sigsby, J. E., F. M. Black. T. A. Bellar, and D. L. Klosterman. 1973. Chemiluminescent method for analysis of nitrogen compounds in mobile source emissions (NO, NO_2, and NH_3). Environ. Sci. Technol. 7:51–54.

Silva, J. A., and J. M. Bremner. 1966. Determination and isotope-ratio analysis of different forms of nitrogen in soils: 5. Fixed ammonium. Soil Sci. Soc. Am. Proc. 30:587–594.

Skjemstad, J. O., and R. Reeve. 1976. The determination of nitrogen in soils by rapid high-temperature Kjeldahl digestion and autoanalysis. Commun. Soil Sci. Plant Anal. 7:229–239.

Smith, K. A. 1977. Gas chromatographic analysis of the soil atmosphere. p. 197–229. In J. C. Giddings et al. (ed.) Advances in chromatography, Vol. 15. Marcel Dekker, Inc., New York.

Smith, M. S., M. K. Firestone, and J. M. Tiedje. 1978. The acetylene inhibition method for short-term measurement of soil denitrification and its evaluation using nitrogen-13. Soil Sci. Soc. Am. J. 42:611–616.

Sörensen, J., J. M. Tiedje, and R. B. Firestone. 1980. Inhibition by sulfide of nitric and nitrous oxide reduction by denitrifying *Pseudomonas fluorescens*. Appl. Environ. Microbiol. 39:105–108.

Sowden, F. J., Y. Chen, and M. Schnitzer. 1977. The nitrogen distribution in soils formed under widely differing climatic conditions. Geochim. Cosmochim. Acta 41:1524–1526.

Stevens, R. K., and J. A. Hodgeson. 1973. Applications of chemiluminescent reactions to the measurement of air pollutants. Anal. Chem. 45:443A–449A.

Stevenson, F. J. 1965a. Amino sugars. In C. A. Black et al. (ed.) Methods of soil analysis, Part 2. Agronomy 9:1429–1436. Am. Soc. of Agron., Madison, Wis.

Stevenson, F. J. 1965b. Amino acids. *In* C. A. Black (ed.) Methods of soil analysis, Part 2. Agronomy 9:1437–1451. Am. Soc. of Agron., Madison, Wis.

Stevenson, F. J. 1981. Organic forms of soil nitrogen. *In* A. L. Page et al. (ed.) Methods of soil analysis, Part 2. Rev. ed. Am. Soc. of Agron., Madison, Wis. (In press.)

Stewart, B. A., and L. K. Porter. 1963. Inability of the Kjeldahl method to fully measure indigenous fixed ammonium in some soils. Soil Sci. Soc. Am. Proc. 27:41–43.

Stewart, B. A., L. K. Porter, and W. E. Beard. 1964. Determination of total nitrogen and carbon in soils by a commercial Dumas apparatus. Soil Sci. Soc. Am. Proc. 28:366–368.

Stewart, B. A., L. K. Porter, and F. E. Clark. 1963. The reliability of a micro-Dumas procedure for determining total nitrogen in soil. Soil Sci. Soc. Am. Proc. 27:377–380.

Stutte, C. A., and R. T. Weiland. 1978. Gaseous nitrogen loss and transpiration of several crop and weed species. Crop Sci. 18:887–889.

Tabatabai, M. A. 1974. Determination of nitrite in soil extracts and water samples by a nitrogen oxide electrode. Commun. Soil Sci. Plant Anal. 5:569–578.

Tabatabai, M. A., and J. M. Bremner. 1972. Assay of urease activity in soils. Soil Biol. Biochem. 4:479–487.

Tam, T. Y., and R. Knowles. 1979. Effects of sulfide and acetylene on nitrous oxide reduction by soil and by *Pseudomonas aeruginosa*. Can. J. Microbiol. 25:1133–1138.

Thomas, R. F., and R. L. Booth. 1973. Selective electrode measurement of ammonia in water and wastes. Environ. Sci. Technol. 7:523–526.

Tiedje, J. M., R. B. Firestone, M. K. Firestone, M. R. Betlach, H. F. Kaspar, and J. Sörensen. 1981. Use of nitrogen-13 in studies of denitrification. *In* K. A. Krohn and J. W. Root (ed.) Recent developments in biological and chemical research with short-lived isotopes. Advances in Chemistry, Am. Chemical Soc., Washington, D.C. (In press.)

Tiedje, J. M., R. B. Firestone, M. K. Firestone, M. R. Betlach, M. S. Smith, and W. H. Caskey. 1979. Methods for the production and use of nitrogen-13 in studies of denitrification. Soil Sci. Soc. Am. J. 43:709–715.

Tiedje, J. M., M. K. Firestone, M. S. Smith, M. R. Betlach, and R. B. Firestone. 1978. Short term measurement of denitrification rates in soils using ^{13}N and acetylene inhibition methods. p. 132–137. *In* M. W. Loutit and J. A. R. Miles (ed.) Microbial ecology. Springer-Verlag New York, New York.

Van Raalte, C. D., and D. G. Patriquin. 1979. Use of the "acetylene blockage" technique for assaying denitrification in a salt marsh. Mar. Biol. 52:315–320.

Vlek, P. L. G. 1979. Ammonia volatilization from flooded rice soils. Soil Sci. Soc. Am. J. 43:630.

Vlek, P. L. G., and E. T. Craswell. 1979. Effects of nitrogen source and management on ammonia volatilization losses from flooded rice-soil systems. Soil Sci. Soc. Am. J. 43:352–358.

Vlek, P. L. G., and J. M. Stumpe. 1978. Effects of solution chemistry and environmental conditions on ammonia volatilization losses from aqueous systems. Soil Sci. Soc. Am. J. 42:416–421.

Wall, L. L., Sr., C. W. Gehrke, T. E. Neuner, R. D. Cathey, and P. R. Rexroad. 1975. Total protein nitrogen: Evaluation and comparison of four different methods. J. Assoc. Off. Anal. Chem. 58:811–817.

Walter, H. M., D. R. Keeney, and I. R. Fillery. 1979. Inhibition of nitrification by acetylene. Soil Sci. Soc. Am. J. 43:195–196.

Watanabe, I., and M. R. de Guzman. 1980. Effect of nitrate on acetylene disappearance from anaerobic soil. Soil Biol. Biochem. 12:193–194.

Weber, C. R. 1966. Nodulating and nonnodulating soybean isolines: II. Response to applied nitrogen and modified soil conditions. Agron. J. 58:46–49.

Wentworth, W. E., E. Chen, and R. Freeman. 1971. Thermal electron attachment to nitrous oxide. J. Chem. Phys. 55:2075–2078.

Wentworth, W. E., and R. R. Freeman. 1973. Measurement of atmospheric nitrous oxide using an electron capture detector in conjunction with gas chromatography. J. Chromatogr. 79:322–324.

Winer, A. M., J. W. Peters, J. P. Smith, and J. N. Pitts, Jr. 1974. Response of commercial chemiluminescent NO-NO$_2$ analyzers to other nitrogen-containing compounds. Environ. Sci. Technol. 8:1118–1121.

Wong, H. K. T., and A. L. W. Kemp. 1977. The determination of total nitrogen in sediments using an induction furnace. Soil Sci. 124:1–4.

Yeomans, J. C., and E. G. Beauchamp. 1978. Limited inhibition of nitrous oxide reduction in soil in the presence of acetylene. Soil Biol. Biochem. 10:517–519.

Yoshinari, T., R. Hynes, and R. Knowles. 1977. Acetylene inhibition of nitrous oxide reduction and measurement of denitrification and nitrogen fixation in soil. Soil Biol. Biochem. 9:177–183.

Yoshinari, T., and R. Knowles. 1976. Acetylene inhibition of nitrous oxide reduction by denitrifying bacteria. Biochem. Biophys. Res. Commun. 69:705–710.

Yung, Y. L., W. C. Wang, and A. A. Lacis. 1976. Greenhouse effect due to atmospheric nitrous oxide. Geophys. Res. Lett. 3:619–621.

Zantua, M. I., and J. M. Bremner. 1975. Comparison of methods of assaying urease activity in soils. Soil Biol. Biochem. 7:291–295.

14 Soil Nitrogen Budgets

J. O. LEGG AND J. J. MEISINGER

Agricultural Research Service,
U.S. Department of Agriculture,
Beltsville, Maryland

I. INTRODUCTION

Several generations of scientists have been challenged by the task of constructing accurate N budgets for various ecosystems. Among the early workers were Lawes et al. (1882), who estimated the N budget for the Broadbalk wheat plots at Rothamsted, England. Lipman and Conybeare (1936) used the N balance approach to prepare an inventory of plant nutrients in the United States. Allison (1955) made an extensive review of N balance experiments and found that quantitative data were usually lacking for some of the items entering the calculations. In his view, the failure of N balance sheets to balance constituted an enigma. In a later review, Allison (1966) noted that marked progress had been made in ascertaining the fate of applied N, which he attributed to the use of N isotopes and improved instrumentation. Trepachev and Pryanishnikov (1976) have also recently reviewed Russian nonlabeled N balance studies.

Nitrogen budgets (balances) have been a valuable tool in expanding our knowledge of the N cycle. They have contributed by identifying mechanisms of N transfer and indicating the size of various N reservoirs. Their main use has been in estimating the net N loss, or the unaccounted for N, in a given agricultural production system. However, to accurately estimate the net N loss and to ascribe this loss to a certain process, an accounting must be made of all of the other major N transformations. For this reason, N budgets have been of inestimable value in contributing to our understanding of the processes of mineralization, immobilization, plant assimilation, leaching, denitrification, etc. Nitrogen budgets also emphasize the fact that all of the biological processes are interrelated; changes in one process are reflected by changes in other processes. The N budget approach is of equal importance when efficient N use by plants is considered. Efficient N utilization is the key to the solution of problems concerned with high crop production, minimal pollution, and energy conservation.

The term N budget is defined herein as the application of mass conservation principles so that N is conserved in the various transformations and biological processes of the system. Nitrogen budget studies generally

Nitrogen in Agricultural Soils—Agronomy Monograph no. 22.

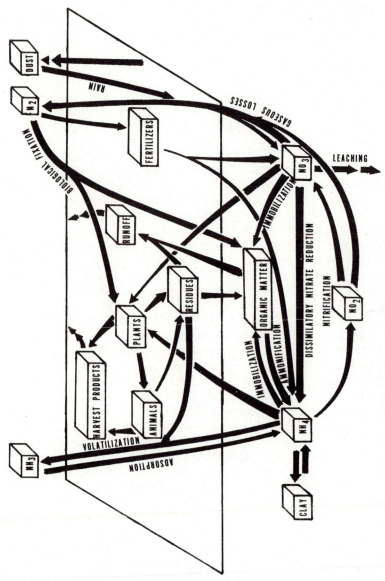

Fig. 1—Schematic representation of N transformations in a general agricultural N cycle.

consider sizes, pathways, and transfer rates among N reservoirs. It should be emphasized that N budgets require a careful definition of the system boundaries including conceptual, spatial, and temporal factors. For example, N balances have been developed from a global scale all the way down to greenhouse pots. Likewise, the time scale may vary from hundreds of years for geological processes down to a single growing season or to a few days or hours for a laboratory experiment. Obviously, with such a broad range in systems, our treatment of N budgets must overlook many applications and must give others only limited attention. We limit our discussion mainly to small-scale agricultural soil-crop systems. Large-scale N budgets are described in Chapter 23 (R. D. Hauck and K. K. Tanji). We place major emphasis on recent research accomplishments that have significantly broadened our understanding of soil N.

II. THE N CYCLE IN RELATION TO N BUDGETS

A. Nitrogen Cycle Diagrams

Diagrams of the N cycle vary widely. For our purposes, a modified form (Frere, 1976) adequately illustrates the main components of soil N budgets (Fig. 1). The overall N cycle may be viewed as a group of smaller interrelated cycles. For example, N is continuously cycling between mineral and organic forms as part of the mineralization-immobilization process. A given N atom may be cycled quite rapidly or may persist for years or centuries (Delwiche, 1977); thus, N does not have a clearly discernible residence time in the soil. The survival of a N-containing compound depends on its chemical availability, the degree to which it is bound on soil colloids, and the ease with which it is broken down by microorganisms.

B. Soil N Equilibrium Concept

Most of the native N that is potentially available to plants is associated with organic matter. Ecosystems generally gain or lose organic N at a diminishing rate until an equilibrium N level is reached (Jenny, 1941). These steady-state levels are influenced by climate, soil, plant, and management variables as discussed in Chapter 1 (F. J. Stevenson). Achieving a steady-state condition may require many years, depending on the degree of change and the environment, as evidenced by the Morrow plot data shown in Fig. 2. Since the system is dynamic, changes in the soil-plant environment (e.g., plant cultural conditions) may lead to a new level of organic N, as shown in Fig. 2 from the Broadbalk experiments at Rothamsted (Jenkinson, 1977). In the continuous wheat (*Triticum aestivum* L.) experiment, the soil of unmanured plots had apparently reached a steady-state condition of about 3 metric tons of N/ha in the surface 23 cm. The soil of the plot receiving 35 metric tons of farmyard manure ha^{-1} year^{-1} has increased in N content and seems to be approaching a new equilibrium value of about 7 metric tons of

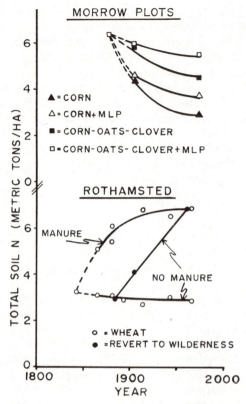

Fig. 2—Nitrogen contained in the topsoil of the Morrow plots and the Broadbalk plots at various times as affected by rotation and other amendments (MLP is manure plus lime plus phosphorus). The dashed lines refer to estimated changes from the beginning of the experiment to the first complete sampling (Jenkinson, 1977; Illinois Agric. Exp. Stn. and Dept. of Agronomy, 1982).

N/ha. Perhaps the most surprising part of Fig. 2 is the rapid increase in soil N for the section that had been in continuous wheat for 40 years and was then allowed to revert to wilderness. After 81 years, the wilderness soil had accumulated as much N as the manured soil. In contrast, annual treatments of up to 144 kg of N/ha of commercial fertilizer (not shown in Fig. 2) increased the soil total N content to only about 3.5 metric tons of N/ha over the same period.

The soil N increase of the wilderness section (see Fig. 2) required a mean accumulation rate of 49 kg of N ha^{-1} year^{-1} from sources such as precipitation, dry deposition, and biological fixation. This sizeable annual N increment indicates that the mixed herbaceous vegetation, which continuously covered this section, was an efficient means of accumulating and retaining N. The wilderness soil N accumulation cannot be explained simply by the amount of plant residues produced (compared with 35 tons of farmyard manure ha^{-1} year^{-1}). Under constant vegetation the N entering the system is continually being immobilized by plant uptake and by microbial activity brought about by the large amount of carbonaceous material contributed by the plants (Legg, 1975). The build-up of soil N does not continue infinitely, of course, and eventually a steady-state is attained at which N losses equal gains. In cultivated soils fertilized with inorganic N, it is likely that low N soils will exhibit a net gain of N upon cropping, whereas high N soils will exhibit a net loss (Legg & Stanford, 1967). The long-term equilibrium stage of soil N under any particular management system is therefore of major importance relative to N gains or losses.

III. NITROGEN SOURCES IN SOIL-PLANT SYSTEMS

When N inputs and outputs are assessed in any N balance study, it is helpful to begin with a broad overview of N gains and losses (Fig. 1) and the various transformations involved. Figure 1 will also serve as a reference point for the general discussion of the various inputs and outputs that follows.

A. Indigenous Soil Organic N

Our knowledge of the nature and composition of soil organic matter was summarized by Allison (1973) and is also discussed in Chapters 2 (J. L. Young and R. W. Aldag) and 3 (F. J. Stevenson). Usually 95% or more of the N in surface soils is present in organic forms, and the remainder is in mineral forms, including some fixed NH_4^+.

An important component of soil N budgets is the net mineralization of organic N. In most studies this has been estimated by growing a crop that received no fertilizer N or organic amendments. This approach is useful as long as the other inputs, such as precipitation and irrigation, are measured and as long as losses through leaching and denitrification are minimal. If these requirements are met, the N uptake of check plots can be used to estimate N mineralization plus residual mineral N, provided the crop assimilation capacity has not been exceeded. It is beyond the scope of this section to review the many studies involved with estimating N release from indigenous soil organic matter (see Chapt. 17, G. Stanford), but it is sufficient to state that soils differ widely in their ability to mobilize organic N and that a number of environmental factors (temperature, water, etc.) affect the process.

B. Additions Through Crop and Animal Wastes

Crop and animal wastes not only are sources of nutrients but also may decrease nutrient loss by decreasing runoff and wind erosion. Recycling of waste N is considered in greater detail elsewhere (Chapt. 21, J. H. Smith and J. R. Peterson); however, some essential points require discussion under N budgets.

The N content of crop residues produced in the United States annually has been estimated to be about 4 million metric tons (Larson et al., 1978;

Power & Legg, 1978), which is equivalent to 40% of current N fertilizer use on cropland. Crop residues are usually well distributed over the field and form a N source that is slowly available to succeeding crops. Losses of nutrients from crop residues usually are small, normally much less than those from more concentrated materials, such as manure.

An estimated 4.1 to 5.8 million metric tons of N are excreted annually by farm animals in the United States (Van Dyne & Gilbertson, 1978; Yeck et al., 1975). Of the 1.4 to 2.4 million tons of N that can be collected, only about one half becomes incorporated into the soil because of high losses during storage, handling, and spreading (Frere, 1976). On a national scale, the recycling of manure N is not efficient because manures are produced in concentrated areas and because management factors have a large influence on subsequent N losses. However, on a local basis manures can have a major impact on the N economy through NH_3 volatilization and N mineralization (Frink, 1969; Hutchinson & Viets, 1969; Pratt et al., 1976a, b).

C. Additions by Precipitation and Irrigation Water

Additions of N through precipitation are generally of minor consequence on cropland compared with other sources, but with forests, pastures, and rangeland, these inputs are of major importance. The principal forms of N in precipitation are NH_3, N oxides, and organic N. The annual flux of atmospheric N for the United States has been estimated at 5.4 million metric tons (NRC, 1978). There is little information available, however, to support specific quantitative assessments of the nature and sources of N in the precipitation or of the influence of atmospheric movement and chemical transformations on the forms and amounts of N deposited in different localities. The main sources of atmospheric fixed N are combustion of fuels, volatilization of NH_3 from animal wastes and fertilizers, volcanoes, and lightning. The relative importance of the various sources has been reviewed by the National Research Council (1978).

Precipitation N deposited on cropland is difficult to quantify because it is dispersed erratically and indiscriminately over large areas irrespective of use. Largest inputs generally occur in broad areas near major sources such as power generating plants, industrial areas, and large feedlots. Annual precipitation inputs range from <5 kg of N/ha in western desert regions to >30 kg of N/ha near feedlots in the Midwest (NRC, 1978). In Wisconsin, Hoeft et al. (1972) found that the NH_4^+ and organic N in precipitation were highest in the spring and in areas adjacent to barnyards. Nitrate deposition was relatively constant at all locations. Total N additions ranged from 13 to 30 kg ha^{-1} year^{-1}, and an overall state average of about 20 kg ha^{-1} year^{-1} was estimated. More long-term data are needed from many areas to evaluate the importance of N from precipitation with respect to soil N budgets. In general, current estimates of N in precipitation are higher than those obtained previously (Allison, 1965). Recent data from Rothamsted (Jenkinson, 1977) illustrate these long-term trends as follows: 1889–1903, 4.4 kg of N/ha; 1960–1964, 5.4 kg of N/ha; 1969–1970, 8.6 kg of N/ha. This probably reflects increased inputs from sources such as fuel combustion and NH_3 volatilization.

Nitrogen contributed by irrigation water was considered to be of little consequence by Allison (1965). This situation seems to be changing. In N balance studies in the Santa Maria Valley in California, Lund et al. (1978) estimated irrigation N inputs (supplied from ground water) to be 126 kg ha^{-1} $year^{-1}$. The mean NO_3^--N concentration of the ground water was 12.5 mg/liter. Total N inputs from fertilizer and water averaged 679 kg ha^{-1} $year^{-1}$, of which only 30% was utilized by crops, 37% was leached, and 33% remained unaccounted for. Under such management systems, about twice as much N is being leached past the root zone than is currently being applied in the irrigation water; therefore, the ground water will continue to be a significant source of N to the irrigated crops. Lund (1979) has reported a large range in irrigation N inputs that vary from 10 to 145 kg of N ha^{-1} $year^{-1}$. Significant irrigation inputs have also been observed in the Midwest (Saffigna et al., 1977) and in the East (Meisinger, 1976) and point out the need to document this N input for each site.

Crops irrigated with surface waters normally receive little N through irrigation. For example, irrigation inputs in a citrus watershed (Bingham et al., 1971) were estimated at < 3 kg of N ha^{-1} $year^{-1}$ since the waters contained < 0.3 mg of NO_3^--N/liter. In many areas of the United States, however, streams contain > 2 mg of N/liter (NRC, 1978), and this could easily contribute 20 to 40 kg of N ha^{-1} $year^{-1}$.

D. Adsorption of Atmospheric Gases

Ammonia is the primary N compound adsorbed from the atmosphere by soils. Indirectly, atmospheric NH_3 may also be added to soils by foliar plant absorption. Volatile organic N compounds (e.g., amines) may also have significance in localized areas but will not be considered here.

Early work on soil adsorption of atmospheric NH_3 has been reviewed by Malo and Purvis (1964). They found that New Jersey NH_3 concentrations were three times higher than the average atmosphere, with the highest concentrations occurring in the spring. Soil adsorption rates varied with soil type and ranged from about 58 to 228 g of N ha^{-1} day^{-1} under field conditions. This extrapolates to 21 to 83 kg of N ha^{-1} $year^{-1}$. Ammonia deposition estimates of up to 13 kg of N/ha were reported by Jenkinson (1977) for the Rothamsted area. Hanawalt (1969a, b) investigated the factors influencing NH_3 adsorption by New Jersey soils and found that adsorption occurred mainly in the uppermost 0.5 cm of soil. Adsorption was closely related to air permeability and moisture content. Hanawalt (1969b) concluded that NH_3 adsorption by soils was primarily governed by those factors affecting the rate of supply of NH_3 to the soil surface and the fraction of NH_3 molecules actually captured.

The high concentrations of atmospheric NH_3 in New Jersey probably result from industrial activity. In other areas NH_3 evolution from manures is the primary source. Hutchinson and Viets (1969) showed that absorption of NH_3 volatilized from cattle feedyards contributes significantly to the N enrichment of nearby lakes. One lake, about 0.4 km from a large feedyard, absorbed 73 kg of NH_3-N ha^{-1} year^{-1}. Elliott et al. (1971) collected NH_3 plus volatile organic N compounds near a small beef feedlot and pasture. Acid traps placed near the feedlot and 0.8 km away averaged 148 and 16 kg of N ha^{-1} year^{-1}, respectively. Luebs et al. (1973) has reported a 20- to 30-fold increase in the concentration of volatile N compounds in a high density dairy area (1,600 cows/km^2).

The above studies clearly illustrate that substantial amounts of NH_3-N can be transferred from localized industrial and agricultural sites via the atmosphere to the surrounding areas and contribute to the N content of soils and water. Plants may also utilize atmospheric NH_3 by foliar absorption as confirmed by use of ^{15}N-labeled NH_3 (Porter et al., 1972), but the amounts involved under natural conditions are difficult to determine (Hutchinson et al., 1972). Faller (1972) worked with atmospheric NH_3 in long-term plant experiments and showed that the aerial portions of plants can absorb and utilize NH_3 as the only source of N without affecting their normal growth. In a related study, Meyer[1] also found that the foliage of plants evolved small amounts of ^{15}N-labeled NH_3 following application of ^{15}N-labeled fertilizer to the roots.

Micrometeorological methods have recently been used in the field to measure NH_3 exchange between crops and the atmosphere (Denmead et al., 1974, 1976, 1978; Lemon & Van Houtte, 1980). These studies have shown that NH_3 can be moving into or out of the plant canopy at different times, depending on the NH_3 concentration near the soil surface, within the plant canopy, and above the canopy. Factors such as animal manures on the soil surface, plant species, and plant maturity all influence the net gain or loss of NH_3. Evidence is also presented for a small-scale NH_3 cycle within the crop canopy with NH_3 evolved from the soil being absorbed by the crop. Lemon and Van Houtte (1980) also discuss evidence suggesting that plants have a compensation point for NH_3 where they give up NH_3 to the atmosphere when ambient concentrations are low yet absorb it at higher levels.

From the foregoing discussion, it is clear that the contribution of atmospheric gases to soil N budgets may be appreciable in areas where the concentration of NH_3 in the atmosphere is greater than normal. This source should be taken into account when it appears to be an important factor in the environment.

E. Biological N_2 Fixation

Nitrogen fixation has been the primary means whereby a N balance has been maintained in the biosphere over the years, offsetting losses caused by denitrification. A detailed review of the subject is found in Chapter 10 (U. D. Havelka, et al.), and large-scale fixation estimates can be found in Chapter 23 (R. D. Hauck and K. K. Tanji).

---

[1]M. W. Meyer, 1973. Absorption and release of ammonia from and to the atmosphere by plants. Ph.D. Thesis, Univ. of Maryland, College Park.

1. SYMBIOTIC N₂ FIXATION

The N_2 fixation for the *Rhizobium*-legume symbiosis has been estimated to be 140 kg of N ha^{-1} year^{-1} by Burns and Hardy (1975). The actual amounts of N_2 fixed by leguminous crops in specific field situations have been difficult to determine due to the many factors that influence N_2 fixation, such as *Rhizobium*-host relationships, legume species, soil moisture, pH, and nutritional factors. The presence of combined N, particularly NO_3^-, greatly decreases fixation rates.

Annual fixation rates for various legumes are often quoted (e.g., Erdman, 1953). These reports, however, usually do not describe the experimental conditions (available soil N, pH, *Rhizobium* species, moisture, etc.) well enough to allow meaningful estimates at other locations. Past methods of estimating fixation usually compare the total N content of fixing and nonfixing plants or the change in total N content over time (Albrecht, 1920). More recent methods involve the indirect acetylene reduction method for short-term relative comparisons (Bergersen, 1970; Hardy et al., 1968; W. D. P. Stewart et al., 1967) and the ^{15}N labeling of soil N for long-term quantitative estimates (Fried & Middleboe, 1977; Legg & Sloger, 1975; Vallis et al., 1977; Vallis & Henzell, 1969). Better quantitative estimates of symbiotically fixed N_2 under a wide range of conditions are needed for N budget studies.

2. NONSYMBIOTIC N₂ FIXATION

Moore (1966) has comprehensively reviewed nonsymbiotic N_2 fixation, and Burns and Hardy (1975) have estimated this input to range from 2 to 200 kg of N ha^{-1} year^{-1}. Such estimates not only reveal a wide range in fixation values but also reflect the inherent difficulties in measuring nonsymbiotic fixation under natural conditions.

In Moore's opinion, attempts to assign field gains of 10 to 20 kg of N ha^{-1} year^{-1} to nonsymbiotic fixation are pointless unless an accurate balance sheet can be constructed. The Broadbalk continuous wheat N budget (Jenkinson, 1977) probably provides the best long-term estimates; the estimated annual nonsymbiotic inputs were 23 to 29 kg of N ha^{-1} year^{-1} on the zero N plots. These N budget estimates are also supported by acetylene reduction data (Witty et al., 1976), which indicate that a maximum of 5 kg N ha^{-1} year^{-1} could be attributed to rhizosphere fixation, and the remaining 18 to 24 kg N ha^{-1} year^{-1} could be accounted for by blue-green algae.

In contrast to the continuous wheat experiment, N_2 fixation in Broadbalk Wilderness was almost exclusively associated with the rhizosphere of nonleguminous weeds. Legumes had disappeared from the area in 1915, and algal fixation on the soil surface was lacking because of total shading by grass, weeds, and leaf litter. The accumulation of 34 kg of N ha^{-1} year^{-1} in the stubbed area of Broadbalk Wilderness therefore appears to be attributable to nonsymbiotic fixation in the rhizosphere (Witty et al., 1976).

Undoubtedly nonsymbiotic fixation occurs in the field, but there is a lack of quantitative data concerning this type of fixation. Much of the available information is not very useful, but sufficient data are available to suggest that nonsymbiotic fixation may be of agronomic significance under some circumstances. The conditions for optimum nonsymbiotic N_2 fixation in soil have not been clearly defined, but the presence of adequate energy materials, low soil N levels, and adequate mineral nutrients and moisture seem to be the major factors involved.

F. Commercial Fertilizers

Fertilizer N use in the United States has increased about fourfold from 1960 to 1977, from 2.48 to 9.65 million metric tons (ERS, 1977). A large portion of the fertilizer N (39%) is used in the production of corn (*Zea mays* L.). The average rate of application in 1977 for corn harvested for grain was 143 kg of N/ha, with a range of 59 to 185 kg of N/ha among the 18 main corn-producing states. Other major field crops in the United States generally receive much lower rates of N fertilizer. Wheat, for example, is currently fertilized at the average rate of 59 kg of N/ha for those areas actually receiving N fertilizer (64% of total wheat compared with 96% of total corn area). Cotton (*Gossypium hirsutum* L.) is fertilized with N at an intermediate level, about 87 kg of N/ha, to avoid excessive vegetative growth.

Obviously N fertilizers are used differently in different cropping systems, and it is impractical to generalize on the effects of N fertilizer use on soil N budgets. The fate of fertilizer N under specified conditions is discussed in a later section dealing with stable isotopes.

G. Miscellaneous Items

Nitrogen inputs in the seed are generally considered negligible. However, for some crops seed inputs cannot be neglected. For example, at normal seeding rates corn contributes < 0.2 kg of N/ha, wheat contributes about 1.5 kg of N/ha, soybeans (*Glycine max* L.) contribute about 2.5 kg of N/ha, and potatoes (*Solanum tuberosum* L.) contribute from 5 to 9 kg of N/ha. Seed inputs should, therefore, be considered on an individual crop basis.

Another often overlooked source of N is the residual mineral N in the root zone at the beginning of the N balance study. Previous management practices often result in a build-up of soil NO_3^-, especially in those areas receiving low rainfall. This residual NO_3^- is readily available to the succeeding crop and has made a major impact in the first-year N budget on several occasions (Bigeniego et al., 1979; Broadbent & Carlton, 1978; Chapman et al., 1949).

IV. NITROGEN LOSSES FROM SOIL-PLANT SYSTEMS

This section considers a broad-scale view of N losses. Smaller-scale losses are considered in section V D, 1–7.

A. Removal by Crops and Livestock

Estimates of N removed by major U.S. crops in 1977 (Table 1) show that about 2.5 million metric tons of N were harvested as corn, soybeans, and hay and that total N removals exceeded 10 million metric tons. Harvested plant products, therefore, contained about a million tons more N than was applied as N fertilizer in 1977. Most of the N in the harvested products is removed from the farm through the sale of grain, soybeans, livestock, etc.

Nitrogen removed from U.S. farms in 1977 through livestock, including poultry, was about 1 million metric tons (NAS, 1971; USDA, 1977; Yeck et al., 1975). Most of this N goes into the human food chain and is not recycled into agricultural systems, although a portion of the manure produced by farm animals is recycled. The total N removed through livestock operations, i.e., N sold in livestock plus one half of the manure N amounts to about 3.9 million metric tons nationally.

B. Erosion and Runoff

The national loss of N by erosion and runoff from cropland is a serious problem. Current estimates of erosion are in the order of 5 billion metric

Table 1—Estimated removal of N in major crops of the United States in 1977 (NAS, 1971; USDA, 1977, 1978).

Crop	Production, thousands of metric tons	% N†	Total N harvested, thousands of metric tons
Corn, grain	161,486	1.5	2,422
Corn, silage	105,723	0.40	423
Sorghum, grain	20,083	1.42	285
Sorghum, silage	8,486	0.43	36
Oats	10,856	1.84	200
Barley	9,053	1.86	168
Wheat	55,133	2.18	1,202
Rice	4,501	1.26	57
Rye	432	1.81	8
Soybeans	46,711	6.06	2,831
Flaxseed	409	3.71	15
Peanuts	1,670	3.52	59
Sunflower seed	1,248	2.69	34
Cotton lint	3,156	0.18	6
Cotton seed	5,018	3.70	186
All hay	118,893	2.0	2,378
Dry beans	739	2.54	19
Potatoes	15,967	0.34	54
Sweet potatoes	568	0.27	2
Tobacco	877	2.69	24
Sugarbeets	22,784	0.6	137
Sugarcane	25,089	0.32	80
Total			10,626

† % N fresh weight basis.

tons of soil annually, with about 80% lost as waterborne sediments and the remainder lost by wind erosion. It is estimated that one half to three fourths of the eroded soil is from agricultural land (Wischmeier, 1976; Pimentel et al., 1976). If one were to use a value of 3 billion metric tons of soil from cropland, which contains 0.15% N (Willis & Evans, 1977), an estimated 4.5 million metric tons of N would be lost annually. Most of the N lost by soil erosion is in organic forms and does not represent a loss of readily available N. It is a large loss of potentially available N, however, that eventually is deposited in streams, lakes, and oceans with little opportunity to be recycled into agricultural systems.

Excellent research on soil erosion has been carried out on watersheds and smaller areas during the past decade, and the best management practices for specific watershed conditions are being determined. A few examples are discussed herein.

Sediment and nutrient discharges in runoff vary greatly, depending on such factors as topography, soil-plant management practices, infiltration rates, storm patterns, etc. Menzel et al. (1978) pointed out that under Oklahoma conditions, long-term records are needed to compare nutrient and sediment losses due to large yearly variations in storm patterns. The maximum annual discharges were 13 kg/ha of total N and 4 kg/ha of NO_3^--N, with average values about 50% of the maximum values. The annual deposition of N in rainfall averaged 5 kg/ha and thus exceeded the maximum discharge of NO_3^--N in runoff.

Similar losses were reported by Kissel et al. (1976) in Texas. With a grain sorghum-cotton-oats rotation fertilized at recommended N rates, the mean annual loss of NO_3^--N was 3.2 kg/ha, whereas losses of sediment-associated N were about 5 kg ha^{-1} year^{-1}.

The importance of soil cover in nutrient transport by surface runoff was emphasized by Burwell et al. (1975) in a 10-year study in Minnesota on a Barnes loam with 6% slope. The amounts of soluble N and sediment N, respectively, for five soil cover conditions (in kg/ha) were as follows: (i) continuous clean-cultivated fallow, 3.0 and 51.3; (ii) continuous corn, 2.3 and 21.5; (iii) corn in rotation, 1.0 and 13.2; (iv) oats in rotation, 2.6 and 1.9; and (v) hay in rotation, 3.1 and 0.2. The small losses of soluble N in runoff were exceeded by the quantities of NH_3 and NO_3^- contributed by precipitation.

In Saskatchewan, where runoff is caused primarily by snowmelt, the annual N loss from summer fallow in solution and sediment was estimated at 10 kg of N/ha (Nicholaichuk & Read, 1978). The average loss of unincorporated fall-applied N fertilizer by surface runoff was about 4%; loss of N from unfertilized fallow was 2.8 kg/ha.

Various soil conservation practices are effective in reducing N losses in runoff. Schuman et al. (1973) found that a level-terraced watershed planted with corn reduced water, sediment, and N yields when compared with contour-planted corn. Annual water-soluble N losses were low from all watersheds, averaging only 3 kg/ha from contour-planted corn fertilized with 420 kg of N/ha. The majority of N losses were associated with the runoff sediment on both conservation practices. The terraced watershed, however, lost only one-tenth the sediment N compared with the contour-planted watershed. Water-soluble N and sediment N losses in runoff were usually highest at the beginning of the cropping season and decreased progressively throughout the year, reflecting crop canopy establishment, crop nutrient removal, and leaching.

Rainfall simulators have been used on small plots to determine the amounts of soluble N and sediment N in surface runoff (Moe et al., 1967, 1968; Romkens et al., 1973). Moe et al. (1967) simulated high rainfall conditions on plots with 13% slope, with and without surface-applied NH_4NO_3 at 224 kg of N/ha. After 12.8 cm of rainfall was applied, there was less loss of soluble N from fallow soil (15.9 kg/ha) than from sod (33.8 kg/ha), but much more organic N was lost from fallow plots. Only 2% of the applied N was lost from sod plots with an initial moisture content of 12.5% compared with a 14% loss when the initial moisture content was 25.8%. Losses from fallow plots with surface seal were also much greater than from similar plots without surface seal. In a similar experiment, Moe et al. (1968) found that surface runoff losses of urea N were nearly 50% less than those from NH_4NO_3 on both sod and fallow plots. Soil losses, however, were 100 times greater from fallow plots than from sod plots.

The N content of runoff can be measured on a smaller scale by using lysimeters (Chichester, 1977). When N fertilizer was applied to meadow and corn in lysimeters at rates up to 322 and 672 kg ha^{-1} year^{-1}, respectively, N losses in runoff were greatest in the summer when intense rainfall events occurred shortly after fertilizer application. Nitrogen transported in surface water decreased as the amount of effective soil cover increased. Nitrogen losses ranged from 1 kg ha^{-1} year^{-1} for meadow to 10 kg ha^{-1} year^{-1} for clean cultivated corn. This study emphasized the importance of good management, including N applications in balance with crop requirements, soil incorporation of fertilizers, and adequate plant cover to improve fertilizer efficiency and minimize erosion and runoff losses.

The examples of runoff and erosion losses cited above illustrate the variability of such losses, depending on the experimental conditions. In most cases, no attempts were made to obtain complete N balances; however, the measured losses provide a range of values that might be expected in normal agricultural production systems. Runoff losses of soluble N are generally low, except when high rates of N fertilizer are surface applied just before high rainfall events. Results obtained by Timmons et al. (1973) indicate the importance of fertilizer incorporation to minimize N losses in surface runoff. In many cases, the gain of N in the precipitation is greater than the amounts of soluble N in the runoff. It is also apparent that soil moisture content prior to rainfall events, soil sealing, plant cover, and conservation practices such as terracing and contour planting all affect soluble and sediment N losses. Total N losses associated with sediments are usually several-fold greater than soluble N losses in runoff.

C. Leaching Losses

Leaching is often the most important channel of N loss from field soils other than that accounted for in plant uptake (Allison, 1973). Losses occur mainly as NO_3^-, the movement of which is closely related to water movement. The main processes involved in this movement are discussed in detail elsewhere (Chapt. 11, D. R. Nielsen, J. W. Biggar, and P. J. Wierenga; Allison, 1973; NRC, 1978) and will be considered only briefly in this section.

Major losses of N occur when two prerequisites are met: (i) soil NO_3^- content is high, and (ii) water movement is large. There are some agricultural situations where these conditions do not often occur. For instance, NO_3^- generally does not accumulate in native grassland soils, since the NO_3^- is assimilated as rapidly as it is formed. Likewise, NO_3^- concentrations are also low in flooded soils and in acid forest soils. On the other hand, tillage stimulates ammonification of soil organic N and subsequent nitrification, leaves the soil bare for a period of time, and sets the stage for possible NO_3^- loss. Leaching losses are also strongly influenced by seasonal effects, such as water and temperature. In humid temperate regions, mineralization rates are low in winter, but leaching of residual NO_3^- from the previous season often occurs. In the spring, NO_3^- accumulates as nitrification rates increase, and N fertilizers are applied. If heavy rains occur before spring-planted crops are growing vigorously, significant amounts of NO_3^- can be leached below the root zone. Nitrate leaching is least likely to take place during the summer (Allison, 1973) when evapotranspiration usually exceeds precipitation, and plant uptake rates are high.

Leaching losses also occur under subhumid conditions where summer fallow is practiced. In a review of N losses in western Canada, Rennie et al. (1976) concluded that summer fallow practices have allowed deeper penetration of precipitation than was possible before cultivation, resulting in the movement of mineralized soil N into lower horizons. The data suggest that the NO_3^--N profile probably extends downward to at least 12 m over most of Saskatchewan and also that the concentration of NO_3^--N to this depth is relatively uniform. This approximates 600 kg of N below each hectare of cultivated land, equivalent to 9.84 million metric tons, or 21% of the soil N mineralized since cultivation. In a 10-year comparison of NO_3^--N profiles for pasture, wheat-fallow, and 10-year fallow, there was essentially no downward movement of NO_3^- under pasture, but about 500 kg/ha of NO_3^--N was found in the upper 3.6 m of soil in the wheat-fallow rotation. In the continuous fallow, 1,082 kg/ha of NO_3^--N was distributed in the top 3.6 m.

In the United States, Viets (1971) has emphasized the tremendous losses of N when the virgin grassland soils were placed under cultivation. Estimates of such losses range from 30 to 40% of the total N in the upper 30 cm of soil under low erosion conditions to 75% or more where serious erosion has occurred. The distribution of losses among leaching, volatilization, erosion, and denitrification is unknown; however, it has been estimated that crop removal accounted for only 10 to 15% of the loss (Viets, 1971).

B. A. Stewart et al. (1967) studied NO_3^--N contents of 6-m soil profiles taken from fields under various land uses in northeastern Colorado. These data indicate low NO_3^--N levels (100 kg of N/ha or less) under virgin grasslands and alfalfa (*Medicago sativa* L.), intermediate levels (290 kg of N/ha) under dryland wheat-fallow, and high levels (570 kg of N/ha) under irrigated cereals. Profiles under cattle feedlots contained over 1,600 kg of NO_3^--N/ha. In similar studies, Power (1970) found little leaching of fertilizer N applied to bromegrass (*Bromus inermis* Leyss.) compared with corn, in which an accumulation of NO_3^- was found at about 2 m in depth.

In humid areas, NO_3^- can be almost completely removed from the soil profile by leaching, denitrification, or both, or it may accumulate in the soil profile and move downward into the ground water depending on the soil, climate, fertilization, and management practices. Data presented by Hensler and Attoe (1970) illustrate both the movement and accumulation of NO_3^- in Corn Belt soils. After 4 years of different rotations with corn (C), oats (O), and alfalfa-bromegrass meadow (M), the COMM rotation had no accumulation of NO_3^--N at any fertilizer N level in the 0- to 152-cm layer of soil above the control. A CCCC rotation fertilized with 336 kg of N/ha annually accumulated 683 kg of NO_3^--N/ha. Other rotations and N levels gave intermediate results. In all cases, only the corn crop received N fertilizer. Under fallow conditions, there was little retention of fertilizer N in the upper 275 cm of a Plainfield sand in April when N was applied at rates of 112 and 336 kg of N/ha the previous fall.

Gast et al. (1974) studied NO_3^- and Cl accumulation and distribution in a fertilized tile-drained Webster clay loam soil in Minnesota and concluded that there was little loss of NO_3^- from the tile drains. In a later study, Gast et al. (1978) found that NO_3^--N losses through tile lines after 3-years' treatment with 20, 112, 224, or 448 kg of N/ha applied annually to continuous corn were 19, 25, 59, and 120 kg/ha, respectively. In a related 15-year study with a Forman clay loam, MacGregor et al. (1974) concluded that most of the applied fertilizer N was still present in the first 8 m of soil as NO_3^-, but very little remained in the corn rooting zone.

Kolenbrander (1972) systematically studied N leaching losses from lysimeters in relation to soil texture and drainage volume. Losses from soil organic matter ranged from 45 kg of N ha^{-1} year^{-1} on sandy soils to 5 kg of N ha^{-1} year^{-1} on heavy clay soils. The difference was attributed to loss by denitrification. The loss of fertilizer N was calculated to be 3.5% at a level of 60 kg of N/ha on cultivated land but only 1% on grassland at an application level of 250 kg of N/ha. The drainage in these lysimeter experiments was 250 mm/year.

In Ohio, Chichester (1977) found the highest N flux in lysimeter percolate during the winter months when evapotranspiration was low. Since winter percolate volumes were essentially the same for all crop management practices, the critical factor in quantity of N leached was the amount of soluble N remaining in the profile after crop harvest. The amounts of N lost in percolate from meadow lysimeters did not exceed 5% of that applied at the rate of 179 kg of N ha^{-1} year^{-1}. Nitrogen leached from corn-cropped lysimeters, however, often equaled or exceeded the amounts utilized during crop growth at the high (336 kg of N ha^{-1} year^{-1}) rates of application.

The leaching of NO_3^- in irrigated agriculture has received increasing attention in recent years in relation to possible ground water pollution (Ayers & Branson, 1973). The movement of fertilizer nutrients as affected by method of irrigation has been reviewed by Viets et al. (1967). The movement of N from band placement with furrow irrigation has received the greatest attention because of the great variation in redistribution of soluble N both laterally and vertically in the soil. If NO_3^- occurs or is placed below the water level in the furrow, movement will be lateral and downward. The closer the furrows, the greater the downward movement of NO_3^-, and correspondingly less movement will occur laterally into the center ridges. If NO_3^- is above the water level, it will concentrate at the surface and in ridges. This may lead to poor N recovery by the crop because of low root activity in the drier surface soil. With flood and sprinkler irrigation systems, the downward movement of NO_3^- is more uniform in comparison with furrow irrigation.

The amount of excess water passing through the crop root zone varies greatly from one area and from one type of irrigation system to another (McNeal & Pratt, 1978). Where minimal leaching is required for salinity control, sprinkler irrigation systems might percolate about 6 to 20 cm through the soil. In furrow-irrigated fields, however, the amounts of water applied vary appreciably from one point to another, and greater percolation losses occur in the process of supplying sufficient water to all parts of the field. Much of the excess water passes through previously leached soil beneath the irrigation furrow, however, and is relatively ineffective in removing additional N from the root zone. In areas where it is necessary to leach excess salts from the soil periodically for salinity control, considerable loss of NO_3^- may occur if the leaching process does not coincide with periods of low soil NO_3^- content.

Amounts of N leached as a percentage of N inputs for some typical southern California croplands under irrigation have been summarized by McNeal and Pratt (1978). Leaching losses ranged from 13 to 102% and commonly averaged 25 to 50% of the N applied in most cropping situations. They concluded that where excessive losses occurred, grower inputs such as improved water and fertilizer management as well as possible changes in water application method could increase N efficiencies substantially. Several management alternatives for preventing leaching losses of excess N and some management constraints were also presented (McNeal & Pratt, 1978).

D. Denitrification and Other Gaseous Losses

Gaseous losses are covered in detail in Chapters 8 (M. K. Firestone) and 9 (D. W. Nelson) and are discussed here only to highlight some recent work. Denitrification losses from soil-plant systems have usually been obtained by the difference between N additions and recoveries, where leaching losses were measured or controlled.

Gaseous losses of N have also been studied in closed systems in which the evolved gases are collected and determined directly. Stefanson's (1972a, b, c) work illustrates this method and shows a wide range in denitrification rates depending on the water, soil organic C content, and the presence or absence of plants. Greater losses were often observed in the presence of plants, but this observation was not consistent in all experiments (Stefanson, 1972c). Work with an Urrbrae red-brown earth showed that the effect of plant growth on denitrification was greatest at water contents around field capacity. Increasing the soil water on the high C soil increased denitrification losses sharply, with small losses (3%) occurring at water contents of 12 to 20% and large losses (40%) occurring at 20 to 28% moisture. With undisturbed soil cores, denitrification losses were not large, even after 40 days, regardless of N source or the presence of plants. In further work with other soils, Stefanson (1972d) found that the rate of denitrification was greater with NO_3^- sources than NH_4^+ sources.

More recently, direct methods of measuring volatile gases from denitrification have been developed that utilize large quantities of isotopically labeled N in small field plots (Rolston, 1978; Rolston & Broadbent, 1977; Rolston et al., 1976). Denitrification rates as high as 70 kg of N ha^{-1} day^{-1} were measured directly (Rolston & Broadbent, 1977) on a plot maintained near saturation, which received 34 metric tons/ha of manure about 3 weeks before 300 kg of NO_3^--N were applied. Total denitrification from this treatment was about 200 kg of N, with most of the denitrification occurring in the first 5 days. The maximum rate of denitrification in a field plot growing perennial ryegrass was about 10 kg of N ha^{-1} day^{-1}, with most of the 150-kg loss occurring during a 30-day period. For uncropped plots with no C addition and maintained near saturation, the maximum rate was about 2.5 kg of N ha^{-1} day^{-1}, with a total loss of 8 kg occurring over a 15-day period from the 300-kg application of NO_3^--N.

The results obtained by Rolston et al. (1976) represent maximum rates of denitrification under the experimental conditions. Under normal irrigation or rainfall conditions, the soil is poorly aerated for short periods at several times during the growing season. Estimates of denitrification occurring during an entire season were made by Rolston and Broadbent (1977), and a total loss of about 13 kg of N/ha was calculated. This represented about 9% of an application of 150 kg of N/ha, which was of the same order of magnitude as values determined in N balance experiments. This amount is less than that generally observed in other experiments (Hauck, 1971), where fertilizer N loss by denitrification has averaged about 15% by difference measurements. Measurements of denitrification, as carried out by Rolston et al. (1976), are needed on a number of soil types in different field situations to provide better direct estimates of denitrification.

Ammonia volatilization losses have been reviewed by Rolston (1978) and are described in Chapter 9. The amounts involved are highly variable, depending on such factors as type, rate, and method of fertilizer N applied, soil pH and environmental factors such as temperature, moisture, and air exchange conditions. Losses of NH_3 from manures are often quite high (see also Chapt. 21, J. H. Smith and J. R. Peterson), and the half-life of surface-spread manure NH_3-N has been estimated to range from 1.9 to 3.4

days (Lauer et al., 1976). Ammonia volatilization losses from fertilizers may range as high as 70%, but it is possible to minimize NH_3 losses by placement of materials about 10 cm below the soil surface. For example, Hargrove et al. (1977) observed large NH_3 losses (up to 50%) from surface-applied $(NH_4)_2SO_4$ on a calcareous soil, whereas Broadbent and Carlton (1979) reported negligible (but still measurable) losses of labeled N from $(NH_4)_2SO_4$ placed several centimeters below the surface. In some cases, however, it has been the practice to broadcast all fertilizers on the surface of the soil without incorporation (e.g., minimum tillage), and the opportunities for NH_3 volatilization are enhanced. Ammonium fertilizers applied in irrigation water may also result in considerable NH_3 loss. Significant NH_3 losses have also been reported from the flood water of rice, with losses being as high as 50% for urea and 15% for $(NH_4)_2SO_4$ (Vlek & Craswell, 1979). The quantitative estimates of NH_3 volatilization needed for N budgets are not generally available, and this is another area that needs to be documented for each study individually.

Plant gaseous losses of nonelemental N were also detected by Stutte and Weiland (1978) with a pyrochemiluminescent method for the analysis of transpirational vapors condensed in a closed system containing an attached leaf. This method allows estimations of total foliar N loss and the amounts in either reduced or oxidized forms, but specific N forms are not defined. In this and subsequent studies (Stutte et al., 1979; Weiland & Stutte, 1979, 1980; Weiland et al., 1979), significant losses of N were detected for several crop and weed plants, primarily as reduced N forms. As an example, minimal foliar loss of N from a soybean field was estimated at 45 kg of N/ha throughout the 1977 growing season. Such losses followed a diurnal trend, were greatest during early vegetative growth, and increased under high temperatures. Quantitative data of this nature may explain some of the N imbalances where denitrification losses were minimal.

E. Ammonium Fixation

The influence of the NH_4^+-fixation capacity of soils on N gains and losses has not been fully ascertained. However, there is some evidence that the quantities of fixed NH_4^+ do not change greatly over long periods of time when concurrent changes in soil organic N have occurred (Jaiyebo & Bouldin, 1967; Keeney & Bremner, 1964). Fixation can influence $^{15}NH_4^+$-N transformations by reducing NH_3 losses, reducing nitrification rates, etc. In a ^{15}N balance study, Kowalenko (1978) found that 59% of the $^{15}NH_4^+$ applied to the 0- to 15-cm layer of an ammonium-fixing clay loam soil was immediately fixed by clay minerals. About 66% of the recently fixed $^{15}NH_4^+$ was released within 86 days, but the remainder was held tightly throughout a 17-month period. Others (van Praag et al., 1980) have observed a relatively rapid release of recently fixed $^{15}NH_4^+$-N compared with the remineralization of residual ^{15}N as a whole. However, K additions and plant removals of interfering K ions also effect the release of recently fixed NH_4^+ (Allison, 1973; Axley & Legg, 1960; Legg & Allison, 1959; van Praag et al.,

1980). Kowalenko and Cameron (1978) reported that 71 to 96% of recently fixed $^{15}NH_4^+$-N was available to a barley (*Hordeum vulgare* L.) crop over the course of a growing season. Ammonium fixation can thus alter ^{15}N transformations, but it can be accounted for quantitatively in N balance studies by using proper analytical methods (Bremner, 1965).

V. RECENT STUDIES OF N BUDGETS IN SOIL-PLANT SYSTEMS

The preceding section considered N budget mechanisms and relevant data for these processes. However, this information does not lend itself to understanding the N cycle of specific soil-plant systems. Hereafter, more complete agronomic N budgets are discussed.

A. Use of Labeled N in N Budgets

Two general methods are employed in N budget research. One involves the complete total N budget of the system and documents the appropriate total N inputs, outputs, etc. without the use of labeled compounds. The other introduces a specific labeled N input, such as fertilizer N, and calculates the labeled N budget. These two budgeting approaches are not equal. They emphasize different N processes and properties of the system. The total N budget focuses on the total N cycle of the system through the various N additions and losses. A labeled N budget indicates how the labeled N *interacts* with the system by tracing its fate throughout the system. The type of N budget constructed depends on the goals of the study, but it is important to understand that the two approaches are not equivalent.

The main advantage of using ^{15}N is improved sensitivity in tracing a given N input. In practice, this increased sensitivity will be limited by sampling problems, as pointed out by Carter et al. (1967). They compared the recoveries of ^{15}N fertilizer added to 60-cm steel cylinders driven into the soil. Before any ^{15}N was applied, the top 15 cm of soil was removed from all the cylinders, mixed, and then replaced. After labeled N fertilizer was added and sudangrass (*Sorghum vulgare* L.) was grown for 8 weeks, the ^{15}N content of the crop and the soil in each cylinder was determined. The soil ^{15}N was estimated by two methods (i) by compositing seven 1.9-cm cores; and (ii) by removing the entire remaining soil mass, mixing, and subsampling it. The total ^{15}N recoveries for the core sampling method averaged 113% and ranged from 86 to 137%. The other method averaged 100% and ranged from 98 to 101%. The increased variability of the core method is clear, and it is even more striking when one considers that the soil was quite homogeneous before sampling and that seven cores from a 60-cm diam ring is much more intensive sampling than is regularly encountered in soils research. These data clearly illustrate the importance of sampling in labeled N work. The largest sources of error are often associated with the estimation of soil mass per unit volume, crop dry matter, etc., and these sources often determine the accuracy and precision of the final data. The spatial variability of N data is usually larger than P or K data, because many N transformations are linked with water movement or water content. Since the spatial distribution of soil water regimes has been shown to be highly variable (Nielsen et al., 1973) it should not be surprising that the N spatial variability is also large (cf. Broadbent et al., 1980; Pratt et al., 1976c; Rible et al., 1976). The use of labeled N can thus markedly improve tracing sensitivity in fertilizer N balance work, but only if reliable estimates of the size of the various N pools can be obtained, a representative sample collected for analysis, and the experiment performed with adequate replication and the proper local controls.

B. N Balance Methodology

Several authors have reviewed the assumptions, advantages and disadvantages, analytical methodology, and the uses of ^{15}N in soils research (Hauck & Bremner, 1976; Martin & Skyring, 1962; Martin et al., 1963). We agree with these authors, as well as numerous others, who have generally concluded that investigators pay special attention to the total N determination. Ensuring quantitative N recovery is a continuing exercise, even after the adoption of an apparently acceptable version of the Kjeldahl method, as exemplified by the incomplete recovery of NO_3^--N by salicylic acid reported by Craswell and Martin (1975a).

Early balance work was commonly done with greenhouse pots (MacVicar et al., 1950; Walker et al., 1956). They are economical and convenient, allow a large number of treatments, reduce sampling problems, and offer a degree of control over temperature and water. However, they are not easily adapted to measure gaseous losses directly, and the extrapolation of the N balance data to field conditions is usually difficult at best. They have been used primarily in studying the effects of a large number of treatments such as soil amendments, fertilizer sources, and soil physical and chemical properties.

Confined and unconfined field plots of small size (microplots) have also been used by several investigators (Carter et al., 1967; Craswell & Martin, 1975b; Kissel & Smith, 1978; Vallis et al., 1973). These procedures involve applying ^{15}N to a small area (< 1 m^2), growing a crop or incubating the soil, and then sampling the area at various times to follow the labeled N into various mineral and organic pools. The confined microplots are usually constructed by forcing a steel or plastic cylinder into the soil or by building a confining structure around an intact mass of soil. The major disadvantage of the unconfined microplot is the enormous sampling problems produced when the ^{15}N is given unlimited dispersion. The confined microplots obviously limit this dispersion and significantly reduce sampling problems. Vallis et al. (1973) concluded that enclosed microplots are superior for N balance work, although they would not be suitable where lateral subsurface drainage is a factor. Neither of these microplot techniques allows estimation of N leached below the sampling zone or gaseous N losses, unless extensive instrumentation is employed. Their greatest use is in areas of low leaching potential or in short-term studies where leaching is minimal. Their main advantages are that the small areas lower isotope costs, they are field techniques, and they allow complete removal of the soil for sampling.

Field lysimeters have also been used in N balance work (Chichester, 1977; Dinchev & Badzhov, 1969; Zamyatina et al., 1973). They offer the advantages of the confined microplot along with an estimate of N leaching. The main disadvantages are high cost, lack of direct data on gaseous N loss, and restricted soil sampling. However, lysimeters do monitor N transformations under field conditions and are especially useful for those transformations associated with water movement, such as leaching and runoff. The permanent installation of lysimeters also allows long-term studies so that climatic variations can be adequately sampled.

Gas lysimeters have also found use in N balance investigations (Catchpoole, 1975; Craswell & Martin, 1975a; Martin & Ross, 1968; Vlek & Craswell, 1979). These devices extend the pot study approach by confining the atmosphere around the pot and thus allowing a direct estimate of gaseous N loss under the experimental conditions imposed. The instrumentation and analysis associated with gas lysimeters are, of course, more costly, and the final extrapolation of the results to field conditions is often difficult. However, if the purpose is to study the processes and factors affecting gaseous loss or to directly establish such losses, gas lysimeters offer a good experimental technique.

In recent years the trend has been to use several of these methods simultaneously (Catchpoole, 1975; Craswell & Martin, 1975a, b; Rolston et al., 1976, 1979). These approaches usually involve extensive instrumentation of a confined field microplot or the removal of an intact mass of soil to confined environments and subsequent instrumentation. The instrumentation of these plots usually involves sampling of the soil gaseous and liquid phases, sampling of evolved gases, as well as measurement of conventional N pools, such as crop uptake and soil organic N. The main disadvantages of these methods are the high costs, large time investment, and extensive instrumentation. Nevertheless, these comprehensive methods allow the direct estimate of the major N transformations and should allow the construction of a complete N budget by direct measurement.

There is no single best method for constructing N budgets. The proper experimental method will depend on the resources available, the objectives of the study, and the major N transformations of the system under investigation.

C. Problems in N Balance Studies

When the literature on N balance studies was reviewed, two major problems became apparent. One was an inadequate description of the analytical procedures and experimental conditions. Authors often neglected the details of the sampling procedures (e.g., number of cores per plot, number of plants harvested), details of analytical procedures (e.g., digestion time on Kjeldahl, inclusion of NO_3^--N in the Kjeldahl procedure), and details of environmental conditions (e.g., rainfall amounts and intensities). A second problem involves the use of incomplete statistics in reporting N balance data. Investigators are usually content in reporting only means, or point estimates. Point estimates give an incomplete report of the experimental data because data distributions are characterized by two parameters, a mean and a variance. It is important to estimate both parameters in order to make meaningful comparisons within an experiment or between experiments. In the case of replicated trials, it is a relatively simple matter to calculate the N balance for each experimental unit and thus estimate the variance directly. In the case of nonreplicated experiments, one must use indirect methods such as higher order interactions, past experience in replicated trials using similar techniques, or propagation of error approaches. The important point is to report both the means and an estimate of error. For example, a N deficit of 20% is interpreted differently if the associated standard error is 20% ± 6% or 20% ± 16%.

It is sometimes stated that errors in N balances are cumulative and that the final loss estimates are rather crude. This view of the error structure is not completely accurate. Errors in N budgets result from random sources and from bias or systematic sources. The statement that errors accumulate usually applies to the bias component. The statement is an inaccurate description of the random component because it tacitly assumes that the N transformation pools are independent. To assume independent pools would ignore the whole cyclic-interactive nature of N transformations. For example, applied N might be taken up by the crop, which would naturally lower leaching losses; similarly, immobilization of N would lower both crop uptake and leaching losses. This inverse relation is the key that improves the accuracy of the final total N account, assuming that the major transformations are measured. In statistical terms the error structure for the final total N account is given by Eq. [1] (Snedecor & Cochran, 1967):

$$Var(N_T) = Var(N_1) + Var(N_2) + Var(N_3) \dots Var(N_i)$$
$$+ 2\,COV(N_1, N_2) + 2\,COV(N_1, N_3) + 2\,COV(N_2, N_3) \qquad [1]$$
$$+ \dots 2\,COV(N_{i-1}, N_i).$$

This equation states that the variance (Var) of the final total N (N_T) is equal to the sum of the variances of the various N pools plus twice the covariances (COV) of all possible two-way combinations of these pools.

For a simple two-component system, Eq. [1] becomes:

$$Var(N_T) = Var(N_p) + Var(N_s) + 2\,COV(N_p, N_s). \qquad [2]$$

where
    N_T = total N accounted for,
    N_p = total N in plant,
    N_s = total N in soil,

Var = the estimated variance, and
COV = the estimated covariance
     = $(r_{Np, Ns})(S_{Np})(S_{Ns})$.

The covariance is most easily understood by recognizing that it equals the product of the correlation coefficient between N_p and N_s and the standard deviations of N_p and N_s. The data of Carter et al. (1967) point out the importance of the covariance term. For their data, Var(N_p) is 31.7 (C.V. of 12.9%), Var(N_s) is 35.3 (C.V. of 10.7%) for the 0- to 75-cm depth including roots, and the COV(N_p, N_s) is -32.5 ($r = -0.97$). The variance of N_T was only 2.1 (C.V. of 1.5%) due to the high negative correlation between N_p and N_s. If the same data are summarized with N_s measured only to 15 cm to simulate leaching beyond the sampling zone, one finds that the Var(N_s) is 34.6 (C.V. of 16.1%) and the COV(N_p, N_s) is -24.2 ($r = -0.73$). This resulted in a Var(N_T) of 18.1 (C.V. = 5.3%). Thus when the N budget is incomplete, as in omitting the 15- to 75-cm layer of soil, the errors in estimating N_T increase considerably due to the decrease in the correlation between N_p and N_s. The final error in N_T can thus be considerably better than the sum of the random errors of each component.

A more serious statistical problem in the final N_T account involves the case where bias has entered the estimates of the N pools. For example, low recoveries in the Kjeldahl determination will give a biased estimate of the N deficit. Another more obvious cause of bias would be the case where a major N pathway was not included (such as leaching or NH_3 volatilization) and the resulting deficit attributed entirely to denitrification. Bias errors will thus accumulate in the final compartment when this compartment is estimated by difference. These points emphasize the need to minimize bias, underscore the importance of quantitative N recovery, and show the need to document all major N transformations except one if the final deficit is to be attributed to a single avenue of loss.

D. N Balance Studies

In recent years, the number of N balance studies has increased markedly. Allison's (1955) original review involved only about a dozen experiments. His revised article (Allison, 1966) included about 40 studies, whereas more recent studies number in the hundreds. It is beyond the scope of this chapter to make a complete literature review of N balance work; rather, a more in-depth review is made of selected N balance papers that cover a range of agricultural crop systems and methods. Emphasis is given to the work that has been reported since Allison's last review (1966) and also to experiments conducted under field conditions.

1. SPECIFIC SOIL N BALANCE STUDIES

Carter et al. (1967) conducted a replicated field experiment in Alabama using ^{15}N in confined microplots. The pH 5.9 fine sandy loam soil contained 0.045% N. Leaching was restricted by movable covers suspended over the

Table 2—Recovery of labeled N (Carter et al., 1967, Part 1) from confined
microplots after 2 months.

Treatment		Recovery of ^{15}N				
			Soil			
Timing	Crop†	Plant tops	Roots plus 0–15 cm	15–30 cm	30–60 cm	Total‡
				%		
NaNO$_3$						
Seeding	F	--	15.0	65.1	19.5	99.6 ± 0.9
Seeding	S	42.0	25.3	29.7	1.5	98.5 ± 0.5
24-day	S	38.0	27.5	33.5	1.0	100.0 ± 0.5
24-day	S§	51.7	13.1	27.3	5.9¶	98.0 ± 0.9
(NH$_4$)$_2$SO$_4$						
Seeding	F	--	52.6	37.6	7.2	97.4 ± 0.9
Seeding	S	53.7	37.6	6.4	1.5	99.2 ± 0.5
24-day	S	40.8	56.1	1.4	1.1	99.4 ± 0.5
24-day	S§	51.0	43.3	1.5	1.7¶	97.5 ± 0.9

† F = fallow; S = sudangrass.
‡ ± Values are one standard error of the mean.
§ Plots were uncovered and received 20.8 cm of water compared with 11.4 cm for the others.
¶ Includes ^{15}N in 60- to 75-cm depth.

Table 3—Recovery of labeled N (Carter et al., 1967, Part 2) from confined
microplots after various times.

Treatment		Recovery of ^{15}N after 2 months			
		Fallow total	Cropped to sudangrass		
N rate	Soil pH		Plant tops	Soil	Total
kg/ha			%		
NaNO$_3$					
336	4.5–5.0	91	44	49	93
	5.5–6.0	90 (77)†	42	53	95
	6.5–7.0	93	45	49	95
672	4.5–5.0	89	23	69	92
	5.5–6.0	90	24 (21)‡	70 (50)‡	94 (71)‡
	6.5–7.0	92	24	70	94
(NH$_4$)$_2$SO$_4$					
336	4.5–5.0	95	41	53	94
	5.5–6.0	93 (95)§	49 (51)§	44 (42)§	94 (93)§
	6.5–7.0	93	50	43	93
672	4.5–5.0	95	29	67	96
	5.5–6.0	92 (91)§	33 (30)‡	62 (53)‡	95 (83)‡
	6.5–7.0	92	36	59	96
	L.S.D. 0.05	2	--	--	2

† Fallow plots sampled after 10 months.
‡ Plots not covered during rainfall received 20.0 cm of water; all others received 17.1 cm of water.
§ Plots receiving nitrification inhibitor.

plots. In the first experiment, sudangrass was seeded and N applied at 280 kg of N/ha as $NaNO_3$ or $(NH_4)_2SO_4$ either at seeding or 24 days after planting. The resulting fertilizer N recoveries (Table 2) were all very high, averaging 99%, which led the investigators to conclude that leaching losses beyond 60 cm and gaseous N losses were small under these experimental conditions. The data (Table 2) point out the greater downward movement of fertilizer N when no crop is grown, when $NaNO_3$ is the N source, and when the plots are unprotected. The following year a similar field experiment was conducted, except that the soil was sampled to 90 cm. Two N sources, three pH levels, and two N rates were studied on cropped (sudangrass) and uncropped cylinders. Leaching was again restricted, and supplementary treatments, such as a nitrification inhibitor, sheltered and unsheltered cylinders, and harvest time beyond 2 months, were included. Labeled N recovery averaged 93% (Table 3) but ranged from 88 to 96%. Cropping improved total recoveries, especially for $NaNO_3$ at low pH levels. The $(NH_4)_2SO_4$ gave greater recoveries than the $NaNO_3$, especially for the fallowed cylinders at low pH levels. The superiority of the NH_4^+ recovery was also greater at the high N rate. The ^{15}N losses were not attributed to leaching, since small amounts were recovered from the deep soil layers. The authors concluded that ^{15}N was probably lost in gaseous form, because efforts were taken to minimize sampling and analytical errors and to prevent leaching. The nitrification inhibitor did not significantly alter total ^{15}N recovery (Table 3) despite evidence that it effectively blocked nitrification. After 10 months, a total of 23% of the labeled N was unaccounted for, with a 10% loss during the first 2 months and an additional 13% loss during the next 8 months. The reduced rate of loss was attributed to the lower temperatures experienced during the eight intervening fall, winter, and spring months. Uncovered plots lost an average of 17% more ^{15}N than corresponding covered plots. Labeled N was found in the lowest soil layer of the uncovered plots, indicating that leaching was a contributing factor. This increased loss is surprising since only 30 mm more rain fell on the uncovered plots, but the time of this extra rainfall and the preceding soil conditions were not given. When the overall results of the first (essentially complete recovery) and the second experiment (about 7% loss) are compared, it is apparent that the controlling factors governing fertilizer N losses were quite elusive and transitory. The authors concluded that slow denitrification losses from subsoil microsites could account for the greater NO_3^--N losses and would also be in harmony with the proposition that the poorly aerated soil volume could be greatly influenced by small environmental changes, such as short-period changes in soil water content. The possibility also exists for leaching of NO_3^- beyond the sampling zone through large pores.

2. SMALL GRAIN N BALANCES

A fertilizer N budget was recently reported by Campbell and Paul (1978) for spring wheat grown to maturity in Saskatchewan using small confined field microplots (Campbell et al., 1977a). This involved two replications of seven N rates in combination with natural rainfall and rainfall plus

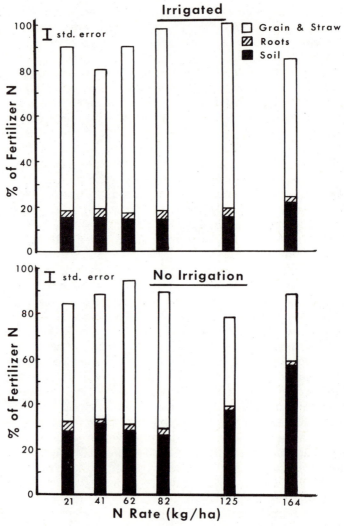

Fig. 3—Fate of fertilizer N applied to wheat under irrigated and dryland conditions (Campbell & Paul, 1978).

supplemental irrigation. The ^{15}N was added 10 cm below the soil surface as $K^{15}NO_3$ at seeding to soil with a loam texture that had previously been cropped to wheat. Labeled N loss by leaching was not a major factor under these experimental conditions, since only 4% of the residual fertilizer N was located in the 60- to 90-cm depth and none was found below this depth, even for the irrigated treatments. Labeled N unaccounted for averaged 13% across N rates for the nonirrigated treatments compared with 10% for the irrigated treatments (Fig. 3). Part of this loss can be attributed to random error, which amounted to 6% using a 99% probability level. Therefore, a minimum of 4 to 7% of the ^{15}N loss can be attributed to denitrification. Irrigation treatment did not significantly influence this loss due to careful irrigation management, which avoided excess water additions (no water addition was greater than about 30 mm). Crop ^{15}N utilization (grain plus straw) was higher on the irrigated treatments (71 compared with 50%), which was attributed to greater dry matter production (Campbell et al., 1977b). Roots accounted for 2 to 4% of the fertilizer N across treatments. The percent recovery of fertilizer N in aboveground plant material declined at the excessive N rates (two high N additions) of the nonirrigated treatments. The irrigated soil contained about 15% of the fertilizer N, whereas the nonirrigated soil contained 30 to 55%. The authors did not determine residual ^{15}NO$_3^-$-N, so it is impossible to state what portion of this labeled soil N was immobilized and what portion was inorganic. However, the low total NO$_3^-$-N levels in the irrigated treatments, excluding the highest N rate (Campbell et al., 1977a), indicate that most of the labeled soil N was in organic form. This study shows that about 13% ± 6% of the fertilizer N was lost from nonirrigated dryland wheat and about 10% ± 6% from irrigated wheat. The crop was a good sink for fertilizer N, with N assimilation closely related to dry matter production. The soil also contained significant levels of fertilizer N, with the quantity of residual N markedly increasing when N rates exceeded crop assimilation capacity.

A field lysimeter experiment reported by Zamyatina (1971) and Zamyatina et al. (1973) utilized 1 m^2 by 1.5 m deep lysimeters containing 20 cm of topsoil, 80 cm of subsoil, and a bottom layer of quartz sand (Zamyatina et al., 1968). Two Spodosols, each replicated four times, were investigated. One had a clay texture, pH of 4.3, and total N content of 0.100%; the other was a sandy loam with pH 4.6 and 0.076% N. Both soils were limed 10 months before seeding (final pH was not reported). Labeled (NH$_4$)$_2$SO$_4$ was applied at 90 kg ha^{-1} year^{-1} for 3 years to the 7- to 15-cm soil layer just before seeding, but no labeled N was added during the fourth year. Oats (*Avena sativa* L.) or barley was grown and soil samples were collected after the first and third growing seasons. The first-year N budget (Table 4) showed no strong effect of soil texture, with both soils immobilizing about 44% of the added N, primarily in the topsoil. Plants utilized about 36% of the ^{15}N, and drainage losses were negligible. From 18 to 22% of the initial ^{15}N application was not recovered. In the combined data for 1967 and 1968 (Table 4), plant recovery increased markedly for both soils compared with that in 1966, whereas immobilized ^{15}N decreased. Labeled N also appeared in the drainage water from the sandy soil; however, total ^{15}N recovery was about the same as in the first year. Fourth-year uptake of residual ^{15}N was only 3% from the clay loam and 1% from the sandy loam soil. The small fourth-year ^{15}N uptake points out the rapid stabilization of the immobilized ^{15}N (cf. Broadbent, 1980; Riga et al., 1980). Overall N losses ranged from 5 to 22% and averaged 16% under the conditions of this field experiment. The high rate of ^{15}N immobilization in 1966, resulting in low plant assimilation, compared with lower immobilization rates in 1967 and 1968 was not explained by the authors. One possible explanation is that increased organic C was available to the soil microflora the first year as a result of soil disturbance during lysimeter filling or from liming, and this C dominated initial ^{15}N transformations overriding soil textural effects and

Table 4—Labeled N balance sheet for small grain crops grown in field lysimeters as reported by Zamyatina et al. (1968, 1971, 1973).

Year	N added	Recovery of ^{15}N† Plants‡	Soil 0–20 cm	Soil 20–120 cm	Soil Total	Drain water	Grand total	Loss
	kg/ha	%						
			Clay loam soil					
1966	90	34	29	15	44	<1	78	22
1967 & 1968	180	67	20	8	28	<1	95	5
1966–1968	270	56	23	10	33	1	89	11
			Sandy loam soil					
1966	90	38	29	15	44	<1	82	18
1967 & 1968	180	50	21	1	23	8	81	19
1966–1968	270	46	24	6	30	5	81	19

† Standard error of a mean is about ± 1% throughout.
‡ Plant tops plus roots.

plant uptake. In subsequent years the microbial effects subsided, and plant uptake and soil textural effects became more pronounced. The data, therefore, offer a good illustration of the transient nature of soil ^{15}N transformations.

Another field ^{15}N balance with oats and barley that used filled lysimeters 1 m^2 by 1.2 m deep over three growing seasons has been reported (Koren'kov et al., 1975). Labeled NH_4^+-N was added to an acid sandy loam soil at 90 kg of N/ha, and the average ^{15}N recovery values over 3 years were as follows: 46% recovered in crop uptake; 31% recovered in soil (0 to 120 cm); 4% recovered in lysimeter drainage water; and 19% was unaccounted for losses. The results were generally the same for both $(NH_4)_2SO_4$ and aqua ammonia. The first-year ^{15}N balance showed an enhanced soil immobilization compared with the later 2 years of the study.

Another ^{15}N budget with cereals using field drainage lysimeters 1 m^2 by 40 cm deep was reported by Dinchev and Badzhov (1969) and by Badzhov and Ikonomova (1971). The study involved two crops and two N sources for the wheat (Table 5). The N was applied at the rate of 100 kg of N/ha, with one third at planting and two thirds as spring topdress for the wheat, and all applied at planting for the corn crop. When $^{15}NO_3^-$ and $^{15}NH_4^+$ sources were compared with wheat, the NO_3^- source lost more N in the drainage water and less was immobilized in the soil. This apparently resulted from the high mobility of NO_3^-, its susceptibility to denitrification, and its lower assimilation into the soil organic N pool (cf. Jansson, 1963; Riga et al., 1980). The end result was that almost 8% more N was lost from the NO_3^- source (34 vs. 26%), although large N losses occurred with both sources. Crop uptake also influenced N losses, with wheat accumulating 43% of the ^{15}N compared with 11% with corn. A larger percentage of the ^{15}N was recovered in the soil when cropped to corn, but the overall loss was about 20% greater for corn than for wheat. This large crop effect may be due to the N timing in relation

Table 5—Average fertilizer N budget for three soils receiving 100 kg of N/ha as labeled $(NH_4)_2SO_4$ or KNO_3 (Dinchev & Badzhov, 1969).[†]

N source[‡]	Crop	Recovery of labeled N					
		In crop		Lysimeter water	Soil 0–40 cm	Grand total	Unaccounted for losses
		Tops	Roots				
				%			
NO_3^-	Wheat	38.5	1.9	2.0	23.2	65.7	34.3
NH_4^+	Wheat	41.2	2.3	0.5	30.3	74.3	25.7
NH_4^+	Corn	10.5	0.6	1.2	40.9	53.2	46.8
Estimated SE of mean§		±1.3	±0.2	±0.1	±1.1	±1.3	±1.3

† These data have been averaged over the three soils reported.
‡ N sources were $K^{15}NO_3$ for NO_3^- and $(^{15}NH_4)_2 SO_4$ for NH_4^+.
§ Error estimated from the soil by N source and soil by crop interactions.

to crop N demand. The wheat crop was fertilized in phase with crop demand, whereas the corn crop received all of its fertilizer N at planting, and several weeks elapsed before rapid crop uptake. Other studies have shown that gaseous ^{15}N losses (Riga et al., 1980) and crop ^{15}N recovery (Olson et al., 1979; Riga et al., 1980; Zardalishvili et al., 1976) are strongly influenced by the time of N application. The overall N losses in this study may seem excessive, but in view of the heavy-textured soils, the shallow 40-cm lysimeters, and the periodic presence of excess water (significant leaching losses), the magnitude of the losses is not surprising. In fact, during drainage events, a 40-cm deep lysimeter should have a surface soil water potential of only about −0.04 bar, and this would promote poor aeration and denitrification losses. These data should not be considered atypical, however, since they are indicative of the ^{15}N transformations occurring in soils with a high water table or a subsoil textural discontinuity.

A series of ^{15}N balance experiments with wheat have been reported from Australia by Craswell and Martin (1975a, 1975b) and Craswell and Strong (1976). They used a black clay soil (74% clay) with a pH of 8.5 and 0.100% N. Under greenhouse conditions, Craswell and Martin (1975a) compared cropped (wheat) vs. fallow pots, $Ca(^{15}NO_3)_2$ vs. $(^{15}NH_4)_2SO_4$, and 56 vs. 63% water contents. They found high (97%) total ^{15}N recoveries for all treatments. These were unexpected results, because the high water contents (<0.1 bar tension) and lack of drainage should have promoted denitrification. Apparently a lack of readily available C restricted denitrification despite the presence of an active crop root system and the fact that Craswell and Martin (1974) had shown that this soil has a good potential for denitrification under more anaerobic conditions. To investigate the small losses of ^{15}N, a gas lysimeter experiment was also conducted. No significant N_2 or N_2O losses were detected; however, a small loss of labeled N was detected as NH_3 despite the fact that NO_3^- was the source of ^{15}N. The authors suggested that this loss might be the result of the plant itself losing NH_3 during maturity and senescence (cf. Hooker et al., 1980; Jansson, 1963).

In other wheat experiments, using packed undrained columns placed in field microplots on the same clay soil, Craswell and Martin (1975b) showed that 43% of the labeled N was found in the grain plus straw, about 14% in the soil organic matter (including roots), and 10% in the soil mineral phase after an addition of about 112 kg of N/ha as $Ca(^{15}NO_3)_2$. Total recovery was 67%, which significantly differed from 100%. Similar low total recoveries were recorded for fallow columns. The authors attributed much of this loss to a single 84-mm rainfall event that occurred 8 weeks after planting and saturated the soil. Lack of drainage in the columns, the presence of NO_3^-, and the nearly saturated water contents thus combined to produce sizeable denitrification losses. The importance of these short periods of high rainfall was underscored in subsequent experiments with sheltered columns where a total of 94% (L.S.D. for comparison with 100% was 3.6%) of the labeled N was recovered from both cropped and fallowed columns. In a third experiment using fallowed field microplots that were unprotected from rain and received 188 mm of rainfall (100 mm coming in four events) over a 16-week period, a total of 98% of the ^{15}N (L.S.D. for comparison with 100% was 2.4%) was recovered. The authors thus concluded that denitrification begins only under rather saturated water conditions in this soil, but under such conditions, significant N losses occur rather quickly.

Craswell and Strong (1976) continued their ^{15}N balance work with wheat by pressing 11-cm diam by 60-cm deep cylinders into the black clay soil. The soil cores were open ended and were subjected to two water regimes (255 vs. 444 mm) after labeled $Ca(NO_3)_2$ was applied at 96 kg of N/ha to the 15- or 45-cm depth. Results from the periodic harvests (Fig. 4) show high total ^{15}N recoveries on the cropped series with an average loss of 7% after 17 weeks. The water regime did not greatly affect the total recoveries on the cropped cores but did affect the distribution of the labeled N between soil and crop (see Fig. 4). Labeled N recovered in the aboveground crop ranged from 40 to 65%, and about 15% was immobilized in the soil. Fallowed treatments showed significant ^{15}N losses after 17 weeks, with the largest losses occurring on the high-water treatments and shallow placements. Leaching losses of ^{15}N were only significant on the fallowed treatments receiving high rainfall and deep N placement, and these amounted to about 12%. A 67-mm rainfall event occurred during week 9, but it apparently caused little ^{15}N loss on cropped cores. The wheat crop was an effective means of conserving N in this experiment, as evidenced by the difference in total recoveries between the cropped and fallowed treatments. This N conservation role was present only at high-water levels. The authors speculate that the N conserved in cropped cylinders might be the result of removal of mineral N, lower soil water contents, or both. Deep placement of N was also a means of reducing gaseous N losses.

This series of studies with the same soil and crop, using several methods, illustrates the elusive nature of soil ^{15}N; losses may be large or small under similar macroenvironmental conditions. The final labeled N transformations are the result of a complex combination of temperature, water, available C, and plant-growth conditions that interact over time to produce the final ^{15}N budget.

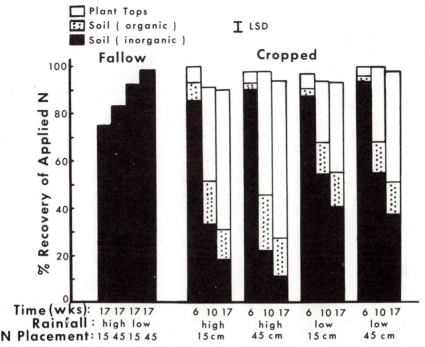

Fig. 4—Fate of fertilizer N applied to fallow or cropped (wheat) soil cores receiving 255 mm (low) or 444 mm (high) of rainfall and two depths of N placement (15 or 45 cm deep) at three harvest times (Craswell & Strong, 1976).

3. N BALANCES WITH CORN

Nitrogen balances with corn have historically used nonlabeled materials in long-term experiments. These studies generally have not provided a completely satisfactory N account due to inadequate measurements of one or more of the N pools, including indirect estimates of grain N removal, failure to collect initial and final soil samples, failure to analyze for organic N or to include NO_3^- in the total N analysis, failure to include control plots, and unsatisfactory soil sampling (small number of cores or only shallow depths sampled). Therefore, these N accounts should be considered as first approximations rather than accurate estimates. The N budget for a N rate experiment spanning 17 years of cropping under western Iowa conditions on a Moody silty clay loam soil was reported by Jolley and Pierre (1977). Their data indicated that 56 kg of N ha^{-1} year^{-1} was adequate for maximum yields and that at this rate the N added was accounted for as 18% in the soil NO_3^--N pool, 18% in soil organic N, 38% in harvested grain, and about 26% unaccounted for. At the two higher rates of 112 and 168 kg of N ha^{-1} year^{-1}, the grain removed 17 and 11% of the applied N, whereas that remaining in the soil increased to 57 and 52%, respectively. Unaccounted for N losses averaged 32% for the two higher rates. Power et al. (1973) reported 7 years of data from a North Dakota study comparing four N sources, each at two rates on a Parshall fine sandy loam having a pH of 6.5. For the 55 kg of N/ha input, about 67% was recovered in the aboveground dry matter and 10% as soil mineral N to the 3.6-m depth. At the 110 kg of N/ha rate, about 53% was recovered by the crops and 29% as soil NO_3^--N. The remaining N, which averaged 21% across all corn treatments, represented the N tied up in relatively stable organic forms plus the gaseous losses of N through NH_3 volatilization and denitrification. In a 3-year study on a Plano silt loam in Wisconsin, Olsen et al. (1970) observed that with an annual rate of 112 kg of N/ha, about 46% was removed in the crop, and 26% was present in the soil as mineral N to the 1.2-m depth. At the 336 kg of N/ha rate, the crop removed only 19% of the N, and the soil contained 69% in mineral form. Thus, an overall average of 21% of the added N was transformed into gaseous losses, incorporated into soil organic N, or leached below 120 cm. These studies show that NO_3^--N accumulates in soils quite readily when N rates exceed the point of crop response.

Field-scale ^{15}N balances with corn are uncommon, because large plots increase costs. However, Chichester and Smith (1978) conducted such a study utilizing four monolith lysimeters at Coshocton, Ohio. The lysimeters covered 8 m^2, were 2.4 m deep, and extended into the underlying parent rock. The soil was a moderately well-drained Keene silt loam on a 6% slope with a slowly permeable mottled horizon at about 65 cm, which restricted root development (Kelly et al., 1975). Previous rotation and fertility treatments resulted in differing organic N and pH status (see Table 6). In 1972, $Ca([15]NO_3)_2$ was added at 336 kg of N/ha, no N was added in 1973, and 178 kg of N/ha of unlabeled N was added in 1974. The ^{15}N contents of the runoff, percolate, and crop were determined, and after 3 years the soil mineral

Table 6—Fate of ^{15}N over 3 years after a one-time addition of labeled $CaNO_3$ at 336 kg of N/ha to corn grown on large monolith lysimeters (Chichester & Smith, 1978).

Tillage system[†]	Soil data		Water transported		Crop uptake		Soil, 0–240 cm			
	Total N	pH	Runoff[‡]	Leached	Grain	Stover[§]	Organic	Inorganic	Total[¶]	Lost
	kg/ha					kg of N/ha[#]				
MT	3,262	6.5	<1	110	75	26	98	9	318	18
			(<1)	(33)	(22)	(8)	(29)	(3)	(95)	(5)
CT	3,106	6.1	4	108	82	24	74	7	299	37
			(1)	(32)	(25)	(7)	(22)	(2)	(89)	(11)
MT	2,752	5.2	<1	106	70	24	41	7	248	88
			(<1)	(32)	(21)	(7)	(12)	(2)	(74)	(26)
CT	2,624	5.8	7	114	67	18	32	19	257	79
			(2)	(34)	(20)	(5)	(10)	(6)	(77)	(23)

† MT = mulch tillage, corn stover residue on surface; CT = conventional tillage, plow, bare soil surface.
‡ Includes ^{15}N in water and sediment phase of runoff, but sediment amounted to <1 kg/ha.
§ Includes N in cobs.
¶ Includes an estimate of ^{15}N in roots and crowns of the last crop, which was <1 kg of N/ha.
# Values in parentheses are the percentages of the original 336 kg of N/ha input.

and organic ^{15}N contents were estimated by core samples taken to 2.4 m. The 3-year ^{15}N data (Table 6) show that the mulch tillage lost little ^{15}N in the runoff, whereas conventionally tilled lysimeters lost 1 to 2%. Leaching losses accounted for about one third of the ^{15}N; most of these losses occurred between November and May, with the largest occurring the first year. About 29% of the total ^{15}N was taken up by the crop, and most of this was removed in the grain. This low percentage recovery resulted from the large ^{15}N addition and subsequent movement of the excess ^{15}N below the root zone after the first year. Larger amounts of ^{15}N were traced into the soil organic fraction on the lysimeters previously receiving high fertility. The mulch tillage also accumulated more organic ^{15}N, especially in the surface 0- to 15-cm layer. Levels of labeled inorganic N were generally low after 3 years except in one lysimeter. Estimated gaseous ^{15}N losses ranged from about 6 to 26% and were strongly related to the previous management practices and, more specifically, to ^{15}N immobilization. In this study the largest ^{15}N sinks were leaching (33%) and crop uptake (29%). Immobilization plus denitrification amounted to about 35%, with denitrification losses being inversely related to ^{15}N immobilization. These values resulted from a large one-time application of ^{15}N in a mobile form to a soil-plant system with a restricted root zone in a humid climate where about 45% of the annual precipitation percolated through the soil.

Zardalishvili and Kvaratsyelia (1973) have also reported the results of a ^{15}N study in shallow lysimeters using $^{15}NH_4^{15}NO_3$ at 150 kg of N/ha. After one growing season, 35% of the N fertilizer was accounted for in the crop, 31% was accounted for in the 40-cm deep soil profile, 6% was accounted for in the leachate, and 28% was unaccounted for. Other work with $(^{15}NH_4)_2SO_4$ applied to irrigated corn in microplots in the Midwest reported gaseous ^{15}N losses of about 20% (Olson, 1980).

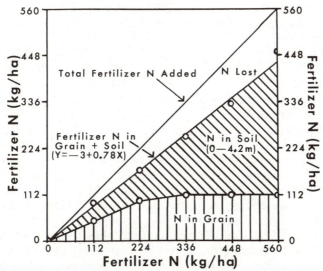

Fig. 5—Average annual fertilizer N balance over a 3-year period for irrigated corn grown on a sandy loam soil fertilized with $(NH_4)_2SO_4$ (Broadbent & Carlton, 1978, 1979).

Another labeled N balance with corn was reported by Broadbent and Carlton (1978). They used large (270 m²) irrigated plots and six levels of ¹⁵N depleted $(NH_4)_2SO_4$ applied below the soil surface, one-third preplant and two-thirds sidedressed. The Hanford sandy loam soil contained 0.02% N and had a slowly permeable layer at about 3 m (Carlton & Hafez, 1973). Total labeled N uptake was determined annually, although only the grain was removed from the plots. After 3 years of ¹⁴N additions, the soil was sampled to 4.2 m to determine its labeled organic and inorganic N content. The average annual fertilizer N balance (Fig. 5) indicated that about 42% of the fertilizer N could be accounted for in the grain until N rates of about 250 kg/ha were reached. Thereafter, grain fertilizer N removal was nearly constant at about 110 kg of N ha⁻¹ year⁻¹. The fertilizer N remaining in the soil ranged from about 44 kg/ha (39%) to nearly 350 kg/ha (62%). This fertilizer-derived soil N occurred primarily as organic compounds at the low rates and as NO_3^- at the high rates. The total fertilizer N recovered in grain and soil averaged about 78% across all rates as shown by the linear regression in Fig. 5. The unaccounted for N losses were attributed to denitrification, since little labeled N was found below 4.2 m and very little NH_3 volatilization was measured (Broadbent & Carlton, 1978, 1979). This study indicated that a fairly consistent proportion of fertilizer N was denitrified annually in this irrigated sandy loam soil. Furthermore, fertilizer N added in excess of crop requirements accumulated in the soil, predominately in inorganic forms. In a practical fertilizer management situation, this excess soil N should be taken into account when future fertilizer applications are made if crop production is to be maximized and environmental impact minimized.

Krauter (1975) and Broadbent and Krauter (1974) have also described a field experiment involving successive crops of irrigated wheat and corn on a Yolo fine sandy loam that received 112 kg of N/ha of ¹⁵N-enriched $(NH_4)_2SO_4$ at planting. Soil core samples were taken after the final crop to estimate labeled organic and inorganic N. The final ¹⁵N account was as follows: 55% was utilized by both crops; 30% was soil organic N; 3% was soil inorganic N; and 12% was unaccounted for. Since very little labeled N occurred below 120 cm, the authors concluded that leaching was minimal, and they attributed the 12% loss to denitrification.

Another large-plot (27 m²) ¹⁵N-depleted fertilizer balance was reported from Nebraska by Bigeniego et al. (1979). The study was conducted on a Sharpsburg silty clay loam containing about 0.2% N and investigated two application times (planting vs. sidedressing), each at three rates of N. Due to high levels of residual NO_3^--N and the high organic N status of this Mollisol, there was little response to N beyond the initial increment of 56 kg of N/ha. With 56 kg of N/ha, the planting and sidedress application methods resulted in labeled crop N recoveries of 65 and 83%; labeled soil NO_3^--N to 150 cm accounted for 4 and 14%, bringing the total fertilizer N account to 69 and 96% for the planting and sidedressing treatments, respectively. Plant recoveries for the two high rates (112 and 168 kg of N/ha) averaged 58% for both application methods. However, the residual soil NO_3^--N averaged 17% for the planting treatment and 34% for the delayed application. Total labeled N recovered was about 16% larger for sidedressing (92 vs. 76%) at the two highest rates of N. The fertilizer N not accounted for was lost by denitrification, by incorporation into the soil organic N pool, or by both. The labeled N could not be traced into the soil organic N pool, however, due to high isotope dilution. Nevertheless, it was evident that applying the N in phase with crop demand resulted in better fertilizer utilization and larger residual mineral N levels.

4. RICE N BALANCES

Nitrogen transformations and balances with paddy rice (*Oryza sativa* L.) are much more complex than conventional systems, because flooded soil offers a unique set of chemical (oxidation-reduction status), physical (puddled soil), and microbial (aerobic vs. anaerobic) conditions, all of which influence N transformations and the ultimate fate of N. See Chapters 8 and 12 (W. H. Patrick, Jr.) also. These conditions generally result in a large potential for N loss through denitrification, NH_3 volatilization, and leaching beyond the root zone.

Rice N budgets generally use large pots in order to confine the flood water. Broadbent and Mikkelsen (1968) used ^{15}N-enriched fertilizers on a flooded Sacramento clay (pH 6.8) in the greenhouse to study the effects of placement and timing, N source, and N rate (30 vs. 60 mg of N/kg of soil). Their data indicated that overall N losses were greater with urea than with $(NH_4)_2SO_4$, especially at the higher N rates. Ammonium sulfate lost about 12% of the labeled N at both N rates, whereas urea lost about 14% at the low rate and 23% at the higher rate. The greater losses with urea might be caused by NH_3 volatilization, especially for the higher rates of N that were unincorporated. Vlek and Craswell (1979) and Mikkelsen et al. (1978) have demonstrated that significant NH_3 losses can occur with continuously flooded rice, especially from urea sources. However, the extent of such losses is governed by the flood water pH, which in turn is controlled by the CO_2 status as related to aquatic photosynthesis, soil respiration, and chemical reactions as well as fertilizer management practices such as time, placement, and rate of N application. Broadbent and Mikkelsen's data also show that placement affected N losses, with banded applications at planting having the lowest losses, topdressing losing the most, and a split application losing intermediate quantities.

The effect of various water regimes and rice straw additions was investigated by Yoshida and Padre (1975), who added labeled $(NH_4)_2SO_4$ to 10-kg pots of Maahas clay soil. They investigated three water regimes: continually submerged, upland (field capacity), and alternating wet and dry (four wet-dry cycles). After 4 weeks virtually all of the NH_4^+-N had been oxidized to NO_3^--N except in the flooded treatment, where no nitrification had occurred. The ^{15}N budgets for the soil not receiving rice straw were as follows: flooded—plant uptake 52%, soil residual 19% (total 71%); upland —plant uptake 26%, soil residual 36% (total 62%); and fluctuating water— plant uptake 28%, soil residual 27% (total 55%). With rice straw additions, the total recoveries increased an average of 8%; however, plant utilization decreased an average of 13%, and residual soil N increased an average of 21% due to the initial immobilization and slow remineralization. The water regime is thus a major factor in determining the fate of applied N, because it impacts the oxidation-reduction status and microbial transformations.

Yoshida and Padre (1977) also conducted a field ^{15}N balance experiment with submerged rice using 1.25-m^2 plots isolated within larger plots. The small plots were treated with 50 kg of N/ha as ($^{15}NH_4)_2SO_4$ at various stages of crop development. At maturity the ^{15}N distributions were as follows: broadcast at planting—15% in crop, 45% in soil (total 60%); incorporated at planting—32% in crop, 52% in soil (total 84%); broadcast at maximum tillering—32% in crop, 36% in soil (total 68%); broadcast at panicle initiation—42% in crop, 38% in soil (total 80%); broadcast at booting—52% in crop, 33% in soil (total 85%); and broadcast at flowering—50% in crop, 34% in soil (total 84%). Both total ^{15}N recovery and crop ^{15}N uptake were enhanced by soil incorporation and applications at later stages of rice development. However, grain yields decreased when N applications were delayed beyond panicle initiation. These data illustrated the importance of applying N in harmony with crop demand to increase both fertilizer N recovery and grain yields.

Koyama et al. (1977) determined the ^{15}N budget for field-grown paddy rice in 0.8-m^2 microplots on a clay soil containing 0.23% N. They compared ^{15}N-labeled NH_4^+, NO_3^-, and urea applied as a small basal dressing, followed by four topdressings up to crop ripening. The NH_4^+-N sources produced rice ^{15}N uptakes of 47%, residual soil ^{15}N values of 26%, and unaccounted for losses of 27%. The NO_3^--N sources gave ^{15}N uptakes of 20%, residual soil ^{15}N values of 2%, and unaccounted for losses of 78%. Urea gave ^{15}N uptakes of 45%, with 6% remaining in the soil and 49% lost. The N losses in this study represent the combined processes of denitrification, NH_3 volatilization, and leaching below the 60-cm sampling depth. Similar findings of NH_4^+ superiority over NO_3^- sources have been observed by Datta et al. (1971), but these investigators were able to trace the N losses to evolved N_2 in their gas lysimeters.

Patrick and Reddy (1976) have also reported a field ^{15}N balance for paddy rice using large replicated plots with Al barriers. The soil was a Crowley silt loam having a pH of 5.8 and containing 0.08% N. Identically treated microplots were also established nearby and were harvested several times during the growing season to follow the fate of ^{15}N over small time increments. The ^{15}N balance data (Table 7) showed crop uptakes of 50 to 60%, with the largest amounts recovered in split applications. Crop utilization effects dominated the total ^{15}N recoveries, with split application losing the least ^{15}N. The missing ^{15}N represents the combined effects of denitrification, NH_3 volatilization, leaching below 15 cm, and transport around the barriers. The authors attributed most of this ^{15}N deficit to nitrification and subsequent denitrification. Microplot samples indicated that the major ^{15}N transformations occurred soon after application, with relatively little change in soil ^{15}N status beyond 62 days. Residual ^{15}N (straw and labeled soil N) transformations indicated that about 81% remained in organic form, with the crop utilizing about 6% and about 13% being unaccounted

Table 7—Distribution of labeled N applied to continuously flooded rice for three different fertilization schemes (Patrick & Reddy, 1976; Reddy & Patrick, 1978).

(NH₄)₂SO₄ fertilization method				Plant uptake			Soil N§	Grand total¶	Lost
Time†	Soil placement	¹⁵N added	Source of ¹⁵N‡	Grain	Straw	Total			
		kg/ha					%#		
14	7 cm	100	(NH₄)₂SO₄	31	18	49±2	26±3	75±4	25±4
		44	Residual	3	2	5±1	83±2	88±7	12±7
26	Surface	100	(NH₄)₂SO₄	32	19	51±7	27±3	78±9	22±9
		46	Residual	4	2	6±1	84±1	90±11	10±11
26 & 53	Surface	100	(NH₄)₂SO₄	37	24	61±1	24±4	85±4	15±4
		48	Residual	4	2	6±1	77±2	83±7	17±7

† Days after planting.
‡ (NH₄)₂SO₄ was applied the first year. Residual ¹⁵N was the previous years soil ¹⁵N plus straw ¹⁵N. Residual uptake was the average of nonfertilized and fertilized (nonlabeled fertilizer) plots.
§ Soil plus roots, roots amounted to about 4%.
¶ Inorganic soil ¹⁵N and ¹⁵N in floodwater was <1%.
# Percent of ¹⁵N input, either (NH₄)₂SO₄ or residual material.

for (Reddy & Patrick, 1978). Over the span of two growing seasons, an average of about 26% of the added ¹⁵N was not accounted for. The high ¹⁵N recoveries under waterlogged conditions can be attributed to the continuous flooding and the fertilizer management practices, which included deep placement of early N, split applications, NH₄⁺ fertilizer source, and N rates that could be readily assimilated by the crop.

5. GRASSLAND SYSTEMS

Several recent studies have utilized ¹⁵N to determine N cycling and ¹⁵N balance in grassland ecosystems. Generally these studies have involved labeled microplots on established pastures or native grasslands.

With an open microplot system, Henzell (1971) reported ¹⁵N recoveries of about 60 to 80% over a 7-year period with mown rhodesgrass (*Chloris gayana* Kunth) fertilized with several solid N fertilizers. However, the pathways of loss could not be identified. In other experiments with open microplots, Vallis et al. (1973) applied (¹⁵NH₄)₂SO₄ to a Mollisol and found 36% accumulated in the rhodesgrass tops, 28% in the litter plus roots to 28 cm, and about 19% in the soil to 74 cm. Total recovery was 83% ± 3% after 8 weeks of growth, and since little ¹⁵N was found deep in the soil or beyond the microplot the 17% loss most likely represented gaseous evolution. For a solodic soil, total recoveries of K¹⁵NO₃ after 43 weeks were 77% ± 4%, with 17% accounted for in the tops and litter, 30% in the roots to 30 cm, and 30% in the soil. Labeled N was found in the lowest soil depth and to 60 cm beyond the microplot, which indicated that the ¹⁵N losses represent leaching, surface runoff, and gaseous losses. Labeled urea added to a Spodosol at 5 kg of N/ha gave essentially complete recovery (98% ± 7%)

Table 8—Fertilizer N balance for a rhodesgrass pasture receiving 150 kg of N/ha
as $^{15}NH_4$ $^{15}NO_3$ (Catchpoole, 1975).

| Time after ^{15}N addition | Plant uptake | | Soil recovery | | | | Runoff water | Grand total | ^{15}N lost | Cumu- lative rainfall | Runoff |
	Tops	Stubble and litter	0– 15	15– 30	30– 60	Total					
Weeks					%					mm	
4	1	7	19	5	3	27	1	35	65	466	330 ± 30
8	6	5	18	4	4	26	3	40	60	629	410 ± 26
12	20	7	16	3	2	21	1	49	51	883	720 ± 15
16	17	6	15	3	2	20	1	44	56	960	610 ± 78
40	9	4	14	1	1	16	12	41	59	1,053	730 ± 31
L.S.D.† 0.05	8	5	8	5	5	11	12	15‡	15‡	--	--

† The L.S.D. is for comparisons between times within a column.
‡ This is a minimum estimate of error for total ^{15}N since it applies to the soil plus plant recoveries but does not include the error in the runoff sector.

after 8 weeks of growth, with plants accumulating 35%, roots to 30 cm accounting for 16%, and about 47% found in the soil. This high urea recovery contrasts sharply with the results of Henzell (1972), who lost 60% of the N from unlabeled urea in the same field. However, the two budgets are not equivalent, because Henzell's study included grazing animals, higher rates of N (374 kg of N/ha), and a longer time period (208 weeks).

Catchpoole (1975) traced ^{15}N in a rhodesgrass pasture using 21-cm diam cylinders driven 60 cm into a clay-textured soil containing 0.33% N. The soil had a pH of 5.9 and was on a 5 to 8% slope. The monthly ^{15}N budgets (Table 8) indicated large early losses, with total recoveries averaging about 42%. Others have also noted the rapid initial transformation of ^{15}N (Jansson, 1963; Patrick & Reddy, 1976; Riga et al., 1980). These early losses were attributed to leaching and denitrification as enhanced by the rainfall pattern (Table 8). Low crop uptake (only 7%) during the first 4 weeks also contributed to the N losses by leaving a large mineral N pool in the heavy-textured soil. Early leaching losses were evidenced by ^{15}N enrichment in the 75- to 125-cm depth. A supplementary gas lysimeter experiment was also conducted with field-moist cores, but gaseous ^{15}N evolution did not reach detectable levels. Cores with higher water levels, however, did show losses over 4 weeks. The percentage ^{15}N losses, as NH_3 plus N_2, were 4, 12 and 28% for cores at field capacity, with the water table at 5 to 8 cm below the soil surface, and with the water table at the soil surface, respectively. Despite these losses, the grass in the gas lysimeters utilized nearly 50% of the ^{15}N and was thus a much better competitor for mineral N than the field crop. Surface runoff losses were quite variable (note L.S.D. in Table 8), and one plot lost 47% of its ^{15}N via runoff. Excluding this plot, about 2% was lost in runoff. This loss is relatively low considering that over one half of the precipitation ran off the plots. However, from the standpoint of ^{15}N balance, the runoff losses were significant and should be determined for such soil-plant systems. These data illustrate the potential for large ^{15}N losses under wet conditions. Once N is applied, it is subject to a variety of loss processes that are controlled by several interacting factors that are difficult to document under field conditions and equally difficult to reproduce under controlled conditions.

Coastal bermudagrass (*Cynodon dactylon* L.) was grown in two 61-cm diam cylinders by Kissel and Smith (1978) to trace the fate of 560 kg of N/ha applied as Ca ($^{15}NO_3$)$_2$. The field study was conducted on a Houston black clay (a Vertisol) with a pH of about 7.8. Coastal bermudagrass tops accounted for 49% of the fertilizer N; and about 34% was in organic forms (roots plus soil organic N). Only 6% was in the mineral form, and overall recoveries totaled about 90%. Thus, denitrification losses were not > 10%. The large immobilization observed in this study was possibly caused by the recent plowing of the site, which stimulated microbial turnover, although much of the soil organic N could have originated from nonrecovered roots since there was a good association between harvested root N and soil organic N. A residual study the next year showed that 17% of the immobilized ^{15}N mineralized. Keeney and MacGregor (1978) used very small plastic cylinders, 7.5 by 10 cm deep, in a short-term study of the fate of ^{15}N-labeled urea applied at 300 kg of N/ha (to simulate excreted urine). The study was conducted on an acid (pH 4.7) silt loam soil growing ryegrass (*Lolium perenne* L.) and white clover (*Trifolium repens* L.) and included ($^{15}NH_4$)$_2$SO$_4$ and K$^{15}NO_3$ for comparative purposes. Samples obtained after 1 day gave complete recoveries for all sources, but after 22 days the ^{15}N recoveries were 89, 61, and 70% for urea, (NH$_4$)$_2$SO$_4$, and KNO$_3$, respectively. Some NO$_3^-$ was leached beyond 30 cm early in the experiment after 23 mm of rainfall, but recovery from urea was much higher than might be expected from an earlier study (Simpson, 1968). The apparent lack of NH$_3$ volatilization may have been due to NH$_3$ absorption by the plant canopy and subsequent recycling (Denmead et al., 1976). Urea showed the largest tendency to accumulate in the soil and plant organic forms, whereas the NO$_3^-$ source showed the least tendency, and the NH$_4^+$ source was intermediate.

Clark (1977) did not present a complete ^{15}N budget for his 5-year study of the internal N cycling in shortgrass prairie. However, no perceptible losses occurred from the soil-plant system after the first sampling. Changes in the distribution of ^{15}N in the plant and soil compartments suggested that the N requirement of blue grama (*Bouteloua gracilis* L.) was met by internal translocation, rapid mineralization of organic N from plant and microbial origin, and a slow mineralization of more stable organic N compounds. All three pathways are important mechanisms in shunting N back to the plant and thus form a rather tight internal N cycle under native conditions.

The field examples cited above provide rather precise estimates of ^{15}N budgets in grassland systems under specific conditions. Such estimates have necessarily been made on a small scale due to the cost of ^{15}N-enriched fertilizers. However, an important component of grassland systems—grazing animals—is not taken into account. In nonlabeled experiments, Power (1977) reported that 3% of the applied fertilizer N (90 kg of N ha^{-1} year^{-1} for 11 years) was removed by beef animals, and 18% was unaccounted for. At least part of the 18% loss can be attributed to animals via NH$_3$ volatilization from excreted N (Woodmansee, 1978, 1979). Woodmansee et al. (1978) have also discussed other factors influencing the N budget of grassland ecosystems.

6. ORGANIC AMENDMENTS

Partial N balances with organic amendments are frequently reported because of the long-standing interest in animal manures and the more recent revival of interest in industrial and municipal by-products. Additional complications arise with organic N sources due to increased C availability, variability arising from initial composition differences, composition changes during storage, spatial variability produced by nonuniform application, changes in soil bulk density, and the increased importance of NH_3 volatilization as an avenue of gaseous N loss. In addition, it has not been feasible to utilize ^{15}N with these N sources because of nonuniform labeling and cost limitations. These N balances have therefore utilized the nonlabeled approach and have usually involved large single doses of organic N or small rates over long periods of time. It is a frequent observation that these N balances have rather large deficits (Wallingford et al., 1975; Pratt et al., 1976a, b).

Pratt et al. (1976a, b) reported a N balance for dairy manure additions in the Chino-Corona Basin of California on a well-drained Hanford sandy loam with no textural discontinuities. The manure treatments were soil incorporated just before establishing winter barley or summer sudangrass. At the end of the study, soil samples were taken to 4.5 m at 10 locations within each plot, which allowed estimation of NO_3^- and Cl contents to within 20% of the true mean (Pratt et al., 1976c). A Cl balance was constructed from the manure and irrigation inputs and from the crop removals. The Cl data allowed an estimation of the fraction of the applied water that leached. Nitrogen leaching losses were estimated from this leaching fraction and the NO_3^- concentrations below the root zone. Both types of manure gave increased crop N removals (Table 9) with increasing manure rates, except for the highest rates, where salts affected crop growth. Crop N removal from 21 metric tons/ha of liquid manure was about equal to that removed from 79 metric tons/ha of dry manure. Irrigation had little effect on crop N removals, but it increased leaching losses. About 180 kg of N ha^{-1} year^{-1} was lost from the check plots, which can be attributed to mineralization of soil organic N. Manured plots generally increased their soil organic N content, but dry manure treatments had the greatest increases per unit of applied N. The net effect of the remaining N cycle processes (Table 9) indicated a net N loss, which is most likely associated with denitrification and NH_3 volatilization. These losses increased as manure rates increased, and liquid manure had greater losses than dry manure at comparable loading rates. At the low and moderate rates of dry manure, these losses were very low (1–11%). Pomares-Garcia and Pratt (1978) observed similar low losses (about 9%) from pots treated with dry manure and ^{15}N in a greenhouse experiment using a light-textured soil. The N loss data do not show an irrigation effect, indicating that water input per se was of secondary importance to available

Table 9—Average annual N balance for manured plots cropped to winter barley and summer sudangrass after four growing seasons (Pratt et al., 1976a).

Treatment		Added in irrigation and manure	Removed in forage crops	Estimated N loss by leaching§	Change in soil organic N¶	Net effect of other N processes#	Net effect, as % of N added
Manure†	Rate‡						
		——————— kg of N ha⁻¹ year⁻¹ ———————					%
		114 cm of irrigation††					
0	0	85	212	53	−180	0‡‡	--
Dry	40	698	379	135	154	30 ± 20§§	4
Dry	79	1,356	460	314	435	147 ± 20	11
Dry	158	2,627	350	519	1,010	748 ± 20	28
Liquid	21	1,007	468 ·	231	74	234 ± 20	23
Liquid	42	1,928	465	483	191	789 ± 20	41
		142 cm of irrigation††					
0	0	107	175	111	−179	0‡‡	--
Dry	40	743	362	260	110	11 ± 20§§	1
Dry	79	1,378	419	425	436	98 ± 20	7
Dry	158	2,650	317	548	1,118	667 ± 20	25
Liquid	21	1,029	414	354	−1	262 ± 20	25
Liquid	42	1,951	406	578	142	825 ± 20	42

† Dry = open air storage from dairy corral; liquid = fresh manure collected daily.
‡ Metric tons per hectare on an air-dry basis.
§ Estimated from Cl balance on same plots (see text).
¶ Soil N at end of experiment less that estimated at start, assuming values in next column are zero for check plots (see footnote below). Positive values are increases.
# This is the net difference between N losses of denitrification, NH_3 volatilization, and runoff; and N inputs in rainfall, nonsymbiotic fixation, and the change in soil mineral N. Positive data are net losses, see footnote ‡‡ for check plots.
†† Avg. yearly addition; 114 cm ≅ evapotranspiration; 142 cm ≅ 1.33 × evapotranspiration.
‡‡ The net effects on check plots were assumed to be zero, i.e., runoff and gaseous losses equaled inputs from rainfall, nonsymbiotic fixation, and change in soil mineral N.
§§ Approximate SE of mean, estimated from the N loss × irrigation interaction.

C in influencing gaseous N loss in this light-textured soil. These data show the large influence of available C and N on subsequent N transformations. The liquid manure was more biologically active per unit of applied N, resulting in greater mineralization and crop uptake, greater leaching, greater gaseous losses, and less soil organic N accumulation. The less active dry manure offered a mirror image of these transformations. Gaseous losses were small at modest manure rates but increased to major proportions at high loading rates even in this well-drained soil.

Rolston et al. (1978, 1979) reported a field ¹⁵N balance on uncropped 1-m² plots that had received 300 kg of N/ha as Ca ($^{15}NO_3$)$_2$ and were maintained at two water levels (about −0.01 and −0.06 bar). The well-drained Yolo loam soil received either a manure (34 metric tons/ha) or a control (no manure) pretreatment. The plots were heavily instrumented to monitor soil temperature, soil water, and gaseous evolution, and the soil was also core sampled at the end of the 115-day experiment. The summer experiment (23°C soil temperature) showed that the plot receiving high water and manure lost 72 to 78% of the ^{15}N as N$_2$ plus N$_2$O, as determined by direct measurement and mass balance difference, respectively. The remaining labeled N was accounted for mainly through leaching (about 13%) and as organic forms (about 8%). The other water treatment receiving manure lost 16 to 27% of the ^{15}N as N$_2$O plus N$_2$ for the direct and mass balance estimates, respectively. The remaining ^{15}N was predominately in mineral form (about 53%). About 21% occurred in organic forms, and no leaching losses were observed. By comparison, the control plots lost very little gaseous ^{15}N (averaged about 4%), but the high-water control lost about 85% of its ^{15}N through leaching; the other control retained about 87% of the ^{15}N in mineral forms. About 8% of the ^{15}N was incorporated into organic forms in both control plots. The ^{15}N balance for the manured plots in the winter experiment (8°C soil temperature) showed much smaller gaseous N losses, averaging about 11% by direct measurement and 16% by mass balance across water levels. Leaching accounted for virtually all the remaining ^{15}N in the high-water plot, whereas residual mineral N dominated the ^{15}N account in the other water treatment. The uncropped control plots showed virtually no directly measured gaseous loss in the winter, with leaching removing 97% of the ^{15}N on the wettest plot and mineral N accounting for the major portion of the ^{15}N on the low water control plot. These data indicate the large gaseous N loss that can occur with manures and the interacting biological and physical factors (available C, water, temperature) that affect ^{15}N transformations and the final ^{15}N budget.

7. FOREST ECOSYSTEMS

Studies of N budgets in forest ecosystems are made difficult by the longevity and size of forest trees; nevertheless, several reports in which both conventional and isotopic methods were used have provided considerable information on the major pools of N, nutrient cycling, and N balance in such systems (e.g., Gessel et al., 1973; Grier et al., 1974; Knowles, 1975; Miller et al., 1976; Nömmik, 1973; Overrein, 1972).

Gessel et al. (1973) reported results of a 20-year study of N regimes in Douglas-fir (*Pseudostuga taxifolia* L.) forests of the Pacific Northwest under a variety of soil and climatic conditions. A summary of N distributions showed that about 85% of the total N in the ecosystem was in the soil, 10% was in vegetation, and 5% was in the forest floor litter. Annual transfer of N between components of the ecosystem at the Thompson research area was estimated to be as follows, in kg of N/ha: inputs (mainly precipitation), 1.1; uptake by the forest, 38.8; total return to the forest floor, 16.4; leaching from forest floor, 4.8; and leaching beyond the rooting zone, 0.6. This represents an essentially closed system, since leaching and gaseous losses were minimal due to the low inputs and small amounts of mineralized N present at any given time. There is an apparent redistribution of N in the system as the forest develops, with a gradual increased accumulation of N in the forest and forest floor, and a subsequent decrease in soil N content. This indicates that as the forest matures, it receives an increasing portion of N from the forest floor and a decreasing portion from the soil. Gessel et al.

(1973) estimated the annual needs of an actively growing Douglas-fir forest to be about 50 to 70 kg of N/ha and indicated that if total soil N falls below 5,000 kg/ha, the forest may not have an adequate N supply.

Bormann et al. (1977) reported a 13-year study of the N budget for an aggrading hardwood forest on a northeastern U.S. watershed. The main inputs were 6.5 kg of N/ha from precipitation and an estimated 14.2 kg of N/ha from biological fixation. Of the 83.5 kg of N/ha added to the inorganic pool within the ecosystem, only about 5% was lost from the system in streamflow. As in the Douglas-fir ecosystem, recycling is involved in maintaining a relatively small mobile pool of N within the living biomass of the hardwood forest.

Early use of ^{15}N in forest ecosystems has been reviewed by Hauck (1968), and more recent work has been described by Knowles (1975). Overrein (1968) began a series of ^{15}N tests to determine leaching and volatilization losses from ^{15}N-labeled NO_3^-, NH_4^+, and urea during periods up to 40 months. Lysimeters were constructed using intact 30-cm forest soil cores 40 cm deep, in plastic cylinders provided with covers to permit gas sampling. After 12 weeks, the total ^{15}N leaching losses at the 250-kg rate were 2, 22, and 92% for urea, NH_4^+, and NO_3^-, respectively (Overrein, 1969). Ammonia volatilization losses from NH_4^+ and NO_3^- sources were not significant at any rate of application, and there was no evidence of denitrification from any source at rates of 250 kg of N/ha or less. After 40 months, the soil profile was sampled, and essentially complete recovery of ^{15}N was attained with NH_4^+ and NO_3^- sources applied at 100- and 250-kg N/ha rates, whereas unaccounted losses from urea were 15 and 20% at the respective rates (Table 10). Leaching losses from urea were negligible at the 100-kg N/ha rate but increased to 7% with 250 kg of N/ha, whereas comparable leaching losses from the NO_3^- source ranged from 80 to 90%. The highest internal recovery of ^{15}N within this system was from the NH_4^+ source at the 100-kg N/ha rate, where only 3% was lost by leaching after 40 months. Usually 90% or more of the ^{15}N remaining in the soil and humus layers after 40 months was in forms not extractable with $4.0N$ or $1.0N$ KCl.

Other experiments have shown that ^{15}N from urea accumulates mainly in the organic layers, whereas that from NH_4^+ resides mainly in the mineral

Table 10—^{15}N balance after a 40-month lysimeter study of a forest soil with three N sources applied at 100 and 250 kg of N/ha (Overrein, 1972).

	N source and rate (kg of N/ha)					
	NH_4^+-N		NO_3^--N		Urea	
Sources of ^{15}N	100	250	100	250	100	250
	————————————— % —————————————					
Gaseous losses, 12 weeks	0	0	0	0	<1	<1
Plant material	5	16	2	1	13	9
Litter and humic layer	29	34	9	4	68	53
Mineral soil	66	27	6	8	5	11
Leachate	3	17	82	90	<1	7
Total recovery	103	94	99	103	86	80

soil of forest ecosystems (Knowles, 1975). These data agree with the data in Table 10 by Overrein (1972). Other workers generally have not measured gaseous and leaching losses, and thus the N balance is less complete.

Current ^{15}N budgets for forest ecosystems are not entirely satisfactory since microplot measurements have been used in a system that is exceptionally macroscale. In view of the current interest in forest fertilization to meet increased demands for forest products, it is evident that a need exists for macroplot experiments with ^{15}N.

E. Summary

The preceding labeled and conventional N balances have both common and contrasting general features. Some of the common qualitative features are slow mineralization of ^{15}N once it is incorporated into organic forms, preferred use of NH_4^+ by soil microbes, rapid transformation of mineral ^{15}N soon after addition, accumulation and/or transport of N out of the soil-crop system when N rates exceed crop assimilation capacity, significant incorporation of ^{15}N fertilizer into the soil organic N pool, significant loss of N in gaseous forms, and the rather efficient use of N by adapted crops that are fertilized with moderate rates of N supplied in harmony with the crops' demand for N. When attempts are made to quantitatively summarize these data for the N cycle processes, the contrasting features become apparent. One approach is to consider only the ranges in recovery as reported by Kundler (1970): incorporation into soil organic matter, 10 to 40%; recovery by the crop, 30 to 70%; losses in leachate, 5 to 10%; and gaseous losses, 10 to 30%. Other reported ranges (Hauck, 1973) were 50 to 75% for crop recovery and 0 to 40% for denitrification losses, although hypothetical values were 15% for denitrification losses, 55% for crop removal, and 30% for immobilization (Hauck, 1971). These preceding values are useful for generalizations but can rarely be applied to specific agricultural systems. This should not come as a surprise if one considers the complex N transformations occurring in nature. The final N balance sheet is the product of many physical-chemical-biological processes, all interacting with the environment and each other over time. Seemingly innocuous changes in the environment often have major impacts on N transformations, and conversely, rather large environmental changes often have negligible impact. Several of the preceding N balances contain these anomalies and serve to emphasize the inadequacy of reporting averages, medians, or ranges for the N transformations in different agricultural systems or even for the same system over different times and environmental conditions.

Future challenges to soil scientists engaged in N balance research will most likely center on a more detailed description of the systems' physical, chemical, microbial, and biological properties over rather short time periods to better understand the controlling factors governing the fate of N. This approach provides insight into the proper way to combine these factors into a broader general framework to expand our understanding and increase our ability to predict and manage the fate of N under other conditions.

VI. APPLICATIONS OF N BALANCES TO SOIL AND CROP PROBLEMS

A. General Aspects

The N balances discussed above resulted from investigation of N transformations from a research point of view, i.e., N pathways, N fluxes, and the size of various N pools, using small-scale experimental units. Others investigate N management on a field scale. The latter covers a wide range of complexity, from those constructed with actual field data collected within the study area to those constructed from casual estimates of various transport processes. We will not attempt to review all of these N budgets but rather will indicate some of the recent work of this type covering a range of production systems. Bouldin et al. (1975) and Johnson et al. (1976) have reported a N balance for a rural watershed in upstate New York that was devoted primarily to dairy production. Schuman et al. (1975) studied N losses from two watersheds under continuous corn in southwest Iowa. Nitrogen budgets involving vegetable crops have been published by Cameron et al. (1978), Meisinger (1976), Lund (1979), and Saffigna et al. (1977). These N studies have provided valuable insight into the hydrological and geological connections between agricultural N and the larger scale N cycle of the study area.

B. Nitrogen Budgets Applied to Environmental Problems

Mass balance investigations are a natural approach in studying environmental aspects of the agricultural N cycle because they emphasize a basic principle that everything is interconnected in nature. Nitrogen balance principles have been used extensively in California for a number of years to investigate various soil- and crop-related problems. This section will review these studies because they represent a range of N balance methodologies applied to diverse agronomic conditions.

Early investigations in southern California employed conventional research approaches, such as the lysimeter studies of Chapman et al. (1949), which were begun in 1933. These studies investigated N transformations under four different field crop systems involving legumes and nonlegumes, each at three levels of N fertilization. Although the specific results of these studies can be faulted for their lack of direct applicability to field conditions, nonreplication, and artificially deep surface soil, they did establish certain base line data and some relevant general conclusions, such as the quantity of N added in precipitation, the effect of N fertilization on reducing symbiotic N_2 fixation, the ready availability of residual soil NO_3^-, and the ease with which soil NO_3^- is lost in drainage water. The major conclusion was that "the greatest N economy will be achieved where available N is maintained at the lowest possible point consistent with satisfactory crop yields." These studies were extended over 20 years, and the long-term results showed a lack of significant gaseous N loss under the minimal irrigation scheme and well-drained soil conditions of the study (Broadbent & Chapman, 1950; Pratt et al., 1960; Pratt & Chapman, 1961).

In a study of the N balance for the Grover City-Arroyo Grande Basin by Stout and Burau (1967), deep soil samples were collected from various land use areas representing both intensive and extensive agriculture and an area receiving sewage effluent. These preliminary samplings found elevated NO_3^- levels under lands used for both agriculture and human waste disposal. Stout and Burau then combined the N balance approach with the assumption of steady-state soil organic matter levels (Jenny, 1941) to estimate the quantity of N leached from various land use areas. A N budget applied to a steady-state soil organic matter system results in several simplifying conditions because N mineralization is balanced by organic N returns from crop residues and immobilization. This simplifying approximation has also been used by other investigators (Adriano et al., 1972b; Cameron et al., 1978; Fried et al., 1976; Meisinger, 1976; Pratt et al., 1972; Tanji et al., 1977) to reduce the soil N compartment to an input-output relation, i.e.:

$$N_{inputs} - N_{outputs} = \text{Change in soil N storage.}$$

At steady-state the change in soil N equals zero, so that inputs equal outputs. However, a word of warning should be sounded about this assumption. Steady-state conditions often require years to achieve, as is evident from the Morrow plot data (Fig. 2). If a steady-state condition is incorrectly assumed and the total soil N pool decreases only 0.001% annually in the surface 30 cm, a net annual production of about 40 kg of N/ha would occur. This amount is comparable with N losses commonly observed for denitrification or leaching; yet, it would be virtually impossible to detect experimentally without a long-term study. Therefore, imposing a steady-state condition on a soil N balance study should be done only after *careful justification*. The steady-state approximations will be approached most closely when one management system has been imposed over many years and the time increment for the N budget also covers several years. Fried et al. (1976) have discussed the application of this steady-state concept in some detail. The study of Stout and Burau (1967) ushered in an era of extensive N balance investigations using deep soil samples that were aimed at documenting the N leaching losses more accurately.

A 3-year N balance was applied to a citrus watershed (Bingham et al., 1971) that contained shallow (about 75 cm deep), light-textured (sandy loam), well-drained soils. The entire area was underlain by impervious bedrock so that water moved through the soil to the bedrock and then moved laterally until it surfaced into streams. Bingham and colleagues viewed the watershed much like a macrolysimeter (Bingham et al., 1971; Davis & Grass, 1966; Grass et al., 1966). Annual rainfall was about 25 cm, which fell irregularly during the nongrowing season between November and April. Irrigation is therefore necessary, and about 90 cm of water was applied annually. The hydrological data summary (Fig. 6) indicates that water inputs exceed drainage all year, with the difference between them

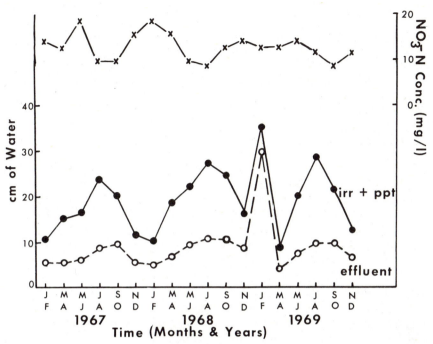

Fig. 6—Average bimonthly water inputs, irrigation plus precipitation (irr + ppt), water outputs (effluent), and NO₃⁻-N concentrations from a 389-ha citrus watershed over 3 years
(Bingham et al., 1971).

(i.e., the estimated evapotranspiration) being greatest in summer and least
in winter. The large rainfall event in early 1969 is also quite evident. The effluent curve mimics the water input curve for these shallow soils. Nitrate N
levels in the effluent averaged about 12.7 mg of NO_3^--N/liter and indicate a
significant NO_3^- enrichment since incoming irrigation water contained
<0.5 mg of NO_3^--N/liter. The common fertilizer practice was to apply
about 145 kg of N/ha through the irrigation system. The N budget for these
citrus groves was estimated (in kg of N/ha) as follows: fertilizer, 145;
rainfall plus irrigation plus NH_3 deposition, 5; harvested crop removal, 50
(33% of inputs); effluent NO_3^--N, 65 (43% of inputs); and unaccounted for
losses, 35 (24% of inputs). Unaccounted for N represents denitrification,
NH_3 volatilization, and any net immobilization into the soil organic N or
the standing citrus crop. The annual water budget showed a 115-cm input
that could be partitioned into 62 cm of evapotranspiration, 20 cm of runoff,
and 33 cm of leaching losses. If 65 kg of N/ha are dissolved in 33 cm of
drainage, the NO_3^--N concentration would be about 19.7 ppm. The NO_3^--N
concentration of the winter drainage approaches this value on several occasions, indicating that the source of N in these winter months is predominantly soil drainage, whereas other periods represent some combination of drainage, surface runoff, and irrigation return flow. A Cl balance in
the same watershed also suggests salt leaching during the winter and accumulation during the summer (Bingham et al., 1971; Grass et al., 1966).

These data indicated a need for N fertilizer management practices that minimize leaching losses but maintain production. Practices such as foliar fertilization and split soil applications at N rates that are consistent with crop needs, as determined through leaf analysis, have been suggested (Bingham et al., 1971; Embleton & Jones, 1974, 1978; Jones & Embleton, 1969). Embleton and Jones (1978) have demonstrated the value of foliar applications and moderate soil fertilization in reducing the NO_3^- pollution potential under citrus.

Several studies have combined a Cl balance and a N balance, with Cl serving as a microbially inert tracer for NO_3^--N. The Cl and NO_3^--N concentrations of deep soil cores can then be used to estimate the mass of NO_3^--N leached below the root zone, provided certain conditions are met. Pratt et al. (1978) have reviewed this approach along with its attendant assumptions, and Olson (1978) and Stewart (1978) have critiqued this method. These methods offer alternative estimations of NO_3^--N leaching and improvement of field N balance calculations, providing the underlying assumptions are met and representative soil samples can be collected. Pratt et al. (1976c) and Rible et al. (1976) have discussed the large spatial variability of NO_3^--N concentrations below the root zone, which presents a major problem when these methods are applied even on an experimental plot basis. Highly accurate N balances should not be expected with these approaches without an intensive sampling program.

Pratt and Adriano (1973) have summarized their earlier N balance work that showed that soil drainage characteristics, N input rate, and crop removal all influence the fate of applied N (Table 11). For example, with freely draining soils at moderate N rates (< 160 kg of N/ha) crop N removal exerts a major influence, with the excess N mainly appearing below the root zone. At higher N rates in freely drained soils, the crop effect is reduced, and the excess N appears below the root zone as well as in gaseous losses (denitrification and NH_3 losses). In soils with restricted drainage, the crop influence is again reduced, and gaseous losses dominate as the major pathway in removing excess N. Lund et al. (1974) have continued to study the soil drainage effects and have shown that high subsoil clay contents, hardpans, and textural discontinuities are inversely related to the NO_3^--N content below the root zone. These profile characteristics can affect both denitrification and drainage volume, although the denitrification aspects probably played the major role in Lund's study area, which received most of its N from manure.

A major effort was also made by Rible et al. (1973, 1974) to extend the combined Cl and N balance approaches to over 90 locations in California with a wide range of soils, crops, and N management systems. This effort involved collecting soil samples to 15 m, determining NO_3^- and Cl contents, classifying subsoil texture, preparing water and N management histories at each site, and applying the mass balance equations developed by Pratt et al. (1972). The sites included vegetable crops, field crops, tree crops, and nonagricultural areas. The NO_3^--N concentration of the soil water samples was above 20 mg/liter in about 70% of the sites sampled (Rible et al., 1979). This study indicated several problems when research methodologies were

Table 11—Summary of Cl and N balance data reported by Pratt and Adriano (1973), Pratt et al. (1972), and Adriano et al. (1972a, b) from deep soil cores.

Soil drainage[†]	Input[‡]	Crop removal[§]	Leached from root zone[¶]	Unaccounted for N[#]	Crop
			kg of N ha⁻¹ year⁻¹		
FD	117	39 (33)	80 (67)	−2 (−)	Citrus
FD	161	40 (25)	130 (80)	−9 (−)	Citrus
FD	137	90 (66)	35 (25)	12 (9)	Asparagus
FD	333	40 (12)	140 (41)	155 (47)	Citrus
FD	437	67 (15)	310 (71)	60 (14)	Strawberries, barley
FD	485	140 (29)	120 (25)	225 (46)	Asparagus
RD	171	40 (23)	40 (23)	90 (53)	Citrus
RD	360	121 (34)	50 (14)	190 (53)	Potatoes, cereal crops

† FD = freely drained, open porous soil; RD = restricted drainage, clay layer, hardpan, etc.

‡ Inputs include fertilizer additions and irrigation water.

§ Values in parentheses are the output as a percentage of the input.

¶ Estimated from Cl and N data from deep cores as per Pratt et al. (1978). These data are probably accurate to ±20% for the kg of N ha⁻¹ year⁻¹ data.

Net effect of denitrification losses, NH_3 volatilization, and changes in the soil organic N pool (net immobilization being positive, net mineralization being negative).

applied to grower conditions, e.g., uncertainties in site history, spatial variability of soil characteristics, geological sources of NO_3^- unrelated to current land use, nonsteady-state soil organic matter conditions, and spatial variability of NO_3^--N and Cl in deep soil cores. The spatial variability was particularly troublesome, e.g., in only 20% of the sites was the true NO_3^--N concentration mean estimated to be within 20% of the observed sample mean. Because of these uncertainties in data quality, the association between N leaching estimates and N inputs, water inputs, and various soil properties was poorly defined. The data from the most reliable sites, however, showed a significant direct relationship between N leached and N inputs, but the correlation between NO_3^--N concentration and N inputs was not significant. Rible and co-workers also found that soil profile characteristics had an impact on both NO_3^--N concentration and the quantity of N leached.

Lund (1979) reported studies involving a combined Cl and N balance on seven vegetable crop sites in the Santa Maria Valley. Eight or nine 100-m² areas were studied intensively within each field, and careful samples were taken of the soils, crops, and irrigation waters at each site. Records of each grower's fertilization and irrigation practices were also collected. These field N balances differed from Rible's earlier work in the greater degree of onsite documentation of N and Cl sources and sinks and a more detailed soil characterization. This study again revealed a large spatial variation in soil NO_3^--N and Cl concentrations and also a temporal variation. The final N budgets for all the fields included both positive and negative values in the unaccounted for N. The negative values occurred at sites with legumes (beans) in the crop system and indicated a net mineralization of N from the soil. Data from three of the best documented sites (Table 12) indi-

Table 12—Estimated N budgets from combined Cl and N balances on Mollisols growing vegetable crops (Lund, 1979).

Soil description[†]	Crops grown	Water budget		NO_3-N concentration in soil water	N balance sheet				
		Irrigation	Drainage		Fertilizer N	Irrigation water N	Crop	Leached	Lost[‡]
		— cm —		µg/ml	— kg of N ha^{-1} year^{-1} —				
1.4-3.3% OM, loam to silty clay loam subsoil	Broccoli[§] Cauliflower	88	16	136	665[§]	15	125 (18[¶])	270 (40)	285 (42)
2.2-3.9% OM, sandy loam to silty clay loam subsoil	Artichokes	52	12	73	290	140	50 (12)	95 (22)	285 (66)
1.5-3.2% OM, sandy loam to silty loam subsoil	Lettuce[§] Broccoli Cauliflower Celery	101	64	40	475[§]	145	140 (23)	290 (47)	190 (31)

[†] OM = organic matter.
[‡] Lost = N denitrified, NH₃ volatilization, plus net N immobilization in soil organic matter.
[§] Usually grew two crops per year.
[¶] Numbers in parentheses = percentage of total inputs.

cate large N losses. Only about 17% of the N inputs could be accounted for in the harvested crops, with the remainder being leached (about 38%) or lost through volatile pathways (about 45%). These large gaseous N losses were also supported by direct field denitrification estimates (Ryden & Lund, 1980). The NO_3^--N concentration in the soil water below the root zone was also high and represented the combined effects of water and N management at each site. When all seven sites were considered together, the NO_3^--N concentrations were most strongly related to water drainage, with lower concentrations being associated with higher drainage volumes. This probably resulted from greater dilution and greater denitrification. The relations for the mass of N leached across all seven sites showed that both NO_3^--N concentration and leaching volume were directly associated with N leaching, but the relationships were not strong. The relatively high losses in Table 12 were attributed to high organic matter content, fine textures, and frequent irrigation of the Mollisols.

Labeled and conventional N budgets representing California's large area of annual grassland have been studied with filled lysimeters using different rates and times of N fertilizer applied to grass, legumes, or grass-legume mixtures (Jones et al., 1974, 1977). Fertilizer N recoveries from annual preplant applications of 100 kg of N/ha to pure-grass stands were 33% from crop uptake, 60% from soil plus root recoveries, 10% from leaching losses, and negligible amounts from gaseous losses. Comparable results from delayed (3 months after planting) N additions were 59% from crop uptake, 24% from soil plus root recovery, 3% from leaching losses, and 14% from gaseous losses. Leaching plus gaseous losses from pure legume stands were greater than the above losses (30 and 18% for preplant and delayed applications, respectively). When 500 kg of N/ha was applied to either grasses or legumes, the leaching and gaseous losses increased considerably. A comparison of unfertilized grass-legume mixtures with fertilized (112 kg of N/ha) pure grass stands showed lower leaching losses and substantially greater forage and total N yields with mixtures. These studies have shown that the N needs of annual grasslands can be efficiently met through modest N fertilizer additions or through legume-grass mixtures.

Nitrogen balance principles also contribute a basic foundation to N models. These models have been developed for several levels of sophistication, including simple input-output descriptions of steady-state systems covering long time periods (Fried et al., 1976; Tanji et al., 1977), nonsteady-state models covering annual time steps (Tanji et al., 1979), and the more elaborate simulation models covering time steps of less than a day (Tanji & Mehran, 1979). The most sophisticated models find application as research tools that summarize and describe the complex soil-plant system of experimental plots on a day-to-day basis. The less complex models find use as management tools over large areas and longer time periods. To illustrate the use of one of these models, the total N data from the Kearney experimental site (Broadbent & Carlton, 1978, 1979) were summarized over 3 years using a nonsteady-state description (Tanji et al., 1979). The nonsteady-state condition was utilized because of inorganic N accumulation at the high N rate and the very low organic N status of the soil (about 0.02% N, Broadbent & Carlton, 1978). For the nonsteady-state situation the excess inorganic N can be estimated from the N balance equation:

$$N_{EXIN} = N_F - N_{HC} - N_D + (N_{NA} - N_{OL}) + N_{SON}$$

where

N_{EXIN} = the excess inorganic N of the system;

N_F = the N fertilizer addition;

N_{HC} = the total N in harvested grain, as reported by Broadbent and Carlton (1979);

N_D = the N denitrified, estimated as 22% of the total inorganic N, as per Fig. 5 from the labeled N data of Broadbent and Carlton (1978, 1979);

N_{NA} = the N from natural additions of rainfall, irrigation and fixation, estimated as 8 kg of N/ha from Tanji et al. (1977);

N_{OL} = the N lost in processes such as erosion, NH_3 volatilization, etc., estimated as negligible; and

N_{SON} = the change in soil organic N through mineralization and subsequent return of organic N residues. Mineralization was estimated at 40 kg of N/ha and N residues estimated from the stover N returned as per Broadbent and Carlton (1978, 1979).

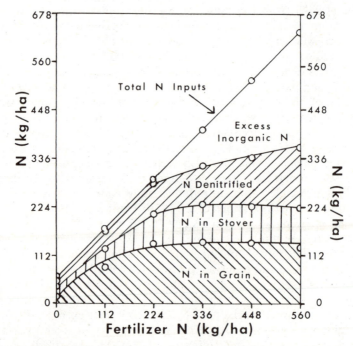

Fig. 7—Average annual distribution of total N inputs on an irrigated sandy loam soil as reported by Broadbent and Carlton (1978, 1979) and summarized by the use of a N model described by Tanji et al. (1979).

The initial inorganic N content of the soil was estimated as 40 kg of N/ha, and the returned residues were mineralized at 4% per year in subsequent years. The fate of the annual N addition, averaged over the 3 years (Fig. 7), clearly illustrates two major conclusions reached by Broadbent and Carlton: (i) maximum N use efficiency can be achieved at the same fertilizer rate that is required to obtain maximum grain yields, and (ii) optimal production of corn is consistent with minimum pollution hazard through careful management of fertilizer and irrigation. These conclusions arise largely from soil NO_3^--N that accumulates with excessive fertilizer N additions; therefore, the key in minimizing this excess inorganic N is to adjust N fertilizer rates to reflect both the crop N requirement and the soils N supply status (N supplied from residual inorganic N and N mineralization).

C. Summary

These studies have attempted to evaluate agriculture's role in the overall N cycle through extensive use of N balances in several different agronomic systems. They were selected to illustrate a wide range of N balance methodologies, including conceptual models, field experimental plots, lysimeter studies, field N balances, and watershed budgets. No single approach is superior to the others, because each has its own purpose and attendant advantages and disadvantages. For example, if long-term effects are of primary interest for a single management system over a large area, a steady-state conceptual model can provide good first estimates of the N transformations with a small resource expenditure. If short-term information is needed on the effects of different N and irrigation systems, then instrumented field plots might be appropriate. If the questions deal with an existing large area N problem, a watershed budget may identify the major sources and sinks of N. The appropriate N balance method for a given problem will thus depend on the questions being investigated, the resources available, the final population to which the data will be applied, and the desired accuracy of the results.

VII. SUMMARY AND CONCLUSIONS

The fate of N in soil-plant systems has held the attention of agricultural scientists for many years. Early investigators could only guess at the possible pathways of N loss, and N balance sheets thus constituted a formidable enigma. As analytical procedures improved and new techniques became available (especially labeled N), this enigma was greatly reduced. The amount of data pertaining to soil N budgets has increased greatly during the past decade. As the area under consideration decreases, N budget precision improves, and ultimately with instrumented microplots and isotopic tracers, highly precise measurements have been attained under a variety of conditions. In fact, several recent studies have been successful in constructing a complete ^{15}N budget by direct measurement and have thus eliminated inexplicable losses. However, it has also been pointed out that the use of isotopic N does not guarantee better results. This is evident from several recent studies in which poor experimental design, inadequate sampling, inadequate chemical analysis, incomplete environmental description, and incomplete statistical summaries have seriously decreased the value of the data.

Every N budget study involves the entire N cycle to some degree. Furthermore, the final N budget of a particular system is the product of many N transformations carried out through physical, chemical, and biological agents interacting with each other and the environment over time. It is therefore impossible to make meaningful quantitative generalizations of the various N cycle processes without carefully describing the complete ecosystem and its interacting environment. This does not mean, however, that N budget studies have not contributed greatly to our general understanding of the fate of N. Several general qualitative conclusions can be drawn from published N balance studies.

External N sources in soil-plant systems vary greatly depending on the degree of N intensiveness. Under native conditions, N inputs from precipitation, gaseous adsorption, and biological fixation dominate. Precipitation plus gaseous adsorption inputs vary from nil to over 50 kg of N ha^{-1} year^{-1}, depending on the location as related to nearby sources of N. It appears that N inputs from these two sources have increased over the past 50 to 75 years. Likewise, nonsymbiotic fixation can supply anywhere from nil to 100 kg of N ha^{-1} year^{-1} depending on the amount of C, sunlight, and nutrients. But whatever the specific value of these N inputs, it is apparent that they can effectively supply modest levels of N over long periods of time for a particular forest, lake, or agricultural system. As the demand for N output increases, the N inputs should also increase, if the level of N within the system is to be maintained. These additional inputs might be accomplished through biological fixation and the addition of crop residues, animal manures, or fertilizer N. Obviously the quantities added will depend on the rates of fertilizer or manure and the inclusion of legumes in the cropping system. These rates and cropping system decisions are based on economic, agronomic, and social considerations. Over the past 40 years, the United States has met the increasing demand for N outputs largely through increased mineral fertilizer N applications. Accompanying this expansion has been a decreased dependence on biological fixation and use of animal wastes. However, the serious energy situation facing modern agriculture will probably slow this trend so that biologically fixed N and by-product N will be utilized more effectively.

Nitrogen budgets constructed for smaller scale units (fields and experimental plots) have clearly shown that the major mechanisms by which N is removed from soils are crop uptake, leaching, and gaseous losses. Erosion losses are usually small compared with the above outputs, but they can be significant under certain conditions. The magnitude of leaching losses will vary from nil to over 50%, depending on the soil NO_3^--N content, water content, rooting depth, water addition, and pattern of water movement (macropore vs. complete displacement). Leaching losses are usually most serious during the cool winter and spring months in humid climates, although significant losses are also possible from smaller water additions moving through macropores, especially for shallow-rooted crops. Nitrogen losses via denitrification also vary greatly depending on NO_3^--N level, available C, temperature, and oxygen or moisture status. The magnitude of these losses is quite unpredictable. Large N deficits have been observed after short periods of high water content or after prolonged conditions of field capacity water contents; yet other studies have reported little or no gaseous loss under seemingly ideal denitrification conditions. The unpredictable nature of these losses has not been adequately explained and illustrates our incomplete understanding of the interacting factors that trigger large denitrification events or that combine to produce low level denitrification losses in macroscopically well-aerated soil. Ammonia volatilization losses have been shown to be quite large for surface applications of compounds containing NH_4^+-N, especially at pH values above 7 and when weather conditions favor drying. Soil incorporation usually greatly reduces such losses. Studies with labeled N have allowed soil scientists to follow the internal transformation of fertilizer N. Such studies have revealed the rapid stabilization of fertilizer N once it is converted to organic forms and the preferential incorporation of NH_4^+-N over NO_3^--N into soil organic N.

Nitrogen balance studies have also shown that leaching and gaseous N losses are greatest when N inputs exceed crop assimilation capacities. This situation leaves excess NO_3^--N in the soil and permits leaching or gaseous loss after the growing season. These N losses can affect the environment beyond the agricultural sector, emphasizing the fact that the soil-plant N cycle is intimately connected to the surrounding large-scale N cycle.

This discussion has emphasized the loss mechanisms of soil N, but N balance studies have also clearly shown that N can be used quite effectively by crops. When N inputs do not exceed crop assimilation capacity and the N is applied in phase with crop demand, the N is utilized very efficiently.

In the final analysis, soil N budget principles provide a sound basis for estimating N gains, losses, and transformations for any agricultural management system. Understanding these processes and the factors influencing them will allow maximum efficiency in the utilization of N resources and concurrently minimize hazards associated with N loss.

LITERATURE CITED

Adriano, D. C., P. F. Pratt, and F. H. Takatori. 1972a. Nitrate in the unsaturated zone of an alluvial soil in relation to fertilizer nitrogen rate and irrigation level. J. Environ. Qual. 1: 418–422.

Adriano, D. C., F. H. Takatori, P. F. Pratt, and O. A. Lorenz. 1972b. Soil nitrogen balance in selected row crop sites in Southern California. J. Environ. Qual. 1:279–283.

Albrecht, W. A. 1920. Symbiotic nitrogen fixation as influenced by the nitrogen in the soil. Soil Sci. 9:275–327.

Allison, F. E. 1955. The enigma of soil nitrogen balance sheets. Adv. Agron. 7:213–250.

Allison, F. E. 1965. Evaluation of incoming and outgoing processes that affect soil nitrogen. *In* W. V. Bartholomew and F. E. Clark (ed.) Soil nitrogen. Agronomy 10:573–606. Am. Soc. of Agron., Madison, Wis.

Allison, F. E. 1966. The fate of nitrogen applied to soils. Adv. Agron. 18:219–258.

Allison, F. E. 1973. Soil organic matter and its role in crop production. Elsevier Scientific Publishing Co., New York.

Axley, J. H., and J. O. Legg. 1960. Ammonium fixation in soils and the influence of potassium on nitrogen availability from nitrate and ammonium sources. Soil Sci. 90:151–156.

Ayers, R. S., and R. L. Branson. 1973. Nitrates in the upper Santa Ana River Basin in relation to groundwater pollution. California Agric. Exp. Stn. Bull. 861.

Badzhov, K., and E. Ikonomova. 1971. ^{15}N for studying nitrogen transformations in soil, nitrogen nutrition of plants, and in assessing available nitrogen in soil. p. 21–32. *In* Nitrogen-15 in soil-plant studies. Proc. of a Research Coordination Meeting, Sofia, Bulgaria. 1–5 Dec. 1969. Int. Atomic Energy Agency, Vienna.

Bergersen, F. J. 1970. The quantitative relationship between nitrogen fixation and the acetylene-reduction assay. Aust. J. Biol. Sci. 23:1015–1025.

Bigeniego, M., R. D. Hauck, and R. A. Olson. 1979. Uptake, translocation and utilization of ^{15}N-depleted fertilizer in irrigated corn. Soil Sci. Soc. Am. J. 43:528–533.

Bingham, F. T., S. Davis, and E. Shade. 1971. Water relations, salt balance, and nitrate leaching losses of a 960-acre citrus watershed. Soil Sci. 112:410–418.

Bormann, F. H., G. E. Likens, and J. M. Melillo. 1977. Nitrogen budget for an aggrading northern hardwood forest ecosystem. Science 196:981–983.

Bouldin, D. R., A. H. Johnson, and D. A. Lauer. 1975. The influence of human activity on the export of phosphorus and nitrate from Fall Creek—transport in streams. p. 61–120. *In* K. S. Porter (ed.) Nitrogen and phosphorus, food production, waste and the environment. Ann Arbor Science Publications, Ann Arbor, Mich.

Bremner, J. M. 1965. Total nitrogen. *In* C. A. Black et al. (ed.) Methods of soil analysis, Part 2. Agronomy 9:1149–1178. Am. Soc. of Agron., Madison, Wis.

Broadbent, F. E. 1980. Residual effects of labeled N in field trials. Agron. J. 72:325–329.

Broadbent, F. E., and A. B. Carlton. 1978. Field trials with isotopically labeled nitrogen fertilizer. p. 1–41. *In* D. R. Nielsen and J. G. MacDonald (ed.) Nitrogen in the environment, Vol. I. Nitrogen behavior in field soil. Academic Press, Inc., New York.

Broadbent, F. E., and A. B. Carlton. 1979. Field trials with isotopes—plant and soil data for Davis and Kearney sites. p. 433–465. *In* P. F. Pratt (principal investigator) Nitrate in effluents from irrigated lands. Final report to the National Science Foundation, University of California, Riverside.

Broadbent, F. E., and H. D. Chapman. 1950. A lysimeter investigation of gains, losses, and balance of salts and plant nutrients in an irrigated soil. Soil Sci. Soc. Am. Proc. 14:261–269.

Broadbent, F. E., and C. Krauter. 1974. Nitrate movement and plant uptake of N in a field soil receiving ^{15}N-enriched fertilizer. p. 236–239. *In* Proc. of the Second Annual NSF-RANN Trace Contaminants Conference, Asilomar, Pacific Grove, Calif. 29–31 Aug. 1974. Univ. of California, Berkeley.

Broadbent, F. E., and D. S. Mikkelsen. 1968. Influence of placement on uptake of tagged nitrogen by rice. Agron. J. 60:674–677.

Broadbent, F. E., R. S. Rauschkolb, K. A. Lewis, and G. Y. Chang. 1980. Spatial variability of nitrogen-15 and total nitrogen in some virgin and cultivated soils. Soil Sci. Soc. Am. J. 44:524–527.

Burns, R. C., and R. W. F. Hardy. 1975. Nitrogen fixation in bacteria and higher plants. Springer-Verlag New York, New York.

Burwell, R. E., D. R. Timmons, and R. F. Holt. 1975. Nutrient transport in surface runoff as influenced by soil cover and seasonal periods. Soil Sci. Soc. Am. Proc. 39:523–528.

Cameron, D. R., R. DeJong, and C. Chang. 1978. Nitrogen inputs and losses in tobacco, bean and potato fields in a sandy loam watershed. J. Environ. Qual. 7:545–550.

Campbell, C. A., D. R. Cameron, W. Nicholaichuk, and H. R. Davidson. 1977a. Effects of fertilizer N and soil moisture on growth, N content, and moisture use by spring wheat. Can. J. Soil Sci. 57:289–310.

Campbell, C. A., H. R. Davidson, and F. G. Warder. 1977b. Effects of fertilizer nitrogen and soil moisture on yield, yield components, protein content and N accumulation in the aboveground parts of spring wheat. Can. J. Soil Sci. 57:311–327.

Campbell, C. A., and E. A. Paul. 1978. Effects of fertilizer N and soil moisture on mineralization, N recovery and A-values, under spring wheat grown in small lysimeters. Can. J. Soil Sci. 58:39–51.

Carlton, A. B., and A. A. R. Hafez. 1973. N-14 nitrogen fertilizer tracing experiments in the San Joaquin Valley (1972–1973). Calif. Agric. 27(11):10–13.

Carter, J. N., O. L. Bennett, and R. W. Pearson. 1967. Recovery of fertilizer nitrogen under field conditions using nitrogen-15. Soil Sci. Soc. Am. Proc. 31:50–56.

Catchpoole, V. R. 1975. Pathways for losses of fertilizer nitrogen from a Rhodes Grass pasture in Southeastern Queensland. Aust. J. Agric. Res. 26:259–268.

Chapman, H. D., G. F. Liebig, and D. S. Rayner. 1949. A lysimeter investigation of nitrogen gains and losses under various systems of covercropping and fertilization, and a discussion of error sources. Hilgardia 19:57–128.

Chichester, F. W. 1977. Effects of increased fertilizer rates on nitrogen content of runoff and percolate from monolith lysimeters. J. Environ. Qual. 6:211–216.

Chichester, F. W., and S. J. Smith. 1978. Disposition of ^{15}N-labeled fertilizer nitrate applied during corn culture in field lysimeters. J. Environ. Qual. 7:227–233.

Clark, F. E. 1977. Internal cycling of ^{15}nitrogen in shortgrass prairie. Ecology 58:1322–1333.

Craswell, E. T., and A. E. Martin. 1974. Effect of moisture content on denitrification in clay soil. Soil Biol. Biochem. 6:127–129.

Craswell, E. T., and A. E. Martin. 1975a. Isotopic studies of the nitrogen balance in a cracking clay. I. Recovery of added nitrogen from soil and wheat in the glasshouse and gas lysimeter. Aust. J. Soil Res. 13:43–52.

Craswell, E. T., and A. E. Martin. 1975b. Isotopic studies of the nitrogen balance in a cracking clay. II. Recovery of nitrate ^{15}N added to columns of packed soil and microplots growing wheat in the field. Aust. J. Soil Res. 13:53–61.

Craswell, E. T., and W. M. Strong. 1976. Isotopic studies of the nitrogen balance in a cracking clay. III. Nitrogen recovery in plant and soil in relation to the depth of fertilizer addition and rainfall. Aust. J. Soil Res. 14:75–83.

Datta, N. P., N. K. Banergee, and D. M. V. Prasada Rao. 1971. A new technique for study of nitrogen balance sheet and an evaluation of nitrophosphate using ^{15}N under submerged conditions of growing paddy. p. 631–638. In J. S. Kanwar et al. (ed.) Int. Symp. on Soil Fertility Evaluation, New Delhi, India. Indian Society of Soil Science, New Delhi, India.

Davis, S., and L. B. Grass. 1966. Determining evapotranspiration from a 1,000-acre lysimeter. Trans. ASAE 9:108–109.

Delwiche, C. C. 1977. Energy relations in the global nitrogen cycle. Ambio 6:106–111.

Denmead, O. T., J. R. Freney, and J. R. Simpson. 1976. A closed ammonia cycle within a plant canopy. Soil Biol. Biochem. 8:161–164.

Denmead, O. T., R. Nulsen, and G. W. Thurtell. 1978. Ammonia exchange over a corn crop. Soil Sci. Soc. Am. J. 42:840–842.

Denmead, O. T., J. R. Simpson, and J. R. Freney. 1974. Ammonia flux into the atmosphere from grazed pasture. Science 185:609–610.

Dinchev, D., and K. Badzhov. 1969. Balance-sheet of the nitrogen introduced into the soil, drawn with the aid of labelled isotopes. (In Bulgarian.) Soil Sci. Agrochem. 4(4):45–52.

Economic Research Service. 1977. 1978 Fertilizer situation. USDA, Economic Research Service, U.S. Government Printing Office, Washington, D.C.

Elliott, L. F., T. M. McCalla, L. N. Mielke, and T. A. Travis. 1971. Ammonium, nitrate, and total nitrogen in the soil water of feedlot and field soil profiles. Appl. Microbiol. 28:810–813.

Embleton, T. W., and W. W. Jones. 1974. Foliar-applied nitrogen for citrus fertilization. J. Environ. Qual. 3:388–391.

Embleton, T. W., and W. W. Jones. 1978. Nitrogen fertilizer management programs, nitrate pollution potential, and orange productivity. p. 275–295. In D. R. Nielsen and J. G. MacDonald (ed.) Nitrogen in the environment, Vol. 1. Nitrogen behavior in field soil. Academic Press, Inc., New York.

Erdman, L. W. 1953. Legume inoculation: What it is; what it does. USDA Farmer's Bull. no. 2003.

Faller, V. N. 1972. Sulphur dioxide, hydrogen sulphide, nitrous gases and ammonia as sole source of sulphur and nitrogen for higher plants. (In German.) J. Plant Nutr. Soil Sci. 131:120–130.

Frere, M. H. 1976. Nutrient aspects of pollution from cropland. p. 59–90. In Control of water pollution from cropland. Vol. II. An overview. U.S. Government Printing Office, Washington, D.C.

Fried, M., and V. Middleboe. 1977. Measurement of amount of nitrogen fixed by a legume crop. Plant Soil 47:713–715.

Fried, M., K. K. Tanji, and R. M. Van De Pol. 1976. Simplified long term concept for evaluating leaching of nitrogen from agricultural land. J. Environ. Qual. 5:197–200.

Frink, C. R. 1969. Water pollution potential estimated from farm nutrient budgets. Agron. J. 61:550–553.

Gast, R. G., W. W. Nelson, and J. M. MacGregor. 1974. Nitrate and chloride accumulation and distribution in fertilized tile-drained soils. J. Environ. Qual. 3:209–213.

Gast, R. G., W. W. Nelson, and G. W. Randall. 1978. Nitrate accumulation in soils and loss in tile drainage following nitrogen applications to continuous corn. J. Environ. Qual. 7:258–261.

Gessel, S. P., D. W. Cole, and E. P. Steinbrenner. 1973. Nitrogen balances in forest ecosystems of the Pacific Northwest. Soil Biol. Biochem. 5:19–34.

Grass, L. B., S. Davis, and E. Shade. 1966. Watershed lysimeter study of salt balance using Colorado River water to irrigate residual soils. Trans. ASAE 9:528–529, 533.

Grier, C. C., D. W. Cole, C. T. Dyrness, and R. L. Fredriksen. 1974. Nutrient cycling in 37- and 450-year-old Douglas-fir ecosystems. p. 21–34. In R. H. Waring and R. L. Edmonds (ed.) Integrated research in the coniferous forest biome. Coniferous Forest Biome Bull. No. 5, Univ. of Washington, Seattle.

Hanawalt, R. B. 1969a. Environmental factors influencing the sorption of atmospheric ammonia by soils. Soil Sci. Soc. Am. Proc. 33:231–234.

Hanawalt, R. B. 1969b. Soil properties affecting the sorption of atmospheric ammonia. Soil Sci. Soc. Am. Proc. 33:725–729.

Hardy, R. W. F., R. D. Holsten, E. K. Jackson, and R. C. Burns. 1968. The acetylene-ethylene assay for N_2 fixation: Laboratory and field evaluation. Plant Physiol. 43:1185–1207.

Hargrove, W. L., D. E. Kissel, and L. B. Fenn. 1977. Field measurements of ammonia volatilization from surface applications of ammonium salts to a calcareous soil. Agron. J. 69:473–476.

Hauck, R. D. 1968. The use of ^{15}N in forest soil-fertilizer nitrogen investigations. p. 64–71. In Forest fertilization, theory and practice. Tennessee Valley Authority, Muscle Shoals, Ala.

Hauck, R. D. 1971. Quantitative estimates of nitrogen-cycle processes: Concepts and review. p. 65–80. In Nitrogen-15 in soil-plant studies. Proc. of a Research Coordination Meeting, Sofia, Bulgaria. 1–5 Dec. 1969. Int. Atomic Energy Agency, Vienna.

Hauck, R. D. 1973. Nitrogen tracers in nitrogen cycle studies—past use and future needs. J. Environ. Qual. 2:317–327.

Hauck, R. D., and J. M. Bremner. 1976. Use of tracers for soil and fertilizer nitrogen research. Adv. Agron. 28:219–266.

Hensler, R. F., and O. J. Attoe. 1970. Rural sources of nitrates in water. In V. L. Snoeyink (ed.) Nitrate and water supply: Source and control. Proc. of the 12th Sanitary Engr. Conf., Urbana, Ill. 11–12 Feb. 1970. Univ. of Illinois, Urbana.

Henzell, E. F. 1971. Recovery of nitrogen from four fertilizers applied to Rhodes grass in small plots. Aust. J. Exp. Agric. Anim. Husb. 11:420–430.

Henzell, E. F. 1972. Losses of nitrogen from a nitrogen-fertilizer pasture. J. Aust. Inst. Agric. Sci. 38:309–310.

Hoeft, R. G., D. R. Keeney, and L. M. Walsh. 1972. Nitrogen and sulfur in precipitation and sulfur dioxide in the atmosphere in Wisconsin. J. Environ. Qual. 1:203–208.

Hooker, M. L., D. H. Sander, G. A. Peterson, and L. A. Daigger. 1980. Gaseous N losses from winter wheat. Agron. J. 72:789–792.

Hutchinson, G. L., R. J. Millington, and D. B. Peters. 1972. Atmospheric ammonia: Absorption by plant leaves. Science 175:771–772.

Hutchinson, G. L., and F. G. Viets, Jr. 1969. Nitrogen enrichment of surface water by absorption of ammonia volatilized from cattle feedlots. Science 166:514–515.

Illinois Agric. Exp. Stn. and Dept. of Agronomy. 1982. The Morrow plots: A century of learning. Illinois Agric. Exp. Stn. Bull. 775.

Jansson, S. L. 1963. Balance sheet and residual effects of fertilizer nitrogen in a 6-year study with N15. Soil Sci. 95:31–37.

Jaiyebo, E. O., and D. R. Bouldin. 1967. Influence of fertilizer and manure additions and crop rotations on nonexchangeable ammonium content of two soils. Soil Sci. 103:16–22.

Jenkinson, D. S. 1977. The nitrogen economy of the Broadbalk experiments. I. Nitrogen balance in the experiments. Rothamsted Exp. Stn. Report for 1976, part 2 (1977):103–109.

Jenny, H. 1941. Factors of soil formation. McGraw-Hill Book Co., New York.

Johnson, A. H., D. R. Bouldin, E. A. Goyette, and A. M. Hedges. 1976. Nitrate dynamics in Fall Creek, New York. J. Environ. Qual. 5:386–391.

Jolley, V. D., and W. H. Pierre. 1977. Profile accumulation of fertilizer-derived nitrate and total nitrogen recovery in two long term nitrogen-rate experiments with corn. Soil Sci. Soc. Am. J. 41:373–378.

Jones, M. B., C. C. Delwiche, and W. A. Williams. 1977. Uptake and loss of ^{15}N applied to annual grass and clover in lysimeters. Agron. J. 69:1019–1023.

Jones, M. B., J. E. Street, and W. A. Williams. 1974. Leaching and uptake of nitrogen applied to annual grasses and clover-grass mixtures in lysimeters. Agron. J. 66:256–258.

Jones, W. W., and T. W. Embleton. 1969. Development and current status of citrus leaf analysis as a guide to fertilization in California. p. 1669–1671. *In* H. D. Chapman (ed.) Proc. of the First Int. Citrus Symp., Vol. III, Riverside, Calif. 16–26 March 1968. Univ. of California, Riverside.

Keeney, D. R., and J. M. Bremner. 1964. Effect of cultivation on the nitrogen distribution in soils. Soil Sci. Soc. Am. Proc. 28:653–656.

Keeney, D. R., and A. N. MacGregor. 1978. Short-term cycling of ^{15}N-urea in a ryegrass-white clover pasture. N.Z. J. Agric. Res. 21:443–448.

Kelly, G. E., W. M. Edwards, L. L. Harrold, and J. L. McGuiness. 1975. Soils of the North Appalachian experimental watershed. USDA Misc. Pub. no. 1296, U.S. Government Printing Office, Washington, D.C.

Kissel, D. E., C. W. Richardson, and Carl Burnett. 1976. Losses of nitrogen in surface runoff in the Blackland Prairie of Texas. J. Environ. Qual. 5:288–292.

Kissel, D. E., and S. J. Smith. 1978. Fate of fertilizer nitrate applied to coastal bermudagrass on a swelling clay soil. Soil Sci. Soc. Am. J. 42:77–80.

Knowles, R. 1975. Interpretation of recent ^{15}N studies of nitrogen in forest systems. p. 53–65. *In* B. Bernier and C. H. Winget (ed.) Forest soils and forest land management. 1973. Proc. Fourth North American Forest Soils Conf., Quebec, Canada. Les Presses de l'Universite Laval, Quebec.

Kolenbrander, C. J. 1972. Eutrophication from agriculture with special reference to fertilizers and animal waste. p. 305–327. *In* Effects of intensive fertilizer use on the human environment. FAO Soils Bull. no. 16, Rome.

Koren'kov, D. A., L. I. Romanyuk, N. M. Varyushkina, and L. I. Kirpaneva. 1975. Use of the stable ^{15}N isotope to study the balance of fertilizer nitrogen in field lysimeters on sodpodzolic sandy loam soil. Sov. Soil Sci. 7:244–248.

Kowalenko, C. G. 1978. Nitrogen transformations and transport over 17 months in field fallow microplots using ^{15}N. Can. J. Soil Sci. 58:69–76.

Kowalenko, C. G., and D. R. Cameron. 1978. Nitrogen transformations in soil-plant systems in three years of field experiments using tracer and non-tracer methods on an ammoniumfixing soil. Can. J. Soil Sci. 58:195–208.

Koyama, T., M. Shibuya, M. Tokuyasu, T. Shimomura, T. Ide, and K. Ide. 1977. Balance sheet and residual effects of fertilizer nitrogen in Saga paddy field. p. 289–296. *In* Proc. Int. Seminar on Soil Environment and Fertility Management in Intensive Agric. Soc. Sci. Soil Manure, Japan.

Krauter, C. F. 1975. The uptake, storage and loss of nitrogen in the field using ^{15}N labeled fertilizers. Ph.D. Thesis. Univ. of California, Davis. Univ. Microfilms. Ann Arbor, Mich. (Diss. Abstr. B 36:18).

Kundler, P. 1970. Utilization, fixation, and loss of fertilizer nitrogen. (In German.) Albrecht-Thaer-Arch. 14:191–210.

Larson, W. E., R. F. Holt, and C. W. Carlson. 1978. Residues for soil conservation. p. 1–16. *In* W. R. Oschwald (ed.) Crop residue management systems. ASA Spec. Pub. no. 31, Madison, Wis.

Lauer, D. A., D. R. Bouldin, and S. D. Klausner. 1976. Ammonia volatilization from dairy manure spread on the soil surface. J. Environ. Qual. 5:134–141.

Lawes, J. B., J. H. Gilbert, and R. Warrington. 1882. On the amount and composition of the rain and drainage waters collected at Rothamsted. J. R. Agric. Soc. Engl. Ser. 2, 17:241–279, 311–350, and 18:1–71.

Legg, J. O. 1975. Influence of plants on nitrogen transformations in soils. p. 221–227. *In* M. K. Wali (ed.) Prairie: A multiple view. The Univ. of North Dakota Press, Grand Forks, N.D.

Legg, J. O., and F. E. Allison. 1959. Recovery of N^{15}-tagged nitrogen from ammonium fixing soils. Soil Sci. Soc. Am. Proc. 23:131–134.

Legg, J. O., and C. Sloger. 1975. A tracer method for determining symbiotic nitrogen fixation in field studies. p. 661–666. *In* E. R. Klein and P. D. Klein (ed.) Proc. of the Second Int. Conf. on Stable Isotopes, Oak Brook, Ill. 20–23 Oct. 1975. U.S. Energy Res. and Development Admin., U.S. Dep. of Commerce, Springfield, Va.

Legg, J. O., and G. Stanford. 1967. Utilization of soil and fertilizer N by oats in relation to the available N status of soils. Soil Sci. Soc. Am. Proc. 31:215–219.

Lemon, E., and R. Van Houtte. 1980. Ammonia exchange at the land surface. Agron. J. 72: 876–883.

Lipman, J. G., and A. B. Conybeare. 1936. Preliminary note on the inventory and balance sheet of plant nutrients in the United States. N. J. Agric. Exp. Stn. Bull. 607.

Luebs, R. E., K. R. Davis, and A. E. Laag. 1973. Enrichment of the atmosphere with nitrogen compounds volatilized from a large dairy area. J. Environ. Qual. 2:137–141.

Lund, L. J. 1979. Nitrogen studies for selected fields in the Santa Maria Valley. A. Nitrate leaching and nitrogen balances. p. 355–415. *In* P. F. Pratt (principal investigator) Nitrate in effluents from irrigated lands. Final Report to the National Science Foundation, Univ. of California, Riverside.

Lund, L. J., D. C. Adriano, and P. F. Pratt. 1974. Nitrate concentrations in deep soil cores as related to soil profile characteristics. J. Environ. Qual. 3:78–82.

Lund, L. J., J. C. Ryden, R. J. Miller, A. E. Laag, and W. E. Bendixen. 1978. Nitrogen balances for the Santa Maria Valley, p. 395–413. *In* P. F. Pratt (ed.) Proc. Natl. Conf. on Management of Nitrogen in Irrigated Agric. Sacramento, Calif. 15–18 May 1978. Univ. of California, Riverside.

MacGregor, J. M., G. R. Blake, and S. D. Evans. 1974. Mineral nitrogen movement into subsoils following continued annual fertilization for corn. Soil Sci. Soc. Am. Proc. 38:110–113.

MacVicar, R., W. L. Garman, and R. Wall. 1950. Studies on nitrogen fertilizer utilization using N^{15}. Soil Sci. Soc. Am. Proc. 15:265–268.

Malo, B. A., and E. R. Purvis. 1964. Soil absorption of atmospheric ammonia. Soil Sci. 97: 242–247.

Martin, A. E., E. F. Henzell, P. J. Ross, and K. P. Haydock. 1963. Isotopic studies on the uptake of nitrogen by pasture grasses. I. Recovery of fertilizer nitrogen from the soil:plant system using Rhodes grass in pots. Aust. J. Soil Res. 1:169–184.

Martin, A. E., and P. J. Ross. 1968. A nitrogen balance study using labelled fertilizer in a gas lysimeter. Plant Soil 28:182–186.

Martin, A. E., and G. W. Skyring. 1962. Losses of nitrogen from the soil:plant system. p. 19–34. *In* A review of nitrogen in the tropics with particular reference to pastures: A symposium. Bull. no. 46 Commonwealth Bureau Past. Field Crops, Hurley, Berkshire, England.

McNeal, B. L., and P. F. Pratt. 1978. Leaching of nitrate from soils. p. 195–230. *In* P. F. Pratt (ed.) Proc. Natl. Conf. on Management of Nitrogen in Irrigated Agric. Sacramento, Calif. 15–18 May 1978. Univ. of California, Riverside.

Meisinger, J. J. 1976. Nitrogen application rates consistent with environmental constraints for potatoes on Long Island. Search Agric. (Geneva, N.Y.) 6:1–18.

Menzel, R. G., E. D. Rhoades, A. E. Olness, and S. J. Smith. 1978. Variability of annual nutrient and sediment discharges in runoff from Oklahoma cropland and rangeland. J. Environ. Qual. 7:401–406.

Mikkelsen, D. S., S. K. De Datta, and W. N. Obcemea. 1978. Ammonia volatilization losses from flooded rice soils. Soil Sci. Soc. Am. J. 42:725–730.

Miller, R. E., D. P. Lavender, and C. C. Grier. 1976. Nutrient cycling in the Douglas-fir type —silvicultural implications. p. 359–390. *In* America's renewable resource potential— 1975: The turning point. Proc. 1975 Annual Convention, Washington, D.C. 28 Sept.– 2 Oct. 1975. Soc. of Am. Foresters, Washington, D.C.

Moe, P. G., J. V. Mannering, and C. B. Johnson. 1967. Loss of fertilizer nitrogen in surface runoff water. Soil Sci. 104:389–394.

Moe, P. G., J. V. Mannering, and C. B. Johnson. 1968. A comparison of nitrogen losses from urea and ammonium nitrate in surface runoff water. Soil Sci. 105:428–433.

Moore, A. W. 1966. Non-symbiotic nitrogen fixation in soil and soil-plant systems. Soils Fert. 29:113–128.

National Academy of Sciences. 1971. Atlas of nutritional data on United States and Canadian feeds. National Academy of Sciences, Washington, D.C.

National Research Council. 1978. Nitrates: An environmental assessment. National Academy of Sciences, Washington, D.C.

Nicholaichuk, W., and D. W. L. Read. 1978. Nutrient runoff from fertilized and unfertilized fields in Western Canada. J. Environ. Qual. 7:542–544.

Nielsen, D. R., J. W. Biggar, and K. T. Erb. 1973. Spatial variability of field-measured soil-water properties. Hilgardia 42:215–259.

Nommik, H. 1973. Assessment of volatilization loss of ammonia from surface-applied urea on forest soil by ^{15}N recovery. Plant Soil 38:589–603.

Olsen, R. J., R. F. Hensler, O. J. Attoe, S. A. Witzel, and L. A. Peterson. 1970. Fertilizer nitrogen and crop rotation in relation to movement of nitrate nitrogen through soil profiles. Soil Sci. Soc. Am. Proc. 34:448–452.

Olson, R. A. 1978. Critique of an approach to measuring leaching of nitrate from freely drained irrigated fields. p. 257–265. *In* D. R. Nielsen and J. G. MacDonald (ed.) Nitrogen in the environment, Vol. I. Nitrogen behavior in field soil. Academic Press, Inc., New York.

Olson, R. V. 1980. Fate of tagged nitrogen fertilizer applied to irrigated corn. Soil Sci. Soc. Am. J. 44:514–517.

Olson, R. V., L. S. Murphy, H. C. Moser, and C. W. Swallow. 1979. Fate of tagged fertilizer nitrogen applied to winter wheat. Soil Sci. Soc. Am. J. 43:973–975.

Overrein, L. N. 1968. Lysimeter studies on tracer nitrogen in forest soil: I. Nitrogen losses by leaching and volatilization after addition of urea-^{15}N. Soil Sci. 106:280–290.

Overrein, L. N. 1969. Lysimeter studies on tracer nitrogen in forest soil: II. Comparative losses of nitrogen through leaching and volatilization after the addition of urea-, ammonium-, and nitrate-^{15}N. Soil Sci. 107:149–159.

Overrein, L. N. 1972. Isotope studies on nitrogen in forest soil. II. Distribution and recovery of ^{15}N-enriched fertilizer nitrogen in a 40-month lysimeter investigation. Medd. Nor. Skogforsoeksves. 30:308–324.

Patrick, W. H., and K. R. Reddy. 1976. Fate of fertilizer nitrogen in a flooded rice soil. Soil Sci. Soc. Am. J. 40:678–681.

Pimentel, D., E. C. Terhune, R. Dyuson-Hudson, S. Rochereau, R. Samis, E. A. Smith, D. Denman, D. Reifschneider, and M. Shepherd. 1976. Land degradation: Effects on food and energy resources. Science 194:149–155.

Pomares-Garcia, F., and P. F. Pratt. 1978. Recovery of ^{15}N-labeled fertilizer from manured and sludge-amended soil. Soil Sci. Soc. Am. J. 42:717–720.

Porter, L. K., F. G. Viets, Jr., and G. L. Hutchinson. 1972. Air containing N-15 ammonia: Foliar absorption by corn seedlings. Science 175:759–761.

Power, J. F. 1970. Leaching of nitrate nitrogen under dryland agriculture in the Northern Great Plains. p. 111–122. *In* Relationship of agriculture to soil and water pollution. Proc. of the First Annual Agric. Pollution Conf., Rochester, N.Y. Nov. 1969. Cornell University Press, Ithaca, N.Y.

Power, J. F. 1977. Nitrogen transformations in the grassland ecosystem. *In* J. K. Marshall (ed.) The belowground ecosystem: A synthesis of plant-associated processes. Dep. of Range Science, Colorado State Univ., Ft. Collins.

Power, J. F., J. Alessi, G. A. Reichman, and D. L. Grunes. 1973. Recovery, residual effects, and fate of nitrogen fertilizer sources in semi-arid region. Agron. J. 65:765–772.

Power, J. F., and J. O. Legg. 1978. Effect of crop residues on the soil chemical environment and nutrient availability. p. 85–100. *In* W. R. Oschwald (ed.) Crop residue management systems. ASA Spec. Pub. no. 31, Madison, Wis.

Pratt, P. F., and D. C. Adriano. 1973. Nitrate concentrations in the unsaturated zone beneath irrigated fields in Southern California. Soil Sci. Soc. Am. Proc. 37:321–322.

Pratt, P. F., and H. D. Chapman. 1961. Gains and losses of mineral elements in an irrigated soil during a 20-year lysimeter investigation. Hilgardia 30:445–467.

Pratt, P. F., H. D. Chapman, and M. J. Garber. 1960. Gains and losses of nitrogen and depth distribution of nitrogen and organic carbon in the soil of a lysimeter investigation. Soil Sci. 90:293–297.

Pratt, P. F., S. Davis, and R. G. Sharpless. 1976a. A four year field trial with animal manures. I. Nitrogen balances and yield. Hilgardia 44:99–112.

Pratt, P. F., S. Davis, and R. G. Sharpless. 1976b. A four year field trial with animal manures. II. Mineralization of nitrogen. Hilgardia 44:113–125.

Pratt, P. F., W. W. Jones, and V. E. Hunsaken. 1972. Nitrate in deep soil profiles in relation to fertilizer rates and leaching volume. J. Environ. Qual. 1:97–102.

Pratt, P. F., L. J. Lund, and J. M. Rible. 1978. An approach to measuring leaching of nitrate from freely drained irrigated fields. p. 223–256. *In* D. R. Nielsen and J. G. MacDonald (ed.) Nitrogen in the environment, Vol. I. Nitrogen behavior in field soil. Academic Press, Inc., New York.

Pratt, P. F., J. E. Warneke, and P. E. Nash. 1976c. Sampling the unsaturated zone in irrigated field plots. Soil Sci. Soc. Am. J. 40:277–279.

Reddy, K. R., and W. H. Patrick. 1978. Residual fertilizer nitrogen in a flooded rice soil. Soil Sci. Soc. Am. J. 42:316–318.

Rennie, D. A., G. J. Racz, and D. K. McBeath. 1976. Nitrogen losses. p. 325–353. *In* Proc. of Western Canada Nitrogen Symp. 20–21 Jan. 1976, Calgary, Alberta, Canada. Alberta Agric., Edmonton, Alberta.

Rible, J. M., K. M. Holtzclaw, and P. F. Pratt. 1973. Nitrate in the unsaturated zone. p. 37–66. *In* P. F. Pratt (principal investigator) Nitrate in effluents from irrigated lands. Annual Report to the National Science Foundation, 1973. Univ. of California, Riverside.

Rible, J. M., K. M. Holtzclaw, and P. F. Pratt. 1974. Nitrate concentrations in the unsaturated zone beneath irrigated land. p. 271–292. *In* P. F. Pratt (principal investigator) Nitrate in effluents from irrigated lands. Annual Report to the National Science Foundation, 1974. Univ. of California, Riverside.

Rible, J. M., P. A. Nash, P. F. Pratt, and L. J. Lund. 1976. Sampling the unsaturated zone of irrigated lands for reliable estimates of nitrate concentration. Soil Sci. Soc. Am. J. 40:566–570.

Rible, J. M., R. F. Pratt, and L. J. Lund. 1979. Nitrates in the unsaturated zone of freely drained fields. p. 297–320. *In* P. F. Pratt (principal investigator) Nitrate in effluents from irrigated lands. Final report to the National Science Foundation, Univ. of California, Riverside.

Riga, A., V. Fischer, and H. J. van Praag. 1980. Fate of fertilizer nitrogen applied to winter wheat as $Na^{15}NO_3$ and $(^{15}NH_4)_2SO_4$ studied in microplots through a four-course rotation: 1. Influence of fertilizer splitting on soil and fertilizer nitrogen. Soil Sci. 130:88–99.

Rolston, D. E. 1978. Application of gaseous-diffusion theory to measurement of denitrification. p. 309–335. *In* D. R. Nielsen and J. G. MacDonald (ed.) Nitrogen in the environment, Vol. I. Nitrogen behavior in field soil. Academic Press, Inc., New York.

Rolston, D. E., and F. E. Broadbent. 1977. Field measurement of denitrification. USEPA Res. Report Ser. EPA-600/2-77-23, Ada, Okla.

Rolston, D. E., F. E. Broadbent, and D. A. Goldhamer. 1979. Field measurement of denitrification: II. Mass balance and sampling uncertainty. Soil Sci. Soc. Am. J. 43:703–708.

Rolston, D. E., M. Fried, and D. A. Goldhamer. 1976. Denitrification measured directly from nitrogen and nitrous oxide gas fluxes. Soil Sci. Soc. Am. J. 40:259–266.

Rolston, D. E., D. L. Hoffman, and D. W. Toy. 1978. Field measurement of denitrification: I. Flux of N_2 and N_2O. Soil Sci. Soc. Am. J. 42:863–869.

Romkens, M. J. M., D. W. Nelson, and J. V. Mannering. 1973. Nitrogen and phosphorus composition of surface runoff as affected by tillage method. J. Environ. Qual. 2:292–295.

Ryden, J. C., and L. J. Lund. 1980. Nature and extent of directly measured denitrification losses from some irrigated vegetable crop production units. Soil Sci. Soc. Am. J. 44:505–511.

Saffigna, P. G., D. R. Keeney, and C. B. Tanner. 1977. Nitrogen, chloride, and water balance with irrigated Russet Burbank potatoes in a sandy soil. Agron. J. 69:251–257.

Schuman, G. E., R. E. Burwell, R. F. Piest, and R. G. Spomer. 1973. Nitrogen losses in surface runoff from agricultural watersheds on Missouri Valley loess. J. Environ. Qual. 2:299–302.

Schuman, G. E., T. M. McCalla, K. E. Saxton, and H. T. Knox. 1975. Nitrate movement and its distribution in the soil profile of differentially fertilized corn watersheds. Soil Sci. Soc. Am. Proc. 39:1192–1197.

Simpson, J. R. 1968. Losses of urea nitrogen from the surface of pasture soil. Int. Congr. Soil Sci., Trans. 9th (Adelaide) 2:459–466.

Snedecor, G. W., and W. G. Cochran. 1967. Statistical methods. 6th ed. Iowa State University Press, Ames.

Stefanson, R. C. 1972a. Soil denitrification in sealed soil-plant systems. I. Effect of plants, soil water content and soil organic matter content. Plant Soil 37:113–127.

Stefanson, R. C. 1972b. Soil denitrification in sealed soil-plant systems. II. Effect of soil water content and form of applied nitrogen. Plant Soil 37:129–140.

Stefanson, R. C. 1972c. Soil denitrification in sealed soil-plant systems. III. Effect of disturbed and undisturbed soil samples. Plant Soil 37:141–149.

Stefanson, R. C. 1972d. Effect of plant growth and form of nitrogen fertilizer on denitrification from four South Australian soils. Aust. J. Soil Res. 10:183–195.

Stewart, B. A. 1978. Critique of an approach to measuring leaching of nitrate from freely drained irrigated fields. p. 267–273. *In* D. R. Nielsen and J. G. MacDonald (ed.) Nitrogen in the environment, Vol. I. Nitrogen behavior in field soil. Academic Press, Inc., New York.

Stewart, B. A., F. G. Viets, Jr., G. L. Hutchinson, W. D. Kemper, F. E. Clark, M. L. Fairbourn, and F. Strauch. 1967. Distribution of nitrates and other water pollutants under fields and corrals in the middle South Platte Valley of Colorado. USDA, ARS41-134, U.S. Government Printing Office, Washington, D.C.

Stewart, W. D. P., G. P. Fitzgerald, and R. H. Burris. 1967. In situ studies on N_2 fixation using the acetylene reduction technique. Proc. Natl. Acad. Sci. U.S.A. 58:2071–2078.

Stout, P. R., and R. G. Burau. 1967. The extent and significance of fertilizer buildup in soils as revealed by vertical distributions of nitrogenous matter between soils and underlying water reservoirs. p. 283–310. *In* N. C. Brady (ed.) Agriculture and the quality of our environment. Am. Assoc. for Adv. Sci. Pub. no. 85, Washington, D.C.

Stutte, C. A., and R. T. Weiland. 1978. Gaseous nitrogen loss and transpiration of several crop and weed species. Crop Sci. 18:887–889.

Stutte, C. A., R. T. Weiland, and A. R. Blem. 1979. Gaseous nitrogen loss from soybean foliage. Agron. J. 71:95–97.

Tanji, K. K., F. E. Broadbent, M. Mehran, and M. Fried. 1979. An extended version of a conceptual model for evaluating annual nitrogen leaching losses from cropland. J. Environ. Qual. 8:114–120.

Tanji, K. K., M. Fried, R. M. Van De Pol. 1977. A steady state conceptual nitrogen model for estimating nitrogen emissions from cropped lands. J. Environ. Qual. 6:155–159.

Tanji, K. K., and M. Mehran. 1979. Conceptual and dynamic models for nitrogen in irrigated croplands. p. 555–646. *In* P. F. Pratt (principal investigator) Nitrate in effluents from irrigated lands. Final Report to the National Science Foundation, Univ. of California, Riverside.

Timmons, D. R., R. E. Burwell, and R. F. Holt. 1973. Nitrogen and phosphorus losses in surface runoff from agricultural land as influenced by placement of broadcast fertilizer. Water Resour. Res. 9:658–667.

Trepachev, Ye. P., and D. N. Pryanishnikov. 1976. Methods of studying the soil nitrogen balance in long term experiments. Sov. Soil Sci. 8:219–231.

U.S. Department of Agriculture. 1977. Agricultural statistics. USDA, U.S. Government Printing Office, Washington, D.C.

U.S. Department of Agriculture. 1978. Crop production. 1977. Annual Summary, Crop Reporting Board, Economics, Statistics, and Cooperatives Serv., Washington, D.C.

Vallis, I., and E. F. Henzell. 1969. A method for measuring nitrogen fixation. CSIRO Aust. Div. Trop. Past. Rep. 1968–1969, p. 71–72.

Vallis, I., E. F. Henzell, and T. R. Evans. 1977. Uptake of soil nitrogen by legumes in mixed swards. Aust. J. Agric. Res. 28:413–425.

Vallis, I., E. F. Henzell, A. E. Martin, and P. J. Ross. 1973. Isotopic studies on the uptake of nitrogen by pasture plants. V. ^{15}N balance experiments in field microplots. Aust. J. Agric. Res. 24:693–702.

Van Dyne, D. L., and C. B. Gilbertson. 1978. Estimating U.S. livestock and poultry manure and nutrient production. USDA, Economics, Statistics, and Cooperatives Serv. Pub. ESCS-12. Washington, D.C.

van Praag, H. J., V. Fischer, and A. Riga. 1980. Fate of fertilizer nitrogen applied to winter wheat as $Na^{15}NO_3$ and $(^{15}NH_4)_2SO_4$ studied in microplots through a four-course rotation: 2. Fixed ammonium turnover and nitrogen reversion. Soil Sci. 130:100–105.

Viets, F. G., Jr. 1971. Water quality in relation to farm use of fertilizer. BioScience 21:460–467.

Viets, F. G., Jr., R. P. Humbert, and C. E. Nelson. 1967. Fertilizers in relation to irrigation practice. *In* R. M. Hagan et al. (ed.) Irrigation of agricultural lands. Agronomy 11:1009–1023. Am. Soc. of Agron., Madison, Wis.

Vlek, P. L. G., and E. T. Craswell. 1979. Effect of nitrogen source and management on ammonia volatilization losses from flooded rice-soil systems. Soil Sci. Soc. Am. J. 43:352–358.

Walker, T. W., A. F. Adams, H. D. Orchiston. 1956. Fate of labeled nitrate and ammonium nitrogen when applied to grass and clover grown separately and together. Soil Sci. 81:339–351.

Wallingford, G. W., L. S. Murphy, W. L. Powers, and H. L. Manges. 1975. Denitrification in soil treated with beef feed-lot manure. Commun. Soil Sci. Plant Anal. 6:147–161.

Weiland, R. T., and C. A. Stutte. 1979. Pyro-chemiluminescent differentiation of oxidized and reduced N forms evolved from plant foliage. Crop Sci. 19:545–547.

Weiland, R. T., and C. A. Stutte. 1980. Concomitant determination of foliar nitrogen loss, net carbon dioxide uptake, and transpiration. Plant Physiol. 65:403–406.

Weiland, R. T., C. A. Stutte, and R. E. Talbert. 1979. Foliar nitrogen loss and CO_2 equilibrium as influenced by three soybean (*Glycine max*) postemergence herbicides. Weed Sci. 27:545–548.

Willis, W. O., and C. E. Evans. 1977. Our soil is valuable. J. Soil Water Conserv. 32:258–259.

Wischmeier, W. H. 1976. Cropland erosion and sedimentation. p. 31–57. *In* Control of water pollution from cropland, Vol. II. USDA, ARS-H-52, U.S. Government Printing Office, Washington, D.C.

Witty, J. F., J. M. Day, and P. J. Dart. 1976. The nitrogen economy of the Broadbalk experiments. II. Biological nitrogen fixation. Rothamsted Exp. Stn. Report for 1976, part 2 (1977):111–118.

Woodmansee, R. G. 1978. Additions and losses of nitrogen in grassland ecosystems. BioScience 28:448–453.

Woodmansee, R. G. 1979. Factors influencing input and output of nitrogen in grasslands. p. 204–235. *In* N. R. French (ed.) Perspectives in grassland ecology: Results and applications of the US/IBP grassland biome study. Springer-Verlag New York, New York.

Woodmansee, R. G., J. L. Dodd, R. A. Bowman, and F. E. Clark. 1978. Nitrogen budget of a shortgrass prairie ecosystem. Oecologia 34:363–376.

Yeck, R. G., L. W. Smith, and C. C. Calvert. 1975. Recovery of nutrients from animal wastes —an overview of existing options and potentials for use in feed. p. 192–194, 196. *In* Proc. Third Int. Symp. on Livestock Wastes—1975. Managing livestock wastes, Urbana, Ill. Am. Soc. of Agric. Engr. Pub. Proc. no. 275, Am. Soc. of Agric. Engr., St. Joseph, Mich.

Yoshida, T., and B. E. Padre. 1975. Effect of organic matter application and water regime on the transformation of fertilizer nitrogen in a Philippine soil. Soil Sci. Plant Nutr. 21:281–292.

Yoshida, T., and B. C. Padre. 1977. Transformation of soil and fertilizer nitrogen in paddy soil and their availability to live plants. Plant Soil 47:113–123.

Zamyatina, V. B. 1971. Nitrogen balance studies using ^{15}N-labelled fertilizers based on nitrogen-15 studies in the USSR. p. 33–45. *In* Nitrogen-15 in soil-plant studies. Proc. of a Research Coordination Meeting, Sofia, Bulgaria. 1–5 Dec. 1969. Int. Atomic Energy Agency, Vienna.

Zamyatina, V. B., N. I. Borisova, N. M. Varushkina, S. V. Burtzeva, and L. I. Kirpaneva. 1968. Investigations on the balance and use of ^{15}N-tagged fertilizer nitrogen by plants in soils. Int. Congr. Soil Sci., Trans. 9th (Adelaide) 2:513–521.

Zamyatina, V. B., N. M. Varushkina, L. I. Kirpaneva, V. J. Porshneva, Yu. Semyonov. 1973. Transformation and balance of fertilizer nitrogen. p. 178–188. *In* D. A. Koren'kov et al. (ed.) Use of the stable isotope ^{15}N in agricultural research. (In Russian.) Trans. of Primenenie stabilnogo izotopa ^{15}N v issledovaniykh po zemledeliyu. Kolos, Moscow.

Zardalishvili, O. Yu., N. A. Aladashvili, and V. G. Tetruashuili. 1976. Gaseous nitrogen losses and plant uptake of fertilizer nitrogen. Sov. Soil Sci. 8:19–22.

Zardalishvili, O. Yu., N. T. Kvaratsyelia. 1973. Balance of nitrogen on eroded brown forest soils of Western Georgia. p. 144–145. *In* D. A. Koren'kov et al. (ed.) Use of the stable isotope ^{15}N in agricultural research (In Russian.) Trans. of Primenenie stabilnogo izotopa ^{15}N v issledovaniykh po zemledeliyu. Kolos, Moscow.

15 Crop Nitrogen Requirements, Utilization, and Fertilization

R. A. OLSON AND L. T. KURTZ

University of Nebraska, Lincoln, Nebraska, and
University of Illinois, Urbana, Illinois

I. INTRODUCTION

Effective management of N presents a greater challenge to the farm operator than does that of any other fertilizer nutrient. No other nutrient requires as much attention and no other brings greater rewards for wise management.

Nitrogen can enter or leave the soil-plant system by more routes than any other nutrient. Nitrogen is subject to losses via NH_3 volatilization, denitrification, and leaching and it may be augmented by rainfall and biological fixation. These interchanges are important processes in the N cycle and they operate under both natural and cultivated conditions. In some situations, small amounts of gasses containing N escape to the atmosphere from plant leaves. Nitrogen is also contained in water of guttation excreted from foliage and in exudations from roots. Because of gains and losses through natural processes, calculations of the net balance of soil N by a mass balance approach are never more than approximations.

In contrast to most other plant nutrients, no mechanism for long-term storage of plant-available fertilizer N exists in soils. Although NH_4^+-N is held against leaching in the cation exchange complex, it is readily transformed microbially to NO_3^- which is subject to leaching and denitrification.

Nutrients, such as P, K, Ca, and Mg, are often applied in amounts greater than will be utilized by crops in the current season. These nutrients have high residual values and soil levels can be "built up." In contrast, N fertilizer has relatively little "carryover" so that the N-supplying capacity of the soil cannot be permanently increased by massive applications of fertilizer N. Consequently, fertilizer N is applied on a crop-by-crop basis rather than for a rotation or a crop sequence.

On an overall basis, far more N than any other nutrient is removed in crops and supplied in fertilizer. Nitrogen is the nutrient present in greatest amount in the grain of widely grown cereal crops. While many root, fruit, and some vegetable crops remove less N than other nutrients, areas in these crops are small in comparison to grains so that N dominates in fertilizer sales.

While many soils are so well endowed with P, K, and some other nutrients that they can be cropped for years before severe deficiencies arise, N usually becomes deficient within a few seasons. Few soils, with the exception of some peats and mucks, supply enough N for yields that are accepted as adequate in contemporary agriculture. Nitrogen fertilizer is almost universally needed for high levels of production of nonleguminous crops.

II. PLANT USE OF N

Nitrogen is without doubt the most spectacular of all essential nutrients in its effect on plant growth. A sufficiency of N imparts a dark green, luxuriant appearance to rapidly growing field crops. Certainly it is no accident of nature that the all-important protein components of plants coincide with a copious supply of N in the root zone.

A. Functions of N in Plant Growth

Plant life could not exist on earth in the absence of N. Major essential roles of N in plant growth include: (i) component of the chlorophyll molecule, (ii) component of amino acids, the building blocks of proteins, (iii) essential for carbohydrate utilization, (iv) component of enzymes, (v) stimulative to root development and activity, and (vi) supportive to uptake of other nutrients.

The basic raw materials, water, CO_2, and solar energy, would be of no value for the synthesis of carbohydrates in the absence of *chlorophyll*, a complex photoreceptor which contains four N atoms per molecule (Fig. 1). Chlorophyll is involved in accepting energy from solar radiation and in carrying out the photosynthetic process by which plant materials are formed. When the amount of N is inadequate, the leaf will be light green or yellow, thereby expressing a recognized symptom of N deficiency.

Since all amino acids contain N in their molecular configurations and since proteins are assembled from component amino acids, N is necessary for protein synthesis. Most plant proteins contain around 16% N and plants, themselves, are dependent on production of proteins for their propagation. A typical example is the cereals which possess threshold protein levels below which grain will not form.

A certain level of N must be present in plant cells for optimum utilization of carbohydrates produced during photosynthesis. Under deficiency conditions excessive deposition of carbohydrates takes place in vegetative cells with consequent thickening of the cell wall, limited formation of protoplasm, reduced succulence, and reduced growth. A growing crop must have a continuous free energy input for synthesizing macromolecules from simple precursors and for active transport of ions and other synthesis materials throughout the plant. Carrier of this free energy is adenosine triphosphate (ATP), another indispensable N-containing compound.

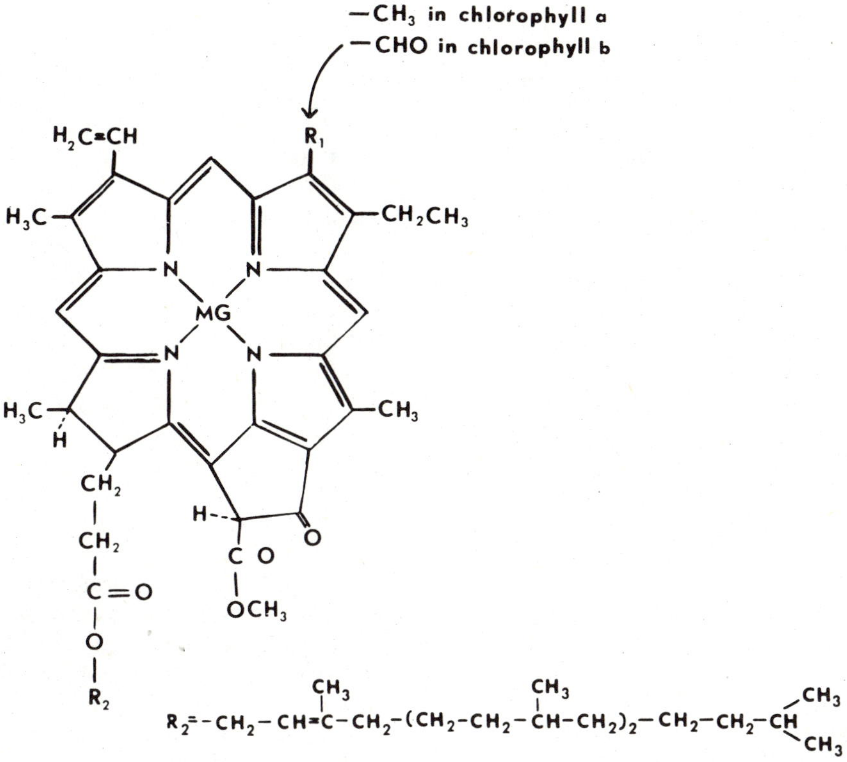

Fig. 1—Structural formula of the chlorophyll molecule.

The many enzymes associated with plant growth are all complex proteins containing N (Stryer, 1975). Restrictions in growth result from N shortages for the production of the molecules involved in the various enzyme systems. These complex proteins largely determine the pattern and rate of chemical transformations in plant cells. Hormone molecules, which are proteins, steroids, or derivatives of amino acids, serve as coordinators of activity of different plant cells.

The role of N in plant growth and associated root development is well documented (Keller & Smith, 1967; Kmoch et al., 1957; Holt & Fisher, 1960; Olson et al., 1964a). Total root mass as well as rooting depth are enhanced by optimal N availability. Extension of roots facilitates absorption of water and other nutrients required for growth. Classic examples include alleviation of drouth effects where deep subsoil moisture is present (Keller & Smith, 1967; Linscott et al., 1962). Uptake of fertilizer P is enhanced particularly when NH_4^+ has been applied. Yield increases from fertilizer N, as well as from other applied nutrients, are accomplished with very little increase in water consumption. Since increases in yield are far out of proportion to the increase of water consumed, water efficiency is notably enhanced (Viets, 1962; Olson et al., 1964b). This factor is of particular importance for regions of limited moisture availability.

B. Uptake, Translocation, and Storage of N

1. UPTAKE

Much of the soil-derived N enters the plant as NO_3^-. Chemical and biological processes make it the most prevalent ionic species of N throughout the rooting zone of well-drained soils. Consequently, the majority of crops have, during their evolution, developed mechanisms for handling NO_3^-. Most common crops also readily absorb NH_4^+ and, if any preference exists, it is usually in favor of NH_4^+ early and NO_3^- late in the season. This sequence corresponds to events in the soils of the temperate region. In early growth stages, roots are largely in the surface layer and the NH_4^+ form predominates because nitrification is limited by low temperatures. As the soil becomes warmer, nitrification proceeds, the root system extends, and the amount and uptake of NO_3^- predominates over NH_4^+. In corn, early growth is best with a combination of NO_3^- and NH_4^+ (Schrader et al., 1972; Warncke & Barber, 1973) with NH_4^+ being used preferentially for synthesis of amino acids and protein (Schrader et al., 1972).

Ammonium is the main form of N available to crops grown under conditions of poor soil aeration, such as submerged rice culture. Under these conditions, nitrification is limited and, moreover, any NO_3^- that is formed or added as fertilizer is subject to rapid loss through denitrification (see Chapter 12). In addition, foliar absorption of NH_3 from the atmosphere can represent a significant contribution to the N nutrition of crops (Hutchinson et al., 1972; Porter et al., 1972). In agricultural regions, NH_3 is more prevalent in the atmosphere than other mineral N species as evidenced by the partition of N compounds in rainfall (Viets & Hageman, 1971; Olson et al., 1973; Tabatabai & Laflen, 1976).

The actual process of N uptake by plants requires movement of ionic species of N to root surfaces for absorption. Most N movement occurs as NO_3^- in the convective flow of soil water to plant roots in response to transpiration in the above-ground portion of the crop. Since attraction between NO_3^- and soil colloids is negligible, NO_3^- is mobile and is readily carried to plant roots by mass flow. In contrast, attraction between NH_4^+ and soil colloids is substantial and its movement in and with soil water is much less. When potential uptake exceeds the supply from mass flow, the concentration of N species at the root surface is lowered and the process of diffusion begins. Although diffusion is of minor importance in most cropping situations on well-drained soils, notable exceptions occur. Where soil characteristics result in localized concentrations of roots (for example, strong peds may cause roots to clump and to follow larger voids) and distances to root surfaces are short, movement by diffusion can be substantial (Barley, 1970). An example of a situation in which diffusion is very important occurs in submerged rice culture (FAO/IAEA, 1970a; Patrick & Mikkelsen, 1971) when high utilization efficiency of NH_4^+ is attained from fertilizer incorporated in the surface few centimeters of soil where most of the roots are concentrated. Crop root geometry adjusts to zones of nutrient concentration with extra proliferation in those zones (Olson & Dreier, 1956a; Miller & Ohlrogge, 1958; Russel & Shone, 1972), enhancing possibilities for diffusional movement.

Nitrate that has moved with percolating water into the lower part of the rooting profile becomes important in dry seasons. When surface horizons dry out, NO_3^- in the lower depths of the rooting profile is the source of N for the crop during the later growth stages. Little or no NH_4^+ exists at those greater soil depths.

Aside from the moisture factor, rate of NO_3^- uptake is controlled by its concentration in the soil solution and by plant metabolism. Uptake rates from nutrient solutions by members of the grass family reach a maximum at a concentration of 0.2 mM of NO_3^- (Huffaker & Rains, 1978) or at 0.3 mM in combinations of NO_3^- and NH_4^+ (Warncke & Barber, 1973). Uptake of NO_3^- is specific with limited influence by most other anions (Rao & Rains, 1976), an exception being the depressive action of phosphate (Wallace & Mueller, 1957). Major soil cations are generally positive in their effects by masking negative charges on the cell wall and allowing NO_3^- to migrate closer to the plasmamembrane uptake sites (Elzam & Epstein, 1965), with K^+ proving more complementary than Ca^{2+} or Na^+ (Wright & Davison, 1964). Ammonium, however, is depressive in some plants (Wallace & Mueller, 1957). When both NH_4^+ and NO_3^- are present in nutrient solution, assimilation of NO_3^- into organic N is retarded and NH_4^+ preferentially utilized by corn (Schrader et al., 1972).

A preponderence of evidence supports the concept of active (energy dependent) uptake of NO_3^-. The presence of the APTase enzyme on the plasmalemma with APT indicates a metabolic system involved in active transport (Lott & Hodges, 1970). Transmembrane electropotential measurements in tissues of oats and pea seedlings supply further indication of an active influx system (Higinbotham et al., 1967). Although the nature of the binding site is not clearly understood, absorption is influenced by Mo, and Viets and Hageman (1971) suggested that a molybdo-protein carrier is formed at the root cell surface. Contributing to the difficulties in understanding the absorption process is the considerable efflux of NO_3^- that occurs from roots simultaneously with uptake (Morgan, 1970).

2. TRANSLOCATION AND STORAGE

Following uptake by roots, a series of independent transformations occur (Pate, 1971). Nitrate may be stored in the roots, reduced, and synthesized into amino acids by root tissues, or transported across root cells and deposited in the xylem for movement into the shoots. Amino acids synthesized in roots may, likewise, be stored there or transported to the shoots. Subsequently, portions of both NO_3^- and amino acids may be temporarily stored in the stem and petiole cells and some may move into the leaves where further storage or reduction of NO_3^- occurs. Eventually, amino acids from any of these accumulation regions may be deposited in the phloem for translocation to reproductive tissue, to younger regions of the shoot, or back to the roots. Carrier proteins are involved in the transport of molecules across membranes adding to the essentiality of N for the movement of products of photosynthesis to points of storage or use in the plant.

The rate at which NO_3^- is reduced to NH_3 within the plant appears to be controlled by NO_3^- reductase, which then has further control over the rate of NO_3^- uptake. The enzyme itself is continuously synthesized and degraded with its steady state concentration related to the plant's energy status (Aslam et al., 1973). Light also has been found to increase NO_3^- reductase induction, possibly indirectly through increased permeability of tissue and rate of uptake of NO_3^- (Beevers et al., 1965).

The site of NO_3^- reduction and consequently the predominant N form in the xylem fluids varies among plant species. Some bean types are known to effect most of the reduction in their roots (Raven & Smith, 1976) while cereals seemingly accomplish much of the reduction in shoot and leaf cells (Beevers & Hageman, 1969; Raven & Smith, 1976). Some species like cocklebur have essentially all the N in the xylem in the NO_3^- form while many woody species have only organic forms. In either case, quantities are influenced by the amount of root growth and the concentration of NO_3^- in the nutrient solution. Concentration of NO_3^- in the growth medium also has considerable bearing on the plant part where reductase activity occurs (Wallace & Pate, 1967). At a NO_3^- level of 5–10 mg/liter, virtually all reductase activity occurred in the root; but above 100 mg/liter, very rapid synthesis of large amounts of reductase took place in the shoot with a noticeable decline of reductase activity in the roots.

Substantial evidence exists that ATP has a controlling influence on NO_3^- transport serving as an energy source for two types of transport systems (Berger, 1973).

Ultimately, the processes of uptake, reduction, and translocation in cereals result in incorporation of the N into amino acids and protein in the vegetative parts of the plant with subsequent breakdown of the vegetative protein into amino acids during the reproductive stage and their translocation, synthesis, and storage into grain proteins in the kernel. The complete process of protein synthesis, starting with the N uptake phase, has been shown to continue from germination to physiological maturity of corn although at a declining rate after silking (Hanway, 1962; Chevalier & Schrader, 1977).

The rate of N uptake by field crops is very rapid during the grand period of vegetative growth (Fig. 2; Olson, 1978), i.e., approximately 4 kg/ha daily in corn and wheat and 6 kg/ha in soybeans for achieving the respective grain yields. Since more total energy input is required for producing protein than starch, soybeans cannot match cereals in grain yield potential. Related, however, to their protein production capability is the longer period of N uptake by soybeans.

Comparative uptake and utilization of N under field conditions is strongly influenced by environmental conditions throughout the growing season including the position of any available N in the rooting zone in relation to the available water supply and consequent root activity. Thus, in dryland farming, enhanced grain protein and yield of winter wheat have

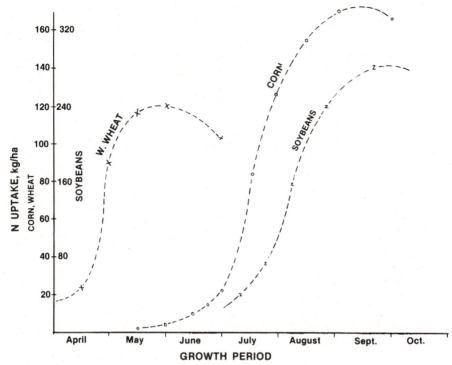

Fig. 2—Average rates of accumulation of N in the above-ground crop of nonirrigated wheat and irrigated corn and soybeans grown at several locations in Nebraska with average yields (kg/ha) of 2,360 for wheat, 9,450 for corn, and 3,375 for soybeans (Olson, 1978).

been associated with available NO_3^- in the deeper portions of the rooting zone, after water had been depleted from upper horizons prior to the grain filling stage (Smika & Grabouski, 1976). Likewise, NO_3^- was not taken up by corn from greater depths of the profile until relatively late in the season after roots had arrived and become active there. Furthermore, such NO_3^- was transported rather directly to the grain for protein synthesis with less vegetative involvement than N taken up earlier from the more shallow depths (Gass et al., 1971). It is increasingly evident that in efficient crop production, a compromise must be reached between the magnitude of vegetative growth and grain yield. A sufficient leaf area for maximum photosynthesis of carbohydrate is essential for top yields, but this does not mean that the largest cereal genotypes produce the most grain. The high yields of the short, stiff-stemmed straw genotypes of wheat and rice supporting the Green Revolution in developing countries confirm this fact. Likewise, earlier planting of corn in temperate regions normally results in shorter plants, but larger grain yields than later planting. In the same context, delayed summer sidedressing of N usually results in shorter plants, greater yields, higher grain/stover ratios, and more economic production than application at planting time (Olson et al., 1964a).

With adequacy of other inputs, increasing levels of N result in higher percentages of protein in the cereal grains (Finney et al., 1957; Schlehuber & Tucker, 1967; Johnson et al., 1973; Olson et al., 1976). The recognized negative relationship between grain yield and protein percentage (Terman et al., 1969) exists under conditions where sufficient N is present for expression of maximum grain yield, but not enough for maximum protein synthesis, the situation that is generally most economical for farmers.

3. GENETIC EFFECTS ON CROP UTILIZATION OF N

Differences have been recognized among species and within genotypes of species in their abilities to absorb and utilize available N from soil in soil-plant systems or other nutrient medium. How much these differences may be related to the previously noted differential capacities for reducing NO_3^- in root cells is not known. In any case, a greater grain protein percentage is obtainable in wheat than in rice or corn and, likewise, greater protein is obtainable in corn grain than in grain sorghum. Interestingly, currently available, high-yielding varieties of grain sorghum under identical N management do not yield as well as corn, but take up more total N from the soil. The major reason for this disparity is the fact that grain sorghum translocates much less of its N from vegetative tissue to grain. Thus, the stover of grain sorghum contains about 50% more total N than does corn stover (Perry & Olson, 1975).

Not only are there differences in magnitude of protein accumulation among species, but major disparities exist in the quality of the proteins for human nutrition. Although low in total protein, the essential amino acid distribution in rice is more favorable than that of any other food grain, particularly with respect to the lysine component. Plant breeders in recent times have made concentrated efforts toward altering protein composition in the major cereals, and this in turn could influence management practices for deriving maximum utilization efficiency of applied N. Perhaps the first breakthrough giving promise in this direction was the discovery of a corn mutant having substantially higher-than-normal lysine concentration in the endosperm. This increased lysine is attributable to a recessive gene designated opaque-2 (Mertz et al., 1964) which, when incorporated into inbred lines, has invariably resulted in lower yields of grain with distinctly higher contents of lysine, histidine, arginine, and aspartic acid (Nelson, 1969). Studies on effects of applied N to the opaque-2 derivatives generally have demonstrated greater total protein production and increases in the less desirable zein component at the expense of the essential amino acids. A high lysine sorghum type (Singh & Axtell, 1973), a barley mutant with greater than normal lysine content (Munck et al., 1970), and similar findings in wheat (Johnson et al., 1978) have more recently been recognized. The way is opened for much enlightening research involving use of these genetic variants in combination with N management practices to improve grain protein quality. Although tedious and expensive research will be required, this approach should help in the endeavors to improve protein nutrition at the end of this century.

4. LOSS OF N FROM THE PLANT

Season-long measurements of the total N contents of grain crops express the rapid accumulation of N in the above ground vegetation prior to grain-filling, after which a distinct decline is observed (Fig. 2; Doneen, 1934; Miller, 1939; Olson & Rhoades, 1953; Boatwright & Haas, 1961; Herron et al., 1963; Rumberg & Sneva, 1970). Various explanations have been given for this loss including translocation back to the soil through the roots, abscission of leaves, and dropping of flowering parts after blooming. A recent suggestion is that much of the disappearance may be due to volatilization of NH_3 or N oxides from the plant following senescence (Daigger et al., 1976). Whatever the cause, the loss is of significant proportions and a better understanding of the processes involved might lead to greater efficiency in N use.

C. Biochemical Pathways of N in the Plant

A first step in N utilization in plants is reduction of NO_3^- through NO_2^- to NH_4^+. This process, which may be represented by the following equation, requires considerable energy:

$$NO_3^- + 2e \xrightarrow[\text{reductase}]{\text{nitrate}} NO_2^- + 6e \xrightarrow[\text{reductase}]{\text{nitrite}} NH_4^+.$$

Reduction of NO_3^- to NO_2^- is the rate-limiting step in the transformation. This step, which is catalyzed by reduced nicotinamide adenine dinucleotide (NADH)–nitrate oxidoreductase (nitrate reductase), occurs in the cytoplasm. Generation of NADH requires carbohydrate oxidation and, thus, indirectly requires utilization of solar energy.

Because the reaction is controlled by NO_3^- reductase and since the rate of NO_3^- reduction influences subsequent reactions in the plant, much effort has been expended to characterize the enzyme. It is known to be a metal protein that contains both Fe and Mo. This enzyme increases in amount when the plant is supplied with NO_3^- and also during the periods when temperature, moisture, and light are favorable for plant growth. Conversely, reduced levels of enzyme activity are found when the plant is under water or temperature stress. The enzyme is believed to have a specific requirement for NADH as its electron donor within the plant.

This first process of utilization of N in the plant, NO_3^- reduction, involves photosynthate and glycolytic metabolism. This interrelationship (Fig. 3) has been represented by Viets and Hageman (1971). In the presence of light and CO_2, sugars and starches, along with phosphorylated glycerates, are produced in the chloroplasts and moved into the cytoplasm where NADH from the oxidation of 3-phosphoglyceraldehyde is used as the source of energy for the reduction of NO_3^-.

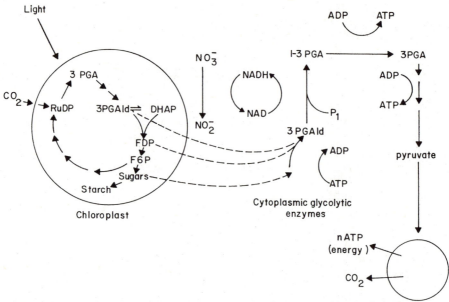

Fig. 3—A simplified metabolic scheme illustrating the interrelationship between photosynthate, glycolytic metabolism, and nitrate reduction. Abbreviations are: RuDP, ribulose diphosphate; 3PGA, 3-phosphoglyceric acid; 3 PGAld, 3-phosphoglyceraldehyde; DHAP, dihydroxyacetone phosphate; FDP, fructose diphosphate; F6P, fructose-6-phosphate; NAD and NADH, oxidized and reduced nicotinamide adenine dinucleotides, respectively; and ADP and ATP, adenosine diphosphate and adenosine tri-phosphates, respectively (Viets & Hageman, 1971).

The second step in the utilization of N in the plant is the reduction of NO_2^- to NH_3. This reaction is catalyzed by NO_2^- reductase, considered to be ferrodoxin–nitrite oxido-reductase. In green tissue, both ferrodoxin and NO_2^- reductase are believed to be located in the chloroplasts and as a consequence, NO_2^- reduction is closely linked to solar energy. In the absence of chlorophyll, for example, in roots or other nongreen plant parts, NO_2^- reduction proceeds by a different system with carbohydrate oxidation as the source of energy.

After NH_3 is produced, it is assimilated rapidly and only traces of NH_3 are normally found in most plants. In fact, greater than trace amounts in plant tissues are often toxic. A major pathway for assimilation of NH_3 is its incorporation into glutamic acid to form glutamine, a reaction catalyzed by glutamine synthetase. Figure 4 shows NH_3 assimilation by incorporation with glutamic acid in the first equation and regeneration of glutamic acid in the second (Noggle & Fritz, 1976). One molecule of glutamine, thus formed, may then react with α-keto glutaric acid and the enzyme glutamic acid synthase to form two molecules of glutamic acid. Thus, double the amount of glutamic acid has been generated and is useable for further assimilation of NH_3.

Incorporation of NH_3 into an amino acid may be followed by a transamination reaction in which the amino group ($-NH_2$) can be transferred to

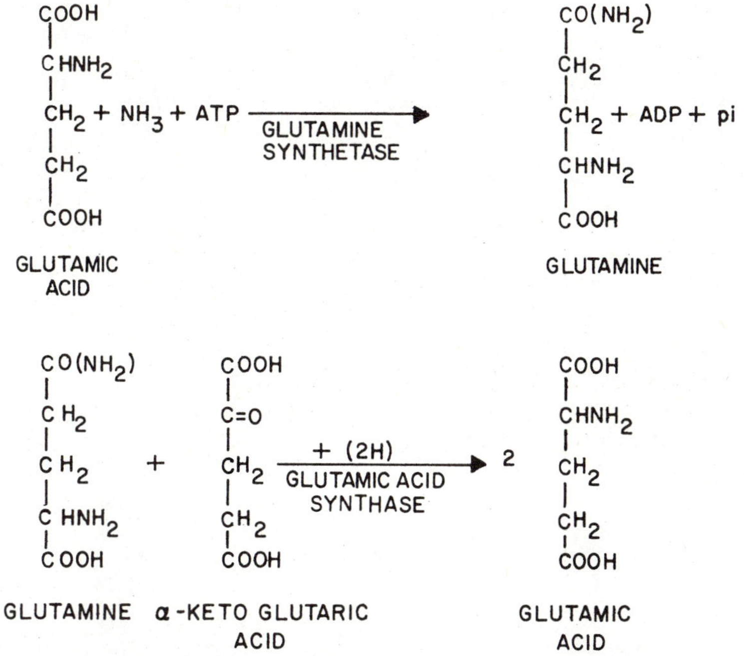

$$
\begin{array}{c}
\text{COOH} \\
| \\
\text{CHNH}_2 \\
| \\
\text{CH}_2 + \text{NH}_3 + \text{ATP} \xrightarrow[\substack{\text{GLUTAMINE} \\ \text{SYNTHETASE}}]{} \\
| \\
\text{CH}_2 \\
| \\
\text{COOH}
\end{array}
\qquad
\begin{array}{c}
\text{CO(NH}_2) \\
| \\
\text{CH}_2 \\
| \\
\text{CH}_2 + \text{ADP} + \text{pi} \\
| \\
\text{CHNH}_2 \\
| \\
\text{COOH}
\end{array}
$$

GLUTAMIC ACID   GLUTAMINE

$$
\begin{array}{c}
\text{CO(NH}_2) \\
| \\
\text{CH}_2 \\
| \\
\text{CH}_2 \\
| \\
\text{CHNH}_2 \\
| \\
\text{COOH}
\end{array}
+
\begin{array}{c}
\text{COOH} \\
| \\
\text{C=O} \\
| \\
\text{CH}_2 \\
| \\
\text{CH}_2 \\
| \\
\text{COOH}
\end{array}
\xrightarrow[\substack{\text{GLUTAMIC ACID} \\ \text{SYNTHASE}}]{+ (2H)}
2
\begin{array}{c}
\text{COOH} \\
| \\
\text{CHNH}_2 \\
| \\
\text{CH}_2 \\
| \\
\text{CH}_2 \\
| \\
\text{COOH}
\end{array}
$$

GLUTAMINE   α-KETO GLUTARIC ACID   GLUTAMIC ACID

Fig. 4—Assimilation of NH_3 by incorporation into glutamic acid to form glutamine. Additional glutamic acid is then synthesized (Noggle & Fritz, 1976).

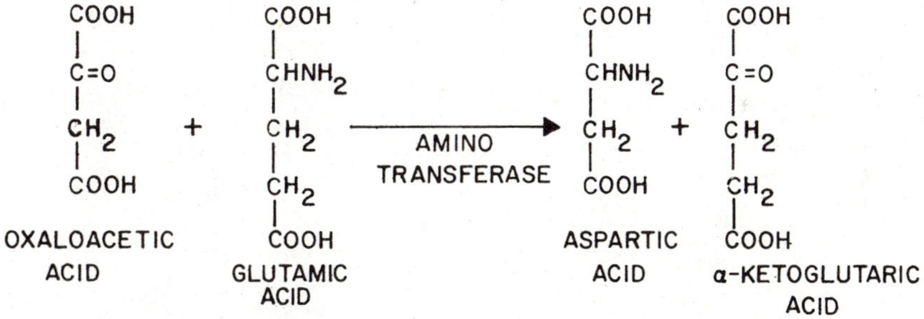

$$
\begin{array}{c}
\text{COOH} \\
| \\
\text{C=O} \\
| \\
\text{CH}_2 \\
| \\
\text{COOH}
\end{array}
+
\begin{array}{c}
\text{COOH} \\
| \\
\text{CHNH}_2 \\
| \\
\text{CH}_2 \\
| \\
\text{CH}_2 \\
| \\
\text{COOH}
\end{array}
\xrightarrow[\substack{\text{AMINO} \\ \text{TRANSFERASE}}]{}
\begin{array}{c}
\text{COOH} \\
| \\
\text{CHNH}_2 \\
| \\
\text{CH}_2 \\
| \\
\text{COOH}
\end{array}
+
\begin{array}{c}
\text{COOH} \\
| \\
\text{C=O} \\
| \\
\text{CH}_2 \\
| \\
\text{CH}_2 \\
| \\
\text{COOH}
\end{array}
$$

OXALOACETIC ACID   GLUTAMIC ACID   ASPARTIC ACID   α-KETOGLUTARIC ACID

Fig. 5—A transamination reaction by which $-NH_2$ is transferred to another metabolite, e.g., oxaloacetic acid, to form additional amino acids or other amino compounds (Noggle & Fritz, 1976).

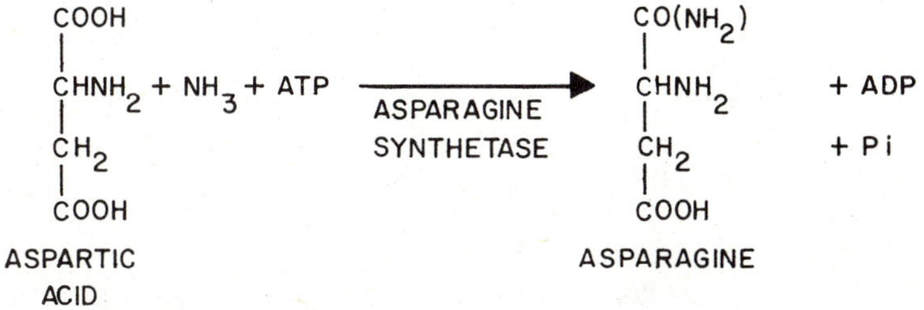

$$
\begin{array}{c}
\text{COOH} \\
| \\
\text{CHNH}_2 + \text{NH}_3 + \text{ATP} \\
| \\
\text{CH}_2 \\
| \\
\text{COOH}
\end{array}
\xrightarrow[\substack{\text{ASPARAGINE} \\ \text{SYNTHETASE}}]{}
\begin{array}{c}
\text{CO(NH}_2) \\
| \\
\text{CHNH}_2 \qquad + \text{ADP} \\
| \\
\text{CH}_2 \qquad\quad + \text{Pi} \\
| \\
\text{COOH}
\end{array}
$$

ASPARTIC ACID   ASPARAGINE

Fig. 6—An alternate pathway for assimilation of NH_3 by combination with an amino acid to form an amide (Noggle & Fritz, 1976).

another metabolite thus forming other amino acids or amino compounds. Transamination reactions are catalyzed by enzymes known as amino trans-ferases, an example (Fig. 5) being the formation of aspartic acid from oxaloacetic acid and glutamic acid (Noggle & Fritz, 1976).

A few additional pathways for assimilation of NH_3 are known. These include a reaction of NH_3 and CO_2 to form carbamyl phosphate, which in turn is converted to the amino acid, arginine. Still another example is the biosynthesis of amides by combination of NH_3 with an amino acid. In this way aspartic acid is converted to the amide, asparagine (Fig. 6).

Amino acids are considered to be the building blocks of protein and they are assembled in specific sequences to form different proteins. Thus, a single protein molecule may contain from 50 to 1,000 monomer amino acids. Not all amino acids in plants are combined into proteins, and 20–40 different ones are found in the free state in various plant species. Proportions of different amino acids in both the free and combined states are characteristics of plant species and sometimes of cultivars within species. Amino acids are also synthesized in the plant into a variety of complex nitrogenous compounds involving life functions.

D. Genetic Effects on Biochemical Pathways

Although the N or protein percentages in crops marketed commercially have not changed appreciably through the years, corn hybrids and cultivars of grains have been selected or bred that contain high or low levels of protein. For example, corn hybrids have been developed that are consistently higher in N and protein than the normal commercial hybrid even when N fertilizer beyond that required for maximum yield is supplied to both.

Since NO_3^- reduction appears to be the step which limits the rate of synthesis of amino acids and proteins, it is sometimes regarded as a reaction which should be closely related to plant growth and yield. With this reasoning, the activity of NO_3^- reductase should be a good indicator of growth rate. Consequently, differences in NO_3^- reductase activity among cultivars and hybrids have been studied as a possible indicator of potential yield or protein production. Genetic differences in levels of NO_3^- reductase have been reported among corn inbreds as well as among hybrids and cultivars of other crops such as sorghum and wheat (Hageman et al., 1976).

In some cases, NO_3^- reductase activity of hybrids or cultivars has been found to be correlated with yield of grain or protein, but usually a general relationship rather than a close correlation has been obtained. A number of plant characteristics and weather factors doubtlessly influence the relationship. Drought stress, for example, is known to greatly retard plant processes. Improvement of crops by selection and breeding for enhancement of enzyme levels and similar biochemical characteristics may appear promising, but difficult to achieve.

III. NITROGEN IN CROP PRODUCTION

A. Nitrogen Levels in Crops Associated with Deficiency, Sufficiency, and Excess

Efforts toward establishing N deficiency/sufficiency levels by plant analysis have been complicated by a number of variables associated with plant growth and by the transitory nature of N itself. Among the more important are variations in N requirements among species and among varieties within a given species. There are also variations in N concentration among parts of a given plant, rapid changes in N concentrations of plant parts with stage of growth, differences in concentration imposed by climatic variables, varied N concentrations with deficiency or excess of another nutrient in the plant, and changing N levels in plant parts due to disease or pest attacks.

The higher protein and N concentrations of leguminous plant tissue over nonlegumes are well recognized, and they are important considerations in animal and human nutrition, particularly in less developed regions of the world where grain crops constitute most of the diet. Furthermore, varieties

Table 1—Nitrogen contents of irrigated corn plants sampled at early silk. Mead Field Laboratory, Nebraska, 1967 (R. A. Olson, unpublished data).

Plant part	N treatment, kg/ha		
	0	84	168
		%	
Whole plant	1.33	1.82	1.93
Top leaf	2.89	3.29	3.31
Third leaf down	2.82	3.24	3.28
Ear leaf	2.53	2.92	3.09
Top internode	0.93	1.06	1.12
Middle internode	0.48	0.81	1.02
Top node	1.21	1.38	1.36
Middle node	1.00	1.26	1.52
	Grain yield at harvest, kg/ha		
	6,489	9,260	10,269

Table 2—Influence of soil water level on concentration and total uptake of N by Starr millet (Bennet et al., 1964).

Soil water level[†]	Starr millet		
	Yield	N concentration	Total N uptake
	kg/ha	%	kg/ha
M₁	15,187	2.26	343
M₂	20,832	1.99	415
M₃	23,332	1.58	400

† M_1, no irrigation; M_2 and M_3 were irrigated when water in the top 61 cm of soil was reduced to 65% and 30%, respectively, of the available soil water capacity.

Table 3—Plant N concentrations associated with deficiency, sufficiency, and excess in several important agricultural crops.

Crop	N measured	Plant part	N content at designated nutritional status of crop†				Reference
			Deficient	Low	Sufficient	Very high, possibly excessive	
			%				
Alfalfa *Medicago sativa* L.	Total	Top 15 cm at early bloom	<4.0	4.0-4.5	4.5-5.0	>5.0	Jones (1967)
Bermudagrass *Cynodon dactylon* (L.) Pers.	Total	Whole tops 4-5 weeks after clipping	<1.5	1.5-2.5	2.5-3.0	>3.0	Burton (1954)
Bromegrass *Bromus inermis* Leyss.	Total	First cutting for hay at early flower stage	1.0-1.5	1.5-2.0	2.0-2.5	>2.5	Russell et al. (1954)
Corn *Zea mays* L.	Total	Ear leaf at silk	<2.25	2.26-2.75	2.76-3.5	>3.5	Jones (1967)
Cotton *Gossypium hirsutum* L.	Total	Upper mature leaves	<2.5	2.5-3.0	3.0-4.5	>4.5	Sabbe et al. (1972); Sabbe & MacKenzie (1973)
Grain sorghum *Sorghum bicolor* L. Moench	Total	Third leaf below head at bloom stage	<2.5	2.5-3.0	3.0-4.0	>4.0	Lockman (1972)
Potato *Solanum tuberosum* L.	NO_3^-	Petiole of 4th leaf from growing tip	<0.8	0.8-1.0	1.0-1.2	>1.2	Lorenz et al. (1964)
Millet *Pennisetum glaucum* L.	Total	Whole tops, 4-5 weeks after clipping	<1.9	1.9-2.2	2.5-3.5	>3.5	Clapp & Chambles (1970)
Rice *Oryza sativa* L.	Total	Most recently fully extended leaf at maximum tillering	<2.4	2.4-2.8	2.8-3.6	>3.6	Mikkelsen & Hunziker (1971)
Soybeans *Glycine max* (L.) Merr.	Total	Upper fully developed trifoliate leaves just prior to pod set	<4.0	4.0-4.25	4.25-5.5	>5.5	Anonymous (1971)‡
Sugar beet *Beta saccharifera*	NO_3^-	Petioles of recently matured leaves	<0.1	0.1-0.2	0.2-0.3	>0.3	Ulrich (1950)
Sugarcane *Saccharum officinarum* L.	Total	Blades 3, 4, 5, 6	<1.0	1.0-1.5	1.5-2.7	>2.7	Schmehl & Humbert (1964)
Wheat *Triticum aestivum* L. Winter	Total	Total above-ground plant at head emergence from the boot	<1.25	1.25-1.75	1.75-3.0	>3.0	Ward et al. (1973)
Spring			<1.5	1.5-2.0	2.0-3.0	>3.0	

† Ranges presented in many cases are adjustments or additions to original authors' values to fit the format employed here.

‡ ... Univ. Plant Analysis Lab. Unpublished data.

and hybrids within a given species vary in N uptake, utilization, and storage as previously detailed. Thus, any expression of sufficiency of N will of necessity express a range for even a designated species.

A given genotype of most species will have a maximum N concentration during early growth. As the season progresses and the plant increases in size, N percentage declines with dilution in the more rapidly accumulating carbohydrate products of photosynthesis. Thus, the total plant N concentration of corn at physiological maturity is commonly no more than one-third of that in the late seedling stage. A substantial change in leaf N concentration can occur within a few days during the grain filling period while N is being translocated from vegetative parts to the seeds. Even greater disparity exists among plant parts at the same growth stage. Table 1 illustrates the differences in N concentrations among designated parts of the corn plant and also the effect of N fertilization in accentuating those differences. Leaves contained three or more times the percent N of internodal tissues, nodes contained moderately more than internodes, and N concentration in all three parts declined from the top of the plant downward.

Climatic factors have a major impact on plant N concentration. Temperature and moisture are especially important since they control the rate of dry matter production. Very favorable growth conditions are associated with maximum growth rate and attendant dilution of N in plant tissues. The data of Table 2 are indicative. These results with two cuttings of millet were obtained by supplying supplemental irrigation to maintain available water at three different levels in the top 61 cm of soil (Bennet et al., 1964). On the other extreme are the high concentrations of NO_3^- that accumulate in plants which are terminated in growth by drouth and which are no longer capable of assimilating NO_3^- that continues to be absorbed by the roots. Plants that are damaged by diseases or pests will also show higher nutrient N concentrations than adjacent healthy plants.

Deficiency of another element that restricts crop growth normally results in elevated N concentrations. Conversely, supplying a deficient nutrient will dilute the N level in the larger mass of dry matter produced. Also, any nutrient excess that reduces growth may cause the N percentage to increase.

These several variables acting alone or in consonance can influence markedly the level of N likely to occur in a given sample of crop tissue. For this reason the values presented in Table 3 cover rather wide ranges and even wider ranges are possible under unusual conditions.

B. Amounts of N in Crops and Distribution within the Crop

1. TOTAL N

The quantities of N found in different crops vary greatly with species and environments in which the crops are produced. Yield level variations account for a portion of the differential while varied genotype accumulation and storage characteristics explain much of the remainder. Substantial-

Table 4—Approximate total N content and distribution in good yields of major
harvested crops. (Derived in part from Anonymous, 1972).

Crop	Plant parts	Yield	Total N†
		kg/ha	
Alfalfa	Total forage	18,000	510
Bermuda (coastal)	Total forage	22,500	565
Brome—nonirrigated	Total forage	7,000	175
irrigated	Total forage	10,000	250
Corn	Grain	10,000	150
	Stover	9,000	80
Cotton	Lint + seed	1,690 + 2,530	105
	Stalks, leaves, burrs	5,000	70
Potatoes (Irish)	Tubers	56,000	170
	Vines	5,000	115
Rice	Grain	7,900	85
	Straw	10,000	40
Sorghum	Grain	9,000	135
	Stover	5,000	60
Soybeans	Grain	2,800	180
	Straw	5,400	75
Sugar beets	Roots	68,000	140
	Tops	36,000	145
Sugarcane	Stalks	112,000	180
	Tops and trash	50,000	225
Wheat	Grain	5,400	110
	Straw	6,000	45

† Substantial variation from these values can occur depending on soil N status and
fertilization effected, i.e., total N of the end product continues to increase with added
N beyond that required for maximum yield.

ly larger N quantities are generally found in the harvest of forage legumes
than in representative yields of cereal crops (Table 4). Grain legumes, like
soybeans, also contain large amounts of N in the harvested product, but
with both leguminous types a portion of the N has come from symbiotic
fixation of atmospheric N_2 during the crop year.

Some of the grass crops harvested for hay remove large quantities of N
and require high N fertilizer rates for maximum yields. In sugarcane
production, relatively little N is removed in the stalks from which sugar is
extracted. A huge loss occurs, however, when the tops and leaf trash on the
cane field are burned to facilitate stalk harvest. Sugar beets also become
heavy soil N depleters when the tops are removed as does corn and other
grain crops when harvested for silage. Table 4 gives some examples of N
removals in harvested crops.

More precise differentiation of N into plant parts has been determined
for corn and soybeans (Hanway, 1962; Hanway & Weber, 1971) and for
small grains (Doneen, 1934; Boatwright & Haas, 1961). However, few data
have been reported on the amounts of N remaining in the roots at harvest
due to the difficulty in obtaining representative samples of this component.
Attempts at measuring the total weight of roots below the soil surface have
given values in the range of 2,000–5,000 kg/ha for some of the grain crops.
As a general rule, total N in roots at harvest probably is about half of that in

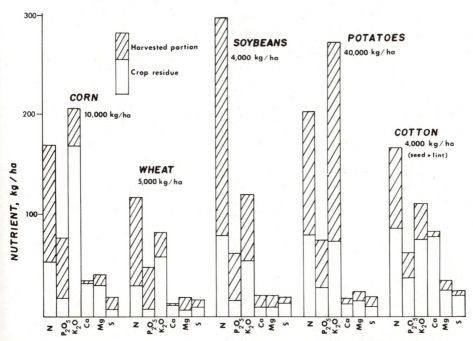

Fig. 7—Nutrients contained in harvested portion and residues of good yields of major agricultural crops (Olson, 1978).

the above ground forage component. Legume roots are a different case and the notably greater amount of N which they contain constitutes most of the benefit to grain crops following a legume which has been harvested for hay throughout its growth period.

The distribution of N between the harvested portion and crop residue, presented graphically in Fig. 7, demonstrates why N is depleted more rapidly than any other element with intensive cropping. Even if residues are returned to the soil in grain crop farming, the 110–170 kg/ha removed still represents a major replacement cost and accounts for N being the foremost fertilizer expense item.

2. TOXIC NO$_3^-$ ACCUMULATION

The distribution pattern described above applies under conditions of satisfactory water supply for normal growth and nominal rate of fertilizer N application. Most of the N in the crop is in combined organic forms at physiological maturity. When a grain crop like corn, oats, or grain sorghum is killed or severely damaged by drouth, excess NO$_3^-$ is likely to accumulate in plant parts. This occurs, in part, because roots continue absorbing NO$_3^-$ after the topmost part of the plant is dead. The highest concentrations are found in the lower stalk, but some of this NO$_3^-$ will disappear in the fermentation that occurs during ensiling. If the NO$_3^-$-N concentration is in excess of 0.2% NO$_3^-$-N in either silage or the directly pastured forage, death

Table 5—The NO_3^--N contents of foodstuffs collected from grocery store shelves in Sweden in December 1971 (Schupan, 1972).

Kind of material	No. of samples	Range	NO_3^--N, mg/kg of fresh weight
Cereals	3		0
Maize frozen	1		13
Tomato	2		13
Cauliflower	2		14
Potato	1		15
Sweet pepper	5		16
Borecole (not fertilized)	2		17
Cabbage	3		17
Mushroom (not fertilized)	3		17
Green pea frozen	4		19
Cucumber	2		24
Red cabbage	3		24
Brussels sprouts frozen	4		27
Applepuree (baby food)	3		28
Mixed vegetables (baby food)	4		35
Carrotpuree (baby food)	3		37
French bean frozen	3		59
Carrotdrink (baby food)	5		60
Broccoli frozen	3	(55–100)	78
Borecole	2		103
Isberg lettuce	7	(78–115)	104
Spinach with ham (baby food)	4		107
Spinach frozen	16	(146–541)	313
Dill frozen	2		490
Lettuce	16	(109–1,222)	554

of animals due to methemoglobinemia is likely (Garner, 1958). Even half that amount can cause reduced milk production of cows and give evidence of vitamin A deficiency in the animals.

Most of the NO_3^- toxicity problems experienced by farmers have been with feeding of corn residues to animals. Conditions most often associated with excessive NO_3^- accumulation in corn vegetation include drouth, high temperature, low P or K levels, and excessive K fertilization (Kurtz & Smith, 1966). Liberal N fertilization, whether from an organic or inorganic source, must be involved as well.

Some of the leaf crops grown for human consumption have also been found to accumulate excessive NO_3^- and are possibly responsible for methemoglobinemia, particularly in babies (see Chapt. 16, D. R. Keeney). Crops known to accumulate excessive NO_3^- in the forage with overly liberal fertilization include spinach, lettuce, and dill. Rarely is this a problem with adults because the leaf crops constitute only a fractional part of the diet. In contrast, the infant has a much smaller diversity of foods and a less developed digestive tract. Table 5 presents results of a Swedish foodstuffs survey of fresh, frozen, and canned goods obtained from grocery store shelves (Schupan, 1972). Large variations existed among samples from different sources, but the mean values for the high accumulators demonstrate the magnitude of the problem.

3. TOXIC NO_2^- ACCUMULATION

Excessive levels of NH_3 in soils are known to be toxic to crop plants as well as many of the soil microflora. In many cases the interpretation may not have been exactly correct because, as shown by Birch and Eagle (1969), high levels of NH_3 were not toxic to seedlings in the absence of NO_2^-. The toxicity of NO_2^- to plants is well documented and is usually associated with acid conditions (Maze, 1911; Curtis, 1949; Bingham et al., 1954). At neutral to alkaline conditions, these researchers found no toxic effects of the NO_2^-, which suggests that toxicities might be expected in strongly acid soils due to differential response of *Nitrosomonas* and *Nitrobacter*, the latter being quite intolerant of strong acidity. However, due to chemical instability, NO_2^- is seldom present in acid soils (see Chapter 9). The soil NO_2^- level responsible for toxicity varies with plants from as little as 2 mg/kg to as much as 100 mg NO_2^--N/kg soil (Curtis, 1949; Bingham et al., 1954; Oke, 1966).

Under normal circumstances, free NO_2^- is not found in healthy plant tissues. The plant root appears to have a defensive mechanism for ridding itself of any accumulated NO_2^- through the Van Slyke reaction: $R\text{-}NH_2 + HNO_2 \rightarrow ROH + H_2O + N_2$ whereby an alcohol, water, and elemental N are produced as an amine is destroyed (Maze, 1911; Mevius & Dikussar, 1930). However, with certain abnormal conditions for growth, for example, after herbicide treatments of susceptible plants, NO_2^- has been found to accumulate in plant tissues and to contribute to the observed toxicity symptoms and eventual plant death from herbicide action (Klepper, 1974, 1975). Accumulation occurred because of the inhibition of the normal process of NO_2^- reduction. Herbicide-treated plants, further, have been noted to emit NO_x gasses to the atmosphere in proportion to the amount of herbicide employed (Klepper, 1978) with a concurrent dissipation of the accumulated NO_2^-.

The herbicide action is probably not greatly different from the combination of conditions that causes NO_2^- to accumulate in human systems from ingestion of high levels of NO_3^-. It is not the NO_3^-, but the reaction of the derived NO_2^- with hemoglobin that produces the methemoglobin responsible for the toxicity (Jaffe & Heller, 1964). Where leaf crops like canned spinach have caused the malady, the responsible factor was usually the conversion of NO_3^- to NO_2^- during storage (Phillips, 1968).

C. Influence of Fertilizer N on Crop Quality

The predominant positive impact of fertilizer N on crop quality is in its enhancement of the protein content of grain and forage crops. Recent literature on fertilization of cereals shows increasing protein content of the grain with increasing N rate applied above some rate which increases yield at the expense of protein. Protein increase from that point onward is nearly a straight line function with rate of N until a protein maximum for the variety is reached, a rate well in excess of that required for maximum yield.

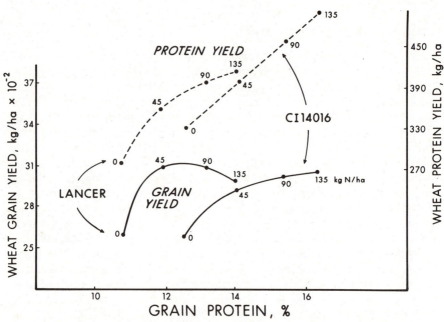

Fig. 8—Grain yield and protein responses of two winter wheat varieties to applied N in 6 experiments in Nebraska in 1969–70. Numbers on each of the curves represent rate of N applied (Johnson et al., 1973).

All cultivars do not respond similarly to N fertilizer. For example, in the work of Johnson et al. (1973), the experimental wheat variety 'CI 14016' consistently contained around 2% more protein than Lancer at all N rates. Maximum protein content of CI 14016 was 16.3% and was still increasing at the highest N rate of 135 kg/ha. In contrast, Lancer had about the same maximum grain yield, but a protein content of only 11.8% attained with an N rate of only 45 kg/ha (Fig. 8).

Overstimulation of vegetative growth is a hazard for cereals grown in drier regions with high N fertility levels. This effect on vegetative growth is more apparent with N than with other nutrients. Excessive vegetative growth uses available soil water at the expense of grain yield, but if grain does form, the shriveled kernels are higher than normal in protein percentage. Weather conditions of the High Plains of the United States makes this occurrence rather common in winter wheat producing areas.

Studies on time of N application to winter wheat indicate comparable yield results from spring and fall applications in most years (Olson & Rhoades, 1953; Schlehuber & Tucker, 1967). Protein percentage tends to become progressively higher as time of N application is delayed, providing sufficient rainfall is received to carry fertilizer N to the root system. With foliar treatments, N application may be delayed so that protein, but not yield, is increased (Finney et al., 1957). Protein increases ranging from 2 to 4% have, thus, been noted from spraying urea solution on winter wheat in three increments to a modest total rate of 50 kg N/ha, with the greatest

Table 6—Effect of foliar fertilization on grain yield and protein of four crops (R. A. Olson, unpublished data).

Crop and fertilizer†	Grain yield	Grain protein
kg N + P + K + S/ha	kg/ha	%
Winter wheat		
0 + 0 + 0 + 0	4,995	15.8
60 + 15 + 30 + 6	4,928	16.4
60 + 0 + 30 + 6	4,725	17.8
60 + 15 + 0 + 6	4,995	17.2
L.S.D. (0.05)	n.s.	
Corn		
0 + 0 + 0 + 0	10,080	8.3
50 + 10 + 12 + 5	10,017	9.3
50 + 0 + 12 + 5	9,765	9.7
50 + 10 + 0 + 5	10,143	9.6
50 + 10 + 12 + 0	10,017	9.5
L.S.D. (0.05)	n.s.	
Grain sorghum		
0 + 0 + 0 + 0	8,370	7.8
40 + 8 + 10 + 4	8,235	9.3
40 + 0 + 10 + 4	8,640	9.7
40 + 8 + 0 + 4	8,303	9.8
40 + 8 + 10 + 0	8,573	9.5
L.S.D. (0.05)	n.s.	
Soybeans		
0 + 0 + 0 + 0	3,105	37.8
80 + 8 + 24 + 4	2,430	42.4
80 + 0 + 24 + 4	2,093	41.8
80 + 8 + 0 + 4	2,098	43.4
80 + 8 + 24 + 0	2,090	42.8
L.S.D. (0.05)	270	

† Fertilizer solutions prepared with urea, potassium polyphosphate, K_2SO_4, Na_2SO_4, and a wetting agent and applied to the total amounts indicated in three equal foliar applications at 10-day intervals beginning at the start of grain filling. Mead Field Laboratory, Nebraska, 1976.

benefit being derived from the treatment made at early blooming stage (Sadaphal & Sas, 1966). Similarly, the data in Table 6 give evidence of about 2% increases in the protein concentration of three cereals receiving moderate N treatment by successive foliar applications during the grain filling period, but no grain yield increases resulted from such late applications. Soybeans, on the other hand, suffered serious yield loss due to foliar burn, but increased 5% in protein content. Grain of all four crops treated with the foliar combinations was distinctly lower in moisture than the controls at harvest, indicating dessication of foliage by the applied salts.

Applications of economic rates of N for corn just before the grand period of growth have also been more effective for maximizing protein yields than have earlier treatments. It is especially evident in this crop that a large N supply that stimulates heavy vegetative growth early is not especially efficient in producing and storing proteinaceous materials in the grain.

The literature is quite consistent in reporting that fertilizer N applied to rice during the growing season rather than before, or at planting, enhances

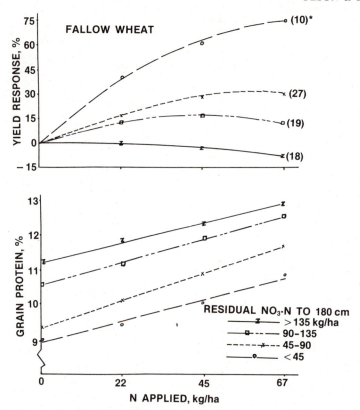

Fig. 9—The impact of residual NO_3^--N in soil on grain yield response of fallow wheat to fertilizer N and on grain protein content. *Number of field experiments involved: a total of 74 in Nebraska in 1962–68 (Olson et al., 1976).

total N uptake and utilization for grain protein (FAO/IAEA, 1970b; Beachell et al., 1972; DeDatta et al., 1972). It is frequently observed that N applied early in the season is responsible for increased plant height and lodging as well as less grain and total protein than are later applications. With submerged rice culture, applications shortly before primordial initiation have proved optimum for maximum protein production with a given nominal N rate on representative soils with sufficient N reserves to get the plant well started. In the case of upland rice, however, such delayed applications may be disappointing if followed by an extended period of limited rainfall.

The quantity of residual mineral N in the rooting zone significantly influences crop responses to applied fertilizer N. Data obtained in a fallow-cropping area in Nebraska is illustrated in Fig. 9. When residual soil NO_3^--N levels were > 135 kg/ha, grain yields were depressed by all increments of fertilizer N. On the other hand, if the residual N was < 45 kg/ha, the yield response curve rose sharply up to 67 kg/ha of fertilizer N. Large differences in protein content of the grain existed without applied N for the different residual NO_3^- levels, but contrary to the yield effects these differences persisted as parallel straight line functions across all fertilizer N rates. Protein level probably could not have been maintained equivalent across these residual ranges with any reasonable rate of applied N. Apparently some modest level of residual NO_3^- must exist in the lower rooting zone and be utilized late during the growing season for maximum protein production. Since water is extracted from the upper rooting zone early in the season, applied fertilizer N is likely to remain in the surface soil throughout the year of application. Further, as with corn, it is probable that N taken up late from the lower rooting zone will be channeled more directly to grain with less immobilization in vegetative parts.

Availability of N from various sources will not explain all the variability of protein of cereals from field to field, but adding other parameters assists materially. For example, a multiple regression combining four factors, i.e., residual soil NO_3^-, available water to 1.5 m at planting, maximum air temperature for the 15- to 20-day period before maturity, and precipitation during 40 to 45 days before maturity, explained 96% of the variation in protein concentrations in wheat grain (Smika & Greb, 1973).

Nitrogen carriers do not vary in utilization efficiency of applied N for protein production if each carrier is applied in accordance with its limitations. Submerged rice culture is an exceptional situation where NO_3^- carriers are inappropriate. Although delayed application of the NO_3^- source was helpful in investigations by FAO/IAEA (1970b) postponed application did not compensate completely. Variations in the denitrification potentials among the 15 sites were suggested when radically better results were obtained at some locations when applications were delayed rather than applied at transplanting. A companion greenhouse study showed microbial denitrification to be responsible for the disparities since irradiation sterilization of the soils before potting eliminated differences in denitrification losses of N among the soils.

Closely related to the protein content of wheat grain is baking quality (loaf volume) of flour produced from the grain. Greater loaf volume is associated with increased protein content of most good baking quality wheats (Aitken & Geddes, 1939; Finney & Barmore, 1948). It has been demonstrated that increased protein coming from N fertilization has the same beneficial effect (Fajersson, 1968; Ramig & Rhoades, 1963).

As was noted previously (II-D), increased protein in cereal grains from applied N does not necessarily indicate greater nutritional quality. The prolamin fraction (zeins) of the protein increases at a more rapid rate than the glutelins, globulins, and albumins as N fertilization is increased. Lysine, the most limiting essential amino acid in cereals, is highest in the globulins and albumins. Thus, dilution of grain protein by increased zein results in a lower nutritional quality per unit of protein for human and monogastric animal consumption. Generally, the increased total protein in corn, barley, and wheat resulting from N fertilization has a lower percentage of lysine although total lysine in kg/ha is usually greater because of the greater yield (Price, 1950; Nelson, 1969; Gustafsson, 1969). Other essential amino acids also decline with increasing prolamin as noted in rice (Juliano, 1972). These short-comings in grain protein quality are not of concern for ruminant animals which possess the capacity for synthesizing their own essential amino acids so that the enhanced protein from fertilization represents higher quality feed irrespective of amino acid distribution. Nor is the decline in lysine with increasing protein axiomatic. A mutant of barley selected in Denmark has shown increased lysine almost in proportion to elevation in protein from increasing fertilizer N rate (Ingversen et al., 1973). Similar observations have been made with certain genotypes of wheat (Hucklesby et al., 1971) and barley (Zoschke, 1970).

Beyond the protein benefit to cereals and forages, a number of different quality features in other crops are attributed to fertilizer N. Greater leafiness and succulence of leaf crops accompanying generous N fertilization increases their acceptance in the market place. An associated hazard exists requiring care in producing these features, viz., that free NO_3^- in the leaf tissue be kept within levels that are acceptable from a toxicity standpoint. Adequate N is essential as well for most other vegetable and fruit crops to meet standards of appearance and taste required in the market by the consumer.

Fertilizer N can also have adverse effects on crop quality. In the small grains, lodging caused by too much N can be responsible for shriveled kernels, and result in abnormally high protein in the grain. Kernels of lodged grain are likely to sprout in the head and give rise to high alpha-amylose activity which is deleterious to baking quality of the flour (Fajersson, 1968). Diseases, like rust and mildew on wheat, are also detrimental to baking quality and are often associated with excessive applications of N fertilizer. Higher protein from excess N is damaging in barley used for malting purposes and also in soft wheats used for pastry four. Decreased shelf life of several fruits and vegetables, declines in certain amino acids, and loss of flavor of some commodities have also been noted with excessive N fertilization (Schupan, 1972).

D. Impact of Applied Fertilizer N on Crop Utilization of Other Nutrients

Fertilizer N has long been recognized to have maximum stimulative action on plant growth by reason of the numerous N functions related to growth. Not the least among these functions is the role of applied N in facilitating crop uptake of other essential nutrients. The most thoroughly documented example among the nutrients is the increase in fertilizer P uptake from soil treated with fertilizer N (Miller & Ohlrogge, 1958; Robertson et al., 1954; Olson & Dreier, 1956b; Grunes et al., 1958; Blair et al., 1971). A precise mode of action is not clearly defined, but the physical association of NH_4^+ with phosphate is necessary for the phenomenon to occur. Contentions are made for the involvement of both physiological and physical/chemical phenomena including root proliferation in the zone of placement to assist the phosphate intake. In any case, the interaction of the two ionic species is responsible for more ready availability to the plant of monoammonium phosphate than other common P carriers, (Olsen et al., 1950; Olson et al., 1956) and explains the advantages of the long-accepted practice of having a low N/P ratio in starter fertilizer formulations. Recent investigations have given further evidence that concurrent placement of anhydrous NH_3 with an ammonium phosphate solution in soil may significantly enhance uptake of the fertilizer P compared with separate placement of the materials (Murphy et al., 1978).

Placement of some fertilizer N in a band with Zn carriers in calcareous soils has been found beneficial to the Zn uptake by plants (Viets et al., 1957; Pumphrey et al., 1963). In this case, soil acidification and preservation of the applied Zn in a more soluble form in the restricted soil zone of Zn placement is considered the predominant factor involved. Enhanced root proliferation effected by N in that zone is probably a contributing factor. Similar synergistic action to that of N on Zn has been observed with N on the uptake of other heavy metals by crops on alkaline soils. In the case of Fe, a shortage of N has been noted to limit production of riboflavin which must be present for normal translocation of Fe to shoots from the roots (Wallace, 1971).

Aside from its impact on crop root action in absorbing companion fertilizer nutrients from a point of common placement in the soil, applied N is also responsible for a crop taking up a greater total quantity of several other nutrients from the soil. This can be largely attributed to the greater total yield of crop even though the concentration of a particular nutrient may have been reduced considerably because of dilution with the higher yield. The data of Table 7 for corn are indicative of this increased requirement for other nutrients as yield is enhanced by increments of fertilizer N (Barber & Olson, 1968). Dilutions in P, Zn, and Mn concentrations were imposed by N increments, but they were more than compensated for in total uptake by the yield increases. Concentrations of K were essentially constant across N rates, but S and Cu were increased in concentration so much that their total uptakes were disproportionate to the increased yields. On the other hand, some crops, like grasses grown for forage, have notably higher K concentration with increasing rate of N applied (Russell et al., 1954). Note in Fig. 10 that 270 kg/ha elevated the K concentration in bromegrass around 50% all season despite the fact that yields were increased by 7 times the control. However, N did not affect the concentrations of P, Ca, and Mg in this experiment.

Table 7—Influence of increments of applied N on grain yield of irrigated corn and on nutrients contained in the aboveground crop (Barber & Olson, 1968).

N applied	Crop yield		Nutrients contained in crop						
	Grain	Stover	N	P	K	S	Zn	Mn	Cu
			kg/ha						
0	5,229	4,613	78	14.2	148	10.2	0.22	0.15	0.016
84	8,379	5,175	127	19.9	200	12.7	0.20	0.19	0.019
169	9,828	5,850	183	20.3	235	15.1	0.23	0.19	0.032
253	9,765	6,863	200	22.1	245	17.1	0.26	0.23	0.046

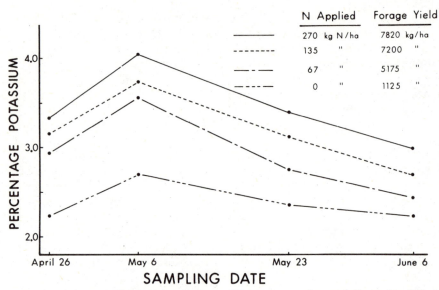

Fig. 10—Percentage K in bromegrass forage at different harvest dates as influenced by N ferti-lizer. Each value is a mean of fall- and spring-applied NH₄NO₃ and fall-applied anhydrous NH₃ (Russell et al., 1954).

E. Efficient Use of N Fertilizer

Because of the transitory nature of N in soil, its tendency for loss from the soil, and its potential for becoming a pollutant of air and water, ferti-lizer N should receive more care in its overall management than any other of the primary and secondary plant nutrients. Nitrogen can also have a more deleterious impact on the chemical properties of many soils than other major fertilizer nutrients due to its tendency to accelerate soil acidification processes.

Acidity induced by the fertilizer N must be neutralized sooner or later if soil conditions and productivity are to be maintained. The acidifying effect of fertilizer N varies among carriers. Some N compounds, such as KNO_3, tend to increase soil pH, while others such as $(NH_4)_2SO_4$, decrease soil pH quite markedly. A generally accepted factor has been that 1.8 kg of effective lime is required to neutralize each kg of NH_4^+-N applied in fertilizer. Exact amounts of acidity developed can also be expected to vary with the propor-tions of N lost by leaching and denitrification and by removal by crops. Some authors (Kurtz & Smith, 1966) have estimated that the neutralization factor should be as large as 3.6 kg of lime/kg of fertilizer N, but this does not allow for N-removal by crops. Fortunately, the lime requirement of soils can be determined independently by soil tests so that fertilizer-induced soil acidity is automatically corrected in a good soil management program.

1. RATES OF APPLICATION

A consideration of the dynamics of N in soils might indicate that the recommendation for fertilizer N should be a net amount resulting from

Table 8. Nitrogen recommendations for grain crops in Nebraska based on residual NO$_3^-$ in the rooting profile to 180 cm at planting. (Derived from Anonymous, 1973–74).

Soil NO$_3^-$ to 180 cm, kg/ha	Crop and yield objective				
	Corn, with yield objective in kg/ha (bu/a)				
	1,350 (50)	6,300 (100)	9,450 (150)	12,600 (200)	15,750 (250)
	Recommended N rate, kg/ha				
< 56	55	110	180	250	325
112	0	56	125	190	270
168	0	0	65	135	215
225	0	0	10	80	160
280	0	0	0	20	100
338	0	0	0	0	45
>338	0	0	0	0	0
	Grain sorghum, with yield objective in kg/ha (bu/a)				
	2,025 (30)	4,050 (60)	6,075 (90)	8,100 (120)	10,125 (150)
	Recommended N rate, kg/ha				
Soil organic matter content = 1%					
< 56	10	56	90	125	170
112	0	0	35	65	110
168	0	0	0	10	56
>168	0	0	0	0	0
Soil organic matter content = 2%					
< 56	0	35	65	100	145
112	0	0	10	45	90
168	0	0	0	0	35
>168	0	0	0	0	0
Soil organic matter content = 3%					
< 56	0	10	45	80	125
112	0	0	0	20	65
168	0	0	0	0	10
>168	0	0	0	0	0

	Winter wheat, with N recommendation by regions					
	Western	West south central	Central	East south central	North-eastern	South-eastern
	Recommended N rate, kg/ha					
< 15	80	100	90	90	90	90
56	65	80	65	65	65	65
112	45	45	45	35	35	35
168	20	35	20	20	0	0
>168	0	0	0	0	0	0

several debits and credits. Residual fertilizer NO$_3^-$ in the soil along with nonfertilizer sources, such as decomposition of soil organic matter and crop residues, would be credits. In some cases, contributions would also come from animal manures and from symbiotic fixation. Crop removals and losses from leaching, denitrification, and erosion would be deducted and a balance obtained. Intuitively, the net result of these gains and losses would be adjustments to make to the N content of the expected mature crop as an estimate of the amount of fertilizer N to be applied.

In actual practice, the major determinant of the appropriate rate of fertilizer N is the crop and its probable yield. The amount of mineral N in the soil can also be important and is often taken into account particularly in the deeper soils of cooler, drier climates. Analyses to determine soil levels of residual mineral N are frequently used in such areas as a partial basis for fertilizer recommendations. Chapter 17 treats the topic of soil testing for N in detail.

Crops vary widely in their requirements for fertilizer N. Perennial forage grasses, such as Bermuda and brome, generally remove more N than cereal crops (Fisher & Caldwell, 1958; Russell et al., 1954; Pesek et al., 1971). Corn, with sufficient water for optimum growth and maximum yield, requires N rates that would cause serious lodging and yield loss with hard red winter wheats on most soils. However, soft wheats as grown in the Pacific Northwest region of the United States and in the lowlands of Europe require N rates comparable to those for high corn yields.

Table 8, prepared for Nebraska conditions, serves as an example of fertilizer N rates that take into account the projected yields of major crops as well as levels of mineral N and organic matter in the soils.

In more humid regions where soils are less likely to contain residual inorganic N, fertilizer N recommendations are often made by multiplying the expected yield of the crop by a factor that has been found applicable for that area. In the East North Central region of the United States, for example, a fertilizer N recommendation for corn may be 1.2 × bu. expected yield. The multiplier can be adjusted for the economic optimum return by taking into account the relative price of corn and the cost of fertilizer N (*Illinois Agronomy Handbook,* Anonymous, 1978). The N-requirement of high-yielding crops is often large in comparison to the variations in the N-supplying abilities of the soils of the area. In such situations, inorganic and organic N from the soil may be regarded as reasonably constant and may not be routinely considered separately as adjustments to the fertilizer N requirement.

Efficiency in crop use of fertilizer N generally decreases with increasing application rates. Correspondingly, the economic return from each added unit of N declines so that fertilizer N is not recommended beyond the point where the cost of the last fertilizer increment is equal to price return of the additional yield produced.

2. TIME OF APPLICATION

For cereals, maximum efficiency of fertilizer N is obtained by the latest possible application that is compatible with the stage of development that permits ready uptake (Olson et al., 1964a; Welch et al., 1971; Herron et al., 1971; Hucklesby et al., 1971; Miller et al., 1975). This is due, at least in part, to the presence of the late-applied N in quantity throughout the period of grain formation without having been partially consumed for production of unnecessary vegetative growth. Furthermore, the opportunities for N losses by leaching, runoff, volatilization, and denitrification are reduced because an active root system is present for absorbing the fertilizer N when it is applied.

Late N application is not desirable for crops like sugarbeets, however, which need a large N supply through the early portion of the growing season for developing the extensive leaf surface area and maximum photosynthetic capability. But in the final quarter of its growth, the beet crop should be under conditions of mild N deficiency to assure maximum sugar yield. High levels of N late in the season promote continued vegetative growth at the expense of sugar storage in the roots (Rhoades & Harris, 1954; Alexander et al., 1954). Late N treatment is also ineffective for seed production of cool season grasses like brome because the seed head potential is determined at the very earliest stages of growth in the spring (Newell, 1973).

Depending on soil and climatic conditions, N application may need to be divided into two or more increments during the growing season for maximum fertilizer efficiency. Seasonal increments are especially likely to be necessary with humid region cropping, and with irrigation of sandy soils. Thus, the practice of metering N solutions into the irrigation water has proved effective (Fischbach, 1964) and is commonplace for corn grown on sandy lands under center pivot irrigation systems.

The chemical carrier of fertilizer N has a significant influence on the effective time of application. Nitrate sources can be subject to severe losses from leaching or denitrification in the winter and early spring under humid climatic conditions, but are ideal for application just prior to heavy crop demand for N due to the ready mobility of NO_3^-. Anhydrous NH_3, on the other hand, is virtually nonleachable and not subject to denitrification during wet conditions between fall and early spring so long as soil temperatures are below approximately 5°C. But it is not nitrified fast enough to move rapidly with water into the crop rooting zone if applied just before heavy demand for N by the crop.

3. METHOD OF APPLICATION

The major portion of the N used in the USA is applied by the four following methods: (i) surface broadcast; (ii) incorporation by an attachment to plow, cultivator, planter or disk; (iii) injection of liquids and gases through knives by pumping; and (iv) distribution in irrigation water. From the standpoint of fertilizer-N efficiency the effectiveness of each method varies with soil and climatic conditions and character of the N carrier.

Surface broadcasting has its shortcomings; especially in sod or orchard situations where it is not possible to immediately incorporate the fertilizer by some tillage procedure. Given drying conditions, especially on soils of high pH, urea and some of the NH_4^+ salts can lose significant quantities of the N as NH_3 by volatilization to the atmosphere (Volk, 1959; Ernst & Massey, 1960; Meyer et al., 1961; Fenn & Kissel, 1973). Tillage or planting equipment that affords immediate incorporation with the soil alleviates this hazard.

The injection of liquids and gases, especially, anhydrous NH_3, requires careful adaptation to soil water conditions. Volatilization losses can occur if the soil is either too dry or too wet. Furthermore soil hardening can result if the soil is too wet at the time of application because the puddling effected by tillage is further aggravated by the dispersing action of the NH_4^+. Even with favorable water, injection must be at a sufficient depth to minimize losses. Several factors including soil texture, soil pH, knife spacing, and application rate interact to determine the most desirable depth (Jackson & Chang, 1947).

Application through irrigation water (fertigation) affords the unique opportunity of supplying N throughout the season according to crop needs, a special advantage in sprinkler irrigation of sandy soils. A hazard associated with sprinkler irrigation is the possibility of nonuniform distribution caused by excessive wind. With irrigation by flooding methods, differential penetration of the treated water may occur due to variations in microrelief and soil permeability across a field.

Particular care must be taken with all application methods to insure that appreciable concentrations of applied N do not contact planted seed. Solid forms of fertilizer N are usually highly soluble salts and with limited soil moisture they can have a severe plasmolyzing effect on the seed, delaying if not preventing germination. As little as 10–15 kg N/ha placed with the seed can reduce stands of some crops. Anhydrous NH_3 (Colliver & Welch, 1970) and high concentrations of NH_3 coming from diammonium phosphate and hydrolyzing urea granules (Olson & Dreier, 1956a) are particularly toxic to germinating seeds and young seedlings.

IV. INFLUENCE OF CLIMATE AND CROPPING SYSTEMS ON N USE

A. Climate and N Fertilization

Fertilization practices are often a reflection of the weather conditions that are likely to be encountered in the area. This is particularly true with N since its behavior in soil is greatly influenced by moisture and temperature. Some generalized limitations of climate on the use of fertilizer N for broad areas of the USA were outlined by Nelson and Uhland (1955).

As noted earlier, N is most effective when applied shortly before the period of most rapid growth and greatest demand by the crop. Sidedressing avoids losses due to wet weather prior to planting. A discussion of the reasons for sidedress application and the amounts of N to be applied is given in *Cornell Recommends for Field Crops* (Anonymous, 1979). On the other hand, sidedressing is not dependable where excessively wet or droughty weather is likely in the early season, since such conditions may cause an undesirable delay in application (Olson et al., 1964a). In the North Central area, the predominant N carrier is anhydrous NH_3, which is usually injected into the soil in the early spring before the crop is planted, although sidedressing after planting is also practiced to some extent. In areas where winters are cold and the soil is frozen into the spring, anhydrous NH_3 is sometimes applied in late fall to avoid the peak labor period during the early spring season.

In much of the western United States, evapotranspiration exceeds annual rainfall so that leaching is much more limited. Nevertheless, even in dry regions, inorganic N is often found at greater soil depths than would be predicted from average weather data. During exceptional storms or unusually wet seasons, greater than normal movement must occur. Also, Thomas (1970) and Thomas and Swoboda (1970) have discussed the possibilities that NO_3^- may penetrate more deeply than expected in some soils because of anion exclusion effects.

Although leaching may not be a problem in colder, drier regions, N losses still occur by denitrification. Recent work in the Prairie Provinces of Canada showed fall applications of N were usually not as effective in producing yield responses as spring applications. For example, in one series of 20 field experiments during a 4-year period in Alberta, fall applications were only 55% as efficient as spring. Losses from different carriers as measured by ^{15}N-labeling were 41%, 30%, and 16% from KNO_3, urea, and $(NH_4)_2SO_4$, respectively. No evidence for leaching could be found by sampling to the depth of 120 cm and analyzing for $^{15}NO_3^-$ (Malhi & Nyborg, 1978).

Since many of the cultivated soils in drier regions are irrigated, leaching occurs to varying degrees depending on the farm operators' skill in water management. Denitrification may also be severe when soils are over-irrigated.

In irrigated areas of the USA, anhydrous NH_3 is the most frequently used N carrier and applications are either made before or sidedressed after planting. Nitrogen solutions are also used, sometimes as sidedressed applications or applied in water with sprinkler irrigation. Other than for the use of fertigation, overall N management with irrigated soils in dry regions is little different than that employed in humid regions of comparable temperatures.

Nitrogen fertilization for soils in the dry tropics or in the dry seasons in the tropics is much like fertilization in drier areas of the USA. Fertilizer may be applied at longer periods of time before the crop is grown without loss by leaching. When N is applied after a rain and before a dry period is likely, the fertilizer should be incorporated since it will not be utilized effectively in the dry surface soil. As in dry regions under temperature climates, N must be used judiciously in relation to water available in the dry tropics. In seasons of poor rainfall, application of N fertilizer at planting may stimulate early growth and hasten the onset of moisture stress, whereas application at midseason may afford a positive yield response. It is important to avoid the use of $(NH_4)_2SO_4$ as the primary N source on low buffer capacity tropical soils because of its strong acidifying action (Russell, 1968).

In the humid tropics or in wet seasons in the tropics, rainfall is so great and waterlogged conditions are so frequent that losses due to denitrification and leaching are extensive. Under these conditions, N fertilizer is best applied by parcelling it out in small doses throughout the rainy periods to assure that some N will be present for the crop at all times. At present, fertilizer N recommendations in the humid tropics are usually based on field trials in a particular region and on the cropping system employed. Work reported by Nye (1950) suggests that the soil C/N increases with bush fallow and that N response is determined by that ratio.

Nitrogen fertilization of submerged soils for lowland rice requires special care (see Chapt. 12, W. H. Patrick). When the crop is continually flooded, urea and NH_4^+ or NH_3 carriers are preferred since much denitrification and loss of N is likely to occur with NO_3^- carriers. The practice in many areas is to apply a portion, perhaps one-fourth, of the N fertilizer at planting time with the remainder applied broadcast on the water approximately at the time of primordial initiation in the crop (FAO/IAEA, 1970b).

B. Nitrogen Carriers and Cropping Systems

For most crops and soils, the choice of the N carrier is more economic than agronomic. The ranking of anhydrous NH_3 as the leading carrier is due to its being the least expensive form on the market rather than to its superior characteristics as a plant nutrient. The fact that all kinds of fertilizer N are normally transformed to NO_3^- in the soil means that the initial form is not overly important in most situations.

V. NITROGEN FERTILIZATION IN THE FUTURE

There seem to be few agronomic reasons to expect that practices in the use of N fertilizer will change greatly in the next decade or so. Present practices seem to be well adapted to the agricultural conditions and systems. Imposition of unexpected economic constraints could, of course, arise and have marked effects (see Chapt. 20, E. R. Swanson).

Increasing cost of energy and shortages of natural gas have been discussed in connection with fertilizer use. While it seems that energy prices will certainly continue to increase, the manufacture of fertilizer for food production will probably not be greatly curtailed. As energy prices increase, food prices will probably also increase. Studies of the effect of energy prices on fertilizer use have shown that energy conservation for most other uses will likely be extreme before fertilizer supplies will be greatly affected. The fact that N-supplies in the USA in 1977–78 exceeded consumption and caused the shut down of some NH_3 plants lends credence to this expectation. Nevertheless, increased consumption and higher N costs seem fairly definite for the future. Fluctuations in production and use on the international scene affect the domestic market and are difficult to predict (Brown, 1979).

Crops that are presently being grown will likely continue to be grown on the same scale as at present. Minimum and conservation tillage practices will probably increase and these usually require somewhat higher levels of fertilizer N.

Since no process for fixing atmospheric N appears ready to displace the Haber process for making NH_3, this product is likely to continue to be the major carrier in extensive cropping systems in the forseeable future. Hydrogen, from sources other than petroleum, may eventually be used more extensively for NH_3 synthesis. More strict regulations on the use and transportation of NH_3 may develop for safety reasons, in which case, urea will probably be used more extensively. Regardless of which is the major carrier, useage of nitrification inhibitors is likely to increase.

Ten years from now, extensive, large scale agriculture is likely to be operating very much like it is now and it is even more likely that N fertilizer will continue to be used in increasing amounts. Economic factors, rather than agronomic developments, will probably prompt the changes in cropping systems and N useage.

LITERATURE CITED

Aitken, T. R., and W. F. Geddes. 1939. The relation between protein content and strength of gluten-enriched flours. Cereal Chem. 16:223–230.

Alexander, J. T., C. C. Schmer, L. P. Orleans, and R. H. Cotton. 1954. The effect of fertilizer applications on leaf and yield of sugar beets. Proc. Am. Soc. Sugar Beet Technol. 8:370–379.

Anonymous. 1972. Remember the plant food content of your crops. Better crops with plant food. 56(1):1–2. Potash Institute of North America.

Anonymous. 1973–74. Neb Guides G73-37, G74-112, and G74-174. Coop. Ext. Serv., Inst. Agr. and Nat. Res., University of Nebraska, Lincoln.

Anonymous. 1978. Illinois Agronomy Handbook 1979–80. p. 32–39. Circ. 1165. University of Illinois, Urbana.

Anonymous. 1979. Cornell recommends for field crops. p. 22–23. New York College of Agriculture at Cornell University, Ithaca, New York.

Aslam, M., R. C. Huffaker, and R. L. Travis. 1973. The interaction of respiration and photosynthesis in induction of nitrate reductase activity. Plant Physiol. 52:137–141.

Barber, S. A., and R. A. Olson. 1968. Fertilizer use on corn. p. 163–168. In L. B. Nelson, M. H. McVickar, R. D. Munson, L. F. Seatz, S. L. Tisdale, and W. C. White (ed.) Changing patterns in fertilizer use. Soil Sci. Soc. Am., Inc., Madison, Wis.

Barley, K. P. 1970. The configuration of the root system in relation to nutrient uptake. Adv. Agron. 22:159–201.

Beachell, H. M., R. S. Khush, and B. O. Juliano. 1972. Breeding for high protein content in rice. p. 419–428. In Rice breeding. Int. Rice Res. Inst., Los Baños, Philippines.

Beevers, L., and R. H. Hageman. 1969. Nitrate reduction in higher plants. Ann. Rev. Plant Physiol. 20:495–522.

Beevers, L., L. E. Schrader, D. Flesher, and R. H. Hageman. 1965. The role of light and nitrate in the induction of nitrate reductase in radish cotyledons and maize seedlings. Plant Physiol. 40:691–698.

Bennet, O. I., D. Boss, D. A. Ashley, V. J. Kilmer, and E. D. Richardson. 1964. Effects of soil moisture regime on yield, nutrient content, and evapotranspiration for three annual forage species. Agron. J. 56:195–198.

Berger, E. A. 1973. Different mechanisms of energy coupling for the active transport of proline and glutamine in Escherichia coli. Proc. Nat. Acad. Sci. U.S.A. 70:1514–1518.

Birch, P., and D. Eagle. 1969. Toxicity of seedlings to nitrite in sterilized composts. J. Hortic. Sci. 44:321–330.

Bingham, F. T., H. D. Chapman, and A. L. Pugh. 1954. Solution culture studies of nitrite toxicity to plants. Soil Sci. Soc. Am. Proc. 18:305–308.

Blair, G. J., C. P. Mamaril, and M. H. Miller. 1971. Influence of nitrogen source on phosphorus uptake by corn from soils differing in pH. Soil Sci. Soc. Am. Proc. 63:235–238.

Boatwright, C. O., and H. J. Haas. 1961. Development and composition of spring wheat as influenced by nitrogen and phosphorus fertilization. Agron. J. 53:33–36.

Brown, J. 1979. Nitrogen—round and round she goes. Fert. Prog. 10(6):16–18.

Burton, G. W. 1954. Coastal Bermuda grass. Georgia Agric. Exp. Stn. Bull. New Ser. 2.

Chevalier, P., and L. E. Schrader. 1977. Genotypic differences in nitrate absorption and partitioning of N among plant parts in maize. Crop Sci. 17:897–901.

Clapp, J. G., Jr., and D. S. Chambles. 1970. Influence of different defoliation systems on the regrowth of pearl millet, hybrid Sudan grass and two sorghum—Sudan grass hybrids from terminal, axillary, and basal buds. Crop Sci. 10:345-349.

Colliver, G., and L. F. Welch. 1970. Toxicity of preplant anhydrous ammonia to germination and early growth of corn: I. Field studies. Agron. J. 62:341-346.

Curtis, D. 1949. Nitrite injury on avocado and citrus seedlings in nutrients solution. Soil Sci. 68:441-450.

Daigger, L. A., D. H. Sander, and G. A. Peterson. 1976. Nitrogen content of winter wheat during growth and maturation. Agron. J. 68:815-818.

DeDatta, S. K., W. N. Obcemea, and R. K. Jana. 1972. Protein content of rice grain as affected by nitrogen fertilizer and some triazines and substituted ureas. Agron. J. 64:785-788.

Doneen, L. D. 1934. Nitrogen in relation to composition, growth, and yield of wheat. Washington Agric. Exp. Stn. Bull. 296.

Elzam, O. E., and E. Epstein. 1965. Absorption of chloride by barley roots: Kinetics and selectivity. Plant Physiol. 40:620-624.

Ernst, J. W., and H. F. Massey. 1960. The effects of several factors on volatilization of ammonia formed from urea in the soil. Soil Sci. Soc. Am. Proc. 24:87-90.

Fajersson, F. 1961. Nitrogen fertilization and wheat quality. Agri. Hort. Genet. 19:1-195.

Food and Agriculture Organization, and International Atomic Energy Agency. 1970a. Fertilizer management practices for maize: Results of experiments with isotopes. Tech. Rep. Ser. no. 121, Int. Atomic Energy Agency, Vienna. p. 78.

Food and Agriculture Organization, and International Atomic Energy Agency. 1970b. Rice fertilization. Tech. Rep. Ser. no. 108, Int. Atomic Energy Agency, Vienna. p. 177.

Fenn, L. B., and D. E. Kissel. 1973. Ammonia volatilization from surface applications of ammonium compounds on calcareous soils: I. General theory. Soil Sci. Soc. Am. Proc. 38:606-610.

Finney, K. F., and M. A. Barmore. 1948. Loaf volume and protein content of hard winter and spring wheats. Cereal Chem. 25:291-312.

Finney, K. F., J. W Meyer, F. W. Smith, and H. C. Fryer. 1957. Effect of foliar spraying of Pawnee wheat with urea solutions on yield, protein content, and protein quality. Agron. J. 49:341-347.

Fischbach, P. E. 1964. Irrigate, fertilize in one operation. Nebr. Agric. Exp. Stn. Q. 8(Summer):15-17.

Fisher, F. L., and A. G. Caldwell. 1958. Coastal Bermuda grass as an irrigated hay crop. Progress Report 2035. Texas Agric. Exp. Stn.

Garner, G. B. 1958. Learn to live with nitrates. Mo. Agric. Exp. Stn. Bull. 708.

Gass, W. B., G. A. Peterson, R. D. Hauck, and R. A. Olson. 1971. Recovery of residual nitrogen by corn (Zea mays L.) from various soil depths as measured by ^{15}N tracer techniques. Soil Sci. Soc. Am. Proc. 35:290-294.

Grunes, D. L., H. R. Haise, and L. O. Fine. 1958. Proportionate uptake of soil and fertilizer phosphorus by plants as affected by nitrogen fertilization. II. Field experiments with sugar beets and potatoes. Soil Sci. Soc. Am. Proc. 22:49-52.

Gustafsson, A. 1969. A study of induced mutations in plants. p. 9-31. In Proc. Symp. Induced Mutations in Plants. Pullman, Wash., 19-27 June 1969. Int. Atomic Energy Agency, Vienna.

Hageman, R. H., R. L. Lambert, M. Loussaert, M. Dalling, and L. A. Klepper. 1976. p. 103-134. In Genetic improvement of seed proteins. Workshop Proc., Nat. Acad. Sci., Washington, D.C.

Hanway, J. J. 1962. Corn growth and composition in relation to soil fertility: II. Uptake of N, P and K and their distribution in different plant parts during the growing season. Agron. J. 54:217-222.

Hanway, J. J., and C. R. Weber. 1971. Accumulation of N, P and K by soybean (Glycine max. (L.) Merrill) plants. Agron. J. 63:406-408.

Herron, G. M., A. F. Dreier, A. D. Flowerday, W. L. Colville, and R. A. Olson. 1971. Residual mineral N accumulation in soil and its utilization by irrigated corn. Agron. J. 63:322-327.

Herron, G. M., D. W. Grimes, and J. T. Musick. 1963. Effects of soil moisture and nitrogen fertilization of irrigated grain sorghum on dry matter production and nitrogen uptake at selected stages of plant development. Agron. J. 55:393-396.

Higinbotham, N., B. Etherton, and R. J. Foster. 1967. Mineral ion contents and cell transmission electropotentials of pea and oat seedling tissue. Plant Physiol. 42:37-46.

Holt, E. C., and F. L. Fisher. 1960. Root development of coastal Bermuda grass with high nitrogen fertilization. Agron. J. 52:593–596.

Hucklesby, D. P., C. M. Brown, S. E. Howell, and R. H. Hageman. 1971. Late spring application of nitrogen from efficient utilization and enhanced production of grain protein of wheat. Agron. J. 63:274–276.

Huffaker, R. C., and D. W. Rains. 1978. Factors influencing nitrate acquisition by plants; assimilation and fate of reduced nitrogen. Vol. 2. p. 1–43. In D. R. Nielsen and J. G. MacDonald (ed.) Nitrogen in the environment. Academic Press, New York.

Hutchinson, G. L., R. J. Millington, and D. B. Peters. 1972. Atmospheric ammonia: Absorption by plant leaves. Science 175:771–772.

Ingversen, J. A., J. Anderson, H. Doll, and B. Køie. 1973. Nuclear techniques for seed protein improvement. Panel Proc. Ser.: 193–198. Int. Atomic Energy Agency, Vienna.

Jackson, M. L., and S. C. Chang. 1947. Anhydrous ammonia retention by soils as influenced by depth of application, soil texture, moisture content, pH value and tilth. Agron. J. 39: 623–633.

Jaffe, E. R., and P. Heller. 1964. Methemoglobinemia in man. Prog. Hematol. 4:48.

Johnson, V. A., A. F. Dreier, and P. H. Grabouski. 1973. Yield and protein responses to nitrogen fertilizer of two winter wheat varieties differing in inherent protein content of their grain. Agron. J. 65:259–263.

Johnson, V. A., P. J. Mattern, K. D. Wilhelmi, and S. L. Kuhr. 1978. Seed protein improvement in common wheat: Opportunities and constraints. p. 23–32. In Seed protein improvement by nuclear techniques. Int. Atomic Energy Agency, STI/PUB/479. Vienna.

Jones, J. B., Jr. 1967. Interpretation of plant analysis for several agronomic crops. p. 49–58. In Soil testing and plant analysis. Special Publ. Ser. no. 2. Soil Sci. Soc. Am., Madison, Wis.

Juliano, B. O. 1972. Physicochemical properties of starch and protein in relation to grain quality and nutritional value of rice. p. 389–405. In Rice breeding. Int. Rice Res. Inst., Los Banos, Philippines.

Keller, W. D., and G. E. Smith. 1967. Ground water contamination by dissolved nitrate. Geol. Soc. Am., Spec. Pap. 90:48–59.

Klepper, L. 1974. A mode of action of herbicides: inhibition of the normal process of nitrite reduction. Nebr. Agric. Exp. Stn. Res. Bull. 259. p. 42.

Klepper, L. 1975. Inhibition of nitrite reduction by photosynthetic inhibitors. Weed Sci. 23: 188–190.

Klepper, L. 1978. Nitric oxide (NO) and nitrogen dioxide (NO_2) emissions from herbicide-treated soybean plants. Atmos. Environ. 13:537–542.

Kmoch, H. G., R. E. Ramig, R. L. Fox, and F. E. Koehler. 1957. Root development of winter wheat as influenced by soil moisture and nitrogen fertilization. Agron. J. 49:20–25.

Kurtz, L. T., and G. E. Smith. 1966. Nitrogen fertility requirements. p. 195–235. In W. H. Pierre, S. R. Aldrich, and W. P. Martin (ed.) Advances in corn production: Principles and practices. Iowa State Univ. Press, Ames, Iowa.

Linscott, D. L., R. L. Fox, and R. C. Lipps. 1962. Corn root distribution and moisture extraction in relation to nitrogen fertilization and soil properties. Agron. J. 54:185–189.

Lockman, R. B. 1972. Mineral composition of grain sorghum plant samples. Part III: Suggested nutrient sufficiency limits at various stages of growth. Commun. Soil Sci. Plant Anal. 3:295–304.

Lorenz, O. A., K. B. Tyler, and F. S. Fullmer. 1964. Plant analysis for determining the nutritional status of potatoes. p. 226–240. In C. Bould (ed.) Plant analysis and fertilizer problems. Am. Soc. Hortic. Sci.

Lott, J. N. A., and T. K. Hodges. 1970. In situ localization of ATPase in plant root. Plant Physiol. 46(Suppl.):23.

Malhi, S. S., and M. Nyborg. 1978. The fate of fall-applied nitrogen in northern Alberta as determined by the ^{15}N-technique. Int. Congr. Soil Sci., Trans. 11th (Edmonton, Canada) I:370.

Maze, P. 1911. Recherches sur la formation de l' aide nitreux dans la cellule vegetale et animale. C. R. Acad. Sci., Paris. 153:357.

Mertz, E. T., L. S. Bates, and O. E. Nelson. 1964. Mutant gene that changes protein composition and increases lysine content of maize endosperm. Science 145:279–280.

Mevius, W., and J. Dikussar. 1930. Nitrite als Stickstoffguellen fur hohere Planzen. Jahrb. Wiss. Bot. 73:633.

Meyer, R. D., R. A. Olson, and H. F. Rhoades. 1961. Ammonia losses from fertilized Nebraska soils. Agron. J. 53:241–244.

Mikkelsen, D. S., and R. R. Hunziker. 1971. A plant analysis survey of California rice. Agrichem. Age 14:18–22.

Miller, E. C. 1939. A physiological study of the winter wheat plant at different stages of development. Kans. Agric. Exp. Stn. Bull. 47.

Miller, H. F., J. Kavanaugh, and G. W. Thomas. 1975. Time of N application and yields of corn in wet, alluvial soils. Agron. J. 67:401–404.

Miller, M. H., and A. J. Ohlrogge. 1958. Principles of nutrient uptake from fertilizer bands: 1. Effect of placement of nitrogen fertilizer on the uptake of band-placed phosphorus at different soil phosphorus levels. Agron. J. 50:95–97.

Morgan, M. A. 1970. Direct and indirect effects of calcium and magnesium on a proposed nitrate absorption mechanism. Ph.D. Thesis. North Carolina State Univ. Xerox Univ. Microfilms, Ann Arbor, Mich. (Diss. Abstr. 36:160; Order no. 71-10290).

Munck, L., K. E. Karlsson, A. Hagberg, and B. Eggum. 1970. Gene for improved nutritional value in barley and seed protein. Science 168:985–987.

Murphy, L. S., D. R. Leikam, R. E. Lomond, and P. J. Gallagher. 1978. Dual application of N and P—better agronomics and economics? Fert. Solutions 22(4):8–20.

Nelson, L. B., and R. E. Uhland. 1955. Factors that influence loss of fall applied fertilizers and their probable importance in different sections of the United States. Soil Sci. Soc. Am. Proc. 19:492–496.

Nelson, O. E. 1969. The modification by mutation of protein quality in maize. p. 41–54. In New approaches to breeding for improved plant protein. Panel Proc. Ser. Int. Atomic Energy Agency Panel of 17–24 June 1968. Rostanga, Sweden.

Nelson, O. E., E. T. Mertz, and L. S. Bates. 1965. Second mutant gene affecting the amino acid pattern of maize endosperm proteins. Science 150:1,469.

Newell, L. C. 1973. Smooth bromegrass. p. 254–262. In M. E. Heath, D. S. Metcalfe, and R. F. Barnes (ed.) Forages, the science of grassland agriculture. 3rd ed. Iowa State Univ. Press, Ames. Iowa.

Noggle, R. N., and G. J. Fritz. 1976. Introductory plant physiology. p. 277–279. Prentice-Hall. Englewood Cliffs, N.J.

Nye, P. H. 1950. The relation between nitrogen responses and previous soil treatment and the carbon/nitrogen ratios in soils of the Gold Coast Savannah areas. Int. Congr. Soil Sci., Trans. 4th (Amsterdam) I:246–249.

Oke, O. 1966. Nitrite toxicity to plants. Nature (London) 212:528.

Olsen, S. R., W. R. Schmehl, F. S. Watenabe, C. O. Scott, W. A. Fuller, J. V. Jordan, and R. Kunkel. 1950. Utilization of phosphorus by various crops as affected by source of material and placement. Colo. Agric. Exp. Stn. Tech. Bull. 42.

Olson, R. A. 1978. The indispensable role of nitrogen in agricultural production. p. 1–31. In P. F. Pratt (ed.) Natl. Conf. Manage. of Nitrogen in Irrigated Agric. U.S. Nat. Sci. Found., USEPA, and Univ. of California, Riverside.

Olson, R. A., and A. F. Dreier. 1956a. Fertilizer placement for small grains in relation to crop stand and nutrient efficiency in Nebraska. Soil Sci. Soc. Am. Proc. 20:19–24.

Olson, R. A., and A. F. Dreier. 1956b. Nitrogen, a key factor in fertilizer phosphorus efficiency. Soil Sci. Soc. Am. Proc. 20:509–514.

Olson, R. A., A. F. Dreier, G. W. Lowrey, and A. D. Flowerday. 1956. Availability of phosphate carriers to small grains and subsequent clover in relation to: 1. Nature of soil and method of placement. Agron. J. 48:106–111.

Olson, R. A., A. F. Dreier, C. Thompson, K. Frank, and P. H. Grabouski. 1964a. Using fertilizer nitrogen effectively on grain crops. Nebr. Agric. Exp. Stn. Bull. 479. p. 10–14.

Olson, R. A., K. D. Frank, E. J. Deibert, A. F. Dreier, D. H. Sander, and V. A. Johnson. 1976. Impact of residual mineral N in soil on grain protein yields of winter wheat and corn. Agron. J. 68:769–772.

Olson, R. A., and H. F. Rhoades. 1953. Commercial fertilizers for winter wheat in relation to the properties of Nebraska soils. Nebr. Agric. Exp. Stn. Res. Bull. 172.

Olson, R. A., E. C. Seim, and J. Muir. 1973. Influence of agricultural practices on water quality in Nebraska: A survey of streams, ground water, and precipitation. Water Resour. 9:301–311.

Olson, R. A., C. A. Thompson, P. H. Grabouski, D. D. Stukenholtz, K. D. Frank, and A. F. Dreier. 1964b. Water requirement of grain crops as modified by fertilizer use. Agron. J. 56:427–432.

Pate, J. S. 1971. Movement of nitrogenous solutes in plants. p. 165–187. In Nitrogen-15 in soil plant studies. Int. Atomic Energy Agency STI/PUB/278. Vienna.

Patrick, W. H., Jr., and D. S. Mikkelsen. 1971. Plant nutrient behavior in flooded soil. p. 187–215. *In* R. A. Olson, T. J. Army, J. J. Hanway, and V. J. Kilmer (ed.) Fertilizer technology and use. Soil Sci. Soc. Am., Inc., Madison, Wis.

Perry, L. J., and R. A. Olson. 1975. Yield and quality of corn and grain sorghum grain and residues as influenced by fertilization. Agron. J. 67:816–818.

Pesek, J., G. Stanford, and N. L. Case. 1971. Nitrogen production and use. p. 263–264. *In* R. A. Olson, T. J. Army, J. J. Hanway, and V. J. Kilmer (ed.) Fertilizer technology and use. Soil Sci. Soc. Am., Inc., Madison, Wis.

Phillips, W. E. J. 1968. Changes in nitrate and nitrite contents of fresh and processed spinach during storage. J. Agric. Food Chem. 16:88–91.

Porter, L. K., F. G. Viets, Jr., G. L. Hutchinson. 1972. Air containing nitrogen-15 ammonia: foliar absorption by corn seedlings. Science 175:759–780.

Price, S. A. 1950. The amino acid composition of whole wheat in relation to its protein content. Cereal Chem. 27:73–74.

Pumphrey, F. H., F. E. Koehler, R. R. Allmaras, and S. Roberts. 1963. Method and rate of applying zinc sulfate for corn on zinc-deficient soils in western Nebraska. Agron. J. 55:235–238.

Ramig, R. E., and H. F. Rhoades. 1963. Interrelationships of soil-moisture level at planting time and nitrogen fertilization on winter wheat production. Agron. J. 55:123–127.

Rao, K. P., and D. W. Rains. 1976. Nitrate absorption by barley: 1. Kinetics and energetics. Plant Physiol. 57:55–58.

Raven, J. A., and F. A. Smith. 1976. Nitrogen assimilation and transport in vascular land plants in relation to intracellular pH regulation. New Phytol. 76:415–431.

Rhoades, H. F., and L. Harris. 1954. Cropping and fertilization practices for the production of sugar beets in western Nebraska. Proc. Am. Soc. Sugar Beet Technol. 8:71–80.

Robertson, W. K., A. J. Ohlrogge, and D. M. Kinch. 1954. Phosphorus utilization by corn as affected by placement and nitrogen and potassium fertilization. Soil Sci. 77:219–226.

Rumberg, C. B., and F. A. Sneva. 1970. Accumulation and loss of nitrogen during growth and maturation of cereal rye (*Secale cereale*). Agron. J. 62:311–313.

Russell, E. W. 1968. The place of fertilizers in food crop economy of tropical Africa. The Fertilizer Soc., London.

Russell, J. S., C. W. Bourg, and H. F. Rhoades. 1954. Effect of nitrogen fertilizer on the nitrogen, phosphorus, and cation contents of bromegrass. Soil Sci. Soc. Am. Proc. 18:292–296.

Russell, R. S. 1977. Plant root systems. p. 82–86. McGraw-Hill Book Co. (UK) Limited, Maidenhead, Berkshire, England.

Sabbe, W. E., J. L. Keogh, R. Maples, and L. H. Hileman. 1972. Nutrient analysis of Arkansas cotton and soybean leaf tissue. Arkansas Farm Res. 21:2.

Sabbe, W. E., and A. J. MacKenzie. 1973. Plant analysis as an aid to cotton fertilization. p. 299–313. *In* L. M. Walsh and J. D. Beaton (ed.) Soil testing and plant analysis (Revised ed.) Soil Sci. Soc. Am., Madison, Wis.

Sadaphal, M. N., and N. P. Das. 1966. Effect of spraying urea on winter wheat (*Triticum aestivum*). Agron. J. 58:137–141.

Schlehuber, A. M., and B. B. Tucker. 1967. Culture of wheat. *In* K. S. Quisenberry and L. P. Reitz (ed.) Wheat and wheat improvement. Agronomy 13:117–179.

Schmehl, W. R., and R. P. Humbert. 1964. Nutrient deficiencies in sugar crops. p. 415–450. *In* H. B. Sprague (ed.) Hunger signs in crops. David McKay Co., New York.

Schrader, L. C., D. Domska, P. U. Jung, Jr., and L. A. Peterson. 1972. Uptake and assimilation of ammonium-N and nitrate-N and their influence on the growth of corn (*Zea mays* L.). Agron. J. 64:690–695.

Schupan, W. 1972. Effects of the application of inorganic and organic manures on the market quality and on the biological value of agricultural products. p. 198–224. *In* Effects of intensive fertilizer use on the human environment. Soils Bull. FAO, Rome.

Singh, R., and J. D. Axtell. 1973. High lysine mutant gene (*hl*) that improves protein quality and biological value of grain sorghum. Crop Sci. 13:535–539.

Smika, D. E., and P. H. Grabouski. 1976. Anhydrous ammonia applications during fallow for winter wheat production. Agron. J. 68:919–922.

Smika, D. E., and B. W. Greb. 1973. Protein content of winter wheat grain as related to soil and climate factors in the semi-arid central Great Plains. Agron. J. 65:433–436.

Stryer, L. 1975. Biochemistry. W. H. Freeman & Co., San Francisco, Calif.

Tabatabai, M. A., and J. M. Laflen. 1976. Nitrogen and sulfur content and pH of precipitation in Iowa. J. Environ. Qual. 5:108–112.

Terman, G. L., R.E. Ramig, A. F. Dreier, and R. A. Olson. 1969. Yield-protein relationships in wheat grain as affected by nitrogen and water. Agron. J. 61:755–759.

Thomas, G. W. 1970. Soil and climatic factors which affect nutrient mobility. p. 1–20. *In* O. P. Engelstad (ed.). Nutrient mobility in soils: Accumulation and loss. Spec. Publ. no. 4. Soil Sci. Soc. Am., Madison, Wis.

Thomas, G. W., and A. R. Swoboda. 1970. Anion exclusion effects on chloride movement in soils. Soil Sci. 110:163–166.

Ulrich, A. 1950. Critical nitrate levels of sugar beets estimated from analysis of petioles and blades with special reference to yields and sucrose concentrations. Soil Sci. 69:291–309.

Viets, F. G., Jr. 1962. Fertilizers and the efficient use of water. Adv. Agron. 14:223–264.

Viets, F. G., Jr., L. C. Boawan, and C. L. Crawford. 1957. The effect of nitrogen and type of nitrogen carrier on plant uptake of indigenous and applied zinc. Soil Sci. Soc. Am. Proc. 21:197–201.

Viets, F. G., Jr., and R. H. Hageman. 1971. Factors affecting the accumulation of nitrate in soil, water, and plants. USDA Agric. Handbook no. 413. p. 63.

Volk, G. M. 1959. Volatile loss of ammonia following surface application of urea to turf or bare soils. Agron. J. 51:746–749.

Wallace, A. 1971. Root excretions in tobacco plants and possible implications on the iron nutrition of higher plants. p. 138. *In* Regulation of the micronutrient status of plants by chelating agents and other factors. Edwards Bros. Inc., Ann Arbor, Mich.

Wallace, A., and R. T. Mueller. 1957. Ammonium and nitrate nitrogen absorption by rough lemon seedlings. Proc. Am. Soc. Hortic. Sci. 69:183–188.

Wallace, W., and J. S. Pate. 1967. Nitrate assimilation in higher plants with special reference to the cocklebur (*Xanthium pennsylvanicum* Wallr.). Ann. Bot. 31:213–228.

Ward, R. C., D. A. Whitney, and D. G. Westfall. 1973. Plant analysis as an aid in fertilizing small grains. p. 329–348. *In* L. M. Walsh and J. D. Beaton (ed.) Soil testing and plant analysis (Revised ed.). Soil Sci. Soc. Am., Madison, Wis.

Warncke, D. D., and S. A. Barber. 1973. Ammonium and nitrate uptake by corn (*Zea mays* L.) as influenced by nitrogen concentration and NH_4^+/NO_3^- ratio. Agron. J. 65:950–953.

Welch, L. F., D. L. Mulvaney, M. G. Oldham, L. V. Boone, and J. W. Pendleton. 1971. Corn yields with fall, spring and sidedress nitrogen. Agron. J. 63:119–123.

Wright, M. J., and R. L. Davison. 1964. Nitrate accumulations in crops and nitrate poisoning in animals. Adv. Agron. 16:197–247.

Zoschke, M. 1970. Effect of additional nitrogen nutrition at later growth stages on protein content and quality in barley. p. 345–356. *In* Proc. Symp. Improving Plant Protein by Nuclear Techniques. Int. Atomic Energy Agency. Vienna.

16 Nitrogen Management for Maximum Efficiency and Minimum Pollution[1]

DENNIS R. KEENEY

University of Wisconsin
Madison, Wisconsin

I. INTRODUCTION

It is clear that agriculture must evolve towards conserving nonrenewable energy resources and minimizing adverse environmental impacts. Of the essential plant nutrients which can realistically be managed, N undoubtedly has the greatest potential environmental and health impact. Further, while small relative to total U.S. energy use, N fertilizer manufacturing has the largest energy requirement for any single facet of production agriculture. The objectives of this chapter are twofold: (i) to consider the impacts of N in the environment, and (ii) to examine various management systems for conservation of N (and, hence, minimization of pollution) in agro-ecosystems. Much of the discussion will draw on principles detailed in other chapters.

II. N REQUIREMENTS FOR FOOD AND FIBER

The National Research Council (NRC, 1978) estimated yearly worldwide biological N fixation on agricultural lands at 89×10^6 tons (10^{12} g) and industrial fertilizer N production at 49×10^6 tons in 1976. Atmospheric inputs from combustion and lightning would add to this total. In contrast, utilizing a daily protein-N requirement of 4–6 g/capita per day, the annual protein-N uptake requires only $6-9 \times 10^6$ tons N/year (Bolin & Arrhenius, 1977), yet malnutrition is a fact of life in many parts of the world. The disparity results from inefficient food distribution, wastage during storage, losses in the conversion of plant proteins to animal proteins, low quality proteins, and the fact that production of food calories also depends on the plant N supply (Bolin & Arrhenius, 1977). Indeed, Sukhatme (1977) presents convincing arguments that, on average, per capita protein supply exceeds needs by about 60% for almost all of the developing countries; inadequate energy intake thus becomes a major factor in protein malnutrition.

[1] Contribution from The College of Agricultural and Life Sciences, University of Wisconsin-Madison, Madison, WI 53706.

If food production must increase by 88% in year 2000 to adequately feed the anticipated 6–7 billion population and this must be done on only 10% more arable land than present (Chancellor & Ross, 1976), then agricultural utilization of N (biological N fixation and fertilizers) would have to at least double (to about 280×10^6 tons of N/year) because the additional lands and land currently used will likely require higher rates of N to sustain yields. Long-term alternatives include utilizing more legume and cereal grains directly for human consumption, relying less on animal protein (Pimentel et al., 1975; Kaul, 1977; Sukhatme, 1977), and increasing the efficiency of use of our N resources. The first alternative involves major societal changes, especially in the developed nations; the second will be discussed further in this chapter.

III. ADVERSE HEALTH AND ENVIRONMENTAL IMPACTS OF N

A number of potentially adverse impacts of N on health and the environment can be identified (Table 1). Some have been known for decades; others have only recently been recognized. All can be, and usually are, caused by one or more anthropogenic (human) manipulations of the N cycle. Specific control strategies are often difficult and may well have secondary effects on other phases of the N cycle.

A. Nitrogen and Human Health

1. METHEMOGLOBINEMIA

Nitrate per se is relatively nontoxic to humans; acute NO_3^- poisoning in an adult requires a single oral ingestion of 1–2 g of NO_3^--N, far above normal exposure limits (Fassett, 1973; NRC, 1978). Rather, the adverse effect of high NO_3^- is actually due to NO_2^-; NO_3^- can be reduced to NO_2^- in

Table 1—Potential adverse environmental and health impacts of N.

Impact	Causative agents
Human health	
Methemoglobinemia in infants	Excess NO_3^- and NO_2^- in waters and food.
Cancer	Nitrosamines from NO_2^-, secondary amines.
Respiratory illness	Peroxyacyl nitrates, alkyl nitrates, NO_3^- aerosols, NO_2^-, HNO_3 vapor in urban atmospheres.
Animal health	
Environment	Excess NO_3^- in feed and water.
Eutrophication	Inorganic and organic N in surface waters.
Materials and ecosystem damage	HNO_3 aerosols in rainfall.
Plant toxicity	High levels of NO_2^- in soils.
Excessive plant growth	Excess available N.
Stratospheric ozone depletion	Nitrous oxide from nitrification, denitrification, stack emissions.

the intestine of some animals and in the human infant during the first few months of life (Comly, 1945; Shuval & Gruener, 1972). Nitrite is rapidly absorbed into the blood where it oxidizes the Fe of hemoglobin to the ferric state, forming methemoglobin which cannot function in oxygen transport. Lethal effects generally do not occur until over 50% of the blood hemoglobin is oxidized. The health, and possibly the heredity of the individual, may affect susceptibility (NRC, 1978).

The large majority of cases of acute NO_3^- toxicity have been reported from households with a private well water supply. Conversely, most of the ingested NO_3^- is from vegetables and of NO_2^- from saliva (NRC, 1978). The current U.S. Public Health Service drinking water standard of 10 mg/liter of NO_3^--N, which was suggested by the comprehensive review of Walton (1951), has withstood several critical examinations. Recently, the National Research Council (NRC, 1978) concluded that this standard "affords reasonable protection to the majority of newborns against methemoglobinemia derived from NO_3^--contaminated water supplies."

2. N-NITROSO COMPOUNDS

Secondary amines can react with nitrosating agents [primarily nitrous acid anhydride (N_2O_3)] in vivo or in the environment to form nitrosamines (Eq. [1]–[3]).

$$NO_2^- + H^+ \rightleftharpoons HNO_2 \tag{1}$$

$$2HNO_2 \rightleftharpoons N_2O_3 + H_2O \tag{2}$$

$$\begin{array}{c} R \\ \diagdown \\ \diagup \\ R \end{array} NH + N_2O_3 \rightarrow \begin{array}{c} R \\ \diagdown \\ \diagup \\ R \end{array} N - N = 0 \tag{3}$$

This reaction is favored by low pH (the pKa for 50% HNO_2 ionization is 3.36) and the pH for maximum rate of formation of nitrosamines is considered to be about 3.5–4 (Mirvish, 1970, 1975). While nitrosamines have been detected in the atmosphere, little is known about the mechanisms of formation (NRC, 1978).

A number of N-nitroso compounds have been shown to induce tumors in test animals on almost every vital tissue; they are also mutagenic and teratogenic (Magee, 1971, 1977; Shank, 1975; Crosby and Sawyer, 1976; MRC, 1978). While no definitive human studies of carcinogenicity are available, and no direct, scientifically documented cause-and-effect data have yet been gathered, some circumstantial evidence relating exposure to NO_3^- or NO_2^- to the incidence of cancer is available (NRC, 1978).

3. RESPIRATORY ILLNESS

High levels of NO_3^- aerosols and peroxyacetyl nitrates have long been known to cause lung irritation and respiratory illness to humans in urban areas (NRC, 1978). However, since the major sources of NO_x are mobile and stationary power sources, this topic is beyond the scope of a book on soil N.

B. Animal Health

While the effects of NO_3^-/NO_2^- toxicity in livestock are generally similar to those observed for humans, much higher doses are usually required and the response varies depending on the rate and type of food consumed, as well as the rate of consumption (Crawford et al., 1966). While the incidence of NO_3^- poisoning in livestock has been considered to be quite high (NRC, 1972), this conclusion is not supported by the literature (Deeb & Sloan, 1975).

C. Environmental Impacts

1. EUTROPHICATION

Surface waters, particularly lakes and estuaries, will have increased biological productivity when nutrient input, particularly N and P, increases. Such changes may be desirable (e.g., an increase in fish production in low productivity water bodies). However, overenrichment (eutrophication) brings about many generally undesirable changes including proliferation of algae and aquatic macrophytes, dissolved oxygen depletion in the bottom water, and a decrease in water clarity (Brezonik, 1969). Most low-productivity lakes are commonly believed to be P- rather than N-limited, while the reverse is often true for eutrophic lakes. A survey of 49 U.S. lakes (Miller et al., 1974) showed P to be limiting to algal growth in 35 of these lakes. While the reasons are not clearly understood, the productivity of estuaries is quite often limited by N (Goldman, 1976).

The N cycle in aquatic systems is in many ways more complex than in terrestrial systems (Brezonik, 1972; Keeney, 1973), and the role of N in eutrophication can seldom be specifically quantified. Most N reaches surface waters from nonpoint sources (primarily from surface runoff and in some cases from ground water). Even if these sources are minimized, N fixation by algae along with N in atmospheric precipitation, may provide sufficient N that eutrophication continues unchecked.

2. MATERIALS AND ECOSYSTEM DAMAGE DUE TO ACID RAIN

Acid pH (<5.7) rainfall occurs over large areas of northern Europe and the northeastern USA (NRC, 1978), and the trend with time is toward a decrease in pH and an expansion of the affected area. At least in the USA, the acidity is largely due to strong mineral acids (H_2SO_4 and HNO_3 with a minor contribution from HCl). The relative contribution from HNO_3 has increased with time. For example, Likens (1976) found the contribution of

HNO_3 to rainfall acidity at Hubbard Brook, New Hampshire increased from 15% in 1964 to 30% in 1974. Nearly all of this HNO_3 arises from combustion of fossil fuels and subsequent oxidation of NO_x (NRC, 1978).

The material damage and ecological effects of acid rain are difficult to document. Effects could include acidification of soils with loss of exchangeable cations and a decline in soil fertility (Norton, 1976; McFee et al., 1976); vegetative damage (Tamm & Cowling, 1976); acidification of lakes with subsequent effects on algae, macrophytes, invertebrates, and fish (Likens, 1976); and increased mobility of metals (Gorham & Gordon, 1963).

3. TOXICITY OF NO_2^- TO PLANTS

While significant quantities of NO_2^- seldom occur in soils (Chapter 2, J. L. Young and R. W. Aldag) elevated levels of NO_2^- have been observed in alkaline soils treated with urea or NH_3 fertilizers (see Chapt. 7, E. L. Schmidt). Depression in plant growth due to NO_2^- has been documented (Court et al., 1962); toxicity seems to develop in soils containing > 10 mg/kg of NO_2^--N and is enhanced by acidic pH (Bingham et al., 1954). Since acid soil conditions preclude accumulation of NO_2^-, phytotoxicity from NO_2^- is not a major agronomic problem.

4. EXCESSIVE N IN TERRESTRIAL SYSTEMS

Several problems, depending on the nature of the crop, can occur if excess available N is present in agronomic systems. High NO_3^- may accumulate in forage or silage (Wright & Davison, 1964) or crop quality may suffer (Viets, 1965). Much less is known about the effects of excess N inputs to nonmanaged terrestrial ecosystems; major ecological effects such as stimulation of grass and shrub growth would be expected.

5. EFFECTS ON STRATOSPHERIC OZONE

In the past decade there has been increased concern over human activities that could deplete the stratospheric ozone (O_3) layer which shields the biosphere from harmful exposures to UV radiation (Crutzen & Ehalt, 1977; NRC, 1978). Several potential threats to the O_3 layer exist, including water vapor and NO_x from supersonic transports, chlorofluorocarbons (freons), and production of N_2O from nitrification and denitrification.

In the stratosphere, O_3 is formed by photolysis of molecular oxygen by far UV radiation; this does not occur in the lower atmosphere since these wavelengths do not penetrate below the stratosphere. There are several natural sinks and sources of O_3 and, thus, the O_3 shield varies little over time (NRC, 1978).

One of the sinks for O_3 is NO (NRC, 1978). Nitrous oxide is a major source of NO. It is transported to the stratosphere by turbulent diffusion and undergoes photolysis by absorbing UV radiation which produces N_2 and singlet oxygen [O('D)]. Singlet oxygen, which also is produced in the photolysis of O_3, reacts with N_2O to produce predominantly N_2 and O_2 (Eq. [5]), but also some NO (Eq. [6]).

$$NO + O_3 \rightarrow NO_2 + O_2 \qquad [4]$$

$$N_2O + O('D) \rightarrow N_2 + O_2 \qquad [5]$$

$$N_2O + O('D) \rightarrow 2NO \qquad [6]$$

Large increases in stratospheric N_2O could lead to increased exposure of the biosphere to UV and appreciable trapping of outgoing planetary IR radiation (Crutzen, 1974). Thus, concern has been expressed of the possibility of significant O_3 depletion due to greatly increased N_2O from nitrification and denitrification resulting from increased anthropogenic use of fixed N (Crutzen, 1974; CAST, 1976; Sze & Rice, 1976; Hahn & Junge, 1977; Crutzen & Ehalt, 1977; Johnston, 1977). Combustion of fossil fuel also produces N_2O (Pierotti & Rasmussen, 1976; Weiss & Craig, 1976), but this appears to be a minor source compared to terrestrial nitrification and denitrification (NRC, 1978).

An increase in solar UV radiation at the earth's surface would have several effects; mortality from skin cancer would increase and unpredictable effects would likely occur at the ecosystem level and to the climate (NRC, 1978).

Prediction of the potential impact on the O_3 layer due to increases in the cycling of fixed N is fraught with uncertainties since accurate values for several critical variables and transformations are not known. These include the fate of anthropogenically fixed N over the short term (1–2 decades), the fraction of the denitrification products as N_2O, and the atmospheric residence time of N_2O. The most recent analysis (NRC, 1978) indicates a "most probable" range of O_3 depletion of 1.5–3.5% sometime in the 21st or 22nd century. This impact is much smaller than earlier estimates and led to the conclusion that "the current value to society of these activities that contribute to global N fixation far exceeds the potential cost of any moderate . . . postponement of action to reduce the threat of future ozone depletion by N_2O." The report did emphasize, however, that this issue should be reexamined as new information becomes available.

D. Perspectives

While several real and potential health and environmental impacts of N exist, agriculture and forestry practices have major effects only on those impacts which involve excess NO_3^- in drinking water, eutrophication, and perhaps on O_3 depletion. An intelligent risk assessment analysis of these impacts is exceedingly difficult, especially when human health and social goals and values are involved.

The growing worldwide demands for food will involve a continued increase in the use of fixed N in agriculture. Much of this increase must come from fertilizer. However, given current estimates of limited fossil fuel reserves and exponential growth of energy use, N fertilizers will inevitably increase in cost and perhaps in scarcity. Thus, a "self-correcting" scenario can be envisaged, since agronomic efficiency of N utilization must increase leading to lower accumulation of N in undesirable environmental pools.

The long-term evolution of N use in agriculture likely will involve increased reliance on biological N fixation and better management of wastes (including use of urban and industrial wastes if they are available and environmentally acceptable). Nitrogen fertilizer will continue to play an important role in many farm systems, but it will be managed as a scarce resource. The remainder of this chapter will discuss the principles and practices of managing N to maximize its efficiency of use concurrent with minimizing potential environmental impacts.

IV. TRENDS IN ANTHROPOGENIC N FIXATION

While current and projected industrial N fixation can be determined with some accuracy, biological N fixation estimates are subject to much uncertainty. The degree of uncertainty will increase as the degree of aggregation increases, i.e., from local to national to global scales.

A. Worldwide

Industrial N fertilizer consumption has risen markedly since the 1950's (Table 2). Since 1970, a large portion of this increase has been in Russia, East Europe, and Asia. In 1965, for example, North America and West Europe accounted for 60% of the world consumption; in 1980, their projected demand is 37%. This trend will continue. The UNIDO estimates a total demand of about 165×10^6 tons N by year 2000 and they predict a surplus of available production of about 1.4×10^6 tons N in the developed countries and a deficit of about 2.8×10^6 tons N in developing countries by year 2000.

Table 2—World fertilizer consumption and future demand.[†]

Year	N. America	W. Europe	Russia and E. Europe	Japan and Oceania	Asia	Africa and Latin America	Total
			million metric tons N				
			Consumption				
1960	2.6	3.0	1.6	0.6	1.3	0.6	9.7
1965	4.6	4.3	3.3	0.8	2.4	1.2	16.4
1970	7.1	6.0	6.8	1.1	5.8	1.9	28.7
1974	8.8	7.4	10.1	1.0	8.4	2.9	38.6
			Future demand				
1980	12.3	10.0	16.9	1.1	15.3	4.6	60.2
1985	15.5	12.4	23.6	1.2	22.0	6.5	81.2
1990	19.1	15.0	31.4	1.4	29.9	8.7	105.5
2000	27.5	21.0	50.5	1.6	49.5	14.1	164.2

† From estimates by the United Nations Industrial Development Organization (UNIDO), supplied by W. C. White, The Fertilizer Institute, Washington, D.C.

Table 3—Global N fixation rates for terrestrial systems.†

System	N_2 fixed	Fixed N
	kg ha^{-1} year^{-1}	× 10^6 tons
Agricultural		
Legumes	140	35
Rice	30	4
Other crops	5	5
Grasslands	15	45
Forest	10	40
Other	2	10
Total		139

† Burns and Hardy (1975).

Table 4—Global sources of fixed N, 1976.†

Source	Amount, × 10^6 tons N/year
Biological fixation	
Agricultural lands	89
Other terrestrial	49
Ocean	
Pelagic	20–120
Sediments	10
Industrial fixation	
Fertilizer	49
Other industrial NH_3	21
Combustion	21
Wood burning	10–200
Lightning	30
Total	299–589

† NRC (1978).

Biological N fixation is more difficult to quantify and no estimates of trends are available. Global terrestrial N fixation has recently been estimated by Burns and Hardy (1975) at about 139 × 10^6 tons N/year, with about 89 × 10^6 tons occurring on managed agricultural lands (Table 3). Increases can be expected as more legumes are used and as improvements in biological N fixation are made (Hardy & Havelka, 1975).

A recent estimate of the global sources of fixed N (NRC, 1978) is given in Table 4. Great uncertainty exists in estimates of ocean N fixation and potential atmospheric inputs from wood burning. Using the lower total estimates (299 × 10^6 tons N/year), 1976 anthropogenic N fixation (agricultural lands, industrial, combustion, wood burning) contribution to the worldwide fixed N is about 190 × 10^6 tons N/year or 63% of the total. Agricultural activities account for about 46% of the global sources of fixed N, but fertilizers are only about 16% of the estimated worldwide N fixation.

B. United States

While industrial N fixation does not dominate the total N budget globally, it can be relatively large on the local and regional scales. An ex-

Table 5—Inputs and major transfers of fixed N for the United States (excluding Alaska and Hawaii) in 1975.[†]

Source or transfer	Amount	Amount due to human activities
	$\times 10^6$ tons N/year	
Biological N fixation		
Tilled land		
Nonleguminous fixation	0.7	--
Leguminous fixation	4.4	--
Pasture and range	2.9	--
Forest and federal lands	3.8	--
Other land	0.2	--
Total	12.0	4.3
Other inputs of fixed N		
Fertilizers	9.4	9.4
Other industrial fixation	3.1	3.1
Precipitation	5.4	2.7
Total	17.9	15.2
Total inputs	29.9	19.2
Major internal transfers of N		
Uptake by crops	16.8	16.8
Human wastes	1.3	1.3
Animal wastes	5.3	5.3

[†] NRC (1978).

ample is the U.S. N budget (NRC, 1978) given in Table 5. It is important that double-counting be avoided when an inventory such as Table 5 is being compiled; if the actual transfer routes of N within the system are not considered, overestimation of inputs may result. Crop residues and human and animal wastes may be important local sources, but they represent internal N transfers previously accounted for. Many earlier summaries of U.S. sources of N (Stanford et al., 1970; Viets & Hageman, 1971; Parr, 1973; Frere, 1976) have erred in treating wastes as sources rather than as transfers. No net transfer of mineralization from soil organic N was included in Table 3, on the assumption that an equivalent amount of N is immobilized as is mineralized. About 19.5×10^6 tons of N (65% of the total) originates for human activities with about 31% from fertilizer. Nearly 34% of the anthropogenic fixed N reappears as wastes.

V. SOURCES OF N POLLUTION

The impacts of N on health or the environment can be on a local or watershed (as is usually the case for ground- and surface-water pollution), regional (acid rain) or global (O_3 depletion) scale. It is imperative to define the extent of these impacts before control measures are adopted. For example, a theoretical consideration of N fertilizer impacts on ground water in Illinois (Klepper, 1978) indicated that if controls on fertilizer use were desired, they would be more effective regionally than statewide.

A. Point Sources

Point sources of N (e.g., sewage outfalls and industrial effluents) can cause localized but intense pollution problems. While engineering technology to control these sources is generally available, it is becoming increasingly cost-effective to apply many of these wastes to land, where they then become potential nonpoint sources (Brink, 1975; Chapt. 21, J. H. Smith and J. R. Peterson). Since these wastes can have agronomic benefit, their use in agriculture becomes a local N input to be managed just as any other N source.

B. Nonpoint Sources

Because of their dispersed nature, nonpoint sources have been, and will always be, difficult to define quantitatively and to control. There are numerous nonpoint sources of N, including atmosphere (aerosols, dust, and

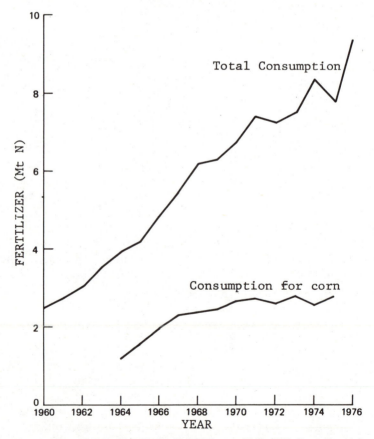

Fig. 1—Consumption of N fertilizer in the United States, 1960-1976. Reproduced from NRC (1978) with permission of the National Academy of Sciences, Washington, D.C.

precipitation); runoff from farmyards and feedlots, and agricultural, urban, and forested lands; and deep percolation from crop and pasture lands, and septic tank disposal fields. Nitrogen inputs to surface and ground waters are largely from nonpoint sources (Cooke & Williams, 1970; Loehr, 1974; McElroy et al., 1975). The National Commission on Water Quality (NCWQ, 1975) estimated that 5.1×10^6 tons of N would enter navigable U.S. streams in 1977, with 93% of this from nonpoint sources. They also estimated that 82% of the nonpoint source N would be from agricultural activities. Denitrification could be considered the major nonpoint source of N_2O to the atmosphere, as opposed to more identifiable point sources of NO_x such as auto exhausts and stationary combustion emissions.

The routes, rates of transport, and reactions of N must be known to evaluate nonpoint source N pollution. For example, while surface runoff may carry considerable N to surface waters, subsurface flow may add considerable NO_3^--N (Stewart et al., 1975). If runoff is lessened through improved conservation and tillage practices, more NO_3^- may be leached through the profile or be denitrified. Ground water recharge rates may be extremely slow, and the effect of leached NO_3^- may thus occur over extended periods.

A comprehensive consideration for sources and control of nonpoint N pollution is beyond the scope of this chapter. It is perhaps more informative to consider some N mass balances at the regional or ecosystem level to define the potential N pollution in agricultural systems where man is managing the N cycle to his advantage.

VI. SOME EXAMPLES OF AGRICULTURAL N POLLUTION

A. U.S. Corn Belt

The five major corn (*Zea mays* L.) producing states (Iowa, Illinois, Indiana, Minnesota, and Ohio) have nearly 60% of the total U.S. harvested corn area (13.7×10^6 ha in 1973). Nationwide, fertilizer use on corn increased parallel to total U.S. consumption until about 1967, but has leveled off while total U.S. fertilizer use increased (Fig. 1). The average N fertilizer rates for corn in the Corn Belt increased from about 60 kg N/ha in 1964 to about 120 kg N/ha in 1970, and have remained about constant since (NRC, 1978).

Studies by Welch (1972), Boone and Welch (1972), and Welch (1977, Univ. of Illinois, Urbana, pers. comm.) have shown that, since about 1965, more fertilizer N has been added to Illinois harvested cropland than is removed in harvested crops (Fig. 2). Concurrent with this increased fertilizer N use, the NO_3^--N concentration in Illinois rivers increased, although at least in the Kaskaskia River, the upward trend has not continued (Fig. 3).

Tile-drained fields can discharge considerable NO_3^--N, particularly if N fertilizer rates are above that recommended for optimum crop yields (Baker et al., 1975; Miller, 1979). Meek et al. (1969) and Gilliam et al.

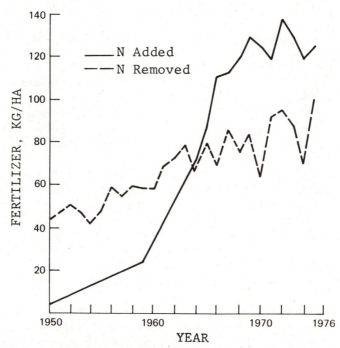

Fig. 2—Annual average rates of addition of fertilizer N and removal of crop N for corn production in Illinois. Reproduced from NRC (1978) with permission of the National
Academy of Sciences, Washington, D.C.

(1979) demonstrated that raising the water table by drainage control enhanced denitrification and lessened considerably the NO_3^--N in tile drainage
waters.

The positive correlation between fertilizer use and surface water NO_3^--
N concentrations in Illinois streams (Commoner, 1970) sparked considerable debate and national interest on the possible links between high NO_3^-
waters and fertilizer use. In 1971, the Illinois Pollution Control Board conducted numerous public hearings on the problem. There are several possible
sources of NO_3^--N to the surface and ground waters of Illinois, including
fertilizer, mineralized organic soil N, sewage, animal manures, and the
atmosphere. However, only fertilizer and mineralized N were important
sources in the watersheds examined. Aldrich[2] pointed out that during the
period of rapid NO_3^--N increase in the water, land use also was changing to
soybeans (*Glycine max* L.) at the expense of land planted to small grains
and meadow. This greater land area in cultivation would result in more soil
N being mineralized, along with less scavaging of residual N, especially in
the spring.

The statistical relationship between various factors such as land use or
fertilizer application rate and NO_3^- in surface and ground waters of various
Illinois watersheds has been examined in several studies (Taylor, 1973;

[2] S. R. Aldrich. 1972. Supplemental statement. *In* The matter of plant nutrients, R 71-15.
Illinois Pollut. Control Board, Chicago.

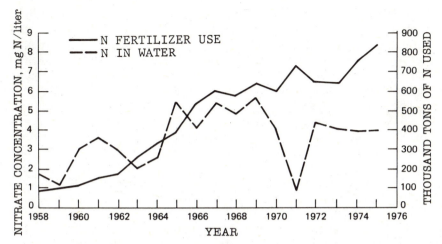

Fig. 3—Trends in the use of N fertilizer in Illinois and in the average NO₃⁻-N concentrations in the Kaskaskia River at Shelbyville, Illinois. Reproduced from NRC (1978) with permission of the National Academy of Sciences, Washington, D.C.

Abrams & Barr, 1974; Gentzsch et al., 1974; Klepper, 1978). Each of these studies found a strong positive correlation between fertilizer use and NO_3^--N levels; however, all lacked accurate data for both parameters, and due to the interrelationships of many variables (e.g., fertilizer use and land in crops), it was impossible to isolate the effects of any single factor. Gentzsch et al. (1974) found that poorly drained soils with matric horizons at shallow depths in Southern Illinois were low in NO_3^--N, but that the amount of live-stock and the rate of fertilizer applied were positively related to NO_3^--N in soils and well waters in the other soils examined.

The Illinois Pollution Control Board ultimately concluded that the NO_3^--N levels in certain Illinois streams, particularly in the east central area, were significantly increased by the use of fertilizer N. They also determined, however, that individual farm management was intimately involved, since well-managed farms likely contributed less leachable NO_3^--N than farms that produce average yields.

B. Irrigated Agriculture

Much of the increase in the yield and quality of agricultural crops has been due to improved cultural practices. Soil water is frequently the limiting plant growth factor, even in humid climates. Thus, it is not surprising that irrigation has expanded rapidly, especially in arid U.S. regions. In 1890, about 1.6 million ha were irrigated in the U.S. By 1965, this had increased to 16 million ha (Law & Witherow, 1971) and presently is greater than 22 million ha (Law & Witherow, 1971; Irrig. J., 1975). Irrigated lands accounted for 24% of the total value of U.S. crop production in 10% of the crop area in 1965 (Law & Witherow, 1971).

Residual NO_3^- in the soil profile is leached with the irrigation water (see Chapt. 11, D. R. Nielsen and J. W. Biggar). High rates of leaching and nitrification in the permeable soils commonly irrigated, and relatively high N fertilizer application rates (due to good economic returns for most irrigated crops) combine to make NO_3^- leaching a serious problem on many irrigated soils (Branson et al., 1975; Pratt, 1978; Rauschkolb, 1978).

The impacts of irrigated agriculture on NO_3^- levels in surface and ground waters have been evaluated most extensively in California (Pratt, 1978). Nitrogen fertilizer use in California has experienced the marked increase typical of the U.S. California agriculture is typified by its wide diversity, in comparison to the corn-soybean emphasis of the Corn Belt. Table 6 illustrates the increase in N fertilizer application rates that has occurred from 1950 to the present; California agriculture used considerable N fertilizer even in 1950.

Ayers and Branson (1973) and Ayers (1978) evaluated the N balance in the Santa Ana River Basin of California; the same methodology was used by Miller and Smith (1976) to evaluate the N balance in the southern San Joaquin Valley. The NO_3^--N concentration of the Santa Ana River increased from an average of about 2 mg/liter in 1930 through 1950 to about 6 mg/liter in 1969. In summary, the mass balances indicated a considerable input of fertilizer N to the valleys. Manure N was considered as an additive input; the analyses indicated that disposal of animal wastes was a considerable problem, especially in localized areas.

Table 6—Changes in harvested area, area fertilized with N, and rate of application of N for California crop categories since 1950.[†]

Crop category	Surveyed area		Fertilized area	Rate of fertilization
	1,000 ha		%	kg N/ha
		1950		
Agronomic crops	2,662		29	61
Fruits and nuts	573		58	133
Vegetables	272		81	103
Total	3,507‡		38	86
		1960		
Field crops	1,810		70	108
Forage crops	1,039		22	52
Fruits and nuts	597		77	121
Vegetables	290		90	157
Total	3,736‡		61	105
		1973		
Field crops	1,587		84	139
Forage crops	958		50	74
Fruits and nuts	575		82	126
Vegetables	330		89	202
Total	3,450‡		74	131

† D. R. Nielsen, Univ. of California, Davis, pers. comm. (1975). Forage and field crops were combined into one category in 1950.

‡ For comparative purposes, range and turf are not included in these totals.

Branson et al. (1975) report that the average soil water NO_3^--N concentrations in the Santa Ana Basin at a depth of 3–6 m were 405, 326, and 290 mg/liter for corral, pasture, and cropland, respectively. At the top of the water table, the respective NO_3^--N concentrations were 251, 326, and 198 mg/liter. Adriano et al. (1972a) found NO_3^--N at 3.3- to 15-m depths below nine row crop sites in the Basin to range from 36 to 122 mg/liter in the saturation extract. Consideration of the N balance at these sites showed losses by denitrification to range from zero to as much as 52% of the total N input. The high NO_3^- concentration below the root zone indicated that considerable leaching losses would occur, but the rate of NO_3^- movement through soil was slow. They estimated that from 10 to 50 years would be required before current fertilization practices would affect the NO_3^--N concentration of the ground water. Current fertilization practices on vegetables supply about 135 kg N ha^{-1} year^{-1} more than is removed by the crops. Well waters pumped from deeper aquifers at these sites currently average >20 mg/liter.

The effect of overfertilization and subsequent NO_3^- leaching was identified in some subwatersheds of the Santa Ana Basin. One of these is a 388-ha watershed of sandy loam soil planted largely to citrus (Davis & Grass, 1966; Bingham et al., 1971). This watershed was developed in 1960, and is underlain by impervious granite. Water moves out of the watershed as surface flow at only one point, and the only inputs are precipitation and irrigation. Hence, it functions as a macrolysimeter. During 1967 to 1970, about 45% of the water added percolated through the soil. This percolate removed 51, 62, and 78 kg of NO_3^--N/ha in 1967, 1968, and 1969, respectively. The NO_3^--N concentration of this drainage water ranged from 37 to 81 mg/liter. Essentially, all of this N was from fertilizer. This left a balance of 20–30 kg/ha unaccounted for of the 130 kg/ha annual N application. For the total upper Santa Ana Basin, about 120 kg/ha of N in excess of crop needs were applied to irrigated vegetable crops in 1969 (Ayers & Branson, 1973). Pratt et al. (1972) estimated that from 0 to 100 kg/ha of NO_3^--N were leached below 30 m in four citrus fields in the Basin, with NO_3^--N concentrations in the saturation extracts of the subsoil ranging up to 80 mg/liter. Nightingale (1972) reported similar levels in Fresno County. In further research (Adriano et al., 1972b; Pratt & Adriano, 1973) from 21 to 123 mg/liter of NO_3^--N was found in the soil water under various irrigated vegetables in the Santa Ana Basin. Nitrate-N leached beyond the root zone ranged from 25 to 912 kg ha^{-1} year^{-1}, while from 0 to 960 kg ha^{-1} year^{-1} was apparently denitrified.

There are several monitoring studies of irrigation drainage which indicate little or no increase in NO_3^--N due to agricultural activities. This includes the Imperial Valley (Meek et al., 1969) and Coachella Valley (Bower et al., 1969) in California, and the upper Rio Grande River system of New Mexico and Texas (Bower & Wilcox, 1969). Ideal conditions for denitrification are probably largely responsible for the low NO_3^- in the drainage.

Tile drain monitoring in southern California (Devitt et al., 1976) showed that from 0 to 350 kg/ha of NO_3^--N were removed from tile-drained irrigated lands. A general positive relationship between NO_3^- in the tile effluent and N fertilizer rates existed. As with the free drainage situation, the presence of a clay layer generally resulted in less NO_3^- in tile drainage, indicating denitrification.

Determination of residual NO_3^- in the profiles of irrigated croplands in Colorado (Ludwick et al., 1976) showed that about 150–175 kg/ha of NO_3^--N was in the upper 90 cm of soil after cropping (see also Stewart, 1970). A gradual accumulation of NO_3^- in the soil profile occurred over time, apparently due to overfertilization.

In Nebraska, Muir et al. (1973) found that NO_3^--N of the ground water increased, on the average, from 2.5 mg/liter in 1961 to 3.1 mg/liter in 1971. This occurred while fertilizer N use quadrupled and irrigation increased 50%. Stream N levels also increased, but were related largely to animal and human population increases.

The California data indicate that the NO_3^--N levels in ground water under irrigated croplands are 25–30 mg/liter, and only when application rates exceed those which are efficiently utilized by the crops does the leaching of N become excessive. In fact, for many crops, with good agronomic practices and profitable production, about 20 mg/liter of NO_3^--N in drainage effluents may be the best achievable. More efficient water use by modification of current irrigation practices will likely result in higher concentrations of NO_3^- in the drainage waters, although total fertilizer N losses could well be lessened. Table 7 illustrates this assumption (Branson et al., 1975). This theoretical analysis does not include losses of N by denitrification. At present, denitrification cannot be satisfactorily monitored or modeled under field conditions. However, attainment of lower NO_3^- levels than those presently observed, or lessening of denitrification losses, will require adoption of better management practices than presently are used.

Lowered rates of irrigation is one management practice that would seem desirable from at least two standpoints; less total N leached and less denitrification from the drier soil. However, with present irrigation methods, salinity buildup in the irrigation return flow is a major problem in the arid regions. Salts from irrigation water, animal wastes, and fertilizers are concentrated by evaporation and transpiration, and additional dissolved salts enter the return flow from percolation through the soil. As a result, the salinity of a stream will increase with each irrigation diversion. The resulting accumulation of soluble salts in the root zone can markedly decrease

Table 7—Average NO_3^--N concentrations with various combinations of volume of drainage water (in surface cm/year) and leached NO_3^-.†

Drainage volume	Amount of N leached, kg/ha				
	22.4	44.8	56	89.7	112.1
12.7	18.0	35.0	44.0	70.0	88.0
25.4	8.8	18.0	22.0	35.0	44.0
50.8	4.4	8.8	11.0	18.0	22.0
76.2	2.9	5.9	7.3	12.0	14.6
101.6	2.2	4.4	5.5	8.8	11.0
127.0	1.8	3.5	4.4	7.0	8.8

† Branson et al. (1975).

crop yields, and as irrigation water quality declines, more water must be used to leach salts from the root zone to prevent adverse effects on crops. A certain portion of the irrigation water, commonly termed the *leaching requirement*, must pass below the root zone. The salt discharge from irrigated lands is poorly understood, but this degradation of water quality should be minimized (Branson et al., 1975). Modification of irrigation practices to minimize salt loads can be accomplished by lowering the leaching fraction. This causes more of the salts to form insoluble precipitates, minimizes the use of the irrigation water, and reduces the amount of return flow (Law & Skogerboe, 1972; Branson et al., 1975). Other innovative irrigation practices to minimize leaching are discussed later.

Specialty (vegetable and root) crops are difficult to manage with respect to efficient N utilization. They are usually irrigated, and often require high rates of N fertilizers. An example is commercially grown potatoes (*Solanum tuberosum* L.). Potatoes are commonly grown on coarse-textured soils. They are shallow-rooted (Lesczynski & Tanner, 1976), which lessens the amount of effective soil volume, and hence, available N and water, compared to a deep-rooted crop. Also, although potato tubers are low in N content (1.1–1.6% N, dry wt basis), their high dry matter yield plus the N in the tops results in a high N requirement (Saffigna et al., 1976). Further, potatoes should not be subject to water stress even for short periods or tuber malformations will result. Therefore, irrigation is usually essential.

A recent monitoring study of ground water NO_3^--N and Cl^- (Saffigna & Keeney, 1977) in the central Wisconsin sand plain region which has about 20% of the area under sprinkler irrigation (and 25% of this area in potatoes) showed that the NO_3^--N concentrations of ground water were significantly above background and ranged from 4 to 23 mg/liter. The ratio of Cl^-/NO_3^--N was relatively constant. The data long with intensive field research suggested that much of the N and Cl^- were derived from fertilizers (Saffigna et al., 1977).

Meisinger (1976) estimated the N balance sheet of potato fields in Long Island, New York. Many wells in this area are high in NO_3^--N (10–15 mg/liter), presumably from N fertilizers, although septic tanks may also be a major source. His analysis indicated that at >155 kg fertilizer N/ha, NO_3^--N pollution resulted. With an average recharge of 50% of precipitation (about 550 mm), 55 kg N/ha excess of fertilizer N over N uptake would result in 10 mg/liter of NO_3^--N in the recharge water. Consideration of average grower yields indicated that about 170 kg/ha of fertilizer N would result in environmentally acceptable NO_3^--N levels in the water without seriously depleting yields given average weather and management. The present fertilizer rate in this area is about 225 kg N/ha.

C. Livestock Operations

In the past, wastes from domestic animals were carefully husbanded fertilizers. However, as food demands have increased and agriculture has become mechanized, commercial fertilizers have been increasingly relied on to meet nutrient needs. With improved technology the price of fertilizers declined to less than the costs of handling and spreading manure on the land.

Part of the protein consumed by animals is undigested or only partly digested, and appears in the fecal matter. The remaining N is digested through protein metabolism and a varying amount of the digested protein, depending on the efficiency of conversion, is excreted in the urine (Nye, 1973). From 70 to 75% of the N added in feeds is normally recovered in the wastes; some also is lost from the animal to the atmosphere.

Much of the N excreted by farm animals is not collected, that is it falls on pasture or holding areas, while with feeding operations such as caged poultry, nearly all of the N is collectable. Frere (1976) cites evidence that of the collectable manure N, about half is lost during storage, handling, and spreading. He estimates that of the manure N produced in the USA, about 2.2×10^6 tons is collected and only 1.1×10^6 tons are ultimately available for crop growth. About 40 to 50% of the organic N in manure is normally mineralized during the first cropping year (Frere, 1976). The additional organic N will be released in subsequent years, but not necessarily at the time when needed by the plants.

Under warm, moist conditions, urea hydrolyzes rapidly to NH_3 and CO_2. The pH in the feedlot surface layer rapidly rises to between 8.5 and 9.9 (Stewart, 1970; Adriano et al., 1974) and NH_3 is evolved. The amount of NH_3 and volatile bases lost from feedlots depends on evaporation, stocking rates, and cleaning operations. Stewart (1970) reported NH_3 losses from urine ranging from 25 to 90% of the added N, while Adriano et al. (1974) found that N losses from a mixture of urine and feces approached 50%. Elliot et al. (1971) measured losses of 148 kg NH_3-N and 21 kg volatile organic N per ha. This N is transported by wind away from the immediate vicinity of the feedlot, and can cause odor nuisances. If deposited in waters, it can cause significant N loadings (Hutchinson & Viets, 1969; Luebs et al., 1974).

The high NH_3 and high pH conditions at the surface of a feedlot initially retards nitrification (Stewart, 1970). Considerable NO_2^- may be formed at the onset of nitrification, and Stewart (1970) found 10-92 mg/kg of NO_2^--N 4 weeks after treatment of soil columns with cattle urine. Volatile amines, including secondary amines, have been identified in acid traps adjacent to a cattle feedlot (Mosier et al., 1973). The combination of secondary amines and NO_2^- makes the formation of nitrosamines such as dimethylnitrosamine (DMNA) possible (Ayanaba et al., 1973a). Nitrosamine formation has been noted in sewage (Ayanaba et al., 1973b), but DMNA formation in cattle feedlots apparently is insignificant (Mosier & Torbit, 1976).

Numerous agronomic studies, some dating back to the early 1900's, have shown relatively little difference in increases in crop yields when N is added as properly handled and applied manure or as chemical fertilizers, although manure N is less rapidly available than fertilizer N (Peterson et al., 1971; Magdoff, 1978). The use of manure to enrich croplands seldom is profitable in terms of replacement of commercial fertilizers; instead, this practice, while desirable for addition of organic matter, usually must be regarded as an expense associated with the waste generating enterprise (Clawson, 1971).

The amount of the NO_3^- in the deep percolation under feedlots depends on the amount of NO_3^- formed, the rate of infiltration of water, and the extent of denitrification. Active feedlots often have a low infiltration rate, because of the puddled conditions of the soil. When a lot is taken out of use, the soil surface dries, nitrification is rapid, and significant amounts of NO_3^- may be leached. Significant NO_3^- leaching may also occur when feedlots are established on coarse-textured soils, when stocking rates are low, or when manure is removed frequently. Organic compounds and NH_4^+ also can leach through the profile under feedlots. Reduced conditions conducive to denitrification have been found under a number of active feedlots (Elliot et al., 1972; Mielke et al., 1974). Stewart et al. (1968) compared the NH_4^+ and NO_3^- concentrations in ground water beneath four Colorado feedlots with those in adjacent irrigated fields, and concluded that the feedlots were not a significant source of inorganic N to ground waters.

Land disposal of livestock wastes can also cause contamination of surface waters (Miner & Willrich, 1970). Application of animal wastes to snow-covered or frozen land, in particular, results in high concentrations of N in spring runoff (Frere, 1976). To overcome this problem, wastes can be stored in pits or lagoons until soil and climate conditions are suitable for application.

D. Grasslands

A major portion of the world's animal products are produced on grassland ecosystems. These ecosystems encompass a wide diversity of environments, productivity and degree of management. They range from low productivity rangelands in semi-arid and arid regions, moderate productivity sheep pastures in the hill country farming as practiced in Scotland and New Zealand, the high productivity, heavily fertilized (200–400 kg N ha^{-1} year^{-1}) permanent grasslands of England and Northern Europe, and moderate to high productivity improved grass-legume [ryegrass (*Lolium perenne*)-white clover (*Trifolium repens*) or subterranean clover (*Trifolium subterraneum*)] grazed pastures such as found in northwest U.S., New Zealand, and Australia. These systems, particularly those which do not require high N fertilizer use, likely will be of increasing importance as sources of high quality protein and of wool in a resource-limited economy.

The potential for water pollution from low to moderate productivity systems is likely minimal due to their generally N-deficient status and low N turnover rates (Dahlman et al., 1967; Whitehead, 1970; Porter, L. K., 1975). Woodmansee (1978), however, points out the high potential for NH_3 volatilization from the feces and urine of large animals which may lead to a degrading N balance in the western U.S. grasslands which are moderately or heavily grazed.

The heavily fertilized permanent grasslands of Europe offer a relatively high potential for NO_3^- loss by leaching and denitrification (Woldendorp et al., 1965). While rapid uptake of N by the grass during periods of favorable growth minimizes leaching losses, considerable loss can occur in spring rainy periods. For example, Henkens (1977) estimates annual losses of 192 kg N/ha by denitrification, 38 kg N/ha by NH_3 volatilization and 44 kg N/ha by leaching from an intensively stocked (4 cow/ha) Holland dairy farm receiving 400 kg/ha of N fertilizer annually.

In theory, the intensely grazed grass-legume pasture should be an almost ideal agricultural ecosystem. Fertilizer N inputs are minimal, soil organic N is accumulating, little fossil fuel energy is needed to manage the system, and the tight N cycle assumed for grasslands should prevent NO_3^- leaching (Walker, 1956; Whitehead, 1970). However, O'Conner (1974) expressed concern over the high NO_3^--N levels in drainage waters from pastoral agriculture in New Zealand. Baber and Wilson (1972) found an average of 25 mg NO_3^--N/liter in ground water draining an area of intensive pastoral agriculture on porous pumice soils in the North Island, New Zealand.

Recent research (Ball et al., 1979) has shown that significant leaching of NO_3^--N from pastured ryegrass-white clover pastures is likely. The problem occurs in the urine-affected areas. These areas receive up to 60 g N/m^2, largely as urea, and the high pH-high NH_3 conditions are conducive to rapid nitrification. Since considerably more NO_3^--N is present than can be effectively utilized by the herbage, leaching will occur with excess rainfall. No quantitative data on leaching losses are available for these types of systems; mass balance considerations would indicate a loss of 50–100 kg N ha^{-1} $year^{-1}$ by leaching which, when coupled with product removal, NH_3 volatilization and N transfer to shade, water, and milking areas, could also put these systems in a negative N balance despite the fact that biological N fixation averages nearly 200 kg N ha^{-1} $year^{-1}$ (Hoglund et al., 1979; Ball, R., personal communication, DSIR-Grasslands, Palmerston North, New Zealand). Relatedly, Jones et al. (1977) also found that considerable NO_3^--N was leached from unfertilized grass-clover mixtures in a lysimeter study in California. Most of the leaching occurred during the rainy period in winter.

E. Tropical Agriculture

While N fertilizers are used in most of the agricultural systems of the developed world to maintain available soil N for stable continuous crop production, this approach has not yet been of great success in the humid tropics. In the tropics, N inputs come either from blue-green algae in paddy rice (*Oryza sativa* L.) production or from symbiotic and nonsymbiotic N fixation during the natural vegetative phase of the shifting cultivation systems (Nye & Greenland, 1960, 1964; Sanchez, 1977; Greenland, 1977). Greenland (1977) argues that the shifting cultivation system cannot be highly productive over extended periods and that increased biological N_2 fixation must preclude an increase in productivity in tropical agriculture.

Very little information is available on pathways and magnitudes of N losses in tropical agriculture. The combination of erosion and high oxidation rates can lead to rapid loss of soil N during cultivation (Nye & Greenland, 1960, 1964; Sanchez et al., 1973; Bartholomew, 1977). Since much of the soil organic N in tropical soils may be in the surface 5–10 cm, erosional losses can be substantial. For example, Greenland (1977) reports unpublished results of D. R. Lal, International Institute of Tropical Agriculture, Ibadan, Nigeria, showing very little erosion loss under minimum or zero tillage, but losses as high as 100 kg N ha^{-1} year^{-1} with conventional tillage. Denitrification losses have not been quantified, but Greenland (1977) points out that in many cases conditions are excellent for high rates of denitrification. Losses also can occur during burning of the vegetation (Bartholomew, 1977).

As Bartholomew (1977) points out, a unique aspect of shifting cultivation in tropical agriculture is the magnitude of N changes which occur. During the first 2 years of the native vegetation phase, gains of 200–350 kg N ha^{-1} year^{-1} in the soil and vegetation are common. Despite the fact that apparently large amounts of N are available for crop uptake, crop yields may be low and crops sometimes respond to N fertilization even the first year after cropping. Bartholomew (1977) hypothesizes that large amounts of NO$_3$⁻-N are leached below the root zone during cropping, but are returned to the surface by the deep-rooted natural vegetation.

Tropical cropping systems offer high potential for NO$_3$⁻-N leaching losses (Date, 1973). Table 8 (from Bartholomew, 1977) gives some estimates of potential downward NO$_3$⁻-N movement in the humid tropics. There are several reports of high NO$_3$⁻-N loss by leaching. For example, in Nigeria, Jones (1968) noted NO$_3$⁻-N increases in the subsoil of from 92 to 560 kg/ha after arable cropping over that in subsoils of fallow plots. While, as Bartholomew (1977) points out, this NO$_3$⁻-N can be utilized to some extent by the fallow vegetation. The high potential for leaching losses represents not only a localized pollutant source, but also an inefficient use of N. Innovative management systems are needed to reduce this loss of N.

The large diversity of soils and climates in the tropics leads to extremely complicated management decisions. Many of the aspects involved are covered in the comprehensive treatise edited by Bornemisza and Alvarado (1975).

Table 8—Some estimates of the net vertical movement of NO$_3$⁻-N in tropical soils.†

Location	Rainfall	Drainage	Net NO$_3$⁻-N movement
		mm/year	
Senegal	660	128	660
Senegal	660	118	590
Ivory Coast	1,569	845	4,225
Ivory Coast	1,758	828	4,140
Nigeria	460	153	538
Nigeria	1,381	96	724
Nigeria (Savanna)	506	125	266

† See Bartholomew (1977) for original reference.

Table 9—Soil, climatic and management factors that affect crop yields.

Soil and site properties	Soil water supply in root zone	Weather	Short-term management	Long-term management
Slope, drainage, texture, structure, permeability, infiltration, presence of restricting layer, depth of topsoil	Preseason—precipitation in fall and spring, previous cropping	Precipitation distribution	Previous crop	Tile or surface drainage
		Temperature	Tillage, ground cover	Soil conservation practices
	Growing season —rainfall, evapotranspiration, rooting depth, weeds	Storms	Variety of crop	Irrigation system
Availability of plant nutrients		Weather during critical growth periods	Plant density	
			Planting date	Land available, land purchases
Water retention in the root zone		Date of killing frost	Timing and rate of application of fertilizers	Cropping system
Soil aeration (need for artificial drainage)				Liming
			Weed and insect control	Machinery and equipment
Susceptibility to flooding				

VII. FACTORS AFFECTING CROP YIELDS AND USE OF N

Any agricultural management system for maximizing crop uptake of N must, directly or indirectly, use the system management approach; i.e., the impact of the total system on crop yields must be considered. There are a number of excellent reviews of the many factors that affect crop yields, including their response to N fertilizers (e.g., Viets, 1965; Pierre et al., 1966; Black, 1967; Ingestad, 1977; Frissel, 1977; Barber, 1974; and Chapter 15). Table 9 summarizes these factors; they can be quantified only in reference to specific crops and regions. Largely unpredictable weather factors provide the context in which farmers with variable amounts of management skills, available labor and capital must make the short- and long-term management decisions (including N fertilization) shown in Table 9. The greatest need is to predict accurately the N dose-response relationship for a given crop on a given farm. Table 10 summarizes factors involved in the on-farm response of crops to N fertilizer.

An important factor in fertilizer dose-response relationships is the rate at which the plant accumulates N; corn, for example, typically accumulates much of its N in the period beginning about 1 month after emergence (Hanway, 1960). Loss of N through leaching or denitrification before this period can result in environmental pollution and reduced yields. The use of crops or varieties characterized by rapid uptake of N during the early growth period, and timing of fertilizer applications to coincide with the rapid uptake period, can result in more efficient fertilizer use. Similarly, fertilizer placement to encourage deeper and more dense root growth could lead to increased uptake of N before it is leached from the root zone.

In some cases, the type of N fertilizer used can affect the response curve, although usually it makes little difference what form is chosen (Viets,

Table 10—Factors that affect the yield response of nonleguminous crops to N fertilizers.

Factor	Related variables
Availability of soil organic N and of inorganic N in the root zone	Soil aeration and temperature; amount and type of organic N in root zone; previous crop, wastes and residues; leaching and/or denitrification; previous N fertilization
Rate of accumulation of N by plants	Weather, soil and management factors, genetics
Rooting depth and density	Genetics, weather, soil properties, fertilizer placement
Final yield of N in crop	Genetics, weather, management, availability of other nutrients
Fertilizer formulation	Type and timing of application, slow release or inhibitor characteristics
Availability of fertilizer N:	
Immobilization	NH_4^+-fixing clay minerals, C/N ratio of residues
Volatilization of NH_3	Application of anhydrous NH_3 to soils which are too dry, moist or sandy, or at too high a rate; application of urea or low pressure solutions to soil surface without incorporation
Denitrification	Soil aeration, temperature, pH, availability of organic C, presence of NO_3^-
Leaching of NO_3^-	Rate and amount of soil water movement, plant uptake, denitrification
Competitive plant uptake by weeds, interrow crops	Weeds, immobilization
Application in the fall	Spring soil temperature, rainfall

1965). Other factors that can reduce the efficiency of use of fertilizer N include immobilization of N due to addition of C-rich residues; poor application practices, such as surface application of NH_3 solutions or urea; poor timing of application; or fall applications of anhydrous NH_3.

The number of variables that can affect crop yields is large, and many of the factors discussed here can have substantial impacts on the efficiency of crop use of N fertilizer. Because of numerous uncertainties and an inability to separate the influences of the multiple factors acting simultaneously, it is often not possible to apply the experimentally derived relationships between fertilizer applications and crop yield response to predict quantitatively the effect of fertilizer use under any set of specific on-farm conditions (Swanson, 1957).

VIII. CONTROL OF N POLLUTION FROM CROPLANDS

Many crop production practices have the potential to create adverse environmental impacts with respect to N. When the natural capacity of a system to cycle N is exceeded in a given locality, the result may be the accumulation of excessive N in the ground or surface waters, or the loss of large amounts of NH_3 or N_2O to the atmosphere. This N comes from the various soil pools, exogenous N in residues of plant or animal origin, or fertilizers. As discussed earlier, knowledge of the sources of N and of the physical and chemical processes of transport and transformation is usually inadequate to determine quantitatively the specific sources of excess N or its potential to cause adverse environmental impacts, even in some of the most intensively studied crop systems. Further, it is difficult to apply knowledge gained at one site to problems at other locations.

Techniques for minimizing N losses and maximizing N efficiency are many and diverse; they include established practices of efficient agricultural management, innovative applications of agricultural technology, regulations limiting fertilizer applications, and possibly even changes in fundamental patterns of land use and crop production. The sections which follow briefly examine several possible approaches, the conditions under which each might be applied, and the implications of such applications for environmental quality and for agricultural production.

A. Agricultural Best Management Practices

Practices which decrease soil erosion and surface runoff will also reduce the amount of N lost from croplands by these routes. Such often used and widely appreciated practices, thus, may have beneficial impacts on environmental quality as related to N.

1. SOIL CONSERVATION

Contouring, strip cropping, terracing, and the use of crop residues or vegetative cover protect soil from wind and water erosion. They, thus, also limit the movement of soil N (especially the particulate organic fraction) from the land and protect surface water from an added N burden. Such measures do not necessarily serve to control other N-related problems, such as leaching or denitrification. In fact, they could increase these impacts. Practices which maintain good soil quality will reduce the need for fertilizer N and maintain crop yields, which in turn, could mitigate potential problems of excess N. Because of differences in climate, terrain, and soil types, the most effective techniques for erosion control vary widely from one area to another. Bandel et al. (1975) noted that N deficiency symptoms on corn were more marked with no-till than conventional tillage, while Tyler and Thomas (1977) found greater NO_3^- leaching with no-till than conventional tillage. Different tillage systems will need evaluation for their impacts on N utilization.

2. WATER CONSERVATION

In irrigated crop systems, the rate of movement of NO_3^- to ground water (by leaching) and to surface waters (by return flow) is a function of the volume of water moving through the soil and its NO_3^- concentration. By increasing the efficiency of water use, it may be possible to reduce substantially the mass emission of NO_3^--N leaving the root zone (Pratt, 1976; Smika et al., 1977; Smika & Watts, 1978), although the concentration of

NO_3^- in the smaller volume of effluent may be quite high (Devitt et al., 1976; Table 7).

The irrigation systems commonly used are flooding, sprinkler, and furrow (Taylor & Ashcroft, 1972). Flood irrigation involves leaching under saturated soil conditions, which is regarded as less efficient in terms of solute leaching than unsaturated flow (Nielsen et al., 1972). Flood irrigation also offers more opportunity for denitrification. Furrow irrigation can lead to inefficient N use, since this method involves lateral and upward water movement into the row between the furrows. Nitrate is carried with this water, and may accumulate in the dry surface soil layer of the row where it is not available to the plants (Taylor & Ashcroft, 1972).

Two irrigation methods have recently been developed which offer promise with respect to reducing the quantity of water needed. These are subsurface application and drip irrigation. Both methods involve application of small amounts of water to soils at frequent intervals, and permit additions of fertilizer with the irrigation water. Subsurface irrigation requires special soil conditions (a highly permeable surface soil underlain at 2–3 m by a relatively impermeable subsoil). This method has resulted in severe salinity problems at some locations (Taylor & Ashcroft, 1972). Drip irrigation involves application of water under pressure through buried mains and laterals to surface placed emitters, which release the water at a slow rate (Marsh et al., 1975). This method is reported to be less expensive (in terms of operating costs) and easier to operate than other irrigation methods, but to have problems with clogging of the emitters, and salt accumulation.

While minimal irrigation techniques seem to have potential for controlling N losses, the economic feasibility of the more careful (and often labor-intensive) management required has not been evaluated for most situations.

3. CROPPING SEQUENCE

The practice of planting different crops on the same field in successive years continues to be quite common, although the cropping sequences have changed considerably with the trend to more specialized farming. For example, the Midwest U.S. farm before 1950 often was a self-contained unit, with some livestock and poultry. Forage crops as well as grain were needed for the livestock and often much of the feed produced remained on the farm. In these systems, rotations of corn, small grains (for grain and for a cover crop for the meadow seeding) and meadow were common, with a typical sequence consisting of 2–4 years of meadow, and 1–4 years of corn. Little N fertilizer was required, as N was adequately supplied to the corn crop (which required less N as it yielded less than the corn crops grown today) by manure and legume residues. Ideally, crop rotation practiced in this way provided more economic stability since the farmer's income was not predicated entirely on one or two crops. Erosion, crop diseases, and pests also were minimized. Further, the deep-rooted legume had the potential for scavenging NO_3^- from the profile (Schertz & Miller, 1972).

As the result of economic considerations, and the general compatibility of the crops, continuous cropping or sequences such as the corn-soybean rotation are increasingly common. Soybeans can scavenge residual fertilizer N from the corn crop, as well as use mineralized soil N and symbiotically fixed N. Johnson et al. (1974) estimated that soybeans removed 195,000 metric tons of N from cultivated land in Illinois, which was about 56% as much N as was removed by corn. Their calculations (Johnson et al., 1975) indicate that on corn fields receiving 224 kg N/ha, about 40% of the N removed by soybeans came from residual N fertilizer. A more effective cropping sequence for immobilizing residual N might be one which utilized an early season, rapid-growing grass or cereal crop. However, to get an economic return from these crops requires a full growing season.

While considerable data are available for corn and corn-soybean sequences, long-range, comprehensive evaluation of the rate and extent of NO_3^- leaching and fertilizer N efficiency is needed for various other sequences. Systems using grass or grain crops such as adapted sorghum (*Sorghum vulgare* Pers.) or sudangrass (*Sorghum sudanensis*), along with double cropping and intercropping should also be considered (Nelson et al., 1977). Adoption of such rotations would clearly have impacts on other farming practices, which also must be evaluated.

4. COVER CROPS

Cover crops are sometimes planted after the principal crop has been removed to control erosion. These crops can also take up residual inorganic N. When the crop is incorporated at the time of seedbed preparation in the spring, this N is at least partly available for subsequent crops. Welch (1974)[3] showed that a rye crop, planted after the fall corn harvest, effectively scavenged residual N. However, adverse weather will often prevent planting or growth of the cover crop. It also prevents fall plowing, and the costs of establishment presently discourage the use of such cover crops strictly for N conservation. Another possible approach is to interseed a low-cost crop such as oats (*Avena sativa* L.) late in the corn season. However, problems with seedling sowing and establishment would seem a priori to preclude such an approach as a worthwhile venture on a wide scale. Furthermore, the fate of the organic N in the plowed-down cover crop is uncertain. For example, some of the N might not be mineralized until too late in the growing season to be utilized by following crops.

5. WATERSHED MANAGEMENT

In some cases system management of an entire watershed may be possible. An excellent example is the Hula Basin, Israel (Avnimelech et al., 1978). In this case 2,000 ha of flooded soils were artificially drained. Extensive nitrification resulted, endangering the water quality of Lake

[3] L. F. Welch. 1974. Cover crops to immobilize nitrogen. p. 52–56. *In* Nitrogen as an environmental quality factor—technical, social, and economical considerations. 1974–1975 Progress Report to the Rockefeller Foundation, submitted by the Coun. on Environ. Qual., College of Agric., and Agric. Exp. Stn., Univ. of Illinois, Urbana.

Kinneret (Sea of Galilee). An interdisciplinary research project spanning several years resulted in implementation of management techniques to minimize leaching and maximize denitrification and crop uptake of N. Another example, although not yet in the implementation stage, is the studies on the Fall Creek watershed, New York (Porter, K. S., 1975), where impacts and sources of N were identified, methods for and consequences of control were assessed, and system analyses of these factors were conducted.

B. Improved Management of Nutrient Systems

In general, adverse impacts of N occur when N inputs greatly exceed amounts which can be efficiently used by crops (Stanford, 1973). This statement has been supported by a large number of investigations involving application of N fertilizer to nonleguminous crops at rates ranging from those which would be expected to show marked yield response to rates far above what the crops could be expected to utilize (and usually much greater than typical application rates). Figure 4 (Broadbent & Carlton, 1978) demonstrates this concept. Yields increased at levels of up to 180 kg N/ha; at those input rates, little fertilizer N was retained in the soil. At the 360 kg N/ha rate, they estimated that a residual of at least 162 kg N/ha was removed by leaching, denitrification, or both. In a similar study at a different site, crop yields were maximized at N application rates of about 220 kg/ha, and residual inorganic N in the soil increased at application levels above that value (Broadbent & Carlton, 1978).

If fertilizer applications were always at the same efficient dose-response levels, losses of N from N-fertilized croplands would be minimal.

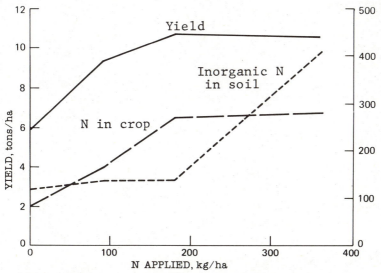

Fig. 4—Yield, crop uptake of N, and residual N in soil after 1975 crop harvest, Davis site, 60 cm irrigation (Broadbent and Carlton, 1978). Reprinted by permission of Academic Press, New York.

However, maximum economic efficiency of production usually requires above-optimal inputs of N fertilizer (cf. Chapt. 20, E. R. Swanson). Also, N recovery by agronomic crops is seldom more than 70%, and the average value is probably nearer to 50% (Allison, 1955; Viets, 1965; Chapt. 15, L. T. Kurtz and R. A. Olson). In addition, the optimum rate of fertilizer application is influenced by a multitude of site-specific conditions, and by the weather.

In general, information is inadequate to determine with precision the optimal rates of fertilizer N input even on a single farm. One possible way of minimizing such pollution is to restrict N fertilizer applications to rates that are significantly below the economic optimum. It may also be possible to use other management practices which can increase the efficiency of crop utilization of fertilizer N, or reduce losses of the unused portion to the environment. The impacts of many of these techniques on crop production and on N-related environmental quality have yet to be evaluated in full-scale field studies.

1. SOIL AND TISSUE TESTING

The efficiency of N fertilizer use can be further improved by even more emphasis than at present on effective field testing and educational programs. Such programs have traditionally been conducted by agricultural extension personnel, oftentimes in conjunction with privately owned soil- and tissue-testing laboratories.

The inherent difficulties in use of soil tests for predicting the amount of soil N available for a crop have been discussed previously (see Chapter. 15, L. T. Kurtz and R. A. Olson). With annual crops such as corn, tissue tests for N also are difficult to interpret (Jones & Eck, 1973), and tissue analysis has the added disadvantage that it is often too late in the growing season to correct a deficiency that might be revealed. Pierre et al. (1977a, b) have reported good success in predicting the N requirement of corn following corn by use of the N content of corn grain, and considerable success has been obtained in the use of tissue tests for prediction of N needs in commercial fruit production (Emble et al., 1974). However, until more reliable soil tests are developed, fertilizer N recommendations for annual crops (especially long season crops such as corn) will require considerable skilled judgment on the part of the extension agronomist and the farmer.

Efforts should be made to provide farmers with as much information as possible to avoid using more N fertilizer than required. Continued research on soil and tissue testing for predicting crop needs of N is essential. Much of this research must of necessity be closely related to on-farm conditions and be of long duration so that results can be meaningfully interpreted.

2. TIMING OF APPLICATIONS

The most logical approach to increasing N fertilizer efficiency and hence lessening environmental impacts of N is to supply the N as it is needed by the crop, i.e., to match applications closely to the N uptake curve throughout the growing season (Scarsbrook, 1965). Theoretically, this could be accomplished with slow-release fertilizers, but these require further development before their use is practical. Alternatively, fertilizer N can be supplied in a number of applications to match crop needs. However, each appliction increases operating costs, and the use of field equipment is usually limited to early in the growing season. However, with crops irrigated by sprinkler or drip irrigation, N can be supplied in the irrigation water (Mulliner & Frank, 1975; Smika & Watts, 1978).

Application of fertilizer N in the fall in cool climates for the next growing season has long been a controversial subject (Larsen & Kohnke, 1946; Nelson & Uhland, 1955; Welch et al., 1966; 1971; Stevenson & Baldwin, 1969; Beauchamp, 1977). Fall application has the advantages to the farmer and fertilizer industry of spreading out the labor and equipment demand; oftentimes soil conditions also are more favorable in the fall. Field experiments comparing the relative efficiency (usually expressed as yield response) of fall- and spring-applied fertilizer N have given results ranging from no difference (Chalk et al., 1975; Beauchamp, 1977; Hendrickson et al., 1978a) to large yield advantages with spring application (Nelson & Uhland, 1955; Welch et al., 1966; 1971; Stevenson & Baldwin, 1969; Warren et al., 1975). This divergence of findings is likely due to year-to-year climatic differences, and to the many other factors which may affect plant yields.

It is generally recommended that fall application of NH_4^+- or NH_4^+-producing forms of fertilizer not be done until the soil temperature is < 10°C; that fall fertilization not be practiced on coarse textured or poorly drained soils; and that perhaps 10–20% more fertilizer N be used in fall than in spring application. However, studies have shown that nitrification will continue, although at reduced rates, to freezing and will commence when the soil warms (Sabey, 1968). Hence a warm, wet spring could result in extensive loss of fertilizer N through leaching, denitrification, or both. Thus, fall fertilization is often discouraged as an environmentally and economically unsound practice (NRC, 1978).

If nitrification of the fertilizer can be delayed through use of slow-release forms or nitrification inhibitors, fall fertilization would be more acceptable. Beauchamp (1977) found no advantage from fall-applied slow-release forms of N as compared to urea. Warren et al. (1975) obtained evidence that a nitrification inhibitor greatly lessened losses of N from fall-applied anhydrous NH_3. However, in a warmer climate (Georgia), Boswell et al. (1976) found that nitrification control was not sufficient to prevent N losses from fall application.

Split applications of fertilizer N are often used on row crops and grasslands (Castle & Reid, 1968; Reid & Castle, 1970; Welch et al., 1971; Hanson et al., 1978). Usually one, and sometimes two sidedress applications are made. This practice has the advantage that fertilizer rates can be adjusted to take into account weather conditions and anticipated yields and runoff losses may be reduced (Langdale et al., 1979), but incorporation of the fertilizer is often required and root damage may result from the later applications (Kurtz & Smith, 1966).

3. SLOW-RELEASE FERTILIZERS

By supplying N in a slowly available form, the amount of N available for volatilization, leaching, or denitrification at a given time will be reduced. Organic sources of N (soil organic N, manure and other organic wastes, or plowed-under legume crops) are in this category. Inorganic fertilizers which take time to solubilize are also available. Organic formulations (such as substituted ureas) and inorganic preparations (such as plastic-coated pellets) have been developed (Parr, 1967; Hadas et al., 1975; Hagin & Cohen, 1976; Verstraeten & Livens, 1977). These products, as well as organic N, have the disadvantage that they are not especially suited to many crops which require considerable N over a short period (Parr, 1973). A sulfur-coated urea (SCU) product has been developed by the Tennessee Valley Authority (Prasad et al., 1971). This material has received extensive agronomic evaluation in recent years with generally mixed results in terms of yield response when compared to conventional inorganic N fertilizers. Since SCU likely will cost 25–50% more per unit of N than unaltered urea (Parr, 1973), a substantial benefit would be needed to justify its use. Most of the yield comparisons with SCU have been conducted at near to optimal fertilization rates, whereas the benefits of a slow-release material would be more pronounced at suboptimal rates. The general experience has been that SCU is especially effective for crops grown on coarse-textured soils under severe leaching conditions (Oertli, 1975; Liegel & Walsh, 1976; Allen et al., 1978). Sanchez et al. (1973) found SCU to be 29–59% more effective than urea or $(NH_4)_2SO_4$ with short-statured, high-yielding rice, and that the benefits were sufficient to overcome the higher cost of SCU. Greater carry-over of fertilizer N from SCU than from inorganic N fertilizers for the following crop almost always is observed (Oertli, 1975; Cox & Addiscott, 1976; Liegel & Walsh, 1976; Beauchamp, 1977; Allen et al., 1978). Dalal and Prasad (1975) noted less volatilization of NH_3 from surface-applied SCU than from urea.

It would appear that slow-release formulations have their main application to rice, high value crops, and turf, where the slow-release characteristics can be used to economic benefit. They might prove cost-effective on soils where leaching is a problem. An alternative to the use of only one slow-release material is to blend materials of varying solubility to provide a product which would have a specific N release rate (Prasad et al., 1971; Parr, 1967, 1973; Verstraeten & Livens, 1977). Further research and development would seem warranted in this area.

4. FERTILIZER PLACEMENT

Some benefits may be gained from more effective placement of N fertilizers. For example, if urea or low-pressure NH_3 solutions are applied to the soil surface, significant losses of NH_3 may result (Kurtz & Smith, 1966; Simpson, 1968) unless rain or irrigation water moves the N into the soil. Usually, however, little benefit is gained from deep (> 15 cm) placement of volatile fertilizers such as anhydrous NH_3, and some damage due to high salts, NO_2^-, or unfavorable pH can occur if urea or NH_3 is placed too close to the germinating seed (Isensee & Walsh, 1971; Passioura & Wetselaar, 1972). Some data are available on the depth to which crops will utilize N. For example, corn and small grains will recover N to at least a 60-cm depth (Dancer & Peterson, 1969; Daigger & Sander, 1976). The effects of depth of placement need further evaluation. Deep placement might be desirable with minimum-tillage practices to minimize denitrification. Saffigna et al. (1976) and Hendrickson et al. (1978a) noted that solid N fertilizers placed in certain positions in the hill of potatoes were less susceptible to leaching than in other locations; Hendrickson et al. (1978a) also found that the high osmotic potential produced in the fertilizer band effectively inhibited nitrification in a sandy soil.

5. IMPROVED USE OF PRESENT TECHNOLOGY

Numerous instances exist in which the efficiency of use of fertilizer N can be improved without the introduction of new technology. For example, nonuniform application of broadcast fertilizer N will limit yields; low rates of application in one area are not compensated for by excessive rates in other areas (Voss, 1976)[4]. Other examples include placement and timing of applications (discussed previously) as well as improved management of all factors which can be manipulated to increase crop yields.

6. TECHNIQUES TO MINIMIZE LEACHING ON COARSE-TEXTURED SOILS

Several techniques to reduce water infiltration and/or leaching have been tested on experimental plots. For example, Snyder et al. (1974) examined the effect of a plastic mulch and a water-repellent siliconate applied over a fertilizer band to reduce nutrient leaching in a coarse-textured soil. They found the siliconate to be as effective as the plastic mulch and much superior to an organic mulch. These techniques and others would require further development and economic evaluation before they could be recommended for field crops.

7. FOLIAR APPLICATIONS

Ammonia and urea are effectively absorbed by plant leaves (Wittwer et al., 1963; Scarsbrook, 1965; Viets, 1965; Porter et al., 1972); urea is preferred in foliar treatments because of its rapid penetration into leaves. This practice has been used for many years on high value truck crops, citrus, and deciduous orchards to supplement the soil N supply. It is most convenient and economical with crops that are irrigated by sprinkler systems or that require frequent spray treatments with micronutrients or pesticides. The practice has not been widely adopted for field crops such as corn and wheat, primarily because of economics. Also, severe leaf burn occurs if the concentra-

---

[4] R. D. Voss. 1976. Urea use in current crop management systems—effect of nonuniform applications. p. 193. Agron. Abstr.

tion of urea in the spray solution is $>1-3\%$, and *biuret* (a phytotoxic condensation product of urea that commonly occurs in manufactured urea fertilizers) must be $<0.25\%$ of the total N (Jones & Embleton, 1965; Viets, 1965). Low-biuret urea fertilizers often are difficult to obtain.

The maximum amount of N usually applied at any one time by foliar spray is 10–20 kg/ha. Since corn, for example, requires as much as 2–3 kg N ha^{-1} day^{-1} during its maximum uptake phase, foliar feeding would be needed once or twice a week. This is not economical at present. It would also be quite difficult to apply the solution to corn or other row crops by conventional means (tractor-sprayer equipment) by mid-season.

Research on California citrus crops has shown that yields are comparable with equivalent foliar applications or soil applications of N, and oftentimes N application can be reduced several fold by use of foliar application (Jones & Embleton, 1965; Jones et al., 1970).

Foliar application of urea leads to some loss of N as NH_3, at a rate depending on the time of day the urea is applied and on the weather (Viets, 1965). Further research, development, and educational efforts on foliar application of N with high-value crops seem warranted. At present, this approach appears to offer little promise as a general practice to increase fertilizer efficiency in most field crops. In addition, the limited supply of low-biuret urea has restricted the expansion of the practice even in cropping systems where its value is established, such as California citrus.

8. CARBON SUPPLEMENTS

In most cropped systems, the return of crop residues is the major source of maintanance of reasonable organic matter levels in soils. The incorporation of large amounts of organic C into soils may minimize leaching of residual NO_3^- from fertilizer applications by immobilization of inorganic N in the biomass of soil microflora. Corn stalk residues (including cobs) are commonly tilled into the soil under current practices. In high-yielding corn, this residue may amount to as much as 5,000–10,000 kg/ha of dry matter, with a C/N ratio of 40:1 to 60:1. This high C/N ratio is quite favorable for immobilization of N (Bartholomew, 1965), although the residue is not incorporated until plowing, which may occur at varying times after harvest. In the fall, immobilization of N by incorporated crop residues would be minimal because of cool temperatures.

Steelman[5] evaluated the advantages of incorporation of additional organic C as ground-up corn cobs or paper in amounts ranging from 4.4 to 35 metric tons of dry matter/ha. As would be expected, there were significant decreases in the concentrations of NO_3^- in the soil. Corn yields were not depressed except at the highest level of supplementation. However, this approach is clearly not feasible for the great expanse of cropped soils. The greatest limitation is the lack of a ready supply of organic C materials available at low cost to the farmer.

---

[5] R. J. Steelman. 1974. A method of lowering the amount of nitrate-nitrogen available for leaching into surface waters. M.A. Thesis, Washington Univ., St. Louis, Mo.

9. UREASE AND NITRIFICATION INHIBITORS

In recent years, there has been considerable interest in and research on chemicals which retard the rate of hydrolysis of urea or the rate of nitrification of NH_3. The theory behind this approach is that any means of retaining the N in the NH_4^+ or urea form reduces loss of NO_3^- by leaching or denitrification and, in the case of urea, also reduces loss by volatilization (Bremner & Douglas, 1973).

While considerable success has been achieved in inhibiting urease activity in laboratory studies (Bundy & Bremner, 1973), this technique has not yet been demonstrated to be effective in the field. Urease inhibition suffers from the added disadvantage that in sandy soils urea could be leached beyond the root zone (Gasser, 1964).

The formulation of fertilizers containing nitrification inhibitors offers promise for improving the efficiency of use of fertilizer N. The only highly effective compound currently approved for sale in the USA is nitrapyrin, the common name for 2-chloro-6-(trichloromethyl)pyridine. Nitrapyrin is the active ingredient in N-Serve, which is patented and marketed by Dow Chemical Company (Mullison & Norris, 1976). The compound inhibits the activity of *Nitrosomonas* (Goring, 1962). Laboratory and field studies have shown that this compound is quite effective in inhibiting nitrification of anhydrous NH_3 for up to 6 weeks in the field (e.g., Hughes & Welch, 1970; Briggs, 1975; Hendrickson et al., 1978a, b). Its chief metabolic breakdown product, 6-chloropicolinic acid, is nontoxic (Mullison & Norris, 1976). The compound is volatile, and is soluble only in certain organic solvents and in anhydrous NH_3. Hence, it is somewhat difficult to handle, but can be formulated with anhydrous NH_3 and applied quite easily. It degrades by chemical and biochemical pathways in the soil and by photolysis when exposed to light, and leaves no residue beyond the cropping year (Mullison & Norris, 1976).

Positive yield responses due to use of nitrapyrin have been observed (e.g., Swezey & Turner, 1962; Huber et al., 1974; Warren et al., 1975; Prasad, 1976). The increase in yield is usually attributed to lessening of N losses through leaching or denitrification, but reductions in plant diseases (such as stalk rot on corn) apparently due to predominance of NH_4^+ over NO_3^- nutrition also appears to be a factor (Warren et al., 1975). However, several negative yield response experiments also have been reported [Hendrickson et al., 1978a with potatoes; Parr et al., 1971 with ryegrass and cotton (*Gossypium hirsutum* L.); Nielsen et al., 1967 and Goh and Young, 1975 with wheat (*Triticum aestivum* L.); Osborne, 1977 with ryegrass]. It appears that the unpredictable responses obtained involve a combination of factors, including soil type, weather, and crop. More research is needed to evaluate methods of applying these compounds and to evaluate other possible inhibitors such as carbon disulfide and sodium trithiocarbonate (Ashworth et al., 1975). As experience is gained with these and with other biological inhibitors, significant progress can be expected in tailoring the rate of nitrification to crop needs. However, it remains to be seen whether these materials will be cost-effective, or will significantly diminish NO_3^- pollution problems on croplands.

C. Limitations on Rates of Fertilizer Applications

A control option that has been the subject of considerable research attention and public policy debate in recent years is the proposal to restrict N fertilizer application rates to levels below the recommended economic optimum rates. Implicit in such a proposal is a judgment that some sacrifice of efficiency in crop production might be acceptable as a cost of maintaining or improving surface or ground water quality. Applications of fertilizer N could be restricted, in theory, by a number of policy mechanisms, including the imposition of regulatory limits, zoning land by allowable fertilizer N input rates, taxing N fertilizer, selling rights to use N fertilizer, or subsidizing farmers for losses in production that might result from reduced N fertilizer use.

The possibility of restricting N fertilizer applications has been most widely studied in Illinois. In 1972 and 1975, the Illinois Pollution Control Board (1972, 1975)[6,7] considered the imposition of restrictions, and in both cases rejected the proposal on the ground that an adequate scientific basis does not exist which could support the selection of any particular limits.

There are many problems associated with regulating N fertilizer use (Taylor & Swanson, 1974; Chapt. 20, E. R. Swanson). Enforcement would be difficult and expensive at best, and would increase the level of bureaucracy, which has traditionally been resisted by the autonomous-minded farmer. Taxation would need to be tied continually to farm income, since this approach assumes that the economic optimum rate would automatically lessen. However, in reality this is not true, because grain prices and income fluctuate. Further, grain prices are not reliably known at the beginning of the season. Zoning land or subsidizing farmers for yield losses involve numerous debatable judgmental decisions, and the latter penalizes those with a high level of management. Selling rights to N fertilizer use encourages transfer of wealth to those most able to afford to bid on N fertilizer rights.

D. Fundamental Changes in Agriculture

In addition to the improved management practices and regulatory approaches discussed earlier, some basic shifts in patterns of agricultural production can be envisioned which would be likely to reduce N-related environmental pollution. None of the changes is likely to be pursued solely for

---

[6] Illinois Pollution Control Board. 1972. Opinion of the Board in the matter of plant nutrients, R 71-75, Illinois. Pollut. Cont. Board, Chicago, Ill.

[7] Illinois Pollution Control Board. 1975. In the matter of plant nutrients, R 71-75. Opinion and order of the Board, 1 Oct. 1975, Illinois. Pollut. Cont. Board, Chicago, Ill.

the sake of control of N losses, and in most cases there is little research to indicate the quantitative effects of any of these measures on particular environmental problems. Nevertheless, both the environmental implications and the possible economic and social consequences of such changes deserve to be examined.

1. SYMBIOTIC N FIXATION BY CORN AND OTHER CEREALS

The possibility exists that the genetic information for the control of nitrogenase can be transferred from plant species that fix N to those that do not, and research to accomplish this is being pursued in several laboratories (Hardy & Havelka, 1975; Dobereiner, 1977). While such a breakthrough could be the ideal solution to the problem of cost and production of N fertilizer in the future, its success is very speculative at present (Hardy & Havelka, 1975; Evans & Barber, 1977). Further, widespread N fixation by cereals would increase the fixed N used in agriculture, perhaps compounding pollution problems.

Another approach currently being studied is the use of N_2-fixing free-living bacteria in the root rhizosphere to produce fixed N for the plant. Again, many problems exist, not the least of which are the high energy requirement for N fixation and the vast numbers of bacteria required. A plant-algal symbiosis in rice fields with the weed *Azolla* and the algae *Anabaena* has been reported in which from 60 to 140 kg N/ha is fixed (Hardy & Havelka, 1975). This system currently is being evaluated further.

2. INCREASED N FIXATION BY LEGUMES

Considerable research effort is currently being expended to increase N fixation by legumes, mainly through genetic engineering of *Rhizobium* (Hardy & Havelka, 1975; Evans & Barber, 1977; Stewart, 1977). If this effort succeeds, the result might well be major shifts in U.S. agriculture in the coming decades toward less dependence on fertilizer N. The farming systems which evolve will have to be evaluated with respect to their potential for environmental pollution.

3. CHANGES IN CROP PATTERNS

A shift to increased area in legumes, and decreased area of grains, would reduce the amount of N fertilizer used. However, the environmental consequences of possible shifts in cropping patterns is open to speculation. Major alterations in the relative production of forage crops of different kinds could occur only in conjunction with changes in patterns of livestock production, land use, and possibly in the human diet. Similarly, improvements in the technology for conversion of plant protein to food protein could also have large impacts on dietary habits, and thus, on the markets for specific crop products. The environmental and social consequences of such adjustments in the agricultural system are very difficult to project.

4. "ORGANIC" FARMING

Commoner (1977) has suggested that farms which combine production of field crops and livestock and return animal manures to croplands are environmentally preferable to farms that produce only grain, because of the combined effects of reduced use of chemical fertilizers and pesticides, and the recycling of animal waste to the soil. Some commercial farms currently rely exclusively on organic sources of N. A study comparing a group of 14 large-scale organic farms (farms which used conventional machinery, but no commercial fertilizers or pesticides) with 14 conventional farms in the Midwest USA was conducted over 3 years (Lockeretz et al., 1978). The average corn yield on organic farms was not statistically different from that of conventional farms, but the organic farms had less corn land (24% of the area in corn compared to 37% for the conventional farms). Yields of wheat, soybeans, and oats were also similar. Calculated soil loss by erosion was less, and average organic matter somewhat higher for the organic farms. Gross income on organic farms was 90% of that on conventional farms, but because of lower costs of production, the net return per ha was the same over the 3-year period for both groups. Commoner (1977) suggests that these results show that organic farming techniques, supplemented by relatively low rates of additions of inorganic fertilizer, could attain production rates equivalent to those of conventional farms while minimizing fertilizer-related nutrient losses. At this time, however, there are almost no data available on leaching and other losses of N from croplands farmed organically, and quantitative assessments of the environmental impacts of a shift toward organic techniques cannot be made. Furthermore, Lockeretz et al. (1978) point out that the organic farms are adding less K and P than is being removed, and that eventually these soil reserves will be sufficiently depleted to require fertilizer additions. A similar situation possibly exists for N, although N balances for these systems are not available. The long-term effect would likely be reduced crop production unless even larger shifts to legume-based agriculture is made. Large scale introduction of legume-based agriculture in grain producing areas would require changes in technology, economics and the human diet. Recently, reports on organic farming were prepared by a USDA study team (USDA, 1980) and by the Council for Agricultural Science and Technology (CAST, 1980).

5. REDUCED AREA IN CULTIVATION

Aldrich[8] has suggested that the total amount of land in cultivation may be as important a variable as fertilizer application rates in explaining NO_3^- levels in Illinois surface waters. Cultivation increases the rate of mineralization of soil organic N, and may thus contribute to NO_3^- leaching even in the absence of added fertilizer. However, the limited data available (Klepper, 1978) suggest that the mineralization of soil organic N is probably not the major source of NO_3^- in Illinois surface waters, and that the area planted in

---

[8] Aldrich, Op. cit.

crops that are heavily fertilized, rather than total cultivated acreage, may be the most critical land-use variable.

Any substantial reduction in cultivated acreage would almost certainly be accompanied by a reduction in total crop production. About 85% of the corn grown in Illinois is used as livestock feed; major changes in cattle feeding practices (e.g., increased use of pasture and range instead of feedlots), a reduction in the size of cattle populations, and more reliance on imported meat would probably be associated with any large reductions in crop area.

IX. SUMMARY

Human activities, in particular fossil fuel combustion, fertilizer N use, and enhanced biological N fixation, have increased the mass flow of fixed N through some pathways of the biogeochemical N cycle. As a result, excessive accumulations of N have occurred or may occur in some compartments with adverse effects ranging from localized to global in scale. While agriculture has a major role in altering many of the flows of N, the lack of an adequate scientific information base limits understanding of the exact magnitude of risks and the consequence of possible control measures.

Most productive agricultural ecosystems rely on energy-intensive practices; a major portion of this energy is in the production of N fertilizers. It seems likely that energy-related rather than environmentally related concerns will have the major influence on agricultural N management in the coming decades.

Consideration of some example agricultural ecosystems show that, in many cases, the efficiency of N use can be improved. Many techniques to improve this efficiency, such as erosion control, improved irrigation, increased use of legume crops, cover crops, intercropping, matching available N to crop needs, slow-release N fertilizers, and nitrification inhibitors are already available. Their effective utilization involves application of systems analysis approaches to farming. Other techniques will be developed as the social and economic pressures for efficient N management increase. Possible innovative techniques include improved crop varieties, nonleguminous symbiotic N fixation, and basic changes in cropping systems. Many of these changes involve major social and economic shifts.

LITERATURE CITED

Abrams, L. W., and J. L. Barr. 1974. Corrective taxes for pollution control: An application of the environmental pricing and standards systems to agriculture. J. Environ. Econ. Manag. 1:12–19.

Adriano, D. C., A. C. Chang, and R. Sharpless. 1974. Nitrogen loss from manure as influenced by moisture and temperature. J. Environ. Qual. 3:258–261.

Adriano, D. C., P. F. Pratt, and F. H. Takatori. 1972. Nitrate in unsaturated zone of an alluvial soil in relation to fertilizer nitrogen rate and irrigation level. J. Environ. Qual. 1:418–422.

Adriano, D. C., F. H. Takatori, P. F. Pratt, and O. A. Lorenz. 1972b. Soil nitrogen balance in selected row-crop sites in Southern California. J. Environ. Qual. 1:279–283.

Allison, F. E. 1955. The enigma of soil nitrogen balance sheets. Adv. Agron. 7:213–250.

Allen, S. E., G. L. Terman, and H. G. Kennedy. 1978. Nutrient uptake by grass and leaching losses from soluble and S-coated urea and K. Agron. J. 70:264–268.

Ashworth, J., G. G. Briggs, and A. A. Evans. 1975. Field injection of carbon disulphide to inhibit nitrification of ammonia fertilizer. Chem. Ind. (London) 1975 (17):749–750.

Avnimelech, Y., S. Dasberg, A. Harpaz, and I. Levin. 1978. Prevention of nitrate leakage from the Hula basin, Israel: A case study of watershed management. Soil Sci. 125:233–239.

Ayanaba, A., W. Verstraete, and M. Alexander. 1973a. Formation of dimethylnitrosamine, a carcinogen and mutagen, in soils treated with nitrogen compounds. Soil Sci. Soc. Am. Proc. 37:565–568.

Ayanba, A., W. Verstraete, and M. Alexander. 1973b. Microbial contribution to nitrosamine formation in sewage and soil. J. Natl. Cancer Inst. 50:811–813.

Ayers, R. S. 1978. A case study—Nitrates in the upper Santa Ana River Basin in relation to groundwater pollution. p. 355–368. In P. F. Pratt (ed.) Proc. Natl. Conf. Manage. of Nitrogen in Irrigated Agriculture. 15–18 May 1978. Dep. Soil and Environ. Sci., Univ. of California-Riverside.

Ayers, R. S., and R. L. Branson. 1973. Nitrates in the upper Santa Ana River Basin in relation to groundwater pollution. Calif. Agric. Exp. Stn. Bull. 861, Univ. of California, Riverside.

Baber, H. T., and A. T. Wilson. 1972. Nitrate pollution of groundwater in the Waikato region. J. N.Z. Inst. Chem. 56:179–183.

Baker, J. L., K. L. Campbell, H. P. Johnson, and J. J. Hanway. 1975. Nitrate, phosphorus, and sulfate in subsurface drainage water. J. Environ. Qual. 4:406–412.

Ball, R., D. R. Keeney, P. W. Theobald, and P. Nes. 1979. Nitrogen balance in urine affected areas of a New Zealand pasture. Agron. J. 71:309–314.

Bandel, V. A., S. Dzienia, G. Stanford, and J. O. Legg. 1975. N behavior under no-till vs. conventional corn culture. I. First-year results using unlabeled N fertilizer. Agron. J. 67:782–786.

Barber, S. A. 1974. Nitrogen efficiency. Part 2: The midwest. Fert. Solutions 5(6):58–60.

Bartholomew, W. V. 1965. Mineralization and immobilization of nitrogen in the decomposition of plant and animal residues. In W. V. Bartholomew and F. E. Clark (ed.) Soil nitrogen. Agronomy 10:285–306. Am. Soc. of Agron., Madison, Wis.

Bartholomew, W. V. 1977. Soil nitrogen changes in farming systems in the humid tropics. p. 27–42. In A. Ayanaba and P. J. Dart (ed.) Biological nitrogen fixation in farming systems of the tropics. John Wiley & Sons, New York.

Beauchamp, E. G. 1977. Slow release N fertilizers applied in fall for corn. Can. J. Soil Sci. 57:487–496.

Bingham, F. T., H. G. Champman, and A. L. Pugh. 1954. Solution-culture studies of nitrite toxicity to plants. Soil Sci. Soc. Am. Proc. 18:305–308.

Bingham, F. T., S. Davis, and E. Shade. 1971. Water relations, salt balance and nitrate leaching losses of a 960-acre citrus watershed. Soil Sci. 112:410–418.

Black, C. A. 1967. Soil-plant relationships. Wiley & Sons, New York.

Bolin, B., and E. Arrhenius. 1977. Nitrogen—An essential life factor and a growing environmental hazard. Report from Nobel Symp. no. 38. Ambio 6:96–105.

Boone, L. V., and L. F. Welch. 1972. The more nitrogen in corn, the less in our water supply. Ill. Res. 14(4):5–7.

Bornemisza, E., and A. Alvarado (ed.). 1975. Soils management in tropical America. Soil Sci. Dep., North Carolina State Univ., Raleigh. 563 p.

Boswell, f. C., L. R. Nelson, and M. J. Bitzer. 1976. Nitrification inhibitor with fall-applied vs. split nitrogen applications for winter wheat. Agron. J. 68:737–740.

Bower, C. A., J. R. Spencer, and L. V. Weeks. 1969. Salt and water balance, Coachella Valley California. Am. Soc. Civil Eng. Proc., J. Irrig. Drain. Div. 95:55–64.

Bower, C. A., and L. V. Wilcox. 1969. Nitrate content of the upper Rio Grande as influenced by nitrogen fertilization of adjacent irrigated lands. Soil Sci. Soc. Am. Proc. 33:971–972.

Branson, R. L., P. F. Pratt, J. D. Rhoades, and J. D. Oster. 1975. Water quality in irrigated watersheds. J. Environ. Qual. 4:33–40.

Bremner, J. M., and L. A. Douglas. 1973. Effects of some urease inhibitors on urea hydrolysis in soils. Soil Sci. Soc. Am. Proc. 37:225–226.

Brezonik, P. L. 1969. Eutrophication: The process and its modeling potential. p. 68–116. In H. D. Putnam (ed.) Modeling the eutrophication process. Proc. Wkshp., Dep. Environ. Eng. Sci., Univ. of Florida, Gainesville.

Brezonik, P. L. 1972. Nitrogen: Sources and transformations in natural waters. p. 1–50. *In* H. E. Allen and J. R. Kramer (ed.) Nutrients in natural waters. Wiley-Interscience, New York.

Briggs, G. G. 1975. The behaviour of the nitrification inhibitor "N-Serve" in broadcast and incorporated applications to soil. J. Sci. Food Agric. 26:1083–1092.

Brink, N. 1975. Water pollution from agriculture. J. Water Pollut. Control Fed. 47:789–795.

Broadbent, F. E., and A. B. Carlton. 1978. Field trials with isotopically labeled nitrogen fertilizer. p. 1–41. *In* D. R. Nielsen and J. G. MacDonald (ed.) Nitrogen in the environment, Vol. 1. Nitrogen behavior in field soil. Academic Press, New York.

Bundy, L. G., and J. M. Bremner. 1973. Effects of substituted *P*-benzoquinones on urease activity in soils. Soil Biol. Biochem. 5:847–853.

Burns, R. C., and R. W. F. Hardy. 1975. Nitrogen fixation in bacteria and higher plants. Springer-Verlag, New York.

Castle, M. E., and D. Reid. 1968. The effects of single compared with split applications of fertilizer nitrogen on the yield and seasonal production of a pure grass sward. J. Agric. Sci. 70:383–389.

Chalk, P. M., D. R. Keeney, and L. M. Walsh. 1975. Crop recovery and nitrification of fall and spring applied anhydrous ammonia. Agron. J. 67:33–37.

Chancellor, W. J., and J. R. Goss. 1976. Balancing energy and food production: 1975–2000. Science 192:213–218.

Clawson, W. J. 1971. Economics of recovery and distribution of animal waste. J. Animal Sci. 32:816–820.

Comly, H. H. 1945. Cyanosis in infants caused by nitrates in well water. J. Am. Med. Assoc. 129:112–116.

Commoner, B. 1970. Threats to the integrity of the nitrogen cycle: Nitrogen compounds in soil, water, atmosphere and precipitation. p. 70–95. *In* S. f. Singer (ed.) Global effects of environmental pollution. Springer-Verlag, New York.

Commoner, B. 1977. Cost-risk-benefit analysis of nitrogen fertilization: A case history. Ambio 6:157–161.

Cooke, G. W., and R. J. B. Williams. 1970. Losses of nitrogen and phosphorus from agricultural land. Water Treat. Exam. 19:253–274.

Council for Agricultural Science and Technology (CAST). 1976. Effect of increased nitrogen fixation on stratospheric ozone. Report no. 53, Counc. for Agric. Sci. Technol., Ames, Iowa.

Council for Agricultural Science and Technology (CAST). 1980. Organic and conventional farming compared. Report no. 84, Counc. for Agric. Sci. Technol., Ames, Iowa.

Court, M. N., R. C. Stephens, and J. S. Wais. 1962. Nitrite toxicity arising from the use of urea as a fertilizer. Nature (London) 194:1263–1265.

Cox, D., and T. M. Addiscott. 1976. Sulphur-coated urea as a fertiliser for potatoes. J. Sci. Food Agric. 27:1015–1020.

Crawford, R. F., W. K. Kennedy, and K. L. Davison. 1966. Factors influencing the toxicity of forages that contain nitrate when fed to cattle. Cornell Vet. 56:3–17.

Crosby, N. E., and R. Sawyer. 1976. N-Nitrosamines: A review of chemical and biological properties and their estimation in foodstuffs. Adv. Food Res. 22:1–71.

Crutzen, P. J. 1974. Estimation of possible variations in total ozone due to natural causes and human activities. Ambio 3:201–210.

Crutzen, P. J., and D. H. Ehalt. 1977. Effects of nitrogen fertilizers and combustion on the stratospheric ozone layer. Ambio 6:112–117.

Dahlman, R. C., J. S. Olson, and K. Doxtader. 1967. The nitrogen economy of grassland and dune soils. p. 54–82. *In* Proc. Biol. and Ecol. of Nitrogen. Natl. Acad. Sci., Washington, D.C.

Daigger, L. A., and D. H. Sander. 1976. Nitrogen availability to wheat as affected by depth of nitrogen placement. Agron. J. 68:524–526.

Dalal, R. C., and M. Prasad. 1975. Comparison of sulphur-coated urea and ammonium sulphate amended with N-Serve as sources of nitrogen for sugarcane. J. Agric. Sci. 85:427–433.

Dancer, W. S., and L. A. Peterson. 1969. Recovery of differentially placed NO_3-N in a silt loam by five crops. Agron. J. 61:893–895.

Date, R. A. 1973. Nitrogen, a major limitation in the productivity of natural communities, crops and pastures in the Pacific area. Soil Biol. Biochem. 5:5–18.

Davis, S., and L. B. Grass. 1966. Determining evapotranspiration from a 1,000-acre lysimeter. Trans. Am. Soc. Agric. Eng. 9:108–109.

Deeb, B. S., and K. W. Sloan. 1975. Nitrates, nitrites, and health. Bull. 750, Univ. of Illinois, Agric. Exp. Stn., Urbana, Ill.

Devitt, D., J. Letey, L. J. Lund, and J. W. Blair. 1976. Nitrate-nitrogen movement through soil as affected by soil profile characteristics. J. Environ. Qual. 5:283–288.

Dobereiner, J. 1977. Biological nitrogen fixation in tropical grasses—possibilities for partial replacement of mineral N fertilizers. Ambio 6:174–177.

Elliot, L. F., T. M. McCalla, L. N. Mielke, and T. A. Travis. 1972. Ammonium nitrate and total nitrogen in the soil water of feedlot and field soil profiles. J. Appl. Microbiol. 28:810–813.

Elliot, L. F., G. E. Schuman, and F. G. Viets, Jr. 1971. Volatilization of nitrogen containing compounds from beef cattle areas. Soil Sci. Soc. Am. Proc. 35:752–755.

Embleton, T. W., W. W. Jones, and R. L. Branson. 1974. Leaf analysis proven useful in increasing efficiency of nitrogen fertilization of oranges and reducing nitrate pollution potential. Commun. Soil Sci. Plant Anal. 5:436–440.

Evans, H. J., and L. E. Barber. 1977. Biological nitrogen fixation for food and fiber production in foods. Science 197:332–339.

Fassett, D. W. 1973. Nitrates and nitrites. p. 7–25. In Toxicants occurring naturally in foods. Nat. Acad. Sci., Washington, D.C.

Frere, M. H. 1976. Nutrient aspects of pollution from cropland. p. 59–90. In Control of water pollution from cropland, Volume II—An overview. Report no. EPA-600/2-75-02b ARS-H-5-2, USEPA and USDA, Washington, D.C.

Frissel, J. J. 1977. Application of nitrogen fertilizers: Present trends and projections. Ambio 6:152–156.

Gasser, J. K. R. 1964. Urea as a fertilizer. Soils Fert. 27:175–180.

Gentzsch, E. P., E. C. A. Runge, and T. R. Peck. 1974. Nitrate occurrence in some soils with and without natric horizons. J. Environ. Qual. 3:89–93.

Gilliam, J. W., R. W. Skaggs, and S. B. Weed. 1979. Drainage control to diminish nitrate loss from agricultural fields. J. Environ. Qual. 8:137–142.

Goh, K. M., and A. W. Young. 1975. Effects of fertilizer nitrogen and 2-chloro-6-(trichloromethyl)pyridine (N-Serve) on soil nitrification, yield, and nitrogen uptake of "Arawa" and "Hilgendorf" wheats. N.Z. J. Agric. Res. 18:215–225.

Goldman, J. C. 1976. Identification of nitrogen as a growth-limiting factor in wastewaters and coastal marine waters through continuous culture algal assays. Water Res. 10:97–104.

Gorham, E., and A. G. Gordon. 1963. Some effects of smelter pollution upon aquatic vegetation near Sudbury, Ontario. Can. J. Bot. 41:371–378.

Goring, C. A. I. 1962. Control of nitrification by 2-chloro-6-(trichloromethyl)-pyridine. Soil Sci. 93:211–218.

Greenland, D. J. 1977. Contribution of microorganisms to the nitrogen status of tropical soils. p. 13–26. In A. Ayanaba and P. J. Dart (ed.) Biological nitrogen fixation in farming systems of the tropics. John Wiley & Sons, New York.

Hadas, A., U. Kafkafi, and A. Peled. 1975. Initial release of nitrogen from urea form under field conditions. Soil Sci. Soc. Am. Proc. 39:1103–1105.

Hagin, J., and L. Cohen. 1976. Nitrogen fertilizer potential of an experimental urea formaldehyde. Agron. J. 68:518–520.

Hahn, J., and C. Junge. 1977. Atmospheric nitrous oxide: A critical review. Z. Naturforsch. 32:190–214.

Hanson, C. L., J. F. Power, and C. J. Erickson. 1978. Forage yield and fertilizer recovery by three irrigated perennial grasses as affected by N fertilization. Agron. J. 70:373–375.

Hanway, J. J. 1960. Growth and nutrient uptake by corn. Extension Service Pamphlet 277, Iowa State Univ. Extension, Ames.

Hardy, R. W. F., and U. D. Havelka. 1975. Nitrogen fixation research: A key to world food? Science 188:633–643.

Hendrickson, L. L., D. R. Keeney, L. M. Walsh, and E. A. Liegel. 1978a. Evaluation of nitrapyrin as a means of improving nitrogen efficiency in irrigated sands. Agron. J. 70:699–703.

Hendrickson, L. L., L. M. Walsh, and D. R. Keeney. 1978b. Effectiveness of nitrapyrin in controlling nitrification of fall- and spring-applied anhydrous ammonia. Agron. J. 70:704–708.

Henkens, C. H. 1977. Agro-ecosystems in the Netherlands, Part II. p. 79–97. In M. J. Frissel (ed.) Cycling in mineral nutrients in agricultural ecosystems. Agro-Ecosystems, Vol. 4, Special Issue, Elsevier, The Netherlands.

Hoglund, J. H., J. R. Crush, J. L. Brock, and R. Ball. 1979. Nitrogen fixation in pasture. XII. General discussion. N.Z. J. Exp. Agric. 7:45–51.

Huber, D. M., H. L. Warren, and D. W. Nelson. 1974. Effect of nitrification inhibitors on yields of corn and wheat. p. 1–6. *In* Proc. Indiana Plant, Food Agric. Chem. Conf., 15–18 Nov. 1974. Purdue Univ., West Lafayette, Ind.

Hughes, T. D., and L. F. Welch. 1970. 2-chloro-6(trichloro-methyl) pyridine as a nitrification inhibitor for anhydrous ammonia applied in different seasons. Agron. J. 62:821–824.

Hutchinson, G. L., and F. G. Viets, Jr. 1969. Nitrogen enrichment of surface water by absorption of ammonia volatilized from cattle feedlots. Science 166:514–515.

Ingestad, T. 1977. Nitrogen and plant growth: Maximum efficiency of nitrogen fertilizers. Ambio 6:146–151.

Irrigation Journal. 1975. Survey issue. Irrig. J. 25:15–22.

Isensee, A. R., and L. M. Walsh. 1971. Influence of banded fertilizer on the chemical environment surrounding the band. I. Effect on pH and solution nitrogen. J. Sci. Food Agric. 22:105–109.

Johnson, J. W., L. F. Welch, and L. T. Kurtz. 1974. Soybeans' role in nitrogen balance. Ill. Res. 16:6–7.

Johnson, J. W., L. F. Welch, and L. T. Kurtz. 1975. Environmental implications of N fixation by soybeans. J. Environ. Qual. 4:303–306.

Johnston, H. S. 1977. Analysis of the independent variables in the perturbation of stratospheric ozone by nitrogen fertilizers. J. Geophys. Res. 82:1767–1772.

Jones, E. 1968. Nutrient cycle and soil fertility on red ferrallitic soils. Int. Congr. Soil Sci., Trans. 9th (Adelaide) III:419–427.

Jones, J. B., Jr., and H. U. Eck. 1973. Plant analysis as an aid in fertilizing corn and grain sorghum. p. 349–364. *In* L. M. Walsh and J. D. Beaton (ed.) Soil testing and plant analysis, 2nd ed. Soil Sci. Soc. Am., Madison, Wis.

Jones, M. B., C. C. Delwiche, and W. A. Williams. 1977. Uptake and losses of ^{15}N applied to annual grass and clover in lysimeters. Agron. J. 69:1019–1033.

Jones, W. W., and T. W. Embleton. 1965. Urea foliage sprays. Calif. Citrogr. 50(9):334, 355–359.

Jones, W. W., T. W. Embleton, S. B. Boswell, G. E. Goodall, and E. L. Barnhart. 1970. Nitrogen rate effects on lemon production, quality, and leaf nitrogen. J. Am. Soc. Hortic. Sci. 95:46–49.

Kaul, A. K. 1977. Protein resources and production. Ambio 6:141–145.

Keeney, D. R. 1973. The nitrogen cycle in sediment-water systems. J. Environ. Qual. 2:15–29.

Klepper, R. 1978. Nitrogen fertilizer and nitrate concentrations in tributaries of the Upper Sangamon River in Illinois. J. Environ. Qual. 7:13–22.

Kurtz, L. T., and G. E. Smith. 1966. Nitrogen fertility requirements. p. 195–236. *In* W. H. Pierre, S. R. Aldrich, and W. P. Martin (ed.) Advances in corn production: Principles and practices. Iowa State Univ. Press, Ames.

Langdale, G. W., R. A. Leonard, W. G. Fleming, and W. A. Jackson. 1979. Nitrogen and chloride in small upland Piedmont watersheds. II. Nitrogen and chloride transport in runoff. J. Environ. Qual. 8:57–63.

Larsen, J. E., and H. Kohnke. 1946. Relative merits of fall- and spring-applied nitrogen fertilizers. Soil Sci. Soc. Am. Proc. 11:378–383.

Law, J. P., Jr., and G. V. Skogerboe. 1972. Potential for controlling quality of irrigation return flows. J. Environ. Qual. 1:140–145.

Law, J. P., Jr., and J. L. Witherow. 1971. Irrigation residues. J. Soil Water Conserv. 26:54–56.

Lesczynski, D. B., and C. B. Tanner. 1976. Seasonal variation of root distribution of irrigated field-grown Russet Burbank potatoes. Am. Potato J. 53:69–78.

Liegel, E. A., and L. M. Walsh. 1976. Evaluation of sulfur-coated urea (SCU) applied to irrigated potatoes and corn. Agron. J. 68:457–463.

Likens, G. E. 1976. Acid precipitation. Chem. Eng. News 54:29, 44.

Lockeretz, W., G. Shearer, R. Klepper, and S. Sweeney. 1978. Field crop production on organic farms in the midwest. J. Soil Water Conserv. 33:130–134.

Loehr, R. C. 1974. Characteristics and magnitude of nonpoint sources. J. Water Pollut. Control Fed. 46:1849–1872.

Ludwick, A. E., J. O. Reuss, and E. J. Langin. 1976. Soil nitrates following four years continuous corn and as surveyed in irrigated farm fields of central and eastern Colorado. J. Environ. Qual. 5:82–86.

Luebs, R. E., K. R. Davis, and A. E. Lang. 1974. Diurnal fluctuation and movement of atmospheric ammonia and related gases from diaries. J. Environ. Qual. 3:265–269.

Magdoff, F. R. 1978. Influence of manure application rates and continuous corn on soil-N. Agron. J. 70:629–632.

Magee, P. N. 1971. Toxicity of nitrosamines: Their possible human health hazards. Food Cosmet. Toxicol. 9:207–218.

Magee, P. N. 1977. Nitrogen as a health hazard. Ambio 6:123–125.

Marsh, A. W., R. L. Branson, S. Davis, C. D. Gustafson, and F. K. Aljibury. 1975. Drip irrigation. Leaflet 2740, Div. Agric. Sci., Univ. of California, Davis.

McElroy, A. D., S. Y. Chin, J. W. Nebgen, A. Aleti, and A. E. Vandegraft. 1975. Water pollution from nonpoint sources. Water Res. 9:675–681.

McFee, W. W., J. M. Kelly, and R. H. Beck. 1976. Acid precipitation: Effects on soil base pH and base saturation of exchange sites. p. 727–735. *In* L. S. Dochinger and T. A. Seliga (ed.) Proc. First Int. Symp., Acid Precipitation and the Forest Ecosystem, 12–15 May 1975, Columbus, Ohio. U.S. For. Serv. Gen. Tech. Report NE-23.

Meek, B. D., L. B. Grass, and A. J. MacKenzie. 1969. Applied nitrogen losses in relation to oxygen status of soils. Soil Sci. Soc. Am. Proc. 33:575–578.

Meisinger, J. J. 1976. Nitrogen application rates consistent with environmental constraints for potatoes on Long Island. Search Agric. (Geneva, N.Y.) 6:1–19.

Mielke, L. N., N. P. Swanson, and T. M. McCalla. 1974. Soil profile conditions of cattle feedlots. J. Environ. Qual. 3:14–17.

Miller, M. H. 1979. Contribution of nitrogen and phosphorus to subsurface drainage water from intensively cropped mineral and organic soils in Ontario. J. Environ. Qual. 8:42–48.

Miller, R. J., and R. B. Smith. 1976. Nitrogen balance in the Southern San Joaquin Valley. J. Environ. Qual. 5:274–278.

Miller, W. E., T. E. Maloney, and J. C. Greene. 1974. Algal productivity in 49 lake waters as determined by algal assays. Water Res. 8:667–679.

Miner, J. R., and T. L. Willrich. 1970. Livestock operations and field-spread manure as sources of pollutants. p. 231–240. *In* T. L. Willrich and G. E. Smith (ed.) Agricultural practices and water quality. Iowa State Univ. Press, Ames.

Mirvish, S. S. 1970. Kinetics of dimethylamine nitrosation in relation to nitrosamine carcinogenesis. J. Natl. Cancer Inst. 44:633–639.

Mirvish, S. S. 1975. Formation of N-nitroso compounds—chemistry, kinetics, and in vivo occurrence. Toxicol. Appl. Pharmacol. 31:325–351.

Mosier, A. R., C. E. Andre, and F. G. Viets, Jr. 1973. Identification of aliphatic amines volatilized from cattle feedyard. Environ. Sci. Technol. 7:642–644.

Mosier, A. R., and S. Torbit. 1976. Synthesis and stability of dimethyl nitrosamine in cattle manure. J. Environ. Qual. 5:465–468.

Mulliner, H. R., and K. D. Frank. 1975. Anhydrous ammonia application in irrigation water versus mechanical and its effect on corn yields. Trans. Am. Soc. Agric. Eng. 18:526–538.

Mullison, W. R., and M. G. Norris. 1976. A review of the toxicological, residual, and environmental effects of nitrapyrin and its metabolite 6-chloropicolinic acid. Down Earth 32:22–27.

Muir, J., E. Seim, and R. A. Olson. 1973. A study of factors influencing the nitrogen and phosphorus contents of Nebraska waters. J. Environ. Qual. 2:466–470.

National Commission on Water Quality (NCWQ). 1975. Staff draft report. U.S. Govt. Printing Off., Washington, D.C.

National Research Council (NRC). 1972. Accumulation of nitrate. Natl. Acad. Sci., Washington, D.C.

National Research Council (NRC). 1978. Nitrates: An environmental assessment. Nalt. Acad. Sci., Washington, D.C.

Nelson, L. B., and R. E. Uhland. 1955. Factors that influence loss of fall applied fertilizers and their probable importance in different sections of the United States. Soil Sci. Soc. Am. Proc. 19:492–496.

Nelson, L. R., R. N. Gallaher, M. R. Holmer, and R. R. bruce. 1977. Corn forage production in no-till and conventional tillage double-cropping systems. Agron. J. 69:635–638.

Nielsen, D. R., R. D. Jackson, J. W. Cary, and D. D. Evans (ed.). 1972. Soil water. Am. Soc. Agron., Soil Sci. Soc. Am., Madison, Wis.

Nielsen, K. F., F. G. Warder, and W. C. Hinman. 1967. Effect of chemical inhibition of nitrification on phosphorus absorption by wheat. Can. J. Soil Sci. 47:65–71.

Nightingale, H. I. 1972. Nitrates in soil and ground water beneath irrigated and fertilized crops. Soil Sci. 114:300–311.

Norton, S. A. 1976. Changes in chemical processes in soils caused by acid precipitation. p. 711–724. *In* L. S. Dochinger and T. A. Seliga (ed.) Proc. 1st Int. Symp., Acid Precip. and the For. Ecosys., 12–15 May 1975, Columbus, Ohio. U.S. For. Serv. gen. tech. Report NE-23.

Nye, J. C. 1973. Animal wastes. p. 95–110. *In* Nitrogenous compounds in the environment. Hazardous Materials Advisory Committee, Science Advisory Board, Report no. EPA-SAR-73-001, USEPA, Washington, D.C.

Nye, P. H., and D. J. Greenland. 1960. The soil under shifting cultivation. Tech. Commun. no. 51, Commonwealth Bureau of Soils, Harpenden, U.K.

Nye, P. H., and D. J. Greenland. 1964. Changes in the soil after clearing the tropical forest. Plant Soil 21:101–112.

O'Connor, K. F. 1974. Nitrogen in agrobiosystems and its environmental significance. N.Z. Agric. Sci. 8:137–148.

Oertli, J. J. 1975. Efficiency of nitrogen recovery from controlled-release urea under conditions of heavy leaching. Agrochimica 19:326–335.

Osborne, G. F. 1977. Some effects of the nitrification inhibitor [2-chloro-6(trichloromethyl) pyridine] on the use of fertilizer nitrogen and the growth of two wheat varieties. Aust. J. Exp. Agric. Animal Husb. 17:645–651.

Parr, J. F. 1967. Biochemical considerations for increasing the efficiency of nitrogen fertilizers. Soils Fert. 30:207–213.

Parr, J. F. 1973. Chemical and biochemical considerations for maximizing the efficiency of fertilizer nitrogen. J. Environ. Qual. 2:75–84.

Parr, J. F., B. R. Carroll, and S. Smith. 1971. Nitrification inhibition in soil: I. A comparison of 2-chloro-6(trichloromethyl)pyridine and potassium azide formulated with anhydrous ammonia. Soil Sci. Soc. Am. Proc. 35:469–473.

Passioura, J. B., and R. Wetselaar. 1972. Consequences of banding nitrogen fertilizers in soil. II. Effects on the growth of wheat roots. Plant Soil 36:461–473.

Peterson, J. R., T. M. McCalla, and G. E. Smith. 1971. Human and animal wastes as fertilizers. p. 557–596. *In* Fertilizer technology and use, 2nd ed. Soil Sci. Soc. Am., Inc., Madison, Wis.

Pierotti, D., and R. A. Rasmussen. 1976. Combustion as a source of nitrous oxide in the atmosphere. Geophys. Res. Lett. 3:265–267.

Pierre, W. H., S. R. Aldrich, and W. P. Martin (ed.). 1966. Advances in corn production: Principles and practices. Iowa State Univ. Press, Ames. 476 p.

Pierre, W. H., L. Dumenil, and J. Henao. 1977a. Relationship between corn yield, expressed as a percentage of maximum, and the N percentage in grain. II. Diagnostic use. Agron. J. 69:221–226.

Pierre, W. H., L. Dumenil, V. D. Jolley, J. R. Webb, and W. D. Shrader. 1977b. Relationship between corn yield, expressed as a percentage of maximum, and the N percentage in grain. I. Various N-rate experiments. Agron. J. 69:215–220.

Pimentel, D., W. Dritschilo, J. Krummel, and J. Kutzman. 1975. Energy and land constraints in food protein production. Science 190:754–761.

Porter, K. S. (ed.). 1975. Nitrogen and phosphorus: Food production, waste and the environment. Ann Arbor Science, Ann Arbor, Mich. 372 p.

Porter, L. K. 1975. Nitrogen transfer in ecosystems. p. 1–30. *In* E. A. Paul and A. D. McLaren (ed.) Soil Biochemistry. Vol. 4. Marcel Dekker, New York.

Porter, L. K., F. G. Viets, Jr., and G. L. Hutchinson. 1972. Air containing nitrogen-15 ammonia: Foliar absorption by corn seedlings. Science 175:759–761.

Prasad, M. 1976. Nitrogen nutrition and yield of sugarcane as affected by N-Serve. Agron. J. 68:343–346.

Prasad, R., G. B. Rajale, and B. A. Lakhdive. 1971. Nitrification retarders and slow-release nitrogen fertilizers. Adv. Agron. 23:337–383.

Pratt, P. F. 1976. Irrigation for minimal nitrate pollution. p. 99–101. *In* Proc. 2nd Symp., Res. Appl. to Natural Needs. Vol. VI. Coping with man-made and natural hazards. Natl. Sci. Found., Washington, D.C.

Pratt, P. F. (ed.). 1978. Proceedings, National Conference on Management of Nitrogen in Irrigated Agriculture. Dep. Soil Environ. Sci., 15–18 May 1978, Univ. of California, Riverside. 442 p.

Pratt, P. F., and D. C. Adriano. 1973. Nitrate concentrations in the unsaturated zone beneath irrigated fields in southern California. Soil Sci. Soc. Am. Proc. 37:321–322.

Pratt, P. F., W. W. Jones, and V. E. Hunsaker. 1972. Nitrate in deep soil profiles in reaction to fertilizer rates and leaching volume. J. Environ. Qual. 1:97–102.

Rauschkolb, R. S. 1978. Overview of nitrogen in irrigated agriculture. p. 53–60. *In* P. F. Pratt (ed.) Proc., Natl. Conf. Manage. of Nitrogen in Irrigated Agric. Dep. Soil Environ. Sci., 15–18 May 1978, Univ. of California, Riverside.

Reid, D., and M. E. Castle. 1970. The effects of the date of applying anhydrous ammonia on a solid nitrogen fertilizer on the spring growth from a pure perennial ryegrass sward. J. Agric. Sci. 75:523–532.

Sabey, B. R. 1968. The influence of nitrification suppressants on the rate of ammonium oxidation in Midwestern USA field soils. Soil Sci. Soc. Am. Proc. 32:675–679.

Saffigna, P. G., and D. R. Keeney. 1977. Nitrate and chloride in ground water under irrigated agriculture in central Wisconsin. Ground Water 15:170–177.

Saffigna, P. G., D. R. Keeney, and C. B. Tanner. 1977. Nitrogen, chloride and water balance with irrigated Russet Burbank potatoes in central Wisconsin. Agron. J. 69:251–257.

Saffigna, P. G., C. B. Tanner, and D. R. Keeney. 1976. Nonuniform infiltration under potato canopies caused by interception, stemflow and hilling. Agron. J. 68:336–342.

Sanchez, P. A. (ed.). 1977. A review of soils research in tropical Latin America. Tech. Bull. 19, N. C. Agric. Exp. Stn., Raleigh, N.C. 197 p.

Sanchez, P. A., A. Gavidia, G. E. Ramirez, R. Vergara, and F. Minguillo. 1973. Performance of sulfur-coated urea under intermittently flooded rice culture in Peru. Soil Sci. Soc. Am. Proc. 37:789–792.

Scarsbrook, C. E. 1965. Nitrogen availability. *In* W. V. Bartholomew and F. E. Clark (ed.) Soil nitrogen. Agronomy 10:481–502. Am. Soc. Agron., Madison, Wis.

Shank, R. C. 1975. Toxicology of N-nitroso compounds. Toxicol. Appl. Pharmacol. 31:361–368.

Shertz, D. L., and D. A. Miller. 1972. Nitrate-N accumulation in the soil profile under alfalfa. Agron. J. 64:660–664.

Shuval, H. I., and N. Gruener. 1972. Epidemiological and toxicological aspects of nitrates and nitrites in the environment. Am. J. Publ. Health 62:1045–1052.

Simpson, J. R. 1968. Losses of urea nitrogen from the surface of pasture soils. Int. Congr. Soil Sci., Trans. 9th (Adelaide) II:459–465.

Smika, D. E., D. F. Heermann, H. R. Duke, and A. R. Batchelder. 1977. Nitrate-N percolation through irrigated sandy soils as affected by water management. Agron. J. 69:623–626.

Smika, D. E., and D. G. Watts. 1978. Residual nitrate-N in fine sand as influenced by N fertilizer and water management practices. Soil Sci. Soc. Am. J. 42:923–926.

Snyder, G. H., H. Y. Ozaki, and N. C. Hayslip. 1974. Water repellent soil mulch for reducing fertilizer nutrient leaching. II. Variables governing the effectiveness of a siliconate spray. Soil Sci. Soc. Am. Proc. 38:678–681.

Stanford, G. 1973. Rationale for optimum nitrogen fertilization for corn production. J. Environ. Qual. 2:159–166.

Stanford, G., C. B. England, and A. W. Taylor. 1970. Fertilizer use and water quality. Publication ARS-41-168, Agric. Res. Serv., USDA, Washington, D.C.

Stevenson, C. K., and C. S. Baldwin. 1969. Effect of time and method of nitrogen application and source of nitrogen on the yield and nitrogen content of corn (*Zea mays* L.). Agron. J. 61:381–384.

Stewart, B. A. 1970. Volatilization and nitrification of nitrogen from urine under simulated cattle feedlot conditions. Environ. Sci. Technol. 4:479–582.

Stewart, B. A., F. G. Viets, Jr., and G. L. Hutchinson. 1968. Agriculture's effect on nitrate pollution of groundwater. J. Soil Water Conserv. 23:13–15.

Stewart, B. A., D. A. Woolhiser, W. H. Wischmeier, J. H. Caro, and M. H. Frere. 1975. Control of water pollution from cropland, Volume I: A manual for guideline development. Report no. EPA-600/1-75-026a and ARS-H-5, USEPA, USDA, Washington, D.C.

Stewart, W. D. P. 1977. Present-day nitrogen-fixing plants. Ambio 6:166–173.

Sukhatme, P. V. 1977. Nitrogen in malnutrition. Ambio 6:137–140.

Swanson, E. R. 1957. Problems of applying experimental results to commercial practice. J. Farm Econ. 39:382–389.

Swezey, A. W., and G. O. Turner. 1962. Crop experiments on the effect of 2-chloro-6-(trichloromethyl)pyridine for the control of nitrification of ammonium and urea fertilizers. Agron. J. 54:532–535.

Sze, N. D., and H. Rice. 1976. Nitrogen cycle factors contributing to N_2O production from fertilizers. Geophys. Res. Lett. 3:343–346.

Tamm, C. O., and E. B. Cowling. 1976. Acidic precipitation and forest vegetation. p. 845–856. *In* L. S. Dochinger and T. A. Seliga (ed.) Proc. 1st Int. Symp., Acid Precip. For. Ecosys., 12–15 May 1975, Columbus, Ohio. U.S. For. Serv. Gen. Tech. Report NE-23.

Taylor, C. R. 1973. An analysis of nitrate concentrations in Illinois streams. Ill. Agric. Econ. 13:12–19.

Taylor, C. R., and E. R. Swanson. 1974. Economic impact of imposing per acre restrictions on use of nitrogen fertilizer in Illinois. p. 1–5. Illinois Agric. Econ., July 1974, Univ. of Illinois, Urbana.

Taylor, S. H., and G. L. Ashcroft. 1972. Physical edaphology: The physics of irrigated and nonirrigated soils. W. H. Freeman & Co., San Francisco.

Tyler, D. D., and G. W. Thomas. 1977. Lysimeter measurements of nitrate and chloride losses from soil under conventional and no-tillage corn. J. Environ. Qual. 6:63–66.

U.S. Department of Agriculture. 1980. Report and recommendations on organic farming. Prepared by USDA study team, July 1980. 310-944/96. U.S. Government Printing Office, Washington, D.C.

Verstraeten, L. M. J., and J. Livens. 1977. Various slow-release nitrogen fertilizers and their combinations as a source of nitrogen for winter wheat. Soil Organic Matter Studies I: 359–364. Int. Atomic Energy Authority, Vienna, Austria.

Viets, F. G., Jr. 1965. The plant's need for and use of nitrogen. *In* W. V. Bartholomew and F. E. Clark (ed.) Soil nitrogen. Agronomy 10:503–549. Am. Soc. Agron., Madison, Wis.

Viets, F. G., Jr., and R. H. Hageman. 1971. Factors affecting the accumulation of nitrate in soil, water and plants. Agric. Handbook 413, Agric. Res. Serv., USDA, Washington, D.C.

Walker, T. W. 1956. The nitrogen cycle in grassland soils. J. Sci. Food Agric. 7:66–72.

Walton, G. 1951. Survey of literature relating to infant methemoglobinemia due to nitrate-contaminated water. Am. J. Publ. Health 41:986–996.

Warren, H. L., D. M. Huber, D. W. Nelson, and O. W. Mann. 1975. Stalk rot incidence and yield of corn as affected by inhibiting nitrification of fall-applied ammonium. Agron. J. 67:655–660.

Weiss, R. F., and H. Craig. 1976. Production of atmospheric nitrous oxide by combustion. Geophys. Res. Lett. 3:751–753.

Welch, L. F. 1972. More nutrients are added to soil than are hauled away in crops. Ill. Res. 14(1):3–4.

Welch, L. F., P. F. Johnson, J. W. Pendleton, and L. B. Miller. 1966. Efficiency of fall-versus spring-applied nitrogen for winter wheat. Agron. J. 58:271–274.

Welch, L. F., D. L. Mulvaney, M. G. Oldham, L. V. Boone, and J. W. Pendleton. 1971. Corn yields with fall, spring, and sidedress nitrogen. Agron. J. 63:119–123.

Whitehead, D. C. 1970. The role of nitrogen in grassland productivity. Bull. 48, Commonw. Agric. Bur., Farnham Royal, Bucks, U.K. 202 p.

Wittwer, S. H., M. J. Bukovac, and H. B. Tukey. 1963. Advances in foliar feeding of plant nutrients. p. 429–448. Fertilizer technology and usage. Soil Sci. Soc. Am., Madison, Wis.

Woldendorp, J. W., K. Dilz, and G. J. Kolenbrander. 1965. The fate of fertilizer nitrogen on permanent grassland soils. Proc. Gen. Meeting Europe Grassland Federation, First, Wageningen. p. 53–68.

Woodmansee, R. G. 1978. Additions and losses of nitrogen in grassland ecosystems. Bio-Science 28:448–453.

Wright, M. J., and K. L. Davison. 1964. Nitrate accumulation in crops and nitrate poisoning in animals. Adv. Agron. 16:197–217.

17   Assessment of Soil Nitrogen Availability

GEORGE STANFORD[1]

U.S. Department of Agriculture,
Science & Education Administration
Beltsville, Maryland

I. INTRODUCTION

Before the dramatic upsurge in N fertilizer use began in about 1945, the level of crop production had become increasingly dependent on the capacity of the soil to supply N. Because of the limited supply and high cost of N fertilizers, farmers were dependent almost entirely on biological N fixation and judicious use of manures and crop residues to sustain crop yields.

By 1945, N fertilizer use in the USA had risen to about one-half million metric tons, 80% or more of which was used in states other than those comprising the cornbelt and the major wheat-producing areas (USDA, 1966). By 1976, N fertilizer use had risen to about 10.4 million metric tons, of which about 55% was applied to corn (*Zea mays* L.) and wheat (*Triticum aestivum* L.), representing approximately a 40-fold increase in N use for those crops since 1945 (USDA, 1977). In contrast, the average N use for all other crops grown in the USA increased approximately 10-fold during this period.

In response to the rapidly changing N situation, agronomists and soil scientists throughout the world began developing methods for assessing the N-supplying capacities of soils as an aid in predicting N fertilizer needs. Research conducted on this problem up to about 1970 has been summarized and evaluated by Bremner (1965), Harmsen and Van Schreven (1955), Harmsen and Kolenbrander (1965), Jenkinson (1968), Stanford and Legg (1968), Keeney and Bremner (1966a) and Dahnke and Vasey (1973). Most of the earlier studies emphasized developing methods of soil N evaluation based on short-term incubation under controlled laboratory conditions and calibration with yield responses to N in the field and greenhouse (for examples, see Olson et al., 1960; Stanford & Hanway, 1955; Fitts et al., 1953; Clement & Williams, 1962; Cook et al., 1957; Saunder et al., 1957; Synghal et al., 1959; Cooke & Cunningham, 1958; Gallagher & Bartholomew, 1964).

In a few laboratories, services for recommending N fertilizer use based on short-term N mineralizations were instituted (Hanway & Dumenil, 1955; Olson et al., 1960). These ventures were relatively short lived, however, for two important reasons: (i) it became increasingly apparent that the contribution of N mineralized during the cropping season often was obscured by

---

[1] Deceased 28 January 1981.

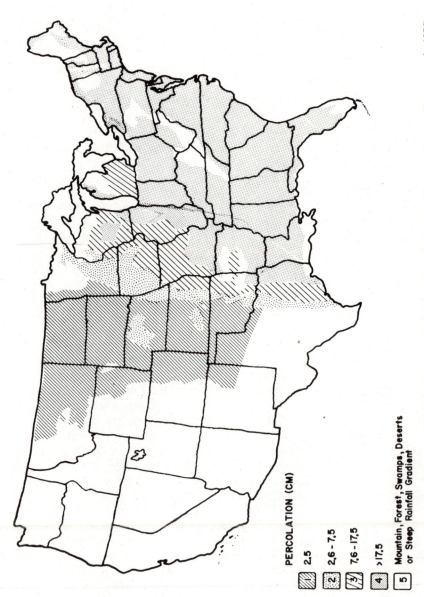

PERCOLATION (CM)

1 2.5

2 2.6 - 7.5

3 7.6 - 17.5

4 >17.5

5 Mountain, Forest, Swamps, Deserts
   or Steep Rainfall Gradient

Fig. 1—Average annual potential percolation below the root zone in well-drained soils planted to corn (Stewart et al., 1975).

the presence of varying amounts of residual mineral N accumulated in the soil from previous fertilization, and (ii) the low cost of fertilizer N discouraged development and use of N soil tests aimed at achieving more effective use of N.

Renewed interest in developing improved methods for assessing soil N availability was stimulated, in part, by concern over environmental consequences of applying more N than required by the crop, and by economic considerations. In recent years, a limited number of field and laboratory studies have centered on developing more definitive methods of assessing soil N availability, involving mineralizable as well as residual mineral N, and more rigorous systems of predicting optimal N requirements of crops (e.g., Reuss & Geist, 1970; Stanford & Smith, 1972; Carter et al., 1974; Oyanedel & Rodriguez, 1977).

II. ESTIMATING RESIDUAL MINERAL NITROGEN IN SOILS

Hundreds of field experiments have been conducted throughout the world during the past 30 years with the main objective of determining the nature and magnitude of crop yield responses to N fertilization under a broad range of soil, climatic, and management conditions. In large part, results of such experiments have provided the basis for making N fertilizer recommendations to growers.

With rising use of N fertilizers, the importance of residual mineral N as a source of plant-available N became increasingly evident, particularly in large areas of the western USA and Canada where the likelihood of NO_3^- removal from the root zone by percolating water is minimal (see Area 1, Fig. 1).

In Fig. 1, *annual potential percolation of water* is defined as the average amount of water that would move below the root zone (120 cm) under corn in four major land resource areas (Stewart et al., 1975). Area 5 was omitted from the simulation study for technical reasons as discussed by Stewart et al. (1975). Despite the broad generalizations involved, Fig. 1 provides a useful basis for evaluating the potential significance of residual mineral N in relation to N fertilizer use in different regions of the United States. Detailed consideration of this problem would require, however, that rainfall distribution, temperature regimes, estimates of rooting depths of crops, soil-water-crop relations, and other factors be taken into account for specific agricultural situations at the local or individual farm level.

In the previous agronomy monograph entitled *Soil Nitrogen,* Harmsen and Kolenbrander (1965) discussed the influence of varying climatic patterns and water regimes, in major agricultural regions of the world, on soil inorganic N fluctuations. Other early reviews (e.g., Bremner, 1965; Scarsbrook, 1965) minimized mineral N accumulation in soils and its significance in interpreting biological and chemical indexes of soil organic N availability. In the intervening years, the importance of evaluating residual mineral N in soils has become widely recognized, particularly in Area 1 (Fig. 1) and portions of Area 5 where average annual percolation is < 2.5 cm (Dahnke & Vasey, 1973; Smith, 1977). In Areas 2, 3, and 4 (Fig. 1), much less has been learned regarding the significance of residual mineral N, although published data (van der Paauw, 1962, 1963; Peterson & Attoe, 1965; White et al., 1958; White & Pesek, 1959; Jolley & Pierre, 1977) indicate that carryover of mineral N to succeeding crops may vary substantially from year to year even in subhumid and humid climates. For example, van der Paauw (1963) observed significant residual effects of N applied to the previous crops when winter rainfall from November to February was 20 cm or less. Pronounced residual effects on yield of rye (*Secale cereale* L.) occurred following 10–15 cm of winter rainfall (van der Paauw, 1962, 1963). Expressing the residual effects of previous applications as percentages of yields obtained with current N applications gave residual values in the range of 10–25% when winter rainfall did not exceed 20 cm. Similar results have been reported in Britain (Cooke, 1967).

Olson et al. (1960), however, soon questioned the general applicability of N fertilizer recommendations based on measurement of soil organic N mineralization during short-term incubation. They found, for example, that correlation of mineralization values with wheat grain yield response to applied N following fallow was lower (60 fields; $r = 0.46$) than for wheat grown in a continuous cropping system (60 fields; $r = 0.64$). The difference was attributed to substantially greater and more variable accumulations of NO_3^- in the root zone during fallow than under continuous cropping.

Comprehensive field and laboratory studies with corn, sorghum [*Sorghum bicolor* (L.) Moench], and small grains in Nebraska (Herron et al., 1968, 1971) and in western USA and Canada (Hunter & Yungen, 1955; Leggett, 1959; Soper & Huang, 1963; Spencer et al., 1966; Young et al., 1967; Stewart et al., 1967; Geist et al., 1970; Soper et al., 1971; Onken & Sunderman, 1972; Ludwick et al., 1976) consistently demonstrated the influence of above-optimal N fertilization, for both irrigated and nonirrigated crops, on residual NO_3^- accumulation. An example for corn is given in Fig. 2 (Herron et al., 1968). This 3-year experiment was conducted on a deep loess-derived Sharpsburg silty clay (fine, montmorillonitic, mesic Typic Argiudoll) that had been in alfalfa (*Medicago sativa* L.) for several years. Each year the optimal rate of N applied as NH_4NO_3 was approximately 84 kg/ha. Even with no N applied (not shown in Fig. 2) appreciable amounts of NO_3^--N were found in the 180-cm profile at the end of the 1966 season (115–140 kg/ha). The substantial accumulations of NO_3^- when N rates exceeded crop needs are clearly evident in Fig. 2. Other striking examples of NO_3^- accumulations resulting from above-optimal N applications have been reported by Jolley and Pierre (1977), Adriano et al. (1972), and Ludwick et al. (1976). Herron et al. (1971) concluded that such effects were particularly evident in years when adverse environmental factors or management practices limit yields. The Nebraska studies also showed a substantially greater accumulation of residual NO_3^- with sidedressing than with preplant application of N fertilizer (Herron et al., 1968, 1971).

As expected, the depth of sampling required to adequately assess the quantity of residual NO_3^- available or accessible to plant roots is dependent

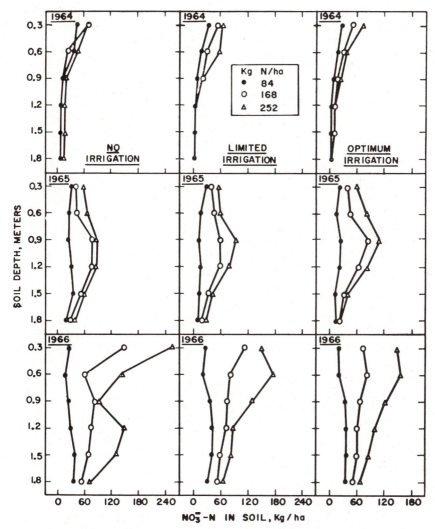

Fig. 2—Nitrate-N at 30-cm intervals in Sharpsburg silty clay loam profiles after corn harvest, as affected by irrigation and amounts of applied N (Herron et al., 1968).

on effective rooting depth (Soper & Huang, 1963; Herron et al., 1968; Soper et al., 1971; Reuss & Rao, 1971; Onken & Sunderman, 1972; Kerbs et al., 1973; Ludwick et al., 1977). In a Sharpsburg soil (Fig. 2), sampling to 180 cm was required to evaluate the plant-available reserve of residual NO_3^-. From this type of soil, corn roots took up ^{15}N-labeled KNO_3 with equal facility from deep (140 cm) and surface-soil layers (Herron et al., 1968). For practical reasons, it would be desirable to sample to the minimal depth required to establish a suitable relation between N uptake by the crop and amount of residual mineral N in the soil (Smith, 1977). Nebraska studies showed that amounts of NO_3^- in the surface soil (30 cm) often were well correlated with amounts in the 180-cm profile for treatments within a given irrigation regime and for a given year. However, upon combining years and irrigation practices, percentages of the total profile NO_3^- in the surface soil varied from 10 to 48 (Herron et al., 1968).

In contrast, Onken and Sunderman (1972) in a 3-year N rate experiment on irrigated grain sorghum consistently found 40% or more of the residual fertilizer NO_3^-, from the previous crop, in the top 30 cm of clay loam soils. From multiple regressions of grain yield on residual NO_3^- and other variables (combining data for years and locations), it was evident that measuring residual NO_3^- to a depth of 61 cm gave no better relationship ($r = 0.95$) than was obtained from the surface 15- or 30-cm layers ($r = 0.94$ and 0.96, respectively). These remarkably consistent results, which differ from those of Herron et al. (1968) and Soper et al. (1971), were attributed to the very slow water infiltration rate of the clay loam soils used (0.2 cm/hour), low rainfall (46 cm/year, avg), and high evaporation potential.

Based on results from 22 N-rate field experiments with barley on fallowed and nonfallowed soils of the Chernozemic order (Canadian soil classification system), Soper et al. (1971) showed that the relationship between N uptake by barley (*Hordeum vulgare* L.) and quantity of NO_3^- in the soil improved with successive increases in depth of soil sampling down to 61 cm. Coefficients of determination (r^2) for N uptake vs. soil NO_3^- were 0.32, 0.64, 0.84, and 0.82, respectively, with sampling depths of 15, 30, 61, and 91 cm. Based on results of this kind, the amount of NO_3^- in the 60-cm layer is used in Manitoba and other western Canadian provinces as a basis for predicting N fertilizer needs of cereal grains. Measurement of NO_3^- as a guide to N fertilization for grain crops is being done routinely or is being developed in the Prairie provinces of Canada (Nyborg et al., 1976; Campbell, 1978), the Great Plains, and other dryland or irrigated areas of the western USA, e.g., North Dakota (Young et al., 1967); South Dakota (Carson, 1975); Montana (Sims & Jackson, 1971); Colorado (Geist et al., 1970); Nebraska (Herron et al., 1971); Texas (Onken & Sunderman, 1972; Hipp & Gerard, 1971); eastern Washington (Leggett, 1959), and in other western states in Areas 1 and 5 of Fig. 1 (Smith, 1977).

Control of N supplied to sugar beets (*Beta vulgaris* L.) is of critical importance because percent sucrose in the beets and recovery of sucrose during processing is reduced sharply by excessive N uptake. A large number of comprehensive studies have emphasized the importance of residual mineral N derived from excessive N fertilization as a varying and often major source of N for irrigated sugar beets in Washington (James, 1971; James et al., 1971; Roberts et al., 1972), in Colorado (Reuss & Rao, 1971; Ludwick et al., 1976), in Idaho (Carter et al., 1974, 1975, 1976), in Montana (Halvorson & Hartman, 1975); and in Utah (James et al., 1977).

Because the criteria for assessing optimal N needs of sugar beets are more clearly definable than for crops that do not respond adversely to excessive N, considerable progress is being made in controlling use of N on this crop (Westfall et al., 1977, 1978). Similar opportunities exist for sugarcane (*Saccharum officinarum* L.) which responds adversely to excessive use of N (Stanford, 1963; Stanford & Ayres, 1964).

Assessment of available soil N supply from residual NO_3^- measurements alone is inadequate if significant differences exist among soils in the capacity to mineralize soil organic N. Although this has been recognized for sugar beets in Areas 1 and 5 of Fig. 1 (Roberts et al., 1972; Carter et al., 1974; Stanford et al., 1977), the possible utility of estimating both mineralizable and residual mineral N has been largely ignored in most of the studies cited earlier involving corn, wheat, and other crops. This aspect of the problem of assessing soil N availability will be considered later in this chapter.

In Areas 2, 3, and 4 (Fig. 1), relatively few field studies have been designed to assess the contributions of residual mineral N to succeeding crops. White et al. (1958), using oats as the test crop, estimated first-year residual effects of N fertilizer applied to corn from regressions of N yield of oats (*Avena sativa* L.) on N applied. Quantities of residual N varied from almost nil to 49% of the previous-year application to corn. Soil sampling for NO_3^- determinations was restricted to the 53-cm depth, and in most cases, the NO_3^- was concentrated below the plow layer (about 15–53 cm). With oat experiments showing substantial residual effects of N previously applied to corn (White et al., 1958; White & Pesek, 1959), regressions of total N content of mature oats on quantity of residual NO_3^- were highly significant and accounted for 75% or more of the variations in N uptake. According to the authors, an even closer relation might have resulted, in certain instances, with soil sampling to greater depth.

The importance òf residual NO_3^- in Areas 2 and 3 may be greater than commonly realized, particularly in Area 2 where estimated average percolation ranges between 2.5 and 7.5 cm annually. Results from long-term N-rate experiments with corn reported from Minnesota (MacGregor et al., 1974) suggest that a moderate N application rate (e.g., 100–150 kg/ha) might have resulted in significant NO_3^- accumulations within the root zone, during years of below-normal rainfall, in the course of 10–15 years of continuous cropping. Since the study was designed to compare long-term accumulations of NO_3^- below the rooting depth, however, annual variations in residual NO_3^- were not measured. Similarly, at the end of a 10-year experiment with N applied annually to continuous corn, Nelson and MacGregor (1973) found substantial and varying amounts of NO_3^- in the upper 150-cm layer of soil with above-optimal N application. Annual rainfall during the 10-year period ranged from 51 to 82 cm (avg 67 cm). Presumably, variations among years in amounts of residual NO_3^- in the root zone also were appreciable, although these were not reported. Even more definitive evidence of the need for assessing residual mineral N in Area 2 was reported by Jolley and Pierre (1977).

Evidence of residual mineral N carryover in the root zone between succeeding crops in Area 3 was observed in comprehensive field studies conducted in Wisconsin by Olsen et al. (1970). In an experiment on a well-drained, loess-derived Plano silt loam soil (fine-silty, mixed, mesic Typic Argiudoll), corn was grown for 3 years with annual N applications of 0, 112, and 336 kg/ha as NH_4NO_3. The highest N rate was above-optimal and, by the end of the third corn crop, resulted in an accumulation of about 700 kg N/ha in the zone accessible to roots (150 cm). With the near-optimal N application (112 kg/ha per year), net accumulation of NO_3^--N attributable to fertilizer was about 100 kg/ha. The magnitude of residual NO_3^- carry-over varied among years owing to differences in annual precipitation. At the end of a 4-year experiment on another deep loess-derived soil (Rozetta silt loam; fine-silty, mixed, mesic Typic Hapludalf), Olsen et al. (1970) also observed substantial NO_3^- accumulations following corn grown in corn-meadow rotations when N applications exceeded crop requirement for attained yields. The foregoing results are in marked contrast to those reported by Boswell and Anderson (1970) in Georgia where, with early winter N application, most of the residual or unused NO_3^- had moved below the effective root zone (60- to 90-cm layer) during the following year even with continuous cropping to rye (*Secale cereale* L.) and millet (*Pennisetum typhoidum* L.). Cumulative amounts of rainfall were 124 and 154 cm on the two sites, considerably higher than amounts recorded in the Iowa, Minnesota, and Wisconsin experiments.

Factors that need to be considered in assessing residual mineral N as a guide to N fertilization include (i) scope and intensity of soil sampling, (ii) depth of sampling, and (iii) time of sampling. Items (i) and (ii) are related to the lateral and vertical variability in NO_3^- distribution in the field (Biggar, 1978) and on effective rooting depth, as discussed earlier, and are not amenable to broad generalizations. The relation of spatial variability to minimal sampling requirements differs among soils, climates, and management systems each of which must be evaluated locally (Reuss et al., 1977; Ludwick et al., 1977; James, 1978). Generally soil should be sampled shortly before planting or very early in the growing season to reflect the available NO_3^- supply. Autumn sampling preceding the next year's crop is undesirable because of possible over-winter NO_3^- loss by leaching (James, 1978) or accumulation by mineralization (Ludwick et al., 1977).

III. INCUBATION METHODS FOR MEASURING MINERALIZATION OF SOIL ORGANIC NITROGEN

The importance of measuring residual mineral N, under appropriate conditions, as a guide to N recommendations is well established. Workers actively involved in this aspect of soil N evaluation are cognizant of the need for assessing mineralizable soil organic N as well as residual NO_3^- in areas where soil organic N differs appreciably among soils (Dahnke & Vasey, 1973; Smith, 1966; Roberts et al., 1972; Carter et al., 1974; Geist et al., 1970). Evidently, however, none of the numerous biological or chemical methods that have been proposed for estimating the organic N supplying capacities of soils is deemed sufficiently reliable to warrant its routine use in soil testing laboratories. Bremner (1965), Dahnke and Vasey (1973), and Campbell (1978) have reviewed many of the published studies involving assessment of soil organic N availability by means of biological and chemical methods.

A. Short-term Incubation Methods

1. AEROBIC INCUBATION

Based on a critical review of more than 30 different incubation procedures, Bremner (1965) concluded that the reliability and reproducibility of methods for measuring soil N mineralization determine, in large part, their suitability for assessing the potential ability of soils to provide N for crop growth. A satisfactory method should provide for rigorous standardization with respect to methods of sampling, drying, grinding and sieving, storing, and incubating the soils (Bremner, 1965). Moreover, total mineral N, (exchangeable NH_4^+ + NO_2^- + NO_3^-)-N, should be measured, since some soils will not nitrify all of the NH_4^+ produced during incubation. To ensure NH_4^+ recovery, Keeney and Bremner (1967) proposed extraction of soils with $2N$ KCl following incubation. Control of water content during incubation was regarded as a major problem for soils having a wide range in water-holding capacities. Keeney and Bremner (1967) attained optimal water content by adding a constant level of water to a soil-sand mixture (10 g of soil mixed with 30 g of washed sand was added to a bottle containing 6 ml of water). This method appears to meet essential requirements for reliably measuring short-term (14 days) N mineralization (Bremner, 1965). Various investigators have used the method with apparent success (Robinson, 1968a, b; Ryan et al., 1971; Herlihy, 1972; Baerug et al., 1973) judging from the relatively high correlations often obtained between mineralized N and N uptake by plants grown in the greenhouse, yield of crops, or other indexes of soil N availability.

The importance of measuring both NH_4^+ and NO_3^- following incubation is clear from the work by Vlassak (1970), Nommik (1976), Verstraeten et al. (1970a, b), and Geist (1977) who found that NH_4^+ was the dominant end product of mineralization in forest soils. Cooke (1967), in a review of the methods used in Great Britain, emphasized the importance of measuring NH_4^+ and NO_3^- (see, for example, Gasser & Kalembasa, 1976). With Hawaiian soils (Stanford et al., 1965) variations in the extent to which NH_4^+ was nitrified during short-term incubation required determination of total mineral N as an index of soil N availability. Where experience shows that NH_4^+ accumulation is negligible, however, measurement of NO_3^- production may be sufficient (Smith, 1966; Rixon, 1969).

2. ANAEROBIC INCUBATION

Incubation of soils under anaerobic (waterlogged) conditions and measurement of NH_4^+ released has attracted considerable attention because of its simplicity as compared to most aerobic incubation procedures. Waring and Bremner (1964) found a high correlation between amounts of NH_4^+ produced anaerobically and amounts of NH_4^+ + NO_3^- + NO_2^- mineralized during aerobic incubation. Using 25 Iowa soils, Keeney and Bremner (1966a) found that amounts of NH_4^+ released anaerobically during a 7-day incubation at 40°C were better correlated with N uptake by ryegrass (*Lolium perenne* L.) (r = 0.71–0.82) than were amounts released anaerobically during 14-day incubation at 30°C (r = 0.38–0.68). The varying correlation coefficients (r) reflect different pretreatments of soils, e.g., field moist, air dried, and air dried followed by varying storage times. Aerobic incubation correlated well with N uptake by ryegrass when soils were air dried and stored (r = 0.75), but the anaerobic method was superior for field-moist or air dried (not stored) conditions. Hanway and Ozus (1966) also found that NH_4^+ released during anaerobic incubation was better correlated with N uptake by ryegrass (r = 0.85) than the NO_3^- produced during aerobic incubation (r = 0.77) using 24 field-moist soils. Similar results were reported by Kadirgamathoiyah and MacKenzie (1970). It should be noted that the soils used by Keeney and Bremner (1966a) and by Hanway and Ozus (1966) contained varying amounts of mineral N before cropping, which presented certain problems in attempting to correlate values for organic N mineralization with N uptake by plants in the greenhouse. These and similar experiments will be considered further in Section V-A, where the simultaneous evaluation of initial mineral (residual) and mineralizable N in relation to N uptake by the crop is discussed.

Other investigators have compared aerobic and anaerobic methods for N mineralization (Cornforth, 1968; Robinson, 1967; Robinson, 1968a; Smith, 1966; Ryan et al., 1971; Gasser & Kalembasa, 1976; Baerug et al., 1973; Herlihy, 1972; Smith & Stanford, 1971; Geist, 1977). Smith (1966) found no correlation between NH_4^+ released by anaerobic incubation and N uptake by orchardgrass (*Dactylis glomerata* L.), but found a relatively good relation with aerobic production of NO_3^-. Similar results were reported by Robinson (1967) who suggested modifying the method of Waring and Bremner (1964) to provide for steam distillation of the filtered extract, rather than the soil plus extract, in order to avoid release of NH_4^+ by alkaline hydrolysis of soil organic matter. Sahrawat and Ponnamperuma (1978) also observed appreciably greater NH_4^+ release upon distilling the soil plus extract as compared to the extract alone, particularly with soils containing 3% organic matter or more.

Another recent modification of the anaerobic procedure involves incubation of soils in 2N KCl instead of water (Olsen et al., 1974). Interestingly, with 20 of the 35 soils studied, the average amount of NH_4^+-N formed during incubation in water was higher (36.7 ppm) than with KCl incubation (25.8 ppm). For the remaining 15 soils, average amounts of NH_4^+-N produced anaerobically in water and 2N KCl solution, respectively, were 21.6 and 23.8 ppm. Before incubation, the average NO_3^--N content of all soils was 43 ppm. Average amounts of NO_3^--N remaining in the 35 soils following incubation in water and KCl, respectively, were 2 and 31 ppm. One possible explanation for the higher accumulation of NH_4^+ in the water than in the KCl system for certain soils might be dissimilatory reduction of NO_3^- to NH_4^+ (Stanford et al., 1975a). This type of NO_3^- dissimilation would be most significant in soils containing relatively high contents of NO_3^- and readily decomposable energy sources (Stanford et al., 1975b; Buresh & Patrick, 1978; Caskey & Tiedje, 1979).

In agreement with Keeney and Bremner (1966a) and Hanway and Ozus (1966), Cornforth (1968) found that both aerobic incubation and anaerobic incubation gave values that correlated highly with N uptake by corn in the greenhouse (r = 0.93 and 0.89, respectively). Similarly, Gasser and Kalembasa (1976) found a very high correlation (r = 0.98) between N mineralized anaerobically (7 days, 40°C) and aerobically, and these indexes correlated equally well with N uptake by ryegrass (r = 0.93).

Comparing arable and ley soils, Baerug et al. (1973) found consistently higher correlation of N uptake by ryegrass in greenhouse culture with aerobic incubation than with anaerobic incubation (14 days, 30°C). Although correlation of NH_4^+ released anaerobically with N uptake by ryegrass was much poorer with ley soils (r = <0.2) than with arable soils (r = >0.6), similar relations (r = >0.6) were observed for ley and arable soils based on aerobic incubation.

Smith and Stanford (1971) modified the method of Waring and Bremner (1964) by first extracting initial mineral N with $0.01M$ $CaCl_2$ followed by incubating the soil residue in "minus-N" nutrient solution at 35°C for 14 days. The NH_4^+ produced during incubation was recovered by distillation of the extract obtained by centrifuging and washing the soil residue with $0.01M$ $CaCl_2$ (Stanford & DeMar, 1970). These modifications precluded reduction of NO_3^- to NH_4^+ (Olsen et al., 1974) and production of NH_4^+ by hydrolysis of soil organic N compounds during distillation (Robinson, 1967; Sahrawat & Ponnamperuma, 1978). Using 39 soils representing several important agricultural regions of the United States, Smith and Stanford (1971) compared amounts of NH_4^+ released during anaerobic incubation, as described above, with amounts of $(NH_4^+ + NO_3^- + NO_2^-)$-N produced during aerobic incubation. The amounts of NH_4^+ produced anaerobically in five aridic calcareous soils were much less than the amounts of $(NH_4^+ + NO_3^- + NO_2^-)$-N released during aerobic incubation. Reasons for the anomalous behavior of various western calcareous soils are unknown. For the remaining 34 soils, the amounts of mineral N released during 4 weeks of anaerobic and aerobic incubation were similar and relatively well correlated (r = 0.93). The average amounts of N released (34 soils) with anaerobic and aerobic incubations, respectively, were 40 and 45 ppm. Based on 2-week incubations, the degree of correlation was somewhat less (r = 0.86) and average amounts of mineral N released, respectively, were 22 and 29 ppm.

Soil N availability indexes based on anaerobic incubation procedures may be uniquely suited to estimating soil N supplying abilities of flooded rice soils (Subbiah & Bajaj, 1962; Lin et al., 1973; Dolmat et al., 1980). Sims et al. (1967) incubated 50-g samples of water-submerged soils (6 or 12 days, 35°C) obtained from rice-producing areas in Arkansas. After incubation, the system was extracted with $1N$ NaCl–$0.1N$ HCl solution, filtered, and NH_4^+ was recovered from the filtrate by distillation. With 19 clay soils, amounts of NH_4^+ produced during 6-day anaerobic incubation correlated well (r = 0.95) with rice grain yields in the greenhouse and the relation was improved by including initial NH_4^+ content of the soils. There was no advantage to extending the incubation period to 12 days despite the accompanying larger releases of NH_4^+. Corresponding relationships for 42 silt loam soils were significant, but much less pronounced, owing to the very narrow range in their N-supplying capacities. Initial NO_3^--N contents of the soils were low, ranging narrowly between 0.4 and 16 ppm. Hence, NO_3^- reduction during incubation probably contributed insignificant amounts of NH_4^+.

In a subsequent study with 90 silt loam soils, Sims and Blackmon (1967) compared the method described above with the method of Waring and Bremner (1964) modified only with respect to temperature (40°C) and incubation time (6 days). Results of the two methods were well correlated (r = 0.92). Thus, the relation of NH_4^+ produced during incubation and N uptake by flooded rice (*Oryza sativa* L.) in the greenhouse was similar for both incubation procedures (r = approx. 0.70). Nitrogen uptake by rice grown under flooded conditions on 39 soils of the Philippines correlated well (r = 0.82 and 0.84) with NH_4^+ produced during anaerobic incubation (1 week, 40°C; or 2 weeks, 30°C) (Dr. K. L. Sahrawat, personal communication). According to Dolmat et al. (1980), rough rice yields from the control plots of 31 N-rate field experiments in Louisiana gave higher correlations with amounts of soil NH_4^+ released anaerobically (r = approx. 0.6) than with amounts of mineral N formed during aerobic incubation (r = approx. 0.4) indicating a superiority of anaerobic mineralization as an index of N availability for rice. Walker (1979)[1] recently developed a model for estimating the N requirement of flooded rice based on measurement of soil N supplying ability by anaerobic incubation together with a knowledge of the N requirement of the crop.

Recent studies have demonstrated the value of anaerobic incubation (Waring & Bremner, 1964) for assaying relative soil N availability as a guide to forest fertilization (Shumway & Atkinson, 1978; Powers, 1980). Considerable progress is being made in developing suitable procedures for relating soil N released on incubation to yield potential or growth response of trees to applied N under a wide range of soil and climatic conditions.

B. Potentially Mineralizable Soil Nitrogen

Amounts of soil N mineralized in short-term incubations are dependent on methods used in pretreating the soil before incubation (Bremner, 1965; Harmsen & Van Schreven, 1955). More recent demonstrations of the differential effects of drying, storing, freezing, or other means of handling field-moist soils on short-term N mineralization values have been presented (Keeney & Bremner, 1966a; Hanway & Ozus, 1966; Robinson, 1968a, b; Baerug et al., 1973; Selmer-Olsen et al., 1971). To obtain comparable values in a given laboratory or among laboratories, it is necessary to rigorously standardize methods of sample pretreatment as well as incubation conditions (Bremner, 1965).

---

[1] C. F. R. Walker. 1979. Simple model for estimating the nitrogen requirement of rice. M.S. Thesis. Pontificia Universidad Catholica de Chile.

Even with rigorous controls, results of short-term incubation do not necessarily reflect the potential, long-term N supplying capacities of soils (Stanford & Smith, 1972). Nitrogen mineralized during short time periods under aerobic conditions may be heavily weighted by the N derived from decomposition of recently incorporated residues and microbial tissues relative to that from the mineralizable fractions of soil organic N (Stanford et al., 1974). The presence of crop residues with high C/N ratios also affects the net mineralization that occurs in short-term incubations (Chichester et al., 1975). Hence, the relative significance of N derived from the various mineralizable sources may well differ with short- and long-term measurements. These considerations may help to explain why attempts at relating short-term N mineralization data to N uptake by a particular crop or a succession of crops grown on different soils under uniform conditions have met with varying degrees of success.

Determining the long-term mineralization capacities of soils is laborious, expensive, and time-consuming. The reliability of results derived from extended incubations in static, closed systems is questionable. Stanford and Smith (1972) proposed a system of incubation that provides for periodic removal of mineralized soil N by leaching with $0.01M$ $CaCl_2$.

The soil (mixed with sand or exfoliated vermiculite) is placed in a leaching tube with glass wool pads placed above and below the sample. The initial mineral N is removed by leaching with $0.01M$ $CaCl_2$, after which suction (0.6–0.7 bar) is applied to provide optimal water content. The sample is stoppered and incubated at 35°C. Following each period of incubation, the soil is leached and treated as described above (Stanford & Smith, 1972;

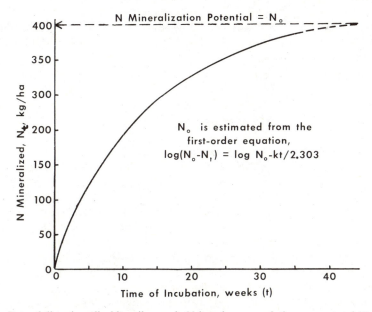

Fig. 3—Potentially mineralizable soil organic N based on cumulative amounts of N mineralized during consecutive incubations.

Stanford et al., 1974). The series of incubations is continued for as long as is deemed necessary to describe adequately the relation between cumulative N mineralization, N_t, and time of incubation, as depicted in Fig. 3. The N mineralization potential, N_o, is estimated from the cumulative amounts of N mineralized under optimal conditions of soil water and temperature, based on the assumption that N mineralization obeys first-order kinetics, i.e., $\log (N_o - N_t) = \log N_o - kt/2.303$. The accepted value of N_o is that which results in best fit for the linear relation between $\log (N_o - N_t)$ vs. t (Stanford & Smith, 1972).

The potentially mineralizable N, N_o, was determined for 39 soils obtained from 15 states ranging from California, Oregon, and Washington to North and South Carolina; and from North Dakota, Minnesota, and Montana to Texas, Alabama, Mississippi, and Georgia. Five soil orders were represented (Entisols, Alfisols, Aridisols, Ultisols, and Mollisols), and management histories varied widely among sites within each order. Estimated values for N_o ranged from 18 to 305 ppm, and amounts of N, (NH_4^+ + NO_3^- + NO_2^-)-N, mineralized during the first 2-week incubation ranged from 2.1 to 74.2 ppm. The regression of N_o (ppm) on initial 2-week mineralization values, $N_{2\,weeks}$ (ppm), was as follows: $N_o = 40.8 + 3.98 N_{2\,weeks}$ ($r^2 = 0.74$). On average, N mineralized in 2 weeks was about 25% of N_o. The reliability of $N_{2\,weeks}$ for predicting N_o is questionable (Stanford et al., 1974). In addition, Smith and Young (1975), Smith et al. (1977), and Oyanedel and Rodriguez (1977) have reported values of N_o for a wide range of soils.

As developed in Section V, N_o is a definable soil characteristic which may be of value in estimating N supplying capacities of soils under specified environmental conditions. Moreover, N_o provides a common basis for evaluating various chemical and biological availability indexes under a broad range of soil conditions and for making quantitative estimates of N mineralization in the field.

IV. CHEMICAL INDEXES OF SOIL ORGANIC NITROGEN AVAILABILITY

Efforts to develop suitable chemical indexes continue to be stimulated by the need for more reliable and rapid methods of assessing soil N availability. Although progress has been made since the reviews by Bremner (1965) and Dahnke and Vasey (1973), none of the proposed chemical indexes have been put to general use. The possible utility of any chemical index for a broad range of soils depends on the degree to which it correlates with reliable biological measurements of soil N availability, e.g., N uptake, crop yield, or mineralizable N. Thus, as emphasized by Bremner (1965) and by Hanway and Ozus (1966), special attention must be given to developing sound biological criteria for calibrating and interpreting empirical indexes of soil N availability. Evaluation of various chemical indexes, based on the widely differing biological assays used by different investigators, is most difficult and, often, impossible.

Chemical methods for assessing soil organic N range from those in-volving relatively mild extraction by boiling water (Keeney & Bremner, 1966a), boiling $0.01M$ CaCl$_2$ (Stanford, 1968b), or autoclaving in $0.01M$ CaCl$_2$ (Stanford & DeMar, 1969, 1970) to determination of total soil N. Be-tween these extremes are found numerous extraction procedures involving use of mineral acids, bases, oxidants, and chelating reagents (e.g., sodium pyrophosphate) at different concentrations and temperatures.

A. Intensive Extraction Methods

Increasing severity of extraction with removal of substantial propor-tions of the soil often results in high correlations between total extractable N or distillable NH$_4^+$ and total soil N. For example, Geist and Hazard (1975) found a very close relation ($r = 0.99$) between total organic N (45 soils) and the amount of NH$_4^+$ released by boiling soils in $4.5N$ NaOH. The distilled NH$_4^+$-N comprised about 20% of the total N.

The acid-hydrolyzable N fractions of 27 soils, obtained by 12-hour re-fluxing with $6N$ HCl, also correlated highly with, and comprised about 75% of, the total soil N ($r = 0.99$) (Keeney & Bremner, 1966b). Such drastic treatment completely removes the fractions of soil organic N that are sus-ceptible to mineralization (Fig. 4) (Stanford, 1968a). For a given extraction time, boiling acid extractions of increasing intensity, with respect to acid concentration, gradually reduce the amounts of N mineralized from the ex-tracted residues. However, the amounts of acid-hydrolyzable N removed far exceeded the accompanying reductions in N mineralized during 11 weeks under anaerobic conditions, even with $1N$ acid (Fig. 4).

Stanford (unpublished data) found that the amount of distillable NH$_4^+$ released from 60 widely different soils by the Walkley-Black organic matter method (K$_2$Cr$_2$O$_7$–H$_2$SO$_4$ oxidation) constituted about 60% of the total or-ganic N, and the correlation with total N was high ($r = 0.98$). Even with ex-traction of lesser proportions of the soil organic N, Keeney (1965) found

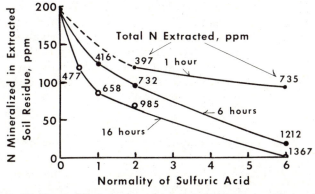

Fig. 4—Effect of boiling an Elliott clay loam for 1, 6, and 16 hours in sulfuric acid on amounts of N extracted and N mineralized during 11 weeks of anaerobic incubation of the extracted soil residues.

that total N was closely related to amounts of NH_4^+ produced during alkaline permanganate extraction ($r = 0.97$) by the method of Subbiah and Asija (1956) and NH_4^+ released by the NaOH-microdiffusion method ($r = 0.98$) of Cornfield (1960). It is beyond the scope of this chapter to review the numerous other studies in which close relations were found between total soil N and amounts extracted by various reagents, nor would this be fruitful. After all, the goal is to devise a chemical index that correlates highly with some reliable measure of biological N availability. As has been demonstrated repeatedly, total soil N does not constitute a reliable index of soil N availability for a broad range of soils.

The possibility that a particular fraction or fractions of the total N extracted from soils by $6N$ boiling HCl might constitute a useful index of N availability has been examined by Keeney and Bremner (1964, 1966b), Kadirgamathoiyah and MacKenzie (1970), Cornforth (1968), Freney and Simpson (1969), and Osborne (1977). Campbell (1978) has comprehensively discussed the biological significance of such fractions in respect to soil N availability. The various organic N fractions, except nonexchangeable NH_4^+, were partially depleted by cultivation, but only small changes occurred in the relative proportions of nonhydrolyzable and various forms of hydrolyzable N (total, NH_4^+, hexosamine, amino acid, and unidentified) (Keeney & Bremner, 1964). Prolonged incubation of 27 soils also led to a marked decrease in all fractions of the $6N$ HCl-hydrolyzable and the nonhydrolyzable portion, but not the nonexchangeable NH_4^+ fraction. Losses by incubation or cropping had little effect on the percentage distribution of the different forms of N (Keeney & Bremner, 1966b; Osborne, 1977) and none of the more or less discrete organic fractions provided a satisfactory index of soil N availability.

On the other hand, Cornforth (1968) observed that the proportions of acid hydrolyzable and nonhydrolyzable N in 10 Trinidad soils often changed markedly after four successive croppings in the greenhouse. Nitrogen uptake by corn (4 crops, roots + tops) correlated with hydrolyzable and nonhydrolyzable N fractions of uncropped soils as follows: total hydrolyzable ($r = 0.85$), NH_4^+ ($r = 0.84$), amino acid ($r = 0.85$), hydroxy amino acid ($r = 0.90$), and nonhydrolyzable ($r = 0.73$). Although relations with N uptake were similar for total hydrolyzable N and the fractions comprising it, the correlations were no better than those occurring between N uptake and total soil organic N ($r = 0.88$). Such findings question the feasibility of devising useful chemical indexes of soil N availability based on intensive extraction procedures that remove far greater amounts of soil N than conceivably are readily susceptible to mineralization.

B. Extraction Methods of Intermediate Intensity

Various modifications of the alkaline $KMnO_4$ method, involving measurement of the NH_4^+-N released from soils by alkaline hydrolysis and oxidation of organic N compounds, have been widely investigated (Stanford, 1978). Bremner (1965) suggested that indexes of soil N availability obtained by alkaline $KMnO_4$ distillation were of limited value. Subsequent studies involving this method have shown inconsistent relations with biological indexes of N availability (Prasad, 1965; Keeney, 1965; Keeney & Bremner, 1966a; Stanford & Legg, 1968; Cornforth, 1968; Jenkinson, 1968; Herlihy, 1972; Stanford, 1978). Usually, the best correlation with biological assays of N availability have been obtained in studies involving a narrow range of soil conditions (Cornforth, 1968; Herlihy, 1972). For heterogeneous soil groups, however, such relationships often have been no better than those existing between total soil organic N and biological indexes (Keeney, 1965; Keeney & Bremner, 1966a; Stanford & Legg, 1968; Jenkinson, 1968; Stanford, 1978).

Stanford (1978) determined the amounts of NH_4^+ released when a large number of soils were steam-distilled with various concentrations of NaOH (0.125, 0.25, and 0.625N) alone, or in combination with differing amounts of $KMnO_4$. Consistently, relatively low correlations were found between N_o and amounts of NH_4^+ released by steam-distilling soils in the NaOH solutions (r values ranged from 0.65 to 0.75). Although there was marked variation in amounts of NH_4^+ released by the various combinations of NaOH and $KMnO_4$ during steam distillation, the correlations with N_o ranged narrowly for 39 soils (r = 0.79 to r = 0.85). The relations between N_o and NH_4^+ release attributable to alkaline oxidation (obtained by subtracting NH_4^+ released by NaOH alone from that released by the combined reagents) were less precise for predicting N_o than similar relations based on acid $KMnO_4$ extraction (Stanford & Smith, 1978). Correlations between soil NH_4^+ distilled from NaOH alone and from the combined reagents consistently were high (r = 0.9), which may indicate that the NH_4^+ released by NaOH hydrolysis and $KMnO_4$ oxidation was derived from the same soil organic N sources. Under certain conditions the alkaline $KMnO_4$ method provides a reasonably suitable index of soil N availability, as reported by various investigators (Tamhane & Subbiah, 1962; Herlihy, 1972). In such instances, a procedure involving steam distillation (Keeney, 1965; Herlihy, 1972; Stanford, 1978) offers the advantage that little time is required (5–6 min for distillation and titration of each sample).

Oxidative release of NH_4^+ from soils by acid chromate or dichromate digestions of intermediate intensity recently has been proposed, for the first time, as an index of soil N availability (Nommik, 1976). Using the humus (A_o horizon) and underlying mineral soil layers (5 cm) of acid forest soils, he measured the amounts of MgO-distillable NH_4^+ produced during 2-hour boiling in 25 ml of sodium chromate-phosphoric solution containing 4.7 g $Na_2CrO_4 \cdot 4H_2O$ and 1.6g H_3PO_4. The distillable NH_4^+ was closely related to N mineralized during incubation (9 weeks, 20°C), and the relations were similar for the humus and mineral layers (r = 0.92 and 0.88, respectively). Moreover, the corresponding regressions of N mineralized on NH_4^+-N released by acid oxidation were similar (slopes of regression were 0.79 and 0.78, respectively). Interestingly, Nommik (1976) also measured the CO_2 liberated during acid oxidation, and the ratio of C released to N mineralized varied widely (C/N = 10 to 40). However, the predictability of N mineralized, based on multiple regressions of N mineralized on N and C released during chemical oxidation, was no better than that achieved by simple regression of N minerlized on NH_4^+-N released oxidatively. As observed by Nommik (1976) and others, (Verstraeten et al., 1970a; Vlassak, 1970; Geist, 1977) very little nitrification occurred during aerobic incubation of acid forest soils.

Little attention has been given to evaluating neutral aqueous $Na_2P_4O_7$ as an extractant of mineralizable soil N. Amounts of organic N extracted from two soils by boiling for 5 hours with various concentrations of $Na_2P_4O_7$ solution, in relation to amounts of N mineralized anaerobically (10 weeks) from the extracted residues, are depicted in Fig. 5A. In the concentration range of 0–0.1N, removal of organic N was accompanied by a sharp decline in N mineralization of the extracted residues. A high proportion of the N removed by more intensive extraction, however, evidently was derived from organic N fractions that were not readily susceptible to mineralization. The distillable NH_4^+ fraction of the $Na_2P_4O_7$-extractable organic N,

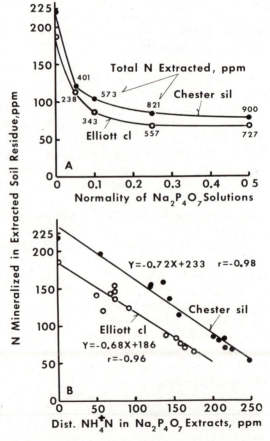

Fig. 5—Effect of extracting soils with aqueous sodium pyrophosphate solutions on N mineralized in the extracted soil residues, in relation to (A) total N extracted by 5-hour boiling in different concentrations of $Na_2P_4O_7$, and (B) distillable NH_4^+-N extracted by $Na_2P_4O_7$ treatments of varying intensities.

obtained with varying intensities of extraction involving differing reagent concentrations and varying duration and temperature of extraction, was linearly related to the quantity of N mineralized anaerobically (10 weeks) from the extracted soil residues (Fig. 5B). The similar slopes of the linear regressions may indicate that the distillable NH_4^+ fraction is uniquely related to soil N mineralization. Further study of this approach to assessing soil N availability may be worthwhile. However, it should be noted that the cumulative amounts of distillable NH_4^+-N removed from 14 soils by several consecutive extractions with boiling $0.01M$ $CaCl_2$, Y, were closely related to and similar to the amounts of distillable NH_4^+ removed by exhaustive extraction with boiling $0.5N$ $Na_2P_4O_7$, X, i.e., $Y = 1.6 + 1.1X$; $r = 0.98$ (Stanford, 1968b). The latter results indicate that the source of distillable NH_4^+ was the same for these two extractants, despite the vastly different quantities of total organic N removed.

Attempts at evaluating soil N availability using procedures that involve room-temperature extraction or boiling with dilute mineral acids (HCl or H_2SO_4) have met with limited success (Gallagher & Bartholomew, 1964; Prasad, 1965; Smith, 1966; Quinn & Salomon, 1966; Cornforth, 1968; Robinson, 1968a; Herlihy, 1972; Stanford, 1978; Fox & Piekielek, 1978b). Even relatively mild acid treatments remove substantial proportions of relatively inert, biologically resistant organic N fractions (Fig. 4).

Alkaline hydrolysis methods of intermediate intensity used in recent years include (i) *room-temperature extraction* with $0.5N$ Na_2CO_3 (Gallagher & Bartholomew, 1964); $1N$ NaOH (Cornfield, 1960; Keeney & Bremner, 1966a; Cornforth, 1968); $0.1N$ $Ba(OH)_2$ (Keeney & Bremner, 1966a; Jenkinson, 1968; Herlihy, 1972); and (ii) *boiling* in $0.25N$ $NaHCO_3$–$0.25N$ Na_2CO_3 solution or other bases, e.g., NaOH, MgO, $Ba(OH)_2$, and $Ca(OH)_2$ (Prasad, 1965; Jenkinson, 1968) or in NaOH solutions of various concentrations (Keeney, 1965; Stanford, 1978). In general, amounts of NH_4^+ or total organic N released by the foregoing methods were rather poorly related to biological measurements of N availability, presumably due to the relatively large amounts of N extracted even with low concentrations of reagent.

C. Relatively Mild Extraction Methods

It is manifestly impossible to devise a chemical extraction procedure that simulates the action of microorganisms in releasing plant-available forms of soil N. Nevertheless, relatively mild extraction methods have been suggested in which high proportions of N may be derived from the biologically active pool.

Using ^{14}C-labeled soil, Jenkinson (1968) found that the organic substances extracted by boiling water, cold dilute HCl, or cold dilute $Ba(OH)_2$ were highly labeled, whereas those removed by more drastic treatment were only lightly labeled. Incubation studies indicated that the ^{14}C-labeled portion of the soil organic matter, the biomass, constituted the principal source of mineralizable C (Jenkinson, 1968). Legg et al. (1971), using ^{15}N-labeled soil, found that the degree of labeling markedly decreased in successive extracts obtained by autoclaving the soil in $0.01M$ $CaCl_2$ (Stanford & DeMar, 1969). Moreover, they found that the ratio of labeled to total N in the N released during successive 2-week incubations (40 weeks total) consistently was about twice the ratio of labeled to total N in the soil mass. Legg et al. (1971) concluded that the labeled portion of soil organic N was twice as susceptible to mineralization as the indigenous soil N. In contrast, the ratio,

$$\frac{\text{tagged N (extract)/total N (extract)}}{\text{tagged N(soil)/total N(soil)}}$$

was about one for a 16-hour autoclave extraction, indicating almost equal susceptibility to extraction of labeled and indigenous soil N. Although these results emphasize the difficulty of interpreting extractable soil N in terms of plant availability, they do not rule out the possibility of developing reliable empirical chemical indexes useful in predicting N mineralization potentials of soils. Mild extractants that remove relatively greater amounts of N from the active biomass than from the more difficultly extractable or mineralizable forms have been studied extensively.

The hot-water extraction method proposed by Livens (1959a), and modified by Keeney and Bremner (1966a) dissolves a relatively small proportion (1–2%) of the total soil N. Amounts extracted by 60-min boiling in water correlated as well with N uptake by ryegrass as did amounts of N mineralized by aerobic or anaerobic methods (Keeney & Bremner, 1966a), in general agreement with findings of Jenkinson (1968), Gasser and Kalembasa (1976), Verstraeten et al. (1970a, b), Lathwell et al., (1972), and Kadirgamathoiyah and MacKenzie (1970).

Livens (1959a, b) concluded that the alkali-distillable NH_4^+ fraction of the hot-water N extract was more closely related to mineralizable soil N than was the total extracted N. Despite this observation, Verstraeten et al. (1970a), have continued measuring the total N extracted by boiling water. Stanford (1968b) fractionated the extracts obtained by 16-hour boiling of soils in $0.01M$ $CaCl_2$ and found that the NaOH-distillable fraction correlated much better ($r = 0.95$) with N mineralized during anaerobic incubation than did α-amino N ($r = 0.72$) or the N in the unidentified residue ($r = 0.55$). Similar results were found upon fractionating extracts obtained by autoclaving soils in $0.01M$ $CaCl_2$ for 16 hours (Stanford & DeMar, 1969). In the latter study, the distillable NH_4^+-N fraction comprised from 20 to 43% of the total N extracted. Correlations of the various fractions with mineralizable N were: Distillable NH_4^+, $r = 0.94$; acid-hydrolyzable distillable NH_4^+, $r = 0.69$; α-amino, $r = 0.72$; unidentified N, $r = 0.66$; and the sum of fractions, $r = 0.76$.

Smith and Stanford (1971) and Stanford and Smith (1976) found a relatively good relation between distillable NH_4^+ determined by the autoclave method and mineralizable N, for a broad range of soils, when calcareous aridisols from the western United States were excluded. Moreover, these same calcareous soils mineralized much less N under anaerobic conditions than under aerobic conditions (Smith & Stanford, 1971). Upon eliminating these soils the correlation between N mineralization potential, N_o, and the chemical index was relatively good (r = 0.87) for a large number of widely differing soils (Stanford & Smith, 1976).

The soil biomass probably constitutes the major source of plant-available N amenable to chemical dissolution or hydrolysis by relatively mild extraction procedures (Jenkinson, 1968). In support of this concept, polysaccharide contents of such extracts often correlated highly with amounts of N mineralized during incubation. Jenkinson (1968) found a close relation between the "glucose", derived from polysaccharides as determined by the anthrone method, and mineralized N, for three groups of soils (r = 0.81, 0.95, and 0.97). Sugars, i.e., "glucose equivalent," extracted from 15 soils with $0.1N$ Ba(OH)$_2$ by the method of Jenkinson (1968) correlated highly (r = 0.98) with N released by aerobic and anaerobic mineralization (Gasser & Kalembasa, 1976). Herlihy (1972) also found a good relationship, using the same "glucose" extractant, for 16 variously treated soils (r = 0.87). Using a more rigorous extractant, 5% trichloracetic acid, Robinson (1968a) found a very poor correlation between "glucose" and mineralized N. Stanford and DeMar (1970) reported that "glucose" and distillable NH$_4^+$ in extracts of 13 soils, obtained by autoclaving in $0.01M$ CaCl$_2$, were closely related (r = 0.98). Using 39 widely differing soils from different agricultural regions, Smith and Stanford (1971) found that "glucose" extracted by autoclaving in $0.01M$ CaCl$_2$ ranged from 200 to 2,500 ppm and correlated significantly with mineralizable N (r = 0.85). As suggested earlier, the close association often observed between "glucose" and mineralizable N, using mild extractants, may indicate that these constituents are derived from a common source, i.e., the biomass. This view is strengthened by the finding that, although differing amounts of "glucose" were extracted from soils by various mild extractants, e.g., dilute Ba(OH)$_2$ (Jenkinson, 1968), boiling water (Keeney & Bremner, 1966a), or dilute NaHCO$_3$ (MacLean, 1964), the values thus obtained were highly correlated (r > 0.90).

It appears, therefore, that greatest emphasis in the search for a suitable chemical index of soil N availability should be directed toward evaluating relatively mild extractants. The possibility that some fraction or fractions of the N in such extracts might provide a better index than their total N content should be considered (Stanford, 1968a, 1968b; Stanford & DeMar, 1970).

The amount of N extracted by water or $0.01M$ CaCl$_2$ depends on the temperature and duration of extraction. For example, Stanford and Smith (1976) found, on average, for 212 calcareous and noncalcareous soils, that the amount of distillable NH$_4^+$ obtained by 16-hour autoclaving ($0.01M$ CaCl$_2$) was about 2.5 times that obtained by 16-hour boiling in this solution, and the amounts were well correlated (r = 0.93). In contrast, Verstraeten et al. (1970a), found a relatively low correlation (r = 0.51) between amounts of organic N extracted by distilled water at room temperature (30 min, 20°C) and by boiling (60 min) using 35 cultivated and meadow soils. Four to five times more organic N was extracted at 100°C than at room temperature, and correlation coefficients, r, relating these values to N mineralized aerobically (30 days, 30°C), respectively, were 0.55 and 0.83. These results suggest that water extraction at room temperature is more selective in removing mineralizable organic N fractions of the biomass than is the more intensive extraction by boiling in water for 60 min.

Another relatively mild procedure consisting of a 15-min, room-temperature extraction with $0.01M$ NaHCO$_3$ was proposed by MacLean (1964). The total organic N extracted by this method from 24 soils was closely related to short-term N uptake by oats ($r = 0.85$) measured by the method of DeMent et al. (1959) and longer-term N uptake by ryegrass in a greenhouse ($r = 0.76$). Smith (1966) also obtained encouraging results using the MacLean (1964) method in connection with a greenhouse study of 60 soils (30 soil types) cropped to orchardgrass. Jenkinson (1968) found that NaHCO$_3$-extractable organic N and "glucose" were more closely related to mineralized N, using a heterogeneous group of 14 soils, than was the total N extracted by Ba(OH)$_2$ or boiling water. Recently, Fox and Piekielek (1978a) modified the method of MacLean (1964) as follows: (i) extraction was done exactly as prescribed by MacLean (1964), i.e., 5g of soil were shaken in 100ml of $0.01M$ NaHCO$_3$ for 15 min and filtered; and (ii) the relative organic matter contents of the extracts were estimated from UV absorbance (260 nm wavelength). In a 2-year study, the N-supplying capability of eight sites, as determined from N uptake by corn grown in field plots, was closely related to UV absorbance ($r = 0.87$). This r-value is higher than that found for the relation between NaHCO$_3$-extractable N and soil N supplying capability ($r = 0.77$). The corresponding relation for total organic N extracted by boiling in $0.01M$ CaCl$_2$ was very good ($r = 0.86$). Using UV absorbance to estimate relative organic matter contents of extracts obtained by various mild extraction procedures merits further study since this approach is less time consuming than any yet proposed.

Critical tests of the general utility of any chemical index of N availability must be based on evaluations involving a broad range of soils. Moreover, the relative N-supplying abilities of all soils used in such evaluations must be determined by a single reliable and reproducible biological procedure. The N mineralization potential, N_o, is considered to meet these requirements since it is derived from long-term incubation. Biological indexes based on short-term incubation are less meaningful because results are influenced by method of sample pretreatment to a greater extent than is usually found with chemical extractions (Keeney & Bremner, 1966a). With long-term incubation, the transitory effects of recent crop residues or method of sample handling are minimized (Stanford et al., 1974).

A suitable chemical index should provide a reliable basis for predicting the quantity of potentially mineralizable N, N_o, for a broad range of soils. Prediction of N_o from distillable NH$_4^+$ obtained by 16-hour boiling or autoclaving in $0.01M$ CaCl$_2$ was reported by Stanford and Smith (1976). For 275 soils, the regression of N_o on autoclave distillable NH$_4^+$-N (N_i) was as follows:

$$N_o = 4.1(\pm 1.0)\, N_i + 6.6\, (r = 0.87).$$

For calcareous aridisols of Idaho and California, significantly larger regression coefficients were obtained (5.5–6.7).

A more broadly applicable chemical index for predicting N_o recently was developed by Stanford and Smith (1978). The method involves a relatively mild, room-temperature extraction of soils for 1 hour with $0.05N$ $KMnO_4$ or $0.1N$ $KMnO_4$ in $1N$ H_2SO_4, followed by steam distillation of the extract with NaOH to measure NH_4^+ released oxidatively. Amounts of NH_4^+ extracted by $1N$ H_2SO_4 (1 hour, room-temperature) from 62 soils essentially were unrelated to N_o ($r = 0.3$). The relation between N_o and amounts of NH_4^+-N released oxidatively (HOx-H), however, was relatively close as shown by the following regressions:

$$N_o = 2.1(HOx-H) + 1.6 \quad (r = 0.89) \tag{1}$$

and

$$N_o = 3.1(HOx-H) - 8.5 \quad (r = 0.89). \tag{2}$$

Equations [1] and [2], respectively, are based on extraction with solutions of 0.1 and $0.05N$ $KMnO_4$ in $1N$ H_2SO_4. Note that (HOx-H) denotes the amount of NH_4^+ extracted by acid plus oxidant (HOx) minus the amount extracted by $1N$ acid alone (H).

From the slopes of these regressions, it is evident that amounts of NH_4^+-N released oxidatively by reactions of soils with acid $0.1N$ and $0.05N$ $KMnO_4$, respectively, were about one-half and one-third of N_o. Regressions for calcareous ($n = 19$) and noncalcareous soils ($n = 43$) did not differ significantly. Moreover, the regression slope relating N_o to oxidative NH_4^+-N release for a broad range of calcareous and near neutral aridisols from Idaho, based on extraction with acid $0.1N$ $KMnO_4$, did not differ significantly from that obtained in Eq. [1] above (unpublished data, this laboratory). Thus, the proposed method is more broadly applicable for predicting N_o than is the method involving autoclaving in $0.01M$ $CaCl_2$ (Stanford & Smith, 1976). Moreover, results indicated that the method, as described, releases NH_4^+ by oxidation of soil organic N fractions that are readily susceptible to mineralization.

V. INTERPRETING CHEMICAL AND BIOLOGICAL ASSAYS OF SOIL NITROGEN AVAILABILITY

Early investigations raised serious doubts regarding the feasibility of interpreting and utilizing laboratory measurements of soil N availability under the widely varying soil and climatic conditions encountered in the field. Harmsen and Van Schreven (1955) concluded that reliable interpretations can be expected only when dealing with a single soil type, climatic zone, or farming system. Moreover, they stressed that interpretations must vary from year to year, even with the foregoing restrictions. Finally, they concluded that determinations of the N supplying capabilities of soils can never reach the same accuracy as is achieved with assessing the P and K status of soils.

Table 1. Relation of N uptake (Y) to initial mineral N (X_1) and N mineralization (X_2) as determined by regression and correlation analyses.

Source	Index of soil N availability	Regression equation†	Correlation coefficients‡					
			$r_{y1.2}$	$r_{y2.1}$	r_{12}	r_{y1}	r_{y2}	R
Stanford & Legg (1968), 12 soils	N mineralized, 40 days (aerobic)	$Y = 0.60X_1 + 0.78X_2 + 15.5$	0.95**	0.87**	0.32	0.88**	0.66*	0.97**
Hanway & Ozus (1966), 24 soils	N mineralized (anaerobic), 1 week, 40C	$Y = 0.98X_1 + 0.27X_2 + 25.5$	0.89**	0.78**	0.18	0.82**	0.59**	0.93**
	N mineralized (aerobic), 2 weeks, 35C	$Y = 0.87X_1 + 0.30X_2 + 30.2$	0.78**	0.60**	0.44*	0.82**	0.67**	0.89**
Keeney (1965), 25 soils	N mineralized (anaerobic), 1 week, 40C	$Y = 0.75X_1 + 0.23X_2 + 2.0$	0.92**	0.83**	0.42*	0.88**	0.73**	0.96**
	N mineralized (aerobic), 2 weeks, 30C	$Y = 0.84X_1 + 0.46X_2 - 1.6$	0.90**	0.66**	0.30	0.88**	0.56**	0.94**
	60-minute boiling in water (total N in extract)	$Y = 0.79X_1 + 0.29X_2 - 2.6$	0.90**	0.72**	0.39	0.88**	0.66**	0.95**

* Significant at 5% level.
** Significant at 1% level.
† Y = N uptake by plants, mg/pot: One crop of oats (Stanford & Legg); corn followed by 3 ryegrass cuttings (Hanway & Ozus); and 3 ryegrass cuttings (Keeney). X_1 = initial mineral N, ppm; X_2 = N mineralized, ppm.
‡ $r_{y1.2}$ = correlation of Y and X_1, independent of X_2; $r_{y2.1}$ = correlation of Y and X_2, independent of X_1. Remaining r-values are based on simple, 2-factor correlations; R = multiple correlation coefficient.

A. Under Controlled Conditions

Greenhouse pot cultures have been widely used as a first step in comparing and calibrating chemical and biological indexes of soil N availability. Usually, soil N uptake by the plants provides a more reliable basis for such evaluations than does plant growth in the zero-N pots or yield response to applied N. Even under ideal and controlled plant growth conditions, however, a common difficulty encountered in establishing meaningful relationships between soil N measurements and greenhouse results is that soils may vary appreciably in amounts of mineral nitrogen present before cropping. The importance of this factor was first emphasized particularly by Attoe (1964), Peterson and Attoe (1965), and Smith (1966). The problem, of course, arises from the fact that mineral N already present in the soils is readily available to plants, whereas N measured by short-term incubations or chemical methods is merely an index of the relative amounts of N that will be supplied from soil organic sources. Smith (1966) demonstrated, using multiple regression analyses, that initial NO_3^- (X_1) and organic N released by incubation or chemical extraction (X_2) were not additive in their effects on N uptake by plants (Y), as can be seen by comparing regression coefficients in the following equation:

$$Y = 0.73\,X_1 + 0.16\,X_2 - 8.7\,(R = 0.88).$$

In this example, X_2 represents the NO_3^--N released during 2-week aerobic incubation using a modified version of the Stanford and Hanway method (1955).

Evaluating N availability indexes is particularly complex when successive crops are grown in the greenhouse (Hanway & Ozus, 1966; Keeney & Bremner, 1966a; Smith, 1966; Baerug et al., 1973; Lathwell et al., 1972). Typically, the relation between initial mineral N and N uptake becomes poorer with successive crops, while the relation of uptake to biological and chemical indexes improves. Hence, the cumulative N taken up by a succession of crops in the greenhouse may be treated as a function of initial mineral N and relative soil organic N availability.

Much can be learned regarding the relative contributions of these two N sources using multiple regression analysis provided these independent variables are not well correlated. This is illustrated in Table 1. Note that the multiple R was relatively high in all cases, except one. As Smith (1966) observed, the regression coefficients for X_1 and X_2 differ markedly. With one exception, the partial correlation coefficient, $r_{y2.1}$, significantly exceeded r_{y2}. In general, the independent variables, X_1 and X_2, were poorly correlated.

Interestingly, the results in Table 1 indicate that short-term anaerobic mineralization, as an index of soil organic N availability, was superior to short-term aerobic mineralization. Hanway and Ozus (1966) and Keeney and Bremner (1966a) arrived at similar conclusions, but in a different way.

They did not include N uptake data for the first crop, since results were highly correlated with initial mineral N (r was 0.93 in both studies). Despite the significant contributions of initial mineral N to succeeding crops or cuttings, total N uptake after the first crop was used alone in evaluating indexes of organic N availability. Although the approach illustrated in Table 1 is more objective, it should be emphasized that, in either case, the relations are empirical in the sense that indexes instead of actual estimates of soil N mineralization were employed.

A less empirical method of relating N uptake to initial mineral N and N mineralization was reported by Stanford et al. (1973b) in connection with a study of 39 soils of widely diverse origin. The greenhouse experiment involved application of ^{15}N-labeled N fertilizer, using sudangrass [*Sorghum sudanense* (Piper) Stapf] as the test crop. Soil water content was adjusted to field capacity, daily or twice-daily, throughout the experiment. Amounts of N mineralized during the 6-week period of plant growth, assuming optimal soil water contents, were estimated from the soil N mineralization potentials (Stanford & Smith, 1972) based on average weekly temperatures (Stanford et al., 1973a). The regression of A-values, Y, (Fried & Dean, 1952), determined from ^{15}N data, on initial mineral plus estimated mineralized N, X, was as follows (mg/kg of soil):

$$Y = 0.95X - 2.6 \, (r = 0.94).$$

The A-value denotes the quantity of soil-derived N that is equivalent in availability to the applied N, as determined from ^{15}N analyses of the plants. This study demonstrates the feasibility of estimating reliably the amounts of N actually mineralized in the course of a cropping period in the greenhouse.

B. Under Field Conditions

As suggested earlier, the main obstacle in predicting N needs of the crop under field conditions is the variation in attainable yields associated with unforeseeable climatic and cropping conditions. In irrigated agriculture, this factor is minimized. In other situations records of previous yields, climate (e.g., rainfall, temperature, and evapotranspiration), and soil and crop management may be used to estimate the probability of attaining given levels of crop production.

1. RESIDUAL NITRATE

In areas where N fertilizer recommendations are guided by residual soil NO_3^- measurements, the interpretation of these values in terms of N fertilizer requirement is usually based on (i) expected yield, (ii) estimates of N needed for different yield levels, and (iii) the level of NO_3^- in the soil. In general, the availability of residual NO_3^- to the crop is considered to be about equal to that from applied fertilizer. A simple example illustrates the state of the art. In some areas, it is assumed that 0.9 kg of N is required to produce 27.2 kg of wheat (2 lb N/1 bushel of wheat). Thus, if the grower expects to produce 2,018 kg/ha (30 bushels/acre), approximately 67 kg/ha (60 lb N/acre) will be required. If the level of NO_3^--N in the upper 60-cm layer of soil is 45 kg/ha (40 lb/acre), then 1,345 kg/ha (20 bushels/acre) is the expected yield without additional N. Hence, to reach the projected attainable yield (2,018 kg/ha), an additional 20–25 kg N/ha of fertilizer N is needed. This general approach to estimating N needs is being used for a number of crops in areas where residual NO_3^- is measured as a guide to N fertilizer use. Basically, N recommendations are estimated from N requirements of the crop for expected attainable yield and the amount of residual NO_3^- in the root zone.

A general understanding of the relation between crop yield response to applied N and residual NO_3^- can be obtained by the method shown in Fig. 6 (Dahnke et al., 1977) which gives an approximation of the "critical" soil NO_3^- level. Below this level, the probability of an economic yield response to applied N is high (Nelson & Anderson, 1977). A statistical method for determining the positions of the vertical and horizontal lines in the graph has been described (Cate & Nelson, 1971; Nelson & Anderson, 1977). This method of relating percentage yield to residual NO_3^- is informative and useful under conditions where mineralizable N varies little among experimental sites. With substantial variations in seasonal N mineralization, however, it would be necessary to combine this source of available N with residual NO_3^-.

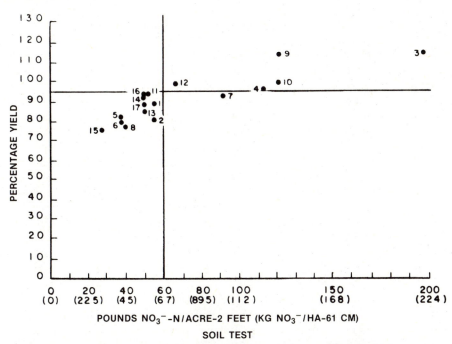

Fig. 6—Scatter diagram of percentage yield vs. NO_3^--N soil test value for 17 potato trials, 1971–1975 (Dahnke et al., 1977).

2. MINERALIZABLE SOIL NITROGEN AND RESIDUAL MINERAL NITROGEN

Soils differ widely in their capacities to mineralize soil organic N under optimal conditions. In the field, the factors that influence rate of N mineralization are seldom optimal. Although the dominant factors usually are soil temperature and available water supply, other conditions associated with soil properties and farming practice, e.g., cultivation, other essential nutrients, liming, and cropping system also are influential (Campbell, 1978).

Reuss and Geist (1970) and Geist et al. (1970) were among the first to attempt an accounting of both available soil mineral N and N released from soil organic matter in predicting yield response to N or N uptake by the crop. As an approximation, they considered that fertilizer N, residual mineral N, and an estimated fraction of the N released from organic matter by chemical extraction are used with equal facility by the crop. They expressed the relationship by a second degree polynomial equation as follows:

$$Y = B_0 + B_1(X_1 + aX_2) + B_2(X_1 + aX_2)^2$$

where B_0, B_1, and B_2 are regression coefficients; Y = yield or other parameter; X_1 = fertilizer + residual mineral N; X_2 = an index of soil organic N availability; and a is the estimated fraction of X_2 released (Reuss & Geist, 1970).

Geist et al. (1970) used this approach to estimate the optimal N requirements of malting barley in 34 rate-of-N field experiments and obtained relatively good relationships between grain yield or protein yield and the sum of residual mineral N, fertilizer N, and the adjusted organic N index, aX_2. Apparently, estimates of total soil N supply $(X_1 + aX_2)$ were dominated by X_1, because there was no discernible difference in the relative effectiveness of various indexes of soil organic N availability. In fact, yield predictions involving NaOH-distillable N, which correlates highly with total soil N (Geist & Hazard, 1975), were no better than with anaerobic N mineralization. Such a result probably reflects a relatively broader range of residual mineral N and fertilizer N (X_1) than of potentially available organic or total soil organic N (X_2). However, their approach cannot be carefully evaluated since only statistical data are given in the paper. Using the same statistical methods, Geist (1977) later concluded that anaerobic N mineralization was a better index than total soil N.

The importance of considering both initial mineral N (N_i) and mineralizable N (N_m) in relation to sugar percentage in sugar beets has been demonstrated (Roberts et al., 1972). Based on multiple regressions, the relative amounts of available organic N as indicated by NH_4^+ released during 16-hour autoclaving (Stanford & DeMar, 1969) contributed 40% as much as N_i to the decline in percent beet sugar. Early in the season (23 June) N_m had no discernible effect on the NO_3^- content of the petioles, relative to N_i; but at a later date (26 July), the relative contribution of N_m was 80% of N_i as revealed by multiple regression analyses.

A more direct approach to estimating N needs of the crop may be described as follows (Stanford, 1973):

$$N_c = N_i + N_m + N_f. \qquad [3]$$

In this equation, N_c = the uptake of N associated with a specified maximum or attainable economic yield; N_i and N_m, respectively, denote the measured initial quantity of mineral N in the profile and the estimated N mineralized during the cropping season; and N_f denotes the amount of N fertilizer needed. Plant utilization or fractional recoveries of N_f, N_i, and N_m may differ according to Eq. [4]. Solving for N_f gives Eq. [5]. Assuming, as a first approximation, that $e_i = e_m = e_f$, the N fertilizer requirement may be estimated as shown in Eq. [6].

$$N_c = e_i N_i + e_m N_m + e_f N_f. \qquad [4]$$

$$N_f = (N_c - e_i N_i - e_m N_m)/e_f. \qquad [5]$$

$$N_f = (N_c/e_f) - (N_i + N_m). \qquad [6]$$

Using an approach similar to Eq. [5], Carter et al. (1974, 1975) estimated the N fertilizer requirements of sugar beets. They estimated the optimal N content of sugar beets to be about 5.5 kg/metric ton (field weight). Thus, expected yield, multiplied by 5.5 defines the total requirement, N_c. In their studies, N_i signifies the quantity (kg/ha) of NO_3^--N in the soil zone accessible to roots before planting, and N_m denotes the amount of N mineralized in the root zone (about 40 cm) during a 3-week incubation at a temperature of 30°C. Empirical utilization coefficients for N_f, N_i, and N_m were estimated from earlier field experimental data. A summary of some of their results is presented in Fig. 7. Here, variations in Y/Y_{max} for 1968 mainly reflect residual effects of NO_3^- from previous applications of N to potatoes, as well as effects of varying rates applied to the 1968 sugar beet crop. The 1969 results reflect residual effects of N from the preceding potato and sugar beet crops and a moderate application (56 kg N/ha) to the 1969 crop. The value, N_T, denotes the estimated total amount of N available to the crop from the various plots ($N_T = e_f N_f + e_i N_i + e_m N_m$). It should be noted that average N_m at this site ranged narrowly between 207 and 216 kg/ha during the course of this 3-year study. In contrast, N_i of early season samples varied widely among plots and years: 75–192 kg/ha in 1968; and 49–105 kg/ha in 1969.

Over a broader range of field conditions, the varying temperature and soil water regimes profoundly affect the quantity of N mineralized during the growing season. Stanford et al. (1977) attempted to take these variables into account in predicting uptake of N by sugar beets from control plots (no N applied) using 32 irrigated experiments in southern Idaho. On the control plots in 1971 and 1972, fresh-root and sucrose yields, respectively, ranged from 23 to 82 and from 4 to 13 metric tons/ha. Total N uptake at harvest ranged from 80 to 420 kg/ha, and N%TDM (tops and roots) ranged from

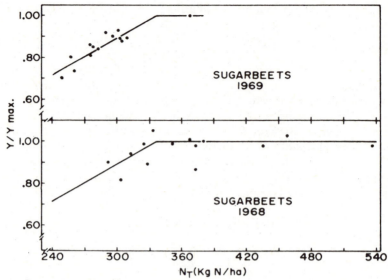

Fig. 7—Effect of available N(N$_T$) on the root yield potential of sugar beets (Carter et al., 1974).

0.7 to 2.1. Amounts of NO$_3^-$-N in the root zone varied between 20 and 240 kg/ha, and mineralization potentials, N$_o$, to the 45-cm depth ranged between 600 and 1,070 kg/ha. Estimated amounts of N mineralized, N$_m$, (6 months, April through September), derived from monthly average estimates of temperature and available soil water, ranged widely between 100 and 230 kg/ha. Expressed as a percentage of N$_o$, these values of N$_m$ encompass the relatively narrow range of 15.8–22.3 which, in turn, reflects the narrow ranges in average monthly temperatures and water regimes among these irrigated sites.

The method used in estimating the average monthly effects of temperature and soil water content on N mineralization is described in the foregoing study (Stanford et al., 1977). The ranges shown above, serve to indicate that there was considerable variation in the N supplying capabilities of the soils. Since the study concerns plots that received no N fertilizers, the relation between N uptake (Y), initial mineral N (N$_i$), and estimated N mineralized during the season (N$_m$) can be expressed as follows:

$$Y = 0.73 \, (N_i + N_m) + 5.3.$$

The coefficient, 0.73, is similar to the value used by Carter et al. (1974) to denote fractional fertilizer N recovery (0.65).

The feasibility of estimating N mineralized under varying temperature and soil water regimes in the field, under fallow and without irrigation, was demonstrated by Smith et al., (1977) using microplots on eight soil types. Soil water contents were measured monthly. Temperature fluctuations recorded at a single site were used for calculations at the eight sites, since the microplots were not widely separated (within 40 km of each other). Cumulative measured and calculated amounts of N mineralized over a period of several months were reasonably well correlated ($r = 0.76$–0.96) and the regression slopes did not differ significantly from unity. Thus, the study supports the validity of estimating temperature and soil water effects on N mineralization as proposed by Stanford et al. (1973a) and Stanford and Epstein (1974).

Oyanedel and Rodriguez (1977) determined the N mineralization potentials of 23 soils in the Central Zone of Chile, following the method proposed by Stanford and Smith (1972). Values of N_o ranged widely from 100 to 520 ppm, and mineralization rate constants, k, derived from the expression $\log (N_o - N_t) = \log N_o - kt/2.303$ were similar to those reported by Stanford and Smith (1972), i.e., 0.054 ± 0.012 week^{-1}. Based on average monthly temperature under irrigated wheat, the average fraction of available water storage capacity (field capacity—wilting point) present each month (June through December), and the N requirement of wheat for a specified attainable yield, they estimated the optimal amounts of fertilizer N needed to satisfy the soil N deficits for two representative soil groups. The effect of temperature was estimated from the relation given by Stanford et al. (1973a). The effect of soil moisture was estimated from the relation between relative N mineralized, N_t/N_{to}, and the average relative soil water content, θ/θ_o (Cavalli & Rodriguez, 1975), i.e., $N_t/N_{to} = -13.8 + 1.11$ θ/θ_o, in which θ_o denotes percent soil water at field capacity and θ is the average percent water for the period of measurement; and N_t and N_{to}, respectively, denote the N mineralized under the corresponding water regimes. A similar relation was reported by Stanford and Epstein (1974). The exponential equation, expressing N mineralized, N_t, as a function of potentially mineralizable N, N_o, soil temperature (T = absolute temperature), and relative soil water content, θ/θ_o, as derived by Oyanedel and Rodriguez (1977) is shown below:

$$N_t = N_o \exp 2.3(7.71 - 2758/T)(1.11\,\theta/\theta_o - 0.138).$$

They prepared a nomograph that expresses this relationship (Fig. 8). In the example given, with N_o = 200 ppm, average weekly temperature = 10°C, and weekly average relative water content = 0.8, the estimated N_t = 1.3 ppm or 2.6 kg/ha per week.

To further test the validity of the foregoing equation, Prado and Rodriguez (1978) conducted rate-of-N experiments at five field locations with wheat. They estimated N fertilizer requirement, N_f, using Eq. [5] discussed earlier in this section:

$$N_f = (N_c - e_iN_i - e_mN_m)/e_f.$$

In this equation, N_i represents the quantity of residual NO_3^--N to a depth of 40 cm; N_m, or N_t, estimated from the equation given in the previous paragraph (Oyanedel & Rodriguez, 1977) ranged from 70 to 122 kg/ha; N_c for wheat was determined from N uptake-yield relations in the five field experiments. The fractional recovery of fertilizer N, e_f, also was derived from the

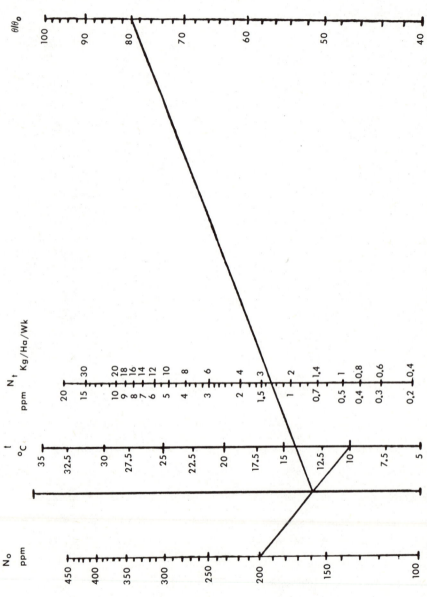

Fig. 8—Nomograph for estimating the amount of N mineralized (N_t) from potentially mineralizable soil nitrogen (N_o), average weekly temperature (t, °C), and weekly average relative moisture content (θ/θ_o) (Oyanedel and Rodriguez, 1977).

field experiments and ranged from 0.55 to 0.64 (average 0.51). The coefficient, e_i, was assumed to be equal to e_f. Fractional uptake of N_m (e_m) was estimated to be 0.40. Predicted and observed quantities of N_f required for maximum yield were close (within 10%) and correlated highly ($r = 0.93$).

Various workers have attempted to devise mathematical models that describe the relation between N inputs and outputs (see Chapt. 19, K. K. Tanji). Future progress in developing suitable models undoubtedly will contribute to a broader understanding of the N system as a whole. Ultimately, the net supply of N made available to crops is a resultant of the complex biochemical processes that determine rate of N turnover, mineralization-immobilization relationships, and other N transformations. These aspects are treated in other chapters of this volume.

LITERATURE CITED

Adriano, D. C., F. H. Takatori, P. F. Pratt, and O. A. Lorenz. 1972. Soil nitrogen balance in selected row-crop sites in southern California. J. Environ. Qual. 1:279–283.

Attoe, O. J. 1964. Tests for available soil nitrogen. p. 392–401. In V. Sauchelli (ed.) Fertilizer nitrogen, its chemistry and technology. Reinhold Publ. Corp., New York.

Baerug, R., I. Lyngstad, A. R. Selmer-Olsen, and A. Øien. 1973. Studies on soil nitrogen. I. An evaluation of laboratory methods for available nitrogen in soils from arable and ley-arable rotations. Acta Agric. Scand. 23:173–181.

Biggar, J. W. 1978. Spatial variability of nitrogen in soils. p. 201–222. In D. R. Nielsen and J. G. MacDonald (ed.) Nitrogen in the environment: Vol. 1, Nitrogen behavior in field soil. Academic Press, New York.

Boswell, F. C., and O. E. Anderson. 1970. Nitrogen movement comparisons in cropped versus fallowed soils. Agron. J. 62:499–503.

Bremner, J. M. 1965. Nitrogen availability indexes. In C. A. Black (ed.) Methods of soil analysis, Part 2. Agronomy 9:1324–1345. Am. Soc. of Agron., Madison, Wis.

Buresh, R. J., and W. H. Patrick, Jr. 1978. Nitrate reduction to ammonium in anaerobic soil. Soil Sci. Soc. Am. J. 42:913–918.

Campbell, C. A. 1978. Soil organic carbon, nitrogen, and fertility. p. 173–271. In M. Schnitzer and S. V. Kahn (ed.) Soil organic matter; developments in soil science 8. Elsevier Sci. Pub. Co., New York.

Carson, P. L. 1975. Recommended nitrate-nitrogen tests. In W. C. Dahnke (ed.) Recommended chemical soil test procedures for the north-central region. North Dakota Agric. Exp. Stn. Bull. 499.

Carter, J. N., M. E. Jensen, and S. M. Bosma. 1974. Determining nitrogen fertilizer needs for sugar beets from residual nitrate and mineralizable nitrogen. Agron. J. 66:319–323.

Carter, J. N., D. T. Westermann, M. E. Jensen, and S. M. Bosma. 1975. Predicting nitrogen fertilizer needs for sugar beets from residual nitrate and mineralizable N. J. Am. Soc. Sugar Beet Technol. 18:232–244.

Carter, J. N., D. T. Westermann, and M. E. Jensen. 1976. Sugar beet yield and quality as affected by nitrogen level. Agron. J. 68:49–55.

Caskey, W. H., and J. M. Tiedje. 1979. Evidence for clostridia as agents of dissimilatory reduction of nitrate in soils. Soil Sci. Soc. Am. J. 43:931–936.

Cate, R. B., Jr., and L. A. Nelson. 1971. A simple statistical procedure for partitioning soil test correlation data into two classes. Soil Sci. Soc. Am. Proc. 35:658–659.

Cavalli, I., and J. Rodriguez S. 1975. Effect of moisture in nitrogen mineralization of nine soils of Santiago province. Cienc. Invest. Agraria 2:101–111.

Chichester, F. W., J. O. Legg, and G. Stanford. 1975. Relative mineralization rates of indigenous and recently incorporated ¹⁵N-labeled nitrogen. Soil Sci. 120:455–460.

Clement, C. R., and T. E. Williams. 1962. An incubation technique for assessing the nitrogen status of soils newly ploughed from leys. J. Soil Sci. 13:82–91.

Cook, F. D., F. G. Warder, and J. L. Daughty. 1957. Relationship of nitrate accumulation to yield response of wheat in some Saskatchewan soils. Can. J. Soil Sci. 37:84–88.

Cooke, G. W. 1967. The control of soil fertility. Crosby Lockwood Ltd., London.

Cooke, G. W., and R. K. Cunningham. 1958. Soil N. III. Mineralizable nitrogen determined by an incubation technique. J. Sci. Food Agric. 6:324–330.

Cornfield, A. H. 1960. Ammonia released on treating soils with N sodium hydroxide as a possible means of predicting the nitrogen-supplying power of soils. Nature (London) 187: 260–261.

Cornforth, I. S. 1968. The potential availability of organic nitrogen fractions in some West Indian soils. Exp. Agric. 4:193–201.

Dahnke, W. C., D. C. Nelson, L. J. Swenson, and M. C. Thoreson. 1977. Relationships of potato yields and quality to soil test values and fertilizer applications. Farm Res. 34:21–27.

Dahnke, W. C., and E. H. Vasey. 1973. Testing soils for nitrogen. p. 97–114. In L. M. Walsh and J. D. Beaton (ed.) Soil testing and plant analysis. Soil Sci. Soc. Am., Inc., Madison, Wis.

DeMent, J. D., G. Stanford, and C. M. Hunt. 1959. A method for measuring short-term nutrient adsorption by plants: III. Nitrogen. Soil Sci. Soc. Am. Proc. 23:371–374.

Dolmat, M. T., W. H. Patrick, Jr., and F. J. Peterson. 1980. Relation of available soil nitrogen to rice yield. Soil Sci. 129:229–237.

Fitts, J. W., W. V. Bartholomew, and H. Heidel. 1953. Correlation between nitrifiable nitrogen and yield response of corn to nitrogen fertilization on Iowa soils. Soil Sci. Soc. Am. Proc. 17:119–122.

Fox, R. H., and W. P. Piekielek. 1978a. A rapid method for estimating the nitrogen-supplying capability of a soil. Soil Sci. Soc. Am. J. 42:751–753.

Fox, R. H., and W. P. Piekielek. 1978b. Field testing of several nitrogen availability indexes. Soil Sci. Soc. Am. J. 42:747–750.

Freney, J. R., and J. R. Simpson. 1969. The mineralization of nitrogen from some organic fractions in soil. Soil Biol. Biochem. 1:241–251.

Fried, M., and L. A. Dean. 1952. A concept concerning the measurement of available nutrients. Soil Sci. 73:263–271.

Gallagher, P. W., and W. V. Bartholomew. 1964. Comparison of nitrate production and other procedures in determining nitrogen availability in southeastern coastal plain soils. Agron. J. 56:179–184.

Gasser, J. K. R., and S. J. Kalembasa. 1976. Soil nitrogen. IX. The effects of leys and organic manures on the available-N in clay and sandy soils. J. Soil Sci. 27:237–249.

Geist, J. M. 1977. Nitrogen response relationships of some volcanic ash soils. Soil Sci. Soc. Am. J. 41:996–1000.

Geist, J. M., and J. W. Hazard. 1975. Total nitrogen using a sodium hydroxide index and double sampling theory. Soil Sci. Soc. Am. Proc. 39:340–343.

Geist, J. M., J. O. Reuss, and D. D. Johnson. 1970. Prediction of nitrogen fertilizer requirements of field crops. II. Application of theoretical models to malting barley. Agron. J. 62: 385–389.

Hanway, J., and L. Dumenil. 1955. Predicting nitrogen fertilizer needs of Iowa soils. III. Use of nitrate production, together with other information, as a basis for making nitrogen fertilizer recommendations for corn in Iowa. Soil Sci. Soc. Am. Proc. 19:77–80.

Hanway, J. J., and T. Ozus. 1966. Comparisons of laboratory and greenhouse indexes of nutrient availability in soils. Iowa Res. Bull. 544.

Halvorson, A. D., and G. P. Hartman. 1975. Manure good source of N for beets. Montana Farmer—Stockman 62(9):1–4.

Harmsen, G. W., and G. J. Kolenbrander. 1965. Soil inorganic nitrogen. In W. V. Bartholomew and F. E. Clark (ed.) Agronomy 10:43–92. Am. Soc. of Agron., Madison, Wis.

Harmsen, G. W., and D. A. Van Schreven. 1955. Mineralization of organic nitrogen in soil. Adv. Agron. 7:299–398.

Herlihy, M. 1972. Some soil and cropping factors associated with mineralization and availability of nitrogen in field experiments. Ir. J. Agric. Res. 11:271–279.

Herron, G. M., A. F. Dreier, A. D. Flowerday, W. L. Colville, and R. A. Olson. 1971. Residual mineral N accumulation in soil and its utilization by irrigated corn (Zea mays L.). Agron. J. 63:322–327.

Herron, G. M., G. L. Terman, A. F. Dreier, and R. A. Olson. 1968. Residual nitrate nitrogen in fertilized deep loess-derived soils. Agron. J. 60:477–482.

Hunter, A. S., and J. A. Yungen. 1955. The influence of variations in fertility levels upon the yield and protein content of field corn in eastern Oregon. Soil Sci. Soc. Am. Proc. 19:214–218.

Hipp, B. W., and C. J. Gerard. 1971. Influence of previous crop and nitrogen mineralization on crop response to applied nitrogen. Agron. J. 63:583–586.

James, D. W. 1971. Soil fertility relationships of sugar beets in central Washington: Nitrogen. Washington Agric. Exp. Stn. Tech. Bull. 68.

James, D. W. 1978. Diagnostic soil testing for nitrogen availability. Utah Agric. Exp. Stn. Bull. no. 497.

James, D. W., F. J. Francom, R. F. Wells, and D. V. Sisson. 1977. Control soil fertility for high sugar beet yield and quality. Utah Agric. Exp. Stn. Bull. 496.

James, D. W., A. W. Richards, W. H. Weaver, and R. L. Reeder. 1971. Residual soil nitrate measurement as a basis for managing nitrogen fertilizer practices for sugar beets. J. Am. Soc. Sugar Beet Technol. 16:313–322.

Jenkinson, D. S. 1968. Chemical tests for potentially available nitrogen in soil. J. Sci. Food Agric. 19:160–168.

Jolley, V. D., and W. H. Pierre. 1977. Profile accumulation of fertilizer-derived nitrate and total nitrogen recovery in two long-term nitrogen-rate experiments with corn. Soil Sci. Soc. Am. J. 41:373–378.

Kadirgamathoiyah, S., and A. F. MacKenzie. 1970. A study of soil nitrogen organic fractions and correlation with yield response of Sudan-sorghum hybrid grass on Quebec soils. Plant Soil 33:120–128.

Keeney, D. R. 1965. Identification and estimation of readily mineralizable nitrogen in soils. Ph.D. Thesis. Iowa State Univ. Diss. Abstr. 26:5624–5625.

Keeney, D. R., and J. M. Bremner. 1964. Effect of cultivation on the nitrogen distribution in soils. Soil Sci. Soc. Am. Proc. 28:653–656.

Keeney, D. R., and J. M. Bremner. 1966a. Comparison and evaluation of laboratory methods of obtaining an index of soil nitrogen availability. Agron. J. 58:498–503.

Keeney, D. R., and J. M. Bremner. 1966b. Characterization of mineralizable nitrogen in soils. Soil Sci. Soc. Am. Proc. 30:714–718.

Keeney, D. R., and J. M. Bremner. 1967. Determination and isotope-ratio analysis of different forms of nitrogen in soils. VI. Mineralizable nitrogen. Soil Sci. Soc. Am. Proc. 31:34–39.

Kerbs, L. D., J. P. Jones, W. L. Thiessen, and F. P. Parks. 1973. Correlation of soil test nitrogen with potato yields. Commun. Soil Sci. Plant Anal. 4:269–278.

Lathwell, D. J., H. D. Dubey, and R. H. Fox. 1972. Nitrogen-supplying power of some tropical soils of Puerto Rico and methods for its evaluation. Agron. J. 64:763–766.

Legg, J. O., F. W. Chichester, G. Stanford, and W. H. DeMar. 1971. Incorporation of ^{15}N-tagged mineral nitrogen into stable forms of soil organic nitrogen. Soil Sci. Soc. Am. Proc. 35:273–276.

Leggett, G. E. 1959. Relationships between wheat yield, available moisture, and available nitrogen in eastern Washington dry land areas. Washington Agric. Exp. Stn. Bull. 609.

Lin, C. F., A. H. Chang, and C. C. Tseng. 1973. The nitrogen status and nitrogen supplying power of Taiwan soils. J. Taiwan Agric. Res. 22:186–203.

Livens, J. 1959a. Contribution to a study of mineralizable nitrogen in soil. (In French). Agricultura 7:27–44.

Livens, J. 1959b. Studies concerning ammoniacal and organic soil nitrogen soluble in water. (In French). Agricultura 7:519–532.

Ludwick, A. E., J. O. Reuss, and E. J. Langin. 1976. Soil nitrates following four years continuous corn and as surveyed in irrigated farm fields of central and eastern Colorado. J. Environ. Qual. 5:82–86.

Ludwick, A. E., P. N. Soltanpour, and J. O. Reuss. 1977. Nitrate distribution and variability in irrigated fields of northeastern Colorado. Agron. J. 69:710–713.

MacGregor, J. M., G. R. Blake, and S. D. Evans. 1974. Mineral nitrogen movement into subsoils following continued annual fertilization of corn. Soil Sci. Soc. Am. Proc. 38:110–113.

MacLean, A. A. 1964. Measurement of nitrogen supplying-power of soils by extraction with sodium bicarbonate. Nature (London) 203:1307–1308.

Nelson, L. A., and R. L. Anderson. 1977. Partitioning of soil test-crop response probability. p. 19–37. In T. R. Peck, J. T. Cope, Jr., and D. A. Whitney (ed.) Soil testing: Correlating and interpreting the analytical results. ASA Spec. Publ. 29. Am. Soc. of Agron., Madison, Wis.

Nelson, W. W., and J. M. MacGregor. 1973. Twelve years of continuous corn fertilization with ammonium nitrate or urea nitrogen. Soil Sci. Soc. Am. Proc. 37:583–586.

Nommik, H. 1976. Predicting the nitrogen-supplying power of acid forest soils from data on the release of CO_2 and NH_3 on partial oxidation. Commun. Soil Sci. Plant Anal. 7:569–584.

Nyborg, M., J. Neufeld, and R. A. Bertrand. 1976. Measuring crop available nitrogen. p. 102–117. *In* Proc. Western Canada Nitrogen Symp. Calgary, Alberta, 20–21 Jan. 1976. Alberta Agric., Edmonton, Alberta.

Olsen, R. J., R. F. Hensler, O. J. Attoe, S. A. Witzel, and L. A. Peterson. 1970. Fertilizer nitrogen and crop rotation in relation to movement of nitrate nitrogen through soil profiles. Soil Sci. Soc. Am. Proc. 34:448–452.

Olsen, A. R. S., A. Øien, R. Baerug, and I. Lyngstad. 1974. Studies on soil nitrogen. II. Anaerobic incubation of soil in potassium chloride solution. Acta Agric. Scand. 24:217–221.

Olson, R. A., M. W. Meyer, W. E. Lamke, A. D. Woltemath, and R. E. Weiss. 1960. Nitrate production rate as a soil test for estimating fertilizer nitrogen requirements of cereal crops. Int. Congr. Soil Sci., Trans. 7th (Madison, Wis.) VII:463–470.

Onken, A. B., and H. D. Sunderman. 1972. Applied and residual nitrate-nitrogen effects on irrigated grain sorghum yield. Soil Sci. Soc. Am. Proc. 36:94–97.

Osborne, G. J. 1977. Chemical fractionation of soil nitrogen in six soils from southern New South Wales. Aust. J. Soil Res. 15(2):159–165.

Oyanedel, C., and J. Rodriguez S. 1977. Estimation of N mineralization in soils. Cienc. Invest. Agraria. 4:33–44.

Peterson, L. A., and O. J. Attoe. 1965. Importance of soil nitrates in determination of need for and recovery of fertilizer nitrogen. Agron. J. 57:572–574.

Powers, R. F. 1980. Mineralizable soil nitrogen as an index of nitrogen availability to forest trees. Soil Sci. Soc. Am. J. 44:1314–1320.

Prado, O., and J. Rodriguez. 1978. Nitrogen fertilizer requirement estimates of wheat. Cienc. Invest. Agraria 5:29–40.

Prasad, R. 1965. Determination of potentially available nitrogen in soils—A rapid procedure. Plant Soil 23:261–263.

Quinn, J. G., and M. Salomon. 1966. Hydrolysis of soil nitrogen by strong acids. Nature (London) 211:664–665.

Reuss, J. O., and J. M. Geist. 1970. Prediction of the nitrogen requirements of field crops. I. Theoretical models of nitrogen release. Agron. J. 62:381–384.

Reuss, J. O., and P. S. C. Rao. 1971. Soil nitrate nitrogen levels as an index of nitrogen fertilizer needs of sugarbeets. J. Am. Soc. Sugar Beet Technol. 16:461–470.

Reuss, J. O., P. N. Soltanpour, and A. E. Ludwick. 1977. Sampling distribution of nitrates in irrigated fields. Agron. J. 69:588–592.

Rixon, A. J. 1969. The influence of annual and perennial irrigated pastures on soil fertility as shown by the yield and quality of a subsequent wheat crop. Aust. J. Agric. Res. 20:243–255.

Roberts, S., A. W. Richards, M. G. Day, and W. H. Weaver. 1972. Predicting sugar content and petiole nitrate of sugar beets from soil measurements of nitrate and mineralizable nitrogen. J. Am. Soc. Sugar Beet Technol. 17:126–133.

Robinson, J. B. D. 1967. Anaerobic incubation of soil and the production of ammonium. Nature (London) 214:534.

Robinson, J. B. D. 1968a. A simple available soil nitrogen index. I. Laboratory and greenhouse studies. J. Soil Sci. 19:269–279.

Robinson, J. B. D. 1968b. A simple available soil nitrogen index. II. Field crop evaluation. J. Soil Sci. 19:280–290.

Ryan, J. E., J. L. Sims, and D. E. Peaslee. 1971. Laboratory methods for estimating plant available nitrogen in soil. Agron. J. 63:48–51.

Sahrawat, K. L., and F. N. Ponnamperuma. 1978. Measurement of exchangeable NH_4^+ in tropical rice soils. Soil Sci. Soc. Am. J. 42:282–283.

Saunder, D. H., B. S. Ellis, and A. Hall. 1957. Estimation of available nitrogen for advisory purposes in southern Rhodesia. J. Soil Sci. 8:301–312.

Scarsbrook, C. E. 1965. Nitrogen availability. *In* W. V. Bartholomew and F. E. Clark (ed.) Soil nitrogen. Agronomy 10:481–502. Am. Soc. of Agron., Madison, Wis.

Selmer-Olsen, A. R., A. Øien, R. Baerug, and I. Lyngstad. 1971. Pretreatment and storage of soil samples prior to mineral nitrogen determination. Acta Agric. Scand. 21:57–63.

Shumway, J., and W. A. Atkinson. 1978. Predicting nitrogen fertilizer response in unthinned stands of Douglas-fir. Commun. Soil Sci. Plant Anal. 9:529–539.

Sims, J. L., and B. G. Blackmon. 1967. Predicting nitrogen availability to rice. II. Assessing available nitrogen in silt loams with different previous year crop history. Soil Sci. Soc. Am. Proc. 31:676–680.

Sims, J. L., J. P. Wells, and D. L. Tackett. 1967. Predicting nitrogen availability to rice: I. Comparison of methods for determining available nitrogen to rice from field and reservoir soils. Soil Sci. Soc. Am. Proc. 31:672–675.

Sims, J. R., and G. D. Jackson. 1971. Rapid analysis of soil nitrate with chromotropic acid. Soil Sci. Soc. Am. Proc. 35:603–606.

Smith, C. M. 1977. Interpreting inorganic nitrogen soil tests: Sample depth, soil water, climate, and crops. p. 85–98. *In* T. R. Peck, J. T. Cope, Jr., and D. A. Whitney (ed.) Soil testing: Correlating and interpreting the analytical results. ASA Spec. Publ. 29. Am. Soc. of Agron., Madison, Wis.

Smith, J. A. 1966. An evaluation of nitrogen soil test methods for Ontario Soils. Can. J. Soil Sci. 46:185–194.

Smith, S. J., and G. Stanford. 1971. Evaluation of a chemical index of soil nitrogen availability. Soil Sci. 111:228–232.

Smith, S. J., and L. B. Young. 1975. Distribution of nitrogen forms in virgin and cultivated soils. Soil Sci. 120:354–360.

Smith, S. J., L. B. Young, and G. E. Miller. 1977. Evaluation of soil nitrogen mineralization potential under modified field conditions. Soil Sci. Soc. Am. J. 41:74–76.

Soper, R. J., and P. M. Huang. 1963. The effect of nitrate nitrogen in the soil profile on the response of barley to fertilizer nitrogen. Can. J. Soil Sci. 4:350–358.

Soper, R. J., G. J. Racz, and P. I. Fehr. 1971. Nitrate nitrogen in the soil as a means of predicting the fertilizer nitrogen requirements of barley. Can. J. Soil Sci. 51:45–49.

Spencer, W. F., A. J. MacKenzie, and F. G. Viets, Jr. 1966. The relationship between soil tests for available nitrogen and nitrogen uptake by various irrigated crops in the western states. Soil Sci. Soc. Am. Proc. 30:480–485.

Stanford, G. 1963. Sugarcane quality and nitrogen fertilization. Hawaii. Plant. Rec. 56:289–333.

Stanford, G. 1968a. Effect of partial removal of soil organic nitrogen with sodium pyrophosphate or sulfuric acid solutions on subsequent mineralization of nitrogen. Soil Sci. Soc. Am. Proc. 32:679–682.

Stanford, G. 1968b. Extractable organic nitrogen and nitrogen mineralization in soils. Soil Sci. 106:345–351.

Stanford, G. 1973. Rationale for optimum nitrogen fertilization in corn production. J. Environ. Qual. 2:159–166.

Stanford, G. 1978. Evaluation of ammonium release by alkaline permanganate extraction as an index of soil nitrogen availability. Soil Sci. 126:244–253.

Stanford, G., and A. S. Ayres. 1964. The internal nitrogen requirement of sugarcane. Soil Sci. 98:338–344.

Stanford, G., A. S. Ayres, and M. Doi. 1965. Mineralizable soil nitrogen in relation to fertilizer needs of sugarcane in Hawaii. Soil Sci. 99:132–137.

Stanford, G., J. N. Carter, and S. J. Smith. 1974. Estimates of potentially mineralizable soil nitrogen based on short-term incubations. Soil Sci. Soc. Am. Proc. 38:99–102.

Stanford, G., J. N. Carter, D. T. Westermann, and J. J. Meisinger. 1977. Residual nitrate and mineralizable soil nitrogen in relation to nitrogen uptake by irrigated sugar beets. Agron. J. 69:303–308.

Stanford, G., and W. H. DeMar. 1969. Extraction of soil organic nitrogen by autoclaving in water. I. The NaOH-distillable fraction as an index of nitrogen availability in soils. Soil Sci. 107:203–205.

Stanford, G., and W. H. DeMar. 1970. Extraction of soil organic nitrogen by autoclaving in water. II. Diffusible ammonia, an index of soil nitrogen availability. Soil Sci. 109:190–196.

Stanford, G., and E. Epstein. 1974. Nitrogen mineralization water relations in soils. Soil Sci. Soc. Am. Proc. 38:103–106.

Stanford, G., and J. Hanway. 1955. Predicting nitrogen fertilizer needs of Iowa soils. II. A simplified technique for determining relative nitrate production in soils. Soil Sci. Soc. Am. Proc. 19:74–77.

Stanford, G., and J. O. Legg. 1968. Correlation of soil nitrogen availability indexes with nitrogen uptake by plants. Soil Sci. 105:320–326.

Stanford, G., and S. J. Smith. 1972. Nitrogen mineralization potentials of soils. Soil Sci. Soc. Am. Proc. 36:465–472.

Stanford, G., and S. J. Smith. 1976. Estimating potentially mineralizable soil nitrogen from a chemical index of soil nitrogen availability. Soil Sci. 122:71–76.

Stanford, G., and S. J. Smith. 1978. Oxidative release of potentially mineralizable soil nitrogen by acid permanganate extraction. Soil Sci. 126:210–218.

Stanford, G., M. H. Frere, and D. H. Schwaninger. 1973a. Temperature coefficient of soil nitrogen mineralization. Soil Sci. 115:321–323.

Stanford, G., J. O. Legg, and S. J. Smith. 1973b. Soil nitrogen availability evaluations based on nitrogen mineralization potentials of soils and uptake of labeled and unlabeled nitrogen by plants. Plant Soil 39:113–124.

Stanford, G., J. O. Legg, S. Dzienia, and E. C. Simpson, Jr. 1975a. Denitrification and associated nitrogen transformations in soils. Soil Sci. 120:147–152.

Stanford, G., J. O. Legg, and T. E. Staley. 1975b. Fate of ^{15}N-labeled nitrate in soils under anaerobic conditions. p. 667–673. *In* E. R. Klein and P. D. Klein (ed.) Proc. 2nd Int. Conf. on Stable Isotopes, 20–23 Oct. 1975, Oak Brook, Ill. Sponsored by Div. of Biol. and Med. Res., Argonne Natl. Lab., Argonne, Ill. U.S. Energy and Dev. Admin.

Stewart, R. A., F. G. Viets, Jr., G. L. Hutchinson, W. D. Kemper, F. E. Clark, M. L. Fairbourne, and F. Strauch. 1967. Distribution of nitrates and other pollutants under fields and corrals in the middle South Platte Valley of Colorado. USDA-ARS 41-134.

Stewart, B. A., D. A. Woolhiser, W. H. Wischmeier, J. H. Caro, and M. H. Frere. 1975. Control of water pollution from cropland, Vol. 1. USDA Report no. ARS-H-5-1.

Subbiah, B. V., and G. L. Asija. 1956. A rapid procedure for estimation of available nitrogen in soils. Curr. Sci. 25:259–260.

Subbiah, B. V., and J. C. Bajaj. 1962. A soil test procedure for assessment of available nitrogen in rice soils. Curr. Sci. 31:196.

Synghal, K. N., J. A. Toogood, and C. F. Bentley. 1959. Assessing nitrogen requirements of some Alberta soils. Can. J. Soil Sci. 39:120–128.

Tamhane, R. V., and B. V. Subbiah. 1962. Correlation of soil tests with pot and field trials in the evaluation of soil fertility. Soil Sci. Plant Nutri. (Tokyo) 8:5–14.

U.S. Department of Agriculture. 1966. Consumption of commercial fertilizers and primary plant nutrients. Statistical Bull. no 375. Statistical Reporting Service and Crop Reporting Board, Washington, D.C.

U.S. Department of Agriculture. 1977. The 1978 fertilizer situation. Publication FS-8. Econ. Res. Serv., Washington, D.C.

van der Paauw, F. 1962. Effect of winter rainfall on the amount of nitrogen available to crops. Plant Soil 16:361–380.

van der Paauw, F. 1963. Residual effect of nitrogen fertilizer on succeeding crops in a moderate marine climate. Plant Soil 19:324–331.

Verstraeten, L. M. J., K. Vlassak, and J. Livens. 1970a. Factors affecting the determination of available soil nitrogen by chemical methods. I. Comparison of extractable with mineralized nitrogen. Soil Sci. 110:299–305.

Verstraeten, L. M. J., K. Vlassak, and J. Livens. 1970b. Factors affecting the determination of available soil nitrogen by chemical methods—comparison of extractable carbon with mineralized nitrogen. Soil Sci. 110:365–370.

Vlassak, K. 1970. Total soil nitrogen and nitrogen mineralization. Plant Soil 32:27–32.

Waring, S. A., and J. M. Bremner. 1964. Ammonium production in soil under waterlogged conditions as an index of nitrogen availability. Nature (London) 201:951–952.

Westfall, D. G., M. G. Barnes, and N. E. Pence. 1977. Interpretation and practical field utilization of the relationships between brei nitrate and sugar content of sugar beets. J. Am. Soc. Sugar Beet Technol. 19:307–315.

Westfall, D. G., M. A. Henson, and E. P. Evans. 1978. The effect of soil sample handling between collection and drying on nitrate concentration. Commun. Soil Sci. Plant Anal. 9: 169–185.

White, W. C., L. Dumenil, and J. Pesek. 1958. Evaluation of residual nitrogen in soils. Agron. J. 50:255–259.

White, W. C., and J. Pesek. 1959. Nature of residual nitrogen in Iowa soils. Soil Sci. Soc. Am. Proc. 23:39–42.

Young, R. A., J. L. Ozbun, A. Bauer, and E. H. Vasey. 1967. Yield response of spring wheat and barley to nitrogen fertilizer in relation to soil and climatic factors. Soil Sci. Soc. Am. Proc. 31:407–410.

18 The Effects of Pesticides on Nitrogen Transformations in Soils

C. A. I. GORING AND D. A. LASKOWSKI

The Dow Chemical Company U.S.A.
Midland, Michigan

I. INTRODUCTION

Prior to the development of technologically advanced food and fiber production systems, soils had evolved to natural states of equilibrium with their associated plant, microbial, and animal populations. Each soil was a unique expression of evolutionary events involving the interaction of climate, topography, organisms, and parent material. Additional change occurred at a very slow rate.

Modern crop production techniques have abruptly altered the status of agricultural soils. Tillage, mineral fertilizers, soil amendments, crops, irrigation, pesticides, and harvest practices have had a substantial impact. Some of these inputs have accelerated the rates of long-term change, and there is concern that the changes are detrimental to biological populations.

Alterations in soils caused by crop production are not necessarily undesirable. Intensive agriculture produces more food and feed and uses less land. It also may be indirectly beneficial to many types of animals that share in the increased bounty. On the other hand, some of the faunal populations are frequently damaged or destroyed when soils are converted to agricultural uses. The same is true for some of the plant populations. Whether or not this is undesirable depends on whether the organisms destroyed are pests that reduce yield or beneficial organisms essential for good soil fertility.

This is the framework within which this chapter examines the influence of pesticides on N transformations. Do pesticides have an impact on the microorganisms responsible for N transformations in soil, and if so, are the effects short term, long term, or both, and are they desirable, undesirable, or immaterial?

II. SOIL VARIABILITY

It is important to appreciate the wide range of natural soil variations in order to judge the significance of the impact of pesticides on soil behavior.

Soils are highly variable spatially, physically, chemically, and biologically. Unique profiles occur that differ greatly in pH, texture, structure, and organic matter content. The horizons within each profile may vary greatly in the amount and type of minerals and organic residues, in the arrangements and groupings of particles, and in pH.

Soil organic matter may vary considerably in composition and stability. Carbon/nitrogen ratios, functional group content, rates of formation and degradation, and relative amounts of individual constituents all can vary from soil to soil and within the same soil under different cropping conditions.

Microbiological populations and processes in soils are also highly variable, including the organisms and processes responsible for N transformations.

Rates of N mineralization and immobilization may vary three- to five-fold over the range of moisture contents that ordinarily occur in soils. The processes occur much more rapidly at high than at low temperatures and cease when the soil is frozen. As much as a 25-fold difference in the rate of hydrolysis of urea has been observed.

Rates of nitrification are equally variable. The process is much more rapid at high than at low pH's and oxygen tensions. Little or no nitrification takes place in dry or extremely wet soils. Optimum moisture is between one-half and two-thirds moisture-holding capacity. From 2 to 35°C, rates may increase as much as 50- to 100-fold.

Rates of denitrification are profoundly influenced by temperature, moisture, oxygen status, organic matter, and other factors and under some conditions can be extremely rapid, with complete loss of nitrate in 2 to 3 days. Rates can increase as much as 100-fold in soils at 450% water-holding capacity (WHC) compared with soils at 50% WHC and in soils at 30°C compared with soils that are just above freezing.

The numbers per gram of soil of the symbiotic N_2-fixing organisms (*Rhizobium*) fluctuate widely, ranging from as little as one to more than a million. Survival of the organisms is highly dependent on many soil, crop, and climatic factors. They are sensitive to very acid conditions. They survive low temperatures but are easily killed at 40°C and above. They also die rapidly as the soil dries, although some cells will escape death and go into a resting condition. Leguminous roots stimulate the organisms to multiply rapidly. Most spectacular results are observed in soils with naturally low populations that are inoculated with a strain well suited to the plant being grown.

Nonsymbiotic N_2 fixation is accomplished by many kinds of microorganisms in soil, including aerobic, facultative, anaerobic, and photosynthetic bacteria and blue-green algae. The process occurs under a wide range of environmental conditions in many different soil and water habitats. Fixation may vary from insignificant to very substantial amounts of N ha^{-1} year^{-1}. It is generally favored by higher temperatures and is greatest under moist tropical conditions where there is a large turnover of soil organic matter.

III. BEHAVIOR OF PESTICIDES

If a pesticide is to have any significant short-term effect on N transformations in soil, it must impact at normally used rates on the organisms and processes involved and must move via leaching or volatilization through a significant portion of the soil profile. If a pesticide is to have any significant long-term impact, it must accumulate in the soil at equilibrium concentrations sufficiently high to affect N transformations. Since the actions of pesticides on N transformations are inextricably linked with their behavior in soils, it is important that the relationships between their properties and their behavior be appreciated.

Some principles of pesticide behavior in soil have emerged from the myriad of investigations that have been conducted. Sorption of pesticides is, for the most part, directly related to the organic matter content of the soil and the hydrophobicity of the pesticide (Hamaker & Thompson, 1972). Sorption coefficients between soil water and soil organic C (K_{oc}) vary from < 1 to > 200,000 (Kenaga & Goring, 1980). Most pesticides have coefficients < 20,000. A few have coefficients > 20,000 because of their extremely low water solubility. DDT, the pesticide having the lowest water solubility, has a coefficient of 238,000. Carboxylic acid pesticides have the highest water solubilities and lowest sorption coefficients, ranging downward from about 50 for 2,4,5-T to < 1 for dicamba.

The potential leachability of a pesticide is inversely related to its sorption coefficient. Pesticides with K_{oc}'s < 50 can be considered readily leachable; those > 20,000 can be considered nonleachable. Even readily leachable pesticides are rarely leached from the soil profile because of the combined effects of decomposition and evapotranspiration (Hamaker, 1975).

The relative tendency of pesticides to volatilize from water can be estimated from their distribution coefficients between air and water (K_w). These are calculated from Henry's Law and vary from about 2 to over 2 $\times$ 10^8 at 20°C (Goring, 1967; Hamaker, 1972b). Diffusion through air and water in soil is equally important at about a K_w of 20,000. Diffusion of pesticides through soil air is generally considered to be significant when K_w's fall < 20,000. Soil fumigants have K_w's ranging from as low as 1.8 for carbon disulfide to as high as 163.8 for 1,2-dibromo-3-chloro propane. Chlorinated hydrocarbon insecticides are surprisingly volatile because of their low water solubilities relative to their vapor pressures. DDT has a K_w of about 303. Many pesticides have K_w's < 20,000.

The relative diffusibility of a pesticide through soil is a function of its K_{oc} and K_w values and can be calculated for any soil if these values and the organic C and water content of the soil are known (Goring, 1972). The initial volatilization of a pesticide from the soil surface is directly related to its vapor pressure, and longer term volatilization is directly related to its diffusability through soil (McCall & Swann, 1978). Pesticides with high vapor pressures, low water solubilities, or both are most likely to diffuse through the air phase and be rapidly lost from soil as a result of volatilization.

Rates of degradation of pesticides in soil are a function of temperature and moisture (Goring et al., 1975; Hamaker, 1972a), degree of sorption (Briggs, 1976), extent of occlusion in the soil organic matrix (Goring et al., 1975; Hamaker & Goring, 1975), in some instances, initial concentration (Goring et al., 1975; Hamaker, 1972a), and intrinsic capacity of the pesticide to resist degradation (Briggs, 1976; Crosby, 1970; Kaufman, 1970; Kearney & Plimmer, 1970). Equations that will accurately express rates of degradation can be very complex. First-order kinetics are ordinarily used as a simple approximation.

Rates vary widely with half-lives ranging from <2 weeks to >6 months and, in a few instances, 1 to 5 years (Goring et al., 1975). Most pesticides have half-lives of <6 months. Pesticides that have half-lives >6 months, such as the chlorinated hydrocarbon insecticides, are usually strongly sorbed and recalcitrant to degradation.

Finally, the impact of pesticides on organisms in soil appears to be primarily related to the concentration of the pesticide in the soil water. This would certainly be true for soil microorganisms.

Significant accumulation in soil from yearly applications of most pesticides will not occur since half-lives for degradation are usually <6 months (Goring et al., 1975; Hamaker, 1966), and for many there are substantial losses from volatilization. For these pesticides any impact on N transformations is likely to be short term.

A few pesticides, specifically the chlorinated hydrocarbon insecticides that have half-lives ranging from about 1 to 10 years, may accumulate in soil to equilibrium levels in the range of 2 to 15 times the annual rate of application (Hamaker, 1966). These pesticides could have significant long-term impacts if the rates of application required to affect N transformations are <2 to 15 times annual use rates.

IV. EFFECTS OF PESTICIDES ON N TRANSFORMATIONS

Data on the effects of pesticides on N transformations are summarized in Tables 1 through 5. Common names are given for the pesticides. Trade names and structures are readily available (Kenaga & Morgan, 1978; Mullison et al., 1979; Worthing, 1979). Recommended rates of application per year were gleaned from several sources (Martin, 1972; Martin & Worthing, 1974, 1977; Mullison et al., 1979; Worthing, 1979). Rates of application required for >25% inhibition were also estimated. The data for these estimates was obtained from both "in vivo" soil studies and "in vitro" studies in other types of media. The estimates are probably low because they do not take into account the potential for reduced pesticide impact as a result of uneven distribution, interception by foliage, and split applications. Furthermore, estimates made from in vitro studies do not take into account the potential for reduced pesticide impact as a result of sorption by soil solids. Frequently, the studies conducted did not involve enough rates of application to permit estimation of the margin between recommended rates and rates required for at least 25% inhibition.

A. Mineralization/Immobilization

Virtually no information is available on the effects of pesticides on N immobilization in soil. Table 1 summarizes the data on N mineralization. For most of the pesticides, >25% inhibition was not observed for any of the rates tested, even though they were, for the most part, substantially greater than the recommended rates of application. In some instances, the rates evaluated were in the same range as the recommended rates of application.

Some pesticides inhibited the hydrolysis of urea at relatively low rates of application (chloranil, diuron, fenuron, linuron, monuron, and phenyl mercuric acetate). Bundy and Bremner (1973a) have shown that a variety of quinones are good inhibitors of urease activity. The urea herbicides could possibly be metabolic substitutes for urea that interact readily with urease but are not easily degraded by the enzyme. The activity of phenyl mercuric acetate can probably be attributed to its capacity to react with a wide variety of enzymes in a nonspecific way.

A considerable number of insecticides (carbofuran, DDT, fenitrothion, fensulfothion, gamma HCH, heptachlor, isobenzan, and terbuphos), fumigants and biocides (chloropicrin, dazomet, and DD), fungicides (benomyl), and herbicides (aminotriazole, chlorthiamid, glyphosate, lenacil, metribuzin, MSMA, and terbacil) were reported to stimulate mineralization of N. Stimulation by fumigants and biocides has usually been attributed to destruction of a part of the soil population and release of mineral N from the decomposing tissues (Jenkinson & Powlson, 1970). Insecticides probably act in the same way.

Mineral N may also be released from N-containing herbicides, such as aminotriazole, during their decomposition, but the reason for increased mineralization in the presence of the herbicide MSMA is unknown. Increased mineralization in the presence of the fungicide benomyl may have been caused by release of N during its decomposition or could be due to its toxicity to earthworms.

Increased N mineralization frequently occurs at recommended rates of application of insecticides, fumigants, and biocides but at rates much higher than those recommended for herbicides.

B. Nitrification

Reports of stimulation of nitrification by pesticides are rare. Most pesticides either inhibit nitrification or have no effect. Investigators generally study either the disappearance of NH_4^+ or the appearance of NO_3^-. In rare instances, the appearance and disappearance of NO_2^- are also studied.

Table 2 summarizes the data on inhibition of nitrification. Greater than 25% inhibition was not observed at any of the rates tested for a considerable number of pesticides. The rates evaluated were usually much greater than recommended rates of application but in some instances were in the same range.

For another group of pesticides, minimal rates required for >25% inhibition could be estimated. These rates were generally much greater than recommended rates but not in all instances. For example, fumigants and biocides (such as allyl alcohol, carbon disulfide, chloropicrin, dazomet, DBCP, DD, ethylene dibromide, metham-sodium, and methyl bromide) inhibit nitrification at rates at or below recommended rates. Carbon disulfide is particularly active and inhibits nitrification at rates far below recommended rates.

The dithiocarbamate fungicides (ferbam, maneb, nabam, zineb, and ziram) appear to inhibit nitrification at rates near the upper end of recommended rates probably because they release carbon disulfide when they decompose in soil. It is unlikely that the repeated low rates of application typical of their use as foliage fungicides will significantly inhibit nitrification, but a massive single application might cause inhibition. Phenyl mercuric acetate has long been recognized as a highly active antimicrobial compound and has been shown by Bundy and Bremner (1973b) to be a moderately active nitrification inhibitor. The fungicide etridiazole is a potent inhibitor similar in activity to nitrapyrin (Huber, 1980).

The activity of the herbicide aminotriazole (1H-1,2,4-triazol-3-amine) was not surprising since it is a close analogue of the nitrification inhibitor 4-amino-1,2,4-triazole, commonly called ATC. The high activity of diallate was unexpected.

Some herbicides, such as barban, methabenzthiazuron, metobromuron, monolinuron, and simazine, appear to inhibit conversion of NO_2^- to NO_3^- at rates that are much lower than the rates required to inhibit conversion of NH_4^+ to NO_2^- and in the same range as the rates normally recommended for weed control.

C. Denitrification

Relatively few pesticides have been studied for inhibition of denitrification. Table 3 summarizes the data. Most of the chemicals evaluated did not inhibit denitrification even at rates much greater than recommended rates. Notable exceptions were phenyl mercuric acetate and metham-sodium, which inhibited denitrification at rates equal to or substantially below recommended rates.

Nitrapyrin was reported to inhibit denitrification at 50 ppm in a sand/bark medium or a solution culture of *Pseudomonas*. It did not inhibit denitrification at 50 ppm in soil. This rate of application is far above the recommended rate of nitrapyrin for inhibition of nitrification.

D. Symbiotic N₂ Fixation

Table 4 summarizes the data on the effects of pesticides on inhibition of growth of *Rhizobium* or on nodulation. Inhibition was not observed for most of the pesticides at any of the rates tested, but as before, these rates for some of the pesticides were no higher than those ordinarily recommended.

A wide variety of herbicides were low in toxicity to *Rhizobium* in agar or liquid culture. Some examples are chlorpropham, 2,4-D, dalapon, 2,4-DB, fluometuron, linuron, maleic hydrazide, and simazine. However, nodulation was inhibited at low rates of application by many of these herbicides. Included were alloxydim-sodium, chloramben, chlorpropham, linuron, nitralin, prometryne, simazine, trifluralin, and vernolate. Such inhibition was attributable to the deleterious effect of the chemicals on the growth and development of the plant root system and not to any direct impact on *Rhizobium* (Greaves et al., 1978; Dunigan et al., 1972; Kapusta & Rouwenhorst, 1973; Misra & Gaur, 1974). It is likely that any chemical applied to soil at dosages that are toxic to plant roots will inhibit nodulation even though the chemical may not be directly toxic to *Rhizobium*. For example, the nitrification inhibitor nitrapyrin did not reduce survival of *Rhizobium meliloti* on alfalfa seeds or in soil at dosages that caused substantial changes in the morphology of nodules and root tips (McKell & Whalley, 1964). The fungicide fenaminosulf and the insecticide gamma HCH both inhibited nodulation at rates of application that damaged plant roots.

Some biocides were reported to be toxic to *Rhizobium* at recommended rates of application. Examples were chloropicrin applied to soil (Ebbels, 1967) and phenyl mercuric acetate applied to seed (Jakubisiak & Golebiowska, 1963). A variety of other fungicides and insecticides used for seed treatment, including captan, chloranil, chlorpyrifos, dieldrin, endrin, gamma HCH, isobenzan, and thiram, were not toxic to *Rhizobium* at recommended rates of application.

E. Nonsymbiotic N₂ Fixation

The data on the effects of pesticides on nonsymbiotic N_2 fixation and N_2-fixing organisms are summarized in Table 5. With some notable exceptions, high application rates were required for inhibition. However, dinoseb appeared to be inhibitory to nonsymbiotic fixation at low rates. A variety of herbicides, including EPTC, molinate, monuron, propanil and trifluralin, were highly inhibitory to blue-green algae in liquid or sand culture at rates of application similar to recommended rates. The herbicide siduron was toxic to *Azotobacter* in soil at low concentrations.

V. AGRONOMIC IMPLICATIONS

It is clear from the foregoing studies that even though many pesticides applied at recommended rates will not affect N transformations on a short-term basis, some will, and the agronomic implications of their impact will have to be taken into consideration.

Insecticides, fumigants and biocides enhance mineralization apparently because they destroy a part of the soil population that then decays to release mineral N. Repeated treatment with such pesticides apparently does not

(text continued on page 711)

Table 1—Inhibition of N mineralization by pesticides.

Pesticide	Recom- mended rates of applica- tion/year	Estimated rates of application for >25% inhibition[†]	Substrate	References
	——— kg/ha ———			
		Fungicides		
Anilazine	2–10	25–1,000	Soil	Dubey & Rodriguez (1970) Mazur and Hughes (1975)
Benomyl	0.25–1	>75[‡]	Soil	Mazur & Hughes (1975)
Chloranil	<2	<50	Urea in soil	Bundy & Bremner (1973a)
Maneb	0.5–15	25–1,000	Soil	Dubey & Rodriguez (1970) Mazur & Hughes (1975)
Phenyl mercuric acetate	0.2–20	10–100	Urea in soil	Bremner & Douglas (1971, 1973)
		10–100	Soil	van Faassen (1973)
Quintozene	1–20	>1,000	Soil	Caseley & Broadbent (1968)
		Herbicides		
Alachlor	2–3	>3.4	Alfalfa in soil	Lewis et al. (1978)
Aminotriazole	2–10	>87.5[‡]	Soil	van Schreven et al. (1970)
Bensulide	2.5–20	>6.7	Alfalfa in soil	Lewis et al. (1978)
Bentazone	1–2	>100	Soil	Marsh et al. (1978)
Cacodylic acid	1–20	>1,000	Peptone in soil	Bollen et al. (1977)
Chloramben	0.5–4	>3.4	Alfalfa in soil	Lewis et al. (1978)
Chloroxuron	6–14	>100	Soil	Odu & Horsfall (1971)
Chlorthal-dimethyl	6–14	>11.8	Alfalfa in soil	Lewis et al. (1978)
Chlorthiamid	5–14	>100[‡]	Soil	Davies & Marsh (1977) Marsh & Davies (1978)
Dalapon	1–20	>127.5	Soil	van Schreven et al. (1970)
Dichlorprop	1–2	>25.5	Soil	van Schreven et al. (1970)
Dinitramine	0.4–0.8	>0.6	Alfalfa in soil	Lewis et al. (1978)
Dinoseb	0.75–12	>3.4	Alfalfa in soil	Lewis et al. (1978)
Diuron	0.5–30	5–10	Urea in soil	Cerville et al. (1977)
		>2.2	Alfalfa in soil	Lewis et al. (1978)
EPTC	3–6	>2.2	Alfalfa in soil	Lewis et al. (1978)
Fenuron	2–30	5–10	Urea in soil	Cerville et al. (1977)
Fluometuron	1–2	>2.2	Alfalfa in soil	Lewis et al. (1978)
Glyphosate	0.7–5.6	>100[‡]	Soil	Marsh et al. (1977)
Ioxynil	0.25–1	>13.2	Soil	van Schreven et al. (1970)
Lenacil	0.5–2	>100[‡]	Soil	Marsh & Davies (1978)
Linuron	0.5–2.8	5–10	Urea in soil	Cerville et al. (1977)
		>2.8	Alfalfa in soil	Lewis et al. (1978)
		>133	Soil	Marsh et al. (1974); Grossbard (1971)
MCPA	0.25–2.5	>6.7	Soil	Grossbard (1971)
Mecoprop	1.5–3	>10	Soil	van Schreven et al. (1970)
Metribuzin	0.25–0.75	>2.2	Alfalfa in soil	Lewis et al. (1978)
		>100[‡]	Soil	Marsh et al. (1977)
Monuron	5–30	5–10	Urea in soil	Cerville et al. (1977)
MSMA	2–3	>1,000[‡]	Peptone in soil	Bollen et al. (1977)
Naptalam	0.5–7	>6.7	Alfalfa in soil	Lewis et al. (1978)
Nitralin	0.25–1	>10	Urea in soil	Lewis et al. (1978)
Paraquat	0.5–3	>25	Peptone in soil	Tu & Bollen (1968a, b)
Picloram	0.02–3	>1,000	Urea in soil	Goring et al. (1967)
		>10	Peptone in soil	Tu & Bollen (1969)
		>4.8	Soil	van Schreven et al. (1970)

(continued on next page)

Table 1—Continued.

Pesticide	Recom-mended rates of applica-tion/year	Estimated rates of application for >25% inhibition[†]	Substrate	References
	——— kg/ha ———			
Profluralin	0.75–1.5	>1.7	Alfalfa in soil	Lewis et al. (1978)
Prometryne	0.5–1.5	>3.4	Alfalfa in soil	Lewis et al. (1978)
Pyrazon	1–4	>100	Soil	Davies & Marsh (1977)
Simazine	0.5–20	>133	Soil	Marsh et al. (1974); Grossbard (1971)
2,4,5-T	0.25–8	>100	Soil	Marsh & Davies (1978)
Terbacil	0.5–6	>100[‡]	Soil	Marsh & Davies (1978)
Triallate	1–3	>6.7	Soil	Grossbard (1971)
Trifluralin	0.5–4	>100[‡]	Soil	Davies & Marsh (1977)
		>1.4	Peptone in soil	Lewis et al. (1978)
Insecticides				
Aldrin	0.5–5	200–1,000	Soil	Eno (1958)
Carbofuran	0.25–10	>500[‡]	Soil	Ross (1974)
Chlordane	1–10	>200	Soil	Eno (1958)
DDT	0.5–30	>500[‡]	Peptone in soil	Bollen et al. (1954); Eno (1958); Ross (1974); Wilson & Choudri (1946)
Demeton	<1	>5	Soil	Eno (1958)
Dieldrin	0.5–5	>2.5	Soil	Eno (1958)
Fenitrothion	1–3	>500[‡]	Soil	Ross (1974)
Fensulfothion	1–5	>500[‡]	Soil	Ross (1974)
Gamma HCH	0.2–2	>1,000[‡]	Peptone in soil	Bollen et al. (1954); Eno (1958)
		>50[‡]	Soil	Jaiswal (1967); Jaiswal et al. (1971); Raghu & McRae (1967)
Heptachlor	0.2–0.5	>9[‡]	Soil	Eno (1958); Jaiswal (1967)
Isobenzan	0.2–1	>6.6[‡]	Soil	Jaiswal (1967); Jaiswal et al. (1971)
Terbuphos	1	>100[‡]	Soil	Laveglia & Dahm (1974)
Nitrification inhibitors				
Nitrapyrin	0.25–1	>10	Urea in soil	Bremner & Bundy (1976); Bundy & Bremner (1974a); Goring (1962)
Soil Fumigants				
Chloropicrin	200–400	>220[‡]	Soil	Rovira (1976)
Dazomet	400	>400[‡]	Soil	Ebbels (1971)
DD (1,3-D)	100–800	>400[‡]	Soil	Ebbels (1971); Lebbink & Kolenbrander (1974)
Methan-sodium	100–400	>300[‡]	Soil	Lebbink & Kolenbrander (1974)
Methyl bromide	400–800	>220	Soil	Rovira (1976)

† 1 ppm or 1 pound/acre was assumed to be equivalent to approximately 1 kg/ha.
‡ Apparent stimulation of N mineralization at or below the indicated dosages.

Table 2—Inhibition of nitrification by pesticides.

Pesticide	Recommended rates of application/year	Estimated rates of application for >25% inhibition in soil† Ammonium oxidation (Nitrite oxidation)	References
		——————— kg/ha ———————	
		Fungicides	
Anilazine	2–10	100–250 (15–60)	Dubey & Rodriguez (1970); Mazur & Hughes (1975)
Benomyl	0.25–1	10–100	Foster & McQueen (1977); Gowda et al. (1977); Mazur & Hughes (1975)
Captafol	2–10	>10	Atlas et al. (1978)
Captan	2–10	50–100	Agnihotri (1971); Wainwright & Pugh (1973)
Dichloran	2–10	10–100	Caseley & Broadbent (1968)
Etridiazole	4.5–9	<25	Turner (1979)
Fenaminosulf	0.1–10	10–20	Agnihotri (1973)
Fentin acetate	0.25–0.5	>10	Barnes et al. (1971); Barnes et al. (1973)
Ferbam	1–15	4–40	Jaques et al. (1959)
Folpet	0.5–2	>10	Atlas et al. (1978)
Maneb	0.5–15	15–25	Dubey & Rodriguez (1970); Jaques et al. (1959); Mazur & Hughes (1975)
Nabam	25–100	<100	Chandra & Bollen (1961)
Phenyl mercuric acetate	0.2–20	10–50	Bundy & Bremner (1973b); Bundy & Bremner (1974b); van Faassen (1973); Gaur & Misra (1977)
Quintozene	1–20	>1,000	Caseley & Broadbent (1968)
Thiram	2–5	25–100	Radwan (1965); Wainwright & Pugh (1973)
Zineb	1–15	2.5–25	Jaques et al. (1959)
Ziram	1–15	2.5–25	Jaques et al. (1959)
		Herbicides	
Alachlor	2–3	>12	Thorneburg & Tweedy (1973)
Allidochlor	4–6	>8	Otten et al. (1957)
Allyl alcohol	100–200	100–200	Höflich (1968); Koike (1961)
Ametryne	2–4	100 (10–100)	Dubey (1969)
Aminotriazole	2–10	2–10	Domsch & Paul (1974); Otten et al. (1957); van Schreven et al. (1970)
Asulam	1–4	>20	Ratnayake & Audus (1978)
Atrazine	1–4	>250	Bartha et al. (1967); Domsch & Paul (1974); Eno (1962); Thorneburg & Tweedy (1973)
Barban	0.5–2	100–250 (1–2)	Bartha et al. (1967); Domsch & Paul (1974)
Bensulide	2.5–20	>6.7	Lewis et al. (1978)
Bentazone	1–2	50–100	Marsh et al. (1978)
Bifenox	1–2	15–30	Turner (1979)
Bromacil	2–15	>100	Horowitz et al. (1974); Pancholy & Lynd (1969)
Bromoxynil	0.5–1	50–100	Debona & Audus (1970); Ratnayake & Audus (1978)

(continued on next page)

Table 2—Continued.

Pesticide	Recommended rates of application/year	Estimated rates of application for >25% inhibition in soil† Ammonium oxidation (Nitrite oxidation)	References
		kg/ha	
Bromoxynil octanoate	0.5-1	>38	Ratnayake & Audus (1978)
Buturon	0.5-1.5	>5	Domsch & Paul (1974)
Cacodylic acid	1-20	1,000-2,000	Bollen et al. (1977)
Carbetamide	2	>40	Ratnayake & Audus (1978)
Chloramben	0.5-4	>3.4	Lewis et al. (1978)
Chlorfenac	2-20	(200-300)	Mayeux & Colmer (1962)
Chloroxuron	1-6	>500	Domsch & Paul (1974); Grossbard & Marsh (1974); Odu & Horsfall (1971)
Chlorpropham	2-4	20-40	Bartha et al. (1967); Domsch & Paul (1974); Hale et al. (1957); Otten et al. (1957); Teater et al. (1958)
Chlorthal-dimethyl	6-14	>11.8	Lewis et al. (1978)
Chlorthiamid	5-40	250-500	Debona & Audus (1970); Domsch & Paul (1974); Marsh & Davies (1978)
2,4-D	0.25-4	25-100	Domsch & Paul (1974); Koike & Gainey (1952); Shaw & Robinson (1960); Smith et al. (1946); Teater et al. (1958)
Dalapon	1-20	50-100 (>700)	Davies & Marsh (1977); Domsch & Paul (1974); Mayeux & Colmer (1962); Otten et al. (1957); van Schreven et al. (1970)
Desmetryne	0.5-1.5	>0.5	Domsch & Paul (1974)
Diallate	1.5-4	2.5-5	Domsch & Paul (1974)
Dicamba	0.1-8	50-100	Ratnayake & Audus (1978)
Dichlobenil	2.5-20	>1,000	Debona & Audus (1970); Domsch & Paul (1974)
Dichlorprop	1-2	>25.5	van Schreven et al. (1970)
Difenzoquat	0.5	>5	Atlas et al. (1978)
Dinitramine	0.4-0.8	>0.6	Lewis et al. (1978)
Dinoseb	0.75-12	>5	Domsch & Paul (1974); Lewis et al. (1978)
Diphenamid	4-6	>250	Bartha et al. (1967)
Diquat	0.5-2	>5	Domsch & Paul (1974)
Diuron	0.5-30	20-500 (<150)	Bartha et al. (1967); Corke & Thompson (1970); Domsch & Paul (1974); Dubey (1969); Grossbard & Marsh (1974); Horowitz et al. (1974); Lewis et al. (1978)
DNOC	1-10	>10	Domsch & Paul (1974)
Endothal	2-6	>1,000	Debona & Audus (1970)
EPTC	3-6	100-250	Bartha et al. (1967); Lewis et al. (1978)
Fenuron	2-30	100-250	Bartha et al. (1967)

(continued on next page)

Table 2—Continued.

Pesticide	Recommended rates of application/year	Estimated rates of application for >25% inhibition in soil† — Ammonium oxidation (Nitrite oxidation)	References
		——— kg/ha ———	
Fluometuron	1–2	250–500	Grossbard & Marsh (1974); Horowitz et al. (1974); Lewis et al. (1978)
Glyphosate	0.7–5.6	>100	Marsh et al. (1977)
Ioxynil	0.25–1	50–100	Debona & Audus (1970); Domsch & Paul (1974); Ratnayake & Audus (1978); van Schreven et al. (1970)
Ioxynil octanoate	0.25–1	50–100	Ratnayake & Audus (1978)
Lenacil	0.5–2	>100	Domsch & Paul (1974); Marsh & Davies (1978)
Linuron	0.5–2.8	100–250 (<150)	Bartha et al. (1967); Corke & Thompson (1970); Domsch & Paul (1974); Freyer & Kirkland (1970); Grossbard & Marsh (1974); Lewis et al. (1978); Marsh et al. (1974)
MCPA	0.25–2.5	>5	Domsch & Paul (1974); Freyer & Kirkland (1970)
Mecoprop	1.5–3	>10	Domsch & Paul (1974); van Schreven et al. (1970)
Methabenzthiazuron	2.0–3.5	>3.5 (3.5)	Domsch & Paul (1974)
Metobromuron	1.5–2.5	100–500 (2.5)	Domsch & Paul (1974); Grossbard & Marsh (1974)
Metoxuron	2.5–5	100–500	Grossbard & Marsh (1974)
Metribuzin	0.25–0.75	100–250	Lewis et al. (1978); Marsh et al. (1977)
Molinate	2–4	50–100	Turner (1979)
Monalide	4	>10	Domsch & Paul (1974)
Monolinuron	0.5–3	>5 (5)	Domsch & Paul (1974)
Monuron	5–30	50–250	Bartha et al. (1967); Grossbard & Marsh (1974); Hale et al. (1957); Otten et al. (1957)
MSMA	2–3	1,000–2,000	Bollen et al. (1977)
Naptalan	0.5–7	>6.7	Lewis et al. (1978)
Neburon	2–4	>10	Domsch & Paul (1974); Horowitz et al. (1974)
Nitralin	0.5–1.5	>1.1	Lewis et al. (1978)
Oxadiazon	1–5	25–50	Ratnayake & Audus (1978)
Paraquat	0.5–3	500–1,000	Debona & Audus (1970); Domsch & Paul (1974); Thorneburg & Tweedy (1973); Tu & Bollen (1968a); Tu & Bollen (1968b)

(continued on next page)

Table 2—Continued.

Pesticide	Recommended rates of application/year	Estimated rates of application for >25% inhibition in soil†	
		Ammonium oxidation (Nitrite oxidation)	References
		——— kg/ha ———	
Pebulate	4–6	100–250	Bartha et al. (1967)
Pendimethalin	1	>5	Atlas et al. (1978)
Phenmediphan	1	>2	Domsch & Paul (1974)
Picloram	0.02–3	20–1,000	Debona & Audus (1970); Dubey (1969); Goring et al. (1967); Thorneburg & Tweedy (1973); Tu & Bollen (1969); van Schreven et al. (1970)
Profluralin	0.75–1.5	>1.7	Lewis et al. (1978)
Prometryne	0.5–1.5	10–>100 (10–100)	Domsch & Paul (1974); Dubey (1969); Lewis et al. (1978)
Propanil	1–4	10–50	Bartha et al. (1967); Corke & Thompson (1970); Debona & Audus (1970); Turner (1979)
Propham	2.5–5	100–250	Bartha et al. (1967)
Pyrazon	1–4	100–250	Davies & Marsh (1977); Domsch & Paul (1974)
Simazine	0.5–20	>250 (5–10)	Bartha et al. (1967); Domsch & Paul (1974); Eno (1962); Freyer & Kirkland (1970); Gaur & Misra (1977); Horowitz et al. (1974); Marsh et al. (1974); Nayyar et al. (1970); Thorneburg & tweedy (1970)
Sulfallate	3–6	6–24	Otten et al. (1957); Teater et al. (1958)
Swep	4–6	100–250	Bartha et al. (1967)
2,4,5-T	0.25–8	100–250	Marsh & Davies (1978)
2,3,6-TBA	10–20	250–500	Debona & Audus (1970)
TCA	7.5–30	>50	Domsch & Paul (1974); Otten et al. (1957)
Terbacil	0.5–6	>100	Marsh & Davies (1978); Ratnayake & Audus (1978)
Terbutryne	1–2	>10	Domsch & Paul (1974)
Thiobencarb	4–5	>25	Atlas et al. (1978)
Triallate	1–3	>5	Domsch & Paul (1974); Freyer & Kirkland (1970)
Trifluralin	0.5–4	100–200	Davies & Marsh (1977); Lewis et al. (1978); Ratnayake & Audus (1978); Thorneburg & Tweedy (1973)
Vernolate	1.5–3	>250	Bartha et al. (1967)

(continued on next page)

Table 2—Continued.

Pesticide	Recom-mended rates of applica-tion/year	Estimated rates of application for >25% inhibition in soil†	References
		Ammonium oxidation (Nitrite oxidation)	
	———————	kg/ha ———————	
		Insecticides	
Aldicarb	0.5–5	50–500	Lin et al. (1972)
Aldrin	0.5–5	>250	Bartha et al. (1967); Eno (1958); Martin et ai. (1959); Narain & Datta (1973); Shaw & Robinson (1960)
Azinphosmethyl	0.25–2	>12	Thorneburg & Tweedy (1973)
Carbaryl	0.25–2	100–250	Bartha et al. (1967)
Carbofuran	0.25–10	>500	Lin et al. (1972); Ross (1974); Turner (1979)
Carbophenothion	0.5–2	>500	Lin et al. (1972)
Chlordane	1–10	>250	Eno (1958); Martin et al. (1959); Shaw & Robinson (1960)
Chlorpyrifos	0.5–1	250–500	Lin et al. (1972)
DDT	0.5–30	>250	Bartha et al. (1967); Bollen et al. (1954); Eno (1958); Martin et al. (1959); Ross (1954); Wilson & Choudri (1946)
Dieldrin	0.5–5	>250	Bartha et al. (1967); Martin et al. (1959); Narain & Datta (1973)
Disulfoton	0.25–4	>10	Thorneburg & Tweedy (1973)
Endrin	0.1–0.7	>250	Bartha et al. (1967); Martin et al. (1959)
Ethoprophos	1–10	>11.2	Elliot et al. (1974)
Fenitrothion	1–3	100–500	Ross (1974)
Fensulfothion	1–5	>500	Ross (1974)
Fonofos	1–2	250–500	Lin et al. (1972)
Gamma HCH	0.2–2	100–1,000	Bollen et al. (1954); Eno (1958); Gaur & Misra (1977); Jaiswal (1967); Jaiswal et al. (1971); Martin et al. (1959); Narain & Datta (1973)
Heptachlor	0.2–0.5	>100	Jaiswal (1967); Martin et al. (1959); Narain & Datta (1973); Shaw & Robinson (1960)
Isobenzan	0.2–1	5–250	Bartha et al. (1967); Jaiswal (1967); Jaiswal et al. (1971); Srivastava (1966)
Malathion	0.2–20	100–250	Bartha et al. (1967)
Methoxychlor	1–10	>250	Bartha et al. (1967)
Oxamyl	0.2–6	>6	Elliot et al. (1977)
Parathion	0.1–10	>250	Bartha et al. (1967); Eno (1958)
Phorate	1–2	>250	Bartha et al. (1967)
Propoxur	0.5–1	50–100	Lin et al. (1972)
Terbuphos	1	>100	Laveglia & Dahm (1974)
Toxaphene	2–6	>20	Martin et al. (1959)
Triazophos	0.3–2	>8	Elliot et al. (1977)

(continued on next page)

Table 2—Continued.

Pesticide	Recommended rates of application/year	Estimated rates of application for >25% inhibition in soil[†] Ammonium oxidation (Nitrite oxidation)	References
		kg/ha	
Trichlorfon	<5	>500	Lin et al. (1972)
Trichloronate	<5	50–500	Lin et al. (1972)
		Nematicides	
Fenamiphos	1–20	500–1,000	Lin et al. (1972)
		Nitrification inhibitors	
Nitrapyrin	0.25–1	0.05–20 (>50)	Goring (1962)
		Soil fumigants	
Carbon disulfide	100–400	<15[‡] 0.1–0.5[§]	Ashworth et al. (1975); Ashworth et al. (1977); Bremner & Bundy (1974)
Chloropicrin	200–400	100–200	Munnecke & Ferguson (1960); Rovira (1976); Stark et al. (1939); Tam & Clark (1943); Winfree & Cox (1958)
Dazomet	400	100–200	Chandra & Bollen (1961); Ebbels (1971); Smith & Weeraratna (1975)
DBCP	10–125	25–100[‡] 1–2 (>32)[§]	Good & Carter (1965); Goring & Scott (1976); Koike (1961) Goring & Scott (1976)
DD (1,3-D)	100–800	100–200 (<200)[‡] 2–4 (16–32)[§]	Abd-El-Malek et al. (1967); Ebbels (1971); Elliot et al. (1974); Elliot et al. (1977); Elliot & Mountain (1963); Good & Carter (1965); Goring & Scott (1976); Koike (1961); Lebbink & Kolenbrander (1974); McCants et al. (1959); Tam (1945); Wolcott et al. (1960) Goring & Scott (1976)
Ethylene dibromide	25–200	25–50[‡] 0.5–1 (>32)[§]	Good & Carter (1965); Goring & Scott (1976); Koike (1961); McCants et al. (1959); Thiagalingam & Kanehiro (1971); Thiegs (1955) Goring & Scott (1976)
Metham-sodium	100–400	100–200	Koike (1961); Munnecke & Ferguson (1960); Lebbink & Kolenbrander (1974)
Methyl bromide	400–800	400–800	McCants et al. (1959); Munnecke & Ferguson (1960); Rovira (1976); Thiagalingam & Kanehiro (1971); Thiegs (1955); Winfree & Cox (1958)

† 1 ppm or 1 pound/acre was assumed to be equivalent to approximately 1 kg/ha.
‡ Field conditions.
§ Closed containers.

Table 3—Inhibition of denitrification by pesticides.

Pesticide	Recom-mended rates of applica-tion/year	Estimated rates of application for >25% inhibition in soil[†]	References
	——— kg/ha ———		
	Fungicides		
Captafol	2–10	>10	Atlas et al. (1978)
Captan	2–10	<100	Bollag & Henninger (1976)
Ferbam	1–15	>100	Mitsui et al. (1964)
Folpet	0.5–2	>10	Atlas et al. (1978)
Maneb	0.5–15	50–100	Bollag & Henninger (1976); Mitsui et al. (1964)
Nabam	25–100	50–100	Bollag & Henninger (1976); Mitsui et al. (1964)
Phenyl mercuric acetate	0.2–20	<5	Henninger & Bollag (1976)
Thiram	2–5	>100	Mitsui et al. (1964)
Zineb	1–15	>100	Mitsui et al. (1964)
Ziram	1–15	>100	Mitsui et al. (1964)
	Herbicides		
Ametryne	2–4	>100	Bollag & Henninger (1976)
Atrazine	1–4	>100	Bollag & Henninger (1976)
2,4-D	0.25–4	>100	Bollag & Henninger (1976)
Diuron	0.5–30	>100	Bollag & Henninger (1976)
Endothal	2–6	>100	Bollag & Henninger (1976)
Linuron	0.5–2.8	>100	Bollag & Henninger (1976)
Propham	2.5–5	>100	Bollag & Henninger (1976)
Siduron	2–12	>100	Bollag & Henninger (1976)
Simazine	0.5–20	>100	Bollag & Henninger (1976)
	Insecticides		
Carbaryl	0.25–2	>100	Bollag & Henninger (1976)
Carbofuran	0.25–10	>100	Bollag & Henninger (1976)
DDT	0.5–30	>100	Bollag & Henninger (1976)
Diazinon	0.2–5	>100	Bollag & Henninger (1976)
Endrin	0.1–0.7	>100	Bollag & Henninger (1976)
HCH	0.5–2	>40	Mitsui et al. (1964)
Methoxychlor	1–10	>100	Bollag & Henninger (1976)
Parathion	0.1–10	>100	Eno (1958)
	Nitrification inhibitors		
Nitrapyrin	0.25–1	50 (*Pseudomonas* sp. liquid culture)	Henninger & Bollag (1976)
		50 (Sand/bark medium)	Mills & Pokorny (1978)
		>50	Bremner & Blackmer (1980); Henninger & Bollag (1976); Mitsui et al. (1964)
	Soil fumigants		
Metham-sodium	100–400	50–200	Mitsui et al. (1964)

† 1 ppm or 1 pound/acre was assumed to be equivalent to approximately 1 kg/ha.

Table 4—Inhibition of symbiotic N-fixing organisms by pesticides.

Pesticide	Recommended rates of application/year	Estimated rates of application for >25% inhibition†	Organism and/or test system	References
	——— kg/ha ———			
Fungicides				
Anilazine	2–10	>15	*R. Japonicum*—Agar	Kapusta & Rouwenhorst (1973)
Captan	--	> Recommended rates	Seed treatment	Diatloff (1970)
	2–10	< Recommended rates	*Rhizobium*—Agar	Afifi et al. (1969)
Chloranil	--	> Recommended rates	Seed treatment	Diatloff (1970)
Dichlone	--	< Recommended rates	*Rhizobium*—Agar	Afifi et al. (1969)
Fenaminosulf	0.1–10	>500	*Rhizobium*—Liquid culture	Karanth & Vasantharajan (1974)
		<100‡	Nodulation	
Phenyl mercuric acetate	0.2–20	< Recommended rates	*Rhizobium*—Agar	Afifi et al. (1969)
	--	> Recommended rates	Seed treatment	Diatloff (1970); Misra & Gaur (1974)
		< Recommended rates	Seed treatment	Jakubisiak & Golebiowska (1963)
Thiram	2–5	< Recommended rates	*Rhizobium*—Agar	Afifi et al. (1969)
	--	> Recommended rates	Seed treatment	Tu (1977); Diatloff (1970); Jakubisiak & Golebiowska (1963)
Herbicides				
Alachlor	2–3	>7.5	*R. Japonicum*—Agar	Kapusta & Rouwenhorst (1973)
Allidochlor	4–6	>20	*R. Japonicum*—Agar	Kapusta & Rouwenhorst (1973)
Alloxydim-sodium	2–4	2–4‡	Nodulation and nitrogenase activity	Greaves et al. (1978)
Atrazine	1–4	>15	*R. Japonicum*—Agar	Kapusta & Rouwenhorst (1973)
Chloramben	0.5–4	>15	*R. Japonicum*—Agar	Kapusta & Rouwenhorst (1973)
		5.6–16.8‡	Nodulation	Dunigan et al. (1972); Olumbe & Veatch (1969)
Chloroxuron	1–6	>100	*Rhizobium*—Liquid culture	Kaszubiak (1966)
Chlorpropham	2–4	<15	*R. Japonicum*—Agar	Kapusta & Rouwenhorst (1973)
		>33.6	*R. Japonicum*—Soil	
		16.8–33.6‡	Nodulation	

(continued on next page)

Table 4—Continued.

Pesticide	Recommended rates of application/year	Estimated rates of application for >25% inhibition†	Organism and/or test system	References
		——— kg/ha ———		
Chlorthal-dimethyl	6–14	>15	R. Leguminosarum—Soil	Fields et al. (1967)
		>52.5	R. Japonicum—Agar	Kapusta & Rouwenhorst (1973)
2,4-D	0.25–4	25–100	R. Trifolii—Agar	Fletcher (1956)
		>100	Rhizobium—Liquid culture	Kaszubiak (1966)
Dalapon	1–20	>4,000	Rhizobium—Liquid culture	Kaszubiak (1966)
2,4-DB	1.5–3	25–300	R. Trifolii—Agar	Fletcher (1956)
		>300	Rhizobium—Liquid culture	Kaszubiak (1966)
Dicamba	0.1–8	>6	R. Japonicum—Agar	Kapusta & Rouwenhorst (1973)
Dinoseb acetate	0.75–12	>100	Rhizobium—Liquid culture	Kaszubiak (1966)
Fluometuron	1–2	>300	Rhizobium—Liquid culture	Kaszubiak (1966)
Linuron	0.5–2.8	1.1–5.6‡	Nodulation	Dunigan et al. (1972)
		>7.5	R. Japonicum—Agar	
		>50	Rhizobium—Liquid culture	Kaszubiak (1966)
MCPA	0.25–2.5	25–100	R. Trifolii—Agar	Fletcher (1956)
		>300	Rhizobium—Liquid culture	Kaszubiak (1966)
MCPB	1.7–3.4	50–300	R. Trifolii—Agar	Fletcher (1956)
Metobromuron	1.5–2.5	100–1,000	Rhizobium—Liquid culture	Kaszubiak (1966)
Monolinuron	0.5–3	100–1,000	Rhizobium—Liquid culture	Kaszubiak (1966)
Naptalam	0.5–7	>26	R. Japonicum—Agar	Kapusta & Rouwenhorst (1973)
Nitralin	0.5–1.5	0.8–4.2‡	Nodulation	Dunigan et al. (1972)
		>7.5	R. Japonicum—Agar	Kapusta & Rouwenhorst (1973)
		>16	Rhizobium—Liquid culture	
		>11.2	R. Japonicum—Soil	
		<1.1‡	Nodulation	
Picloram	0.02–3	>1,000	R. Phaseoli—Agar	Goring et al. (1967)
Prometryne	0.5–1.5	1.1–8.4‡	Nodulation	Dunigan et al. (1972)
		4,000–10,000	Rhizobium—Liquid culture	Kaszubiak (1966)
Propachlor	3.5–5	>20	R. Japonicum—Agar	Kapusta & Rouwenhorst (1973)
Pyrazon	1–4	500–10,000	Rhizobium—Liquid culture	Kaszubiak (1966)

(continued on next page)

Table 4—Continued.

Pesticide	Recommended rates of application/year	Estimated rates of application for >25% inhibition†	Organism and/or test system	References
	——— kg/ha ———			
Simazine	0.5–20	4,000–10,000	*Rhizobium*— Liquid culture	Kaszubiak (1966)
		<2‡	Nodulation	Misra & Gaur (1974)
2,4,5-T	0.25–8	25–100	*R. Trifolii*—Agar	Fletcher (1956)
TCA	7.5–30	1,000–10,000	*Rhizobium*— Liquid culture	Kaszubiak (1966)
Trifluralin	0.5–4	>10	*R. Japonicum*— Agar	Kapusta & Rouwenhorst (1973);
		1.7–4.2‡	Nodulation	Dunigan et al. (1972)
Vernolate	1.5–3	>15	*R. Japonicum*— Agar	Kapusta & Rouwenhorst (1973);
		2.8–14‡	Nodulation	Dunigan et al. (1972)

Insecticides

Pesticide	Recommended rates of application/year	Estimated rates of application for >25% inhibition†	Organism and/or test system	References
Aldrin	0.5–5	>2.5	*R. Japonicum*— Agar	Kapusta & Rouwenhorst (1973)
Azinphos-methyl	0.25–2	>15	*R. Japonicum*— Agar	Kapusta & Rouwenhorst (1973)
Carbaryl	0.25–2	>12.5	*R. Japonicum*— Agar	Kapusta & Rouwenhorst (1973)
		>28	Nodulation	
Chlordane	1–10	>10	Nodulation	Simkover & Shenefelt (1951)
Chlorpyrifos	--	> Recommended rates	Seed treatment	Tu (1977)
DDT	0.5–30	100–1,000	Nodulation	Eno (1958); Gaur & Pareek (1969); Pareek & Gaur (1970); Wilson & Choudri (1946)
Diazinon	0.2–5	>2.5	*R. Japonicum*— Agar	Kapusta & Rouwenhorst (1973)
Dieldrin	--	> Recommended rates	Seed treatment	Diatloff (1970)
	0.5–5	>25	*R. Japonicum*— Agar	Kapusta & Rouwenhorst (1973)
		>500	Nodulation	Selim et al. (1970)
Dimethoate	--	> Recommended rates	Seed treatment	Diatloff (1970)
Disulfoton	0.25–4	>7.5	*R. Japonicum*— Agar	Kapusta & Rouwenhorst (1973)
		>16.8	*R. Japonicum*— Soil	
		>16.8	Nodulation	
		>50	*Rhizobium*— Liquid culture	Kaszubiak (1966)

(continued on next page)

Table 4—Continued.

Pesticide	Recommended rates of application/year	Estimated rates of application for >25% inhibition[†]	Organism and/or test system	References
		kg/ha		
Endrin	--	> Recommended rates	Seed treatment	Diatloff (1970)
	0.1–0.7	>12	Nodulation	Taha et al. (1972)
Fensulfothion	1–5	>70	*R. Japonicum—* Agar	Kapusta & Rouwenhorst (1973)
Gamma HCH	--	> Recommended rates	Seed treatment	Diatloff (1970); Tu (1977);
	0.2–2	1–40[‡]	Nodulation	Eno (1958); Misra & Gaur (1974); Selim et al. (1970); Simkover & Shenefelt (1951)
		>20	*R. Japonicum—* Agar	Kapusta & Rouwenhorst (1973)
Isobenzan	--	> Recommended rates	Seed treatment	Diatloff (1970)
Malathion	0.2–20	>15	*R. Japonicum—* Agar	Kapusta & Rouwenhorst (1973)
Methoxychlor	1–10	>2.5	*R. Japonicum—* Agar	Kapusta & Rouwenhorst (1973)
Trichlorphon	<5	>9	Nodulation	Taha et al. (1972)
Nitrification inhibitors				
Nitrapyrin	0.25–1	10–20[‡]	Nodulation	McKell & Whalley (1964)
Plant growth regulators				
Maleic hydrazide	2–5	280–2,300	*Rhizobium—* Liquid culture	Nickell & English (1953)
Soil fumigants				
Chloropicrin	200–400	<300	Nodulation	Ebbels (1967)
Dazomet	400	300	Nodulation	Ebbels (1967)
DD (1,3-D)	100–800	300	Nodulation	Ebbels (1967)

† 1 ppm or 1 pound/acre was assumed to be equilent to approximately 1 kg/ha.
‡ Inhibition of nodulation due to the phytotoxicity of the chemical to the plant.

Table 5—Inhibition of nonsymbiotic N-fixing organisms or fixation by pesticides.

Pesticide	Recommended rates of application/year	Estimated rates of application for > 25% inhibition[†]	Organism and/or test system	References
	——— kg/ha ———			
		Fungicides		
Captafol	2–10	>10	Soil—Acetylene reduction	Atlas et al. (1978)
Folpet	0.5–2	>10	Soil—Acetylene reduction	Atlas et al. (1978)
		Herbicides		
Aminotriazole	2–10	>20	Algae—Sand	DaSilva et al. (1975)
		88–875	*A. Chroococcum* —Soil	van Schreven et al. (1970)
Chloramben	0.5–4	1,000–7,000	*Azotobacter*— Agar	Babak (1968)
Chlorbufam	1–4	1,000–7,000	*Azotobacter*— Agar	Babak (1968)
Chlorthal-dimethyl	6–14	>50	*Azotobacter*— Soil	Fields et al. (1967)
2,4-D	0.25–4	100–200	*A. Chroococcum*— Liquid culture	Colmer (1953)
		2,000–3,000	*A. Agile*—Liquid culture	
Dalapon	1–20	13–128	*A. Chroococcum* —Soil	van Schreven et al. (1970)
Dichlorprop	1–2	26–255	*A. Chroococcum* —Soil	van Schreven et al. (1970)
Dinoseb	0.75–12	3–6	Soil—Acetylene reduction	Vlassak et al. (1976)
Diphenamid	4–6	1,000–7,000	*Azotobacter*— Agar	Babak (1968)
Diquat	0.5–2	<20	Algae—Sand	DaSilva et al. (1975)
EPTC	3–6	<0.1	Algae—Liquid culture	Ibrahim (1972)
Ioxynil	0.25–1	13.2–132.3	*A. Chroococcum* —Soil	van Schreven et al. (1970)
Linuron	0.5–2.8	10	Algae—Sand	DaSilva et al. (1975)
MCPA	0.25–2.5	20	Algae—Sand	DaSilva et al. (1975)
Mecoprop	1.5–3	10–100	*A. Chroococcum* —Soil	van Schreven et al. (1970)
Molinate	2–4	1–10	Algae—Liquid culture	Ibrahim (1972)
Monuron	5–30	1,000–7,000	*Azotobacter*— Agar	Babak (1968)
Neburon	2–4	1,000–7,000	*Azotobacter*— Agar	Babak (1968)

(continued on next page)

Table 5—Continued.

Pesticide	Recom-mended rates of applica-tion/year	Estimated rates of application for >25% inhibition[†]	Organism and/or test system	References
		——— kg/ha ———		
Paraquat	0.5–3	<20	Algae—Sand	DaSilva et al. (1975)
Pebulate	4–6	1,000–7,000	Azotobacter—Agar	Babak (1968)
Picloram	0.02–3	>1,000	A. Chroococcum Agar C-Butricum—Liquid culture	Goring et al. (1967)
		>48	A. Chroococcum —Soil	van Schreven et al. (1970)
Propanil	1–4	1	Algae—Liquid culture	Ibrahim (1972)
Siduron	2–12	<10	Azotobacter—Soil	Fields & Hemphill (1968)
Trifluralin	0.5–4	0.1–1	Algae—Liquid culture	Ibrahim (1972)
Insecticides				
Aldrin	0.5–5	>25	Azotobacter—Soil	Eno (1958)
Gamma HCH	0.2–2	>9[‡]	Soil—Kjeldahl analysis	Jaiswal (1967)
		>6[‡]	Soil—^{15}N analysis	Raghu & MacRae (1967)
Heptachlor	0.2–0.5	>9[‡]	Soil—Kjeldahl analysis	Jaiswal (1967)
Isobenzan	0.2–1	>6.6[‡]	Soil—Kjeldahl analysis	Jaiswal (1967)
Malathion	0.2–20	>100	Algae—Sand	DaSilva et al. (1975)
Plant growth regulators				
Maleic hydrazide	2–5	100–1,000	A. Vinelandii—Agar	Nickell & English (1953)
Soil fumigants				
DD (1,3-D)	100–800	>200 <200	Azotobacter—Soil Clostridium—Soil	Abd-El-Malek et al. (1967)

[†] 1 ppm or 1 pound/acre was assumed to be equivalent to approximately 1 kg/ha.
[‡] Apparent stimulation of nonsymbiotic N_2 fixation at or below the indicated dosages.

continue to increase available soil N once the susceptible population is destroyed (Jenkinson & Powlson, 1970). In some instances, N mineralization may be decreased. For example, paraquat-treated barley straw (*Hordeum vulgare* L.) decays more slowly in soil (Grossbard & Cooper, 1974). Repeated applications of pesticides, especially dithiocarbamates, to a soil in which sugarcane (*Saccharum officinarum* L.) was growing was reported to inhibit mineralization of N and cause N deficiency (Dubey, 1970). The simultaneous application of urea fertilizer and pesticides that inhibit urea hydrolysis could result in decreased losses of NH_3 from the surface of the soil. All of these potential effects could influence fertilizer N recommendations.

Effects on nitrification also need to be considered. Russell and Hutchinson (1909) were among the first to observe the temporary decrease in the nitrifying capacity of soils by disinfectants, but 34 years elapsed before it was clearly shown that the resulting shift to NH_4^+ nutrition could greatly enhance the growth of at least one crop, pineapple (*Ananas comosus* L. Merr.) (Tam, 1945, Tam & Clark, 1943). Conditions for optimum response have already been outlined (Nightingale, 1937), and the practical importance of increased growth response of some crops to NH_4^+ nutrition in fumigated soils emphasized (Nightingale, 1948).

Not all crop responses to NH_4^+ nutrition in disinfected soil are positive (Davidson & Thiegs, 1966). A number of crops, including tomatoes (*Lycopersicon esculentum* Mill.), potatoes (*Solanum tuberosum* L.), and tobacco (*Nicotiana tabacum* L.), are highly sensitive to excessive NH_4^+ nutrition, although the situation can be mitigated by applying enough NO_3^- fertilizer to ensure a mixed NH_4^+-NO_3^- diet.

Inhibition of nitrification has other significant agronomic impacts. Losses of N from leaching and denitrification may be reduced. Nutrient uptake and plant composition may be altered. Plant disease may be enhanced or reduced. All of these agronomic effects need to be considered when pesticides that are known to inhibit nitrification at recommended rates are utilized. Similar consideration must be given to pesticides that have the potential to reduce denitrification.

It is undesirable to inhibit N_2 fixation with pesticides. Much of the N returned to the soil from the atmosphere is due to symbiotic and nonsymbiotic N_2 fixation. Fortunately, most pesticides have little, if any, influence on N_2-fixing organisms at recommended rates. When there is a potential for an undesirable impact, it can often be anticipated and minimized. For example, pesticides that are toxic to legumes at recommended rates not only damage the growth of the plants but also impair nodulation and subsequent symbiotic N_2 fixation (Garcia & Jordan, 1969; Hamdi & Tewfik, 1969; Karanth & Vasantharajan, 1974; McKell & Whalley, 1964; Olumbe & Veatch, 1969; Rolston et al., 1976). Their use on legumes should be avoided.

Pesticides that are used to protect legume seed from attack by fungi and insects might also be toxic to *Rhizobium* and reduce the effectiveness of seed inoculation. However, rates and methods of application of pesticide

Table 6—Potential cumulative levels of chlorinated hydrocarbon insecticides in soil in comparison with rates of application required to inhibit N transformations.

Chlorinated hydrocarbon insecticides	Estimated average half-lives in soil	Potential cumulative levels in soil‡	Estimated rates of application required to inhibit various N transformation processes or organisms in soil				
			Mineralization of N	Nitrification	Denitrification	Symbiotic N_2-fixing organisms	Nonsymbiotic N_2-fixing organisms
	Years†	kg/ha	kg/ha				
Aldrin	0.7	0.8–7.9	200–1,000	>250	--	>2.5	>25
chlordane	8.1	12.2–122	>200	>250	--	>10	--
DDT	10.5§	7.9–471	>500	>250	>100	100–1,000	--
Dieldrin	1.0	1.0–9.9	>2.5	>250	--	>500	--
Endrin	11.8	1.8–12.4	--	>250	>100	>12	--
HCH	3.3	2.6–10.5	--	--	>40	--	--
Gamma HCH	1.6	0.6–5.6	>1,000	100–1,000	--	1–40	>9
Heptachlor	6.8	2.0–5.1	>9	>100	--	--	>9
toxaphene	12.6	37.4–112.2	--	>20	--	--	--

† Calculated from Hamaker and Thompson (1972).
‡ Calculated from Hamaker (1966) and recommended annual rates of application (Table 1–5).
§ Estimate for DDT was made for aerobic conditions.

and inoculant can often be chosen so as to ensure that nodulation occurs (Diatloff, 1970). Strains of *Rhizobium* that are resistant to the pesticide and effective in causing nodulation might also be available (Gillberg, 1970; Hofer, 1958; Kaszubiak, 1968).

Data obtained thus far indicate that most pesticides are unlikely to cause significant reductions in the nonsymbiotic N_2-fixing population. Furthermore, the amount of N contributed to most agricultural soils by nonsymbiotic N_2 fixation is usually low (see Chapt. 10, R. W. Hardy and U. D. Havelka).

Certain organisms (notably blue-green algae) in flooded soils can, however, fix substantial amounts of N (Chapt. 10). some herbicides are highly toxic to blue-green algae (Ibrahim, 1972). The use of such herbicides in rice (*Oryza sativa* L.) may be undesirable. On the other hand, certain insecticides seem to stimulate N_2 fixation in flooded soils (Raghu & MacRae, 1967) possibly by stimulating the growth of bluegreen algae.

As previously discussed, persistent pesticides such as the chlorinated hydrocarbon insecticides could affect N transformations on a long-term basis if cumulative levels in soil exceeded the rates required for inhibition.

Table 6 gives estimates of the potential cumulative levels of chlorinated hydrocarbon insecticides calculated from estimated half-disappearance times in soil (Hamaker & Thompson, 1972) and recommended annual rates of application. The calculations were made (Hamaker, 1966) assuming that half-disappearance times approximated half-lives. Table 6 also summarizes estimated rates of application required to inhibit N transformations.

The data in Table 6 suggest that in most instances the chlorinated hydrocarbon insecticides would not impact on N transformations even at the highest potential cumulative levels. For some of the chemicals and types of N transformations, data were either not available or the impact on N transformations was measured at rates of application below the highest potential cumulative levels. In a few instances (for example, the data for DDT and gamma HCH on nodulation), the rates required for inhibition were in the same range as the potential cumulative levels.

VI. ENVIRONMENTAL AND REGULATORY IMPLICATIONS

Pesticide use in agriculture is subject to state and federal regulations designed to protect public health and environmental quality. A current concern of the regulatory agencies is the potential adverse effects of pesticides on soil microbial processes, especially those involving N transformations. An assessment of such effects requires the development of suitable test procedures.

Although there has been some progress in standardizing test methods (Atlas et al., 1978; Cooper et al., 1978; Goh & Edmeades, 1978), the question arises as to how reliable and valid these methods are for determining the effect of pesticides on N transformations in soil since the microorganisms involved are variable in number and highly responsive to changes in environmental conditions. Another difficulty is that it is unlikely that any test procedures will be able to predict adaptive responses of these organisms to the environment after they are no longer being impacted by the pesticide.

Still another problem in interpreting test results is the fact that the pesticide may have both adverse and beneficial effects on the soil microbial population. For example, inhibition of nitrification and denitrification can usually be considered beneficial. Chemicals designed to inhibit these processes have and are being developed. On the other hand, if such chemicals also inhibit biological N_2 fixation, their value must be judged in terms of the balance between their adverse and beneficial effects.

Continuing research is needed to better understand and quantify the extent to which pesticides influence N transformations in soil, including short-term and long-term studies. Continued effort is needed to standardize procedures and to ascertain which transformations are so responsive to changes in test conditions that suitable standardized tests cannot readily be devised.

At the present time, it seems inappropriate to exert regulatory control on pesticide use based on their impact on soil microbial processes. This is because of the highly variable nature and continuously changing populations of the organisms involved and their adaptive responses, the difficulty of developing valid test procedures, and the high cost/benefit ratio of pesticide use in terms of overall soil productivity.

LITERATURE CITED

Abd-El-Malek, Y., Y. M. Monib, M. N. Zayed, and M. Abd-El-Nasser. 1967. The effect of DD nematicide on soil microorganisms. Folia Microbiol. 13:270–274.

Afifi, N. M., A. A. Moharram, and Y. A. Hamdi. 1969. Sensitivity of *Rhizobium* species to certain fungicides. Arch. Microbiol. 66:121–128.

Agnihotri, V. P. 1971. Persistence of captan and its effects on microflora, respiration and nitrification in a forest nursery soil. Can. J. Microbiol. 17:377–383.

Agnihotri, V. P. 1973. Effect of dexon on soil microflora and their ammonification and nitrification activities. Indian J. Exp. Biol. 11:213–216.

Ashworth, J., G. G. Briggs, and A. A. Evans. 1975. Field injection of carbon disulfide to inhibit nitrification of ammonia fertilizer. Chem. Ind. (London) 17:749–750.

Ashworth, J., G. G. Briggs, A. A. Evans, and J. Matula. 1977. Inhibition of nitrification by nitrapyrin, carbon disulfide and trithiocarbonate. J. Sci. Food Agric. 28:673–683.

Atlas, R. M., D. Pramer, and R. Bartha. 1978. Assessment of pesticide effects on non-target soil microorganisms. Soil Biol. Biochem. 10:231–239.

Babak, N. M. 1968. Sensitivity of *Azotobacter* to some antibiotics and herbicides. (In Russian.) Mikrobiologiya 37:338–344.

Barnes, R. D., A. T. Bull, and R. C. Poller. 1971. Behavior of triphenyltin acetate in soil. Chem. Ind. (London) 7:204.

Barnes, R. D., A. T. Bull, and R. C. Poller. 1973. Studies on the persistence of the organotin fungicide fentin acetate (triphenyltin acetate) in the soil and on surfaces exposed to light. Pestic. Sci. 4:305–317.

Bartha, R., R. P. Lanzilotta, and D. Pramer. 1967. Stability and effects of some pesticides in soil. Appl. Microbiol. 15:67–75.

Bollag, J. M., and N. M. Henninger. 1976. Influence of pesticides on denitrification in soil and with an isolated bacterium. J. Environ. Qual. 5:15–18.

Bollen, W. B., H. E. Morrison, and H. H. Crowell. 1954. Effect of field and laboratory treatments with BHC and DDT on nitrogen transformations and soil respiration. J. Econ. Entomol. 47:307–312.

Bollen, W. B., L. A. Norris, and K. L. Stowers. 1977. Effect of cacodylic acid and MSMA on nitrogen transformations in forest floor and soil. J. Environ. Qual. 6:1–3.

Bremner, J. M., and A. M. Blackmer. 1980. Mechanisms of nitrous oxide production in soils. p. 279–291. In P. A. Trudinger et al. (ed.) Biogeochemistry of ancient and modern environments. Australian Academy of Science, Canberra, Australia.

Bremner, J. M., and L. G. Bundy. 1974. Inhibition of nitrification in soils by volatile sulfur compounds. Soil Biol. Biochem. 6:161–165.

Bremner, J. M., and L. G. Bundy. 1976. Effects of potassium azide on transformations of urea nitrogen in soils. Soil Biol. Biochem. 8:131–133.

Bremner, J. M., and L. A. Douglas. 1971. Inhibition of urease activity in soils. Soil Biol. Biochem. 3:297–307.

Bremner, J. M., and L. A. Douglas. 1973. Effects of some urease inhibitors on urea hydrolysis in soils. Soil Sci. Soc. Am. Proc. 37:225–226.

Briggs, G. G. 1976. Degradation in soils. p. 41–54. In K. I. Beynon (ed.) Persistence of insecticides and herbicides. Proc. British Crop Protection Council Symp., Reading, England. 22–24 March 1976. British Crop Protection Council, London.

Bundy, L. G., and J. M. Bremner. 1973a. Effects of substituted p-benzoquinones on urease activity in soils. Soil Biol. Biochem. 5:847–853.

Bundy, L. G., and J. M. Bremner. 1973b. Inhibition of nitrification in soils. Soil Sci. Soc. Am. Proc. 37:396–398.

Bundy, L. G., and J. M. Bremner. 1974a. Effect of urease inhibitors on nitrification in soils. Soil Biol. Biochem. 6:27–30.

Bundy, L. G., and J. M. Bremner. 1974b. Effects of nitrification inhibitors on transformations of urea nitrogen in soils. Soil Biol. Biochem. 6:369–376.

Caseley, J. C., and F. E. Broadbent. 1968. The effect of five fungicides on soil respiration and some nitrogen transformations in Yolo fine sandy loam. Bull. Environ. Contam. Toxicol. 3:58–64.

Cerville, S., P. Nannipieri, G. Giovanni, and A. Perna. 1977. Effect of soil on urease inhibition by substituted urea herbicides. Soil Biol. Biochem. 9:393–396.

Chandra, P., and W. B. Bollen. 1961. Effects of nabam and mylone on nitrification, soil respiration, and microbial numbers in four Oregon soils. Soil Sci. 92:387–396.

Colmer, A. R. 1953. The action of 2,4-D upon the Azotobacter of some sugarcane soils. Appl. Microbiol. 1:184–187.

Cooper, S. L., G. I. Wingfield, R. Lawley, and M. P. Greaves. 1978. Miniaturized methods for testing the toxicity of pesticides to microorganisms. Weed Res. 18:105–107.

Corke, C. T., and F. R. Thompson. 1970. Effects of some phenylamide herbicides and their degradation products on soil nitrification. Can. J. Microbiol. 16:567–571.

Crosby, D. G. 1970. The nonbiological degradation of pesticides in soils. p. 86–94. In Pesticides in the soil—ecology, degradation and movement. Int. Symp., East Lansing, Mich. 25–27 Feb. 1970. Michigan State Univ., East Lansing, Mich.

DaSilva, E. J., L. E. Henriksson, and E. Henriksson. 1975. Effect of pesticides on blue-green algae and nitrogen-fixation. Arch. Environ. Contam. Toxicol. 3:193–204.

Davidson, J. H., and B. J. Thiegs. 1966. The effect of soil fumigation on nitrogen nutrition and crop response. Down Earth 22(1):7–12.

Davies, H. A., and J. A. P. Marsh. 1977. The effect of herbicides on respiration and transformation of nitrogen in two soils. II. Dalapon, pyrazone and trifluralin. Weed Res. 17:373–378.

Debona, A. C., and L. J. Audus. 1970. Studies on the effects of herbicides on soil nitrification. Weed Res. 10:250–263.

Diatloff, A. 1970. The effects of some pesticides on root nodule bacteria and subsequent nodulation. Aust. J. Exp. Agric. Anim. Husb. 10:562–567.

Domsch, K. H., and W. Paul. 1974. Simulation and experimental analysis of the influence of herbicides on soil nitrification. Arch. Microbiol. 97:283–301.

Dubey, H. D. 1969. Effect of picloram, diuron, ametryne, and prometryne on nitrification in some tropical soils. Soil Sci. Soc. Am. Proc. 33:893–896.

Dubey, H. 1970. A nitrogen deficiency disease of sugar cane probably caused by repeated pesticide applications. Phytopathology 60:485–487.

Dubey, H. D., and R. L. Rodriguez. 1970. Effect of dyrene and maneb on nitrification and ammonification and their degradation in tropical soils. Soil Sci. Soc. Am. Proc. 34:435–439.

Dunigan, E. P., J. P. Frey, L. D. Allen, Jr., and A. McMahon. 1972. Herbicidal effects of the nodulation of *Glycine max* (L.) Merrill. Agron. J. 64:806–808.

Ebbels, D.L. 1967. Effect of soil fumigants on *Fusarium* wilt and nodulation of peas (*Pisum sativum* L.). Ann. Appl. Biol. 60:391–398.

Ebbels, D. L. 1971. Effects of soil fumigation on soil nitrogen and on disease incidence in winter wheat. Ann. Appl. Biol. 67:235–243.

Elliot, J. M., C. F. Marks, and C. M. Tu. 1974. Effects of the nematicides DD and Mocap on soil nitrogen, soil microflora, and populations of *Pratylenchus penetrans,* and flue-cured tobacco. Can. J. Plant Sci. 54:801–809.

Elliot, J. M., C. F. Marks, and C. M. Tu. 1977. Effects of certain nematicides on soil nitrogen, soil nitrifiers, and populations of *Pratylenchus penetrans* in flue-cured tobacco. Can. J. Plant Sci. 57:143–154.

Elliot, J. M., and W. B. Mountain. 1963. The influence of spring and fall application of nematicides on *Pratylenchus penetrans* and quality of flue-cured tobacco grown with various forms of nitrogen. Can. J. Soil Sci. 43:18–26.

Eno, C. F. 1958. Insecticides and the soil. J. Agric. Food Chem. 6:348–351.

Eno, C. F. 1962. The effect of simazine and atrazine on certain of the soil microflora and their metabolic processes. Soil Crop Sci. Soc. Fla. Proc. 22:49–56.

Fields, M. L., R. Der, and D. D. Hemphill. 1967. Influence of DCPA on selected soil microorganisms. Weeds 15:195–197.

Fields, M. L., and D. D. Hemphill. 1968. Influence of siduron and its degradation products on soil microflora. Weeds 10:417–420.

Fletcher, W. W. 1956. Effect of hormone herbicides on the growth of *Rhizobium trifolii.* Nature 177:1244.

Foster, M. G., and D. J. McQueen. 1977. The effects of single and multiple application of benomyl on non-target soil bacteria. Bull. Environ. Contam. Toxicol. 17:477–484.

Freyer, J. D., and K. Kirkland. 1970. Field experiments to investigate long-term effects of repeated applications of MCPA, tri-allate, simazine and linuron. Report after six years. Weed Res. 10:133–158.

Garcia, M. M., and D. C. Jordan. 1969. Action of 2,4-DB and Dalapon on the symbiotic properties of *Lotus corniculatus* (Birdsfoot trefoil). Plant Soil 30:317–334.

Gaur, A. C., and K. C. Misra. 1977. Effect of simazine, lindane and ceresan on soil respiration and nitrification rates. Plant Soil 46:5–15.

Gaur, A. C., and R. P. Pareek. 1969. Effect of dichlorodiphenyl trichloro-ethane (DDT) on leghemoglobin content of root nodules of *Phaseolus aureus* (green gram). Experientia 25: 777.

Gillberg, B. O. 1970. On the effects of some pesticides on *Rhizobium* and isolation of pesticide resistant mutants. Arch. Microbiol. 75:203–208.

Goh, K. M., and D. C. Edmeades. 1978. Field measurements of symbiotic nitrogen fixation in an established pasture using acetylene reduction and a ^{15}N method. Soil Biol. Biochem. 10:13–20.

Good, J. M., and R. L. Carter. 1965. Nitrification lag following soil fumigation. Phytopathology 55:1147–1150.

Goring, C. A. I. 1962. Control of nitrification by 2-chloro-6-(trichloromethyl) pyridine. Soil Sci. 93:211–218.

Goring, C. A. I. 1967. Physical aspects of soil in relation to the action of soil fungicides. Annu. Rev. Phytopathol. 5:285–318.

Goring, C. A. I. 1972. Agricultural Chemicals in the Environment: A quantitative viewpoint. p. 793–863. *In* C. A. I. Goring and J. W. Hamaker (ed.) Organic chemicals in the soil environment, Vol. 2. Marcel Dekker, Inc., New York.

Goring, C. A. I., J. D. Griffith, F. C. O'Melia, H. H. Scott, and C. R. Youngson. 1967. The effect of Tordon® on microorganisms and soil biological processes. Down Earth 22(4): 14–17.

Goring, C. A. I., D. A. Laskowski, J. W. Hamaker, and R. W. Meikle. 1975. Principles of pesticide degradation in soil. p. 135–172. *In* R. Haque and V. H. Freed (ed.) Environmental dynamics of pesticides. Plenum Press, New York.

Goring, C. A. I., and H. H. Scott. 1976. Control of nitrification by soil fumigants and N-Serve® nitrogen stabilizer. Down Earth 32(3):14–17.

Gowda, T. K. S., V. R. Rao, and N. Sethunathan. 1977. Heterotropic nitrification in the simulated oxidized zone of a flooded soil amended with benomyl. Soil Sci. 123:171–175.

Greaves, M. P., L. A. Lockhart, and W. G. Richardson. 1978. Measurement of herbicide effects on nitrogen fixation by legumes. p. 581–586. In Proc. Vol. 2, 1978. British Crop Protection Conference—Weeds, London. 20–23 Nov. 1978. British Crop Protection Council, London.

Grossbard, E. 1971. The effect of repeated field applications of four herbicides on the evolution of carbon dioxide and mineralization of nitrogen in soil. Weed Res. 11:263–275.

Grossbard, E., and S. L. Cooper. 1974. The decay of cereal straw after spraying with paraquat and glyphosate. p. 337–343. In Proc. 12th British Weed Control Conference, London. 18–21 Nov. 1974. British Crop Protection Council, London.

Grossbard, E., and J. A. P. Marsh. 1974. The effect of seven substituted urea herbicides on the soil microflora. Pestic. Sci. 5:609–623.

Hale, M. G., F. H. Hulcher, and W. E. Chappell. 1957. The effects of several herbicides on nitrification in a field soil under laboratory conditions. Weeds 5:331–341.

Hamaker, J. W. 1966. Mathematical prediction of cumulative levels of pesticides in soils. Adv. Chem. Ser. 60:122–131.

Hamaker, J. W. 1972a. Decomposition: Quantitative aspects. p. 253–340. In C. A. I. Goring and J. W. Hamaker (ed.) Organic chemicals in the soil environment, Vol. 1 Marcel Dekker, Inc., New York.

Hamaker, J. W. 1972b. Diffusion and volatilization. p. 341–397. In C. A. I. Goring and J. W. Hamaker (ed.) Organic chemicals in the soil environment, Vol. 1. Marcel Dekker, Inc., New York.

Hamaker, J. W. 1975. The interpretation of soil leaching experiments. p. 115–133. In R. Haque and V. H. Freed (ed.) Environmental dynamics of pesticides. Plenum Press, New York.

Hamaker, J. W., and C. A. I. Goring. 1975. Turnover of pesticide residues in soils. p. 219–243. In D. D. Kaufman et al. (ed.) Bound and conjugated pesticide residues. ACS Symp. Ser. 29, Am. Chem. Soc., Washington, D.C.

Hamaker, J. W., and J. M. Thompson. 1972. Adsorption. p. 49–143. In C. A. I. Goring and J. W. Hamaker (ed.) Organic chemicals in the soil environment, Vol. 1. Marcel Dekker, Inc., New York.

Hamdi, Y. A., and M. S. Tewfik. 1969. Effect of the herbicide trifluralin on nitrogen fixation in Rhizobium and Azotobacter and on nitrification. Acta Microbiol. Pol. 1(18):53–58.

Henninger, N. M., and J. M. Bollag. 1976. Effect of chemicals used as nitrification inhibitors on the denitrification process. Can. J. Microbiol. 22:668–672.

Hofer, A. W. 1958. Selective action of fungicides on Rhizobium. Soil Sci. 86:282–286.

Höflich, G. 1968. The effect of various chemical substances on nitrification, some soil microorganisms, and leaching of nitrogen. (In German.) Albrecht-Thaer-Arch. 12:691–699.

Horowitz, M., T. Blumenfield, G. Herzlinger, and N. Halin. 1974. Effects of repeated applications of ten soil-active herbicides on weed population, residue accumulation, and nitrification. Weed res. 14:97–109.

Huber, D. M. 1980. Nitrification inhibitors—a timely innovation? Solutions (March–April) 24(2):86, 88, 90, 92.

Ibrahim, A. N. 1972. Effect of certain herbicides on growth of nitrogen-fixing algae and rice plants. Symp. Biol. Hung. 11:445–448.

Jaiswal, S. P. 1967. Effect of some pesticides on soil microflora and their activity. J. Res. Punjab Agric. Univ. 4:223–226.

Jaiswal, S. P., A. K. Verna, and C. N. Babu. 1971. Studies on the effect of some agrochemicals on nitrogen transformation in soil. 2. Ammonium nitrogen carrier. J. Res. Punjab Agric. Univ. 8:447–450.

Jakubisiak, B., and J. Golebiowska. 1963. Influence of fungicides on Rhizobium. Acta Microbiol. Pol. 12:196–202.

Jaques, R. P., J. B. Robinson, and F. E. Chase. 1959. Effects of thiourea, ethyl urethane and some dithiocarbamate fungicides on nitrification in Fox sandy loam. Can. J. Soil Sci. 39: 235–243.

Jenkinson, D. S., and D. S. Powlson. 1970. Residual effects of soil fumigation on soil respiration and mineralization. Soil Biol. Biochem. 2:99–108.

Kapusta, G., and D. L. Rouwenhorst. 1973. Interaction of selected pesticides and Rhizobium japonicum in pure culture and under field conditions. Agron. J. 65:112–115.

Karanth, N. G. K., and V. N. Vasantharajan. 1974. Phytotoxicity of Dexon (p-dimethyl-aminobenzene diazo sodium sulfonate) towards the legume *Crotalaria juncea.* Proc. Indian Nat. Sci. Acad. Part B 40:576–585.

Kaszubiak, H. 1966. The effect of herbicides on *Rhizobium.* I. Susceptibility of *Rhizobium* to herbicides. Acta Microbiol. Pol. 15:357–364.

Kaszubiak, H. 1968. The effect of herbicides on *Rhizobium.* II. Adaptation of *Rhizobium* to afalon, aretit and liro-betarex. Acta Microbiol. Pol. 17:41–50.

Kaufman, D. D. 1970. Pesticide metabolism. p. 73–86. *In* Pesticides in the soil—ecology, degradation and movement. Int. Symp., East Lansing, Mich. 25–27 Feb. 1970. Michigan State Univ., East Lansing, Mich.

Kearney, P. C., and J. R. Plimmer. 1970. Relation of structure to pesticide decomposition. p. 65–72. *In* Pesticides in the soil—ecology, degradation and movement. Int. Symp. Michigan State Univ., East Lansing, Mich.

Kenaga, E. E., and C. A. I. Goring. 1980. Relationship between water solubility, soil sorption, octanol-water partitioning, and concentration of chemicals in biota. p. 78–115. *In* J. C. Eaton et al. (ed.) Aquatic toxicology. ASTM Special Tech. Pub. 707, Am. Soc. for Testing and Materials, Philadelphia.

Kenaga, E. E., and R. W. Morgan. 1978. Commercial and experimental organic insecticides. Entomological Soc. of Am. Spec. Pub. 78-1, College Park, Md.

Koike, H. 1961. The effects of fumigants on nitrate production in soil. Soil Sci. Soc. Am. Proc. 25:204–206.

Koike, H., and P. L. Gainey. 1952. Effects of 2,4-D and CADE, singly and in combination on nitrate and bacterial content of soils. Soil Sci. 74:165–172.

Laveglia, J., and P. A. Dahm. 1974. Influence of AC 92,100 (Counter®) on microbial activities in three Iowa surface soils. Environ. Entomol. 3:528–533.

Lebbink, G., and G. J. Kolenbrander. 1974. Quantitative effect of fumigation with 1,3-dichloropropene mixtures and with metham-sodium on the soil nitrogen status. Agric. Environ. 1:283–292.

Lewis, J. A., G. C. Papavixas, and T. S. Hora. 1978. Effect of some herbicides on microbial activity in soil. Soil Biol. Biochem. 10:137–141.

Lin, S., B. R. Funke, and J. T. Schulz. 1972. Effects of some organophosphate and carbamate insecticides on nitrification and legume growth. Plant Soil 37:489–496.

Marsh, J. A. P., and H. A. Davies. 1978. The effect of herbicides on respiration and transformation of nitrogen in two soils. III. Lenacil, terbacil, chlorthiamid, and 2,4,5-T. Weed Res. 18:57–62.

Marsh, J. A. P., H. A. Davies, and E. Grossbard. 1974. The effect of a single field application of very high rates of linuron and simazine on carbon dioxide evolution and transformation of nitrogen within soil. p. 53–57. *In* Proc. 12th British Weed Control Conference, London. 18–21 Nov. 1974. British Crop Protection Council, London.

Marsh, J. A. P., H. A. Davies, andE. Grossbard. 1977. The effect of herbicides on respiration and transformation of nitrogen in two soils. I. Metribuzin and glyphosate. Weed Res. 17: 77–82.

Marsh, J. A. P., G. I. Wingfield, H. A. Davies, and E. Grossbard. 1978. Simultaneous assessment of various responses of the soil microflora to bentazone. Weed Res. 18:293–300.

Martin, H. 1972. Pesticide manual. 3rd ed. British Crop Protection Council, London.

Martin, H., and C. R. Worthing. 1974. Pesticide manual. 4th ed. British Crop Protection Council, London.

Martin, H., and C. R. Worthing. 1977. Pesticide manual. 5th ed. British Crop Protection Council, London.

Martin, J. P., R. D. Harding, G. H. Cannell, and L. D. Anderson. 1959. Influence of five annual field applications of organic insecticides on soil biological and physical properties. Soil Sci. 87:334–338.

Mayeux, J. V., and A. R. Colmer. 1962. The effect of 2,3,6-trichlorophenylacetic and 2,2-dichloropropionic acids on nitrite oxidation. Appl. Microbiol. 10:206–210.

Mazur, A. R., and T. D. Hughes. 1975. Nitrogen transformations in the soil as affected by the fungicides benomyl, dyrene, and maneb. Agron. J. 67:775–758.

McCall, P. J., and R. L. Swann. 1978. Nitrapyrin volatility from soil. Down Earth 34(3):21–27.

McCants, C. B., E. O. Skogley, and W. G. Woltz. 1959. Influence of certain soil fumigation treatments on the response of tobacco to ammonium and nitrate forms of nitrogen. Soil Sci. Soc. Am. Proc. 23:466–469.

McKell, C. M., and R. D. B. Whalley. 1964. Compatibility of 2-chloro-6 (trichloromethyl) pyridine with *Medicago sativa* inoculated with *Rhizobium meliloti.* Agron. J. 56:26–28.

Mills, H. A., and F. A. Pokorny. 1978. The influence of nitrapyrin on N retention and tomato growth in sand-bark media. J. Am. Soc. Hortic. Sci. 103:662–664.

Misra, K. C., and A. C. Gaur. 1974. Influence of simazine, lindane, and ceresan on different parameters of nitrogen fixation by groundnut. Indian J. Agric. Sci. 44:837–840.

Mitsui, S., I. Watanabe, M. Homma, and S. Honda. 1964. The effect of pesticides on the denitrification in paddy soil. Soil Sci. Plant Nutr. 10:107–115.

Mullison, W. R., R. W. Bovey, A. P. Burkhalter, T. D. Burkhalter, H. M. Hull, D. L. Sutton, and R. E. Talbert. 1979. Herbicide handbook. 4th ed. Weed Sci. Soc. of Am., Champaign, Ill.

Munnecke, D. E., and J. Ferguson. 1960. Effect of soil fungicides upon soil-borne plant pathogenic bacteria and soil nitrogen. Plant Dis. Rep. 44:552–555.

Narain, P., and N. P. Datta. 1973. Influence of nitrification inhibitors and some insecticides on nitrification of urea under aerobic and waterlogged conditions. Agrochimica 18:79–89.

Nayyar, V. K., N. S. Randhawa, and S. L. Chopra. 1970. Effect of simazine on nitrification and microbial population in a sandy-loam soil. Indian J. Agric. Sci. 40:445–451.

Nickell, L. G., and A. R. English. 1953. Effect of maleic hydrazide on soil bacteria and other organisms. Weeds 2:190–195.

Nightingale, G. T. 1937. The nitrogen nutrition of green plants. Bot. Rev. 3:85–174.

Nightingale, G. T. 1948. The nitrogen nutrition of green plants. II. Bot. Rev. 14:185–221.

Odu, C. T. I., and M. A. Horsfall. 1971. Effect of chloroxuron on some microbial activities in soil. Pestic. Sci. 2:122–125.

Olumbe, J. W. K., and C. Veatch. 1969. Organic matter—amiben interaction on nodulation and growth of soybeans. Weed Sci. 17:264–265.

Otten, R. J., J. E. Dawson, and M. M. Schreiber. 1957. Effects of several herbicides on nitrification in soil. Proc. Northeast. Weed Control Conf. 11:120–127.

Pancholy, S. K., and J. Q. Lynd. 1969. Bromacil interactions in plant bioassay, fungi cultures and nitrification. Weed Sci. 17:460–463.

Pareek, R. P., and A. C. Gaur. 1970. Effect of dichlorodiphenyl trichloroethane (DDT) on symbiosis of *Rhizobium* sp. with *Phaseolus aureus* (green gram). Plant Soil 33:297–304.

Radwan, M. A. 1965. Persistence and effect of TMTD on soil respiration and nitrification in two nursery soils. Forest Sci. 11:152–159.

Raghu, K., and I. C. MacRae. 1967. The effect of the gamma isomer of benzene hexachloride upon the microflora of submerged rice soils. II. Effect upon nitrogen mineralization and fixation, and selected bacteria. Can. J. Microbiol. 13:621–627.

Ratnayake, M., and L. J. Audus. 1978. Studies on the effects of herbicides on soil nitrification. Pestic. Biochem. Physiol. 8:170–185.

Rolston, M. P., A. C. P. Chu, and I. R. P. Fillory. 1976. Effect of paraquat on the nitrogen-fixing activity of white clover. N.Z. J. Agric. Res. 19:47–49.

Ross, D. J. 1974. Influence of four pesticide formulations on microbial processes in a New Zealand pasture soil. II. Nitrogen mineralization. N.Z. J. Agric. Res. 17:9–17.

Rovira, A. D. 1976. Studies on soil fumigation. 1. Effects of ammonium, nitrate and phosphate in soil and on the growth, nutrition and yield of wheat. Soil Biol. Biochem. 8:241–247.

Russell, E. J., and H. B. Hutchinson. 1909. The effect of partial sterilization of soil on the production of plant food. J. Agric. Sci. 3:111–144.

Selim, K. G., S. A. Z. Mahmoud, and M. T. El-Mokadem. 1970. Effect of dieldrin and lindane on the growth and nodulation of *Vicia faba*. Plant Soil 33:325–329.

Shaw, W. M., and B. Robinson. 1960. Pesticide effects in soil on nitrification and plant growth. Soil Sci. 90:320–323.

Simkover, H. G., and R. D. Shenefelt. 1951. Effect of benzene hexachloride and chlordane on certain soil organisms. J. Econ. Entomol. 44:426–427.

Smith, N. R., V. T. Dawson, and M. E. Wenzel. 1946. The effect of certain herbicides on soil microorganisms. Soil Sci. Soc. Am. Proc. 10:197–201.

Smith, M. S., and C. S. Weeraratna. 1975. Influence of some biologically active compounds on microbial activity and on the availability of plant nutrients in soils. II. Nitrapyrin, dazomet, 2-chlorobenzamide and tributyl-3-chlorobenzyl ammonium bromide. Pestic. Sci. 6:605–615.

Srivastava, S. C. 1966. The effects of Telodrin on nitrification of ammonia in soil and its implication on nitrogen nutrition of sugarcane. Plant Soil 25:471–473.

Stark, F. L., J. B. Smith, and F. L. Howard. 1939. Effect of chlorpicrin fumigation on nitrification and ammonification in soil. Soil Sci. 48:433–442.

Taha, S. M., S. A. Z. Mahmoud, and S. H. Salem. 1972. Effect of pesticides on *Rhizobium* inoculation, nodulation and symbiotic N-fixation of some leguminous plants. Symp. biol. Hung. 11:423–429.

Tam, R. K. 1945. The comparative effects of a 50-50 mixture of 1,3-dichloropropene and 1,2-dichloropropane (DD mixture) and of chloropicrin on nitrification in soil and on the growth of the pineapple plant. Soil Sci. 59:191–205.

Tam, R. K., and H. E. Clark. 1943. Effect of chloropicrin and other soil disinfectants on the nitrogen nutrition of the pineapple plant. Soil Sci. 56:245–261.

Teater, R. W., J. L. Mortensen, and P. F. Pratt. 1958. Effect of certain herbicides on rate of nitrification and carbon dioxide evolution in soil. J. Agric. Food Chem. 6:214–216.

Thiagalingam, K., and Y. Kanehiro. 1971. Effect of two fumigating chemicals, 2-chloro-6-trichloromethyl pyridine and temperature on nitrification of added ammonium in Hawaiian soils. Trop. Agric. (Trinidad) 48:357–364.

Thiegs, B. J. 1955. Effect of soil fumigation on nitrification. Down Earth 11(1):14–15.

Thorneburg, R. P., and J. A. Tweedy. 1973. A rapid procedure to evaluate the effect of pesticide on nitrification. Weed Sci. 21:397–399.

Tu, C. M. 1977. Effects of pesticide seed treatments on *Rhizobium japonicum* and its symbiotic relationship with soybeans. Bull. Environ. Contam. Toxicol. 18:190–199.

Tu, C. M., and W. B. Bollen. 1968a. Effect of paraquat on microbial activities in soils. Weed Res. 8:28–37.

Tu, C. M., and W. B. Bollen. 1968b. Interaction between paraquat and microbes in soils. Weed Res. 8:38–45.

Tu, C. M., and W. B. Bollen. 1969. Effect of Tordon® herbicides on microbial activities in three Willamette Valley soils. Down Earth 25(2):15–17.

Turner, F. T. 1979. Soil nitrification retardation by rice pesticides. Soil Sci. Soc. Am. J. 43:955–957.

van Faassen, H. G. 1973. Effects of mercury compounds on soil microbes. Plant Soi. 38:485–487.

van Schreven, D. A., D. J. Lindenbergh, and A. Koridan. 1970. Effect of several herbicides on bacterial populations and activity and the persistence of these herbicides in soil. Plant Soil 33:513–532.

Vlassak, K., K. A. H. Heremans, and A. R. van Rossen. 1976. Dinoseb as a specific inhibitor of nitrogen fixation in soil. Soil Biol. Biochem. 8:91–93.

Wainwright, M., and G. J. F. Pugh. 1973. The effect of three fungicides on nitrification and ammonification in soil. Soil Biol. Biochem. 5:577–584.

Wilson, J. K., and R. S. Choudri. 1946. Effects of DDT on certain microbiological processes in the soil. J. Econ. Entomol. 39:537–538.

Winfree, J. P., and R. S. Cox. 1958. Comparative effects of fumigation with chlorpicrin and methyl bromide on mineralization of nitrogen in everglades peat. Plant Dis. Rep. 42:807–809.

Wolcott, A. R., F. Maciak, L. N. Shepherd, and R. E. Lucas. 1960. Effects of Telone® on nitrogen transformations and on growth of celery in organic soil. Down Earth 10(1):10–14.

Worthing, C. R. 1979. The pesticide manual. A world compendium. British Crop Protection Council, London.

19 Modeling of the Soil Nitrogen Cycle

KENNETH K. TANJI

University of California
Davis, California

I. INTRODUCTION

The behavior of N in the soil-plant-water system is dynamic and involves numerous interactions (see Chapt. 11, 14, 22, and 23, this book). For economic and environmental reasons, N use efficiency by crop plants and N losses through runoff, leaching, and denitrification need to be examined more analytically and quantitatively. One approach to integrating and synthesizing a large body of knowledge on soil N transformations and transport is mathematical modeling and computer simulation.

Mathematical and simulation modeling may have quite diverse objectives: testing existing or new concepts and hypotheses, obtaining greater conceptual understanding of complex problems, obtaining more quantitative evaluation or prediction of observed phenomena or experimental data, identifying research needs, and helping develop guidelines for best management practices.

This chapter focuses on systems-level models and their application to field studies; it does not cover process-level models for batch and column studies (see Chapter 11, this book). Section II of this chapter provides some exposure to computers, modeling principles, and simulation. Section III surveys representative N models and provides in-depth descriptions and appraisals of selected N simulation models. Section IV contains philosophical and technical critique as well as research needs.

II. COMPUTERS AND SIMULATION MODELS

With the invention of the Harvard Mark I in 1944, the first automatic calculating machine, ENIAC in 1946, the first electronic digital computer, and Univac I in 1951, the first commercially available general-purpose digital computer, there has been an exponential growth in the development and marketing of digital computers. This explosive development in computers followed two lines: larger and larger computer systems, starting from Univac I to the IBM 360's in the 1960's, and now CRAY-I, the "Super Computer," as well as smaller and smaller computers, starting with mini-computers in 1965, microcomputers in 1970, and now microprocessors

(Davis, 1977). This rapid development can be attributed to advances in electronic circuitry. The first-generation computers used vacuum tubes, the second-generation computers used a circuit board with transistors and diodes, and the third-generation computers used microminiature circuits, including the "chip." The pocket electronic calculator contains a single microelectronic chip of about 25 mm^2 that performs mathematical functions (McWhorter, 1976).

Concurrent with computer development and applications, a computer science jargon (such as FORTRAN and BASIC, hardware and software, batch operations and time sharing, library functions and subroutines, flow chart and do loop, and debugging and calibration) has evolved. Thus, in the last 2 decades, there has been a bewildering array of new developments both in computers and in terminology. This section summarizes computers and simulation modeling techniques. For more extensive treatments, see Readings from Scientific America (1975), Spencer (1974), Barden (1977), and McGlynn (1976).

A. Computers and Programming

1. COMPUTER SYSTEMS

A computer is nothing more than a very fast electronic calculating machine. What a computer can do or cannot do needs to be clarified. One of the most widespread myths is that computers can think. That is absolutely incorrect. Since, unlike the human mind, a computer does not have creative abilities, it can carry out only what it is instructed to do, nothing more. A computer may be likened to an electronic slave: man does the thinking; the computer does the work. Moreover, no matter how well or how correct the computer is instructed, the quality of the output data is influenced to a large degree by the quality of the input data. If the input data is in error, incomplete, or fed in the wrong sequence, it results in GIGO (garbage in, garbage out).

A question frequently asked is "do computers make mistakes?" The answer is a qualified no, because more often than not, errors are introduced by human beings. Some computer malfunction is always a probability, but that is readily detected by computer-center personnel. Computers are, however, susceptible to electrical disturbance, ambient temperature and humidity, dust, salty air, etc. and need to be protected against such elements.

Computing systems may be broadly categorized into computers, minicomputers, and microcomputers. Computers by size, speed, and cost could formerly be clearly categorized, but with improvements in electronic circuitry and market competition, such criteria may no longer apply. For example, the distinction between minicomputers and microcomputers is becoming less clear-cut.

A minicomputer system is made up of hardware and software. Hardware consists of the physical units, such as the chassis housing the computer and input/output devices (teletype, card reader, magnetic tape, etc.). Software consists of the programs and subroutines that perform routine calculations and compiler program to convert programs written in symbolic language, e.g., FORTRAN, into machine language that the computer understands.

The four essential functional units of a minicomputer system are the Central Processing Unit (CPU), the memory, input/output controllers, and the secondary storage controllers (Barden, 1977). The CPU performs the four basic arithmetic functions (addition, subtraction, multiplication, division); the program (instructions), and data are stored in the memory; and there is usually a secondary storage unit, such as a magnetic tape unit. The minicomputer software includes machine language, a combination of binary ones and zeros. Minicomputers in general are dedicated single-user computers, unlike the larger-scale computing systems, which can operate with more than one program.

In 1971, a smaller and less expensive version of minicomputers, known as the microcomputer, was developed. The heart of the microcomputer is the microprocessor, a small (25-mm^2) CPU on an integrated circuit chip that performs various standard functions and operations (Freiberger & Chew, 1978). A microcomputer is an inexpensive general-purpose computer that is sold as a personal computer, such as the Radio Shack TRS-80. These microcomputers may be fitted with input/output devices and software programs similar to those of minicomputers.

A typical medium-sized computer system such as the Burroughs B6700 system at the University of California at Davis has dual processors with each processor independent of the other. The core storage consists of 16 and 64 K modules for a total of 256 K words. The input/output processor is capable of handling up to six simultaneous input or output operations. The disk storage contains 610 million bytes, whereas the magnetic tape storage consists of four nine-track and one seven-track magnetic-tape drives. Other standard peripherals include two card readers, two printers, and one card punch attached to the system. The data communication processor is capable of supporting up to 256 lines, 32 of which are attached in support of remote batch computing and time sharing. The B6700 runs under the control of the Master Control Program, which permits 20 to 30 jobs to be processed at the same time. Programming languages available are FORTRAN, ALGOL, BASIC, COBOL, and PL/1. Applications software are available in support of statistical analysis, linear programming, and numerical analyses.

2. PROGRAMMING

A program is a set of instructions for solving a given problem with a computer. The formulation and testing of a computer program is initiated with a problem definition (Spencer, 1974), an analysis to determine exactly what problem is to be solved, how it is to be solved, and what programming language to use.

Flow charting is a pictorial representation of the sequence of logical and arithmetic operations needed to solve the problem. It not only aids in writing the program but also provides a means of presenting the program logic to users. Coding is the writing of the program. A single flowchart symbol may result in several computer instructions. Symbolic (source) language is used to simplify the programming task. A translator converts the source program into machine language (object program) that is obeyed by a computer. In addition to FORTRAN and BASIC, N cycle modelers also use CSMP (Continuous System Modeling Program), a special programming package developed by IBM.

In the process of coding a program, it is very common to make mistakes that need to be detected and corrected (debugged) in the testing stage of program execution. After the program is working properly on the computer, it is highly desirable to document it and develop a user's manual.

B. Development and Application of Systems Simulation Models

1. STEPS IN DEVELOPING SYSTEMS MODELS

Although no hard and fast rules exist for developing models, Fig. 1 presents a guideline for the development and application of large-scale systems. If we take a problem situation, the initial step is to specify modeling objective(s), which is followed by system identification. The physical nature of the system is examined, and the system of interest is isolated for study. Within this system of interest, various processes and conditions affect the item of interest (state variable). Available information and data are evaluated and analyzed preparatory to mathematical model formulation. Since most systems are not continuously monitored, spatially and temporally, the modeler views the behavior of a system through small portholes.

Riley (1970) points out four general classes of mathematical models. Mathematical representation of a system may be achieved by either lumped parameter or distributed parameter models. The former is sometimes known as the "black box" type of model, and the latter views a problem analytically in both space and time. The mathematical representation of system behavior may be described in terms of deterministic or stochastic processes. The former gives cause and effect relations, and the latter deals with probability. Because natural systems are rather complex, a mathematical representation is usually a conceptual simplification.

Computer synthesis involves formulation of a computer model for the mathematical model, including the programming step. After the computer program has been debugged, a systems model usually requires calibration, a fine tuning of model coefficients and parameters to get a better fit between calculated and observed data. The calibrated model is then tested with an independent set of data to determine its level of fitness between computed output and observed data. After the model has been validated, it is ready for computer runs commensurate with modeling objectives.

The results obtained are analyzed, and if they do not meet the quality desired, one or more of the steps may need to be reexamined and modified as indicated by the feedback loops. If the computer outputs meet the desired standards, the outputs are summarized and interpreted.

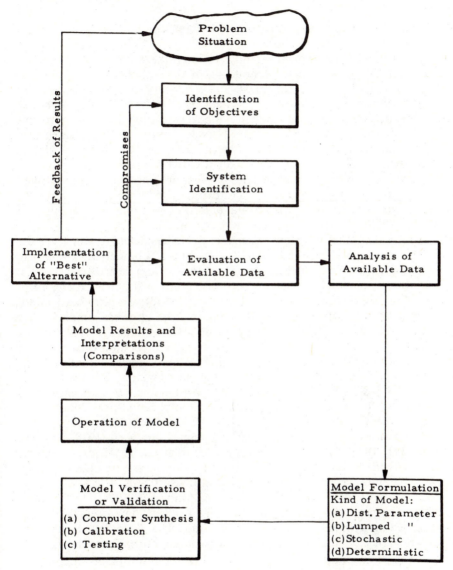

Fig. 1—Steps in the development and application of a simulation model (Riley, 1970).

III. REPRESENTATIVE NITROGEN MODELS

This section contains a survey of representative dynamic N simulation models and evaluation of selected ones. These models were formulated with specific objectives in mind, and examples are given later. The more general model objectives were cited earlier in Section I.

A. Survey of Dynamic N Simulation Models

Tanji and Gupta (1978) recently reviewed the state of the art and research needs for N simulation modeling of the soil-plant-water system in croplands. The physical, biological, and chemical processes in this system are described mathematically by mechanistic models, empirical models, or combinations of both. The simulation models are integrating the up to now isolated bodies of knowledge on soil N cycle and providing increased conceptual understanding. However, these models simulate processes such as denitrification, for which little or no quantitative field data are available, or they include parameters and coefficients that are not commonly measured in the field and have to be estimated or extrapolated from other sources. As more model testing with experimental data is carried out, the credibility of such modeling efforts will become more acceptable.

Table 1 lists representative dynamic N models, some of which are described and evaluated in more detail later. These N models vary widely in scope. Some were designed to simulate N losses from larger-scale systems, such as watersheds and irrigation projects (Frere et al., 1975; Frere, 1976; Donigian & Crawford, 1976; Shaffer et al., 1977), or farm-sized multi-cropped units (Duffy et al., 1975). Some of these large-scale models have been verified with smaller-scale experiments (Dutt et al., 1972; Reddy et al., 1979a, b). Most of these models, however, simulate N and water fluxes in smaller-scale systems, such as field plots and lysimeters. Many of the models are verified with cornfield experiments (Watts;[1] Davidson et al., 1978a; Tanji & Mehran, 1979; Tillotson[2]). Most models consider fertilizer N as an input, but some also consider the land application or soil incorporation of crop residues (Beek & Frissel, 1973; van Veen;[3] Rao et al., 1981), animal manures (Beek & Frissel, 1973; Frissel & van veen, 1978; Hagin & Amberger, 1974; Reddy et al., 1979a, b; Rao et al., 1981), or sewage effluents (Selim & Iskandar, 1978, 1981). Many modelers use FORTRAN IV for programming language, though several use CSMP (Beek & Frissel, 1973; Frissel & van Veen, 1981; Hagin & Amberger, 1974; Duffy et al., 1975). The programs and sample input and output data are available as published material or on request from the modelers. Complete documentation (reference manual) is available on some models from the authors.

The physical processes usually modeled in the soil-plant-water system are infiltration, redistribution, evaporation, transpiration, and seepage of soil water. Other upper-boundary conditions considered are snowmelt (Donigian & Crawford, 1976), surface runoff (Shaffer et al., 1977; Frere et al., 1975; Donigian & Crawford, 1976), heat flux (Beek & Frissel, 1973; Frissel & van Veen, 1978; Hagin & Amberger, 1974; Watts & Hanks, 1978), or air flow (Reddy et al., 1979b). Other lower-boundary conditions considered are semiinfinite soil depth and assumption of unit-hydraulic gradi-

---

[1] D. G. Watts. 1975. A soil-water-nitrogen-plant model for irrigated corn on coarse textured soils. Ph.D. Thesis. Utah State Univ., Logan.

[2] W. R. Tillotson. 1979. Simulation of soil-water and nitrogen fluxes under cropped conditions. I. Model theory. M.S. Thesis. Utah State Univ., Logan.

[3] J. A. van Veen. 1977. The behaviour of nitrogen in soil, a computer simulation model. Ph.D. Thesis. Univ. of Amsterdam, published by ITAL-Wageningen, The Netherlands.

Table 1—Listing of dynamic N simulation model and subsequent modification and/or applications.

References	Brief description and comments
Dutt et al. (1972)	One of the earliest digital simulations of N in soil-water-plant systems. Designed for use by USBR for irrigation projects. Moisture flow program verified with soil moisture tension in cropped lysimeter. Biological-chemical program considers N processes empirically (regression or algebraic); does not consider denitrification. Nitrogen transformation subroutine validated with data from incubation type of data. Model simulated N flows in lysimeters cropped with barley (*Hordeum vulgare* L.) and milo (*Sorghum bicolor* L. Moench), and fertilized with ammonium sulfate and potassium nitrate. Model validation includes N uptake, residual soil organic N, and nitrates in lysimeter effluents.
Shaffer et al. (1977)	Dutt et al. (1972) model extended to tile-drained croplands and incorporated into a large irrigation return flow model to handle N as well as dissolved mineral constituents. Model allows user to select the degree of sophistication in simulation or to bypass certain subroutines. Zero-order denitrification and transition-state nitrification added. Model verified for salinity but not N.
Gupta et al. (1978a); Shaffer & Gupta (1981)	Shaffer et al. (1977) model simulated waste water applications in field test sites in Apple Valley, Minn., and Suisun Marsh, Calif. In the former, predicted seasonal average nitrate of drainage water was compared with measured averages for low and high application rates. In the latter, salinity simulation was validated but not N.
Beek & Frissel (1973)	First CSMP N simulation model. Used water and heat flow CSMP model of deWit and van Keulen (1972). Nitrogen modeling focused on decomposition of organic matter in soils, including added straw and manure, in the absence of growing plant. Denitrification not considered. Organic matter subdivided into proteins, sugars, cellulose, and lignin and treated separately for mineralization. Simulated microbial biomass and nitrification with moisture, temperature, and pH as environmental variables. Results are not verified.
Hagin & Amberger (1974)	This CSMP model is similar to that of Beek and Frissel (1973). Major differences include restricting nitrification to the top 10-cm layer of soil, modeled oxygen transfer in soils, and allow denitrification only when oxygen concentration is below a certain level, and the presence of growing plants that required modeling evapotranspiration. Presented simulation for Mediterranean and Central European conditions but not tested.
van Veen[3]; Frissel & van Veen (1978)	Modified Beek and Frissel's model (1973) by subdividing a single complex model into submodels of mineralization, nitrification, denitrification, ammonia volatilization, ammonium clay fixation, and nitrate leaching. Nitrogen transformations based on modeled microbial activities, e.g., C/N ratio no longer used but uptake of C by biomass regulating decomposition. This second generation model was tested on greenhouse and field experiments involving straw, fertilizer, or both in the absence of plants. Verification data presented on inorganic N and microbial biomass with time.
Frissel & van Veen (1981)	This third stage of development of the CSMP model involved simplifying the substrate components to be more in line with those measured by soil scientists, i.e., rapidly decomposable added material such as carbohydrates, slowly decomposable added material such as cellulose, rapidly decomposable N-containing substances such as proteins and amino sugars, rapidly decomposable microbial debris products, resistant microbial debris products, and old organic matter. Model verification carried out with greenhouse and field data show third-stage model to be the most accurate.

(continued on next page)

Table 1—Continued.

References	Brief description and comments
Mehran & Tanji (1974)	Proposed irreversible first-order kinetics for nitrification, denitrification, mineralization, immobilization, and plant uptake and reversible first-order kinetics for ammonium-ion exchange. Model verified with published incubation data.
Gupta et al. (1978b)	Simulated field soil water cycle in cornfield plots. Extended Nimah and Hanks' model (1973) by modeling root distribution factor and radial distance between roots. Evaluated spatial variability of soil hydraulic conductivity (Simmons et al., 1979). Various subroutine options available. Model validation included profile water contents over growing season.
Tanji & Mehran (1979); Tanji et al. (1981)	Coupled water flow submodel (Gupta et al., 1978b) with N transformations (Mehran & Tanji, 1974) and transport. Model includes checks on daily mass balance of water and N. Sensitivity analysis carried out on mineralization, simulations verified with cornfield data including plant uptake, residual inorganic N, and profile nitrate contents.
Frere et al. (1975)	Formulated agricultural chemical transport model (ACTMO) for storm events in agricultural watersheds. Large-scale model consists of hydrological, chemical, and erosional models. Nitrogen processes considered are mineralization with temperature and moisture variables, plant uptake in proportion to water uptake, fertilizer inputs, and vertical movement and lateral outflow. Model tests for carbofuran, sediment, and water discharges but not N.
Frere (1976)	This simplified regional-scale computer model considers nitrate leaching losses from fall or spring ammonium fertilizer applications before roots have reached full extension. Considers only nitrification of added fertilizer and not initial soil N at time of fertilization or denitrification. Presented results for eastern and central USA.
Duffy et al. (1975)	CSMP simulation model to evaluate effects of Corn Belt farm management practices and weather on nitrates in tile-drain effluents. Modeled nitrification, mineralization-immobilization, and denitrification, with site-specific empirical relationships. Validated simulations on 62-ha farm in Illinois with corn (*Zea mays* L.), wheat (*Triticum aestivum* L.), and soybean (*Glycine max* L. Merr.) croppings with water table depths, tile flow, and nitrates in soil profile and tile effluent.
Vithayathil et al. (1979)	This is a reference manual for FIELD, the Duffy et al. (1975) model reprogrammed into FORTRAN IV. It includes additional verified simulations of tile flow and nitrate contents for 1970 to 1974 as well as sensitivity analysis on the nitrate concentration in the tile effluent for 1972. Results of sensitivity analysis for 10% change of the variables indicate water content in the unsaturated region is by far most sensitive. Less significant factors are nitrates in the soil profile and amount of precipitation.
Watts[1]; Watts & Hanks (1978)	Water flux simulated by Nimah and Hanks (1973) and soil temperature by Hanks et al. (1971). N processes modeled are temperature-dependent mineralization, nitrification, and urea hydrolysis as well as plant uptake. Model validated with cornfield experiments in coarse-textured central Great Plains soil, including profile soil water content and nitrate and ammonium concentrations as well as N plant uptake.

(continued on next page)

Table 1—Continued.

References	Brief description and comments
Tillotson[2]; Tillotson et al. (1980)	Water flow by Nimah and Hanks' (1973) model. Nitrogen transformations including urea hydrolysis simulated by Wagenet et al. (1977); mineralization and immobilization not considered but ammonia volatilization is considered. Plant uptake described by N flux into the root (Nye & Tinker, 1977) with root uptake absorption coefficient (Warncke & Barber, 1973). Model validated with corn lysimeter data.
Donigian & Crawford (1976)	Formulated agricultural runoff model (ARM) for watershed. Model simulated rainfall and snowmelt runoffs and nutrient and other water quality contributions to stream channels from surface and subsurface sources. N transformation model similar to Mehran and Tanji (1974) but with temperature corrections on rate constants. Model validated for rainfall runoff, sediments, and some pesticide residues.
Reddy et al. (1979a, b)	Donigian and Crawford (1976) model extended to handle animal waste loadings. Modified mineralization simulation with considerations for potentially mineralizable N (Stanford & Smith, 1972) and short-term rate kinetics. Added ammonia volatilization from animal wastes that is dependent on temperature, air flow rate, and cation exchange capacity. Mineralization and volatilization validated with laboratory data.
Davidson et al. (1978a); Davidson et al. (1978b); Davidson & Rao (1978)	Two models developed. Simplified management model: assumes homogeneous profile, displacement of resident soil water and solutes ahead of infiltrating water at field capacity, first-order kinetics for N transformations (Mehran & Tanji, 1974). Water extraction related to PET and Molz and Remson (1970) model, and N uptake by empirical Michaelis-Menten type of model regulated by water uptake and available nitrate and ammonium. Model tested on literature N uptake by corn including Watts.[1] Refined research model: considers simultaneous transport of water and N on multilayered soils, first-order kinetics for N transformation, modification of Molz and Remson extraction model with root growth and root length distribution simulated and constraints on actual transpiration, and plant uptake of N by Michaelis-Menten kinetics. Model currently being validated on Watts[1] and other data sources.
Rao et al. (1981)	Davidson et al. (1978a, b) model extended to consider N behavior in cropped lands receiving crop residues and animal manures. Soil organic N and C subdivided into readily mineralizable and slowly mineralizable fractions (phases). Computer program is modular, and up to four subroutines may be modified or replaced.
Selim & Iskandar (1978, 1981)	Davidson et al. (1978a) research model modified for application of waste water on croplands. Soil organic N excluded, ammonium exchange described by Freundlich isotherms. The 1978 paper describes model and sensitivity analyses on N transformation rate coefficients, rate of N uptake, and amounts and scheduling of waste water applications. The 1980 paper considered water table in lower boundary carried out additional sensitivity analyses on rates of nitrification, exchange and plant uptake, and modified denitrification rate by degree of water saturation. Model validated with lsyimeter experiment receiving treated municipal sewage effluent on grasses. Also validated simulation model comprising water flow from Selim and Iskandar (1978) and N flow from Tanji and Mehran (1979).

ent for most models and static or quasi-dynamic water table for some (Shaffer et al., 1977; Duffy et al., 1975; Selim & Iskandar, 1981). Most of the models simulate unsaturated water flow in soils, but a few simulate saturated flow for the underdrained or water table systems. The more simplified water flow simulation models assume homogeneous soil profiles and describe soil water movement empirically (Frere et al., 1975; Frere, 1976; Duffy et al., 1975; Davidson & Rao, 1978). Most models, however, simulate unsaturated water flow by a partial differential equation describing transient nonsteady flow that includes a root water extraction term. Such models are solved numerically, mainly by the implicit finite-difference scheme.

The biological and chemical processes usually modeled are N transformations and plant uptake. Most models consider nitrification, denitrification, mineralization, and immobilization, but a few do not consider one or more of these, such as denitrification (Dutt et al., 1972; Beek & Frissel, 1973; Frere et al., 1975; Watts & Hanks, 1978; Tillotson[2]) or mineralization-immobilization (Frere et al., 1975; Frere, 1976; Selim & Iskandar, 1978; Tillotson[2]). Other biological, chemical, or physical processes affecting soil N that are modeled are urea hydrolysis (Dutt et al., 1972; Watts;[1] Tillotson[2]), ammonia volatilization (van Veen;[3] Reddy et al., 1979b), microbial biomass (Beek & Frissel, 1973; Frissel & van Veen, 1978, 1981), NH_4^+ exchange or adsorption (Mehran & Tanji, 1974; Donigian & Crawford, 1976; Davidson et al., 1978a; Selim & Iskandar, 1981), NH_4^+ clay fixation (van Veen)[3], and biological N_2 fixation (Duffy et al., 1975).

Nitrogen transformations and plant uptake are described by empirical or mechanistic models. Empirical models include multiple regression (Dutt et al., 1972) or algebraic expressions (Frere et al., 1975; Duffy et al., 1975; Watts & Hanks, 1978; Reddy et al., 1979a, b). Many models involve chemical kinetics, mainly first order, or in one model, zero order for denitrification (Shaffer et al., 1977), and Michaelis-Menten type of kinetics. Many models take into account effects of environmental conditions such as temperature on rate constants so that it partially regulates N transformations, but others ignore this factor (Mehran & Tanji, 1974; Frere et al., 1975; Duffy et al., 1975; Tanji & Mehran, 1979; Davidson et al., 1978a; Selim & Iskandar, 1978; Tillotson et al., 1980). A few models consider the effects of pH on N transformations (Beek & Frissel, 1973; Hagin & Amberger, 1974; Reddy et al., 1979b) as well as soil aeration or water saturation status (Hagin & Amberger, 1974; Selim & Iskandar, 1981; Frissel & van Veen, 1978).

Solute transport through the soil is described by simple displacement or convective-diffusion models. The former includes solute transport by discrete soil plates (depths), which may or may not include mixing of invading and resident soil water (Dutt et al., 1972; Frere et al., 1975; Duffy et al., 1975; Davidson & Rao, 1978). The convective-diffusion transport models used in the CSMP models take discrete soil plates too, but diffusion is also considered. In the FORTRAN models the transport is usually described by differential equations containing sink/source terms so that N transformations and transport are calculated simultaneously. In CSMP models, in contrast to the FORTRAN models, N transformations are calculated separately, followed by transport.

Attempts have been or are being made to validate simulation models. Testing of larger-scale models usually involves comparison of N concentrations in runoffs and, infrequently, soil N. Testing of smaller-scale models usually involves comparison of soil water contents, inorganic soil N, uptake in the harvested crop, and N in collected seepage waters. van Veen[3] and Frissel and van Veen (1978) validated microbial populations and organic matter decomposition. Reddy et al. (1979b) validated NH_3 volatilization. Rao et al. (1981) are currently validating denitrification using Rolston's field data from the University of California at Davis (P. S. C. Rao, personal communication). Validation is usually incomplete in those modeling efforts because direct measurements are not available for all of the modeled processes or computer outputs.

As pointed out earlier, dynamic N simulation modeling is providing increased insight into a complex problem and pointing out gaps in our knowledge. As more comprehensive field data become available, these models can be subjected to further testing. It is of interest that except for two models (Duffy et al., 1975; Shaffer et al., 1977), all models assume one-dimensional flow. The need for two-dimensional modeling is critical for specific site conditions. It is of interest also that except for one model (van Veen)[3], simulations on microbial biomass are not validated for those models using Michaelis-Menten kinetics. Since simulations involve only a few crop plants, grains, and grasses, a need exists to model other crops or cropping systems. As the modeling approaches become sophisticated and refined, the cost of simulation has been increasing. A need exists for achieving computational efficiency and lower costs for computer runs. The greatest deficiency, however, appears to be an adequate field data base.

B. Evaluation of Selected Simulation Models

The models appraised in this section were selected in part on the basis of their contribution to N modeling efforts, completeness in the documentation of the computer simulation model, and response to a questionnaire distributed to the modelers by this writer. Several of the models noted in the survey of simulation models should be considered here, but the modelers did not provide the information requested and, therefore, are unfortunately excluded.

The computer simulation models considered in this section include a CSMP model that focuses on microbial degradation of organic matter components (van Veen;[3] Frissel & van Veen, 1981) as well as another CSMP model that simulates, in particular, the N emissions through tile drainage (Duffy et al., 1975; Vithayathil et al., 1979). FORTRAN models include a simplified management model to simulate N leaching losses from field plots (Davidson & Rao, 1978; Rao et al., 1981) as well as a large-scale model that is used mainly for simulating the quality of return flows from irrigation projects (Shaffer et al., 1977). In addition, appraisals are given on three research-oriented FORTRAN models that simulate N losses from experimental field plots and lysimeters (Davidson et al., 1978a; Tanji & Mehran, 1979; Tillotson[2]).

The objectives of those modeling efforts and their intended or actual applications were noted in Table 1. The following gives a more complete description and evaluation of the representative simulation models.

1. WATER FLOW SIMULATION

To simulate soil N fluxes, it is first necessary to simulate soil water fluxes as illustrated by Fig. 2. The outputs from the water flow model serve as inputs to the N flow model, so that the concentrations (mass vol⁻¹) and masses (conc × vol) of N can be calculated. Simulations of water and N fluxes are usually carried out in comparatively small temporal and spatial increments. Time scale is usually daily or less, and spatial scale is usually 30 cm or less. Some of the major factors affecting the field soil water cycle include water application and precipitation, estimation of actual from

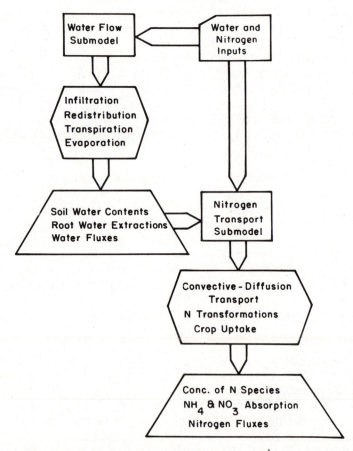

Fig. 2—A smplified block diagram of the water flow and N transport submodels in the dynamic N simulation model (Tanji & Mehran, 1979).

potential evapotranspiration, soil hydraulic properties, water uptake by plants, and water table conditions (Gupta et al., 1978b).

The governing partial differential equation for the one-dimensional nonsteady flow of water in the root zone (Nimah & Hanks, 1973) can be expressed as

$$\frac{\partial \theta}{\partial t} = \frac{\partial}{\partial z}\left[K(\theta)\ \frac{\partial H}{\partial z}\right] + S(z,t) \qquad [1]$$

where θ is the volumetric water content, t is time, z is the vertical distance from the soil surface, K is the hydraulic conductivity, H is the hydraulic head (comprising of pressure head h and gravity head z), and S is the sink term for root water extraction. Another version of the nonlinear partial-differential equation as used by Shaffer et al. (1977) and Davidson et al. (1978a) takes into account soil water diffusivity D. Equation [1] may be conceptually described as follows: the product of hydraulic conductivity and hydraulic gradient (terms within the brackets) gives flux, the first group of terms in the RHS of Eq. [1] gives rate of change of flux with soil depth, and the sum of the two groups of terms gives the rate of change of water content with respect to time.

This equation requires a knowledge of the functional relationships between K and θ. Taking K to be pressure dependent, i.e., replacing $K(\theta)$ by $K(h)$, simplifies the lower boundary conditions and allows consideration for layered cases.

Nimah and Hanks (1973) input data for h vs. θ and K vs. θ, whereas Davidson et al. (1978a) estimate $K(\theta)$ by

$$K(\theta) = \exp[a\theta^b + c] \qquad [2]$$

where a, b, and c are -3.347, -0.62, and 10.175, respectively, for Eustis f.s. soil, and Gupta et al. (1978b) estimate K from h using the following polynomial

$$\ln K(h) = a_0 + a_1 \ln h + a_2 (\ln h)^2 \qquad [3]$$

and, in turn, the measured θ and h are related by

$$\theta = b_0 + b_1 \ln h + b_2 (\ln h)^2 \qquad [4]$$

with coefficients that vary in value for site-specific field locations and soil layers.

Equation [1] needs to be transformed so that only one variable exists, and this is done with $C(\theta) = \partial\theta/\partial h$, where $C(\theta)$ is the soil water–differential capacity. By the chain rule of calculus,

$$\frac{\partial \theta}{\partial t} = \frac{\partial \theta}{\partial h} \cdot \frac{\partial h}{\partial t} = C(\theta)\ \frac{\partial h}{\partial t}. \qquad [5]$$

Substituting Eq. [5] into [1] with $K(h)$ instead of $K(\theta)$ yields

$$C(\theta)\frac{\partial h}{\partial t} = \frac{\partial}{\partial z}\left[K(h)\frac{\partial H}{\partial z}\right] - S(z). \qquad [6]$$

Equation [6] is subject to the following initial and boundary conditions:

(a) $h = h_0(z)$	$0 < z < Z$	$t = 0$	
(b) $h_d \leq h \leq 0$	$z = 0$	$t > 0$	
(c) $h = P(t)$	$z = 0$	$t > 0$	
(d) $q = K(h)$	$z = Z$	$t > 0$	[7]
(e) $h = 0$	$z = Z$	$t > 0$	
(f) $q = Q_z$	$z = Z$	$t > 0$	

where h_0 is the initial pressure head, z is the distance to the lower boundary from the surface, h_d is the dry limit of h, $P(t)$ is the depth of ponding, q is the volumetric flux, and Q_z is the prescribed flux. For Eq. [7], (a) is the initial condition, (b) is for the Neuman condition at the upper boundary for values of h between saturation and dry limit, (c) is for the infiltration condition, (d) is the unit hydraulic gradient at the lower boundary for a semi-infinite depth, (e) is for the static water table, and (f) is for the dynamic water table. This is a generalized form of the soil water flow model applicable to several lower-boundary conditions (Gupta et al., 1978b) which is solved by the finite difference method.

The sink term S for water extraction is defined by Nimah and Hanks (1973) to be

$$S(z,t) = \frac{[H_r + (\text{RRES} \cdot z) - h(z,t) - s(z,t)] \cdot \text{RDF}(z) \cdot K(\theta)}{\Delta x \cdot \Delta z} \qquad [8]$$

where H_r is the effective water potential in the root at $z = 0$, RRES is a root resistance term equal to $1 + Rc$ in which Rc is a flow coefficient in the plant root system that is assumed to be 0.05, $s(z,t)$ is the solute potential, $\text{RDF}(z)$ is the proportion of total active roots in Δz, and Δx is the distance between plant roots that is assumed to be unity. The value of H_r is iteratively calculated until plant root extraction over the soil profile is equal to the potential transpiration (PT), provided H_r is higher than the potential below which wilting will occur (H_w). So H_r is bounded on the wet end by $H_r = 0.0$ and the dry end by H_w in this extraction model. The daily PT is taken from input data.

Davidson et al., (1978a) use the Molz and Remson (1970) extraction model, defined as

$$S(z,\theta,t) = ET[D(\theta)\,RDF(z,t)]/\int_0^L D(\theta)\,RDF(x,t)\,dz, \qquad [9]$$

where ET is the volumetric evapotranspiration rate and L is the depth to the bottom of the root zone. RDF is obtained by Davidson et al. (1978a) from an empirical model

$$RDF(z) = a[\exp(-bz)][\sin(cz)] \qquad [10]$$

with $a = 3.384$, $b = -0.035$, and $c = 0.0314$ as an illustrative case. Equation [9] also distributes the ET demand over the entire root zone according to the product $[D(\theta)RDF(z,t)]$. The ET demand is constrained by

$$ET = PET,\ AW \geq 0.2\ TAW$$
$$ET < PET,\ AW < 0.2\ TAW \qquad [11]$$

where

$$TAW = \int_0^L (\theta_{FC} - \theta_{15})dz. \qquad [12]$$

Here, the ET demand is set equal to potential evapotranspiration (PET) when available water (AW) is $\geq 20\%$ of the total available water (TAW), and when AW is $<20\%$ of TAW, ET is decreased linearly to zero. TAW is defined as that water contained in the soil profile between "field capacity" (θ_{FC}) and 15-bar soil water contents (θ_{15}). The ET in Eq. [9] was adjusted by multiplying it by an appropriate crop coefficient to account for changes in crop water requirement during the growing season.

On the other hand, Gupta et al. (1978b) modified Eq. [8] to

$$S(z,t) = K(h)\{[H_r - h)/R(z) - RDF(z)]/\Delta z \Delta x\} \qquad [13]$$

where $R(z)$ is the radius of water flow to roots. This modification calculates for $RDF(z)$ and Δx rather than taking measured input data on $RDF(z)$ or assuming Δx to be unity as in Nimah and Hanks (1973). The value of RDF is calculated as the ratio of root length density L_v in Δx and total L_v in the root zone. L_v is empirically defined by

$$L_v = a[1 - b\exp(-ct)] \qquad [14]$$

with $a = 3.5$, $b = 1.0$, and $c = 0.02$ for corn grown in Yolo soil.

The parameter Δx is obtained from

$$\Delta x = (1/\pi L_v)^{1/2}. \qquad [15]$$

The PT, when the soil moisture is not limited, is estimated from PET and leaf area index (LAI) by the Ritchie and Burnett method (1971)

$$PT = PET(a + b\,LAI)^{1/2}, \qquad [16]$$

with $a = -0.21$ and $b = 0.7$ over the range of $0.1 \leq$ LAI ≤ 2.7 for corn. When LAI > 2.7, PT = PET; and when LAI < 1.0, PT is considered neglible. The actual rate of transpiration (T) is computed by assuming linear decrease

$$T = \text{PT}(\text{SWS}/\text{AW}) \tag{17}$$

where SWS is the actual soil water stored at any given time in the root zone, and AW is the available water between "field capacity" and wilting point taken to be 10 bars.

The Shaffer et al. (1977) model for evaluating root water extraction uses the Blaney and Harris equation (Jensen, 1973)

$$U = k \cdot T \cdot P(2.54/100) \tag{18}$$

where U is the total extraction rate due to consumptive use for a given period, k is the consumptive-use coefficient for a particular crop for the period, T is the mean temperature in °F, and P is the percent of annual daylight hours occurring in the time period. The following computes extraction on a daily basis:

$$S_j^i = \left[\frac{U}{\Delta x}\right]\left[\frac{\text{KP} \cdot \text{DEL}}{\text{DAYS IN PERIOD}}\right] \tag{19}$$

where S_j^i is consumptive use for the ith unit time and jth soil depth node, KP is the fraction of total extraction in the foot of soil in which node j occurs, and DEL is the length of Δx expressed in feet. The values of k, T, P, and KP are supplied as input data. The Shaffer et al. (1977) model has an irrigation scheduling subroutine in conjunction with the above CONUSE subroutine.

The CSMP models from The Netherlands (Beek & Frissel, 1973; Frissel & van Veen, 1978, 1981) and Washington University (Duffy et al., 1975) define water flow by

$$v = D(d\theta/dx) + K \tag{20}$$

where v is the flow rate and other parameters have been previously defined. In these CSMP models, the soil profile is divided into soil plates or compartments like the Dutt et al. (1972) model. Frissel and his associates take this plate to be 4 cm thick, and Duffy et al. (1975) take it at 15 cm, whereas Dutt et al. (1972) divides the profile into 10 plates irrespective of length. Since the flow rate of water in the CSMP models is calculated in the soil at the boundary of two adjacent plates (compartments), i and i-1, the average diffusivity ($D_{i-1/2}$) and average hydraulic conductivity ($K_{i-1/2}$) are calculated from

$$D_{i-1/2} = [D(\theta_{i-1}) + D(\theta_i)]/2 \tag{21}$$

$$K_{i-1/2} = [K(\theta_{i-1}) + K(\theta_i)]/2, \tag{22}$$

and the flow rate of water from plate i into i-1 is computed by

$$V_i = D_{i-1/2} [(\theta_{i-1} - \theta_i)/l] + K_{i-1/2} \qquad \text{for } i = 2, \ldots, L \qquad [23]$$

where l is the thickness of the plate, and L is the length of the soil column or is the index of the layer containing the water table for the Duffy et al. (1975) model.

For the simplified management model, Davidson and Rao (1978) consider "piston displacement." The depth to which the wetting front (d_{wf}) will advance during infiltration is estimated from

$$d_{wf} = [I(\theta_f - \theta_i)], \; \theta_f > \theta_i \qquad [24]$$

where I is the amount of infiltrating water, and θ_f and θ_i are the θ in the wetted zone behind the wetting front and the initial θ, respectively. It is assumed that θ_i is completely displaced by the wetting front, so that the depth of solute front (d_{sf}) for a nonadsorbing solute will be

$$d_{sf} = I/\theta_f. \qquad [25]$$

After the cessation of infiltration, drainage or redistribution takes place to an assumed θ of field capacity, θ_{FC}. The infiltrated water, after the adjustment for ET losses, is distributed in the soil profile to a depth, dx, by successive approximations to satisfy the following conditions:

$$(I - 2ET) = \int_0^{dx} [\theta_{FC} - \theta(z)] \, dz, \; dz \le d_{sf}. \qquad [26]$$

The factor 2 for ET reflects the assumption that the redistribution period is 2 days. Daily ET is provided and assumed to be constant over the period of simulation. Equation [9] is used for root water extraction, with RDF defined by

$$RDF(z) = \exp[-az] - b \qquad [27]$$

with $a = 0.005$ and $b = 0.471$ for millet (*Coix lacryma-jobi* L.) grown on Eustis f.s. soil.

Vithayathil et al. (1979) relate ET to open-pan evaporation (OET) by the ratio R of ET to OET, which varies with the amount of crop cover, percent available moisture and stress factor (SF), and the season:

$$ET = OET \cdot R \cdot SF, \qquad [28]$$

and extraction is distributed from the root zone in a specified pattern that is read in.

For the upper boundary (at the soil surface), all the models have provisions for the infiltration of rainfall and applied irrigation water at soil saturation θ. Most models assume that evaporation from the soil is negligible during the infiltration process. The Vithayathil et al. (1979) and Shaffer et al. (1977) models consider surface runoff when the infiltration rate is less than the applied rate or when head is >2 cm.

The CSMP models of Beek and Frissel as well as Duffy et al. (1975) simulate water flow and θ in the first plate of the soil column separately from lower plates. The latter calculates for θ by

$$\theta_i(t) = \theta_{i(t-1)} + \frac{v_i(t) - V_{i+1}(t) - E_i(t)}{1} \qquad [29]$$

where E_i is the ET; and the former calculates for rate of flow in the first plate by

$$V_i = D_{i-1/2}(\theta_s - \theta)/D_{i-1/2} + K_{i-1/2} \qquad [30]$$

where $D_{i-1/2}$ and $K_{i-1/2}$ are the average diffusivity and conductivity at the soil surface at saturated $\theta(\theta_s)$.

Attempts are being made to consider the spatial variability of soil properties. One such effort, advanced by Simmons et al. (1979) for soil hydraulic conductivity, is

$$K = \alpha^2 K_m \exp[\beta(\theta - \theta_s)] \qquad [31]$$

where α is the scale factor, K_m is the scale mean K, β is a location-dependent parameter, and saturated $K(K_s)$ is given by

$$K_s = \alpha^2 K_m. \qquad [32]$$

For the lower boundary (bottom of the root zone), most modelers assume a semiinfinite depth and unit hydraulic gradient. In the presence of a water table and if drainage is being calculated, saturated water flow is also simulated. Shaffer et al. (1977) handles drainage with a drainout subprogram that computes the position of the water table and drain discharge as a function of time:

$$\frac{\partial y_{x,t}}{\partial t} = \frac{K(\theta_s) D_a}{s} \frac{\partial^2 y_{s,t}}{\partial x2}, \qquad [33]$$

and the shape of the water table is simulated by a fourth-degree parabola:

$$y_{x,0} = (8y_{m,0}/L^4)(L^3x - 3L^2X^2 + 4LX^3 - 2X^4), \qquad [34]$$

where x and y are the vertical and transverse space coordinates, s is the specific yield, D_a is the average saturated thickness, $y_{x,0}$ is the initial water table height above the drain at position x, $y_{m,0}$ is the initial water table height at a point midway between drains, and L is the horizontal drain spacing. This drainage model basically computes stream tube volumes following the streamlines to the drain. On the other hand, Vithayathil et al.

(1979) assume water to flow vertically down to deep drainage and horizontally to the tile in equal proportions from the water table and below.

For additional details on the specificity of the lower and upper boundaries, the reader is referred to the original reports and publications. The mathematical modeling approaches given have some similarities, with differences in the details. The more refined models attempt to simulate the field soil water cycle mechanistically and as completely as possible, but the less refined ones require substantial input data on soil properties such as relations between K and θ and h, root growth and effective root distribution, and plant water use, because they are not mathematically incorporated into the model. The extent to which these models are validated is considered later.

2. SOIL N TRANSFORMATIONS AND PROCESSES

Tanji and Gupta (1978) have reviewed efforts on modeling soil N transformations up to about 1977. In simulation models, the transport and transformations of N in the soil are calculated simultaneously, but since not all of the N species are mobile, the N transformations and other processes are considered independently for clarity.

Table 2 summarizes N modeling efforts into empirical, kinetic, and Michaelis-Menten type of models.

Dutt et al. (1972) described N transformations by an empirical rate equation of the general form

$$Y = c + b_1 x_1 + b_2 x_2 + \ldots + b_n x_n \qquad [35]$$

where Y is the transformation rate (dependent variable), X_n is the basic transformed parameter (independent variable), b_n is the regression coefficient, and c is a constant (y-intercept). For example, the nitrification rate is defined as

$$Y = 4.64 + 0.00162\, T(\mathrm{NH_4\text{-}N}) + 0.00162 \ln_{10} (\mathrm{NH_4\text{-}N})$$
$$- 2.51 \ln_{10} (\mathrm{NO_3\text{-}N}) \qquad [36]$$

where T is temperature in °C. Similar regression equations were obtained for urea hydrolysis, mineralization, and immobilization. Denitrification and $\mathrm{NH_3}$ volatilization were not considered, but in a later N-transformation subroutine (Shaffer et al., 1977) a zero-order kinetic model was incorporated for denitrification:

$$dN/dt = -k_0. \qquad [37]$$

Mehran and Tanji (1974) suggested irreversible first-order kinetics for nitrification, denitrification, mineralization, immobilization, and plant uptake:

$$dN/dt = -k_1 N, \qquad [38]$$

as well as reversible first-order kinetics for ion exchange of NH_4^+. These rate models are coupled and solved simultaneously:

Table 2—Mathematical modeling approaches for soil N transformations and other processes.

Models	Modeling approaches
1) Dutt et al. (1972)	1) Nitrification, mineralization-immobilization, and urea hydrolysis modeled by regression equation; NH_4^+ sorption by equilibrium cation exchange equation; N plant uptake proportional to water uptake.
2) Shaffer et al. (1977)	2) Same as (1) except denitrification by zero-order kinetics.
3) Beek & Frissel (1973)	3) Growth of nitrifier and ammonifier by Michaelis-Menten kinetics; NH_4^+ oxidation by first-order kinetics with environmental variables; mineralization of proteins, sugars, cellulose, lignin and living biomass by first-order kinetics; immobilization by first-order kinetics including considerations for microbial biomass and C/N ratio; NH_3 volatilization by diffusion; NH_4^+ clay fixation by equilibrium model.
4) Frissel & van Veen (1978)	4) Revised version of (3): growth of *Nitrosomonas* and *Nitrobacter* by Michaelis-Menten kinetics including considerations of oxygen levels; denitrification by a physical-biological model involving oxygen diffusion; mineralization-immobilization considers organic matter grouped into fresh applied organic matter such as animal manures, straw and waste water, and resistant biomass residues; NH_3 volatilization by physicochemical model including dissociation of NH_4OH; NH_4^+ clay fixation by reversible first-order kinetics.
5) Frissel & van Veen (1981)	5) Revised version of (4) and (3); mineralization-immobilization described by coupled N and C flows among six pools.
6) Mehran & Tanji (1974)	6) Nitrification, denitrification, mineralization, immobilization, and N plant uptake by irreversible first-order kinetics; NH_4^+ exchange by reversible first-order kinetics.
7) Tanji & Mehran (1979)	7) Same as (6) except N uptake by variable NO_3^- and NH_4^+ absorption coefficient.
8) Duffy et al. (1975)	8) Nitrification, denitrification, and net mineralization by empirical models; N_2 fixation by soybean; N uptake proportional to water uptake.
9) Vithayathil (1979)	9) Same as (8) with revised denitrification and net mineralization rates.
10) Davidson et al. (1978a)	10) Nitrification, mineralization, and immobilization by first-order kinetics and modified by soil water pressure head; denitrification by first-order kinetics considering pressure head, water content and organic matter content; NH_4^+ sorption by linear partition model; NH_4^+ and NO_3^- uptake by Michaelis-Menten kinetics.
11) Davidson & (1978b)	11) Same as (10) but simplified; NH_4^+ exchange by reversible first-order kinetics.
12) Tillotson et al. (1980)	12) Nitrification and urea hydrolysis by first-order kinetics; NH_3 volatilization by first-order kinetics from $(NH_4)_2CO_3$ formed from urea hydrolysis; NH_4^+ sorption by linear partition model; NH_4^+ and NO_3^- plant uptake involving diffusion to roots.

$$d[N_c]/dt = -\sum_{i=1}^{n} K_i[N_c] + \sum_{j=1}^{m} K_j[N_m], \qquad [39]$$

where N_c is the concentration of N species of interest and N_m is the concentration of other N species, K_i and K_j are the respective first-order rate constants for $i = 1, \ldots, n$ sink and source mechanisms. For instance, the rate of change in solution NH_4^+ concentration with time is given by

$$d[NH_4]_S/dt = -(K_1 + K_{se} + K_4 + KK_6)[NH_4]_S$$
$$+ K_{es}[NH_4]_e + K_6[OrgN]_i. \qquad [40]$$

Figure 3 aids in ascertaining the sink/source mechanisms in Eq. [40]. In a more recent N subroutine, Tanji and Mehran (1979) described NH_4^+ and NO_3^- plant uptake by a variable absorption scheme:

$$Uptake = \lambda^I C^I S(z,t)/\theta, \qquad [41]$$

where λ^I is the NH_4^+ absorption coefficient, which can be set to unity if uptake is assumed to be proportional to root water extraction, C^I is the concentration of solution NH_4^+, S is the rate of extraction, and θ is soil water content. A similar equation was used for absorption of NO_3^- by plant roots.

Davidson et al. (1978a) also invoked first-order kinetics for N transformations but considered some of the transformation rate constants to be dependent upon environmental and other factors. For example, the rate constant for nitrification ($\overline{K_i}$) was empirically adjusted for water suction (h):

$$k_1 = \overline{K_1} f_1(h). \qquad [42]$$

Similar h-dependency was assumed for mineralization. The effects of θ and mineralizable organic N on denitrification were also considered.

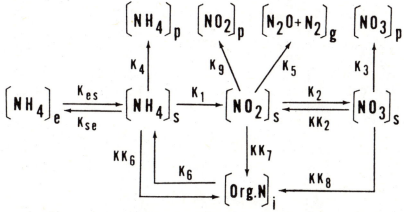

Fig. 3—Possible transformations of soil N. Subscripts K and KK denote rate constants; subscripts $e, s, p, i,$ and g refer to exchangeable, solution, plant, immobilized, and gaseous phases, respectively (Mehran & Tanji, 1974).

Davidson et al. (1978a) described adsorption-desorption of NH_4^+ by a simple linear partition model

$$S = Rc \qquad [43]$$

where S and c are the adsorbed and solution NH_4^+, and R is a partition coefficient. Tillotson et al. (1980) used Eq. [43] for both NH_4^+ and urea sorption.

Davidson et al. (1978a) simulated N uptake with a Michaelis-Menten type of model:

$$q_A = q_N^{max} \left[\frac{A(z,t)}{K_m = A(z,t) + B(z,t)} \right], \qquad [44]$$

where q_A and q_N^{max} are the actual and maximum rates of NH_4^+ uptake, K_m is the value of $(A + B)$ when $q_N = 0.5\, q_N^{max}$, and A and B are the concentration of NH_4^+ and NO_3^-. A similar model is used for NO_3^- uptake. The total N uptake is satisfied by uptake of both NH_4^+ (q_A) and NO_3^- (q_B).

Tillotson et al. (1980) considered three factors in nutrient supply to plants: external concentration at the root surface, some measure of the nutrient uptake rate, and the growth rate of the plant. Flux of nutrients into the soil surface is defined by

$$I = 2\pi rc\alpha \qquad [45]$$

where I is the rate of nutrient uptake per unit length of root, r is the root radius, $2\pi r$ is the circumference of root, c is the nutrient concentration in the soil solution, and α is the root uptake coefficient. The values of α, obtained from their study with corn, were

$$\ln_{10} \alpha_{NH_4} = -6.81 - 0.63 \ln_{10} (NH_4^+) - 116.8\, (NO_3^-) \qquad [46]$$

$$\ln_{10} \alpha_{NO_3} = -6.80 - 0.61 \ln_{10} (NO_3^-) - 127.1\, (NH_4^+). \qquad [47]$$

The root radius is assumed to be constant, and root growth is simulated as in the Davidson et al. (1978a) model.

In addition, Tillotson et al. (1980) have considered the volatilization of NH_3 from the decomposition of ammonium carbonate produced by hydrolysis of urea in alkaline soils. A first-order rate model was invoked with a rate constant of 0.3 hour^{-1}.

Duffy et al. (1975) simulated N transformations in an empirical and site-specific manner (Table 2). Their model is the only model that deals with symbiotic N_2 fixation. The rate of N fixation by soybean (N_f) is described by

$$N_f = kf \cdot r_g \qquad [48]$$

where kf is a constant (0.011 mg N cm^{-3}), and r_g is the rate of root growth, which in turn is obtained from

$$r_g = k_r \exp \left\{ -\left[\frac{55 - (t - t_p)}{35}\right]^2 \right\}, t \geq t_p \qquad [49]$$

where t_p is the day of planting, and k_r for soybean is assumed to be roughly 0.8 of the 2.5 cm day^{-1} for corn.

Although the presence of a growing plant is not considered in their modeling efforts, Frissel and his associates (Beek & Frissel, 1973; van Veen;[3] Frissel & van Veen, 1978, 1981) have contributed significantly to soil N simulation modeling, particularly in the areas of decomposition of organic matter and the application of Michaelis-Menten and chemical kinetics. In their most recent article, Frissel and van Veen (1981) traced their development of N models in four stages. In the first stage, it was believed that the C/N ratio of the organic material controlled mineralization and immobilization. In the second stage, differences in the rate of decomposition of organic components in plant residues were considered and these components included amino sugars, proteins, cellulose, and lignin fractions as well as microbial biomass (Fig. 4). At this second stage, the C/N ratio was no longer assumed to control organic matter decomposition but, rather, C uptake by the growing microbial biomass. In the third stage, N transformations were incorporated into a multilayer soil profile model with some simplifications. And in the fourth stage, the soil organic matter fractions were reaggregated into C and N pools that corresponded better with the soil organic matter fractions commonly measured by agronomists.

It is not possible to describe here all of the N transformations modeled by Frissel and his associates because of their sheer numbers. Instead, gener-

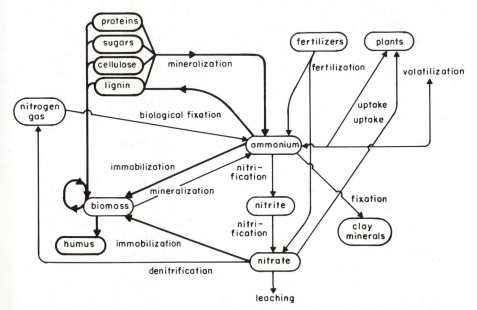

Fig. 4—Soil N transformations and components of soil organic matter (Frissel & van Veen, 1978).

alized equations are given. One of the basic considerations in their CSMP models is enzyme kinetics and microbial growth. The relationship between growth of microbes and consumption of nutrient substrate (van Veen)[3] is given by

$$-\frac{d[S]}{dt} = \frac{1}{Y_g} \cdot \left(\frac{dG}{dt}\right)_g \qquad [50]$$

where S is the nutrient substrate (N, C, etc.), Y_g is the gross growth yield factor, i.e., amount of biomass synthesized per unit of S consumed, G is the microbial biomass produced, and $(dG/dt)_g$ is the gross growth rate of biomass. The Monod exponential growth of microbes is defined by

$$(dG/dt)_g = \mu_g G \qquad [51]$$

where μ_g is the gross specific microbial growth rate that is obtained from

$$\mu_g = \mu_g^{max} \cdot (S/k_s + S) \qquad [52]$$

where μ_g^{max} is the maximum gross specific growth rate and K_s is the half-saturation constant for $\mu_g = \frac{1}{2} \mu_g^{max}$. Microbial biomass is also subject to death, so that the net growth is described by

$$(dG/dt)_d = (dG/dt)_g - (dG/dt)_d = \mu_g G - pG \qquad [53]$$

where the subscripts n, g, and d denote net growth, gross growth, and death, respectively, and p is the death rate constant.

Frissel and his associates consider some of the environmental factors affecting the N transformations (Table 2). More specifically, the correction (reduction) factors for less than optimal rates are incorporated into Eq. [52] after μ_g^{max}. Because of synergistic as well as antagonistic interactions among these reduction factors, temperature is usually considered as a primary factor, and the minimum value of other reduction factors is taken (Frissel & van Veen, 1981). For example:

$$\text{Reduction Effect} = \text{TCOF} \cdot \text{WCOF} \cdot \text{PHF} \qquad [54]$$

where TCOF, WCOF, and PHF are coefficients for temperature, θ, and pH, respectively. If their respective values were 0.8, 0.6, and 0.5, Eq. [54] would yield 0.24 and 0.5, if only the coefficient with minimum value is taken. They consider that the above two options are extreme and instead take TCOF and the minimum value of WCOF and PHF, which yields 0.40.

Equations [50] to [54] are invoked for all the microbially mediated N processes (Table 2). Other soil N processes not involving microbes include NH_3 volatilization, NH_4^+-clay fixation, and nitrate leaching. The last item will be covered later. Biological N_2 fixation and plant uptake were not modeled by Frissel and associates, and they are included for the sake of completeness in Fig. 4.

Volatilization of NH_3 is described by physicochemical processes, and it is assumed that such gaseous losses occur only from the surface soil depth. The concentration of NH_3 in the soil-solution phase (NH_3^s) is calculated from the dissociation of NH_4OH

$$NH_4OH \rightleftharpoons NH_3^s + H_2O \qquad [55]$$

and gaseous NH_3 (NH_3^g) from

$$NH_3^s \rightleftharpoons NH_3^g. \qquad [56]$$

Temperature effect on the chemical equilibrium constants for Eq. [55] and [56] are considered. The NH_3^g formed is lost to the atmosphere ($NH_3^{g,v}$) according to Fick's law

$$NH_3^{g,v} = D_{NH_3^g} \cdot NH_3^g \cdot \frac{AFV/BD}{x} \qquad [57]$$

where $D_{NH_3^g}$ is the diffusion rate constant taken as 1.6×10^4 cm² day⁻¹, AFV is the air-filled porosity of the soil, BD is the soil bulk density, and x is the distance to the soil surface from the point of NH_3^g production.

Fixation of NH_4^+ by clay minerals is described by a reversible first-order rate model

$$NH_4^s \underset{K_s}{\overset{K_f}{\rightleftharpoons}} NH_4^f \qquad [58]$$

where NH_4^s and NH_4^f are, respectively, solution and fixed NH_4^+, and K_f and K_s are, respectively, the rate constants for fixation and desorption, with $K_f \gg K_s$ and a prescribed maximum limit or extent of fixation.

For nitrification, Frissel and his associates assume that NH_4^+ and NO_2^- are growth-limiting substrates for *Nitrosomonas* and *Nitrobacter,* respectively. Equations [50] to [53] are used to calculate the gross specific growth rate, gross growth rate, death rate, and net growth rate. The oxidation rate of NH_4^+ is equal to the production rate of NO_2^-, and similarly, the oxidation rate of NO_2^- is equal to the production rate of NO_3^-.

As indicated in Fig. 4, the general theme of their model is that soil organic matter may be characterized into several C and N pools. The growth of biomass is controlled by the availability of C added to the soil. If no C is added to the soil, the microbial population remains constant. The four pools, proteins, carbohydrates, cellulose and hemicellulose, and lignin, are decomposed in decreasing order. Decomposition of these organic matter fractions is a result of the increase in biomass growth. Immobilization of N is proportional to the growth rate of the biomass. Mineralization is assumed to occur simultaneously and independently of immobilization. It is assumed that the production of NH_4^+ occurs only from the mineralization of protein-containing organic matter. A criterion used to assess mineralization-immobilization is the C/N ratio; i.e., if this ratio is <20 to 30, net mineralization will occur, and if it is greater, net immobilization will occur. van Veen[3] points out that the C/N ratio criterion can be as high as 40 to 50, with net mineralization occurring if the organic matter contains such resistant compounds as cellulose and lignin. Such a phenomenon is said to be accounted for with the C and N pools described in Fig. 4.

Most recently, Frissel and van Veen (1981) revised the mineralization-immobilization portion of the soil N-cycle modeling. Figure 5 shows six pools of C, three of which contain N, and the C and N flow pathways as well as the type of rate model used. Shown also is the production of CO_2 in the decomposition of organic matter. In this scheme, as soon as part of the C is used for additional growth of biomass, a corresponding quantity of N,

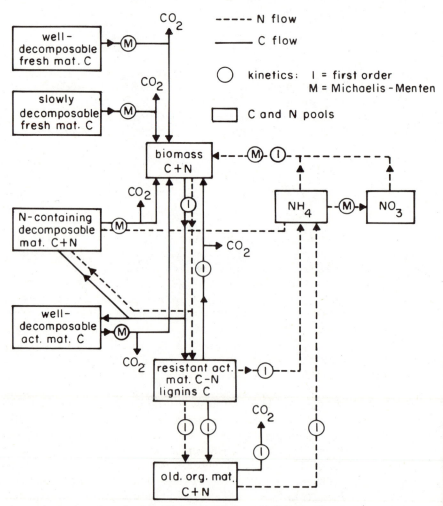

Fig. 5—Model for mineralization and immobilization of soil N coupling C and N transformations (Frissel & van Veen, 1981).

depending on the C/N ratio of the substrate and biomass, is contributed to the NH_4^+ pool.

For illustrative purposes, the mathematical modeling for Pool 3, N-containing decomposable organic matter, is examined. The rate of loss of C from Pool 3 (dC/dt) is given by

$$\left(\frac{dC}{dt}\right)_3 = \left[CNB \cdot \left(\frac{dB}{dt}\right) \right] / Y_3 \qquad [59]$$

where CNB is the C/N ratio of the biomass, $(dB/dt)_3$ is the net rate of increase in biomass, and Y_3 is the efficiency coefficient for the growth of biomass feeding on Pool 3. The rate of CO_2 production $(dCO_2/dt)_3$ is obtained from

$$\left(\frac{dCO_2}{dt}\right)_3 = \left[CNB \cdot \left(\frac{dB}{dt}\right)_3 (1 - Y_3) \right] / Y_3 \qquad [60]$$

and the rate of assimilation of C by the biomass from

$$\left(\frac{dC}{dt}\right)_{biomass} = CNB \cdot \left(\frac{dB}{dt}\right)_3 \qquad [61]$$

The mineralization of N from Pool 3 is obtained from

$$\left(\frac{dN}{dt}\right)_3 = \left(\frac{dC}{dt}\right)_3 / CNP \qquad [62]$$

where CNP is the C/N ratio of protein in Pool 3. The rates of C and N inputs into Pool 3 from the biomass pool are

$$\left(\frac{dC}{dt}\right)_{biomass} = CNP \cdot FR \cdot dB_D \qquad [63]$$

and

$$\left(\frac{dN}{dt}\right)_{biomass} = FR \cdot dB_D, \qquad [64]$$

respectively, where FR is the fraction of biomass debris products transferred into Pool 3 and dB_D is the total decay rate for all of the biomass. Similar types of equations are used to describe the other C and N fluxes shown in Fig. 5.

For denitrification, Frissel and his associates modeled the O_2 status in the soil, assuming that denitrification can take place at some minimal O_2 concentration level, and the rate of production of N_2O and N_2 is proportional to the NO_3^- consumed by heterotrophs, i.e.,:

$$-dNO_3/dt = \mu_g \cdot Y_{NO_3} \cdot NO_3 \qquad [65]$$

where μ_g is the gross growth rate of biomass and Y_{NO_3} is the efficiency factor for use of NO_3^-. van Veen[3] describes a complex and involved model in CSMP statements for the gaseous diffusion of O_2 through air-filled and water-filled pores in soils as well as microbial consumption of O_2 in the decomposition of organic matter. The air-filled pores are assumed to exist as discrete vertical cylinders in a given soil layer that are interconnected between layers. The rate of flow of O_2 (DO2C) from layer I to $I+1$ is given in CSMP language as

$$DO2C(I) = [DIFO2(I) - DIFO2(I+1)]/VOL(I)/BD - CONSO2(I) \quad [66]$$

where DIFO2 is the diffusion rate, VOL is the volume of the soil layer, BD is the bulk density, and CONSO2 is the rate of consumption of O_2 by microbes, which is defined by

$$CONSO2(I) = GRB * YO2 \quad [67]$$

where GRB is the growth rate of the heterotrophs and YO2 is the efficiency factor for consumption of O_2. The concentration of soil O_2 decreases if the rate of O_2 diffusion is less than the rate of O_2 consumption. The minimum O_2 concentration level at which denitrification can proceed is taken as 0.2 mg liter^{-1} and denitrification is also allowed to proceed until the NO_3^- concentration is reduced to 0.001 mg of NO_3^--N/g of dry soil. The diffusion and consumption of NO_3^- are modeled similarly to Eq. [66] and [67]. Further details are in van Veen[3].

It has been shown that the mathematical modeling for soil N processes is quite diverse in scope and intensity. As indicated in Table 2, all models except one or two consider nitrification, denitrification, mineralization, immobilization, plant uptake, and NH_4^+ exchange, but most models do not consider NH_3 volatilization, urea hydrolysis, and biological N_2 fixation. Many models use first-order kinetics for biological as well as physicochemical processes, some use a combination of Michaelis-Menten and chemical kinetics, and a few use nonmechanistic or empirical approaches. Several models account for environmental variables such as temperature, pH, and O_2 concentration, though most do not. For the microbially mediated N transformations, only one model considers microbial biomass. It appears that the scope and intensity of N modeling are dictated in part by site-specific conditions and availability of field data.

3. TRANSPORT OF MOBILE N SPECIES

Many models consider the reactivity of soil N simultaneously with transport of mobile soil N. The basic mathematical model (Bresler, 1973) for solute transport is

$$\frac{\partial(\theta C_i)}{\partial t} = \frac{\partial}{\partial z}\left[D(v,\theta)\frac{\partial C_i}{\partial z}\right] - \frac{\partial}{\partial z}\left(\frac{qC_i}{\partial z}\right) \pm \phi_i \quad [68]$$

where C_i is the concentration of the solute N species i, θ is the volumetric soil water content, t is time, z is the vertical space coordinate, D is the dispersion coefficient, v is the average pore water velocity, q is the volumetric water flux, and ϕ_i is the sink/source term(s) for solute species i. In some models, D is the apparent diffusion coefficient that is not v- or θ-dependent.

In the case of NO_3^-, it is affected by one source mechanism (nitrification) and several sink mechanisms (denitrification, plant uptake, immobilization), for which Eq. [39] was given as a generalized expression. Coupling of water and N flow with these soil N processes (Tanji & Mehran, 1979) results in

$$\frac{\partial(\theta C_i)}{\partial t} = \frac{\partial}{\partial z}\left[D_{(v,\theta)}\frac{\partial C_i}{\partial z}\right] - \frac{\partial}{\partial z}\left(\frac{zC_i}{\partial z}\right) + \sum_{i=1}^{n} K_i[C_i] + \sum_{j=1}^{n} K_j[C_i]. \qquad [69]$$

Ammonium concentrations are affected by several sink and source mechanisms; these are considered in Eq. [40]. If the adsorption and desorption of NH_4^+ is described by a linear partition model (Eq. [43]), then Eq. [68] will have the following form (Davidson et al., 1978a):

$$\frac{\partial(\theta C_i)}{\partial t} + \frac{\partial(\varrho S)}{\partial t} = \frac{\partial}{\partial z}\left[D_{(v,\theta)}\frac{\partial C_i}{\partial z}\right] - \frac{\partial}{\partial z}\left(q\frac{C_i}{\partial z}\right)$$

$$[70]$$

$$+ \sum_{i=1}^{n} K_i(C_i) + \sum_{j=1}^{n} K_j(C_j)$$

where ϱ is the soil bulk density and S is the adsorbed NH_4^+. The above mathematical models are solved by the finite-difference method similar to that of soil water flow (Eq. [1]).

Models that simulate water flow in discrete soil plates consider the generalized solute flow equation given by Eq. [68], but the sink/source mechanisms are usually calculated independently. The reactivity of N is calculated independently within a given soil plate over a given time interval, and the new concentration values of N are used in transport of solutes from one plate to another. Beek and Frissel (1973) assume that NO_3^- is the only mobile N species and that the leaching of nitrate (m_i^t) is caused by mass (M_i^{mass}), dispersive (m_i^{disp}), and diffusive (m_i^{diff}) fluxes, i.e.,

$$m_i^t = m_i^{mass} + m_i^{disp} + m_i^{diff} \qquad [71]$$

where

$$m_i^{mass} = q_i\, C_{i-1} \qquad [72]$$

$$m_i^{disp} = |q_i|\, D^{disp}\, (C_{i-1} - C_i/l) \qquad [73]$$

$$m_i^{diff} = |q_i|\, D^{diff} + D^{disp}\, \alpha\left(\frac{\theta_{i-1} + \theta_i}{2}\right)\, |q_i|\, \frac{l}{2} \qquad [74]$$

in which D^{disp} and D^{diff} are the dispersion and diffusion coefficients, respectively, l is the distance between midpoints of adjacent soil plates i and $i-1$, the absolute value of volumetric water flux q_i is taken to handle either downward or upward flow, and α is tortuosity factor. Duffy et al. (1975) use the same model (Eq. [71]–[74]), but Dutt et al. (1972) consider only Eq. [72]. In a more recent model, Frissel and van Veen (1981) have simplified the transport model to

$$dC/dt = [D_A \cdot (C_i - C_{i+1})]/l + L_e[C_i + C_{i+1})/2] \qquad [75]$$

where D_A is the apparent diffusion coefficient and L_e is the leaching efficiency factor.

Davidson and Rao (1978), in their simplified N-management model, take a different approach by integrating N transformations and transport over the whole soil profile. First, Eq. [68] is simplified to

$$\frac{\partial(\theta C_i)}{\partial t} = - \frac{\partial q C_i}{\partial z} \pm \phi_i. \qquad [76]$$

Applying Eq. [75] to NO_3^- yields

$$\frac{\partial(\theta C_3)}{\partial t} = - \frac{\partial(q C_3)}{\partial z} + k_1 \theta C_2 \qquad [77]$$

where C_3 is NO_3^-, C_2 is NH_4^+, and k_1 is the reaction rate constant for nitrification. Equation [77] in the integral form over soil depth z yields

$$\frac{\partial}{\partial t} \int_0^L \theta C_3 dz = \int_0^L - \frac{\partial q C_3}{\partial z} dz + \int_0^L k_1 \theta C_2 dz. \qquad [78]$$

If we assume that k_1 is independent of depth and take

$$T_2 = \int_0^\infty \theta C_2 dz \qquad [79]$$

$$T_3 = \int_0^\infty \theta C_3 dz \qquad [80]$$

where T_2 and T_3 are the total mass of NH_4^+ and NO_3^- in the soil profile, integration of Eq. [78] gives

$$dT_3/dt = \Delta_q C_3 + k_1 T_2 \qquad [81]$$

where (dT_3/dt) is the change in the total mass of NO_3^- in the profile with time and $\Delta_q C_3$ is the net solute flux. If $\Delta_q C_3$ is assumed to be zero, i.e., no change in storage of soil N for a specific time period, Eq. [81] further reduces to a first-order rate model (Eq. [38]). A similar mathematical modeling approach is taken for NH_4^+.

4. COMPUTER SIMULATION MODELING

The mathematical models described in the previous section need to be formulated into computer models (Fig. 1) before they are solved or executed with computers. The details are too extensive for each of the computer models to be described completely here. Instead, an attempt is made to provide the reader with an overall perspective on the general nature and computer systems used, the capabilities of these models, and the extent of model evaluation.

Figure 6 is a typical flow diagram of a water-N simulation model from Vithayathil et al. (1979). The inputs include water, N, and climatic data, and the outputs include water and N data. To simulate such outputs from the inputs, it is necessary to interface and integrate the processes affecting various water and N variables.

Table 3 summarizes some general information on selected models. It points out the programming language and computer system used and the nature of the system simulated. Some notion is given on the size of computer memory needed to simulate such problems and the typical processor time and total costs. The majority of these models are medium-sized programs requiring medium-sized computing systems, except for the Shaffer et al. (1977) model, which requires a large computing system. The computing time varies widely with the refinement and complexity of the calculations, length of time and number of spatial nodes simulated, and extent of printouts. The Rao et al. (1981) model is a coarsely tuned model that requires the least computing needs, and hence, is the least costly. The costs shown may be somewhat misleading for some of the models. For example, the Gupta et al. (1978b) model, when run on unprocessed primary data, costs about $50, but once the input data have been prepared and stored on the disk, it costs less than $20 for water flow simulation. The Tanji and Mehran (1979) model accepts the water output data, and the actual costs of simulating N transformations and transport on a daily basis for 132 days is $12.70. Hence, the combined cost of simulating a season's growth of a cornfield varies from about $31 to $63, depending on whether or not the water flow input data have been prepared in advance. The cost of computer runs is minimal when compared with the costs of obtaining field input data or data for testing these models.

The Dutt et al. (1972) model, as shown in Fig. 7, examines a flow line for physical, biological, and chemical processes. This model is comprised of a Moisture-Flow Program and a Biological-Chemical Program, interfaced with a Program INTFACE to match the outputs from the first into the second program because the Δx's are not the same. This model simulates, on a daily basis, water and N flows, and the printouts may range from daily to monthly levels. The model has been tested with a lysimeter experiment. Table 4 compares computed and measured N in barley and milo. These simulation results are considered noteworthy in that a N-simulation model was validated for probably the first time. This model is well documented,

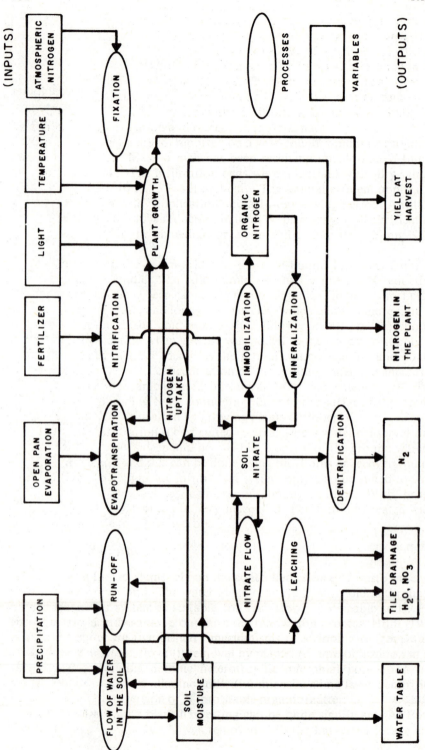

Fig. 6—Flow diagram showing the processes and variables in a systems level simulation model (Vithayathil et al., 1979).

Table 3—General nature of selected simulation models and their computing needs.

Computer models	Programming language	Computer system	System simulated	Typical storage needed	Typical processor time	Typical total cost
Dutt et al. (1972)	FORTRAN IV	CDC 6400	N leaching losses from cropped lysimeter for a year	~70 K words	--	--
Shaffer et al. (1977)	FORTRAN IV	CDC CYBER 74-28	Return-flow salinity and N from irrigation project for a year	70–200 K words	7.5 minutes	$62.50
Beek & Frissel (1973)	CSMP/360	IBM 360	Organic matter decomposition over 45 days in 70-cm profile	--	0.25 minute/day simulated	--
Frissel & van Veen (1978)	CSMP Level 3	IBM 360	Organic matter decomposition in 70-cm profile over 150 days	--	--	--
Duffy et al. (1975)	CSMP/FORTRAN IV	IBM 360	Tile drainage from 62-ha farm for a year	--	--	--
Vithayathil et al. (1979)	FORTRAN IV	IBM 360/65	Tile drainage from 62-ha farm for a year	90 K Bytes	1 minute	$ 9.00
Davidson et al. (1978a)	FORTRAN IV	AMDAHL 470 V/6-11	Nitrogen flow for corn grown in lysimeter for 97 days	256 Bytes	~30 minutes	--
Rao et al. (1981)	FORTRAN IV	AMDAHL 470 V/6-11	Nitrogen flow in cornfield plots for a season	51 Bytes	0.2 minute	$ 1.78
Gupta et al. (1978b)	FORTRAN IV	Burroughs B6700	Water flow in cornfield for 132-day season	16.5 kiloword-minutes	4.6 minutes	$18.51
Tanji & Mehran (1979)	FORTRAN IV	Burroughs B6700	Nitrogen flow in cornfield for 132-day season	11.5 kiloword-minutes	2.2 minutes	$12.70
Tillotson et al. (1980)	FORTRAN IV	Burroughs B6700	Salinity and N flow in cropped lysimeter for 158 days	64.3 kiloword-minutes	5.9 minutes	$24.37

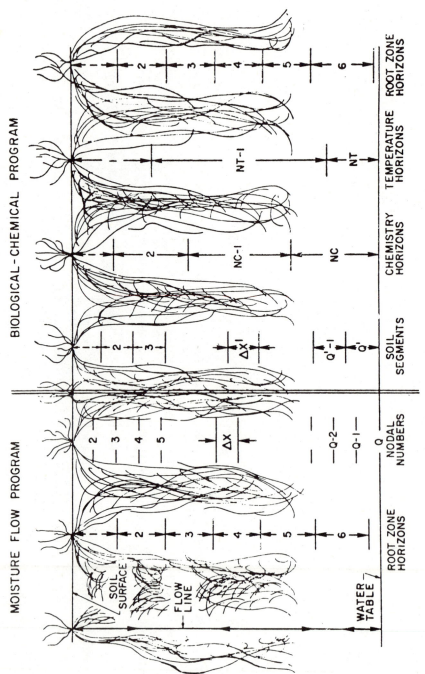

Fig. 7—Spatial division of the soil-plant-water system along a flow line (Dutt et al., 1972).

Table 4—Calculated and observed N uptake by barley and milo (Dutt et al., 1972).

Soil type	Fertilizer	Crops	Lysim-eter	Straw	Grain	Roots	Total ob-served	Total pre-dicted
					g of N/lysimeter			
Panoche c.l.	$(NH_4)_2SO_4$	Barley	2	0.35	0.94	0.53	1.82	2.20
			3	0.34	1.00	0.53	1.87	2.20
Panoche f.s.l.	KNO_3	Barley	5	0.43	1.14	0.65	2.22	2.24
			6	0.38	1.12	0.57	2.07	2.24
Panoche c.l.	$(NH_4)_2SO_4$	Milo	2	0.28	0.77	0.42	1.47	0.86
			3	0.38	0.67	0.57	1.62	0.86

with a user's manual, complete program listing, definition of variable names, and sample input and output data.

The Dutt et al. (1972) model was later incorporated into a larger model by Shaffer et al. (1977) that is being used by the U.S. Bureau of Reclamation to evaluate the water quality of return flows from present and future irrigation projects. The dissolved mineral constituents have been validated but not N. Figure 8 schematically describes the irrigation return flow model. A unique feature is the option to bypass any of the submodels given in Fig. 8 or subroutines within the submodel. This model has been widely used to evaluate the potential environmental impacts of proposed irrigation projects. Shaffer et al. (1977) published a 227-page report documenting this model.

Beek and Frissel (1973) were the first to formulate a CSMP model. Figure 9 is a schematic representation of their model with a focus on microbial decomposition of organic matter (Fig. 4) and the leaching of NO_3^-. A complete listing of the program has been published, along with some simulations on a hypothetical problem.

This model was later verified with experimental data by van Veen.[3] The second model consisted of some modifications in the treatment of organic matter decompositions. The CSMP program, with adequate comment, is available from Frissel. The details on the theory and assumptions are found in van Veen's doctoral dissertation. van Veen[3] and Frissel and van Veen (1978) tested the model on two experiments—a greenhouse soil incubation study and a field plot study. Despite some deviations between simulated and experimental data, the model appears to simulate trends quite well (Frissel & van Veen, 1978). The verification on biomass population is a noteworthy achievement.

Frissel and van Veen (1981) further modified the soil N cycle in a third model (Fig. 5). Using the same experimental data as above, they compared the simulation results on mineral N in the greenhouse pot experiments (Fig. 10) using the three N models (Table 1). Figure 11 shows calculated and measured mineral N at three soil depths in the field experiment using the third model (Fig. 11). The simulation improved substantially over the earlier models.

A second CSMP model was formulated and tested by Duffy et al. (1975). This model was later translated into FORTRAN IV. Figure 12 shows the predicted flow and NO_3^- concentrations in the tile outflows from

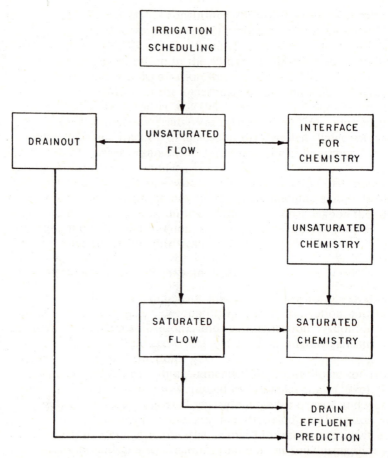

Fig. 8—Schematic representation of the USBR irrigation return flow salinity and nutrient model (Shaffer et al., 1977).

the corn root zone from a typical field. Additional simulations have been obtained by Vithayathil et al. (1979) in the same area for 1973 to 1974 with similar results. Vithayathil et al. (1979) also carried out sensitivity analysis on various perturbed factors (Table 5). Each factor was changed 10% from its nominal value, and the effects on NO_3^- concentrations are noted. For example, an R value of 5 indicates that a 10% change in the factor would produce a 50% change in the output. As shown in Table 5, water content in the unsaturated zone of the soil profile caused the largest change, followed by NO_3^- concentration in the profile, and precipitation.

Davidson et al. (1978a) formulated a simplified management model as well as a refined research model. Davidson and Rao (1978) tested the simplified model against experimental data from NaNagara et al. (1976) and Watts and Hanks (1978). Table 6 shows simulation results on N uptake for three growth periods and cumulative for days 34 to 97. The gross simplifications in the model (Eq. [75]–[80]) no doubt contributed to the deviations be-

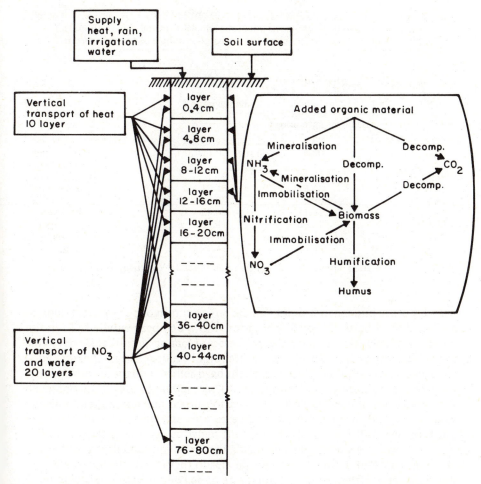

Fig. 9—General structure of a CSMP model to describe the behavior of N in soils (Beek & Frissel, 1973).

tween measured and calculated data. Table 7 gives a more extensive testing of the management model. The unaccounted for N for measured data is that necessary to obtain a mass balance in the field measurements. The simulated results appear to be in better agreement in the second corn experiment than the first. Rao et al. (1981) extended this model to include inputs of crop residue and animal manure. Documentation for this extended model is available from these workers.

Davidson et al. (1978a) present simulation results for their research model in Fig. 13 and 14 for profile θ and NO_3^- during three irrigation events. Table 8 compares simulated and measured N uptake for corn. Calculated results from Model III are based on the simplified management model (Table 1), and those from Models I and II are based upon more refined considerations. Model I assumes mass flow of NO_3^- into roots with

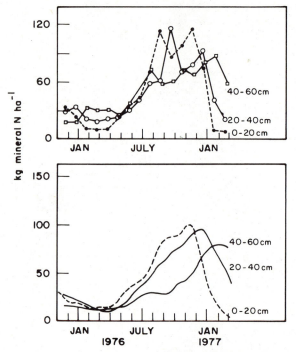

Fig. 11—Measured (*upper figure*) and calculated (*lower figure*) values of mineral N in the soil of a field experiment using the third model (Frissel & van Veen, 1981).

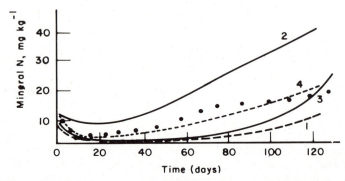

Fig. 10—Mineral N in soil of a pot experiment. *Dots*: observed values. *Curve 1*: calculated, first model, without priming. *Curve 2*: calculated, first model, with priming. *Curve 3*: calculated, second model. *Curve 4*: calculated, third model. In the second and third models, the biomass controls the transfer rates, so priming is automatically included (Frissel & van Veen, 1981).

water extraction, and Model II considers diffusion as well as mass flow. As given in Table 8, substantial improvements in calculated N uptake are noted with the research model. The mathematical description of the computer model along with a complete program listing has been published by Davidson et al. (1978a).

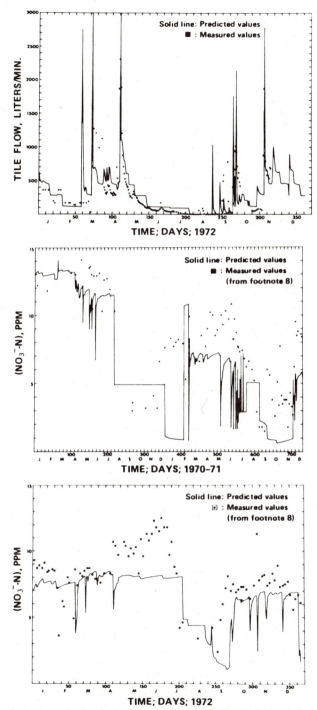

Fig. 12—Simulations on tile flow and NO_3^- concentrations in a Midwestern tile-drained farm (Duffy et al., 1975).

Table 5—Relative sensitivity coefficients for output variable CNORZ (nitrate
concentration in the tile effluent) for 10% change in the factors
(Vithayathil et al., 1979).

	Relative sensitivity coefficient (R_1, R_2, R_3, R_4) for the four quarters and for the whole year				
Perturbed factors	R_1 First quarter (1 Jan.– 31 March)	R_2 Second quarter (1 April– 30 June)	R_3 Third quarter (1 July– 30 Sept.)	R_4 Fourth quarter (1 Oct.– 31 Dec.)	R_T Whole year
Water content in unsaturated region	23.0	41.1	20.2	63.3	36.9
Nitrate concentration in the soil profile	7.8	6.4	4.5	2.4	5.3
Precipitation	1.6	2.9	12.0	4.6	5.3
Initial nitrate concentration in the cornfield	2.7	2.8	1.1	1.4	2.0
Rate of N uptake	0.0	0.0	1.9	5.2	1.8
Amount of deep drainage	1.4	0.7	2.1	1.2	1.35
Evapotranspiration	0.7	0.7	1.8	1.4	1.2
Net mineralization rate	0.0	0.2	1.4	2.8	1.1
Flow rate of nitrate in the crop area	0.5	0.3	1.2	2.1	1.0
Surface runoff	0.1	0.2	2.2	0.5	0.8
Amount of N fertilizer	0.0	0.0	0.1	2.4	0.6
Fraction of land under tile influence	0.8	0.4	0.7	0.6	0.6
Flow rate of nitrate in the tile area	0.1	0.1	1.8	0.4	0.6
Rate of denitrification	0.0	0.2	0.9	0.9	0.5
Rate of nitrification	0.0	0.0	0.1	1.7	0.5

Table 6—Comparison of calculated and measured N uptake by corn grown under
field conditions (Davidson & Rao, 1978).

	Nitrogen uptake		
Growth period	Measured[†]	Calculated	% error
days		mg of N/plant	
34–49	1,435	1,928	+35%
49–76	1,101	1,948	+77%
76–97	1,496	683	−54%
Total (34–97)	4,002	4,559	+14%

† Measured data was taken from NaNagara et al. (1976).

The Gupta et al. (1978b) model is unique in that it contains several subroutine options for simulating field soil water depending on availability of
data base. Figure 15 contains simulation on profile θ for the third irrigation
(Rooting depth of 1 m) in three irrigation treatments, and Fig. 16 contains
simulated daily seepage at the 20-cm depth for the whole growing season.
Water flow simulation was verified mainly with profile θ, as indicated in
Fig. 17. It is noted that simulated values on θ are within the standard deviation of measured values. Table 9 summarizes a consideration on spatial

Table 7—Comparison of N balance sheet prepared from measured data (Watts[1]) and that calculated for corn grown under three irrigation management schemes (Davidson & Rao, 1978).

	Low irrigation (0.8 ET)		Medium irrigation (1.15 ET)		High irrigation (1.5 ET)	
	Measured	Cal-culated	Measured	Cal-culated	Measured	Cal-culated
			kg of N ha⁻¹			
Initial mineral N	58	58	55	55	47	47
Fertilizer N added	250	250	250	250	250	250
Mineralized from organic N	84	86	84	77	84	76
Plant uptake	178	200	160	206	153	186
Mineral N in profile at harvest	38	132	44	39	26	6
Leaching loss	60	62	73	136	120	182
Unaccounted	116	--	112	--	83	--

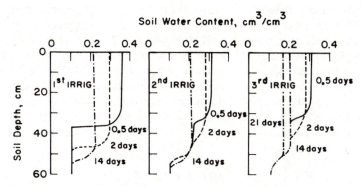

Fig. 13—Simulated soil water content distributions in a deep uniform loam soil profile during infiltration and redistribution following three irrigation events (Davidson et al., 1978a).

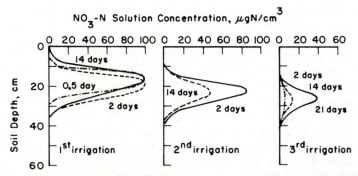

Fig. 14—Simulated NO₃⁻-N solution concentrations in the soil profile at selected times following three irrigation events corresponding to Fig. 13 (Davidson et al., 1978a).

Table 8—A comparison of measured N uptake by corn grown under field conditions and that predicted by three simulation models (Davidson et al., 1978b).

Growth period	Measured N uptake[†]	Calculated N uptake		
		Model I	Model II	Model III
days		mg of N/plant		
34–49	1,435	1,097	1,254	1,928
49–76	1,593	1,101	2,000	1,948
76–97	974	1,496	1,278	683
Total (34–97)	4,002	3,693	4,533	4,559
% Error		−7.7	+13.3	+13.9

† From NaNagara et al. (1976).

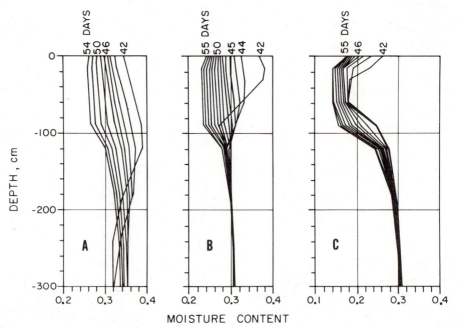

Fig. 15—Illustration of moisture content (cm³/cm³) simulation for the third irrigation cycle for three different treatments: A (5/3 ET), B (3/3 ET), and C (1/3 ET) (Gupta et al., 1978b).

Table 9—Evaluation of spatial variability in soil hydraulic conductivity (Tanji & Mehran, 1979).

	100-cm irrigation (5/3 ET)		20-cm irrigation (1/3 ET)	
Scale factor α	0.404	2.364	0.404	2.364
Saturated K, cm day⁻¹	2.0	63.3	2.0	63.3
Soil water after preirrigation, cm	105	105	105	105
Irrigation during cropping, cm	101.9	101.9	20.1	20.1
Transpiration, cm	48.2	34.2	35.6	22.8
Evaporation, cm	6.4	6.4	6.7	6.7
Seepage, cm	76.8	118.7	22.6	56.3
Soil water after cropping, cm	70.2	42.3	60.2	39.3

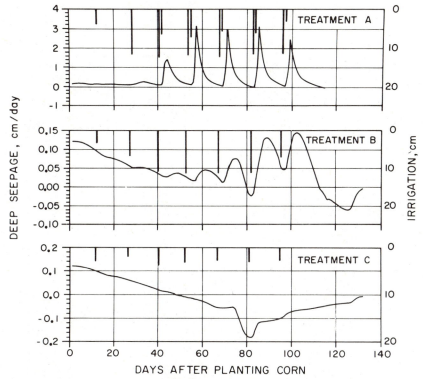

Fig. 16—Daily seepage at the 20-cm depth. Note that the vertical scale for each treatment is different and that seepage downward is considered positive (Gupta et al., 1978b).

variability in soil hydraulic conductivity (Eq. [31]). A scale factor of 0.404 is greater than the mode but less than the average, and 2.364 represents about 10% of the highest saturated conductivities. Figure 18 shows the simulated effects of scaling conductivity on water content over the growing season. This water flow simulation model has been fully documented (Gupta et al., 1978b).

Tanji and Mehran (1979) coupled the above water flow model to a N-transformation and transport model. Table 10 shows the results of simulating four N fertilizer treatments, for which reasonable agreements were obtained. Additional validation of the model was obtained by com-

Table 10—Comparison between measured and computed results for the 5/3 ET irrigation regime (Tanji & Mehran, 1979).

Application rate kg of N ha^{-1}	Plant uptake kg of N ha^{-1}		Residual inorganic N kg of N ha^{-1}		Calculated leaching losses
	Measured	Calculated	Measured	Calculated	
0	77	57	117	69	70
90	155	121	137	91	77
180	214	202	134	89	123
360	283	260	295	260	300

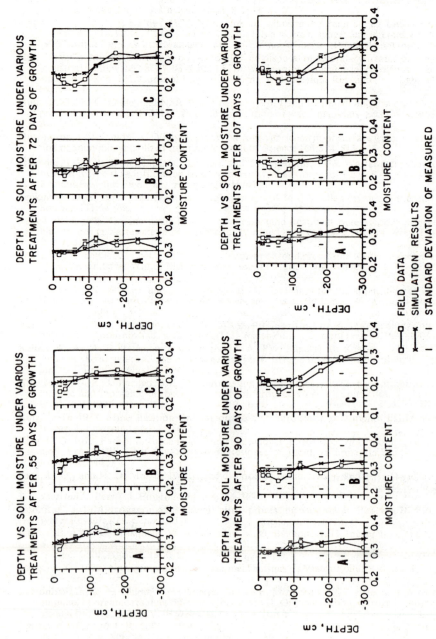

Fig. 17—Measured and predicted soil moisture over the 1975 corn growing season (Gupta et al., 1978b).

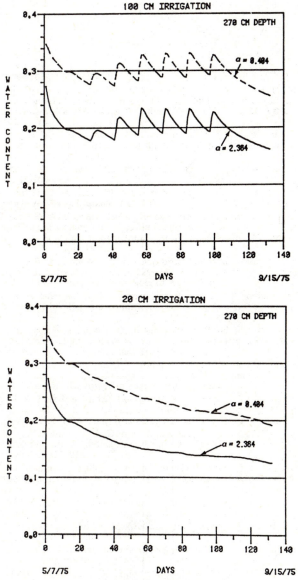

Fig. 18—Simulated daily changes in water contents (cm³/cm³) at the 270-cm soil depth for wet (100-cm) and dry (20-cm) irrigation treatments with two scale factors for hydraulic conductivity (Tanji & Mehran, 1979).

paring profile NO_3^- concentrations, for example Fig. 19. Similar results were obtained for the 0-, 180-, and 360-kg N/ha treatments.

The above illustrative simulations and validated simulations are indicative of the capabilities of dynamic N simulation models. Progress has been considerable despite the relative scarcity of experiments suitable for model testing.

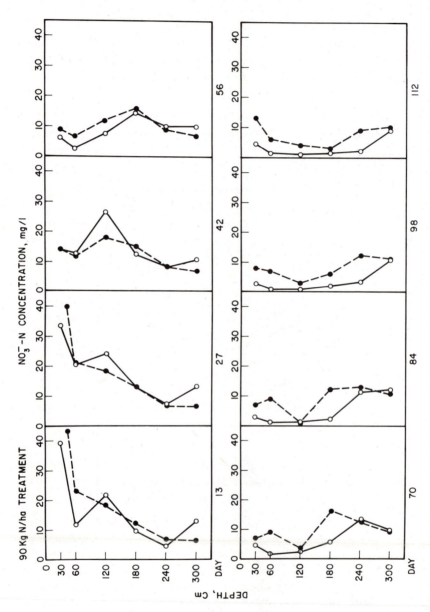

Fig. 19—Measured (*solid line*) and simulated (*broken line*) NO$_3^-$ concentration profiles at different days for the 90-kg N/ha treatment of the 5/3 ET irrigation regime (Tanji & Mehran, 1979).

IV. TECHNICAL AND PHILOSOPHICAL CRITIQUE

This section appraises both the technical and philosophical aspects of soil N modeling. A general critique precedes specific critiques on the N models cited in this chapter. My comments, of course, are but one person's opinion and may not reflect the diverse opinions held by modelers and non-modelers. I happen to be an experimentalist as well as a modeler.

A. General Critique

Agronomists dealing with the soil N cycle have suggested management options to improve crop production and N use efficiency. Today, however, the questions being asked about N emissions from croplands and their environmental impacts are so broad and complex that even knowledgeable agronomists must rely heavily on experience and intuition. The factors and processes contributing to or affecting N losses are so numerous and interactive that without resorting to some kind of modeling it would be extremely difficult to integrate and synthesize the large body of existing information and data, let alone those that are unknown. However, mass-balance and dynamic simulation modeling are not a panacea for all problem-solving situations. These modeling efforts have both negative as well as positive features.

If one examines the negative aspects, it should be clearly recognized that a model is only a substitute for the real system. The one and only complete model of a natural system is the system itself (Tanji & Gupta, 1978). The real world complexities are simplified for modeling. At times, the simplifications are so gross (as in large-scale mass-balance models) that the results are not very meaningful. Computer science and numerical analyses have advanced to such levels that nonspecialists in those fields have insufficient understanding of the capabilities and utilities of models. Dynamic modelers generally describe their work in sophisticated mathematical terms and computer jargon that are difficult to follow (Tanji, 1981). Frequently, adequate documentation of the model, including user's manual, is not provided, so that even modelers have difficulties deciphering computer models developed elsewhere. Another frustrating aspect of computer modeling is that a computer program written and debugged on one computer system is not operational when fed into a second computer system. This problem occurs even if the computers are identical systems because of local modifications made at the respective computer centers.

Many models are developed for a site-specific problem and are not amenable to more generalized problem situations. If the model is to be applied to another problem situation, both the mathematical and computer models usually have to be modified. Although much progress has been made in soil N modeling, most such studies have been research oriented and do not lend themselves to providing practical management guidelines (Tanji, 1980). On occasion, an oversell by modelers has contributed to greater skepticism by the nonmodelers. To achieve greater acceptance and credibility, more rigorous and extensive model testing should be carried out. Input data and experimental data to test these models, however, are often found to be incomplete, inadequate, or both. A need exists for closer collaboration among experimentalists and modelers.

On a more positive note, because problems related to N are exceedingly complex and difficult to answer, we should take advantage of combining the power of the human mind and computers. There is no doubt that the advent of computers and modeling techniques have had and will continue to provide significant advances. Systems analysis and modeling do help organize and integrate information and data. Such efforts foster interdisciplinary inputs, which are critical for the larger-scale problems. Systems modeling forces one to get a clearer definition of the problem and provides increased insight and conceptual understanding. Mathematical modeling and simulation are serving a useful means of quantitatively evaluating cause and effect relations that are not well defined or quantified as well as testing the more fully known concepts and hypotheses. Modeling has frequently identified gaps in data or knowledge. The more refined models are computing parameters over time and space that complement or add to observed measurements, increasing the value of these limited observations. It is clear that the use of computers for data processing and modeling will escalate with time. Agronomists should become familiar with computers so as not to be left off the mainstream of researching and managing our soil resources.

B. Specific Critique

The theoretical basis for the dynamic simulation of water and N fluxes in the soil-plant-water system is the equation of continuity (Eq. [1] and [68]). These models are process oriented in that physical, biological, and chemical processes are simulated. Simplifying assumptions are, however, made in the more coarsely tuned computer models. Such models, strictly speaking, do not mathematically simulate all of the processes considered or all of the required model parameters and coefficients and, instead, rely heavily on input data. For instance, the rooting depth as a function of time and space can be assumed to be invariant (Dutt et al., 1972; Shaffer et al., 1977) or can be given as input data rather than simulated (Watts & Hanks, 1978). On the other hand, greater efforts are made in the more finely tuned models to define and simulate the processes. In the case of plant roots, Davidson et al. (1978a) and Gupta et al. (1978b) mathematically simulate root growth over small intervals of time and space. One simulation model (Davidson & Rao, 1978) has been so simplified that it approaches some of the more refined mass-balance models (e.g., Tanji et al., 1979). However, the latter is capable only of simulating growing season or annual time increments, whereas the former has the resolution to simulate daily values. In contrast to mass-balance models, which aggregate time and space, all of the dynamic models numerically compute at daily or smaller time increments for several months to a year.

Each of the dynamic models was formulated with rather specific objectives in mind and for specific local conditions. Most of the models have been tested against only one experimental data base, and only a few have been applied to more than one data base, e.g., Davidson and Rao (1978) on data of NaNagara et al. (1976) and Watts.[1] There are only a few cases where two or more different models have been applied to a common data base, e.g., Watts and Hanks (1978) and Davidson and Rao (1978) on data of Watts,[1] or Frissel and van Veen (1978, 1980) on data of van Veen.[3] Application of a given model to a different experiment or different set of experimental conditions usually requires some modifications because of the more or less unique specificity of the models.

Some attempts are being made to develop computer models that have more generality. The model of Shaffer et al. (1977) has options to bypass various subroutines. The model of Gupta et al. (1978b) allows the user to select the particular subroutine option or to add the users' subroutine. Soil N modeling is progressing to a stage where a certain group of modelers rely on first-order kinetics, whereas others prefer a combination of Michaelis-Menten and first-order kinetics or empirical approaches. There is a potential in the future for some unification or standardization of models into subgroups with similar mathematical modeling approaches.

Because dynamic soil N-cycle modeling started less than a decade ago, critical appraisals of the various modeling efforts have not yet fully evolved. As pointed out earlier, several of the N models have been incorporated into larger-scale system models (Frere et al., 1975; Donigian & Crawford, 1976; Shaffer et al., 1977; Reddy et al., 1979a, b). The results simulated in these large-scale models have not been validated in the field. Some have been tested on smaller-scale studies such as lysimeters (Dutt et al., 1972) and incubation experiments (Reddy et al., 1979a, b). Except for the model of Frere et al. (1975), the modelers chose to scale up small-scale models to larger regional models without modifications. There is a problem here in that the available data base is not in conformance with the model sophistication. Other problems identified (Tanji, 1980) in these large-scale models include little or no consideration of N transformations in surface drainage waters and deficiencies in subsurface hydrology modeling that contributes the base flow in surface waterways. The model of Duffy et al. (1975) for farm-size systems requires extensive model calibration with field data. The simulations for 5 years, however, show good conformity with observed data. It is the only model that considers biological N_2 fixation. Vithayathil et al. (1979) evaluated that model extensively, including sensitivity analyses of many model parameters. The simplified model of Davidson and Rao (1978) lends itself to large-scale simulation but has not been linked to a large-scale hydrology model. Instead, it is currently being used to obtain the more practical N management guidelines under small-scale experimental conditions. This model has been extended by Rao et al. (1981) to consider animal manures and crop residues and is the only model in which rates and extent of denitrification are being verified.

For the more finely tuned research models applicable to small-scale experimental plots, considerable advances have been made. Frissel and van

Veen (1981) made extensive and comparative simulations of their three levels of soil N simulations (Fig. 10). Their models are unique in that microbial activities are explicitly simulated and the calculated microbial populations have been verified. The approach taken to model denitrification is also noteworthy. However, their models do not consider the presence of growing plants. Other soil N modelers have taken the suggestion of Mehran and Tanji (1974) by using first-order kinetics. Some have extended such a model to include considerations of temperature and other environmental parameters. Watts and Hanks (1978) validated their model with cornfield data, invoking some empiricism to obtain better fit between simulated and observed data. The research-oriented model of Davidson et al. (1978a) is being widely applied. It has been tested mainly with experiments involving coarsely textured soils with little or no soil organic matter. Its performance under other soil conditions has yet to be explored. The model of Tanji and Mehran (1979) has been evaluated extensively, including considerations on spatial variability of soil hydraulic conductivity as well as experimental verification. However, the mineralization and denitrification rate constants obtained from sensitivity analyses have not been directly confirmed. Selim and Iskandar (1981) modified the model of Davidson et al. (1978a) to simulate waste water applications on grasses. The simulated results are in good agreement with the experimental lysimeter data. Selim and Iskandar (1980) have also made some comparative modeling studies by linking their water flow submodel with Tanji and Mehran's (1979) N flow submodel. Tillotson et al. (1980) tested their model with lysimeter data. They have taken a unique approach for NO_3^- and NH_4^+ plant uptake that is different from the Michaelis-Menten type of kinetics used by Davidson et al. (1978a). Most of these dynamic models use numerical approximations to solve water and N flows. Only a few models have provisions to check on the calculated mass balance of water and N (Davidson et al., 1978a; Tanji & Mehran, 1979; Tillotson et al., 1980).

It is clear that significant advances have been made in the dynamic simulation of soil N. These modeling efforts are providing a better conceptual understanding of this complex system. Validation of simulation modeling is severely limited, however, because of the scarcity of available experiments to test these models. There is a need for more coordination between experimentalists and modelers.

LITERATURE CITED

Barden, W., Jr. 1977. Minicomputers and microcomputers. Howard W. Sams & Co., Indianapolis, Ind.

Beek, J., and M. J. Frissel. 1973. Simulation of nitrogen behavior in soils. Centre for Agricultural Publishing and Documentation, Wageningen, The Netherlands.

Bresler, E. 1973. Simultaneous transport of solute and water under transient, unsaturated flow conditions. Water Resour. Res. 9:975–986.

Davidson, J. M., and P. S. C. Rao. 1978. Use of mathematical relationships to describe the behavior of nitrogen in the crop root zone. p. 291–319. *In* P. F. Pratt (ed.) Proc. Natl. Conf. on Management of Nitrogen in Irrigated Agriculture, Sacramento, Calif., 15–18 May 1978. Univ. of California, Riverside.

Davidson, J. B., D. A. Graetz, P. S. C. Rao, and H. M. Selim. 1978a. Simulation of nitrogen movement, transformation, and uptake in plant root zone. EPA-600/3-78-029.

Davidson, J. M., P. S. C. Rao, and R. E. Jessup. 1978b. Critique of computer simulation modeling for nitrogen in irrigated croplands. p. 131-143. *In* D. R. Nielsen and J. G. Mac-Donald (ed.) Nitrogen in the environment, Vol. I. Academic Press, Inc., New York.

Davis, R. M. 1977. Evolution of computers and computing. Science 195:1096-1102.

deWit, C. T., and H. van Keulen. 1972. Simulation of transport processes in soils. Centre for Agricultural Publishing and Documentation, Wageningen, The Netherlands.

Donigian, A. S., Jr., and N. H. Crawford. 1976. Modeling pesticides and nutrients on agricultural lands. EPA-600/2-76-043.

Duffy, J., C. Chung, C. Boast, and M. Franklin. 1975. A simulation model of biophysicochemical transformations of nitrogen in tile-drained corn belt soil. J. Environ. Qual. 4: 477-486.

Dutt, G. R., M. J. Shaffer, and W. J. Moore. 1972. Computer simulation model of dynamic bio-physicochemical processes in soils. Tech. Bull. 196, Agric. Exp. Stn., Univ. of Arizona.

Freiberger, S., and P. Chew. 1978. Personal computing and microcomputers. Hayden Book co., Rochelle Park, N.J.

Frere, M. H. 1976. Simulation of potential percolation and nitrate leaching, Appendix B. p. 149-175. *In* B. A. Stewart (coordinator). Control of Water Pollution from Cropland, Vol. II. An overview. USDA-ARS, EPA-ORD.

Frere, M. H., C. A. Onstad, and H. N. Holtan. 1975. ACTMO, an agricultural chemical transport model. USDA-ARS-H-3. U.S. Government Printing Office.

Frissel, M. J., and J. A. van Veen. 1978. A critique of "computer simulation modeling for nitrogen in irrigated croplands." p. 145-162. *In* D. R. Nielsen and J. G. MacDonald (ed.) Nitrogen in the Environment, Vol. I. Academic Press, Inc., New York.

Frissel, M. J., and J. A. van Veen. 1981. Simulation model for nitrogen immobilization and mineralization. p. 359-381. *In* I. K. Iskandar (ed.) Modeling wastewater renovation by land disposal. John Wiley & Sons, Inc., New York.

Gupta, S. C., M. J. Shaffer, and W. E. Larson. 1978a. Review of physical/chemical/biological models for prediction of percolate water quality. p. 121-132. *In* H. L. McKim (coordinator) Int. Symp. on the State of the Knowledge in Land Treatment of Wastewater, vol. 1, Hanover, N.H. 20-25 Aug. 1975. U.S. Army, Cold Regions Res. and Engineering Lab., Hanover, N.H.

Gupta, S. K., K. K. Tanji, D. R. Nielsen, J. W. Biggar, C. S. Simmons, and J. L. MacIntyre. 1978b. Field simulation of soil-water movement with crop water extraction. Water Science and Engineering Paper no. 4013, Dep. of Land, Air and Water Resources, Univ. of California, Davis.

Hagin, J., and A. Amberger. 1974. Contribution of fertilizers and manures to the N- and P-load of waters. A computer simulation. Final Rep. to Deutsche Forschungsgemeinschaft from Technion–Israel Institute of Technology, Soil & Fertilizer Division, Haifa.

Hanks, R. J., D. D. Austin, and W. T. Ondrechen. 1971. Soil temperature estimation by a numerical method. Soil Sci. Soc. Am. Proc. 35:665-667.

Jensen, M. E. (ed.). 1973. Consumptive use of water and irrigation water requirements. Tech. Comm. Rep. on Irrigation Water Requirements, Irrigation and Drainage Div., Am. Soc. Civil Engineers.

McGlynn, D. R. 1976. Microprocessors. John Wiley & Sons, Inc., New York.

McWhorter, F. W. 1976. The small electronic calculator. Sci. Am. 234:88-98.

Mehran, M., and K. K. Tanji. 1974. Computer modeling of nitrogen transformations in soils. J. Environ. Qual. 3:391-395.

Molz, F. J., and I. Remson. 1970. Extraction-term models of soil moisture use by transpiring plants. Water Resour. Res. 6:1346-1356.

NaNagara, T., R. E. Phillips, and J. E. Leggett. 1976. Diffusion and mass flow of nitrate-nitrogen into corn roots under field conditions. Agron. J. 68:67-72.

Nimah, M. N., and R. J. Hanks. 1973. Model for estimating soil, water, plant and atmospheric interrelations: I. Description and sensitivity. Soil Sci. Soc. Am. Proc. 37:522-527.

Nye, P. H., and P. B. Tinker. 1977. Solute movement in the soil-root system. Blackwells Sci. Publ., Oxford.

Rao, P. S. C., J. M. Davidson, and R. E. Jessup. 1981. Simulation of nitrogen behaviour in the root zone of cropped land areas receiving organic wastes. p. 81-95. *In* M. J. Frissel and J. A. van Veen (ed.) Simulation of nitrogen behaviour of soil-plant systems. Centre for Agric. Pub. & Documentation, Wageningen, The Netherlands.

Readings from Scientific American. 1975. Computers and computation. W. H. Freeman and Co., San Francisco, Calif.

Reddy, K. R., R. Khaleel, M. R. Overcash, and P. W. Westerman. 1979a. A nonpoint source model for land areas receiving animal wastes: I. Mineralization of organic nitrogen. Trans. ASAE 22:863–872.

Reddy, K. R., R. Khaleel, M. R. Overcash, and P. W. Westerman. 1979b. A nonpoint source model for land areas receiving animal wastes: II. Ammonia volatilization. Trans. ASAE 22:1398–1405.

Riley, J. P. 1970. Computer simulation of water resource systems. p. 49–274. *In* J. P. Riley (ed.) Systems analysis of hydrologic problems, Utah Water Res. Lab, Utah State Univ.

Ritchie, J. T., and E. Burnett. 1971. Dry land evaporative flux in a subhumid climate: II. Plant influences. Agron. J. 63:56–62.

Selim, H. M., and I. K. Iskandar. 1978. Nitrogen behavior in land treatment of wastewater: A simplified model. p. 171–179. *In* H. L. McKim (coordinator) Int. Symp. on the State of Knowledge in Land Treatment of Wastewater, Vol. 1, Hanover, N.H. 20–25 Aug. 1978. U.S. Army Cold Regions Res. and Engineering Lab., Hanover, N.H.

Selim, H. M., and I. K. Iskandar. 1981. A model for predicting nitrogen behavior in slow and rapid infiltration systems. p. 479–507. *In* I. K. Iskandar (ed.) Modeling wastewater renovation by land disposal. John Wiley & Sons, New York.

Shaffer, M. J., and S. C. Gupta. 1981. Hydrosalinity models and field validation. p. 137–181. *In* I. K. Iskandar (ed.) Modeling wastewater renovation by land treatment. John Wiley & Sons, Inc., New York.

Shaffer, M. J., R. W. Ribbens, and C. W. Huntly. 1977. Prediction of mineral quality irrigation return flow. V. Detailed return flow salinity and nutrient simulation model. EPA-600/2-77-179E, EPA. Ada, Okla.

Simmons, C. S., J. W. Biggar, and D. R. Nielsen. 1979. Scaling field measured soil water properties. Hilgardia 47(4):77–174.

Spencer, D. D. 1974. Computers in action. Hayden Book Co., Rochelle Park, N.J.

Tanji, K. K. 1980. Problems in modeling nonpoint sources of nitrogen in agricultural systems. p. 165–183. *In* M. R. Overcash and J. M. Davidson (ed.) Environmental impact of nonpoint source pollution. Ann Arbor Science Publishers, Ann Arbor, Mich.

Tanji, K. K. 1981. Approaches and philosophy in modeling. p. 20–41. *In* I. K. Iskandar (ed.) Modeling wastewater renovation by land treatment of wastewater. John Wiley & Sons, Inc., New York.

Tanji, K. K., M. Mehran, and S. K. Gupta. 1981. Water and nitrogen fluxes in the root zone of irrigated maize. p. 51–66. *In* M. J. Frissel and H. van Veen (ed.) Simulation of nitrogen behaviour of soil-plant systems. Centre for Agricultural Publishing and Documentation, Wageningen, The Netherlands.

Tanji, K. K., F. E. Broadbent, M. Mehran, and M. Fried. 1979. An extended version of a conceptual model for evaluating annual nitrogen leaching losses from croplands. J. Environ. Qual. 8:114–120.

Tanji, K. K., and M. Mehran. 1979. Conceptual and dynamic models for nitrogen in irrigated croplands. p. 555–646. *In* P. F. Pratt (principal investigator) Nitrate in effluents from irrigated lands. Final Report to the National Science Foundation, Univ. of California, Riverside.

Tanji, K. K., and S. K. Gupta. 1978. Computer simulation modeling for nitrogen in irrigated croplands. p. 79–130. *In* D. R. Nielsen and J. G. MacDonald (ed.) Nitrogen in the environment, Vol. I. Academic Press, Inc., New York.

Tillotson, W. R., C. W. Robbins, R. J. Wagenet, and R. J. Hanks. 1980. Soil water, solute and plant growth simulation. Utah Agric. Exp. Stn. Bull. 502, Utah State Univ.

Vithayathil, F. J., H. A. Neuwrith-Hirsch, C. Chung, J. Duffy, M. Franklin, and C. Boast. 1979. Reference manual for FIELD, a program for the simulation of nitrogen flow in a tile-drained Corn Belt agricultural field. Report to the National Science Foundation, Project no. AEN 73-07848 A03, Washington Univ., St. Louis, Mo.

Wagenet, R. J., J. W. Biggar, and D. R. Nielsen. 1977. Tracing the transformations of urea fertilizer during leaching. Soil Sci. Soc. Am. Proc. 41:896–902.

Warncke, D. D., and S. A. Barber. 1973. Ammonium and nitrate uptake by corn (*Zea mays* L.) as influenced by NH_4/NO_3 ratio. Agron. J. 65:950–953.

Watts, D. G., and R. J. Hanks. 1978. A soil-water-nitrogen model for irrigated corn on sandy soils. Soil Sci. Soc. Am. Proc. 42:492–499.

20 Economic Implications of Controls on Nitrogen Fertilizer Use

EARL R. SWANSON

University of Illinois
Urbana, Illinois

I. INTRODUCTION

Interest in examining economic consequences of public policy for controlling N fertilizer use stems from environmental and health concerns. Chapter 16 (D. R. Keeney) of this monograph presents a summary of the environmental hazards associated with N levels in air, water, and soil. A number of assessment studies have been made of the role that N plays in environmental quality (National Research Council, 1972, 1978; Porter, 1975; USEPA, 1973; Viets & Hageman, 1971).

The primary purpose of imposing controls on N fertilizer use is to achieve an improvement in water quality and to protect the ozone layer. Interest in the effect of N fertilizer on potential destruction of ozone is of more recent origin than the effect on NO_3^--N levels in water (National Research Council, 1978, p. 225–336).

Control of fertilizer use by public agencies may also be initiated to attain desired crop production goals during a national emergency. However, the market for N fertilizer is normally relied upon to control allocation and use. Thus, the amounts applied by producers depend on the price of N fertilizer, the expected price of the crop, and the expected relationship between crop yield and rate applied per hectare. This decentralized type of control of fertilizer use in crop production is not, however, applicable to achievement of desired levels of environmental quality because, in general, there are no market prices reflecting the values of varying levels of environmental quality.

In order to evaluate the consequences of improving environmental quality through controls on N fertilizer, a comprehensive framework is needed which includes both the economic impacts of adjustments in crop production *and* the effects on environmental quality (Langham, 1971; Mishan, 1971; Randall, 1972; Gros & Swanson, 1976). Although environmental quality effects are amenable to certain types of economic analysis (e.g., National Research Council, 1978, p. 567–594), in this chapter only the economic consequences of controlling N fertilizer use related to agricultural production are considered. Nevertheless, brief mention should be made of the fate of N in applied fertilizer. The N not taken up by the crop may be re-

Nitrogen in Agricultural Soils—Agronomy Monograph no. 22.

tained in the soil, leached, or volatilized. Estimates of the relationship of N fertilizer use and water quality have often been based on budgets. Soil N budgets are treated in Chapter 14 (J. O. Legg and J. J. Meisinger) of this monograph.

Some researchers have used regression analysis to explain variations in NO_3^--N levels in surface waters (Abrams & Barr, 1974; Horner, 1977; Klepper et al., 1974; Klepper, 1978; Singh & Sekhon, 1976; Taylor, 1973). Because a large number of other factors interact with N fertilizer, neither the budget nor the regression approach has precisely defined the role of N fertilizers in altering water quality. Biological theory has not developed to the point where definitive guidance can be provided for the variables to be included. Further, there has not been any systematic collection of data on those variables, other than N fertilizer, presently considered good candidates for explaining the variations in NO_3^--N level in surface water (e.g., livestock, organic matter in soil, and drainage patterns).

It is likely that actions will be proposed to reduce fertilizer's contribution to a perceived NO_3^--N problem in water or N_2O emissions from soils even though these contributions are not well defined. The Illinois Pollution Control Board did, in fact, conduct hearings in 1971 on regulations that would have controlled N fertilizer applications. The Board decided that there was an insufficient basis for establishing regulations (Illinois Pollution Control Board, 1972). Various ways of controlling NO_3^--N pollution are currently being considered in order to implement Section 208 of the Federal Water Pollution Control Act. Different methods of control would, of course, produce different effects; this chapter presents the economic impact that selected N-related policies would have on farm income, food costs, and level and location of crop production.

II. ALTERNATIVE METHODS OF CONTROL

In our present economic system, which relies primarily on markets to allocate scarce resources among competing alternative uses, there is no provision for the level of NO_3^--N in water to be taken into account in the decision-making process of producers. Ability of the environment, surface or subsurface water in this case, to absorb NO_3^--N without adverse environmental consequences is not infinite and the market mechanism does not automatically allocate this limited capacity among the various sources. Because of the diffuse nature of NO_3^--N sources, identification of the persons or other entities liable for the elevated NO_3^- content cannot easily be made. In such cases, reducing NO_3^--N levels in water may be accomplished by some type of public policy to intervene in the system. If the system is viewed as one consisting of quantities and prices or costs, then controls take two general forms. They may be directed toward changing the level of a physical variable in the system (direct) or a price or cost (indirect). The latter type is indirect in that it relies on economic incentives to achieve a desired level of NO_3^--N in the water.

Public policy alternatives focused on physical variables to be considered in this chapter are: (i) per hectare restrictions on commercial N fertilizer rates, (ii) restrictions on NO_3^--N concentrations in leachate or effluent, (iii) water treatment to reduce NO_3^--N content, and (iv) restrictions on the N balance at the farm level. These alternatives may be called regulatory or direct.

Also discussed are the following public policy alternatives that alter the economic incentives: (i) an excise tax, (ii) an effluent charge, (iii) a market for rights to use N fertilizer, and (iv) an information or education program to improve the base for decisions on N fertilizer levels. These eight control methods, four direct and four indirect, are not mutually exclusive; a combination of two or more could be adopted, and, of course, other alternatives could be developed.

Although some information is available on the problem of intraseason timing of application as well as carryover effects (Hills et al., 1978; Ogg, 1978; Pearson et al., 1961; Stauber & Burt, 1973; Stauber et al., 1975; Taylor & Frohberg, 1978), the economic analyses of alternative methods of control that have been considered up to this point have not included these considerations.

III. ECONOMIC FRAMEWORK

The choice of the economic framework for tracing the impact of N fertilizer controls depends on the unit of analysis. However, a crucial part of any framework is the crop yield response to applications of N fertilizer. This topic is discussed in Chapter 15 (L. T. Kurtz and R. A. Olson), and in section XI of this chapter. With either direct or indirect cntrols an attempt is made to alter the rate of fertilizer application and information on the consequent effect on yield is a prerequisite to assessment of economic impacts.

If controls are to be imposed only in a local region (e.g., a watershed) it may be unnecessary to analyze the impact within the context of a national model. In an Illinois study, Klepper et al. (1974) concluded that the problem of NO_3^--N exceeding the U.S. Public Health Service standard of 10 mg/liter was not a statewide problem, but was concentrated in certain areas. This indicates that selective restrictions on N fertilizer use rather than statewide restrictions may be appropriate.

If controls are instituted at the national level or in a major crop-producing state, the economic analysis of the impacts on such variables as farm income and food costs must realistically be accomplished within a national model. Even in rather short periods for adjustment to controls, it is important that substitution possibilities be taken into account. Such possibilities exist in both consumption and production. For example, various livestock rations may be formulated with differing consumption requirements for crops using N fertilizer. On the production side, rotations may change, different forms of fertilizer might be used, timing of applications may be altered, etc. Along with such substitutions in consumption and production, the locational pattern of crop production may vary to follow the law of

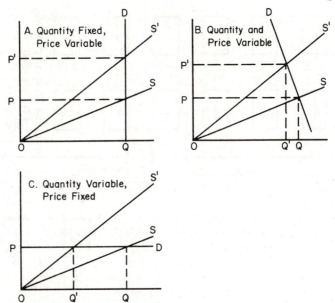

Fig. 1—Effect of marginal cost increases under various demand assumptions. D = demand, P = price, Q = quantity, and S = supply.

comparative advantage. If the full range of substitution possibilities is not considered, the impacts of controls are apt to be overestimated. In general, each possibility for substitution has a buffering effect on the system.

The nature of the demand assumptions is crucial to interpretation of the results of any analysis designed to evaluate the impact of controls. These demand assumptions influence the price and production effects of programs that control the use of N fertilizer. As the price and production effects vary, the consequences of N controls on total revenues (gross income) will also vary. Figure 1 illustrates the difference in the demand concept among models, and provides a framework for classifying the five models used.

For each of the three demand patterns in Fig. 1, a control on the use of N fertilizer increases the marginal cost of crop production. Marginal cost, which is the increase in total production cost occurring when output is increased by one unit, is based on variable cost. In all three types of demand patterns, supply (marginal cost function) increases in a similar way from S to S' when a control is imposed. The effect on price and quantity, however, varies because the demand assumptions are different.

In Fig. 1A, the demand is a fixed quantity, or perfectly inelastic with respect to price. In this situation only the price changes; quantity remains constant at $0Q$. Thus, the full effect of a marginal cost increase is reflected in price. Before the control is imposed, the total revenue (quantity times price) is $0Q \times 0P$; after the supply curve rises to S', total revenue increases to $0Q \times 0P'$. The national models employed by Heady et al. (1972), Mayer and Hargrove (1971), Nagadevara et al. (1975), and Taylor and Swanson (1975) use this type of demand assumption.

A model of Corn Belt agriculture (Swanson & Taylor, 1977; Taylor & Frohberg, 1977) has six crops and 17 land resource areas. The demand assumptions for two major crops, corn and soybeans, are illustrated in Fig. 1B. In this pattern, consumption is reduced as incresed marginal costs are reflected in increased prices. For crops other than corn and soybeans the inelastic demand of Fig. 1A is used. Note that an increase in the marginal cost from S to S' may cause total revenue either to increase or decrease, depending on the slope of the demand curve. In Fig. 1B, the total revenue increases after N fertilizer controls are imposed, from $0Q \times 0P$ to $0Q' \times 0P'$. The demand assumption presented in Fig. 1B is preferred for analysis of controls having economic effects at the regional and national levels, but often the increased complexity of implementing this concept leads to the use of the concepts in Fig. 1, A and C.

Individual farm and watershed models usually have crop demands that are perfectly elastic (Fig. 1C). For example, see Klepper et al. (1977); Onishi and Swanson (1974); Onishi et al. (1974); Palmini et al. (1978); Schneider and Day (1976); Walker and Swanson (1974). Kasal (1976) analyzed a river basin with this type of model. As marginal costs increase from S to S' there is no increase in price, and total revenue decreases from $0Q \times 0P$ to $0Q' \times 0P$. The decrease in income will be overestimated if a large fraction of the production is affected by the controls because crop price increases occur that offset, to some degree, the decreased production.

The different demand concepts underlying the models used should be kept in mind as the economic effects of the various N control alternatives are presented in later sections of this chapter.

Linear programming has been the principal analytical tool used by economists in the assessment of the consequences of control programs. This method uses an optimization technique that permits many variables to be considered simultaneously. A solution is determined by maximizing or minimizing a linear criterion function, subject to a set of linear equalities or inequalities. In economic applications the criterion function is usually net returns (income) in a maximization problem or costs in a minimization problem. Various constraints, such as the amount and location of land resources of differing quality, are taken into account. Applied to individual farms, the method permits a more detailed analysis of the interrelationships among enterprises and consideration of a wider range of technical alternatives than is possible with conventional, less formal methods such as partial budgeting. At the regional or national level, the method allows an analysis of the influence that markets and regional variations in soil productivity and climate have on the location of agricultural production.

The analytic capability of linear programming is especially important for estimating the effects of N-related control measures. Restrictions on N fertilizer will, for example, have "ripple" effects throughout the agricultural economy by altering the competitive advantage of crops that depend in varying degrees on N fertilizer. As a result of changes in competitive positions among crops, land use patterns, crop prices, farm income, and food costs change. The crops disadvantaged by N controls are replaced to some extent by those crops not directly affected by the controls. The possibility of substituting different kinds of grain in a producing region to meet feed and food requirements tends to moderate the economic impact of public policy intervention.

IV. PER-HECTARE RESTRICTIONS ON N FERTILIZER USE

A. National Analyses

National linear programming models have been used to estimate various impacts of this type of control. All of the studies reported in this section used a cost-minimization model with fixed demand requirements (Fig. 1A). In the following we sample the results of a number of these studies to suggest the general nature of the conclusions. Even though the basic structure of the economic models used in these studies is similar, the details and focus of the analyses may differ, thus causing the estimated impacts to vary from study to study.

Mayer and Hargrove (1971) used a national linear programming model to analyze the consequences of limitations on N fertilizer use. Limiting fertilizer use on all farms in the nation had effects on farm prices and incomes similar to a policy of limiting total cropland. Limiting N fertilizer rates to approximately 1969 levels (123 kg/ha) over the period 1970 to 1980 would not have raised farm prices and incomes or the farm portion of consumer food costs *if* government land retirement programs were relaxed. Continuation of this policy after 1980 without reduction in exports could cause a rapid increse in prices and food costs. If N fertilizer had been banned in 1970, Mayer and Hargrove estimated that all cropland would be used and that total food costs would have increased by about $11 billion, or about 12%. Net farm income would have increased by $0.5 billion, or about 3%.

Taylor and Swanson (1974, 1975) used a national linear programming model to study two sets of N fertilizer restrictions—one set applicable only in Illinois and another set nationwide. Although the restrictions apply only to corn, sorghum, and wheat, the areas planted to other crops may also be affected in the process of meeting the demand at minimum cost. Both analyses determined the following: first, the hectares necessary to meet high and low export demand predictions when N restrictions of 168, 112, and 56 kg/ha are imposed; and second, some additional effects of these restrictions.

The low export demand level corresponds approximately to the national exports in calendar year 1970 for the eight crops in the model. The high demand level, which is a little greater than actual exports for the 1973 crop year, represents the following increases above the low export levels: 40% for corn and sorghum, 15% for soybeans and cottonseed, and 93% for small grains.

In general, restricting use of N fertilizer will lower yields. Therefore, if domestic and export demands are to be met, the number of hectares planted to some crops must be increased. When low export levels are considered, nationwide restrictions require a far greater increase in total crop area than do restrictions only in Illinois. The benchmark solution with no restrictions in Illinois requires a total of 86.9 million ha to meet the low export demand, while the 56-kg/ha restriction only in Illinois needs 87.2 million ha nationwide, an increase of <1%. However, if the restriction of 56 kg/ha is enforced throughout the entire United States, the acreage needed for crops increases 7.5%. A similar comparison using the high export figures was made. With a 56-kg/ha limit set throughout the United States, the area needed to meet high export levels increases by 8.0% over the no-restriction levels. Such an increase would have serious environmental consequences in terms of increased erosion and sedimentation, thus increasing off-site damages and possibly lowering water quality by increasing P levels. These adverse effects should be taken into account in developing policy for N control.

B. Regional Analyses

A linear-programming model of the Corn Belt by Taylor and Frohberg (1977) provides an example of regional analysis. This model includes the production of six crops, namely corn, soybeans, wheat, oats, hay, and pasture, which are economically important to the Corn Belt. In this model the Corn Belt includes all of Illinois and Iowa, together with parts of Indiana, Kansas, Michigan, Minnesota, Missouri, Nebraska, Ohio, South Dakota, and Wisconsin. Solutions to the Corn Belt model are obtained by maximizing the sum of consumer's and producer's surpluses minus the variable costs of producing the six crops. Consumer's surplus is measured by comparing the amount consumers are willing to pay for food with the amount they actually pay. Producer's surplus, or the return to the fixed resource (land), represents net return to the landowner. Estimates of demand and supply as functions of price are required to implement the concepts of consumer's and producer's surpluses.

Corn and soybeans are of major economic importance to the Corn Belt, which produces 70% of the nation's corn and 60% of the soybeans. Therefore, the demand for these two crops is introduced into the model in the manner indicated in Fig. 1B, and both prices and quantities for these two crops are determined within the model. In contrast, the demands for the other four crops are treated as fixed quantities (Fig. 1A).

The Corn Belt is divided into 17 land resource areas, each with 11 land capability units that reflect variations in the suitability of soil for crops. Crop production alternatives in these resource situations can vary in several ways; by crop rotation, with an average of 11 rotations per unit; by conservation practice, namely, contouring, terracing, and no practice (straight row); and by tillage methods, namely, fall plow, spring plow, and chisel plow. Rotations, rather than single crop activities, were included to reflect the influence of the previous crop on fertilizer and pest management requirements of the current crop.

This model was used to assess the effects of reducing the average rate of N fertilizer application from 157 kg/ha to a maximum of 56 kg/ha. Imposing this limit reduces consumer's surplus by $3.3 billion and increases net return to land by $2.0 billion, leaving a net decrease of $1.3 billion. Net return to land increases as a result of the fertilizer restriction, because land becomes more scarce and, hence, more valuable as a source of crop production. The average annual net return to Corn Belt land changes from $215 to $274/ha, an increase of about 27%.

The average corn price rises from $96.85 to $121.26 per metric ton, an increase of 25%. But the N load, that is, the total N added minus the N removed in harvested crops, decreases by 47%. Whether such a reduction in the N load would meet or surpass water quality goals is not, of course, indicated by these results. In any event, a N fertilizer reduction of about 67% in the entire Corn Belt would have a substantial economic impact on farmers and consumers.

C. State Analyses

Nagadevara et al. (1975) studied the impacts of a number of environmental restrictions including limitations on N fertilizer when applied only in one state (Iowa), but analyzed within the context of a national model. A similar approach had been used for limitations on N fertilizer by Mayer and Hargrove (1971) and Taylor and Swanson (1973). If a single state applies restrictions, such action decreases the competitive advantage of crops dependent upon N fertilizer. Because the demand requirements for that crop can be met by production from other states, farm income in the state with restrictions is depressed.

V. RESTRICTIONS ON NO_3^--N CONCENTRATION IN LEACHATE OR EFFLUENT

Rather than limiting the amount of N fertilizer used, this method of control would require that the NO_3^--N content of the leachate or effluent be equal to or below a specified standard reflecting the environmental and health requirements. The difficulties in implementing this approach with available monitoring technology would be substantial. Further, this approach would not take into account possible atmospheric impacts.

Examples of this approach are provided by Onishi and Swanson (1974) and Schneider and Day (1976). Both studies used profit-maximizing linear-programming models of small areas, in the Onishi-Swanson case a watershed and in the Schneider-Day case a single farm. Thus, no aggregate implications of controls were evaluated.

Upper limits of the concentration of NO_3^--N in the leachate below the root zone were set at 10 mg/liter and 20 mg/liter in the Onishi and Swanson (1974) study. An equation for assessing the potential NO_3^--N concentration in water leaching below the root zone was developed by Stout and Burau (1967). Application of this equation indicates that corn fertilized at the rate of 112 kg or less of N/ha does not release any NO_3^--N into ground water because the N uptake by the corn grain is greater than the amount of N supplied. Theoretically, no N leaches into ground water because the equation assumes an equilibrium between N application and uptake; in reality, some N is released in the leachate even at the lower rates. Accordingly, an adjustment was made in the equation by assuming that the amount of N available in a given area for a crop is the sum of the amount applied plus the amount already in the soil. The amount already in the soil is estimated by calculating the N taken up by the crop if no N fertilizer is applied. The total amount of N thus calculated was inserted in the Stout and Burau (1967) equation to estimate the potential NO_3^--N concentration in the leachate below the root zone. This concentration was estimated to be 70 mg/liter under conditions in which net income was maximized. Restrictions at 10 mg/liter reduced net farm income by approximately 50% from the net farm income level with no restriction.

The Schneider-Day study considered a number of potential environmental quality problems on 16 farms. The nutrient budget approach was used and one of the control alternatives considered was placing a limit on the excess of applied N over plant uptake of N. This upper limit was chosen to be 1.68 kg/ha. Applying a rainfall-dilution equation at Wisconsin levels results in approximately 1.0 mg/liter of NO_3^--N in ground water. In only 2 of the 16 farms did this constraint limit net returns.

Singh et al. (1978) provide a method for determining the level of N fertilizer application rate that will maximize economic return and not exceed a permissible loss of N from the soil-root zone. Their calculations are based on experiments conducted at the Punjab Agricultural Farm in Ludhiana, India. A loss of 60 kg N/ha was set as the maximum permitted from an environmental standpoint. Two optimum fertilizer rates were compared, one without and one with this constraint. A small increment of corn yield would need to be sacrificed to meet the standard for permissible pollution, but the constraint did not reduce economic returns from wheat.

VI. TREATMENT OF WATER TO REDUCE NO_3^--N CONTENT

Rather than intervention in the form of fertilizer N controls, the option of water treatment has been considered. This control policy, as well as an effluent charge, was considered by Horner (1975, 1977) in his study of the Westlands Water District in the San Joaquin Valley of California. Removal of NO_3^--N from subsurface drainage water by bacterial denitrification was included in the linear-programming model as an alternative to a penalty to the farmer on the NO_3^--N concentration of water draining from the district. The results of this study will be discussed in Section IX.

VII. RESTRICTIONS ON THE N BALANCE
AT THE FARM LEVEL

The nutrient budget approach (Fried et al., 1976) provides an accounting framework by which the consequences of forcing particular N balances may be estimated. This budgeting approach assumes that the emission of NO_3^- to the ground water from an area such as a farm is the difference between the N input into the soil and that removed. In principle, a N balance may be calculated for a farm based on farming practices and crop yields and this balance may provide an indicator for a control method. A positive balance implies that NO_3^- is entering ground water; a negative balance implies that little or no NO_3^- is entering ground water.

A study by Walker and Swanson (1974) assumes that controls on nitrogen fertilizer use would focus on maintaining a particular N balance. Effects of such a control program on rotations and farm income were studied by varying the N-balance over a range. A 119-ha cash-grain farm in central Illinois was chosen for the analysis. The analysis used a linear-programming model for a 5-year planning period. The N accounting system recognized two sources of N: commercial N fertilizer and N added by legume crops (soybeans, alfalfa, and clover). The N from these sources either remains on the farm or is removed in harvested crops. The balance is positive if the sum of the two sources is greater than the amount removed in harvested crops, and negative if the sum is less. Pollution of water with NO_3^--N becomes more likely as the N balance becomes increasingly positive. Crop production alternatives include corn, soybeans, wheat, alfalfa, and sweet clover, each of which is produced by different methods and in different rotations. The resource constraints are land, labor, and capital. Land is considered homogeneous and the size of the farm remains constant because no land is purchased or sold. The constraint on labor is the amount of labor available in each month of the year. Capital is limited to operating expenses only; no long-term investment credit is considered.

With the N balance required to be zero, the cropping system with the maximum farm income above direct costs is continuous corn with N fertilizer application rate of 224 kg/ha. This system required purchase of 26.6 metric tons of N fertilizer. If, in addition to meeting the zero N balance requirement, N fertilizer were restricted to half of that amount (13.3 metric tons), a rotation of 60% corn and 40% soybeans results in maximum returns. The banning of N fertilizer results in a highest-return system of continuous soybeans, not likely to be a viable system because of pest problems. It should be remembered that this individual farm analysis assumes demend of the nature of Fig. 1C. If a large number of producers were involved in such a control program, the aggregate effects on price and their subsequent effect on the relative profitability of crops would alter the optimal cropping system. This is why national and regional models are needed to analyze policies applicable to large groups of farms.

VIII. AN EXCISE TAX ON N FERTILIZER

The preceding controls have been direct in the sense that they were applied to physical variables in the system. Even though there were economic consequences in terms of income and cost effects, there was no direct manipulation of prices or costs to achieve the desired NO_3^--N concentration in water via actions of producers responding to such altered prices and costs. The excise tax is our first example of a control which relies on producers to respond to prices and/or costs modified by a tax to achieve desired results.

A method referred to as an environmental pricing and standards system has been suggested by Baumol and Oates (1971). In brief, the method involves a corrective tax which would be set to achieve a publicly determined standard, in this case, NO_3^--N in water.

Abrams and Barr (1974) used a national linear-programming model to estimate the tax rate on N fertilizer use in Illinois needed to achieve various water quality standards in the Illinois waters affected. The focus was on Illinois because, at that time, it was the only state formally considering N fertilizer regulations. The estimated tax required to achieve the 10 mg/liter of NO_3^--N was found to range from about 2.5 to 5.0 cents/kg of N in fertilizer. The required tax rose sharply to about 33 cents/kg of N to achieve a standard of 5 mg/liter NO_3^--N.

Taylor (1975a, 1975b) investigated, with the use of a national linear-programming model, the income and crop pattern shifts within the United States resulting from the imposition of excise taxes of 6.6 cents, 13.2 cents, and 26.4 cents/kg of N in fertilizer in Illinois. Again this focus on Illinois was due to the active consideration by that state of controls.

The response of farmers to the N fertilizer tax depends in part on the relationship of crop yield to fertilizer application. Because yields decrease at an increasing rate, a given increase in fertilizer price with or without a tax causes the application rate to be reduced more at high than at low yield levels. Thus, the initial tax of 6.6 cents/kg reduces the amount of N applied to corn in Illinois by 17%, calculated from a situation with no tax; an additional 6.6 cents (a total of 13.2 cents) results only in an additional 8% reduction. The highest tax considered, 26.4 cents/kg, reduces N use on corn by 32%.

Even the smallest of the three tax rates has an appreciable effect. Illinois crop farmers suffer an immediate 5% decline in income, a 40% drop in fertilizer applications, and a reduction of 1.3 million ha of corn. However, this loss, in terms of area, is more than compensated for by an increase of 1.6 million ha of soybeans.

Crop acreage changes in the rest of the Corn Belt (in this case considered to be Indiana, Iowa, Missouri, and Ohio) are also of interest. With an Illinois tax of 6.6 cents, the Corn Belt exclusive of Illinois increases corn area by 0.9 million ha when Illinois decreases its area by 324,000 ha of corn. This shift means that part of the feed grain requirement must be met from production outside of the Corn Belt, and this may aggrevate the problem of NO_3^--N in the waters of those areas.

IX. AN EFFLUENT CHARGE

Effluent charges as a means of pollution control are usually associated with point sources. In the case of a drain pipe carrying waste from a single factory, the technology exists for measuring at least some of the characteristics of the effluent, thus providing a basis for a charge to the factory owner. For nonpoint sources, even if the pollution parameters of the effluent can be measured, the difficulty of identification of the relative contribution of various farms may prevent the allocation of the charge among farmers. Nevertheless, efficiencies of the effluent charge system may warrant development of technology to permit such an allocation.

In Section VI, reference was made to a study by Horner (1975, 1977) which considered water treatment as an alternative. Systems were compared in terms of total or social cost, regardless of its incidence, of reducing the NO_3^- to a 2 mg/liter level of NO_3^--N in irrigation subsurface drainage water. Three systems were compared in the study (Horner, 1975). One system was assumed to operate with no standard on the NO_3^--N concentration in the drainage water. The second system taxed farmers according to the contribution of their cropping system to the NO_3^--N concentration in the drainage water. The third system treated the drainage water to meet the 2 mg/liter NO_3^--N standard. Costs of the two control methods, effluent charges and water treatment, included the reduction of farm income under each of the two control methods. This reflects the changes in cropping practices under controls as compared with the system of no controls. Treatment costs are also included for the treatment alternative. Total costs, both public and private, necessary to achieve the 2-mg/liter standard were about 1.5 times greater for the treatment alternative than for the effluent charge alternative (Horner, 1975, p. 37). The results of this study confirm the hypothesis of Baumol and Oates (1971) that the social costs of meeting a standard are minimized through a system of pricing and standards.

Jacobs and Casler (1979) compared a system of effluent taxes with regulation to control discharges of P from cropland. They concluded that effluent taxes resulted in lower social costs than regulation requiring a uniform reduction in P discharge for all producers.

X. A MARKET FOR RIGHTS TO USE N FERTILIZER

The concept of a market for rights to purchase commercial N fertilizer is basically simple, although the operational details may be complex. On the basis of the water quality standard specified for the year, a public agency decides how much N fertilizer is to be used that year. In the form of coupons or certificates issued annually, rights to purchase a given quantity of fertilizer are sold on the open market, with purchasers bidding for these rights. The procedure might start with the agency asking a representative sample of users to indicate the quantity they would order at various prices. With this information the agency can then decide what price to set to ensure that approximately the number of rights the agency wants to issue will be sold.

After the initial disposition of rights by the agency, individual users and nonusers can buy and sell rights among themselves. Nonusers, such as environmental groups, can influence the amount of fertilizer used by trying to change the number of rights through the political system or by buying rights and then not using them.

Using a national model, Taylor (1975a) examined the effects of imposing rights for five different quantities of N fertilizer in Illinois only. These five quantities (784, 471, 305, 203, and 180 thousand metric tons) were selected to correspond to the assumed decline in the amount of N used when excise taxes of 6.6, 13.2, 19.8, and 26.4 cents/kg of N are imposed. For this analysis the Corn Belt was defined to include Illinois, Indiana, Iowa, Missouri, and Ohio. The impact of the market-rights program was assessed in terms of three areas: Illinois, the other Corn Belt states, and those states outside of the Corn Belt. The results underline the national corn area redistribution that occurs when a control method is introduced in only one state. Although the precise changes in the various regions are determined by the interdependent relationships in the model, several general trends may be noted.

As might be expected, the imposition of a market for N fertilizer rights in a given state (Illinois in this case) disadvantages that state in corn and sorghum production. For the smallest quantity of rights, 180,000 metric tons, the area of these crops in Illinois drops by about 67%, from 4.6 to 1.5 million ha. Because a fixed quantity of corn and sorghum must be produced nationally in this model, increases in areas of these crops occur both in the other Corn Belt states (an added 2.4 million ha) and in states outside of the Corn Belt (an added 1.6 million ha). As a result of a market for rights in Illinois, a total of 0.9 million additional ha are needed nationally to produce the required corn and sorghum.

In addition to the shifts in location of corn and sorghum production, small grain areas in the various regions also change. The pattern for small grains parallels that for corn and sorghum—a decrease in Illinois and an increase outside of Illinois. However, the shifts are not as dramatic as those for corn and sorghum, because on a national scale Illinois' small grain production is of less importance than its corn and sorghum. Also, small grain crops are not affected as much by the restrictions on available N fertilizer. As in the case of corn and sorghum, more total area is required to meet the national needs for small grain. At the most restricted level of rights, 2.3 million additional ha would be required.

The location of soybean production is changed substantially by the market for rights in Illinois. In general, the shifts are in the opposite direction of those for the other crops considered. With the smallest quantity of rights, soybean area in Illinois increases almost fourfold. This increase is accompanied by a reduction in the other states within the Corn Belt and outside of the Corn Belt. Because Illinois soybean yields are higher than the yields in other states, 1.2 million fewer ha are needed nationally to meet the demand than when there is no market for rights.

Less productive land will be used to meet the demand, thereby increasing the unit cost of production. An 8% cost increase occurs as we go to the smallest quantity of rights. In the long run, the increased costs must be covered by the price of these grains. Soybean costs are not affected, but wheat, because of its N requirement, follows roughly the same pattern as corn.

XI. INFORMATION PROGRAMS

If farmers are applying more N fertilizer than necessary to meet their economic goals, then an educational program could be an effective way to convince them to voluntarily reduce the amount applied. This would be an adjustment in the direction of reducing the NO_3^--N pollution potential in water and the N_2O impact on the stratosphere. Farmers would save money in the long run and environmental quality would improve. But are most farmers actually applying too much fertilizer?

One approach to this question is to use experimental data and compare the economic optimum with the rates used by farmers. A critical deficiency in this approach is the likely differences between experimental response functions and those characteristic of commercial farms. A number of attempts have been made to describe response functions likely to be found on commercial farms (FAO, 1966; Ibach & Adams, 1968).

In assessing the effect on rate of N fertilizer application of an information program which appeals to the economic self-interest of the producer, the algebraic form of the response function may play a critical role. The optimal level is usually sensitive to the form of the function (Dillon, 1977; Swanson, 1963; Heady & Dillon, 1961). A recent resurgence of interest in the von Liebig law as a description of the response function (Cate & Hsu, 1978) may cause a reexamination of the matter of choice of function.

Taylor and Swanson (1973) approached the issue of the economically optimal amount of N fertilizer to apply to corn in two ways. First, by comparing experimental response functions with the results of a survey of yields on commercial farms. Second, they compared experimental results with response functions, using the Spillman function, that were based on the judgments of agronomists familiar with both experimental data and farm practices (Ibach & Adams, 1968). While the experimental response of corn yields to N appeared to be consistently higher than the response experienced by commercial farmers, the economically optimal fertilization rates for commercial farms were only slightly lower than the optimal rates for the experimental situations. Apparently, any reduction in fertilizer use in this situation will reduce profits of an individual farmer. Consequently, an educational program would be ineffective in reducing the N load (the total amount of N added minus the amount removed by crops). In fact, if farmers were better informed about response functions and price ratios, an educational program could possibly increase the N load, because in terms of economic returns, more Illinois farmers apparently apply too little rather than too much N. We cannot, of course, generalize these results to other areas or to other crops.

Calculating the economic optimum level of N fertilizer application from experimental results may be useful in determining if those farmers identified as the better managers are applying more than an economic optimum. The response functions of these farmers are more apt to be similar to those based on experimental data. In two studies (Swanson et al., 1973; Frohberg & Taylor, 1975) experimental data were used with decision models which took into account the risks involved in the returns from corn production. In general, the optimal application rates as determined from these models were greater than those used by average farmers, and in most cases in excess of those used by farmers following management programs economically justifying higher levels of N fertilizer application.

XII. SUMMARY AND CONCLUSIONS

Substantial effort was expended in the 1970's to evaluate the likely economic consequences of various controls on N fertilizer use. Our knowledge of the production, price, and income consequences is substantially greater than the environmental and health consequences of such controls.

Improvements in health and the quality of the environment are the reasons for examining the effects of various control measures. Therefore, for a complete evaluation of these measures public policy makers should place the production, farm income, and food cost consequences of controls side by side with expected benefits. The combined information will then form the basis from which policy decisions can be made.

In this chapter, two general types of controls were considered—those that are regulatory in nature and those that rely on economic incentives and improved information. The administrative costs involved with the regulatory approach are usually larger than those of policies which use economic incentives.

There is evidence that the problem of elevated NO_3^--N levels in water is a rather localized one. If this is an accurate description of the problem, and if controls are to be imposed in a rather small region, it is important that the economic analysis of impacts takes into account the economic losses likely to be incurred by producers in that area as well as the gains for those producers outside of the area. If the controls are for large regions, or for the nation, the economic analysis should be done in a framework that allows the pattern of cropping and adoption of technologies to adapt to the new situation of controls. It also is important that demand considerations be introduced into the calculations in such a way that the impact of changed production on price and, hence, income may be evaluated.

Empirical results presented in this chapter are illustrative of various analyses, but a quantitative comprehensive summary statement is not possible. We may conclude, however, that initial reductions from present levels of N fertilizer use on corn and sorghum to approximately 110 kg/ha would require a very minor additional area of cropland. However, if export demand is high and if this limit is halved, there would be severe pressures on the land resource. When fertilizer use is initially reduced in the range of 10 to 25%, the effects on cropping patterns, area requirements, and income are moderate. The impact becomes incresingly severe, however, when the use of N fertilizer is further restricted.

Finally, any analysis of the economic impact of controls on the use of N fertilizer as a part of an environmental assessment must ideally be done in the context of the total crop production system and other public policies presently affecting agriculture. Interventions which cause alterations in the levels of N fertilizer use are likely to cause other adjustments in the system that relate to erosion and sedimentation, the impact of other plant nutrients on environmental quality, and the use of pesticides.

The interactions of controls on N fertilizer with other public policies affecting agriculture, such as price and income policies and soil conservation, are outside of the scope of this chapter. The reader may find the following analyses to be of interest in pursuing this aspect of the interactions of control of N fertilizer with other policies: Farris and Sprott, 1971; Casler, 1972; Farris and Sprott, 1972; Lacewell and Masch, 1972; Taylor and Frohberg, 1977; USEPA, 1978.

ACKNOWLEDGMENTS

Comments and suggestions of George C. Casler, Wesley D. Seitz, and C. Robert Taylor are gratefully acknowledged.

LITERATURE CITED

Abrams, L. W., and J. L. Barr. 1974. Corrective taxes for pollution control: An application of the environmental pricing and standards system to agriculture. J. Environ. Econ. Manage. 1:296–318.

Baumol, W. J., and W. E. Oates. 1971. The use of standards and prices for protection of the environment. Swed. J. Econ. 73:42–54.

Casler, G. L. 1972. Economic and policy implications of pollution from agricultural chemicals: Comment, Am. J. Agric. Econ. 54:534–535.

Cate, R. B., and Y. T. Hsu. 1978. An algorithm for defining linear programming activities using the law of the minimum. North Carolina Agr. Exp. Stn. Tech. Bull. no. 253. p. 29.

Dillon, J. L. 1977. The analysis of response in crop and livestock production. 2nd ed. Pergamon Press, New York. p. 213.

Farris, D. E., and J. M. Sprott. 1971. Economic and policy implications of pollution from agricultural chemicals. Am. J. Agric. Econ. 53:661–662.

Farris, D. E., and J. M. Sprott. 1972. Economic and policy implications of pollution from agricultural chemicals: Reply. Am. J. Agric. Econ. 54:536.

Fried, Maurice, Kenneth K. Tanji, and Ronald M. Van De Pol. 1976. Simplified long term concept for evaluating leaching of nitrogen from agricultural land. J. Environ. Qual. 5: 197–200.

Food and Agriculture Organization of the United Nations (FAO). 1966. Statistics of crop response to fertilizers. Rome. 112 p.

Frohberg, K. K., and C. R. Taylor. 1975. The influence of risk arising from weather variability on the optimal nitrogen fertilization level of corn. Ill. Agric. Econ. 150:23–26.

Gros, J. G., and E. R. Swanson. 1976. A framework for evaluating public policy on the use of agricultural chemicals. J. Environ. Manage. 4:403–411.

Heady, E. O., and J. L. Dillon. 1961. Agricultural production functions. Iowa State University Press, Ames. p. 667.

Heady, E. O., H. C. Madsen, K. J. Nicol, and S. H. Hargrove. 1972. Agricultural and water policies and the environment: An analysis of national alternatives in natural resource use, food supply capacity, and environmental quality. CAED Rep. no. 40T. Center for Agric. and Rural Dev., Iowa State Univ.

Hills, F. J., F. E. Broadbent, and M. Fried. 1978. Timing and rate of fertilizer nitrogen for sugarbeets related to nitrogen uptake and pollution potential. J. Environ. Qual. 7:368–372.

Horner, G. L. 1975. Internalizing agricultural nitrogen pollution externalities: A case study. Am. J. Agric. Econ. 57:33–39.

Horner, G. L. 1977. An economic analysis of nitrogen reduction in subsurface drainage water. California Agric. Exp. Stn., Giannini Foundation Res. Rep. no. 323. 90 p.

Ibach, D. B., and J. R. Adams. 1968. Crop yield response to fertilizer in the United States. USDA Econ. Res. Serv. Stat. Bull. 431. p. 295.

Illinois Pollution Control Board. 1972. Opinion of the Board in the matter of plant nutrients. R-71-15. Springfield, Ill. p. 42.

Jacobs, James J., and George L. Casler. 1979. Internalizing externalities of phosphorous discharges from crop production to surface water: Effluent taxes versus uniform reductions. Am. J. Agric. Econ. 61:309–317.

Kasal, James. 1976. Trade-offs between farm income and selected environmental indicators. USDA Tech. Bull. 1550. Econ. Res. Service. p. 27.

Klepper, R., R. Parker, G. B. Shearer, and B. Commoner. 1974. Fertilization application rates and nitrates in Illinois surface waters. Illinois Inst. for Environ. Qual. Document no. 72-38. Springfield, Ill. p. 167.

Klepper, R., W. Lockeretz, B. Commoner, M. Gertler, S. Fast, D. O'Leary, and R. Blobaum. 1977. Economic performance and energy intensiveness on organic and conventional farms in the Corn Belt: A preliminary analysis. Am. J. Agric. Econ. 59:1–12.

Klepper, R. 1978. Nitrogen fertilizer and nitrate concentrations in tributaries of the Upper Sangamon river in Illinois. J. Environ. Qual. 7:13–22.

Lacewell, R. D., and W. R. Masch. 1972. Economic incentives to reduce chemicals used in commercial agriculture. South. J. Agric. Econ. 4:203–208.

Langham, M. R. 1971. A theoretical framework for viewing pollution problems. South. J. Agric. Econ. 3:1–8.

Mayer, L. V., and S. H. Hargrove. 1971. Food costs, farm incomes, and crop yields with restrictions on fertilizer use. CAED Rep. no. 38. Center for Agric. and Econ. Dev., Iowa State Univ.

Mishan, E. J. 1971. The post-war literature on externalities. J. Econ. Lit. 9:1–28.

Nagadevara, S. V., E. O. Heady, and K. J. Nicol. 1975. Implications of application of soil conservancy and environmental regulations in Iowa within a national framework. CAED Rep. no. 57. Center for Agric. and Rural Dev. Iowa State Univ., Ames. 136 p.

National Research Council. 1961. Status and methods of research in economic and agronomic aspects of fertilizer response and use. Pub. 918, Natl. Acad. of Sci., Washington, D.C. 89 p.

National Research Council. 1972. Accumulation of nitrate. Natl. Acad. of Sci., washington, D.C. 106 p.

National Research Council. 1978. Nitrates: An environmental assessment Natl. Acad. of Sci., Washington, D.C. 723 p.

Ogg, C. 1978. The welfare effects of erosion controls, banning pesticides, and limiting fertilizer application in the Corn Belt: Comment. Am. J. Agric. Econ. 60:559.

Onishi, H., and E. R. Swanson. 1974. Effect of nitrate and sediment constraints on economically optimal crop production. J. Environ. Qual. 3:234–238.

Onishi, H., A. S. Narayanan, T. Takayama, and E. R. Swanson. 1974. Economic evaluation of the effect of selected crop practices on nonagricultural uses of water. Water Resour. Center Res. Rep. no. 79, Univ. of Illinois at Urbana-Champaign. p. 52.

Palmini, D. J., C. R. Taylor, and E. R. Swanson. 1978. Potential impact of selected agricultural pollution controls on two counties in Illinois. Agric. Econ. Res. Rep. 154. Dep. of Agric. Econ., Univ. of Illinois, Urbana-Champaign. p. 41.

Pearson, R. W., H. V. Jordan, O. L. Bennett, C. E. Scarsbrook, W. E. Adams, and A. W. White. 1961. Residual effects of fall and spring applied nitrogen fertilizer on crop yields in the southeastern United States. USDA Tech. Bull. 1254. Washington, D.C. p. 19.

Porter, Keith S. (ed.). 1975. Nitrogen and phosphorus, food production, waste and the environment. Ann Arbor Science Publ., Inc. Ann Arbor, Mich. 372 p.

Randall, A. 1972. Market solutions to externality problems: Theory and practice. Am. J. Agric. Econ. 54:175–183.

Schneider, R. R., and R. H. Day. 1976. Diffuse agricultural pollution: The economic analysis of alternative controls. Water Resour. Center Tech. Rep. WIS WRC 76-02. Univ. of Wisconsin, Madison. 100 p.

Singh, B., and G. S. Sekhon. 1976. Nitrate pollution of ground water from nitrogen fertilizer and animal wastes in the Punjab, India. Agric. Environ. 3:57–68.

Singh, B., C. R. Biswas, and G. S. Sekhon. 1978. A rational approach for optimizing application rates of fertilizer nitrogen to reduce potential nitrate pollution of natural waters. Agric. Environ. 4:57–64.

Stauber, M. S., and O. R. Burt. 1973. Implicit estimate of residual nitrogen under fertilized range conditions in the North Great Plains. Agron. J. 65:897–901.

Stauber, M. S., O. R. Burt, and F. Linse. 1975. An economic analysis of nitrogen fertilizer of grasses when carry-over is significant. Am. J. Agric. Econ. 57:463–471.

Stout, P. R., and R. G. Burau. 1967. The extent and significance of fertilizer buildup in soils as revealed by vertical distributions of nitrogenous matter between soils and underlying water reservoirs. p. 283–310. In N. C. Brady (ed.) Agriculture and the quality of our environment. Am. Assoc. for the Adv. of Sci., Washington, D.C.

Swanson, E. R. 1963. The static theory of the firm and three laws of plant growth. Soil Sci. 95: 338–343.

Swanson, E. R., C. R. Taylor, and L. F. Welch. 1973. Economically optimal levels of nitrogen fertilizer for corn: An analysis based on experimental data, 1966–1971. Ill. Agric. Econ. 13(2):16–25.

Swanson, E. R. 1975. Problems and policy alternatives-agriculture. p. 87. In Wesley D. Seitz (ed.) Proc. Wrkshp. on Non-Point Sources of Water Pollut., Allerton House, 20–21 March 1975, College of Agric., Special Pub. no. 37, Univ. of Illinois, Urbana-Campaign.

Swanson, E. R., and C. R. Taylor. 1977. Potential impact of increased energy costs on the location of crop production in the Corn Belt. J. Soil Water Conserv. 32:126–129.

Swanson, E. R., C. R. Taylor, and P. J. van Blokland. 1978. Economic effects of controls on nitrogen fertilizer. Illinois Agr. Exp. Stn. Bull. 757. College of Agric., Univ. of Illinois, Urbana-Champaign. p. 37.

Taylor, C. R. 1973. An analysis of nitrate concentrations in Illinois streams. Ill. Agric. Econ. 13(1):12–19.

Taylor, C. R. 1975a. A regional market for rights to use fertilizer as a means of achieving water quality standards. J. Environ. Econ. Manage 2(1):7–17.

Taylor, C. R. 1975b. The nitrate controversy: Three proposed policies and their economic effects. Ill. Res. 17:6–7.

Taylor, C. R., and E. R. Swanson. 1973. Experimental nitrogen response functions, actual farm experience and policy analysis. Ill. Agric. Econ. 13(2):26–32.

Taylor, C. R., and E. R. Swanson. 1974. Economic impact of imposing per acre restrictions on use of nitrogen fertilizer in Illinois. Ill. Agric. Econ. 14(2):1–5.

Taylor, C. R., and E. R. Swanson. 1975. The economic impact of selected nitrogen restrictions on agriculture in Illinois and the United States. Agric. Econ. Res. Rep. 133, Dep. of Agric. Econ., Univ. of Illinois at Urbana-Champaign. p. 37.

Taylor, C. R., and K. K. Frohberg. 1977. The welfare effects of erosion controls, banning pesticides, and limiting fertilizer application in the Corn Belt. Am. J. Agric. Econ. 59: 25–36.

Taylor, C. R., and K. K. Frohberg. 1978. The welfare effects of erosion controls, banning pesticides, and limiting fertilizer applications in the Corn Belt: Reply. Am. J. Agric. Econ. 60:560–561.

U.S. Environmental Protection Agency. 1973. Nitrogenous compounds in the environment. Report no. EPA-SAB-73-001. Hazardous Materials Advisory Comm., Sci. Advisory Board, Washington, D.C.

U.S. Environmental Protection Agency. 1978. Alternative policies for controlling nonpoint agricultural sources of water pollution. Report no. EPA-600/5-78-005. Environmental Res. Lab., Off. of Res. & Dev., Athens, Ga.

Viets, F. G., and R. H. Hageman. 1971. Factors affecting the accumulation of nitrate in soil, water, and plants. Agr. Handbook 413. USDA-ARS.

Walker, M. E., Jr., and E. R. Swanson. 1974. Economic effects of a total farm nitrogen balance approach to reduction of potential nitrate pollution. Ill. Agric. Econ. 14(2):21–27.

21

Recycling of Nitrogen Through Land Application of Agricultural, Food Processing, and Municipal Wastes[1]

J. H. SMITH

Snake River Conservation Research Center
Agricultural Research Service, USDA
Kimberly, Idaho

J. R. PETERSON

Research & Development Laboratory
Metropolitan Sanitary District of Greater Chicago
Cicero, Illinois

I. INTRODUCTION

Crop residues, animal, municipal, and more recently, food processing wastes are among the many materials applied to soil as fertilizer or for disposal. Before waste treatment systems were developed, most food processing and municipal wastes were discharged to rivers, lakes, and oceans, often severely polluting these waters and wasting fertilizer nutrients. Many innovations in waste management have been developed and considerable research is underway on the beneficial use of wastes for crop growth while giving cost-effective disposal.

The objective of this chapter is to summarize and evaluate available information on the value and management of agricultural, food processing, and municipal wastes as they are applied to land for N fertilization of growing crops.

II. AGRICULTURAL WASTES

A. Crop Residues

1. CROP RESIDUE DECOMPOSITION AND NITROGEN MINERALIZATION

Crop residues are a valuable natural resource. Approximately 363 million metric tons of crop residues are produced annually in the USA and

[1] Contribution from the Agricultural Research Service, USDA, and the Metropolitan Sanitary District of Greater Chicago (CRIS no. 5704-20790-002).

Table 1—Annual crop residue yields and N contents in the USA (Larson et al., 1978).

Residue	Total yield	Nitrogen content	
	10⁶ metric tons	%	10⁶ metric tons
Corn (*Zea mays* L.) and			
Sorghum (*Syricum granum* L.)	154	1.1	1.69
Cereal grain†	96	0.7	0.67
Soybeans (*Glycine max* L.)	53	2.2	1.17
Rice (*Oryza sativa* L.)	4.7	0.6	0.03

† Wheat (*Triticum aestivum*), barley (*Hordeum vulgare*), oats (*Avena sativa*), and rye (*Secale cereale*).

these residues contain approximately 4 million metric tons of N (Larson et al., 1978). Most crop residues are returned to the soil following harvest.

Yield and N content figures provide perspective on crop residue value and management problems associated with crop residues in the USA (Table 1). The N content of residues varies widely. For example, wheat straw N may range from 0.2 to 1.0% or higher.

Nitrogen transformations in decomposing crop residues in soil are reasonably well understood. Bartholomew (1965) prepared a comprehensive review of the subject and this section is intended to describe and evaluate subsequent research developments (See also Chapt. 6, S. L. Jansson and J. Persson).

When crop residues are incorporated into soil, the N requirements for a maximum decomposition rate depend upon many factors. The environmental factors were discussed in some detail by Bartholomew (1965) and will not be dealt with here. One point, however, that should be considered is the fact that environmental conditions are frequently less than optimum in the field, resulting in slower decomposition rates than those observed in the laboratory. Under these conditions, less N is required for crop residue decomposition in the field than in the laboratory. We shall emphasize field results.

After crop residues are incorporated into the soil the initial phases of decomposition proceed rapidly. In a field experiment with wheat straw, Smith and Douglas (1971) observed 20% straw weight loss in 10 weeks in the fall while soil temperatures decreased to 4°C. During 13 months 75% straw weight loss occurred. Jenkinson (1971) surveyed the literature on plant residue decomposition and reported that the proportion of crop residues decomposed under different climatic conditions with different plant materials was remarkably similar. Excluding acid soils, approximately one-third of the residue remained after 1 year and one-fifth remained after 5 years. He reported that even fresh green plant materials behave in this way, contrary to the widespread belief that such residues decompose rapidly and completely in the soil.

Previous cropping history and soil fertility have considerable impact on crop residue decomposition and N mineralization. The N requirements of subsequent crops also can be important. Smith and Douglas (1968) noted that while residual N from previous cropping had little influence on wheat straw decomposition in the field, straw buried in plots increased in N percentage in relation to previous N fertilization. Further, yields of non-N responsive Pinto beans (*Phaseolus vulgaris* L.) were not influenced by straw applications, while sugar beet (*Beta vulgaris* L.) yields decreased due to addition of straw residues. Soil and crop analyses showed that N immobilization was essentially completed in the fall. The addition of 7.5 kg N/metric ton of straw, with N applied in the spring for growing the sugar beet crop, compensated for straw-immobilized soil N (Smith et al., 1973).

Studies in Iowa to determine the long-term effects of crop residue additions on soil organic matter and on N immobilization and mineralization, showed that after the addition of 16 tons/ha per year of cornstalk residues for 11 years, soil organic N increased by 32%. Alfalfa residue addition increased soil organic N by 41%. Calculated break-even point values for maintaining soil organic matter and N at the original levels indicated that 6 metric tons/ha per year of either alfalfa or corn residues were required (Larson et al., 1972).

Many laboratory experiments have been run in which researchers attempted to develop threshold values for N immobilization and release. Generally, crop residues containing 1.5% N or more will decompose at "normal" rates without bringing about a net immobilization of soil N. Crop residues containing <1.5% N are considered deficient in N and will lead to depletion of available forms of soil N during decomposition. Such residues are often amended with N to compensate for net immobilization and to hasten crop residue decomposition. In most cases, this is an uneconomical, wasteful practice based on a false premise. Nitrogen addition seldom accelerates plant residue decomposition (Allison, 1973). Nitrogen addition did not change the amount of straw remaining in soil after 3 years cropping under dryland conditions in Montana and Nebraska (Greb et al., 1974).

Nitrogen fertilizer use, soil fertility levels, and crop productivity have increased since many of the original laboratory experiments on N immobilization were performed. As a result, the impact of N on crop residue decomposition has changed. Field experiments by Brown and Dickey (1970) on nonirrigated soils in Montana showed that 95% of the buried wheat straw decomposed in 18 months and the residue increased from 0.3 to 1.2% N without added N. Similar results were found by Smith and Douglas (1971) in irrigated soils in Idaho, who found that 70% of the applied N (to bring the straw N to 1.5%) was lost in the first few weeks of incubation and that straw N was mineralized at N concentrations <1.0% (See Fig. 1). Under field conditions, with irrigation or rainfall, leaching as well as the numerous factors that limit decomposition rates apparently decrease the amount of N required for maximum decomposition. Evidently, decomposition proceeds at some rate that is not limited by N when N percentages in the crop residues are below the theoretical equilibrium value of 1.5%.

Several reports are available on the influence of above-ground crop residues on soil organic matter and N transformations, but few deal with roots. Powers (1968) reported that the effect of N content of bromegrass (*Bromus inermis* L.) on the rate of decomposition was similar to that of

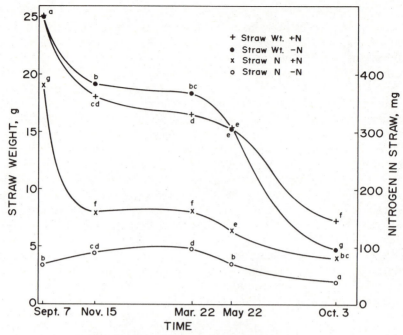

Fig. 1—Weight of Nugaines wheat straw remaining in soil at indicated dates, and N content of straw. Points associated with different letters are different at the 0.05 significance level (from Smith and Douglas, 1971).

most other crop residues. Roots containing 0.8% N immobilized some soil N for a time while roots containing 1.4% N caused little or no N immobilization. Smith (1966) using ^{14}C-labeled plant parts of corn, wheat, and soybeans (*Glycine max* L.), showed that roots decomposed more slowly than tops. He also reported that corn roots decomposing in soil decreased the decomposition of soil organic matter by almost 50%. This protective effect would also decrease the rate of soil N mineralization and, therefore, corn may be grown continuously when adequately fertilized.

With other low N plant materials such as wood chips, sawdust, or wood bark, the decomposition rate and N mineralization is limited by composition. Wood products contain less soluble carbohydrates and higher lignin concentration than most crop plants, and thus, have lower decomposition rates. Most of the wood products investigated by Allison (1965) contained an average of 0.1% N, and the decomposition rates of most of the 28 wood species investigated were not influenced by N addition. These wood materials would not be expected to mineralize enough N to supply growing crops for a long time after incorporation into soil and would only slowly decompose in the soil.

The importance of rice straw (*Oryza sativa* L.) as a crop residue and its impact on soil fertility worldwide have been largely overlooked until recently. In the past 10 years, several definitive publications pointing out the differences between rice straw and wheat straw management have appeared.

While rice culture in the USA produces a relatively insignificant 4.7 million metric tons of residue annually, the remainder of the world probably produces more rice residues than any other cultivated crop residue. In Asia, excluding mainland China, the annual production of rice straw is approximately 454 million metric tons (Tanaka, 1973). Rice straw contains an average of 0.5% N and a total of approximately 2.3 million metric tons of N. Rice straw is used for a variety of purposes including building materials and furnishings, paper, fuel, animal feed, culture media, and composts, but its major use is for incorporation in soil to provide plant nutrients. Rice culture included both lowland (flooded) and highland (nonflooded) rice. This distinction is made because there is a striking difference between straw decomposition and N requirements under these two conditions.

Williams et al. (1968) determined that under lowland field conditions, net immobilization of N, as determined by yield responses, occurred at 0.54% N in the original straw. Straw with higher N content increased yields on N deficient soils, while straw with lower N content decreased yields. They reported that under upland conditions, N requirements for rice straw decomposition were similar to that of wheat straw.

Rao and Mikkelsen (1976) in laboratory experiments found that rice straw immobilized about half as much N as wheat straw. Sain and Broadbent (1977) enclosed rice straw in nylon bags and buried the bags in the soil. Decomposition of the buried rice straw was influenced very little by N additions. Yoneyama and Yoshida (1977a, 1977b, 1977c) conducted a series of experiments using ^{15}N to trace N transformations during rice straw decomposition. They reported that the N percentage in the rice straw increased during the first 2 weeks because of N immobilization while further increases in percentage of N resulted from C loss. The mature rice plants grown on soils enriched with ^{15}N-labeled rice straw derived approximately 6% of their N from the straw. Sampling with stage of growth showed that the percentage of N in the plants derived from straw was relatively constant. These authors also observed that N mineralization and immobilization by plant residues appears to be simultaneous, beginning at an early stage of decomposition even when the N content of the residue is low and net mineralization is not detectable.

Thus, N addition to rice straw for decomposition under lowland conditions is neither necessary nor desirable and, as found by Williams et al. (1968), sometimes N may suppress rice straw decomposition or cause yield depression. Most rice straw contains adequate N for decomposition and will provide part of the N needed to grow succeeding rice crops.

Krantz et al. (1968) have summarized the practical considerations in handling crop residues in the field. Early crop residue incorporation into moist soil was regarded as a key factor in residue management. Further, fertilizing the growing crop at the time N is needed by the crop is much more efficient than fertilizing the crop residue. When N is added to crop residues, part of the N is immobilized, but has little effect on decomposition rates. Nitrogen immobilization by decomposing crop residues decreases N use efficiency because losses occur in each subsequent turnover. A portion of the immobilized N is remineralized during the immediate cropping year and becomes available to growing crops. The balance of the immobilized N mineralizes at about the same rate as the native soil N in subsequent years.

2. DENITRIFICATION

Four factors must be present simultaneously for denitrification to occur in the field (see Chapt. 8, M. K. Firestone). They are (i) anoxic conditions, (ii) an organic energy source, (iii) presence of denitrifying microorganisms, and (iv) presence of NO_3^-.

Crop residues decomposing in soil may promote conditions contributory to denitrification. This depends primarily on the nature of the crop residue. Most crop residues, when mixed with warm moist soil, will decompose rapidly enough to deplete O_2 and, thus, lower the redox potential in the immediate area. Soil water content and texture will probably determine whether or not the threshold conditions for denitrification will be reached. Crop residues low in N that are not fertilized with N will not yield NO_3^- during the stage of rapid decomposition and NO_3^- in the soil at the time of residue incorporation will probably be mostly immobilized. Therefore, denitrification is unlikely in the presence of low N crop residue, providing N fertilizer has not been added. Approximately 85% of the crop residues produced in the USA are low enough in N that denitrification of the residue N will be minimal. One exception is soybean straw, which contains an average of 2.2% N. Incorporating this residue in a warm, moist soil will cause rapid decomposition, and nitrification is likely. With irrigation or rainfall, the redox potentials will be lowered and some denitrification could occur. How frequently this will happen in the field is open to speculation. The same conditions would develop with plowdown of green manure crops, alfalfa, bean or pea straw, or other high N crop residues.

Crop residues decomposing in soil can either decrease or increase ground water pollution by NO_3^-. For example, Smith and Douglas (1971) showed that approximately 70% of the N added to straw in the field was leached out of the straw into the soil. Nitrogen not immobilized will be susceptible to leaching. If the field is irrigated or rainfall is greater than the soil water holding capacity, the applied N may leach below the root zone and ultimately reach the ground water causing pollution. The potential for pollution is similar to that of low rate fertilizer N as described by Schuman et al. (1975).

Low N crop residues in soil may also decrease pollution potential by acting as a sink and immobilizing excess fertilizer N or mineralized soil N, retaining it in a nonmobile form until later when it will become available to growing crops.

High N crop residues, returned to the soil when N is not needed by growing crops, may pose a pollution hazard. Soybean straw, edible bean straw, pea (*Pisum sativum* L.) residue, or alfalfa (*Medicago sativa* L.) when incorporated into soil will decompose rapidly and the N will be subject to nitrification. This NO_3^-, like any fertilizer N that is applied at the wrong time, can leach and pollute ground water (Burwell et al., 1976; Letey et al., 1977). Crop residues can and should be managed to minimize N pollution hazards.

B. Animal Manures

1. ANIMAL MANURE DECOMPOSITION AND NITROGEN MINERALIZATION

Domestic animals in the USA produce over 1 billion metric tons of fecal wastes each year. Liquid effluents amount to over 360 million metric tons. Used bedding, paunch manure from slaughtering facilities, and dead carcasses make the total annual animal waste production close to 2 billion metric tons (Wadleigh, 1968). For centuries, animal wastes have been used as fertilizers for crop production. With the advent of low-cost commercial fertilizers, the use of animal manures became less cost effective. Since 1973, energy shortages have increased commercial fertilizer costs and made manure use more economical.

Modern animal feeding practices, where thousands of animals are concentrated in small areas, have made manure handling difficult. Waste must be stored so that its application corresponds with favorable weather, crop, and land conditions. Nutrient loss, especially NH_3, can occur during storage and can cause some social and economic problems. However, with careful management, the benefits outweigh the problems, making animal manure a valuable resource (Elliott & Swanson, 1976).

Many pathways are followed by N during animal waste use and conversion (Fig. 2). The processes affecting these conversions are physical, chemical, and biological. Microorganisms play a major role in N conversions in decomposing manure in the feedlot and soil.

The N forms in fresh excreta are mainly urea, NH_3, organic, and from birds, uric acid, with virtually no NO_3^-. Manure composition will vary with feed ration composition and with the animal species. For example, the N content of fresh chicken manure is much higher than that of the other animals (Table 2). Even though these variabilities exist, many manure decomposition pathways are common to all types of manure (Fig. 2).

When manure and urine are deposited, their composition changes immediately. For instance, urine contains large quantities of urea, which is converted rapidly to CO_2 and NH_4^+ by the enzyme urease. The N in urine comprises 21% (hog), 35.5% (horse), 52% (cattle), and 63% (sheep) of the total N excreted (Elliott & McCalla, 1973).

Mathers et al. (1972) determined the N concentration in manure from 23 Texas high plain feedlots. The N percentage ranged from 1.16 to 1.96% and averaged 1.34%. The way manure is handled affects its chemical content. As decomposition proceeds, the mineral percentage in the manure increases, and the mineralized N is converted to NO_3^-. When excessive rates of manure are added to soil, (100 metric tons/ha or more) NH_3 tends to accumulate in the soil, temporarily slowing NO_3^- formation.

Mathers et al. (1972, 1975) and Mathers and Stewart (1970), in a series of experiments where varying rates of manure were applied to soil, found that manure applications up to 22 metric tons/ha supplied enough N for growing corn or sorghum without excess NO_3^- in the soil profile. Higher

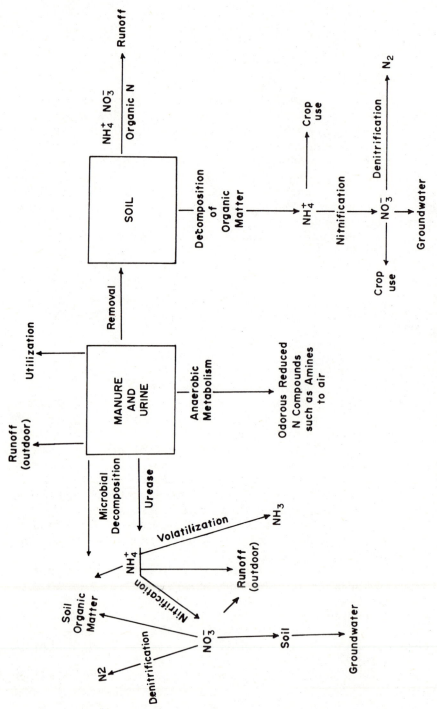

Fig. 2—Manure decomposition pathways (from Elliot and McCalla, 1973).

Table 2—Average moisture and N content of animal manures (from Elliott and McCalla, 1973).

	Moisture in fresh manure	N in fresh manure	Total dry solids		N/animal per day
	%	kg/ metric ton	kg/animal per day	%	g
Dairy cattle	79 1.0	10	3.7	4.6	172
Beef cattle	80 0.9	9	4.1	4.5	186
Swine	75 1.6	16	0.32	6.4	20
Horse	60 0.7	7	6.4	1.8	118
Sheep	65 1.7	17	0.5	4.6	21
Broiler	80 5.2	52	0.03	6.9	2.2
Hen	75–80 4.3	43	0.03	6.2	1.7

rates caused NO_3^- accumulation with up to 1,200 kg of NO_3^--N/ha in 180 cm of soil when 224 metric tons/ha manure were applied.

Nitrate pollution of surface or ground water is one of the major concerns in concentrated livestock holding areas or in soils receiving large amounts of manure (Gilmour et al., 1977). Mielke et al. (1974) described three layers in the soil profile under cattle feedlots. The surface layer is manure overlaying a mixed layer of organic and mineral material. Below this is the top of the mineral soil. The two upper layers are relatively impermeable when moist and allow little downward movement of NO_3^-. These authors reported higher NO_3^- concentrations to 1.7 m in one feedlot profile than in adjacent cropland, but little NO_3^- occurred in another location in the same feedlot. Some ammonification and nitrification occurs in feedlots. Adriano et al. (1971) found higher NO_3^- concentrations under California dairy corrals than under irrigated pastures or cropland at the 3- to 6-m depths. Marriott and Bartlett (1972) in Pennsylvania applied up to 85 metric tons/ha of manure and found relatively high NO_3^- concentrations in the soil to depths of 1.5 m when more than 17 metric tons of manure/ha were applied.

The literature on manure application to soils is voluminous. Pertinent recent references concerned with the probability of high soil NO_3^- concentrations accumulating with large applications of animal manure include: Webber and Lane (1969); Olsen et al. (1970); Concannon and Genetelli (1971); Hensler et al. (1971); Mathers and Stewart (1971); Peterson et al. (1971); Carlile (1972); Kimble et al. (1972); MacMillan et al. (1972, 1975); Murphy et al. (1972); Weeks et al. (1972); Bielby et al. (1973); Walter et al. (1974); Meek et al. (1974); Stewart and Chaney (1975); Loehr et al. (1976); and Pratt et al. (1976).

Although wastes have been used for centuries as nutrient sources for plant growth, a rational basis for their use has not been developed. Application rates have been based on experience and little information has been available about decomposition rates or nutrient availability. In recent years, renewed interest in using manure for its fertilizer value has led to several experiments to obtain information on decomposition rates and to develop a rational basis for using manure. Pratt et al. (1973) proposed a theoretical basis for matching manure applications with crop needs. This theory is being tested by them and others to evaluate and refine estimates of N availability.

Organically combined N must be mineralized before it can become available to plants. Therefore, the mineralization rate is the key to the application rate of any given material. The yearly mineralization rates are expressed as a series of fractional mineralizations of any given application, or the residual of that application. These will be referred to hereafter as the decay series. For example, the decay series, 0.30, 0.10, 0.05, means that for any given application, 30% is mineralized the first year, 10% of the unmineralized residual is mineralized the second year, and 5% of the residual is mineralized the third and all subsequent years. The same series is applied individually to each yearly application of organic N (Pratt et al., 1973).

Constant annual manure applications that will supply enough N for the present crop will ultimately cause excessive fertilization. Therefore, to meet annual crop N requirements with manure applications, decreasing amounts should be applied each year. For example, based on 1.5% N, 30% moisture, and a 0.35, 0.15, 0.10, 0.05 decay series, to supply 200 kg of available N/ha each year, it would be necessary to add 54 metric tons the 1st year, and decrease the rate to <27 metric tons for the 20th year (Pratt et al., 1973).

Decomposition patterns of manures differ, requiring development of several decay series. The decay series 0.90, 0.10, 0.05 is typical of organic wastes such as chicken manure containing N, primarily in the rapidly mineralized urea and uric acid forms. The N in these materials is nearly as available as inorganic N. The decay series 0.75, 0.15, 0.10, 0.05 represents materials in which about 50% of the N is present as urea or uric acid and the balance is slowly mineralizable organic compounds. Fresh wastes from dairy or beef cattle are in this category. Other decay series are used to represent more stable organic materials such as accumulated dried dairy or beef cattle manure found in corrals (Pratt et al., 1973).

Table 3—Ratio[†] of yearly N input to annual N mineralization rate of organic wastes at constant yearly mineralization rate for five decay series for various times after initial application (Pratt et al., 1973).

Decay series	Typical material[‡]	Time, years							
		1	2	3	4	5	10	15	20
		N input/mineralization rate							
0.90, 0.10, 0.05	Chicken manure	1.11	1.10	1.09	1.09	1.08	1.06	1.05	1.04
0.75, 0.15, 0.10, 0.05	Fresh bovine waste, 3.5% N	1.33	1.27	1.23	1.22	1.20	1.15	1.11	1.06
0.40, 0.25, 0.06	Dry corral manure, 2.5% N	2.50	1.82	1.74	1.58	1.54	1.29	1.16	1.06
0.35, 0.15, 0.10, 0.05	Dry corral manure, 1.5% N	2.86	2.06	1.83	1.82	1.72	1.40	1.23	1.13
0.20, 0.10, 0.05	Dry corral manure, 1.0% N	5.00	3.00	2.90	2.44	2.17	1.38	1.13	1.04

† This ratio equals kg of N required to produce 1 kg of available N annually.
‡ The N content is on a dry weight basis.

The decay series can be used to obtain a constant N output each year or a particular N output in the current year by adjusting the manure applications. A constant annual N output can be obtained by using Table 3 which presents the ratio of yearly inputs to the annual N mineralization rate, at a constant yearly mineralization rate for five decay series for 20 years. In this case the application for any specific year can be obtained by multiplying the ratio for that year times the yearly mineralization rate desired. For example, if a constant yearly mineralization rate of 100 kg N/ha were desired, using a decay series of 0.40, 0.25, 0.06, the input rates would be 250, 182, 154, and 109 kg/ha of manure N each year, respectively, for the 1st, 2nd, 5th, and 20th years. Nitrogen input rates for years not listed will be between those listed. Using these ratios, the required amounts of any of the five manures can be calculated if the N and water contents are known (Pratt et al., 1973).

The decay series used in Table 3 were largely the results of the combined judgments of Pratt, Broadbent, and Martin, except the decay series 0.40, 0.25, 0.06, which was taken from a field trial in the Coachella Valley, in which the availability of the N was compared with the availability of inorganic sources. These series should be tested further for verification or modification.

Nitrogen losses in manure handling are an important consideration when determining the fertility value of the manure. Turner (1976) evaluated N loss in various storage, treatment, and handling systems (Table 4). Nitrogen losses in storage and handling ranged from 10 to 84%. Management systems can be used that will increase N losses when large quantities of manure are available to spread on limited land areas or can be minimized in cases where efficient N use for growing crops is desired.

Turner (1976) developed a less complicated approach to manure N availability than the decay series previously discussed. His availability coefficient A indicates that portion of the total soil-incorporated manure N that is, or becomes available. The value of A is influenced by the portion of the total N that is already in the inorganic form when the manure is incorporated, the amount of organic N in the current-year application that is miner-

Table 4—Nitrogen remaining after accounting for storage, treatment and handling losses (Turner, 1976).

Manure storage, treatment, and handling system	N remaining, %
Oxidation ditch, anaerobic lagoon, irrigation or liquid spreading	16
Anaerobic lagoon, irrigation or liquid spreading	22
Deep pit storage, liquid spreading	34
Aerobic lagoon, irrigation or liquid spreading	40
Open lot surface storage, solid spreading	40
Roofed storage in manure pack, solid spreading	65
Fresh manure, directly field spread: (time between application and incorporation)	
1-4 days; warm, dry soil	65
7 days or more; warm, dry soil	50
1-4 days; warm, wet soil	85
7 days or more; warm, wet soil	70
7 days or more; cool, wet soil	90

Table 5—Availability coefficient A for the first 5 application years (Turner, 1976).

Type of manure	Availability coefficient for the year of application				
	1	2	3	4	5
Poultry, fresh	0.75	0.80	0.85	0.90	0.93
Poultry, aged, covered	0.60	0.75	0.80	0.84	0.87
Dairy, fresh	0.50	0.65	0.70	0.74	0.78
Dairy, liquid manure tank (1–3 weeks storage)	0.42	0.54	0.60	0.64	0.68
Dairy, anaerobic lagoon storage	0.30	0.38	0.45	0.50	0.53
Beef, feedlot stockpiled	0.35	0.45	0.50	0.53	0.55

alized during the year, and the amount of residual organic N from previous applications that is mineralized during the current year. Estimated values for A (Table 5) are based on the assumption that manure will be applied each year on the same land for an indefinite number of years. The availability coefficient, therefore, approaches 1.0 with time.

In the case of poultry waste and fresh bovine waste, the initial year recovery rates proposed by Turner (1976) appear to be lower than those of Pratt et al. (1973). This may be partly explained by the climatic differences between the Pacific Northwest and California. This also points to the necessity of obtaining data from other areas to enable more precise evaluation of manure fertilization requirements in areas of widely differing climates.

Other approaches have been made to estimate the amount of available N from manure. Walter et al. (1974) prepared a computer model for predicting the soil profile NO_3^- distribution resulting from heavy application rates of anaerobic liquid dairy waste to coarse-textured soil. Gilmour et al. (1977) introduced a half-life concept of manure decomposition. The half-life is the time required for 50% of the manure incorporated into the soil to decompose. The half-life increases with stage of decomposition and time as the more resistant materials remain in the soil. They described this relationship with the equation:

$$t_{1/2} = \frac{0.693}{k \times (M_2/M_1) \times 0.933^{-[(\text{Annual heat units}/365) - (T_2 - T_1)]}}$$

where

$t_{1/2}$ = manure decomposition half-life in days;

k = reaction velocity constant at optimum temperature (27°C);

M_1 = soil moisture for optimum mineralization (% of water-holding capacity, WHC);

M_2 = actual soil moisture (% of WHC);

T_2 = experimental reference temperature, °C; and

T_1 = 5°C = temperature at which annual heat units = 0.

The factor 0.933 applies when the velocity of the reaction is decreased 50% for each 10°C temperature decrease.

This mathematical model was developed to predict C loss during organic matter decomposition in soil, but because C loss and N mineralization are closely related in waste materials with narrow C/N ratio, as in animal manures, it should also be useful for predicting N mineralization. Additional testing in the field will determine how useful mathematical models may be for predicting N availability from manures.

Meek et al. (1975) prepared guidelines for manure use and disposal in the western USA. The first objective of manure application should be designed to minimize pollution and maximize plant nutrient conservation.

2. DENITRIFICATION

Fresh manure contains very little NO_3^--N, therefore, the potential for N loss through denitrification is nearly zero initially. However, soon after manure is incorporated into the soil, decomposition begins, and with large applications of manure, the NO_3^- concentration in the soil sometimes increases to high levels. Denitrification of NO_3^- from animal manures is affected by the same factors that were discussed earlier in this chapter and in Chapter 8 (M. K. Firestone).

Larson and Gilley (1976) stated that considerable N will be lost by denitrification in moist soils after application of manures, sewage sludges, and sewage effluents. Rolston and Broadbent (1977) found that denitrification was greater in manured plots than in similar nonmanured cropped field plots, probably because of an abundant energy source. Carlile (1972) studied the transformation and movement of N from slurry manure applied to soil. He developed a system for denitrifying soil solutions containing NO_3^- by adding energy materials. Two such materials that worked successfully were methanol and milk, either of which removed up to 90% of the NO_3^- in 1 hour retention time in his anaerobic filter. There is an interest in denitrifying excess NO_3^- under field conditions without the use of additives.

Guenzi et al. (1978) conducted greenhouse investigations in which they studied nitrification and denitrification with 45- and 90-metric ton/ha manure rates and with ^{15}N-labeled N sources. They determined that oxygen levels in the soil were decreased, but not completely depleted by manure decomposition. However, anaerobic microsites were developed which permitted denitrification to occur. They showed that nitrification and denitrification were proceeding simultaneously. In field experiments, Concannon and Genetelli (1971) showed that soluble organic matter from manure applied to soil was leached 90 cm or deeper. This organic matter provided energy for denitrification. Murphy et al. (1972) reported greater NO_3^- concentrations in fields that had received one manure application than after two annual treatments. They attributed this to leaching of organic materials to greater depths and in greater concentrations with additional manure applications, resulting in enhanced denitrification.

Kimble et al. (1972) attempted to estimate denitrification by measuring leached NO_3^- and the NO_3^-/Cl ratios in plots receiving 66 metric tons/ha of manure. The ratio decreased with depth to 182 cm, indicating that denitrification had occurred. Meek et al. (1974) showed that with 180 metric tons/ha manure, soluble organic C leached to 80 cm or deeper. Nitrate was leached to considerable depths in the soil and denitrified, and the disappearance of NO_3^- correlated with the appearance of soluble manganese. They

Table 6—Denitrification coefficient for four drainage conditions (Turner, 1976).

Degree of soil drainage	Denitrification coefficient
Excessive or somewhat excessive	--
Well to moderately drained	0.85
Somewhat poorly to poorly drained	0.70
Very poorly drained	0.60

also measured redox potentials and correlated them with NO_3^- loss by denitrification.

This research has provided background for greater quantification of nitrification and denitrification of manure N in soil. Further, Turner (1975, 1976) developed a denitrification coefficient. This coefficient indicates that portion of the available inorganic N remaining in the soil after accounting for denitrification loss. Aeration, one of the main factors in denitrification, is related to soil drainage. Therefore, excessively drained soils show little or no denitrification potential and this potential increases as soil drainage decreases to the very poorly drained soils (Table 6).

Nitrogen loss from denitrification following manure applications to land will probably be associated primarily with systems where manure is applied at high rates. In summarizing available data, manure application rates below 20 metric tons/ha will probably produce fairly low NO_3^- concentrations in soil and low concentrations of leached organic material below the manure mixing layer. Therefore, denitrification losses would be expected to be less than at high disposal rates (100 or more metric tons/ha) where high concentrations of leached organic materials, high NO_3^- concentrations, and low redox potentials resulting from intensive microbiological activity would maximize denitrification. This is a favorable factor in decreasing ground water and soil pollution from NO_3^- in manure disposal sites and under corrals.

3. NITROGEN POLLUTION POTENTIAL FROM ANIMAL MANURES

During decomposition of manure, NO_3^- is the main end product from the nitrogenous portions of the manure, and NO_3^- may accumulate to very high concentrations in the soil. Several unfavorable conditions may develop in soil and ground water as a result of high NO_3^- concentrations. High NO_3^- in the soil may result in high to very high NO_3^- concentrations in forage, feed, or food. High NO_3^- concentrations can be toxic to animals or humans consuming the products grown on land. Nitrate concentrations in the ground water may increase above the limits set by the U.S. Public Health Service (see Chapt. 16, D. R. Keeney). The movement of NO_3^- through the soil profile varies with soil, climate, and management (Chapt. 11, D. R. Nielsen and J. W. Biggar). Generally, NO_3^- will move through porous, well-drained soils with water from either rainfall or irrigation. Although the rate of movement varies, Adriano et al. (1971) suggested that with 18 to 25 cm of drainage water per year, NO_3^- may reach a water table at a 30-m depth in 10–50 years. Nitrate does not readily move through tight soils, uncleaned animal yards that have become sealed, or in low rainfall nonirrigated areas.

Stagnant water or poorly aerated soil layers encourage denitrification and much of the NO_3^- is lost rather than moved into the ground water (Meek et al., 1975).

We will discuss a few animal waste disposal related experiments that point out the problems associated with NO_3^- leaching. Bielby et al. (1973) treated lysimeters with slurry manure to provide 560, 1,120, and 1,685 kg N/ha, and found that in 2 years of cropping to corn, 26.6, 26.2, and 24.0% of the added N leached out of the lysimeters. Concannon and Genetelli (1971) applied manure to field plots at rates of 0, 33, 66, and 99 metric tons/ha and found that the 66 and 99 metric ton rates produced NO_3^--N concentrations above 10 mg/liter in the soil solution for August or October applications, but not for June applications. Marriott and Bartlett (1972) found that manure applications containing 250–1,460 kg N/ha produced NO_3^--N concentrations in the soil solution ranging from 3 to 140 mg/liter.

In evaluating soil and water pollution potential from manure applications, the greatest need is to determine leaching losses for various manure rates for specific cropping sequences, water management systems, soils, and climates. When soils are loaded with large amounts of manure, denitrification is the key mechanism in preventing or decreasing NO_3^- leaching beyond the root zone. Pratt et al. (1976) indicated that organic loading was more important than water management in creating O_2-deficient systems necessary for denitrification. They also reported that the NO_3^- concentration in the unsaturated zone and the amount of NO_3^- leached were linearly related to available N in the soil and also to the amount of dry or liquid manure added. High organic loading favors denitrification and adds a buffering effect to NO_3^- leaching from high manure applications. In most well-aerated soils, this buffering probably is not sufficient to prevent large leaching losses and high NO_3^- concentrations in drainage waters from large manure applications. Conversely, in heavy, poorly drained soils, there is evidence that denitrification will greatly decrease or eliminate NO_3^- leaching losses when large amounts of animal manure are applied. When annual applications cease, the lower soil profile becomes aerated, denitrification stops, and NO_3^- leaches from the profile until the added organic source is depleted. In a humid area with 125 cm of annual precipitation, Fordham and Giddens (1974) found that feedlots remained wet on the surface most of the time, which was conducive to denitrification, and they found only 3 mg/liter NO_3^--N at the 120-cm depth under the feedlot.

Of nuisances related to manures, odor is probably the most readily noticeable, but the least definable and the most difficult to control. Although manure odors can be characterized chemically, odor nuisance judgements are subjective. The odor of fresh livestock manure is inoffensive to most people, but odors produced by anaerobic bacterial activity during fermentation of wet manure can be offensive (McCalla et al., 1977).

Ammonia from manures is usually not as offensive as anaerobically produced volatile non-N compounds. Stewart (1970) in soil column studies showed that 25% of the applied urine-N was lost as NH_3 from wet and 90% from dry soil surfaces. Peters and Reddell (1976) reported a 10–20% loss of total N as volatilized NH_3 from heavy manure applications on soil columns.

Adriano et al. (1974a) found that manure application rate did not affect NH₃ loss, but showed volatile losses approaching 50% of manure N. McCalla et al. (1970) in a laboratory experiment showed up to 90% of surface-applied urine N was lost by volatilization of NH_3.

Hutchinson and Viets (1969) measured NH_3 concentrations in the air downwind from a 90,000 cattle feedlot. They postulated potential water pollution of nearby lakes with NH_3. Luebs et al. (1973) measured NH_3 concentrations in the air in a high concentration dairy area of California and calculated that a lake in that area could absorb up to 190 kg N/ha of lake surface annually. In a 6-m-deep lake this would constitute enough N to make an NH_3 concentration of 0.75 mg/liter. This is in excess of the U.S. Environmental Protection Agency (USEPA, 1973) recommended permissable NH_3 level for public water systems of 0.5 mg/liter. Rainfall in these areas also contributes additional NH_3-N to soil and water and increases the pollution potential from high concentrations of livestock. These observations suggest that large livestock enterprises, with their associated pollution problems from odors and NH_3, should not be located in highly populated areas.

III. FOOD PROCESSING WASTES

A. The Nature and Composition of Food Processing Wastes

Food processing wastes may be liquid as in the cases of effluents from processing potatoes, sugar beets, vegetables, fruits, meats, or dairy products or they may be solid or semisolid materials such as soil, rocks, peel wastes, pulps, or paunch manure. They vary in composition from very low nutrient liquid wastes from dairy washing operations to very high nutrient concentrations from whey or meat processing. Almost all food processing waste effluents may be used for irrigating agricultural land to supply water and nutrients for growing crops. Waste water containing a high salt or Na concentration that would damage crops and soil cannot be used for irrigation.

The amount of vegetables, fruits, and meats processed in the USA annually was estimated by Hunt et al. (1976), and the processed product, the waste water produced, and the resulting biochemical oxygen demand (BOD) load are shown in Table 7. The total N added by irrigation with these waste

Table 7—Wastes from processed vegetables, fruits, and animal slaughtering (Hunt et al., 1976).

Product	Amount processed	Waste water	BOD
	10^6 metric tons	10^6 m³	10^6 kg
Vegetables	12.8	185	232
Fruits	10.9	129	131
Meat	32.5	327	557
Total	56.2	641	920

waters is significant, ranging from 6,500 metric tons N/year for vegetable processing wastes, somewhat less from fruit processing, and 2,400 metric tons annually from meat-processing wastes.

1. VEGETABLE PROCESSING WASTES

Large volumes of water are used for washing, transporting, blanching, and cooking vegetables. Each of these processes extracts some vegetable constituents, enriching the water with organic material and plant nutrients. As water is used and recycled to lower level steps in the vegetable processing, its nutrient and organic concentration increases and its quality decreases until the water can no longer be used for processing. Then it is discarded. Table 8, compiled from several publications on waste water quality, shows NO_3^-, total N, and chemical oxygen demand (COD) concentrations in the waste waters from processing several vegetables. Nitrate-N concentrations are generally low (< 3 mg/liter) in vegetable-processing waste waters. However, the organic N in the waste water will be converted to NO_3^--N during treatment or in the soil when it is used for irrigation. When using vegetable processing waste water for irrigation, the organic N must be considered as the main N component and evaluated as fertilizer for crops and for its soil and ground-water pollution potential.

The total N content of the vegetable-processing waste waters reported in Table 8 range from 6 to 66 mg/liter with tomato (*Lycopersicon esculentum* Mill.) and brussels-sprouts (*Brassica oleracea* Gemmifera) processing producing low N waste water and potato (*Solanum tuberosum* L.), corn,

Table 8—Nitrogen and COD composition of vegetable processing wastes.

Crop	NO_3^--N	Total N	COD
	In wastewater		
	mg/liter		
Snap beans[†]	0.9	31.2	176
Sweet corn[†]	1.7	61.9	1,043
Brussels-sprouts[†]	0.4	5.7	15
Beets[†]	2.9	66.4	854
Peas[†]	0.1	44.7	707
Tomatoes[†]	0.6	6.3	95
Cabbage[†]	1.6	31.3	229
Tomatoes[‡]	0.4	6.8	47
Corn[‡]	trace	27.3	316
Potatoes[§]	0.9	55.0	1,680
	In solid waste		
	% N		
Tomato[¶]	--	2.33	--

[†] Shannon et al., 1968.
[‡] Stanley Assoc., 1977; values are biochemical oxygen demand, BOD.
[§] Smith et al., 1977.
[¶] Timm, H., N. B. Akesson, M. O'Brien, W. J. Flocker, and G. York. 1976. Soil and crop response to variable loading of canning wastes. Presented at Am. Soc. of Agric. Eng. Meeting, Davis, Calif., June 1976.

cabbage (*Brassica oleracea* Capitata), and beet (*Beta vulgaris* L.) processing relatively high N waste waters. The COD concentration of the waste water is also included because in some waste materials, the COD and N concentrations have some interesting relationships. For example, potato-processing waste waters have a COD/N ratio in the range of 25:1 to 30:1 representing from 3.3 to 4.0% N in the organic waste materials. Other vegetable processing wastes may have predictable COD/N ratios, but these data are not available.

Applying food processing waste water to land can supply large amounts of N. Smith et al. (1975, 1978) and Smith (1976) showed that potato processors applied from 160 to 490 cm of waste water annually, which supplied from 1,080 to 2,200 kg N/ha. Potatoes are processed most of the year and large amounts of waste water are discharged from the processing plants. Daily discharge ranges from 1.9 to 19 million liters. This long processing season often results in excessive N applications to the land used for waste water discharge. Other vegetables such as peas, green beans, sweet corn, tomatoes, and brussels-sprouts are processed for a much shorter season each year. Waste water from these processing operations is discharged to the land for only a few months, therefore, their fertilizer potential is much less than that from potato processing. The actual fertilizer N obtained from processing these other vegetables is not known, but can be calculated from the general data in Table 8 if the amount of water available from their processing is known.

Another factor related to N fertilization with waste water from vegetable processing is the efficiency of conversion from the organic to available N. De Haan et al. (1973) stated that waste water applications must be adjusted so that plant nutrient as well as purification requirements are met. To meet plant nutrient requirements, the availability of plant nutrients in the waste water must be known. They developed a "relative efficiency index" for N and other nutrients utilizing potato starch waste. The "relative efficiency index" of potato starch waste N compared to commercial fertilizer N was as follows: potatoes, 0.5; beets, 0.5; cereals, 0.2; and grass, 0.8. These values need further verification in other climatic areas and with other soils, waste waters, and crops than those in The Netherlands, but the concept is good.

Some vegetable processing operations generate a large quantity of solid or semisolid waste. In the potato processing industry, several kinds of solid waste are generated. Waste water from the processing plants is usually passed through a primary clarifier, where the settleable solids are removed from the clarifier underflow and concentrated on a vacuum filter. These wastes contain from 6 to 15% dry matter with a N concentration similar to that of the soluble COD that passes through the filter. The filter cake is usually fed to livestock as a substantial part of the fattening ration. Some potato processors do not use a clarifier, but apply the solid wastes directly to the fields. In this case, organic C and N loading is increased by the additional solid materials resulting in a higher rate of N addition than with clarified waste effluent alone. Additional land should be used to assimilate the additional nutrients. Other solid wastes include substandard products that must be discarded, soil, rock, and mud from transporting and washing the potatoes. These waste materials are fed to livestock or are discarded on land or in landfills as appropriate. Other solid wastes encountered in vegetable processing include peels from tomatoes and other vegetables, and pomace or pulp from tomato juice processing, or other similar materials. These are generally disposed of on land and contribute substantially to N fertilization because they contain relatively high N concentrations. For example, Timm et al. (see footnote Table 8) reported that tomato wastes contain 2.3% N on a dry weight basis.

2. SUGAR BEET PROCESSING LIQUID WASTES

Sugar beet processing waste water is being evaluated for irrigation use by Smith and others in Idaho (Smith & Hayden, 1980). Preliminary data are presented in Table 9. In Idaho, sugar beets are processed mostly during the winter season. Organic materials in the waste water pass through the soil more readily than do potato processing wastes and the N also readily leaches through the soil. Additional research is underway to improve cleanup and utilization of the sugar beet waste waters. The concentration of waste constituents in the waste water depends in part on the type of processing operation; however, the N rates in all these applications are high, but when divided by two or three seasons of fertilization they do not give excessively high N fertilization rates.

3. FRUIT PROCESSING WASTES

Fruit processing waste water data are rather scarce, but those few available data indicate that except for grape wastes, these wastes are usually low in N (Table 10).

Solid wastes from the fruit processing industry consist of fruit peels, pits, pomace, seeds, and stems. These wastes as applied to land for disposal are high in water. The N concentrations range from 0.77 to 1.37% and when applied to land, will supply from 1.2 to 1.8 kg N/metric ton of the wet

Table 9—Sugar beet processing plant waste water applied October 1975 to December 1977 (Smith et al., 1977; and unpublished data).

Processing	Waste water applied	COD applied		Nitrogen applied	
	cm	mg/liter	kg/ha	mg/liter	kg/ha
		Twin Falls, Idaho			
Sugar beets	142	3,580	42,930	104	1,330
Juice	--	1,990	--	53	--
		Rupert, Idaho			
Sugar beets	111	1,616	20,830	65	717
		Nampa, Idaho			
Sugar beets	170	1,758	24,238	55	719
Juice	--	660	--	14	--

Table 10—Liquid fruit processing waste N and COD in wastewater.

Crop	NO₃⁻-N	Total N	COD
		mg/liter	
Apple products†	trace	2.2	170
Pear (*Pyrus communis* L.)†	trace	2.6	1,230
Grapes (*Vitus* sp.)‡	2.0	49.9	909
Citrus (*Citrus* sp.)§	3.0	7.8	150

† Stanley Assoc., 1977; values are BOD.
‡ Shannon et al., 1968.
§ Koo, 1974.

Table 11—"Solid" wastes from fruit processing (Reed et al., 1973).

Crop	Water	Total N
	%	
Peach-pear	86.7	0.77
Peach [*Prunus persica* (L.) Batsch]	85.1	0.75
Pear	87.2	1.37
Mixed fruit†	88.9	0.85

† Mixed fruit was peach, pear, plum, grape, and cherry.

fruit processing waste material (Table 11). From 50 to 200 metric tons/ha of the wet waste material can be applied to meet the fertilization needs of crops grown on these lands.

4. MEAT AND DAIRY PROCESSING WASTES

Meat processing liquid wastes are generally low in NO₃⁻-N, but they may increase in NO₃⁻ during conventional waste water treatment (Table 12). The N and COD concentrations in meat packaging waste water are relatively high and probably inversely related to water use efficiency in processing plants. The N and COD concentrations in waste water are highest for slaughter wastes and lowest for meat cutting wastes (Table 12). Meat processing waste water is generally easily treated in conventional waste treatment systems and is well suited for use on irrigated cropped land.

Dairy manufacturing wastes are extremely variable in composition (Table 12). Whey has the highest NO₃⁻, total N, and COD concentrations. While whey is perhaps best suited for use as a livestock feed rather than land application, considerable whey is disposed of by irrigation. The high N and COD concentration pose some special problems when irrigating with whey. Extremely high N fertilization can result. The high COD causes rapid microbial growth and can produce odor problems and anaerobic conditions in and on the soil which damage growing crops. Dilution of whey with water is usually necessary for irrigation to decrease the severity of the problems in the field.

Other dairy manufacturing waste waters contain much less N and COD than whey and should be well suited for irrigation with no greater problems than balancing the N content of the waste water with the cropping require-

Table 12—Meat and dairy processing waste N and COD concentrations.

Source	NO$_3$-N	Total N	COD
		mg/liter	
In meat packing wastes			
Catch basin effluent†	--	124	4,180
Extended aeration influent‡	0.4	92	1,630
Extended aeration effluent‡	2.6	10	122
Slaughter waste, nontreated§	0	69	2,029
Slaughter waste, treated§	0.4	10	139
Custom meat cutting§	0	20	139
In dairy manufacturing wastes			
Whey§	7.8	685	53,225
Fresh milk packaging§	0	24	2,290
Curds (waste not whey)§	0	18	725
Condensed milk§	2.5	1	96
Ice cream§	0	1	21

† Tarquin et al., 1974.
‡ Witherow, 1975.
§ J. H. Smith, Kimberly, Idaho, USDA-ARS, unpublished data from 24 meat and dairy processing plants in Idaho.

ments. Nitrogen efficiency factors for meat and dairy processing waste water N are not yet available.

Paunch manure is probably the major waste from livestock slaughtering that would be used for fertilizer. According to Baumann (1971) paunch manure contains approximately 2% total N on a dry weight basis. This material can be applied to fields and would be expected to produce an almost immediate fertilizer benefit. Values for availability of paunch manure N should be similar to those of fresh livestock manure discussed earlier. There are no solid wastes associated with dairy manufacturing that would normally be used for fertilizer.

B. Irrigating Agricultural Land

Application systems for food processing waste water fall into three different categories:

1) *Irrigation*—Irrigating agricultural land to produce crops is the system most often used and has the advantage of conserving and using at least part of the nutrients contained in the waste water. The water may be applied by surface methods or by sprinkling. Surface application lessens NH$_3$ volatilization compared to sprinkling and can be used on some soils even in midwinter, as the heat in the water thaws the soils and maintains infiltration. Sprinkling increases volatile NH$_3$ losses and creates some aerosol problems without utilizing the temperature advantages of surface application.

2) *High-rate Infiltration*—When large volumes of waste water are applied to sandy or gravelly soil where infiltration and percolation

rates are very high, many nutrients are lost by percolation and plant nutrients are used inefficiently.

3) *Overland Flow*—This is a method of water application where soils are relatively impermeable and the water is purified to some extent by contact with the growing vegetation, the organic matter laying on the surface, and by limited contact with the soil. This method has been studied by Gilde et al. (1971).

The most desirable waste water application method will be determined by several factors that must be evaluated at each site when the systems are designed. The system used will influence nutrient recovery and the utilization of N. Two concepts of waste handling in relation to the nutrient content are the disposal concept (the main consideration in infiltration-percolation systems), and the recycling–reuse concept. The recycling–reuse concept is becoming much more important with developing energy shortages and increased fertilizer cost. Many food processing waste treatment and disposal systems were designed to utilize the maximum rate of water and nutrients. Frequently, there was no other consideration than that a crop could be grown to keep the area looking acceptable and to avoid waste water applications that would create nuisance situations such as ponding and objectionable odor. High land cost and proximity, and the necessity for landleveling and water retention or recycling on the disposal and treatment site have led to the concept of applying as much waste water as possible. In many cases, N loading is extremely excessive and pollution of soil and ground water and the production of high NO_3^- forage results.

C. Nitrogen Loading and Utilization on Land

When food processing waste water is applied to agricultural land for treatment and disposal, the first limiting factor in the system design is usually the N application rate. Reasonable predictions of acceptable loading rates can be made for a given site if soil conditions, type of crop to be grown, depth to water table, frequency and intensity of rainfall, and similar pertinent information is available (Loehr, 1974). Nitrogen concentrations in food processing waste water varies widely, as discussed earlier. When designing or evaluating a food processing waste water irrigation field it is necessary to obtain data on N concentrations in the waste water and the amount of water being applied.

Loading rates for food processing waste water irrigation are available for a few cases. Smith et al. (1975) studied waste disposal at five potato processing waste water irrigation systems in Idaho and calculated N application rates for the fields. Nitrogen applications ranged from 800 to 2,200 kg/ha annually. These values are higher than the grass crops grown on the fields can be expected to utilize and will increase soil NO_3^- and possibly pollute ground water under the fields. Adriano et al. (1974b, 1975) measured N fertilization and utilization at two sites in lower Michigan. At one site vegetables, fruit, and occasionally meats were processed for 20 years before the study was initiated and the waste water was applied to land that grew quackgrass (*Agropyron repens* (L.) Beauv.) that was clipped but not harvested. Average annual N application was 365 kg/ha. At the other site, dairy products were processed and the waste also applied to a field that grew quackgrass. Annual N application was 359 kg/ha. These application rates were not excessive if the grass had been harvested and removed from the fields.

While N loading should be a primary consideration in designing food processing waste water irrigation systems, the question of organic loading should also be considered. If the limits of N application are the amount of N that can be utilized by crops, then organic matter application rates will seldom be excessive. Organic matter applications in waste water should be limited to the amount that will decompose during the time between applications. The exception to this is wintertime irrigation with warm water when the organic matter accumulates. Jewell (1976) and Jewell and Loehr (1975) in field and laboratory experiments showed that under favorable conditions, soils that are conditioned to receive waste water containing organic matter can utilize high loading rates. In laboratory experiments, they found that at 26°C removal efficiency was nearly 100% at a vegetable processing waste water application rate of 19,000 kg COD/ha per day. Field sampling of waste water at two vegetable processing plant spray irrigation fields indicated that loadings up to 9,000 kg COD/ha per day were removed with >99% efficiency. While these removal rates may not be widely obtainable, the reports showed that food processing wastes are readily treated in the field and that the capacity of the soil to assimilate these types of organic wastes is very large.

The data reported earlier showed that most food processing waste water contains highly variable concentrations of NH_4^+ and organic N. During waste treatment or decomposition in soil, the organic N is converted to NH_4^+. The capacity of soils to absorb NH_4^+ is generally considered to be quite high. However, in cases where large amounts of NH_4^+ or readily mineralized organic wastes are applied to soil, the exchange capacity of the soil may be exceeded and NH_4^+ may migrate. Lance (1972) developed a method for calculating potential NH_4^+ migration in soil and related it to the NH_4^+ adsorption ratio (AAR). This relationship is similar to the Na adsorption ratio that is used extensively in evaluating Na movement in soils. The AAR is calculated as:

$$AAR = \frac{NH_4^+}{[\frac{1}{2}(Ca^{2+}) + \frac{1}{2}(Mg^{2+})]^{1/2}}$$

where concentrations are expressed in meq/liter. As this ratio increases the exchangeable NH_4^+ percentage increases. In high infiltration rate soils, the adsorption of NH_4^+ is not a very important N removal factor because the soils are soon saturated with NH_4^+. But in agricultural soils that have a high CEC, most of the NH_4^+-N can be removed from the waste water used for irrigating crops. The NH_4^+ will subsequently be converted to NO_3^- in the aerated soil between waste water irrigations.

D. Nitrification and Denitrification

The organic materials contained in the waste water from vegetable and fruit processing are mostly water soluble, readily decomposable, and have a relatively low molecular weight. Before application, much of the waste water is filtered leaving only the soluble and reflocculated organic materials. This is evident in the extremely rapid decomposition rates observed by Jewell (1976) with vegetable processing waste water organic materials. As these organic materials rapidly decompose, the organic N is also rapidly converted to NH_4^+ and then to NO_3^-. Ammonification is seldom a rate-limiting step and crops grown on the waste water treated fields will usually have adequate or excessive N. The waste water organic matter is usually adequate to high in N and decomposition is seldom slowed by N deficiency.

Excessive amounts of N are often added in food processing waste water, indicating that denitrification must also be considered. Smith et al. (1976) determined the potential for denitrification in a field irrigated with potato processing waste water. About 2,400 kg N/ha was applied during 2 years of irrigation. The saturated zone in the soil rose from below 140 cm to the 65-cm depth. Anaerobic conditions were measured at the 65-cm depth by platinum electrodes. The soluble organic material that leached to 65 cm provided energy for the denitrifying microorganisms, and denitrification removed nearly all of the NO_3^-. In another experiment with a deep water table (>25 m), Smith et al. (1978) showed that denitrification could be enhanced by irrigating with high organic waste water at strategic times during the warm season. This lowered redox potentials and promoted denitrification. All irrigations lowered redox potentials, but only irrigations with high organic waste water during warm weather lowered the redox potential sufficiently that denitrification occurred. These studies showed that denitrification can be managed and used as a tool to regulate leaching and loss of N to the ground water in waste water treatment and disposal fields.

E. Pollution Potential

Irrigating with food processing waste water in many cases applies N greatly in excess of that required for growing crops and management becomes the key to pollution control. Adriano et al. (1974b, 1975) measured N leaching in the study cited previously and found that when the grass grown on the fields treated with waste water was not harvested and removed, most of the applied N was leached. For fields receiving 365 kg and 359 kg N/ha per year, 76 and 69% of the added N was leached, respectively. Much of the leaching loss probably represented N that had been returned to the field in unharvested quackgrass.

In contrast to Adriano's experience, Smith et al. (1976) showed that organic matter from potato processing waste water applied to soil the previous winter decomposed as soil temperatures increased in the summer, releasing N that was utilized by growing grass. The excess N was denitrified as it leached into the anaerobic zone near the water table. In this case, a large excess of N was disposed of without polluting the ground water. When irrigating with relatively high organic waste water, it is possible to manage the soil redox potentials to develop occasional low redox conditions that will denitrify any excess NO_3^- when needed, even in soil without a high water table. Temporary artificial water tables develop in the soil at soil particle size phase changes such as a change from silt loam to gravel, and these can be made anaerobic rather readily by irrigating with water with a high oxygen demand.

In a well-managed system in California, Meyer (1974) reported NO_3^- buildup in the soil at depths to 90 cm when irrigating with waste water from fruit and vegetable processing. He was able to grow a winter cereal crop in addition to the summer vegetation and remove most of the residual N from the soil profile that had accumulated during waste water irrigation. This decreased NO_3^- leaching and ground water pollution.

Timm et al. (see footnote Table 8) applied large quantities of tomato processing waste solids to fields at rates from 448 to 1,792 metric tons/ha. This applied 1,461–5,844 kg N/ha which created a lodging problem when growing barley. Excessive NO_3^--N accumulated (up to 8,700 $\mu g/g$ in the growing crop), with severe potential for NO_3^- leaching through the soil and into the ground water.

IV. MUNICIPAL WASTES

A. Sewage Effluent

In an effort to provide clean rivers and lakes the U.S. Congress passed Public Law 92-500 in 1972 requiring fishable and swimmable waters by 1983. The cost of doing this with conventional waste water treatment methods will be very high. This law created a renewed interest in using soil to renovate waste water. Secondary benefits of this system include water and plant nutrient conservation. Basically there are three methods of using land for renovating waste water: (i) high-rate infiltration, (ii) overland flow, and (iii) crop irrigation.

1. HIGH-RATE INFILTRATION

High-rate infiltration relies on the physical, chemical, and biological properties of the soil profile to remove impurities from waste water. An intensive study of this method was started at Phoenix, Arizona in 1967 (Bouwer et al., 1974b). This site is in the Salt River Valley with a fine loamy sand (0–0.9 m) underlain with layers of sand, gravel, boulders, and traces of clay 75 m deep, where there was an impermeable clay layer. The static water table was at 3 m.

Secondary effluent was applied to the infiltration bed for 10–30 days followed by a 10- to 20-day drying period. The maximum hydraulic loading was 122 m/year using a 20- to 30-day effluent loading and a 10- and 20-day drying period for summer and winter, respectively. However, Bouwer et al. (1974a) found that a loading of 91 m effluent/year resulted in removal of 30% of the applied N. The effluent contained 20–40 mg NH_4^+-N, 0–3 mg NO_2^--N, 0–1 mg NO_3^--N, and 1–6 mg organic N/liter which resulted in a N addition of about 28,000 kg/ha. The wet-dry cycle used was 10 days wet and 10–20 days of drying. Oxygen and organic C were the limiting factors for denitrification. The effluent from the infiltration basin was suitable for unrestricted crop irrigation and recreation in Arizona.

Satterwhite et al. (1976) reported on a year-round rapid infiltration system at Fort Deven, Mass. which has been receiving unchlorinated Imhoff effluent since 1942. The infiltration beds were underlain with silty sand to sandy gravel with 10 to 15% silt and clay. The annual effluent application was 27.1 m with a 2-day application and a 14-day drying period. A total-N balance showed a 60–80% reduction in total N, primarily by denitrification. The ground water in the immediate area contained from 10 to 20 mg NO_3^--N/liter.

In comparing the warm-arid and the cold-humid locations from the previously noted studies, the cold-humid site had a greater N reduction. This may possibly have been caused by higher organic C in the Imhoff effluent, lower loading rate, and long drying cycles, which allowed more time for N mineralization at the cold-humid site. Bouwer et al. (1974a) showed 80–90% denitrification when glucose was added to the system to supply organic C. In both of these locations the ground water quantity and quality were affected. The NO_3^- concentration of the ground water was increased, but this impact was ameliorated by lateral NO_3^- movement at the Fort Deven site and pumping NO_3^- containing water to the surface for crop irrigation and nutrient utilization by growing plants at Phoenix.

High-rate infiltration is best suited to areas where water conservation is essential and the soil is deep and permeable. Control of the ground water at the site is necessary and this can be done using tile drains or recovery pumps (Reed, 1972). This high-rate infiltration system may also be suited to some seasonal operations such as canners or summer camps. With proper management, this system can be very successful. The N in this treated water is readily available to growing crops.

2. OVERLAND FLOW

Overland flow systems rely on the controlled release of waste water effluent onto sloping land. The water ideally should flow in sheets through a grass cover. This system is suitable for very slowly permeable soils. Precision grading is advisable to optimize the renovation process. Hunt (1972) stated that there are five primary mechanisms of stripping N from waste water during overland flow: (i) removal by plants, (ii) immobilization in the cells of the expanding aerobic heterotrophic microbial population, (iii) gaseous loss of NH_3 on alkaline soils, (iv) micosite-slime layer denitrification, and (v) the conversion of N into stable organic matter.

The N removal by plant harvesting can be quite significant. At Paris, Texas, 42% of the applied N was removed with the harvested forage crop

(Hunt, 1972). However, in a sparse pine forest little N was removed by the vegetation and no improvement was realized on frozen ground (Sopper, 1968). The immobilization of N by the expanding microbial population and the conversion of N into stable organic matter may be significant during the startup of an overland flow site, but when the system reaches steady-state these two mechanisms are of minor significance.

Slime layer denitrification tends to occur on wet areas where the waste water is repeatedly applied. This process is similar to a trickling filter with its slime-layer on the media.

Raw waste water from Melbourne, Australia is treated in the winter by overland flow through Italian ryegrass (*Lolium perenne* L.). The retention time of the field is about 2 days and the ryegrass is grazed during the summer when the site is not being treated. Seabrook (1975) estimated 60% total N removal with this system. The raw waste water contained 14.3 mg organic N, 35 mg NH_4^+-N, 0.75 mg NO_2^--N, and 0 NO_3^--N/liter. Daily application during the winter was about 1.9 cm.

The overland flow system can handle more effluent than the crop irrigation system, and the treated water goes to surface streams with some evaporation and percolation. The N removal is primarily by plant uptake and denitrification. Overland flow can be a viable means of waste water renovation if the system is managed properly and the waste water application rate does not exceed the removal capacity of the soil and crop produced.

3. CROP IRRIGATION

Crop irrigation maximizes both water renovation and plant nutrient conservation. The effluent is generally applied to crops by overhead sprinkler or flooding at rates of 0.6–2.4 m/year.

The success of waste water renovation will be dependent on the crop selected. Long season high-yielding sod crops utilize more N than annual crops. Hook and Kardos (1977) reported reed canarygrass (*Phalaris arundinacea* L.) receiving effluent daily at a rate of 5 cm/week provided 57–71% removal by forage harvest and the concentration of NO_3^--N in the ground water was generally below 10 mg/liter. An effluent ratio of NH_4^+-N/NO_3^--N of 2:1 resulted in less NO_3^--N loss to recharge water than when the sewage treatment plant produced effluent having a NH_4^+-N/NO_3^--N ratio from 1:2 to 1:5. At Hanover, New Hampshire Iskandar et al. (1976) observed the NH_4^+-N remained in the top 45 cm of the soil profile during the winter. In the summer, almost all the NH_4^+ was nitrified to NO_3^-.

In a study by Clapp et al. (1977) in Minnesota, forages utilized N throughout the growing season better than did corn. Over the entire growing season, less NO_3^- and NH_4^+-N were present in the soil water (60- and 125-cm depth) when 415 kg N/ha from effluent than when 341 kg N/ha from inorganic fertilizer was applied.

Irrigation of trees with sewage effluent appears most promising on a young, fast growing forest because N uptake is greatest during this growth time. Sopper (1968) reported the renovation efficiency of the forest decreases with time because there is no annual harvest. During the early stages of effluent application, the N content of the tree leaves, the soil biomass, and the soil organic matter increases; however, this condition eventually reaches a new steady state, and the N utilization and water renovation decreases unless the trees are harvested. Sopper (1968) reported a mixed hardwood forest site in Pennsylvania which was capable of renovating sewage effluent for ground water recharge. Hook and Kardos (1977) reported no N renovation was observed on this same site during 1973 after 8 years of year-round operation.

The selection of crops to be irrigated with sewage effluent is limited. Baier and Fryer (1973) reported stone fruits, citrus, grapes, sugar beets, potatoes, avocados, apples, melons, and squash were of lower quality when fertilized with sewage effluent, as compared with chemical fertilization and irrigation. They attributed this, in part, to the continuous addition of N throughout the growing season. Since the N is only needed for vegetative plant growth, this problem could have been reduced if an alternate water source were used during the period of fruit set. Day and Tucker (1959) reported that sewage effluent–irrigated wheat and oats produced 263 and 249% more pasture forage, respectively, than did check plots that received only pump water. Barley was more sensitive than wheat to the detrimental effects of sewage effluent. Corn yields may be reduced from effluent irrigation because the effluent N concentration is not high enough during the critical period of corn growth, unless supplemented by N fertilizer. Hook and Kardos (1977) observed the greater N leaching with corn because the highest rainfall occurs in the spring and fall when the cropped land is not irrigated. They also found that double-cropping or no-till planting can reduce the movement of NO_3^- to the ground water. This system of using the plant to hold the NO_3^- in the surface soil has the added benefit of reducing soil erosion. The management of a no-till system is more difficult than conventional tillage and with some soils periodic deep tillage is needed to maintain the infiltration capacity.

Treated sewage effluents can be effectively used as a source of N for growing crops if the crop N requirements are considered and N is either supplemented or withheld as the need develops. In addition, sewage effluent will be further purified, ground water can be recharged, and excess waste water disposed by this efficient method.

B. Sewage Sludge

1. SLUDGE PROPERTIES

Sewage sludge is obtained from waste water treatment. The N content of the sewage sludge depends on the nature of the waste water, type and extent of waste water treatment, type of sludge stabilization, and age of the sludge. Table 13 gives N concentrations of the anaerobically digested sludge from three sewage treatment plants with primary and waste-activated secondary waste water treatment.

Table 13—Total and NH_3-N concentrations in anaerobically digested sludge from three waste water treatment plants (Peterson et al., 1973; McCalla et al., 1977).

Water treatment plant	Capacity	Total N	NH_3-N
	m³/day	——— % dry wt basis ———	
Hanover Park	2.3×10^4	5.2	2.4
Calumet	9×10^5	5.6	3.6
West-Southwest	3×10^6	6.9	3.3

The Hanover Park treatment plant is small with primarily domestic sewage while the Calumet and West-Southwest treatment plants are large with about a 3:2 ratio of domestic to industrial waste water. The total N concentration of the digested sludge at the West-Southwest Plant is higher because approximately two-thirds of the primary sludge is not anaerobically digested, but is diverted to drying beds.

King (1977) reported a threefold total N change in sludge composition over a 6-month period at Chatham, Ontario. McCalla et al. (1977) observed no significant seasonal differences in the N concentration of the digested sludge at the Metropolitan Sanitary District of Greater Chicago (MSDGC) West-Southwest waste water treatment plant. The greater diversity of industries using the larger plant may explain the lack of a significant seasonal N flux in this sludge.

Sommers (1977) reported on the chemical composition of sewage sludge from 150 treatment plants located in six states (Table 14). These data were skewed, therefore, the discussion will be based on the range and

Table 14—Concentrations of organic C, total N, NH_3-N, and NO_3^--N in sewage sludge (Sommers, 1977).

Component	Type[†]	Sample Number	Range	Median	Mean
			——————— % ———————		
Organic C	Anaerobic	31	18–39	26.8	27.6
	Aerobic	10	27–37	29.5	31.7
	Other	60	6.5–48	32.5	32.6
	All	101	6.5–48	30.4	31.0
Total N	Anaerobic	85	0.5–17.6	4.2	5.0
	Aerobic	38	0.5–7.6	4.8	4.9
	Other	68	0.1–10.0	1.8	1.9
	All	191	0.1–17.6	3.3	3.9
			——————— mg/kg ———————		
NH_4^+-N	Anaerobic	67	120–67,600	1,600	9,400
	Aerobic	33	30–11,300	400	950
	Other	3	5–12,500	80	4,200
	All	103	5–67,600	920	6,540
NO_3^--N	Anaerobic	35	2–4,900	79	520
	Aerobic	8	7–830	180	300
	Other	--	--	--	--
	All	43	2–4,900	140	490

† Type of sludge treatment.

median values only. The NH_4^+-N constituted 28% of the total N present in anaerobically digested sludges. King (1977) reported 29% of the total N in anaerobic sludge was NH_4^+-N, ranging from 15 to 59%.

The organic C concentration in the sludges of unspecified treatments was the highest and the N concentration was the lowest. The resulting C/N ratio was 18. The C/N ratios of the anaerobic and aerobic sludges were 6.4 and 6.1, respectively. Peterson et al. (1973) reported 21% of the volatile solids were hexane-soluble materials. Varanka et al. (1976) reported that 19% of the total organic fraction of anaerobically digested sludge was fats, oils, and waxes. The other organic C materials include polysaccharides and proteinaceous constituents (Sommers, 1977).

The presence of NO_3^--N in sewage sludge is a function of sludge age and moisture content. Nitrate is not present in anaerobic sludge freshly drawn from a digester. However, during storage in an open lagoon the solid fraction settles, and the liquid fraction, which contains most of the NH_4^+, is oxygenated by wave action and nitrification can occur.

2. SLUDGE NITROGEN REACTIONS IN SOIL

When sewage sludge is applied to land, the following physical and microbiological processes occur: decomposition, ammonification, nitrification, immobilization, volatilization, and denitrification. Additions of sludge to the soil cause a rapid growth of zymogenous microorganisms that thrive on the sludge organic fraction.

The biological activity of a sludge-amended Blount silt loam soil (fine, illitic, mesic Aeric Ochraqualfs) was studied by Varanka et al. (1976). The soil had been receiving liquid digested sewage sludge for 6 years and was in continuous corn production. The total accumulative sludge application (dry wt) ranged from 0 to 369 metric tons/ha. Table 15 presents the concentrations of the major organic matter fractions found in the soil during the sixth year of sludge application and in the sludge used in 1973 and 1974.

The soil concentration of fats, waxes, and oils increased most with these sludge additons. This fraction is rather stable in soils. The total organic C nearly doubled with the addition of 180 metric tons organic matter from the sludge. About 50% of this C was evolved during the 6 years.

Soils were collected on three dates in 1974 for bioassays. Liquid sludge was first applied (15 July) after the second sampling (Varanka et al., 1976). Their data showed no clear proof of restrictions in microbial populations and enzyme activities. The percentage of denitrifiers and the protease and amylase activities increased as a result of sludge additions and invertase and urease activities were unaffected by sludge treatments. The actinomycetes population increased with rate of sludge additions after the 15 July sludge application. Also, the soil was drier at the last sampling (29 August). This favored actinomycetes growth over fungi and bacteria. The CO_2 evolution was always positively correlated with the amount of sludge applied.

The decomposition rate of digested sludge in soil was studied by Miller (1974) at 1/3 bar and saturated soil water contents and at application rates of 90 and 224 metric tons dry sludge/ha for 1, 3, and 6 months. Incubation

Table 15—Major organic matter fractions in percentage of total organic matter in the surface (0–15 cm depth) of Blount silt loam, the sludge-mended soil, and two samples of digested sludge collected in 1973 and 1974 (Varanka et al., 1976).

	Soil		Sludge	
	Nonamended	Amended with 369 metric tons/ ha sludge		
Fractions			1973	1974
		% of total organic matter		
Fats, waxes, and oils	1.48	8.68	19.75	19.07
Resins	4.03	5.30	3.82	8.20
Water-soluble polysaccharides	7.79	5.47	3.22	14.38
Hemmicellulose	15.45	13.15	4.04	5.97
Cellulose	4.98	3.86	3.45	3.22
Lignin-humus	29.58	31.11	16.78	14.51
Protein	32.41	29.81	24.11	39.60
Total recovered	95.72	97.38	75.17	104.95
Total organic matter†	2.48	5.43	44.76	38.53

† Organic C determined by Walkley-Black method; Total organic matter (% of dry soil) = % C × 1.724.

temperatures were programmed for normal Ohio diurnal and seasonal variations. The sludge used had a C/N ratio of 8. The soils used were Paulding clay (very-fine, illitic, nonacid, mesic Typic Haplaquepts), Celina silt loam (fine, mixed, mesic Aquic Hapludalfs), and Ottokee sand (mixed, mesic Aquic Udipsamments). Miller used the rate of CO_2 evolution to measure the decomposition rate of sludge. At the high application rate, decomposition depended largely on soil texture or chemical properties. On the sandy soil, moisture did not affect the rate of sludge decomposition, but on the clay soil decomposition stopped when the soil was saturated. Soil temperature had a major influence on the rate of sludge decomposition.

Mann and Barnes (1956) observed mineralization of 38% of organic C from digested sludge applied at an annual rate of 18.8 metric tons/ha over 9 years. Premi and Cornfield (1969) found that at low sludge application rates (114 kg/ha of NH_4^+-N from sludge) rapid mineralization occurred, but with higher sludge application rates there was a lag phase before nitrification began. Also, mineralization of the soil organic N was stimulated at the lower application rates. In their 8-week study, mineralization of N from the solid portion of the sludge was negligible. Molina et al. (1971) noted that only the NH_4^+ fraction was immediately available to the plants. They observed inhibited germination of corn and soybeans if freshly digested, high NH_3 sludge was applied. Aging the sludge before application to the soil or avoiding immediate planting after a liquid sludge application eliminated this inhibition.

Addition of digested sludge with a C/N ratio of 8 to most soils will rapidly increase soil NO_3^- levels, and if very high sludge application rates are used, this excess NO_3^- may leach to ground water. To reduce the rate of sludge N oxidation, a high C material can be added to the sludge. Agbim et al. (1977) used wood and bark which they mixed at various ratios with sludge to regulate the rate of CO_2 evolution. The greatest rate of CO_2 evolution was with sludge alone, and as the ratio of either wood or a wood-bark mix was increased, the rate of CO_2 evolution decreased. The net results are that by proper management of this sludge-wood mixture, sludge can be applied at higher application rates, the soil organic content will be increased, the crop will have a more uniform supply of NO_3^-, and the excess NO_3^- will be less apt to leach.

Composted sludge and wood chips were studied in soil by Tester et al. (1977) in a laboratory incubation experiment. The evolved CO_2 and NH_3 were measured, as well as the remaining C and N fractions in the soil. In 64 days of incubation, 16% of the sludge-compost C was evolved and NH_3 volatilization was minimal, regardless of soil texture. Approximately 6% of the compost N in the loam sand treatment had mineralized after 54 days. Evidence of interlattice NH_4^+ fixation was noted in the silt loam and silty clay soils. Ryan and Keeney (1975) reported that when waste water sludge was applied to the soil surface, 11–60% of the applied NH_3-N was lost by volatilization. These losses decreased as the clay content of the soil increased. Volatilization rate increased with increasing rate of sludge application and with repeated application of sludge.

Gilmour and Gilmour (1979) developed a computer simulation model of sludge decomposition. The model used first-order kinetics and considered temperature and water content changes during decomposition. Sludge was considered to have a rapid and a slow decomposing phase. Model output showed sludge half-life to be 13–51 months depending on temperature and moisture regimes. The month of sludge addition to the soil had little effect on sludge half-life. Monthly temperature and moisture data gave better values of half-life than annual average data. When "optimum" data were inserted into the model, half-life was reduced by a factor of 2 to 7 and they concluded that laboratory studies under "optimum" conditions would greatly underestimate half-life under field conditions.

Peterson et al. (1973)[2] studied NH_3 volatilization and NH_4^+ fixation by sludge fertilized calcareous strip-mined spoil material (pH 7.8). During this 2-week laboratory experiment, fixed, exchangeable, and water-soluble NH_4^+ were measured. The highest liquid sludge treatment was equivalent to 89.6 metric tons/ha dry solids (1,622 kg NH_4^+-N/ha). This spoil material had a 28% clay content. The dominant clay mineral was illite (54%). Kaolinite, chlorite, and vermiculite were estimated to be 28, 8, and 11% of the total clay fraction, respectively. The surface area of the clay fraction was 138 m^2/g which indicated a small quantity of expandable clay minerals. This experiment indicated that 35% of the added NH_4^+ was fixed by the clay minerals. After 2 weeks of incubation, up to 14.7% of the added NH_4^+ was water-soluble and 8.7% was exchangeable. By difference, the NH_3 volatilization was calculated to be about 50%, regardless of sludge application rate, with most of this loss occurring during the first week. No NO_2^- or

---

[2] J. R. Peterson, J. Gschwind, L. Papp-Vary, and R. L. Jones. 1973. Ammonia volatilization and ammonium fixation by sludge fertilized strip-mined spoil material. Agron. Abstr. p. 180.

NO_3^- was detected in the soil after the first or second week of the experiment.

Keeney et al. (1975) estimated that 15–20% of the sludge N is mineralized the first year and, after initial application, about 6, 4, and 2% of the remaining N is released for the 3 subsequent years.

King (1973) found that in 18 weeks 22% of the surface-applied and 38% of the incorporated sludge N was mineralized. Surface application of sludge resulted in 36% of the applied N being lost and incorporation resulted in a 22% loss in 18 weeks. Only a small part of this loss was NH_3 volatilization. King concluded that denitrification was the major pathway of N loss. The large addition of organic materials provides the energy source required for denitrification and causes an O_2 depletion with the rise in microbial activities. Using this concept, weekly or biweekly addition of liquid sludge on cropland should increase the denitrification rate just as Bouwer et al. (1974b) observed with waste water.

The Illinois Environmental Protection Agency[3] has proposed NH_3 volatilization rates of 50% of the sludge NH_4^+ on sandy and nonsandy soils and a 75% volatilization rate on clay-type soil if the sludge is left on the soil surface. They estimated a 50% loss on sandy soils and a 20% loss on silt and clay-type soils if the sludge is applied by shallow incorporation, such as discing. If the sludge is applied by deep incorporation, such as plowing, the volatilization rate was estimated at 50% on sandy soils and zero on finer textured soils.

To conserve the sludge N, the liquid sludge should be promptly incorporated into the soil. Conversely, to apply the maximum amount of sludge without exceeding the capacity of the crop-soil system and losing NO_3^- to the aquifer, the sludge should be aged, dewatered, and surface applied. The NH_3 volatilization rate will approach 50% with NH_3 surface application and with sludge incorporation the gaseous NH_3 losses will decrease as the depth increases and the soil texture becomes finer. The mineralization rate of the organic N in the sludge is greater at warmer soil temperatures.

3. PLANT RESPONSES TO SLUDGE NITROGEN

To the agronomist, the use of sewage sludge on land must result in an increase in crop quality and quantity. To this end a great deal of research has been done. Kelling et al. (1977b) reported optimum rye and sorghum-sudan (*Sorghum bicolor* × *S. Sudanese*) forage yield with the application of 7.5 metric tons/ha sludge solids on Plano silt loam (fine-silty, mixed, mesic Typic Argiudolls) and with 15 metric tons/ha on Warsaw sandy loam (fine-loamy over sandy or sandy-skeletal, mixed, mesic Typic Argiudolls) during the year of application. Yield of the first crop following the application of 30- and 60-metric tons/ha sludge was depressed, possibly because of excess soluble salts present in the sludge. The residual effects of the sludge were

[3] Design criteria for municipal sludge utilization on agricultural land—draft copy—2nd ed. Tech. Policy WPC-3. Illinois EPA, Springfield. 1977. 35 p.

observed in the crop yield for at least 3 years. Over the 4-year period, 50% of the available N was recovered by the crops at the lower sludge treatment rates and 14% of the available N was recovered at the 60-metric ton/ha application rate.

Coker (1966) observed grass-clover yield increases of 15.2 kg/ha dry forage, for sludge applied at a low rate (70 kg N/ha), and 11 kg dry forage/ kg N when applied at a high rate (132 kg N/ha). Stewart et al. (1975) recovered 12% of the sludge N in the corn crop with an application of 400 kg sludge N/ha. Higher rates of sludge application resulted in a less efficient utilization of the sludge N by the corn.

Hinesly et al. (1979) studied the residual effect of digested sludge on corn. Sludge had been applied for 5 years with a maximum total application of 61.1 metric tons/ha of sludge solids for an average of 12 metric tons/ha per year. From 1974 to 1977, corn was grown on these plots without adding more sludge. In 1976 and 1977 significantly higher corn yields were obtained with the highest sludge application. During the years of sludge application, corn yields were higher at all rates of sludge application compared to inorganic fertilizer. The total N and organic C concentration of the 0- to 15-cm soil depth increased with increasing sludge application and the residual total N and organic C remained stable 4 years after the sludge application stopped.

If high sludge application rates are necessary for the operation of a sludge disposal site and if the climate is warm and humid, coastal bermudagrass (*Cynodon dactylon* L. Pers.) with a winter rye (*Secale cereale* L.) cover crop may prove to be the best user of sludge N. King and Morris (1972) reported 63% of the sludge-applied N was used by this cropping combination. The sludge supplied 724 kg N/ha per year. The N concentration of the forage increased with increasing rates of sludge application. The longer the growing season of the crop, the less chance NO_3^- has to leach below the root zone.

The renewed interest in sludge use on land was not primarily because of its fertilizer potential, but rather as an alternative to other methods of sludge disposal. Ecologically, all methods have good and bad points. In the mid-1970's, because of the oil embargo and threats of future oil shortages, a great deal of the organic fertilizer stockpile was reduced. Depending on the quality of the waste, some sources, such as animal manures, were completely cleared. Sewage sludges in the smaller cities were often used for their fertilizer value. In the large cities this also occurred, but owing to the size of the stockpiles, the distance, and the nature of some of the sludges, complete removal of all sludges has not occurred.

4. POLLUTION

This discussion will be limited to N. The application of sludge on land may cause N enrichment of surface and ground waters. If rates and time of application of sludge on cropland are properly managed, the amount of N reaching surface and ground water can be held to a reasonable level.

At high sludge application rates, better management is necessary to hold surplus N in the soil and in the root zone. The practices of double cropping (King & Morris, 1972; Hook & Kardos, 1977) adding wood products (Agbim et al., 1977), composting with carbonaceous wastes (Epstein et al., 1978), or intermittent applications of liquid sewage sludge with dry cycles between application (Bouwer et al., 1974a) can effectively immobilize or volatilize a part of the sludge N added to the soil. The ideal situation is to retain N in the root zone.

Kelling et al. (1977a) reported improved soil aggregation and increased infiltration with sludge application to the soil. Runoff water from a sludge-treated soil slightly increased in organic N but organic N and sediment load were decreased. Peterson et al. (1979) observed no change in total N concentration of runoff water after 5 years of sludge application in two watersheds. A decrease of NH_4^+-N in one watershed and no change in the other was observed after 5 years. An increase from 0.09 to 1.61 mg (NO_2^- + NO_3^-)-N/liter in one watershed and from 0.04 to 0.88 mg (NO_2^- + NO_3^-)-N/liter in the other watershed was observed after 5 years of sludge application.

Atmospheric NH_3 concentration over a sludge-fertilized field was studied by Beauchamp et al. (1978). They observed the maximum NH_3 flux occurred about midday. The *half-life* or time necessary to lose 50% of the ammoniacal N (NH_3-N and exchangeable NH_4^+-N) in the sludge was 3.6 days in May and 5.0 days in October. During the 5-day experiment in May, 60% of the 150 kg/ha ammoniacal N applied in the sludge was volatilized, while during the 7-day experiment in October 56% of the 89 kg/ha ammoniacal N applied was volatilized. The gaseous NH_3 may be absorbed by the surrounding vegetation, surface water, or soil (Denmead et al., 1976).

V. SUMMARY

Crop residues produced annually in the USA amount to approximately 208 million metric tons, which contain 3.56 million metric tons of N. Most of the residues are returned to the soil to supply plant nutrients and organic matter. Turnover in the organic waste materials depends upon their original N percentage and a number of other factors with N release from low N residues being slow and from higher N residues relatively rapid. Rice residues produced outside the USA amount to 454 million metric tons annually; they contain 2.3 million metric tons N. In lowland rice culture, the rice residues containing approximately 0.5% N, have adequate N for optimum decomposition rate, and will supply some of the N needed for the succeeding crop of rice. With the exception of highly acid soils, most organic wastes decompose rapidly with 33% remaining after 1 year and 20% after 5 years.

Animal manure production in the USA is approximately 2 billion tons annually. Manure decomposes fairly rapidly in soil releasing some of the contained N at relatively predictable rates depending upon composition. Several models have been developed for manure decomposition and N release. Large amounts of manure may be applied for disposal or small amounts for growing crops, the latter is preferred. Pollution of soil and water will probably result from high manure applications, but is not likely if the rates are below 20 metric tons/ha.

A wide variety of vegetables, fruits, meats, and dairy products, amounting to an estimated 56 million metric tons/year are processed that produce large amounts of solid and liquid waste materials containing about 35,000 metric tons N. Most of these waste materials are suited for use on land for fertilization and irrigation of crops. The organic matter in the wastes decomposes rapidly in soil, releasing the N. Most food processing wastes contain high enough N concentrations that they readily serve as sources of fertilizer nutrients. Decomposition is seldom limited by N deficiency in food processing wastes. Overfertilization and potential for pollution of soil and ground water can be a problem with many food processing wastes. Dairy manufacturing and meat packing wastes may need pretreatment or special handling because of unusually high COD and N concentrations in the waste materials to avoid soil and ground-water pollution and other problems such as odors.

Municipal waste effluents are being used extensively for irrigation and fertilization of agricultural crops or for reclamation on land. Distribution methods include high-rate infiltration, overland flow, and crop irrigation. The choice of methods of water distribution is site dependent. Some of the methods are capable of recharging surface and ground water, and providing water and nutrients for growing plants. Nitrogen in these previously treated waste waters is usually NH_4^+ or NO_3^- and is readily available to growing plants. Nitrogen fertilization can be efficient and turnover rates rapid with irrigation using municipal waste water. Nitrogen mineralization in sewage sludge is relatively slow compared with municipal effluents and other waste materials.

LITERATURE CITED

Adriano, D. C., A. C. Chang, and R. Sharpless. 1974a. Nitrogen loss from manure as influenced by moisture and temperature. J. Environ. Qual. 3:258–261.

Adriano, D. C., A. E. Erickson, A. R. Wolcott, and B. G. Ellis. 1974b. Certain environmental problems associated with longterm land disposal of food processing wastes. p. 222–244. In Process Manage. Agric. Wastes Proc., 6th Cornell Agric. Waste Manage. Conf.

Adriano, D. C., L. T. Novak, A. E. Erickson, A. R. Wolcott, and B. G. Ellis. 1975. Effects of long-term land disposal by spray irrigation of food processing wastes on some chemical properties of the soil and subsurface water. J. Environ. Qual. 4:242–248.

Adriano, D. C., P. F. Pratt, and S. E. Bishop. 1971. Fate of inorganic forms of nitrogen and sale from land-disposal manures from dairies. p. 243–246. In Livestock waste management and pollution abatement. Am. Soc. of Agric. Eng., St. Joseph, Mich.

Agbim, N. N., B. R. Sabey, and D. C. Markstrom. 1977. Land application of sewage sludge: V. Carbon dioxide production as influenced by sewage sludge and wood waste mixtures. J. Environ. Qual. 6:446–451.

Allison, F. E. 1965. Decomposition of wood and bark sawdust in soil, nitrogen requirements, and effects on soil. USDA, Bull. 1332. 58 p.

Allison, F. E. 1973. Soil organic matter and its role in crop production. Elsevier Scientific Publ. Co., New York. 637 p.

Baier, D. C., and W. B. Fryer. 1973. Undesirable plant responses with sewage irrigation. Am. Soc. of Chem. Eng. J. Irrig. and Drain. Div. (June). p. 133–141.

Bartholomew, W. V. 1965. Mineralization and immobilization of nitrogen in the decomposition of plant and animal residues. *In* W. V. Bartholomew and F. E. Clark (ed.) Soil nitrogen. Agronomy 10:285–306. Am. Soc. Agron., Madison, Wis.

Baumann, D. J. 1971. Dehydration of cattle rumen and whole blood. p. 313–322. *In* Proc. 2nd Natl. Symp. on Food Processing Wastes. Denver, Colo. 23–26 March 1971.

Beauchamp, E. G., G. E. Kidd, and G. Thurtell. 1978. Ammonia volatilization from sewage sludge applied in the field. J. Environ. Qual. 7:141–146.

Bielby, D. G., M. H. Miller, and L. R. Webber. 1973. Nitrate content of percolates from manured lysimeters. J. Soil Water Conserv. 28:124–126.

Bouwer, H., J. C. Lance, and M. S. Riggs. 1974a. High-rate land treatment II: Water quality and economic aspects of the Flushing Meadows project. J. Water Pollut. Control Fed. 46:844–859.

Bouwer, H., R. C. Rice, and E. D. Escarcega. 1974b. High-rate land treatment I: Infiltration and hydraulic aspects of the Flushing Meadows project. J. Water Pollut. Control. Fed. 46:834–843.

Brown, P. L., and D. D. Dickey. 1970. Losses of wheat straw residue under simulated field conditions. Soil Sci. Soc. Am. J. 34:118–121.

Burwell, R. E., G. E. Schumann, K. E. Saxton, and H. G. Heinemann. 1976. Nitrogen in subsurface discharge from agricultural watersheds. J. Environ. Qual. 5:325–329.

Carlile, B. L. 1972. Transformation, movement, and disposal of nitrogen from animal manure wastes applied to soils. Unpublished Ph.D. Dissertation, Washington State Univ., Pullman, 70 p. (Livestock and the Environ.: EPA-600/2-77-092, May 1977 Abstract no. 2448-A8).

Clapp, C. E., D. R. Linden, W. E. Larsen, G. C. Marten, and J. R. Nylund. 1977. Nitrogen removal from municipal wastewater effluent by crop irrigation system. p. 139–150. *In* R. C. Loehr (ed.) Land as a waste management alternative. Proc. 1976 Cornell Agric. Waste Manage. Conf. Ann Arbor Sci. Publ., Ann Arbor, Mich.

Coker, E. G. 1966. The value of liquid sewage sludge I. The effect of liquid sewage sludge on growth and composition of grass-clover swards in south-east England. J. Agric. Sci. 67:91–97.

Concannon, T. J., Jr., and Emil J. Genetelli. 1971. Groundwater pollution due to high organic manure loadings. p. 249–253. *In* Livestock waste management and pollution abatement. Am. Soc. of Agric. Eng. Proc., St. Joseph, Mich.

Day, A. D., and T. C. Tucker. 1959. Production of small grain pasture forage using sewage effluent as a source of irrigation water and plant nutrients. Agron. J. 51:569–572.

De Haan, F. A. M., G. J. Hoogeveen, and F. Reim Vis. 1973. Aspects of agricultural use of potato starch wastewater. Neth. J. Agric. Sci. 21:85–92.

Denmead, O. T., J. R. Freney, and J. R. Simpson. 1976. A closed ammonia cycle within a plant canopy. Soil Biol. Biochem. 8:161–164.

Elliott, L. F., and T. M. McCalla. 1973. The fate of nitrogen from animal wastes. p. 86–110. *In* Conf. Proc. Nitrogen in Nebraska Environ.

Elliott, L. F., and N. P. Swanson. 1976. Land use of animal wastes. p. 80–90. *In* Land application of waste materials. Soil Conserv. Soc. Am. Ankeny, Iowa.

Epstein, E., D. B. Keane, J. J. Meisinger, and J. O. Legg. 1978. Mineralization of nitrogen from sewage sludge and sludge compost. J. Environ. Qual. 7:217–221.

Fordham, H. W., and J. Giddens. 1974. Soil pollution from feedlots in Georgia. Georgia Agric. Res. 15(4):17–19.

Gilmour, C. M., F. E. Broadbent, and S. M. Beck. 1977. Recycling of carbon and nitrogen through land disposal of various wastes. p. 171–174. *In* L. F. Elliott and F. J. Stevenson (ed.) Soils for management of organic wastes and wastewaters. Am. Soc. of Agron., Madison, Wis.

Gilmour, J. T., and C. M. Gilmour. 1979. A simulated model for sludge carbon decomposition in soil. J. Environ. Qual. 9:194–199.

Gilde, L. C., A. S. Kester, J. P. Law, C. H. Neeley, and D. M. Parmelee. 1971. A spray irrigation system for treatment of cannery wastes. J. Water Pollut. Control Fed. 43:2011–2025.

Greb, B. W., A. L. Black, and D. E. Smika. 1974. Straw buildup in soil with stubble mulch fallow in the semi-arid great plains. Soil Sci. Soc. Am. J. 38:135–136.

Guenzi, W. D., W. E. Beard, F. S. Watanabe, S. R. Olsen, and L. K. Porter. 1978. Nitrification and denitrification in cattle manure-amended soil. J. Environ. Qual. 7:196–202.

Hensler, R. F., W. H. Erhardt, and L. M. Walsh. 1971. Effect of manure handling systems on plant nutrient cycling. p. 254-257. *In* Livestock Waste Management and Pollution Abatement. Am. Soc. of Agric. Eng. Proc., St. Joseph, Mich.

Hinesly, T. D., E. L. Ziegler, and G. L. Barrett. 1979. Residual effects of irrigated corn with digested sewage sludge. J. Environ. Qual. 8:35-38.

Hook, J. E., and L. T. Kardos. 1977. Nitrate relationships in the Penn State "Living Filter" system. p. 181-198. *In* R. C. Loehr (ed.) Land as a waste management alternative. Proc. 1976 Cornell Agric. Waste Management Conf., Ann Arbor Sci. Publ., Ann Arbor, Mich.

Hunt, P. 1972. Microbiological responses to the land disposal of secondary-treated municipal-industrial wastewater. p. 77-93. Chap. 5. *In* Wastewater management by disposal on the land. Spec. Rep. 171. Corps of Eng., U.S. Army, Cold Regions Res. and Eng. Lab., Hanover, N.H.

Hunt, P. G., L. C. Gilde, and N. R. Francingues. 1976. Land treatment and disposal of food processing wastes. p. 112-135. *In* Land application of waste materials. Soil Conserv. Soc. Am. Ankeny, Iowa.

Hutchinson, G. L., and F. G. Viets, Jr. 1969. Nitrogen enrichment of surface water by absorption of ammonia volatilized from cattle feedlots. Science 166:514-515.

Iskandar, I. K., R. S. Sletten, D. C. Leggett, and T. F. Jenkins. 1976. Wastewater renovation by a prototype slow infiltration land treatment system. Corps. of Eng., U.S. Army, Cold Regions Res. and Eng. Lab., Hanover, N.H. 44 p.

Jenkinson, D. S. 1971. Studies on the decomposition of 14-C labelled organic matter in soil. Soil Sci. 111:64-70.

Jewell, W. J. 1976. Organic assimilation capacities of land treatment systems receiving vegetable processing wastewaters. *In* 31st Annu. Purdue Univ. Ind. Waste Conf. West Lafayette, Ind. 38 p.

Jewell, W. J., and R. C. Loehr. 1975. Land treatment of food processing wastes. Paper no. 75-2513. Presented at Winter Meetings Am. Soc. of Agric. Eng., Chicago, Ill. Dec. 1975. 38 p.

Keeney, D. R., K. W. Lee, and L. M. Walsh. 1975. Guidelines for the application of wastewater sludge to agricultural land in Wisconsin. Tech. Bull. no. 88, Dep. of Natural Resour., Madison, Wis.

Kelling, K. A., A. E. Peterson, and L. M. Walsh. 1977a. Effect of wastewater sludge on soil moisture relationship and surface runoff. J. Water Pollut. Control Fed. 49:1698-1703.

Kelling, K. A., A. E. Peterson, L. M. Walsh, J. A. Ryan, and D. R. Keeney. 1977b. A field study of the agricultural use of sewage sludge: I. Effects on crop yield and uptake of N and P. J. Environ. Qual. 6:339-345.

Kimble, J. M., R. J. Bartlett, J. L. McIntosh, and K. E. Varney. 1972. Fate of nitrate from manure and inorganic nitrogen in a clay soil cropped to continuous corn. J. Environ. Qual. 1:413-415.

King, L. D. 1973. Mineralization and gaseous loss of nitrogen in soil applied liquid sewage sludge. J. Environ. Qual. 2:356-358.

King, L. D. 1977. Fate of nitrogen from municipal sludges. p. 161-165. *In* Disposal residues on land. Proc. Natl. Conf. on Disposal of Residues on Land. Information Transfer, Inc., Rockville, Md.

King, L. D., and H. D. Morris. 1972. Land disposal of liquid sewage sludge: I. The effect on yield, in vivo digestibility, and chemical composition of coastal bermudagrass (*Cynodon dactylon* L. Pers.). J. Environ. Qual. 1:325-329.

Koo, R. C. J. 1974. Irrigation of citrus with citrus processing wastewater. Pub. no. 28, Florida Water Resour. Res. Center. 73 p.

Krantz, B. A., F. E. Broadbent, W. A. Williams, K. G. Baghott, K. H. Ingebretsen, and M. E. Stanley. 1968. Research with nitrogen fertilizer emphasizes. . .Fertilize crop—not crop residue. Calif. Agric. 22(8):6-8.

Lance, J. C. 1972. Nitrogen removal by soil mechanisms. J. Water Pollut. Control Fed. 44:1352-1361.

Larson, W. E., C. E. Clapp, W. H. Pierre, and Y. B. Morachan. 1972. Effects of increasing amounts of organic residues on continuous corn: II. Organic carbon, nitrogen, phosphorus, and sulphur. Agron. J. 64:204-208.

Larson, W. E., and J. R. Gilley. 1976. Soil-climate-crop considerations for recycling organic wastes. Trans. Am. Soc. of Agric. Eng. 19(1):85-89, 96.

Larson, W. E., R. F. Holt, and C. W. Carlson. 1978. Residue for soil conservation. *In* Proc. Symp. Crop Residue Manage.: I. Effect on Soil. Am. Soc. of Agron.

Letey, J., J. W. Blair, Dale Devitt, L. J. Lund, and P. Nash. 1977. Nitrate-nitrogen in effluent from agricultural tile drains in California. Hilgardia 45:289-319.

Loehr, R. C. 1974. Agricultural waste management problems processes approaches. Academic Press, Inc., New York. 576 p.

Loehr, R. C., T. B. S. Prakasam, E. G. Srineth, T. W. Scott, and T. W. Bateman. 1976. Design parameters for animal waste treatment systems. Nitrogen control. EPA-600/2-76-190. 144 p.

Luebs, R. E., K. R. Davis, and A. E. Laag. 1973. Enrichment of the atmosphere with nitrogen compounds volatilized from a large dairy area. J. Environ. Qual. 2:137–141.

MacMillan, K., T. W. Scott, and T. W. Bateman. 1972. A study of corn response and soil nitrogen transformations upon application of different rates and sources of chicken manure. p. 481–494. In Waste Management Res., Proc. 1972 Cornell Agric. Waste Manage. Conf., Ithaca, N.Y.

MacMillan, K. A., T. W. Scott, and J. W. Bateman. 1975. Corn response and soil nitrogen transformations following varied applications of poultry manure treated to minimize odor. Can. J. Soil Sci. 55:29–34.

Mann, H. H., and T. W. Barnes. 1956. The permanence of organic matter added to the soil. J. Agric. Sci. 48:160–163.

Marriott, L. F., and H. D. Bartlett. 1972. Contribution of animal waste to nitrate nitrogen in soil. p. 435–440. In Waste Manage. Res. Proc. 1972 Cornell Agric. Waste Manage. Conf.

Mathers, A. C., and B. A. Stewart. 1970. Nitrogen transformations and plant growth as affected by applying large amounts of cattle feedlot wastes to soil. p. 207–214. In Relationship of Agric. to Soil and Water Pollut. Proc. Cornell Agric. Waste Manage. Conf. Rochester, N.Y. 19–21 Jan. 1970.

Mathers, A. C., and B. A. Stewart. 1971. Crop productions and soil analyses as affected by applications of cattle feedlot waste. p. 229–231. In Proc. Int. Symp. on Livestock Waste Manage. and Pollut. Abatement. Am. Soc. of Agric. Eng., St. Joseph, Mich.

Mathers, A. C., B. A. Stewart, J. D. Thomas, and B. J. Blair. 1972. Effects of cattle feedlot manure on crop yields and soil conditions. Texas Agric. Exp. Stn., Res. Center Tech. Rep. no. 11:13.

Mathers, A. C., B. A. Stewart, and J. D. Thomas. 1975. Residual and annual rate effects of manure and grain sorghum yields. p. 252–350. In Conf. Proc. 3rd Int. Symp. on Livestock Wastes. Am. Soc. of Agric. Eng., St. Joseph, Mich.

McCalla, T. M., L. R. Frederick, and G. L. Palmer. 1970. Manure decomposition and fate of breakdown products in soil. p. 241–255. In T. L. Willrich and G. E. Smith (ed.) Agric. practices and water quality. Iowa State Univ. Press, Ames, Iowa.

McCalla, T. M., J. R. Peterson, and C. Lue-Hing. 1977. Properties of agricultural and municipal wastes. p. 11–43. In L. F. Elliott (ed.) Soils for management of organic wastes and wastewaters. Am. Soc. of Agron., Madison, Wis.

Meek, B. D., A. J. McKenzie, T. J. Donovan, and W. F. Spencer. 1974. The effect of large applications of manure on movement of nitrate and carbon in an irrigated desert soil. J. Environ. Qual. 3:253–258.

Meek, B. L., W. Chesnin, W. Fuller, R. Miller, and D. Turner. 1975. Guidelines for manure use and disposal in the Western Region, USA. Washington State Univ., College of Agric. Bull. 814. 18 p.

Meyer, J. L. 1974. Cannery wastewater for irrigation and groundwater recharging. Calif. Agric. 28(8):12.

Mielke, L. N., N. P. Swanson, and T. M. McCalla. 1974. Soil profile conditions of cattle feedlots. J. Environ. Qual. 3:14–17.

Miller, R. H. 1974. Factors affecting the decomposition of anaerobically digested sewage sludge in soil. J. Environ. Qual. 3:376–380.

Molina, J. A., O. C. Braids, T. D. Hinesly, and J. B. Cropper. 1971. Aeration-induced changes in liquid digested sewage sludge. Soil Sci. Soc. Am. Proc. 35:60–63.

Murphy, L. S., G. W. Wallingford, W. L. Powers, and H. L. Manges. 1972. Effects of solid beef feedlot wastes on soil conditions and plant growth. p. 449–464. In Waste Manage. Res. Proc. 1972. Cornell Agric. Waste Manage. Conf.

Olsen, P. J., R. F. Hensler, and O. J. Attoe. 1970. Effect of manure application, aeration, and soil pH on soil nitrogen transformations and on certain soil test values. Soil Sci. Soc. Am. Proc. 34:222–225.

Peters, R. E., and D. L. Reddell. 1976. Ammonia volatilization and nitrogen transformations in soils used for beef manure disposal. Trans. Am. Soc. of Agric. Eng. 19:945–952.

Peterson, J. R., C. Lue-Hing, and D. R. Zenz. 1973. Chemical and biological quality of municipal sludge. p. 26–37. In W. E. Sopper and L. T. Kardos (ed.) Recycling treatment municipal wastewater and sludge through forest and cropland. Penn State Univ. Press, University Park, Penn.

Peterson, J. R., T. M. McCalla, and G. E. Smith. 1971. Human and animal wastes as fertilizers. p. 557–595. *In* R. A. Olson et al. (ed.) Fertilizer technology and use. 2nd ed. Soil
Sci. Soc. Am., Madison, Wis.

Peterson, J. R., R. I. Pietz, and C. Lue-Hing. 1979. Water, soil, and crop quality of Illinois
coal mine spoils amended with sewage sludge. p. 359–368. *In* W. E. Sopper and S. N.
Kerr (ed.) Symp. Wastewater and Sludge Recycling on Forest Land and Disturbed Land.
Penn State Univ. Press, University Park, Penn.

Powers, J. F. 1968. Mineralization of nitrogen in grass roots. Soil Sci. Soc. Am. Proc. 32:
673–674.

Pratt, P. F., F. E. Broadbent, and J. P. Martin. 1973. Using organic wastes as nitrogen
fertilizer. Calif. Agric. 27(6):10–13.

Pratt, P. F., S. Davis, and R. G. Sharpless. 1976. A four-year field trial with animal manures.
I. Nitrogen balances and yields. II. Mineralization of nitrogen. Hilgardia 44:99–125.

Premi, P. R., and A. H. Cornfield. 1969. Incubation study of nitrification of digested sewage
sludge added to soil. Soil Biol. Biochem. 1:1–4.

Rao, D. N., and D. S. Mikkelsen. 1976. Effect of rice straw incorporation on rice plant growth
and nutrition. Agron. J. 68:752–755.

Reed, S. 1972. Wastewater management by disposal on the land. Cold Regions Res. and Eng.
Lab., Corps of Eng., U.S. Army. Spec. Rep. 171, Hanover, N.H. p. 2–34.

Reed, A. D., W. E. Wildman, W. S. Seyman, R. S. Auyers, J. D. Prato, and R. S. Rauschkolb. 1973. Soil recycling of cannery wastes. Calif. Agric. 27(3):6–9.

Rolston, D. E., and F. E. Broadbent. 1977. Field measurement of denitrification. EPA-600/2-
77-233. 75 p.

Ryan, J. A., and D. R. Keeney. 1975. Ammonia volatilization from surface-applied wastewater sludge. J. Water Pollut. Control Fed. 47:386–393.

Sain, P., and F. E. Broadbent. 1977. Decomposition of rice straw in soils as affected by some
management factors. J. Environ. Qual. 6:96–100.

Satterwhite, M. B., G. L. Stewart, B. J. Condike, and E. Vlach. 1976. Rapid infiltration of
primary sewage effluent at Fort Deven, Mass. Corps of Eng., U.S. Army, Cold Res. and
Eng. Lab., Hanover, N.H. 34 p.

Schuman, G. E., T. M. McCalla, K. E. Saxton, and H. T. Knox. 1975. Nitrate movement and
its distribution in the soil profile of differentially fertilized corn watersheds. Soil Sci. Soc.
Am. Proc. 29:1192–1197.

Seabrook, B. L. 1975. Land application of wastewater in Australia. The land treatment system
at Werribee, Victoria. USEPA-430/9-75-017, MCD 16. 54 p.

Shannon, S., M. T. Vittum, and G. H. Gibbs. 1968. Irrigating with wastewater from processing plants. New York State Agric. Exp. Stn. Res. Circ. no. 10, 9 p.

Smith, J. H. 1966. Some inter-relationships between decomposition of various plant residues
and loss of soil organic matter as measured by carbon-14 labeling. p. 223–233. *In* The
use of isotopes in soil organic matter studies. Pergamon Press. Rep. FAO/IAEA Tech.
Meet. Brunswick-Volkenrode. 1963.

Smith, J. H. 1976. Treatment of potato processing wastewater on agricultural land. J.
Environ. Qual. 5:113–116.

Smith, J. H., and C. L. Douglas. 1968. Influence of residual nitrogen on wheat straw decomposition in the field. Soil Sci. 106:456–459.

Smith, J. H., and C. L. Douglas. 1971. Wheat straw decomposition in the field. Soil Sci. Soc.
Am. Proc. 35:269–272.

Smith, J. H., C. L. Douglas, and M. J. LeBaron. 1973. Influence of straw application rates,
plowing dates, and nitrogen applications on yield and chemical composition of sugarbeets. Agron. J. 65:797–800.

Smith, J. H., R. G. Gilbert, and J. B. Miller. 1976. Redox potentials and denitrification in a
cropped potato processing wastewater disposal field. J. Environ. Qual. 5:397–399.

Smith, J. H., R. G. Gilbert, and J. B. Miller. 1978. Redox potentials in a cropped potato processing wastewater disposal field with a deep water table. J. Environ. Qual. 7:571–574.

Smith, J. H., C. W. Robbins, J. A. Bondurant, and C. W. Hayden. 1977. Treatment of potato
processing wastewater on agricultural land: Water and organic loading, and the fate of
applied plant nutrients. p. 769–781. *In* R. C. Loehr (ed.) Land as a Waste Management
Alternative. Proc. 1976 Cornell Agric. Waste Manage. Ann Arbor Sci. Publ., Ann Arbor,
Mich.

Smith, J. H., C. W. Robbins, J. A. Bondurant, and C. W. Hayden. 1978. Treatment and disposal of potato processing wastewater by irrigation. USDA-SEA, Conserv. Res. Rep.
22. 43 p.

Smith, J. H., and C. W. Hayden. 1980. Irrigating with sugarbeet processing wastewater. J. Am. Soc. Sugar Beet Technol. 20(5):484–502.

Smith, J. H., C. W. Robbins, and C. W. Hayden. 1975. Plant nutrients in potato processing wastewater used for irrigation. p. 159–165. *In* Proc. 26th Ann. Pacific Northwest Fertilizer Conf., 15–17 July 1975, Salt Lake City, Utah.

Sommers, L. E. 1977. Chemical composition of sewage sludge and analysis of their potential use as fertilizer. J. Environ. Qual. 6:225–232.

Sopper, W. E. 1968. Wastewater renovation for reuse: Key to optimum use of water resources. Water Res. 2:471–480.

Stanley Associates Engineering Ltd. 1977. Review of treatment technology in the fruit and vegetable processing industry in Canada. EPX 3-WP-77-5. p. 176–178.

Stewart, B. A. 1970. Volatilization and nitrification of nitrogen from urine under simulated cattle feedlot conditions. Environ. Sci. Technol. 4:579–582.

Stewart, B. A., and R. L. Chaney. 1975. Wastes: Use or discard? p. 160–166. *In* Proc. 30th Annu. Meeting. Soil Conserv. Soc. of Am. 10–13 Aug. 1975, San Antonio, Texas.

Stewart, N. E., E. G. Beauchamp, C. T. Corke, and L. R. Webber. 1975. Nitrate nitrogen distribution in corn land following applications of digested sewage sludge. Can. J. Soil Sci 55:287–294.

Tanaka, A. 1973. Methods handling the rice straw in various countries. Int. Rice Comm. Newsl. 22(2):120 Food and Agric. Organ., Bangkok, Thailand.

Tarquin, A., H. Applegate, F. Rizzo, and L. Jones. 1974. Design considerations for treatment of meatpacking plant wastewater by land application. p. 107–113. *In* Proc. 5th Natl. Symp. on Food Processing Wastes. Syracuse, N.Y., 13–15 Jan. 1969. EPA-660/2/-74-058.

Tester, C. F., L. J. Sikora, J. M. Taylor, and J. F. Parr. 1977. Decomposition of sewage sludge compost in soil: 1. Carbon and nitrogen transformation. J. Environ. Qual. 6:459–463.

Turner, D. O. 1975. On-the-farm determination of animal waste disposal rates for crop production. Conf. Proc. 3rd Int. Symp. on Livestock Wastes. Am. Soc. of Agric. Eng., St. Joseph, Mich.

Turner, D. O. 1976. Guidelines for manure application in the Pacific Northwest. Coop. Ext. Serv. College of Agric., Wash. State Univ., Pullman, Wash., EM 4009. 25 p.

USEPA. 1973. Water quality criteria. 1972. A report of the committee on water quality criteria. Environ. Studies Board, Natl. Acad. of Sci., Natl. Acad. of Eng., Washington, D.C. 594 p.

Varanka, M. W., Z. M. Zabiocki, and T. D. Hinesly. 1976. The effect of digester sludge on soil biological activity. J. Water Pollut. Control Fed. 48:1728–1740.

Wadleigh, C. H. 1968. Wastes in relation to agriculture and forestry. USDA Misc. Pub. 1065. U.S. Govt. Printing Off., Washington, D.C. 112 p.

Walter, M. F., G. D. Bubenzer, and J. C. Converse. 1974. Movement of manurial nitrogen in cool, humid, climates. Paper no. 74-2018. 67th Annu. Meeting. Am. Soc. of Agric. Eng., Oklahoma State Univ., Stillwater. 21 p. (Livestock and the Environ. EPA 600/2-77-092. May 1977, Abstr. no. 1712-A1).

Webber, L. R., and T. H. Lane. 1969. The nitrogen problem in the land disposal of liquid manure. p. 124–130. *In* Animal Waste Manage. Proc. Cornell Univ. Conf. on Agric. Waste Manage. Syracuse, N.Y.

Weeks, M. E., M. E. Hill, S. Karczmarczk, and A. Blackmer. 1972. Heavy manure applications: benefit or waste? p. 441–447. *In* Waste Manage. Res. Proc. 1972 Cornell Agric. Waste Manage. Conf.

Williams, W. A., D. S. Mikkelsen, K. E. Mueller, and J. E. Ruckman. 1968. Nitrogen immobilization by rice straw incorporated in lowland rice production. Plant Soil 28:49–60.

Witherow, J. L. 1975. Small meat-packer waste treatment systems II. Purdue Univ. Ind. 30th Waste Conf., Lafayette, Ind. 942 p.

Yoneyama, T., and T. Yoshida. 1977a. Decomposition of rice residue in tropical soils. I. Nitrogen uptake by rice plants from straw incorporated, fertilizer (ammonium sulfate), and soil. Soil Sci. Plant Nutr. (Tokyo) 23:33–40.

Yoneyama, T., and T. Yoshida. 1977b. Decomposition of rice residue in tropical soils. II. Immobilization of soil and fertilizer nitrogen by intact rice residue in soil. Soil Sci. Plant Nutr. (Tokyo) 23:41–48.

Yoneyama, T., and T. Yoshida. 1977c. Decomposition of rice residue in tropical soils. III. Nitrogen mineralization and immobilization of rice residue during its decomposition in soil. Soil Sci. Plant Nutr. (Tokyo) 23:175–183.

22 Energetics of Nitrogen Transformations

R. F. HARRIS

University of Wisconsin
Madison, Wisconsin

I. INTRODUCTION

Chemical equilibrium and biochemical nonequilibrium thermodynamics provide the ground rules needed to interpret and predict the nature, rate, and extent of N transformations as a function of environmental conditions. In quantitative terms, for a potential reaction represented as

$$\text{reactants (s, l, g, or aq)} \rightleftharpoons \text{products (s, l, g or aq)}$$

where s (solid),[1] l (liquid), g (gas), and aq (aqueous) specify the physical state of the reaction components, the thermodynamic properties of the reaction components provide, under conditions of constant temperature and pressure, the following information (Van der Meer et al., 1980; Caplan, 1976; Stumm & Morgan, 1970):

1) The direction and magnitude of the thermodynamic force driving the reaction toward equilibrium:

$$A = -[\Delta G]$$

$$= -[\Sigma \, \mu \, \text{products} - \Sigma \, \mu \, \text{reactants}]$$

$$= -[\Sigma \, \Delta G_f \, \text{products} - \Sigma \, \Delta G_f \, \text{reactants}]$$

$$= -[(\Sigma \, \Delta G_f^\circ \text{products} - \Sigma \, \Delta G_f^\circ \text{reactants})$$
$$+ RT \ln\{\text{products}\}/\{\text{reactants}\}] \qquad [1]$$

$$= -[\Delta G^\circ + 2.3RT \log Q] = A^\circ - 2.3RT \log Q \qquad [2]$$

$$= -[2.3RT(-\log K + \log Q)] = 2.3RT(\log K - \log Q).$$

2) The rate of the reaction under certain conditions:

$$v = LA - b$$

$$= L(A^\circ - 2.3RT \log Q) - b. \qquad [3]$$

---

[1] Also expressed as c (crystalline) or amorph (amorphous).

Nitrogen in Agricultural Soils—Agronomy Monograph no. 22.

3) The relative activities of the reaction components prevailing under conditions of a specific thermodynamic force/reaction rate:

$$\log \frac{\{products\}}{\{reactants\}} = \log Q = \frac{\Delta G - \Delta G°}{2.3RT} = \frac{-A + A°}{2.3RT}. \qquad [4]$$

4) The relative activities of the reaction components at equilibrium:

$$\log \frac{\{products\}_{eq}}{\{reactants\}_{eq}} = \log Q_{eq} = \log K = \frac{-\Delta G°}{2.3RT} = \frac{A°}{2.3RT}. \qquad [5]$$

5) The amount of heat evolved or absorbed during the reaction (recognizing that for dilute aqueous systems, $\Delta H \cong \Delta H°$, Stumm & Morgan, 1970):

$$\Delta H \cong \Delta H° = \Sigma \Delta H_f° \text{ products} - \Sigma \Delta H_f° \text{ reactants} \qquad [6]$$

where A is the thermodynamic force or affinity of the reaction; ΔG is the Gibbs free energy change of the reaction; μ and ΔG_f are the chemical potential and Gibbs free energy of formation of the reaction components, respectively; $\Delta G_f°$ is the standard free energy of formation of a reaction component; R is the universal gas constant; T is the absolute temperature; {substance} denotes the activity (approximately equal to the concentration for dilute aqueous systems) of the substance; Q is the reaction quotient; K is the activity equilibrium constant for the reaction; $\Delta G°$ and $A°$ are the standard Gibbs free energy change and the standard thermodynamic force of the reaction, respectively; L is the conductance coefficient; b is a constant; ΔH and $\Delta H°$ are the heat and standard heat of the reaction, respectively; and $\Delta H_f°$ is the standard heat of formation of a reaction component. For solids and liquids the standard state is that of the pure element or substance at constant 1 atm pressure; for gases it is the ideal gas at a pressure of 1 atm; for species in aqueous solution it is the hypothetical ideal solution of unit molality; and the standard temperature is 25°C (Alberty & Daniels, 1979; CODATA, 1978; Wilhoit, 1969).

Characterization of the energetic basis of environmental N and other biochemically dominated nutrient transformations is not helped by the confusion of current thermodynamic conventions. Although thermodynamic force is a conceptually simpler and more appropriate (from the standpoint of flow-force relationships) index of the energy drive of a reaction, in line with tradition, the conventional ΔG approach is emphasized for calculation of reaction energetics. Similarly, currently entrenched acceptor/formation-oriented conventions for calculating group transfer energetics will also be retained. For simplicity, unless specified otherwise, the physical state of reaction components is designated as aqueous (water as liquid). Since most non-European chemical and biochemical literature still use the calorie rather than the SI unit of energy (the joule), the calorie is the basic energy unit used in this review. However, most equations are represented in noncommital (RT) form. For such equations, for a standard temperature of 25°C (298.15°K) and recognizing that $R = 8.3143$ J•deg^{-1}•mol^{-1} = 1.9872 cal•deg^{-1}•mol^{-1} (1 cal = 4.184 J) (CODATA, 1978), the term 2.3RT (more precisely, 2.3026RT) = 5.7079 kJ•mol^{-1} = 1.3642 kcal•mol^{-1}. Similarly, recognizing that the Faraday, F, = 96.487 kJ•V^{-1}•mol^{-1} = 23.061 kcal•V^{-1}•mol^{-1} (CODATA, 1978), the term 2.3RT/F = 0.0592 V.

Because of the mechanistic importance of group transfer energetics in biothermodynamics, and the increasing popularity of the "relative electron activity" concept for predicting theoretical equilibrium conditions of redox reactions in environmental chemistry, standardized equations are presented to provide a unifying reference base for comparison of the different thermodynamic approaches currently in use for calculating group transfer energetics. The major emphasis of this review is concerned with quantitative description of the thermodynamic (particularly the biothermodynamic) principles dictating the nature, rate, and extent of environmental N transformations. The latter part of the review uses these principles to evaluate the mechanistic energetics of specific N transformations.

II. BIOENERGETIC PRINCIPLES OF ENVIRONMENTAL NITROGEN TRANSFORMATIONS

A. Nonequilibrium Thermodynamics and Reaction Kinetics

The strength of chemical equilibrium thermodynamics is that the predicted composition of a reaction mixture at equilibrium is dictated solely by the thermodynamic properties of the reactants and products of the reaction and is independent of the mechanisms and kinetics of the reaction; the related weakness is that such equilibrium-based information is of little practical value in the absence of knowledge of the time scale (microseconds or thousands of years?) of the reaction. With the exception of proton transfer reactions, biospheric N transformations tend to proceed at extremely slow (often essentially negligible) rates in the absence of biological catalysts.

The rapidly expanding field of nonequilibrium thermodynamics has made major progress recently toward establishment, in quantitative mechanistic terms, of the dependence of biochemical reaction rates on the thermodynamic forces driving the reactions toward equilibrium (Van der Meer et al., 1980; Van Dam et al., 1980; Caplan, 1976). Because of space considerations and the fact that the current state of the art in nonequilibrium thermodynamics is still largely in the conceptual rather than the rigorously quantitative stage of development, the following interpretation is confined to summarization in highly simplified form of the major concepts relevant to biochemical N transformations.

1. BASIC RELATIONSHIPS BETWEEN THE KINETIC PROPERTIES OF ENZYMES AND THE THERMODYNAMIC FORCE OF THE REACTIONS THEY CATALYZE

For a simple enzyme-catalyzed reaction represented as

$$S + E \rightleftharpoons SE \rightleftharpoons EP \rightleftharpoons E + P, \qquad [7]$$

the rate of reaction, v, is given by (Van der Meer et al., 1980; Cornish-Bowden, 1979; Segel, 1976)

$$v = \frac{\dfrac{V^s_{max}[S]}{K^s_m} - \dfrac{V^p_{max}[P]}{K^p_m}}{1 + \dfrac{[S]}{K^s_m} + \dfrac{[P]}{K^p_m}} \qquad [8]$$

where V^s_{max} and V^p_{max} are the maximum forward and backward velocities, respectively; K^s_m and K^p_m are the half-saturation Michaelis-Menten affinity constants for substrate and product, and $[S]$ and $[P]$ are the concentration of the substrate and product, respectively.

Rearrangement of Eq. [8] for the case where $v = 0$ (i.e., for the system at equilibrium) gives the well-established Haldane relationship

$$\frac{V^s_{max}/K^s_m}{V^p_{max}/K^p_m} = \frac{[P]_{eq}}{[S]_{eq}} = K^c$$

$$\cong \frac{\{P\}_{eq}}{\{S\}_{eq}} = K = \exp(-\Delta G°/RT) = \exp(A°/RT) \qquad [9]$$

where the product and substrate concentrations are assumed to be essentially the same as the corresponding activities (so that K^c, the concentration equilibrium constant, $\cong K$, the activity equilibrium constant).

According to Eq. [9], the kinetic properties of an enzyme must be established within the boundaries dictated by the standard energetics of the reaction catalyzed by the enzyme, a high $A°$ providing maximum flexibility for establishment of the high V^s_{max}/low K^s_m properties necessary for maintaining a high forward velocity under a diverse range of substrate/product conditions (Eq. [8]). Rearrangement of Eq. [9] clarifies, for example, the dependence of the substrate-binding affinity constant on the standard energetics of the reaction

$$\log K^s_m = \log \frac{V^s_{max}K^p_m}{V^p_{max}} - \frac{A°}{2.3RT}. \qquad [10]$$

Under conditions of low substrate concentration, organisms commonly activate enzyme systems of lower K^s_m (Tempest & Neijssel, 1978), and in line with Eq. [10], increase the $A°$ of the reaction by energy transduction (e.g., by energy coupling with exergonic hydrolysis of adenosine triphosphate, ATP), thereby facilitating, through the operation of a lowered K^s_m, maintenance of a high v in the face of decreasing substrate concentration (Eq. [8]). The effect of an energy transduction–induced increase in $A°$ and decrease in K^s_m on the v vs. S relationship is identified specifically by combination and rearrangement of Eq. [8], [9], and [2] (after Van der Meer et al., 1980):

$$\frac{v}{V_{max}^s} = \frac{[S] - [P]/\exp(A°/RT)}{[S] + K_m^s([P]/K_m^p + 1)}$$

$$= \frac{\exp(A/RT) - 1}{\left(\dfrac{K_m^s}{[S] + [P]} + 1\right)\exp(A/RT) + \left(\dfrac{K_m^p}{[S] + [P]} + 1\right)\dfrac{V_{max}^s}{V_{max}^p}} \cdot \qquad [11]$$

As pointed out by Cornish-Bowden (1979), more complex rate equations than Eq. [7], such as those that describe relations of several substrates, lead to more complex Haldane relationships, but for all equations there is at least one relationship of the type identified in Eq. [9] between the kinetic properties of the enzyme and the standard energetics of the reaction catalyzed by the enzyme.

EXAMPLE 1

Energy transduction for reaction rate control under conditions of low substrate concentration: The glutamate dehydrogenase (GDH) system vs. the ATP-requiring glutamine synthetase-glutamate synthase (GS–GOGAT) system for NH_4^+ *assimilation.*

In the presence of relatively high (millimolar) ambient NH_4^+ concentration, most microorganisms use the relatively low affinity (K_m for NH_4^+ in the millimolar range) GDH system for reductive NH_4^+ assimilation; under conditions of low NH_4^+, the ATP-requiring, relatively high affinity (K_m in the micromolar range) GS–GOGAT system is commonly induced (Dalton, 1979):

$\Delta G°'$, kcal

Glutamate dehydrogenase reaction,

$\quad$ NADPH + α-ketoglutarate^{2-} + NH_4^+ + H^+

$\qquad$ = $NADP^+$ + glutamate$^-$ $\qquad\qquad$ −9.9 $\qquad$ [12]

glutamine synthetase–glutamate synthase reaction,

$\quad$ glutamate$^-$ + ATP^{4-} + NH_4^+ = glutamine

$\qquad$ + ADP^{3-} + P_i^- $\qquad\qquad$ −4.7 $\qquad$ [13]

NADPH + α-ketoglutarate^{2-} + glutamine + H^+

$\qquad$ = $NADP^+$ + 2 glutamate$^-$ $\qquad\qquad$ −12.8 $\qquad$ [14]

NADPH + α-ketoglutarate^{2-} + ATP^{4-} + NH_4^+ + H^+

$\qquad$ = $NADP^+$ + glutamate$^-$ + ADP^{3-} + P_i^- $\qquad$ −17.5 $\qquad$ [15]

where NADPH and NADP$^+$ are reduced and oxidized nicotinamide dinucleotide phosphate, respectively; ADP^{3-} is adenosine diphosphate; $\Delta G^{\circ\prime}$ is the standard free energy change of the reaction at pH 7; and Eq. [13] and [14] represent the individual GS and GOGAT reactions, respectively. The thermodynamic force driving the reactions at pH 7 is given by the following:

For the GDH reaction (from Eq. [12] and [2]):

$$A' = A^{\circ\prime} - 2.3RT \log \frac{\{NADP^+\}\{glutamate^-\}}{\{NADPH\}\{\alpha\text{-ketoglutarate}^{2-}\}\{NH_4^+\}}$$

$$= 9.9 + 1.364 \log\{NH_4^+\}$$

$$- 1.364 \log \frac{\{NADP^+\}\{glutamate^-\}}{\{NADPH\}\{\alpha\text{-ketoglutarate}^{2-}\}}. \tag{16}$$

For the GS–GOGAT reaction (from Eq. [15] and [2]):

$$A' = 17.5 + 1.364 \log\{NH_4^+\}$$

$$- 1.364 \log \frac{\{NADP^+\}\{glutamate^-\}\{ADP^{3-}\}\{P_i^-\}}{\{NADPH\}\{\alpha\text{-ketoglutarate}^{2-}\}\{ATP^{4-}\}}. \tag{17}$$

Comparison of Eq. [16] and [17] identifies that the GS–GOGAT system can maintain a higher thermodynamic force and thus a higher reaction rate (Eq. [11]) than the GDH system in the face of low NH$_4^+$ and high relative NADP$^+$/NADPH and glutamate$^-$/α-ketoglutarate^{2-} ratios. Maintenance of a high glutamate$^-$/α-ketoglutarate^{2-} ratio is critical during NH$_4^+$ assimilation since the amino transferase reactions between glutamate$^-$–α-ketoglutarate^{2-} and other keto acid–amino acid pairs tend to show marginally exergonic or endergonic standard energetics. For example:

$$glutamate^- + pyruvate^- = \alpha\text{-ketoglutarate}^{2-} + alanine,$$

$$\Delta G^{\circ\prime} = +1.2 \, kcal$$

$$A' = -1.2 + 1.364 \log \frac{\{glutamate^-\}\{pyruvate^-\}}{\{\alpha\text{-ketoglutarate}^{2-}\}\{alanine\}} \tag{18}$$

so that establishment of a nominal thermodynamic force of 1 to 2 kcal needed to maintain a physiologically acceptable reaction rate for alanine biosynthesis would require, for equimolar activities of pyruvate$^-$ and alanine, a glutamate$^-$/α-ketoglutarate^{2-} ratio of 2 to 3:1 in favor of glutamate$^-$ (Eq. [18]).

2. MULTISTEP REACTIONS: KINETIC IMPLICATIONS OF ENDERGONIC STEPS WITHIN OVERALL EXERGONIC REACTIONS

From a general mechanistic standpoint, biological systems dictate the nature, rate, and extent of reaction processes by control of the amount and activity of the specific enzyme complex catalyzing each step of each process. For a multigroup transfer reaction involving enzyme-bound intermediates, a major issue is the maximum number of groups that can be catalytically transferred in a single step. If there were no limit, then the kinetic properties of the enzyme complex could be established within the boundaries of the gross standard energetics of the reaction process (Eq. [9]). However, for logistic reasons, an enzyme complex appears in general to be restricted energetically to catalysis of a maximum of one or two groups at a time (Buvet, 1977), thereby dictating that in general, multigroup transfer reactions involving enzyme-bound intermediates must proceed mechanistically as a sequence of one- or two-group transfer steps. This restriction, combined with a kinetic-based limitation on the substrate concentration range over which enzymes can effectively function, has major practical implications with respect to the importance of endergonic steps within overall exergonic reactions.[2] In practice, energy transduction is used to overcome the kinetic barriers of such endergonic steps.

EXAMPLE 2

Energy transduction to circumvent endergonic energy barriers: The ATP-requiring nitrogenase system for N_2 reduction to $2NH_4^+$.

Although N_2 reduction to $2NH_4^+$ with a reductant such as H_2 at pH 7 is exergonic (Eq. [22]), the first step in the reaction pathway through diimide and hydrazine (Fig. 1) is highly endergonic (Eq. [19]) (from Eq. [1] and Table A-1a[3]):

	$\Delta G^{\circ\prime}$, kcal	$E_h^{\circ\prime}$, V	
$H_2(g) + N_2(g) = N_2H_2(g)$	$+50.00$	-1.4982	[19]
$H_2(g) + N_2H_2(g) + H^+ = N_2H_5^+$	-20.71	$+0.0349$	[20]
$H_2(g) + N_2H_5^+ + H^+ = 2NH_4^+$	-48.17	$+0.6303$	[21]
$3H_2(g) + N_2(g) + 2H^+ = 2NH_4^+$	-18.88	-0.2777	[22]

where $E_h^{\circ\prime}$ is the standard (pH 7) reduction potential (section III C2b). If we assume that a thermodynamic force of at least 1 kcal is needed at each step to drive the reaction at a reasonable rate, then the steady-state activities of

---

[2] This is a crucial difference between biothermodynamic and equilibrium treatment of reaction energetics, since in equilibrium thermodynamics the presence of an endergonic step within an overall exergonic reaction is irrelevant to the equilibrium composition calculation.

[3] See Appendix.

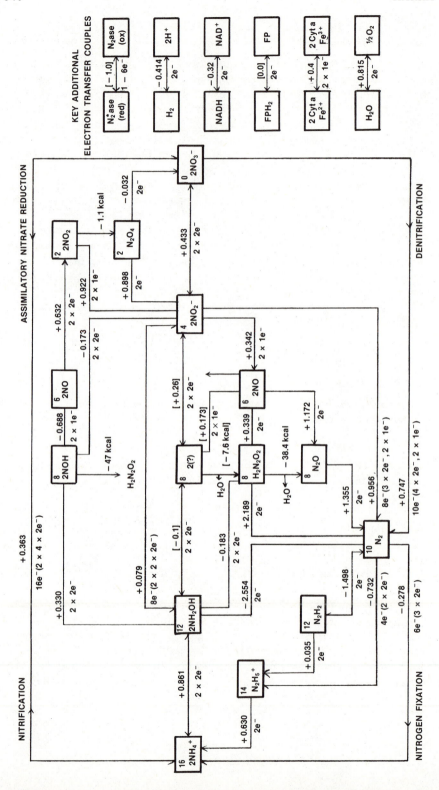

Fig. 1—Energetics of inorganic N metabolism (opposite page).

The N compounds are ranked according to degree of e^- richness, proceeding from ammonium (-3 oxidation state) on the left to nitrate ($+5$) on the right (Latimer, 1952). The number of N-bound e^- (Table 1a) associated with each N compound or doubled N compound is identified in the top left-hand corner of the boxes enclosing the compounds. Redox half-reaction couples are joined by horizontal lines. The $E_h^{\circ\prime}$ of each couple is identified above the line. The numbers of electrons involved in the half reactions are identified below the line as multiples of the $2e^-$ or $1e^-$ mechanistic steps. The $\Delta G^{\circ\prime}$ of non-e^- transfer reactions (joined by vertical lines) are identified in kilocalories. Numbers in square brackets are estimates.

Key biochemical e^- transfer couples are listed on the far right of the diagram. In directional terms, a redox couple may act as an e^- donor and reduce another redox couple as long as the E_h of the e^- donor couple is more negative than that of the couple being reduced. Similarly, a redox couple may act as an e^- acceptor and oxidize another redox couple as long as the E_h of the e^- acceptor is more positive than that of the couple being oxidized. Equations for calculating e^- transfer energetics as a function of $E_h^{\circ\prime}$ are presented in the text (Eq. [65]–[68]) and in Fig. 2.

Full arrows indicate that the reaction is known to proceed mechanistically as shown. Half arrows are used for reactions that are known to proceed but via intermediates (multistep reactions) or only if boosted by energy transduction (e.g., N_2 reduction to N_2H_2; NH_4^+ oxidation to NH_2OH).

N_2H_2 and $N_2H_5^+$ under physiological conditions of 0.78 atm N_2, $10^{-5} M$ NH_4^+ and a reductant equivalent to 10^{-3} atm H_2 would be (from Eq. [19], [20], and [4])

$$\log\{N_2H_2\} = \frac{\Delta G^\prime - \Delta G^{\circ\prime}}{2.3RT} + \log\{H_2\}\{N_2\}$$

$$= \frac{-1 - (+50.0)}{1.364} + \log(10^{-3})(0.78) = -40.5$$

$$\log\{N_2H_5^+\} = \frac{\Delta G^\prime - \Delta G^{\circ\prime}}{2.3RT} + \log\{H_2\}\{N_2H_2\}$$

$$= \frac{-1 - (-20.7)}{1.364} + \log(10^{-3}) + (-40.5) = -29.1.$$

Reactant activities such as $10^{-41}(N_2H_2)$ and $10^{-29}(N_2H_5^+)$ are grossly out of range of the substrate-binding properties of enzymes. Energy transduction with 12 MgATP at $\Delta G^{\circ\prime} = -8.8$ kcal•mol ATP^{-1} hydrolyzed (Stiefel, 1977) would add up to 105.6 kcal (boundary maximum for 100% energy transduction efficiency) of thermodynamic force to the N_2 reduction reaction. If, for example, 47 kcal of this ATP energy were used in effect to boost the $\Delta G^{\circ\prime}$ of the N_2–N_2H_2 reduction step from $+50$ to $+3$ kcal•reaction^{-1}, then the steady-state activity of N_2H_2 would be increased to a more kinetically favorable level of

$$\log\{N_2H_2\} = \frac{-1 - (+3)}{1.364} + \log(10^{-3})(0.78) = -6.$$

Mechanistic aspects of the e^- transfer energetics of biological N_2 reduction are discussed in Example 9, section III C2biii.

3. APPLICABILITY OF THE PHENOMENOLOGICAL FLOW-FORCE RELATIONSHIP TO BIOCHEMICAL REACTIONS

As shown theoretically by Van der Meer et al. (1980) for a simple two-step Michaelis-Menten reaction (Eq. [7]) and supported by experimental data from the multistep oxidative phosphorylation reaction process (Van der Meer et al., 1980), for a system conservative with respect to substrate plus product (i.e., $[S] + [P] = C$, where C is a constant, in Eq. [11]), the reaction rate is linearly dependent on thermodynamic force over a considerable range of force between the equilibrium and saturating substrate condition, i.e., within this range

$$v = LA - b \qquad [23]$$

where L is a proportionality constant (conductance coefficient) and b/L is a (positive) intercept on the A axis of the v vs. A straight line. For the theoretical model, L and b were a function solely of C and the V_{max} and K_m kinetic properties of the enzyme catalyzing the reaction. The applicability of the phenomenological flow-force (Eq. [23]) and related potential Onsager symmetry relationships to biochemical reactions has important conceptual implications with respect to interpretation and prediction of reaction rate control as a function of energy flux coupling between different reactions in biological systems (Van Dam et al., 1980; Caplan, 1976). Within this context, the operational rate of a specific biochemical N transformation catalyzed by an organism will reflect the energy budgeted by the organism to drive the reaction at a rate compatible with the overall energy strategy of the organism for competitive growth and survival.

B. Ecological Considerations

Recognizing that the environmental N cycle exists in a continual state of disequilibrium (powered ultimately by the sun), we may consider the organisms interacting with this cycle to be, in effect, a collection of catalysts organized in such a way as to direct the energy fluxes implicit in the disequilibrium state of the cycle, in accord with the ecological needs of the organisms for self-perpetuation. Evolutionary pressures have maximized energy conservation, especially for chemotrophic organisms that cannot directly tap the sun's energy but must rely on the energy trapped (largely in the form of high energy C-, S-, and N-bound electrons) in the organic and inorganic by-products, excretions, and residues released by photosynthetic organisms.

Biologically catalyzed reactions may be categorized into two broad groups, assimilatory (body-building) and dissimilatory (energy-conserving) reactions. The specific nature and relative balance between these two groups of reactions is under rigorous physiological control, with e⁻ donor conservation tending to be an overriding factor dictating chemotrophic decisions on alternate nutritional pathways. Assimilatory use of N involves the uptake of simple inorganic or organic N monomers, the conversion of these monomers through the amino state into amino acids, purines, pyrimidines and other N-containing intermediary metabolites, and ultimately the condensation of these polymer precursors to form the proteins, nucleic acids, and phospholipids constituting the primary N-containing components of biological cells:

$$N_2 \longrightarrow$$

$$NO_3^- \rightarrow NH_4^+ \rightarrow \text{organic N monomers} \rightarrow \text{organic N polymers.} \qquad [24]$$

Nitrogen substrate selection by chemotrophs to satisfy assimilatory N growth requirements is dominated by the principle of maximum growth with minimum effort and maximized use of e⁻ donor resources. Thus, preformed organic N monomers tend to be preferred over $NH_4^+ > NO_3^- > N_2$. Assimilation of NO_3^- requires diversion of 8e⁻ eq reducing power to change the oxidation state from +5 (NO_3^-) to −3 (NH_4^+); assimilation of N_2 requires only 3e⁻ eq reducing power/mol of NH_4^+ but for mechanistic reasons requires a substantial commitment of high-energy electrons for activation and protection of the nitrogenase catalytic system (as discussed in Example 9 and section V). Apart from the N_2 reduction reaction, most other assimilatory reactions leading to polymer precursor formation are nominally exergonic with respect to physiological e⁻ carriers, such as NAD(P)H–NAD(P)⁺ ($E_h^{o'} = -0.32$ V), universal to biological systems. However, since the objective of an assimilatory reaction is product formation rather than energy conservation, the thermodynamic force of an exergonic assimilatory reaction tends to be used solely to drive the reaction unidirectionally toward product formation.

In contrast to assimilatory reactions, dissimilatory reactions are aimed at trapping the chemical energy from exergonic reactions in a form (such as ATP) that can be used to drive biosynthesis, solute transport, and other endergonic reactions needed for cell growth and maintenance. Thus, dissimilatory pathways reflect a need for maximum energy conservation and compatibility with biological energy-trapping mechanisms (Fig. 2; Thauer et al., 1977; Jones, 1979). As energy substrates, N compounds may act as e⁻ donors, e⁻ acceptors, or both. Since soil environments tend to be e⁻ donor limited, oxidation of diverse reduced forms of N, H, C, S, Fe, and Mn tends to proceed concurrently in soil (unless restricted by thermodynamically unfavorable e⁻ acceptor conditions), whereas the oxidized forms of N and other elements are used sequentially as e⁻ acceptors, in line with their thermodynamic ranking for maximized energy exploitation of limited e⁻ donor resources (Fig. 2; McCarty, 1972; Fenchel & Blackburn, 1979).

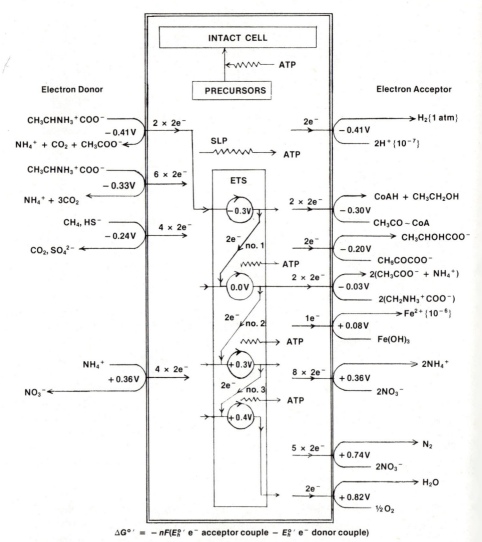

$$\Delta G^{\circ\prime} = -nF(E_h^{\circ\prime}\ e^-\ \text{acceptor couple} - E_h^{\circ\prime}\ e^-\ \text{donor couple})$$

Fig. 2—Relative energetics of dissimilatory N transformations.

Electrons are transferred from an e^- donor couple (left-hand side) through an e^- carrier (center) to an e^- acceptor couple (right-hand side) (cf. Eq. [26]). The H^+ and H_2O components of the half reactions are omitted for simplicity. The $E_h^{\circ\prime}$ is identified within each couple ($E_h^{\circ\prime}$ for $Fe(OH)_3$–Fe^{2+} = -0.27 V). The number of electrons (n) is identified as multiples of the $2e^-$ or $1e^-$ mechanistic steps. In contrast to assimilatory e^- transfer reactions, which characteristically involve unimpeded e^- transfer through the e^- carriers to accomplish an assimilatory goal, dissimilatory e^- transfer reactions are coupled to ATP generation [$E_h^{\circ\prime}$ implications of ATP generation (Example 8) are not included in Fig. 1]. In terms of the $\Delta G^{\circ\prime}$ of a multistep N transformation, ATP conservation is commonly effected with an efficiency of -20 to -25 kcal·mol ATP^{-1}. For the Stickland reaction specifically identified in Fig. 2, one ATP generated by SLP (Example 8) combined with an assumed one ATP generated by ETP via H^+ translocating segment no. 1, gives a thermodynamic efficiency of (from Eq. [70])

$$A_{\Delta G/ATP} = \frac{-4F[-0.03 - (-0.41)]}{2} = \frac{-35.1}{2} = -17.6\ \text{kcal·mol ATP}^{-1}.$$

III. CALCULATION AND INTERPRETATION OF GROUP TRANSFER ENERGETICS

A. General Equations

Most N transformation reactions involve the transfer of a group (X) or groups (nX) from a group donor couple (AX_n–A) to a group acceptor couple (B–BX_n). Such reactions are conveniently represented as a function of coupled group donor-group acceptor half reactions (with the half reactions being referenced, at least in principle, to a standard half reaction couple for energy characterization purposes):

$$AX_n = A + nX \qquad \text{dissociation half reaction}$$

$$nX + B = BX_n \qquad \text{formation half reaction}$$

$$AX_n + B = A + BX_n \qquad \text{complete reaction.} \qquad [25]$$

Biochemical group transfer reactions are commonly represented as (e.g., for Eq. [25])

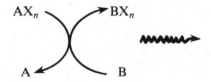

where the notation 〰➤ (away from the reaction) designates exergonic energetics. This convention may be readily expanded to quantify the thermodynamic force driving the reaction as a function of the relative group transfer energetics of the donor and acceptor couples:

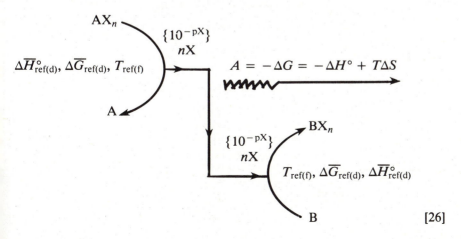

[26]

$$\Delta G = -n(\overline{\Delta G}_{\text{ref(d)}}, \text{group acceptor} - \overline{\Delta G}_{\text{ref(d)}}, \text{group donor})$$

$$= -n(\overline{\Delta G}^{\circ}_{\text{ref(d)}}, \text{group acceptor} - \overline{\Delta G}^{\circ}_{\text{ref(d)}}, \text{group donor})$$
$$+ 2.3RT \log Q \tag{27}$$

$$\Delta G = -n\alpha(T_{\text{ref(f)}}, \text{group acceptor} - T_{\text{ref(f)}}, \text{group donor})$$

$$= -n\alpha(T^{\circ}_{\text{ref(f)}}, \text{group acceptor} - T^{\circ}_{\text{ref(f)}}, \text{group donor})$$
$$+ 2.3RT \log Q \tag{28}$$

$$\Delta H^{\circ} = -n(\overline{\Delta H}^{\circ}_{\text{ref(d)}}, \text{group acceptor} - \overline{\Delta H}^{\circ}_{\text{ref(d)}}, \text{group donor}) \tag{29}$$

where n is the number of groups transferred; $\overline{\Delta G}_{\text{ref(d)}}$ and $\overline{\Delta G}^{\circ}_{\text{ref(d)}}$ are the free energy and standard free energy of dissociation of a group from a group transfer couple and acceptance by a reference couple, respectively; $T_{\text{ref(f)}}$ and $T^{\circ}_{\text{ref(f)}}$ are the potential and the standard potential of a group transfer couple to accept a group relative to the group acceptance potential of reference couple, respectively; α is a proportionality coefficient defining the units of group acceptance potential; $\overline{\Delta H}^{\circ}_{\text{ref(d)}}$ is the enthalpy analog of $\overline{\Delta G}^{\circ}_{\text{ref(d)}}$; $\{10^{-pX}\}$ is the hypothetical activity of the X group in "equilibrium" with each group transfer couple; and ΔS is the entropy change of the reaction.

For calculation of the steady-state activities of the reaction components under conditions of a given thermodynamic force, the group transfer analogs of Eq. [4] and [5] (recognizing that at equilibrium, $\Delta G = 0$) are obtained by rearrangement of Eq. [27] and [28]:

$$\log \frac{\{A\}\{BX_n\}}{\{AX_n\}\{B\}} = \frac{\Delta G - \Delta G^{\circ}}{2.3RT} = \frac{\Delta G}{2.3RT} + \log K$$

$$= [\Delta G + n(\overline{\Delta G}^{\circ}_{\text{ref(d)}}, \text{X acceptor} - \overline{\Delta G}^{\circ}_{\text{ref(d)}}, \text{X donor})]/2.3RT$$

$$= [\Delta G + n\alpha(T^{\circ}_{\text{ref(f)}}, \text{X acceptor} - T^{\circ}_{\text{ref(f)}}, \text{X donor})]/2.3RT \tag{30}$$

$$\log \frac{\{BX_n\}}{\{B\}}$$

$$= \Delta G + n(\overline{\Delta G}^{\circ}_{\text{ref(d)}}, \text{X acceptor} - \overline{\Delta G}_{\text{ref(d)}}, \text{X donor})/2.3RT$$

$$= [\Delta G + n\alpha(T^{\circ}_{\text{ref(f)}}, \text{X acceptor} - T_{\text{ref(f)}}, \text{X donor})]/2.3RT. \tag{31}$$

Group transfer energetics are relevant to a wide range of chemical and particularly biochemical reactions (Stumm & Morgan, 1970; Cornish-Bowden, 1979; Klotz, 1967). From an N transformation standpoint, the H^+ and e^- transfer systems are quantitatively most important.

B. Energetics of Proton Transfer

Proton transfer half reactions are commonly expressed in acid dissociation or ionization form (Eq. [32]); in practice H^+ transfer couples rapidly equilibrate with the water–hydronium couple in aqueous media (Eq. [33]):

$$AH_{n+1} = AH_n + H^+ \qquad\qquad \Delta G_{ioniz}, K_a \qquad [32]$$

$$H^+ + H_2O = H_3O^+$$

$$\overline{AH_{n+1} + H_2O = AH_n + H_3O^+} \qquad\qquad\qquad [33]$$

Because of the convention establishing $\Delta G_f^\circ H^+ = 0$, $\Delta G_f^\circ H_2O \equiv \Delta G_f^\circ H_3O^+$ and $\{H_2O\} \equiv 1$:

$$pH = -\log\{H_3O^+\}_{eq} \equiv -\log\{H^+\}_{eq}$$

$$= pK_a + \log \frac{\{AH_n\}_{eq}}{\{AH_{n+1}\}_{eq}} \qquad\qquad\qquad [34]$$

$$pK_a = -\log K_a = \frac{\Delta G_{ioniz}^\circ}{2.3RT} = \frac{\Delta G_f^\circ AH_n - \Delta G_f^\circ AH_{n+1}}{1.3642} \qquad [35]$$

$$\log \frac{\{AH_n\}_{eq}}{\{AH_{n+1}\}_{eq}} = pH - pK_a. \qquad\qquad\qquad [36]$$

In group transfer terms, pH is the potential of a H^+ transfer couple to accept protons relative to the H^+-accepting potential of the unit activity water–hydronium reference couple. Accordingly, the energetics of H^+ transfer is given by (from Eq. [28], for $\alpha = 2.3RT$)

$$\Delta G = -n\, 2.3RT\,(pH, H^+ \text{ acceptor} - pH, H^+ \text{ donor}) = -1.364\, n\, \Delta pH.$$

The tendency of H^+ transfer couples to equilibrate rapidly through the H_2O–H_3O^+ couple as H^+ carrier (with uncontrollable dissipation of the free energy of the reaction) has led to evolution of a H^+ impermeable membrane barrier with enzymatically controlled H^+ carriers as the basis of the proton motive force mechanism by which biological systems manipulate the free energy of H^+ transfer reactions (Caplan, 1976; Thauer et al., 1977). From a predictive standpoint, the rapidity of H^+ transfer reactions means that Eq. [36] is of indispensable practical value, since it defines (as identified in Tables A-1a, A-1b,[3] and documented extensively in Stumm and Morgan, 1970, and Segel, 1976) the relative distribution of the different ionic forms of N and other compounds prevailing in practice at a given pH in aqueous systems such as soil.

EXAMPLE 3

Distribution of species as a function of pK_a and pH. Calculate the relative distribution of hyponitrous acid, hyponitrite$^-$, and hyponitrite^{2-} at pH 8.

From Table A-1a, the dominant species at pH 8 (species closest to 8 in the pK_a column) is $HN_2O_2^-$. Establishing the activity of $HN_2O_2^-$ at unity for reference purposes, the relative activities of $H_2N_2O_2$ and $N_2O_2^{2-}$ are obtained from pK_a data (Table A-1a) by use of Eq. [36]:

$$\log\{H_2N_2O_2\}_{eq} = pK_{a1} - pH + \log\{HN_2O_2^-\}_{eq} = 7.05 - 8 + 0 = -0.95$$

$$\log\{N_2O_2^{2-}\}_{eq} = pH - pK_{a2} + \log\{HN_2O_2^-\}_{eq} = 8 - 11.54 + 0 = -3.54.$$

C. Energetics of Electron Transfer

The e^- transfer analog of Eq. [25] may be represented in simplified form as

$$A_{red} = A_{ox} + m_A H^+ + ne^- \qquad \Delta H_{e^-(d)}\ \Delta G_{e^-(d)}\ K_{e^-(d)}$$

$$\underline{ne^- + m_B H^+ + B_{ox} = B_{red} \qquad \Delta H_{e^-(f)}\ \Delta G_{e^-(f)}\ K_{e^-(f)}}$$

$$A_{red} + B_{ox} = A_{ox} + B_{red} + (m_A - m_B)\,H^+ \qquad \Delta H \quad \Delta G \quad K \quad [37]$$

where A_{red}–A_{ox} is the e^- donor couple; B_{ox}–B_{red} is the e^- acceptor couple; m_A and m_B are the stoichiometric H^+ coefficients; n is the number of electrons involved in the e^- transfer reaction; ΔH_{e^-}, ΔG_{e^-} and K_{e^-} define the hypothetical energetics of the half reactions; and ΔH, ΔG, and K are the energetics of the complete reaction.

A variety of approaches are available for mass balancing e^- transfer reactions (Stumm & Morgan, 1970; Erickson, 1979; Battley, 1979; Harris & Adams, 1979). For example, for a half reaction expressed as

$$\text{reactants} = \text{products} + mH^+ + ne^-,$$

$$n = \Sigma(H = -1, 0 = +2, \text{ionic charge}) \text{ products}$$

$$-\Sigma(H = -1, 0 = +2, \text{ionic charge}), \text{reactants}.$$

For inorganic N half reactions represented as going to NO_3^- (Table 1a), n gives the N-bound e^- composition of the inorganic N compound (Table 1a; Fig. 1). For organic N materials in which the N is considered to be in the -3 oxidation state and the half reaction is represented as going to CO_2 and NH_4^+ (Table 1b), n gives the C-bound e^- composition of the material (Table 1b; Fig. 3 and 4) and n/a (where a is the number of moles of C in the material) gives the degree of e^- richness of the C in the material (Harris & Adams, 1979); in the bioengineering literature, $n/a = \gamma$ (Table 1b) is called the reductance degree of the material (Erickson, 1979; 1980).

From a semantics standpoint, e^- composition may be expressed in a variety of ways, e.g., equivalents of available e^- (Erickson, 1979); the term e^- equivalent (e^- eq) (Harris & Adams, 1979; Harris, 1981) is used in this review.

Conventions used for characterizing the energetics of environmental e^- transfer reactions reference the e^- transfer couples, at least in principle, to the thermodynamic stability of water [oxygen-hydroxyl (water) or proton-hydrogen as reference couples].

1. OXYGEN-WATER AS ELECTRON TRANSFER REFERENCE COUPLE

Heat of combustion (ΔH_c°) in air has long been used for characterizing the energy content of organic substances and for mass-balancing energy flows in soil (Macfadyen, 1971) and other biological systems (Morowitz, 1968). Emphasis has been placed in recent years on the use of ΔH_c° (Erickson, 1979, 1980; Roels & Kossen, 1978) as a predictive index of microbially mediated transformations in industrial microbiology/bioengineering. The conventional combustion process converts organic N into H_2O, CO_2, and N_2. However, from a biochemical standpoint it is mechanistically more appropriate to consider the terminal N product of organic N oxidation as NH_3 (Erickson, 1980) or NH_4^+ (for pH < 9.2 systems). I have reserved the ΔH_c° term for the classical combustion process and use the terms ΔH_{ox}° and $\Delta G_{ox}^{\circ\prime}$ to characterize the energetics of complete and partial oxidation of N substances with O_2-$2H_2O$ as the reference e^- acceptor in aqueous media (Tables 1a and 1b). The reference reaction is

$$A_{red} + fH_2O = A_{ox} + m_A H^+ + ne^-$$

$$ne^- + nH^+ + \frac{n}{4}O_2 = \frac{n}{2}H_2O$$

$$\overline{A_{red} + \frac{n}{4}O_2 + (f - \frac{n}{2})H_2O = A_{ox} + (m_A - n)H^+} \qquad [38]$$

Recognizing that ΔH_f° and ΔG_f° of $O_2(g)$ and $H^+(aq) = 0$, and $\{H_2O\} \equiv 1$ by convention, from Eq. [38], [6], and [1]

$$\Delta H_c^\circ = [\Delta H_f^\circ A_{ox}(s, l, \text{ or } g)] - [\Delta H_f^\circ A_{red}(s, l, \text{ or } g)$$

$$+ (f - \frac{n}{2})\Delta H_f^\circ H_2O(l)] \qquad [39]$$

$$\overline{\Delta H_{ox}^\circ} = \frac{\Delta H_{ox}^\circ}{n} = \frac{[\Delta H_f^\circ A_{ox}] - [\Delta H_f^\circ A_{red} + (f - \frac{n}{2})\Delta H_f^\circ H_2O(l)]}{n} \qquad [40]$$

Table 1a—Mass balance composition and energy potential of inorganic N and other selected inorganic substances at 25°C.

Substance	Electron transfer half reaction†	ΔH°_{ox}	$\overline{\Delta G}^{\circ\prime}_{ox}$	$\Delta G^{\circ\prime}_{ox}$	pe°_h	$E^{\circ\prime}_h$
		— kcal·e⁻·eq⁻¹ —		kcal·mol⁻¹		V
Ammonium⁺	$NH_4^+ + H_2O = NH_2OH + 3H^+ + 2e^-$	+4.175	+1.050	+2.10	+25.048	+0.8605
	$NH_4^+ + H_2O = NOH(g) + 5H^+ + 4e^-$	−3.116	−5.062	−20.25	+18.817	+0.5955
	$NH_4^+ + H_2O = \tfrac{1}{2}H_2N_2O_2 + 5H^+ + 4e^-$	−11.041	−10.980	−43.92	+14.479	+0.3389
	$NH_4^+ + \tfrac{1}{2}H_2O = \tfrac{1}{2}H_2N_2O_2 + 5H^+ + 4e^-$	−15.206	−15.786	−63.14	+10.956	+0.1305
	$NH_4^+ + H_2O = \tfrac{1}{2}N_2O(g) + 5H^+ + 4e^-$	−9.810	−10.981	−54.90	+17.660	+0.3388
	$NH_4^+ + H_2O = NO(g) + 6H^+ + 5e^-$	−9.386	−10.968	−65.81	+15.071	+0.3394
	$NH_4^+ + 2H_2O = NO_2^- + 8H^+ + 6e^-$	−10.753	−10.430	−83.44	+14.882	+0.3627
	$NH_4^+ + 3H_2O = NO_3^- + 10H^+ + 8e^-$	−3.352	−4.260	−8.52	+21.155	+0.6303
	$2NH_4^+ = N_2H_5^+ + 3H^+ + 2e^-$	−9.232	−11.125	−44.50	+16.123	+0.3326
	$2NH_4^+ = N_2H_2(g) + 6H^+ + 4e^-$	−23.541	−25.198	−75.60	+4.640	−0.2777
	$2NH_4^+ = N_2(g) + 8H^+ + 6e^-$	−17.950	−15.595	−124.76	+12.846	+0.1388
	$3NH_4^+ = N_3^- + 12H^+ + 8e^-$					
Azide⁻	$N_3^- = 1\tfrac{1}{2}N_2(g) + e^-$	−100.118	−102.025	−102.03	−61.010	−3.6091
	$N_3^- + 9H_2O = 3NO_3^- + 18H^+ + 16e^-$	−9.145	−7.849	−125.56	+15.900	+0.4747
Diimide	$N_2H_2(g) = N_2(g) + 2H^+ + 2e^-$	−52.158	−53.345	−106.69	−18.326	−1.4982
	$N_2H_2(g) + 6H_2O = 2NO_3^- + 14H^+ + 12e^-$	−11.260	−10.198	−122.38	+14.469	+0.3728
Dinitrogen	$N_2(g) + H_2O = N_2O(g) + 2H^+ + 2e^-$	+9.800	+12.450	+24.90	+29.904	+1.3549
	$N_2(g) + 2H_2O = H_2N_2O_2 + 2H^+ + 2e^-$	+26.458	+31.675	+63.35	+43.996	+2.1885
	$N_2(g) + 2H_2O = 2NOH(g) + 2H^+ + 2e^-$	+58.158	+55.345	+110.69	+61.347	+3.2150
	$N_2(g) + 2H_2O = 2NO(g) + 4H^+ + 4e^-$	+10.785	+10.345	+41.38	+28.361	+1.2636
	$N_2(g) + 4H_2O = 2NO_2^- + 8H^+ + 6e^-$	+2.669	+3.262	+19.57	+25.502	+0.9564
	$N_2(g) + 4H_2O = 2NO_2(g) + 8H^+ + 8e^-$	+1.982	+3.065	+24.52	+23.024	+0.9479
	$N_2(g) + 4H_2O = N_2O_4(g) + 8H^+ + 8e^-$	+0.274	+2.923	+23.38	+22.920	+0.9417
	$N_2(g) + 6H_2O = 2NO_3^- + 12H^+ + 10e^-$	−3.081	−1.569	−15.69	+21.028	+0.7470

(continued on next page)

Table 1a—Continued.

Substance	Electron transfer half reaction†	$\overline{\Delta H}^\circ_{ox}$	$\overline{\Delta G}^{\circ\prime}_{ox}$	$\Delta G^{\circ\prime}_{ox}$	pe°_h	$E^{\circ\prime}_h$
		kcal·e⁻·eq⁻¹		kcal·mol⁻¹		V
Dinitrogen tetroxide	$N_2O_4(g) + 2H_2O = 2NO_3^- + 4H^+ + 2e^-$	-16.500	-19.535	-39.07	+13.458	-0.0321
Hydrazine⁺	$N_2H_5^+ = N_2(g) + 5H^+ + 4e^-$	-33.635	-35.690	-142.67	-3.618	-0.7316
	$N_2H_5^+ + 6H_2O = 2NO_3^- + 17H^+ + 14e^-$	-11.810	-11.311	-158.36	+13.986	+0.3245
Hydrogen	$H_2(g) = 2H^+ + 2e^-$	-34.158	-28.344	-56.69	0.000	-0.4141
Hydroxylamine	$NH_2OH = ½N_2(g) + H_2O + e^-$	-78.973	-77.695	-77.70	-36.175	-2.5541
	$NH_2OH + 2H_2O = NO_3^- + 7H^+ + 6e^-$	-15.729	-14.257	-85.54	+11.494	+0.1968
Hyponitrous acid	$H_2N_2O_2 + 4H_2O = 2NO_3^- + 10H^+ + 8e^-$	-10.465	-9.880	-79.04	+15.286	+0.3866
Nitric oxide	$NO(g) + 2H_2O = NO_3^- + 4H^+ + 3e^-$	-12.324	-9.512	-28.53	+16.139	+0.4026
Nitrite⁻	$NO_2^- + H_2O = NO_3^- + 2H^+ + 2e^-$	-11.705	-8.815	-17.63	+14.316	+0.4328
Nitrogen dioxide	$NO_2(g) + H_2O = NO_3^- + 2H^+ + e^-$	-23.332	-20.105	-20.10	+13.041	-0.0568
Nitrous oxide	$N_2O(g) + 5H_2O = 2NO_3^- + 10H^+ + 8e^-$	-6.301	-5.074	-40.59	+18.809	+0.5950
Nitroxyl	$NOH(g) + 2H_2O = NO_3^- + 5H^+ + 4e^-$	-18.390	-15.798	-63.19	+10.948	+0.1300
Water	$H_2O \rightarrow ½O_2(g) + 2H^+ + 2e^-$	0.00	0.00	0.00	+20.778	+0.8150

† The physical state of the half-reaction components is aqueous (aq), except for water (l), unless specified otherwise (Table A-1a).

Table 1b—Mass balance composition and energy potential of organic N and other selected organic materials at 25 °C.

Substance	State[†]	Composition[‡] Electron transfer half reaction[§]
Acetaldehyde	aq	$C_2H_4O + 3H_2O \rightarrow 2CO_2 + 10H^+ + 10e^-$
Acetate$^-$	aq	$C_2H_3O_2^- + 2H_2O \rightarrow 2CO_2 + 7H^+ + 8e^-$
Adenine	aq(c)	$C_5H_5N_5 + 10H_2O \rightarrow 5CO_2 + 5NH_4^+ + 5H^+ + 10e^-$
L-Alanine	aq	$C_3H_7O_2N + 4H_2O \rightarrow 3CO_2 + NH_4^+ + 11H^+ + 12e^-$
Allantoin	c	$C_4H_6O_3N_4 + 5H_2O \rightarrow 4CO_2 + 4NH_4^+ + 4e^-$
L-Arginine	c	$C_6H_{14}O_2N_4 + 10H_2O \rightarrow 6CO_2 + 4NH_4^+ + 18H^+ + 22e^-$
L-Asparagine	aq	$C_4H_8O_3N_2 + 5H_2O \rightarrow 4CO_2 + 2NH_4^+ + 10H^+ + 12e^-$
L-Aspartate$^-$	aq	$C_4H_6O_4N^- + 4H_2O \rightarrow 4CO_2 + NH_4^+ + 10H^+ + 12e^-$
Creatine	aq(c)	$C_4H_9O_2N_3 + 6H_2O \rightarrow 4CO_2 + 3NH_4^+ + 9H^+ + 12e^-$
Cyanogen	g	$C_2N_2 + 4H_2O \rightarrow 2CO_2 + 2NH_4^+ + 2e^-$
L-Cysteine	aq(c)	$C_3H_7O_2NS + 4H_2O \rightarrow 3CO_2 + NH_4^+ + HS^- + 10H^+ + 10e^-$
		$C_3H_7O_2NS + 8H_2O \rightarrow 3CO_2 + NH_4^+ + SO_4^{2-} + 19H^+ + 18e^-$
L-Cystine	aq(c)	$C_6H_{12}O_4N_2S_2 + 8H_2O \rightarrow 6CO_2 + 2NH_4^+ + 2S + 2OH^- + 22e^-$
		$C_6H_{12}O_4N_2S_2 + 16H_2O \rightarrow 6CO_2 + 2NH_4^+ + 2SO_4^{2-} + 36H^+ + 34e^-$
Cytosine	aq	$C_4H_5ON_3 + 7H_2O \rightarrow 4CO_2 + 3NH_4^+ + 7H^+ + 10e^-$
Dimethylamine$^+$	aq	$C_2H_8N^+ + 4H_2O \rightarrow 2CO_2 + NH_4^+ + 12H^+ + 12e^-$
Formate$^-$	aq	$CHO_2^- \rightarrow CO_2 + H^+ + 2e^-$
α-D-Glucose	aq	$C_6H_{12}O_6 + 6H_2O \rightarrow 6CO_2 + 24H^+ + 24e^-$
L-Glutamate$^-$	aq	$C_5H_8O_4N^- + 6H_2O \rightarrow 5CO_2 + NH_4^+ + 16H^+ + 18e^-$
L-Glutamine	aq(c)	$C_5H_{10}O_3N_2 + 7H_2O \rightarrow 5CO_2 + 2NH_4^+ + 16H^+ + 18e^-$
Glycine	aq	$C_2H_5O_2N + 2H_2O \rightarrow 2CO_2 + NH_4^+ + 5H^+ + 6e^-$
Guanine	aq(c)	$C_5H_5ON_5 + 9H_2O \rightarrow 5CO_2 + 5NH_4^+ + 3H^+ + 8e^-$
Histidine	aq	$C_6H_9O_2N_3 + 10H_2O \rightarrow 6CO_2 + 3NH_4^+ + 17H^+ + 20e^-$
Hydrogen cyanide	aq	$HCN + 2H_2O \rightarrow CO_2 + NH_4^+ + H^+ + 2e^-$
Hypoxanthine	aq(c)	$C_5H_4ON_4 + 9H_2O \rightarrow 5CO_2 + 4NH_4^+ + 6H^+ + 10e^-$
L-Isoleucine	aq	$C_6H_{13}O_2N + 10H_2O \rightarrow 6CO_2 + NH_4^+ + 29H^+ + 30e^-$
α-Ketoglutarate^{2-}	aq(c)	$C_5H_4O_5^{2-} + 5H_2O \rightarrow 5CO_2 + 14H^+ + 16e^-$
L-Leucine	aq	$C_6H_{13}O_2N + 10H_2O \rightarrow 6CO_2 + NH_4^+ + 29H^+ + 30e^-$
L-Lysine	c	$C_6H_{14}O_2N + 10H_2O \rightarrow 6CO_2 + 2NH_4^+ + 26H^+ + 28e^-$
L-Methionine	aq	$C_5H_{11}O_2NS + 8H_2O \rightarrow 5CO_2 + NH_4^+ + HS^- + 22H^+ + 22e^-$
		$C_5H_{11}O_2NS + 12H_2O \rightarrow 5CO_2 + NH_4^+ + SO_4^{2-} + 31H^+ + 30e^-$
Methylamine$^+$	aq	$CH_6N^+ + 2H_2O \rightarrow CO_2 + NH_4^+ + 6H^+ + 6e^-$
L-Phenylalanine	aq	$C_9H_{11}O_2N + 16H_2O \rightarrow 9CO_2 + NH_4^+ + 39H^+ + 40e^-$
L-Proline	aq	$C_5H_9O_2N + 8H_2O \rightarrow 5CO_2 + NH_4^+ + 21H^+ + 22e^-$
Pyridine	l	$C_5H_5N + 10H_2O \rightarrow 5CO_2 + NH_4^+ + 21H^+ + 22e^-$
Pyruvate$^-$	aq	$C_3H_3O_3^- + 3H_2O \rightarrow 3CO_2 + 9H^+ + 10e^-$
L-Serine	aq	$C_3H_7O_3N + 3H_2O \rightarrow 3CO_2 + NH_4^+ + 9H^+ + 10e^-$
Thiocyanate$^-$	aq	$CNS^- + 6H_2O \rightarrow CO_2 + NH_4^+ + SO_4^{2-} + 8H^+ + 8e^-$
L-Threonine	aq(c)	$C_4H_9O_3N + 5H_2O \rightarrow 4CO_2 + NH_4^+ + 15H^+ + 16e^-$
Thymine	c	$C_5H_6O_2N_2 + 8H_2O \rightarrow 5CO_2 + 2NH_4^+ + 14H^+ + 16e^-$
Trimethylamine$^+$	aq	$C_3H_{10}N^+ + 6H_2O \rightarrow 3CO_2 + NH_4^+ + 18H^+ + 18e^-$
L-Tryptophan	aq	$C_{11}H_{12}O_2N_2 + 20H_2O \rightarrow 11CO_2 + 2NH_4^+ + 44H^+ + 46e^-$
L-Tyrosine	aq	$C_9H_{11}O_3N + 15H_2O \rightarrow 9CO_2 + NH_4^+ + 37H^+ + 38e^-$
Uracil	c	$C_4H_4O_2N_2 + 6H_2O \rightarrow 4CO_2 + 2NH_4^+ + 8H^+ + 10e^-$
Urate$^-$	aq	$C_5H_3O_3N_4^- + 7H_2O \rightarrow 5CO_2 + 4NH_4^+ + H^+ + 6e^-$
L-Valine	aq	$C_5H_{11}O_2N + 8H_2O \rightarrow 5CO_2 + NH_4^+ + 23H^+ + 24e^-$
Xanthine	aq(c)	$C_5H_4O_2N_4 + 8H_2O \rightarrow 5CO_2 + 4NH_4^+ + 4H^+ + 8e^-$
Bacterial biomass	s	$C_4H_7O_{1.5}N(+10\%) + 6.5H_2O \rightarrow 4CO_2 + NH_4^+ + 16H^+ + 17e^-$
		$C_4H_7O_{1.5}N(+10\%) + 6.5H_2O \rightarrow 4CO_2 + \frac{1}{2}N_2 + 20H^+ + 20e^-$
		$C_4H_7O_{1.5}N(+10\%) + 9.5H_2O \rightarrow 4CO_2 + NO_3^- + 26H^+ + 25e^-$

† Letters in parentheses identify the physical state of the substance used for ΔH_{ox}° calculation if different from that used for calculation of the free energy parameters (Tables A-1a, b). Additional source of ΔH_f° data were Wilhoit (1969), Wagman et al. (1968), and Morowitz (1968) and of ΔG_f° data were Thauer et al. (1977) and Burton (1958).

Table 1b—Continued.

Composition‡				Energy properties				
N	C	γ	Equivalent weight	$\overline{\Delta H^\circ_{ox}}$	$\overline{\Delta G^{\circ\prime}_{ox}}$	$\Delta G^{\circ\prime}_{ox}$	pe°_h	$E^{\circ\prime}_h$
%		e⁻ eq· mol C⁻¹	g· e⁻ eq⁻¹	kcal·e⁻ eq⁻¹		kcal·mol⁻¹		V
00.00	54.53	5.00	4.405	−27.398	−26.844	−268.44	+1.100	−0.3491
00.00	40.68	4.00	7.380	−26.084	−25.507	−204.05	+1.205	−0.2911
51.83	44.44	2.00	13.513	−31.063	−30.936	−309.36	−5.399	−0.5265
15.72	40.44	4.00	7.425	−26.494	−26.399	−316.79	+0.843	−0.3298
35.44	30.38	1.00	39.530	−31.765	−34.495	−137.98	−11.509	−0.6808
32.16	41.37	3.67	7.918	−27.774	−27.391	−602.59	−0.573	−0.3727
21.20	36.36	3.00	11.011	−27.110	−27.225	−326.70	−0.346	−0.3656
10.61	36.37	3.00	11.008	−26.608	−26.935	−323.22	−0.133	−0.3530
32.04	36.63	3.00	10.928	−28.630	−28.515	−342.18	−1.875	−0.4215
53.84	46.16	1.00	26.018	−60.349	−54.198	−108.40	−25.951	−1.5352
11.56	29.74	3.33	12.116	−25.982	−27.519	−275.19	+0.605	−0.3783
11.56	29.74		6.731	−26.279	−25.862	−465.52	+2.209	−0.3065
11.66	29.99	3.67	10.923	−26.766	−27.050	−595.10	+0.313	−0.3580
11.66	29.99		7.068	−26.089	−25.759	−875.82	+2.307	−0.3020
37.82	43.24	2.50	11.110					
31.07	53.27	6.00	3.758	−27.315	−26.671	−320.06	+1.227	−0.3415
00.00	26.68	2.00	22.510	−30.328	−28.749	−57.50	−3.796	−0.4316
00.00	40.00	4.00	7.507	−28.006	−28.603	−686.46	−0.189	−0.4253
9.59	41.10	3.60	8.118	−26.257	−26.338	−474.08	+0.693	−0.3271
19.17	41.09	3.60	8.119	−26.283	−26.497	−476.95	+0.577	−0.3340
18.66	32.00	3.00	12.512	−27.419	−27.483	−164.90	−0.535	−0.3767
46.34	39.74	1.60	18.891	−30.472	−31.539	−252.31	−6.716	−0.5526
27.08	46.45	3.33	7.758					
51.83	44.44	2.00	13.513	−41.593	−37.804	−75.608	−10.434	−0.8243
41.17	44.13	2.00	13.611	−29.770	−30.350	−303.50	−4.269	−0.5010
10.68	54.94	5.00	4.373	−26.204	−25.875	−776.24	−1.578	−0.3070
00.00	41.69	3.20	9.005	−26.869	−26.977	−431.64	+0.127	−0.3548
10.68	54.94	5.00	4.373	−26.171	−25.843	−775.29	−1.601	−0.3056
19.16	49.30	4.67	5.221	−26.395				
9.39	40.25	4.40	6.782	−23.901	−24.424	−537.32	+2.874	−0.2441
9.39	40.25		4.974	−24.868	−24.255	−727.65	+4.408	−0.2368
43.69	37.45	6.00	5.345	−27.371	−26.731	−160.39	+1.183	−0.3441
8.48	65.44	4.44	4.130	−26.069	−25.875	−1035.02	+1.635	−0.3070
12.17	52.17	4.40	5.233	−26.439	−26.205	−576.50	+1.251	−0.3213
17.71	75.93	4.40	3.595	−27.014	−26.352	−579.75	+1.142	−0.3277
00.00	41.39	3.33	8.705	−27.628	−27.316	−273.16	+0.054	−0.3695
13.33	34.28	3.33	10.510	−27.983	−28.349	−283.49	−0.959	−0.4143
24.12	20.68	8.00	7.260	−28.118	−24.996	−199.97	+2.455	−0.2689
11.76	40.33	4.00	7.445	−26.256	−27.096	−433.54	+0.478	−0.3600
22.21	47.62	3.20	7.883	−26.378				
23.30	59.93	6.00	3.340	−27.332	−26.706	−480.70	+1.202	−0.3430
13.72	64.69	4.18	4.440	−26.204	−26.063	−1198.89	+1.369	−0.3152
7.73	59.66	4.22	4.768	−26.239	−26.208	−995.92	+1.382	−0.3215
25.00	42.85	2.50	11.210	−26.139				
33.53	35.94	1.20	27.852	−29.391	−32.378	−194.27	−8.790	−0.5890
11.96	51.27	4.80	4.881	−26.190	−25.925	−622.19	+1.483	−0.3092
36.84	39.48	1.60	19.014	−29.191	−31.105	−248.84	−5.523	−0.5338
13.54	46.44	4.25	6.086	−27.610	−27.317	−464.40	+0.341	−0.3696
13.54	46.44		5.173	−27.000	−27.000	−540.00	+0.986	−0.3558
13.54	46.44		4.138	−22.216	−21.913	−547.84	+4.994	−0.1352

‡ The C-bound e⁻ composition is in general identified directly in the half reaction.
§ The physical state of the half-reaction components are $CO_2(g)$, $H_2O(l)$, $N_2(g)$; all other components are aqueous(aq) unless specified otherwise.

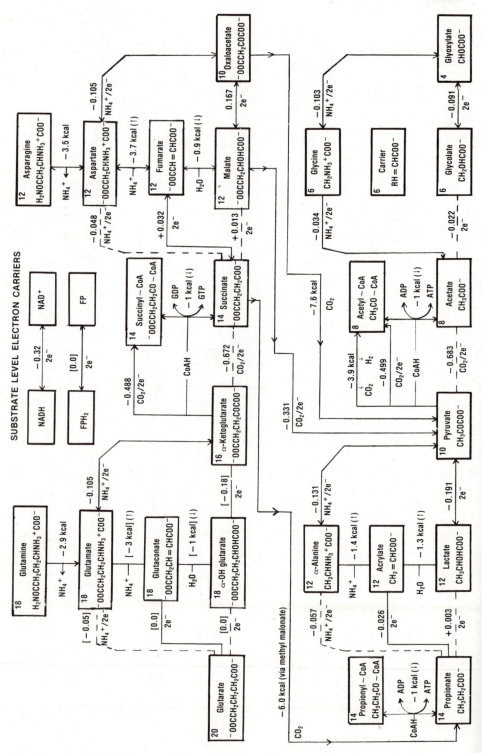

Fig. 3—Energetics of organic N metabolism, with emphasis on glutamate⁻, aspartate⁻, alanine, and glycine (opposite page).

The conventions are analogous to those used in Fig. 1. The number of C-bound e⁻ (Table 1b) associated with each compound is identified in the top left-hand corner of the boxes enclosing the compounds. For reversible non-e⁻ transfer reactions, the direction of the $\Delta G^{\circ\prime}$ is identified with a bracketed arrow. In addition to the full and half arrow convention used in Fig. 1, dotted lines are used for reactions that, although energetically very favorable, do not occur in practice for mechanistic reasons.

$$\overline{\Delta G^{\circ}_{ox}}' = \frac{\Delta G^{0\prime}_{ox}}{n}$$

$$\frac{[\Delta G^{\circ}_f A_{ox} + (m_A - n)\Delta G_f H^+] - [\Delta G^{\circ}_f A_{red} + (f - \frac{n}{2})\Delta G^{\circ}_f H_2O(l)]}{n} \qquad [41]$$

where the physical state of A_{ox} and A_{red} may be s, l, g, or aq in Eq. [40] and [41].

EXAMPLE 4

Derivation of ΔH°_c, $\overline{\Delta H^{\circ}_{ox}}$, and $\overline{\Delta G^{\circ}_{ox}}'$ energy properties. Calculate (i) the ΔH°_c of alanine and (ii) the $\overline{\Delta H^{\circ}_{ox}}$ and $\overline{\Delta G^{\circ}_{ox}}'$ of alanine oxidation to NH_4^+ and CO_2.

i) The combustion half reaction for alanine is

$$C_3H_7O_2N(c) + 4H_2O(l) = 3CO_2(g) + \tfrac{1}{2}N_2(g) + 15H^+ + 15\,e^-. \qquad [42]^4$$

From Eq. [42] and [39] [recognizing that $\Delta H^{\circ}_f N_2(g) = 0$] and Tables A-1a and A-1b:

$$\Delta H^{\circ}_c = [3\Delta H^{\circ}_f CO_2(g)] - [\Delta H^{\circ}_f alanine(c) + (4 - \frac{15}{8})\Delta H^{\circ}_f H_2O]$$

$$= -386.8 \text{ kcal} \cdot \text{mol}^{-1}.$$

ii) The half reaction for alanine oxidation to CO_2 and NH_4^+ in aqueous media at pH 7 is (from Table 1b)

$$C_3H_7O_2N^{\pm}_{(aq)} + 4H_2O(l) = 3CO_2(g) + NH_4^+(aq) + 11H^+ + 12e^-. \qquad [43]$$

---

[4] In practice the heats of formation of organic substances are commonly calculated from experimentally derived combustion data by rearrangement of Eq. [42] (Wilhoit, 1969; Hutchens, 1976b).

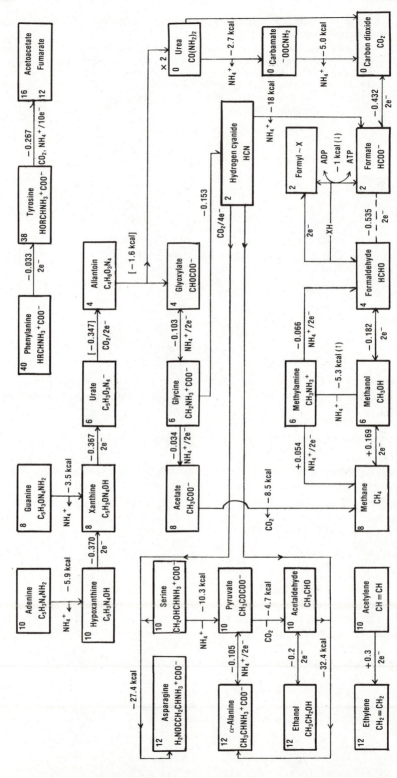

Fig. 4. Energetics of organic N metabolism, with emphasis on pyrimidines, hydrogen cyanide, and methylamine⁺. The conventions are the same as those for Fig. 3.

From Eq. [43] and [40] (recognizing that $\Delta H_f^\circ H^+ = 0$) and Tables A-1a and A-1b:

$$\Delta H_{ox}^\circ = [3\Delta H_f^\circ CO_2(g) + \Delta H_f^\circ NH_4^+]$$

$$- [\Delta H_f^\circ \text{alanine(aq)} + (4 - \frac{12}{2})\Delta H_f^\circ H_2O]$$

$$\overline{\Delta H_{ox}^\circ} = \Delta H_{ox}^\circ/12 = -317.93/12 = -26.494 \text{ kcal} \cdot e^- \text{ eq}^{-1}. \qquad [44]$$

From Eq. [43] and [41] and Tables A-1a and A-1b:

$$\Delta G_{ox}^{\circ\prime} = [3\Delta G_f^\circ CO_2(g) + \Delta G_f^\circ NH_4^+ + (11 - 12)\Delta G_f' H^+]$$

$$- [\Delta G_f^\circ \text{alanine(aq)} + (4 - \frac{12}{2})\Delta G_f^\circ H_2O]$$

$$\overline{\Delta G_{ox}^{\circ\prime}} = \Delta G_{ox}^{\circ\prime}/12 = -316.79/12 = -26.399 \text{ kcal} \cdot e^- \text{ eq}^{-1}. \qquad [45]$$

Comparison of Eq. [44] and [45] shows that $\Delta H_{ox}^\circ \cong \Delta G_{ox}^{\circ\prime}$ for alanine. Similarly, as illustrated in Table 1b, the $\overline{\Delta G_{ox}^{\circ\prime}}$ properties of other organic compounds are a function largely of the enthalpy (rather than the entropy) component of the oxidation reaction (Roels & van Suijdam, 1980). This phenomenon explains why enthalpy energetics may be used to successfully approximate the standard free energy drive of aerobic, organic oxidation reactions.

---

a. Equations Describing Reaction Energetics as a Function of $\overline{\Delta H_{ox}^\circ}$, $\overline{\Delta G_{ox}^{\circ\prime}}$, ΔH_{ox}°, and $\Delta G_{ox}^{\circ\prime}$ —In addition to their intrinsic value for characterizing the energetics of aerobic redox reactions, the $\overline{\Delta H_{ox}^\circ}$, $\overline{\Delta G_{ox}^{\circ\prime}}$, and $\Delta G_{ox}^{\circ\prime}$ energy parameters (Tables 1a and 1b) are applicable to calculation of the heat and free energy/activity relationships of anaerobic and non-e^- transfer reactions through the use of equations such as (analogous to Eq. [27] and [29])

$$\Delta G = \Delta G^{\circ\prime} + 2.3RT \log \frac{\{A_{ox}\}\{B_{red}\}}{\{A_{red}\}\{B_{ox}\}}$$

$$+ 2.3RT(m_A - m_B)(7 - pH) \qquad [46]$$

$$\Delta G^{\circ\prime} = -n(\overline{\Delta G_{ox}^{\circ\prime}} e^- \text{ acceptor} - \overline{\Delta G_{ox}^{\circ\prime}} e^- \text{ donor}) \qquad [47]$$

$$= -(\Sigma \Delta G_{ox}^{\circ\prime} \text{ products} - \Sigma \Delta G_{ox}^{\circ\prime} \text{ reactants}) \qquad [48]$$

$$\Delta H^\circ = -n(\overline{\Delta H_{ox}^\circ} e^- \text{ acceptor} - \overline{\Delta H_{ox}^\circ} e^- \text{ donor}) \qquad [49]$$

$$= -(\Sigma \Delta H_{ox}^\circ \text{ products} - \Sigma \Delta H_{ox}^\circ \text{ reactants}). \qquad [50]$$

EXAMPLE 5

Non-e⁻ transfer reaction energetics as a function of $\Delta G_{ox}^{o'}$: Xanthine catabolism.

Clostridia dissimilate the purine xanthine to glycine, formate, CO_2, and NH_4^+ (Stanier et al., 1976):

$$C_5H_4O_2N_4(8e^-) + 6H_2O + 2H^+ = CH_2(NH_3^+)COO^-(6e^-)$$
$$[-248.84] \qquad [0] \qquad [0] \qquad [-164.90]$$

$$+ HCOO^-(2e^-) + 2CO_2 + 3NH_4^+ \qquad\qquad [51]$$
$$[-57.50] \qquad\quad [0] \qquad [0]$$

where the numbers in parentheses represent the C-bound e⁻ composition, and the numbers in brackets identify the $\Delta G_{ox}^{o'}$ properties (as defined in Table 1b) of the reaction components. From Eq. [51] and [48]:

$$\Delta G^{o'} = -[(-164.90 - 57.50) - (-248.84)] = -26.4 \text{ kcal.}$$

2. PROTON–HYDROGEN AS ELECTRON TRANSFER REFERENCE COUPLE

The convention is to rank e⁻ transfer couples in terms of their potential ($T_{h(f)}$) to accept electrons relative to the e⁻ accepting potential of the unit activity proton–hydrogen couple. The $T_{h(f)}$ of an e⁻ transfer couple (e.g., B_{ox}–B_{red}, Eq. [37]) is given by [recognizing that $\Delta G_f^o H^+$ and $\Delta G_f^o H_2(g) = 0$ and that the activity of the proton–hydrogen reference couple is fixed at unity by convention]

$$T_{h(f)} = T_{h(f)}^o - \frac{2.3RT}{n\alpha} \log \frac{\{B_{red}\}}{\{B_{ox}\}\{H^+\}m_B}$$

$$T_{h(f)}^o = \frac{-(\Delta G_f^o B_{red} - \Delta G_f^o B_{ox})}{n\alpha} \equiv \frac{2.3RT \log K_{e^-(f)}}{n\alpha}. \qquad [52]$$

a. pe_h and pe as Energy Parameters—For $\alpha = 2.3RT$, Eq. [52] becomes

$$pe_h = pe_h^o + \frac{1}{n} \log\{H^+\}m_B - \frac{1}{n} \log \frac{\{B_{red}\}}{\{B_{ox}\}}$$

$$= pe_h^o - \frac{m_B}{n} pH - \frac{1}{n} \log \frac{\{B_{red}\}}{\{B_{ox}\}}$$

$$pe_h^o = \frac{-(\Delta G_f^o B_{red} - \Delta G_f^o B_{ox})}{2.3RT n} \equiv \frac{\log K_{e^-(f)}}{n} \qquad [53]$$

where pe_h is the potential (dimensionless) and pe_h° the standard potential of a couple to accept electrons relative to the e^- accepting potential of the unit activity proton–hydrogen couple. Because of the convention establishing $\Delta G_f^\circ H^+$, $\Delta G_f^\circ H_2(g)$, $\Delta G_f^\circ e^- = 0$, the e^- transfer potential term, pe_h, is mathematically equivalent to the equilibrium thermodynamics term, $pe = -\log\{e^-\}$. However it is important to recognize that whereas the equilibrium thermodynamics pH term may be considered to be operationally equivalent to $-\log\{H_3O^+\}$ in aqueous media (because of the rapid rate of equilibrium between H^+ transfer couples and the H_2O–H_3O^+ couple), N redox couples rarely achieve (and are commonly far removed from) equilibrium with the O_2–$2H_2O$ or $2H^+$–H_2 couples in aqueous media (Stumm & Morgan, 1970), so that for such redox couples the pe_h–pe term is more appropriately considered as a relative e^- transfer potential rather than as an equilibrium activity index.

To date, use of the pe_h–pe concept has been confined largely to construction of equilibrium-based stability diagrams of redox couples (Stumm & Morgan, 1970; McCarty, 1972; Lindsay, 1979) based on equations such as (by analogy with Eq. [30] and [31] for $\alpha = 2.3RT$, as applied to Eq. [37] and [53])

$$\log \frac{\{A_{ox}\}_{eq}\{B_{red}\}_{eq}}{\{A_{red}\}_{eq}\{B_{ox}\}_{eq}}$$

$$= n(pe_h^\circ, e^- \text{ acceptor} - pe_h^\circ, e^- \text{ donor}) + (m_A - m_B)\text{pH}$$

$$\equiv \log K_{e^-(f)}, e^- \text{ acceptor} + \log K_{e^-(d)}, e^- \text{ donor} + (m_A - m_B)\text{pH}$$

$$\log \frac{\{A_{ox}\}_{eq}}{\{A_{red}\}_{eq}} \equiv \log K_{e^-(d)}, e^- \text{ donor} + n(pe_h, e^- \text{ acceptor}) + m_A\text{pH}$$

$$\equiv \log K_{e^-(d)}, e^- \text{ donor} + n(pe_h, e^- \text{ acceptor} + \text{pH})$$

$$+ (m_A - n)\text{pH}. \qquad [54]$$

EXAMPLE 6

Calculate (i) the pe_h° and (ii) the pe_h at 0.2 atm O_2 of the oxygen–water half reaction, and (iii) calculate the activity of NO_3^- required to maintain equilibrium with atmospheric N_2 (0.78 atm) and O_2 (0.2 atm) as a function of pH.

i) From the oxygen–water half reaction (Table 1a) and Eq. [53]:

$$pe_h^\circ \text{ oxygen–water} = -\frac{\Delta G_f^\circ H_2O}{2.3RT \times 2} = \frac{+56.69}{1.3642 \times 2} = +20.778. \qquad [55]$$

ii) For $P_{O_2} = 0.2$ atm (from Eq. [55] and [53]):

$$pe_h \text{ oxygen-water} = +20.778 + \tfrac{1}{2}\log(0.2)^{1/2} - \frac{2}{2}pH$$

$$= +20.603 - pH$$

$$pe_h \text{ oxygen-water} + pH = +20.603. \tag{56}$$

iii) Combination of the N_2-$2NO_3^-$ half reaction as e^- donor with the $\tfrac{1}{2}O_2$-H_2O half reaction as e^- acceptor (Table 1a) gives

	pe_h°	$\log K_{e^-}$
$N_2 + 6H_2O = 2NO_3^- + 12H^+ + 10e^-$	21.028	-210.28
$5(2e^- + 2H^+ + \tfrac{1}{2}O_2) = 5H_2O$	20.778	207.78
$N_2 + H_2O + 2\tfrac{1}{2}O_2 = 2NO_3^- + 2H^+.$		[57]

From Eq. [57] and [54]:

$$\log\{NO_3^-\}_{eq} = [-210.28 + 10(+20.603) + (12 - 10)pH + \log(0.78)]/2$$

$$= -2.18 + pH. \tag{58}$$

According to Eq. [58], NO_3^- is stable with respect to N_2 in aerobic soil as long as pH >2.18, and for a typical soil pH of 7, $\log\{NO_3^-\}_{eq} = -2.18 + 7 = 4.82$. The fact that the level of NO_3^- required to maintain equilibrium with the soil environment is impossibly high ($10^{4.8}M$) demonstrates that the N cycle is far from thermodynamic equilibrium in aerobic soil. Equilibrium-based stability diagrams of N compounds in aquatic and soil environments using the pe and the pe + pH concepts are detailed in Stumm and Morgan (1970) and Lindsay (1979), respectively.

---

b. E_h as Energy Parameter—For $\alpha = F$, Eq. [52] becomes

$$E_h = E_h^\circ + \frac{2.3RT}{nF}\log\{H^+\}^{m_B} - \frac{2.3RT}{nF}\log\frac{\{B_{red}\}}{\{B_{ox}\}}$$

$$= E_h^\circ - \frac{0.0592\, m_B\, pH}{n} - \frac{0.0592}{n}\log\frac{\{B_{red}\}}{\{B_{ox}\}}$$

$$E_h^\circ = \frac{-(\Delta G_f^\circ B_{red} - \Delta G_f^\circ B_{ox})}{nF} \tag{59}$$

where E_h is the potential and E_h° the standard potential of a redox couple to accept electrons relative to the e^- accepting potential of the unit activity proton–hydrogen couple. The relative e^- accepting potential of the redox couple at pH 7, E'_h, is obtained by appropriate modification of Eq. [59]:

$$E'_h = E^{\circ'}_h - \frac{2.3RT}{nF} \log \frac{\{B_{red}\}}{\{B_{ox}\}} = E^{\circ'}_h - \frac{0.0592}{n} \log \frac{\{B_{red}\}}{\{B_{ox}\}}$$

$$E^{\circ'}_h = E^{\circ}_h - 0.4141\, m_B/n$$

$$= \frac{-[\Delta G^{\circ}_f B_{red} - (\Delta G^{\circ}_f B_{ox} + m_B\, \Delta G'_f H^+)]}{nF}$$

$$= \frac{-[\Delta G^{\circ}_f B_{red} - (\Delta G^{\circ}_f B_{ox} - 9.55 m_B)]}{23.061\, n} \tag{60}$$

The $E^{\circ'}_h$ of an e^- transfer couple may also be derived from the $E^{\circ'}_h$ data in Tables 1a, 1b, and Fig. 1 to 4 using the relationship

$$E^{\circ'}_h = \frac{n_1(E^{\circ'}_h)_1 + n_2(E^{\circ'}_h)_2 + \cdots}{n}$$

$$n = n_1 + n_2 + \cdots \tag{61}$$

where n_1, n_2, $\cdots$, and $(E^{\circ'}_h)_1$, $(E^{\circ'}_h)_2$, $\cdots$ are the number of e^- and the $E^{\circ'}_h$, respectively, of the appropriate e^- transfer couples; and the signs of n_1, n_2, $\cdots$ may be positive or negative.

EXAMPLE 7

Derivation of $E^{\circ'}_h$. Calculate the $E^{\circ'}_h$ of (i) the $2NO_2^--N_2$ couple from ΔG°_f data; (ii) the $N_2-N_2H_5^+$ couple from $E^{\circ'}_h$ $N_2-N_2H_2$ and $E^{\circ'}_h$ $N_2H_2-N_2H_5^+$ data; (iii) the alanine-$3CO_2$, NH_4^+ couple as e^- donor from ΔG°_f data; (iv) the NH_4^+, CO_2, acetate-alanine couple from $E^{\circ'}_h$ NH_4^+, pyruvate-alanine and $E^{\circ'}_h$ acetate, CO_2-pyruvate data.

i) The half reaction is (Table 1a)

$$6e^- + 8H^+ + 2NO_2^- = N_2(g) + 4H_2O. \tag{62}$$

From Eq. [62] and [60] and Table A-1:

$$E^{\circ'}_h = -[4\Delta G^{\circ}_f H_2O - (2\Delta G^{\circ}_f NO_2^- + 8\Delta G'_f H^+)]/6F = +0.9564\ \text{V}.$$

ii) From Fig. 1, Table 1a, and Eq. [61]:

$$E^{\circ'}_h = \frac{4E^{\circ'}_h N_2H_2\text{-}2NH_4^+ - 2E^{\circ'}_h N_2H_5^+\text{-}2NH_4^+}{4 - 2} = +0.0349\ \text{V}.$$

iii) The half reaction for the alanine–$3CO_2$, NH_4^+ couple (Table 1b) is written in reduction form for calculation of $E_h^{o\prime}$:

$$12e^- + 11H^+ + NH_4^+ + 3CO_2(g) = C_3H_7O_2N + 4H_2O. \qquad [63]$$

From Eq. [63] and [60] and Tables A-1a and A-1b:

$$E_h^{o\prime} =$$

$$\frac{-[(\Delta G_f^o alanine + 4\Delta G_f^o H_2O) - (3\Delta G_f^o CO_2(g) + \Delta G_f^o NH_4^+ + 11\Delta G_f^o H^+)]}{12F}$$

$$= -0.3298 \text{ V.}$$

iv) From Fig. 3 and Eq. [61]:

$$E_h^{o\prime} = \frac{2E_h^{o\prime} NH_4^+, \text{pyruvate–alanine} + 2E_h^{o\prime} \text{acetate}, CO_2\text{–pyruvate}}{2 + 2}$$

$$= -0.407 \text{ V.}$$

i. General Equations Describing Reaction Energetics as a Function of $E_h^{o\prime}$

Equations for calculating free energy/activity relationships of e^- transfer reactions (Eq. [37]) as a function of $E_h^{o\prime}$ data include (analogous to Eq. [28], [30] and [31] for $\alpha = F$, as applied to Eq. [37] and [60],

$$\Delta G = -nF(E_h' e^- \text{ acceptor} - E_h' e^- \text{ donor})$$
$$+ 2.3RT(m_A - m_B)(7 - pH) \qquad [64]$$

$$\Delta G' = -nF\left[\left(E_h^{o\prime} e^- \text{ acceptor} - \frac{2.3RT}{nF} \log \frac{\{B_{red}\}}{\{B_{ox}\}}\right)\right.$$

$$\left. - \left(E_h^{o\prime} e^- \text{ donor} + \frac{2.3RT}{nF} \log \frac{\{A_{ox}\}}{\{A_{red}\}}\right)\right] \qquad [65]$$

$$\Delta G^{o\prime} = -nF(E_h^{o\prime} e^- \text{ acceptor} - E_h^{o\prime} e^- \text{ donor}) \qquad [66]$$

$$\log \frac{\{A_{ox}\}\{B_{red}\}}{\{A_{red}\}\{B_{ox}\}} = \frac{\Delta G + nF(E_h^{o\prime} e^- \text{ acceptor} - E_h^{o\prime} e^- \text{ donor})}{2.3RT}$$

$$- (m_A - m_B)(7 - pH) \qquad [67]$$

$$\log \frac{\{B_{red}\}}{\{B_{ox}\}} = \frac{\Delta G + nF(E_h^{o\prime} e^- \text{ acceptor} - E_h' e^- \text{ donor})}{2.3RT}$$

$$- (m_A - m_B)(7 - pH). \qquad [68]$$

For example, the $\Delta G^{\circ\prime}$ of reactions involving the $2H^+$-H_2 couple as e^- acceptor or e^- donor (Thauer et al., 1977; Example 2) or the O_2-$2H_2O$ couple as e^- acceptor ($\Delta G_{ox}^{\circ\prime}$, Tables 1a and 1b) may be readily calculated from Eq. [66] recognizing that $E_h^{\circ\prime}$ $2H^+$-H_2 = -0.4141 and $E_h^{\circ\prime}$ O_2-$2H_2O$ = $+0.8150$ (Table 1a).

ii. Energy Coupling Between Exergonic Electron Transfer and Endergonic Dehydrolytic Group Transfer Reactions

Dissimilatory e^- transfer reactions involve the transfer of electrons from an e^- donor to a nicotinamide dinucleotide (NAD$^+$-NADH) or flavo-protein (FAD-FADH$_2$ or FMN-FMNH$_2$) e^- carrier, followed by regeneration of the oxidized component of the e^- carrier via transfer of electrons from the carrier to an e^- acceptor (Fig. 2). Electron transfer energy may be conserved in the form of high-energy phosphate or coenzyme A (CoA) groups (identified as R $\sim$ X, with the energy conserved in R $\sim$ X expressed in terms of the hydrolysis energetics, $\Delta G_{hydrol}^{\circ\prime}$) either during the transfer of electrons from the e^- donor to the NAD$^+$ e^- carrier (substrate level phosphorylation, SLP) and/or during e^- transfer from the carrier (via a membrane-bound e^- transport system, ETS) to the e^- acceptor (e^- transport phosphorylation, ETP) (Fig. 2; Thauer et al., 1977). In $E_h^{\circ\prime}$ terms, the $\Delta E^{\circ\prime}$ between the e^- donor or acceptor and the e^- carrier required to promote such energy conservation is given by

$$\Delta E^{\circ\prime} = \frac{-\Delta G_{hydrol}^{\circ\prime} R \sim X}{\beta n F} \qquad [69]$$

where $\Delta E^{\circ\prime}$ is defined as $E_h^{\circ\prime}$ nominal e^- acceptor $- E_h^{\circ\prime}$ nominal e^- donor, and β is the thermodynamic efficiency of energy transduction (100% efficiency = 1).

The $\Delta G_{hydrol}^{\circ\prime}$ of R $\sim$ X compounds involved in SLP ranges from about -8.5 (acyl $\sim$ CoA) to -10.7 (acetyl $\sim$ P) and -12.3 (P $\sim$ enol pyruvate) kcal$\cdot$mol^{-1} (Thauer et al., 1977). Accordingly, a minimum $\Delta E^{\circ\prime}$ of about 0.25 to 0.35 V is required between a $2e^-$ transfer donor and the NAD$^+$-NADH e^- carrier ($E_h^{\circ\prime}$ = -0.32 V) for SLP. The only redox couples with an $E_h^{\circ\prime}$ in the required -0.5- to -0.65-V range are the aldehyde-carboxy acid and the "carboxylated aldehyde" (keto carboxy acid)-carboxy acid couples (Fig. 3 and 4).

ETP involves the use of e^- transfer energy to create a H$^+$ gradient across the ETS membrane, the resultant proton motive force driving the dehydrolytic formation of ATP (more appropriately considered MgATP under physiological conditions) from ADP and Pi. The $\Delta G_{hydrol}^{\circ\prime}$ of ATP is -7.6 (Thauer et al., 1977) to -8.8 (Stiefel, 1977) kcal$\cdot$mol^{-1}. Thus, from Eq. [69], a minimum $\Delta E^{\circ\prime}$ of about 0.2 to 0.3 V is needed between the energy-transducing components of a $2e^-$ electron transport system to allow generation of ATP at a reasonable rate (Thauer et al., 1977). In $E_h^{\circ\prime}$ terms, the two basic H$^+$-translocating segments of bacterial respiratory systems span between about -0.3 and 0.0 V (NADH dehydrogenase complex, segment no. 1) and between about 0.0 and $+0.3$ V (quinone–cytochrome b complex, segment no. 2); an additional segment, spanning between about $+0.3$ V and $+0.8$ V (high potential cytochrome c–cytochrome oxidase complex, segment no. 3) is exhibited by mitochondrial but not all bacterial respiratory systems (Fig. 2; Jones, 1979; Wikstrom & Krab, 1980).

The efficiency of ATP generation by a redox reaction may be expressed in terms of the ΔG of the uncoupled reaction standardized for production of 1 mol of ATP, $A_{\Delta G/ATP}$ (Eq. [70]), or as the percentage of free energy recovery, η (Eq. [71]):

$$A_{\Delta G/ATP} = \frac{-nF(E_h e^- \text{ acceptor} - E_h e^- \text{ donor})}{n_{ATP}} \qquad [70]$$

$$\eta = \frac{\Delta G_{driven}}{\Delta G_{driving}} \times 100\% = \frac{\Delta G_{hydrol}ATP \times 100\%}{A_{\Delta G/ATP}} \qquad [71]$$

where n_{ATP} is the number of moles of ATP generated by the reaction.

EXAMPLE 8

Substrate level and e^- transport level phosphorylation: Oxidative deamination of α-amino acids.

The first step is oxidative removal of the amino group to form the corresponding keto acid, with NAD^+–NADH as the e^- acceptor (Fig. 3, 4). Because of the endergonic nature of this step (Eq. [72]), low relative activities of the products (the keto acid and NADH) must prevail to provide the needed positive thermodynamic force; the second step generates sufficient energy to accomplish this and also allows (marginally) substrate level energy conservation through acyl $\sim$ CoA to ADP $\sim$ P (Eq. [74] to [76]):

basic components of the reaction (R = CH₃)	$E_h^{\circ\prime} e^-$ donor, V	$\Delta G^{\circ\prime}$, kcal	
$RCHNH_3^+COO^- + NAD^+$			
$= RCOCOO^- + NH_4^+ + NADH$	-0.1310	$+8.72$	[72]
$(RCOCOO^- + NAD^+$			
$= RCOO^- + CO_2 + NADH)$	(-0.6834)	(-16.76)	[73]
$RCOCOO^- + CoAH + NAD^+$			
$= RCO \sim CoA + CO_2 + NADH$	-0.4991	-8.26	[74]

$$RCO \sim CoA + ADP^{3-} + Pi^-$$
$$= RCOO^- + ATP^{4-} + CoAH \qquad\qquad -0.90 \qquad [75]$$

$$RCHNH_3^+COO^- + ADP^{3-}$$
$$+ Pi^- + 2NAD^+$$
$$= RCOO^- + CO_2 + NH_4^+$$
$$+ ATP^{4-} + 2NADH \qquad -0.3248 \qquad -0.44 \qquad [76]$$

$$(RCHNH_3^+COO^- + 2NAD^+$$
$$= RCOO^- + CO_2 + NH_4^+$$
$$+ 2NADH) \qquad\qquad (-0.4072) \quad (-8.04). \quad [77]$$

Equations [73] (Fig. 3) and [77] (Fig. 2) represent Eq. [74] and [76], respectively, without energy conservation. Generation of ATP by SLP requires continual regeneration of NAD^+ via oxidation of NADH by a thermodynamically favorable e^- acceptor. Electron acceptors with $E_h^{o\prime}$ properties in the -0.1- to -0.3-V range such as the $2e^-$ aldehyde–alcohol and keto carboxy acid–hydroxy carboxy acid ($E_h^{o\prime} = -0.1$ to -0.2 V) couples (Fig. 3 and 4) and the $4e^-$ acyl $\sim$ CoA–alcohol ($E_h^{o\prime} \cong -0.3$ V) couple (Fig. 2), are commonly used by obligate and facultative anaerobes for NAD^+ regeneration. Electron acceptors with $E_h^{o\prime}$ properties in the 0.0-V range, such as unsaturated carboxy acid–saturated carboxy acid, α-amino acid–saturated carboxy acid, and amine or alcohol–hydrocarbon couples (Fig. 3 and 4) in addition to acting as e^- sinks for NAD^+ regeneration may also participate in ETP via translocating site no. 1 (Fig. 2). In principle, high $E_h^{o\prime}$ e^- acceptor couples such as N_2O–N_2 ($E_h^o = +1.4$) and $2NO$–N_2O ($E_h^{o\prime} = +1.2$ V) might be expected to generate more ATP via ETP than the oxygen–water ($E_h^o = +0.82$ V) couple; however, in practice such e^- acceptors appear to interact effectively with only one or at most two of the ETS H^+ translocating segments.

For aerobic oxidative deamination of alanine to acetate and CO_2 by a typical soil heterotroph with segments no. 1 and 2 operational for ETP: 1 mol of ATP would be generated by SLP and $2 \times 2 = 4$ mol ATP by ETP (from regeneration of 2NADH with O_2 as the terminal e^- acceptor) (Fig. 2). Accordingly, the thermodynamic efficiency of this dissimilatory reaction would be (from Eq. [70] and [71], respectively)

$$A_{\Delta G/ATP} = \frac{4F[+0.8150 - (-0.4072)]}{5} = -22.5 \text{ kcal} \cdot \text{mol ATP}^{-1}$$

$$\eta = -760/-22.5 = 34\% \quad (\text{for } \Delta G^{o\prime}_{\text{hydrol}}ATP = -7.6 \text{ kcal} \cdot \text{mol}^{-1}).$$

iii. Energy Coupling Between Exergonic Hydrolytic Group Transfer and Endergonic Electron Transfer Reactions

From an energy barrier circumvention standpoint, the exergonic transfer of high potential phosphate or CoA from a group donor to H_2O as group acceptor may be coupled to the e^- donor or the e^- acceptor component of an endergonic e^- transfer reaction to give a corresponding energy boost to the reaction. In e^- transfer potential terms, the $E_h^{\circ\prime}$ of an e^- donor becomes more negative (greater reducing potential), or the $E_h^{\circ\prime}$ of an e^- acceptor becomes more positive (greater oxidizing potential) by a $\Delta E^{\circ\prime}$ of (analogous to Eq. [69]):

$$\Delta E^{\circ\prime} = \frac{\beta n_x(-\Delta G_{hydrol}^{\circ\prime} \, R \sim X)}{nF} \qquad [78]$$

where n_x is the number of high energy groups (maximum of 2 in practice) and n the number of electrons transferred per mole of e^- donor or acceptor.

EXAMPLE 9

Use of hydrolysis energy to increase the reducing potential of an e^- donor: The nitrogenase system.

The Fe protein component (Feprot, N_2aseII or dinitrogenase reductase) of the nitrogenase complex shows an unenergized $E_h^{\circ\prime}$ of -0.29 V (Dalton, 1979; Hageman & Burris, 1980). From Eq. [78], the coupling of the hydrolysis energy of 2Mg ADP $\sim$ P ($\Delta G_{hydrol}^{\circ\prime} \cong -8.2$ kcal: Thauer et al., 1977; Stiefel, 1977) to this $1e^-$ transfer e^- donor would increase its reducing potential to
$E_h^{\circ\prime}$Fe*prot $= E_h^{\circ\prime}$Feprot $- (\beta \times 2 \times 8.2)/(1 \times 23.061)$

$$= -0.29 - 0.71\beta \qquad [79]$$

where $E_h^{\circ\prime}$Fe*prot is the $E_h^{\circ\prime}$ of the energized Fe protein.
From Eq. [79] and [60]:

$$E_h'\text{Fe*prot} = -0.29 - 0.71\beta$$

$$- \frac{2.3RT}{nF}\left(\log \frac{\{MgATP\}^2}{\{MgADP\}^2\{P_i\}^2} + \log \frac{\{Feprot_{(red)}\}}{\{Feprot_{(ox)}\}}\right). \qquad [80]$$

As identified in Eq. [80], a high reducing potential for the energized Fe protein is favored by (i) a high efficiency of energy transduction of phosphate to e^- transfer energy ($\beta = 1$ is 100% efficiency), (ii) a high phosphorylation potential (i.e., high MgATP relative to MgADP, P_i), and (iii) a high e^- donor supply to maintain a high relative activity of reduced to oxidized Fe protein. The importance of the $1e^-$ rather than $2e^-$ transfer characteristic of the Fe protein–2MgATP system with respect to the reducing potential of the energized Fe protein should also be recognized: For a $2e^-$ system coupled to 2MgATP, the $E_h^{\circ\prime}$ would be boosted by only -0.35 V. Alternatively, 4MgATP would have to be coupled to a $2e^-$ transfer system to boost the $E_h^{\circ\prime}$ by -0.7 V (Eq. [78]). Furthermore, n would be 2 rather than 1 in Eq. [80].

Electrons are transferred in $1e^-$ steps from the Fe protein to the MoFe protein concomitant with the hydrolysis of the MgATP, but there is a lag in substrate reduction during the period that the MoFe protein is being reduced to a state adequate for it to transfer electrons to substrates (Hageman & Burris, 1980). This is consistent with the need for a highly negative e^- donor to circumvent the endergonic first $2e^-$ step in the diimide–hydrazine pathway of N_2 reduction to $2NH_4^+$ (Fig. 1; Example 1).

Analogous to the approach used in Example 1, the steady-state activities of the N_2H_2 and $N_2H_5^+$ intermediates, and of the end product NH_4^+, are given by (from Eq. [19] to [22] and [68])

$$\log\{N_2H_2\} = \frac{\Delta G' + 2F(-1.4982 - E_h' N_2^* ase)}{2.3RT} + \log\{N_2\} \qquad [81]$$

$$\log\{N_2H_5^+\} = \frac{\Delta G' + 2F(+0.0349 - E_h' N_2^* ase)}{2.3RT} + \log\{N_2H_2\} \qquad [82]$$

$$\log\{NH_4^+\}^2 = \frac{\Delta G' + 2F(+0.6303 - E_h' N_2^* ase)}{2.3RT} + \log\{N_2H_5^+\} \qquad [83]$$

$$= \frac{\Delta G' + 6F(-0.2777 - E_h' N_2^* ase)}{2.3RT} + \log\{N_2\} \qquad [84]$$

where $E_h' N_2^* ase$ is the operational E_h' of the activated Fe protein–MoFe protein nitrogenase complex as dictated mainly by the E_h' of the activated Fe protein component ($E_h' Fe^* prot$, Eq. [80]).

If we assume a minimum 1-kcal drive requirement for each step and a maximized $E_h' N_2^* ase$ of -1.3 V (representing a high relative level of physiological reductant and ATP, Eq. [80]) with respect to N_2 reduction to N_2H_2, then for physiological activities of $10^{-5} M$ NH_4^+ and 0.78 atm N_2, the steady-state activity of N_2H_2 would be (from Eq. [81])

$$\log\{N_2H_2\}$$

$$= \frac{-1 + 2F[-1.4982 - (-1.3)]}{2.3RT} + \log(0.78) = -7.5. \qquad [85]$$

Such an activity would seem reasonable for a highly labile enzyme-bound intermediate acting as a reactant for a subsequent highly exergonic reaction, as is the case for N_2H_2. Thus, even if the operational $E_h' N_2^*$ase for the N_2H_2–$N_2H_5^+$ and/or $N_2H_5^+$–$2NH_4^+$ steps were substantially higher than -1.3 V, e.g., -0.5 V for each step (such that the gross $E_h' N_2^*$ase for the $6e^-$ transfer would be $-4.6/6 = -0.7667$ V), the resultant driving force/steady-state activity relationships would still be highly favorable for a forward reaction (from Eq. [82], [85], and [83]):

$$\log\{N_2H_5^+\}$$

$$= \frac{-22.6 + 2F[+0.0349 - (-0.5)]}{2.3RT} + \log(10^{-7.5}) = -6 \qquad [86]$$

$$\log\{NH_4^+\}^2$$

$$= \frac{-57.6 + 2F[+0.6303 - (-0.5)]}{2.3RT} + \log(10^{-6}) = -10 \qquad [87]$$

where the sum of the driving forces for the three steps (Eq. [85–87]) constitutes the total driving force for the overall $6e^-$ transfer reaction as calculated by rearrangement of Eq. [84]:

$$\Delta G' = 2.3RT[\log(10^{-5})^2 - \log(0.78)] - 6F[-0.2777 - (-0.7667)]$$

$$= 138.37(-0.7667) + 25.1 = -81 \text{ kcal.} \qquad [88]$$

Whatever the actual mechanism used by the FeMo protein to effect the $6e^-$ reduction of N_2 to $2NH_4^+$ via mandatorily enzyme-bound intermediates, it is apparent that at least the first step requires a very powerful e^- donor and that if the FeMo protein is not pumped up to a sufficiently electronegative level by the Fe protein, then electrons are readily diverted to effect reduction of diverse alternate e^- acceptor couples (of $E_h^{o'}$ properties $\gg -1.5$ V; e.g., N_2O–N_2, $E_h^{o'} = +1.4$ V; C_2H_2–C_2H_4, $E_h^{o'} = +0.3$ V) bound to sites on the FeMo protein different from the N_2-binding site (Hageman & Burris, 1980; Dalton, 1979). In practice, consistent with the extremely high thermodynamic force driving the N_2 reduction reaction (e.g., Eq. [88]), some e^- diversion to the most available alternate e^- acceptor to N_2 in an aqueous medium, the $2H^+$–H_2 couple ($E_h^{o'} = -0.41$ V), occurs even under conditions conducive to maximized N_2 reduction [with an apparent minimum of 1 mol H_2 ($2e^-$ eq) produced/mol of N_2 ($6e^-$ eq) reduced]. However, for many N_2-fixing systems, the electrons in the resultant H_2 gas are commonly released via hydrogenase catalysis and either are recycled back into the nitrogenase complex or (for aerobic systems) are channeled into the ETS for O_2 control with or without ETP (Hageman & Burris, 1980; Emerich & Evans, 1980; Dalton, 1979). The practical implications of such phenomena are discussed in section V.

IV. PATHWAY ENERGETICS OF NITROGEN
TRANSFORMATIONS

Soil N exists largely in the form of organic N polymers linked together by peptide bonds, glycoside bonds, or both. Although hydrolytic cleavage of these bonds is exergonic ($\Delta G^{\circ\prime} = -3$ to -5 kcal·bond^{-1}; Lehninger, 1965), structural accessibility problems frequently make the initial depolymerization process the rate-limiting step in the soil N cycle.

A. Assimilatory Pathways

All forms of inorganic N require reduction to NH_4^+ before immobilization into the organic N precursors of the cell polymers (Eq. [24]).

1. ASSIMILATION OF DINITROGEN

The energetics of N_2 reduction to $2NH_4^+$ via the N_2H_2–$N_2H_5^+$ pathway are evaluated in Examples 2 and 9. The alternate N_2 reduction pathway, via NH_2OH, is unrealistic on thermodynamic grounds ($E_h^{\circ\prime} = -2.6$ V, Fig. 1). Biological N_2 oxidation as a mechanism of N_2 fixation is limited by two sequential energy barriers ($E_h^{\circ\prime} = +1.36$ and 1.17 V, Fig. 1). In practice, the rate of N_2 oxidation in aerobic systems is essentially negligible in the absence of an energy input (e.g., electrical discharge). To date, microorganisms have not developed energy transduction mechanisms to overcome the two initial energy barriers limiting N_2 oxidation. This is understandable from a dissimilatory energy conservation standpoint, since the net energetics of N_2 oxidation to NO_3^- would offer little incentive. However, from an assimilatory N fixation standpoint, the transduction of ATP energy (as in N_2 reduction via N_2H_2–$N_2H_5^+$) or e$^-$ transfer energy (as in NH_4^+ oxidation via NH_2OH) to overcome the endergonic barriers limiting oxidative N_2 fixation appears to be an attractive alternate for aerobic organisms compared with the highly O_2-labile, ATP-driven nitrogenase system used historically by N_2-fixing biological systems (as part of their anaerobic heritage?) to overcome the endergonic barriers limiting reductive N_2 fixation.

2. ASSIMILATORY NITRATE REDUCTION

The pathway from NO_3^- ($+5$ oxidation state) to NH_4^+ (-3) is through NO_2^- ($+3$), an unknown intermediate ($+1$), and NH_2OH (-1) (Fig. 1). The enzymes catalyzing assimilatory NO_3^- reduction are soluble rather than membrane bound (Dalton, 1979), since energy conservation via ETP is not an issue for assimilatory reactions. Nitrate reductase uses NADPH as e$^-$ donor but can also use $FADH_2$ (Dalton, 1979), which is consistent with the $E_h^{\circ\prime} = +0.4$ V for NO_3^- reduction to NO_2^- (Fig. 1). The nitrite reductase complex reduces NO_2^- and NH_2OH to NH_4^+; NAD(P)H is the commonly used e$^-$ donor, although $FADH_2$ may also be involved, and the nitrite reductase for *Azotobacter chroococcum,* in fact, is FAD rather than NAD linked (Dalton, 1979). The accessibility of both steps of the NO_2^- to NH_2OH reduction process to $FADH_2$ suggests that the $E_h^{o\prime}$ of each step is less negative than -0.1 V, thereby precluding nitroxyl as an intermediate (Fig. 1). Similarly, chemical evidence that the most negative $E_h^{o\prime}$ lies on the reduction rather than the oxidation side of the $+1$ intermediate also does not support NOH as the intermediate: In slightly alkaline solution the weak reductant $Fe(OH)_2$ reduces NO_2^- quantitatively to NH_3 in the presence of a threefold excess of reducing agent but without the excess, some N_2O is formed (Latimer, 1952), indicating that under weakly reducing conditions NO_2^- is reduced to the intermediate but that the negative nature of the E_h of the next step retards the reduction process so that the intermediate reaches sufficiently high concentrations to facilitate dehydrative disproportionation (via $H_2N_2O_2$) to N_2O (Fig. 1). Relatedly, the pathway for HNO_2 reduction to NH_4^+ by sulfurous acid is known to proceed via a sequence of $2e^-$ reductions from which $H_2N_2O_2$, N_2O, and NH_3OH^+ can be obtained by appropriate pH adjustment (Latimer, 1952). The identity of the elusive $+1$ intermediate between NO_2^- and NH_2OH has been the subject of considerable speculation. As rationalized by Latimer (1952), dihydroxylammonia $[NH(OH)_2]$ is the most likely candidate on structural grounds. However, whatever the actual identity of the $+1$ intermediate, it is evidently highly unstable with respect to dehydrative disproportionation into N_2O under acid, slightly alkaline, and also physiological conditions (section IV B2).

3. ASSIMILATION OF AMMONIUM INTO ORGANIC NITROGEN POLYMERS

The primary pathway of NH_4^+ immobilization into organic N under the energy- rather than N-limiting conditions characteristic of soil is reductive amination of α-ketoglutarate^{2-} to glutamate$^-$ (Fig. 3) by the GDH system, followed by transamination of other α-keto to α-amino acids using glutamate$^-$ as the NH_4^+-$2e^-$ donor (Example 1). Consistent with the negative $E_h^{o\prime}$ (-0.11 V) of the α-ketoglutarate^{2-}–glutamate$^-$ couple and the endergonic or marginally exergonic energetics of the transamination reactions ($E_h^{o\prime} = -0.10$ to -0.13 V for other keto acid–α-amino acid couples, Fig. 3, 4), the e^- donor for reductive amination is NAD(P)H ($E_h^{o\prime} = -0.3$ V) rather than flavoprotein e^- carriers ($E_h^{o\prime} \cong 0$ to -0.1 V).

The energy barrier for amidation of glutamate$^-$ to glutamine and aspartate$^-$ to asparagine, $\Delta G^{o\prime} = +2.9$ and $+3.5$ kcal$\cdot$mol^{-1}, respectively (Fig. 3), is circumvented by ATP energy transduction (Stanier et al., 1976). The mechanistic role of ATP-linked glutamine synthesis/catabolism in driving NH_4^+ assimilation under NH_4^+-limiting (including N_2-fixing) conditions via the GS–GOGAT system is well established (Dalton, 1979; Example 1). In addition, by virtue of its high-energy amide group ($\Delta G^{o\prime}_{hydrol} = -2.9$ kcal), glutamine serves as an important donor of amino N for biosynthesis of purines and pyrimidines and non–α-amino N in amino acids such as tryptophan, histidine, and arginine (Dalton, 1979; Stanier et al., 1976). The final step in microbial immobilization of N (Eq. [24]), polymer bond formation by dehydrative condensation reactions between the polymer precursors, is endergonic ($\Delta G^{\circ\prime} = +3$ to $+5$ kcal·bond^{-1}; Lehninger, 1965) and requires energy coupling with ATP. In practice, for mechanistic, structural alignment reasons, a minimum of two to three ATP's/bond are required for polymer synthesis (Lehninger, 1965).

B. Dissimilatory Pathways

Dissimilatory reactions are concerned with conservation of energy by SLP and/or ETP mechanisms, with most reactions involving the transfer of high energy electrons from an e$^-$ donor through an e$^-$ carrier system to an e$^-$ acceptor (Section III B).

1. AMMONIFICATION

Ammonium may be liberated from organic N by oxidative, reductive, hydrolytic, or desaturative deamination mechanisms (Fig. 2–4).

a. Amino Acids—The general principles of the oxidative deamination of amino acids are covered in Example 8. In summary, SLP is achieved via acyl~CoA generation during oxidation of the high energy α-keto carboxy acid or aldehyde products of the oxidative deamination reaction. Reoxidation of the resulting NADH by an e$^-$ acceptor is required for maintained SLP. If the $E_h^{\circ\prime}$ of the e$^-$ acceptor is higher than about 0 V, ETP is also possible. From an ammonification standpoint, α-amino acid-saturated acid couples have the potential ($E_h^{\circ\prime} \cong 0.0$ V; Fig. 2–4) as e$^-$ acceptors to facilitate ETP as well as SLP (Fig. 2). For anaerobic organisms capable of assimilatory NO_3^- reduction, dissimilatory NO_3^- reduction to NH_4^+ without ETP (Fig. 2) is more energy efficient than the use of acyl~CoA as an e$^-$ acceptor for NAD$^+$ regeneration.

With O_2 or NO_3^- as e$^-$ acceptor, glutamate$^-$ is readily deaminated oxidatively, followed by complete oxidation to CO_2, the highly positive $E_h^{\circ\prime}$ of O_2–$2H_2O$ and $2NO_3^-$–N_2 providing sufficient thermodynamic force to overcome the endergonic first step of e$^-$ transfer from glutamate$^-$–α-ketoglutarate^{2-} ($E_h^{\circ\prime} = -0.11$ V) to the NAD$^+$/NADH e$^-$ carrier ($E_h^{\circ\prime} = -0.32$ V) (cf. Example 8). In the absence of O_2 and NO_3^-, less endergonic deamination pathways are commonly used. For example, glutamate$^-$ deamination to α-ketoglutarate^{2-} may be coupled in part to α-ketoglutarate^{2-} reduction to α-OH glutarate^{2-} ($E_h^{\circ\prime} = -0.15$ to -0.2 V), probably followed by dehydration to glutaconate^{2-}, and thence decarboxylation to crotonate (in CoA form) to provide a high potential e$^-$ sink ($E_h^{\circ\prime}$ crotonate-butyrate $= -0.02$ V) for the NADH generated during acetyl~CoA formation (Doelle, 1975). Alternatively, the non-e$^-$ transfer mesaconate pathway may be used to transform glutamate$^-$ into pyruvate and acetate ($\Delta G^{\circ\prime}$ only endergonic by 3.6 kcal) (Stanier et al., 1976).

The α-amino group of other amino acids is commonly deaminated via transamination to the α-ketoglutarate^{2-}–glutamate$^-$ couple (cf. Example 1), with the resultant α-keto carboxy acids providing the source of dissimilatory SLP via acyl~CoA production (Example 8). From a reductive standpoint, the $E_h^{o'}$ of α-amino acid–saturated acid couples (0.0 to -0.05 V range, Fig. 2-4) is highly competitive energetically as an e$^-$ sink under anaerobic conditions; however, in practice reductive deamination is restricted to relatively few amino acids (Stanier et al., 1976). Certain clostridia are capable of coupling oxidative deamination of one amino acid (as a source of SLP) to reductive deamination of another amino acid (as a mechanism of NAD$^+$ regeneration and possibly as a source of ETP as well; Thauer et al., 1977) (Fig. 2).

Desaturative deamination of α-amino acids to the unsaturated acid, e.g., aspartate$^-$ $\rightarrow$ fumarate^{2-} + NH$_4^+$, $\Delta G^{o'}$ = $+3.7$ kcal, apparently occurs (Stanier et al., 1976); however, the endergonic nature of the reaction identifies a requirement for a high relative activity of aspartate$^-$ to fumarate^{2-} for a forward reaction. Hydrolytic deamination of α-amino N is likewise endergonic but is also restricted biochemically on mechanistic grounds (Fig. 3 and 4). In contrast, hydrolytic deamination of the amide group of glutamine and asparagine is energetically favorable (Fig. 3), as demonstrated by the need for ATP energy transduction to drive the reverse, amidation, reaction.

Deamination of amino acids may also occur by disproportionation reactions. For example, serine deamination to pyruvate (Fig. 4) catalyzed by serine dehydratase is irreversible, which is in line with its highly exergonic nature:

$$HOCH_2CHNH_3^+COO^- \ (10e^-)$$
$$[-293.49]$$

$$= HCH_2COCOO^- \ (10e^-) + NH_4^+ \ (0e^-) \quad \Delta G^{o'}$$
$$[-273.16] \qquad\qquad [0] \qquad -10.3 \ kcal$$

where the numbers in square brackets are $\Delta G_{ox}^{o'}$ data from Table 1b (cf. Example 5).

Oxidation of phenylalanine to tyrosine ($E_h^{o'}$ = -0.03, Fig. 4) cannot be accomplished by NAD$^+$ and in practice involves FAD as e$^-$ carrier and an energy boost via hydroxylase coupling with O$_2$. Dicarboxy amino acids are susceptible to irreversible decarboxylation reactions, e.g., decarboxylation of aspartate$^-$ to alanine ($\Delta G^{o'}$ = -6.4 kcal) and of glutamate$^-$ to λ-amino butyrate.

b. Other Organic N Compounds—All steps in the aerobic degradation pathway of adenine and guanine through hypoxanthine, xanthine, urate$^-$, and allantoin to urea are exergonic (Fig. 4). Pathway specifics are reviewed in detail by Vogels and van der Drift (1976). Under anaerobic conditions, SLP may be accomplished by non-e$^-$ transfer transformations of purines (Example 5; Stanier et al., 1976; Vogels & van der Drift, 1976). Urea hydrolysis to 2NH$_4^+$ and CO$_2$ is highly exergonic (Fig. 4) but does not appear to be coupled to SLP (Stanier et al., 1976).

Methylamine is oxidatively deaminated to CO_2 via formaldehyde and formate under aerobic conditions (Doelle, 1975), with energetics analogous to oxidative deamination of amino acids (Fig. 4 vs. 3). Under anaerobic conditions, $CH_3NH_3^+$ is readily reduced by methanogens to methane as an e^- acceptor for ETP, which is in line with the relatively high $E_h^{o\prime}$ of $CH_3NH_3^+$–CH_4 (Fig. 4).

2. NITRIFICATION

As identified by the $E_h^{o\prime}$ for NH_4^+ oxidation to NO_2^- (+0.34 V) and NO_2^- oxidation to NO_3^- (+0.43 V) (Fig. 1), the nitrification reaction is obligately aerobic (Fig. 2). Nitrobacters use NO_2^- as an assimilatory e^- donor for CO_2 reduction and a dissimilatory e^- donor for energy generation with O_2 as e^- acceptor. Electrons enter the ETS at the cytochrome a_1 level (Aleem, 1977). SLP is thermodynamically impossible; ETP with segment no. 3 is marginal but evidently occurs as the sole source of energy for nitrobacter. The resultant ATP must not only satisfy normal assimilatory growth and maintenance requirements, but substantial amounts are also needed to raise the energy potential of the NO_2^- electrons ($E_h^{o\prime}$ NO_2^-–NO_3^- = +0.43 V) to a level capable of reducing CO_2 ($E_h^{o\prime}$ CO_2–CH_2O $\cong$ −0.4 V), via ATP-driven reverse e^- transport through the ETS to generate NADH as the primary reductant (Aleem, 1977).

Ammonium oxidation to NO_2^- appears to follow the assimilatory NO_2^- reduction pathway in reverse, with the associated uncertainties regarding the nature of the +1 oxidation state intermediate (Fig. 1; Suzuki, 1974). The first step, NH_4^+ oxidation to NH_2OH ($E_h^{o\prime}$ = +0.86) presents a thermodynamic e^- transfer problem in that the $E_h^{o\prime}$ not only is too high to allow e^- transfer into the ETS ($E_h^{o\prime}$ of high potential cytochrome a $\cong$ +0.4 V) but also is too high for exergonic e^- transfer to the O_2–$2H_2O$ couple ($E_h^{o\prime}$ = +0.82 V). To circumvent this problem, NH_4^+ is reacted directly with O_2 via a hydroxylase reaction, possibly assisted by an energy boost derived from concomitant oxidation of reduced cytochrome P460, the reduced cytochrome P460 being regenerated by $2e^-$ from the terminal step in the NH_4^+-to-NO_2^- oxidation process (Suzuki, 1974). According to this scheme, 4 of the $6e^-$ released during NH_4^+ oxidation to NO_2^- are consumed in effecting NH_4^+ oxidation to NH_2OH, and energy generation (linked to segments no. 3 and possibly 2) is confined to the $2e^-$ transfer step from NH_2OH to the +1 oxidation state intermediate (Suzuki, 1974; Aleem, 1977). Such a scheme is energetically inconsistent with NOH as the +1 intermediate (Fig. 1) but is consistent with the evidence from the assimilatory NO_2^- reduction system that the most negative $E_h^{o\prime}$ in the two-step process of NH_2OH oxidation to NO_2^- lies on the reduction rather than the oxidation side of the +1 intermediate. Specific supporting experimental evidence includes:
1) The hydroxylamine reductase enzyme system appears to be flavoprotein-linked (Aleem, 1977), indicating that both steps are in the > −0.1 V range.
2) With O_2–$2H_2O$ as e^- acceptor, hydroxylamine reductase catalyzes stoichiometric oxidation of NH_2OH to NO_2^-, whereas with e^- car-

riers such as phenazine methosulfate and low potential cytochrome ($E_h^{o'} \cong 0.0$ V) as oxidants, NH_2OH is oxidized to N_2O (degradation product of the unstable $+ 1$ intermediate) rather than to NO_2^- (Suzuki, 1974), indicating that the $E_h^{o'}$ of the $NH_2OH/+1$ intermediate couple is more negative, and the $E_h^{o'}$ of the $+1$ intermediate–NO_2^- couple is more positive, than the e^- carriers.

3) In acid solution, Fe^{3+} is powerful enough to oxidize NH_3OH^+ to the $+1$ intermediate but cannot oxidize the $+1$ intermediate to HNO_2, resulting in production of N_2O (high acid) and N_2 gas (low acid) as degradation products of the unstable intermediate (Latimer, 1952).

Considerable amounts of N_2O may be produced from nitrification reactions (Bremner & Blackmer, 1978). Under conditions conducive to simultaneous production of NH_2OH and NO_2^-, diverse redox and condensation reactions may occur between these two highly unstable compounds (Fig. 1), resulting in variable amounts of N_2O, NO, and/or N_2 (Latimer, 1952; Suzuki, 1974). Under acid conditions (e.g., pH < 5), chemical reactions involving HNO_2, (pK_a HNO_2–NO_2^- = 3.14, Table A-1a) may occur (Chapt. 8, M. K. Firestone). In the presence of O_2, the most likely chemical fate of HNO_2 is self-decomposition into NO_3^- and NO, followed by reaction of the NO with atmospheric O_2 to form NO_3^- via NO_2 and N_2O_4 (Latimer, 1952) (ΔG° calculated from ΔG_f° data in Table A-1a):

$$\Delta G^\circ, \text{kcal}$$

	ΔG°, kcal
$3HNO_2 = 2NO + NO_3^- + H^+ + H_2O$	-2.05
$2NO + O_2 = 2NO_2$	-16.86
$2NO_2 = N_2O_4$	-1.14
$N_2O_4 + H_2O = HNO_2 + NO_3^- + H^+$	-6.63
$2HNO_2 + O_2 = 2NO_3^- + 2H^+$	-26.68

Ammonium, urea, methylamine, purines and pyrimidines, and particularly α-amino acids and amido sulfonic acid (H_2NSO_2OH, sulfamic acid) react with HNO_2 via a van Slyke-type reaction to produce N_2 (Chapt. 8): 1965):

	ΔG°, kcal	
$NH_4^+ + HNO_2 = N_2 + 2H_2O + H^+$	-81.1	
$RNH_3^+ + HNO_2 = N_2 + ROH + H_2O + H^+$	-80.2	[89]

where RNH_3^+ is alanine and ROH is lactic acid in Eq. [89], but similar energetics prevail for other α-amino acid–α-OH acids and amine–alcohol couples (Fig. 3 and 4). From an inorganic reductant standpoint, e^-

reductants such as HSO_3^- readily reduce HNO_2 through the NH_3OH^+ state to NH_4^+; likewise, Sn^{2+} reduces HNO_2 to NH_3OH^+, although under certain conditions, N_2O, NO, and N_2 are the reduction products (Latimer, 1952). Such $1e^-$ reductants as Fe^{2+} and Ti^{3+} reduce HNO_2 to NO but no further, which is in line with the highly negative E_h^o of the NO–NOH couple (Fig. 1; Latimer, 1952).

3. DENITRIFICATION

Denitrification of $2NO_3^-$ to N_2 involves $2NO_2^-$, 2NO, and N_2O as inter-mediates (Fig. 1). Specifics on the organisms, enzymes, e^- carriers, and energy coupling sites involved are reviewed in detail by Thauer et al. (1977). In summary, most of the e^- carriers are membrane bound, con-sistent with the ETP energy conservation objectives of NO_3^- dissimila-tion. Most evidence indicates that each intermediate couples with the ETS for proton motive force generation, in line with thermodynamic pre-diction (Fig. 2), although energy coupling efficiency tends to be much lower than with the O_2–$2H_2O$ e^- acceptor (for which the ETS is primarily de-signed). A common physiological e^- donor is NADH, although electrons may also enter the ETS at the succinate level ($E_h^o{}' \cong 0.0$ V), which is con-sistent with the high potentials of the NO_2^-–NO_3^- (+0.4 V), NO–NO_2^- (+0.3 V), $2NO$–N_2O (+1.2 V) and N_2O–N_2 (+1.4 V) e^- acceptor couples (Fig. 1).

C. Hydrogen Cyanide Metabolism

The general pathway for hydrogen cyanide biosynthesis is via oxidation of glycine (Knowles, 1976). Flavoproteins are considered to be the e^- carriers, which is consistent with the $E_h^o{}'$ of -0.15 V for HCN, CO_2–glycine (Fig. 4). Cyanide assimilation occurs via condensation/rearrange-ment reactions (Knowles, 1976), the atypically high thermodynamic force of these reactions (Fig. 4) suggesting a detoxification rather than an assimila-tory nutrient function. Similarly, the initial hydrolysis step in the oxidative dissimilation of HCN to CO_2 via formaldehyde and formate is highly inef-ficient from a dissimilatory energy conservation standpoint (Fig. 4).

V. EFFICIENCY OF REDUCTIVE DINITROGEN FIXATION

The efficiency of reductive N_2 fixation is dictated by the mechanisms used to meet the basic requirements for a source of (i) relatively high energy electrons ($E_h^o{}' \cong -0.4$ V) for reduction of the N_2; (ii) energy to overcome the activation energy barrier of N_2 reduction; (iii) N_2, free from catalyst-inhibiting O_2; and (iv) materials and energy for construction and mainten-ance of the N_2-fixing machinery (Emerich & Evans, 1980; Postgate & Hill, 1979; Harris, 1978).

Table 2—Efficiency of biological N_2 fixation.

| Assimilatory | Dissimilatory | | | | Reaction | | | | Efficiency | | |
| N reduction | Bio-mass | N_2ase, GS-GOGAT | Growth | Other | Equation† | Eq. no. | $\Delta H°$ | $\Delta G°'$ | $A_{\Delta H \, ox}$ | $A_{\Delta G \, ox}$ | Y_{gluc} |
e⁻ eq glucose·N_2 mol⁻¹							kcal·reaction⁻¹		kcal·N_2 mol⁻¹		mg N·g glucose⁻¹
6	0	0	0	0	$\frac{6}{24}$ $C_6H_{12}O_6$ + N_2 + 1.5H_2O + 2H⁺ = 2NH_4^+ + 1.5CO_2	[90]	−27	−20	168	172	622
6	0	8.8	0	0	$\frac{14.8}{24}$ $C_6H_{12}O_6$ + N_2 + $\frac{8.8}{4}$ O_2 + 2H⁺ = 2NH_4^+ + 3.7CO_2 + 0.7H_2O	[91]	−274	−273	416	424	251
6	0	12.9	0	0	$\frac{18.9}{24}$ $C_6H_{12}O_6$ + N_2 + $\frac{12.9}{4}$ O_2 + 2H⁺ = 2NH_4^+ + 4.7CO_2 + 1.7H_2O	[93]	−389	−390	530	541	197
0	34	0	19.1	0	$\frac{53.1}{24}$ $C_6H_{12}O_6$ + 2NH_4^+ + $\frac{19.1}{4}$ O_2 = $C_8H_{14}O_3N_2$ + 5.3CO_2 + 9.3H_2O + 2H⁺	[94]	−551	−593	1,487	1,519	70
6	34	12.9	19.1	0	$\frac{72.0}{24}$ $C_6H_{12}O_6$ + N_2 + $\frac{32.0}{4}$ O_2 = $C_8H_{14}O_3N_2$ + 10.0CO_2 + 11.0H_2O	[95]	−939	−982	2,017	2,060	52
16	0	0	19.1	0	$\frac{16}{24}$ $C_6H_{12}O_6$ + 2NO_3^- + 4H⁺ = 4CO_2 + 2NH_4^+ + 2H_2O	[96]	−276	−291	448	458	233
16	34	0	19.1	0	$\frac{69.1}{24}$ $C_6H_{12}O_6$ + 2NO_3^- + $\frac{19.1}{4}$ O_2 + 2H⁺ = $C_8H_{14}O_3N_2$ + 9.3CO_2 + 11.3H_2O	[97]	−827	−884	1,935	1,977	54
6	34	24.0	19.1	6	$\frac{89.1}{24}$ $C_6H_{12}O_6$ + N_2 + $\frac{43.1}{4}$ O_2 = $C_8H_{14}O_3N_2$ + 14.3CO_2 + 3H_2 + 12.3H_2O	[98]	−1,211	−1,299	2,496	2,549	42
6	34	12.9	19.1	300	$\frac{372}{24}$ $C_6H_{12}O_6$ + N_2 + $\frac{332}{4}$ O_2 = $C_8H_{14}O_3N_2$ + 85CO_2 + 86H_2O	[99]	−9,339	−9,561	10,419	10,641	10

† The physical states are N_2(g), O_2(g), CO_2(g), H_2(g), H_2O(l), $C_8H_{14}O_3N_2$(s); all other components are (aq).

EXAMPLE 10

Efficiency of biological N_2 fixation: N_2 fixation by bacterial heterotrophs using glucose as assimilatory and dissimilatory e^- donor and O_2 as dissimilatory e^- acceptor.

The assimilatory requirement for N_2 reduction is 6 e^- eq glucose•mol N_2^{-1} (Eq. [90], Table 2). If we assume (i) that the minimum energy requirement for nitrogenase activation is 12 ATP•N_2^{-1} (Example 2) plus 2ATP per 2NH_4^+ to operate the GS–GOGAT system (Example 1), (ii) that the efficiency of ATP generation from glucose oxidation is maximized at 38 ATP•mol glucose^{-1} = 38/24 = 1.583 mol ATP•e^- eq glucose^{-1} (oxidation pathway via glycolysis, the TCA cycle and ETS segments no. 1, 2, and 3), and (iii) that (for simplicity) no SLP is derived from glucose as an assimilatory e^- source for N_2 reduction, then the dissimilatory glucose requirement is 14/1.583 = 8.84 e^- eq glucose•N_2 mol^{-1}, and the total requirement is 6 + 8.84 = 14.84 e^- eq glucose•N_2 mol^{-1} (Eq. [91], Table 2). Reactions [90] and particularly [91] are strongly exothermic and exergonic (Table 2). Under intracellular N_2-fixing conditions of $10^{-5}M$ NH_4^+, 0.78 atm N_2, 0.002 atm O_2, 0.03 atm CO_2, and $10^{-3}M$ glucose, the drive for N_2 fixation (Eq. [92]) is relatively close to the standard value (Eq. [91], Table 2) (from Eq. [91] and [2]):

$$\Delta G' = \Delta G^{\circ\prime} + 2.3RT \log \frac{\{NH_4^+\}^2\{CO_2\}^{3.7}}{\{C_6H_{12}O_6\}^{0.4}\{N_2\}\{O_2\}^{2.2}}$$

$$= -273 + 2.3RT \log \frac{(10^{-5})^2(0.03)^{3.7}}{(10^{-3})^{0.4}(0.78)(0.002)^{2.2}}$$

$$= -282 \text{ kcal•reaction}^{-1}. \qquad [92]$$

Similarly, in practice, the $\Delta G^{\circ\prime}$ of many other microbially catalyzed reactions provides a good index of the operational thermodynamic force driving the reactions toward equilibrium (McCarty, 1972; Thauer et al., 1977).

The efficiency of N_2 fixation may be expressed as a function of diverse dissimilatory or total mass and/or energy parameters. For example, in Table 2 ($\overline{\Delta H}^{\circ}_{ox}$, $\overline{\Delta G}^{\circ\prime}_{ox}$, and equivalent weight data from Table 1b):

$$A_{\Delta H_{ox}} = -n_g \overline{\Delta H}^{\circ}_{ox} \text{ glucose} = 28.006 \, n_g \text{ kcal•}N_2\text{mol}^{-1}$$

$$A_{\Delta G_{ox}} = -n_g \overline{\Delta G}^{\circ\prime}_{ox} \text{ glucose} = 28.603 \, n_g \text{ kcal•}N_2\text{mol}^{-1}$$

$$Y_{gluc} = \frac{1000 \text{ (molecular weight } N_2)}{n_g \text{ (equivalent weight glucose)}} = \frac{28014}{7.507 \, n_g} \text{ mg N•g glucose}^{-1}$$

where $A_{\Delta H_{ox}}$ is the classical heat energy equivalent of glucose, $A_{\Delta G_{ox}}$ is the analogous free-energy equivalent, and n_g is the number of e^- equivalents of glucose (identified directly in the equations, Table 2), required for assimilatory and dissimilatory fixation of N; and Y_{gluc} is the weight-based N yield from glucose. As shown in Table 2 (Eq. [91]), based on current concepts of the minimum biochemical cost of N_2 reduction, the maximum boundary efficiency of biological N_2 fixation is 416 kcal heat energy and 424 kcal free-energy equivalent per N_2 reduced to $2NH_4^+$ or in yield form is 251 mg N•g glucose^{-1}. In comparison, the Haber-Bosch process shows a somewhat higher efficiency of $A_{\Delta H_{ox}}$ = 300 to 350 kcal•N_2^{-1} (Postgate & Hill, 1979; Emerich & Evans, 1980).

For a more typical bacterial ATP-generating efficiency of 26 mol ATP•mol glucose^{-1} = 26/24 = 1.083 mol ATP•e$^-$ eq glucose^{-1} (pathway: glycolysis/TCA/ETS no. 1 and 2), the dissimilatory requirement would be 14/1.083 = 12.9, and the total requirement would thus become 6 + 12.9 = 18.9e$^-$ eq glucose•N_2 mol^{-1} (Eq. [93], Table 2). Bacterial growth-linked N_2 fixation imposes an additional 34e$^-$ eq assimilatory requirement (Table 2) combined with an additional 19.1e$^-$ eq dissimilatory requirement (20.7 mol ATP•207 g cells^{-1} = 20.7/1.083 = 19.1e$^-$ eq glucose•N_2 mol cells^{-1}; Thauer et al., 1977; Harris, 1981), giving a total requirement of 53.1e$^-$ eq glucose• N_2 mol^{-1} for NH_4^+-assimilating growth (Eq. [94], Table 2) and a total requirement for N_2-fixing growth of 53.1 + 18.9 = 72.0e$^-$ eq glucose•N_2 mol^{-1} (Eq. [95], Table 2). Reactions for NO_3^--assimilation and NO_3^--assimilating growth are given for comparison (Eq. [96] and [97], respectively, Table 2). The energetics of the reactions in Table 2 may be calculated from energy of formation (Eq. [1]), $\overline{\Delta H_{ox}^\circ}$ or ΔH_{ox}° (Eq. [49] and [50]), $\overline{\Delta G_{ox}^{0\prime}}$ or $\Delta G_{ox}^{0\prime}$ (Eq. [47] and [48]), or $E_h^{o\prime}$ (Eq. [66]) data. For example, recognizing that (from Table 1b) the ΔH_{ox}° of $C_8H_{14}O_3N_2$–$8CO_2$, N_2 = 2 × 20 × –27.000 = –1080 kcal, then the ΔH° for N_2-fixing bacterial growth (Eq. [95], Table 2) is given by (from Eq. [50]):

$$\Delta H^\circ = -[1080 - 72.03(-28.006)] = -937 \text{ kcal•reaction}^{-1}.$$

For comparison with the bioengineering literature (Erickson, 1979, 1980), the energetic efficiency, η_{e^-} of NH_4^+-assimilating growth (Eq. [94], Table 2) is derived as

$$\eta_{e^-} = (n_b/n_g) \times 100 = (34/53.1) \times 100 = 64\%$$

or in enthalpy terms, η_H, as ($\overline{\Delta H_{ox}^\circ}$ data from Table 1b)

$$\eta_H = \frac{n_b \, \overline{\Delta H_{ox}^\circ} \text{ biomass} \times 100}{n_g \, \overline{\Delta H_{ox}^\circ} \text{ glucose}} = \frac{34 \times -27.610 \times 100}{53.1 \times -28.006} = 58\%$$

where n_b is the C-bound e$^-$ composition of a N_2 mole of bacterial biomass (Table 1b).

The efficiency of N_2 fixation decreases dramatically for growth-linked systems (Table 2). Increased assimilatory e^- costs caused by storage of poly-hydroxybutyrate (PHB; equivalent weight = 4.8 g•e^- eq^{-1}) or glycogen due to imbalanced e^- donor uptake vs. use (e.g., energy-limited growth caused by inadequate O_2 or N-limited growth resulting from nitrogenase inhibition by excess O_2) may result in further substantial decreases in N_2 fixation efficiency. For example, with 50% (wt/wt) PHB storage, efficiency declines to 72 + (207 ÷ 4.8) = 115e^- eq glucose•N_2 mol^{-1} = 32 mg N•g glucose^{-1}. Increased dissimilatory e^- costs may be caused by shunting of electrons through the nitrogenase system to reduce protons to H_2 (Emerich & Evans, 1980). The dissimilatory cost depends on whether the H_2 is evolved, recycled through the nitrogenase system or reoxidized via ETS with or without ETP. For example, for 3 mol of H_2 generated/mol of N_2 reduced, the increased cost ranges from about 11e^- eq (H_2 recycled or reoxidized via ETS no. 1 and 2) to about 17e^- eq glucose (H_2 reoxidized with no ETP, or lost) (Eq. [98], Table 2). Another phenomenon contributing to increased dissimilatory e^- costs is e^- dissimilation for O_2 control (nitrogenase protection) rather than energy conservation (Postgate & Hill, 1979). For example, the efficiency of N_2-fixing azotobacters decreases progressively with increasing O_2 concentration, commonly becoming about 72 + 300e^- eq glucose•N_2 mol^{-1} = 10 mg N•g glucose^{-1} at high O_2, with an associated massive output of heat (Eq. [99], Table 2).

VI. APPENDIX

A. Selected Values of Thermodynamic Properties for Nitrogen

Enthalpies of formation were obtained largely from CODATA (1978), Wagman et al. (1968), Hutchens (1976b), and Wilhoit (1969) and were updated as indicated by heat of ionization (ΔH°_{ioniz}) data from Christensen et al. (1976). The ΔH°_f for pyrimidines and related compounds were recalculated from the heat of combustion (ΔH°_c) data of Stiehler and Huffman (1935) using ΔH°_f for H_2O and CO_2 from CODATA (1978). Free energies of formation derived from thermodynamic data were based on ΔH°_f and S° values from CODATA (1978), supplemented by other sources as necessary. Most ΔG°_f data were obtained from Wagman et al. (1968) and Hutchens (1976d) and were updated as indicated by pK_a data from Christensen et al. (1976) or Jencks and Regenstein (1976). The choice of which species to use as the reference base for ΔG°_f calculation from pK_a data followed the approach used by Latimer (1952) for inorganic and Burton and Krebs (1953) for organic N.

Table A-1a—Selected values of enthalpy of formation, pK_a and Gibbs free energy of formation of inorganic N and other inorganic substances in the standard state at 25°C.

Substance	Formula	State†	ΔH_f°‡	pK_a	ΔG_f°‡	Source§
			kcal·mol⁻¹		kcal·mol⁻¹	
Ammonia	NH_3	g	−10.98		−3.93	ΔH_f°, S°: 78C. S°(N_2, H_2): 78C
Ammonium	NH_4^+	aq†	−31.85	9.244	−18.99	ΔH_f°, S°: 78C. S°(N_2, H_2, H^+): 78C
Ammonia	NH_3	aq	−19.39		−6.38	ΔH_{ioniz}°, pK_a: 76C (Example 11)
Carbon dioxide	CO_2	g	−94.05		−94.258	ΔH_f°, S°: 78C. S°(C, O_2): 78C
Carbonic acid	H_2CO_3	aq	−167.22	6.363	−148.95	ΔH_f°: 69W. ΔG_f°: 79S
Bicarbonate	HCO_3^-	aq†	−165.39	10.328	−140.27	ΔH_f°: 69W. ΔG_f°: 79S
Carbonate	CO_3^{2-}	aq	−161.84		−126.18	ΔH_f°: 69W. ΔG_f°: 79S
Diimide	N_2H_2	g	+36.00		[+50.00]	ΔH_f°: 76W. ΔG_f°: 76S, 77S
Dinitrogen	N_2	g	0.00		0.00	ΔH_f°, S°: 78C
Dinitrogen tetroxide	N_2O_4	g	+2.19		+23.38	ΔH_f°, ΔG_f°: 68W
Hydrazine	N_2H_4	g	+22.80		+38.07	ΔH_f°, ΔG_f°: 68W
	$N_2H_5^+$	aq†	−2.09	7.961	+19.74	ΔH_{ioniz}°, pK_a: 76C
	N_2H_4	aq	+8.20		+30.6	ΔH_f°, ΔG_f°: 68W
Hydrazoic acid	HN_3	aq	+62.16		+76.9	ΔH_f°, ΔG_f°: 68W
Azide	N_3^-	aq†	+65.96	4.640	+83.23	ΔH_{ioniz}°, pK_a: 76C
Hydrogen	H_2	g	0.00		0.00	ΔH_f°, S°: 78C
Hydroxylamine	NH_3OH^+	aq	−32.8		−15.45	ΔH_{ioniz}°, pK_a: 76C
	NH_2OH	aq†	−23.5	5.948	−7.34	ΔH_f°: 68W. S°: 52L. S°(N_2, H_2, O_2): 78C
Hyponitrous acid	$H_2N_2O_2$	aq†	−15.4	7.05	+6.66	ΔH_f°: 68W. S°: 52L. S°(N_2, H_2, O_2): 78C
Hyponitrite⁻	$HN_2O_2^-$	aq	−12.4	11.54	+16.28	ΔH_f°: 68W. pK_a: 76J
Hyponitrite²⁻	$N_2O_2^{2-}$	aq	−4.1		+32.02	ΔH_f°: 68W. pK_a: 76C

(continued on next page)

Table A-1a—Continued.

Substance	Formula	State†	ΔH_f°‡	pK_a	ΔG_f°‡	Source§
			kcal·mol⁻¹		kcal·mol⁻¹	
Nitric acid	HNO_3	aq	−45.86		−24.72	ΔH_{ioniz}°, pK_a: 76C
Nitrate	NO_3^-	aq†	−49.56	−1.41	−26.64	ΔH_f°: 68W. S°: 78C. $S^\circ[N_2, O_2]$: 78C
Nitric oxide	NO	g	+21.57		+20.69	ΔH_f°, ΔG_f°: 68W
Nitrogen dioxide	NO_2	g	+7.93		+12.26	ΔH_f°, ΔG_f°: 68W
Nitrous acid	HNO_2	aq	−28.5		−13.3	ΔH_f°, ΔG_f°: 68W
Nitrite	NO_2^-	aq†	−26.15	3.143	−9.01	ΔH_{ioniz}°, pK_a: 76C
Nitrous oxide	N_2O	g	+19.6		+24.90	ΔH_f°, ΔG_f°: 68W
Nitroxyl	NOH	g	[+24.0]		[+27.0]	ΔH_f°: 70K. ΔG_f°: 70K, 52L$[E_R^\circ]$
Oxygen	O_2	g	0.00		0.00	ΔH_f°, S°: 78C
Proton	H^+	aq	0.00		0.00	ΔH_f°, S°: 78C
Proton	H^+	aq (pH 7)	0.00		−9.55	$\Delta G_f H^+(pH\ 7) = \Delta G_f^\circ H^+ + 1.3642(-7)$
Water (hydronium)	H_3O^+	aq			−56.69	pK_a: 70S
Water	H_2O	l	−68.315	0.00	−56.69	ΔH_f°, S°: 78C. $S^\circ(H_2, O_2)$: 78C
Hydroxyl	OH^-	aq	−54.98	14.00	−37.59	ΔH_f°: 78C. pK_a: 76C

† Identifies the dominant ionic species at pH 7, as derived from pK_a data (Example 3).

‡ Numbers in square brackets [] are estimates. Numbers underlined were derived by the compiler and are expressed to two decimal places to allow accurate reconversion rather than to represent absolute accuracy. Calculation methods were according to standard procedures (Wagman et al., 1968; Latimer, 1952). The relationship between pK_a and ΔG_f° is, for example (Eq. [35]): $pK_a = (\Delta G_f^\circ AH_n - \Delta G_f^\circ AH_{n+1})/1.3642$. All data are for 298.15°K and zero ionic strength.

§ Source (numbers represent year; see section VII for complete citation): 79S, Sadiq & Lindsay; 78C, CODATA; 77S, Stiefel; 76C Christensen et al.; 76J, Jencks & Regenstein; 76S, Shilov; 76W, Willis et al.; 70S, Stumm & Morgan; 70K, Karapet'yants & Karapet'yants; 69W, Wilhoit; 68W, Wagman et al.; 52L, Latimer.

Table A-1b—Selected values of enthalpy of formation, pK_a, and Gibbs free energy of formation of organic N substances in the standard state at 25°C.

Substance	Formula	State†	$\Delta H_f^\circ\ddagger$	pK_a	$\Delta G_f^\circ\ddagger$	Source§
			kcal·mol⁻¹		kcal·mol⁻¹	
Adenine	$C_5H_5N_5$	c	+22.70		+71.36	ΔH_c°, S°: 35S. ΔH_f°, $S^\circ(H_2O$, etc.): 78C
	$C_5H_6N_5^+$	aq		4.2	+68.59	pK_a (N_1–H⁺): 76C
	$C_5H_5N_5$	aq†			+74.32	Solubility: 68M
	$C_5H_4N_5^-$	aq		9.67	+87.51	pK_a (N_9–H): 76C
L-Alanine	$CH_3CH(NH_2)COOH$	c	−134.5		−88.4	ΔH_f°, ΔG_f°: 76H$_b$
	$CH_3CH(NH_3^+)COOH$	aq	−133.43	2.346	−92.00	ΔH_{ioniz}°, pK_a (α-COOH): 76C
	$CH_3CH(NH_3^+)COO^-$	aq†	−132.7		−88.8	ΔH_{soln}°: 76H$_c$. ΔG_f°: 76H$_d$
	$CH_3CH(NH_2)COO^-$	aq	−121.8	9.855	−75.36	ΔH_{ioniz}°, pK_a(α-NH$_3^+$): 76C
Allantoin	$C_4H_6O_3N_4$	c	−171.60		−106.74	ΔH_c°, S°: 35S. ΔH_f°, $S^\circ(H_2O$, etc): 78C
	$C_4H_6O_3N_4$	aq			[−104]	Solubility: 73W
L-Arginine	$H_2NC(:NH)NH(CH_2)_3CH(NH_2)COOH$	c	−149.0		−57.4	ΔH_f°, ΔG_f°: 76H$_b$
L-Asparagine	$H_2NCOCH_2CH(NH_2)COOH$	c	−188.7		−126.7	ΔH_f°, ΔG_f°: 76H$_b$
	$H_2NCOCH_2CH(NH_3^+)COOH$	aq		2.02	−128.66	pK_a^I(α-COOH): 76J
	$H_2NCOCH_2CH(NH_3^+)COO^-$	aq†	−182.9		−125.9	ΔH_{soln}°: 76H$_c$. ΔG_f°(Aspn. H_2O): 76H$_d$
	$H_2NCOCH_2CH(NH_2)COO^-$	aq		8.8	−113.90	pK_a^I(α-NH$_3^+$): 76J
L-Aspartic acid	$HOOCCH_2CH(NH_2)COOH$	c	−232.6		−174.5	ΔH_f°, ΔG_f°: 76H$_b$
	$HOOCCH_2CH(NH_3^+)COOH$	aq	−228.28	1.99	−175.1	ΔH_{ioniz}°, pK_a (α-COOH): 76C
	$HOOCCH_2CH(NH_3^+)COO^-$	aq	−226.5	3.90	−172.4	ΔH_{soln}°: 76H$_c$. ΔG_f°: 76H$_d$
Aspartate⁻	$^-OOCCH_2CH(NH_3^+)COO^-$	aq†	−225.39		−167.08	ΔH_{ioniz}°, pK_a (β-COOH): 76C
Aspartate²⁻	$^-OOCCH_2CH(NH_2)COO^-$	aq	−216.36	10.00	−153.44	ΔH_{ioniz}°, pK_a (α-NH$_3^+$): 76C
Bacterial biomass	$C_4H_7O_{1.5}N$ (formula weight 103.5 g)	s	[−75.31]		[−35.45] ¶	
Cyanic acid	HCNO	aq	−36.9	3.47	−28.0	ΔH_f°, ΔG_f°: 68W
Cyanate	CNO^-	aq†	−34.42		−23.27	ΔH_{ioniz}°, pK_a: 76C
Cyanogen	C_2N_2	g	+73.84		+71.07	ΔH_f°, ΔG_f°: 68W

Name	Species		State	pK_a		Notes
	$HSCH_2CH(NH_3^+)COOH$		aq	1.71	-82.76	pK_a (α-COOH): 76J
	$HSCH_2CH(NH_3^+)COO^-$		aq†		-80.43	ΔG°_{soln}: 57B
	$HSCH_2CH(NH_2)COO^-$		aq	8.54	-68.78	pK_a (α-NH$_3^+$): 76C
	$^-SCH_2CH(NH_2)COO^-$		aq	10.55	-54.39	pK_a (β-SH): 76C
L-Cystine	$[-SCH_2CH(NH_2)COOH]_2$	-249.6	c		-163.9	$\Delta H^\circ_f, \Delta G^\circ_f$: 76H$_b$
	$[-SCH_2CH(NH_3^+)COOH]_2$		aq	1.65	-164.73	pK_a (α-COOH): 76J
	$C_6H_{13}O_4N_2S_2^+$		aq	2.26	-162.48	pK_a (α-COOH): 76J
	$[-SCH_2CH(NH_3^+)COO^-]_2$		aq†	7.85	-159.40	ΔG°_f: 76H$_d$
	$C_6H_{11}O_4N_2S_2^-$		aq	9.85	-148.69	pK_a (α-NH$_3^+$): 76J
	$[-SCH_2CH(NH_2)COO^-]_2$		aq		-135.25	pK_a (α-NH$_3^+$): 76J
Dimethylamine	$(CH_3)_2NH$	-4.41	g		$+16.35$	$\Delta H^\circ_f, \Delta G^\circ_f$: 68W
	$(CH_3)_2NH_2^+$	-28.80	aq†		-0.83	$\Delta H^\circ_{ioniz}, pK_a$: 76C
	$(CH_3)_2NH$	-16.88	aq	10.76	$+13.85$	$\Delta H^\circ_f, \Delta G^\circ_f$: 68W
α-D-Glucose	$C_6H_{12}O_6$	-302.05	aq		-219.22	ΔH°_f: 69W. ΔG°_f: 57B
L-Glutamic acid	$HOOCCH_2CH_2CH(NH_2)COOH$	-241.3	c		-174.8	$\Delta H^\circ_f, \Delta G^\circ_f$: 76H$_b$
	$HOOCCH_2CH_2CH(NH_3^+)COOH$	-234.74	aq	2.162	-175.95	$\Delta H^\circ_{ioniz}, pK_a$ (α-COOH): 76C
	$HOOCCH_2CH_2CH(NH_3^+)COO^-$	-234.8	aq	4.272	-173.0	ΔH°_{soln}: 76H$_c$. ΔG°_f: 76H$_d$
Glutamate$^-$	$^-OOCCH_2CH_2CH(NH_3^+)COO^-$	-234.43	aq†	9.358	-167.17	$\Delta H^\circ_{ioniz}, pK_a$ (γ-COOH): 76C
Glutamate^{2-}	$^-OOCCH_2CH_2CH(NH_2)COO^-$	-224.85	aq		-154.40	$\Delta H^\circ_{ioniz}, pK_a$ (α-NH$_3^+$): 76C
L-Glutamine	$H_2NCOCH_2CH_2CH(NH_2)COOH$	-197.5	c		-127.3	$\Delta H^\circ_f, \Delta G^\circ_f$: 76H$_b$
	$H_2NCOCH_2CH_2CH(NH_3^+)COOH$		aq	2.17	-129.56	pK_a (α-COOH): 76J
	$H_2NCOCH_2CH_2CH(NH_3^+)COO^-$		aq†		-126.6	ΔG°_f: 76H$_d$
	$H_2NCOCH_2CH_2CH(NH_2)COO^-$		aq	9.13	-114.14	pK_a (α-NH$_3^+$): 76J
Glycine	$CH_2(NH_2)COOH$	-127.35	c		-89.25	$\Delta H^\circ_f, \Delta G^\circ_f$: 76H$_b$
	$CH_2(NH_3^+)COOH$	-124.84	aq	2.351	-92.96	$\Delta H^\circ_{ioniz}, pK_a$ (α-COOH): 76C
	$CH_2(NH_3^+)COO^-$	-123.75	aq†		-89.75	ΔH°_{soln}: 76H$_c$. ΔG°_f: 76H$_d$
	$CH_2(NH_2)COO^-$	-113.14	aq	9.778	-76.41	$\Delta H^\circ_{ioniz}, pK_a$ (α-NH$_3^+$): 76C
Guanine	$C_5H_5ON_5$	-44.15	c		$+11.16$	$\Delta H^\circ_c, S^\circ$: 35S. $\Delta H^\circ_f, S^\circ$(H$_2$O, etc.): 78C
	$C_5H_4ON_5^+$		aq	3.3	$+12.77$	pK_a (N$_7$,-H$^+$): 76J
	$C_5H_5ON_5$		aq†		$+17.27$	Solubility: 68M
	$C_5H_5ON_5^-$		aq	9.2	$+29.82$	pK_a (N$_1$,-C$_2$O group): 76J
	$C_5H_5ON_5^{2-}$		aq	12.3	$+46.60$	pK_a (N$_9$,-H): 76J

(continued on next page)

Table A-1b—Continued.

Substance	Formula	State†	ΔH°_f‡ (kcal·mol⁻¹)	pK_a	ΔG°_f‡ (kcal·mol⁻¹)	Source§
Hydrogen cyanide	HCN	g	+32.3		+29.8	ΔH°_f, ΔG°_f: 68W
	HCN	aq†	+25.6		+28.6	ΔH°_f, ΔG°_f: 68W
Cyanide	CN⁻	aq	+36.0	9.20	+41.15	ΔH°_{ioniz}, pK_a: 76C
Hypoxanthine	$C_5H_4ON_4$	c	−26.69		+18.19	ΔH°_c, S°: 35S. ΔH°_f, $S^\circ(H_2O$, etc.): 78C
	$C_5H_5ON_4^+$	aq		1.845	+18.68	pK_a (N₇–H⁺): 76C
	$C_5H_4ON_4$	aq†			+21.20	ΔG°_{soln}: 57B
	$C_5H_3ON_4^-$	aq		8.855	+33.28	pK_a (N₁–C₂O group): 76C
	$C_5H_2ON_4^{2-}$	aq		12.035	+49.70	pK_a (N₉–H): 76C
L-Isoleucine	$CH_3CH_2CH(CH_3)CH(NH_2)COOH$	c	−152.5		−83.0	ΔH°_f, ΔG°_f: 76H_b
	$CH_3CH_2CH(CH_3)CH(NH_3^+)COOH$	aq	−152.0	2.319	−85.36	ΔH°_{ioniz}, pK_a (α-COOH): 76C
	$CH_3CH_2CH(CH_3)CH(NH_3^+)COO^-$	aq†	−151.6		−82.2	ΔH°_{soln}: 76H_c. ΔG°_f: 76H_d
	$CH_3CH_2CH(CH_3)CH(NH_2)COO^-$	aq	−140.87	9.758	−68.89	ΔH°_{ioniz}, pK_a (α-NH₃⁺): 76C
L-Leucine	$CH_3CH(CH_3)CH_2CH(NH_2)COOH$	c	−153.5		−84.25	ΔH°_f, ΔG°_f: 76H_b
	$CH_3CH(CH_3)CH_2CH(NH_3^+)COOH$	aq	−153.02	2.328	−86.33	ΔH°_{ioniz}, pK_a (α-COOH): 76C
	$CH_3CH(CH_3)CH_2CH(NH_3^+)COO^-$	aq†	−152.6		−83.15	ΔH°_{soln}: 76H_c. ΔG°_f: 76H_d
	$CH_3CH(CH_3)CH_2CH(NH_2)COO^-$	aq	−141.7	9.744	−69.86	ΔH°_{ioniz}, pK_a (α-NH₃⁺): 76C
L-Lysine	$H_2NCH_2(CH_2)_3CH(NH_2)COOH$	c	−162.2			ΔH°_f: 68M.
L-Methionine	$CH_3SCH_2CH_2CH(NH_2)COOH$	c	−181.2		−120.9	ΔH°_f, ΔG°_f: 76H_b
	$CH_3SCH_2CH_2CH(NH_3^+)COOH$	aq		2.28	−123.31	pK_a (α-COOH): 76J
	$CH_3SCH_2CH_2CH(NH_3^+)COO^-$	aq†	−178.4		−120.2	ΔH°_{soln}: 76H_c. ΔG°_f: 76H_d
	$CH_3SCH_2CH_2CH(NH_2)COO^-$	aq		9.21	−107.64	pK_a (α-NH₃⁺): 76J
Methylamine	CH_3NH_2	g	−5.49		+7.67	ΔH°_f, ΔG°_f: 68W
	$CH_3NH_3^+$	aq†	−29.99	10.62	−9.55	ΔH°_{ioniz}, pK_a: 76C
	CH_3NH_2	aq	−16.77		+4.94	ΔH°_f, ΔG°_f: 68W

Compound	Formula	State	ΔH_f°	pK_a	ΔG_f°	Notes
L-Phenylalanine	$(C_6H_5)CH_2CH(NH_2)COOH$	c	−111.6		−50.6	$\Delta H_f^\circ, \Delta G_f^\circ$: 76H$_b$
	$(C_6H_5)CH_2CH(NH_3^+)COOH$	aq		1.83	−52.0	pK_a (α-COOH): 76J
	$(C_6H_5)CH_2CH(NH_3^+)COO^-$	aq†	−108.8		−49.5	ΔH°_{soln}: 76H$_c$, ΔG_f°: 76H$_d$
	$(C_6H_5)CH_2CH(NH_2)COO^-$	aq	−98.13	9.31	−36.80	ΔH°_{ioniz}, pK_a (α-NH$_3^+$): 76C
L-Proline	$C_5H_9O_2N$	c	−125.7		−72.04	ΔH_f°: 68M. S°: 76H$_a$. S°(C etc.): 78C
	$C_5H_{10}O_2N^+$	aq	−125.74	1.952	−76.96	ΔH°_{ioniz}, pK_a (α-COOH): 76C
	$C_5H_9O_2N$	aq†	−125.4		−74.3	ΔH°_{soln}: 76H$_c$, ΔG°_{soln}: 76H$_d$
	$C_5H_8O_2N^-$	aq	−115.09	10.64	−59.78	ΔH°_{ioniz}, pK_a (α-NH$_3^+$): 76C
Pyridine	C_5H_5N	l	+23.89		+42.33	$\Delta H_f^\circ, \Delta G_f^\circ$: 70K
L-Serine	$CH_2(OH)CH(NH_2)COOH$	c	−173.6		−121.6	$\Delta H_f^\circ, \Delta G_f^\circ$: 76H$_b$
	$CH_2(OH)CH(NH_3^+)COOH$	aq	−172.17	2.187	−125.08	ΔH°_{ioniz}, pK_a (α-COOH): 76C
	$CH_2(OH)CH(NH_3^+)COO^-$	aq†	−170.8		−122.1	ΔH°_{soln}: 76H$_c$, ΔG_f°: 76H$_d$
	$CH_2(OH)CH(NH_2)COO^-$	aq	−160.42	9.208	−109.54	ΔH°_{ioniz}, pK_a (α-NH$_3^+$): 76C
Thiocyanic acid	HCNS	aq	+5.27	0.95	+20.85	ΔH°_{ioniz}, pK_a: 76C
Thiocyanate	CNS$^-$	aq†	+18.27		+22.15	$\Delta H_f^\circ, \Delta G_f^\circ$: 68W
L-Threonine	$CH_3CH(OH)CH(NH_2)COOH$	c	−192.9		−131.5	$\Delta H_f^\circ, \Delta G_f^\circ$: 76H$_b$
	$CH_3CH(OH)CH(NH_3^+)COO^-$	aq†			[−123]	ΔG_f°: 57B
Trimethylamine	$(CH_3)_3N$	g	−5.81		+23.65	$\Delta H_f^\circ, \Delta G_f^\circ$: 68W
	$(CH_3)_3NH^+$	aq†	−26.98	9.786	+8.87	ΔH°_{ioniz}, pK_a: 76C
	$(CH_3)_3N$	aq	−18.17		+22.22	$\Delta H_f^\circ, \Delta G_f^\circ$: 68W
L-Tryptophan	$(C_8H_8N)CH_2CH(NH_2)COOH$	c	−99.2		−28.5	$\Delta H_f^\circ, \Delta G_f^\circ$: 76H$_b$
	$(C_8H_8N)CH_2CH(NH_3^+)COOH$	aq		2.38	−30.15	pK_a (α-COOH): 76J
	$(C_8H_8N)CH_2CH(NH_3^+)COO^-$	aq†	−97.8		−26.9	ΔH°_{soln}: 76H$_c$, ΔG_f°: 76H$_d$
	$(C_8H_8N)CH_2CH(NH_2)COO^-$	aq		9.39	−14.09	ΔH°_{ioniz}, pK_a (α-NH$_3^+$): 76J
L-Tyrosine	$(HOC_6H_5)CH_2CH(NH_2)COOH$	c	−160.5		−92.2	$\Delta H_f^\circ, \Delta G_f^\circ$: 76H$_b$
	$(HOC_6H_5)CH_2CH(NH_3^+)COOH$	aq		2.20	−91.60	pK_a (α-COOH): 76J
	$(HOC_6H_5)CH_2CH(NH_3^+)COO^-$	aq†	−154.5		−88.6	ΔH°_{soln}: 76H$_c$, ΔG_f°: 76H$_d$
	$(HOC_6H_5)CH_2CH(NH_2)COO^-$	aq		9.11	−76.17	pK_a (α-NH$_3^+$): 76J
	$(^-OC_6H_5)CH_2CH(NH_2)COO^-$	aq		10.07	−62.43	pK_a (Ar-OH): 76J
Urea	$CO(NH_2)_2$	c	−79.6		−47.1	$\Delta H_f^\circ, \Delta G_f^\circ$: 76H$_b$
	$CO(NH_2)_2$	aq			−48.72	ΔG_f°: 57B

(continued on next page)

Table A-1b—Continued.

Substance	Formula	State†	ΔH_f°‡	pK_a	ΔG_f°‡	Source§
			kcal·mol⁻¹		kcal·mol⁻¹	
Uric acid	$C_5H_4O_3N_4$	c	−148.05		−90.94	ΔH_c°, S°: 35S. ΔH_f°, S°(H₂O etc.): 78C
	$C_5H_4O_3N_4$	aq			−85.84	ΔG_{soln}°: 57B
Urate⁻	$C_5H_3O_3N_4^-$	aq†		5.4	−78.47	pK_a: 76J
Urate²⁻	$C_5H_2O_3N_4^{2-}$	aq		10.3	−64.42	pK_a: 76J
L-Valine	$CH_3CH(CH_3)CH(NH_2)COOH$	c	−147.7		−85.8	ΔH_f°, ΔG_f°: 76Hb
	$CH_3CH(CH_3)CH(NH_3^+)COOH$	aq	−146.88	2.286	−88.42	ΔH_{ioniz}°, pK_a (α-COOH): 76C
	$CH_3CH(CH_3)CH(NH_3^+)COO^-$	aq†	−146.8		−85.3	ΔH_{soln}°: 76Hc, ΔG_f°: 76Hd
	$CH_3CH(CH_3)CH(NH_2)COO^-$	aq	−136.06	9.72	−72.04	ΔH_{ioniz}°, pK_a (α-NH₃⁺): 76C
Xanthine	$C_5H_4O_2N_4$	c	−90.87		−39.79	ΔH_c°, S°: 35S. ΔH_f°, S°(H₂O etc.): 78C
	$C_5H_5O_2N_4^+$	aq		0.8	−34.54	pK_a (N₇-H⁺): 76J
	$C_5H_4O_2N_4$	aq†			−33.45	ΔG_{soln}°: 57B
	$C_5H_3O_2N_4^-$	aq		7.53	−23.18	pK_a (N₁-C₆OH): 76C
	$C_5H_2O_2N_4^{2-}$	aq		11.84	−7.03	pK_a (N₉-H): 76C

† Identifies the dominant ionic species at pH 7 as derived from pK_a data (Example 3).

‡ Numbers in square brackets [] are estimates. Numbers underlined were derived by the compiler and are expressed to two decimal places to allow accurate reconversion rather than to represent absolute accuracy. Calculation methods were according to standard procedures (Wagman et al., 1968; Hutchens, 1976). The relationship between pK_a and ΔG_f° is, for example (Eq. [35]): $pK_a = (\Delta G_f^{\circ}AH_n - \Delta G_f^{\circ}AH_{n+1})/1.3642$. All data are for 298.15°K and zero ionic strength.

§ Source (numbers represent year; see Section VII for complete citation): 78C, CODATA; 76C, Christensen et al.; 76Hₐ to 76H_d, Hutchens; 73W, Weast; 76J, Jencks & Regenstein; 70K, Karapet'yants & Karapet'yants; 69W, Wilhoit; 68W, Wagman et al.; 57B, Burton; 35S, Stiehler & Huffman.

¶ Derived from the theoretically based assumption that $\Delta G_c^{\circ} \simeq \Delta H_c^{\circ}$ (Morowitz, 1968) and experimental $\Delta H_c^{\circ} = -27$ kcal·e⁻ eq cells⁻¹ (Erickson, 1980; Ho & Payne, 1979).

EXAMPLE 11

Derivation of ΔG_f^o from pK_a: (i) Calculate the ΔG_f^o of $NH_3(aq)$ from ΔH_f^o $NH_4^+(aq)$, S^o $NH_4^+(aq)$ and pK_a $NH_4^+-NH_3(aq)$ data; and (ii) calculate the ΔG_f^o of glutamate$^-$ from ΔG_f^o glutamic acid and pK_a glutamic acid-glutamate$^-$ data.

i) The ΔG_f^o $NH_4^+(aq)$ must first be derived from the fundamental ΔH_f^o and S^o properties, starting with calculation of ΔS_f^o $NH_4^+(aq)$. The formation reaction for $NH_4^+(aq)$ is

$$1\frac{1}{2}H_2(g) + H^+(aq) + \frac{1}{2}N_2(g) = NH_4^+(aq).$$

Recognizing that S^o $H_2(g) = 31.207$, S^o $H^+(aq) = 0$, S^o $N_2(g) = 45.770$, and S^o $NH_4^+(aq) = 26.57$ cal•^{o}K^{-1}•mol^{-1} (CODATA, 1978), the standard entropy change of formation, ΔS_f^o, is given by

ΔS_f^o $NH_4^+(aq)$

$= S^o$ $NH_4^+(aq) - [1\frac{1}{2}S^o$ $H_2(g) + S^o$ $H^+(aq) + \frac{1}{2}S^o$ $N_2(g)]$

$= 26.57 - (46.8105 + 0 + 22.885) = -43.1255$ cal•^{o}K^{-1}•mol^{-1}

$T\Delta S_f^o$ $NH_4^+(aq)$

$= [298.15 \, \Delta S_f^o \, NH_4^+(aq)]/1000 = -12.858$ kcal•mol^{-1}.

Recognizing that ΔH_f^o $NH_4^+(aq) = -31.85$ kcal•mol^{-1} (CODATA, 1978), ΔG_f^o $NH_4^+(aq)$ is derived as (Wagman et al., 1968)

ΔG_f^o $NH_4^+(aq) = \Delta H_f^o$ $NH_4^+(aq) - T\Delta S_f^o$ $NH_4^+(aq)$

$= -31.85 - (-12.858) = -18.99$ kcal•mol^{-1}.

The pK_a for NH_4^+ ionization to NH_3 (25°C and zero ionic strength) is (Christensen et al., 1976)

$$NH_4^+(aq) = NH_3(aq) + H^+, \qquad pK_a = 9.244. \qquad [100]$$

From Eq. [100] and [36] (rearranged):

ΔG_f^o $NH_3(aq) = \Delta G_f^o$ $NH_4^+(aq) - 1.3642 \, pK_a = -6.38$ kcal•mol^{-1}.

ii) The ΔG_f^o of glutamic acid (aq) is -173.0 kcal•mol^{-1} (Hutchens, 1976d). The pK_a for glutamic acid ionization to glutamate$^-$ (at 25°C and zero ionic strength) is (Christensen et al., 1976)

$HOOCCH_2CH_2CH(NH_3^+)COO^-$

$$= {}^-OOCCH_2CH_2CH(NH_3^+)COO^- + H^+, \qquad pK_a = 4.272. \quad [101]$$

From Eq. [101] and [36] (rearranged):

ΔG_f° glutamate$^-$

$$= 1.3642\ pK_a + \Delta G_f^\circ \text{ glutamic acid } = -167.17\ \text{kcal}\cdot\text{mol}^{-1}.$$

ACKNOWLEDGMENT

This chapter is a contribution from the College of Agriculture and Life Sciences, University of Wisconsin, Madison, supported in part by the Wisconsin Alumni Research Foundation (project no. 190204) and USDA-SEA Hatch (project no. 2495).

VII. LITERATURE CITED

Alberty, R. A., and F. Daniels. 1979. Physical chemistry. John Wiley & Sons, Inc., New York.

Aleem, M. I. H. 1977. Coupling of energy with electron transfer reactions in chemolithotrophic bacteria. Symp. Soc. Gen. Microbiol. 27:351-381.

Battley, E. H. 1979. Alternate method of calculating the free energy change accompanying the growth of *Saccharomyces cerevisiae* (Hansen) on three substrates. Biotechnol. Bioeng. XXI:1929-1961.

Bremner, J. M. 1965. Total nitrogen. *In* C. A. Black et al. (ed.) Methods of soil analysis, Part 2. Agronomy 9:1148-1178. Am. Soc. of Agron., Inc., Madison, Wis.

Bremner, J. M., and A. M. Blackmer. 1978. Nitrous oxide emission from soil during nitrification of fertilizer nitrogen. Science 199:295-296.

Burton, K. 1957. Free energy data of biological interest. Physiol. Biol. Chem. Exp. Pharmakol. 49:275-298.

Burton, K., and H. A. Krebs. 1953. The free energy changes associated with the individual steps of the tricarboxylic acid cycle, glycolysis and alcoholic fermentation and with the hydrolysis of the pyrophosphate groups of adenosine triphosphate. Biochem. J. 59:94-100.

Buvet, R. 1977. Energetics of coupled biochemical processes and of their chemical models. p. 21-39. *In* R. Buvet et al. (ed.) Living systems as energy converters. Elsevier North-Holland Inc., New York.

Caplan, S. R. 1976. Biothermodynamics. p. 1-37. *In* D. R. Sanadi (ed.) Chemical mechanisms in bioenergetics. ACS Monogr. 172. Am. Chem. Soc., Washington, D.C.

Christensen, J. J., L. D. Hansen, and R. M. Izatt. 1976. Handbook of proton ionization heats and related thermodynamic properties. John Wiley & Sons, Inc., New York.

Committee on Data for Science and Technology. 1978. CODATA recommended key values for thermodynamics 1977. CODATA Bull. 28.

Cornish-Bowden, A. 1979. Fundamentals of enzyme kinetics. Butterworth (Publishers) Inc., Woburn, Mass.

Dalton, H. 1979. Utilization of inorganic nitrogen by microbial cells. Microb. Biochem. 21: 227-266.

Doelle, H. W. 1975. Bacterial metabolism. Academic Press, Inc., New York.

Emerich, D. W., and H. J. Evans. 1980. Biological nitrogen fixation with an emphasis on legumes. p. 117-145. *In* A. S. Pietro (ed.) Biochemical and photosynthetic aspects of energy production. Academic Press, Inc., New York.

Erickson, L. E. 1979. Energetic efficiency of biomass and product formation. Biotechnol. Bioeng. XXI:725–743.

Erickson, L. E. 1980. Biomass elemental composition and energy content. Biotechnol. Bioeng. XXII:451–456.

Fenchel, T., and T. H. Blackburn. 1979. Bacteria and mineral cycling. Academic Press, Inc., New York.

Hageman, R. V., and R. H. Burris. 1980. Electrochemistry of nitrogenase and the role of ATP. Curr. Top. Bioenerg. 10:279–291.

Harris, R. F. 1978. Non symbiotic nitrogen fixation—state of the art. p. 74–82. In Proceedings of the Fertilizer and Aglime Conference, Madison, Wis. 17–18 Jan. 1978. College Agric. Life Sci., Univ. of Wisconsin, Madison.

Harris, R. F. 1981. Effect of water potential on microbial growth and activity. p. 23–95. In J. F. Parr and W. R. Gardner (ed.) Water potential relations in soil microbiology. ASA Spec. Pub. no. 9, Madison, Wis.

Harris, R. F., and S. S. Adams. 1979. Determination of the carbon-bound electron composition of microbial cells and metabolites by dichromate oxidation. Appl. Environ. Microbiol. 37:237–243.

Ho, K. P., and W. J. Payne. 1979. Assimilation efficiency and energy contents of prototrophic bacteria. Biotechnol. Bioeng. XXI:787–802.

Hutchens, J. O. 1976a. Heat capacities, absolute entropies, and entropies of formation of amino acids and related compounds. p. 109–110. In G. F. Fasman (ed.) Handbook of biochemistry and molecular biology. CRC Press, Inc., West Palm Beach, Fla.

Hutchens, J. O. 1976b. Heat of combustion, enthalpy and free energy of formation of amino acids and related compounds. p. 111–112. In G. F. Fasman (ed.) Handbook of biochemistry and molecular biology. CRC Press, Inc., West Palm Beach, Fla.

Hutchens, J. O. 1976c. Heats of solution of amino acids in aqueous solution at 25°C. p. 116–117. In G. F. Fasman (ed.) Handbook of biochemistry and molecular biology. CRC Press, Inc., West Palm Beach, Fla.

Hutchens, J. O. 1976d. Free energies of solution and standard free energy of formation of amino acids in aqueous solution at 25°C. p. 118. In G. F. Fasman (ed.) Handbook of biochemistry and molecular biology. CRC Press, Inc., West Palm Beach, Fla.

Jencks, W. P., and J. Regenstein. 1976. Ionization constants of acids and bases. p. 322–351. In G. F. Fasman (ed.) Handbook of biochemistry and molecular biology. CRC Press, Inc., West Palm Beach, Fla.

Jones, C. W. 1979. Energy metabolism in aerobes. Microb. Biochem. 21:49–84.

Karapet'yants, M. Kh., and M. L. Karapet'yants. 1970. Thermodynamic constants of inorganic and organic compounds. (Translated by J. Schmorak, Israel Program for Scientific Translations.) Ann Arbor-Humphrey Science Publishers, Ann Arbor, London.

Klotz, I. M. 1967. Energy changes in biochemical reactions. Academic Press, Inc., New York.

Knowles, C. J. 1976. Microorganisms and cyanide. Bacteriol. Rev. 40:652–680.

Latimer, W. L. 1952. Oxidation potentials. Prentice-Hall, Inc., Englewood Cliffs, N.J.

Lehninger, A. L. 1965. Bioenergetics. The Benjamin Co., Inc., New York.

Lindsay, W. L. 1979. Chemical equilibria in soils. Wiley-Interscience, New York.

Macfadyen, A. 1971. The soil and its total metabolism. p. 1–13. In J. Phillipson (ed.) Methods of study in quantitative soil ecology: Population, production and energy flow. Blackwell Scientific Publications Ltd., Oxford.

McCarty, P. L. 1972. Energetics of organic matter degradation. p. 91–118. In R. Mitchell (ed.) Water pollution microbiology. Wiley-Interscience, New York.

Morowitz, H. J. 1968. Energy flow in biology: Biological organization as a problem in thermal physics. Academic Press Inc., New York.

Postgate, J. R., and S. Hill. 1979. Nitrogen fixation. p. 191–213. In J. M. Lynch and N. J. Poole (ed.) Microbial ecology: A conceptual approach. John Wiley & Sons, Inc., New York.

Roels, J. A., and N. W. F. Kossen. 1978. On the modelling of microbial metabolism. Prog. Ind. Microbiol. 14:95–203.

Roels, J. A., and J. C. van Suijdam. 1980. Energetic efficiency of a microbial process with an external power input: Thermodynamic approach. Biotechnol. Bioeng. XXII:463–471.

Sadiq, M., and W. L. Lindsay. 1979. Selected standard free energies of formation for use in soil science. p. 386–422. In W. L. Lindsay. Chemical equilibria in soils. Wiley-Interscience, New York.

Segel, I. H. 1976. Biochemical calculations. John Wiley & Sons, Inc., New York.

Shilov, A. E. 1976. Dinitrogen reduction in protic media. p. 42–52. *In* W. E. Newton and C. Nyman (ed.) Proceedings of the First International Conference on Nitrogen Fixation, Pullman, Wash. 3–7 June 1974. Washington State University Press, Pullman.

Stanier, R. Y., E. A. Adelberg, and J. Ingraham. 1976. The microbial world. Prentice-Hall, Inc., Englewood Cliffs, N.J.

Stiefel, E. I. 1977. The mechanisms of nitrogen fixation. p. 69–108. *In* W. Newton et al. (ed.) Recent developments in nitrogen fixation. Academic Press, Inc., New York.

Stiehler, R. D., and H. M. Huffman. 1935. Thermal data. V. The heat capacities, entropies and free energies of adenine, hypoxanthine, guanine, xanthine, uric acid, allantoin and alloxan. J. Am. Chem. Soc. 57:1741–1743.

Stumm, W., and J. J. Morgan. 1970. Aquatic chemistry. Wiley-Interscience, New York.

Suzuki, I. 1974. Mechanism of inorganic oxidation and energy coupling. Annu. Rev. Microbiol. 28:85–101.

Tempest, D. W., and O. M. Neijssel. 1978. Eco-physiological aspects of microbial growth in aerobic nutrient-limited environments. Adv. Microb. Ecol. 2:105–153.

Thauer, R. K., K. Jungerman, and K. Decker. 1977. Energy conservation in chemotrophic anaerobic bacteria. Bacteriol. Rev. 41:100–180.

Van Dam, K., H. V. Westerhoff, K. Krab, R. van der Meer, and J. C. Arents. 1980. Relationship between chemiosmotic flows and thermodynamic forces in oxidative phosphorylation. Biochim. Biophys. Acta 591:240–250.

Van der Meer, R., H. V. Westerhoff, and K. van Dam. 1980. Linear relation between rate and thermodynamic force in enzyme-catalyzed reactions. Biochim. Biophys. Acta 591:488–493.

Vogels, G. D., and C. van der Drift. 1976. Degradation of purines and pyrimidines by microorganisms. Bacteriol. Rev. 40:403–468.

Wagman, D. D., W. H. Evans, V. B. Parker, I. Halow, S. M. Bailey, and R. H. Schum. 1968. Selected values of thermodynamic properties. Technical Note 270-3. U.S. Dep. of Commerce, National Bureau of Standards, U.S. Government Printing Office, Washington, D.C.

Weast, R. C. 1973. Physical constants of organic compounds. p. C75–543. *In* R. C. Weast (ed.) Handbook of chemistry and physics. CRC Press, Inc., West Palm Beach, Fla.

Wikstrom, M., and K. Krab. 1980. Respiration-linked H^+ translocation in mitochondria: Stoichiometry and mechanism. Curr. Top. Bioenerg. 10:51–101.

Wilhoit, R. C. 1969. Thermodynamic properties of biochemical substances. p. 33–81, 305–317. *In* F. D. Brown (ed.) Biochemical microcalorimetry. Academic Press, Inc., New York.

Willis, C., F. P. Lossing, and R. A. Back. 1976. The heat of formation of N_2H_2 and the proton affinity of N_2. Can. J. Chem. 54:1–3.

23   Nitrogen Transfers and Mass Balances

R. D. HAUCK

Tennessee Valley Authority
Muscle Shoals, Alabama

K. K. TANJI

University of California
Davis, California

I. INTRODUCTION

Terrestrial humus, lakes, streams, sea bottoms, and living organisms contain about 0.02% of the earth's N. Much of this N, unlike that in fundamental rocks (which contain about 98% of the earth's total N), is in a continual state of flux. It is this comparatively negligible amount of N flowing within and through complex internal cycles of the biosphere and interacting with the atmosphere (1.98% of earth's N) that is of vital importance to animal and plant life. Of the biosphere N, some forms are readily transformed and flow rapidly through time and space, whereas other forms remain relatively inactive over long time periods, having little short-term impact on the system containing them.

Investigators of ecosystem N cycling are faced with the problem of integrating pieces of information that are obtained on the process level into the totality of N transformations and transfers occurring in the system under study. This integration must be done with respect to time and space and must consider interrelated N processes that occur at different rates for shorter or longer time intervals, that involve different amounts of various forms of N, and that form products that become substrates for other N transformations.

The development of a systematic means for describing and understanding the multiplicity and complexity of N transformations and transfers within and among ecosystems can be approached on several levels of integration. Rowe (1961) discussed a consistent organization of biosphere levels that included the cell, organ, organism, and ecosystem, each object of the hierarchy constituting the environment of the one below and at the same time being part of the structure of the one above. Each level of integration can be studied from the viewpoint of its anatomy (morphology), composition (qualitative and quantitative inventory of objects within it), classification (relationship to its own kind), position in space (microsite to global volume), time, physiology (functional relationship of the object to its

Copyright 1982 © ASA-CSSA-SSSA, 677 South Segoe Road, Madison, WI 53711, USA.
Nitrogen in Agricultural Soils—Agronomy Monograph no. 22.

parts), and ecology (relationship of the object to other objects in its sur-roundings). The levels of integration cannot be mutually exclusive. Thus, for example, the ecology of an organism overlaps the physiology of the eco-system in which it is reacting and being reacted on. Because the hierarchal levels are related to each other in a definable and measurable way, "an understanding of an organization at any level requires attention to the level above and below" (Rowe, 1961).

In any hierarchy of objects, processes occurring within the system or its parts proceed at different rates. The greater the number of different objects and their processes interacting within the system, the more difficult it is to maintain the various time scales in perspective. The residence time of N in any given compartment of the biosphere may vary from short (e.g., seconds) to long (e.g., eons). Dinitrogen may be recycled to the atmosphere within hours after fixation in soil, or it may be locked in storage reservoirs such as coal for 200 million years before release through burning. The time scales for the many and various N cycles are determined by the nature of processes involved, the rates of transfer and mixing of reactants, and the rates of chemical and biological reactions. Bolin (1976) discussed character-istic time scales for the movement of N and related substances in atmos-pheres and waters. For example, vertical mixing of substances in the tropo-sphere may approach completion within a month after addition compared with several months for horizontal mixing. Pronounced seasonal effects on the rates of atmospheric mixings have been observed, e.g., the rate of mix-ing in a given region of the troposphere may be 50 to 100 times greater than, or about equal to, that in a corresponding region of the stratosphere during summer and winter, respectively. Bolin (1976) suggested that when con-sidering large-scale transfers of N, one should also consider those of C, O, P, and perhaps S, which may occur at markedly different rates in processes that affect the cycling of N.

K. K. Tanji (Chapt. 19) discusses the construction and use of models to simulate N flow through agricultural ecosystems. J. O. Legg and J. J. Meisinger (Chapt. 14) discuss N budgets for agricultural soils. Other chapters in this monograph are concerned with processes that cause trans-formations and transfers of N on the cellular and organismal level and with the impact of these processes on the environment. We outline here the major N transformations and transfers within, into, and from various eco-systems and summarize N balance information obtained from some subre-gional, regional, and global studies, keeping in mind the difficulties of interpolating from one level of integration to another or from one time scale to another.

II. N TRANSFORMATIONS AND TRANSFERS

A. General Considerations

Wind and water are the main agents of N transfer over large distances within and among ecosystems. Humans and animals also transport plant-derived N over large distances after grazing or harvest. In the microcosm, diffusion and mass flow are the main processes by which N is transferred within a cell, an organism, or a microsite. Although we discuss here mainly the movement of N within and among macrosystems, we recognize that N transfers and transformations within the microcosms influence the form and amount of N that moves from one object and site to another.

B. N Income, Outgo, and Transfer

1. SINKS AND SOURCES

Terrestrial, lacustrine, and marine ecosystems gain N through biological N_2 fixation, from precipitation, dry fallout, transported sediment, drainage (inflow) waters, and from sources that are the result of animal and human activities. Included among N loss mechanisms are denitrification, leaching, runoff and outflow, NH_3 volatilization, and harvesting. Some transfer processes result in N loss from one ecosystem and N gain by another. However, some N may be transferred within an ecosystem from relatively active N pools to relatively stable sites, e.g., entrapment of NH_4^+-N in clay lattices or immobilization of N in biologically resistant soil organic matter. Obviously, ecosystems differ with respect to the nature and size of their N inputs and removals, and within an ecosystem, all processes whereby N is gained or lost do not occur at the same time. Also, several ecosystems of the same kind within a region may differ in the nature and extent of their N gains and losses. For example, Brezonik and Powers (1973) detected the occurrence of N_2 fixation by blue-green algae in only 15 of 55 Florida lakes; when N_2 fixation was detected, it occurred at markedly different rates at different times within a given lake and not necessarily at the same time among neighboring lakes. Because N supplies, transformations, and transfers are linked, frequent monitoring of all major processes involved in N cycling is requisite for accurately understanding the N economy of a given ecosystem and its surroundings.

A reliable summary of the magnitudes of all N sources and sinks for major ecosystems cannot be prepared from current data, but the discussion that follows includes selected references to the kind of information that is available; these references, in turn, cite additional studies that report measurements of N gains and losses for specific ecosystems.

2. TRANSFERS

a. Lakes—The variety and extent of the measurements that should be made to follow N transfers into and out of contiguous ecosystems can be appreciated by studying a listing of the possible N sinks and sources for a lake and its surroundings. Brezonik and Powers (1973) listed 16 sources of N income and 10 sources of N outgo. The sources are surface waters containing N from agricultural runoff and drainage, animal wastes, marsh drainage, uncultivated and forestlands, storm drainage from urban areas, domestic waste and effluents, industrial waste and effluents, and boating activity wastes; underground waters containing N from uncultivated land, agricultural and urban subsurface drainage, and septic tanks near lake-shores; airborne N from precipitation, aerosols, dust, leaves, and mis-cellaneous debris; and N from N_2 fixation and sediment leaching. Sinks for N include NH_3 volatilization, denitrification, effluent loss, evaporation (aerosol formation), ground water discharge, sediment deposition of detritus, sorption of NH_4^+, insect emergence, and fish and weed harvest.

Because the massive effort needed to obtain reliable data for all N sources and sinks for a given lacustrine ecosystem probably will not be made in the foreseeable future, simple models that simulate N transport into and from lakes have been developed (e.g., see Lee & Sonzogni, 1972). These models consider the size of the lake and its drainage basin, hydrology, land use patterns, and area population, using appropriate in-formation from the literature on nutrient concentrations and N transforma-tion rates.

Quantitative information on natural and cultural (anthropogenic) N sources that contribute to lacustrine N budgets is sparse, but concern over eutrophication has stimulated efforts to measure N and P contributions from various sources. These efforts have resulted in the establishment of approximate N budgets for a few North American and European lakes (for reviews, see Brezonik, et al., 1969; Vollenweider, 1968). As the literature on nutrient export rates as a function of various land use patterns becomes more complete, the accuracy of literature-based estimates of nutrient budgets will increase. The accuracy of these estimates depends not only on the accuracy of the original measurements but also on the applicability of the data to the nutrient budget of a given lake and its surroundings. Never-theless, literature-based models seem to be the only reasonable approach for understanding the N budgets of many lakes under cultural stress (Brezonik & Powers, 1973).

b. Seas—The marine N cycles are no less complex than terrestrial and freshwater cycles. A distinguishing feature of oceanic regimes is that N translocations occur over large geographical areas as compared with those usually occurring on land. Nitrogen cycle processes in the sea appear to be similar qualitatively to those in soils, the main differences between the two environments being the species composition of their micropopulations and macropopulations and the mobility of N dissolved in seawater (Vaccaro, 1965).

Not only is it a formidable task to approximate N budgets for specific locations in seas, it is also difficult to measure specific N transformation rates and transfers. Spacial variability extends both horizontally and vertically. Most N transformations in seas occur in the upper several hundred meters, the uppermost water layers being the euphoric zones (Dug-dale & Goering, 1967). Seas contain highly productive regions where NO_3^- accounts for as much as 30% of the inorganic N taken up. However, in a tropical "desert" zone, as little as 1% of the inorganic N assimilated is NO_3^--N. Turnover rates for inorganic N were found by Hattori and Wada (1973) to vary considerably among regions; for example, residence times of

NO_2^--N were 30 to 60 days in the northern North Pacific, 30 days in the equatorial South Pacific, and 1 day or less in Sagami Bay (Japan) during summer. Residence times for NH_4^+-N and NO_3^--N in subarctic surface waters of the North Pacific were about 15 days and 500 days, respectively. Measurements of turnover rates are complicated by local and seasonal variations in microbial succession, especially in coastal waters. The order and rate of succession may markedly affect N transformation rates; for example, NO_2^- may be produced mainly by NH_4^+ oxidation in one area at a given time and by NO_3^- reduction (carried out by different microorganisms) in another.

Marine N balance accounts show a large apparent loss of residual N, suggesting the occurrence of substantial denitrification (Dugdale & Goering, 1967). However, knowledge of this process in oceans is so sparse that estimates of the annual contribution by oceans to tropospheric N_2O levels range from zero to 40 million metric tons (or 40 Tg).

The actual amount of N_2O transferred from seas to the troposphere is of agricultural interest; the level of tropospheric N_2O affects the amount transferred to the stratosphere where N_2O is involved in the regulation of stratospheric ozone. Mathematical models that attempt to assess the environmental impact of various N inputs to agricultural soils require accurate quantitative information on all major sources of and sinks for N_2O (e.g., see CAST, 1976).

c. Marshlands—Freshwater and salt swamps and marshes trap nutrients moving toward lakes and seas, and they also contribute nutrients to waters. Nitrogen and other nutrients are retained during the growing season and are drained from marshlands during short periods of high water outflows, usually during the spring (Benson, 1965). A study of several marshes in Wisconsin by Lee et al. (1975) showed them to be in approximate N balance, with N outgo during one period being matched with N income during the remainder of the year. Draining marshes to convert them to dry land promotes oxidation and mineralization of organic matter, with release of considerable N in forms susceptible to loss. For salt marshes, N loss through denitrification and tidal outflow appears to balance N gains from precipitation, biological N_2 fixation, and groundwater flow (Valiela & Teal, 1979). Salt marshes actively recycle N as a consequence of an internal demand that generally exceeds N inputs. The internal transformations convert oxidized N forms to NH_4^+ and particulate (proteinaceous) N; these reduced N forms are then exported to coastal water. Oxidized N forms are transported to salt marshes from uplands and in tidal inflow and leave them in reduced forms through tidal outflow. Valiela and Teal (1979) reported a net loss of N through tidal exchange. Their N budget estimate included N gains from bird droppings (insignificant) and N losses through NH_3 volatilization and shellfish harvest (both negligible compared with denitrification and tidal outflow). Gains of N followed the order: ground water flow > biological N_2 fixation > precipitation > bird feces. Net loss of N was slightly greater from denitrification than from tidal outflow (6,940 vs. 6,400 kg of N from 483,800 m^2, equivalent to 143 vs. 132 kg of $N \cdot ha^{-1} \cdot year^{-1}$), but 31,600 kg of N (653 kg·ha^{-1}) was estimated to flow annually from this area and 26,200 kg (542 kg·ha^{-1}) to return through tidal exchange. The authors concluded that the considerable exchange of nutrients between upland soils, marsh, and coastal waters affects the structure and composition of marsh and coastal water ecosystems.

d. Forestlands—Nitrogen enters the forest ecosystem through geologic weathering of parent rock and via meteorological and biological inputs. Chemical and physical weathering of rocks, minerals, and associated organic matter solubilizes N and other nutrients; dissolved and particulate N is added to forests in precipitation, dry fallout, and other aerosols. The biological N inputs are mainly N-containing materials obtained elsewhere and deposited by animals, except in forests where biological N_2 fixation is significant or when N fertilizers are added. Feces of migratory birds and terrestrial animals can add considerable N to local areas within forests, such as in waterfowl marshes and near watering areas of large numbers of mammals (Cooper, 1969). Nitrogen leaves the uncultivated forest in solution or as particulate matter transported by water, as particulates carried by wind, or in gaseous forms after release to the atmosphere by fire or during denitrification.

As with other major ecosystems, generalizations about N transfers and budgets in forests should be made with caution. Forests differ greatly in their N inventories and rates of nutrient cycling, as one would readily observe by comparing a temperate zone forest with a tropical forest, a deciduous forest with a coniferous forest, or an oak/hictory forest (*Quercus* sp./*Carya* sp.) with a mixed deciduous forest. Golley et al. (1975) contrasted N transfers in a tropical moist forest with those of other tropical forest types that differed in the size and composition of their nutrient reservoirs and the size and turnover rates of their biomass components. In forests, vegetation rather than soil can be regarded as the main reservoir of N; because much of the N is stored in woody tissues with low turnover rates, the annual loss of N from undisturbed forests is small. Mature coniferous and deciduous forests are more conserving of N than their younger counterparts (Henderson & Harris, 1975); actively growing secondary vegetation has a higher N concentration and turnover rate than mature vegetation (Golley et al., 1975). Obviously, the greater the transfer of N from vegetation to soil, the greater the opportunity for N transfer from soil to waters. Fertilization and clearcutting tend to increase N export from forests, mainly by increasing runoff and sediment transport (Cole & Gessel, 1965; Bormann et al., 1968). In managed forests, N added as fertilizer may result in long-term gain in total N, but that portion of fertilizer N that is not stored in woody tissue is more susceptible to loss during recycling. At harvest and removal of product, stored N also eventually is transported from the forest. Henderson and Harris (1975) discussed the biogeochemical cycles in three forested landscapes and offered guidelines for studying nutrient cycling dynamics and nutrient transport to streams. They concluded that each of these forests was gaining N, the rate being determined by the rate of industrialization of the surrounding area. Of the N being actively cycled, 2.5% or less was being transported to streams annually. Rain intensity was found to be the most important single factor governing the removal of N from vegetation to soil to waters.

Over the short term, undisturbed forests may have a net gain or loss of N, depending on the extent of biological N_2 fixation. In some forests, biological N_2 fixation may be insignificant (Cooper, 1969), or it may be a forest's largest single source of N (Wollum & Davey, 1975). In general, N transport into waters from undisturbed forests and well-managed forests under cultivation is considered to be small. For a comprehensive review of N management in cultivated forests, see Wollum and Davey (1975).

e. Deserts—Desert soils obtain N mainly from biological N_2 fixation, precipitation, and dry fallout; N is lost mainly through water and wind erosion, denitrification, and NH_3 volatilization. Estimates of annual amounts of N deposited on soils of six desert regions ranged from 3.9 to 11.9 kg•ha^{-1} (for summary, see West, 1978). An annual average deposition of 4 kg of N•ha^{-1} may supply 20 to 50% of the total N taken up by desert plants, depending on the amount of N supplied through biological N_2 fixation. Determinations of N in rain usually include only dissolved N (NH_4^+, NO_2^-, and NO_3^-). Rain over desert regions also may contain appreciable amounts of N as particulates, but quantitative information on the proportion of dissolved to particulate N in desert rain is scarce. About 70% of the atmospheric N contribution to arid soils apparently is derived from dry fallout (June, 1958). However, the net gain of N as dust may be small, depending on how much dust is airborne and carried away by wind. In the western United States, wind is the main erosive agent in arid deserts, whereas water erosion is greatest in semiarid regions (Fletcher et al., 1978). Whether N is lost through erosion or merely translocated depends on the magnitude of the erosive event and the boundaries of the desert. Thus, an erosive event may carry N from the desert floor to a gully (during a rainstorm) or from the desert floor to a nearby mountain (during a windstorm).

In desert soils, compared with soils in humid and subhumid regions, N remains near the surface. Although no less important in other ecosystems, the role of water in determining the extent of N transfers in desert ecosystems is readily apparent. Nitrogen is added directly by rainwater to the desert floor or is washed from one locality to another. Water stress decreases N gain from biological N_2 fixation and N loss via denitrification. The major process leading to N loss from desert soils probably is denitrification (Klubek et al., 1978; Westerman & Tucker, 1978). However, significant loss of N can occur through NH_3 volatilization from alkaline microsites during ammonification. Yaalon (1964) found NH_4^+-N concentrations in rain over desert soils of high pH more than four times as high during warm spring days as during winter, suggesting that NH_3 evolved from these soils after warmer temperatures stimulated ammonification. On the other hand, much of the NH_4^+ formed during mineralization in desert microsites high in N, such as algal crusts, is rapidly nitrified and subject to denitrification (Klubek et al., 1978). The annual loss of N through erosion from western U.S. soils has been reported to be in the range of 0.5 to 30 kg•ha^{-1}, with an average of about 6 to 9 kg•ha^{-1} (Fletcher et al., 1978). A comparable range in values for volatile N losses for desert ecosystems is not available. In their comparison of N gains and losses from four ecosystems, Vlek et al. (1981) reported a N loss via denitrification of 14.2 kg•ha^{-1}•year^{-1} for a desert ecosystem. It is generally assumed that in deserts N gains balance N losses. Because N is concentrated in the shallow surface layers of desert soils or in desert vegetation, erosion (by removing soil relatively high in N content) or fire (by denuding vegetation) can markedly affect the N economy of local desert sites. As much as 70% of the N in the aboveground biomass can be lost as gaseous N from brushfires in semiarid regions (Murray, 1975). Much of this N is believed to be deposited on adjacent ecosystems. Fire is of negligible importance as a mechanism of N loss in arid deserts where vegetative cover is sparse.

f. Grasslands—About 23% of earth's terrestrial surface can be considered grassland. The Great Plains of North America alone comprises 150 million ha of native grasses, unbroken to cultivation. About one third of the world's grasslands can be classified as arid or semiarid. These are regions of increasing concern because of their tendency to approach desert status because of overgrazing, improper management, or cataclysmic natural events that denude the vegetation and erode the soil.

Arid and semiarid grasslands are characterized by a vigorously growing, although often sparse, vegetation that results in the accumulation of plant residues at the soil surface and organic matter in the upper soil layers. Nitrogen and water often are limiting. These grassland regions are extremely variable in rainfall and temperature. They may be highly productive of biomass during years of favorable climate but desertlike in unfavorable years. Because the main storage reservoirs for N are at or near the soil surface (similar to deserts), significant amounts of the actively cycling N in grasslands can readily be transported or lost through surface erosion. Cultivation leading to accelerated oxidation of soil organic matter will increase N cycling and the opportunity for N transfer or loss. Addition of fertilizer N changes sink sizes and N turnover rates resulting in new equilibria between mineralization and immobilization processes. New N inputs, grazing, changes in management practices, and grass species significantly affect the rate and course of N cycling. Power (1977) suggested that the grassland plant-soil system acts as a large sink capable of immobilizing large amounts of N. After equilibration (3 to 5 years) and after biomass demand is satisfied, excess N remains in inorganic, readily transportable forms. In grasslands where excessive loss of N results in decrease in biomass production, the grassland may be subject to the first stage of desertification.

Because of the extremely variable conditions that prevail within a given grassland from year to year and the great differences in N economies that exist among grasslands in different regions, we are reluctant to make a general summary of N transfer data from information that is reported in the world literature. Extensive reviews have been prepared by Bazilevich (1958), Rodin and Bazilevich (1965, 1966, 1968), Dahlman et al. (1969), Woodmansee (1979), and Woodmansee and Adamson (1982). Several mathematical models have been proposed from literature data in order to understand and perhaps design an efficient grassland economy. Jenny (1941) described equations that suggested a geographical consistency in N and other nutrient accumulations, distributions, and equilibria under native grassland conditions. Functional relationships among soil properties, biological components, and ecosystem microenvironments were found to be modified by water regime, temperature, and other independent variables, the influence of which could be predicted. Among the extensions of this work are the models for grasslands proposed by Dahlman et al. (1969), Hunt (1978), McGill et al. (1981), and Van Veen and Paul (1981).

3. RAIN AND RUNOFF

Nitrogen concentrations in rain and surface runoff can be measured with relative ease; therefore, it is not surprising that N contributions by rain and N losses in runoff for various ecosystems are well documented. Considerable data of this type can be found in some of the references cited. Often, however, complementary information on wet and dry particulate fallout, N loss through leaching, or removal of NH_4^+-N through immobilization is not given, which makes difficult the assessment of total N transfers in waters where these processes occur to an appreciable extent.

4. N TRANSFERS FROM PLANT CANOPIES

Recent evidence (reviewed by Wetselaar and Farquhar, 1980) indicates that the total amount of N in plant tops decreases as plants approach maturity. This decrease in N content cannot be explained as N translocation to grain or roots, although storage of N in roots after fruiting is common in perennials. Decrease in N content of plant tops is especially noticeable in annuals containing relatively high concentrations of N. Wetselaar and Farquhar (1980) cited data for several avenues of N loss from living higher plants: leaching of soluble N from plant parts, guttation, root excretion, nitrous acid dismutation reactions near the leaf surface leading to formation of gaseous N, liberation of volatile amines, and NH_3 evolution. Some of these mechanisms do not necessarily lead to N loss from the soil; others may lead to redistribution of N to soils in adjacent areas.

Several recent papers have reported the liberation of NH_3 to the atmosphere from senescing vegetation and from healthy leaf canopies. There is reason to believe also that certain plant species, such as the *Solonacea,* which do not efficiently assimilate NH_4^+-N, apparently are more prone to lose N as NH_3. Söderlund and Svensson (1976) estimated that wildlife, domesticated animals, and coal burning contributed annually 26 to 53 Tg of NH_3-N to the atmosphere. The annual deposition of N from NH_3 to the pedosphere and hydrosphere was estimated to be 87 to 191 Tg. If one assumes a maximum transfer to the atmosphere of 53 Tg of N and a minimum transfer to land and waters of 87 Tg of N, there is an apparent imbalance of at least 34 Tg of N. Fertilizer industry statistics suggest that as much as 10% of industrially fixed N_2 may be lost during NH_3 synthesis and its processing into fertilizer and other industrial products. At the current world levels of NH_3 production, as much as 6 to 9 Tg of industrially fixed NH_3-N could be released annually to the atmosphere, but this amount would increase only slightly the values for NH_3-N evolved to the atmosphere given by Söderlund and Svensson (1976). Lemon and Van Houtte (1980) suggested that as much as 40 to 70 Tg of N could be liberated annually from higher plant leaves. Using a micrometeorological technique, they measured NH_3-N losses in the range of 0.1 to 1.0 mg•m^{-2}•hour^{-1} of leaf surface for young alfalfa plants (*Medicago sativa* L.), corresponding to a N loss of 1 to 10 g•ha^{-1}•hour^{-1}. Losses of NH_3-N sustained at this rate over a substantial part of the growing season would result in a considerable transfer of N from leaves to the atmosphere; calculations based on reported rates of NH_3 evolution suggest that estimates of seasonal N losses as high as 45 kg•ha^{-1} would not be unreasonable.

Leaf canopies can absorb NH_3 in the vapor phase through open leaf stomates and NH_3 trapped in dew. Field measurements of NH_3 flux densities were made within leaf canopies of soybeans (*Glycine max* L.) and quackgrass (*Agropyrum repens* L.) by Lemon and Van Houtte (1980) and of ungrazed mixed pasture (*Trifolium subterraneum* L. and *Lolium rigidum* Goud.), grazed alfalfa pasture, and fertilized maize (*Zea mays* L.) by Denmead et al. (1974, 1976, 1978). These measurements strongly indicate that NH_3 was absorbed within the leaf canopy and released at the top. Whether or not NH_3 is absorbed or evolved depends on ambient NH_3 concentrations, NH_3 flux, stomatal resistance, factors that affect the preceding variables, and unknown internal plant factors that may be species dependent.

Most N balance accounts have not considered possible NH_3 transfer between vegetation and the atmosphere. The apparent discrepancy between NH_3 release from and addition to the biosphere pointed out by Söderlund and Svensson (1976) indicates that NH_3 exchange between the atmosphere and biosphere may be an important N transfer process for all ecosystems. As a result of NH_3 transfer from cultivated plants to uncultivated plants and soils, biological N_2 fixation by cultivated legumes and N fertilizer additions may indirectly contribute to the N supply of unfertilized soils and thus may impact on the quantitative aspects of N transformations occurring in them. Transfers of NH_3 also can occur within undisturbed ecosystems from alkaline microsites to acid ones following litter decay. Ammonia can volatilize from floodwater or irrigation waters that are alkaline or increase in alkalinity to a pH >7.5 after N fertilizer addition. Until recently, little attention has been paid to possible loss of significant amounts of NH_3-N from rice paddy (*Oryza sativa* L.) floodwater, but as much as 50% of the N applied as urea was lost as NH_3 from floodwater overlaying alkaline soils, especially under conditions in which algal blooms, by consuming CO_2, temporarily raise floodwater pH to values >8.5 (IRRI, 1977).

If NH_3 transfers are indeed of the magnitude suggested above, the accuracy of denitrification loss estimates that are based on the calculation of N deficits in experimental systems may be open to question. With current information, we can justifiably only draw attention to the possible significance of NH_3 transfers between the atmosphere and biosphere.

5. N CYCLING IN THE PEDOSPHERE

About 5% of the N cycling annually within the pedosphere interchanges with the atmosphere and hydrosphere (Rosswall, 1976). The remaining 95% interacts solely within soil-microbial-higher plant systems and is not transferred from these systems. Nevertheless, the rates of N transfers among the various components of the pedosphere largely determine the rate of global N cycling.

In the preceding sections, various ecosystems were classified mainly by their types of aboveground vegetation (e.g., trees) or biotic environment (e.g., freshwater). They can be classified also on the basis of their average biomass concentrations or N turnover rates or on the basis of a dominant process. Rosswall (1976) included shrub tundra, subalpine coniferous forest, coniferous forest, oak/birch forest (*Quercus* sp./*Betula* sp.), heath, raised bogs, and many swamps among ecosystems in which NH_4^+ is the predominant form of inorganic N (i.e., in which nitrification rates are low). Included in the type that normally contained both oxidized and reduced forms of inorganic N were temperate deciduous forests on loamy soils, alluvial forests, alder fen (*Alnus*), dry grassland on calcareous soils, tropical savannas, and some tropical forests. Nitrite/nitrate types included moist tropical lowland forests, temperate deciduous forests on calcareous soils, and fertilized meadows. In these groups, elevated soil pH and nitrification activity appear to be directly correlated. Rosswall (1976) estimated the total biomass for seven ecosystems; the amounts ranged from 20 g of dry weight m^{-2} for a Canadian tundra to 240 g$\cdot$m^{-2} for a coniferous forest in Sweden. However, there was a threefold difference in biomass values for two grassland ecosystems and a fourfold difference for tundras in Canada and Sweden. The global microbial biomass averaged 4 g of dry weight$\cdot$m^{-2}. The average N content in microbial biomass ranged from 2.8 to 6.0%, with a mean value of 4.0%, indicating an average annual microbial requirement of 19 g of N$\cdot$m^{-2}. The average residence time of N in microbial biomass is < one tenth and one thousandth of that in plant litter and soil, respectively.

The average residence time for N in world soils has been estimated to be 177 years, ranging from 109 years for a mixed deciduous forest soil to 372 years for tundra mire (for references, see Rosswall, 1976). Unfortunately, comparisons among such values may not be valid because estimates of biomass or N turnover rates for various ecosystems may not have been made using the same method; for example, the estimate of N turnover in the tundra mire cited above was made by measuring change in microbial biomass, whereas the average world estimate was calculated from available data. The calculations, however, were based on several assumptions, one being that the average transfer of N through the microbial biomass is 2.27 times greater than the average microbial N requirement.

Attempts to generalize from empirical data are beset by the problems of data interpretation that are created by extreme variability among soils, climates, and management practices. Extrapolating from nanograms, micrograms, or milligrams of N per meter to global dimensions requires many assumptions about how well the experimental systems reflect the average course of events that occur over large areas and long time intervals (i.e., whether space averaging or time averaging is a valid approach to use). Despite these problems, various attempts have been made to construct N mass balances on the subregional, regional, and global levels. The sections that follow contain a survey of representative N budget models.

III. N MASS BALANCES AND MODELS

Mass balance has also been referred to as a balance sheet or account, material balance, nutrient budget, and as compartmental, conceptual, input-output, or income-outgo models. There are substantial differences among these models regarding modeling objectives, aggregation in space and time, structure, and required inputs. General objectives of modeling may be to (i) obtain better conceptual understanding and increased insight into complex problems, (ii) evaluate available information and ascertain the adequacy of published data for problem evaluation, (iii) test existing as well as new concepts and hypotheses, (iv) estimate by difference an unknown output or input, (v) determine by sensitivity analysis and/or calibration an unknown model coefficient, or (vi) predict observed and/or anticipated phenomena.

A. N Mass Balance Models

Mass balance models examine the inputs to and outputs from systems and usually consider internal storage elements (reservoirs, pools, compartments, sinks, and sources) that are interconnected (fluxes, material flows, flow pathways). The pools have dimensions of $mass \cdot area^{-1}$, whereas the fluxes have dimensions of $mass \cdot area^{-1} \cdot time^{-1}$. The temporal scale of models of interest to agriculturists is typically on a seasonal or an annual basis, whereas the spatial scale varies from the rhizosphere within an experimental field plot to areas of regional and global dimensions. If one assumes that there is no change in the internal storage elements, the model reduces to an input-output ("black box") type. Mass balance models yield outputs for one point in time but can be used to appraise changes as a function of time by comparing two or more time periods, each calculated separately, or by taking the outputs from one time period as the inputs to a succeeding period.

Numerous flowcharts describing the soil N cycle have been drawn (e.g., see Allison, 1965; Bolin & Arrhenius, 1977). Since sufficient data are not available to permit a full understanding of the quantitative aspects of the soil N cycle, a mass balance modeling approach can lead to an interim understanding. Models tend to simplify the real world complexities and take advantage of best estimates and judgments. A major weakness is that they are seldom verified quantitatively, and their validity is usually judged intuitively. On the other hand, if the data base is not adequate for testing mass balance models, it is obvious that it would not be adequate for validating more sophisticated simulation models.

Field plots and farms are the smallest scale units considered here for modeling of N balances. For the small-scale models, agricultural systems are emphasized.

B. Small-Scale Models and N Balances

Comprehensive reviews of N balance data from greenhouse, lysimeter, and field studies have been prepared by Allison (1955, 1965, 1966) and by Legg and Meisinger (Chapt. 14). We discuss here the modeling of some recent data from agricultural experiments.

To assess the contribution of agricultural practices to surface and ground water pollution, Fried et al. (1976) suggested use of a simple conceptual model in which it was assumed that the N pool in intensively cultivated soils maintained a steady-state condition and that the net contribution of soil-derived N was nil (i.e., all leachable N was derived from fertilizer, water, and other N inputs). These assumptions stem from Jenny (1941), who concluded that soil N content stabilizes at some level as determined by crop production and soil management practice. The validity of this concept has been verified by long-term field studies (e.g., Smith, 1942; see also Chapt. 1, F. J. Stevenson) and was tested by Tanji et al. (1977) with a mathematical model consisting of a hydrologic submodel and a N submodel. The hydrologic submodel considers the following water balance:

$$\text{(Irrigation water + Precipitation)} - \text{(Evapotranspiration + Surface runoff}$$
$$+ \text{Deep percolation + Collected subsurface drainage)} = 0. \qquad [1]$$

The N submodel considers the following N balance:

$$\text{(Irrigation water + Precipitation + Applied fertilizer + Fixed } N_2)$$
$$- \text{(Surface runoff + Deep percolation + Collected sursurface drainage}$$
$$+ \text{Harvested crop + Volatilization + Denitrification)} = 0. \qquad [2]$$

Tanji et al. (1977) tested this steady-state model using data obtained from field trials with maize conducted at two sites (at Davis and near Fresno, Calif.). Nitrogen outputs from the Davis plots were found to exceed N inputs; the soil from these plots was not in a steady-state condition. Response to applied N was limited by plant uptake of soil N made available through mineralization of a large pool of soil organic matter (containing about 22,000 kg of $N \cdot ha^{-1}$ in the 3-m profile). At the Kearney Field Station (near Fresno), N outputs equaled inputs, validating the assumption that there was no net change in storage of N in the plant root zone.

Table 1. Comparison between steady-state computations† and measured results from the 1974 Kearney maize plots (Tanji & Mehran, 1979).

Fertilizer applied	Harvested crop coefficient†		Calculated grain yield		Measured grain yield	Calculated mass emission§		Calculated N concentration¶		Measured NO₃⁻-N concentration in soil profile‡		
	A	B	A	B		A	B	A	B	at 1.7 m	at 2.0 m	at 2.3 m
kg of N ha⁻¹			kg of N ha⁻¹ year⁻¹							mg of N liter⁻¹		
112	0.78	0.62	84	67	92	23	41	19	33	4	10	20
224	0.73	0.67	153	140	132	57	69	46	56	34	49	52
336	0.50	0.56	155	173	136	155	136	126	111	86	94	73
448	0.36	0.43	149	176	126	261	234	212	190	102	160	82
560	0.28	0.35	143	179	133	368	332	299	270	197	256	153

† Method A uses a harvested crop coefficient obtained from the rates of measured grain yield to calculated effective N inputs; method B uses a harvested crop coefficient obtained from N isotope data.

‡ From soil sampling at three depths after harvest of the 1974 maize crop.

§ NO₃⁻-N leached beyond root zone.

¶ Average NO₃⁻-N concentration in leachate.

Table 1 reports observed and calculated data for the 1974 Kearney field experiment with maize. Two sets of calculated data are given for grain yield and N lost through leaching. The first set (method A) uses as a harvested crop coefficient the ratio of measured grain yield to effective N inputs to soil (calculated); the second set (method B) uses as a harvested crop coefficient the ratio of ^{15}N in the grain to amount of ^{15}N applied to soil. Other N inputs used in the calculations were from irrigation water and rain, measured to be 2.2 and 2.3 kg of $N \cdot ha^{-1} \cdot year^{-1}$, respectively, and from biological N_2 fixation, estimated to be 3.4 kg of $N \cdot ha^{-1} \cdot year^{-1}$. An estimated denitrification loss of 12 kg of $N \cdot ha^{-1} \cdot year^{-1}$ was used. In general, calculated values were in close agreement with measured values of grain yield when method A was used. For N concentrations in the soil solution, calculated values were in closer agreement with measured values when method B was used.

Recently Frissel (1978) reported on a group study of nutrient cycling in 65 agricultural systems. The framework for this investigation is depicted by a flow chart describing 31 transfer pathways that interconnect the plant, livestock, soil organic matter, and soil mineral pools (Fig. 1). It was concluded that land grazed by livestock and arable land that received little or no N from fertilizer or manure produced for use within the agricultural system a net 0.4 to 20 kg of $N \cdot ha^{-1} \cdot year^{-1}$, crop-livestock agroecosystems produced 40 to 113 kg of $N \cdot ha^{-1} \cdot year^{-1}$, and intensively cultivated and fertilized land usually produced 40 to 180, and sometimes as much as 400, kg of $N \cdot ha^{-1} \cdot year^{-1}$. Difficulties were encountered in estimating N fluxes for all the agricultural systems studied. For the intensively cultivated system, fertilizer provided the major N input; N outgo was provided largely by way of harvested crop. Nitrogen additions as fertilizer and N removals in harvests could be accurately measured. Measurements of N loss via leaching, denitrification, and NH_3 volatilization, and of mineralization-immobilization rates within the soil N pool created the greatest uncertainties about the data.

Data for cropland in the group study discussed above are given in Table 2 (Thomas & Gilliam, 1978). Only average figures are given for each cropping system, thus predictions of N input or output for a specific site cannot justifiably be made, but the average values indicate trends. The smallest N losses were from the grain crop systems, whether losses were by leaching or by surface runoff (inorganic N and organic N in sediments). Leaching losses were almost 30% of the total N input to the systems cropped to potato (*Solanum tuberosum* L.) and cotton (*Gossypium hirsutum* L.). Large N losses occurred in furrow-irrigated cotton following the injection of anhydrous NH_3 to irrigation waters. The percentages of total N from all sources added to each cropping system that were found in crops at harvest were 45, 61, 74, 77, and 95 for systems cropped to cotton, potatoes, soybeans, maize, and wheat (*Triticum aestivum* L.), respectively.

Thomas and Gilliam (1978) suggested that very few agroecosystems in North America are accumulating N in soils. There usually appears to be a small net loss from cultivated soils. They are in agreement with Fried et al. (1976) that the interpretation of N balance data can be simplified to an

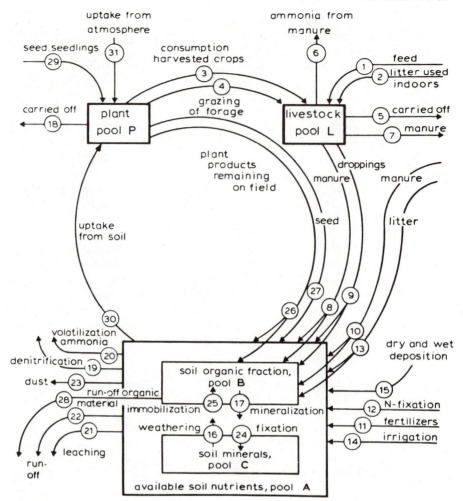

Fig. 1—Flow chart of 31 transfer pathways that interconnect plant, livestock, soil organic matter, and soil mineral pools (Frissel, 1978).

input-output analysis if, over several cropping seasons, N immobilized by soil microorganisms equals N mineralized.

Tanji et al. (1979) extended the N mass balance model of Tanji et al. (1977) to include nonequilibrium or transient conditions. Net mineralization-immobilization of fertilizer N and residual organic N and the availability of residual inorganic soil N were considered in the model. Mineralization of organic N was described by first-order kinetics (Mehran & Tanji, 1974).

To investigate the influence of several model coefficients and constants on N outputs, Tanji et al. (1979) performed sensitivity analyses on the denitrification coefficient, harvested crop coefficient, and mineralization rate constant for organic N in the soil. Sensitivity analysis is carried out by selecting a particular parameter and varying its input value while keeping all

Table 2—N fluxes in selected agroecosystems in the United States
(Thomas & Gilliam, 1978).

	Maize for grain, northern Indiana	Soybeans for grain, northeast Arkansas	Wheat, central Kansas	Potatoes, Maine	Cotton California
			kg of N ha^{-1}		
N inputs					
Fertilizer	112	--	34	168	179
N$_2$ fixation	t†	123	t	t	t
Irrigation water and flooding	10	--	--	--	50
Atmospheric deposition	--	10	6	6	3
Crop residue	41	30	20	65	48
Total inputs	163	163	60	239	280
N outputs from soil					
Net plant uptake	126	120	56	145	127
Denitrification	15	15	5	15	20
Volatilization	t	t	t	t	t
Leaching	15	10	4	64	83
Runoff (inorganic N)	6	3	1	5	50
Runoff (organic N)	10	13	4	10	t
Wind erosion (dust)	--	t	t	t	t
Total outputs	172	161	70	239	280
N inputs − N outputs	−9	2	−10	0	0

† t = trace amounts.

other inputs constant. A parameter is considered to be sensitive if it produces large changes in the outputs. Calculated values were compared with measured values obtained from the maize field trial at Davis, California, discussed earlier in relation to the model before its extension to include transient conditions. Because the soil at Davis contained considerable amounts of organic N, the calculated results were found to be sensitive to the mineralization rate constant, which had an estimated value of 10^{-4} week^{-1} corresponding to a 0.52% mineralization of organic soil N over 52 weeks. This rate of mineralization is substantially smaller than rates reported elsewhere. It is based on the average soil organic N content to a depth of 3 m (N contents ranged from 1,526 ± 303 kg•ha^{-1} in the lower portions of the 3-m profile to 4,231 kg•ha^{-1} in the surface 0.3 m). Detailed discussion of the model's usefulness can be found in the publications cited, including a comparison of calculated and measured values for drainage, N leached, N in harvested grain and stover, and average N content in the soil profile at various depths.

The transient conceptual model, with some modifications, has been applied also to a lysimeter study involving spray irrigation of municipal sewage effluents on grasses (Mehran et al., 1981). The model was modified to adjust for a K deficiency that was observed in the grasses by assuming a Michaelis-Menten type of N uptake that was influenced by the available K/N ratio in the soil. In addition, the model was used as a predictive tool by simulating 5 successive years, using the calculated outputs of residual inorganic and organic soil N as the inputs to the succeeding years.

The results obtained from small-scale N balance models seem to be accepted by other scientists mainly because they are verifiable not only by measurements made at the site for which refinements of the model are particularly applicable but also intuitively, i.e., the results are often similar to those obtained in other experiments and, therefore, are creditable. Uncertainties about the data increase with increase in size of the area under study.

C. Regional Models and N Balances

The N balance account for U.S. agricultural lands reported over 40 years ago by Lipman and Corybeare (1936) agrees in many respects with later estimates of N income-outgo. These investigators made estimates of N and other nutrient gains and losses that were based on the results of 14,500 analyses of top soils (surface 17 cm) of 149 million ha of tilled and nontilled croplands. Input-output rates for biennial and perennial crops were included. Individual estimates of N gains and losses were aggregated to compile a national total (included in Table 3). The N economies of pastures within farms and pastures not on farms were also examined, but data from these studies were not included in the national estimate for croplands (lands from which crops were harvested). It is of interest to note that Lipman and Corybeare conducted their study to assess the depletion of soil fertility so that a "rational land use policy" could be formulated to maintain and enhance soil productively. They concluded that U.S. soils were suffering enormous N losses through erosion and leaching, a situation that would lead to major economic and social stress if soil resources were not used more effectively.

Table 3 contains two additional estimates of N balance for U.S. croplands and two for the total land area of the United States. The estimates of Stanford et al. (1970) for 1947 and 1967 were made mainly to assess the potential contribution of fertilizer N to NO_3^- enrichment of waters. Viets and Hageman (1971) commented on the speculative nature of such N balance accounts, but a comparison of the data from the 1947 and 1967 estimates indicates certain trends: an increase in fertilizer N use, a decrease in soil N loss from erosion, a decrease in soil mineralization rates that correlates with a decrease in leaching of soil N, and an increase in N removal by harvested crops. These trends are consistent with improved soil conservation and soil and crop management practices.

In 1972, the National Academy of Sciences (NAS) published N income-outgo estimates for the total U.S. land surfaces. The main objective of this study was to determine whether measures to increase food and fiber production resulted in NO_3^--N accumulation in soils, waters, foods, and feeds, and "to examine various problems associated with the accumulation of nitrate nitrogen and related nitrogenous compounds in the environment and to recommend courses of action that may mitigate these problems." The NAS concluded that U.S. soils apparently had a net gain of 1.5 Tg of N during 1970, as calculated from the difference between total N income and outgo.

Table 3—Inputs and outputs of N in harvested croplands and total land area in the United States.

Estimated fluxes	For 1930 Lipman and Corybeare (1936)	For 1947 Stanford et al. (1970)	For 1967 Stanford et al. (1970)	For 1970 NAS (1972)	For 1975 NAS (1978)
	Tg† of N year⁻¹ for croplands			Tg of N year⁻¹ for total area	
Inputs of N					
Fertilizer N	0.3	0.7	6.8	7.5	9.4
Symbiotic N_2 fixation	1.5	1.7	2.0	3.6	12.0
Nonsymbiotic N_2 fixation	1.0	1.0	1.0	1.2	12.0
Barnyard manure	0.9	1.3	1.0	--	--
Rainfall	0.8	1.0	1.5	5.6	5.4
Irrigation	<0.1	--	--	--	--
Roots and unharvested portions	1.0	1.5	2.5	--	--
Mineralization of soil organic N	--	--	--	3.1	--
Total inputs	5.5	6.2	14.8	21.0	--
Outputs of N					
Harvested crops	4.2	6.5	9.5	16.8	16.8
Erosion	4.5	4.0	3.0	--	--
Leaching of soil N	3.7	3.0	2.0	--	--
Leaching of fertilizer N	--	--	?	--	--
Denitrification	--	?	?	8.9	--
Volatilization	--	--	--	5.6	--
Total outputs	12.4	13.5	14.5	19.5	--

† Tg = teragram = 10^{12} g = 1 million metric tons.

Some compartments of this N budget were reexamined (NAS, 1978), especially with regard to N_2 fixation. The estimated amount of N_2 fixed biologically in U.S. soils during 1975 (12.0 Tg of N•year⁻¹) was much larger than that estimated for 1970 (4.8 Tg of N•year⁻¹). During 1975, there was an estimated transfer of 29.9 Tg of N as N_2 to fixed N_2 forms within U.S. boundaries through biological, atmospheric, and industrial fixation processes; 65% of this amount was said to be the consequence of human activities (e.g., NH_3 synthesis and legume cultivation).

Lipman and Corybeare (1936) and Stanford et al. (1970) did not include N gains to the available nutrient supply from net mineralization of soil organic matter in their estimates of N inputs. They assumed that rapid oxidation of soil organic matter already had occurred, for the most part, by 1930. Leaching of fertilizer N from 1930 to 1947 was assumed to have been negligible because of the relatively small amounts of fertilizers added during these years. Gaseous losses of N were not considered in the 1930 study; they were considered in the 1947 and 1967 studies, but no estimates were given. Because of obvious knowledge gaps during the preparation of their N balance accounts, Stanford et al. (1970) emphasized the need for more research on denitrification, efficiency of N fertilizer use, and models to aid one's understanding of N transfers and transformations in soil-plant-water systems.

Table 4—Mass balance case studies on diverse regional land uses.

References	Study area, land use, and N problem	Modeling objectives	Findings relative to the soil N pool (see also Table 5)
Ayers & Branson (1973)	1.44×10^3 km² valley floor of the Upper Santa Ana River Basin in southern California. Historical records of land use dating back to 1930 show a shift from citrus and general agriculture to cattle feedlot and urban development. Increasing accumulation of NO_3^- in ground waters is major source of supply water.	To ascertain N pollution potential of ground waters based on 1960 N loading rates to the soil N pool and fate of N.	Major sources of N contributing to the ground water basin are ground water recharge of N-laden municipal sewage effluents, excess N fertilizer applications, applied water containing high NO_3^-, and excessive applications of manures on forage crops. Major sinks of N are uptake of N by vegetation and potential leaching to ground water basin.
Miller & Smith (1976)	17.72×10^3 km² Lower San Joaquin in central California, 10.2×10^3 km² is under irrigation with cotton, barley, grapes (*Vitis*), and irrigated pasture as the major crops. Two centers of population at Fresno and Bakersfield. Localized accumulation of N in ground waters.	To appraise N in the environment in an area of high food and fiber production under semiarid climatic conditions for 1971.	Major inputs of N are fertilizer N, applied waters, and N_2 fixation that are diffuse in origin compared with point loadings of animal manures and municipal wastes. Major outputs are plant uptake and residual soil N, part of which is subject to leaching. Estimated fertilizer and residual soil N removals ranged from a low of 19% removal for young tree crops to a high of 73% for hay crops. Comparison of 1961 vs. 1971 shows yield increases due to higher N application rates rather than higher N use efficiency.

(continued on next page)

Table 4—Continued.

References	Study area, land use, and N problem	Modeling objectives	Findings relative to the soil N pool (see also Table 5)
NAS (1978)	146.4×10^3 km^2 state of Wisconsin, 54% of which is farmland, with population centered in the south central area. Agriculture consists of mixed cash grain, dairy, vegetable crops, and general agriculture. Pollution of some lakes noted.	To evaluate the increased use of N fertilizer and N fluxes through livestock-cropland systems for 1974.	Assessed the soil N pool by subdividing into the larger unavailable soils organic N subpool and the smaller available soil N pool. Unlike the N fertilizer input in the Corn Belt states, the largest single source of N input is N_2 fixation (mainly alfalfa) followed by animal manures. Nitrogen pools and fluxes through animals were coupled with pools and fluxes for crops. Plant uptake was a major sink.
NAS (1978)	116.7×10^3 km^2 Peninsular Florida with population centered in coastal areas. Agriculture dominated by citrus, vegetable crops, and feeder calves. No discernible N water pollution noted.	To analyze for the more important sources and sinks of N in the peninsula averaged over the 1970–1975 period.	Developed submodels for hydrologic, agricultural, and atmospheric flows aggregated for the peninsula. Soil N pool models similar to the Wisconsin study. Unlike Wisconsin, biological N_2 fixation was very low. Major sources of N inputs are atmospheric deposition, manures, and fertilizers. Major sinks are unaccounted-for gaseous losses and plant uptake. Atmospheric submodel indicates production of NO_x and NH_3 greater than return to land. Agricultural submodel indicates leaching and/or volatilization losses high.

The 1972 NAS report included gains of available N through mineralization of soil organic matter and N losses through denitrification. The average release of soil organic N to the available N supply was estimated to be 40 kg of $N \cdot ha^{-1} \cdot year^{-1}$. The estimate of N loss through denitrification was made by revising downward the global estimate made by Delwiche (1970) of 11 kg of $N \cdot ha^{-1} \cdot year^{-1}$, assuming a maximum loss of 15% of the fertilizer N applied annually to U.S. soils and that 90% of the N transported from soils to waters is denitrified. It is emphasized in the 1972 NAS report that the values given "should be considered merely as estimates, any one of which may be in error by a factor of two or more. . . . The data necessary to establish a precise nitrogen budget for the United States are largely lacking, with the exception of inputs in chemical fertilizers." Concern over the potential adverse effects of increased levels of fixed N in the biosphere has stimulated research directed toward quantification of N cycle processes. This research is resulting in the accumulation of more accurate data obtained at local and regional sites; these data can be used to refine the larger space-averaged fluxes. The recent N budget estimates reflect a more accurate data base by giving a better balance between N inputs and outputs than was given by earlier estimates.

Four mass balance case studies for diversified regional land uses are described in Table 4. The N pools and fluxes for the Upper Santa Ana River

Table 5—Estimated annual N fluxes into and from the soil N pool.

Estimated fluxes	Upper Santa Ana River Basin, Calif. for 1960 Ayers & Branson (1973)	Lower San Joaquin Valley, Calif. for 1971 Miller & Smith (1976)	Wisconsin for 1974 NAS (1978)	Peninsular Florida for 1970–1975, avg NAS (1978)
	——— kg of N ha⁻¹year⁻¹ ———			
Inputs into soil				
Precipitation (wet and dry)	9.0	0.4	7.0	10.3
Other atmospheric deposition (NH₃ sorption, etc.)	4.5	3.4	2.7	7.9
Symbiotic N₂ fixation	11.2	26.2	21.7	3.3
Nonsymbiotic N₂ fixation	4.5	3.3	2.7	3.3
Municipal wastes	24.7	2.8	1.9	--
Industrial wastes	0.5	5.7	--	--
Animal manures	28.7	28.7	17.8	14.2
Fertilizers	53.2	67.1	8.7	9.0
Applied waters	37.4	21.0	--	--
Total inputs	173.7	158.6	62.5	44.7
Outputs from soil				
Runoffs	4.7	0.5	1.2	5.0
NH₃ and NO emissions	10.7	12.0	13.0	7.3
Denitrification	16.6	15.9	--	3.0
Plant uptake	88.8	85.5	32.0	5.9
Leaching	--	--	--	--
Residuals (leaching, accumulation, etc.)	52.8	44.8	16.0	23.5
Total outputs	173.6	158.7	62.2	44.7

Basin (Ayers & Branson, 1973) were developed by K. K. Tanji et al., the Wisconsin study (NAS, 1978) by D. R. Keeney, and the Peninsular Florida study (NAS, 1978) by J. J. Messer and P. L. Brezonik.

In addressing the problem of NO_3^- accumulation in ground waters of the Upper Santa Ana River Basin investigation, Ayers and Branson (1973) chose the soil N pool as the focal point. It is this pool that receives N wastes and fertilizers, the fates of which are determined by the amounts recycled through the soil-plant-water system and lost as gases and percolates. A model was constructed that considered the basinwide internal N cycling and the external fluxes. The model consisted of six N pools (soil, substrata, ground water, surface water, land surface, and atmosphere) and 34 N fluxes that were combined into 14 fluxes between the N pools. The estimated annual N fluxes into and out of the soil N pool are given in Table 5. For the Santa Ana River Basin in 1960, Ayers and Branson (1973) estimated that an average of 53 kg of $N \cdot ha^{-1} \cdot year^{-1}$ (30% of N inputs) could have leached, accumulated in the soil, and/or subsequently have been taken up by plants. If one assumes complete nitrification of this N and an average water flux in the basin of 0.24 $m \cdot year^{-1}$, the calculated NO_3^--N concentration in percolates would have averaged 22 $mg \cdot liter^{-1}$. Because all the N in the soil-available N pool does not nitrify during a given year, calculations of maximum possible NO_3^--N concentrations in waters serve to indicate upper limits only. Well waters in the above basin have NO_3^--N concentrations ranging between < 5 and > 25 $mg \cdot liter^{-1}$.

The N fluxes described by the model used by Ayers and Branson (1973) were for the 1960 level of development of the Upper Santa Ana River Basin. In a later unpublished study, Tanji applied this model to the 1970 level of area development (Table 6). Although the amount of fertilizer N used decreased by one half from 1960 to 1970, increases in human and animal wastes apparently increased the potential for NO_3^- enrichment of ground water by 10%.

Table 6—Major annual N fluxed into and from the soil N pool in the Upper Santa Ana River Basin, Calif., for the 1960 and 1970 levels of development (Goldhammer & Tanji, unpublished data).

Major inputs and outputs	1960	1970
	— metric tons $\times 10^3$ —	
Sources		
N fertilizers†	7.4	4.2
Municipal wastewaters‡	3.6	4.3
Animal manures§	2.9	5.2
Sinks		
Plant uptake	12.8	12.3
Gaseous losses¶	3.9	4.7
Potential leaching	7.6	8.4

† 116 kg of N ha⁻¹, 65,600 ha arable land in 1960; 72 kg of N ha⁻¹, 60,750 ha arable land in 1970.
‡ Human population: 658,000 in 1960; 809,100 in 1970.
§ Cows: 71,200 in 1960; 125,200 in 1970. Poultry: 6 million in 1960; 11 million in 1970.
¶ Increases largely due to NH_3 volatilization from cattle manure.

Miller and Smith (1976) used a similar modeling approach for understanding N fluxes in the Lower San Joaquin Valley. The predominant land use in this region is for food and fiber production. Detailed appraisals were made of the amounts of N applied to and removed by 26 different crops during each of 2 years, 1961 and 1971. During this decade, N fluxes involving the soil N pool resulted in an increase of 9 kg of $N \cdot ha^{-1}$, which is equivalent to <0.1% of the total soil N content. During 1971, the average loading rate of N was 159 $kg \cdot ha^{-1} \cdot year^{-1}$. From the mass balance appraisal, the percentage of each N source subject to leaching to ground waters was as follows: residual fertilizer, 45%; irrigation water, 19%; biologically fixed N_2, 13%; industrial wastes, 10%; animal wastes, 6%; municipal wastes, 5%; and rain, 2%. More important than the average loading rate of N is the nature of the loading; for example, a large amount of fertilizer N distributed over a large area may not elevate NO_3^- concentrations in specific waters as much as smaller amounts from other N sources disposed over a smaller area.

The modeling approach and type of N flux data used in the Wisconsin study (NAS, 1978) differed significantly from those used in the California studies. Except for N released from soil organic matter through mineralization ($1\% \cdot year^{-1}$), N inputs were considered equivalent to N outputs (the latter included N export in food and feed). Land used for grain (maize) production increased from 0.6 to 0.9 million ha between 1964 and 1974 but remained about constant at 0.4 million ha for silage production during that period. From 1964 to 1974, N applied as fertilizer increased from an average of 32 $kg \cdot ha^{-1} \cdot year^{-1}$ to 70 to 80 $kg \cdot ha^{-1} \cdot year^{-1}$, with a resultant increase in average grain (maize) yield from 4,700 to 5,500 $kg \cdot ha^{-1}$ (the latter value equivalent to an annual export in grain of about 80 kg of $N \cdot ha^{-1}$).

The statewide average N input for Wisconsin was estimated to be 62 $kg \cdot ha^{-1} \cdot year^{-1}$, of which 16 kg of $N \cdot ha^{-1}$ (26%) was unaccounted-for losses (denitrification, leaching, NH_3 volatilization). The model indicated that the amount of N brought into the state was twice that exported in food and feed and that more efficient use of manure and legume residue would bring N imports and exports into balance. A major problem in the Wisconsin study not encountered in the California case studies was the artificial political boundaries imposed on this state model rather than the natural hydrologic boundaries.

Although much information was gathered for the Peninsular Florida model (NAS, 1978), results were considered to be inconclusive. The input of N into the soil N pool was estimated to be 45 $kg \cdot ha^{-1} \cdot year^{-1}$, which was about 26, 28, and 72% of that for the Upper Santa Ana River Basin, the San Joaquin Valley, and Wisconsin, respectively. During 1973, the N input for citrus, the major crop, was 10 and 123 $kg \cdot ha^{-1}$ from rain and fertilizer, respectively; 42 kg of $N \cdot ha^{-1}$ was removed through harvest. Thus, N inputs for citrus exceeded N output by 91 kg of $N \cdot ha^{-1}$ (68% of the input). The deficit could not be explained by assuming N removal through immobilization in soil organic matter or loss via NH_3 volatilization and denitrification. Leaching of NO_3^- was considered the most probable avenue of N loss, but

NO_3^- concentrations in waters adjacent to citrus groves were not sufficiently high to validate this hypothesis.

For the entire area encompassed by the Peninsular Florida model, N inputs far exceeded N outputs, about 13.2 kg of $N \cdot ha^{-1} \cdot year^{-1}$ (40% of inputs) for cropped plus uncropped soils and 23.5 kg of $N \cdot ha^{-1} \cdot year^{-1}$ (53% of inputs) for croplands. Gaseous N fluxes (e.g., those involving NH_3 and NO_x) were perhaps more important than was realized. The Florida model has raised several questions concerning our knowledge of N cycle processes in acid, coarse-textured soils in high rainfall areas.

Obviously, there is much uncertainty in estimating N fluxes and N pools on a regional level. Because of this, a liberal interpretation of results obtained from regional mass balances for N has been criticized (Viets & Hageman, 1971; NAS, 1978). The California regional models have been criticized for their model structures, incompleteness of N fluxes into various pools, lack of adequate documentation for assumed fluxes, and the manner in which potential N leaching losses were calculated (Viets, 1978; Kohl et al., 1978). Our intent here is neither to defend nor to refute these criticisms but to provide examples of different modeling approaches and to point out that modeling objectives and the specific conditions prevalent at the site under study dictate the modeling approach that is taken.

Despite the uncertainties of N flux values in regional mass balances, regional modeling indicates trends, exposes knowledge gaps, and suggests the relative significance of a particular N cycle process. Thus, the Upper Santa Ana River Basin model indicated the degree of change in N income-outgo between two levels of land development. The Wisconsin model indicated the significance to agriculture of that region of biological N_2 fixation, which contributed almost three times as much N to agricultural soils as did fertilizer applications. The Florida model showed that N inputs far exceeded N outputs and suggested that N losses through leaching and volatilization must be much higher than those that are being experimentally verified.

D. Global Models and N Balances

Even though about 95% of the total N that cycles globally each year cycles solely within soil-microbial-vegetation subsystems (Rosswall, 1976), major changes in N transfers within these subsystems may have significant effects on the global N cycle. Although estimates of global N fluxes are very uncertain, attempts have been made to estimate global N mass balances and N transfers. References to these estimates are given in Table 7, along with our comments on their major findings.

Four recent estimates of N inventory in the global terrestrial system are given in Table 8. The amount of N in rocks, sediments, coal, and clay-fixed NH_4^+ is exceedingly great, but this N is relatively inert and does not readily participate in global N cycling (except when suddenly released, e.g., through coal burning). The more labile N resides in the soil and biomass. In

Table 7—Listing of global mass balances for N with comments on findings and/or conclusions.

References	Comments
Hutchinson (1954)	First attempt at estimating mass of N in geosphere, hydrosphere, and atmosphere from geochemical data. Also estimated mass transfer rate of N from land to oceans via river discharges, as well as N fixation. Qualitatively appraised significance of different N species in gaseous exchange, including potential formation of acidic precipitation and long residence time of N_2 in the atmosphere.
Robinson & Robbins (1970)	Examined NO and NO_2 emissions from anthropogenic and natural sources and their fate in the atmosphere, including reaction of NO with O_3. Concluded that NO_x emissions are relatively unimportant in global circulation, whereas natural biological activities are important, particularly NH_3/NH_4^+ circulation in gaseous and particulate form. Presented detailed fluxes of N species between atmosphere and earth's surface, terrestrial and aquatic, along with residence times.
Delwiche (1970)	First to point out in a global context the increasing fixation of N through industrial and agricultural activities relative to denitrification, and the consequent accumulation of N in terrestrial and aquatic systems. Presented data on energetics of N transformations, energy required in both biological and abiological N fixation, and energy released in all other biological conversions. Refined Hutchinson's global estimates and presented major fluxes between atmosphere and earth's surface.
Burns & Hardy (1975)	Performed comprehensive review on N fixation and indicated its significance in global N cycle. Updated fluxes based on recent findings, in particular, biological N fixation and denitrification. Pointed out two major subcycles, fixation-denitrification and precipitation-volatilization, which act independently in the atmosphere but not in the soil-water pool. Noted that soil-water pool is clearly the dynamic center of global N cycle.
CAST (1976)	Focused on 1974 global N fixation and denitrification in regard to destruction of O_3 in stratosphere by NO and NO_2 formed from N_2O. Estimated increases in fertilizer N usage from 3.5 million tons year^{-1} in 1950 to about 40 million tons year^{-1} in 1974, and total fixed from 174 million tons year^{-1} in 1950 to about 237 million tons year^{-1} in 1974. Also estimated denitrification of 70 to 100 million tons year^{-1} from land surface with about 4 to 6 million tons year^{-1} of N_2O evolved assuming N_2 to N_2O ratio of 16 to 1. Concluded that even though information gaps in N cycle in oceans do not permit accurate estimates, potential regard to O_3 depletion is serious enough to warrant more research.

(continued on next page)

the soil N pool, nonliving organic matter is the main reservoir for N; however, it is the biomass component of total soil organic matter through which N cycles rapidly. The amount of N in plant and animal biomass generally is about an order of magnitude smaller than that in nonbiomass soil components.

Note that rates of N cycling within the global soil-plant subsystem (Table 9) far exceed rates of external global N transfers (Table 10).

Numerous diagrams depicting N flows among the pedosphere, hydrosphere, and atmosphere have been published. One such diagram separates

Table 7—Continued.

References	Comments
Soderlund & Svensson (1976)	The most completely documented mass balance and transfer study to date. Updated estimates on global N pools and fluxes. Concluded that N fixation by fertilizer industry in 1970 was about 26% of present biological terrestrial fixation, that anthropogenic fixation of NO_x was small but would increase due to increased fossil fuel burning to meet higher energy demands, that there may be additional sinks for N_2O in terrestrial and aquatic systems, and that lack of data on denitrification was most acute. Incorporated N fixation by sediments, a flux heretofore overlooked.
Rosswall (1976)	Developed global flowchart for soil N–vegetation system and concluded that internal fluxes between these two far exceeded (95% of the total transfers) external global fluxes. Presented typical N cycling data for desert, deciduous forest, tundra, and grass sward ecosystems.
Delwiche (1977)	Reexamined and refined global N inventory and transfer rates, including detailed fluxes at the process level. Concluded that historic N cycle has been greatly altered by human activities, but the significance of this is difficult to assess.
Delwiche & Likens (1977)	Expressed concern over biological responses to fossil fuel combustion which injects into the atmosphere substances not normally present in the atmosphere and alters the concentrations of other atmospheric constituents. Concluded that deliberate introduction of fixed N into the biosphere by agricultural activity is the largest contributor to potential N enrichment of waters.
NAS (1978)	Conducted analytical review of global and other material balance studies on N cycle and offered critical appraisals in the methodology, estimates, and conclusions derived from such studies. Considered Soderlund and Soderquist model to be the best estimate, but suggested some additional refinements.

Table 8—Recent estimates of N pools in the global terrestrial system.

Compartments	Burns & Hardy (1975)	Soderlund & Svensson (1976)	Delwiche (1977)	NAS (1978)
		Tg† of N		
Plant biomass	1.0×10^5	$1.1 - 1.4 \times 10^4$	--	1.6×10^4
Animal biomass	1.0×10^3	2.0×10^2	--	1.7×10^2
Litter	--	$1.9 - 3.3 \times 10^3$	--	--
Subtotal plants and animals	--	--	8.0×10^3	--
Soil				
Organic matter	--	3.0×10^5	--	3.0×10^5
Microorganisms	--	5.0×10^2	--	5.0×10^2
Subtotal organic N	5.5×10^5	--	1.75×10^6	--
Insoluble inorganic	1.0×10^5	1.6×10^4	--	--
Soluble inorganic	1.0×10^3	?	--	--
Subtotal inorganic N	--	--	1.6×10^5	1.0×10^3
Rocks	1.93×10^{11}	1.9×10^{11}	--	1.9×10^{11}
Sediments	4.0×10^8	4.0×10^8	--	4.0×10^5
Coal	1.0×10^5	1.2×10^5	--	1.2×10^6
Clay-fixed NH_4^+	--	--	--	7.0×10^4

† Tg = teragram = 10^{12} g = 1 million metric tons.

Table 9—Recent estimates of internal fluxes within the global soil-vegetation N cycle.

N fluxes	Rosswall (1976)		Delwiche (1977)		NAS (1978)	
	Tg† of N year⁻¹	kg of N ha⁻¹year⁻¹‡	Tg of N year⁻¹	kg of N ha⁻¹year⁻¹§	Tg of N year⁻¹	kg of N ha⁻¹year⁻¹
Plant uptake	2,530	19.0	1,800	14.8	1,000	8.2
Immobilization	800	6.0	--	--	500	4.1
Litter formation	--	--	1,900	15.6	1,200	9.8
Mineralization	1,730	13.0	1,900	15.6	1,700	13.9

† Tg = teragram = 10^{12} g = 1 million metric tons.
‡ Based on 133×10^9 km² land area.
§ Based on 122×10^9 km² land area.

N flows between the atmosphere and earth's land and ocean surfaces (Söderlund & Svensson, 1976), but in common with other diagrams of global N flows, causative agents for a particular process are grouped together. For example, often no distinction is made among symbiotic, associative, and nonsymbiotic N_2 fixations. However, attempts have been made to consider different levels of the same process in the construction of global N balance accounts. Table 10 contains four recent estimates of external N fluxes to and from global terrestrial systems. Differences in estimates among investigators for a particular N transfer process are obvious. Without exception, investigators of global N balance emphasize the uncertainty of their data and the tenuous assumptions that must be made to complete their estimates. Values for denitrification loss are considered to be the most uncertain. They seldom are based on direct measurements of gaseous N transfers from the biosphere to the atmosphere, and when results of direct measurements are used, extrapolating data obtained from experimental plots to global dimensions requires use of many assumptions that are virtually impossible to validate. Most often, N deficits in N balance accounts are presumed to represent N losses via denitrification. Accurate values are known for the amount of N_2 fixed by industrial processes and N lost as NO_x emissions.

Clearly, the modeling of the global N cycle requires extensive knowledge in many scientific disciplines. Estimates of global mass balances for N and N transfer rates are considered by some investigators to be too inaccurate to be of value. However, proponents of N balance estimates point to their value in sharpening our insight into problems that could result from a greatly changed flow of N between the biosphere and atmosphere. Moreover, as estimates of global N transfer are refined, models of the global N cycle may become increasingly more valuable as tools for predicting the onset of significant changes in N cycling patterns.

The attempt by Hutchinson (1954) to make a systematic analysis of global N balance was motivated by a need for a better understanding of outer space, particularly in regard to exchange of nitrogenous and other gases between the earth's surfaces and the lower atmosphere. Hutchinson's estimate, which was made 3 years before the launching of Sputnik I by the

Table 10—Estimates of external N fluxes for the global terrestrial system.

Fluxes	Burns & Hardy (1975)	Söderlund & Svensson (1976)	Delwiche (1977)	NAS (1978)
		—Tg† of N year^{-1}—		
		Inputs		
Biological N_2 fixation				
Agriculture: Legumes	35	--	--	--
Rice	4	--	--	--
Other crops	5	--	--	--
Grasslands	45	--	--	--
Subtotal agriculture	89	--	--	--
Forest	40	--	--	--
Others	10	--	--	--
Total biological fixation	139	139	99	139
Abiological N fixation				
Industrial (fertilizer N etc.)	30	36	40	70
Combustion (fossil fuels etc.)	20	19	18	61–251
Total abiological fixation	50	55	58	131–321
NO_3^- atmospheric deposition	32	--	79	--
		Outputs		
Denitrification				
N_2O	--	16–69	--	--
N_2	--	91–92	--	--
Total denitrification	140	107–161	--	197–390
		Transfers		
NH_3 volatilization				
Animals and coal burning	--	26–53	53	--
Others (assumed)	--	87–191	22	--
Total NH_3 volatilization	160	113–244	--	18–45
NH_3/NH_4^+ atmospheric deposition				
Wet deposition	--	30–60	--	--
Dry deposition	--	61–126	--	--
Total NH_3/NH_4^+ deposition	73	91–186	65	--
NO_x atmospheric emission	--	--	--	11–33
NO_x atmospheric deposition				
Wet deposition	--	13–30	--	--
Dry deposition	--	19–53	--	--
Total NO_x deposition	200	32–83	250	--
Organic N emission	--	?	--	--
Organic N deposition	--	10–100	--	--
Total atmospheric deposition ($NH_3/NH_4^+/NO_x$)	--	--	--	66–200
Land to river, ocean discharge				
NH_4^+	--	<1	--	--
NO_x	--	5–11	--	--
Organic N	--	8–13	--	--
Subtotal of NH_4^+/NO_x/organic N	--	13–24	130	--
NO_3^-	--	5–11	--	--
Subtotal inorganic N	15	--	220	--
Total river discharge	--	18–33	350	--

Soviet Union, contributed to the data base used by others who were specu-
lating on the possibility of life existing on neighboring planets.

Recent concern over the possible adverse effects on the environment of
increasing levels of fixed N in the biosphere and concern over the accumulation of CO_2, NO_x, and SO_2 in the atmosphere have rekindled interest in estimating global C, N, and S fluxes. Greater food and fiber demands by a
rapidly expanding world population have resulted in nearly a fivefold increase in N fertilizer use during the past 2 decades—from 11.0 Tg of N during 1961 to 53.6 Tg of N during 1980 (Hauck, 1981). Increased coal burning
to meet energy demands is resulting in the release of N, C, and S to the atmosphere. Increased N fertilizer use and coal burning are but two of several
changes in levels of human activities that change the rates of global nutrient
transfers. Concern has been expressed about the greenhouse warming effect
caused by CO_2 accumulation in the atmosphere, the possible reduction of
stratospheric ozone to undesirable levels caused by N_2O released from soils
via denitrification, and the spread of acid precipitation caused by accumulation of SO_2 and, perhaps, NO_x in the atmosphere (Bolin & Arrhenius,
1977). It can be seen that a significantly large increase in N inputs on a
regional or subregional level can measurably affect global N transfers between the earth's land and water surfaces and the atmosphere.

IV. PERSPECTIVE

Studies of N cycle processes continually enlarge in scope. Formerly
pursued mainly at the single-process level, studies of N transformations in
the biosphere now are often multidisciplinary efforts directed toward increasing our understanding of how changes in N transformation and transfer rates impact on subregional, regional, and global environments. Soil remains the focal point of research on N cycle processes. The results of this research are of interest not only to those concerned with food and fiber production but also to limnologists, oceanographers, atmospheric chemists,
ecologists, and other biological and physical scientists who investigate the
effects of N in various environments. More than ever before, investigators
of N in the soil environment think and work at several levels of integration.
For example, studies of nitrification on the subcellular level indicate the involvement of a Cu-containing component of a cytochrome oxidase during
the biological oxidation of NH_3 (NH_4^+) to hydroxylamine. Addition of a
chemical that chelates Cu in the medium surrounding the microbial cell or
colony of cells (microscopic level) may retard NH_3 oxidation, the extent depending on the combined effects of numerous naturally occurring or culturally modified factors operating within microsites that comprise the
macroscopic environment. The main objective of the enzymological investigation of the nitrification process usually is to increase understanding of the
process on the subcellular level. Control of nitrification may be the main
objective of investigations that are made within the microsites of the soil-N
fertilizer reaction zones. Within these microsites, some studies are mainly

concerned with the immobilization of that NH_4^+-N that is not nitrified, the uptake of inorganic N by higher plant roots, the movement of NO_3^--N from the microsite by diffusion or mass flow, or the reduction of oxidized forms of N to N gases that are transferred from the microsite to the atmosphere. The events occurring on the subcellular level rapidly gain added significance on higher levels of integration. As an energy-producing process, the oxidation of NH_3 to NO_2^- is of vital significance to the microorganism. As a process that results in the eventual formation of NO_3^-, nitrification promotes the transfer of N from soil through leaching to ground waters, gaseous evolution, or plant uptake and removal in harvested crops. On the subcellular level, N_2O formation during nitrification apparently is an enzymic response to anoxic conditions of vital significance to the microorganism. On the macrosystem level, N_2O formation during nitrification probably is of negligible economic importance to the farmer (representing an estimated loss of only 1 to 2 kg of $N \cdot ha^{-1} \cdot year^{-1}$). However, if the nitrification process in soils and waters is a significant source of N_2O in the global environment, the amount and rate of N_2O production is of interest to investigators of the ozone regulatory process in the stratosphere, a process that is affected by tropospheric N_2O concentrations, and to those investigating global N mass balances and transfers. On the sub-cellular level, N_2O formation during denitrification is a vital process. On the farm level, it can result in inefficient use of fertilizer N, and on the global level, it can present problems of data interpretation for investigators of global N transfers because the amount of N evolved from this source of N_2O (denitrification) must be separated from the amounts evolved from another biological process (nitrification) and abiological processes such as combustion.

It is clear that studies of N transformation and transfers in the biosphere are of interest to investigators representing various scientific disciplines, but these investigators have different objectives, use different approaches, and focus on different levels of integration. Investigators whose main concern is to accurately measure events that occur at the lower hierarchy of objects (the single-process level) should aid others who use their data with different objectives in mind by clearly reporting the conditions under which the data were obtained and by discussing the limitations of the measurements. Investigators who work at the highest levels of integration should be aware of the errors inherent in space averaging and time averaging data that are obtained from studies at basic levels of integration. We regard consistency of data to be an indication of their validity, especially when similar values for the magnitude of a process are obtained from validly comparable studies made at different levels of integration.

Mass balance estimates for N that are obtained from smaller scale studies are accepted more readily by scientists than are those obtained from regional and global studies. Several reasons for this are suggested. There is less experience among scientists in working with large-scale systems, and thus there is less confidence in judging the validity of estimates for extensive areas. Generally, small-scale systems are more homogeneous with respect to their physical characteristics, chemical reactions, and biological activities than are large-scale systems. For small-scale systems, space averaging is an acceptable practice. The larger the system, the more heterogeneous it is generally, and the greater is the risk of error from space averaging data. Models for N balance in small-scale systems usually require fewer tenuous assumptions than do models for large-scale systems. Measurements in small-scale systems usually are made in greater detail. Most important, the validity of data from small-scale systems is much easier to verify, which leads to credibility.

Because increased understanding of N transfers among the pedosphere, hydrosphere, and atmosphere is being sought by a variety of investigators pursuing different short-term and long-term objectives, new approaches, methods, and instrumentation are being brought to bear on studies of N cycle processes. This increased level of multidisciplinary involvement will result in the refinement of information on all levels of integration.

LITERATURE CITED

Allison, F. E. 1955. The enigma of soil nitrogen balance sheets. Adv. Agron. 7:213–250.

Allison, F. E. 1965. Evaluation of incoming and outgoing processes that affect soil nitrogen. In W. V. Bartholomew and F. E. Clark (ed.) Soil nitrogen. Agronomy 10:573–606. Am. Soc. of Agron., Inc., Madison, Wis.

Allison, F. E. 1966. The fate of nitrogen applied to soils. Adv. Agron. 18:219–258.

Ayers, R. S., and R. L. Branson. 1973. Nitrates in the Upper Santa Ana River Basin in relation to groundwater pollution. Div. of Agric. Sci., Univ. of California, Bull. no. 861.

Bazilevich, N. I. 1958. The minor biological cycle of mineral elements and nitrogen during meadow-steppe and steppe soil formation. (In Russian.) Pochvovedenie 12:9–27.

Benson, D. 1965. How much is a marsh worth? New York State Conservationist Bull. no. L-132, New York State Conservation Dep.

Bolin, B. 1976. Transfer processes and time scales in biogeochemical cycles. Ecol. Bull. (Stockholm) 22:17–22.

Bolin, B., and E. Arrhenius. 1977. Nitrogen—an essential life factor and a growing environmental hazard. Ambio 6:96–105.

Bormann, F. H., G. E. Likens, D. W. Fisher, and R. S. Pierce. 1968. Nutrient loss accelerated by clear-cutting of a forest ecosystem. Science 159:882–884.

Brezonik, P. L., W. E. Morgan, E. E. Shannon, and H. D. Putnam. 1969. Eutrophication factors in north central Florida lakes. Bull. Series 134, Engr. Indust. Exp. Stn., Univ. of Florida, Gainesville.

Brezonik, P. L., and C. F. Powers. 1973. Nitrogen sources and cycling in natural waters. U.S. Environmental Protection Agency Rep. EPA 660/3-73-002, U.S. Government Printing Office, Washington, D.C.

Burns, R. C., and R. W. F. Hardy. 1975. Nitrogen fixation in bacteria and higher plants. Springer-Verlag New York, New York.

Cole, D. W., and S. P. Gessel. 1965. Movement of elements through a forest soil as influenced by tree removal and nutrient additions. p. 95–104. In C. T. Youngberg (ed.) Forest-soil relationships in North America. Oregon State Univ. Press, Corvallis.

Cooper, C. F. 1969. Nutrient output from managed forests. p. 446–463. In Eutrophication: Causes, consequences, correctives. Natl. Acad. of Sci., Washington, D.C.

Council for Agricultural Science and Technology. 1976. Effect of increased nitrogen fixation on stratospheric ozone. CAST Rep. no. 53, Ames, Iowa.

Dahlman, R., J. S. Olson, and K. Doxtator. 1969. The nitrogen economy of grassland and dune soils. p. 54–82. Proc. Conf. on Biology and ecology of nitrogen. Univ. of California, Davis. 1967. Natl. Acad. of Sci., Washington, D.C.

Delwiche, C. C. 1970. The nitrogen cycle. Sci. Am. 223:136–147.

Delwiche, C. C. 1977. Energy relations in the global nitrogen cycle. Ambio 6:106–111.

Delwiche, C. C., and G. E. Likens. 1977. Biological response to fossil fuel combustion products. p. 73–88. *In* W. Stumm (ed.) Global chemical cycles and their alterations by man. Dahlem Konferenzen, Berlin.

Denmead, O. T., J. R. Freney, and J. R. Simpson. 1976. A closed ammonia cycle within a plant canopy. Soil Biol. Biochem. 8:161–164.

Denmead, O. T., R. Nulsen, and G. W. Thurtell. 1978. Ammonia exchange over a corn crop. Soil Sci. Soc. Am. J. 42:840–842.

Denmead, O. T., J. R. Simpson, and J. R. Freney. 1974. Ammonia flux into the atmosphere from a grazed pasture. Science 185:609–610.

Dugdale, R. C., and J. J. Goering. 1967. Uptake of new and regenerate nitrogen in primary productivity. Limnol. Oceanogr. 12:196–206.

Fletcher, J. E., L. Sorensen, and D. B. Porcella. 1978. Erosional transfer of nitrogen in desert ecosystems. p. 171–181. *In* N. E. West and J. Skujins (ed.) Nitrogen in desert ecosystems. Dowden, Hutchinson, and Ross, Inc., Strousburg, Pa.

Fried, M., K. K. Tanji, and R. M. Van De Pol. 1976. Simplified long term concept for evaluating leaching of nitrogen from agricultural land. J. Environ. Qual. 5:197–200.

Frissel, M. J. (ed.). 1978. Cycling of mineral nutrients in agricultural ecosystem. Elsevier Scientific Publ. Co., New York.

Golley, F. B., J. T. McGinnis, R. G. Clements, G. I. Child, and M. J. Dueter. 1975. Mineral cycling in a tropical moist forest ecosystem. Univ. of Georgia Press, Athens.

Hattori, A., and E. Wada. 1973. Biogeochemical cycling of inorganic nitrogen in marine environments with special reference to nitrite metabolism. p. 28–39. *In* E. Ingerson (ed.) Proc. of the Symp. on Hydrogeochemistry and Biogeochemistry. The Clarke Co., Washington, D.C.

Hauck, R. D. 1981. Nitrogen fertilizer effects on nitrogen cycle processes. *In* F. E. Clark and T. Rosswall (ed.) Terrestrial nitrogen cycling processes, ecosystem strategies, and management impacts. Ecol. Bull. (Stockholm) 33:551–562.

Henderson, G. S., and W. F. Harris. 1975. An ecosystem approach to characterization of the nitrogen cycle in a deciduous forest watershed. p. 179–193. *In* B. Bernier and C. H. Winget (ed.) Forest soils and forest land management, Proc. Fourth North American Forest Soil Conf., Laval Univ., Quebec. Aug. 1973. Laval Univ., Quebec.

Hunt, H. W. 1978. A simulation model for decomposition in grasslands. Ecology 58:469–484.

Hutchinson, C. E. 1954. The biochemistry of the terrestrial atmosphere. p. 371–433. *In* G. P. Kuiper (ed.) The earth as a planet. Univ. of Chicago Press, Chicago.

International Rice Research Institute. 1977. Annual report for 1976. IRRI, Los Banos, Philippines.

Jenny, H. 1941. Factors of soil formation. McGraw-Hill Publications, New York.

Junge, C. E. 1958. The distribution of ammonia and nitrate in rainwater over the United States. Trans. Am. Geophys. Union 39:241–248.

Klubek, B., P. I. Eberhardt, and J. Skujins. 1978. Ammonia volatilization from great basin desert soils. p. 107–129. *In* N. E. West and J. Skujins (ed.) Nitrogen in desert ecosystems. Dowden, Hutchinson, and Ross, Inc., Strousburg, Pa.

Kohl, D. H., G. Shearer, and F. Vithayanthil. 1978. Critique of "Nitrogen inputs and outputs: A valley basin study." Nitrogen mass balance studies. p. 183–200. *In* D. R. Nielsen and J. G. MacDonald (eds.) Nitrogen in the environment, Vol. 1. Academic Press, Inc., New York.

Lee, G. F., E. Bently, and R. Adamson. 1975. Effect of marshes on water quality. p. 105–127. *In* A. D. Hasler (ed.) Ecological Studies 10—coupling of land and water systems. Springer-Verlag New York, New York.

Lee, G. F., and W. C. Sonzogni. 1972. Nutrient sources in Lake Mendota. Trans. Wis. Acad. Sci. Arts Lett. LXII:133–164.

Lemon, E., and R. Van Houtte. 1980. Ammonia exchange at the land surface. Agron. J. 72:876–883.

Lipman, J. G., and A. B. Corybeare. 1936. Preliminary note on the inventory and balance sheet of plant nutrients in the United States. New Jersey Agric. Exp. Stn. Bull. 607.

McGill, W. B., H. W. Hunt, R. C. Woodmansee, and J. O. Reuss. 1981. A model for nitrogen cycling in grasslands. Ecol. Bull. (Stockholm) (In press.)

Mehran, M., and K. K. Tanji. 1974. Computer modeling of nitrogen transformations in soils. J. Environ. Qual. 3:391–395.

Mehran, M., K. K. Tanji, and I. K. Iskandar. 1981. A compartmental model for prediction of nitrate leaching losses. Chapt. 13. *In* I. K. Iskandar (ed.) Modeling wastewater renovation by land treatment. USA Cold Regions Res. Engr. Lab., Hanover, N.H.

Miller, R. J., and R. B. Smith. 1976. Nitrogen balance in the Southern San Joaquin Valley. J. Environ. Qual. 5:274–278.

Murray, R. B. 1975. Effect of *Artenisia tridentata* removal on mineral cycling. Ph.D. Thesis. Washington State Univ., Pullman. Univ. Microfilms. Ann Arbor. Mich. (Diss. Abstr. 76: 322).

National Academy of Sciences. 1972. Accumulation of nitrate. Committee on Nitrate Accumulation, Agriculture Board, Div. of Biol. and Agric., National Research Council, M. Alexander, Chairman, Washington, D.C.

National Academy of Sciences. 1978. Nitrates: An environmental assessment. Panel on Nitrates of the Coord. Comm. for Scientific and Technical Assessments of Environmental Pollutants, National Research Council, P. L. Brezonik, Chairman, Washington, D.C.

Power, J. F. 1977. Nitrogen transformation in the grassland ecosystem. p. 195–204. *In* J. K. Marshall (ed.) The belowground ecosystem: A synthesis of plant-associated processes. Colorado State Univ., Fort Collins.

Robinson, E., and R. C. Robbins. 1970. Gaseous atmospheric pollutants from urban and natural sources. p. 50–64. *In* S. F. Singer (ed.) Global effects of environmental pollution. D. Reidel Publ. Co., Dordrecht.

Rodin, L. E., and N. I. Bazilevich. 1965. Dynamics of the organic matter and biological turnover of ash elements and nitrogen in the main types of the world vegetation. Komarov Botanical Inst., Leningrad.

Rodin, L. E., and N. I. Bazilevich. 1966. The biological productivity of the main vegetation types in the Northern Hemisphere of the Old World. For. Abstr. 38:369–372.

Rodin, L. E., and N. I. Bazilevich. 1968. Production and mineral cycling in terrestrial vegetation. Oliver and Boyd, London. (In Russian.)

Rosswall, T. 1976. The internal nitrogen cycle between microorganisms, vegetation, and soil. p. 157–167. *In* Nitrogen, phosphorus and sulfur—global cycles. SCOPE Rep. 7, Ecological Bull. no. 22, Swedish Natural Science Research Council, Royal Swedish Acad. of Sci., Stockholm.

Rowe, J. S. 1961. The level-of-integration concept and ecology. Ecology 42:420–427.

Smith, G. E. 1942. Fifty years of field experiments with crop rotations, manure, and fertilizers. Univ. Missouri Agric. Exp. Stn. Bull. 458.

Söderlund, R., and B. H. Svensson. 1976. The global nitrogen cycle. p. 23–73. *In* Nitrogen, phosphorus and sulfur—global cycles. SCOPE Rep. 7, Ecological Bull. no. 22, Swedish Natural Science Research Council, Royal Swedish Acad. of Sci., Stockholm.

Stanford, E., C. B. England, and A. W. Taylor. 1970. Fertilizer and water quality. ARS 41-168, U.S. Dep. of Agric., U.S. Government Printing Office, Washington, D.C.

Tanji, K. K., F. E. Broadbent, M. Mehran, and M. Fried. 1979. An extended version of a conceptual model for evaluating annual nitrogen leaching losses from croplands. J. Environ. Qual. 8:114–120.

Tanji, K. K., M. Fried, and R. M. Van De Pol. 1977. A steady-state conceptual nitrogen model for estimating nitrogen emissions from cropped lands. J. Environ. Qual. 6:155–159.

Tanji, K. K., and M. Mehran. 1979. Conceptual and dynamic models for nitrogen in irrigated croplands. p. 555–646. *In* Nitrate in effluents from irrigated lands. P. F. Pratt, Principal Investigator, Final Rep. to NSF, Univ. of California, Riverside.

Thomas, G. W., and J. W. Gilliam. 1978. Agro-ecosystems in the U.S.A. p. 182–143. *In* M. J. Frissel (ed.) Cycling of mineral nutrients in agricultural ecosystems. Elsevier Scientific Publ. Co., New York.

Vaccoro, R. 1965. Inorganic nitrogen in sea water. p. 365–408. *In* J. P. Riley and G. Skirrow (ed.) Chemical oceanography, Vol. 1. Academic Press, Inc., New York.

Valiela, I., and J. M. Teal. 1979. The nitrogen budget of a salt marsh ecosystem. Nature (London) 280:652–656.

Van Veen, J. A., and E. A. Paul. 1981. Organic C dynamics in grassland soils. I. Model development. Can. J. Soil Sci., Vol. 61. (In press.)

Viets, F. G. 1978. Critique of "Nitrogen inputs and outputs: A valley basin study." Mass balance and flux of nitrogen as aids in control and prevention of water pollution. p. 173–182. *In* D. R. Nielsen and J. G. MacDonald (ed.) Nitrogen in the environment, Vol. 1. Academic Press, Inc., New York.

Vlek, P. L. G., I. R. P. Fillery, and J. R. Burford. 1981. Accession, transformation, and loss of nitrogen in soils of the arid region. Plant Soil 58:133–175.

Viets, F. G., Jr., and R. H. Hageman. 1971. Factors affecting the accumulation of nitrate in soil, water, and plants. U.S. Dep. of Agric. Handbook no. 413, U.S. Government Printing Office, Washington, D.C.

Vollenweider, R. A. 1968. Scientific fundamentals of the eutrophication of lakes and flowing waters, with particular reference to nitrogen and phosphorus as factors in eutrophication. Organization for Econ. Coop. & Dev. Rep., DAS/CSI 68.27.

West, N. E. 1978. Physical inputs of nitrogen to desert ecosystems. p. 165–170. *In* N. E. West and J. Skujins (ed.) Nitrogen in desert ecosystems. Dowden, Hutchinson, and Ross, Inc., Strousburg, Pa.

Westerman, R. L., and T. C. Tucker. 1978. Denitrification in desert soils. p. 75–106. *In* N. E. West and J. Skujins (ed.) Nitrogen in desert ecosystems. Dowden, Hutchinson, and Ross, Inc., Strousburg, Pa.

Wetselaar, R., and G. D. Farquhar. 1980. Nitrogen losses from tops of plants. Adv. Agron. 33:263–302.

Wollum, A. G., II, and C. B. Davey. 1975. Nitrogen accumulation, transformation, and transport in forest soils. p. 67–106. *In* B. Bernier and C. H. Winget (ed.) Forest soils and forest land management, Proc. Fourth North American Forest Soils Conf., Laval Univ., Quebec. Aug. 1973. Laval Univ., Quebec.

Woodmansee, R. G. 1979. factors influencing input and output of nitrogen in grasslands. p. 117–134. *In* N. R. French (ed.) Perspectives in grasslands ecology. Springer-Verlag New York, New York.

Woodmansee, R. G., and F. J. Adamson. 1982. Nutrient cycles in ecological hierarchies. *In* R. L. Todd (ed.) Proc. Int. Symp. Nutrient cycling in agroecosystems. Ann Arbor Science Publishers, Inc., Ann Arbor, Mich. (In press.)

Yaalon, D. H. 1964. The concentration of ammonia and nitrate in rain water over Israel in relation to environmental factors. Tellus 16:200–204.

SUBJECT INDEX